Lecture Notes in Electrical Engineering

Volume 339

About this Series

"Lecture Notes in Electrical Engineering (LNEE)" is a book series which reports the latest research and developments in Electrical Engineering, namely:

- Communication, Networks, and Information Theory
- Computer Engineering
- Signal, Image, Speech and Information Processing
- Circuits and Systems
- Bioengineering

LNEE publishes authored monographs and contributed volumes which present cutting edge research information as well as new perspectives on classical fields, while maintaining Springer's high standards of academic excellence. Also considered for publication are lecture materials, proceedings, and other related materials of exceptionally high quality and interest. The subject matter should be original and timely, reporting the latest research and developments in all areas of electrical engineering.

The audience for the books in LNEE consists of advanced level students, researchers, and industry professionals working at the forefront of their fields. Much like Springer's other Lecture Notes series, LNEE will be distributed through Springer's print and electronic publishing channels.

More information about this series at http://www.springer.com/series/7818

Kuinam J. Kim
Editor

Information Science and Applications

Springer

Editor
Kuinam J. Kim
Department of Convergence Security
Kyonggi University
Institute of Creative Advanced Technology,
Science and Engineering
Kyonggi-do
Korea, Republic of (South Korea)

ISSN 1876-1100 ISSN 1876-1119 (electronic)
Lecture Notes in Electrical Engineering
ISBN 978-3-662-52613-2 ISBN 978-3-662-46578-3 (eBook)
DOI 10.1007/978-3-662-46578-3

Springer Heidelberg New York Dordrecht London

Softcover reprint of the hardcover 1st edition 2015

Printed on acid-free paper

Springer-Verlag GmbH Berlin Heidelberg is part of Springer Science+Business Media
(www.springer.com)

Contents

Part I

QoS-aware Mapping and Scheduling for Integrated CWDM-PON and WiMAX Network

Siti H. Mohammad[1], Nadiatulhuda Zulkifli[2], Sevia M. Idrus[2]

[1,2]Lightwave Communication Research Group, Faculty of Electrical Engineering, Universiti Teknologi Malaysia, 81310 UTM Johor Bahru, Johor, Malaysia

[1]hasunah@fkegraduate.utm.my
[2]{nadia,sevia}@fke.utm.my

Abstract. Worldwide Interoperability for Microwave Access (WiMAX) has emerged as one of key technologies for wireless broadband access network while Coarse Wavelength Division Multiplexing-Passive Optical Network (CWDM-PON) is one of the potential solutions for future high speed broadband access network. Integrating both networks could enhance the whole network performance by allowing cost-effectiveness, higher capacity, wider coverage, better network flexibility and higher reliability. In this work, scheduling algorithm is proposed as means to maintain the Quality of Service (QoS) requirements of two different media whilst allocating the bandwidth to the subscribers. The NS-2 simulation results demonstrate how network performances of the integrated CWDM-PON and WiMAX networks are improved in terms of delay and throughput.

Keywords: Wireless networks; optical networks; quality of service; scheduling algorithm.

1 Introduction

Coarse Wavelength Division Multiplexing-Passive Optical Network (CWDM-PON) is widely researched given its advantages as a cost-effective choice for short fiber span network transmission up to 100 km [1], capable to support more channels in order to expand the network capacity. Meanwhile, Worldwide Interoperability for Microwave Access (WiMAX) has shown its potential for the next-generation wireless networks technology as it provides both high mobility and large bandwidth. Theoretically, a WiMAX base station can provide broadband wireless access ranging up to 50 km for fixed stations and 5 to 15 km for mobile stations, with a maximum data rate of up to 70 Mbps [2, 3].

Previous researchers have proposed various types of service classes that can be offered in PON system. However, this paper follows the QoS traffic classes proposed by Yang et al. in 2009 [4] as it has the closest network architecture to our work i.e. integrated PON and WiMAX network. Here, CWDM-PON QoS traffic classes can be divided into Expedited Forwarding (EF), Assured Forwarding (AF) and Best Effort (BE). On the other hand, IEEE802.16 standard of WiMAX consists of five QoS classes, which are Unsolicited Grant Service (UGS), real-time Polling Service (rtPS), extended real-time Polling Service (ertPS), non-real-time Polling Service (nrtPS) and Best Effort (BE) [4, 5, 6].

Currently, the integration of optical and wireless technologies to create a single access network has been one of the hot topics among researchers because of the advantages in the mobility and flexibility of the network compared to the fixed network infrastructure such as PON alone. With proven reliability of optical networks at the back end and flexible wireless network at the front end, the integrated network is believed to provide better service to the subscribers exploiting the extended coverage and high bit rates of existing wireless network.

In the research field, many works have been done to explore the potential of integrated optical wireless network such as in [4-10]. So far, it can be seen that a vast majority of previous works on integrated optical wireless have not addressed the issues of fairness and QoS during bandwidth allocation. To our knowledge, no work has been done to integrate the CWDM-PON and WiMAX in order to provide the applications and services for network subscribers while maintaining the QoS over different network channel. Thus, such architecture is proposed for future solution to face the increasing demand from the subscribers.

This study extends the WiMAX scheduling algorithm by Freitag and da Fonseca [9] for the converged network of CWDM-PON and WiMAX, exploiting the existing QoS parameters of the WiMAX network that will be

K.J. Kim (ed.), *Information Science and Applications*,
Lecture Notes in Electrical Engineering 339, DOI 10.1007/978-3-662-46578-3_1

mapped to the three QoS traffic classes of CWDM-PON [4, 5, 6]. The paper is organized as follows. Section 2 describes the network topology. Section 3 presents the proposed QoS mapping and scheduling mechanism. Section 4 discusses the simulation environment used to test the proposed scheduler. Section 5 presents simulation results and analysis. Finally, Section 6 concludes the paper.

2 Network Topology

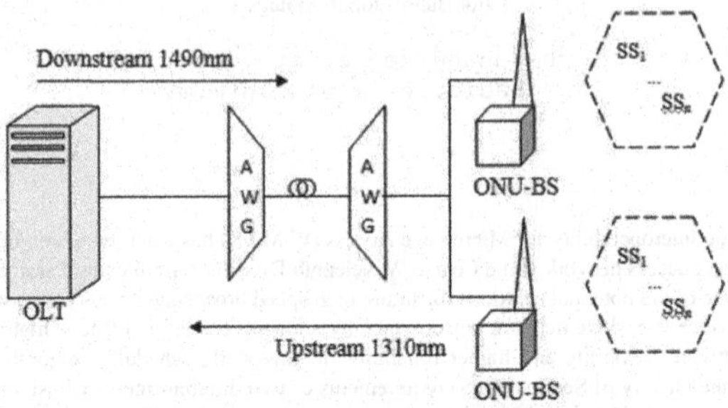

Fig. 1. Proposed integrated CWDM-PON and WiMAX network.

Typically, an integrated CWDM-PON and WiMAX network utilizes the tree topology at the optical backend while Point-to-Multipoint (PMP) topology is applied at the wireless front end. An Optical Line Terminal (OLT) from CWDM-PON is connected to the integrated Optical Network Unit-Base Station (ONU-BS) by two Arrayed Waveguide Gratings (AWGs), adopted from previous CWDM-PON system of Khairi *et al.* in [11]. The OLT is responsible to schedule the up-link traffic among all the ONU-BSs since they share the up-link resources from the AWGs to the OLT, as shown in Figure 1. The ONU-BS is then connected to subscribers station (SS) wirelessly.

3 QoS Mapping and Scheduling Mechanism

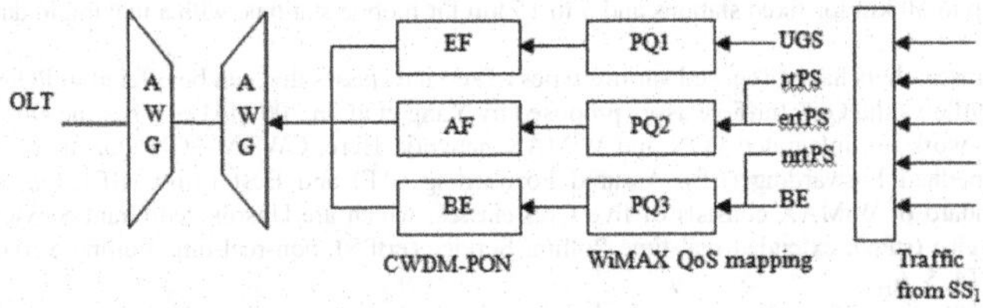

Fig. 2. The proposed QoS mapping of integrated CWDM-PON and WiMAX network.

As the nrtPS service in WiMAX is quite similar to BE, these two service types are combined as Priority Queue 3 (PQ3) and then mapped to BE classes of CWDM-PON in this work. rtPS and ertPS are combined as Priority Queue 2 (PQ2) and are mapped to CWDM-PON AF, while UGS at Priority Queue 1 (PQ1) is mapped to CWDM-PON EF class to be sent to the OLT. Three types of traffic are considered: voice, video, and data, which are associated to PQ1, PQ2, and PQ3 services, respectively. Three queues are used in this proposed scheduler, referred as low priority queue, intermediate queue and high priority queue [9, 10]. The requests are served in a strict priority order from the high priority queue to the low priority queue. The bandwidth requests of the PQ3 service flows are stored in the low priority queue while the bandwidth requests sent by PQ2 connections are held in the intermediate queue. The high priority queue stores the bandwidth request of PQ1 service flows. The proposed mechanism is presented as in Figure 3.

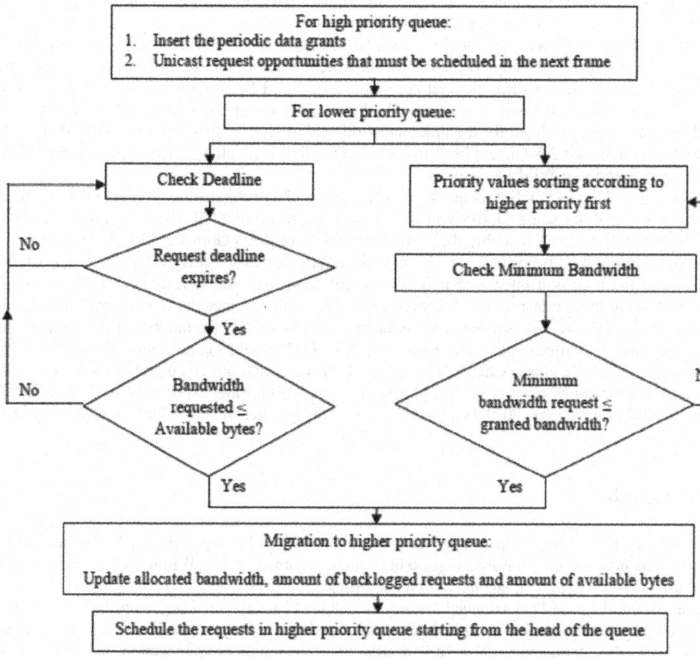

Fig. 3. The flow of the proposed scheduling algorithm.

The data grants are periodically inserted into the high priority queue to guarantee the bandwidth of PQ1 flows. After inserting it, the scheduler checks whether the request in the intermediate queue should be scheduled in the following frame or not, based on the deadline and the minimum bandwidth required. The ONU-BS assigns a deadline to each PQ2 bandwidth request in the intermediate queue to guarantee the maximum delay requirement. Each time the scheduler is executed, the requests which reaching the deadline in the frame following the next one migrate to the high priority queue. According to [9], it would be necessary to know the arriving time of the packets at the SSs queues to calculate the deadline for migration, but the ONU-BS has no access to this information. Thus, the worst possible case is considered, which corresponds to the arrival at the queue immediately after the connection sent the last bandwidth request. Hence, the deadline of a request should be equal to the sum of the arriving time of the last request sent by the connection and its maximum delay requirement. This also applies to the low priority queue when the intermediate queue is empty and there are available slots in the uplink frame.

4 Simulation Setup

The simulation experiment is done in NS-2 to measure the throughput and the delay of the network. PQ1 traffic generator is an "on/off" one with duration of periods exponentially distributed with a mean of 1.2 s and 1.8 s for the "on" and "off" periods, respectively. A new packet for one PQ1 connection is generated every 20 ms with packet size of 66 bytes [12]. According to Alsolami [15], most of the Internet traffics are generated by http, VBR, and FTP. Thus, PQ2 traffic consumes most of the network bandwidth compared to the PQ1 traffic. PQ2 traffic is generated by real MPEG traces [13], with delay requirement of 100 ms. Each connection has its own minimum bandwidth requirement which varies according to the mean rate of the transmitted video. PQ3 traffic is not as sensitive to delay as PQ1 and PQ2; however it is subject to a minimum bandwidth requirement.

For all simulated scenarios, it is assumed that each SS has only one traffic flow to prevent interference between packet scheduling in the SSs and the evaluation of the scheduling mechanism at the BS. The performance of the network is evaluated considering ideal channel condition of CWDM-PON system following the experimental analysis on physical layer. The duplex link was used for both downlink and uplink transmission. The optical characteristics in NS-2 simulation were defined for the link bandwidth and time delay properties. As for CWDM-PON, the channel link capacity is assigned 2.5 Gbps with 2 ms delay. This link is created between the wired CWDM-PON network segment and BS nodes of WiMAX network.

The limitation of CWDM-PON link is mainly contributed by fiber propagation, processing and transmission delays. Our preliminary analysis using Optisystem proves that the received SNR at maximum length 100 km is 15.11 dB (input SNR=27 dB), which is within the allowable SNR limit at 2.5 Gbps i.e. 10.4 dB [14]. With assumption that light velocity inside the fiber is at 2.0×10^{-8} m/s, the light propagation delay is 0.5 ms. Therefore, for network layer analysis of the proposed integrated network, contribution from other delays including device rise/fall time and device processing time; transmission delay is assumed to be three times the fiber propagation delay or 1.5 ms. For each scenario, one ONU-BS is connected to multiple SSs. In this work, the number of BS and SSs are based on the work of Freitag and da Fonseca in [9]. The number of PQ1, PQ2 and PQ3 connections are varied according to the offered traffic load, which are voice, video and data. Due to space limitation, this paper focuses on the effects of increasing voice traffic load to low priority service classes. The scenario for this experiment consists of one BS and 61 SSs. There are 6 PQ2 connections, 20 PQ3 connections, and the number of active PQ1 connections varies from 15 to 35.

5 Results Analysis

Results of the experiment are compared to the simulation results of Freitag *et al.* in [9]. Figure 4 and 5 compare the delay performances of our proposed algorithm, which is named as C-QWi algorithm, with the formerly published algorithm of Freitag *et al.* [9], labeled as QWi algorithm.

From Figure 4, the delay of PQ1 connections are not affected by the increasing connection numbers for the whole connection range. This proves that the scheduler is able to provide data grants at fixed intervals as required by the service. The delay of PQ2 connections show little oscillation as the number of connections increased for all three simulation scenarios because the scheduler has to fulfill the demands of high priority PQ1 traffic class before offering the bandwidth to lower priority PQ2 traffic class.

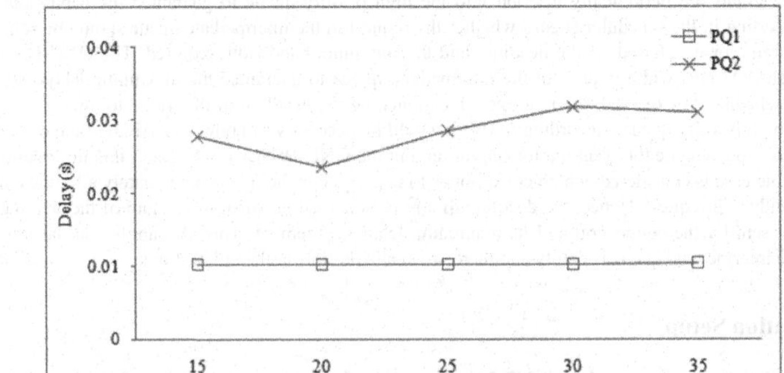

Fig. 4. The delay of high priority classes over increasing PQ1 connections for the proposed C-QWi scheduler.

When comparing both Figures 4 and 5, it can be seen that the service performance is quite similar, except that in between 15 to 30 connections of PQ1, the delay of video service for C-QWi scheduler is much better than the delay of video service for QWi scheduler. In spite of the increment in voice connections, the delay values are considerably lower than the required one, according to WiMAX standard as mentioned in Section 4, which is 100 ms.

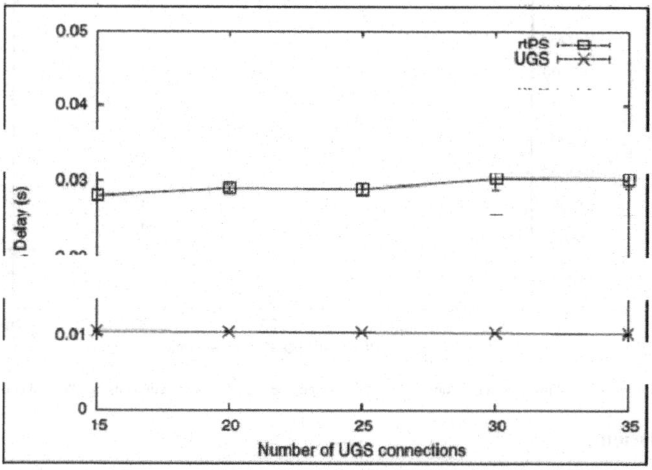

Fig. 5. The delay of high priority classes over increasing UGS connections using QWi scheduler [9].

For throughput performance, the throughput of the lower priority class, PQ3 almost maintains at high bit rate with an increasing number of PQ1 connections (Figure 6). This shows that the C-QWi scheduler can allocate bandwidth to the lower priority classes even if the voice load increases. Large optical link capacity of CWDM-PON at the back end is able to use the excess bandwidth from the higher priority classes to serve the lower priority classes. Thus, it is able to help the network to maintain the throughput of the lower priority class, compared to the 40 Mbps WiMAX network alone with the use of QWi scheduler. This surpasses the throughput performance of nrtPS and BE connections over the number of BE connections for the QWi scheduler, as shown in Figure 7.

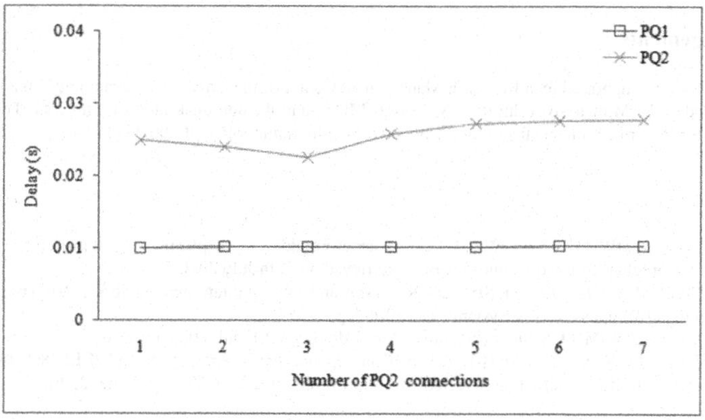

Fig. 6. Average throughput of PQ3 connections over the number of PQ1 connections for the proposed C-QWi scheduler.

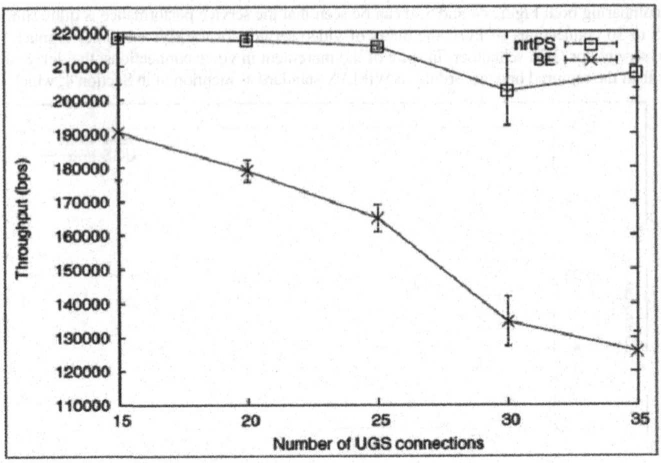

Fig. 7. Throughput of nrtPS and BE connections over the number of UGS connections using QWi scheduler [9].

6 Conclusion

In this study, the QoS-aware scheduling algorithm of integrated CWDM-PON and WiMAX network is concerned. The introduction of CWDM-PON to the WiMAX network enables the exploitation of higher optical link capacity for varied wireless traffic loads. Based on the simulation work, the proposed C-QWi scheme could satisfy the delay requirement for real time traffic, specifically video service, PQ2. The delay of all classes of services were not much affected when the offered load is varied, showing that the optical link delay is negligible, thus would not degrade the performance of the proposed integrated network. 2.5 Gbps CWDM-PON at the back end of the proposed integrated network guarantees high throughput performance of even the lower service classes whilst maintaining high bandwidth provision of higher service classes. The findings show the feasibility of the proposed integrated CWDM-PON and WiMAX network, with significant throughput improvement compared to the WiMAX network alone.

Acknowledgements

This work was supported by a Research Management Center of the University Technology Malaysia. The authors acknowledge the Ministry of Education Malaysia (MOE) and the administration of Universiti Teknologi Malaysia (UTM) for the project financial support through cost centre number R.J130000.7823.4F407.

References

[1] CWDM, DWDM, WDM Carrier-Grade Optical transport Systems. http://www.optelian.com/faq/technology.php. Retrieved on 27th July 2010.
[2] IEEE P802.16Rev2/D2. DRAFT Standard for Local and metropolitan area networks, Part 16: Air Interface for Broadband Wireless Access Systems, Dec. 2007, pp. 2094.
[3] WiMAX Forum. WiMAX System Evaluation Methodology V2.1, Jul. 2008, pp. 230.
[4] K. Yang, S. Ou, K. Guild, and H-H. Chen. Convergence of Ethernet PON and IEEE 802.16 Broadband Access Network, IEEE Journal on Selected Areas in Communications, Vol. 27, Issue: 2, 2009.

[5] I-S. Hwang, J-Y. Lee, C-W. Huang and Z-D. Shyu. Advanced Dynamic Bandwidth Allocation and Scheduling Scheme for the Integrated Architecture of EPON and WiMAX, Tenth International Conference on Mobile Data Management: Systems, Services and Middleware 2009, MDM '09, 2009.

[6] B. Ranaweera, E. Wong, C. Lim and A.Nirmalathas. Quality of Service Assurance in EPON-WiMAX Converged Network, International Topical Meeting on Microwave Photonics 2011 & Asia-Pacific, MWP/APMP Microwave Photonics Conference 2011.

[7] J. Lee, S. H. S. Newa, J. K. Choi, G. M. Lee, and N.Crespi. QoS Mapping over Hybrid Optical and Wireless Access Networks, First International Conference on Evolving Internet 2009, INTERNET '09, 2009.

[8] R. Q. Shaddad, A. B. Mohammad, A. A. M. Al-Hetar, and S. M. Idrus. Performance Analysis of Hybrid Optical-Wireless Access Network Physical Layer, Jurnal Teknologi, 55 (Sains & Kej.), pp. 107-117, 2011.

[9] J. Freitag and N. L. S. da Fonseca. Uplink Scheduling with Quality of Service in IEEE 802.16 Networks, Proceedings in IEEE Global Telecommunications Conference 2007.

[10] J. F. Borin and N. L. S. da Fonseca. Scheduler for IEEE802.16 Networks, IEEE Communications Letters, Vol. 12, No. 4, April 2008.

[11] K. Khairi, Z. A. Manaf, D. Adriyanto, M. S. Salleh, Z. Hamzah, and R. Mohamad, "CWDM PON System: Next Generation PON for Access Network", Proceedings in IEEE 9th Malaysia International Conference on Communications (MICC) 2009, pp. 765-768, 2009.

[12] IEEE Standard for Local and Metropolitan Area Networks – Part 16: Air Interface for Fixed Broadband Wireless Access Systems, IEEE Std., Rev. IEEE Std. 802.16-2004, 2004.

[13] P. Seeling, M. Reisslein and B. Kulapala. Network Performance Evaluation Using Frame Size and Quality Traces of Single-Layer and Two-Layer Video: A Tutorial, IEEE Communications Surveys and Tutorials, vol. 6, no. 2, 2004.

[14] M. Ali and D. Pennickx. Provisioning Method for Bit-Rate-Differentiated Services in Hybrid Optical Networks, Photonic Network Communications, 7:1, pp. 59-76, 2004.

[15] F. J. J. Alsolami. Channel-aware and Queue-aware Scheduling for Integrated WiMAX and EPON, Faculty of Engineering Theses and Dissertations, Electronic Theses and Dissertations, University of Waterloo, Master of Applied Science, 2008.

Resource Management Scheme Based on Position and Direction for Handoff Control in Micro/Pico-Cellular Networks

Dong Chun Lee
R&D Center, Bluwise Inc., Daejeon, Korea
ldch22@hanmail.net

Kuinam J. Kim
Dept. of Convergence Security, Kyonggi Univ., Korea

Jong Chan Lee
Dept. of Computer Information Eng., Kunsan National Univ., Korea
chan2000@knu.ac.kr

Abstract. We propose a handoff control scheme to accommodate mobile multimedia traffic based on the resource reservation procedure using the direction estimation. This proposed scheme uses a novel mobile tracking method based on Fuzzy Multi-Criteria Decision Making (FMCDM), in which uncertain parameters such as Pilot Signal Strength (PSS), the distances between the Mobile Terminal (MT) and the Base Station (BS), the moving direction, and the previous location are used in the decision process using the aggregation function in fuzzy set theory. In performance analysis, our proposed scheme provides a better performance than the previous schemes.

Keywords: Handoff control, FMCDM, position and direction method.

1 Introduction

With the proliferation of various wireless services such as forthcoming Fifth Generation (5G), Wireless Local Area Network (WLAN), and Personal Area Network (PAN), etc, next generation wireless communication systems are considered to support various types of high-speed multimedia traffic with packet switching at the same time. To do that, more upgraded Quality of Services (QoS) and system capacity are needed. Due to the limitations of the radio spectrum, the next generation wireless networks will adopt micro/pico-cellular architectures for various advantages including higher data throughput, greater frequency reuse, and location information with finer granularity. In this environment, because of small coverage area of micro/pico-cells, the handoff rate grows rapidly and fast handoff support is essential [1].

These limitations call for the development of new frameworks and approaches to meet the specific challenge of supporting adaptive QoS in a controlled manner, despite frequent and random mobility of the Mobile Terminal (MT) and the dynamically changing network resource availability. Adaptive QoS issues relate to how handoff must be managed and carefully controlled in order to minimize session

K.J. Kim (ed.), *Information Science and Applications,*
Lecture Notes in Electrical Engineering 339, DOI 10.1007/978-3-662-46578-3_2

dropping due to insufficient resource available in the new cell [2]. As MTs move around, the network must continuously track them down and discover their new locations in order to be able to deliver data to them. Especially wireless radio resources availability varies frequently as MTs move from one access point to another [2-3].

Majority of the previous schemes to support mobility make a reservation for resources in adjacent cells [4-5]. The reserved resource approach offers a generic means of improving the probability of successful handoffs by simply reserving the corresponding resources exclusively for handoff sessions in each cell. The remaining resource can be equally shared among handoff and new sessions. The penalty is the reduction in the total carried traffic load due to the fact that fewer radio resources are granted to new sessions. Also, these techniques cause a waste of resources since it is regardless of the direction of MTs. Therefore, previous methods for predicting and reserving resources for future handoff sessions do not seem to be suitable for mobile multimedia networks. The amount of resources required to successfully perform handoff may vary arbitrarily over a wide range in mobile multimedia networks [6-12].

2 Defining Location in FMCDM

Fig. 1 shows how our scheme divides a cell into many blocks based on the signal strength and then estimates the optimal block stepwise where the MT is located using the FMCDM.

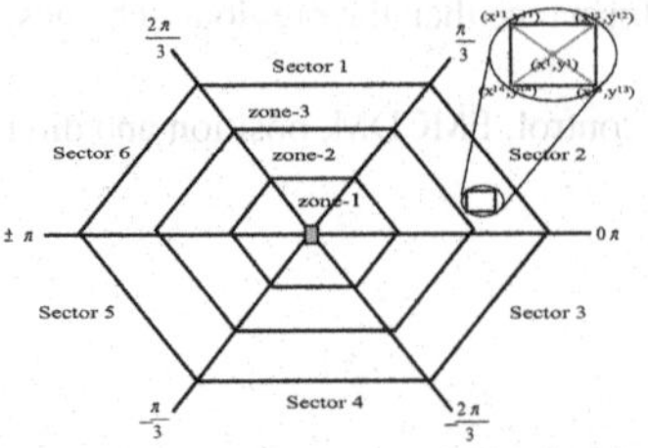

Fig. 1 Sector, Zone and Block

The location of a MT within a cell can be defined by dividing each cell into sectors, zones and blocks and relating these to the signal level received by it at that point. It is done automatically in three phases of sector definition, zone definition and block definition. Then the location definition block is constructed with these results. They are performed at the system initialization before executing the location estimation. The sector definition phase divides a cell into sectors, and assigns a sector number to blocks belonging to each sector. The zone definition phase divides each sector into zones, and assigns a zone number to blocks belonging to each zone. The block definition phase assigns a block number to each block. In order to indicate the location of each block within a cell, 2-dimensional vector (i.e., distance d, angle a) is assigned to each block. After the completion of this phase each block has a set of block information. Three classified zones are used to predict the handoff probability of MTs as shown in the Fig. 1. One cell is consists of zone-1 as a base station coverage, zone-2 as a candidate coverage, and zone-3 as a handoff coverage. However, although a MT moves within the same zone, MT's neighboring cells can be changed. Therefore, to identify them, each zone is partitioned in to six directions which can be easily archived by the pilot signal strength. The sectors are used to know that MT's moving direction is changed. Based on the above location definition method, each block is given his location information which consists of one center point (xi, yi) and four of area point (xij, yij) as shown in Fig. 1. By comparing the location information of each block and the position information of the MT, the position of the MT within a cell is estimated. The set of block information is called the block object. The block object contains the following information: the sector number, the zone number, the block number, the vector data (i.e., distance d, angle a), the maximum and the minimum values of the average Pilot Signal Strength (PSS) for the Line of Sight (LOS) block.

3 Mobile Tracking in FMCDM

3.1 Fuzzy multi-criteria decision parameters

In our paper, the received signal strength, the distance between the MTs and the BS, the previous location, and the moving direction are considered as decision parameters. The received signal strength has been used in many schemes, but it has very irregular profiles due to the effects of radio environments. The distance is considered because it can explain the block allocation plan; however, it may also be inaccurate due to the effect of multi-path fading, etc. It is not sufficient by itself. We consider the previous location. It is normally expected that the estimated location should be near the previous one. Therefore, if the estimated location is too far from the previous one, the estimation may be regarded as inaccurate. We also consider the moving direction. Usually the MT is most likely to move forward, less likely to move rightward or leftward, and least likely to move backward more than one block. The low-speed MT (i.e., a pedestrian) has a smaller moving radius and a more complex moving pattern, while the high-speed MT (i.e., a motor vehicle) has a larger radius and a simpler pattern.

In mobile tracking using FMCDM, the decision function D is defined by combining the degree of satisfaction for multiple evaluation parameters, and the decision is made on the basis of his function. The evaluation parameter can be seen as a proposition. A compound proposition is formed from multiple evaluation parameters with a connective operator, and the total evaluation is performed by totaling the values for the multiple parameters with connective operators. In this method errors in the evaluation parameters impose milder changes on the total evaluation value than in binary logics. This method can also consider multiple inaccurate and insufficient evaluation parameters simultaneously and can compensate for them. This results in the optimal decision.

3.1 Membership function

The membership function with a trapezoidal shape is used for determining the membership degree of the MT because it provides a more versatile degree between the upper and the lower limits than the membership function with a step-like shape. Let us define the membership functions for the PSSs from neighboring BSs. The membership function of PSS_i, $\mu_R(PSS_i)$, is given by Eq. (1). PSS_i is the signal strength received from the BS_i, s_1 is the lower limit, and s_2 is the upper limit. 2 *PSS* values are used for forming the membership function with a trapezoidal shape and for determining the membership degree of estimated signal strength.

$$\mu_R(PSS_i) = \begin{bmatrix} 0, & PSS_i < s_1 \\ 1 - \dfrac{PSS_i - s_1}{|s_2 - s_1|}, & s_1 \le PSS_i \le s_2 \\ 1, & PSS_i > s_2 \end{bmatrix} \tag{1}$$

Now we define the membership function of the distance. The membership function of the distance $\mu_R(D_i)$ is given by Eq. (2), where D_i is the distance between the BS_i and the MT [13]. Also, d_1 is the upper limit, and d_2 is the lower limit.

$$\mu_R(D_i) = \begin{bmatrix} 1, & D_i < d_1 \\ 1 - \dfrac{|D_i - d_2|}{|d_1 - d_2|}, & d_1 \le D_i \le d_2 \\ 0, & D_i > d_2 \end{bmatrix} \tag{2}$$

The membership function of the previous location of the MT $\mu_R(L_i)$ is given by Eq. (3), where L_i is its current location, $E_1, \cdots, E_4$ is the previous location [14]. The previous location is defined by comparing the vector information of previous block and that of the estimated block. These 4 values are used for forming the membership function with a trapezoidal shape and for determining the membership degree of estimated current location as shown in the following figure. In other words, if a current location estimated is too far away from previous location (that is, if the membership degree is too small), it is likely that we have incorrect location estimated.

$$\mu_R(L_i) = \begin{bmatrix} 0; & L_i < E_1 \\ 1 - \dfrac{L_i - E_1}{E_2 - E_1}, & E_1 \le L_i \le E_2 \\ 1, & E_2 \le L_i \le E_3 \\ 1 - \dfrac{L_i - E_3}{E_4 - E_3}, & E_3 \le L_i \le E_4 \\ 0, & L_i > E_4 \end{bmatrix} \tag{3}$$

The membership function of the moving direction $\mu_R(C_i)$ is given by Eq. (4). C_i is the moving direction of the MT, $PSS_1, \cdots, PSS_4$ is the PSS, and o_i the physical difference between the previous location and the current one. These 4 values are used for forming the membership function with a trapezoidal shape and for determining the membership degree of estimated moving direction. The moving direction C_i is defined by comparing the vector information of previous block and that of the estimated block.

$$\mu_R(C_i) = \begin{bmatrix} 0, & C_i < PSS_1 \\ 1 - \dfrac{C_i - PSS_1}{PSS_2 - PSS_1}, & PSS_1 \le C_i \le PSS_2 \\ 1, & PSS_2 \le C_i \le PSS_3 \\ 1 - \dfrac{C_i - PSS_3}{PSS_4 - PSS_3}, & PSS_3 \le C_i \le PSS_4 \\ 0, & C_i > PSS_4 \end{bmatrix} \tag{4}$$

3.2 Location estimation

Most of the FMCDM approaches face the decision problem in two consecutive steps: aggregating all the judgments with respect to all the criteria and per decision alternative and ranking the alternatives according to the aggregated criterion. Also our approach uses this two-steps decomposition [15].

Let J_i (i ∈ {1, 2,..., n} be a finite number of alternatives to be evaluated against a set of criteria K_j (j=1, 2,..., m). Subjective assessments are to be given to determine (a) the degree to which each alternative satisfies each criterion, represented as a fuzzy matrix referred to as the decision matrix, and (b) how important each criterion is for the problem evaluated, represented as a fuzzy vector referred to as the weighting vector.

$$\mu = \begin{bmatrix} \mu_R(PSS_{11}) & \mu_R(D_{12}) & \mu_R(L_{13}) & \mu_R(C_{14}) \\ \mu_R(PSS_{21}) & \mu_R(D_{22}) & \mu_R(L2_{23}) & \mu_R(C_{24}) \\ \mu_R(PSS_{31}) & \mu_R(D_{32}) & \mu_R(L_{33}) & \mu_R(C_{34}) \\ \dots & \dots & \dots & \dots \\ \mu_R(PSS_{n1}) & \mu_R(D_{n2}) & \mu_R(L_{n3}) & \mu_R(C_{nm}) \end{bmatrix} \tag{5}$$

Each decision problem involves n alternatives and m linguistic attributes corresponding to m criteria. Thus, decision data can be organized in $m \times n$ matrix. The decision matrix for alternatives is given by Eq. (5). The weighting vector for evaluation criteria can be given by using linguistic terminology with fuzzy set theory [16]. It is a finite set of ordered symbols to represent the weights of the criteria using the following linear ordering: Very high ≥ high ≥ medium ≥ low ≥ very low. Weighting vector W is represented as Eq. (6).

$$W = (w_i^{PSS}, w_i^{D}, w_i^{L}, w_i^{C}) \tag{6}$$

3.2.1 Sector estimation based on multi–criteria parameters

The decision parameters considered in the Sector Estimation step are the signal strength, the distance and the previous location. The MT is estimated to be located at the sector neighboring to the BS whose total membership degree is the largest. The sector estimation is performed as follows.

Procedure 1: Membership degrees are obtained using the membership function for the signal strength, the distance and the previous location.

Procedure 2: Membership degrees obtained in Procedure 1 for the BS neighboring to the present station are totalized using the fuzzy connective operator as shown in Eq. (7).

$$\mu_i = \mu_R(PSS_i) \cdot \mu_R(D_i) \cdot \mu_R(L_i) \tag{7}$$

We obtain Eq. (8) by imposing the weight on μ_i. The reason for weighting is that the parameters used may differ in their importance.

$$\omega\mu_i = \mu_R(PSS_i) \cdot W_{PSS} + \mu_R(D_i) \cdot W_D + \mu_R(L_i) \cdot W_L , \tag{8}$$

Where W_{PSS} is the weight for the received signal strength, W_D for the distance, and W_L for the location. Also $W_{PSS} + W_D + W_L$ is 1, W_{PSS} is 0.5, and W_D and W_L are 0.3, 0.2, respectively.

Procedure 3: Blocks with the sector number estimated are selected from all the blocks within the cell for the next step of the estimation. Selection is done by examining sector number in the block object information.

3.2.2 Zone estimation based on multi-criteria parameters

The decision parameters considered in the zone estimation step are the signal strength, the distance and the moving direction. From the blocks selected in the sector estimation step, this step estimates the zone of blocks at one of which the MT locates using the following algorithm.

Procedure 1: Membership degrees are obtained using the membership function for the signal strength, the distance and the moving direction.

Procedure 2: Membership degrees obtained in Procedure 1is totalized using the fuzzy connective operator as shown in Eq. (9).

$$\mu_i = \mu_R(PSS_i) \cdot \mu_R(D_i) \cdot \mu_R(C_i) \tag{9}$$

We obtain Eq. (10) by imposing the weight on μ_i.

$$\omega\mu_i = \mu_R(PSS_i) \cdot W_{PSS} + \mu_R(D_i) \cdot W_D + \mu_R(C_i) \cdot W_C \tag{10}$$

Where W_{PSS}, W_D, and W_C are assumed to be 0.6, 0.2, 0.2, respectively.

Procedure 3: Blocks which belong to the zone estimated above are selected for the next step. It is done by examining the zone number of the blocks selected in the sector estimation.

3.2.3 Zone estimation based on multi-criteria parameters

The decision parameters to be considered in the block estimation step are the signal strength, the distance and the moving direction. From the blocks selected in the zone estimation step, this step uses the following algorithm to estimate the block in which the MT may be located.

Procedure 1: Membership degrees are obtained using the membership function for the signal strength, the distance and the moving direction.

Procedure 2: Membership degrees obtained in procedure 1 are totalized using the fuzzy connective operator as shown in Eq. (11).

$$\mu_i = \mu_R(PSS_i) \cdot \mu_R(D_i) \cdot \mu_R(C_i). \tag{11}$$

We obtain Eq. (12) by imposing the weight on μ_i.

$$\omega I_i = \mu_R(PSS_i)\cdot W_{PSS} + \mu_R(D_i)\cdot W_D + \mu_R(C_i)\cdot W_C \qquad (12)$$

Where W_{PSS}, W_D, and W_C are assumed to be 0.6, 0.1, 0.3, respectively.

Procedure 3: The selection is done by examining the block number of the blocks selected in the zone estimation

4 Resource Reservation and Allocation in Direction Prediction

We have a resource management structure to efficiently accommodate multimedia sessions, in which **rs-part** can be reserved only for real-time sessions and temporarily occupied by non-real-time sessions, **ns-part** can be used only for non-real-time sessions and **ss-part** can be reserved by real-time sessions and temporarily occupied by non-real-time sessions depending on the needs. The size of **ss-part** is adapted according to the transmission rate of real-time sessions. A boundary line between two parts is decided on **ss-part**. Namely, the **ss-part** is temporarily occupied by non-real-time sessions during a time interval T_i and is updated at T_{i+1}. The amount of the reserved resources for each session can be adjusted periodically and allocated only to same applications. The basic rule for resource allocation is to allocate the resources corresponding to the minimum transmission rate to all sessions.

Each session needs to have the primary resources, which are a part of reserved resources in real-time sessions and corresponds to Minimum Transmission Rate (MiTR) in non-real-time sessions. If the allocation of the primary resources should not be done, a session request is blocked. If the transmission rate of the primary resources allocated to a real-time session is less than Average Transmission Rate (AvTR), the supplementary resources are allocated and reserved for the session. In case of non-real-time sessions the primary resources corresponding to MiTR are allocated and the supplementary resources are temporarily allocated on the basis of the bursts of real-time sessions.

The BS reserves only the resources corresponding to the minimum transmission rate to the MT. Based on the location and the direction of the MT within a cell, the resource reservation is performed with the following orders: unnecessary state, not necessary state, necessary state, and positively necessary state. If the reservation variable for the MT is changed, the reservation is canceled and the resources have to be released with the reverse order and returned to the pool of available resource.

5 Performance Evaluation

5.1 Simulation Model

The simulation model is based on a B3G system proposed from ETRI [17] as shown in Table 1, which is implemented using MOBILESimulatorV6. The simulation model composed of a single cell, which will keep contact with its six neighboring cells. Each cell contains a BS, which is responsible for the session setup and tear-down of new applications and to serve handoff applications. The moving path and the

MT velocity are affected by the road topology. The moving pattern is described by the changes in moving direction and velocity. In our study we assume that the low speed MTs, the pedestrian, occupy 60% of the total population in the cell and the high-speed MTs, the vehicles, 40%. Vehicles move forward, leftward/rightward and U- Turn. The moving velocity is assumed to have the uniform distribution. The walking speed of the pedestrian is 3~5 *Km/hr*, the speed of the private car and the Taxi 50~120 Km/hr, and the bus 20~90 Km/hr. The speed is assumed to be constant during walking or driving. Fig. 3 shows the blocks and moving path on the road within a square cell. The black circle indicates the branch of the road, and the shaded areas are blocks that the road passes through. Each block is a square and its side is assumed to have the length of 30 m. The time needed for a high speed MT to pass through a block is calculated from the crossing time of a block $BT = r / \upsilon$ where r is the length of a road segment crossing at each block and v is the MT speed. As shown in Fig. 4, the crossing time of a block is dependent on r. In order to reflect more realistic information into our simulation, it is assumed that the signal strength is sampled every 0.5 sec, 0.2 sec, 0.1 sec, 0.1 sec and 0.04 sec for the speed of 10km/h, 20km/h, 50km/h, 80km/h and : 120km/h, respectively.

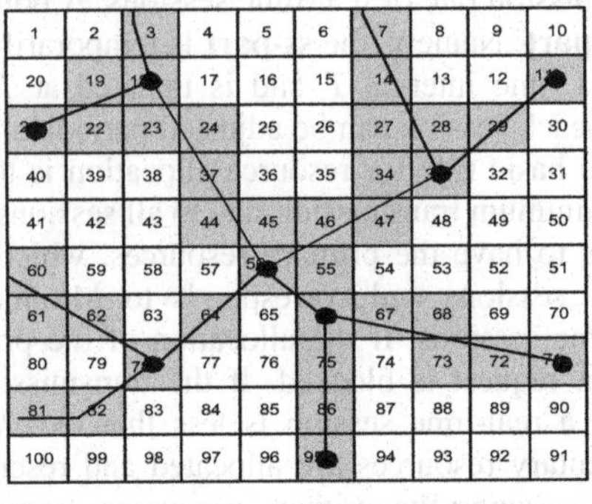

Fig. 3 Blocks and moving path on the road

We consider the following simulation parameters regarding the received signal strength. The mean signal attenuation by the path-loss is proportional to 3.5 times the propagation distance, and the shadowing has a log-normal distribution with a standard deviation of $c = 6dB$.

5.2 Numerical Results

The results of from Fig. 4 to Fig .5 show the comparison between the proposed Directional Reservation scheme and the previous resource reservation schemes that mobile tracking is not applied. Assuming that thirty sessions with MiTR are accepted regardless of service type, Fig. 4 shows the results for an average percentage of resource utilization as a function of sessions arrival with priority to handoff sessions over new sessions, in which the resources occupied by the non-real-time sessions are not considered as an appraised target that is how many non-real-time sessions is multiplexed to the reserved resources by the real-time sessions. The resource utilization for the proposed Directional Reservation scheme is increased up to 25% at average and peak arrivals, than that for Fixed Reservation method and Statistic Reservation method. In Fig. 5, the comparison of transmission delay of the three schemes is plotted against the number of sessions. It is observed that as the number of sessions increase, the proposed Directional Reservation scheme provides a noticeable improvement over the conventional schemes for real-time sessions, while slightly degrading the performance for the non-real-time sessions.

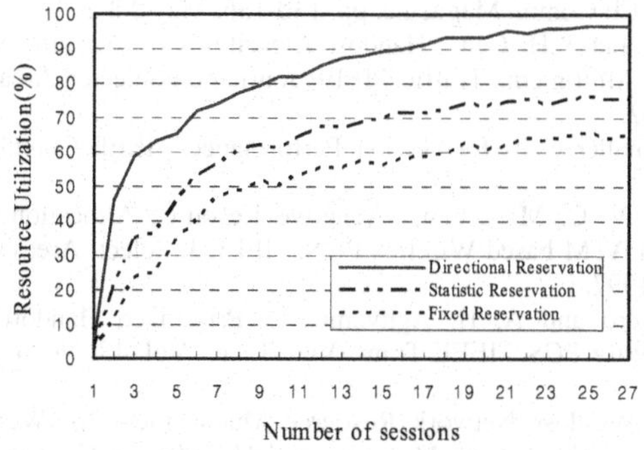

Fig. 4 Comparison of resource utilization

In case that traffic load is over 20, results demonstrate that the delay of the proposed scheme has decreased to about 300ms and 500ms as compared to the Statistic Reservation scheme and Fixed Reservation scheme, respectively.

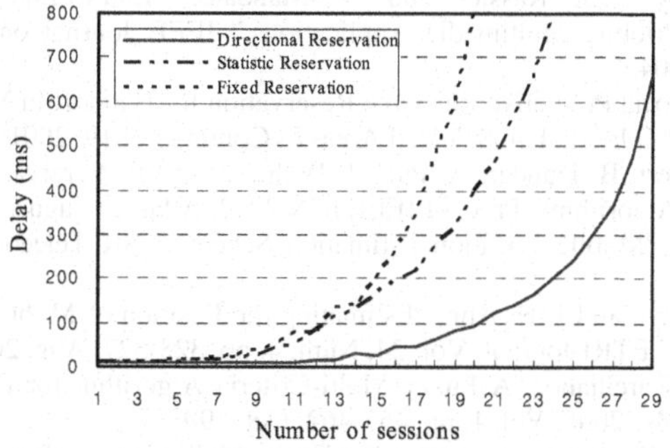

Fig. 5 Comparison of transmission delay

6 Conclusions

This paper is to address the problem of guaranteeing an acceptable level of QoS requirements for MT users as they move from one location to another. This is achieved through reservation variables such as the current location and the moving direction that is presented with a set of attributes that describes the user mobility. In this scheme, MTs are classified according to their reservation variables.

The proposed scheme shows a great improvement of the resource utilization and the dropping rate of handoff sessions. It is because our resource reservation scheme is more adaptive than previous schemes. In the proposed scheme, radio resources are classified as ones having priority to the new sessions and ones having priority to the handoff sessions based on reservation variable. Using this, we improve the dropping rate for the handoff sessions by dynamically adjusting the amount of the reserved resources according to the amount of occupied resources.

References

1. Yu Cheng and Weihua Zhuang, "Diffserv. Resource Allocation for Fast Handoff in Wireless Mobile Internet, IEEE Comm. Magazine, pp. 130-136, May 2002.
2. W. Ju, J.C.L. Liu, and Y.H. Cen, "Handoff Algorithms in Dynamic Spreading WCDMA System Supporting Multimedia Traffic," IEEE Journal on Selected Areas in Comm., vol.21, 2003, pp.1652-1662.
3. W. C. Y. Lee, "Smaller Cells for Greater Performance," IEEE Comm.. Mag., pp. 19–23, Nov. 1991.
4. O. T. W. Yu and V. C. M. Leung, "Adaptive Resource Allocation for Prioritized Call Admission over an ATM-based Wireless PCN," IEEE J. Select. Areas Comm., vol. 15, pp. 1208–1225, Sept. 1997.
5. L. Ortigoza-Guerrero and A. H. Aghvami, "A Prioritized Handoff Dynamic Channel Allocation Strategy for PCS," IEEE Trans. Veh..Tech., Vol. 48, No. 4, pp. 1203–1215, Jul. 1999.
6. B. Shafiq et al, "Wireless Network Resource Management for Web-based Multimedia Document Services," IEEE Comm. Magazine, vol.41, 2003, pp.138-145.
7. W. Mohr, "Further Developments beyond Third Generation Mobile Communications," International Conference on Communication Technology Proceedings, vol.2, 2000.
8. S. Laha et. al., "Evolution of Wireless Data Services: IS-95 to cdma2000," IEEE Comm. Magazine, Oct. 1998.
9. T. Guenkova-Luy, A.J. Kassler and D. Mandato, "End-to-End Quality-of-Service Coordination for Mobile Multimedia Applications," IEEE Journal on Selected Areas in Comm., vol.22, 2004.
10. T. Zhang et al, "Local Predictive Resource Reservation for Handoff in Multimedia Wireless IP Networks," IEEE Journal on Selected Areas in Comm., vol.19, 2001.
11. M. Ergen, S. Coleri, B. Dundar, A. Puri, J. Walrand, and P. Varaiya, "Position Leverage Smooth Handoff Algorithm", Proc. of IEEE ICN 2002, Atlanta, August 2002.
12. J. C. Lee et. al., "Mobile Location Estimation Scheme," SK Telecom. Review, Vol. 9, Dec.1999.
13. J. C. Lee, B. Y. Ryu and J. H. Ahn, "Estimating the Position of Mobiles by Multi-Criteria Decision Making," ETRI Journal, Vol. 24, Num. 4, pp. 323-327, Aug. 2002.
14. C. Naso and B. Turchiano, "A Fuzzy Multi-Criteria Algorithm for Dynamic Routing in FMS," IEEE ICSMC'2008, Vol. 1, pp. 457-462, Oct. 1998.
15. C. H. Yeh and H. Deng, "An Algorithm for Fuzzy Multi-Criteria Decision Making," IEEE ICIPS'2007, pp. 1564-1568, 1997.
16. S. Ku Hwang, "4G Mobile Telecommunications Technology Development Korea," International Forum on Next Generation Mobile Communications, London, May 2012.
17. G. Liu and G. Q. Maguire, Jr., "Efficient Mobility Management Support for Wireless Data Service," in Proc. 45th IEEE Vehicular Technology Conf., pp. 902-906, July 1995.
18. AbdulRahman Aljadhai and Taieb F. Znati, "Predictive Mobility Support for QoS Provisioning in Mobile Wireless Environments," IEEE J. Select. Areas Comm., Vol. 19, No. 10, pp. 1915-1930, Oct. 2001.
19. Dong Chun Lee and J. C. Lee, "Handover Control Method Using Resource Reservation in Mobile Multimedia Networks", IEICE Trans. Comm., VOL.E92–B, NO.8 Aug. 2009

A Pilot Study of Embedding Android Apps with Arduino for Monitoring Rehabilitation Process

Safyzan Salim[1,2,3], Wan Nurshazwani Wan Zakaria[2] and M. Mahadi Abdul Jamil[2]

[1]Communication Technology Section, UniKL-British Malaysia Institute, Bt 8 ¾ Sg Pusu, 53100 Gombak, Selangor, Malaysia,
[2]Department of Electronic Engineering, Biomedical Modeling & Simulation Research Group,
Universiti Tun Hussein Onn Malaysia, Parit Raja 86400, Batu Pahat, Johor
[3]Kolej Kemahiran Tinggi MARA, Ledang, Johor
safyzan@unikl.edu.my, {shazwani, mahadi}@uthm.edu.my

Abstract. This paper proposes the monitoring of a post-stroke rehabilitation activities via Android smart phones. The study focuses on designing the hardware, developing the software and simulating the results in interactive manner. The result is documented for the purpose of post-processing and progressive status tracking. Furthermore, the interactive output may motivate the patients to keep on using this system for rehabilitation. The subject needs to wear a set of sensors over the palm while performing a few basics arm movement. The data will be converted into series of readable data and then transferred to the smart phone via Bluetooth. The experiment demonstrates the capabilities of the sensors to produce information and also the Android Apps in responding such hand movement activities. It is believed that the system offers more information than conventional method and also the ability to improve training quality, results and patients progress. For initial proof of concept, the system will be tested to a healthy normal subject.

Keywords: Post-stroke, hemiparetic-arm, rehabilitation, tilt, android, arduino

1. Introduction

Post-stroke patients in Malaysia always turn to traditional massage and medicine as the alternative treatment [1], [2], [3]. The easy access, relatively cheap, convenient and less hassle compared to the one provided in the hospital that makes them go for the traditional. The question is how to measure the success of the traditional treatment? Will there be any record or data logged for its progress? How about the potential side effects produced by the herbs which are yet to be scientifically proven? [4].

2. Literature Review

Studies have been made in rehabilitation of stroke patients using gyroscopes and accelerometer [5]. They have discussed the better way to covey the data to the computer for post-processing, how to correspond effectively between the sensors and the microcontroller and so on. A problem is raised for this system. The only person able to study the result is the physicist himself, furthermore it is not meant to be used at home.

A web-based system for stroke patients have been developed which allow the result to be transmitted real time from the patient's home to their physicists over the Internet [6]. The chances of packet loss during the therapy may happen due to instability connection, resulting incorrect analysis by the therapists.

Another related contribution was done by R. Ambar, M. S. Ahmad, A. M. Mohd Ali and M. M. Abdul Jamil [7]. We notices that there are a few similarities between these systems with ours. However, it is not comfortable to wear since too many sensor to hook up and a tedious to wear.

A study has been done before, but only limit to a glove and a microcontroller with accelerometer [8]. The analysis done is just based on Excel.

Thus, a wearable sensor which will able to display the result during rehabilitation process, user-friendly and interactive, need to be considered.

The proposed system is secured to a hand glove. The wearable sensor consist of a low-cost single-board microcontroller i.e., to handle data flow from the sensors; tilt sensors: which will detect the hand movement, tilting either left or right. The system applies Bluetooth as a medium to transmit the data to smart phone. The data will be used as the evidence of the progress for rehabilitation process.

This paper reports the development of an affordable interactive wearable device which is capable to capture the data produced during the post-stroke rehabilitation session for monitoring the progress of the patient.

K.J. Kim (ed.), *Information Science and Applications*,
Lecture Notes in Electrical Engineering 339, DOI 10.1007/978-3-662-46578-3_3

3. Experimental Methods

The experimental methods have been composed of three exercises which is part of the rehabilitation treatments. The experiment will determine the ability of post stroke patients to do three movements as in Table 1.

Table 1. Selected Exercises

Experiments	Execution Of Experiments
Right Tilt	Able to tilt the palm clock-wise
Flat	Able to flatten the palm after tilting to right
Left Tilt	Able to tilt the palm counter clock-wise

In this study, the exercises were done by healthy person. The measured person also tries to simulate post-stroke movements during the suggested exercises.

4. Development of The Project

There are four elements of the system: the microcontroller, the tilts sensors, Bluetooth module and a smart phone that running Android OS.

4.1 Arduino Uno

Arduino Uno is a microcontroller board based on the ATmega32. It has 14 digital input/output pins (of which 6 can be used as Pulse Width Modulation outputs), 6 analog inputs, a 16 MHz ceramic resonator, a Universal Serial Bus (USB) connection, a power jack, an In-circuit Serial Programming (ICSP) header, and a reset button. It contains everything needed to support the microcontroller; simply connect it to a computer with a USB cable or power it with a AC-to-DC adapter or battery to get started

4.2 HC06 Bluetooth Module

This Serial Bluetooth brick is easy to use module. It designs for transparent wireless serial connection setup. Serial port Bluetooth module is fully qualified Bluetooth V2.0+EDR (Enhanced Data Rate) 3Mbps Modulation with complete 2.4GHz radio transceiver and baseband. It uses CSR Bluecore 04-External single chip Bluetooth system with CMOS technology and AFH (Adaptive Frequency Hopping Feature). Hope it will simplify your overall design/development cycle. Connect the + pin to 5v and - pin to GND, TX RX pin connect to the serial pins.

4.3 Tilt Sensor Module

The tilt sensor module come with the basic components for operation. Supplying power and it is good to be use. Attach it to object and it will detect whether the object is tilted. Simple usage as it is digital output, so you will know the object is tilted or not by reading the output. It uses SW-460D or SW-520D tilt sensor. The tilt sensor is ball rolling type, not Mercury type.

Comes with a M3 mounting hole for ease of attaching it to any object. On board it provide a tilt switch, high sensitivity and commonly being used for tilt detection. The module comes with power LED and status LED for visual indicator.

4.4 Prototype

The final prototype setup showed the device that was attached to a glove that contains a microcontroller with 2 tilt sensors connected. Once the data been processed in microcontroller, it will then pass to smart phone thru Bluetooth module. Then, with the help of Basic4Android, a simple apps was developed whereas it is able to show the output. The prototype can be seen in Fig. 1.

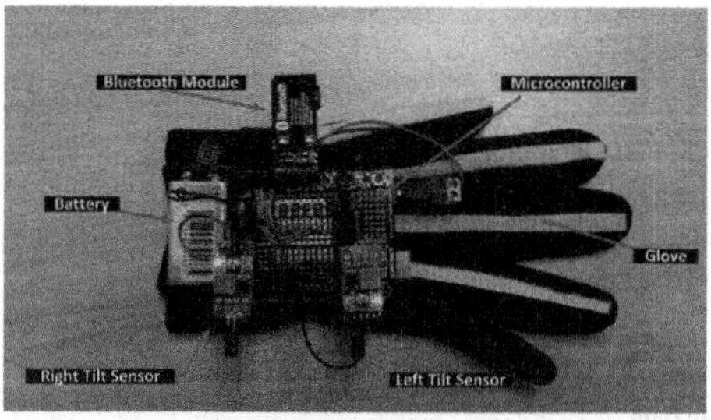

Fig.1. The prototype Wearable Data Acquisition System

5. Results And Discussion

From the experimental works, the followings are the recorded measurement for the developed device which has been segmented into five parts.

5.1 Software Analysis

Before developing the Android Apps, an Arduino sketch has been constructed and tested. The software must be able to detect the tilting movement and document it over Serial Monitor pop-up Window. Series of tests were done in order to get the best setting for tilt sensor especially for *No Movement* mode. Figure 2 document the result of tilt sensors.

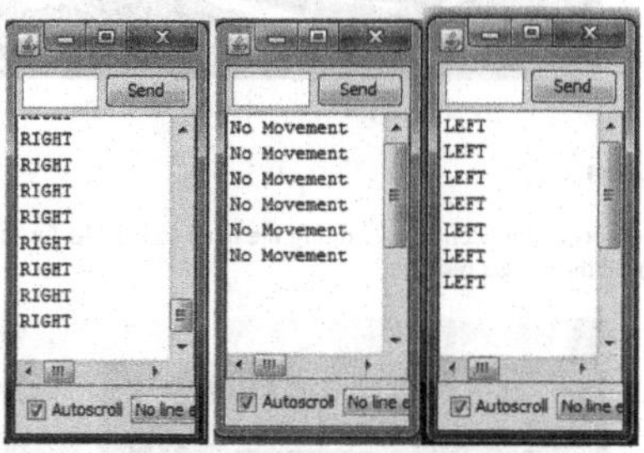

Fig. 2. The responds of tilt sensors

5.2 Android Apps

Figure 3 represent the caption of tilt apps was written using Basic4Android. Once the system is triggered, the system will pick up the signal transmitted from the glove and will respond accordingly to box labelled with ***'a'***, '***b***' or ***'c'***.

Box labelled with '***b***' or ***'c'*** will pop up if the patient able to tilt his hand either to the left or right. Box labelled with 'd' is to check and confirm the data receive from glove. By having such apps, the progress of every treatment will be more interactive and easier.

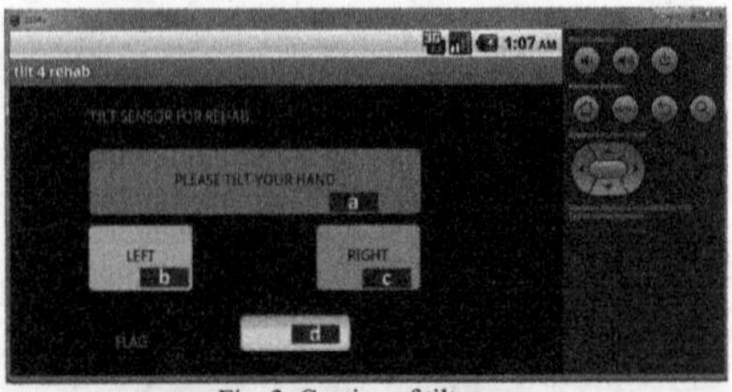

Fig. 3. Caption of tilt apps

5.3 Exercise 1: Tilting to right

Exercise 1 is the activity of tilting the palm to the right. It can be foreseen that such activity responded accordingly as expected. The moment the subject tilt, the apps responded quickly. Figure 4 shows the activity.

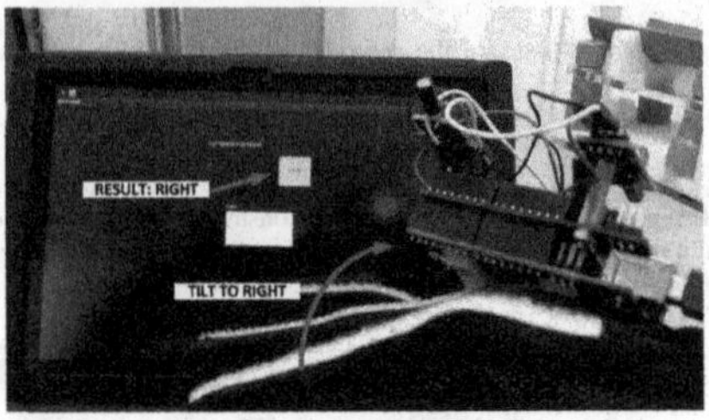

Fig. 4. Basic movement ability: Tilt to right

5.4 Exercise 2: Tilting to left

Figure 5 is the results for the second modeling, i.e., tilting the hand to left. As for this experiment, it is proven that the glove interacted with apps and produced the correct respond.

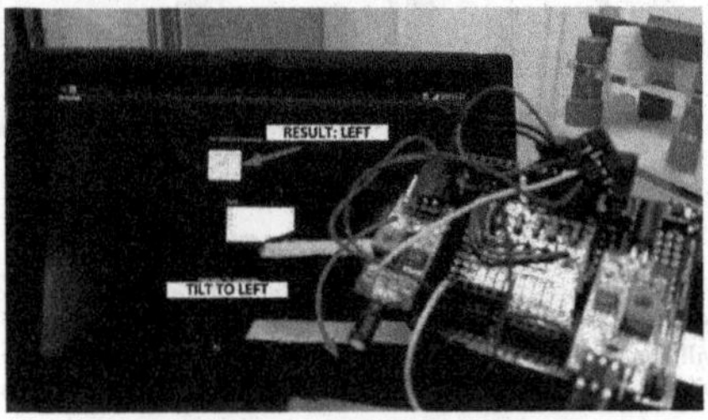

Fig. 5. Basic movement ability: Tilt to left

5.5 Plotting Ability

While the Part 5.3 and Part 5.4 shows the result which able to be understood without any scientific study, the latter shows the clinical results. This system has the ability to graph the motion variation and can be interpreted in terms of clinical. Figure 6 proved the responds.

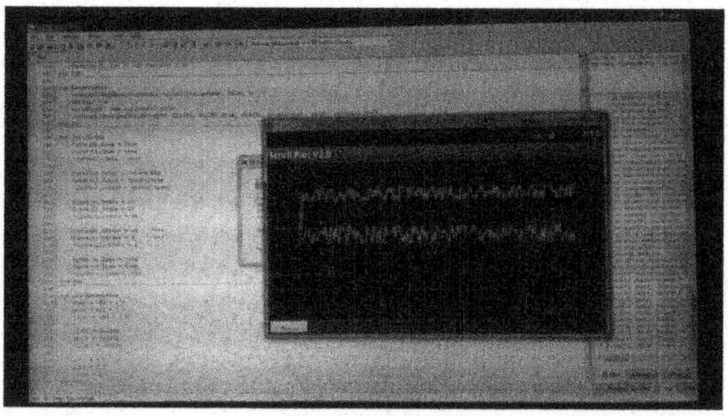

Fig. 6. Plotting the motion variation

6. Conclusion

In this paper, we have presented and evaluated the potential of applying tilts sensors in rehabilitating post stroke patients by introducing the element of open source. Experimental result shows that the prototype system successfully responded accordingly. By having such apps, it may motivate the patients since it works interactively with the glove. Furthermore, the ability to plot the graph can be considered a good achievement. Perhaps later on, the tilts apps will be in the form of games.

Based from the experiments, it can be seen that the application of this device can be extended to other area of physical rehabilitation such as physical movement and gait analysis. Furthermore, such system may also be used in evaluation of athlete's performance thus, the contribution of this study towards Sports Technology cannot be understated.

References

[1] Anuar H.M., Fadzil F., Ahmad N., Abd Ghani N. and Sallehuddin M.S., "Urut Melayu for post-stroke patients: a qualitative study," National Committee for Clinical Research 2010. Available at http://www.nccr.gov.my/view_file.cfm?fileid =25 (accessed 7 Sept 2013).

[2] Wu B., Liu M., Liu H. et al, "Meta-analysis of traditional Chinese patent medicine for ischemic stroke," Stroke, vol. 38, no. 6, 2007, pp. 1973-1979.

[3] Jamal J.A., "Malay traditional medicine an overview of scientific and technological progress," Asia-Pacific Tech Monitor, vol. 23, 2006, pp. 37-49.

[4] Keat W.L. and Siew H.G., "Burden of stroke in Malaysia," International Journal of Stroke, vol. 7, Feb 2012, pp. 165-167.

[5] Safyzan Salim and M. Mahadi Abdul Jamil, "Monitoring of Rehabilitation Process Via Gyro and Accelerometer Sensor", International Colloquium on Sports Science, Exercise, Engineering & Technology, Apr 2014

[6] H. Zheng, R. J. Davies, N. D. Black, "Web-based monitoring system for home-based rehabilitation with stroke patients", Computer-Based Medical Systems, 2005, pp 419 – 424.

[7] R. Ambar, M. S. Ahmad, A. M. Mohd Ali and M. M. Abdul Jamil, "Arduino Based Arm Rehabilitation Assistive Device," Journal of Engineering Technology, vol. 7, 2011, pp. 5-13.

[8] S. Safyzan, A. B. Badri and M. Mahadi Abdul Jamil, "Exploitation of Accelerometer for Rehabilitation Process", IEEE International Conference on Control System, Computing and Engineering, Nov 2013.

A Security Protocol based-on Mutual Authentication Application toward Wireless Sensor Network

Ndibanje Bruce[1], YoungJin Kang[1], Hyeong Rag Kim [2], SuHyun Park[3], Hoon-Jae Lee[3]

[1]Department of Ubiquitous IT, Graduate School of Dongseo University, Sasang-Gu, Busan 617-716, Korea
[2]Departement of IT & Electronics, Pohang University
[3]Division of Computer and Engineering, Dongseo University, Sasang-Gu, Busan 617-716, Korea
bruce.dongseo.korea@gmail.com
rkddudwls55@gmail.com
hrkim@pohang.ac.kr
{subak,hjlee}@dongseo.ac.kr

Abstract. This paper presents an application of security protocol based-on mutual authentication procedure between involved entities that providing data and network security in wireless sensor environment communication. Through WSN, a user can access base station of wireless sensor networks and gets the data. This paper proposes a secure authentication protocol where involved entities are mutual strongly verified before accessing the data. The proposal provides secure features standards such as data integrity, mutual authentication and session key establishment. Furthermore, an extensive analysis shows that the proposed scheme possesses many advantages against popular attacks, achieves better efficiency and can be safeguard to real wireless sensor network applications.

Keywords: authentication protocol, WSN, sensor network, ad hoc

1 Introduction

Data and network security for wireless sensor network are a primary concern due to the easiest deployment area accessibility of the sensor node devices to collect the data transmitted to the base-station traversing some nodes via RF signals and routing schemes. Via a network it is wirelessly possible to reach each one of the sensor nodes ubiquitously as long as a network terminal is accessible. WSNs are widely used in areas such as military, battlefield, homeland security, healthcare, environment monitoring, agriculture and cropping, manufacturing, measurement of seismic activity etc. Every sensor node has some level of computing power, limited storage, and a small communication module to communicate with the outside world over an ad hoc wireless network. For example, in case of healthcare system, sensitive data of patients such as physiological or physical health parameters are sensed, collected and pro-

K.J. Kim (ed.), *Information Science and Applications*,
Lecture Notes in Electrical Engineering 339, DOI 10.1007/978-3-662-46578-3_4

cessed by small sensor nodes that are placed on patient's body. Furthermore, they are often interacting closely in cooperation with the physical environment and the surrounding people, where such exposure increases security vulnerabilities in cases of improperly managed security of the information sharing among different healthcare organizations. Thus far, authentication protocols schemes have been proposed on the link layer [1, 2] and the network layer [3] and they provide sufficient security in the wireless sensor networks. Meanwhile, protocols based on user authentication on the application layer in [4, 5] have been proposed where the secrets are stored into the gateway-node or base station. In that case, when a user wants to access data through the gateway-node or base station he is authenticated using those protocols. This paper deals with mutual authentication security protocol for sensor network where the user and devices perform mutual authentication before session begins. The proposed protocol ensures both devices and data security among wireless networks. The rest of the paper is organized as follows. In Section 2 we present the proposed authentication protocol. The performance analysis is done in Section 3, before concluding in Section 4.

2 The Proposed Scheme

2.1 Design and basic system architecture

The design of security protocol for data and infrastructures security depend on the designer‘s goal in order to minimize the cost (memory and power of sensor nodes) ensuring the efficiency and security of the concerned entities. The basic system architecture is shown in Figure 1 where a user sends a request to the gateway node/base station through the authentication server (AS) for registration phase. After registration process, the user can login to the wireless sensor network to access data where mutual authentication phase is performing to verify the legitimacy of the entities. In the proposed scheme, we establish our security protocol on two main phases: registration phase, authentication phase consisting of login and verification steps. The basic system architecture components consist of user device (UD), using authentication server (AS) and a gateway node/base station (GWn) which acts as a sink where the user can get the wanted data. Also it consists of sensors node to collect data and send to the gateway node. As given in Figure 1, the user login to the device which verifies the user legitimacy and then transmits the request message to GWn. In the process of our scheme, we require to the user to interact directly with the sink node in order to save the sensor nodes energy and memory while they are performing some computation. Table 1 provides a list of some notations and symbols to be used in this paper.

2.2 Protocol communication process : Registration phase

This sub-section details the process of communication where the user can access the data from the sink node through the WSN using the registration and authentication phases. We assume that all components are honest before starting any session, the compromise might occur after or during communication process.

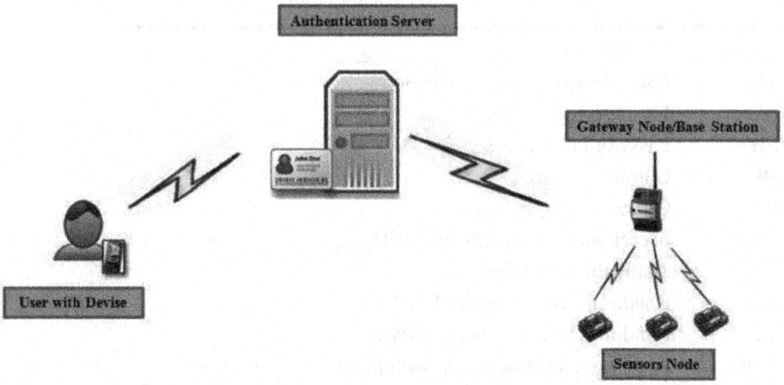

Fig. 1. Design and basic components of the proposed scheme

During the registration phase, the use registers to the *GWn* via the *AS*. The aim of this phase is to negotiate a shared secret key between user device and gateway node for login and authentication phase. The user chooses his *IDud* and *PW*, generates a random number *Xud* and compute *(Xud ⊕PW) ||IDud*. Afterward, the user submits the message to the *AS* for a registration request to the GWn. Up receiving the request message of registration from *UD, AS* checks *IDud (new) =IDud(existing)*, if equal, then reject registration request otherwise, proceed to the next step. *AS* assigns a nonce *Nud* based on user's device. The *AS* also plays the role of third-party to distribute a shared key *SKug* between the *UD* and the *GW*. Then the AS send *Cs= ESKug(Xud ⊕PW) ||IDud|| Nud* to the GWn. While receiving the message from *AS*, the *GWn* generates a secret number *Yg* and compute the following:

- $Ag = h\,(IDud \oplus Yg)$,
- $Pg = g^{h\,(IDud \| (Xud \oplus PW))} \bmod p$

Afterward, the *GWn* personalizes a smart card for the *UD* with the parameters *{Ag, Pg, h (.), g, p, Cs}*. The *GWn* sends the reply message with the above parameters through the *AS* which forward the message to the *UD*. Here, *h (.)* is a collision free one-way function. The *UD* now enters *Xud* into the smart card and it contains *{Ag, Pg, h (.), g, p, Cs}*. The *GWn* store the *IDud* in the table of ids to maintain it for login and authentication steps, this is the end of the registration phase and the procedure flow is illustrated in Figure 2.

2.3 Protocol communication process : Authentication phase

Login Step. This phase is invoked when the *UD* wants to access the data through WSN. In this protocol, the *UD* doesn't communicate with the sensor node. The computation of the mutual authentication consumes a lot of energy's sensor nodes; this why we reserve the mutual authentication phase to the *GWn*. Then, the *UD* input his *IDud* and *PW* and compute the following:

Table 1. Notations and symbols used in this scheme

Notations	Descriptions
ID_{ud}	User identity of login to device
UD	User Device
PW	Password of the user
GWn	Gateway node
ID_G	Gateway ID
Nud	Secret number(as nonce) of UD
AS	Authentication server
Xud	Random secret number of UD
Yg	Random secret number of GWn
EK(m)	Encryption of message m with key *x*
H (.)	One way hash function, e.g., *SHA-1*
$\oplus$,||	Bit-wise *XOR* operation and concatenation function

Step1-LP: Compute $Pg' = g^{h\,(IDud||\,(Xud \oplus PW))}$ *mod p* and verify if *Pg*= *Pg*´if yes, go to the next step if not reject the login request.
Step2-LP: Compute $Vud = g^{h\,(Tud||Nud)}$ *mod p*. Here *Tud* and *Nud* are respectively the timestamp and nonce of the user device. Compute *Qud=h (Vud|| Pg)*
Step3-LP: The user device send the login request message *M1=< Ag, Qud >* to the *GWn*. This is the end of the login step from the *UD* to the *GWn*, the message is sent over a public channel.

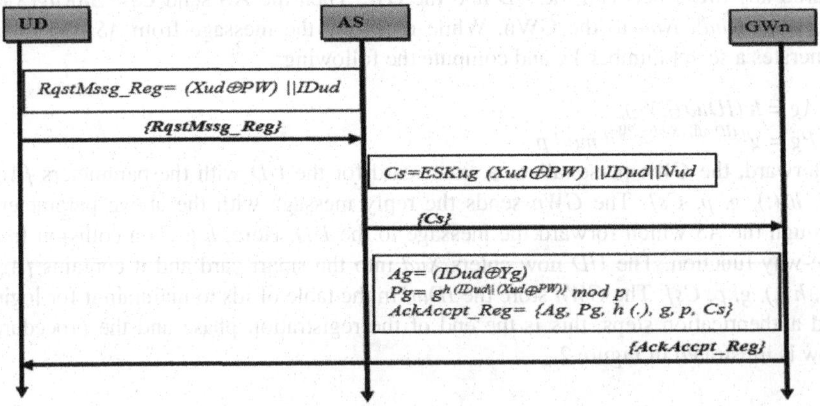

Fig. 2. Registration phase messages flow

Verification step. The verification phase is performed by the *GWn* when he receives the login message request from the *UD* for mutual authentication between the entities as sketched in Figure 3. Upon receiving the login request *M1=< Ag, Qud >* at time *Tg*, the *GWn* authenticates *UD* by the following steps:

Step1-VP: Checks if $(Tg - Tud) \leq \Delta T$ then *GWn* proceeds to the next step, otherwise the step is terminated. Here ΔT shows the expected time interval for the transmission delay and *Tg* is the time stamp of the gateway node.
Step2-VP: From the ids table of the gateway node, the *GWn* verify if *IDud* = *IDud'* if yes, then the gateway node considers it as legitimate user and proceeds to the next step, otherwise, terminates the operations
Step3-VP: The *GWn* generates a nonce *Ng* then calculates *Gg* with the following: $Gg = g^{h\ (Tg||Ng)}\ mod\ p$ and *GWn* compute the session key $SEK = V_{ud}{}^{Yg}\ mod\ p$. Subsequently the gateway node computes $Lg = EKug\ [h\ (IDud)\ ||Tg||ID_G]$ and sends to UD message *M2=<Lg, Ug>* to respond to the login message request in order to process the mutual authentication. Here $Ug = E_{SEK}\ (Gg||Nud)$. After receiving the message *M2* from the *GWn*, the *UD* perform the mutual authentications operations as following:
Step4-VP: The *UD* validates the time *Tg* and check if $(Tud - Tg) \leq \Delta T$ if yes, then continues to the next verification step if not abort.
Step5-VP: From message *M2*, the *UD* decrypts the sub *message* $Ug = DE_{SEK}\ (Gg||\ Nud')$ and check if *Nud'* = *Nud*, also check if $ID_G' = ID_G$ if yes, then continues to the next step if not abort. The *UD* calculate the session key with the knowledge of *Gg* from the decryption of *Ug*, $SEK = Gg^{Nud} mod\ p$.
Step6-VP: After checking every parameter, the *UD* can trust that the *GWn* is the authentic one, and then *UD* sends the last message *M3*, to acknowledge the session key from the gateway. $M3 = E_{SEK}\ (Jud\ ||IDud)$. Here $Jud = h\ ((ID_G)\ ||Ng)$. While receiving the message *M3*, the gateway node performs the followings:
Step7-VP: The *GWn* Computes the session key and decrypts the sub message, obtains Ng' and IDud" .The gateway checks if *Ng'=Ng, IDud"= IDud*, if the conditions are true the *GWn* believes that the *UD* is a legitimate one and it can access the data he wanted otherwise not.
Step8-VP: Furthermore, *UD* and the *GWn* shares the session key *SEK* to perform subsequent operation during a session and the establishment of the session key terminates the authentication phase.

3 Performance Analysis

This section presents the performance analysis evaluation in terms of security and efficiency analysis; we assume that an adversary may intercept *M1, M2,* and *M3*. Also, we assume that an adversary may hack either passwords or steal user device *and* extract secrets, but cannot do both at the same time. As per the current literature, extracting secrets from the smart card memory is quite difficult and some smart card manufacturer companies provide countermeasures against risk of side channel attacks. Based on above assumptions, an attacker may execute certain attacks to breach the proposed protocol.

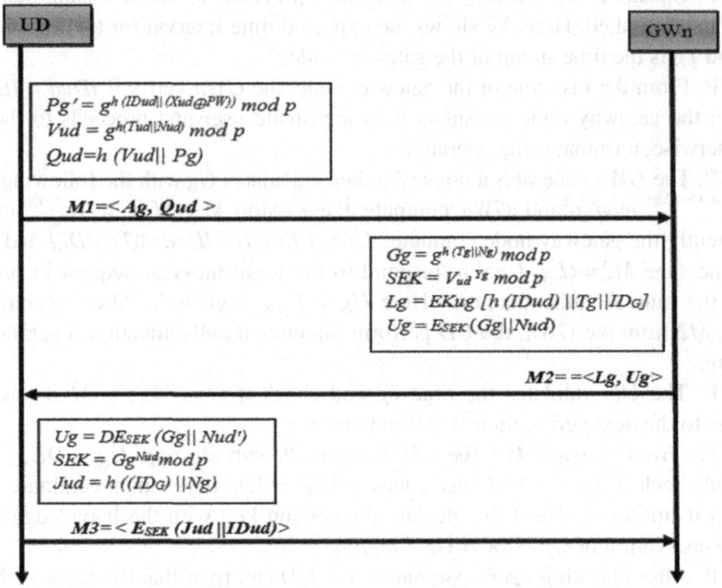

Fig. 3. Authentication phase message flow

3.1 Security Analysis

- *Mutual authentication:* Our protocol provides mutual authentication, between involved entities. For example, in the messages *M2=<Lg, Ug>* and *M3=* E_{SEK} *(Jud ||IDud),* the user device and gateway node achieve the mutual authentication and both them are sure that they are the legitimacies ones.
- *Confidentiality:* The proposed protocol provides adequate confidentiality to their messages (such as, $EKug\ [h\ (ID_{ud})\ ||T_g||ID_G]$, $E_{SEK}\ (J_{ud}\ ||ID_{ud})$). In particular, these messages are confidential from any attacker.
- *Replay Attacks*: As defined in [6], replay attacks occur when the attacker intercepts some of the exchanged message and can re-use them. This protocol is resistant to replay attacks because the authenticities of messages *M1, M2,* are timestamp and nonce based. They are validated by checking the freshness of timestamps $(Tg-Tud) \leq \Delta T$, $(Tud-Tg) \leq \Delta T$ and nonce $(Nud' = Nud, Ng'=Ng)$. Suppose that an attacker intercepts a login request message M1 and attempt to access the WSN by replaying the same message *(M1)*. The verification of this login attempt fails, since the time difference expires (i.e., $(Tg-Tue) \geq \Delta T$). In the same way, he intercepts *M2* or *M3* and try to extracts *< Lg, Ug, Jud>* and attempts to replay one of, the verification request will fail because of the time difference expires again and also, the nonce will show that the message is already used. Hence, our protocol is secure against replaying of messages.

- *Man-in-the-Middle Attacks:* An attacker may attempt a man-in-the-middle (MIMT) attack by modifying the login message *M1=< Ag, Qud >* to *M1*=< Ag*, Qud* >*. Nevertheless, this malicious attempt will not work, as the false IDud* will not be verified at the GWn and the GWn cannot get the original sub-message *{h (Vud|| Pg)}** by decrypting *Qud**. Thus, man-in-the-middle attacks are not applicable our protocol.
- *Session Key Establishment:* The proposed protocol provides session key establishment after the authentication phase. A session key [i.e., $SEK = Gg^{Nud} \bmod p$] is set up between the used device and the gateway node for secure subsequent communication. For each login session, the session key will be different and cannot be replayed after the time expires. Furthermore, the *UD* and *GWn* can securely execute encryptions and decryptions by using of the session key and hence, achieve confidentiality for the subsequent messages.

3.2 Efficiency Analysis

The evaluation of the efficiency is performed in this subsection using H the function to perform one-way hash computation, S as symmetric cryptosystem and MAC the time for performing a MAC. The efficiency analysis gives the output of a statistical estimation of the computational cost, communication cost and node energy cost of our proposed protocol and we make a comparison with existing ones.
From the comparison performance Table 2, our proposed protocol in the term of computation cost, requires 1H and no symmetric cryptosystem whereas in [8, 10] and [11] 6H, 4H and 4H/1S are required, respectively in the registration phase.

Table 2. Comparison performance table

Scheme	Registration		Login and Authentication		
	User	Gateway	User	Gateway	Sensor Node
Das[7]	-	3H	4H	4H	1H
Daojing et al.[8]	1H	5H	5H	5H	1H
Wong et al[9]	-	3H	-	1H	3H
Vaidya et al.[10]	2H	2H	3H	3H	3H
Pardeep et al[11]	1H	3H+1S	4H+2S	3H+2S+ 1MAC	1H+2S+1MAC
Proposed	-	1H	4H+2S	2H+2S	-

Moreover, in the login and authentication phase the proposed protocol requires 6H, and 4S, whereas, [7], [8], [10] and [11] require 9H, 11H, 9H, and 8H/6S/2MAC respectively. The cost of the computation of our proposed protocol is low than others schemes due to the fact that our architecture of the protocol does not allow the user to interact directly with the sensors nodes. In the term of node energy cost, only the user will access the data from the gateway which save the energy of the sensor node. Thus, sensor node's energy consumption in our protocol is much lesser than other protocols.

The performance analysis of the communication cost indicates that, our proposed protocol requires tree messages to fulfill all process of communication through WSN.

4 Conclusion

Secure authentication protocol is an important issue in several different networks. In this paper we proposed a secure authentication protocol through WSN. The evaluation of our protocol showed that it is more robust against many popular attacks and provides many security services and it is efficiency at logical computational costs.

Acknowledgment. This research was supported by Basic Science Research Program through the National Research Foundation of Korea (NRF 2014) funded by the Ministry of Education, Science and Technology. And it also supported by the BB21 project of Busan Metropolitan City.

References

1. A. Perrig, R. Szewczyk, V. Web, D. Culler, J.D. Tygar, SPINS: Security protocol for sensor networks. In Proceeding of the 7th Annual International Conference on Mobile Computing and Networks (MOBBICOM 2001), Rome, Italy, July (2001).
2. P. Kumar, S. Cho, D.S. Lee, Y.D. Lee, H.J Lee, TriSec: A secure data framework for wireless sensor networks using authenticated encryption. Int. J. Marit. Inf. Commun. Sci., 8, 129-135. (2010).
3. C. Karlof, D. Wagner, Secure routing in wireless sensor networks: Attacks and countermeasure. Ad Hoc Network, 293-315. (2003).
4. Z. Benenson, F. Gartner, D. Kesdogan,User Authentication in sensor network (extended abstract). In Proceedings of the Workshop on Sensor Networks Informatik. 34 Jahrestagung der Gesellschaft fur Informatik, Ulm, Germany, September (2004).
5. M. Luk, G. Mezzour, A. Perrig, V. Gligor, MiniSec: A secure sensor network communication architecture. In *Proceeding of the 6th International Conference on Information Processing in Sensor Networks*, *IPSN'07*, Cambridge, MA, USA, 25–27 April 2007.
6. C.C. Chang, C.Y. Lee, Y.C Chiu, Enhanced Authentication scheme with anonymity for roaming service in global mobility networks. *Comput. Commun.* 2009, *32*, 611-618
7. M.L. Das, Two-factor user authentication in wireless sensor networks. *IEEE Trans. Wireless Comm.* 2009, *8*, 1086-1090
8. D. He, Y. Gao, S. Chan, C. Chen, J Bu, An enhanced two-factor user authentication scheme in wireless sensor networks. *Int. J. Ad-Hoc Sensor Wirel. Netw.* 2010, *0*, 1-11
9. K.H.M. Wong, Y Zheng, J. Cao, Wang, S. A dynamic user authentication scheme for wireless sensor networks. In *Proceedings of the IEEE International Conference on Sensor Networks, Ubiquitous, and Trustworthy Computing* (*SUTC'06*), Taichung, Taiwan, 5–7 June 2006
10. B. Vaidya, J.J.P.C, Rodrigues, J.H Park, User authentication schemes with pseudonymity for ubiquitous sensor network in NGN. *Int. J. Commun. Syst.* 2010, *23*, 1201-1222.
11. P. Kumar, A. J. Choudhury , M. Sain , S.G. Lee, H.J. Lee, RUASN: A Robust User Authentication Framework for Wireless Sensor Networks, Sensors, ISSN:1424-8220, 5020-5046; doi: 10.3390/s110505020, November 2011.

A Noninvasive Sensor System for Discriminating Mobility Pattern on a Bed

Seung Ho Cho[1], Seokhyang Cho[*2]

[1]Division of Computer and Media Information Engineering, Kangnam University/ 40 Gangnam-ro, Giheung-gu, Yongin-si, Gyeonggi-do, 446-702, Republic of Korea
[2]School of Information and Communication Engineering, Sungkyunkwan University/ 2066 Seobu-ro, Jangan-gu, Suwon-si, Gyeonggi-do, 440-746, Republic of Korea
[1]shcho@kangnam.ac.kr; [*2]shcho@security.re.kr

Abstract. In this paper, we propose a noninvasive sensor system for discriminating mobility pattern of a resident on the bed without inconvenience. The proposed system consists of a thin and wide film style of piezoelectric force senor, a signal processing board, and data collecting program. There are four different types of motion that were simulated by non-patient volunteers. About 10,000 experimental motions of subjects were performed. Sensor data by human motions were collected and preprocessed by a moving average filter, transformed by FFT, and classified by the *k*-NN algorithm with $k = 1$. The experiment yielded the overall discrimination rate of 89.4%. The proposed system will contribute to differentiating mobility pattern on a bed and distinguishing the physical characteristics of a person.

Keywords: u-Health care, Noninvasive sensor, Mobility pattern, Motion recognition, Ubiquitous computing

1 Introduction

U-Health, as a fusion technology between ubiquitous computer technique and health care, helps to monitor remote patients with wired and wireless networking and provide medical services to them with little constraint of time and space. Recently, with the advent of the u-Health service, the paradigm of health care has shifted from diagnosis to precaution and health care [1].

Most human beings spend about one third of each day sleeping so that the sleeping is likely the most time-consuming activity in their daily lives. Since they sleep mainly on a bed inside a room, the bed may well be the most used furniture at home. That is why we propose, in this paper, a noninvasive sensor system, which can monitor the medical conditions of a human staying in a bed with much less inconvenience than the systems using invasive and touch sensor.

The proposed sensor system consists of three parts, i.e., a piezoelectric force sensor, a signal processing board, and data acquisition program where the piezoelectric force sensor measuring the medical data has been installed under a bed mattress to set the patients free from adhesive sensors. The clinical data collected through the noninva-

K.J. Kim (ed.), *Information Science and Applications*,
Lecture Notes in Electrical Engineering 339, DOI 10.1007/978-3-662-46578-3_5

sive sensor system are used to analyze the sleep patterns and movements of the person on the bed so that the following two pieces of information can be obtained. The first one is about the motion of the person on the bed, i.e., which one of 'sitting up', 'lying down', and 'rolling about', it corresponds to. The second one is whether the person on the bed is a normal person or a patient undergoing rehabilitation.

First, the motions of the person on the bed are considered as three parts: wake up and sitting on the bed, sitting on the bed to lie down, roll over on the bed. Second, the targeted person on the bed should be characterized. The person on the bed should be consider whether normal person or patient under rehabilitation.

This paper is organized as follows. Section 2 compares our experimental result with the recent related experimental results of other groups. Section 3 introduces the elements of the proposed system. Section 4 explains how the experimental environment for monitoring the movements of the person on a bed has been set up and the data is collected. Section 5 explains the significance of experimental results through the pre-/post-analysis of acquired data. Section 6 concludes this paper.

2 Related Works

A. Gaddam et al. [2] monitored the activities of a patient by using the existing home monitoring system together with an intelligent bed having wireless force sensors (manufactured by Tekscan Company [3]) under its legs. Having monitored the existence, the position, and weight of a body for a long period of time, they collected those data to categorize the patient and determine his/her sleep pattern. They further attempted to make a warning given for a patient whose activities deviate from his/her normal behavioral pattern. Their approach made a long-time monitoring of a patient available with a low cost and little abuse of privacy. However, it is hard to recognize the activities of a patient by his/her existence and position. Besides, there is a limit to the accuracy of sleeping time estimates that can be achieved by determining whether a patient is sleeping or not from his/her time of tossing about in the bed.

A. Gaddam et al. [4] conducted another research focusing on the quality of sleeping, which aims to get a more detailed information about the patient's state by using more force sensors to measure the pressure and weight for estimating his/her position and existence. Also, they tried to use each sensor output signal pattern to detect the activities of a patient, like shaking and getting out of the bed, that are not detectable with one sensor. However, the research also showed a limit in distinguishing various activities.

In a research performed by M. Holtzman et al. [5], the pressure sensors were installed not on but below the mattress to measure the breathing and heart beat rates with enhanced user convenience and reduced signal interferences. They quantitatively examined how the intensity, response time, and SNR(signal-to-noise ratio) of the sensor output signals depend on such factors as the material, thickness of the mattress and the position of the sensors, i.e., whether the sensors were installed below or above the mattress. They showed that embedded sensors are valid for use in home environments and could be reliably used to collect physiological data. However, it was shown that resultant low signal levels might require more sophisticated signal extraction algorithms.

M. Borazio and K. Van Laerhoven [6] used a wrist-worn bracelet with luminance sensors and 3-axis accelerator sensor for a long-term monitoring of patients to categorize their sleep patterns and biomedical information into 36 different groups based on the sensor data and video screen captures. They succeeded to identify the sleep cycle, mild spasms, and epileptic. However, since a photometer was used to discern whether it is day or night, numerous errors are expected in real situations where the bracelet is inside the bed sheet or facing downward. In addition, using a camera may cause a serious violation of privacy and wearing a bracelet routinely every day may result in discomfort to the user's daily life.

In connection with increased interest in activity patterns during sleep, S. Cho et al. [7] installed a wireless infrared sensor on the ceiling above vibration sensors located on a bed and used such noninvasive sensors to monitor the sleep postures like lying down, rolling about between sleep, and sitting up on the bed. In this research, they attempted to categorize the activities of patients as a group and for each person, but could get meaningful results from the latter approach. Still a large variability of the vibration sensors can make it difficult to keep the monitoring system tracking human behavioral log in a stable manner.

3 Proposed Sensor System

3.1 Overall System Structure

The proposed sensor system consists of a piezoelectric force senor, a signal processing board, and data collecting program. The voltage signal induced by the movements of the person on the bed is measured by the piezoelectric force sensor and transmitted to the signal processing board. The signal processing board amplifies, filters, and AD-converts the input signal, and then use UART (Universal Asynchronous Receiver/Transmitter) to transmit the AD-converted signal to the PC. The PC executes the data collecting program to exhibit real time graphs and save the collected sensor data.

3.2 Signal Processing Board

This signal processing board amplifies and converts the analog output signal from the piezoelectric force sensor into a digital signal, which is transmitted to the data collecting program on the PC. Fig. 1 shows the total process performed by this signal processing board.

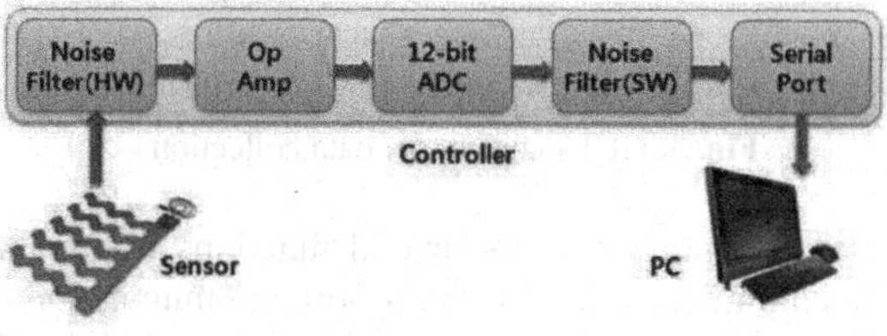

Fig. 1. Flow diagram

The board is an Analog Devices ADuC7026 board fully integrated MCU32 bit ARM7 core, 1 MSPS (Mega-Samples per Second), 12-bit data acquisition system incorporating high performance multichannel ADCs, where the ADC converts an analog input equal to or above 2.5V into 4095. It filters the piezoelectric force sensor output signal to remove the high frequency noise and uses an Op Amp LM 385(Texas Instruments) to amplify the signal by 7 times because most signals have been absorbed into and weakened by the mattress. The amplified analog signal is sampled and quantized 10 times per 10ms and then, the 20% trimmed mean of the 10 quantized numbers (the average of the remaining 6 ones with upper 2 and lower 2 samples discarded) is taken for reducing the noise effect. Thus the board generates one 16-bit quantized data per 10ms (at the sampling rate of 100Hz), forms 4 bytes packets (as depicted in Fig. 2), and sends them to the PC via UART port at the baud rate of 115,200bps. The firmware used in the board was developed by Keil uVision ARM compiler.

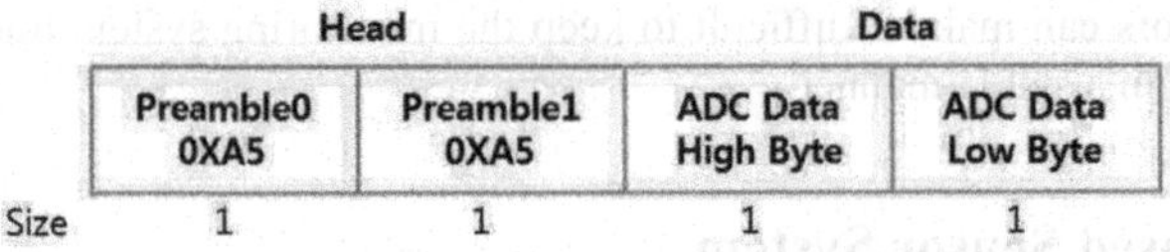

Fig. 2. Structure of a data packet

3.3 Sensor Data Collection

We have developed a GUI based program to collect and save the piezoelectric force sensor output signal as a file in the PC, displaying it in real time just like an oscilloscope(as illustrated in Fig. 3) where the time is measured along the x-axis and the quantized values are measured on the y-axis.

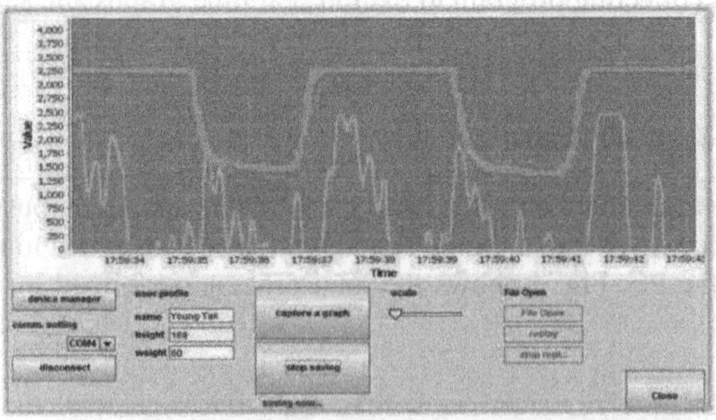

Fig. 3. GUI snapshot for data collection

The data collection program performs several functions where JAVA Comm API 2.0 is used for serial communication. For multi-port communication, we have built an UI to search the communication port. The communication link between PC and the board is connected via the port selected by Windows control panel where the visual tool developed by JAVA and the utility library of JFreeChart and Javacomm are used.

4 Experimental Environment

4.1 Experiment Overview

We installed piezoelectric sensors below the elbow position under the mattress of a bed (Fig. 4) and used them to determine the activities of the person on the bed as one of the three motions, i.e., 'sitting up', 'lying down', and 'rolling about'. Also, based on the 'sitting up' motion, we tried to distinguish whether the person is 'paralytic' or not. We performed the experiments of determining the activities and distinguishing 'paralytic' or not on the 5 normal subjects listed in Table 1. Each subject repeated the three motions and additionally, the motion of 'sitting up' leaning on his/her elbow as if he/she were a paralytic as directed by a paralytic rehabilitation expert. To reduce the signal attenuation due to the absorption, a general hospital mattress other than a normal one was used for our experiment.

Table 1. Information related to the Subject group

Subject	Height(cm)	Weight(kg)	Sex
A	178	73	Male
B	174	75	Male
C	167	71	Male
D	167	73	Male
E	172	72	Male

Fig. 4. Installation of a proposed sensor system

4.2 Experimental Motion Types

Table 2 shows that four kinds of activities, where the last one is what is made for comparison by a normal participant who mimics the way of most paralytics sit up leaning on their elbow on the paralyzed side.

Table 2. Experimental motion types

Types of Motions	Types of Persons	Details
[1] Getting up	Normal person	Getting up in bed with the help of his elbow
[2] Lying down	Normal person	Lying down after sitting on bed
[3] Rolling about	Normal person	Lying in bed, tossing and turning
[4] Getting up	Stroke patient	Getting up in bed with the help of his paralyzed side elbow

5 Recognizing Motions

5.1 Preprocessing

In this study, a moving average filter (MAF) is used to remove the high-frequency noise contained in the sensor output so that the data can be made smooth. The order of the MAF was fixed as 16 because it turned out to yield the highest motion recognition rate after an exhaustive trial and error process.

After the noise cancellation using a MAF, the period of a motion is determined where a motion is regarded as having started or ended if the sensor output signal becomes larger than a threshold and if it shows no variation of magnitude greater than the threshold for a certain period. The threshold was fixed as 20 because it resulted in the most accurate detections of the start and end times through an exhaustive trial and error process.

Figs. 5 (a)/(b) and (c)/(d) show the signals and their FFT(fast Fourier transform) magnitude spectra obtained from the normal/paralysis-like sitting-up motions, respectively [8, 9]. The two spectra show a notable difference only in the low frequency band. Therefore, the sum of the FFT magnitudes in the low frequency band is fed into the feature input of the *k*-NN (Nearest Neighbor) algorithm [10], which determines the motion pattern label (class) of an input sample by the majority rule among the labels of the *k* nearest training samples in the feature space. After an exhaustive trial and error process, we fixed $k = 1$ because it yielded the best classification rate.

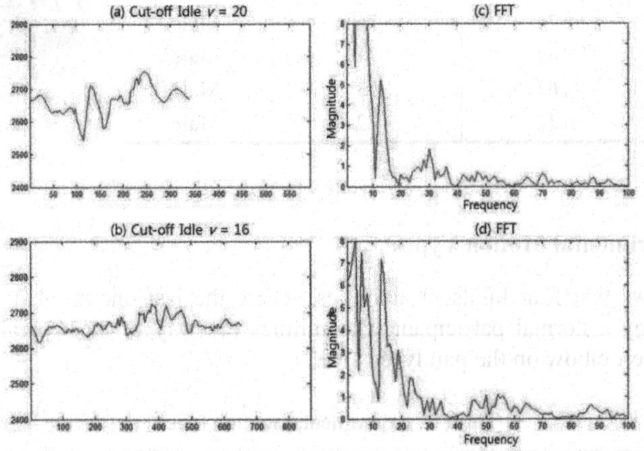

Fig. 5. Extraction of a motion period and FFT (a) a normal person, (b) a stroke patient

5.2 Motion Detecting Experiment

Each of the five participants repeats each motion 8 times so that $5 \times 8 = 40$ motion data can be collected for each of the 4 motions. Among the 40 motion data, 35 ones are randomly selected to be used as the training data (with known class of motion) and the remaining 5 ones are classified into the class of the closest training sample by the *k*-NN algorithm with $k = 1$. This yields 20 classification results. We repeated this experimental analysis 500 times to get a total of 10,000 classification results, for which the overall classification rate of 89.4% was achieved as summarized in Table 3, that is the confusion matrix.

Table 3 lists the numbers of right and wrong classifications together with the classification rates for each of the four motions: 'sitting up', 'lying down', 'rolling about', and 'paralysis-like sitting up'. It shows that the 'rolling about' motion of a normal person can most easily be misclassified as the 'sitting up' motion of a paralytic and that many misclassifications result between the 'lying down' and 'rolling about' motions of a normal person. In contrast, there are very few misclassifications between the 'sitting up' motions of a normal person and a paralytic so that their classification rates are 98.16% and 97.56%, respectively. This implies that our proposed system with the classification scheme is very efficient in distinguishing whether a person on the bed is normal or not.

Table 3. Confusion matrix

Class	**Normal Person**			**Stroke Patient**	**Subtotal**	**Recognition Ratio (%)**
	[1]	[2]	[3]	[4]		
[1] Getting up	2454	45	0	1	2500	98.16
[2] Lying down	194	2046	252	8	2500	81.84
[3] Rolling about	0	166	2001	333	2500	80.04
[4] Getting up	0	0	61	2439	2500	97.56
Subtotal					10000	89.4

6 Conclusion

In this study, we proposed a noninvasive sensor system for monitoring the pattern of a home user on the bed, without causing any inconvenience to his/her daily life unlike the case of using invasive or contact sensors. The proposed sensor system consists of a thin and wide film-type piezoelectric force senor (installed under the mattress), a signal processing board, and data collecting program.

The signals induced by the 10,000 experimental motions of the participants are preprocessed by a MAF, their FFTs are taken, and classified by the *k*-NN algorithm with $k = 1$, which yielded the overall classification rate of 89.4%. While the classification rates for normal 'lying down' and 'rolling about' motions are 81.84% and 80.04%, respectively, those for 'sitting up' motions of a normal person and a paralytic are 98.16% and 97.56%, respectively. This implies that our proposed system may help to recognize the motion and distinguish the physical characteristic (between 'normal' or 'paralytic') of a person on the bed.

This encouraging result leads to our expectation that, with more advanced developments, our monitoring system can be utilized to recognize the behavioral pattern of a person so that a patient, like ones in a stroke rehabilitation program, can have his/her health issue identified to get a medical examination without going to see a doctor.

From the viewpoint of Activities of Daily Living (ADL) that is a standard for evaluating the basic daily activities, our study can be beneficial to the patients [11]. As an example, bed mobility of a patient - one of ADL capability test items as a measure of independent daily life - is now assessed qualitatively in terms of athletic ability by a physician's inspection. So the bed mobility test result is subject to vary with personal perception. Our research result may help to shift the qualitative assessment of bed mobility to the quantitative assessment so that it can lead to establishing a new standard for ADL. We believe that more active researches will be performed in this direction.

References

1. Song, T.: u-Health: Current Status and Policy. Korea Institute of Health and Social Affairs (KIHASA), Republic of Korea (2011)
2. Gaddam, A., Mukhopadhyay, S.C. and Sen Gupta, G.: Necessity of a Bed-Sensor in a Smart Digital Home to Care for Elder-People. Sensors, IEEE (2008) 1340-1343
3. http://www.tekscan.com/force-sensitive-bed-monitoring
4. Gaddam, A., Kaur, K., Sen Gupta, G., and Mukhopadhyay, S.C.: Determination of Sleep Quality of Inhabitant in a Smart Home using an Intelligent Bed Sensing System. In: IEEE Instrumentation and Measurement Technology Conference (I2MTC), Austin (2010) 1613-1617
5. Holtzman, M., Townsend, D., Goubran R., and Knoefel, F.: Validation of Pressure Sensors for Physiological Monitoring in Home Environments. In: '10 IEEE International Workshop on Medical Measurements and Applications Proceedings (MeMeA), Canada (2010) 38-42
6. Borazio, M., Van Laerhoven, K.: Combining Wearable and Environmental Sensing into an Unobtrusive Tool for Long-Term Sleep Studies. In: IHI '12 Proc. of the 2nd ACM SIGHIT International Health Informatics Symposium, ACM, Miami (2012) 71-80
7. Cho, S., Phillips, W.D., Sankar, R., and Moon, B.: A State Preserving Approach to Recognizing Human Behavior using Wireless Infrared and Vibration Sensors. Proc. of IEEE Southeastcon (2012) 1-6
8. Cooley, J.W. and Tukey, J.W.: An Algorithm for the Machine Calculation of Complex Fourier Series. Mathematics of Computation, 19 (1965) 297-301
9. Hyounkyo Oh, et al., Preprocessing in a Noninvasive Sensor System. Proc. of KIPS Spring Conference, 20(1), Republic of Korea (2013) 83-85
10. Cover, T. and Hart, P.: Nearest Neighbor Pattern Classification. IEEE Transactions on Information Theory 13(1) (1967) 21-27
11. Hamilton, BB., Granger CV., Sherwin, FS. et al.: A Uniform National Data System for Medical Rehabilitation, Rehabilitation Outcomes. Analysis and Measurement, Brookes Pub. (1987) 137-147

A Fault-Tolerant Multi-Path Multi-Channel Routing Protocol for Cognitive Radio Ad Hoc Networks

Zamree Che-aron[1], Aisha Hassan Abdalla[1], Khaizuran Abdullah[1], Wan Haslina Hassan[2] and Md. Arafatur Rahman[3]

[1]Department of Electrical and Computer Engineering, International Islamic University Malaysia (IIUM), Kuala Lumpur 53100, Malaysia
one_zamree@hotmail.com, aisha@iium.edu.my, khaizuran@iium.edu.my
[2]Malaysia-Japan International Institute of Technology (MJIIT), Universiti Teknologi Malaysia (UTM), Jalan Semarak, 54100 Kuala Lumpur, Malaysia
wanhaslina@ic.utm.my
[3]Department of Biomedical Electronics and Telecommunications Engineering, University of Naples Federico II, Naples 80138, Italy
arafatur.rahman@unina.it

Abstract. Cognitive Radio (CR) has been proposed as a promising technology to solve the problem of radio spectrum shortage and spectrum underutilization. In Cognitive Radio Ad Hoc Networks (CRAHNs), which operate without centralized infrastructure support, the data routing is one of the most important issues to be taken into account and requires more studies. Moreover, in such networks, a path failure can easily occur during data transmission caused by an activity of licensed users, node mobility, node fault, or link degradation. Also, the network performance is severely degraded due to a large number of path failures. In this paper, the Fault-Tolerant Cognitive Ad-hoc Routing Protocol (FTCARP) is proposed to provide fast and efficient route recovery in presence of path failures during data delivery in CRAHNs. In FTCARP, a backup path is immediately utilized in case a failure occurs over a primary transmission route in order to transfer the next coming data packets without severe service disruption. The protocol uses different route recovery mechanism to handle different cause of a path failure. The performance evaluation is conducted through simulation using NS-2 simulator. The protocol performance is benchmarked against the Dual Diversity Cognitive Ad-hoc Routing Protocol (D2CARP). The simulation results prove that the FTCARP protocol achieves better performance in terms of average throughput and average end-to-end delay as compared to the D2CARP protocol.

Keywords: Multi-path multi-channel routing; Cognitive radio ad hoc network; Fault tolerance; Fast route recovery; Joint path and spectrum diversity

1 Introduction

The recent experiment results conducted by the Federal Communications Commission (FCC) [1] have proved that the static spectrum allocation policy, which allows each

K.J. Kim (ed.), *Information Science and Applications*,
Lecture Notes in Electrical Engineering 339, DOI 10.1007/978-3-662-46578-3_6

wireless service to access fixed frequency bands, poses the spectrum inefficiency problem. Furthermore, due to the rapidly increased demand for wireless services, the radio spectrum is one of the most heavily used and costly natural resources, thus leading to the problem of spectrum scarcity.

The CR technology [2-3], which is being considered as the candidate for the 5th Generation (5G) of wireless communications [4], has been proposed as a promising technology to cope with current spectrum scarcity problem by means of enabling unlicensed users (also called cognitive radio users or Secondary Users (SUs)) to dynamically adjust its operating parameters and thus opportunistically use the licensed spectrum (such as the frequencies licensed for television broadcasting) which is always statistically underutilized by licensed users (also known as Primary Users (PUs)) with condition that interference to the licensed users is below an acceptable threshold level. Cognitive Radio Ad Hoc Network (CRAHN) [5] is a class of Cognitive Radio Networks (CRNs) which applies the CR paradigm to ad hoc scenario.

Data routing is a key function in CRAHNs and faces various significant challenges [6-7] that require in-depth studies. The major purpose of a routing protocol is to exchange up-to-date routing information and determine the appropriate path over which data is transmitted based on routing metrics as well as to discover a new path in case the current path is no longer available. One of the significant challenges on routing in CRAHNs involves the frequent changes of network topology, simply leading to route failures and service outages. The topological changes in CRAHNs occur primarily due to node mobility and intermittent PU activities. Since SUs have to instantaneously vacate the channel that overlaps with a PU's transmission frequency once a PU activity is detected, the channel availability for each SU varies frequently. Therefore, the issues of fault tolerance must be seriously considered in such networks.

In this paper, the Fault-Tolerant Cognitive Ad-hoc Routing Protocol (FTCARP) is proposed. It provides fast and efficient route recovery in presence of path failures during data delivery in CRAHNs. Moreover, the protocol exploits the joint path and spectrum diversity in routing process to offer reliable communication and efficient spectrum usage over the networks. In the proposed protocol, a backup path is utilized in case a failure occurs over a primary transmission route. Different cause of a path failure will be handled by different route recovery mechanism. Providing a backup path in CRAHNs is beneficial because the networks are prone to route breakages resulting from PU activity, node mobility, path fading, signal interference, and node failure.

The remainder of the paper is organized as follows. The overview of the FTCARP protocol and the protocol operations are explained in Section 2. Section 3 describes the simulation configuration and parameters. The simulation results and performance evaluation are presented in Section 4 followed by the conclusion in section 5.

2 Fault-Tolerant Cognitive Ad-hoc Routing Protocol

The Fault-Tolerant Cognitive Ad-hoc Routing Protocol (FTCARP) is a reactive distance-vector routing protocol proposed to deal with path breakages occurring during data transmission in CRAHNs. It shares some common features with D2CARP protocol [8]. The protocol jointly exploits path and spectrum diversity in data routing. By jointly utilizing both diversities, SUs can switch among different paths and channels for data communication in presence of frequency and space varying PU activity. It also takes advantage of the multiple available channels to improve the network performance. In addition, the novel fault-tolerant algorithm is provided to respond to route breakages in a timely manner by creating a backup path for all SUs which are transmitting data packets towards the destination. When the SU gets failure to forward a data packet through the primary route, it instantaneously exploits its backup path for transmitting the next coming data packets without any interruption. The backup path is selected based on the cause of path failure, i.e. PU activity, node mobility or link degradation. For different cause of a path failure, the protocol will call different route recovery mechanism to handle it. By exploiting the fault-tolerant scheme, the data communication still keeps running continually in presence of route breakages. Unlike the D2CARP protocol, in the FTCARP, each SU maintains two routing tables: Primary routing table and Backup routing table.

Even though the D2CARP protocol also provides a joint path and spectrum diversity, only the source node can dynamically switch among different paths and channels for communicating with each other in presence of path failure (i.e. the intermediate nodes can switch among different channels only). This problem is solved in the FTCARP protocol.

2.1 Protocol Operations

This section describes the scenario of primary route discovery, backup path establishment, route maintenance, and route recovery in the FTCARP protocol. Moreover, the explanation of how the control packets are handled is also presented.

Primary Route Discovery

When a source node requires a path towards a destination node for data communication, it broadcasts a Primary Route REQuest (P-RREQ) packet to its neighbors through all its available channels (i.e. not occupied by a PU). For an intermediate node, if the first P-RREQ packet is received, then it creates a primary reverse route pointing to the previous node. Then it records the channel through which the packet has been forwarded as well as rebroadcasting the packet through all its vacant channels. In case it receives an additional P-RREQ packet with the same sequence number from the same node but on different channel, it just records a primary reverse route through that channel without rebroadcasting the packet. In such a way, the intermediate node can establish multi-channel primary reverse routes. Afterwards, the intermediate node will update the record of primary reverse route only if it receives the P-

RREQ packet with a higher sequence number or the same sequence number but lower hop count. The out-of-date P-RREQ packet received by a node will be discarded to prevent the routing loop problem.

If a P-RREQ packet reaches the destination node or an intermediate node which has a valid primary route towards the destination (i.e. an active primary route entry with the same or higher sequence number than the sequence number stored in the P-RREQ packet), it generates a Primary Route REPly (P-RREP). Then it sends it back to the previous node through the same channel that the P-RREQ packet has been transmitted. The destination node will not discard the further P-RREQ packets received from the same node but on different channels. An intermediate node which receives the first P-RREP packet creates a primary forward route pointing to the packet sender through the same channel that the packet has been received. It then forwards the copies of the packet over all its valid primary reverse routes with different vacant channels (i.e. not used by a PU) towards the source of the P-RREQ packet. Afterwards, if an intermediate node receives an extra P-RREP packet from the same sender but on different channel, it establishes a primary forward route for the channel and only forwards the packet on its primary reverse route through that channel. In this fashion, multi-channel primary forward routes will be created. In case an intermediate node receives a fresher or better P-RREP packet (which has a greater sequence number or the same sequence number with smaller hop count), the entry of its primary forward route will be updated.

Once the first P-RREP packet arrives at the source node, a primary forward route for the packet-received channel will be created and then it starts transmitting the data packets destined for the destination node over the primary forward route via that channel. The source node also does not ignore the additional P-RREP packets received from the same node but on different channels. The source and intermediate nodes store active primary forward routes with different available channels (i.e. not occupied by a PU) and randomly select one of them for data communication.

Backup Path Establishment

After the primary path from the source node to the destination node has been established, the data delivery process will be triggered. The process of backup path establishment is performed during data transmission. Only SUs, except the destination node, that involve the data delivery process are able to create the backup path. Each of them is allowed to have only one multi-channel backup path towards a destination node. By ensuring these conditions, when a node on the primary path forwards a data packet, if its backup path towards the destination node has not been created, it broadcasts a Backup Route REQuest (B-RREQ) packet with limited TTL (Time-To-Live) value to its neighbors through all its vacant channels. In the B-RREQ packet, some extra fields are utilized including “Data-Source Node ID”, “Next-Hop Node ID” and “Primary Path Length” field. The “Data-Source Node ID” field contains the ID of the data source. The “Next-Hop Node ID” field stores the ID of the next-hop node over the primary path. Lastly, the “Primary Path Length” field records the hop count of the primary path from the B-RREQ-originating node to the destination node. To avoid creating an ineffective backup path, the B-RREQ packet will be discarded in the following cases: first, it is received by an intermediate node that has the same ID as stored in the "Next-Hop Node ID" field inside the packet; second, it is arrived at an intermediate node whose next hop along the primary path towards the destination is the same as the B-RREQ-originating node; or third, it reaches the B-RREQ-originating node. In addition, the destination node ignores the B-RREQ packet if the hop count stored in the packet is equal to one. In case the B-RREQ packet is not discarded, a node responds to the first received B-RREQ packet by creating a backup reverse route pointing to the previous node through the channel that the packet has been received. The stored backup reverse route entry will be updated only if the node receives the B-RREQ packet with a greater sequence number or the same sequence number but lower hop count. The node is allowed to record only one lastly updated backup reverse route.

After processing the B-RREQ packet, an intermediate node may generate a Backup Route REPly (B-RREP) packet and send it back to the previous node through its backup reverse route if one of the following conditions is reached. First, it has a fresh enough primary path towards the destination and the ID of its next hop over the primary path towards the data source is the same as stored in the "Next-Hop Node ID" field inside the packet. Second, it has a valid primary path towards the destination and the sum of the hop count of the valid primary path and the hop-count value stored in the packet is lower than the value of the "Primary Path Length" field inside the packet. Third, it knows a valid backup path towards the destination and the sum of the hop count of the valid backup path and the hop-count value recorded in the packet is lesser than or equal to the value of the "Primary Path Length" field in the packet. Otherwise, it rebroadcasts the packet to its neighbors via all its available channels. In case the B-RREQ packet reaches the destination node, a B-RREP packet will be generated and forwarded back to the previous node via its backup reverse route. This is only if its ID is the same as recorded in the "Next-Hop Node ID" field inside the packet or the ID of its next hop over the primary path towards the data source is the same as stored in the "Next-Hop Node ID" field of the packet.

The B-RREP packet will be forwarded back along the backup reverse path towards the B-RREQ-originating node. As the B-RREP packet travels back to the B-RREQ source, each intermediate node sets up a backup forward route pointing to the B-RREP sender via the same channel that the packet has come. The backup forward route will be updated if a further B-RREP packet which has a higher sequence number or the same sequence number with a lower hop count is received. After a B-RREP packet arrives at the B-RREQ source, its backup path towards the destination node will be established.

Route Maintenance and Recovery

After a data packet has been successfully forwarded to a next-hop node over the primary path towards the destination, the lifetime of the primary forward route and the primary reverse route is increased in order to maintain the connectivity of primary path.

Additionally, in CRAHNs, a route breakage can be caused by node mobility, node fault, PU activity, etc. To provide fast and efficient route recovery, different cause of a path failure will be resolved by different route recovery mechanism. During data transmission, when a node over the primary path detects a primary route breakage, it first checks the cause of the route failure. If the link breakage results from a PU activity, the node cannot transmit a data packet to the next-hop node via the currently used channel. Subsequently, it suddenly selects another available channel (i.e. not used by a PU) from its primary routing table to deliver next coming data packets to the same next-hop node towards the destination without changing the path direction.

In the situation that a node cannot successfully forward a data packet to its next hop over the primary path due to the mobility or failure of the next-hop node, it instantaneously utilizes its backup path (i.e. its detour) (whose routing entry has been kept in its backup routing table) to deliver the next coming data traffic without severe interruption of data transmission. After the backup path is exploited for data communication, the node establishes a new backup path instead of the currently used one. Subsequently, the node generates a Backup Route ERRor (B-RERR) packet containing the information of the backup path length (determined by hop count) and sends the packet to the source of data. An intermediate node over the primary path responds to the received B-RERR packet by just forwarding the packet to the next hop towards the data source. When the B-RERR packet reaches the source node, it will discover a new primary path for data transmission only if the hop count of the currently used backup path from source to destination stored in the "Backup Path Length" field inside the packet is greater than the hop count of the broken primary path from source to destination. The reason behind this is to prevent producing extremely high end-to-end packet delay.

In case a node detects a link failure during data transmission and no alternative available channel or backup path is found, it generates an N-RERR packet and broadcasts it to all its neighbors. As the N-RERR packet propagates towards the source node, a node that receives the N-RERR packet invalidates all affected routing table entries. When the N-RERR packet arrives at the source node, a new primary route discovery process will be triggered.

3 Simulation Configuration

The efficiency of the FTCARP protocol is evaluated by using NS-2 simulator [9]. A simulation area of 1000 x 1000 m^2 in which 100 movable SUs are located is specified. The distance between two SUs along X- and Y-axis is set to 100 m. There are 10 PUs randomly placed in the simulation area. The PU activities are modeled according to the ON/OFF process with exponential distribution with parameter λ of 100, referred to as PU activity parameter. The ON state represents the period where the channel is occupied by a PU and the OFF state denotes the period where the channel is available for SUs' communications. The UDP (User Datagram Protocol) connection is created for the source (Node 0) and destination node (Node 99). Over the UDP connection, CBR (Constant Bit Rate) traffic with 512 byte data packets at the packet interval of 50 ms is transmitted. The simulation time is set to 400 seconds and the data transmission process is started after 10 seconds. There are 3 non-overlapping channels given for multi-channel data communications. The transmission range of SUs and PUs is set to 150 m. The IEEE 802.11 standard is used for MAC protocol. The two-ray ground reflection model is specified as the radio propagation type.

4 Simulation Results and Performance Evaluation

To evaluate the performance improvement, the D2CARP protocol [8] is considered as a benchmark for comparing with the proposed FTCARP protocol. The protocol performance is evaluated through the simulation with the different number of path failures occurring during data transmission. The NS2 Visual Trace Analyzer [10] is used to analyze the simulation results that are stored in the NS2 trace files. The performance metrics used for evaluation include the average throughput and average end-to-end delay.

In Fig. 1, the average throughput results of FTCARP are compared with that of D2CARP protocol under the different number of path failures. From the simulation results, the D2CARP protocol is clearly not able to cope with the networks encountering high path-failure rate by showing the large performance degradation in terms of the average throughput. It drops considerably when the number of path failures increases. On the other hand, the FTCARP protocol achieves higher average throughput compared to D2CARP. Since the fault-tolerant approach is applied in the FTCARP protocol, a breakage of primary path does not significantly affect the data delivery process. In the network with 7 path failures, the FTCARP achieves a throughput enhancement of about 13.51% over the D2CARP protocol.

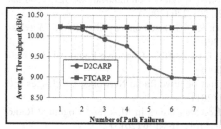

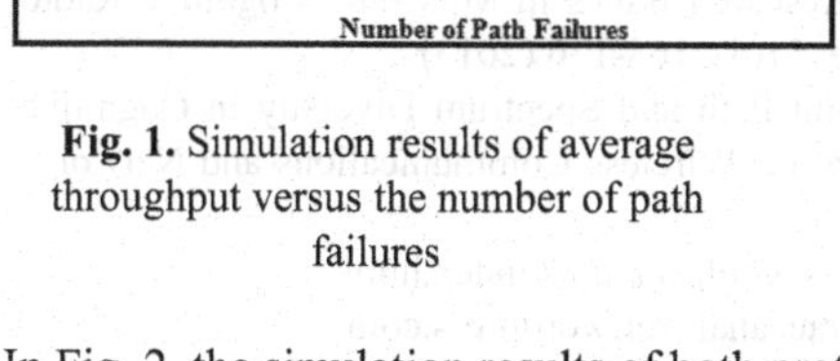

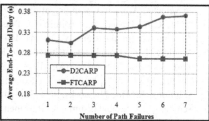

Fig. 1. Simulation results of average throughput versus the number of path failures

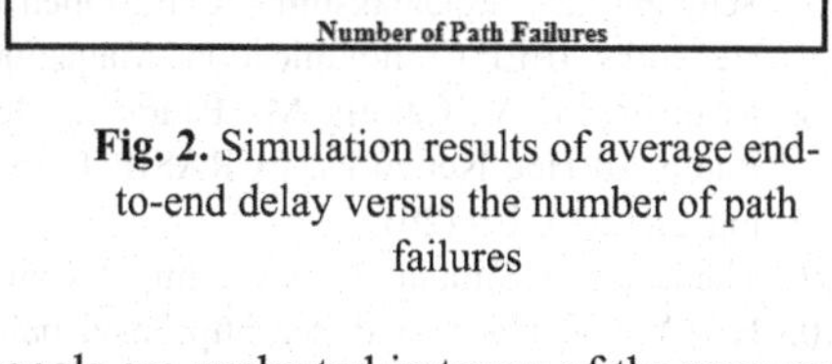

Fig. 2. Simulation results of average end-to-end delay versus the number of path failures

In Fig. 2, the simulation results of both protocols are evaluated in terms of the average end-to-end delay against the increased number of path failures. It is observed that the FTCARP protocol outperforms the D2CARP protocol in all cases due to its fast and efficient route recovery mechanism, whereas the D2CARP has worse performance. In FTCARP, when a primary path breaks during data delivery, a node over the path which encounters the failure immediately switches the transmission route to its backup path. On the other hand, D2CARP suffers from longer delay due to discovering a new transmission path. In the network with 7 path failures, the FTCARP achieves an end-to-end delay enhancement of about 28.36% over the D2CARP protocol.

5 Conclusion

This paper has presented the proposed FTCARP protocol, a fault-tolerant routing protocol for CRAHNs. The protocol jointly exploits the path and spectrum diversity in routing process to provide reliable communications and efficient use of spectrum in the networks. FTCARP can effectively cope with large numbers of path breakages occurring during data transmission by offering the fast and efficient route recovery mechanism. The protocol uses different route recovery mechanism to handle different cause of path failure (i.e. PU activity, node mobility or faulty node). To evaluate the protocol effectiveness, the performance comparison between the proposed FTCARP protocol and the benchmarked D2CARP protocol has been carried out. As compared with the D2CARP, the simulation results have proved that the FTCARP provides higher average throughput and reduces average end-to-end delay in identical scenarios.

References

1. FCC's Spectrum Policy Task Force.: Report of the Spectrum Efficiency Working Group. Technical report, Federal Communications Commission (FCC) (2002)
2. Akyildiz, I. F., Lee, W.-Y., Vuran, M. C., Mohanty, S.: Next Generation/Dynamic Spectrum Access/Cognitive Radio Wireless Networks: A Survey. Computer Networks. 50(13), 2127-2159 (2006)
3. Haykin, S.: Cognitive Radio: Brain-Empowered Wireless Communications. IEEE Journal on Selected Areas in Communications. 23(2), 201-220 (2005)
4. Badoi, C.-I., Prasad, N., Croitoru, V., Prasad, R.: 5G Based on Cognitive Radio. Wireless Personal Communications. 57(3), 441-464 (2011)
5. Akyildiz, I. F., Lee, W.-Y., Chowdhury, K. R.: CRAHNs: Cognitive Radio Ad Hoc Networks. Ad Hoc Networks. 7(5), 810-836 (2009)
6. Cesana, M., Cuomo, F., Ekici, E.: Routing in Cognitive Radio Networks: Challenges and Solutions. Ad Hoc Networks. 9(3), 228-248 (2011)
7. Sengupta, S., Subbalakshmi, K. P.: Open Research Issues in Multi-hop Cognitive Radio Networks. IEEE Communications Magazine. 51(4), 168-176 (2013)
8. Rahman, M. A., Caleffi, M., Paura, L.: Joint Path and Spectrum Diversity in Cognitive Radio Ad-Hoc Networks. EURASIP Journal on Wireless Communications and Networking. 2012(1), 1-9 (2012)
9. The Network Simulator - NS-2, http://www.isi.edu/nsnam/ns/index.html
10. NS2 Visual Trace Analyzer, http://nsvisualtraceanalyzer.wordpress.com

STUDY OF SOUND AND HAPTIC FEEDBACK IN SMART WEARABLE DEVICES TO IMPROVE DRIVING PERFORMANCE OF ELDERS

Chularas Natpratan, Nagul Cooharojananone*

Machine Intelligence and Multimedia Information Technology Lab,
Department of Mathmathetics and Computer Science, Faculty of Science,
Chulalongkorn University, Thailand
ChularasN@gmail.com, Nagul.C@chula.ac.th

Abstract This paper's objective is to study the influence of sound and haptic feedback from smart wearable devices on performance of elderly drivers. The performance is measured with an assumption that those who spend more time to apply brake tend to be more prepared and aware of their surroundings. We create a prototype wearable device and experiment on how it affects drivers. A total of 9 elderly drivers were measured more than 108 times on how long they apply brake while driving. We then perform a paired sample T-test and found performance change from sound feedback to be statistically significant. We then calculate for correlation of the factors and performance. The factors with statistical significance are familiarity with smart devices and gender.

Keywords: Accident, Elderly, Eyesight, Traffic, Wearable device

1 Introduction

In Thailand, more than 40% of traffic accident involves personal cars (cars, trucks, taxi) [1]. The leading cause of these accident (sudden cut-in and tailgate) [1] could have a cause from lack of sufficient reaction to surrounding. This is because instead of looking at the road, the primary task, the driver needs to look at secondary tasks: signs, traffics lights, and surrounding vehicles.

The risk of traffic accident becomes higher, when the duration of secondary tasks is longer. [2],[9],[10] Elders whose senses degraded: eye, ear, skin [3],[8] have a potential to perform secondary tasks longer. This research will focus on people age 40 onwards as this is an age group that starts to show eyesight deterioration [4] and is widely studied by many researches [5,6,7].

This research is conducted under a hypothesis that: with smart wearable device providing haptic and sound feedback, elderly drivers will be more aware of their surroundings, and thus, spending more time to apply brake in their cars when face with red light. We will measure on how long our drivers push the brake.

The research will further unveil the usability and technology acceptance among users, as this has potential to be use under intense and unsafe situation.

*Corresponding author

K.J. Kim (ed.), *Information Science and Applications*,
Lecture Notes in Electrical Engineering 339, DOI 10.1007/978-3-662-46578-3_7

2 Background

2.1 Secondary task

When we observe driving as a main task, we usually consider every other event that causes the attention to shift as distraction or secondary tasks. Many researches were studied to identify and categorize secondary tasks into events inside and outside of the vehicles [11,12]. These categories are then further studied to find the causes (crashes, near misses etc.) and solidify policy making [2]. Hence, the solutions in this paper focus on eyesight. This is because today traffics focus so much on cognitive feedback. Drivers have to shift their vision constantly to look at sign, traffic lights, pedestrians, and surrounding vehicles. This limits drivers' channel of information to only their eyes, and when their eyes deteriorates, their driving skill follows. To help this, we have better traffic light, navigation [4], and raised stripe [5] to provide additional traffic info.

2.2 Elderly limitation

Generally, human eyes can start to deteriorate from the age of 40[4]. The age of 40 is also widely studied by many researched [5,6,7].Though some remain healthy throughout their age, some do take a hit, and most people are unaware of their gradual deterioration. Because of this, sometimes, people realize their senses deteriorate after something happen. However, there has been an attempt to introduce addition feedback for drivers. These additional feedbacks are usually based on three sensories: visual, audio, and haptic. [13,14]

We propose an additional audio and haptic feedback to help drivers in real-road scenario. We will measure the performance of the drivers, with and without the additional feedback as well as conducting a survey on their perceived usefulness.

3 Methodology

3.1 Participants and setting

Participant. For the experiment, we would need participants who are 40 years old or more, have a car, and can drive them fluently and safely. Participants are then given the information about feedback devices and route of the test. This is to ensure the experiment is done under consensus and transparency.

Setting. The setting for this experiment will be the route in which participants will drive their cars. In our case, the route will be along the streets which high amount of traffic lights. This is to make collecting high amount of data more efficient, as we will be collecting data every time our participants brake for red-light.

3.2 Experiment

Prototype Device. In this experiment, we made a prototype with user interfaces of sound and haptic feedback to imitate wearable device. The prototype is consisted of arm strap for phone and an Android smartphone. The Android smartphone then simulate sound and haptic feedback onto the participants' arm. This is the only apparatus on participants' arm (see Fig. 1.).

On an observer side, two apparatuses are needed. First one is another Android Smartphone to trigger sound and haptic feedback. The other apparatus is a stopwatch to measure brake time (see Fig. 1).

After the experiment, we will also ask testers to do a survey. This is to gather background information and see how comfortable test subjects are with driving and smart devices.

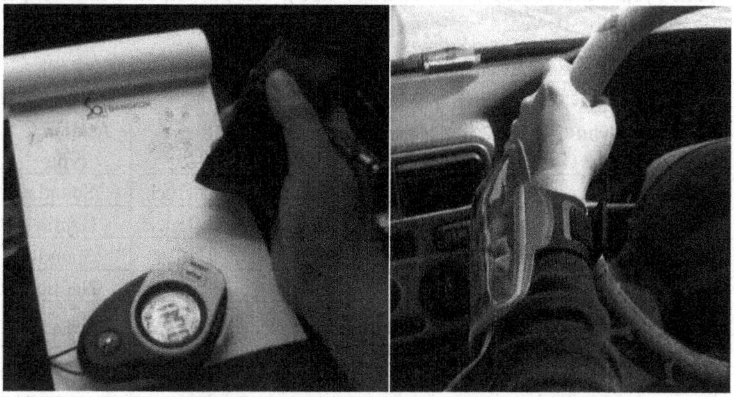

Fig.1. Apparatuses needed during the experiment

Methodology. We schedule with testers for appointments. Generally, it takes an hour per tester to introduce our method, conduct experiment, and work on survey. For our experiment, we have a total of 9 testers.

Once the testers arrive at the testing site, we explain experiment steps, driving route, and user interface of the feedback devices. When everyone is ready, we conduct the following experiment:

1. Let the testers drive normally
2. When approaching red light, at least 200 meter away, the observer choose and trigger one of the three feedbacks (sound, haptic, or none)
3. When the tester starts to apply the brake, the observer start the stopwatch
4. When the tester comes to a stop or re-applies gas, the observer stops the watch.
5. Repeat the experiment with every red-light.

The reason we chose to measure break duration instead of general respond time is because the devices are not meant to induce reaction immediately. As driving safely involve casual deceleration or coasting, the device simulates more as a yellow signal in traffic light rather than red light, which is discussed to affect drivers' behavior. [15]. Instead, we would like to observe how aware the testers are with extra information we have introduced.

4 Result

4.1 Participants

As shown in Table 1, a total of 9 participants were involved in the experiment. Out of those, 5 were male participants and 4 were female. The average age of the participants is 49.1 years old, 5 were age between 41-50 years old (Mean= 43, SD= 2.35) and 4 were older than51years old (Mean= 56.75, SD= 4.35). All participants have experience in driving at least for 10 years.

Table 1. Participants' background information

Testers #	Age	Gender	Driving experience (years)	Familiarity with smart devices	Car Type	Preferred feedback type
1	42	M	20	Not at all	Truck	Sound
2	42	M	10	Not at all	Truck	Haptic
3	41	M	20	Not at all	Car	Sound
4	51	M	15	Familiar	SUV	Haptic
5	47	F	23	Not at all	Car	Haptic
6	43	F	28	Familiar	Car	Haptic
7	59	F	35	Familiar	SUV	Sound
8	56	F	31	Somewhat	Car	Haptic
9	61	M	39	Familiar	SUV	Sound

4.2 Performance result

From the experiment, a total of 108 red-light stopping were recorded. As demonstrated in Table 2, we calculate a paired sample two-tail T-test and found significant changes in performance from both haptic ($p = 0.052$) and statistically significant in sound feedback ($p = 0.003$). On average, 16.9% improvement for sound and 15.5% for haptic feedback. Some participants show a clear sign of improvement on only single type of feedback.

Table 2. Duration of brake apply

Testers #	Mean - control (sec)	Mean - sound (sec)	Mean - haptic (sec)	% improvement - sound	% improvement - haptic
1	8.34	8.77	10.28	21.23	13.49
2	8.53	10.65	9.99	5.06	23.19
3	8.04	9.75	9.13	24.89	17.18
4	9.07	9.90	9.94	8.81	17.21
5	7.88	9.82	9.54	24.69	21.10
6	8.40	9.14	9.84	9.21	9.62
7	7.71	9.62	10.74	29.10	-19.24
8	13.90	17.95	11.23	24.89	39.42
9	10.01	10.50	12.02	4.92	20.08
Paired sample T-Test against control (p-value)		**0.003**	**0.052**		

4.3 Correlation among data

With the data, we now can select outstanding data to set up and test our hypothesis as well as finding correlations. Our points of interests consist of gender, familiarity with smart devices, eyesight, and driving experience.

Hypothesis 1. Female testers' performance are more positively influenced by sound feedback
Hypothesis 2. Testers who are not familiar with smart devices are more positively influenced by sound feedback
Hypothesis 3. Testers who drive a normal car are more positively influenced by sound feedback
Hypothesis 4. Testers who have good eyesight are more positively influenced by haptic feedback
Hypothesis 5. Testers who are less than 50 years old are more positively influenced by haptic feedback
Hypothesis 6. Testers who has more than 20 years of driving experience are more positively influenced by sound feedback

With above hypotheses set up, we calculate for correlation, t-score, and p-value of the hypothesis.

Table 3. Corelation of Hypothesis

Hypothesis	Correlation coefficient	T-Score	Probability (p)
H1: Female and Sound	0.48	1.43	0.10
H2: Familiarity and Sound	0.49	1.48	0.09
H3: Car and Sound	0.49	1.11	0.15
H4: Good eyesight and Haptic	0.37	1.05	0.16
H5: Age > 50 and Haptic	0.20	0.55	0.30
H6: Experience > 20 and Sound	0.18	0.49	0.32

As calculated in Table 3., only two hypotheses are statistically significant at 90%, Gender and Familiarity with device positively influence the performance. For comparison, average of performance improvement for female testers with sound feedback is 21.87% while male average is 13.03%. Moreover, the testers who are not very familiar with smart devices have an average improvement of 20.99% with sound feedback, contrast with testers who are familiar with smart devices at the average of 11.96% in improvement. The contrasts are pretty strong in these factors.

While the other factors do not carry enough certainty to be statistically significant, they do show somewhat promising influence. The testers who drive cars have average improvement with sound feedback of 20.96%, compare to 13.79% improvement of truck and SUV drivers. The testers who have good eyesight have an average of improvement with haptic feedback of 21.82% while the tersest who have eyes-condition average at 10.95%. Testers with age < 50 years old improve with haptic feedback at the average of 18.45% contrast with older testers' average at 12.47%. Lastly, testers who have more than 20 years of experience improve with sound feedback at the average of 18.48% compare to 15.10% of less experience testers.

5 Discussion

5.1 Additional feedback can help in Performance, and potentially safety

Along with other researches, using feedback devices can help improve the performance of driving. In our case, it is to increase duration in which the brake is applied, and our results can support that. As shown in Table 2., the sound feedback improves the average time to apply the brake by 16.98% while the haptic feedback improves by 15.78%. The paired sample two-tail T-test we conduct also supports this with statistically significant p-value of 0.003 for sound and 0.052 for haptic feedback. Moreover, that the sound feedback is more consistent with standard deviation of 9.77, while haptic feedback has value of 15.55.

To translate this to real world scenario, a driver traveling at speed of 40 – 60 Km/second usually 5 – 10 seconds to apply brake. The improvement of 20% has a potential to cover the distance of 10 – 20 meters. Under some scenarios, this improvement could prevent split-second incidences.

5.2 Familiarity with smart devices influent effectiveness of sound feedback

According to the result, testers who are not familiar with smart devices tend to do better with sound feedback. In contrast, those who are familiar with smart devices reveal a higher improvement when they are equipped with haptic feedback.

There could be many reasons behind this factor. One of the very likely reasons is that, the persons who are familiar with smart devices are already familiar with generic notification sound, making the notification sent by our device has less impact. For unfamiliar testers, however, this may be a very alien sound, thus, it grabs their attention better.

5.3 Gender might indirectly influenced performance

As shown in Table 3., female testers have significantly higher improvement with sound feedback compare to male counterpart. This may not be the direct cause, but the combination of habits and equipment usage while driving. Some testers use radio or car console more while they drive. Some testers have louder cars or applied gas heavily. Some testers store their phones in a bag or a purse. All of these habits have potentials to interfere with sound feedback, and thus, might be the reason behind the difference between genders.

5.4 Environment and personal preference

From our experiment, we notice a wide variety of environment the devices will be used in. Inside each person's cars are all different. Some drive with windows down or radio on. Some drive trucks or use manual transmission. This can cause great impact on feedback devices because of the noise. Simply put, we expect testers with noisy cars to do badly with sound feedback, and expect shaky cars to do badly for haptic.

The result, however, is not consistent. Some testers would drive in incredibly shaky cars, such as a truck for public transport, and still have better performance with haptic feedback. Moreover, their personal preferences play almost no part in performance. Some testers would discuss and prefer one feedback type, but can still end up with better performance in another feedback type. Out of 9 testers, only 3 prefer and perform better at the same type. This may link to user friendliness and degree of frustration of the devices.

5.5 Eyesight is not a significant factor

According to our result, the eyesight is not a very influential factor when using our devices. This might mean that the devices are as effective for both who do and do not have eye condition. In fact, only 2 out of 9 testers wear glasses while driving. Some of our testers do not wear glass while driving, and they are familiar with the route.

This aspect is something to consider in future researches. Current technology can already provide routing navigation via sound. The need to use eyes might not be as big of a factor in smart devices usage. In other words, anyone can use the feedback devices with relatively similar effectiveness regardless of their eyesight.

5.6 Age and experience are not strong factors

Similar to eyesight, age and experience do not play a big role according to this experiment. And since the trend of driving becoming less and less complicated, physical condition and skilled execution might not have much impact in performance. Testers who are younger or have less experience with driving perform practically the same as the older and more experienced one.

5.7 Additional user interface factors

Part from the survey we conducted, we also asked for comments from the testers. We then discovered several user interface factors that could potentially affect the result. One of the factors is the alert sound from the feedback devices. We used generic smartphone sound as we feared using any extreme sound would annoy the testers. Instead, it turned out some testers felt the sound was not distinct enough as they are already familiar with generic smartphone sound. The other factor is the haptic feedback itself. As people have different preference, our prototype might vibrate too much or too little.

To sum up, the comment was that our user interface, sound and haptic, feel too similar to that of smartphones, and that might be the reason why those who are familiar with smart devices performed differently. If an opportunity arises, this should be further researched in the future.

6 Conclusion

The objective of this paper is to study the influence of sound and haptic feedback from smart wearable device on driving performance of elderly. The performance is measured with an assumption that those who spend more time to apply brake tend to be more prepared and aware of their surroundings. We create a prototype to simulate sound and haptic feedback and conduct a driving experiment. From the performance result, along with T-test, we found that in general, sound and haptic feedback can improve driving performance. Between the two feedbacks, sound feedback tends to be more consistent and more influential to performance than haptic feedback. We then use the performance result to find the correlation between performance and several other factors to find which have influence on the feedback usage. Two factors came out as statistically significant. One factor is the familiarity of drivers with smart devices. Drivers who are not familiar with smart devices perform better with sound feedback. Another factor is gender. Female drivers can perform better with sound feedback. We believe this is not a direct cause, but rather a tendency to have different driving environment such as using car console and keeping the phone in a purse.

All in all, this paper suggests that smart wearable devices has a potential to improve performance with vary effectiveness depends on the driving environment and personal preference. With upcoming trend of smart devices, these factors should be included into consideration as it may influence the effectiveness of devices and potentially traffic safety.

7 Reference

1. Transport Statistics Sub-Division (2014). Department of Land transport.Statistic of land traffic accident in the area of Royal Thai Police [Excel file]. Retrieved from http://apps.dlt.go.th/statistics_web/st1/accident.xls on July 21, 2014.
2. Ronnie Taib, Kun Yu, Jessica Jung, Anne Hess, Andreas Maier (2013). Human-Centric Analysis of Driver Inattention
3. Piotr Calak (2013). Smartphone evaluation Heuristics for Older Adult.
4. Kazuko Asano, Hideki Nomura, Makiko Iwano, Fujiko Ando, NaoakiraNiino, Hiroshi Shimokata, Yozo Miyake (2005), Relationship between astigmatism and aging in middle-aged and elderly Japanese
5. Saunders H (1981). Age-dependence of human refractive errors. Ophthalmic Physiol Opt 1981;1:159–174.
6. Hayashi K, Masumoto M, Fujino S, Hayashi F (1993). Change in corneal astigmatism with aging. Nippon GankaGakkaiZasshi 1993;97: 1193–1196.
7. Nomura H, Tanabe N, Nagaya S, Ando F, Niino N, Miyake Y, Shimokata H (2000). Eye examinations at the National Institute for Longevity Sciences–Longitudinal Study of Aging: NILS-LSA. J Epidemiol 2000;10:S18–25.
8. KazutakaMitobe, Masafumi Suzuki and Noboru Yoshimura (2012). Development of Pedestrian Simulator for the Prevention of Traffic Accidents Involving Elderly Pedestrians
9. William Consiglio, Peter Driscoll, Matthew Witte, William P. Berg (2003). Effect of cellular telephone conversations and other potential interference on reaction time in a braking response.
10. SeungJun Kim, Jin-Hyuk Hong, Kevin A. Li, Jodi Forlizzi, and Anind K. Dey (2012). Route Guidance Modality for Elder Driver Navigation.
11. Dingus, T.A., Klauer, S.G. (2008). The relative risks of secondary task induced driver distraction.Society of Automotive Engineers, Technical Paper Series 2008-21-0001
12. Horberry, T., Anderson, J., Regan, M.A., Triggs, T.J., Brown, J. (2006). Driver distraction: Theeffects of concurrent in-vehicle tasks, road environment complexity and age on drivingperformance. Accident Analysis & Prevention 38, 185
13. Kim, S., Dey, A.K.(2009). Simulated augmented reality windshield display as a cognitive mappingaid for elder driver navigation. In: Proc. CHI 2009, pp. 133–142
14. Kim, S., Dey, A.K., Lee, J., Forlizzi, J. (2011). Usability of car dashboard displays for elder drivers.In: Proc. CHI 2011, pp. 493–502
15. Chang, M., Messer, C., & Santiago, A. (1985). Timing traffic signal change intervals based on driver behavior. Transportation Research Record, 1027, 20–32.

Indoor WIFI localization on embedded systems

Anya Apavatjrut and Ekkawit Boonyasiwapong

Chiang Mai University,
Chiang Mai, Thailand
anya@eng.cmu.ac.th

Abstract. Localization has recently become increasingly important for ubiquitous computing. Most embedded devices require location-based services for monitoring and tracking purposes. Typically the location services provided by these devices can be achieved using global positioning system (GPS). Although GPS can indicate absolute position based on satellite system, several studies have demonstrated its inaccuracy when deployed indoor due to signal degradation from the obstacles and building structures. In this paper, we have developed a testbed that provided indoor positioning services based on multi-lateration technique. The empirical results have shown that appropriate indoor propagation model can improve the positioning accuracy, several propagation models have been studied and statistical models have been proposed for a site-specific environment.

Keywords: WIFI localization, multi-lateration, propagation model

1 Introduction

Ubiquitous computing usually requires localization for monitoring and tracking. Most devices nowadays can provide location-based services using global positioning system (GPS). The GPS provides an absolute position with a worst case pseudo range accuracy of 7.8 meters at a 95% confidence level [9]. The accuracy is however worsen when deployed indoor as the signal is obstructed by the building structures.

Wireless localization has been proposed as GPS alternative when deployed indoor where localizing with GPS is limited. The principle is to determine relative position of the unknown device based on known location of selected anchor transmitters. There exist several other wireless localization techniques based on both range-based and range-free localization techniques which will later described in section 2.1. However, most of the previous work were developed and their performance were measured based on the simulations. In [13], the authors have realized some of these existing techniques on real testbed and have concluded that real world performance of localization results presents severe limitations due to signal degradation i.e. attenuation, fading, multipath propagation. The results given by the testbed is much worsen from those derived from the simulation. The authors have stated however that accurate channel model may leverage these problems.

K.J. Kim (ed.), *Information Science and Applications*,
Lecture Notes in Electrical Engineering 339, DOI 10.1007/978-3-662-46578-3_8

Thus, in this paper, we aim at focusing on accuracy improvement of localization process by finding appropriate channel models.

The remainder of this paper is organized as follows. In section 2, we surveyed related work in propagation model for WIFI indoor localization. We then discussed the research methodology and presented the overview of system model in section 3.1. Next, the results obtained from the testbed will be discussed in section 3. Finally, we conclude our research work and its perspective in the last section.

2 Literature reviews

2.1 Localization Technique

The principle of wireless localization is to define relative position of an unknown device based on a number of reference nodes with known positions. These reference nodes can be referred to as *beacon* or *anchor* nodes. The localization techniques can be divided into two main categories: range-based and range-free localization [6].

In range-based localization, the localization process makes use of the received signal strength indicator (RSSI) to estimate the distance to the anchor nodes using various channel models. In range-free localization, however, the computation of distance is not required, the position can be determined based on RSSI mapping. By collecting samples of available RSSI at each area during the calibration, the positioning process can delegate an unknown position to be the area which provide the closest samples of RSSI previously collected. Among these localization techniques, multi-lateration and fingerprinting are the most commonly used example of range-based and range-free localization respectively. These techniques were initially presented in [3]. More work on multi-lateration and fingerprinting for WIFI indoor localization can be found in [2, 5].

In our previous work, we have compared in [4] two localization techniques: multi-lateration vs. fingerprinting using real testbed. The results has demonstrated that fingerprinting outperforms multi-lateration positioning technique but required excessive pre-calibration time, resource and computation overhead. The multi-lateration, on the other hand, requires less configuration, however, it provides lower accuracy. However, in the previous work, we deployed a free-space path loss model that neglected the deterioration of signal from the building structure. To obtain a more realistic results, we are interested in this work to determine a more realistic indoor propagation models that take into account the signal corruption due to attenuation, fading and multipath propagation.

2.2 Indoor propagation model

The primary causes of signal degradation are related to attenuation, penetration losses through obstacles as well as multipath propagation caused by reflection, diffraction, and scattering. Usually, the indoor propagation model varies significantly with the environment i.e. building structures, wall material and is usually difficult to be determined. When the signal corrupts, the distance from the reference APs will be computed with error.

In the literature, there are several studies both theoretical and experimental that proposed indoor propagation models. In [11, 8], the statistical model is suggested for indoor radio channel for the simulation and analysis of various indoor communication schemes. In [12], the authors proposed exponential path loss model for different kinds of obstructions between a transmitter and a receiver. We can resume existing indoor propagation model as follows:

Log-Distance Path Loss The signal decays with the distance d. This reduction in the signal level can be modeled using the log-distance path loss modeling [10]:

$$PL(dB) = PL(d_0) + 10n \log \frac{d}{d_0} \tag{1}$$

where d_0 is the close-in reference distance in meters and n is the path loss exponent. The value of n varies depending on propagation environment, i.e., building construction, material, architecture. For example, in free space, n is equal to 2 whereas n takes higher value with the presence of obstruction.

Log-Normal Shadowing The obstructions from the transmitter and the receiver can cause different attenuation, delay and phase shift to the signal. This produces variation of signal levels on a propagation path. The effects of random shadowing can be referred to as log-normal distribution as follows [10]:

$$PL(dB) = PL(d_0) + 10n \log \frac{d}{d_0} + X_\alpha 32 \tag{2}$$

where X_α is a zero-mean Gaussian distributed random variable with standard deviation α attempting to compensate random shadowing resulting from clutter.

Addition of Attenuation Factors In reality, it is hard to describe an exact generic indoor propagation model as the indoor environment differs widely with e.g. material, architecture, size, layouts. These factors effect the characteristics of propagated signal. Thus, empirical model are used to characteristics all the factors that affect environment influences of signal loss. The pathless model can be described as [7]:

$$PL(dB) = PL(d_0) + 10n \log \frac{d}{d_0} - K + \sum_{i=1}^{N_f} FAF_i - \sum_{i=1}^{N_p} PAF_i \tag{3}$$

with K a constant typically obtained via a regression fit to empirical data, FAF_i representing floor attenuation factor for the ith floor traversed by the signal, PAF_i represent i^{th} partition attenuation factor traversed by the signal. The number of floor and partitions that the signal traversed is presented by N_f and N_p respectively. The example values of the parameters of penetration losses

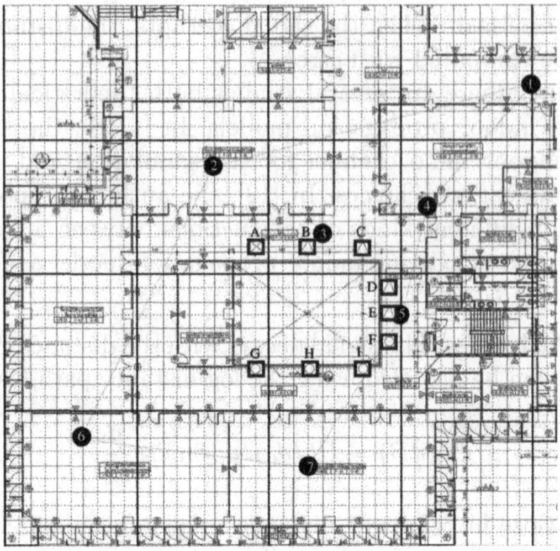

Fig. 1. The layout of the floor plan where we performed the experiments.

for multi-floor and different obstacles are derived from empirical data and be found in [7]. Empirical model for wireless IEEE802.11 WIFI propagation are proposed in [1].

3 Experimental results and evaluation

3.1 Testbed

The experimental results in this paper were derived from a testbed based on Raspberry pi and D-Link wireless adapter DWA-132. Our platform consists of a Raspberry pi module and a remote server. The Raspberry pi computed its location periodically and transferred this coordinate position information to the remote server wirelessly. The remote server provided a real-time graphic interface allowing a user to monitor and visualize the Raspberry pi position within a floor plan of the building.

Our objective is to evaluate the positioning accuracy within a single floor of a building. The floor plan layout where the experiments are performed is shown in figure 1. The position of access points (AP) that are displayed as anchor positions are presented as circle symbols and the experiment is performed on 9 different locations presented as square symbols. Each simulation results is collect every 2 seconds for 10 minutes.

3.2 Indoor Propagation Modeling

In order to determine the propagation model that best described indoor wireless signal transmission. We have studied attenuated pattern of signal coming from

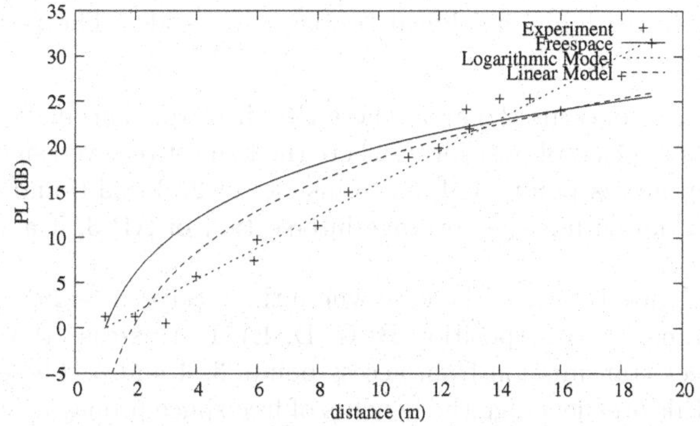

Fig. 2. Attenuation pattern

an access point with two layers of wall obstruction. As wireless signal decays with distance, figure 2 represents the average path loss as a function of the distance from the AP collected during the experiments. Noting that the APs were installed on the ceiling, the distance shown in this figure represent relative distance from the AP collected on the ground not the absolute distance.

From equation 3, we can reformulate the equation as:

$$PL(dB) = PL(d_0) + 10n \log \frac{d}{d_0} + c \tag{4}$$

where n is an attenuation factor and c a constant that characterizes a loss factor due to obstacles inside the building.

By fitting this equation to the experimental results using nonlinear least-squares (NLLS) Marquardt-Levenberg algorithm, we obtained equation 5 with n=2.6 and c=-7.36 that best characterizes signal attenuation in logarithmic pattern. The equation can be rewrite as:

$$PL(dB) = PL(d_0) + 2.6 \cdot 10n \log \frac{d}{d_0} - 7.36 \tag{5}$$

We remarked however that the path loss data collected during the experiment is more likely to decrease in linear manner, hence we propose to investigate also the linear path loss model which can be represented via the following equation:

$$PL(dB) = PL(d_0) + 1.77(d) - 1.82 \tag{6}$$

In the next section, we are investigating the positioning results using the models from equations 5 and 6.

3.3 Path loss model evaluation

Using ranging techniques from equation 5 and 6, we compare in this section the accuracy of trilateration (multi-lateration with 3 anchors) using 3 channel models: free-space, logarithmic and linear propagation model. The positing results fall into 3 categories as follows:

- The first case concerns the area where the the error margin exceeds 10 meters. This is the case of position A where the highest received signal strengths comes from access points 3, 4 and 5 instead of 2, 3 and 6 and lead the positioning to be within the coverage intersection of AP 3, 4 and 5 which is incorrect.
- The second case concerns the case where the receiver is exposed to at least 2 direct transmissions (position B, C, D, E). In this case the receiver can detect direct transmission from access points 3, 4 and 5. The experimental results in this area leverage the benefits of free-space propagation model over logarithmic and linear model (see figure 3).
- The last case represents the case where direct transmission from the APs is limited where at least 2 out of 3 received signals come from the sources which are out of sight, most signals are objected to obstacles or building structures (position G and H), the logarithmic and the linear model defined in equation 5 and 6 gives more accurate results than free-space propagation. (see figure 3)

3.4 Hybrid model

From the experiments in section 3.3 we remarked that the free-space propagation outperforms both logarithmic and linear model when low level of signal obstruction is presence. Otherwise, the other two give more accurate positioning results. The linear model seems to improve the positioning results in lower coverage area. We proposed a hybrid model that aim at combining the benefits between the free-space propagation and the linear model. With an assumption that high received signal strength is likely to come from the access points which is close or is within the line of sight of the transmitter. We assume that when the receiver intercepts a signal strength higher than a certain threshold, the logarithmic path loss model will be deployed, in other circumstances, the linear model is adopted. The path loss threshold used in this experiment is equal to 5 dB. We noticed that the hybrid model can leverage the benefits of the two models and gives more accurate results than any other models in most of the experiments (see figure 3).

3.5 Case study: limited coverage area

The results in section 3.3 has confirmed the benefit of using logarithmic and linear propagation models in the area where direct transmission from the referenced access points is limited. We want to confirm the results by creating a scenario with limited coverage by disable an access point number 3 situated in the middle of the floor plan. The positioning results after redoing the same experiment demonstrated that only 7 out of 9 positions could be determined within error margin of 10 meters using linear and hybrid model (see figure 4). The free-space and logarithmic model can not achieve satisfied error margin in most cases.

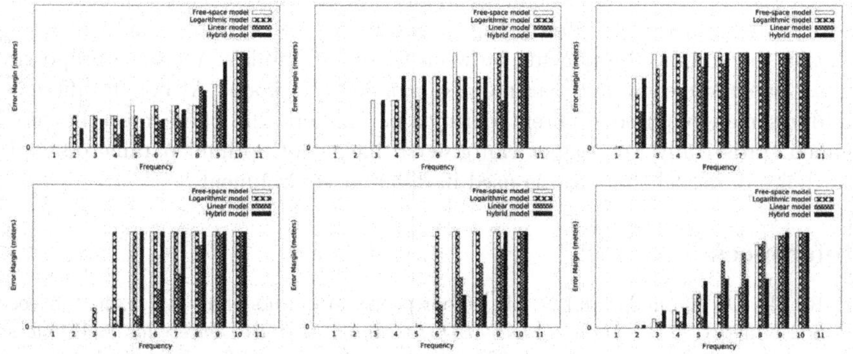

Fig. 3. Frequency distribution of success rate at different error margins for position B, C, D (top) F, G, H (bottom) in full WIFI coverage scenario

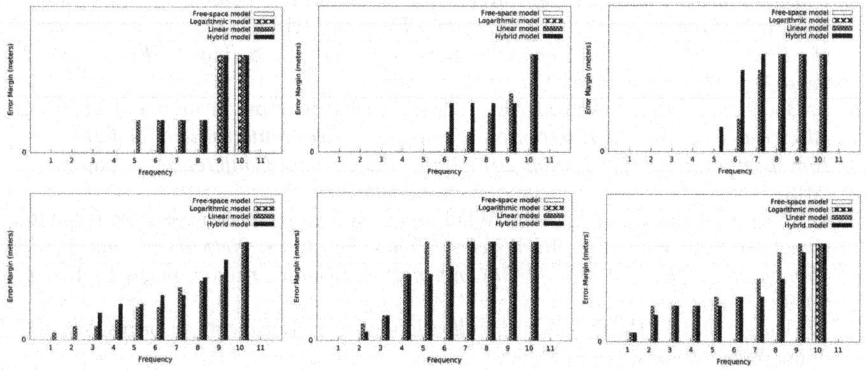

Fig. 4. Frequency distribution of success rate at different error margins for position B, C, D (top) F, G, H (bottom) in limited WIFI coverage scenario

4 Conclusion

In this paper, we proposed to investigate indoor propagation modeling for site-specific environment. The proposed empirical models have taken into account the effects of signal deterioration caused by obstacles and building structures. We have compared the positioning results for four different channel models: free-space, logarithmic, linear and hybrid model using the same dataset using trilalteration technique. By defining appropriate parameters for each model, the results have shown that positioning performance depends tightly on the choices of propagation modeling. Even in different area within the same building, optimal model may differs. We have demonstrated that although logarithmic model has been widely deployed as an indoor channel model, when wireless signal is subject to high level of obstruction, the linear model outperforms the logarithmic model. In order to improve the positioning performance, the hybrid model can be a solution to combine the benefits between these two models. We have proposed in this paper some optimal parameters to define the model. Although different buildings may required different optimal parameters and these must be pre-defined statistically, this paper can give an insight of optimal parameters and the choice of model that can be used in different environment.

References

1. R. Akl, D. Tummala, and X. Li. Indoor propagation modeling at 2.4 ghz for ieee 802.11 networks. In *The sixth IASTED International Multi-Conference on Wireless and Optical Communications, 2006. Wireless networks and Emerging Technologies*, july 2006.
2. T. Bagosi and Z. Baruch. Indoor localization by wifi. In *Intelligent Computer Communication and Processing (ICCP), 2011 IEEE International Conference on*, pages 449 –452, aug. 2011.
3. P. Bahl and V. Padmanabhan. Radar: an in-building rf-based user location and tracking system. In *INFOCOM 2000. Nineteenth Annual Joint Conference of the IEEE Computer and Communications Societies. Proceedings. IEEE*, volume 2, pages 775 –784 vol.2, 2000.
4. S. Boonsriwai and A. Apavatjrut. Indoor wifi localization on mobile devices. In *Electrical Engineering/Electronics, Computer, Telecommunications and Information Technology (ECTI-CON), 2013 10th International Conference on*, pages 1–5, 2013.
5. Q. Chen, G. Huang, and S. Song. Wlan user location estimation based on receiving signal strength indicator. In *Wireless Communications, Networking and Mobile Computing, 2009. WiCom '09. 5th International Conference on*, pages 1 –4, sept. 2009.
6. W. Dargie and C. Poellabauer. *Localization*, chapter Localization, pages 249–266. John Wiley & Sons, Ltd, 2011.
7. A. Goldsmith. *Wireless Communications*. Cambridge University Press, New York, NY, USA, 2005.
8. H. Hashemi, D. Tholl, and G. Morrison. Statistical modeling of the indoor radio propagation channel. i. In *Vehicular Technology Conference, 1992, IEEE 42nd*, pages 338–342 vol.1, 1992.
9. D. of Defense of United States of America and G. NAVSTAR. Global positioning system standard positioning service performance standard, September 2008.
10. T. Rappaport. *Wireless Communications: Principles and Practice*. Prentice Hall PTR, Upper Saddle River, NJ, USA, 2nd edition, 2001.
11. A. A. M. Saleh and R. Valenzuela. A statistical model for indoor multipath propagation. *Selected Areas in Communications, IEEE Journal on*, 5(2):128–137, 1987.
12. S. Seidel and T. Rappaport. A ray tracing technique to predict path loss and delay spread inside buildings. In *Global Telecommunications Conference, 1992. Conference Record., GLOBECOM '92. Communication for Global Users., IEEE*, pages 649–653 vol.2, 1992.
13. G. Zanca, F. Zorzi, A. Zanella, and M. Zorzi. Experimental comparison of rssi-based localization algorithms for indoor wireless sensor networks. In *Proceedings of the Workshop on Real-world Wireless Sensor Networks*, REALWSN '08, pages 1–5, New York, NY, USA, 2008. ACM.

Bacterial Foraging-based Power Allocation for Cooperative Wireless Sensor Networks

Mohammad Abdul Azim[1], Zeyar Aung[2⋆], and Mario E. Rivero-Angeles[3]

[1]Network and Avionic Software Lab, Gyeongsang National University, Korea.
azim@ieee.org
[2]Institute Center for Smart and Sustainable Systems (iSmart),
Masdar Institute of Science and Technology, Abu Dhabi, UAE.
zaung@masdar.ac.ae
[3]National Polytechnic Institute, CIC-IPN, Mexico City, Mexico.
mriveroa@ipn.mx

Abstract. Cooperative communication becomes a popular area of research due to its strength and wide application scope in wireless networking and communications. This technique improves the communication performance largely in capacity enhancement, energy-efficiency, timeliness and contention. Power allocation plays an important role in the cooperative communication paradigm to get the desired performance improvements in the aforementioned aspects. In this paper, we present a bacterial foraging optimization algorithm (BFOA)-based power allocation method for cooperative communications in wireless systems. Comparative measures with non-cooperative approaches are made to justify our proposed method.

Keywords: Wireless Communications, Bacterial Foraging, Constrained Optimization, cooperative Communication, Signal Combining.

1 Introduction

Cooperative communication attracts much attention in wireless communications due to its potential to increase the received signal strength with multiple packets using reduced transmit power. Increased signal strength in turn provides capacity enhancement, energy efficiency, timeliness and reduction in contention.

For example, if the source node S intends to send data to the destination D, the relay node R also sends the same packet received from the source node S. The destination node D therefore combines both the signals received from the source and the relay in order to boost the signal strength. Here, a single relay or a number of relays in the vicinity can contribute to the relaying activity.

In cooperative communications relay node selection and power allocation have ultimate impact on the performance. The best node or the set of best nodes participating as relay(s) influence(s) the performance as the channel condition varies from link-to-link. On the other hand the tradeoff exists as improving the signal strength by injecting packets increases the contention over the networks.

⋆ Corresponding author.

K.J. Kim (ed.), *Information Science and Applications*,
Lecture Notes in Electrical Engineering 339, DOI 10.1007/978-3-662-46578-3_9

In wireless sensor networks (WSNs), distributed approaches dominates the design criteria of all the aspects of networking and selection mechanisms. Considering a homogenous deployment it is not desirable some of the tiny sensors will bear the additional responsibility maintaining and managing the clusters. Centralized environment generally requires explicit messaging for maintenance and therefore undesirable for such homogeneous sensor network deployments.

The **major contributions** of this paper are: (1) Solving the power allocation problem on the cooperative communication paradigm in WSN under constrained receive sensitivity is addressed by utilizing bacterial foraging optimization algorithm (BFOA) in order to minimize the transmit power consumption while achieving the desired received signal strength. (2) The proposed technique provides throughput, energy, timeliness and contention based performance improvements due to the enhancement of the received signal strengths while sending packets with small fraction of transmit power compared to the non-cooperative (NC) communications.

2 Background

Cooperative communications have been increasingly studied in the context of WSN. Reference [16] proposes cooperative communication scheme where sensors in each cluster relay data to the adjacent clusters utilizing the cooperative communication scheme. This technique is only suitable for the cluster based sensor networks. This proposition is not suitable in cases where all the nodes in the network has similar role and no node is superior enough to take additional responsibility to serve as cluster head.

Among the subtopics of the cooperative communications power allocation techniques dominates the interest due to its importance. Reference [14] proposes half transmission power allocation solution (HTPAS) where the transmission power of the relay is set to half of the source's power. Results show the performance is better than the equal power allocation scheme. This simple scheme neglects intelligent allocation of constrained minimization of transmit power.

Equal power allocation (EPA) equally allocate powers to the sender and receiver with a constrained outage probability [5]. The family of simple equal power allocation schemes are almost always inferior to the optimized power allocation scheme while considering the power consumption.

Reference [8] provides a heuristic based approach where the algorithm attempts to minimize the energy cost by utilizing an iterative approach. Reference [11] provides two different heuristic approaches named maximal channel gain (MCG) and least channel correlation (LCC) algorithms under different constrained such as the energy and delay constraints. Hubristic algorithms are rather intuitive approaches. We intentionally avoid designing a hubristic based approach because of the lack of fundamental background behind the algorithms.

Reference [7] solves the power allocation problem in the cooperative communications with decode and forward mechanism where the constraint is defined as the symbol error probability of the radio links. The approximate solution takes a form of water-filling strategy. Unfortunately the water-filling approaches poses the limitation of on convergence in multiple constraints and the improved solution has been proposed [10] for cognitive cooperative communications.

A number of different optimization tools have been utilized in the cooperative communication paradigm. Reference [6] proposes a joint optimization of routing, relay node selection and power allocation under constraint signal to noise ratio by utilizing mixed-integer optimization framework. Reference [4] proposes linear programming approach to solve the multi-source, multi-relay cooperative networks. Reference [15] presents reliable and energy efficient cooperative communication (REEC) technique utilizing simulation annealing based optimization approach to solve the power allocation problem in cooperative communications. Contrarily, we have developed a cooperative power allocation technique for resource constrained WSN based on the BFOA.

3 Proposed Method

This section deals with the underlying theme of the BFOA, the cost function utilized in the BFOA along with the complementary techniques utilized along with our cooperative power allocation scheme. Where the proposed algorithm provides output as Θ:$(P_{Tx}(S)_{dBm}, P_{Tx}(R)_{dBm})$.

3.1 Bacterial Foraging Optimization Algorithm (BFOA)

BFOA is a bio-inspired optimization algorithm similar to ant colony optimization or particle swarm optimization. BFOA is derived from the group foraging behavior of bacteria such as E.coli and M.xanthus. More precisely it is the chemotaxis behavior of bacteria that will perceive chemical gradients in the environment (such as nutrients) and move toward or away from specific signals.

Bacteria decides about the direction of the food based on the gradients of chemicals in the surrounding environment. On the other hand bacteria's secretion attracts and repels each other and control their search direction in the environment. E.coli bacteria uses flagella moves chaotic by tumbling and spinning. It also moves directionally known as swimming. Utilizing (i) chemotaxis (ii) reproduction and (iii) elimination dispersion the bacteria stochastically and collectively swarm toward optima. The detail algorithmic description of BFOA is found in [9].

Mimicking the bacterial behaviour the underlying algorithmic functionalities follow the steps: (a) initializing foraging parameters and variables (b) specifying upper and lower limits of the variables (c) initializing elimination-dispersal, reproduction and chemo-taxis (d) generating bacterial population and positions (e) updating bacteria locations by using tumbling and swimming and (f) executing bacterial reproduction and elimination process.

Let us denote the number of bacteria as S_b, the chemotactic loop limit N_c, the swim loop limit N_s, the reproduction loop limit N_{re}, the number of bacteri for reproduction S_r, the elimination-dispersal loop limit N_{ed}, step-sizes C_i, and the probability of elimination dispersal P_{ed}, respectively.

The chemotactic step is modeled with the generation of a random search direction by $\phi(i) = \Delta(i)/(\sqrt{\Delta(i)^T\Delta(i)})$ also known as tumble, where $\Delta(i)^n$ is the n-dimensional vector with elements having limits of $[-1, 1]$. Each bacterium $\theta^i(j,k,l)$ modifies its position known as swim by $\theta^i(j+1,k,l) = \theta^i(j,k,l) + C(i)\phi(i)$, where C_i is the step-size for each direction $\phi(i)$.

```
BFOA power allocation Θ:(P_Tx(S)_dBm, P_Tx(R)_dBm)
while (Improvement > Threshold) {
    Θ = F(Θ) {
        Initialize: S_b, N_c, N_s, N_re, S_r, N_ed, C_i and P_ed
        Create initial swarm of bacteria as θ^i(j,k,l)∀i where i = 1 : S_b
        Find τ = f(θ^i(j,k,l))∀i
        for l = 1 : N_ed {
            for k = 1 : N_re {
                for j = 1 : N_c
                    for i = 1 : S_b
                        Chemotactic step for bacterium θ^i(j,k,l) controlled by N_s
                    Reproduction with probability S_r
                }
                Elimination-dispersal with probability 0 ≤ P_ed ≤ 1
            }
        }
    }
    τ = ℵ(Θ) {
        p_f = 3.01^i
        if (P̂_Rx(S) + P̂_Rx(R) > P_Rx(Th)) {
            C_BB(P̂_Tx(S), P̂_Tx(R)) = P_Rx(Th) − (P̂_Rx(S) + P̂_Rx(R))
            ℵ = P̂_Tx(S) + P̂_Tx(R) + |log(−C_BB * p_f)|
        }
        else
            ℵ = ∞
    }
```

Fig. 1. BFOA power allocation algorithm.

In each swim the cost is evaluated. If $f(\theta^i(j+1,k,l)) < f(\theta^i(j,k,l))$ the swim is repeated up to N_s times.

The reproduction steps consists of sorting the bacteria population in terms of the objective function. The worse S_r number of the population is destroyed and the best S_r population is duplicated to maintain the population fixed.

In elimination-dispersal step elimination of the bacteria takes place with a probability of P_{ed}, where $0 \leq P_{ed} \leq 1$.

3.2 Cost Function for BFOA

We attempt to utilize the algorithmic strength of BFOA to minimize the energy cost while maintaining the minimum allowable received signal strength utilizing the maximal ratio combining (MRC) scheme of the cooperative sensor networks.

Let, $P_{Tx}(S)$ and $P_{Tx}(R)$ are the powers to be assigned at the transmitters at the source S and relay R. And let $P_{Rx}(Th)$ be the receive sensitivity of the receivers. The designed algorithm therefore assigns Θ:$(P_{Tx}(S)_{dBm}, P_{Tx}(R)_{dBm})$ such that the resultant signal strength by MRC combine at the destination D becomes greater than the receive sensitivity. Alternately the sum of the received signal strengths $P_{Rx}(S)$ and $P_{Rx}(R)$ is greater than $P_{Rx}(Th)$. Let, the $P_{Tx}(min)$ and $P_{Tx}(max)$ are the minimum and maximum allowable level of the transmit power respectively. Then, $P_{Tx}(min) \leq P_{Tx}(S) \leq P_{Tx}(max)$ and $P_{Tx}(min) \leq P_{Tx}(R) \leq P_{Tx}(max)$ becomes the other constrains of the minimization.

$$P_{Tx}(T) = P_{Tx}(S) + P_{Tx}(R) \text{ s.t. } P_{Rx}(S) + P_{Rx}(R) > P_{Rx}(Th) \text{ and } P_{Tx}(Min) < P_{Tx}(S) < P_{Tx}(Max) \text{ and } P_{Tx}(Min) < P_{Tx}(R) < P_{Tx}(Max) \quad (1)$$

Penalty Function To achieve to aforementioned constrained optimization we employ BFOA. Note that the core BFOA provides method for the simple optimization. Making it a constrained optimization several techniques may apply such as primal method, penalty method dual and cutting plane method Lagrangian method. The proposed algorithm employ the logarithmic penalty function method while achieving the constrain in the BFOA.

Let $\hat{P}_{Rx}(S)$ and $\hat{P}_{Rx}(R)$ are estimated received power at D from S and from R respectively. Algorithm approximate the resultant receive power. If this evaluated power becomes less than the acceptable threshold then the basic building block of the penalty cost is set to $C_{BB}(\hat{P}_{Tx}(S), \hat{P}_{Tx}(R)) = P_{Rx}(Th) - (\hat{P}_{Rx}(S) + \hat{P}_{Rx}(R))$. Otherwise, the penalty cost is set to ∞. The resultant cost is defined as $\tau = \hat{P}_{Tx}(S) + \hat{P}_{Tx}(R) + |\log(-C_{BB} * p_f)|$. Here p_f is a multiplication factor that depends on the algorithmic round. In our evaluation we define $p_f = 3.01^i$ for the i_{th} round of the algorithmic run. Note that, $\hat{P}_{Rx}()$ can be modeled as $\hat{P}_{Rx} = \Im(P_{Tx}, d, \gamma)$ as per Friis transmission equation.

3.3 Complementary Techniques

The proposed power allocation technique is the bare allocation of power at the sender and the relay nodes. The complementary techniques therefore are the other mechanisms required to complete the other functionalities of the networking; described in the following subsections.

Routing: The sensor network can be of a flat or a cluster based structure. In homogeneous deployment a simple flat architecture is often preferred. In the flat architecture, the well-known distributed routing technique named greedy forwarding attracts much attention in recent days due to its simplicity and usefulness. The distributed geographic location aware greedy forwarding algorithm is the basis of our routing approach [12]. Further specifying, the Euclidian least reaming distance based routing algorithm employed in our evaluation of the sensor networks as [3].

According to this approach, let the node i intends to forward a packet toward the destination. And let N_i defines the set of the neighbor nodes of node i. Node i calculates $\min(d(N_i, C))$ for the set of N_i and selects the calculated node as the next-hop. Note that $d(j, k)$ defines the Euclidean distance of node j and k.

Access Control: Due to the distributed nature of the WSN architecture, carrier sense multiple access with collision avoidance (CSMA/CA) based simple medium access control (MAC) mechanism is taken as the basis of the access control mechanism. To accommodate the cooperative communications we in fact implemented a modified CSMA/CA in our approach.

Here, each node detecting a busy medium waits for at least twice the transmit time requirement of a single packet before the next attempt to access the medium. Such modification attempts to is to provide priority on relaying packets as the relay nodes priorities cooperatively sending the copy of the packet over its own data packets or the packets it acts as a next hop node to forward. Without such modifications in a simple MAC the same packet occupy unnecessary buffer as multiple copies of same packet remains in multiple nodes over the time.

Relay Selection: Multiple relays can contribute more in the signal boost up with a trade of in the bandwidth inefficiency. In cooperative relay networks n number of relay nodes require $n+1$ different channels. A number of relay selection mechanism therefore introduced in the literature to find the best relay node(s). To achieve such bandwidth advantage we therefore chosen a single relay based cooperative communication scenario in our approach.

In a single relay based approach a number of min-max based relay selection mechanism is devised. Imperfect channel information with min-max algorithm is utilized in our approach while defining the relay node using the method in [13].

According to this approach let a packet is flowing through path consisting of links L_i. For each link L_i there exists a source S and a destination D pair. A potential relay of the link L_i can be of the neighbor nodes in the forwarding path only. Let N_{fi} defines the set of nodes satisfies all the three conditions: (i) N_{fi} are in the radio range of S and D (ii) located closer to the sink compared to the S and finally (ii) closer to the S compared to the D.

Here, the costs of the channels of any pair of nodes ($S \to N_{fi}$, $N_{fi} \to D$ and $S \to D$) are define by the reciprocal to received signal strengths of the pairs. Note that the signal strength can be depicted as a simplified view of the channel condition. Let $C(i, j)$ define the transmission cost from node i and j. The selected relay therefore is the relay satisfies $\max(\min(C(S, N_{fi}), C(N_{fi}, D)))$.

Relaying: The relaying is of amplify and forward approach where a single relay is utilized as an helper node that relay the received packet from the source without any decoding. Upon receiving both the packets from the source and relay the receiver utilize maximal ratio combining (MRC) as signal combining approach to boost the signal strength.

MRC combines the signal in such way that the gain of each channel is made proportional to the root mean square (RMS) signal level and inversely proportional to the mean square noise level in that channel. A detailed performance analysis of single relay amplify forwarding based MRC is in [2], limited to physical layer characterization. Note that this characterization that follows the trend of cooperative communications research where the upper layers are completely ignored. Note that such evaluation is sufficient only for the TDMA like MAC approaches. For the distributed MAC layering approach as CSMA-CA a upper layer consideration is crucial as employed in our evaluation strategies.

4 Experimental Results

We simulate the proposed power allocation technique in Matlab. Where the sensors are deployed randomly and generate packets. Data generated in the field is delivered to the sink in the central location in a multi-hop fashion.

The Chipcon CC2420 dataset [1] values are taken as the transceiver characteristics. Here, the allowable transmit power settings are $P_{dbm} = [0, -1, -3, -5, -7, -10, -15, -25]$. The random backoff as is set to $rand() * 2^{BF-1} * 51.2 * 10^{-6}$ second, where BF denotes the backoff flag. The packets are of 128 byte size. The packet processing time is 2 msec. The retransmission timeout is 3. The physical channel is modelled as Rayleigh fading channel.

A number of N sensors have been deployed in the squared sensor field of length L. During the evaluation the number of the nodes are varied as $N = 100$:

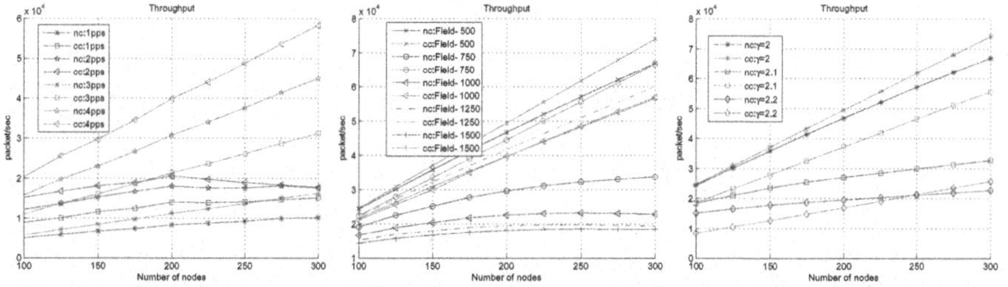

(a) Different packet rates. (b) Different field lengths. (c) Different γ values.

Fig. 2. Throughput performance of the protocols.

50 : 300. Three different aspects of variation in the parameters are considered in this literature: (i) field size and γ remain constant as $L = 1000m$ and $\gamma = 2$ with varying packet generation rate in packet per second (pps) $pps = 1 : 1 : 4$ and (ii) packet generation and γ remain constant as $pps = 4$ and $\gamma = 2$ while the filed length is varied with the length $L = 500m : 250m : 1500m$ and (iii) field size and packet generation rate remain constat as $L = 1000m$ and $pps = 4$ with varying γ with $\gamma = 2 : 0.1 : 2.2$. 200 different random fields are generated in each scenario and averaged out for the results presented. Note that, each deployment assumes a connected network, where each node has at least one neighbor in the forwarding path to send the data to the sink. In case of low density deployments, cooperative communications may not provide sufficient gain in terms of energy efficiency, especially, due to the fact that there may not be a relay node available in the right place. In such scenario no cooperation takes place. Such situation is defined by $d^{\gamma}_{S,D} > d^{\gamma}_{S,R} + d^{\gamma}_{R,D}$.

Figs. 2–5 present performance of the proposed algorithm compared with the NC. The sub-figures (a), (b) and (c) corresponds to the parameter settings of (i), (ii) and (iii) respectively.

Throughput: Fig. 2 presents the throughput performance of the proposed protocol along with the non-cooperative communication. Fig 2(a) presents the throughput performances of the schemes in different packet generation rates. In all the cases the proposed cooperative communications (CC) outperforms the non-cooperative (NC) counterpart. Note that the increasing number of generating packets does not result increasing the throughput in NC as efficiently as NC. In fact in case of high density deployments with 225 nodes or above and with pps 3 or above the throughput starts to fall due to inefficiency of the NC.

Fig 2(b) presents the throughput performance of the protocols with different field sizes. Increasing the field sizes causes the NC performance decreasing. This is due to the fact that increasing the size in turn increase the hop distance a packet needs to flow from the sender to the sink. As the successful packet reception in each hop is lower than that of CC. The multi-hop cases the rate of reception decays exponentially with the increasing of the hop count. For example, if the reception of each hop in case of C and NC is 90% and 80% respectively. After the second-hop and after the third-hop the C reception becomes 81% and 72.9% respectively where the NC becomes 64% and 51.2% respectively.

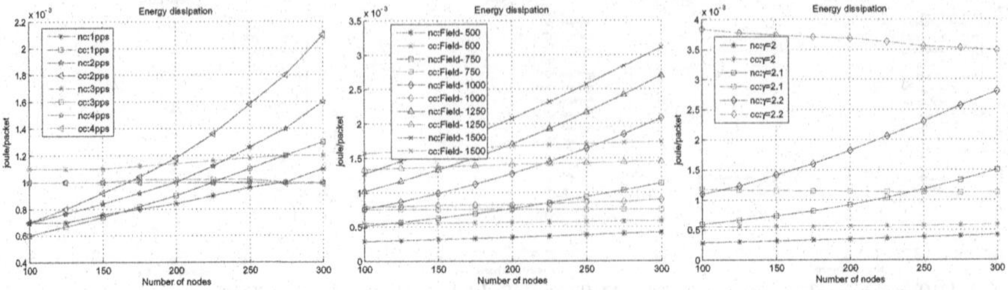

(a) Different packet rates. (b) Different field lengths. (c) Different γs.

Fig. 3. Energy performance of the protocols.

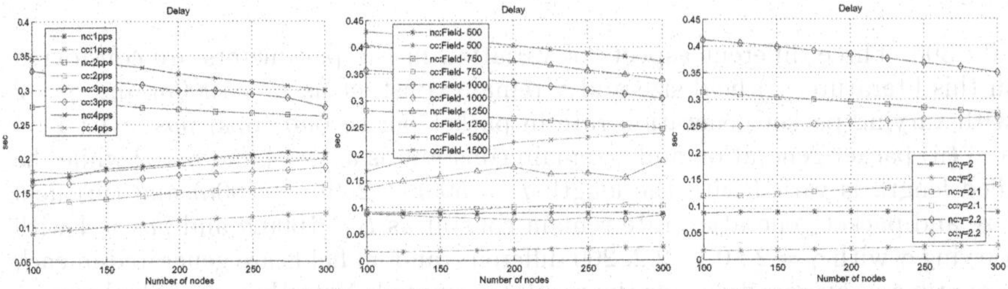

(a) Different packet rates. (b) Different field lengths. (c) Different γ values.

Fig. 4. Delay performance of the protocols.

Fig 2(c) presents the performance of the throughput in different γ. Like the previous cases the CC outperforms the NC. In this aspect there are some exceptions in $\gamma = 2.2$ and with a comparatively less number of nodes in the filed. Note that with low density network there is provision where we do not have available cooperative nodes in a good location.

Energy Dissipation: Fig. 3 presents the energy costs of the CC and NC approaches. Fig. 3(a) shows that increasing of the *pps* increases the energy dissipation of NC abruptly in high density networks. Though the number of packets in the systems are increasing, CC is not affected by implosion of power as much as in case of NC due to its superiority, in terms of capacity improvement. Fig. 3(b) shows the energy dissipation with increased field size. Increasing the filed size causes increasing is energy dissipation in the field due to multi-hop transmissions. But increasing the number of node does not affect the system largely in case of CC. On the other hand the NC not only cannot handle the high data rate but also suffer from high energy dissipation in high density deployment scenarios. Indeed, in some parameter settings the energy performance of the NC outperforms CC (such as in Fig. 3(c)) this is due to the fact that multiple packets (in CC) instead of a single packet (in NC) is transmitted in each link.

Delay: Fig. 4 presents the delay performance of the protocols. Fig. 4(a) shows the delay performance of the communication paradigm with increasing *pps*. The delay increases with increasing of data rates eventually due to higher number

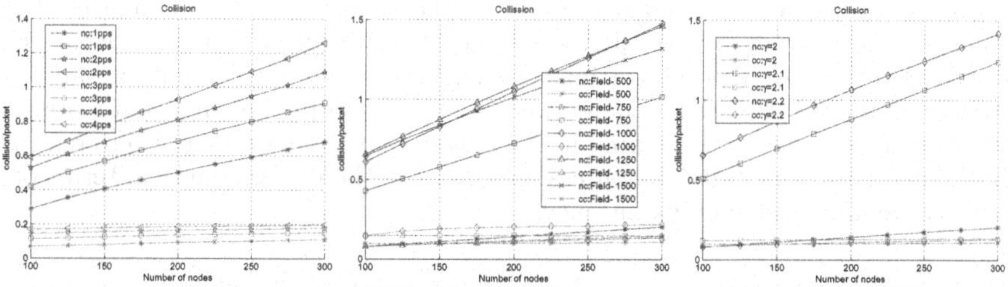

(a) Different packet rates. (b) Different field lengths. (c) Different γ values.

Fig. 5. Collision performance of the protocols.

of collisions. In terms of increasing number of node deployment delay in CC increases due to the same reason i.e., increasing nodes increases the packets in the networks. In case of NC increasing the node results decreasing the delay performance this is due to he far lesser number of packets in sink due to the high packet drops. Note that packets only received at the sink is counted for the delay. Fig. 4(b) shows the performance of delay with increased field size. Eventually increasing the field size increases the delay due to multi-hop functionality. Fig. 4(c) shows the delay performance in different γ. Increasing γ results increasing retransmissions and results increased delay in the networks. In all the cases in Fig. 4(a), Fig. 4(b) and Fig. 4(c) CC outperforms NC due to its signal enhancement and low power transmitter settings.
Collision: Fig. 5 presents the collision performance of the algorithms. The performance measures reveal useful rational behind the improved performance of our proposed CC. Fig. 5(a) reveals that in case of increased node density the proposed CC algorithm does not affected largely due to its low power transmissions. On the other hand with increased *pps* and/or with increased node density NC's collision rate increases abruptly results poor throughput, high energy dissipation (with exceptions) and high delay. In case of varying field size the collision rate jumps in NC due to the shift of single hop to multi-hop presented in Fig. 5(b). Where collision performance of CC remains almost the same. And finally Fig. 5(c) shows that the increased γ results abrupt collision in NC due to the poor receptions and retransmissions.

Different aspects of improvements in terms of performance such as throughput, energy, delay and collisions are investigated and the result clearly demonstrate the superiority of the CC over NC with only minor exceptions.

5 Conclusions and Future Work

To support the cooperative communications this paper presents an optimized power allocation for WSN based on BFOA algorithm. The algorithm attempts to minimize the power allocation with the constrained receive sensitivity. This approach enhances the throughput, energy and delay performance while comparing with the traditional NC communications in the WSN. The underlying reason of the performance enhancements is due to the fact that the CC enhances the received signal strength with MRC combine. Additionally the power allocation results lower transmit power settings consequently holds the contention limited compared to NC where the contention increases abruptly in high data rates. We intend to extend this research in the network coding paradigm where relays XOR messages from multiple sources and improve the performances in terms of both contention and energy-efficiency.

References

1. Chipcon: CC2420 2.4 GHz IEEE 802.15.4 / ZigBee-ready RF Transceiver (2004), http://inst.eecs.berkeley.edu/~cs150/Documents/CC2420.pdf
2. Datta, S.N.: Performance analysis of distributed MRC combining with a single amplify-and-forward relay over Rayleigh fading channels. In: Proc. 2013 IEEE WCNC. pp. 2399–2404 (2013)
3. De, S.: On hop count and Euclidean distance in greedy forwarding in wireless ad hoc networks. IEEE Communications Letters 9(11), 1000–1002 (2005)
4. Farsad, N., Eckford, A.W.: Resource allocation via linear programming for multi-source, multi-relay wireless networks. In: Proc. 2010 IEEE ICC. pp. 1–5 (2010)
5. Goudarzi, H., Pakravan, M.: Equal power allocation scheme for cooperative diversity. In: Proc. 2008 IEEE/IFIP ICI. pp. 1–5 (2008)
6. Habibi, J., Ghrayeb, A., Aghdam, A.G.: Energy-efficient cooperative routing in wireless sensor networks: A mixed-integer optimization framework and explicit solution. IEEE Transactions on Communications 61(8), 3424–3437 (2013)
7. Khabbazibasmenj, A., Vorobyov, S.A.: Power allocation in decode-and-forward cooperative networks via SEP minimization. In: Proc. 3rd IEEE CAMSAP. pp. 328–331 (2009)
8. Li, P., Guo, S., Cheng, Z., Vasilakos, A.V.: Joint relay assignment and channel allocation for energy-efficient cooperative communications. In: Proc. 2013 IEEE WCNC. pp. 626–630 (2013)
9. Passino, K.: Bacterial foraging optimization. Foundations of Computational Intelligence 1, 1–16 (2010)
10. Qi, Q., Minturn, A., Yang, Y.: An efficient water-filling algorithm for power allocation in OFDM-based cognitive radio systems. In: Proc. 2nd IEEE ICSAI. pp. 2069–2073 (2012)
11. Qu, Q., Milstein, L.B., Vaman, D.R.: Cooperative and constrained MIMO communications in wireless ad hocsensor networks. IEEE Transactions on Wireless Communications 9(10), 3120–3129 (2010)
12. Stojmenovic, I.: Position-based routing in ad hoc networks. IEEE Communications Magazine 40(7), 128–134 (2002)
13. Taghiyar, M.J., Muhaidat, S., Liang, J.: Max-min relay selection in bidirectional cooperative networks with imperfect channel estimation. IET Communications 6(15), 2497–2502 (2012)
14. W. Liu, G.L., Zhu, L.: Energy efficiency analysis and power allocation of cooperative communications in wireless sensor networks. Journal of Communications 8(12), 870–876 (2013)
15. Xie, K., Cao, J., Wang, X., Wen, J.: Optimal resource allocation for reliable and energy efficient cooperative communications. IEEE Transactions on Wireless Communications 12(10), 4994–5007 (2013)
16. Zhou, Z., Zhou, S., Cui, S., Cui, J.: Energy-efficient cooperative communication in a clustered wireless sensor network. IEEE Transactions on Vehicular Technology 57(6), 3618–3628 (2008)

Part II

Maintaining a trajectory data warehouse under schema changes with the mobile agent view adaptation technique

Wided Oueslati, Jalel Akaichi
Higher Institute of Management of Tunis

widedoueslati@live.fr

jalelakaichi@isg.rnu.tn

Abstract. With the development of pervasive systems and positioning technology, the analysis of data resulting from moving objects trajectories has attracted a particular interest. Those data are called trajectory data and are stored in a suitable repository called trajectory data warehouse (TDW). TDW view definitions are constructed from heterogeneous mobile information sources schema. Those latter are more and more autonomous and they often evolve by changing their contents and /or their schema then the TDW view definition may become undefined and consequently the analysis process may be affected. For this reason, it is important to achieve view definitions restoring or synchronization following schema changes occurred at the heterogeneous mobile information sources. The goal of this paper is to propose a TDW view definition maintenance based on mobile agent view adaptation system.

Keywords: TDW maintenance, schema changes, mobile agent, view synchronization

1 Introduction

TDW are constructed from TD collected from mobile information sources such as PDA, mobile phones, ipad… The TDW has to be updated and maintained in order to ensure the consistency of information used for analysis purposes. In fact, the trajectory data warehouse management system must update the materialized views when changes of mobile information sources occur. Those changes can be of type data changes or schema changes and both of them may make view definition built among trajectory data warehouse undefined. In previous work, the view redefinition was performed manually by data warehouse developers, until EVE [1,2] project proposed a prototype to automate view definition rewriting by evolving view definitions and view synchronization algorithms [3,4,5]. This latter consists on determining legal rewritings for the affected views referring to rules. These rules

K.J. Kim (ed.), *Information Science and Applications*,
Lecture Notes in Electrical Engineering 339, DOI 10.1007/978-3-662-46578-3_10

enable substitutions retrieval for the affected view definitions components while respecting preference parameters.
The EVE solution has to become more adapted for dynamic, distributed and mobile environement by adopting new techniques like the mobile agents [6,7,8]. For this reason, we propose a system that uses mobile agents to ensure the trajectory data warehouse maintenance under schema changes. Such system permits to avoid the network saturation and to minimize communication coasts.
This paper is organized as follows. In section 2, we present research works related to data warehouse maintenance. In section 3, we determine the view maintenance under schema changes. In section 4, we summarize the work and we propose some extensions to be done in the future.

2 Related works

Research works elaborated in the context of view maintenance can be classified in the following categories:

- View adaptation [9, 10, 11, and 12]: this approach consists in adapting views to changes by adding meta data to materialized views. Those meta data contain structural updates related to materialized view.
- View synchronization (rewriting of views) [12, 13, 14, 15]: many research works were interested in this approach because it is in relation with other problems such as data integration, data warehouse modeling…

In [12] Bel presents an approach for dynamic adaptation of views related to data sources (relations sources) changes. The main idea of this work is to avoid the recompute of views which are defined from several sources. In fact, the key idea is to compute the new view from the old one. Bel presents the view adaptation problem from two different points of views. The first point of view is from the user or from the DW designer or administrator and the second point of view is from data sources. As mentioned above, the views can change their schema independently of data sources. In fact, the user or the DW administrator can bring into play schema changes directly on views by adding an attribute to a view or by modifying the domain of an attribute. In [12], the add of new attribute to a view is the result of new requirements. This operation does not need the rewriting of the view but only the storage of the value of the new attribute. The modification of an attribute domain simulates the creation of new view V' from the old one V. V' must not involve the specific attribute, then an attribute with the same name and having a new type or new definition (domain) is added. The old view V is deleted and the new view V' is renamed as V.

3 View maintenance under schema changes using mobile agents

To ensure data warehouse views maintenance under schema changes, we propose to design a mobile agent's view synchronization system based on EVE solution. In fact, the increase need to decrease the network saturation and to minimize communication costs, have led us to revise the EVE solution to become more adapted for dynamic and distributed environment by adopting new techniques like the mobile agents. In fact, we propose to design a mobile agent's view synchronization system based on EVE solution. This latter has to ensure data warehouse maintenance under schema changes. Our solution decreases the synchronization time due to parallelism permitted by mobile agents and avoids the saturation of the network. The architecture of our system is distributed on four entities which are the mobile Meta Knowledge Base (MKB) agent, the mobile View Knowledge Base (VKB) agent, the mobile detector agent and the mobile synchronizer agent. All those agents know each other via their identifier, names and sites. Thus, any mobile agent of the system can communicate directly with any other mobile agent.

The mobile agent technology imposes itself as a new solution to schema change detection system. In fact the mobile detector agent in contrast to the static one can move from one information source to another to detect changes. The change detection operation consists on comparing for each schema component, two schema versions (schemat-1 , schemat), if they have been found different, it implies that a change has been occurred and it has to be computed. This latter will be sent in parallel to the mobile Meta Knowledge Base (MKB) and the mobile View Knowledge Base (VKB). After receiving messages from mobile detectors about schema changes, the mobile MKB agent must look into all MKB components in order to determine affected knowledge and rules by the indicated schema changes, and then it computes the affected knowledge before sending them to the mobile view synchronizer agent. After receiving messages from mobile detectors about schema changes, the mobile VKB agent must look into all VKB components in order to determine affected views by the indicated schema changes, and then it computes the affected views before sending them to the mobile view synchronizer agent. The role of the mobile view synchronizer (VS) agent is to find legal rewritings for the affected views (the view rewriting is legal when it is compatible with the current information space), then it emits the affected knowledge and views replacement to the mobile MKB agent and the mobile VKB agent respectively. The following figure describes the communication between mobile agents of our proposed system.

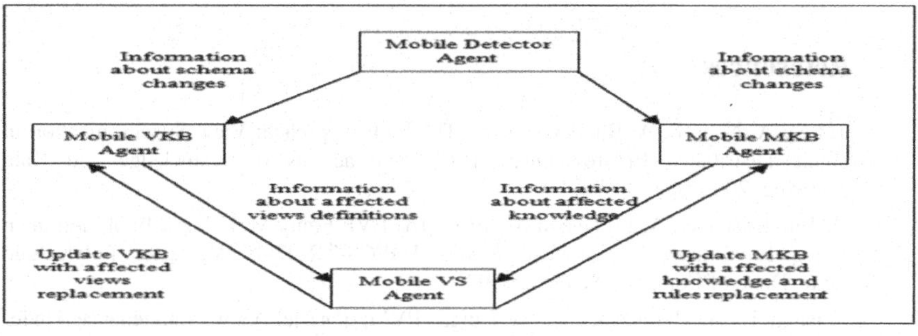

Fig. 1. The mobile agents' communication

The solution that we propose is based on EVE and on a technique resulting from the field of artificial intelligence which is the mobile agents. This new solution permits more autonomy (by separating between schema change detection, affected knowledge and view definition determination and view definition synchronization tasks), parallel schema change detection (by using parallel detector agent instances running over distributed information sources), parallel determination of affected knowledge and view definition determination and parallel view definitions synchronization (by triggering view synchronizer agent instances).
The main functionalities expected from our solution are the change detection at the information sources level, the view synchronization and the updates of the MKB and the VKB. Those functionalities get advantages from the advent of positioning technologies and the emerging mobile devices. In fact, information sources are no longer static but dependant on the carrier mobility. This pushes us to download a detector agent on each mobile device such a PDA.
This kind of agents communicates with our central synchronization system via wireless networks enhanced by a GPS. Furthermore, MKB and VKB may be distributed in various locations according to a classification of domains involving specific information sources. When synchronization is performed, only needed information sources description and a subset of view definitions may be implicated. This obviously increases the whole performance of the system.

5. Conclusion

In this work, we proposed a solution to maintain a trajectory data warehouse under schema changes. Our proposed model based on EVE is enhanced by mobile agent concepts to evolve to a new system whose architecture is distributed on four entities: the Mobile Detector Agent, the Mobile MKB Agent, the Mobile VKB Agent and the Mobile View Synchronizer Agent. Communication between agents is guaranteed by the traditional message sending. In fact, all the agents of the model know each other directly via their identifier, names and sites. As future work we will try to improve the mobile detector agents' intelligence by allowing them to determine automatically knowledge necessary for finding substitutions. This task can be carried out by exchanging information with other mobile detector agents.

References

1 A. J. Lee, A. Nica, E. A. Rundensteiner, "The EVE Approach: View Synchronization in Dynamic Distributed Environments", IEEE Transactions on Knowledge and Data Engineering, 2002, pp.931-954.

2 E. A. Rundensteiner, A. J. Lee and A. Nica, "The EVE Framework: View Evolution in an Evolving Environment", Technical Report WPICS-TR-97-4, Worcester Polytechnic Institute, Dept. of Computer Science, 1997.

3 X. Zhang, E. A. Rundensteiner, L. Ding, "PVM: Parallel View Maintenance Under Concurrent Data Updates of Distributed Sources, in Data Warehousing and Knowledge Discovery", Proceedings, Munich, Germany, 2001, pp. 230–239.

4 E. A. Rundensteiner, A. Koeller, X. Zhang, "Maintaining Data Warehouses over Changing Information Sources", Communications of the ACM. June 2000.

5 A. Gupta, I. S. Mumick, "Maintenance of Materialized Views: Problems, Techniques, and Applications", IEEE Data Engineering Bulletin, 1995.

6 B. Amil. Bougardiye, G. Agrawal, P. Patil, "Mobile Agent Synchronization and Security Issue", International Journal of Engineering Research and Technology, vol 1, Issue 10, 2012.

7 Q. Karma, I. Serguiev Skaia, "Agent Based Framework Architecture for Supporting content Adaptation for Mobile Government", International Journal of Interactive Mobile Technology, vol 7, No 1, 2013, pp. 10-15.

8 P. Biswas, H. Qi, Y. Xu, "Mobile-agent-based Collaborative Sensor Fusion", Information Fusion, July 2008, pp. 399-411.

9 A. Gupta, I. M umick, K. Ross. Adapting Materialized Views after redefinitions SIGMOD, pp 211-222, 1995.

10 A. Nica, E. A. Rundensteiner. View Maintenance after Vew Synchronization. International Database Engineering and application Symposium, pp 213-215, 1999.

11 E. A. Rundensteiner, A. Nica, A. J. Lee. On Preserving Views in Evolving Environments. Proc. Fourth Int Workshop Knowledge Representation Meets Databases. Pp 131- 141. 1997.

12 Z Bellahsene. Schema Evolution in Data Warehouses. Knowledge and Information Systems, 4(3):283–304, 2002.

13 E. A. Rundensteiner, A. Koeller, X. Zhang, A.J. Lee, A. Nica: Evolvable View Environment EVE: A Data Warehouse System Handling Schema and Data Changes of Distributed Sources. Proceedings of the International Database Engineering and Application Symposium (IDEAS'99), Montreal, Canada, April 1999.

14 A. Rajaraman, Y. Sagiv, J. D. Ullman. Answering Queries Using Templates With Binding Patterns. Proc. ACM Symp. Principles Database System, pp 105-112, 1995 .

15 L. V. S. Lakshmanan, F. Sadri, I. N. Subramanian. Schema SQL a Language for Interoperability in Relational Multi-Databases Systems. Proc. 22.nd Int Conf. Very Large Databases, pp 239-250, 1996.

Traffic Analysis in Concurrent Multi-Channel Viewing on P2PTV

Koki Mizutani[1], Takumi Miyoshi[1], and Olivier Fourmaux[2]

[1] Shibaura Institute of Technology, Saitama 337-8570, Japan
[2] UPMC Sorbonne Universités, Laboratoire d'Informatique de Paris 6, Paris 75005, France

Abstract. In recent years, peer-to-peer (P2P) video streaming services (P2PTV) have attracted much attention because of the ability to decrease the load on the content servers by distributing data delivery function to peers. On the other hand, P2P overlay networks are oblivious to the physical network topology and thus may cause undesirable traffic straddling on some Internet service providers (ISPs). To optimize P2PTV traffic, several traffic measurements have been studied and revealed the characteristics of P2PTV traffic. However, these studies did not focus on users' behavior and traffic flow of each user. In this paper, we focus on PPTV that is one of the most famous P2PTV services and collect traffic data when multiple channels are viewed at the same time. Through this measurement, we observed the changes of the number of peers, traffic flows, and packet arrival time. As a result, we found the new characteristics of PPTV such as the transmission state of PPTV by monitoring the variation in the number of new arrival peers. Moreover, we could detect video servers by simultaneously analyzing multi-channel PPTV traffic.

Keywords: P2PTV, traffic analysis, traffic flow, multi-channel

1 Introduction

In recent years, video delivery servers and Internet service providers (ISPs) have been suffering from the increase of traffic load due to the huge demand of multimedia contents. Cisco Systems, Inc. stated that video traffic will be on the increase accounting for 86 percent of all Internet traffic in 2016 [1]. As shown in Fig. 1, peer-to-peer (P2P) video streaming services (P2PTV) have attracted attention as a solution for decreasing the traffic load on content servers due to the nature of P2P communication that can distribute data delivery function to peers. Currently, P2P streaming applications have become getting popular such as SopCast [2], PPTV [3], and PPStream [4]. In the future, P2P traffic will still increase since video streaming services will shift from the client/server model to the P2P model. To distribute the load on the content servers, P2P communication implements a specific peer selection strategy. In P2PTV communication, a host first selects some peers randomly from the peer list that is offered by the server, and then sends request packets to them. As a result, P2PTV may

K.J. Kim (ed.), *Information Science and Applications*,
Lecture Notes in Electrical Engineering 339, DOI 10.1007/978-3-662-46578-3_11

download the video data from farther peers even when there exist some neighbor peers that have the desired video data, and thus establishes inefficient P2P overlay networks. Moreover, it is difficult for ISPs to deduce P2PTV traffic because of the distribution of peers around the world. Therefore, P2P traffic optimization is one of the biggest problems for ISPs.

To optimize P2PTV traffic, several traffic measurements have been studied and revealed its characteristics [5]-[7]. However, these studies are not enough to reveal users' behavior since they only inspected a steady-state traffic in a long duration of time on the same channel. To study P2PTV traffic sufficiently, we need to inspect the behavior of peers and to compare the characteristics of P2PTV among multiple channels. This paper focuses on analyzing traffic characteristics and users' behavior among multiple channels on PPTV, which is one of popular P2PTV applications. We run PPTV applications on multiple PCs to view multiple channels and collect traffic data at the same time. As a result, we show the following characteristics: (1) frequencies in the use of the same peers among multiple channels, (2) the transmission state of PPTV by comparison between the variation in the number of new arrival peers and long duration flows, and (3) the possibility to find super peers that provide stable video streaming by analyzing traffic flows among multiple channels.

2 Related Work

Several conventional studies have investigated the characteristics of P2PTV applications including PPStream, PPTV, SopCast, etc. Hei et al. analyzed the steady state of P2PTV traffic on PPLive (former name of PPTV) with a crawler that can automatically send and receive control packets [5]. They discovered the variation in the characteristics of throughput, users' behavior, and the number of connections on PPLive. They also concluded that users tend to change their channels to others on the desired channel selection. This behavior is similar to

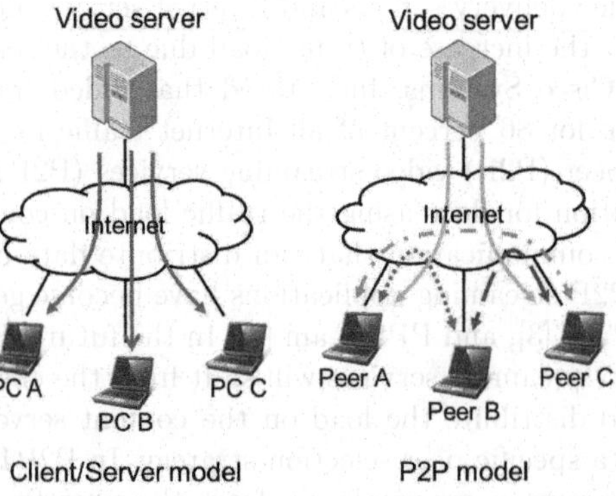

Fig. 1. Network model of video streaming services.

the channel zapping on analog TV. Jia et al. analyzed the characteristics of the traffic on PPStream with crawler [6]. They clarified the geographical characteristics, the variation in the number of peers, and the ratio of sent and received packets. They concluded that PPStream has a hierarchical mesh network topology including super nodes eager to send the video data to many users. Silverston et al. analyzed P2PTV traffic on PPStream and SopCast measuring on multiple points in different locations of the world [7]. They clarified the number of AS hops and the ratio of download and upload data from the viewpoint of measurement locations. They concluded that the P2PTV applications have not been equipped with any traffic localization mechanism because many peers received video data from not own countries but others. In addition, they indicated that P2PTV also has no fairness in sharing the video data because of the differential ratio of download and upload data among peers.

The above studies analyzed P2PTV traffic in a long duration of time and revealed the characteristics of P2PTV traffic, the geographical location of peers, the overlay topology, etc. However, they are not enough to study the characteristic of P2PTV since they did not focus on the characteristics of users' behavior such as traffic per each user. In addition, these studies were measured only on single channel. To reveal and generalize the characteristic of P2PTV traffic, it is essential to investigate P2PTV traffic on multiple channels.

3 Objective of P2PTV Traffic Measurement

These days, various analyses of P2PTV traffic have been performed. However, these studies are not enough to reveal the P2PTV traffic thoroughly considering users' behavior and the extraction of P2PTV characteristics among multiple channels.

In this paper, we present a novel measurement experiment for multi-channel viewing by using multiple measurement PCs. We collect traffic data among multiple channels on P2PTV and analyze the traffic per each user. Based on the collected traffic, we firstly determine *single-channel peers* and *multi-channel peers* among peers that the P2PTV application connects to. A single-channel peer is defined as the peer that appears only in one channel, i.e., the peer is observed on a single channel. A multi-channel peer is defined as the peer that appears in two or more channels, i.e., the peer is observed on multiple channels. On P2PTV, it is not sure that measurement PCs can observe the same peers among multiple channels due to the nature of random peer selection. Therefore, we examine frequencies in the use of the same peers and define peers observed on various channels as multi-channel peers. In our second analysis, we examine the transmission state of P2PTV on a specific channel. This can be done by analyzing the variation in the number of new arrival peers and long duration flows. In particular, if many new arrival peers are intensively found on a channel, we can assume that the measurement PC is trying to search some stable peers at that time. Finally, we specify which peers could be a video server by analyzing traffic from multi-channel peers on each measurement PC. If some measurement

PCs can simultaneously observe peers with the same IP address among multiple channels, we can decide that these peers are video servers deployed for providing stable P2PTV services.

4 Experimental Setting

This section describes our experimental environment. We focused on PPTV that is one of the most famous P2PTV services. We measured traffic for twelve hours (from 12:00 JST to 23:59) on 18th December 2013 on PPTV using ten measurement PCs including six virtual machines. As shown in table 1, we prepared three types of measurement PCs including high-, medium-, and low-performance PCs assuming actual users' environment. Every PC has an Internet connection provided by FLET'S HIKARI NEXT, 100 Mbps optical access service on next generation network (NGN) via Plala HIKARI Mate as an ISP in Japan. For capturing and monitoring traffic, Wireshark [8], a well-known packet sniffer, is installed on every measurement PC. As shown in Fig. 2, we viewed different live streaming contents through 10 popular channels based on the official popularity ranking on PPTV.

In the measurement PCs, we extracted only packets with not less than 1000 Bytes as video packets to cut off control packets and then analyzed only video traffic flows. Note that a flow represents a group of packets consecutively received from the same IP address. If the interval between two consecutive packets from the same IP address is more than 60 seconds, however, we regard them as different flows. Moreover, some flows may continue to exist even after our measurement. Therefore, we excluded some flows that maintained connections within the last 60 seconds of the measurement.

Table 1. Performance of measurement PCs.

Performance	CPU	# of cores	Memory (GB)	OS
High	Intel Core i5-3470S 2.90GHz	4	12	Windows 7
Medium	Intel Core i3-3217U 1.80GHz	2	8	Windows 7
Low	Intel Core i5-3470S 2.90GHz	1	2	Windows 7

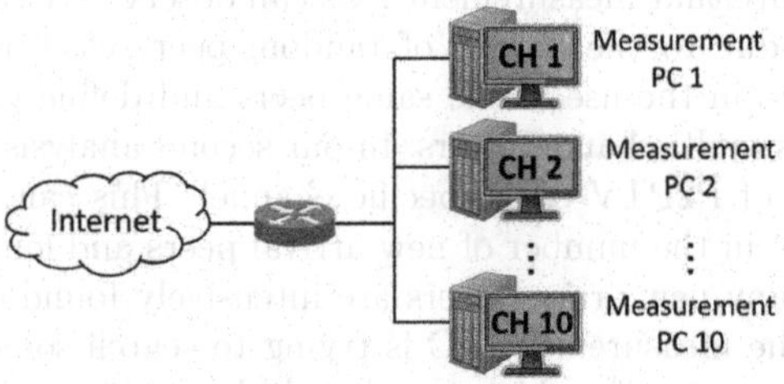

Fig. 2. Experimental environment.

5 Analysis of P2PTV Traffic

In this section, we analyze PPTV traffic measured in our experiment.

5.1 Ratio of multi-channel peers

First, we extracted multi-channel peers that appeared in multiple channels. Then, we calculated the ratio of multi-channel peers to the total number of peers appearing in each channel. As shown in table 2, for instance on channel 1, a measurement PC observed 94 multi-channel peers and 793 peers in total, and therefore the ratio of multi-channel peers accounted for 11.9 percent. Moreover, we categorized channels considering the total number of peers into two groups of channels; popular channel group that has 1000 or more peers and unpopular channel group that has less than 1000 peers. In this experiment, channels 1, 3, 4, 5, 6, and 10 are categorized into unpopular channel group while channels 2, 7, 8, and 9 are categorized into popular channel group. The number of multi-channel peers accounted for 5 to 12 percent of the total number of peers on each channel.

5.2 Analysis of traffic flows

We analyzed traffic flows on channel 1 and 9, which represent the unpopular and popular channel groups, respectively. From Figs. 3 to 6 show the variation in peer numbers, the order of peers arrival, for 45000 seconds. In this experiment, we extracted long duration flows that connected to the same peers for not less than 1200 seconds (20 minutes). Firstly, Figs. 3 and 5 show all traffic flows on channel 1 and 9, respectively. In addition, Figs. 4 and 6 pick up long duration flows from Figs. 3 and 5, respectively. In each of figures, orange lines show traffic flows of single-channel peers, and blue lines show those of multi-channel peers. At the beginning of each measurement, as shown in Figs. 3 and 5, measurement PCs drastically connected to multi-channel peers. Moreover, we found a relationship between the number of new arrival peers and long duration flows. In particular, the comparison between Figs. 3 and 4 indicates that a huge increase of new

Table 2. Statistics of peers and multi-channel peers.

CH#	Total peers	Multi-channel peers	Ratio of multi-channel peers (%)
1	793	94	11.9
3	838	66	7.9
4	680	70	10.3
5	514	56	10.9
6	665	62	9.3
10	544	64	11.8
2	1882	149	7.9
7	1810	94	5.2
8	1516	88	5.8
9	1440	119	8.3

arrival peers is observed on the disappearance of long duration flows from 23000 to 25000 seconds.

To clear the above relationship, we focused on the variation in the number of new arrival peers and long duration flows for 45000 seconds on channel 1 and 9, shown in Figs. 7 and 8, respectively. In each of figures, red lines show the number of new arrival peers, and blue lines show the number of long duration flows. In Fig. 7, a huge increase of new arrival peers is observed while long duration flows disappear from 23000 to 25000 seconds. We conclude that the variation in the number of new arrival peers is a very important point to understand the transmission state of PPTV. We assume that PPTV lost stable peers to download the video data due to the immense decrease of long duration flows and then started searching new stable peers. According to this assumption, we can deduce the transmission state of PPTV by monitoring the variation in the number of new arrival peers. We found the similar tendency in Fig. 8 regardless of the number of peers in channels.

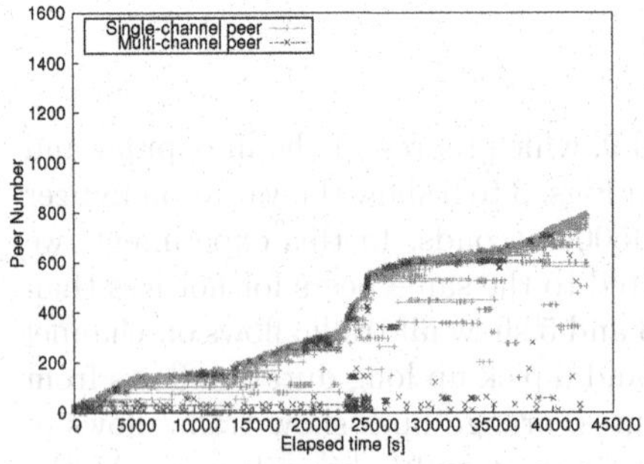

Fig. 3. All traffic flows on channel 1.

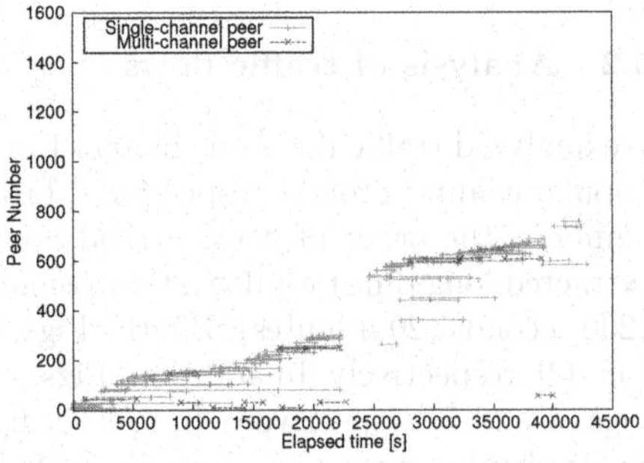

Fig. 4. Long duration flows on channel 1.

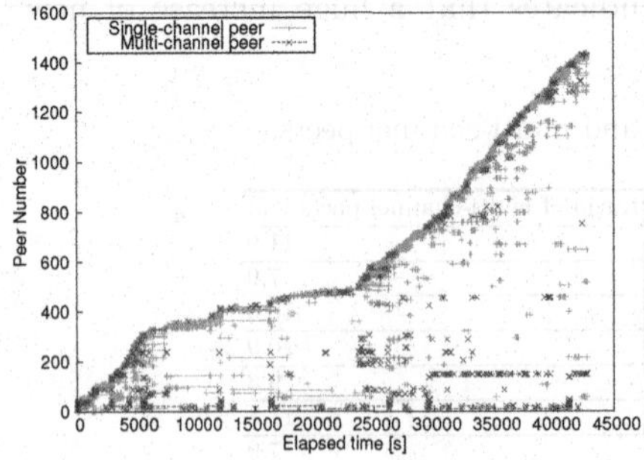

Fig. 5. All traffic flows on channel 9.

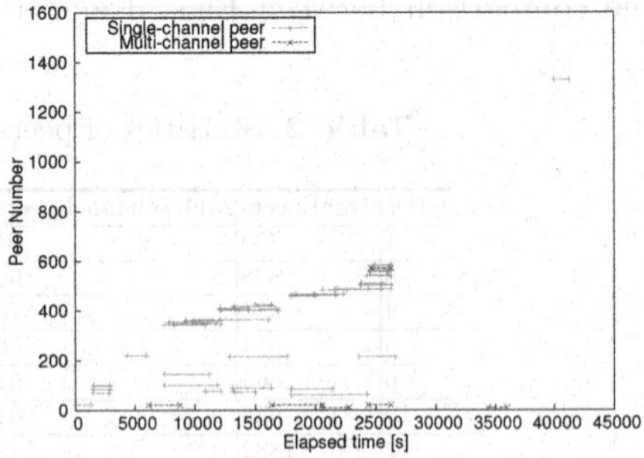

Fig. 6. Long duration flows on channel 9.

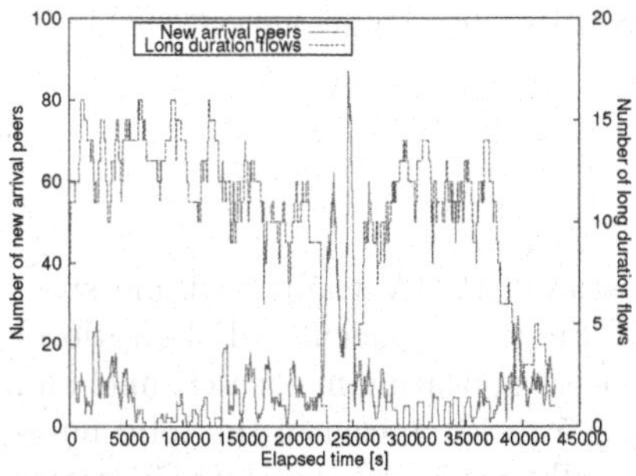

Fig. 7. Relationship between the changes of new arrival peers and long duration flows on channel 1.

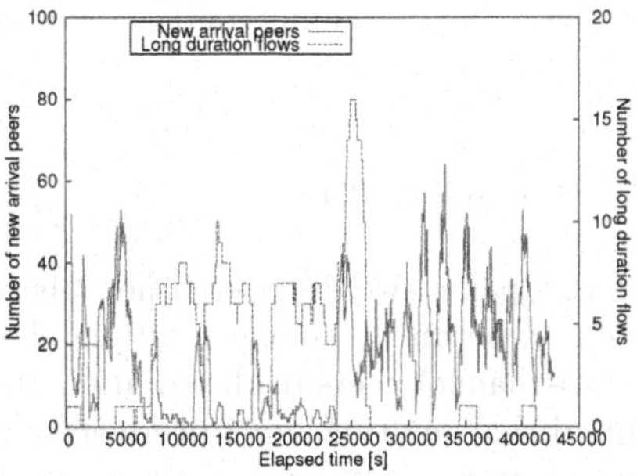

Fig. 8. Relationship between the changes of new arrival peers and long duration flows channel 9.

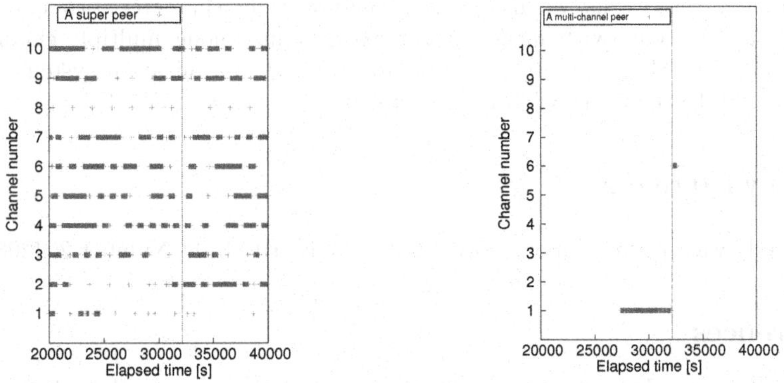

(a) From Super peer

(b) From Multi-channel peer

Fig. 9. Flows from specific peers appearing on each measurement PC.

5.3 Behavior of multi-channel peers

Finally, we try to find the *super peer* among multiple channel. Fig. 9 shows the flows from a multi-channel peer on each channel viewed by each measurement PC. In Fig. 9 (a), measurement PCs simultaneously observed the same peer among multiple channels for a long duration. Thus, we assume that this peer was a video server called super peer, which provides stable video streaming services dedicating itself to sending video packets. On the contrary, in Fig. 9 (b), measurement PCs observed a channel transition of another multi-channel peer. This multi-channel peer was not observed simultaneously by several measurement PCs and disappeared shortly. We assume that this kind of behavior is users because user peers do not have intension to provide stable video streaming services and just select their preferred channels by themselves. According to these analyses, we can simply detect video servers with video traffic measurements on multiple channels.

6 Conclusion

In this paper, we discovered new characteristics of PPTV traffic with measurement experiment among multiple channels. Firstly, we calculated the ratio of multi-channel peers to the total number of peers appearing in each channel and found that multi-channel peers accounted for 5 to 12 percent of the total number of peers observed on each channel. Secondly, we found that huge increases of new arrival peers were observed on the disappearance of long duration flows. According to this characteristic, we can deduce the transmission state of PPTV by monitoring the variation in the number of new arrival peers. Finally, we can also detect super peers by monitoring the same peer observed among multiple channels simultaneously.

For the future works, we are going to analyze traffic characteristics on other P2PTV applications with measurement experiment among multiple channels simultaneously. Moreover, we will configure a traffic optimization system that controls P2PTV traffic considering the transmission state of P2PTV.

Acknowledgment

This study was partly supported by JSPS KAKENHI Grant Number 26420370.

References

1. Cisco Systems, Inc., "Cisco visual networking index: forecast and methodology, 2011-2016," White paper, May 2012.
2. SopCast. http://www.sopcast.org/
3. PPTV. http://www.pptv.com/
4. PPStream. http://www.pps.tv/
5. X. Hei, C. Liang, J. Liang, Y. Liu, and K.W. Ross, "A measurement study of a large-scale p2p iptv system," IEEE Trans. on Multimedia, Vol. 9, Issue 8, pp. 1672-1687, Dec. 2007.
6. J. Jia, C. Li, and C. Chen, "Characterizing ppstream across internet," 2007 IFIP Int'l Conf. Netw. and Parallel Comput. (NPC'07) Workshops, pp. 413-418, Sept. 2007.
7. T. Silverston, O. Fourmaux, K. Salamatian, and K. Cho, "On fairness and locality in p2p-tv through large-scale measurement experiment," IEEE Global Telecommun. Conf. (GLOBECOM 2010), Dec. 2010.
8. Wireshark. http://www.wireshark.org/

Evaluation Method of MANET over Next Generation Optical Wireless Access Network

M.A. Wong[1], N. Zulkifli[2], S.M. Idrus[2] and M. Elshaikh[3]

[1] Department of Engineering Technology, UTeM
williamsu82@yahoo.com
[2] Lighwave Communication Research Group, UTM
{nadia, sevia}@fke.utm,my
[3] School of Computer and Communication Engineering, Unimap
elshaikh@unimap.edu.my

Abstract. Access networks are increasingly shaped by the emerging trend of user applications that called for the need of converged layers to meet the requirement of growing spectrum resources. But currently, majority of the previous works in this domain are still limited to a specific functional layer. Recently discovered, MANETs of access network alone have salient characteristic that is bandwidth constrained where the wireless links of frontend topology will continue to have significantly lower capacity than their wired counterpart. Hence, a unified and hybrid method to establish a performance evaluation of MANET (AODVUU, OLSR and DYMOUM) routing based on IEEE 802.11 mesh topology over passive optical system access network is proposed through simulation in the OMNeT environment. The results show that DYMO has the best performance compared to AODV and OLSR of in terms of throughput and delay as the performance metrics.

Keywords: MANET (Mobile Ad hoc Network) ·AODVUU (Ad-hoc On-Demand Distance Vector) ·OLSR (Optimized Link State Routing) ·DYMOUM (Dynamic MANET On-demand) ·OMNeT ·HOWAN (Hybrid Optical Wireless Access Network)

1 Introduction

Among today's networking solutions, optical networking is considered as the most promising transport technology for access network operations [1] due to its high speed, large capacity and low fiber attenuation [2, 3]. In general, Fiber to the X (FTTX) application can be divided into two basic architectures possible; point-to-point (P2P) [4] and point-to-multipoint (P2MP) which is normally in the market described as Passive Optical Network (PON). Most of the existing work proposed solutions within a specific functional OSI layer such as physical layer or network layer. However, we believe that it is also important to investigate more holistic solutions that consider multiple layers to include more than just one functional layer [5]. For that, an architectural platform that consider cross functional layers e.g. physical and network layers demonstrating a seamless transmission of mobile/ wireless signals over standardized PON topologies has been investigated to provide centralized heterogeneous network with ubiquitous access and mobility [2] such as broadcast access with splitting scheme and WLAN, and this relies on the optical IP architecture.

K.J. Kim (ed.), *Information Science and Applications*,
Lecture Notes in Electrical Engineering 339, DOI 10.1007/978-3-662-46578-3_12

Another promising access solution is the wireless network where WiMAX (Worldwide Interoperability for Microwave Access of standard: IEEE 802.16) [6] is gaining rapid popularity. Aggressive research in this area has continued since then, with prominent studies on routing protocols at network layer such as AODV, DSR and OLSR. Another related architecture study is known as Wireless Optical Broadband Access Network (WOBAN) that uses proactive routing (e.g. OLSR to evaluate the resource usage performance such as capacity and channel availability [7]. The very existing purpose of this wireless mobile access is to address the overloading existing part of the wired infrastructure and spectrum assignments plus operational limitation overloading of existing older wireless technology (e.g. LAN switches) by the improvement of some number of existing wireless access points over fixed optical backbone. For that reason, our proposed work will primarily focus on the cooperation of heterogeneous performance of MANET routing protocols efficiency surrounding optical physical system technologies in OMNeT [1].

To overcome the challenges of introducing scalability and capacity limitations of a single wavelength access network, a multi-lightwave signal network extension over power splitting PON infrastructures and a multi-path routing is applied by exploiting the broadcast multiplex access nature [8, 9] are proposed in this paper requiring current access network infrastructure improvement. Hence, this paper proposes and investigates a novel hybrid network paradigm – Hybrid optical wireless access network (HOWAN) – a combination technology of high-capacity optical access system network and untethered wireless access technologies. The distinguishing feature between HOWAN and WOBAN is HOWAN design is based on a WDM/TDM PON at the optical backhaul with WIFI technology at the wireless front end while WOBAN is using EPON as optical backend with wireless mesh network. Overall, it can be summarized that most dominant optical wireless access networks are typically based on the PON technology such as the Ethernet-PON (EPON), Gigabit-PON (GPON), Long-Reach PON and WOBAN [10, 11].

To our knowledge, none of the existing work in HOWAN has considered this scenario. HOWAN is a novel access architecture that provides wireless at the front end, either a WiFi or WiMAX, while supported by an optical backhaul such as PON [12, 13] with the focus in this article by means of accessing each remote Optical Network Unit/Base Station (ONU/BS) from the optical line terminal (OLT) by frequency division multiplexing (TDM) and dense array wave guide grating (AWG). Specifically, this has brought about the cross layer of HOWAN that also provides the following key features that illustrate its departure from the existing work and mark the contributions of this paper.

- It simplifies the process of monitoring and analysis study by providing a unified interface for accessing application, protocol and system information.
- The generic design of the unified framework was proposed to integrate these disparate specific functional of OSI layer into a cross layer form which should be familiar and further enabled simplifies the process of specifying cross-layer interaction in service-oriented architecture by providing a declarative way to specify how a set of layering improvement of cross-functionality should be composed and adapted in interoperability strategy.
- This cross-level approach offers (i) an improved of very high degree of flexible scalability, to better evaluate with different routing compositions of MANET over passive access network of efficiency, and (ii) towards centralized heterogeneous information system with extensible access control, to include cooperating protocols to find the right set of design level optimizations for a certain use-case which in this article focus on transmission handling. Hence, cross layer HOWAN

is well suited as a rapid prototyping tool for application and system developers. The rest of the paper is organized as follows. Section II is devoted to the related work surrounding the topic of optical framework backhaul system technologies and the extension of wireless mobility routing algorithms. Section III explains framework of the proposed solution of typical HOWAN. Section IV discusses OMNeT preliminary work result of HOWAN. Finally, the paper is concluded in Section V.

2 Background Research

This cross-configuration of the network vision is in line with the Next Generation Optical Access Networks (NGOAN) paradigm including long term evolution access network. [14] It is because of the shared infrastructure and the independent deploying routing rules in the path between the head-end office and the customer. The focus here is to provide reliable elements for access network to be envision in today's optical access networks. At the system level, the components element for next-generation access networks enables broader-band delivery by higher-speed operation, more wavelengths, and more integration.

Prior to the idea of merging and integrating of two existing of different domains, we have proposed our own such platform that combines the capabilities of traffic optical backhaul system and wireless access protocol communication domains, in an attempt to provide a hybrid optical wireless access communication network platform.

The researcher has aimed to design a single integrated control platform simulator which combines the features of both optical backhaul and wireless networks [15] couple with customized interactive nature of traffic transmission such as file downloading (FTP), TCP flow control and streaming video (H.264/AVC with UDP). [16] In this project, we use performance metrics such as UDP traffic throughput and end-to-end data delay as background traffic to study the effects of the MANET protocols over PON. In MANET, mobile nodes have the ability to act as both routers and hosts which might be beneficial to work on analysis of passive optical network that interconnects multiple sub-networks on Tree network. Our goal is to provide an additional source of performance statistics with a unique combination of last mile implementation solutions [17, 10], extension of existing radio coverage and capacity carrying UDP traffic plus passive optical backhaul.

Table 1: MANET Comparison Chart Of AODVUU, OLSR & DYMOUM

Protocol	OLSR	AODVUU	DYMOUM
Category	Table driven/proactive	On demand or Reactive	On demand or Reactive
Shortest Path	Yes	Yes & Fastest Route Discovery	No and can be extend with Shortest Path
Power consumption	Greater than that on demand routing	Grows with increasing mobility of active routes	Grows with increasing mobility of active routes
Resource usage (Bandwidth)	High	Medium	Low

There is a fundamental implementation passive optical network model based on OMNet++ simulation environment [18, 15] and INET framework models which focus on frame network and the physical layer modeling. Given this idea of practicality, this article will be discuss about the viability of creating OMNeT++-based simulator optical access framework with MANET routing deployment over Passive optical Backhaul. We employ and simulate different routing MANET protocols that are DYMOUM, AODVUU, and OLSR as shown in Table 1; driven by point to multipoint optical access system able to support flexible optical splitting mechanism with the joint ability factor to induce better integration and broadcasting transmission-convergence layer [19]. This paper also reports our work on extending the OMNeT++ INET Framework with an omni directional radio model of IEEE 802.11, putting a special emphasis on the implementation of asymmetrical communications because the framework of the backend connection in which downstream data (from an optical backhaul to the end user) flows over non-fiber downlink by wireless access, while upstream data (from the end user to the optical backhaul is sent over fiber link.

Performance simulation based capacity and delay [20] of wireless access based passive optical network access network is undertaken various environments such as in [21] using the OPNET platform. In this project, we analyze a similar situation where the nodes in the MANET send traffic to a common destination that is the centralized office (server) at the passive optical backhaul. We do not intend to dispute or concur with the conclusion drawn by the others author as we are performing the simulations in different environments with single hierarchical optical system backhaul and MANET convergence.

3 Methodology Over HOWAN

The proposed framework of the cross layer solution for HOWAN is adopted from the model of physical layer impairment-aware control and management protocols for optical transport network [22].

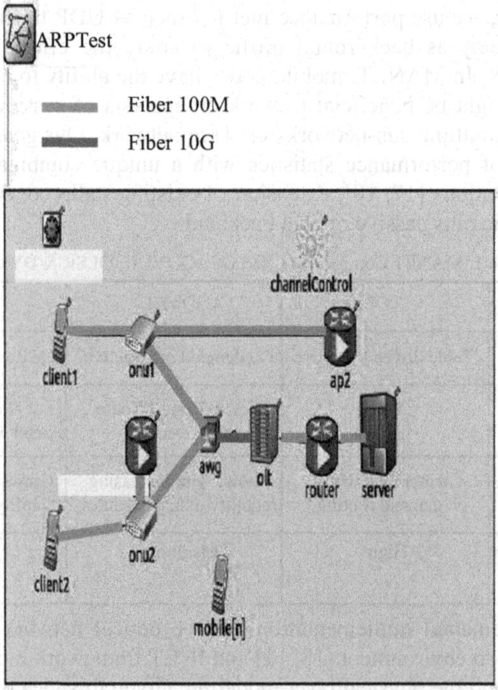

Fig. 1. Framework of Cross Layer Solution for HOWAN

Table 2: Simulation Parameter

Parameter	Values
Number of OLT	1
Number of ONU	2
Number of user nodes	30
Simulation times	200s
Traffic type	UDP
Routing protocol	MANET routing (DYMO, AODV, OLSR)
Carrier Frequency	2.4GHz
WLAN bitrate	2Mbps

The chosen HOWAN conceptual architecture is based on the work in [23] and can be summarized the fundamental foundation is only focus on specific one functional layer of physical system of the design. To support unlimited bandwidth of wired-wireless intensive applications and couple with seamless protocol transparency, cross layer of HOWAN is proposed in the sense that it extends routing channel and configurator signaling and reuses some concepts such as the decision path making without affecting the core functionality of the overall serving access network. In the study for the HOWAN project with emphasis on the fiber optics ecosystem, PON was taken and declared as the best solution in the combination of low complexity and best sensitivity which based on pyramid-type layer model.

The simulation scenario was run for the interval of 50 seconds up to total of 200 seconds in a combination environment of wireless mobility and optical wired immobility of dedicated access. Upfront simultaneous wired-wireless mobility extension and optical backend system bandwidth allocation each are considered independently with 10Gbit/s and 100Mbit/s each, Table 2 show the parameter for the simulation setup. In Figure 1, the significant difference from the referred model is the transmission media since this new updated HOWAN uses both optical and wireless media. The wireless mobility and optical framework aims to combine the properties of the two entities such that wireless mobility is used to simulate the behavior of wireless nodes together with optical wired nodes according to an ad hoc wireless communication protocol.

Our framework assumes a centralized management and control approach where resource and routing allocation process (through Routing and Bandwidth Assignments) is conducted by a single, centralized unit known as the center office (CO). CO is equipped with the traffic engineering where packet is obtained from the network elements. The channel control and the framework configurator are the entity that provides the necessary information to facilitate the resource and routing provisioning process in the central CO. It is the usage of the channel over both time and space.

4 Result Analysis and Discussion

4.1 Random Distribution Graph of Mean End-to-End Delay

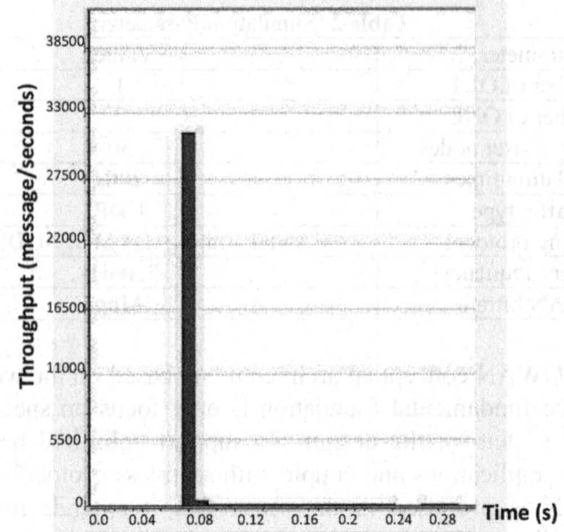

Fig. 2. End-to-End delay DYMOUM routing for UDP (CO) with the mean delay 0.069993

Firstly we see the result of typical proposed access method of the framework as depicted in the Figure 2, the characteristics of DYMOUM protocol routing over Passive Optical Network include showing less delay which means having accuracy in routing information and lower routing overhead. The worst performance in terms of end-to-end delay comes from OLSR protocol routing as shown in Figure 3 due to it is a proactive routing protocol and as such control message are periodically transmitted and constantly floods the wireless end network up to the centralize office optical backhaul and routing traffic to keep its routing tables up to date to all the nodes whether it is active or inactive. As a result, OLSR protocol routing is not bandwidth efficient.

As shown in the Figure 4, the characteristics of AODV protocol routing over Passive Optical Network showing less stability and reliability of route if compared to DYMO protocol routing which means occur increased number route discovery of process and time (TTL [Time-To-Live] value). Due to its nature as a reactive routing protocol, the routing information need to be maintain about the active nodes only regardless whether it is the disconnected nodes or not as compared to OLSR routing protocol.

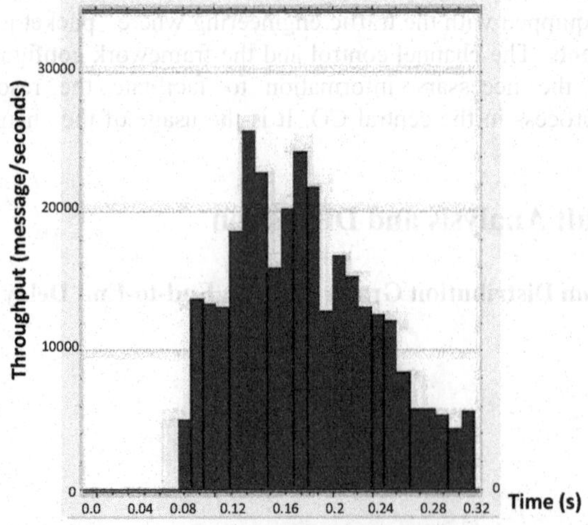

Fig. 3. End-to-End delay OLSR routing for UDP (CO) with the mean delay 0.098408

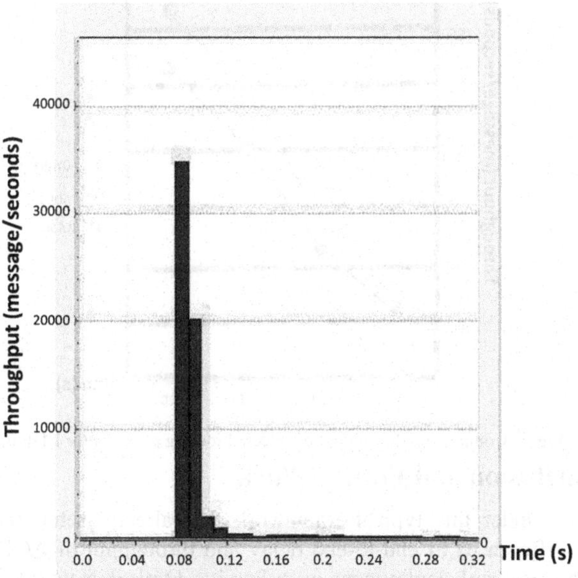

Fig. 4. End-to-End delay AODVUU routing for UDP (CO) with the mean delay 0.08227993

4.2 Throughput Vs Transmission Simulation Time

When End-to-End delay is our main requirement then in high static and dynamic environment, DYMOUM is better in the optical-wireless passive optical network infrastructure system. DYMOUM also has the better number of packets or data successfully transmitted to their final destination as in the Table 3 above despite the increase of mobility time period at 200s because throughput decreases with increase of mobility as shown in Figure 5.

Table 3: Performance Capacity of different network routing in NGOAN

TIME (S) / PROTOCOL	50	100	150	200
DYMOUM	6275	15530	24785	32675
OLSR	5795	11645	20155	27930
AODVUU	6680	14265	24700	31400

It also shows better route process of message or information delivery to final destination. But AODV doesn't show consistent performance if compared to DYMOUM. This is happens because of uncertainty link failure causing full-scale route rediscovery process was initiate thus increasing the routing overhead. The OLSR behave constantly that nodes can easily get routing information and it's easy to establish a session for traffic transmission.

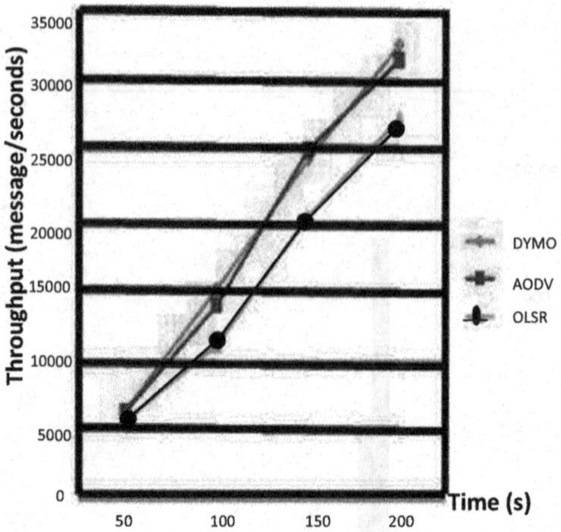

Fig. 5. Comparison throughput of selected routing protocol for UDP (CO)

5 Conclusion and Future Work

In this article, under this typical framework that already setup then examine the performance differences of end-to-end delay and throughput of AODV, DYMO and OLSR routing protocol deployment over Passive Optical Network. Our simulation results shows DYMO deployment over Passive Optical Backhaul is the best scheme in terms of both end to end delay and throughput transmission received. As in AODV deployment over optical backhaul network the switching mechanism during routing is inviting more process switching delay. Thus creating node delay occurs. It shows that throughout this comparison integrations framework compares the routing method efficiency, DYMO can perform well over passive optical backhaul in the way of route discovery and route maintenance due to its nature behavior. The future plan is to study the power consumption and others quality of service such as network access utilization in this particular network.

Acknowledgement

This work was supported by a Research Management Center of the University Technology Malaysia. The authors acknowledge the Ministry of Education Malaysia (MOE) and the administration of Universiti Teknologi Malaysia (UTM) for the project financial support through cost centre number R.J130000.7823.4F407 and University Teknikal Malaysia Melaka (UTeM) for the financial support through SLAB funding to the main author.

References

1. M. Molnár, F. Zhou, B. Cousin "Multicast Routing In Optical Access Networks," in "Optical Access Networks and Advanced Photonics: Technologies and Deployment Strategies, Ioannis P. Chochliouros; George A. Heliotis (Ed.) (2009) pp.162-183.
2. Rodney S. Tucker, "Green Optical Communications-Part I: Energy Limitations In Transport", IEEE Journal Of Selected Topics In Quantum Electronics, Vol. 17, No. 2, March/April 2011, pp. 245-260.
3. Rodney S. Tucker, Green Optical Communications-Part Ii: Energy Limitations In Networks, IEEE Journal Of Selected Topics In Quantum Electronics, Vol. 17, No. 2, March/April 2011, pp. 261-274.
4. B. Batagelj, "Implementation concepts of an optical access network by point–to–point architecture," Electrotechnical Review, Ljubljana, Slovenija, Phil., pp. 259–266, 2010.

5. GreenTouch™: Putting Energy Into Making Networks More Efficient, (Accessed 31st Mei 2012), http://www.greentouch.org/index.php?page=ict-industry combats climate-change.
6. Saba Al-Rubaye, Anwer Al-Dulaimi, Hamed Al-Raweshidy, "Next Generation Optical Access Network Using CWDM Technology," Int. J. Communications, Network and System Sciences, 2009.
7. Chowdhury. P, Mukherjee, B, Sarkar, S, Kramer. G, Dixit, S. "Hybrid Wireless-Optical Broadband Access Network (WOBAN): Prototype Development and Research Challenges" Network, IEEE, May-June 2009.
8. Vaishampayan V.A, Chao Tian, Feuer, M.D, "On the Capacity of a Hybrid Broadcast Multiple Access System for WDM Networks," Symposium on Information Theory Proceedings (ISIT), IEEE International, 2011.
9. Z. Zhao, B.Mosler, T.Braun, "Performance evaluation of opportunistic routing protocols: a framework-based approach using OMNeT++, Conference Proceedings of the 7th Latin American Networking, ACM, October 2012.
10. Yi Zhang, Pulak Chowdhury, Massimo Tornatoe and Biswanath Mukherjee, "Power Efficiency in Telecom Optical Networks", IEEE Comm. Surveys and Tutorials, vol.12, no.4, 4th Quarter, 2010.
11. Bjorn Skubic, Einar in de Betou, Tolga Ayhan and Stefan Dahlfort, "Energy-Efficient Next Generation Optical Access Networks", Topics in Optical Communications, IEEE Comm. Magazine, Jan 2012, pp. 122-127.
12. Pulak Chowdhury, Biswanath Mukherjee, Suman Sarkar, Glen Kramer and Sudhir Dixit, "Hybrid Wireless-Optical Broadband Access Network (WOBAN): Prototype Development and Research Challenges", IEEE Network, May/June 2009, pp. 41-48.
13. Pulak Chowdhury, Massimo Tornatore, Suman Sarkar and Biswanath Mukherjee, "Building a Green Wireless-Optical Broadband Access Network," Journal of Lighwave Technology, IEEE, 2010.
14. Redhwan Q. Shaddad, Abu Bakar Mohammad, Sevia M. Idrus, Abdulaziz M. Al-hetar and Nasir A. Al-geelani, "Emerging Optical Broadband Access Network from TDM PON to OFDM PON," PIERS Proceedings, 2012.
15. Foukalas. Fotis, "Practical implementation of cross-layer design in broadband wireless access networks," International Journal of Mobile Network Design and Innovation, Volume 3, Number 3, ACM January 2010.
16. K. Kim, "A research framework for the clean-slate design of next-generation optical access," Congress on Ultra Modem Telecommunications and Control Systems and Workshops (ICUMT), 2011 3rd International, IEEE, 2011.
17. Qasim, H; Abbas, M; Jameel, T; Naufal, I; Nofal, M, "Performance Analysis of TFRC and UDP over Mobile-IP Network with Computing Flows," Conference on Computational Intelligence, Communication Systems and Networks (CICSyN), 2010 Second International, 28-30 July 2010.
18. K. Kim, "Integration of OMNeT++ hybrid TDM/WDM-PON models into INET framework," in OMNeT++ Workshop, 2011.
19. R. Massin, C. Lamy-Bergot C. J. Le Martret, R. Fracchia "OMNeT++-Based cross-layer simulator for content transmission over wireless ad hoc networks," Journal on Wireless Communications and Networking networks, Volume 2010, April 2010.
20. R Malhotra, Amit.K. Garg "Analysis Of Capacity And Delay In Passive Optical Networks (Pons)" Journal of IJCCR, May 2012.
21. S Lee, Y Jeon, T Lim, K Lee, J Park, "A wireless access network based on WDM-PON for HMIPv6 mobility support, " Springer Science+Business Media, ACM, 28 October 2009.
22. Marinez, "Challenges and Requirements for Introducing Impairment-Awareness into the Management and Control Planes of ASON/GMPLS WDM Networks," IEEE Communications Magazine, Vol. 44, pp. 76-85, Dec 2006.
23. Redhwan Q. Shaddad, Abu Bakar Mohammad, Abdul Aziz M. Al–Hetar, Sevia M. Idrus, "Performance Analysis of Hybrid Optical–Wireless Access Network Physical Layer," Journal of Technology, UTM, 2011.

A Feedback Mechanism Based on Randomly Suppressed Timer for ForCES Protocol

Lijie Cen[1,*], Chuanhuang Li[1], and Weiming Wang[2]

Zhejiang Gongshang University, Hangzhou, Zhejiang Province, China
1011500218@pop.zjgsu.edu.cn, { chuanhuang_li, wmwang }@zjgsu.edu.cn

Abstract. As the shortcomings of closed networks are high lightened, the requirements of next generation network for openness are becoming increasingly intense. Based on the fact that reliable multicast of ForCES protocol plays a significant role in improving the performance of ForCES router, this paper researches the congestion control within the reliable multicast process, and analyzes scalability issues of reliable multicast based on ForCES protocol. Also, this paper proposes a feedback mechanism based on randomly suppressed timer which effectively avoids the ack-implosion problem via analyzing a mathematical model of randomized procedure of multicast feedback process. According to the results of the test, this mechanism enables reliable multicast to better adapt to the instability of network environment.

Keywords: scalability; reliable multicast; congestion; ForCES

1 Introduction

ForCES (Forwarding and Control Element Separation) which specializes in open programmable router architectures and protocols is one of the working groups in the IETF routing field. The difference between ForCES architecture router and traditional router is that the former's control element (CE) communicates with forwarding element (FE) via standard ForCES protocol messages. The ForCES architecture router increases the openness of the network elements and configurability. CE and FE communication of ForCES architecture is important in configuration operations of router, and has impact on router's performance. In order to decrease packet loss and delay of reliable protocol messages that a CE sends to FEs, this paper analyzes reliable multicast scalability issues within the ForCES router and proposes appropriate strategies to improve the performance of the router.

Several multicast congestion control protocols have been proposed, such as Pragmatic General Multicast Congestion Control (PGMCC) [1], TCP-friendly Multicast Congestion Control (TFMCC) [2], Receiver-driven Layered Multicast (RLM) [3], Receiver-driven Layered Congestion control (RLC) [4], Fair Layered Increase/Decrease with Dynamic Layering (FLID-DL) [5], Efficient Joining and Leaving for Receiver Driven Multicast Congestion Control (EJLRDMC) [6], Wave and Equation Based Rate Control (WEBRC) [7] and so on.

K.J. Kim (ed.), *Information Science and Applications*,
Lecture Notes in Electrical Engineering 339, DOI 10.1007/978-3-662-46578-3_13

Currently, there are two primary kinds of reliable multicast transmission protocols based on Acknowledgement (ACK) and Negative Acknowledgement (NACK) [8]. [9] indicates that reliable multicast transmission protocols based on NAK Compression are needed smaller bandwidth, having better scalability than those based on ACK. The reliable multicast of ForCES protocol based on the error detection and recovery of NACK preferably ensures reliability when multicast protocol messages are transmitting.

[10, 11] propose two performance indexes of reliable multicast congestion control: scalability and fairness. Revolving around these two indexes, they also propose several algorithms and protocols.

In order to ensure TCP-friendly fairness, [12] proposes Fair Active Congestion Control (FACC) which falls into single-rate and receiver-driven protocol. According to congestion parameter gathering strategy, this protocol redesigns the filtering algorithm of congestion control parameter to improve robustness and adaptability of the algorithm.

To eliminate "lowest-first" phenomenon, [13, 14] respectively propose Composite Multicast Congestion Control (CMC) and layered multicast (LM) congestion control scheme by particle swarm optimization (PSO) (LM-PSO). These two protocols don't use a single-rate congestion control algorithm, but dynamically adjust each layer of layered multicast. [14] also proposes a feedback suppressed algorithm called "EPS", which can make all receivers send feedback with equal probability.

2 Reliable Multicast Scalability Issues on ForCES Router

The performance of ForCES Router is limited for the following two reasons:

1) In the process of a reliable multicast for ForCES protocol, request messages sent by FEs will show a rapid upward trend with the growing number of FEs, which will lead to ACK implosion, an enormous bandwidth pressure in the transport mapping layer of ForCES, a lot of packet loss, link congestion and reliable multicast performance degradation.

2) When the network environment is unstable, a lot of reliable multicast restoration packets will be lost with the badly available bandwidth of transport mapping layer of ForCES, which results in that each receiver's FE sends a large number of redundant packets to request packet loss restoration, increasing the processing overhead and network bandwidth bottleneck pressure. So it is necessary to take measures to minimize times of sending redundant restoration packets when a CE sends packets to FEs.

3 Feedback Suppressed Strategy Based on Random Delay

3.1 Feedback Suppressed Strategy for FE

In the scenario of reliable multicast of ForCES protocol, a CE starts a cycle timer when multicast transmission starts and periodically sends session packets. Cycle time isn't static, but in a period of time is fixed, which means that the triggered cycle of cycle timer can be set in a CE. After correctly receiving session packets which must be reliable, FEs record timestamps and maximum serial ID numbers session packets contain. When FEs of receivers receive control protocol messages sent by a CE, packet loss is detected according to received session packets. If the packet loss detection is successful, FEs record the serial number of packet loss, generating request packets, starting packet loss suppressed timer. If a FE receives serial numbers of packet loss request sent by other FEs in the group before the timer expires, it checks receiving serial numbers whether it has detected and prepared to send to a CE. If there are repeated request packets, the FE deletes them. Otherwise, a FE sends request packets of packet loss containing serial numbers of packet loss in this side to a CE. At the same time, the FE sends request packets of packet loss to other FEs in the group, indicating the sending ID, starting the restoration timer. If a FE receives corresponding restoration packets before the restoration timer expires, it will cancel the restoration timer. Otherwise, the FE continues to send request packets and start the restoration timer, until it receives all restoration packets. The structure diagram of FE multicast module with feedback suppressed strategy is shown in Fig. 1, and the design flow chart is shown in Fig. 2.

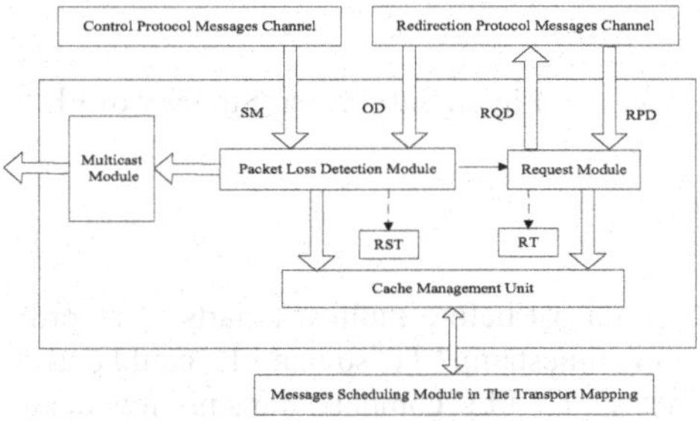

Fig. 1. FE Multicast Module

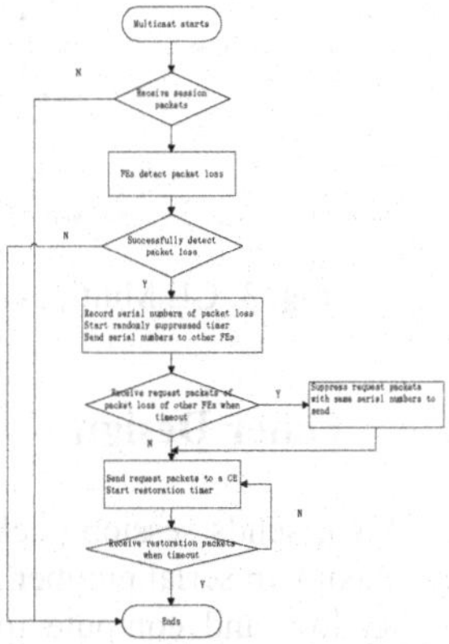

Fig. 2. Suppressed Strategy of FE

3.2 Feedback Suppressed Strategy for CE

When a FE sends feedback packets, it also sends serial numbers of multicast feedback packets to other FEs in the group. Each FE which has received these multicast messages may contain request packets sent by a FE which sent these multicast messages. If FEs use the method of responded restoration that all of FEs received multicast messages send request packets to the FE that sent these multicast messages, FEs in the group should maintain records related to the state and set the timer of packet loss restoration. It will increase FEs' processing overhead, and can't process packets fast, which is not conducive to improve the performance of network. Therefore, it doesn't use the method of FEs to respond each other. Because of a lot of FEs which prepare to receive restoration packets, if a CE sends unicast packets to FEs, a large number of TCP connections need to be established, which will increase the processing overhead of the CE. Though a CE sends fewer restoration packets, they still have a heavy load. Therefore, it uses the method that a CE sends multicast packets. Feedback packets which are sent by FEs of multicast receivers at first time are sent synchronously, and each serial number only responds once. It can reduce unsuccessfully suppressed feedback packets to avoid redundancy. In order to reduce bandwidth usage, feedback packets sent after the second time are sent as a UDP unicast way. The structure diagram of CE multicast module with feedback suppressed strategy is shown in Fig. 3, and the design flow chart is shown in Fig. 4.

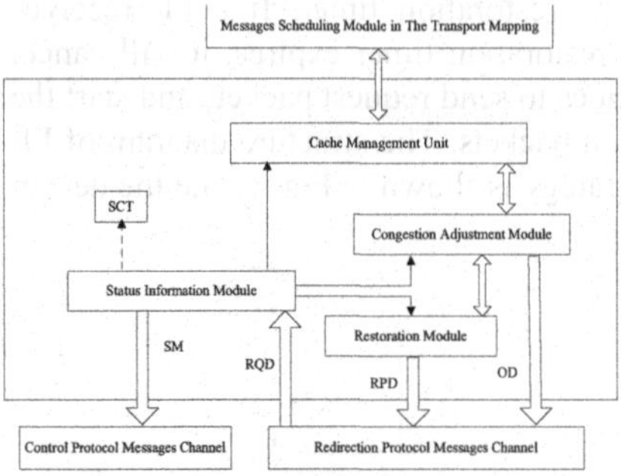

Fig. 3. CE Multicast Module

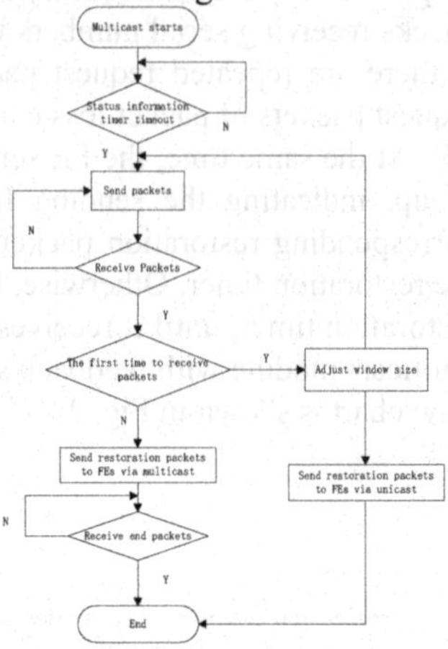

Fig. 4. Suppressed Strategy of FE

4 Timer Design

A CE sends session packets to FEs via unicast before multicast starts. It records the maximum serial number "Si" and sending timestamp "Ti" so that FEs could detect packet loss and compute transmission delay "t". FEs compute transmission delay denoted as "ti" after receiving session packets and set (2ti, 4ti) as the time span of randomly suppressed timer with uniform distribution. The lower limit value "2ti" ensures that FEs accurately and completely detect packet loss, which avoids the packets arriving out of order sending redundant feedback packets. The upper limit value "4ti" ensures that each FE sends serial numbers of multicast feedback packets to other FEs in the group when sending feedback packets, in order to restrain repeated feedback packets. With uniform distribution, feedback packets sent by the CE can smoothly and stably pass through the bottleneck link, which avoids congestion. After randomly suppressed timer expires or feedback packets are restrained by feedback packets sent by other FEs, the FE randomly starts the restoration timer. If restoration packets are not received before the timeout, the FE resets the timer and starts again until it receives restoration packets. The time span of timer is uniformly distributed in (2ti, 4ti), which can avoid packets loss, reduce the time of packet loss restoration, and improve the protocol efficiency.

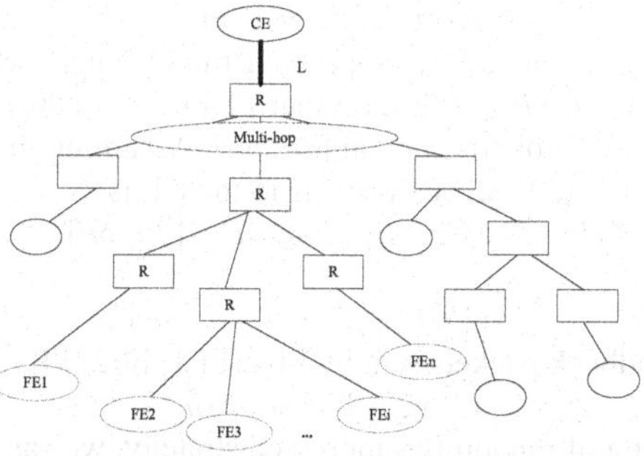

Fig. 5. Topology of ForCES Protocol Multicast

Assume that multicast model is shown in Fig. 5: a CE sends multicast messages to FEs through multiple network paths, and "L" is the bottleneck link. After the multicast starts, the CE sends session packets of unicast to FEs. Assume that the starting time for the multicast process is 0, then the time that session packets arrive at FEs (FE1, FE2, FE3…FEi…FEn) is

$$t1, t2, t3 \ldots ti \ldots tn;$$

then one-way RTT delay recorded by FEs is

$$t1, t2, t3 \ldots ti \ldots tn.$$

Because of network uncertainty, ti can approximately be considered to obey normal distribution. Assume that X(t) = t1, t2, t3 … ti … tn, then

$$fx(t) = \frac{1}{\sqrt{2\pi\delta^2}} \exp\left\{ \frac{-(x-\eta)^2}{2\delta^2} \right\},\ E[X(t)] = \eta . \quad (1)$$

If FEs don't use randomly suppressed strategy, each FE sets ti as the duration packet loss detection timer. The time that a FE sends feedback packets to the CE is

$$3t1, 3t3, 3t3 \ldots 3ti \ldots 3tn.$$

Assume that $3t1 < 3t2 < 3t3 < \ldots < 3ti < \ldots < 3tn$, then it enables

$$X'(t) = 3t1, 3t2, 3t3 \ldots 3ti \ldots 3tn.$$

According to a linear combination of random variables within normal distribution, we can see E[X'(t)] = 3η. It means that the time that feedback packets pass through the bottleneck link "L" at [*3t1*, *3tn*] obeys normal distribution. Assume that the number of multicast packets is *M* and the packet loss rate of arriving each FE is *p*. To simplify the treatment, the average number of packets sent by FEs is *Mp*. The expectation of the number of feedback packets which pass at [*3t1*, *3tn*] is

$$E(NACK) = E[X'(t)] * Mp = 3\eta Mp. \quad (2)$$

According to the characteristics of normal distribution, when the number of FEs *n* increases in the multicast group, η also increases. If it doesn't use randomly suppressed strategy, the load of bottleneck link "L" between a CE and FEs increases with the increasing number of FEs, which has a poor scalability.

After multicast starts, a CE sends session packets to FEs via unicast. If FEs use randomly suppressed strategy under the same multicast model, assuming the starting time of multicast is 0, the time that session packets arrive at FEs (FE1, FE2, FE3…FEi…FEn) is

$$t1, t2, t3 \ldots ti \ldots tn.$$

The length of time of randomly suppressed timer that is set by each FE obeys uniform distribution at [$2ti$, $4ti$]. The time that FEs send feedback packets to the CE after timeout obeys uniform distribution at [$4ti$, $6ti$]. Assuming the time that FEs (FE1, FE2, FE3…FEi…FEn) send feedback packets to the CE is

$$t1', t2', t3' \ldots ti' \ldots tn', ti' \sim [4ti, 6ti].$$

Assume that

$$t1' < t2' < t3' < \ldots < ti' < \ldots < tn'.$$

The number of feedback packets sent by FEs (FE1, FE2, FE3…FEi…FEn) is

$$x1, x2, x3 \ldots xi \ldots xn.$$

Because of the using of randomly suppressed strategy, we can know

$$x1+x2+x3+\ldots+xi+\ldots+xn \leqslant Mp.$$

And we know

$$E(ti) = \frac{4ti + 6ti}{2} = 5ti. \tag{3}$$

Obviously, at the points of time [$5t1$, $5t2$, $5t3$ ... $5t$ ... $5tn$], the number of feedback packets which pass through the bottleneck link "L" is less than or equal to Mp, and it doesn't increase with the number of FEs. So then it ensures the reliable multicast scalability of randomly suppressed strategy.

5 Scalability Test

5.1 Function of Scalability Test

Run test programs of reliable multicast on FEs and a CE. The test program of CE is added to a multicast group via the IGMP protocol. The multicast group which has ID 0xC0000000 and IP 233.4.4.4, includes 10.20.0.229 and 10.20.0.135 two hosts. As shown in Fig. 6, sending conditions of multicast request packets can be got by sending 4000 multicast packets to 0xC0000000.

Fig. 6. Feedback Conditions of Reliable Multicast Packet

Fig. 6 shows that transmission time of feedback packets of FEs is uniformly distributed, there is no redundant packets. It means that the randomly suppressed timer of FE effectively restrains redundant request packet and avoids ACK implosion.

5.2 Performance of Scalability Test

Run test programs of reliable multicast on FEs and a CE. The test program of CE is added to a multicast group via the IGMP protocol. The multicast group which has ID 0xC0000000 and IP 233.4.4.4, includes 10.20.0.59, 10.20.0.190 and 10.20.0.229 three hosts. As shown in Fig. 8, sending conditions of multicast request packets can be got by sending 5000 multicast packets to 0xC0000000.

As shown in Fig. 7, the number of feedback packets arriving at a CE doesn't increase with increasing the number of FEs. Arrival time is uniformly distributed. There are no redundant packets. In the multicast restoration process, it doesn't take more time than when the number of FEs is fewer. It means that the feedback suppressed strategy of reliable multicast for ForCES protocol improves the scalability of multicast transmission protocols.

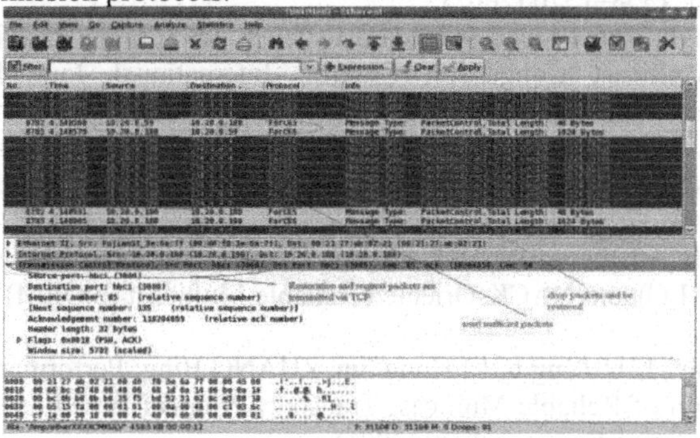

Fig. 7. Sending Conditions of Multicast Restoration Packets

6 Conclusion

The paper analyzes scalability issues of reliable multicast based on the reliable multicast mechanism for ForCES protocol. According to the result of test, a randomly suppressed timer for the FE to restrain feedback packets improves the scalability of reliable multicast for ForCES protocol, and ensures the reliability of multicast.

Acknowledgments. This work was supported in part by a grant from the National Basic Research Program of China (973 Program) (No.2012CB315902), the National Natural Science Foundation of China (No.61402408, 61379120, 61170215), Zhejiang Leading Team of Science and Technology Innovation (No.2011R50010-02, 2011R50010-03, 2011R50010-17), Zhejiang Provincial Key Laboratory of New Network Standards and Technologies (NNST) (No.2013E10012), the Youth Foundation of Zhejiang Gongshang university (QZ13-8).

References

1. L. Rizzo: PGMCC: A TCP-friendly single-rate multicast congestion control

scheme. In: Proceedings of ACM SIGCOMM (2000).
2. Kulatunga, Fairhurst: TFMCC Protocol Behaviour in Satellite Multicast with Variable Return Path Delays. IEEE press (2006).
3. S. McCanne, V. Jacobson, M. Vetterli: Receiver-driven layered multicast. In: Conference proceedings on applications, technologies, architectures, and protocols for computer communications, pp. 117-130, ACM SIGCOMM (1996).
4. L. Vicisano, J. Crowcroft, L. Rizzo: TCP-like congestion control for layered multicast data transfer. In: Proceedings of the Seventeenth Annual Joint Conference of the IEEE Computer and Communications Societies, pp. 996–1003, IEEE press (1998).
5. J. Byers, G. Horn, M. Luby, M. Mitzenmacher, W. Shaver: FLID-DL: congestion control for layered multicast. IEEE Journal on Selected Areas in Communications, vol. 20. pp. 1558–1570 (2002).
6. Karan Singh, Rama Shankar Yadav: Efficient Joining and Leaving for Receiver Driven Multicast Congestion Control. International Journal of Computer Applications, vol. 1, pp. 110–116 (2010).
7. M. Luby, V.K. Goyal, S. Skaria, G.B: Horn. Wave and equation based rate control using multicast round trip time. SIGCOMM Computer Communication Review, vol. 32, pp. 191-204 (2002).
8. V Roca, B Adamson: FCAST: Object Delivery for the Asynchronous Layered Coding (ALC) and NACK-Oriented Reliable Multicast (NORM) Protocols, IETF (2013).
9. WANG Bin, LIU Zeng ji, LI Hong bin, ZHANG Bing: Performance Analysis and Comparison of Reliable Multicast Transport Protocol Based on ACK and NAK. Acta Electronica Sinica, vol. 29 (2001).
10. REN Li yong, LU Xian liang: Reliable Multicast Congestion Control. Computer Applications, vol. 29 (2002).
11. SHI Feng, WU Jian-ping: A Survey on Multicast Congestion Control. Journal of Software, vol. 13, pp. 1441-1449 (2002).
12. WANG Ji-zhou, HUANG Zhi-guo: A TCP Friendly Reliable Multicast Congestion Control Algorithm. Computer and Modernization (2009).
13. MA Hai-yuan, MENG Xiang-ru, MA Zhi-qiang, MA Sen: Composite Multicast Congestion Control by Multi-objective Particle Swarm Optimization. Journal of Applied Sciences, vol. 29, pp. 459-466 (2011).
14. MA Hai-yuan, MENG Xiang-ru, MA Zhi-qiang, LI Jin-liang: Layered multicast congestion control by particle swarm optimization in heterogeneous environment. The Journal of China Universities of Posts and Telecommunications, vol. 18, pp. 85-91 (2011).

Operational Analysis of Interarrival Rate-based Queuing Management Mechanism for High-speed Networks

Mohammed M. Kadhum · Selvakumar Manickam

the date of receipt and acceptance should be inserted later

Abstract With the development of the Internet, it has become essential to operate an effective queue management mechanism that improves Quality of Service (QoS), such as throughput, end-to-end delay, and loss on the Internet. The performance of TCP applications relies on the selection of a queue management mechanism in the network. The current mechanisms used in the Internet have slow response to congestion, which in turn results in large queue size variation and have untimely congestion detection and notification. These consequences degrade the performance significantly because of the high queuing delays and packet loss. This paper presents a proactive queuing management mechanism that would efficiently address the congestion state upon every incoming packet arrival, by deciding the probability with which the packet should dropped or marked. It is designed based on Packet Interarrival Rate and Actual Queue Size parameters. The performance evaluation showed that the proposed mechanism can be adjusted to efficiently control the queue size to some desirable value, which can decrease the queuing delay while maintaining high link utilization and low packet drops.

Keywords TCP/IP protocol · Internet Congestion Control and Avoidance · Network Performance · Queue Management · High-speed Networks

1 Introduction

Due to the high cost of building and operating computer networks, as they require a lot of resources such as human, financial, and material resources, it is necessary to optimize the network utilization, which allows increasing the number of users that utilize the

Mohammed M. Kadhum · Selvakumar Manickam
National Advanced IPv6 Center (NAv6), Universiti Sains Malaysia (USM), 11800 Pulau Pinang, Malaysia
E-mail: {kadhum,selva}@usm.my}

Mohammed M. Kadhum
Telecommunication Research Laboratory School of Computing, Queen's University, K7L 3N6 Kingston, ON, Canada
E-mail: {kadhum@cs.queensu.ca}

K.J. Kim (ed.), *Information Science and Applications,*
Lecture Notes in Electrical Engineering 339, DOI 10.1007/978-3-662-46578-3_14

network resources. The network performance is the key parameter that determines the users' satisfaction. The user goes through the performance of the network when using the network services. It is significant to network owners and operators to enhance the network performance to maximize their profits through minimizing the operation costs, which in turn ensure the users' satisfaction. Network congestion causes degradation in network utilization and performance as mentioned in [3] and [8]. The congestion can be controlled at the router by the interaction between end-to-end congestion control mechanisms and router queue management algorithms, which determines how good the packet arrival rate and the queue lengths. Avoiding or controlling congestion is a critical issue in designing and operating computer network [25]. The future availability of inexpensive buffers and high-speed links and processors will not be able to alleviate the congestion completely because of the mismatching of link speeds, higher offered traffic loads, unforeseeable traffic patterns, large transient loads, and greater degrees of statistical multiplexing [17]. All these issues will continue to act as sources of congestion, as was mentioned by Jain [7]. For that reasons, the study of congestion management in computer networks will go on to be an effective research area. Additionally, as argued by Mahbub et al. [13], with the new increase in network complexities and traffic dynamics, congestion control algorithms employed at the end hosts and in the network routers continue to evolve. Although there is a number of schemes have been proposed for network congestion control, their contributions are still not sufficient enough to manage the router's queue efficiently. According to Durresi et al. [3], the research for new mechanisms will continue as a result of the requirements for congestion control that make it difficult to get a satisfactory solution, and for network strategies that affect the design of a congestion mechanism(s). Thus, a mechanism proposed for a specific network may not work on another network due to different traffic pattern or service requirements. For example, many of schemes developed in the past for best-effort data networks will not work satisfactorily for multiclass IP as mentioned in [21] [14] [15]. According to Barcellos et al. [2] and Molinero-Fernandez et al. [16], Transmission Control Protocol (TCP) dominates more than ninety percent of the Internet traffic such as FTP, Web traffic, electronic mail (e.g., SMTP), and remote terminal (e.g., Telnet). Fifty to seventy percent of TCP traffic is short-lived connections (also called "mice") as asserted in [5]. Ryu et al. [19] concluded that although these applications can tolerate packet delay or packet losses, congestion remains a major problem that leads to poor performance. That is, congestion reduces the network resources' utilization and causes reduction in the performance experienced by the network users. Thus, it is important to minimize the happening of congestion events to improve the resources utilization and to provide users with acceptable performance. If the Internet is developed to become a high performance network that provides widespread services, including real time voice/video, it must understand how congestion occurs and come up with an efficient mechanism to keep the network functioning within its capacity. This paper proposes an innovative proactive mechanism for queuing management which are able to respond to congestion quickly, conveys the congestion notification timely, and manages the queue size according to congestion condition; which results in minimizing the variation of queue size . Consequently, the Internet performance will be improved widely.

The rest of the paper is organized as follows; Section 2 presents the most related work and concepts to the research work presented in this thesis. common multihop routing protocols used for MANETs, namely Dynamic Source Routing (DSR) and Ad Hoc On-Demand Distance Vector (AODV). The proposed algorithm of Interarrival Rate-based Queuing mechanism is described in Section 3. Section 4 shows the process model that implements the proposed algorithm. The parameter settings for the proposed mechanism, in terms of time constant (T) and the packet sliding window size (W_s) parameters, are described in Section 5. Section 6 presents the verification and validation of the proposed mechanism. Results and discussion of the performance evaluation are presented then in Section 7. Finally, conclusions and future work are presented in Section 8.

2 Related Work and Concepts

This section provides the background and some related research on queue management that defines the general framework of this research. To improve the performance of the Internet, by controlling congestion and reducing packet drops, the need for pro-active queue management has widely accepted. Several pro-active queue management mechanisms have been proposed for the Internet. Random Early Detection (RED) [4] is the most popular among these mechanisms and it is recommended by IETF as default mechanism for the Internet routers. RED uses the average queue size, which is collected over long period, to make its control decisions. However, as RED uses the average queue size makes, it reacts to congestion condition slowly, which causes large queue size deviation and improper congestion detection and notification. This would degrade the performance of the Internet seriously [20]. RED algorithm has many problems such as bandwidth unfairness, low throughput under poorly setting parameters, and large queuing delay variance (jitter) because of the fluctuation of the queue level, being unable to handle unresponsive connections, and a high number of consecutive drop. It is our hypothesis that an active queue management mechanism which uses a measure of packet arrival rate with a measure of the queue length for its control decisions will show better ability in realizing the goals of controlling the packet arrival rate to the router, router queue lengths, and network congestion, while achieving a higher performance. Therefore, the proactive queue management mechanism presented in this paper uses the current queue size and the packet interArrival rate for making its control decisions to accomplish the goal of controlling the network congestion.

Several variants of RED have been proposed to overcome the restrictions of RED algorithm. The common goals of these proposals are to stabilize the queue length and to protect the responsive connections. It is typically the same as enforcing fairness among the connections on the link as mentioned in [24] [11] [23]. Some of the proposed algorithms deal with the calculations of the control variable (which is the average queue size) and/or drop/mark probability function; while, the other algorithms are concerned with configuring and setting RED's parameters. Stabilized RED (SRED) [18], Effective RED (ERED) [1], Adaptive RED (ARED) [26], Double Slop RED (DSRED) [27], Multilevel RED (MRED)[10], Hyperbola RED (HRED) [6], LRED [23], Cautious Adaptive RED (CARED)[22], and Nonlinear RED [28].

3 Algorithm of Interarrival Rate-based Queuing Mechanism

It is important to point out that the proposed proactive queue management is not firmly attached to packet forwarding process and its components calculations are not required to be performed in the time critical packet forwarding path. The algorithm works such that the calculation of the average packet interarrival rate (I_r), the current queue size

(Q_{cur}), and the packet dropping/marking probability (P_{ini}), can be carried out in parallel with packet forwarding process. The work can be also calculated by router as a lower priority task as time allows. That is, the algorithm does not weaken the router's ability to process packets. in case of the presence of Explicit Congestion Notification (ECN) [12] packet , one might think that setting a *Congestion Indication* (*CE*) bit in the ECN-capable packet header by algorithm's marking technique, instead of dropping it, adds overhead to the algorithm. Nonetheless, the algorithm is developed to mark as few ECN packets as possible. Thus, the overhead of setting the CE bit in the ECN packet header is kept to a minimum. The router utilizes the proposed mechanism is different from DEC-bit router, for instance, which sets the congestion indication bit in every packet that arrives at the queue whenever the average queue length surpasses a certain threshold [9].

The pseudo-code implementations of the proposed algorithm is presented in Figure 1.

for (each arriving packet (Pkt_i)) {

/* sample the current queue size (Q_{cur}) */

/* compute average packet interarrival rate (I_r) over a period of time T,

wrt the difference of relative transit times for the two successive packets $D(i-1,i)$*/

$I_r = \frac{Pkt(i-1) + (|D(i-1)| - Pkt(i-1)}{T}$

/* compute arithmetic mean of average packet interarrival rate (($\bar{I}_r$)) */

$\bar{I}_r = ((I_r^i / 2).T) + (I_r^i)$

/* compute initial packet dropping/marking probability (P_{ini}) */

$P_{ini} = \frac{(\bar{I}_r - \mu).T) - Q_{cur}}{\bar{I}_r.T}$

/* compute the final packet dropping/marking probability (P_f)*/

$P_f = \frac{P_{ini}}{2.count.P_{ini}}$

/* drop/mark the packet with probability (P_f)*/

dropPacket(Pkt_i, P_f)

markPacket($Pktecn_i$, P_f)}

Fig. 1 Pseudo Code of the Proposed Algorithm

The design and derivation of the algorithm probability function is based on the suppositions of changeable packet interarrival rate with constant acceleration (deceleration). The packet drop/mark probability (P^i) which is given by

$$P_{ini} = \frac{(\bar{I}_r - \mu).T) - Q_{cur}}{\bar{I}_r.T} \tag{1}$$

was calculated as the fraction (the decrease in the queue size due to random packet dropping to the increase in the queue size due to packet arrivals) of queue length increase in consequence of packet arrival over the time constant (T) period that necessary to be randomly dropped (if non-ECN-capable), to help the outgoing transmission link capacity in reducing the packet arrival rate to the queue ($(1 - P^i_{ini}).I_r$ to below the link capacity (μ) and steering the current queue length *(Q_{cur})* to manageable queue size over the time period of *(T)* seconds.

To estimate the interarrival rate after i^{th} packet is received , we calculate the change of interarrival time over the time period of *(T)* seconds, using a sliding window of W_s packets as the average number of bits arrived per unit time over the period of time enclosed by the window of the most recently arrived W_s packets.

$\bar{I_r}$ was the average of the interarrival rate that is given by

$$\bar{I_r} = ((I^i_r/2).T) + (I^i_r) \quad (2)$$

over a time period of T that is calculated as the arithmetic mean of the average packet interarrival rate at the beginning of the interval (I_r) and at the end of the interval $(I^i_r.T) + I^i_r$ presuming that the rate will increase or decrease with the constant acceleration or deceleration of $bits/sec^2$.

3.1 Computational Complexity of the Proposed Algorithm

The proposed algorithm calculates the average packet interarrival rate and the packet drop/mark probability for each arriving packet. The average packet interarrival rate is calculated as in Eq. 2. The packet drop/mark probability is calculated as:

$$P_{ini} = \frac{(\bar{I_r} - \mu).T) - Q_{cur}}{\bar{I_r}.T} \quad (3)$$

$$P_{ini} = 1 - \frac{\mu}{\bar{I}} + \frac{Q_{cur}}{\bar{I_r}.T} \quad (4)$$

The per-packet calculation for the algorithm comprises of 2 addition operations and 1 division operation for the average packet interarrival rate; and 2 addition operations, 1 multiplication operation, and 2 division operations for the packet drop/mark probability. Hence, the proposed algorithm demands 4 addition operations, 1 multiplication operation, and 3 division operations which total to 8 operations for each arriving packet.

3.2 State Storage Space

For the packet interarrival rate estimate process, the algorithm uses a packet sliding window of size W_s packets. This has the disadvantage of requiring twice W_s storage elements for storing the packet sizes and interarrival times of the latest arriving W_s packets rather than 1 storage element required by RED's *EWMA* queue averaging process. Nonetheless, the design of the proposed drop/mark probability function is separated from the packet interarrival rate estimation process and a different packet interarrival rate estimation technique, with lower state requirements such as an $EWMA$ process, can be used to reduce the need for large state storage space.

4 Process Model

The queue module in the router implements the proposed queuing management algorithm. Figure 1 shows the process model implementing the proposed algorithm.

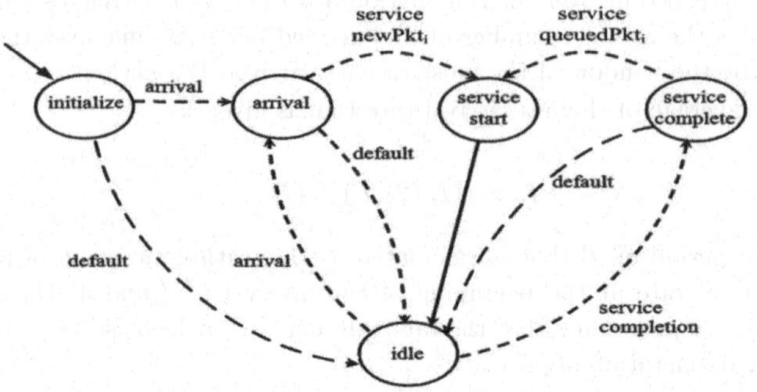

Fig. 2 Finite State Machine Model of the Proposed Algorithm

The algorithm process model, which was created and implemented in ns-2, accepts packets from sources and forwards them to the foreword module after holding each packet for a simulated duration which is referred to as the packet's service time. The service time of each packet is computed as a fraction of the packet's length over the queuing module's service rate.

The process model is comprised of five different states with transitions between them. The initialize state specifies necessary initializations that are performed once only at the beginning of the operation of the queue module. Once the initialization is completed in the initialize state, if a packet arrives at the buffer, a transition is made to the arrival state, otherwise to the idle state. The idle state is where the queue management process would wait for an event to happen to be processed if there is no work to be done or if the server is busy. Once the process enters the arrival state due the arrival of a new packet, either from the initialize or from the idle state, the newly arrived packet is operated upon by the proposed algorithm and it is either randomly dropped or inserted into the buffer. Note that if the packet to be dropped is ECN-capable packet, it will be admitted and then might be marked at the head of the queue, when necessary, to provide fast congestion notification. Therefore, the proposed algorithm is implemented in the arrival state. If the packet is successfully inserted in the queue and the server is busy, a transition is made, from the arrival state, to the idle state to wait for the server to become available. If the packet is inserted into the queue successfully and the server is idle, the process moves from the arrival state to the service-start state. While in the service-start state, the length of the packet at the front of the queue is obtained, its service time is computed, and a self-interrupt is scheduled to signal the completion of the packet's service in the near future. Then the process moves, from the service-start state, to the idle state where it will wait for the next packet to arrive or for the service of the packet that is currently being served to be completed. Once the finishing of a packet's service is perceived by a self-interrupt in the idle state, the process makes a transition from the idle to the service-complete state where the packet at the front of the queue is moved out to the forwarding module. If more packets are waiting in the queue to be served, the process moves from service-complete to service-start state to initiate the processing of the next packet at the head of the queue, as explained previously, before returning to idle state. If, however, the queue is empty, the process moves back from service-complete state to idle state where it will wait for the arrival of a new packet.

5 Parameter Settings

The proposed queuing management mechanism can be adjusted for different network conditions based on the time constant (T) and the packet sliding window size (W_s) parameters.

The time constant (T) specifies the aggressiveness of the random packet dropping/marking process that applied by the the proposed mechanism scheme. The smaller the T parameter, the faster the algorithm will trigger to direct the instantaneous queue length to its optimal value by exercising more packet dropping/marking. On the other hand, the larger the T parameter is, the slower the control and queue size directing.

The packet sliding window size specifies the preciseness with which the average packet interarrival rate is estimated and the sensitivity of the algorithm to rate oscillations. The smaller the window size, the more sensitive the algorithm, and the rate estimation procedure, will be to oscillations in the packet interarrival rate. On the other hand, the larger the window size, the less sensitive the algorithm, and the rate estimation, will be to oscillations in the packet interarrival rate. These two parameters enable tuning the algorithm at different level of sensitivity for early detection and response (reaction) to congestion.

While small T and W_s values enable the algorithm to detect even very short-term traffic oscillations and to react quickly, large values for these parameters make the algorithm less sensitive to short-term traffic oscillations and more moderate in reacting to long-term traffic. The effects of different (T, W_s) parameters settings are investigated based on the numerical results to identify the general relationship between these parameters settings that ensure best performance of the algorithm.

6 Verification and Validation of the Proposed Mechanism

Once the algorithm had been built in ns-2, the verification and validation of the algorithm was made by comparing the results to RED results which obtained from actual simulation runs based on the optimal settings according to the RED guidelines.

All simulations were carried out over a period of 120 seconds which is long enough to cover the transit state of the transmissions. 25 experiments that include different (Min_{th}, Max_{th}) settings were performed using RED with an average queue weight W of 0.002 and a maximum probability $Max_{drop}$of dropping or marking packet of $= 0.1$. From these experiments, it was found that the (Min_{th}, Max_{th}) associated with the pairs (1, 3), (2, 6), (3, 9), (4, 12), (5, 15) had maintained the average queue size below the half of the actual buffer size. Particularly, the highest link utilization obtained was during the experiment when $(Min_{th}, Max_{th}) =$ (5, 15). Hence, this particular experiment is used as a reference experiment for the performance evaluation in order to verify the proposed mechanism.

6.1 Verification

Verification of proposed mechanism was done to ensure that it is programmed and implemented correctly, and it does not have errors, oversights, or bugs. Also, to ensure that he specification and requirements of the model are complete.

The proposed mechanism was run with simple cases; for example, the basic scenario for the purpose of verification and validation of the algorithm included a small window size of 8 packets, a small packet size of 552Bytes, and two TCP sources (one is ECN-capable and the other is non-ECN-capable) sharing a bottleneck link of 300Kbps with 100msec delay that connects a bottleneck router to one destination. The links between the sources and their gateway offer full-duplex connections at a bandwidth of 2Mbps with a propagation delay of 10msec. These simple cases can be easily analyzed and the simulation results can be compared with the analysis. Then, these cases were developed and made more complex to test the proposed mechanism according to the simulation scenario setup shown in Figure 3. The size of the physical buffer was set to 840,000bits which is twice the Bandwidth Delay Product (BDP).

Consistency tests were carried out to ensure that the proposed mechanism produces similar results for input parameters values that have similar effects. Furthermore, to verify its implementation, we used a script to trace and filter the events occurred over the bottleneck link during the run time.

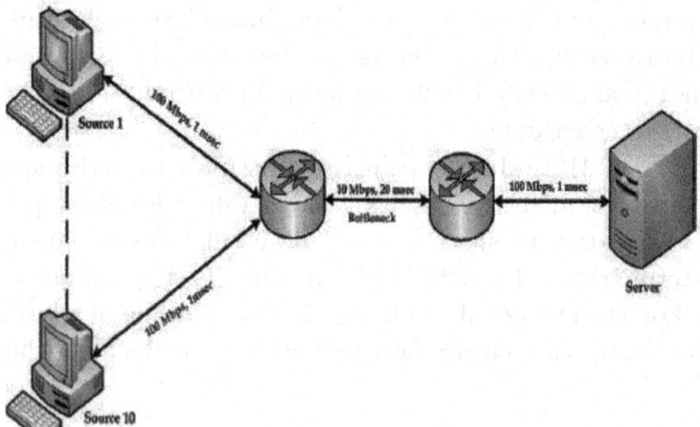

Fig. 3 Simulation Setup

6.2 Validation

Validation of the proposed mechanism was done to ensure that the mechanism meets its intended requirements in terms of the methods employed and the results obtained.

The validity of a simulation model is usually tested for different sets of experimental conditions and for an acceptable range of accuracy related to the purpose for which the model is intended. Therefore, the proposed mechanism was tested for various sets of packet window size (W_s) and time constant (T). A total of 96 experiments were carried out for evaluating the algorithm. The W_s parameter was varied over 8 different values; while, T parameter was varied over 12 different values. The 96 experiments - that represent the different settings of (W_s, T) according to the Cartesian Product - involve the following two parameter sets:

- W_s = {32, 64, 128, 256, 512, 1024, 2048, 4096}, and
- T = {1, 8, 27, 64, 125, 216, 343, 512, 729, 1000, 1331, 1728}

The Graphical Comparisons technique is used to validate the proposed mechanism. Graphs that are presented in next section can be used for the implementation validation of the proposed mechanism.

7 Results and Discussion

In this section, the performance of the proposed queuing management mechanism is evaluated and compared to the RED mechanism in terms of controlling the queue size and the packet arrival rate based on the numerical results obtained from simulations.

7.1 Queue Size Control

This subsection compares the statistical properties of actual queue sizes for the proposed and RED algorithms to disclose the algorithms' abilities in controlling the queue size. Figures 4 shows the actual queue length graphs for the reference RED experiments, while Figure 5 show the reference experiments of the proposed mechanism. From the figures, it is obvious that the proposed mechanism enforces a tighter and smoother control on the current queue size than RED. From Figure 5, it is clear that for small values of packet sliding window (W_s = 32; 64), the proposed mechanism tightly controls the queue size compared to RED and it is much better in keeping the queue size below the half of the actual buffer size. The figure also reveals that for the larger values of (W_s =128; 256; 512; 1024; 2048; 4096), when the T value is increased, the average queue length observed during the experiment increases. This means that the proposed algorithm exercises an aggressive control at small T settings. This is because the smaller the T setting, the shorter the period of time over which the actual queue size has to be directed to the manageable level and therefore, the faster the algorithm will have to diminish the queue size. When the T value is increased, the algorithm becomes less and less aggressive, allowing the queue size to grow at the router. Nevertheless, for smaller values of the W_s settings (32; 64), various T settings do not make much difference regarding the observed average queue length. This is because when W_s settings are small, the effect of W_s dominates the effect of T . Therefore, changing the T setting does not make much difference in the average queue length and this metric is specified more by the W_s setting.

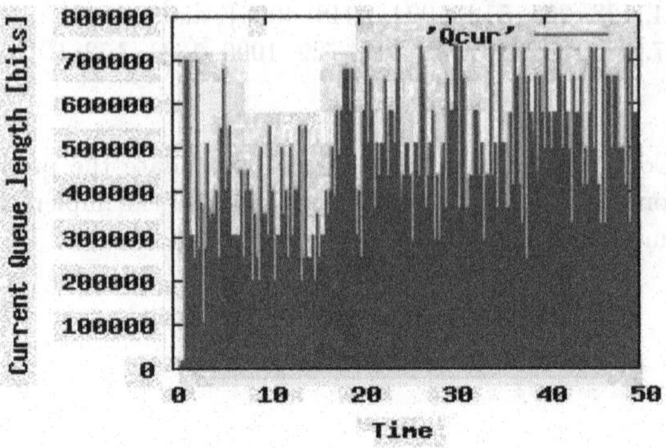

Fig. 4 Actual Queue Size of RED

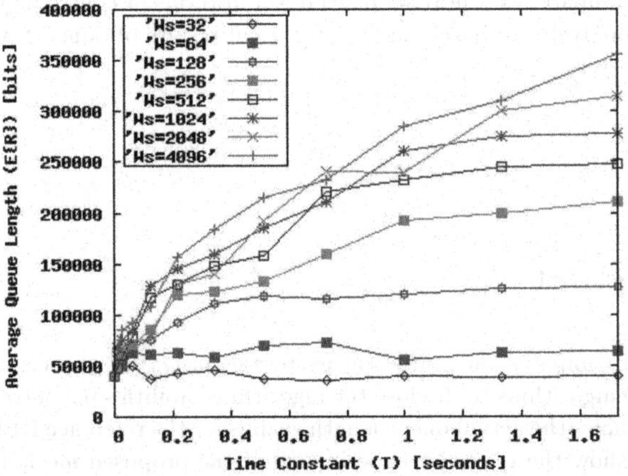

Fig. 5 Average Queue Size of the Proposed Mechanism

7.2 Packet Arrival Rate Control

This section provides a comparison between the proposed and RED mechanisms regarding their ability to control the average packet arrival rate to the buffer below the outgoing transmission link capacity. The average packet arrival rate graphs for different reference experiments of the proposed and RED mechanisms with the different (W_s, T) values are shown in Figures 6 and 7, respectively.

Visual comparison of the average packet arrival rate graphs for the proposed mechanism to those for RED shows that proposed mechanism performs better in controlling the average packet arrival rate near the outgoing transmission link capacity . The size and frequency of the sudden temporary decreases in the packet arrival rate, due to random drop/mark of packets, are much smaller for experiments of the proposed mechanism compared to RED. This verifies that the proposed mechanism is better capable of avoiding global synchronization which causes this type of sudden periodic temporary decreases in the packet arrival rate.

Figure 7 illustrates the average packet arrival rate arrived at the router buffer during the experiments for the full set of 96 experiments of the proposed mechanism performed using the aforementioned various (W_s, T) settings. Figure 6 reveals that for small T values (1, 8, 27, 64, 125 milliseconds) when T is increased, the average packet arrival rate increases as well. For very small values of T setting, the proposed mechanism practices very aggressive control at the queue, preventing packet arrival rate from growing significantly. Increasing the T setting moderates the control practiced by the proposed mechanism which in turn permits the packets to continue to flow into the buffer at a higher average rate. For larger T values (216, 343, 512, 729, 1000, 1331, 1728 milliseconds), the average packet arrival rates obtained are all close to the outgoing transmission link capacity. This is because, even if the T value is large enough to theoretically allow achieving very large packet arrival rates at the queue, by reducing the random packet droppings/markings, packet drops due to buffer overflows will decrease the packet arrival rate to a value close to the outgoing transmission link capacity. Thus, the average arrival rate achieved for various large values of T settings are all around the outgoing transmission link capacity.

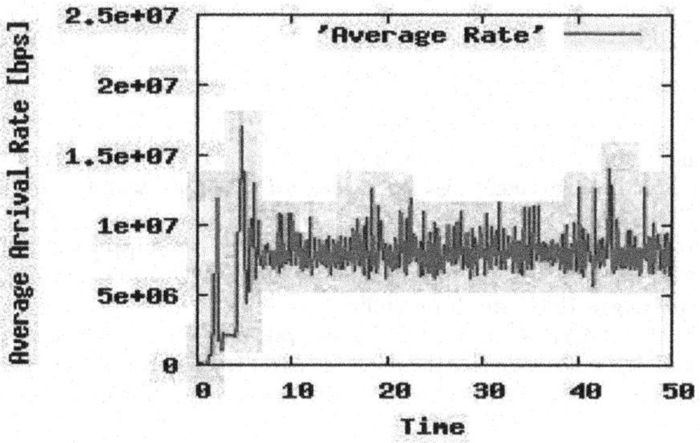

Fig. 6 Average Packet Arrival Rate of RED

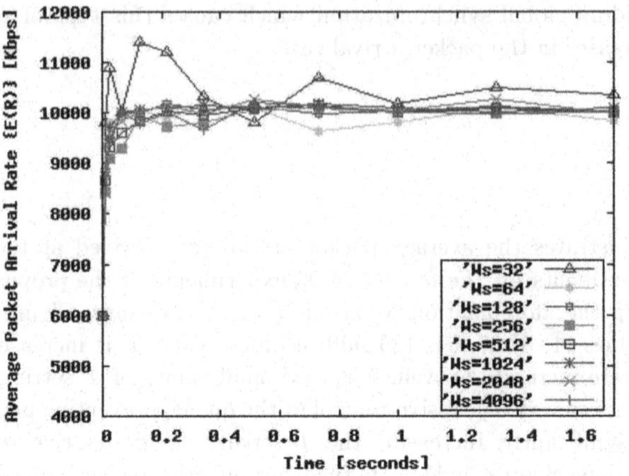

Fig. 7 Average Packet Arrival Rate of the Proposed Mechanism

8 Conclusion

This paper presented an innovative queuing management mechanism that aims to limit the queuing delay by enforcing a maximum allowed queue size, sending congestion notification timely by dropping or marking packets, and to achieve the maximum link utilization possible with minimum packet drops. To achieve this purpose, the proposed mechanism combines two measures; the Current Queue Size and the Packet Interarrival Rate. This combination outperformed the single measure used by Random Early Detection (RED), which is the Average Queue Size. The performance evaluation based on the numerical simulations results showed that the proposed mechanism offers a superior performance to that of RED. The evaluation confirmed that the proposed queuing management mechanism can be set with various (W_s, T) settings to efficiently control the actual queue size below some required optimal value to decrease the queuing delay whereas maintaining high link utilization and low packet drop ratio.

Acknowledgements This research work is funded by Universiti Sains Malaysia (USM), Short-term Grant No. 304/PNAV6312117.

References

1. Babek Abbasov and Serdar Korukoglu. Effective red: An algorithm to improve red's performance by reducing packet loss rate. *Journal of Network and Computer Applications*, 32(3):703 – 709, 2009.
2. M.P. Barcellos and A. Detsch. Congestion control with ecn support in poll-based multicast protocols. In *Computers and Communications, 2005. ISCC 2005. Proceedings. 10th IEEE Symposium on*, pages 724–729, June 2005.
3. A. Durresi, M. Sridharan, Chunlei Liu, M. Goyal, and R. Jain. Congestion control using multilevel explicit congestion notification in satellite networks. In *Computer Communications and Networks, 2001. Proceedings. Tenth International Conference on*, pages 483–488, 2001.
4. Sally Floyd and V. Jacobson. Random early detection gateways for congestion avoidance. *IEEE/ACM Transactions on Networking*, 1(4):397–413, Aug 1993.

5. Liang Guo and I. Matta. The war between mice and elephants. In *Network Protocols, 2001. Ninth International Conference on*, pages 180–188, Nov 2001.
6. Liujia Hu and AD. Kshemkalyani. Hred: a simple and efficient active queue management algorithm. In *Proceedings of the 13th International Conference on Computer Communications and Networks, 2004. ICCCN 2004.*, pages 387–393, Oct 2004.
7. R. Jain. Congestion control in computer networks: issues and trends. *Network, IEEE*, 4(3):24–30, May 1990.
8. M.M. Kadhum and S. Hassan. The effect of ecn on short tcp sessions. In *Telecommunications and Malaysia International Conference on Communications, 2007. ICT-MICC 2007. IEEE International Conference on*, pages 708–712, May 2007.
9. M.M. Kadhum and S. Hassan. Fast congestion notification mechanism for ecn-capable routers. In *International Symposium on Information Technology, 2008. ITSim 2008*, volume 4, pages 1–6, Aug 2008.
10. Jahon Koo, Byunghun Song, Kwangsue Chung, Hyukjoon Lee, and Hyunkook Kahng. Mred: a new approach to random early detection. In *Proceedings of the 15th International Conference on Information Networking*, pages 347–352, 2001.
11. Srisankar Kunniyur and R. Srikant. Analysis and design of an adaptive virtual queue (avq) algorithm for active queue management. *SIGCOMM Comput. Commun. Rev.*, 31(4):123–134, August 2001.
12. I.K.-K. Leung and J.K. Muppala. Packet marking strategies for explicit congestion notification (ecn). In *Performance, Computing, and Communications, 2001. IEEE International Conference on.*, pages 17–23, Apr 2001.
13. H. Mahbub and J. Raj. *High Performance TCP/IP Networking: Concepts, Issues, and Solutions.* U.S ed., Prentice-Hall, 2003.
14. A. Matrawy and I. Lambadaris. A survey of congestion control schemes for multicast video applications. *Communications Surveys Tutorials, IEEE*, 5(2):22–31, Fourth 2003.
15. A. Matrawy, I. Lambadaris, and Changcheng Huang. Comparison of the use of different ecn techniques for ip multicast congestion control. In *Universal Multiservice Networks, 2002. ECUMN 2002. 2nd European Conference on*, pages 74–81, 2002.
16. P. Molinero-Fernandez and N. McKeown. Tcp switching: exposing circuits to ip. *Micro, IEEE*, 22(1):82–89, Jan 2002.
17. S.A Nor, S. Hassan, O. Ghazali, and M.H. Omar. Enhancing dccp congestion control mechanism for long delay link. In *International Symposium on Telecommunication Technologies (ISTT), 2012*, pages 313–318, Nov 2012.
18. T.J. Ott, T. V. Lakshman, and L.H. Wong. Sred: stabilized red. In *Proceedings of the Eighteenth Annual Joint Conference of the IEEE Computer and Communications Societies.IEEE INFOCOM '99*, volume 3, pages 1346–1355 vol.3, Mar 1999.
19. S. Ryu C. Rump C. Qiao. Advances in active queue management (aqm) based tcp congestion control. *Telecommunication Systems*, 25:317–351, 2004.
20. Zhen qiu ZHAN, Jie ZHU, and Yi-Jin SUN. Further results on local stability of lrc-red algorithm in internet. *The Journal of China Universities of Posts and Telecommunications*, 19(5):99 - 103, 2012.
21. L. Rizzo. Pgmcc: A tcp-friendly single-rate multicast congestion control scheme. In USA New York, editor, *in Proceedings of the Conference on Applications, Technologies, Architectures, and Protocols for Computer Communication*, volume 30, 2000.
22. Mohit P. Tahiliani, K.C. Shet, and T.G. Basavaraju. Cared: Cautious adaptive {RED} gateways for tcp/ip networks. *Journal of Network and Computer Applications*, 35(2):857 - 864, 2012. Simulation and Testbeds.
23. Chonggang Wang, Bin Li, Y. Thomas Hou, Kazem Sohraby, and Yu Lin. Lred: a robust active queue management scheme based on packet loss ratio. In *INFOCOM 2004. Twenty-third AnnualJoint Conference of the IEEE Computer and Communications Societies*, volume 1, pages -12, March 2004.
24. M. Welzl. *Network Congestion Control: Managing Internet Traffic.* John Wiley & Sons, 2005.
25. Naixue Xiong, Athanasios V. Vasilakos, Laurence T. Yang, Cheng-Xiang Wang, Rajgopal Kannan, Chin-Chen Chang, and Yi Pan. A novel self-tuning feedback controller for active queue management supporting {TCP} flows. *Information Sciences*, 180(11):2249 - 2263, 2010.
26. Yue-Dong Xu, Zhen-Yu Wang, and Hua Wang. Ared: a novel adaptive congestion controller. In *Machine Learning and Cybernetics, 2005. Proceedings of 2005 International Conference on*, volume 2, pages 708–714 Vol. 2, Aug 2005.

27. Bing Zheng and M. Atiquzzaman. Dsred: improving performance of active queue management over heterogeneous networks. In *IEEE International Conference on Communications, 2001. ICC 2001*, volume 8, pages 2375–2379 vol.8, 2001.
28. Kaiyu Zhou, Kwan L. Yeung, and Victor O.K. Li. Nonlinear red: A simple yet efficient active queue management scheme. *Computer Networks*, 50(18):3784 – 3794, 2006.

E-Government project enquiry framework for a continuous improvement process, status in Malaysia and Comoros

Said Abdou Mfoihaya[1], Mokhtar Mohd Yusof[2]

Faculty of Information and Communication Technology-Universiti Teknikal Malaysia Melaka
comoransaid@live.com[1], mokhtaryusof@utem.edu.my[2]

Abstract. Information and Communication Technology (ICT) transformed the way we live today, both in the private and public sectors. Technologies such as e-Commerce and banking systems show advancement in the private sector while the public sector is characterized by the reinvention of public services processes and delivery. However, many obstacles have been realized in developing and least developed countries leading into failure of many e-Government projects. This study aims to propose a hypothetical e-Government projects enquiry (EPE) framework that can serve as a basis for investigating e-Government projects implementation against best practices, techniques, and methodologies in this area. The study focused on three levels of enquiry namely the strategy level, the operational level, and the technical level. On the basis of a sound study on the Malaysian and Comorian e-Government systems status, we found that the proposed framework is noteworthy for questioning current e-Government environments performance and guiding the implementation of new e-Government projects initiatives.

Keywords: EPE framework, process reengineering, Customer Relationship Management (CRM), Public Private Partnership (PPP).

1 Introduction

The United Nations` 2014 e-Government survey claimed that there are significant opportunities to transform the public administration into a powerful instrument of sustainable development through e-Government and innovation [1]. Therefore, e-Government can be considered as a key enabler of a nation`s development in several areas. Since its birth two decades ago, many models and frameworks have been developed including those presented in [4]. However, they focused on developing particular structures for implementing e-Government initiatives. On the other hand, the EPE framework does not only offer the opportunity of building new structures but, in addition, it allows oriented enquiry process based on well established domains, hence, reducing the complexity of dealing with independent and complex structures. In most countries all over the world e-Government initiatives continue to emerge [1]. At the same time, there are many cases showing the failure of numerous e-Government projects especially in developing countries. In [2] the authors argued that while the historical development of e-Government is showing the prominent role of ICT in creat-

K.J. Kim (ed.), *Information Science and Applications*,
Lecture Notes in Electrical Engineering 339, DOI 10.1007/978-3-662-46578-3_15

ing efficient and transparent Government there are considerable failures that occurred in many countries worldwide. We proposed the EPE framework as a tool for understanding the current situation on e-Government projects and guiding its improvement process.

2 Literature background

2.1 E-Government from challenges to failure

According to [3], e-Government progress can be classified as total failure: when the initiatives were adopted but abandoned or never implemented, partial failure: when the initiatives were implemented but the major goals were not attained, and Success: when the major goals were attained without undesirable outcomes. [7] Revealed that 35% of the e-Government projects are total failure while 50% are partial failure in developing countries. And only 15% of these projects are completed successfully. This high failure rate is the result of a number of influencing factors. [5] Mentioned different factors affecting the implementation of e-Government projects in developing countries. They include top management, technology, project management, organizational, complexity, size, and process factors. To implement successful e-Government projects [6] suggested that people, organizations, processes, and technology which represent 80% and 20% respectively need to be considered. The table below depicts the major challenges in this area.

Table 1. Major Challenges of e-Government projects implementation in developing countries

Barrier	Description	Sources
Political influence	Willingness of political authorities to adopt e-Government and employees resistance from change.	ITU (2008), Nugi Nkwe (2012), Ebrahim and Irani (2005), Kayani et al (2011), N. Qaisar and H. G. A. Khan (2010)
Human resources	Lack of skilled manpower from the development to implementation of e-Gov implementation	Ashaye and Irani (2013),E. Lau (2003),Q. Li and E. O. Abdalla (2014), Nugi Nkwe (2012), Ebrahim and Irani (2005), N. Qaisar and H. G. A. Khan (2010), E. Rakhmanov (2009)
Funding capability	Willingness of Governments to provide proper funds for the implementation of e-Government initiatives through the development of ICT infrastructures, training programs, etc.	E. Lau (2003), Nugi Nkwe (2012), Ebrahim and Irani (2005), N. Qaisar and H. G. A. Khan (2010), E. Rakhmanov (2009), Monga, A. (2008), A. M. Odat (2012)
Organizational Capacity	Lack of strategic leadership programs, project management skills, cooperation, and partnership.	Ashaye and Irani (2013), E. Lau (2003), Q. Li and E. O. Abdalla (2014), Nugi Nkwe (2012), Ebrahim and Irani (2005), A. M. Odat (2012)
Security & Privacy	Protection of Government sensitive data and information. This applies also to personal data.	Monga, A. (2008), E. Rakhmanov (2009), Kayani et al (2011), Ebrahim and Irani (2005), Nugi Nkwe (2012), Q. Li and E. O. Abdalla (2014), Ashaye and Irani (2013)

2.2 Strategic planning and modeling

Strategic planning is the process of defining the direction and strategy of an organization for making decisions on resources allocation to pursue the strategy and extend the mechanisms of control to guide the implementation of the strategy for the future of the organization. In [8] the author argued that the process of strategic planning involves four basic components: vision & mission, philosophy & policy, strategies & goals, and programmes & objectives. Combined together these components are critical factors for sound management of e-Government projects one of the biggest challenges of e-Government implementation as mentioned in [5]. According to Gounder [9] there are six critical elements for Government transformation: being customer centered, collaborating for success, working across silos, coping with change, and technology infrastructure. The same study has strongly recommended that e-Government programs should focus on customer needs, an issue which has been addressed in [13]. Although [13] focused on the business area, it contains important concepts useful for developing strong customer relationship. Additional concepts that a prominent role in e-Government implementation include public private partnership [12], [14] and customer relationship management [10,11], [15]. While the PPP represents contracts signed between the Government body and private entities that call private partners to supply technologies and deliver desired services in association with the Government to empower the public services sector, the CRM is a strategy that aims at maximizing the access, benefits and citizens satisfaction in the e-Government domain.

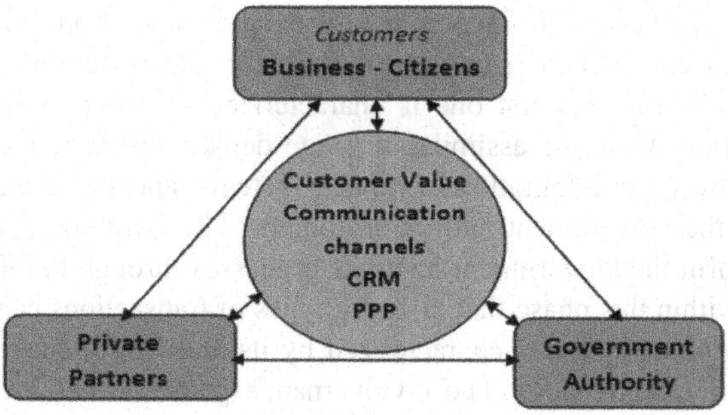

Fig. 1. E-Government stakeholders' interaction model

2.3 Reengineering Government processes

Government process reengineering is highly influenced by several factors and conditions not only within the Government and its institutions but also within the community and society in general [16]. According to [17], making decisions on people during e-Government implementation is a complex issue that requires a plan. Although people resists toward implementing e-Government, this technology offers a number of advantages [16]. It emancipates public servants and citizens from heavy works, offers public service quality, allows communication among people, and enforces laws. For a successful reinvention of Government processes, [3] suggested that Governments should make use of internal experts who can understand social and cultural patterns of people. Until recently, business process management (BPM), business process reengineering (BPR), and enterprise resource planning (ERP) were mainly used in the private sector. However due to the emerging relationships between private and public sectors and their role in process transformation, these concepts saw their integration in the public sector. BPM has been used primarily in the private sector early in 1980s and later from 1990s public sector organizations started using it to constantly adapt their business processes with emerging business environments and operations [18]. According to [20], BPR encompasses many horizons from a radical change of the current work environment or structure to a complex transformation of organizational, human, and technological dimensions. On the other hand, ERP is traditionally designed for private organizations. But due to its potential benefits, the public sector began to integrate it in the public services delivery process [21]. ERP allows ubiquitous on demand network access to a number of shared devices over a computer network in a minimum resources usage. This characteristic of ERP is nowa-days an important aspect especially for interactions between Government agencies and business companies.

2.4 E-Government models and frameworks

E-Government models are well established structures that define the stages of implementing e-Government while its frameworks identify and integrate together the key elements for the implementation of e-Government projects.

The *Jungwoo Lee* model discussed in [30] is an exciting study that combines key features and concepts of earlier models. Five major stages were identified. The presenting stage which is the first one is characterized by the presentation of simple online information while the assimilation stage depicts the implementation of basic computing abilities, the integration of related systems, and the interaction that takes place between the Government and its customers. The third stage, reforming stage, attempts to reform the Government business processes through the information technologies. It is within this phase that different types of transactions can take place. The next phase, morphing stage, is characterized by its abilities of shaping all processes and services for effectiveness. The e-Governance phase is the last the stage. This model represents the highest level of e-Government development in which major change in government and governance are observed.

On the other hand, the *Wimmer`s holistic reference framework for e-Government* [31] integrates of four principal perspectives or layers. The first layer defines the strategy and the basic roles which are referred to as organizational requirements in [31]. The second layer consists of public services, processes and their workflows. It is on this layer where the basic steps taken for each process are also defined. The third layer is termed as interaction. At this level, the service performance, the relationship between components such as people, data, processes, and public services have to be defined and established. Last, the information technology layer determines the information technology components and their implementation.

3 The EPE framework

We proposed in this study a hypothetical framework for driving the process of enquiring e-Government projects against best practices. This framework consists of four major stages or layers as explained below:

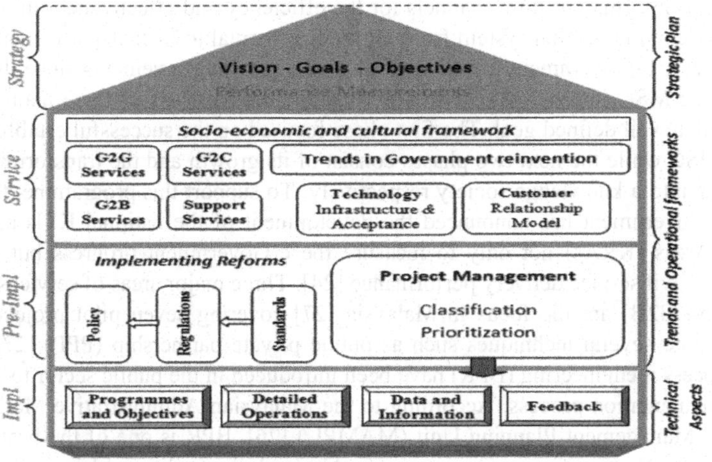

Fig. 2. Proposed e-Government projects enquiry (EPE) framework

The Strategy stage: The strategy stage or layer specifies the overall orientation of the programme by defining the vision, goals, and objectives of the project. In addition, this layer gives the opportunity to define the success measurement standards. That is, a clarified strategic plan can be developed early in this phase.

The Service stage: The second stage of the framework is concerned with a structure defining all services (public, administrative, and support), socio-economic and cultural factors, the infrastructure development, and the interaction mechanism between all project stakeholders. Furthermore, this stage allows the reengineering of Government processes.

Pre-implementation stage: For successful implementation of e-Government initiatives, one of the important steps is to establish a base of reforms which target both people`s and administrative changes. This can be achieved through reform initiatives such as policy enforcement, regulations, and standards. This layer enables also the application of technical skills in project management.

Implementation stage: As long as the environment is already set up, it is now time to proceed with more detailed programmes and objectives, the key operations or steps needed, and identification of the type of data and information. Also, to secure the effectiveness of the system, setting a feedback strategy is more important. This stage is characterized with the physical implementation of the new initiatives.

4 Status of e-Government in Malaysia and Comoros

Malaysia: According to [22], e-Government initiatives began in 1996 with the creation of the Multimedia Super Corridor (MSC). They include the construction of the telecommunication infrastructure, which in turn introduced the Government Integrated Telecommunications Network (GITN) enabling the integration of networks and the creation of information flow channels for the efficiency and effectiveness in the public sector communication system from intra-departmental to inter-departmental orientations [23]. E-Government in Malaysia has seen great improvements since the creation of the MSC in 1996 [25]. The MSC vision 2020 consists of three main phases each with a well defined goal. The first phase focused on the successful establishment of the MSC while the next two phases consist of its growth and the transformation of Malaysia into a knowledge society respectively. To support this programme, the Malaysian Government has announced the development of the national KPIs and Key Result Areas (KRAs) not only to measure the e-Government progress but also to boost the civil service delivery performance [24]. Three major areas of services (G2G, G2C, and G2B) are the focus in Malaysia [27] covering seven pilot projects [29]. Furthermore several techniques such as public private partnership (PPP) [27], business process reengineering (BPR) have been introduced in the public sector to support the implementation process. According to the Malaysian Administrative Modernization and Management Planning Unit (MAMPU) [26], BPR is one of the best strategies to strengthen systems and procedures. Although the traditional bureaucratic machinery is gradually fading through a series of initiatives to reinvent the Government, [28] suggested that Malaysian ICT should incorporate modernization, changes, and reforms to challenge the internal Governance framework in Government operations.

Comoros: While important initiatives are being implemented in developed and developing countries including Malaysia, we observed that e-Government in Comoros has a poor status. According to [19], few Comoros Government institutions developed websites and only about 50% of these websites satisfied the web presence phase of the e-Government implementation models. They deal with providing basic online information of Government activities. In addition [19] mentioned the telecommunication infrastructure status and human capital as the key limitations slowing down the implementation of e-Government initiatives in Comoros. Since 2008, the Comoros Government started the development of an e-Government strategy focused on the transformation of the Comoros public finance sector by the year 2019. Other initiatives were considered in this strategy including the corruption control and the public schools network. In the light of this strategy, the public finance strategic planning document was developed in 2009 by the Comoros Government, setting the goals and objectives of the transformation. In order to implement these initiatives, we suggest that government authorities should develop an implementation roadmap that takes in consideration the best practices in this field. This roadmap should also highlights the current limitations for the purpose of creating plans for the future. In this regard, Comoros presents various opportunities to transform its public sector, first, through the current ICT infrastructure development under the RCIP 4 programme, second, through a learning process from existing e-Government environments.

5 Research Method

The aim of this study is to propose an enquiry framework of e-Government projects. Through an intensive structured review, our investigation focused on three key perspectives. The strategic plan perspective enabled us to discover the foundation of e-Government implementation while the trends and operational frameworks helped us to learn and analyze relevant practices necessary for implementing e-Government initiatives. Finally, through the technical aspects perspective, we could explore important technical skills in the field. The choice of this approach was done according to the nature of the study which required a long exploration, description, and analysis of the e-Government principles and concepts. The same approach allowed us to study the current e-Government environment in Malaysia and Comoros. The nature of the data used in this study includes research articles, open data, and country reports.

6 Analysis and discussion

Throughout this study we found that there is a need for e-Government investigation methodology in order to support its implementation process. We also found that there is a wide gap between different e-Government environments and between new e-Government implementation strategies and the current situation in developing and least developed countries. We suggest that more customizable structures need to be developed for the study of the current environment to measure its performance in a continuous process. We developed a hypothetical framework that can serve as the basis of a questioning process for e-Government projects. Based on the actual findings, our framework is more likely to address several issues especially on methods, measurement, and improvement programmes. When tested with the current status in Malaysia and Comoros (see Fig. 3), the enquiry process is characterized by ''WHAT'' questions for Malaysia while for the case of Comoros the process is highly dominated with ''HOW'' questions. This is showing the applicability of the framework in different situations. However, the application of the framework to analyze real world environments is necessary to confirm its strengths.

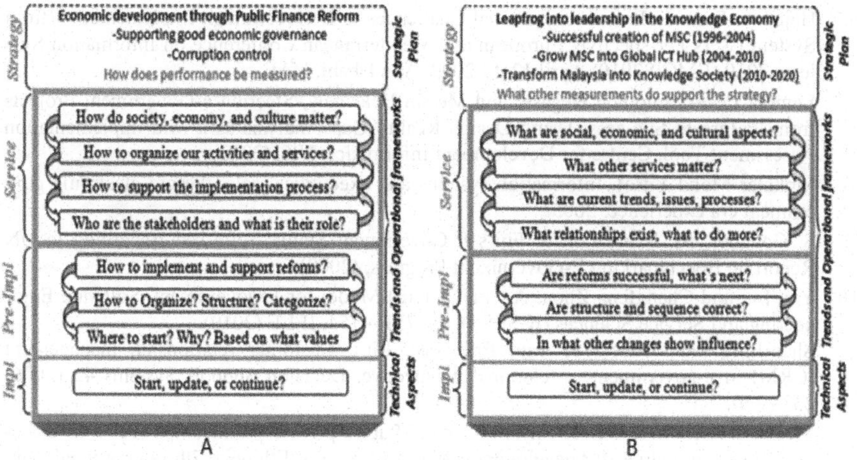

Fig. 3. Enquiry process sample in Comoros (A) and Malaysia (B)

7 Conclusion

In this study, we discussed several aspects which were rarely mentioned in existing models and frameworks of e-Government while their role is significant for the success of e-Government implementation especially in developing and least developed countries where failure in this area is creating a burden for others to carry out. As a result, we developed a framework consisting of components necessary for the implementation of e-Government projects which are the basis of the enquiry process that can be used to guide the implementation of these projects. This study can serve as an input to conduct further research works. Also, it can be used by managers of e-Government projects to better support their activities. Furthermore, the framework developed in this study needs to be applied in real world situations in order to confirm its strengths. This is a limitation of our study which can be addressed in further research works.

8 References

1. United Nations e-Government survey 2014, e-Government for the future we want, available at http://unpan3.un.org/egovkb/Portals/egovkb/Documents/un/2014-Survey/E-Gov_Complete_Survey-2014.pdf, accessed on October 28, 2014.
2. Darwin et al., E-Government: Historical Evolution and Current Trends. Canadian Journal on Data Information and Knowledge Engineering. Vol. 3 No. 2, July 2012.
3. Richard Heeks, Most e-Government-for-Development Projects Fail How Can Risks be reduced? IDPM (2003).
4. Shahkooh et al., A Foresight based Framework for E-government Strategic Planning. Journal of software, vol. 4, no. 6, august 2009.
5. H. S. A. Nawi, A. A. Rahman, O. Ibrahim, Government ICT Project Failure Factors: Project Stakeholders' Views. Journal of information systems research and innovation, (2012).
6. Hajed Al-Rashidi, Examining Internal Challenges to E-Government Implementation from System Users Per-spective. European and Mediterranean Conference on Information Systems 2010 (EMCIS2010) April 12-13 2009, Abu Dhabi, UAE.
7. Lemma Lessa, Solomon Negash, and Mesfin Belachew, Steering e-Government Projects from Failure to Suc-cess: Using Design-Reality Gap Analysis as a Mid-Implementation Assessment Tool, Centre for Development Informatics, UK, (2012).
8. Mokhtar Mohd Yusof, Information systems and executives` role, the pre-electronic government era experience, 2005.
9. S. Gounder, An Evaluation of Fiji's E-Government Status: Assessed According to UN Report on Benchmarking E-Government Progress, (2009).
10. Yan Liu and Changfeng Zhou, A Citizen Trust Model for E-government, Software Engineering and Service Sciences (ICSESS), pp. 751 – 754, IEEE (2010).
11. Shan-Ling Pan, Chee-Wee Tan, Eric T.K. Lim: Customer relationship management (CRM) in e-government: a relational perspective. Decision Support Systems 42 (2006) 237–250.
12. Mesfin Belachew and R K Shyamasundar.: Public private partnerships (PPP) in the e-government initiatives for developing nations: the case of Ethiopia. 7th International Conference on Theory and Practice of Electronic Governance. pp. 42-45. ACM (2013).

13. Alexander Osterwalder.: the business model ontology a proposition in a design science approach. UNIVERSITE DE LAUSANE (2004).
14. W. Lance Bennett and Philip N. Howard.: Evolving Public-Private Partnerships: A New Model for e-Government and e-Citizens. Daryl M. West, "Global E-Government, (2006).
15. Li Bo and Li Hui.: Enterprises as customers: customer-centered e-government system based on CRM in China local government. Management and Service Science. pp. 1-4. IEEE (2009).
16. D. Osborne, reinventing government: what a difference a strategy makes, 7h Global Forum on Reinventing Government, 2007.
17. LUO Guanghua.: e-government, people and social change: a case study in china. EJISDC (2009) 38, 3, 1-23.
18. K Ortbach et al.: A Dynamic Capability-based Framework for Business Process Management: Theorizing and Empirical Application. 2012 45th Hawaii International Conference on System Sciences.
19. K. A. Mohamed.: Evaluation of the status of the e-government in Comoros. DIPLO (2013).
20. Abraham van der Vyver and Jayantha Rajapakse.: E-Government Adoption and Business Process Re-Engineering in Developing Countries: Sri Lankan and South African Case Studies. International Journal of Innovation, Management and Technology, Vol. 3, No. 6, December 2012.
21. Trevor Clohessy and Thomas Acton.: Enterprise Resource Planning for e-Government in the Cloud. ICTIC 2013. Conference of Informatics and Management Sciences March, 25. - 29. 2013.
22. Alias et al.: Evaluating e-Government Services in Malaysia Using the EGOVSAT Model. International Confer-ence on Electrical Engineering and Informatics 17-19 July 2011, Bandung, Indonesia.
23. S. M. Alhabshi.: E-government in Malaysia: Barriers and Progress. International Institute of Public Policy and Management, University Malaya, Kuala Lumpur. Volume 18, No. 3, October 2008.
24. Z. Zakaria et al.: Key Performance Indicators (KPIs) in the Public Sector: A Study in Malaysia. Asian Social Science. Vol. 7, No. 7; July 2011.
25. R Hussein et al.: The influence of organizational factors on information systems success in e-Government agen-cies in Malaysia. EJISDC (2007) 29, 1, 1-17.
26. The Malaysian Administrative Modernisation and Management Planning Unit (MAMPU), www.mampu.gov. my/web/en/business-process-reengineering.
27. Maniam Kaliannan, Halimah Awang, and Murali Raman.: Public-Private Partnerships for E-Government Ser-vices: Lessons from Malaysia. International Journal of institutions and economies. Vol. 2, No. 2, October 2010, pp. 207-220.
28. Mohsin Bin HJ Ahmad and Raha Binti Othman.: implementation of electronic Government in Malaysia: The status and potential for better service to the public. Public sector ICT management review. October 2006- March 2007, vol. 1. No. 1.
29. Roslind Kaur.: Malaysian e-Government Implementation Framework. Faculty of Computer Science and Information Technology University of Malaya. May 2006.
30. J. Lee, 10 year retrospect on stage models of e-Government: A qualitative meta-synthesis, Government Information Quarterly 27 (2010) 220–230.
31. MARIA A. WIMMER, Integrated Service Modelling for Online One-stop Government, Volume 12 (3): 149–156.

Objective Non-intrusive Conversational VoIP Quality Prediction using Data mining methods

Sake Valaisathien, Vajirasak Vanijja

IP Communications Laboratory, School of Information Technology,
King Mongkut"s University of Technology Thonburi, Bangkok, Thailand
sake.online@mail.kmutt.ac.th, vachee@sit.kmutt.ac.th

Abstract. Nowadays, there is a growth in the number of applications running on the Internet involving real-time transmission of speech and audio streams. Among these applications, Voice over Internet Protocol (VoIP) has become a widespread application based on the Internet Protocol (IP). However, its quality-of-service (QoS) is not robust to network impairments and codecs. It is hard to determine conversational voice quality within real-time network by using ITU-T standards, PESQ and E-model. In this research, three data mining methods: Regression-based, Decision tree and Neural network were used to create the prediction models. The datasets were generated from the combination of PESQ and E-model. The statistical error analysis was conducted to compare accuracy of each model. The results show that the Neural network model proves to be the most suitable prediction model for VoIP quality of service.

Keywords: VoIP, QoS, E-model, PESQ, Data mining

1 Introduction

Voice over IP (VoIP) is a communication technology carrying voice data packets across the Internet Protocol (IP) network. However, IP is a best-effort service that does not guarantee quality of service (QoS) for real-time applications such as VoIP. The impairment factors such as packet loss, one-way delay and codec types can degrade voice quality. In order to meet a commercial service agreement, an efficient speech quality measurement is a necessary requirement.

In general, speech quality is measured through Mean Opinion Score (MOS) defined in ITU-T P.800 [6]. It is a five-point scale, varied on a 5 (Excellent) to 1 (Poor). By the ITU-T standard, the MOS was originally obtained from asking the opinion of people about the speech quality and this is known as a subjective method. The method can be a listening test (one-way) or conversational test (interactive). The terminology, MOS-LQ refers to listening quality, whereas MOS-CQ refers to conversational quality [16]. The major problems of the subjective method are time consuming and expensive. Therefore instead of the real human testing, objective methods have been developed.

K.J. Kim (ed.), *Information Science and Applications*,
Lecture Notes in Electrical Engineering 339, DOI 10.1007/978-3-662-46578-3_16

The objective methods are divided into two types that are intrusive and nonintrusive methods. Intrusive methods are more accurate by comparing a reference signal to a degraded signal. PESQ is one of the most successful intrusive methods from ITU-T [1]. Since it requires a hard-in-obtain reference signal, PESQ is unsuitable for real-time quality monitoring tasks. Moreover, PESQ results only in listening quality which is unusable for conversational applications such as IP telephony. Nonintrusive methods typically calculate quality by observing the environment parameters. The E-model, a transmission planning tool, can evaluate a speech quality from several factors such as network impairments and codec types [2]. It provides conversational quality by including an end-to-end delay in the calculation. However the E-model is inefficient when some parameters are missing. Also, it normally has less accuracy than PESQ. To combine the advantages of both PESQ and E-model, the new conversational prediction model was proposed in other research called PESQ/E-model [3].

In order to study other objective non-intrusive VoIP quality evaluation methods, some of data mining techniques were presented. In [3],[14], the regression-based model proved to be more capable than the E-model when lacking parameters for the prediction. Though the nonlinear regression model is powerful, it is too static that each model exists for each condition. Therefore, it is difficult to use in a dynamic network such as the Internet. With its learning ability, the neural network model has been studied for quality prediction and monitoring purposes [4,5],[15]. Nevertheless, those works did not include a comparison with other methods.

In this paper, several data mining models used to create prediction models of objective, non-intrusive, conversational voice quality for VoIP network were conducted and the accuracy results of each model were compared. The datasets were generated from the PESQ/E-model which proved more accurate than the individual PESQ or E-model [3]. Along with the experiment, the various parameters were scoped to impairment factors which are packet loss rate, one-way delay and voice codecs of G.711 [11], G.729 [12] and Speex [13] were considered.

2 ITU-T objective voice quality evaluation methods

ITU-T voice quality evaluation of objective methods can be divided into two types, intrusive and non-intrusive method. The details of each method can be described as follow.

2.1 Intrusive methods

This objective methods reduces difficulty of the subjective methods. The Intrusive method commonly takes reference signals and degraded signals into account. The ITU-T Recommendation P.862 [1], known as PESQ (Perceptual Evaluation of Speech Quality) is the one most widely used. Its measurement compares the reference voice with the output signal passed through a network. Therefore, the hard-to-obtain reference signal in large and complex network makes PESQ inefficient. Since PESQ does not consider the effect of delay impairments, it can be used only to one-way communications. The quality score called "Mean Opinion Score Listening Quality Objective" or MOS-LQO.

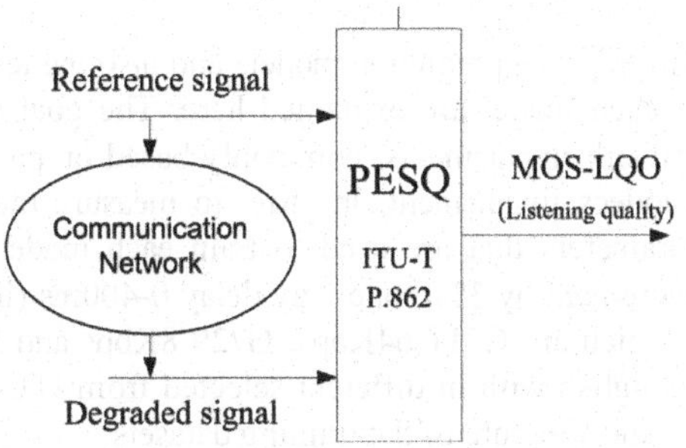

Fig. 1. PESQ diagram

2.2 Non-intrusive methods

The most practical widespread method used to objectively, non-intrusively, and conversationally measure voice quality is ITU-T G.107 recommended the E-model [2]. By not requiring any reference signal, the computational model combines the various impairment parameters into a total value, Rating-factor or R ranging from 0 to 93.2. Involving the effect of delays, the primary output represents the conversational score that can be calculated from the equation (1).

$$R = R_o - I_s - I_d - I_{e\text{-}eff} + A \qquad (1)$$

Where R_o is the basic signal-to-noise ratio (SNR), I_s denotes the impairments happening with the speech signal, I_d is the impairments caused by delay, and $I_{e\text{-}eff}$ is the impairments caused by the low bit rate of codecs. The advantage factor A can be used for the compensation when there are other advantages of accessing network to the user. Most of parameters used in the E-model are pre-calculated and are defined in ITU-T recommendation G.113 [7].

According to [8], the E-model can neglect some parameters not related to the packet-switching network like VoIP and be simplified into the equation (2) below.

$$R = R_o - I_d - I_{e\text{-}eff} \qquad (2)$$

Finally, R can be converted into a MOS scale by equation (3). While it has the effect of delay impairments, the output MOS is equivalent to the conversational quality called MOS- Conversation Quality Objective (MOS-CQO).

$$\begin{aligned} &(R \le 0): MOS = 1 \\ &(0 < R < 100): MOS = 1 + 0.035R + R(R-60)(100-R)\cdot(7\times10^{-6}) \\ &(R \ge 100): MOS = 4.5 \end{aligned} \qquad (3)$$

3 Experiment

3.1 Experimental testbed

Datasets preparation to train the prediction models and also the testing dataset used to validate reliability of each model are explained here. The goal of this dataset is to cause the effect of network impairments, commonly based on packet loss or network one-way delay, and codecs impairment. In order to measure the variation of voice quality, the factor parameters that are used to train each model consist of random packet loss 0-30% (increment by 3%), one-way delay 0-400ms (increment by 40ms.) and the three codecs which are G711 64Kbps, G729 8Kbps and Speex 11Kbps. The voice samples are in English with in different selected from ITU-T P.50 speech corpus [9]. Table 1 shows the structure of the training datasets.

Table 1. The structure of training datasets

Codec	Packet Loss (%)	Delay (ms.)	MOS-CQO
G.711,G.729,Speex	0-30	0-400	1-4.5

The following figure describes an overall modeling process in the experiment. Firstly, the voice samples [9] inserted into the VoIP simulator suffered packet loss, and PESQ measured voice quality into MOS-LQO. Secondly, the obtained MOS-LQO were converted into the conversational quality scores by including one-way delay impairments as the method, PESQ/E-model, described in [3]. Thirdly, the obtained conversational scores served as the actual MOS-CQO were formed with each condition into the training dataset as in the Table 1. Then the training dataset was inserted into the four training models for linear regression, nonlinear regression, decision tree and neural network model. Each model was tuned by using 10-fold cross validation. The testing dataset was used to obtain the prediction of MOS-CQO. Finally, the predicted and actual MOS-CQO were used to evaluate the accuracy of each model and resulted in a statistical result. The overall process is shown in the Fig. 2.

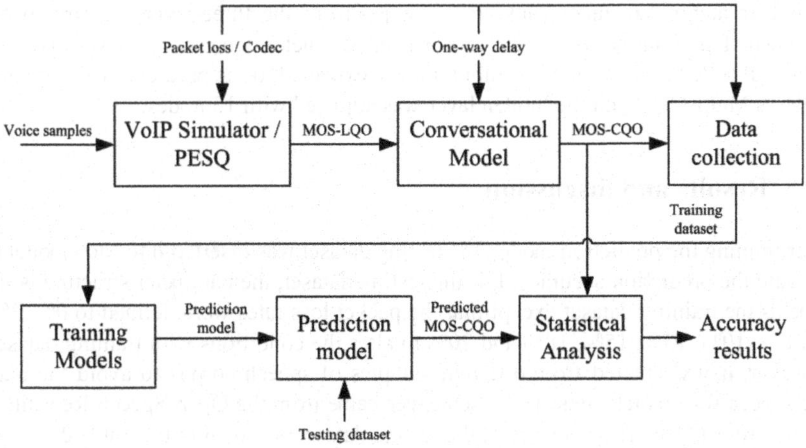

Fig. 2. Experiment testbed

3.2 Modeling methods

This section describes several data mining models used in this paper in conducting the voice quality prediction experiment.

Regression-based model. The model has been widely used in statistical analysis to study the relationship between a dependent variable and one or more independent variables. Equation (8) shows the nonlinear regression (NR) function of conversational voice quality representing MOS-CQO as a dependent variable and impairment factors as independent variables: ρ denoted packet loss probability and δ denoted one-way delay. In this study, the training dataset fitted well with the 3rd order polynomial surface function (cubic function) with the goodness of fit (R^2) between 0.93-0.95.

$$MOS\text{-}CQO = a\rho^3 + b\delta^3 + c\rho^2 + d\delta^2 + e\rho^2\delta + f\rho\delta^2 + g\rho + h\delta + i\rho\delta + j \quad (4)$$

Decision tree model. In the classification method, REPTree (RT) is one of decision tree models. Missing values are dealt with using the C4.5 method of fractional instances. This model helps decrease the complexity of the decision tree model by reducing error from variance by pruning. In this experiment, the tree size is 595 was built.

Neural network model. By using feed-forward architecture for neural network model, a multilayer perceptron (MLP) utilized back propagation. That is a supervised algorithm as the learning technique. It is different from other mathematic models like E-model. This model can be able to learn and retrain the new relationship of voice quality and impairment factor. This is a great benefit for VoIP network that various parameter factors are inconstant. In this experiment the three-layer network model was used. The input nodes are the impairment parameter consists of packet loss rate, delay and codec types while the output node is expected to the predicted voice quality which is MOS-CQO and the hidden layer was adjusted with 15 nodes.

4 Results and discussion

After training the prediction model, the testing dataset was inserted into each model to evaluate the prediction accuracy. For the testing dataset, the parameter structure is the same as the training dataset except that the packet loss rates were defined to 0%, 3%, 5%, 8%, 10%, 13%, 15%, 18% and 20% to alter the conditions with training dataset. However, it was created from different samples of speech corpus to avoid the bias. The speech files which consist of 8 sentences came from the Open Speech Repository (OSR) project [10]. The structure of the testing dataset is shown in the Table 2.

Table 2. The structure of testing datasets

Codec	Packet Loss (%)	Delay (ms.)
G.711,G.729,Speex	0, 3, 5, 8, 10, 13, 15, 18, 20	0-400

As the result, the correlation of the predicted MOS-CQO from each model by using the testing dataset as the conditions and actual value from the conversational model, PESQ/E-model were obtained. Linear regression turned out the lowest value which is 0.895. With 0.95 confident intervals, that made this model unreliable to use further. Hence, the IP network condition has a nonlinear relationship with the quality score. However, the neural network model gave the highest value at 0.989 of positive relationship. Moreover, it resulted to 0.1268 Root Mean Square Error (RMSE) being the lowest error among the other models such as decision tree and nonlinear regression. The results are shown in the Table 3.

Table 3. Models comparison

Statistical analysis	Liner regression	Nonlinear regression	Decision tree	Neural Network
Correlation coefficient	0.8951	0.9826	0.9859	0.9890
Root mean squared error	0.4213	0.1514	0.1364	0.1268

In the codecs perspective, Speex gave the minimum errors followed by G.729 and then G.711 respectively. For all tested codecs, the neural network model can be suitable prediction model giving a lowest error. The result is depicted in Fig. 2.

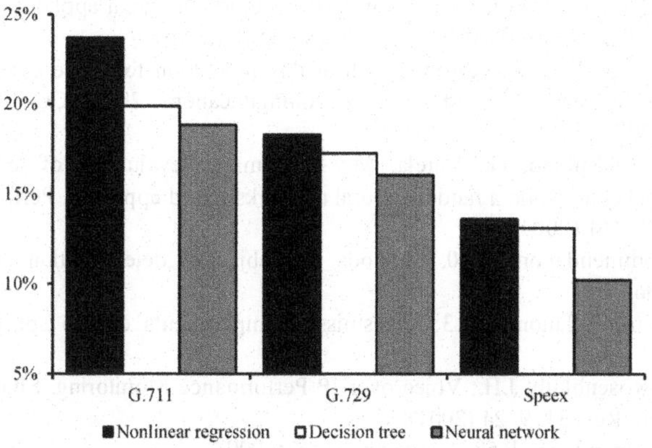

Fig. 3. The error rate of the trained models for each codecs

5 Conclusions and future works

This paper briefly reviews the recommendation of the ITU-T voice quality measurement methods and uses the several data mining models that are the regression-based, decision tree and neural network to create the voice quality prediction model. With the training dataset created from PESQ/E-model, the accuracy is improved from the ITU-T methods. As the result, the nonlinear regression model has a good prediction result, but it is not a suitable model for an adaptation of network due to its inflexibility. For example, if there are multiple codecs, multiple models are formed. However, the neural network and decision tree can overcome the problem. Finally, the result shows that the neural network has the closest relation with PESQ/E-model and the least errors after the accuracy testing. For Speex, every model output its prediction more precise than the other codecs.

Since the scope of this experiment is only limited to packet loss, one-way delay and a few codecs, other impairments such as modern codecs, speaker's ages, language or gender are required to improve the study in modeling of voice quality evaluation.

Acknowledgements. We are grateful to J.H. Chan for providing scientific guidance and also to T. Triyason for great technical support. Thank you, Mr. John F. Lawry for editing.

6 References

1. ITU-T Recommendation P.862: Perceptual evaluation of speech quality (PESQ), an objective method for end-to-end speech quality assessment of narrowband telephone networks and speech codecs, (2001)
2. ITU-T Recommendation G.107 The E-model a computational model for use in transmission planning, (2014)

3. Sun, L., Ifeachor, E.C.: Voice quality prediction models and their application in VoIP networks. IEEE Transactions on Multimedia. 8, 809–820 (2006).
4. Sun, L., Ifeachor, E.C.: Perceived speech quality prediction for voice over IP-based networks. IEEE International Conference on Communications, 2002. ICC 2002. pp. 2573–2577 vol.4 (2002).
5. Mohamed, S., Rubino, G., Varela, M.: Performance evaluation of real-time speech through a packet network: a random neural networks-based approach. Performance Evaluation. 57, 141–161 (2004).
6. ITU-T Recommendation P.800. Methods for subjective determination of transmission quality, (1996)
7. ITU-T Recommendation G.113: Transmission impairments due to speech processing, (2007)
8. Cole, R.G., Rosenbluth, J.H.: Voice over IP Performance Monitoring. SIGCOMM Comput. Commun. Rev. 31, 9–24 (2001).
9. ITU T Recommendation P.50: Artificial voices, (1999)
10. The Open Speech Repository, http://www.voiptroubleshooter.com/open_speech
11. ITU-T Recommendation G.711. Pulse code modulation (PCM) of voice frequencies, (1988)
12. ITU-T Recommendation G.729: Coding of speech at 8 kbit/s using conjugate-structure algebraic-code-excited linear prediction (CS-ACELP), (2012)
13. Speex: a free codec for free speech, http://www.speex.org/
14. Goudarzi, M., Sun, L., Ifeachor, E.: Modelling Speech Quality for NB and WB SILK Codec for VoIP Applications. 2011 5th International Conference on Next Generation Mobile Applications, Services and Technologies (NGMAST). pp. 42–47 (2011).
15. Radhakrishnan, K., Larijani, H.: A Study on QoS of VoIP Networks: A Random Neural Network (RNN) Approach. Proceedings of the 2010 Spring Simulation Multiconference. pp. 114:1–114:6. Society for Computer Simulation International, San Diego, CA, USA (2010).
16. ITU-T Recommendation P800.1: Mean Opinion Score (MOS) terminology (2006)

A Hybrid Incentive-based Peer Selection Scheme for Unstructured Peer-to-Peer Media Streaming Systems

Victor II Romero[1,2] and Cedric Angelo Festin[1]

[1] Department of Computer Science
College of Engineering
University of the Philippines-Diliman
Diliman, Quezon City, Philippines
http://dcs.upd.edu.ph/

[2] Division of Natural Sciences and Mathematics,
University of the Philippines Visayas Tacloban College,
Magsaysay Blvd., Tacloban City, Philippines
{vmromero,cmfestin}@up.edu.ph

Abstract. The success of peer to peer file sharing protocols such as BitTorrent and Gnutella make peer to peer networks an attractive alternative for implementing media streaming services. However, due to the continuous nature of the content, the key components of such conventional file sharing applications are unable to suffice to the more stringent requirements of media streaming. In this study, we propose an alternative to one of its key components that is the incentive based peer selection routine. The primary motivation of our proposed scheme is the utilization of peer contribution information on both local and global context for a more informed peer selection process.

Keywords: Incentive Mechanisms, Peer-to-Peer, Media Streaming

1 Introduction

The peer-to-peer paradigm provides an alternative approach to building distributed network applications. In P2P, participating nodes called *peers* simultaneously assume client and server roles so that they do not only consume but also contribute resources such as CPU time, disk space, and network bandwidth; as a result, such systems are inherently self scaling allowing them to support its growing population without need for additional infrastructure. The success of P2P in traditional file sharing protocols such as BitTorent[1] motivated its use in the more challenging area of P2P media streaming.

In traditional P2P file sharing, the sequence of piece acquisition does not affect end user experience as long as the file is eventually completed. Media streaming, on the other hand, necessitates that pieces arrive in some sort of order to support a continuous and seamless playback. P2P streaming systems are mainly classified as either tree-based or mesh-based depending on the structure

K.J. Kim (ed.), *Information Science and Applications*,
Lecture Notes in Electrical Engineering 339, DOI 10.1007/978-3-662-46578-3_17

of its underlying overlay network[5]. In tree-based systems the structure of the overlay is well defined, peers maintain parent-child relationships which govern the flow of data diffusion in the network. The static distribution paths allow tree-based systems to operate with minimal signaling overhead but make them highly susceptible to service interruption caused by peer transience. In addition, the potentially high number of leaf peers in trees results to inefficient utilization of aggregate system resources. Mesh-based systems, on the other hand, impose no static organization on the underlying overlay network. Participants in such systems establish peering relationships in a dynamic fashion motivated by the availability of data making them inherently more resilient to failures. Furthermore, since data may be retrieved from multiple sources, mesh-based systems are also able to more efficiently use available resources.

The strengths of the peer-to-peer paradigm rests on the assumption that participants cooperate and contribute. However, it has been shown in literature that peers, being autonomous entities, behave strategically – maximizing utility while minimizing cost[3]. At the most extreme case, called *free riding*, selfish peers may choose to not contribute anything in the network. The selfish tendencies of peers at higher concentrations could cause the system to cease functioning as a useful entity, making them a primary concern. To preserve system functionality, it is therefore necessary to put in place a mechanism that motivates peer cooperation. Incentive mechanisms achieve this goal by providing incentives to peers relative to their contribution to the network. As a result, the limited aggregate system resources are allocated to peers that are not only capable, but also willing to propagate received services to the rest of the network. In this study, we characterize incentive mechanisms as either *local* or *global* depending on the scope of contribution information used to introduce bias in the peer selection process. Leveraging this concept, we propose a hybrid incentive based peer selection scheme which uses peer contribution information on both contexts to: (1) provide service differentiation among peers from different peer classes, (2) deter free riding, and (3) meet specific goals unique to P2P media streaming. The rest of the document is outlined as follows: Section 2 presents the works that primarily motivated this study. Section 3 and 4 discusses the proposed scheme, and the simulation module and experiments used to assess its performance. Section 5 analyzes the simulations results, and Section 6 concludes the study.

2 Related Works

2.1 BitTorrent

BitTorrent[1] is the precursor to existing mesh-based peer to peer media streaming systems. It functions by dividing hosted files into smaller *pieces*, and allowing participating peers to exchange completed pieces with each other. As a result, data in the network is replicated faster, and the download process becomes highly resilient to peer transience. To aid in increasing data availability in the network, BitTorrent peers use the *rarest piece first* selection policy [4], wherein the cardinality of all pieces available in the local swarm is computed and the least replicated one is chosen for download. BitTorrent implements an *optimistic tit-for-tat* peer selection scheme so as to motivate cooperation among peers. Local peers unchoke the highest locally contributing peers allowing them to pull desired content. To allow the local peer to search for potential future exchange partners, at least one random peer is optimistically unchoked regardless of its contribution.

2.2 Incentive Schemes for P2P Media Streaming

Pulse[8] is an unstructured P2P live streaming system designed to scale to large populations. It uses *pairwise feedback driven tit-for-tat* as its primary incentive mechanism and *excess based altruism* as its secondary to constantly update exchange connections. Peer contribution value is maintained locally, corresponding to the number of unique blocks received. A pulse peer maintains two types of exchange connections, namely *missing* and *forward.* Missing connections refer to those which are close to the local peer's current buffer position and are therefore more capable in supporting the continuity of the stream while the forward connections are those to which the local peer altruistically allocates its excess bandwidth. During the peer selection routine, the local peer populates the missing list with the peers that have contributed during the last epoch. Peers in the said list are then sorted based on contribution and a predefined number of top contributing peers are selected as *missing connections* for the next epoch. The remainder of the peers are added to the *forward* list and are sorted based on a historical score which is derived from the number of unique blocks received from the remote peer in the entire duration of the streaming session when it was not in the local peer's missing connections. Depending on the amount of excess bandwidth available, the local peer then allocates a variable number of exchange connections to the top scoring peers referring to them as its *forward connections.* Simulation results indicate that when resources are globally scarce, Pulse's incentive mechanism allows high contributing peer to stream content while starving those with less contribution. In addition to this, the *tit-for-tat* selection cause peers to favor associations with peers belonging to the same bandwidth classes.

A distributed incentive mechanism for mesh based P2P video streaming is presented in [7]. In their approach, the authors use a peer's global contribution as basis for their peer selection scheme. The computation and maintenance of global contribution information is distributed across participating peers in the network. When a new peer joins the the streaming session, the *tracker* node is tasked with the selection of the said peer's bank set - the subset of peer responsible for keeping track of a peer's contribution. Each time a peer i sends new data to a remote peer j, it also sends a message to both peers' bank sets indicating the amount of data recently uploaded. The receiving peer j's bank set then sends a message to the sending peer i's bank set to confirm the update, after which i's contribution value is increased. Using the global contribution information of remote peers, the local peer can then set priorities for the transmission of data to peers, favoring those with higher contribution. Simulations indicate that the said incentive mechanism is effective in deterring free riding.

3 Proposed Approach

We finally present the proposed hybrid incentive based peer selection scheme. The goal of the approach is to be able to utilize peer contribution information in both the local and global contexts, and capitalize on their individual strengths so as to: (1) enable peers to directly reciprocate received services using the local contribution information maintained, and (2) straightforwardly deter free riding in the network through a more encompassing representation of a peer's contribution in the network. We assume that the download process, similar to the implementation of the BitTorrent protocol, is organized in fixed time intervals called *epochs.* Throughout the duration of each, remote peers in a local swarm may hold strictly one of the two states: (1) *choked* or (2) *unchoked.* An unchoked peer is one that is allowed to pull content from the local peer while a choked peer is prohibited. To determine which peers will be unchoked, each epoch is preceded by a peer selection process that selects a predefined number of peers (default: 4) from the set of interested peers. The proposed approach comprises of two steps, namely, (1) the *exploitation* phase and (2) the *retention and exploration* phase. The exploitation phase is designed to take advantage of the known beneficial exchange connections. During this phase the local peer assigns to each interested peer i a primary score defined in Equation 1, which is the product of the remote peer's usability and contribution during the most recent epoch.

$$score_{1,i} = usability_i * epoch_i \tag{1}$$

A remote peer's usability refers to the ratio between the number interesting pieces owned by the remote peer and the number of missing pieces in the local peer's current buffer. The score assigned in the first phase of the peer selection scheme indicates a remote peer's ability to support the continuity of the local peer's stream. Remote peers are then sorted based on $score_{1,i}$ and the top most 3 peers with non zero scores are selected.

$$usability_i = \frac{interesting_i}{missing} \tag{2}$$

Remaining interested peers are subjected to the second phase of the peer selection process where they are assigned a secondary score defined in Equation 3. The numerator part of the equation assigns weights to a remote peer's local and global contribution. By adjusting the value of β it is possible to vary the priority given to either contribution contexts. When the β is set to either 0 or 1, the peer selection scheme functions solely based on the global or local contribution, respectively.

$$score_{2,i} = \frac{\beta * score_{local,i} + (1-\beta)score_{global,i}}{\sum_{j=1}^{m}[\beta * score_{local,j} + (1-\beta)score_{global,j}]} \tag{3}$$

The denominator normalizes the secondary scores so that their sum equates to unity. A weighted random function is then used to determine the next remote peer to be unchoked. This process is repeated until all remaining exchange connections have been allocated. The local and global scores, defined in Equations 4 and 5 are also normalized to avoid one overwhelming the other. An α value is also used to balance between a remote peer's most recent and historical contribution.

$$score_{local,i} = \frac{\alpha * epoch_i + (1-\alpha)historical_i}{\sum_{j=1}^{m}[\alpha * epoch_j + (1-\alpha)historical_j]} \tag{4}$$

$$score_{global,i} = \frac{global_i}{\sum_{j=1}^{m} global_j} \tag{5}$$

4 Simulations & Experiments

To test the effectiveness of the proposed peer selection scheme, we create an event-driven simulation module for peersim[2]. The simulated network consists of 100 peers composed of 1 primary source and 99 sharing peers. To determine how the effects of the approach are manifested in a heterogeneous network, the sharing peers are grouped into 3 classes, each with different bandwidth characteristics as detailed in Table 1. The stream is 100MB in size and has a constant SBR of 384 kbps. We use a similar data representation scheme as that in BitTorrent and divide the stream into smaller *pieces* that are 256 KB in size. A piece represents the smallest amount of stream data that can be processed by the player components and has a play-out time of 5.333 seconds. To facilitate better shareability of content in the network, each piece is further subdivided into 16 smaller units called *blocks* that represent the smallest amount of data that can be transmitted between peers. To continuously drive the data diffusion process, each peer upon completion of a piece updates all peers in its local swarm on the availability of new data. Notified peers then reevaluate the interestingness of the local peer thus pointing them to new data sources. The buffer is represented by a *buffer head* which indicates the unowned piece closest to the current playback time , and *buffer size* which when interpreted with the buffer head indicates the current window of interesting pieces. Unlike traditional file sharing where the buffer essentially encompasses the entire file, our buffer size is significantly reduced(default:35) in order to account for the continuous nature of media streaming. The buffer sliding policy, similar to that in [8], depends on a set value of *buffer tolerance* (default: 3), i.e., the number of pieces that must exist in the current buffer before it is allowed to slide to the right and update the buffer head position. An unchoked peer uses a piece scheduling algorithm to determine which piece to pull from the unchoking peer. We use the scheme proposed in [9] with the demarcation point set to 1; as a result, the unowned piece closest to the current playback pointer is pulled only when no other pieces are available. Otherwise, the next piece to be downloaded is selected using the rarest piece first policy.

We set up 5 experiments with varying values for α and β, as detailed in table 2, to encompass the cases wherein the peer selection process is biased by purely local contribution, purely global contribution, and their hybrid. In experimental

Peer class	Count	uplink	downlink
A	33	64	512
B	33	384	1500
C	33	1984	6440

Table 1: Peer class distribution in the simulated network

setup I, setting the β value to 1 indicates that only local contribution will be used to introduce bias in the selection process. In addition to this, setting α value to 1 further restricts the selection routine so that only the most recent contribution of peers is considered. Experimental setup II takes into account the remote peer's accumulated historical contribution in addition to its contribution during the most recent epoch. Experimental setup III uses the local score computed in setup II along with the remote peer's global contribution. Experimental setup IV uses only global contribution for the second phase of the selection process. Experimental Setup V is similar to Setup IV but skips the first phase of the selection process, and is therefore purely global in nature. Each setup is subjected to varying degrees of free riding, ranging from 0 to 70% in 10% increments.

Setup	α value	β value	skip
I	1	1	false
II	0.5	1	false
III	0.5	0.5	false
IV	0	0	false
V	0	0	true

Table 2: Experimental Setup

For statistical consistency, all experiments are subjected to 30 runs and on varying `RANDOM_SEED` values - the global seed used for all random events in the simulation module which includes the selection of peers that comprise one's local swarm, and the assignment of free riding peers. At the end of each run, a report is generated which details relevant statistics such as *completion time*, *interruption time*, and *stream start time*.

5 Results

Simulation results reveal that the proposed peer selection scheme cause the system to provide varying qualities of service to participating peer relative to their contribution. As seen in Figure 1(setup III), peers from classes B and C experience better streaming performance in terms of stream completion time, stream start time, and stream interruption time. When the network is fully cooperative (malicious probability = 0), it can be observed that class A peers, being the least capable, experience the most inferior streaming experience. Compared to class B peers which acquire all pieces of the streamed content at an average time of 31.284 minutes, class A peers conclude their download process 6.848 minutes later representing 21.89% increase in duration. There is less difference in completion time between class B and class C peers with the latter finishing 2.784 minutes earlier despite the significant increase in bandwidth capacity. We attribute this phenomenon to the lack of available content, as described in [6], such that class C peers are unable to exploit their download bandwidth superiority. A similar trend can be observed with regards stream interruption time and stream start time. Class A peers are typically interrupted and averages 6.919 interruptions while majority of class B and class C peers do not experience any interruption. Class A peers also start processing stream pieces 33.3 seconds later than class B peers while class C peers do so 19.74 seconds earlier than class B peers. It can also be observed from the figure that even when free riding peers are introduced to the network, cooperative peers are able maintain their respective streaming qualities relative to their contribution to the network. Interestingly, free riding peers experience minimal interruptions as compared to class A peers. However, it must also be noted that the former are gravely penalized in terms of stream start time which indicates that requests for exchange connections by free riders are entertained only when no other peer in the local swarm is interested. Incidentally, during this point in time there is high data availability which causes free riders to experience the *late comer's advantage*. The ability of the scheme to suppress the effects of free riding in the system is consistent across all degrees of tested free riding concentrations.

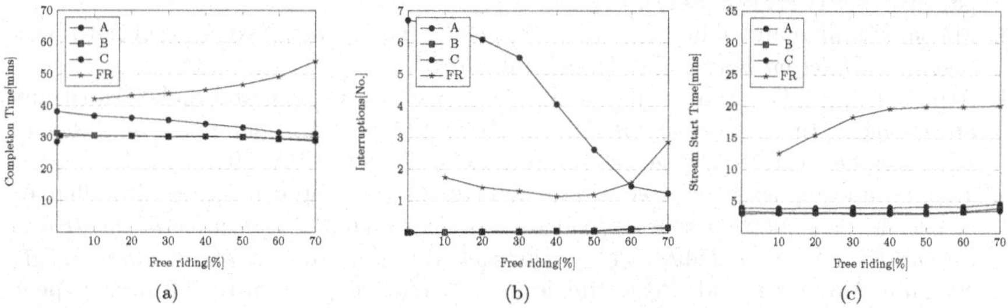

Fig. 1: System Performance (Setup III)

To conclude our analysis we look at how the system reacts to the different α and β configurations, and how they are manifested in different peer classes under varying free riding concentrations. It is evident that a particular setup may be beneficial for a certain peer class while not necessarily so for another. Class A peers for example, can take advantage of the weaker service differentiation in Setup I while higher bandwidth peers, on the other hand, perform better when their contribution is more appropriately represented in the network. However, as free riding becomes more prevalent, all contributing peers, regardless of class, develop affinity towards stronger service differentiation as it allows them to preserve the streaming performance observed in the the fully cooperative scenario.

6 Conclusion

In this study we proposed a hybrid incentive based peer selection scheme which takes into account a peer's local and global contribution. We also presented an evaluation function that is capable of emulating different incentive forms which includes purely global, purely local and their hybrid. We conclude based on simulation results that our proposed scheme effectively provides quality of service relative to their contribution in the network, which persists through all tested free riding concentrations. In addition, by further degrading the performance of lower bandwidth peers, the proposed scheme further reduces stream interruption of higher bandwidth peers.

References

1. Bittorrent - deliverying the world's contents. [Online; last accessed 24-Nov-2014 from `http://www.bittorrent.com/`].
2. Peersim - a p2p simulator. [Online; last accessed 24-Nov-2014 from `http://peersim.sourceforge.net/`].
3. Michal Feldman and John Chuang. Overcoming free-riding behavior in peer-to-peer systems. *SIGecom Exch.*, 5(4):41–50, July 2005.
4. Arnaud Legout, G. Urvoy-Keller, and P. Michiardi. Rarest first and choke algorithms are enough. In *Proceedings of the 6th ACM SIGCOMM Conference on Internet Measurement*, IMC '06, pages 203–216, New York, NY, USA, 2006. ACM.
5. Jiangchuan Liu, Sanjay G. Rao, Bo Li, and Hui Zhang. Opportunities and challenges of peer-to-peer internet video broadcast. In *In (invited) Proceedings of the IEEE, Special Issue on Recent Advances in Distributed Multimedia Communications*, 2007.
6. Nazanin Magharei and Reza Rejaie. Understanding mesh-based peer-to-peer streaming. In *Proceedings of the 2006 International Workshop on Network and Operating Systems Support for Digital Audio and Video*, NOSSDAV '06, pages 10:1–10:6, New York, NY, USA, 2006. ACM.
7. A. Montazeri and B. Akbari. Mesh based p2p video streaming with a distributed incentive mechanism. In *Information Networking (ICOIN), 2011 International Conference on*, pages 108–113, Jan 2011.
8. F. Pianese, D. Perino, J. Keller, and E.W. Biersack. Pulse: An adaptive, incentive-based, unstructured p2p live streaming system. *Multimedia, IEEE Transactions on*, 9(8):1645–1660, Dec 2007.
9. Yipeng Zhou, Dah Ming Chiu, and John C.S. Lui. A simple model for analyzing p2p streaming protocols. *2012 20th IEEE International Conference on Network Protocols (ICNP)*, 0:226–235, 2007.

A Transmission Method to Improve the Quality of Multimedia in Hybrid Broadcast/Mobile Networks

Hyung-Yoon Seo,[1] Byungjun Bae[2] and Jong-Deok Kim,[1*]
[1] Department of Electrical and Computer Enginneering, Pusan National University Busan, Republic of Korea
[2] Electronics and Telecommunications Research Institute Daejeon, Republic of Korea

tanyak@pusan.ac.kr[1], 1080i@etri.re.kr[2], kimjd@pusan.ac.kr[1*]

Abstract. This paper proposes the method through the mobile communication network when broadcast contents occur receiving errors in the Hybrid Broadcast/Mobile Network. Generally, the characteristic of the mobile communication network is pay network and peer-to-peer mobile. Therefore, it is necessary to reduce the amount of retransmission data in order to reduce the load of users and networks for recovery. This paper proposes the method to recovery for Hybrid DMB System which combines the T-DMB of the major mobile TV standard with the mobile communication network. The proposed method utilizes the Reed-Solomon techniques based on cross-layer. The proposed method transmits the additional information for recovery when errors are detected. It can reduce by ¼ the resource compared with the retransmission of the original MPEG2-TS.

Keywords: cross-layer, error correcting, Hybrid Transmission Method, Reed-Solomon

1 Introduction

As the digital technology develops, the boundary of broadcasting and communication crumbles down due to the active convergence of the broadcasting network and the communication network. The convergence of the said networks means that the boundary between them has become blurred. Therefore, broadcasting contents can be transmitted not only through a broadcasting network, but also through a communication network and vice versa. As a result, a convergence of the services, such as VOD (Video-On-Demand), data broadcasting, and Internet broadcasting services is created.

Hybrid Mobile Networks can be configured to various forms. Hybrid DMB System that combines T-DMB [1] with wireless mobile communication network (3G, LTE, Wi-Bro, Wi-Fi, etc.) is actively researched in Korea, and this paper is based on the Hybrid DMB System.

* Corresponding Author

K.J. Kim (ed.), *Information Science and Applications*,
Lecture Notes in Electrical Engineering 339, DOI 10.1007/978-3-662-46578-3_18

T-DMB incurs errors during the transmission due to the nature of wireless networks. T-DMB uses Reed-Solomon (204, 188) [2] for the error correction, and can only correct the errors of 8 bytes. Thus, problems occur where they are more than 8 bytes to be corrected.

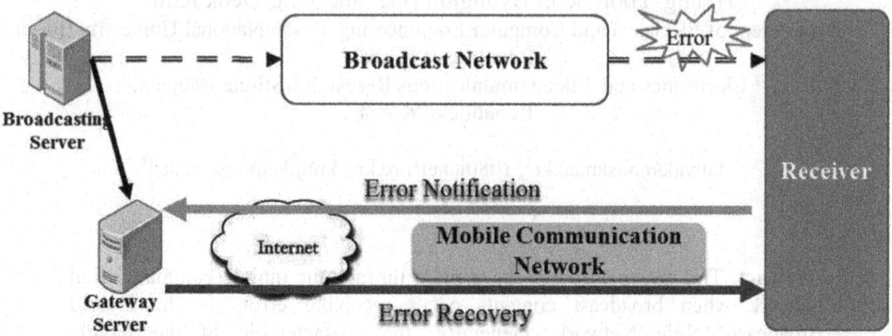

Fig. 1. The concept of error recovery through mobile communication network in Hybrid Broadcast/Mobile Communication Network

Hybrid DMB System uses the wireless mobile communication network and the T-DMB. However, a wireless mobile communication network has limited bandwidth and is shared by multiple users, so wireless mobile communication networks are highly sensitive to the load.

This paper proposes a transmission method that is transmitted through wireless mobile communication network and that minimizes the transmission load to improve the quality of multimedia service that is transmitted through the T-DMB network.

2 Analysis of the Error Characteristic of T-DMB

This paper analyses the real T-DMB streams [3] of three broadcasting networks in Korea: KBS, KNN, and myMBC.

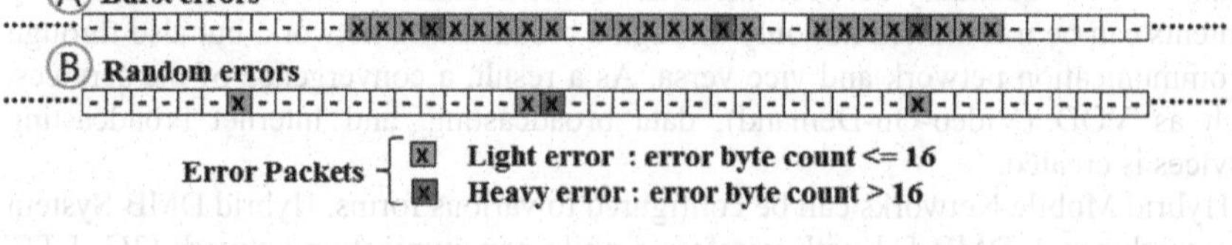

Fig. 2. T-DMB Error Characteristics

Figure 2 shows the characteristic of the error when analyzing the T-DMB stream. The incidence of errors is different, but the errors are distinguished into two characteristics: burst error that occurs continuously (A of Figure 2) and random errors that do not occur continuously (B of Figure 2). Also, most of the MPEG2-TS packets that have errors are analyzed to be Byte Error (BE) which has the error portion of the MPEG2-TS packet.

Based on the results of the analyzed packets that had BE in this paper, the BE was distinguish to be Light Error (LE) and Heavy Error (HE) as shown Figure 2 and Table 1. LE means that the number of error bytes of MPEG2-TS packets is smaller than 16, which is 85-90% of the total errors. HE means that the number of error bytes of MPEG2-TS packets is more than 16, which is 5-10% of the total errors.

Table 1. Error Type

Error Type	Error Byte Count (EC)	Error Ratio (%)
LE	EC $\leq$ 16	85 - 90
HE	EC > 16	10 - 15

3 Proposed Transmission Method for Error Recovery

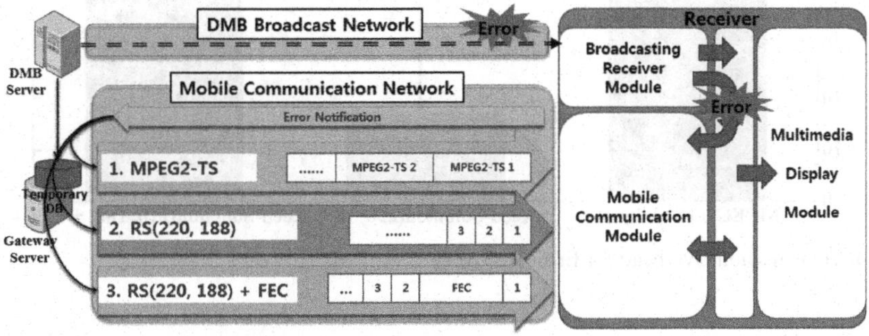

Fig. 3. Transmission Method for Error Recovery

Figure 3 shows the three transmission methods for the recovery. The DMB receiver requests for error recovery through the wireless mobile communication network to the Gateway Server when errors are detected. The Gateway Server has three transmission methods for the recovery: first, the original MPEG2-TS; second, the RS (220, 188); and third, the RS (220, 188) + FEC.

The first transmission method aims to retransmit the original MPEG2-TS packet that has errors. The second transmission method can recover LE which is 85-90% of the total errors according to the analyzed results in this paper but cannot recover the HE. Thus, this paper uses the third transmission method which is RS (220, 188) + FEC for the recovery so that it can recover not only LE but also HE.

4 Performance Evaluation of Transmission Method

This paper proposes the transmission methods through the wireless mobile communication networks. This paper compares the load of each transmission method because wireless mobile communication networks are highly sensitive to the load.

Figure 4 shows the transmission load for error recovery. The transmission method of the original MPEG2-TS has 100% the load. The transmission method of RS (220, 188) reduces load by 83% compared with the transmission method of the original MPEG2-TS, but there is a problem that it cannot recover the HE. The transmission method of RS (220, 188) + FEC can recover not only LE but also HE, and this method has reduced load by 57% compared with the original MPEG2-TS. However, this method has higher load than that of RS (220, 188).

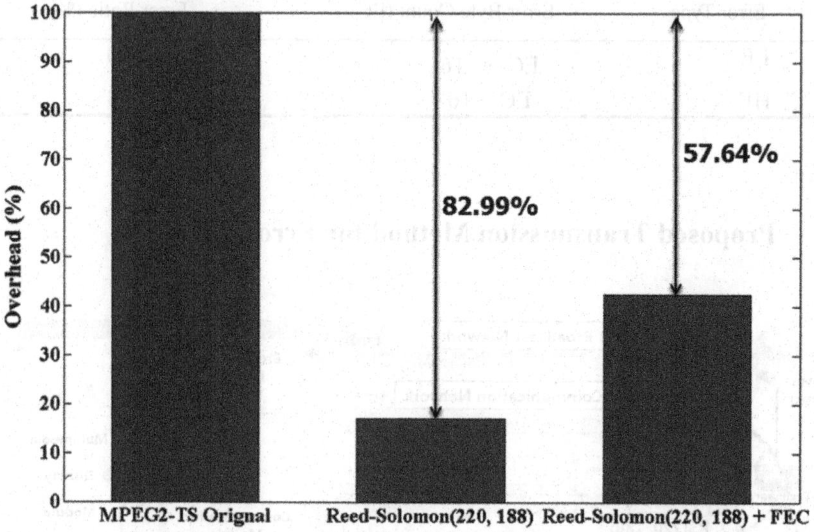

Fig. 4. Transmission Overhead for Error Recovery

5 Conclusion

This paper analyzed the characteristics of the error of the real T-DMB stream. This paper proposed a transmission method that reflects the characteristics of error to improve the quality of multimedia, and this transmission method could reduce the transmit load significantly.

Acknowledgments.

This work was supported by the ICT R&D program of MSIP/IITP. [14-000-02-002, Development of Service and Transmission Technology for Convergent Realistic Broadcast]

References

1. ETSI EN 300 401, "Radio Broadcasting Systems: Digital Audio Broadcasting (DAB) to Mobile, Portable and Fixed Receivers," v.1.4.1, June, 2006.
2. ETSI EN 300 744, Digital Video Broadcasting (DVB); Framing Structure, Channel Coding and Modulation for Digital Terrestrial Television, July 1999.
3. Hyung-Yoon Seo, Byungjun Bae, and Jong-Deok Kim, "An Efficient Transmission Scheme of MPEG2-TS over RTP for a Hybrid DMB System," ETRI Journal Vol. 35 No. 4 pp. 655-665 Aug. 2013.

A Failureless Pipelined Aho-Corasick Algorithm for FPGA-based Parallel String Matching Engine

HyunJin Kim

School of Electronics and Electrical Engineering, Dankook University, 152, Jukjeon-ro, Suji-gu, Yongin-si, Gyeonggi-do, Republic of Korea,
hyunjin2.kim@gmail.com
WWW home page: http://sites.google.com/site/dankooksoc

Abstract. This paper proposes a failureless pipelined Aho-Corasick (FPAC) algorithm that generates the failureless pipelined deterministic-finite automaton (DFA). The failureless pipelined DFA generated by the FPAC algorithm does not store the failure pointers for reducing hardware overhead. Moreover, by sharing common prefixes, the information for storing states can be compressed. Because the pipeline register stores the state in each stage, the failureless pipelined DFA can perform multiple state transitions in parallel. Therefore, throughput can be increased with multiple homogeneous DFAs. In the experiments with cost-effective FPGAs, the implementation of the proposed FPAC algorithm shows high performance and low hardware overhead compared to several FPGA-based string matching engines.

Keywords: Aho-Corasick algorithm, deterministic-finite automaton, field-programmable gate array, string matching

1 Introduction

The string matching engine is the essential device for the network security [1], bio informatics, image pattern recognition, etc. Due to the requirements of high performance and increased number of target patterns, a dedicated pattern matching system should have high throughput. Each target pattern can be composed of multiple characters. DFA is popularly adopted in the string matching due to the deterministic time of state transitions.

The traditional Aho-Corasick (AC) algorithm in [2] constructs DFA for several patterns. In the DFA, transitions between states can be performed in a fixed number of clock cycles; therefore, it is guaranteed that throughput is unchanged for the worst-case situation. In addition, common prefixes are shared, so that hardware overhead can be reduced. However, for the deterministic state transition time, the failure pointers towards the longest matched suffix should be added for each state, which is great burden in the real implementation. Moreover, the hardware complexity of the combinational part in the implementation increases with the number of states in a DFA.

K.J. Kim (ed.), *Information Science and Applications*,
Lecture Notes in Electrical Engineering 339, DOI 10.1007/978-3-662-46578-3_19

FPGA is one of suitable devices that can be used in the string matching due to its high flexibility and programmability. The DFA generated by the original AC algorithm, however, is not suitable for the FPGA-based string matching because of a large numbers of states in a DFA and failure pointers in each state [3]. On the other hand, a parallel string matching engine adopts the parallel failureless Aho-Corasick (PFAC) algorithm in [4], where multiple homogeneous DFAs without the failure pointers are processed by multiple threads in the graphic processing unit (GPU), respectively. However, the application of the original PFAC algorithm to the FPGA-based string matching is not be suitable because the number of homogenous DFAs can be proportional to the maximum pattern length in a set of patterns.

In this paper, the failureless pipelined Aho-Corasick (FPAC) algorithm is proposed. In the proposed FPAC algorithm, in order to reduce hardware overhead, the generated DFA does not store the failure pointers. In addition, common prefixes are shared in the DFA, so that the information about states can be compressed. For each stage, pipeline registers are inserted; therefore, the state transitions in multiple stages can be performed in a pipelined fashion. Because only one output state exists in each pipelined stage, the implementation for generating the identification of the longest matched pattern can be simplified. In addition, throughput can be increased with multiple homogeneous failureless pipelined DFAs. In our experiments with the cost-effective FPGA, hardware overhead is reduced by 16.1% on average. In addition, the maximum operating frequency is increased by 26% on average, compared to several FPGA-based string matching methods.

2 Previous Works and Motivations

There are several previous works that can be implemented using an FPGA. Firstly, the content-addressable memory (CAM) can be emulated using logic resources in an FPGA [5]. By comparing an input sequence with patterns using logic comparators in each row, the identification numbers for multiple matched patterns can be generated, where the identification number for the longest matched pattern can be easily outputted by the priority encoder. In the emulation, because all target patterns are implemented with logic comparators, a large number of combinational logic cells are required. In addition, the operating frequency is low due to the slow priority encoding to generate the identification number for the longest matched pattern.

On the other hand, in the implementation of a DFA using an FPGA, there are two methods: the memory-based and general string matching engines. The memory-based string matching adopts memory macro blocks in an FPGA, where target patterns are mapped onto the memory blocks [3, 6]. In the memory-based string matching, therefore, a small number of logic elements are used in the FPGA. Because the target resource of the proposed algorithm is the logic element, the proposed algorithm is categorized into the general string matching.

In the DFA-based general string matching engine implemented in the FPGA, combinational logic elements are used to construct state transitions. In addition, registers are required to store current state. By describing the failure pointers towards the initial state as default, the combinational logic for decoding the state transitions can be simplified. As the number of target patterns, *T*, increases, the number of states, *n*, could grow. In addition, the number of the failure pointers towards the non-initial states increases because the matched suffixes can be frequent for a large number of states; therefore, the DFA construction with failure pointers increases hardware complexity.

The PFAC algorithm can reduce the complexity of combinational logic for a DFA by removing the failure pointers towards non-initial states [4]. By applying the multiple initialized failureless DFAs in parallel, throughput can be enhanced dramatically, compared to the case with a single DFA for the traditional AC algorithm. However, for the FPGA-based string matching, the previous PFAC algorithm requires high hardware overhead for implementing multiple DFAs. In addition, because the initialized DFAs are required for the parallel string matching, the enhanced throughput is not guaranteed for the worst-case situation.

The proposed FPAC algorithm is motivated from the failureless DFA. In the PFAC algorithm that requires multiple failureless DFAs, hardware overhead is proportional to the number of the failureless DFAs in FPGA-based string matching. In addition, the majority of the DFA implementation is the combinational logic, which causes unbalanced usage of the available hardware resource in an FPGA. Therefore, in the proposed FPAC algorithm, the failureless pipelined DFA is constructed by adding the pipelined registers between stages. Because an input sequence is applied into multiple stages at the same time, the string matching can be performed in parallel. In addition, the unbalanced resource usage can be amortized with the inserted registers.

3 Proposed FPAC algorithm

3.1 Failureless Pipelined DFA

Fig. 1 shows the concept of the failureless pipelined DFA generated by the proposed FPAC algorithm. In Fig. 1, input sequence is provided for the input of each stage in parallel. In each stage, state transitions are performed in parallel. The dotted arrows are the pipelined failure pointers towards the initial state of the next stage. In the failureless pipelined DFA, there is no failure pointer toward non-initial states. The normal arrows mean the pipelined goto functions, where the destination of a goto function is the state in the next stage. In addition, the pipelined failure pointers towards the added initial state are applied for each stage. A pipelined failure pointer crosses $\mathbf{stage}_i$ and $\mathbf{stage}_{i+1}$, when the source and destination of the failure pointers exist in $\mathbf{stage}_i$ and $\mathbf{stage}_{i+1}$, respectively. When $\mathbf{i} > 0$ and the current state is the initial state for $\mathbf{stage}_i$, the next state is also the initial state for $\mathbf{stage}_{i+1}$. Because there is no matched prefix for pattern sets in the initial state in $\mathbf{stage}_i$, no patterns are matched in the next state. When the current state is in the last stage, the next state is the initial state in the $\mathbf{stage}_0$. In this case, when the maximum target pattern length is l, states are rotated from $\mathbf{stage}_0$ to $\mathbf{stage}_l$ in the failureless pipelined DFA. In the example of Fig. 1, states are rotated from $\mathbf{stage}_0$ to $\mathbf{stage}_4$ because the maximum pattern length is four of **bbab**.

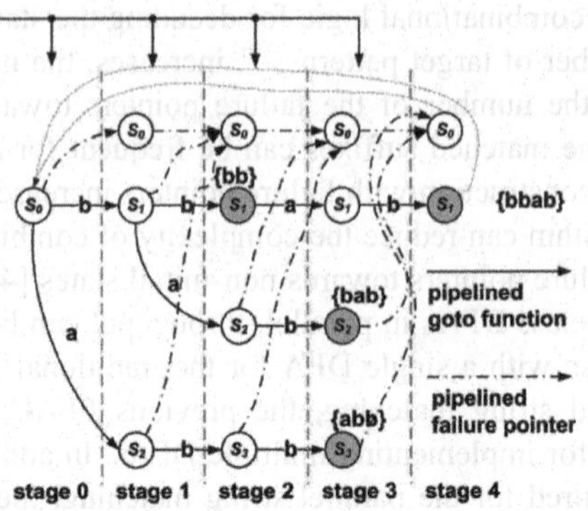

Fig. 1. Example of a failureless pipelined DFA for patterns **abb**, **bab**, **bb**, and **bbab**.

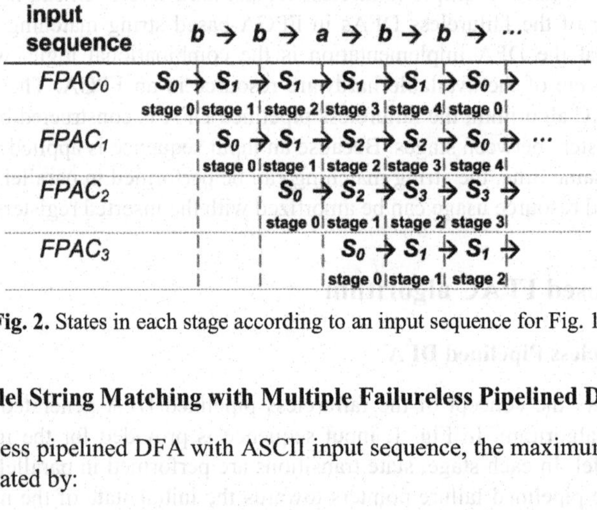
input sequence
b → b → a → b → b → ...
FPAC$_0$
S_0 → S_1 → S_1 → S_1 → S_1 → S_0 →
stage 0 stage 1 stage 2 stage 3 stage 4 stage 0
FPAC$_1$
S_0 → S_1 → S_2 → S_2 → S_0 → ...
stage 0 stage 1 stage 2 stage 3 stage 4
FPAC$_2$
S_0 → S_2 → S_3 → S_3 →
stage 0 stage 1 stage 2 stage 3
FPAC$_3$
S_0 → S_1 → S_1 →
stage 0 stage 1 stage 2

Fig. 2. States in each stage according to an input sequence for Fig. 1.

3.2 Parallel String Matching with Multiple Failureless Pipelined DFAs

For a failureless pipelined DFA with ASCII input sequence, the maximum throughput can be calculated by:

$$\text{Throughput}=8 \times F_{max}, \quad (1)$$

where F_{max} means the maximum operating frequency in the implementation. When F_{max} is 200 MHz, the maximum throughput is 1.6 Gbps, which is not sufficient for 10/100 Gbps networking environments. In order to increase the maximum throughput, multiple homogeneous failureless pipelined DFAs can be adopted for the parallel string matching. If each DFA has different input sequence, the amount of input data to be processed at a time increases; therefore, when n failureless pipelined DFAs are adopted, it is expected that the multiplied maximum throughput can be calculated by:

$$\text{Throughput} = n \times 8 \times F_{max}. \quad (2)$$

Table 1. Characteristics of target patterns in rule sets.

rule name	*num(patterns)*	*num(bytes)*	*min(l)*	*max(l)*	*avg(l)*
backdoor	955	8,875	1	94	9.3
chat	49	431	1	38	8.8
deleted	615	7,399	1	72	12.0
exploit	243	1,906	1	109	7.8
policy	114	1,154	1	114	10.1

4 Experimental Results

4.1 Experimental Environments

The proposed FPAC algorithm was evaluated using the C++ library and Boost graph library (BGL), where an executable program was created [7]. The executable program generated a HDL code for the proposed FPAC algorithm. In the experiments, the Xilinx's FPGA and its tool chain were adopted [8]. The HDL code was synthesized using XST in 64-bit ISE 14.4 of Xilinx. The operating system was 64-bit Microsoft Windows 7 that ran on the Intel(R) Core(TM) I7-2620M CPU @2.70 GHz with 8 Gbytes physical memory and a solid-state disk (SSD). The target device was the Spartan-3E xc3s500e and xc3s1200e. Even though there were several high performance FPGA device families, it was considered that the price of high performance devices was too high for the general string matching engine. Therefore, because the Spartan-3E FPGA family was suitable for high volume and cost-sensitive applications, it was adopted in our experiments. In the target FPGA, there were many CLBs (Configurable Logic Blocks), where four slices were located in a CLB. In a slice, there were two 4-input LUTs (Look-Up Tables) and two flip-flops (FFs). Hardware overhead was described with the numbers of used slices, slice FFs, and 4-input LUTs.

Five sets of target patterns denoted as *backdoor*, *chat*, *deleted*, *exploit*, and *policy* were extracted from Snort v2.8 rules [9]. Table 1 summarizes the characteristics of the five sets of target patterns, where *num(patterns)* and *num(bytes)* mean the number of target patterns and total sum of characters of target patterns in each set, respectively. In addition, *min(l)*, *max(l)*, and *avg(l)* are the minimum, maximum, and average pattern lengths.

For the apples-to-apples comparisons, several previous works introduced in **Previous Works and Motivations** section were adopted. Firstly, a primitive string matching engine, *primitive*, with multiple pattern matchers was evaluated. In the primitive string matching engine, each pattern matcher compared a pattern with input sequence with logic comparators [5]. If the pattern was matched with the input sequence, its own identification number was provided. In this case, the number of pattern matchers was the same as the number of patterns. In addition, by using a priority encoder, only one identification number for the longest matched pattern was provided. Secondly, the DFA constructed by the traditional AC algorithm, *aho*, was evaluated. If multiple patterns were matched in an output state, one identification number for the longest

Table 2. Hardware overhead and F_{max} according to rule sets.

items	*backdoor*	*chat*	*deleted*	*exploit*	*policy*
target device	1200e	500e	500e	500e	500e
slices	5,191	309	4,319	1,351	820
used/total	60%	7%	93%	29%	17%
slice FFs	6,914	403	5,986	1,766	1,168
used/total	40%	4%	64%	19%	12%
4-input LUTs	8,985	551	7,510	2,366	1,453
used/total	52%	6%	81%	25%	16%
F_{max}(MHz)	128	192	128	134	147

pattern in the state was provided; therefore, the priority encoder for each output state was not required. Thirdly, the DFA constructed by the PFAC algorithm, *pfac*, was evaluated. Even though the application of the previous PFAC algorithm was focused on the acceleration with GPU, the HDL code was generated to evaluate the application of the PFAC algorithm to FPGA. Even though multiple failureless DFAs should be implemented, only one failureless DFA was implemented in the experiments in order to know the effectiveness of the inserted pipeline registers in the proposed algorithm. All previous works were evaluated with the generated Verilog HDL codes. After compiling the HDL codes, the maximum operating frequency and hardware overhead were summarized.

4.2 Experimental Data and Discussion

Considering the experimental environments mentioned above, experiments were performed. Table 2 shows the hardware overhead and F_{max} according to the rule set. The device utilization was estimated after the synthesis using Xilinx's XST. The logic utilization was proportional to the total sum of characters of target patterns. Considering the structure of the CLB with 4-input LUTs and storage elements, the ratio of combinational logic to registers was large. From the maximum operating frequency in Table 2, it was concluded that F_{max} decreased as hardware overhead increased, which meant that as the number of target patterns increased, hardware structure became more complex. Considering both the synthesis report and structure of the generated DFA, the critical path existed in the priority encoder for generating the identification number of the longest matched pattern. In the implementation of one failureless pipelined DFA by the proposed FPAC algorithm, the throughput was ranged from 1.024 Gbps from 1.536 Gbps. Therefore, when n failureless pipelined DFAs were implemented, it was expected that the throughput can be ranged from $n \times 1.024 \sim n \times 1.536$ Gbps.

For the five target rule sets mentioned above, three previous works, the proposed FPAC algorithm, *proposed*, was compared to *primitive*, *aho*, and *pfac* in terms of hardware overhead and F_{max} in Table 3, where the proposed algorithm provided better synthesis results with small number of used slices and F_{max}.

Table 3. Comparision with previous works in terms of hardware overhead and F_{max}.

items	algorithm	*backdoor*	*chat*	*deleted*	*exploit*	*policy*
slices	*primitive*	5,736	444	4,394	1,893	1,299
	aho	N/A	308	N/A	N/A	1,011
	pfac	N/A	348	N/A	1,509	953
	proposed	5,191	309	4,319	1,351	820
slice FFs	*primitive*	1,708	361	1,204	1,124	1,098
	aho	N/A	355	N/A	N/A	1,043
	pfac	N/A	348	N/A	1,594	1,044
	proposed	6,914	403	5,986	1,766	1,168
LUTs	*primitive*	10,627	656	8,050	3,061	1,687
	aho	N/A	543	N/A	N/A	1,787
	pfac	N/A	620	N/A	2,676	1,683
	proposed	8,985	551	7,510	2,366	1,453
F_{max} (MHz)	*primitive*	78	137	81	97	123
	aho	N/A	206	N/A	N/A	154
	pfac	N/A	145	N/A	126	131
	proposed	128	192	128	134	147

For two rule sets with many target patterns, *backdoor* and *deleted*, due to the non-sufficient memory of 8 GBytes, the generated HDL codes by *aho* and *pfac* were not synthesized, Moreover, the HDL code of *exploit* by *aho* was also not synthesized. The generated HDL codes by *aho* and *pfac* were too complex to obtain the synthesis results in the experimental environments. Therefore, it was concluded that the hardware complexity in the next state encoding was rapidly increased with the number of total characters in target patterns in *aho* and *pfac*. For *chat* and *policy* patterns, the maximum operating frequency of *aho* was slightly higher compared to the cases of the proposed algorithm. Because the critical path of *proposed* was in the priority encoder, it was thought that F_{max} become lower due to the delay of the priority encoder.

Even though there were failing pointers in *chat* rule set, *aho* provided low hardware overhead and higher F_{max} over *pfac*. However, for *policy*, *pfac* required lower hardware overhead than *aho*. Because almost failure pointers went towards states near the initial state, a small number of failure pointers were repeated for each state, which could be helpful to obtain better synthesis results for *aho* in *chat* rule set.

Only *primitive* and *proposed* succeeded the synthesis for all five rule sets. Compared to *primitive*, the proposed algorithm required a large number of slice FFs, which was used for storing states in pipeline registers and identification numbers for output states. Especially, one-hot encoding was adopted to implement an FSM with a large number of states in an FPGA, which increased the number of used slice FFs.

Even though the number of used slice FFs increased, the number of used slices was not greatly affected because the required numbers of LUTs were greater. Please be reminded that there were two LUTs and two slice FFs in a slice. Therefore, the number of used slices of *proposed* was smaller than that of *primitive*. In addition, compared to *primitive*, the hardware overhead in terms of used slices was decreased by 1.7%-36.9%. The maximum operating frequency was increased by 19.1%-63.9%.

5 Conclusion

The proposed failureless pipelined Aho-Corasick algorithm reduces hardware overhead by adopting simple combinational logic in each stage. In addition, the maximum operating frequency increases due to the simplicity of the combinational part between stages. For all target rule sets in the experiments, the HDL codes for the proposed algorithm are well synthesized. In addition, the hardware implementations show better results in terms of both hardware overhead and the maximum operating frequency. By implementing the proposed algorithm using cost-effective FPGAs, a string matching engine can be parallelized with high updatability. Considering the advantages of the proposed FPAC algorithm and efficient implementation results, it is concluded that the FPAC algorithm can be practically adopted in the string matching engine using an FPGA.

Acknowledgement

This work was partly supported by the ICT R\&D program of MSIP/IITP, Republic of Korea. [14-000-05-001, Smart Networking Core Technology Development] and Basic Science Research Program through the National Research Foundation of Korea(NRF) funded by the Ministry of Education. (NRF-2014R1A1A2A16055699).

References

1. Lin, P.-C., Lin, Y.-D, Lee, T.-H., and Lai, Y.-C.: Using string matching for deep packet inspection. IEEE Computer. 41, 23-28 (2008)
2. Aho, A.V., Corasick, M. J.: Efficient string matching: an aid to bibliographic search. Communication ACM. 18, 333- 340 (1975)
3. Tan, L., Sherwood, T.: A high throughput string matching architecture for intrusion detection and prevention. Proc. 32nd IEEE/ACM Int'l Symp. Comp. Arch. 112-122 (2005)
4. Lin, C.-H. et al.: Accelerating pattern matching using a novel parallel algorithm on GPUs. IEEE Trans. Comp. 62, 1906-1916 (2013)
5. Sourdis, I., Pnevmatikatos, D.: Fast, large-scale string match for a 10Gbps FPGA-based network intrusion detection system. Field Programmable Logic and Application, Lecture Notes in Comp. Sci. 2778, 880-889 (2003)
6. Kim, H., Kim, H.-S., Kang, S.: A memory-efficient bit-split parallel string matching using pattern dividing for intrusion detection systems. IEEE Trans. Parallel and Distributed Systems. 22, 1904- 1911 (2011)
7. Siek, J. G., Lee, L.-Q. Lumsdaine, A.: Boost graph library: user guide and reference manual. The Pearson Education (2001)
8. Xilinx Inc. Available: http://www.xilinx.com.
9. Snort, ver.2.8, Network Intrusion Detection System. Available: http://www.snort.org.

Revised P2P Data Sharing Scheme over Distributed Cloud Networks

Wonhyuk Lee[1], TaeYeon Kim[2], Seungae Kang[3], HyunCheol Kim[3]*

[1] Korea Institute of Science & Technology Information
Daejon, Korea, 305-806
livezone@kisti.re.kr

[2] Electronics and Telecommunications Research Institute
Daejon, Korea, 305-700
tykim@etri.re.kr

[3] Dept. of Computer Science, Namseoul University
Cheonan, Korea, 331-707
{sahome,hckim}@nsu.ac.kr

Abstract. Distributed cloud networks can be seen as a cooperative network composed of millions of hosts spread around the world and it is also a distributed shared resource. In the P2P technology, as opposed to the existing client/server concept, devices are actively connected with one another in order to share the resources and every participant is both a server and a client at the same time. In P2P distributed cloud networks, peers are able to directly share and exchange information without the help of a server. This results in a prompt and secure sharing of network resources and data handling. However, flooding algorithm that is used in distributed P2P network generated query message excessively. Our objective in this paper is to proposes a presents a restricted path flooding algorithm that can decrease query message's occurrence to solve P2P network's problems. It includes concepts as well as systematic procedures of the proposed scheme for fast path flooding in distributed P2P distributed cloud networks.

1 Introduction

The Internet can be seen as a cooperative network composed of millions of hosts spread around the world and it is also a distributed shared resource. The current network status requires, more so than ever, a scheme in which bandwidths are more efficiently managed and network capacities are fully utilized as well as application programs that can send packets more safely to farther distances [1]. Traditional client/server models with a vertical hierarchical structure generate problems such as vulnerability against malicious attacks, inability to be expanded, and bottleneck issues in local networks. Therefore, various researches were conducted on the P2P (Peer-to-

* Corresponding author

K.J. Kim (ed.), *Information Science and Applications*,
Lecture Notes in Electrical Engineering 339, DOI 10.1007/978-3-662-46578-3_20

peer network) model, which is a horizontal network model in which all hosts that form a network can act as a client and the server at the same time.

The existing client/server-based model falls short of satisfying such requirements. For example, it cannot react against malicious attacks or resource bottleneck effects in focused structures and there are complexity and high-cost issues when attempting to mitigate the original structure [2]. In the P2P technology, as opposed to the existing client/server concept, devices are actively connected with one another as shown in Fig. 2 in order to share the resources and every participant is both a server and a client at the same time. When the peers that exist in a physical network are registered in the P2P service, a virtual network among the registered peers, namely a P2P overlay network, is formed.

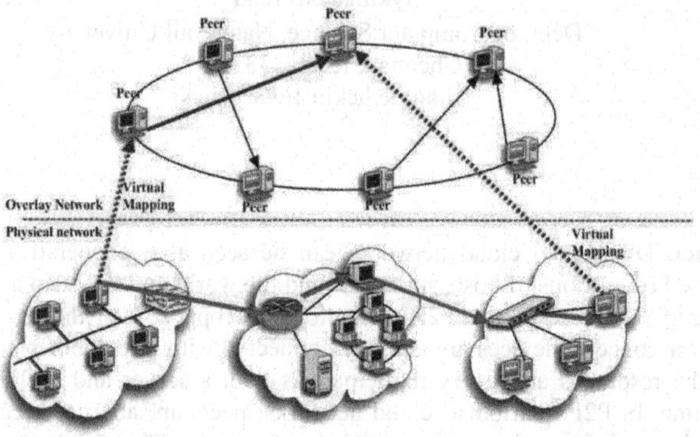

Fig. 1. Traffic Concentration Problem in Client-Server Networks

In the P2P technology, as opposed to the existing client/server concept, devices are actively connected with one another in order to share the resources and every participant is both a server and a client at the same time. When the peers that exist in a physical network are registered in the P2P service, a virtual network among the registered peers, namely a P2P distributed cloud network, is formed. In the P2P distributed cloud network, peers are able to directly share and exchange information among themselves without the help of the server. Through this, a safer and a faster network resources sharing and data handling are made possible. However, the P2P distributed cloud network has a drawback in that it induces an inefficient operation of network bandwidth because it consumes such a vast amount of network bandwidth. In order to overcome these difficulties, this paper proposes a new flooding algorithm that enhances the overall performance of the network by reducing the excessive query message traffic that occurs in the distributed P2P network, which is a type of P2P distributed cloud network [3][4][5].

The remainder of this paper is organized as follows. Section 2 describes the characteristics of each P2P distributed cloud network type and it primarily focuses on the flooding method in the distributed P2P network along with a possible solution. Section 3 compares the existing flooding method with the proposed flooding method through simulation in order to analyze and contrast their quantified performances.

Section 4 concludes this paper by summarizing some key points made throughout and assessing the representation of analyzed results.

2 Restricted Path Flooding

2.1 Method

In the existing flooding algorithm, when a host wants to send a query message or relay a received query message to other peers, it chose the method that sends the message to all its adjacent peers. Such method is useful in that it is able to promptly distribute specific information in a vast network. However, the drawback of the flooding method is that it induces large traffics within the network. Not only this overloaded the entire network but it also creates bottleneck issues [6][7]. So the distributed P2P network uses the TTL value to control the area to which a query message is delivered to.

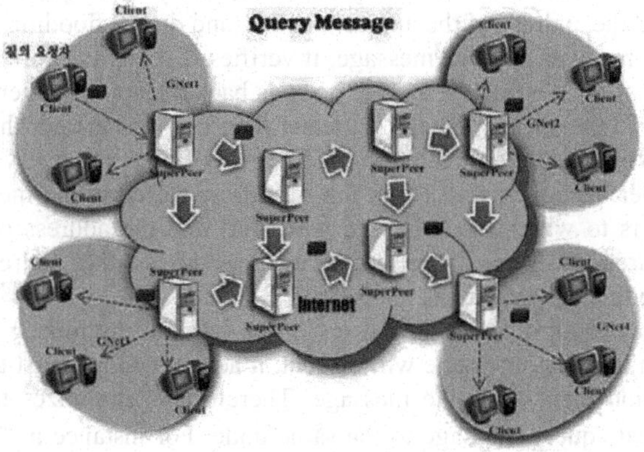

Fig. 2. Restricted Path Flooding Scheme

Nonetheless, finding the appropriate TTL value is not a simple task. If the TTL value is too large, it will result in an unnecessary overloading and when it's too small, the appropriate target node will not be found because there are too little a number of nodes to send to. Another problem with the flooding method is that there is a high probability that duplicates of a query message could be sent in a P2P network with high connectivity. Performing a flooding on duplicate messages will cause the number of nodes that receive the duplicate messages to continuously increase. Also, since messages will only be sent to similar nodes, it will bring about a difficulty in detecting the target node. To prevent this, the P2P that uses the flooding-style search method is designed so that it discards any duplicate message and does not execute the flooding again.

Nevertheless, since duplicate messages will occur regardless of flooding, the feasibility of this technique is very low. In order to mitigate that problem, this research paper proposes the restricted path flooding algorithm. Fig. 2 represents the proposed Restricted Path Flooding algorithm. In the restricted path flooding technique, the P2P network is, first, designed to have an distributed cloud structure and the query-generating peers are made to send query messages to only the superpeer The superpeer does a flooding on the query message that includes the list of the adjacent peers' address.Upon receiving the query message, the nodes check if its adjacent nodes exist in the list of addresses included in the query message and it will not send the query message the nodes that exist in the address list. This method is utilized so that a duplicate query message is not delivered to nodes in the distributed P2P network.

2.2 Algorithm

Let's assume that a temporary node generated a query message in order to request for data resources. The node that sends a query message does a flooding on the query message that it generated to its adjacent nodes where it first configures a TTL value suitable for the network environment, generates a query message that includes the query details and the address of the adjacent nodes, and does a flooding on it.

When a node receives a query message, it verifies if it has the corresponding data resources. When it locates such resource, it sends back a response otherwise it checks if the TTL value in the query message is greater than 1. If it is greater than 1, it checks if the query message had been received before and it floods the query message only when it had never been received. Here, the S_address stored in the message (the address of the peers to which this message was sent) and the address of the adjacent peers are compared and when there is a redundant address, that address is removed from the list of addresses that it has to send to.

After having compared all the addresses in the list and optimizing the list of addresses to which its query message will be sent, it adds its address list to the previous S_Address list and then sends the message. Therefore it minimizes the chances of sending a duplicate query message to the same node. For instance in Fig. 3 (a), let's say Node A in the network sends a query message to request for data.

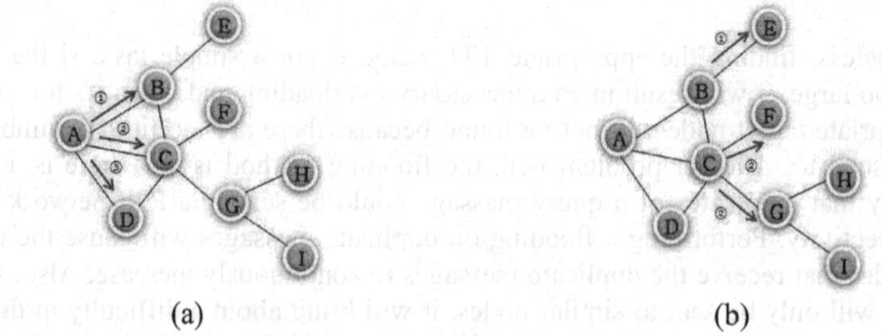

(a) (b)

Fig. 3. Sample Network (a) and Restricted Flooding Network (b)

When Nodes B, C, and D receive the query message and they do not have the data requested by Node A, these nodes will then flood. Fig. 3 (b) shows when the restricted path flooding algorithm is used.

3 Performance Analysis

3.1 Network Topology

In order to evaluate the performance of the proposed algorithm, we generated three different types of network topology and compared the performance for each. The most important element when considering network topology is the number of adjacent nodes. This number refers to the average number of nodes adjacent to each node.

Each of the simulated network topologies was structured to have an average of 4 adjacent nodes. The first network topology is 2-dimensional mesh topology. In this mesh-type topology, which is composed of 10x10 nodes, all nodes within the 2- dimensional mesh topology excluding the 4 nodes in the corners have 4 adjacent nodes. A typical adjacent node of this topology has about 3:96 nodes. The second network topology is a topology created by the gt-itm topology generator. This network topology has the form of a 2-phase hierarchical random graph type.

The first phase forms a connective graph with the 10 generated nodes. In this connective graph, nodes are connected with one another with probability p and each node is composed of 10 vertices randomly scattered. The second phase uses each of the generated nodes, composed of vertices, as the root of the new graph in order to generate a new random graph that consists of 10 nodes.

3.2 Performance Analysis

The latest results show that the current dispersion P2P network is in the form of power-law type-node dispersion with a reducing number of adjacent nodes due to the power function. The power-refraction dispersion index of -2:2 are used to represent the node dispersion of the Internet and the power-refraction dispersion index of -2:3 represent the node dispersion type of an unstructured P2P network.

A network topology generator known as Inet was used and the network topology obtained by this has an average number of adjacent nodes of approximately 4. First, this simulation compared the amount of traffic and the number of query messages generated by the flooding algorithm and the restricted path flooding algorithm in each network topology defined by the topology up to 50 unit times and compared the amount of traffic, the number of query messages, and the time elapsed until the arrival at the arbitrary target node.

Examining the results obtained through the simulation, it can be seen that the proposed restricted path flooding algorithm generates a fewer number of query messages and less traffic than the standard flooding algorithm. The difference in the number of query messages generated by flooding and the restricted path flooding is smaller for a lower TTL value and an increase in the TTL value could result in a maximum 58%

reduction in the number of generated query messages. For the power-law dispersion, the reason why there is no change in the number of generated query messages and the amount of traffic after a certain amount of time is because all the nodes have been reached.

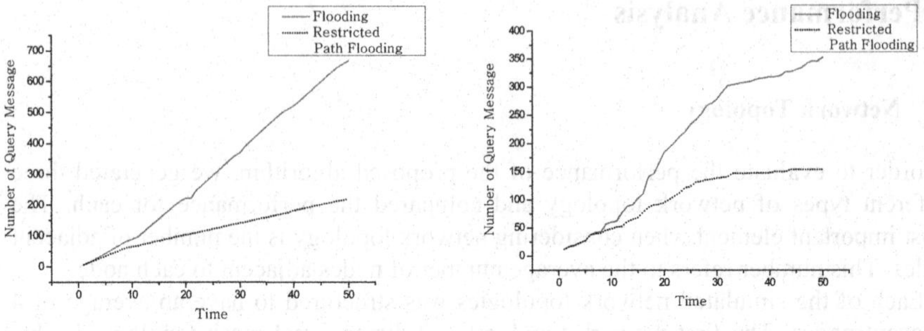

(a) (b)

Fig. 4. Number of Query Message in 2-Dimensional Mesh Topology (a) and Random Topology (b)

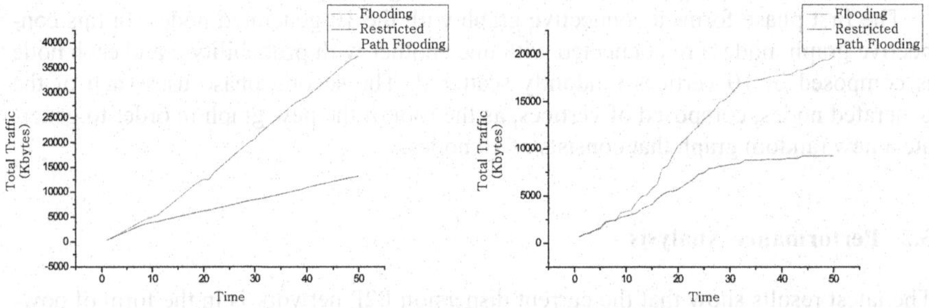

(a) (b)

Fig. 5. The amount of Traffic in 2-Dimensional Mesh Topology (a) and Random Topology (b)

Fig. 4 and Fig. 5 compare the number of query messages created, the amount of generated traffic, and the time it takes for the flooding algorithm and the restricted path flooding algorithm to arrive at the tentative target node in each network topology defined in the topology.

When a delivery of a query message to the target node is successfully made, it can be seen that the difference in the number of query messages generated among flooding and restricted path flooding is not significant. Also, the power-law dispersion in which one node is connected to many nodes showed only 14% of decrease. Therefore, the longer the time it takes for a query message to get to the target node, the higher efficiency of the restricted path flooding algorithm than the flooding algorithm.

5 Conclusions

The Internet is replacing the traditional telephone network as the ubiquitous network infrastructure. Internet customers are increasing at an exponential rate and will continue to increase in the near future. As the Internet became vaster, accommodating an enormous number of users who swarmed to the online world, it created issues such as the bottleneck problem between the local network and the backbone caused by the network overloading between a relatively smaller number of servers and many clients. Such phenomenon resulted in the emergence of the P2P network, thereby resolving the network difficulties. This paper takes sharing way of incomplete data in order to improve performance in decentralized P2P network. Such attempt could bring about a P2P network with an improved performance.

Acknowledgement This work was supported by the ICT R&D program of MSIP/IITP, Republic of Korea. [14-000-05-001, Smart Networking Core Technology Development]

References

1. Min Yang, Yuanyuan Yang, ``An Efficient Hybrid Peer-to-Peer System for Distributed Data Sharing``, IEEE Transactions on Computers, Vol. 59, Issue 9, pp. 1158-1171, 2010.
2. Dharanipragada J., Haridas H., "Stabilizing Peer-to-Peer Systems Using Public Cloud: A Case Study of Peer-to-Peer Search``, International Symposium on Parallel and Distributed Computing, pp. 135-142, 2012.
3. Miller K., Wolisz A., "Transport Optimization in Peer-to-Peer Networks``, Parallel, Distributed and Network-Based Processing, pp. 567-573, 2011.
4. Min Yang, Yuanyuan Yang, ``Peer-to-Peer File Sharing Based on Network Coding ``, International Conference on Distributed Computing Systems, pp. 168-175, 2008.
5. Y. Chawathe, S. Ratnasamy, L. Breslau, N. Lanham, and S. Shenker, "Making Gnutella-like P2P systems Scalable``, ACM SIGCOMM'03, Aug. 2003.
6. C. Lv, P. Cao, E. Cohen, K. Li, and S. Shenker, "Search and Replication in Un-structured Peer-to-Peer Networks``, ICS, 2002.
7. B. Yang and H. Garcia-Molina, "Improving Search in Peer-to-Peer Networks``, ICDCS, 2002.

LQDV Routing Protocol Implementation on Arduino Platform and Xbee module

Ho Sy Khanh and Myung Kyun Kim*

School of Electrical Engineering, University of Ulsan
mkkim@ulsan.ac.kr

Abstract. So far today, most of routing protocols in ad-hoc network have been evaluated by simulation. However, the simulation usually does not reflect the impacts of real environment on the performance of the routing protocols. In this paper, we have implemented Link Quality Distance Vector (LQDV) routing protocol which is a reliable routing protocol in static wireless networks on Arduino platform. We also have evaluated the performance of our LQDV protocol implementation on a real environment. Furthermore, we introduce a simple way to help developers to create implementations based on Arduino platform and Xbee module.

Keywords: Wireless Network, Reliable Routing Protocol, ETX Link Quality Estimator, Arduino and Xbee.

1 Introduction

In all over areas in the research work, simulation is very useful tool for researchers. Ad-hoc network is not an exception. However, simulation does not guarantee protocol work well in practice. So, after simulation, it is necessary to create an implementation to verify that it can work well in the real environment. However, implementation is much more difficult than simulation. Creating implementation requires developers to design not only routing protocol but also all components of system. LQDV [1] was proven by simulation to be a reliable routing protocol in the high load static wireless network. Therefore, in this paper we implement LQDV on the real devices based on Arduino platform and Xbee module to validate the performance of LQDV routing protocol in the real environment. Furthermore, we introduce a simple way to help developers to create implementations based on Arduino platform and Xbee module.

The rest of this paper is organized as follows. Section 2 introduces LQDV routing protocol. Section 3 describes routing implementation design. Section 4 shows the results. Finally, section 5 presents conclusions.

K.J. Kim (ed.), *Information Science and Applications*,
Lecture Notes in Electrical Engineering 339, DOI 10.1007/978-3-662-46578-3_21

2 LQDV Protocol Overview

The LQDV routing protocol is a modification of AODV [2]. LQDV differs from AODV in three aspects. First, LQDV uses HETX [1] link metric instead of minimum hop-count. Second, to reduce the number of control packet overhead, if an intermediate node receives RREQ not for the first time, it will wait for other RREQs. After a period of waiting time, it rebroadcasts RREQ with the least HETX value. Third, when a destination receives RREQ, it also waits for other RREQs. After a period of waiting time, the destination will send RREP along path with the least HETX value. The route discovery process of LQDV at intermediate node is shown as follows [1].

Route Request Packet Processing of LQDV at intermediate node B

```
When B receives RREQ[id, hetx] packet from A
If it is the first RREQ packet of path id then
    B creates a route entry such that RE[id, pHop:= A, srcHETX := hetx + HETX(A,B)]
    B updates RREQ as RREQ[id, hetx := hetx + HETX(A,B)]
    B rebroadcasts RREQ and set the timer ΔdelayRREQ to collect multiple RREQs
else if a RE[id, pHop, srcHETX] exists then
    if (hetx+HETX(A,B) <srcHETX) then
        B updates the route entry such that RE[id, pHop:=A, srcHETX:= hetx+HETX(A,B)]
        B updates and keeps RREQ: RREQ[id, hetx := hetx + HETX(A,B)]
          When ΔdelayRREQ becomes 0, B rebroadcasts RREQ
    else
          B drops the new RREQ packet
    end if
end if
```

3 LQDV Protocol Implementation

Our implementation was developed on Arduino platform and Xbee module as shown in Fig. 1. Arduino is an open-source platform used for building electronics projects. Arduino consists of both a physical programmable circuit board and Integrated Development Environment (IDE) [3].

Fig. 1. Implementation Arduino board.

The Arduino platform has become quite popular with people just starting out with electronics. The Arduino IDE uses a simplified version of C++, making it easier to learn to program. Arduino provides a standard form factor that breaks out the functions of the micro-controller into a more accessible package [3]. Xbee module is a wireless network interface running the IEEE 802.15.4 standards. Each node consists of Arduino Mega2560 board (ATmega2560-16AU Atmel 8-bit Microcontrollers – MCU 256 KB Flash Memory, 8 KB SRAM, 4 KB EEPROM, Clock Speed 16 MHz) and Xbee module. Specifications of Xbee module is shown in Table 1.

Table 1. Parameters of Xbee module.

Xbee Specifications	Value
RF Data Rate	250,000 bps
Media Access	IEEE 802.15.4
Receiver Sensitivity	-92 dBm (1% packet error rate)
Transmission Power	-10 dBm
Operating Frequency	ISM 2.4 GHz
Serial Interface Data Rate	1200 bps - 250 kbps
Number of Channels	16 Direct Sequence Channels
Addressing Options	PAN ID, Channel and Addresses

3.1 Software Layering Architecture

Fig.2 shows the network architecture. LQDV was implemented as sublayer between the network and MAC layer. LQDV maintains three tables. The routing table contains the information of routes to destinations. The PROBE table contains the information about its neighbors and link quality metric between it and these neighbors. The seen table contains information of route discovery, including source address and route ID. The seen table is used for determining whether the received RREQ is for the first time or not.

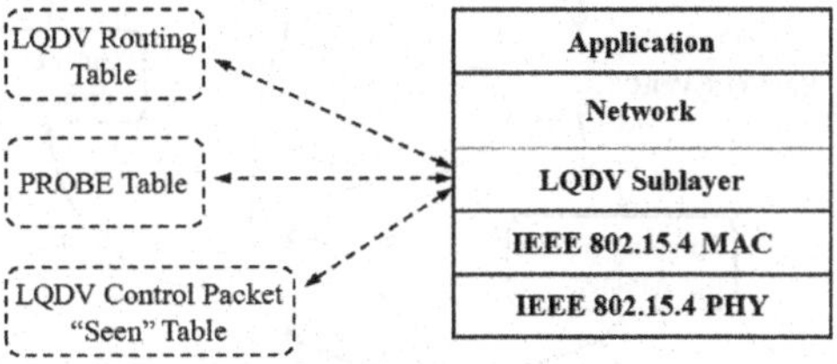

Fig. 2. Software layering architecture of our implementation.

LQDV control and data packets have the same packet header. The header is shown in Fig.3. The LQDV uses four type of control message: RREQ, RREP, RERR, and PROBE. The structure of each message is shown in Fig 4.

2 bytes	2 bytes	1 byte	1 byte	1 byte
Destination Address	Source Address	TTL	Type	Length

Fig. 3. LQDV protocol header.

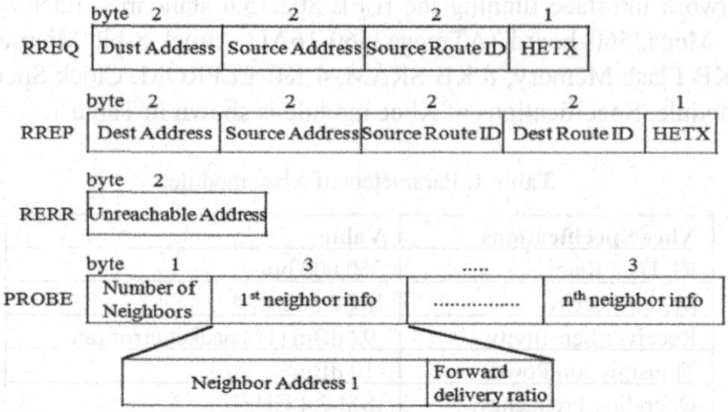

Fig. 4. Structure of control messages.

All messages is attached a layer 3 header before passing down layer 2.

3.2 Programming Model

Because microcontroller in Arduino has only one thread, it needs to check events in an infinitive loop. All events and structure of main program is depicted in Fig. 5.

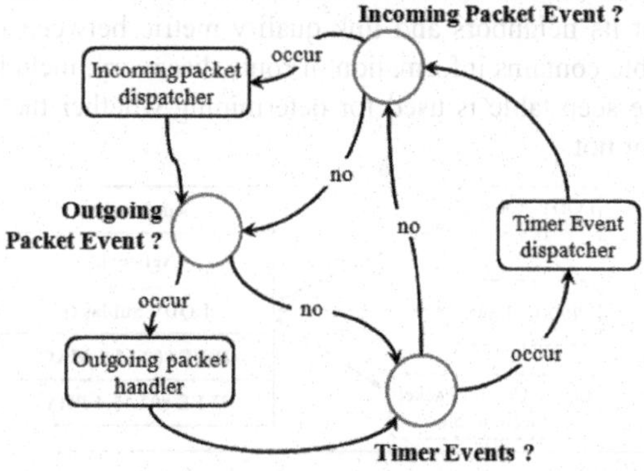

Fig. 5. Programming model of our implementation.

- **Incoming packet event:** This event is detected when a packet is available at serial port. If this event occurs, the function of incoming packet dispatcher will be called. The function performs a switch on the type of packet, contained in the IP header of the received packet, and calls the appropriate routine to handle the packet.
- **Outgoing packet event:** This event is detected when the network layer receives a requirement of sending a data packet from application. If this event occurs, the function of outgoing packet handler will be called. This function performs route discovery if a active route to destination is available on routing table, whereas it sends the data packet to Xbee module.

- **Time events:** Due to limitation of timer on hardware, soft timer is implemented by using timestamp. Each timer contains a timestamp. The timestamp is set to current time plus time of delay. The expiration of a timer is detected when timestamp is smaller than or equal to checking time. All timers will be checked and handled according to type of time event if occurs.

3.3 LQDV Routing Function Implementation

The processing steps of route discovery and data transmission in our LQDV protocol implementation is depicted in Fig. 6.

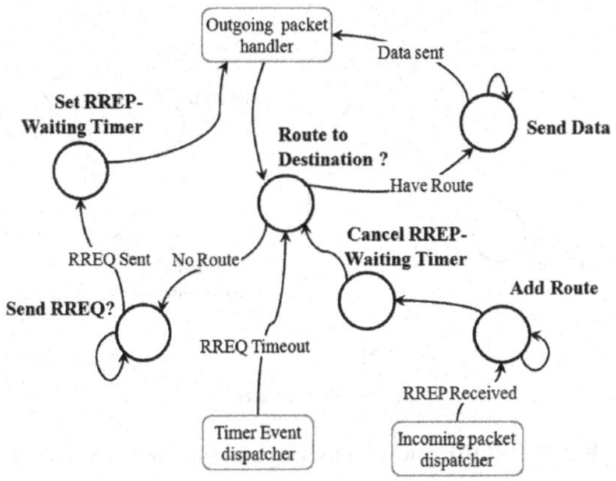

Fig. 6. LQDV route discovery and data transmission flow.

- **Outgoing Data Packet Handling and Route Discovery (RREQ/RREP)**

  When receiving a data packet from application layer, the node will check whether there is route to destination or not. If existed, the node will send data packet, whereas, the node perform route discovery. Right after sending RREQ, the node will set a RREP-waiting timer for waiting RREP. When the timer expires, if node has not received any RREP, rediscovery of route will be performed. If the node receives RREP before the RREP-waiting timer expires, the node will cancel the RREP-waiting timer and send data packet in buffer to Xbee module.

- **Implementing Packet Forwarding**
  When a data packet arrives at LQDV host, it is examined and a decision is made as to whether packet is passed up the application layer (if the host is indeed the destination) or forwarded. In the later case, a lookup in the LQDV routing table is performed to determine the next-hop address. And then, the packet is sent to Xbee module.
- **Implementing Link Quality Measurement**
  To measure quality of links between a node and its neighbors, each node periodically broadcast PROBE messages. A timer is set for this work. When the timer ex-

pires, a PROBE packet which contains the number of PROBE packet received from each neighbors (forward delivery ratio) is broadcasted and the node also reset the timer. When receiving a PROBE packet, a node handles this packet as depicted in Fig.7. The node first updates information of the neighbor which sent PROBE packet in PROBE table. And then it updates the PROBE timer. If the PROBE timer expires, this means the node has not received any PROBE packet for a period of time and therefore, the link between it and that neighbor is considered to break. The calculation of HETX values is carried out when a window timer event happens. As mentioned earlier, the time axis is divided into time windows. A window timer event occurs when a node moves from this window to next window. Each time this event happens, the node calculates HETX value and update to PROBE table.

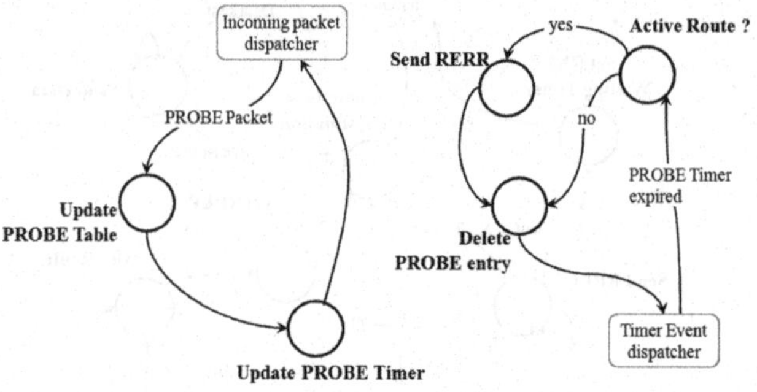

Fig. 7. PROBE packet handling and link failure detection.

4 Experimental Evaluation

We performed an outdoor experimentation in real environment to evaluate the performance of the protocol. Our experimentation was implemented on a static 8-node network of approximately 10m*30m. We set up two simultaneous flows. Xbee module is configured with lowest power level (-10 dBm), 16 bit address, non-beacon enabled mode. Each data packet consists of 13 bytes of 802.15.4 frame overhead, 100 bytes of data payload and 2 bytes of frame check sequence: 115 bytes in total.

Fig.8 shows the average of packet delivery ratio depending on the input data rate for each flow. The result shows that when traffic load increases, the packet delivery ratio is reduced a little bit slightly. But, after 16 packets per second, the packet delivery ratio has decreased sharply. That is because of limitation of Xbee module. That proves that this routing protocol is reliable in high traffic load network.

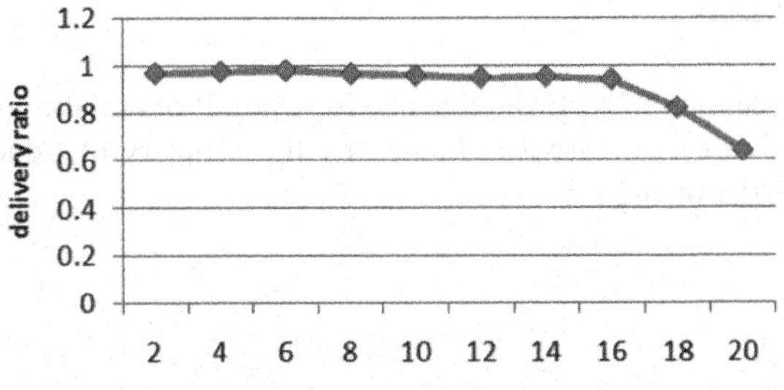

Fig. 8. Packet delivery ratio according to traffic rate.

The Fig.9 shows the average number of link failure times and the average of end-to-end HETX value. When traffic load increases, average HETX increases a little bit due to impact of traffic load. This implies that choosing route after link failure is correct and does not impact by route rediscovery. That is because of advantage of HETX metric. We can also see that, when traffic load increases, the number of link failures changes slightly. This means the path which is chosen after rediscovery is reliable.

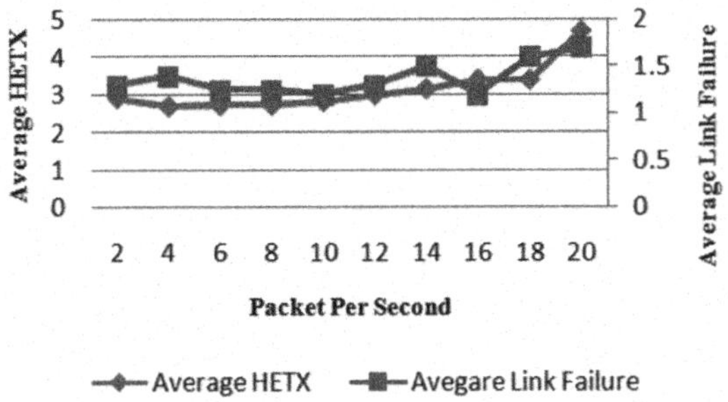

Fig. 9. Average HETX value and number of link failures according to traffic rate.

5 Conclusion

In this paper, we described the design for LQDV routing protocol implementation. The result shown that LQDV routing protocol can work well and be reliable in real environment. We also introduced a simple way using Arduino and Xbee module for routing protocol implementation. We hope that this way help researcher easily create implementations.

Acknowledgement

This research was supported by Basic Science Research Program through the National Research Foundation of Korea(NRF) funded by the Ministry of Education, Science and Technology(2010-0024245).

Reference

1. Dinh Duong Mai, Anh Tai Tran and Myung Kyun Kim, "Impact of route request packets to the stability of routes on static wireless networks", International Journal of Software Engineering and Its Applications Vol.8, No.6 (2014), pp.51-66.
2. C. E. Perkins and E. M. Royer, "Ad-hoc on-demand distance vector routing", Proceedings of IEEE WMCSA(1999).
3. http://www.arduino.cc/

A Virtualization and Management Architecture of Micro-Datacenter

Byungyeon Park[1], Wonhyuk Lee[1], TaeYeon Kim[2], HyunCheol Kim[3,*]

[1] Korea Institute of Science & Technology Information
Daejon, Korea, 305-806
{bypark,livezone}@kisti.re.kr

[2] Electronics and Telecommunications Research Institute
Daejon, Korea, 305-700
tykim@etri.re.kr

[3] Dept. of Computer Science, Namseoul University
Cheonan, Korea, 331-707
hckim@nsu.ac.kr

Abstract. With the cloud computing, storage and computing resources are moving to remote resources such as virtual servers and storage systems in large DCs(Data Centers), which raise many performance and management challenges. We consider distributed cloud systems, which deploy micro DC that are geographically distributed over a large number of locations in a wide-area network. In this article, we also argue for a micro DC-based network model that provides higher-level connectivity and logical network abstraction that are integral parts of wellness applications. We revisit our previously proposed logical network models that are used to configure the logical wellness network.

1 Introduction

Cloud computing is becoming a promising platform and have developed rapidly recently. In contrast to traditional enterprise IT solution, it enables enterprises to procure computing resources on demand basis and delegate management of all the resources to the cloud service provider. With emerging software defined networking (SDN) paradigm, it provides a new chance to fast service provisioning in the cloud with the network through standard programmable interfaces [1][2]. With the cloud computing, datacenters (DCs) promote on-demand provisioning of computing resources and services. Storage and computing resources (i.e., IT resources) are moving to remote resources such as virtual servers and storage systems in large datacenters, which raise many performance and management challenges [3][4].

With increasing interest in the concept of wearable computing and the popularization of hand-held devices, those movable devices are becoming start point of wellness information [5]. As shown in Fig. 1 and Fig. 2 we consider distributed cloud systems, which deploy micro DC that are geographically distributed over a large number of

* Corresponding author

K.J. Kim (ed.), *Information Science and Applications*,
Lecture Notes in Electrical Engineering 339, DOI 10.1007/978-3-662-46578-3_22

locations in a wide-area network. This distribution of micro DC over many locations in the network may be done for several reasons, such as to locate resources closer to mobile devices, to reduce bandwidth costs, to increase availability, etc [6][7]. Micro DCs are interconnected via high-speed optical wavelength division multiplexing (WDM) links. In this article, we also argue for a micro DC-based network model that provides higher-level connectivity and logical network abstraction that are integral parts of wellness applications. We revisit our previously proposed logical network models that are used to configure the logical wellness network.

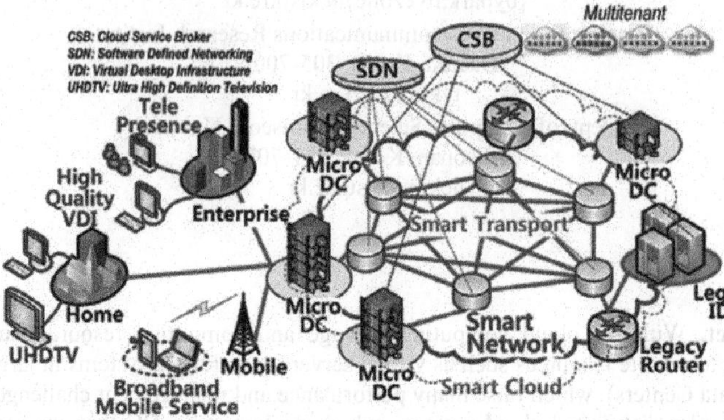

Fig. 1. Micro DC-based Smart Cloud Network Overall Architecture (Target Network)

2 Logical Network Management Schema

2.1 Topology Schema

As the topology class which expresses the configuration of logic network, it has trunk class as a subordinate class. Trunk class consists of two classes; Vlan class which contains logical information for management of Carrier Ethernet trunk and interface class which physical information. The UML diagram of topology schema, as shown in Fig. 2, it shows specific parameters which are used in the class. Each parameter has been defined using enterprise MIB information. The logic network has been configured based on the definition of PBB-TE technology.

2.2 Performance Monitoring Schema

The performance monitoring class has the following subordinate classes; Throughput, Frame Loss and Frame Delay. It consists of Frame Loss Class, Fame Loss Ratio and Fame Delay Variation. Fig. 10 shows a complete view of the performance monitoring schema. In fact, it has five classes to monitor logic network performance. Each class has been prepared based on the performance monitoring technology which is defined in ITU-T Y.1731. In terms of a UML diagram of the performance monitoring schema, specific parameters have been used. Each parameter has been defined using enterprises MIB information.

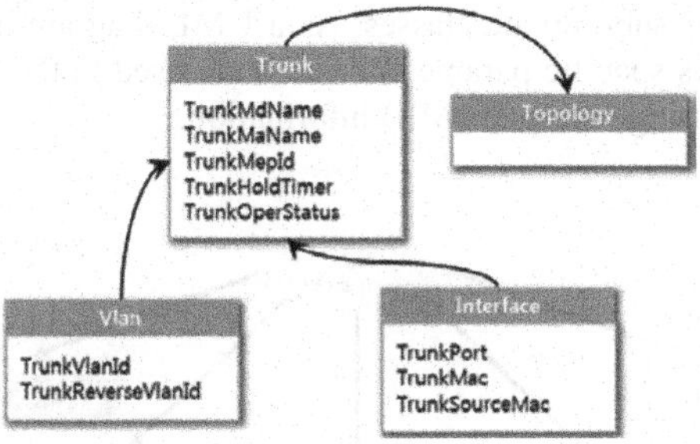

Fig. 2. Topology UML Schema

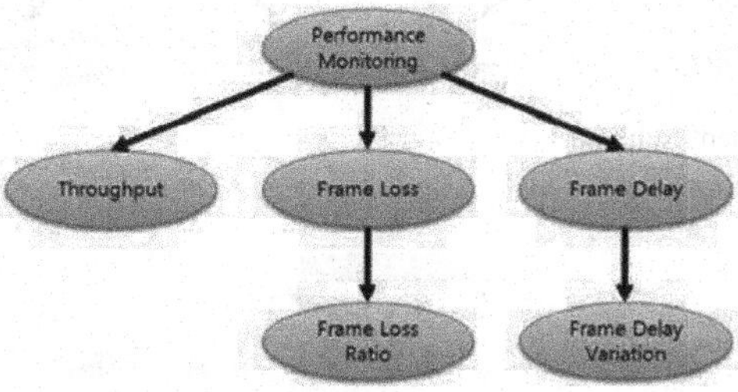

Fig. 3. Performance Monitoring Schema

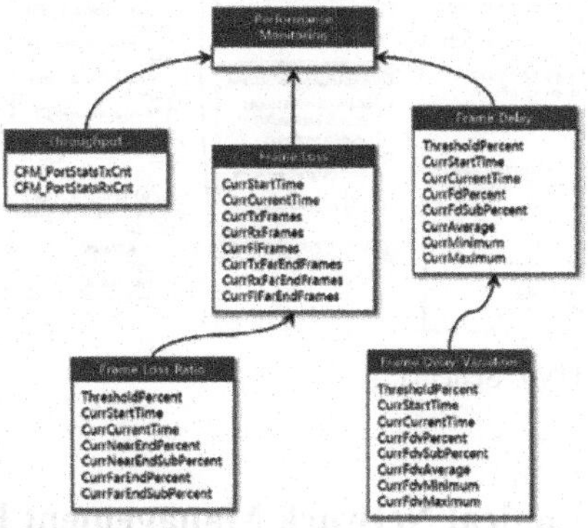

Fig. 4. Performance Monitoring UML Schema

2.3 Fault Management Schema

The fault management schema which is aimed to manage logic network faults has the following subordinate classes; Fault Verification, Power, Fault Notification, Fault Isolation and Fault Detection. Power Class includes Power Status Class which can analyze current power supply status and control the warning process and Warning Threshold Class. Fault Isolation Class has Linktrace Messages Class which can figure out network faults by analyzing LTM/LTR messages and Linktrace Relay Class. The Fig. 5 shows a complete view of the fault management schema. It consists of five super classes and four subordinate classes. As a UML diagram of the fault management schema, it shows specific parameters which are used in the class. Each parameter has been defined using enterprise MIB information.

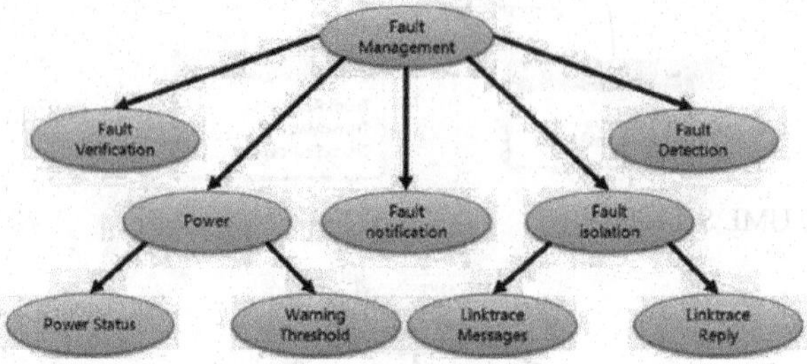

Fig. 5. Fault Management Schema

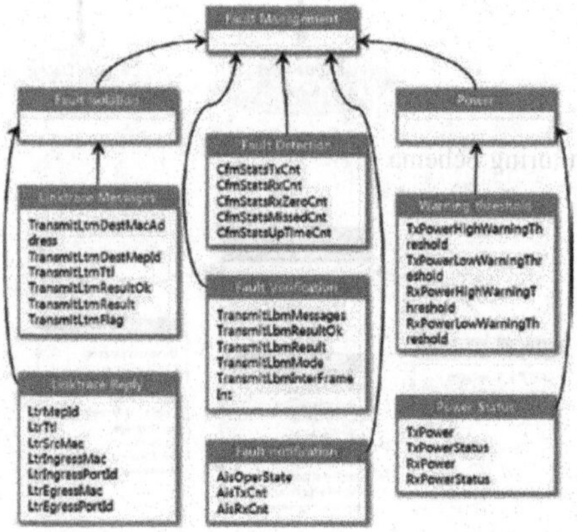

Fig. 6. Fault Management UML Schema

3 Development of Logical Network Management Framework

The logic network management framework established in this paper consists of three databases. It is comprised of DC_info, OOO_topo and OOO_inter databases. System information database table which should be created first after the establishment of framework is DC_info. It contains μ-DC configuration, performance, fault management table names as well as basic information on μ-DC equipment.

μ-DC Configuration Information Database is a database table in which logic network configuration information is stored. This table is automatically created when μ-datacenter equipment is added. It is automatically named in the form of 'OOO_topo.'

For example, Fig. 8 shows μ-DC equipment in Daejeon so that it is named 'DJN_topo.'

mysql> desc mers_info;

Field	Type	Null	Key	Default	Extra
ipAddr	varchar(15)	NO	PRI	NULL	
sysName	text	NO		NULL	
sysDescr	text	YES		NULL	
sysLocation	text	YES		NULL	
sysUpTime	varchar(20)	YES		NULL	
topoTable	varchar(10)	NO		NULL	
perfoTable	varchar(10)	NO		NULL	
faultTable	varchar(10)	NO		NULL	
interTable	varchar(10)	NO		NULL	

9 rows in set (0.00 sec)

Fig. 7. μ-DC Information Table

mysql> desc mers_info;

Field	Type	Null	Key	Default	Extra
ipAddr	varchar(15)	NO	PRI	NULL	
sysName	text	NO		NULL	
sysDescr	text	YES		NULL	
sysLocation	text	YES		NULL	
sysUpTime	varchar(20)	YES		NULL	
topoTable	varchar(10)	NO		NULL	
perfoTable	varchar(10)	NO		NULL	
faultTable	varchar(10)	NO		NULL	
interTable	varchar(10)	NO		NULL	

9 rows in set (0.00 sec)

Fig. 8. μ-DC Information Table

μ-DC Interface Information Database is a database table in which μ-DC interface information is stored. It is also automatically created when μ-DC equipment is added just like topology table. It is automatically named in the form of 'OOO_inter.' For example; Fig. 8 shows μ-DC equipment in Daejeon so that it is named 'DJN_inter.'

mysql> mysql> desc DJN_inter;

Field	Type	Null	Key	Default	Extra
ifIndex	varchar(4)	NO	PRI	NULL	
Descr	text	YES		NULL	
Mtu	int(11)	YES		NULL	
Speed	bigint(20)	YES		NULL	
PhysAddress	varchar(20)	YES		NULL	
OperStatus	varchar(10)	YES		NULL	
InOctets	bigint(20)	YES		NULL	
InUcastPkts	bigint(20)	YES		NULL	
InNUcastPkts	bigint(20)	YES		NULL	
InErrors	bigint(20)	YES		NULL	
OutOctets	bigint(20)	YES		NULL	
OutUcatPkts	bigint(20)	YES		NULL	
OutNUcastPkts	bigint(20)	YES		NULL	
OutErrors	bigint(20)	YES		NULL	

14 rows in set (0.01 sec)

Fig. 9. μ-DC Interface Table

μ-DC Configuration Information is the most important part in management framework. The logic network in which μ-DC equipment is configured across the nation can be precisely understood. Fig. 10 shows μ-DC in Seoul. It can be understood at a sight that it consists of μ-DC equipment and trunk in Daejeon and Gwangju. In addition, MAC address, Vlan ID and port number which are essential in configuring PBB-TE can be checked.

```
mysql> mysql> desc DJN_inter;
+----------------+--------------+------+-----+---------+-------+
| Field          | Type         | Null | Key | Default | Extra |
+----------------+--------------+------+-----+---------+-------+
| ifIndex        | varchar(4)   | NO   | PRI | NULL    |       |
| Descr          | text         | YES  |     | NULL    |       |
| Mtu            | int(11)      | YES  |     | NULL    |       |
| Speed          | bigint(20)   | YES  |     | NULL    |       |
| PhysAddress    | varchar(20)  | YES  |     | NULL    |       |
| OperStatus     | varchar(10)  | YES  |     | NULL    |       |
| InOctets       | bigint(20)   | YES  |     | NULL    |       |
| InUcastPkts    | bigint(20)   | YES  |     | NULL    |       |
| InNUcastPkts   | bigint(20)   | YES  |     | NULL    |       |
| InErrors       | bigint(20)   | YES  |     | NULL    |       |
| OutOctets      | bigint(20)   | YES  |     | NULL    |       |
| OutUcatPkts    | bigint(20)   | YES  |     | NULL    |       |
| OutNUcastPkts  | bigint(20)   | YES  |     | NULL    |       |
| OutErrors      | bigint(20)   | YES  |     | NULL    |       |
+----------------+--------------+------+-----+---------+-------+
14 rows in set (0.01 sec)
```

Fig. 10. μ-DC Topology Information

Fig. 11 reveals the interface information of μ-DC equipment. As L2 switch equipment, μ-DC equipment has several interfaces. It includes MAC address per interface, MTU size, current operating status and port number which is essential in configuring a logic network.

MERS-SEL Interface Information

ifIndex	Descr	Mtu	Speed	PhysAddress	OperStatus	InOctets	InUca
64	1000Gbic850Sx Port 1/1 Name	1950	1000000000	0:24:43:97:70:0	up	22119063	2
65	1000Gbic850Sx Port 1/2 Name	1950	1000000000	0:24:43:97:70:1	up	6424	
66	1000Gbic Port 1/3 Name	1950	1000000000	0:24:43:97:70:2	up	942171299	3815
67	1000Gbic Port 1/4 Name	1950	1000000000	0:24:43:97:70:3	up	2524115896	749
68	1000Gbic Port 1/5 Name	1950	0	0:24:43:97:70:4	down	0	
69	1000Gbic Port 1/6 Name	1950	0	0:24:43:97:70:5	down	0	
70	1000Gbic Port 1/7 Name	1950	0	0:24:43:97:70:6	down	0	
71	1000Gbic Port 1/8 Name	1950	0	0:24:43:97:70:7	down	0	
72	1000Gbic Port 1/9 Name	1950	0	0:24:43:97:70:8	down	0	
73	1000Gbic Port 1/10 Name	1950	0	0:24:43:97:70:9	down	0	

Fig. 11. μ-DC Interface Information

4 Conclusions

This paper has investigated a framework in which a logic network is described and managed by particular application based on science & technology research network resource specification. As a result, a schema through which topology, performance and fault information can be systematically managed in accordance with international standards has been completed. In addition, database has been created based on the schema which has been designed in accordance with international standards, and a network management framework through which a logic network can be managed has been built. Then, information has been brought from the current Carrier Ethernet equipment and provided to an administrator.

However, a further study needs to be conducted on the construction of a management framework which can reveal topology, performance and fault information in a more dynamic manner using the collected information. Even though the current management framework shows the configuration of a logic network in a static manner using a table, the network could be operated more effectively once a management framework just like a weather map is built.

Acknowledgement This work was supported by the ICT R&D program of MSIP/IITP, Republic of Korea. [14-000-05-001, Smart Networking Core Technology Development]

References

1. Banikazemi M.; Olshefski D., Shaikh A., Tracey, J., Guohui Wang, "Meridian: an SDN platform for cloud network services," Communications Magazine, IEEE, Vol. 51, Issue 2, pp. 120–127, 2013.
2. Alicherry M., Lakshman T.V., "Network aware resource allocation in distributed clouds," INFOCOM, pp. 963-971, 2012.
3. Mon-Yen Luo, Jun-Yi Chen, "Software Defined Networking across Distributed Datacenters over Cloud," International Conference on Cloud Computing Technology and Science (CloudCom), pp. 615-622, 2013.
4. M. F. Bari, R. Boutaba, R. Esteves, L. Z. Granville, M. Podlesny, M. G. Rabbani, Z. Qi, and M. F. Zhani, "Data Center Network Virtualization: A Survey," Communications Surveys & Tutorials, IEEE, Vol. 15, pp. 909-928, 2013.
5. Sunyoung Kang, Hyuncheol Kim, Seungae Kang, "Virtual private network for wellness sports information," Multimedia Tools and Applications, Online First Article, Springer, 2014.
6. M.-Y. Luo and J.-Y. Chen, "Towards Network Virtualization Management for Federated Cloud Systems," International Conference on Cloud Computing, IEEE, 2013.
7. Khethavath P., Thomas J., Chan-Tin E., Hong Liu, "Introducing a Distributed Cloud Architecture with Efficient Resource Discovery and Optimal Resource Allocation", IEEE, Services (SERVICES), pp. 386-392, 2013.
8. Shicong Meng, Ling Liu, Ting Wang, "State Monitoring in Cloud Datacenters", Transactions on Knowledge and Data Engineering, IEEE, Vol. 23, Issue 9, pp. 1328-1344, 2011.
9. Hassan, Mohammad Mehedi, Song, Biao, Hossain, M.Shamim, Alamri, Atif, "QoS-aware Resource Provisioning for Big Data Processing in Cloud Computing Environment," International Conference on Computational Science and Computational Intelligence(CSCI), pp. 107–112, 2014.
10. Haiying Shen, Guoxin Liu, "An Efficient and Trustworthy Resource Sharing Platform for Collaborative Cloud Computing," Transactions on Parallel and Distributed Systems, IEEE, Vol. 25, Issue 4, pp. 862–875, 2014.

Multi-disjoint Routes Mechanism for Source Routing in Mobile Ad-hoc Networks

Baidaa Hamza Khudayer[*], Mohammed M. Kadhum[‡] and Wan Tat Chee[†]

[*]Department of Information Technology, Al Buraimi University College, Oman
[*‡†]National Advanced IPv6 Center (NAv6), University Sains Malaysia, Pulau Penang, Malaysia
[‡]School of Computing, Queens's University, Canada

baidaa@buc.edu.om, kadhum@nav6.usm.my, kadhum@cs.queensu.ca, tcwan@cs.usm.my

Abstract— The capabilities of Mobile Ad hoc Networks (MANETs) pave the way for wide range of potential applications personal, civilian, andmilitary environments as well as emergency operations.The characteristics of MANETS impose serious challenges in deploying these networks efficiently. The limitation of transmission range of mobile nodes emerge the use of multi-hop to exchange and route data between nodes across the network. Routing is the most important process that can affect the overall performance in MANETs. A variety of routing protocols have been developed for MANETs. These protocols present poor performance under different conditions due to their strategies in learning about routes to destinations. In this paper, we present an innovative Multi-disjoint Routes mechanism that is based on source routing concept to improve routing in MANTEs. The simulation evaluation results showed that the proposed routing mechanism outperforms the common routing protocols, DSR and AODV.

Index Terms—MANET; Multipath Routing; Source Routing; DSR; AODV.

1 INTRODUCTION

Mobile Ad-hoc Networks (MANET) can be deployed in areas where the network communication infrastructure is not available due to a natural disaster such as floods, earthquakes, hurricanes, and war [6]. MANET should be useful for military troops (Fig.1) where there is a need to establish a network of mobile nodes in a hostile area where wired networking is impossible or not sufficient [3-5]. MANET is also beneficial in many place or activities. For example, MANETs can be set up in congress meetings where the attendees want to share information; or in a shopping complex where shoppers can exchange data including fashion, prices, and advertisements; or in an airport where users can be updated on the departures/arrivals schedule and the weather conditions, and so

K.J. Kim (ed.), *Information Science and Applications*,
Lecture Notes in Electrical Engineering 339, DOI 10.1007/978-3-662-46578-3_23

on. MANET is also important in domains such as demotic where several robots can communicate with each other to perform different tasks [3] [6].

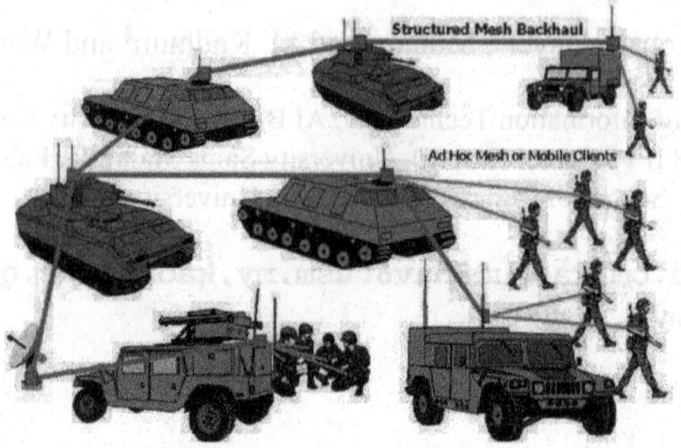

Fig. 1. Example of a Mobile Ad-hoc Network in a military action

The roles of nodes in MANETs include the discovery and maintenance of routes for delivering data to intended destination(s). The nature of nodes in MANETs - in that the deviate randomly and unpredictably in both speed and direction - poses serious challenges in routing data while managing mobility. Nodes mobility has significant negative impact on routing protocols as it causes frequent topology changes that a require dynamic and efficient response to such changes in order to maintain routes between source and destination. Therefore, the performance of MANETs depends mainly on the efficiency of the routing protocol and how it reacts to the sudden changes in the network topology [7] [8].

In dynamic environments such as MANETs, the overall network capacity and performance can be increased by utilizing "multihopping" concept, which allows exchanging data among mobile nodes across the network. Consequently, a mobile node can convey data on behalf of a sender node to the desired destination. Thus, in addition to that nodes in MANETS have can be a sender or receiver, they also can be a data forwarder. This means that destination can receive data from source that even if the source node is not within the transmission range of the destination. While packets will travel across several wireless nodes to reach the destination, this can help saving energy as a mobile node needs to reach the destination by sending data to next hop only. However, for effective network, nodes that contribute to the forwarding progression should be distributed well in the forwarding area. As stated by Frodigh et al. [15], the multihopping mechanism help conserving transmit energy resources, reducing interferences, and increasing the overall network throughput.

The aim of the research work presented in this paper is to improve routing in MANETs. Therefore, we propose an effective multipath routing mechanism that is based on source routing concept to establish multi-disjoint routes for data delivery. The rest of the paper is organized as follows; Section 2 presents the most common multihop routing protocols used for MANETs, namely Dynamic Source Routing (DSR) and Ad Hoc On-Demand Distance Vector (AODV). The proposed multi-disjoint routes mechanism is described in Section 3.Section 4 shows the scenario operation of the proposed routing mechanism. The performance evaluation is presented then in Section 5. The results are discussed in Section 6. Finally, conclusions and future work are presented in Section 7.

2 COMMON MULTIHOP ROUTING PROTOCOLS IN MANETS

In this section, the most common in practice multihop routing protocols developed for MANETs are presented.

2.1 Dynamic Source Routing (DSR)

DSR [1] [2] is an on-demand routing protocol developed to limit the bandwidth consumption used for periodically updating routing tables in the table-driven routing. DSR does not require neighbor detection where periodic hello messages are sent by a node to inform neighbors of its presence. As a source routing protocol, DSR injects the complete path information (list of nodes) in the header of every packet designated to a particular destination. Thus, the intermediate nodes do not waste resources for routing decisions by maintaining up-to-date routing information to forward data.

DSR uses two strategies [10]; Route Discovery and Route Maintenance. Each of them operates entirely when there is a demand for a route. Route Discovery is the strategy that a source node S uses to obtain a route to a particular destination node D. This strategy is performed only when node S has a data packet need to send to node D and that node S does not have a route to node D in its cache. As illustrated in Fig. 2, Route Discovery is performed by the source node S where it broadcasts a Route Request (RREQ) to all nodes within its transmission range. Similarly, this RREQ is then propagated (flooded) through the entire MANET.

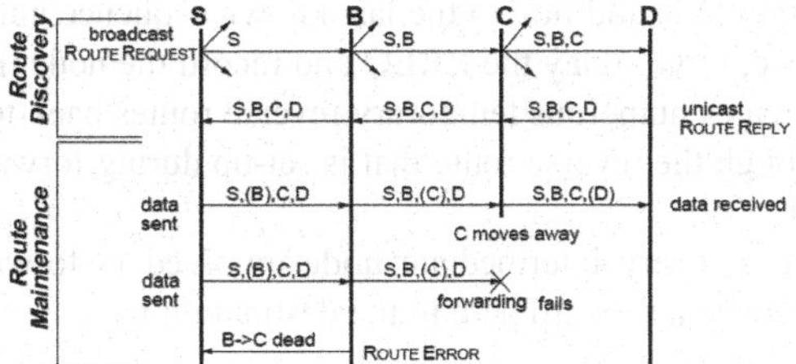

Fig. 2. Route discovery and route maintenance in DSR

In response to the RREQ, Route Reply (RREP) will be unicastly send back to S when the destination node is found [11]. RREP is sent either by the destination itself or any other node that has or knows the way to the destination. Routing information carried by RREQ is used reversely in sending the RREP packet. The cost of the RREQ propagation during the route discovery process is reduced by having each node stores the learned or heard sources routes in its cache.

Route Maintenance is required when node S detects that the route to the destination D is no longer active (valid) to transfer data due to link breakage which might be caused by node mobility (node C went out of the range of node B). In this case, node B notifies node S via a Route Error (RERR) packet [12]. Node S might then utilize any other available route to node D from its cache; otherwise, node S will initiate another Route Discovery to find a new route to node D. Route maintenance is used only when node S needs to send a data packet to node D. DSR does not require periodic messages of any kind at any level.

DSR supports unidirectional and asymmetric routes, which means that node S sends a packet to node D through a route and receives a packet D from another route.

Considerable routing overhead is involved due to the source-routing mechanism employed in DSR [12]. This routing overhead is directly proportional to the path length. The performance of DSR is not stable when the network size increases with respect to mobility, pause time, and transmission rate of nodes [16].

2.2 Ad Hoc On-Demand Distance Vector (AODV)

AODV [9] keeps the desirable features of DSR in that routes should be maintained only among mobile nodes that need to exchange data. AODV uses hop-by-hop routing, sequence numbers, and periodic hello messages in addition to the basic on-demand strategy of Route Discovery and Route Maintenance used by DSR. In AODV, nodes have to remember only the next hop towards the destination rather than remembering the entire route, as in DSR.

As illustrated in Fig. 3, when the source node S is willing to send data to a specific destination D, it looks for a route by broadcasting RREQ to all of its neighbors within its transmission range. Each neighbor receives the RREQ checks on whether it has a route to the destination D. if so, it will send RREP that includes the number of hops towards the destination D in addition to the last known sequence number for D stored in its cache; otherwise, it will relay the RREQ and record the node that it received the RREQ from. This creates numerous temporary reverse routes back to the source node S. RREP travels through the reverse route that is set-up during forwarding the RREQ, as depicted in Fig. 4.

As shown in Fig. 4, every intermediate node involved in forwarding the RREP back to node S will create a forward path to the destination D.

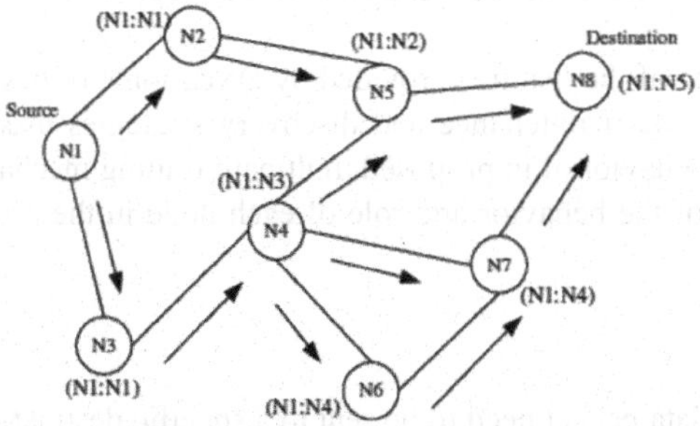

Fig. 3. RREQ Propagation during route discovery

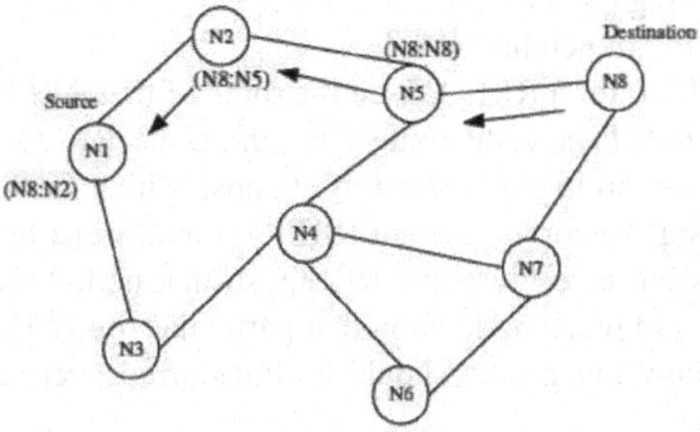

Fig. 4. Path of the RREP to the source node S

In AODV, for a node to indicate its existence, it periodically broadcasts a HELLO message to neighbors within its transmission range only. If no HELLO message is received from the next hop involved in an active route within a certain period, it is assumed the next hop is no longer exists. This basic strategy helps in detecting link breakage. A RERR is send back to the source node S when link breakage is detected, and then a new route discovery process will be initiated.Although AODV protocol is considered simple and doesn't require much computation process and has less traffic for communication over existing links, compared to DSR, however, intermediate nodes bring about inconsistent routes and make stale entries if their cache has old destination sequence number and source sequence number. Also, heavy control overhead can be caused by several RREPs in response to one RREQ. Moreover, periodic hello messages used in AODV consume more bandwidth than necessary.

It is important to mention that the aforementioned common protocols suffer from high latency in finding a route to destination and their excessive RREQ propagation (flooding) strategy can result in network clogging.

3 MULTI-DISJOINT ROUTES MECHANISM

To take advantage of the features provided by DSR, some of desirable features and procedures of the route maintenance and discovery strategies used by DSR are utilized for the nodes behavior of in proposed multipath routing mechanism.

The description of the behavior and role of each node in the proposed mechanism is as follows:

Source Node S

When there is a data packet need to be sent to a specific destination node D, node S will:

- Generate a RREQ packet with single ID, the address of the node, and the timestamp (time of sending the RREQ).
- Broadcasts the RREQ to neighbors.
- Once received a RREP for a RREQ, used the route of that RREP to send data.
- If more than one RREP (several routes) for the same RREQ, distribute the data over these routes according to their timestamps, which is RTT of sending the RREQ and receiving the corresponding RREP – the destination node D has to update on the time when it received the RREQ, so that node S is aware of the time that its RREQ takes to reach node D on that particular route. This would help node S in deciding on how much data should be transferred over each of the multiple routes

Receiver Node M

Node M can be any node in the network. M is allowed to send or forward more than one RREP for one RREQ. Once it receives a RREQ packet, it acts as follows:

- If M is the destination node, send a RREP packet back to S.
- If the destination node is within the transmission range of node M, then send the received RREQ to the destination only (to prevent the unnecessary RREQ propagation through the entire network, which consequently reduces the overhead of traffic load).
- Otherwise, add M address in the RREQ's route table and broadcast the updated RREQ.

Destination Node D

Destination node D decides which routes are going to be used for data delivery between node S and node D. Initially, the number of routes should be defined. To improve the overall network performance, four routes number is considered the best choice. Thus, the destination node D is configured to send maximum of four RREP to only four broadcasts of the same RREQ. Therefore, node D should assign four different numbers for every RREP. The cache of node D then will include all accepted disjoint routes to source node S. Destination node D acts as follows:

- Once a RREQ packet arrives, examine the route cache for a route back to source node S.
- If a route found, use it to send back RREP packet.
- Otherwise, D believes that the RREQ arrives through the best and the shortest route; it then stores this route and refers to it to perform the following decisions for the next 400msec when another RREQ arrives (as the ideal TTL for RREQ when it first sent is 500msec, within which, if no RREP received, that RREQ will be re-transmitted):

1. Number each route based on their arrivals.
2. Make the first numbered route (route #1) as a reference route.
3. Compare route #1with the next in the list routes.
4. Removing of the route that has one or several common nodes with route #1.
5. Update the routes list number.
6. Make route #2 as the reference and repeat the processes 2 to 6 until all routes become node disjoint routes and so on.

- For each of the disjoint established routes, record it in the cache and send RREP packet to source node S unicastly.

For route maintenance in the presence of link failure during forwarding a packet, when a receiver (intermediate) node M notices that the next hop on the route to destination node D is unreachable, it verify its cache on whether there is an alternative route to node D. If a route is found, then node M will use that route to deliver the data and update its cache. Otherwise, it drops the packet and notifies the source node S with RERR packet. Node M is not required to initiate route discovery on behalf of source node S.

4 SCENARIO OPERATION OF THE PROPOSED ROUTING MECHANISM

In this section, we show how the proposed multi-disjoint routes mechanism works. Assume mobile nodes are connected and communicate with each other according to the scenario shown in Fig. 5. Source node S has data to send to a specific destination D. Node S then initiates a RREQ and broadcasts it to neighbor nodes (node C, node B, and node A). These nodes then will rebroadcast to their neighbors and so on, until reaches the destination node D, as depicted in Fig. 6.

Fig. 5. Scenario of a MANET operation

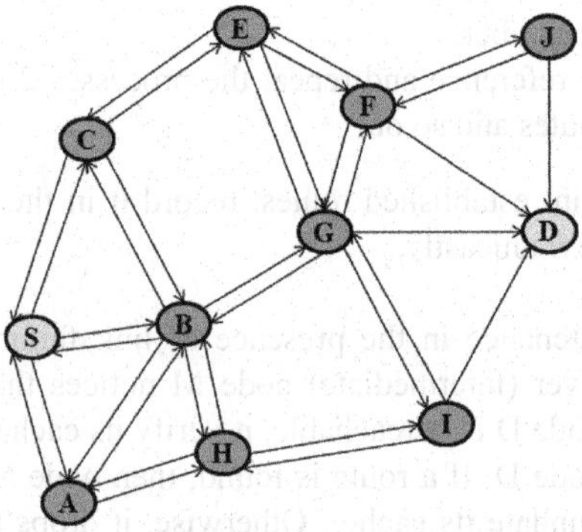

Fig. 6. RREQ propagation in the network

Assume that the first RREQ from source node S is arrived at destination node D through the rout {S-B-G-D}. Node D will consider this route as the best and the shortest route and will send back a RREP packet to node S immediately. Node D then will make this route as a reference route and number it "route#1".

Assume that within the next 400msec multiple copies of the same RREQ arrive through the following routes:

route#2 = {S-C-E-G-D}
route#3 = {S-B-G-F-D}
route#4 = {S-B-H-I-D}
route#5 = {S-A-H-I-D}
route#6 = {S-C-E-F-J-D}
route#7 = {S-C-B-G-D}

route#8 = {S-A-H-I-G-D}
route#9 = {S-B-G-I-D}
route#10 = {S-A-B-G-D}
route#11 = {S-C-E-F-I-D}

Node D will then compare route#1 with all other routes to eliminate the routes with common nodes.

Reference → route#1 = {S-B-G-D}

route#2 = {S-C-E-G-D}	remove
route#3 = {S-B-G-F-D}	remove
route#4 = {S-B-H-I-D}	remove
route#5 = {S-A-H-I-D}	keep and update #
route#6 = {S-C-E-F-J-D}	keep and update #
route#7 = {S-C-B-G-D}	remove
route#8 = {S-A-H-I-G-D}	remove
route#9 = {S-B-G-I-D}	remove
route#10 = {S-A-B-G-D}	remove
route#11= {S-C-E-F-I-D}	remove

Consequently, node D will cache the resulting three disjoint routes - route#1 = {S-B-G-D}, route#5 = {S-A-H-I-D}, and route#6 = {S-C-E-F-J-D} - and then unicastly send RREP to source node S using each of the resulting routes, as illustrated in Fig. 7.

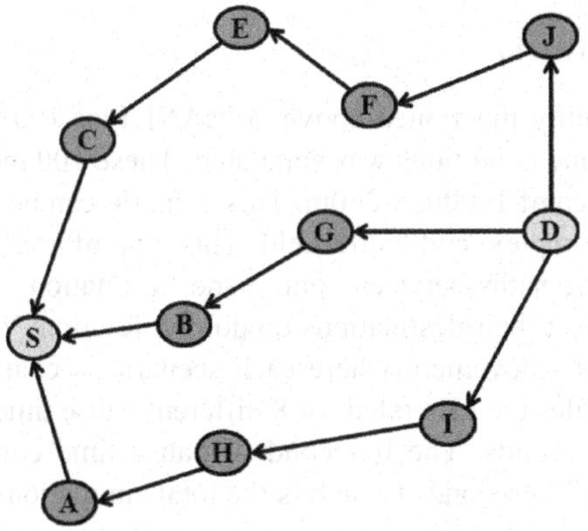

Fig. 7. Disjoint routes of RREP back to node S

Source node S will then use the learnt routes to send its data as shown in Fig. 8 The data distribution over these routes is done based on delay time of each route, regardless the number of hops contained in each route.

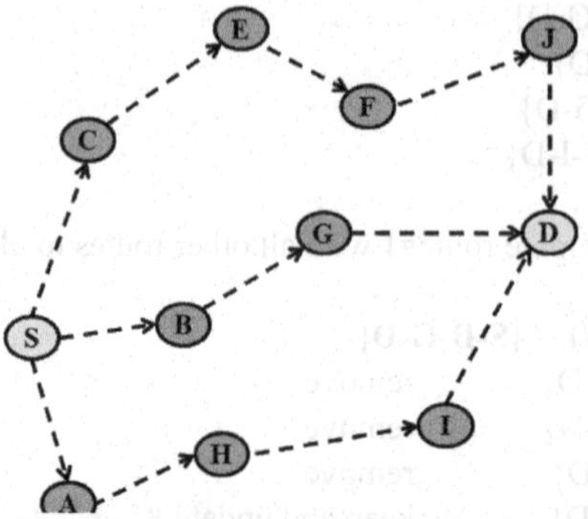

Fig. 8. Data transfer over the learnt disjoint routes

5 PERFORMANCE EVALUATION

The goal of the performance evaluation is to examine the ability of the proposed routing mechanism to improve the performance of MANETs by establishing reliable routes that guarantee successful delivery of data packets. The performance then compared to the common routing protocols DSR and AODV. The evaluation is done according to the most common scenarios and settings published in the Recent Trends in Networks and Communications [14].

5.1 Evaluation Scenario

To examine the ability mentioned above, a MANET of 100 mobile nodes with a variety of workloads and conditions was simulated. These 100 mobile nodes are moving in a rectangular area of 1000m x 250m. This scenario can be potentially happened in airport, shopping complex, and battle field. This type of area is chosen to so as to impose the use of long paths between source and destination. The simulation starts with 30 nodes that select their destinations randomly. Random Waypoint (RWP) [13] model is used for node movement where each scenario is characterized by a pause time. The movement files are generated for 8 different pause times: 0, 20, 40, 80, 160, 320, 640, and 1280 seconds. The 0 second of pause time corresponds to nonstop movement while the 1280 seconds (which is the total simulation time) corresponds to stationary mobile node with no movement. As the performance can be very sensitive to node mobility, the movement scenario files are generated with 80 different mobility patterns (10 scenarios for each of the pause time settings). The nodes move with a speed that is distributeduniformly between 1 m/sec and 8m/sec. However, this paper report the numerical results from simulations based on the average speed of 4 m/sec. The data traffic type generated by sources to is constant bit rate (CBR) with sending rates of 8 packets/sec and packet size of 512 bytes. The transmission range of each mobile node is 250m. The research work presented in this paper does not consider congestion or interference that might encountered by a packet travelling over an active path.

5.2 Evaluation Metrics

In studying the performance of the proposed routing mechanism, and comparing to DSR and AODV protocols, the following three evaluation metrics were chosen:

- Packet Delivery Ratio (PDR):
  PDR is the ratio of the number of packets sent by a source node S to the number of packets received by its intended destination D. it helps in measuring the completeness and correctness of our proposed routing mechanism in regards to packet loss and throughput desired.

- Normalized Routing Load (NLR):
  NLR is the total number of routing packets transmitted for a packet successfully delivered to the destination node D. Each transmission of the packet by a node on the route to node D is counted as one transmission. It helps in examine the efficiency of our proposed routing mechanism and how well it works in dynamic environment like MANETs.

- Route Optimality (RO):
  RO is the difference between the lengths of the shortest route exists to the destination node D and the route that a packet actually passed through towards its destination. It examines the ability of our proposed routing mechanism to efficiently select the shortest path when the network topology changes.

6 Results and Discussion

This section presents the performance evaluation of the proposed routing mechanism based on the numerical results obtained from the simulations. The performance is measured and compared to DSR and AODV according to the aforementioned performance evaluation metrics.

6.1 Packet Delivery Ratio (PDR)

Here, we discuss the results of the simulations and highlight the relative performance of the proposed routing mechanism, DSR and AODV protocols in terms of PDR. Fig. 9 provides an assessment of how these routing protocols are functioning under different mobility patterns.

From the observed PDR shown in Fig. 9, it is noticeable that all of the three routing methods are able to deliver a large amount of data packets at large pause time (when the node mobility is low). Regardless of the pause time rate, the proposed routing mechanism managed to deliver over 88% of the data packets, outperforming DSR and AODV. When the pause time rate is high (higher mobility), AODV performs somewhat poorly compared to DSR, delivering about 77% of the data packets. The rest of data packets are dropped due to stale entries in nodes routing tables, resulting from broken routes. DSR also fails to reach higher percentages of PDR compared to the proposed routing mechanism under higher node mobility. Neither DSR nor AODV are capable of reacting adaptively to topology changes for higher node pause time.

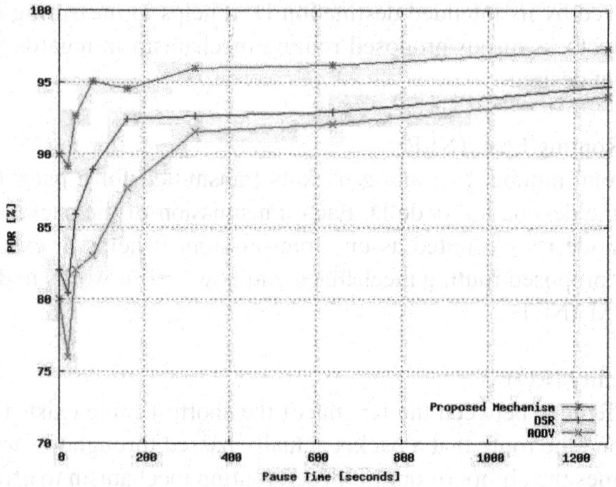

Fig. 9. PDR observed for the proposed routing mechanism compared to DSR and AODV

6.2 Normalized Routing Load (NRL)

The routing overhead curves shown in Fig 10 verified the character these routing methods. From Fig. 10, it is clear that DSR and AODV protocols routing methods enforce massively different amounts of overhead compared to the proposed routing mechanism, which has the least overhead.

For all of the routing methods, the NRL is decreasing with higher node mobility, and it isincreasing as the mobility increases. The proposed routing mechanism offers lower overhead due to its stable multiple routes provided for delivering data. AODV requires more than 10 times the overhead of the proposed mechanism when the pause time is 0 compared to that of DSR, about 2 times of the proposed routing mechanism. This is because of that DSR can limit routing overhead of RREQ packets by using non-propagating RREQ technique and storing forwarded and overheard packets.

AODV presents high routing overhead because of its periodic updates and unnecessary traffic generation in response to route requests, and also because of its route discovery process that broadcast the RREQs to every node in the network. Due to the nature of on-demand routing in DSR and AODV, when the number of nodes sending data increases, there will be more required routes to the corresponding destinations, which need to be maintained, and therefore the routing overhead will increase accordingly. However, as the network continues functioning, NRL for both DSR and AODV drops significantly because they can use the routing information learnt from previous route discoveries.

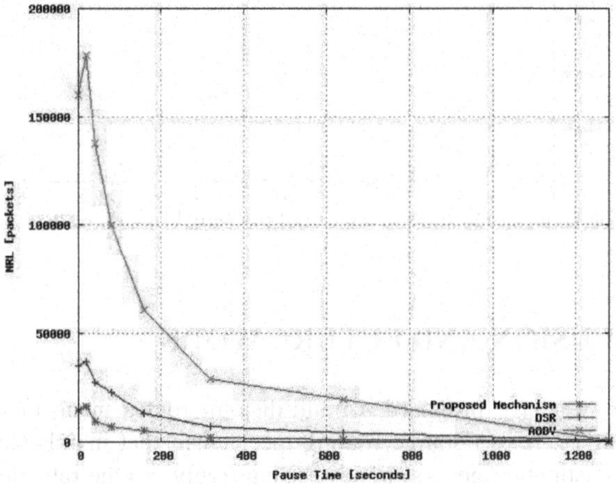

Fig. 10. Comparison of NRL wrt pause time.

6.3 Route Optimality (RO)

Fig. 11 shows the difference between the length of the shortest route available at the nodes (intermediate or destination) and the length of route that a packet used to reach an intermediate node or its destination. The lower difference points out that the packet used a shortest route, whilehigher difference signifies that the packets travelled over route with more nodes (hops).

Both of the proposed routing mechanism and DSR use close-to-optimal routes, while AODV has a noticeable long tail. This means that AODV takes up to 5 more nodes than the shortest route for some packets. With respect to node mobility rate and regardless of the pause time, the proposed routing mechanism does well and show very small difference regarding the optimality of routing compared to DSR and AODV. DSR still performs better than AODV in finding the close-to-optimal routes. AODV has a major difference in the length of the routes it uses compared with the shortest possible routes available. AODV uses routes that are so close to the shortest routes when pause time is very low compared to higher pause time value.

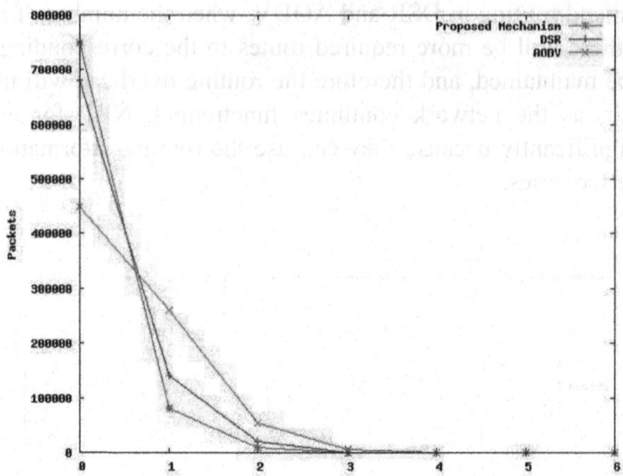

Fig. 11. Difference between the shortest route and the actual route a packet used to reach the destination

7 CONCLUSION AND FUTURE WORK

In this paper, we have proposed a multi-disjoint routes mechanism to improve routing in MANETs. It is a source routing mechanism that mainly depends on the behavior of the destination node(s) in advertising (replying) the reliable and efficient routes for the source nodes to use to send their data packets. Upon the receiving of a route request (RREQ), the destination node(s) will learn and analyze the routes through which RREQs have arrived, and then send reply (RREP) for the disjoint routes only. In turn, the source node will use these routes to send its application's data via multiple routes. This would guarantee the delivery of data and ensure higher throughput. In addition, nodes have some intelligence that would help them recognizing the destination node when it iswithin their transmission range. If so, then these nodes will unicast RREQs to destination only, preventing RREQs from propagation in the entire networks that would consequently increase the traffic and overhead and waste network resources. The performance of the proposed routing mechanism was evaluated using ns-2 simulator. Simulation results showed that the proposed mechanism outperforms the common routing protocols, DSR and AODV. For different conditions and scenarios, it offers greater PDR, which means that it is provide high throughput, and less packet drops. In addition to the route optimality, NRL results showed that the proposed mechanism has the lowest overhead compared to DSR and AODV, which is credited to it efficiency and reliability in selecting the best shortest routes. For further research, we plan to develop a prediction algorithm to detect link failure and topology changes to improve the performance and utilization of resources in MANETs.

8 REFERENCES

[1] David B. Johnson and David A. Maltz. Dynamic source routing in ad hoc wireless networks. In Mobile Computing, edited by Tomasz Imielinski and Hank Korth, chapter 5, pages 153–181. Kluwer Academic Publishers, 1996.

[2] David Johnson, David Maltz, Yih-Chun Hu: The Dynamic Source Routing Protocol for Mobile Ad Hoc Networks for IPv4, RFC 4728.

[3] Dalton, N.S.; Dalton, R.C., "The Theory of Natural Movement and its Application to the Simulation of Mobile Ad Hoc Networks (MANET)," Fifth Annual Conference on Communication Networks and Services Research (CNSR '07), vol., no., pp.359-363, 14-17 May 2007.

[4] Fischer, J.; Grossmann, M.; Felber, W.; Landmann, M.; Heuberger, A., "A measurement-based path loss model for wireless links in mobile ad-hoc networks (MANET) operating in the VHF and UHF band," IEEE-APS Topical Conference on Antennas and Propagation in Wireless Communications (APWC), vol., no., pp.349-352, 2-7 Sept. 2012.

[5] Ding Junxia, "Simulation and Evaluation of the Performance of FSR Routing Protocols Based on Group Mobility Model in Mobile Ad Hoc," International Conference on Computational Intelligence and Software Engineering (CiSE), vol., no., pp.1-4, 10-12 Dec. 2010.

[6] Soundararajan, S.; Bhuvaneswaran, R.S., "Multipath load balancing & rate based congestion control for mobile ad hoc networks (MANET)," Second International Conference on Digital Information and Communication Technology and it's Applications (DICTAP), vol., no., pp.30-35, 16-18 May 2012.

[7] William Stallings, 2010. Data and Computer communications, 9th Edition, Prentice Hall PTR (ISBNISBN-10: 0131392050).

[8] Wong, K.D.; Kwon, T. J.; Varma, V., "Towards commercialization of ad hoc networks," IEEE International Conference on Networking, Sensing and Control, vol.2, no., pp.1207-1211 Vol.2, 2004.

[9] C. Perkins, E. Royer and S. Das: Ad hoc On-demand Distance Vector (AODV) Routing, RFC 3561.

[10]Tarique, M.; Tepe, K.E.; Naserian, M.," Energy saving dynamic source routing for ad hoc wireless networks," Third International Symposium on Modeling and Optimization in Mobile, Ad Hoc, and Wireless Networks, 2005. WIOPT 2005. vol., no., pp.305-310, 3-7 April 2005.

[11]Al-Mekhlafi, Z.G.; Hassan, R., "Evaluation study on routing information protocol and dynamic source routing in Ad-Hoc network," 7th International Conference on Information Technology in Asia 2011 (CITA 11), vol., no., pp.1-4, 12-13 July 2011.

[12]Mitra, P.; Poellabauer, C.; Mohapatra, S., "On Improving Dynamic Source Routing for Intermittently Available Nodes in MANETs," Fourth Annual International Conference on Mobile and Ubiquitous Systems: Networking & Services, 2007. MobiQuitous 2007, vol., no., pp.1-5, 6-10 Aug. 2007.

[13]Bettstetter, C.; Resta, G.; Santi, P., "The node distribution of the random waypoint mobility model for wireless ad hoc networks," IEEE Transactions on Mobile Computing, vol.2, no.3, pp.257-269, July-Sept. 2003.

[14]Meghanathan, N., Boumerdassi, S., Chaki, N., Nagamalai, D. (eds.) Recent Trends in Networks and Communications. CCIS, vol. 90. Springer, Heidelberg (2010).

[15]M. Frodigh, P. Johansson, and P. Larsson. Wireless ad hoc networking: the art of networking without a network. Ericsson Review, No.4:248–263, 2000.

[16]Baidaa Hamza Khudayer, Mohammad M. Kadhum, Reliability of Dynamic Source Routing in Heterogeneous Scalable Mobile Ad Hoc Networks, IEEE International Conference on Communication, Networks and Satellite – COMNESAT 2014, pp.71-79, Jakarta, Indonesia, 4-5 November 2014.

Passive Network Monitoring using REAMS

Amir Azodi, David Jaeger, Feng Cheng, Christoph Meinel

Hasso Plattner Institute (HPI)
University of Potsdam
14482 Potsdam, Germany
amir.azodi@hpi.de, david.jaeger@hpi.de,
feng.cheng@hpi.de, christoph.meinel@hpi.de

Abstract. As computer networks grow in size and complexity, monitoring them becomes more challenging. In order to meet the needs of IT administrators maintaining such networks, various *Network Monitoring Systems (NMS)* have been developed. Most NMSs rely solely on active scanning techniques in order to detect the topology of the networks they monitor. We propose a passive scanning solution using the logs produced by the systems within the networks. Additionally, we demonstrate how passive monitoring can be used to develop a holistic knowledge graph of the network landscape.

Index terms— Network Monitoring, Network Graph, Event Processing, Event Normalization

1 Introduction

A primary benefit of NMSs is that they allow the administrators to monitor the systems across their network more effectively by being able to observe events of interest, whether they are related to the security, availability, quality of service, or other aspects of keeping a network and its available services operating in an optimal manner. Often administrators need to target systems matching a specific criteria within their network's landscape, for instance when trying to patch newly discovered security vulnerabilities, issuing software updates, or detecting hardware anomalies. License management and regulatory compliance is another area where operating a NMS can be beneficial. Interestingly, although the concept of a generic and automated NMS — using logical network topology graphs — has been around for quite some time[4], software vendors have largely opted for providing vendor specific NMSs. As a result, even for simple tasks such as getting the system uptime information of a particular host, network administrators have had to deal with different NMSs or even rely on simpler methods (i.e. locally/remotely reading a host's system logs). Many software vendors provide their own means of monitoring their systems within a network. Therefore these NMSs are often not self-contained and can only monitor systems specific to a subset of the network.

K.J. Kim (ed.), *Information Science and Applications*,
Lecture Notes in Electrical Engineering 339, DOI 10.1007/978-3-662-46578-3_24

1.1 Network Monitoring

In investigating NMSs we observe that performance has traditionally been (and to a large extent still is) one of the main challenges and is a major barrier facing the development of more effective NMSs. As a result, much research has been directed at developing more efficient methods and algorithms to discover and map network topologies and available services. Some implementations of these systems have shown promising results[7]. However, the proposed systems generally rely on low level network scanning tools to actively discover network objects. The problem with this approach is that most of the scanning techniques would not work within the IPv6 address space — due to the vastness of the address space available in IPv6 (i.e. 128bits) — and therefore are inevitably going to be outdated and ineffective.

Some workarounds exist and others may be developed in the future that would allow for more efficient and effective scanning of IPv6 networks but they would likely prove ineffective in many cases (such as when scanning blindly) due to the fundamental principals of the new protocol. Other approaches using Ethernet and Physical layer protocols are somewhat immune to this shift in the IP protocol. However, they remain limited in their effectiveness due to other constraints such as a lack of access to the information contained within higher level network protocols, which contain the majority of useful information. Deep packet inspection can provide some insight into the information in those layers but the performance overhead would be prohibitively high when dealing with larger networks. In addition, active scanning techniques are often limited by the honesty of the hosts they scan. This is not a major limitation when dealing with trusted and secure network objects. However, in a security context, this will result in a view of the network which does not contain perhaps relevant information about unknown or external hosts interacting with the network under observation.

The traditional approach for monitoring systems across a network has been to deploy client software (agent) to run on the network objects across the monitored network. This approach has the advantage of providing precise and accurate information about the monitored systems. In return the NMS (the server component) would be able to build a bird's-eye view of the network to be used by network administrators. Another advantage of such a system is that they often provide a gateway into the network objects they monitor (through the agent component), allowing the administrators to modify the network object's configuration and to remotely execute commands on it should they need to. Such access is useful in cases when a service provided by the network object needs to be restarted, reconfigured, new software needs to be installed, etc.

Though the client server approach has many benefits, it also has some shortcomings. For instance, many of the network objects within a network may not have the capability and/or the capacity to run external software (i.e. monitoring agents) or they may be unsupported by the NMS vendor. This would exclude them from the NMS's network landscape view. Systems such as Routers, Switches, Firewalls and Gateways may all hold important information that would contribute to the security and/or availability of a network and its services, provided that the network administrators would be notified of important events or incidents in addition to being able to pro-actively monitor the network's status.

In response to these needs, several protocols and systems have been designed for monitoring network systems. Simple Network Management Protocol (SNMP)[5] is one such protocol. Devices that support it can be monitored using SNMP. The SNMP design is based on a Manager and Agents. The manager component is usually part of a NMS while the agents often come pre-installed with the firmware of network objects. When an event occurs on a system with an SNMP agent and an SNMP trap is triggered, the incident is reported to the SNMP manager. The NMS would then decide on what to do with the event. For example, if a server's case fan has suddenly stopped working, the administrators could receive an email to alert them of the fact. This would allow them to replace the damaged fan before the temperatures in the system could rise and cause more severe damage to other components such as the hard drives or the CPUs.

Although SNMP is widely implemented in the more advanced — and more expensive — networking equipment, many devices often lack support for the protocol. Additionally, there are instances when proprietary software is introduced in the SNMP implementation which breaks compatibility even between managers and agents from the same vendor[14]. Nagios XI [8] is an example of a NMS which heavily relies on active network scanning techniques and on its agents deployed across a network to provide a highly versatile and feature rich platform for administrators to monitor their networks. Nagios is a good example of the strengths of active monitoring mechanisms. In fact most NMS systems rely on agents deployed on systems across the network. Given that this is the most popular and accurate way to map and monitor a network, alternative approaches to doing so have not been pursued and developed to their fullest extent. As stated above, we focus on passive network monitoring (i.e. with the aid of deployable agents) in order to further advance its capabilities with the overall aim that they can be used alongside active techniques to help create a more complete view of networks for NMSs. Figure 1, explains the logical separation of the two approaches.

2 Event Processing and Information Extraction

Security Information and Event Management (SIEM) systems monitor networks and systems primarily using the events (logs) produced by those systems. SIEM systems can offer an in depth view of the systems they monitor without needing agents. Just like a plain central log server which many companies operate, a SIEM system collects events in a central repository and processes and analyses the events without negatively affecting the performance of the systems it is monitoring. There can be challenges with regards to scalability as more systems are monitored, but there are existing event processing solutions capable of scaling

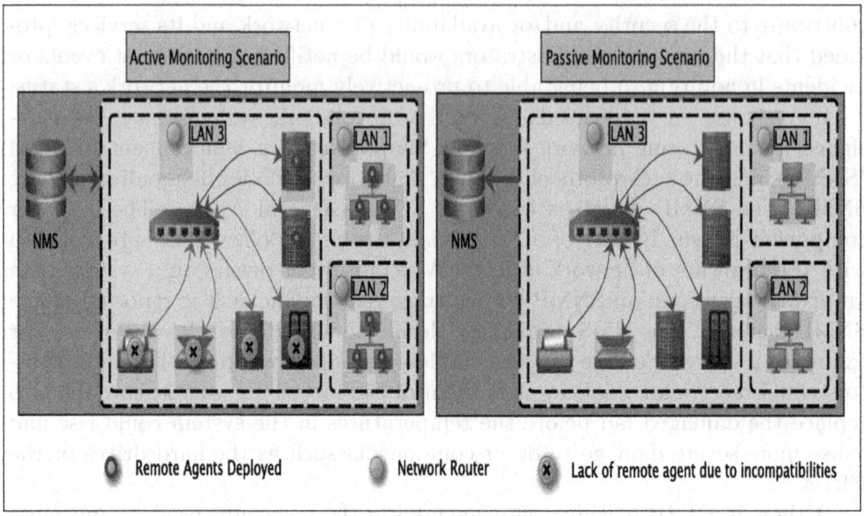

Fig. 1. This figure demonstrates the logical difference between Active and Passive network monitoring mechanisms.

up to a point where this is less of a concern. Real-time Event Analytics and Monitoring System (REAMS)[1] (previously SAL) is a research system with the aim of combining Intrusion Detection System (IDS) capabilities as well as SIEM system functionality in a centralized and scalable way. The fundamental capabilities needed to achieve its stated goals are delivered through an *Event Processing Engine* capable of uniquely identifying the type of the individual events it receives in (near) real-time. We use REAMS as an event processing engine to help with building a live network graph for the purposes of monitoring the operations of a network. The process of extracting information from events can be logically subdivided into five steps. Determining valuable sources of information, preprocessing the incoming events (including event normalization), event sanitization, extracting relevant information, and context information injection.

Information Sourcing Sourcing the information necessary to build a network graph can be a challenge. Today, the vast majority of networks use some form of a SIEM system or a log server to gather the events their systems produce in a central place. This is often mandatory and dictated by regulatory bodies. One reason is that when a system is compromised by an attacker, one of the first steps the attacker would take would be to wipe or manipulate the system's log files in order to prevent being discovered. Given that basic remote logging capabilities are an integral part of any system, almost any device can be configured to use a central log server. For the purposes of this paper, we focus on using information gathered by a SIEM system (namely REAMS), but the approach taken can be similarly applied to other SIEM systems or log servers.

Event Processing The quality of the information extracted from an event depends greatly — if not entirely — on the event format. In [2] the authors argue that many events generated across a network can not be harvested for useful information. In many cases this is due to the format of the event. For instance, in some cases only the event ID is recorded and special — often proprietary — software is needed to decode it into a consumable format while in other cases it might be due to the poor quality of the event itself. For example, extracting useful information from the event in Listing 1.1 would prove difficult.

Listing 1.1. An example of a poorly modeled event

```
Debug: it works, Storage Error
```

As stated, in order to access events, we use a SIEM system (i.e. REAMS). Today, there are many commercial SIEM systems and log servers available[15][10][16][12]. These systems generally process the events they receive in one of two ways. The first way is to simply process the structured portion of the event and store the event message (often the more important portion) in its simple form. Network administrators are then able to run text searches over the event's message field in order to investigate incidents. Additionally, the structured part of the event allows for filtering (e.g. temporal) of the events.

The second and more advanced method of processing events includes an event normalization stage. Systems using this process effectively convert the format of the events they process into a single unified event format. This allows for correlation engines to operate on a single event format instead of many different ones which would greatly reduce the complexities of processing the events. Given that operating on normalized and unified events is considerably more efficient for automated systems, we focus on using one such system (i.e REAMS). REAMS operates on one global event stream. Events produced across the network are aggregated into a *global* event stream before being processed. By making the event processing step atomic, the system can use parallel processing to boost the efficiency of its event normalization step. Once the event has been received by the system, it passes through a text analysis step which creates a quasi unique template from it. The template is then used to find other events which the system knows of (using proximity searching techniques), in order to find a matching tokenizer for the event. The tokenizer is then used to extract information in multiple steps from the Wrapper, Header and Message portions of the event.[1] Figure 2 is the full stack of network related information REAMS can extract from the events it processes.

[1] **Wrapper** is defined as an event format that encapsulated the original event. e.g. GELF[17]
Header is the portion of the event which is common across all events of a particular format. e.g. Syslog[9] header
Message is the core information that is to be persisted using the event. e.g. user x logged into system y.

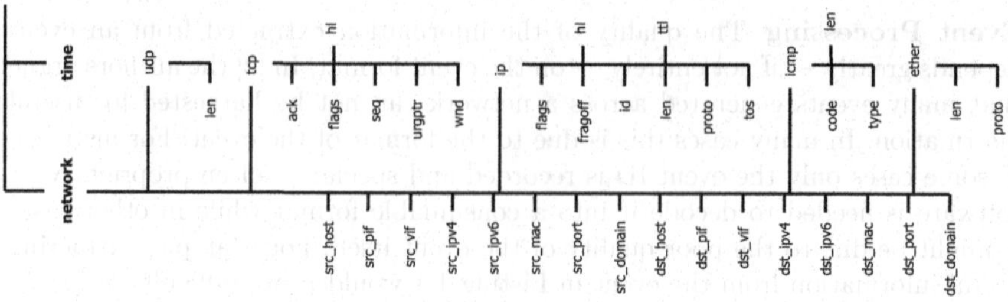

Fig. 2. Network discovery relevant fields extracted from processed events.

Event Sanitization Logically not every processed event contains all the network layer information supported by REAMS. However, it is trivial to filter and use only events that contain information that can populate (at least a subset of) the network related fields supported by REAMS. In Section 3 we use the network related information we extract to generate a logical network graph. It is therefore important — in the interest of clarity and performance — for our system to be able to drop those events which are of little or no use. We achieve this by dropping events that do not match a white list of rules or alternatively by specifying rules that match the events we wish to drop and subsequently doing so. Using REAMS, we have developed a list of events that once matched, should be ignored. These events are then not considered when the view of the network is constructed.

The Extraction Process In order for REAMS to know about the semantics of the events we are interested in, some rules need to be written. REAMS can understand and parse the different events it receives as long as it has a rule for it in its knowledge base. The extraction process specifically depends on the tokenizer component of the rule. For our examples we developed a tokenizer capable of processing a test set of Cisco ASA firewall events.

Adding Context Information In [1] the authors demonstrate how extra *knowledge* can be added to events being processed. In cases where the event instance does not contain some important information, a prepared rule is able to append more information to the event. This can help with events produced by some applications in which only an event ID is provided which needs to be resolved for its true meaning. In such cases the knowledge base of the event processor would be able to resolve event IDs to respective message and process the message instead.

3 Using Processed Events to Generate a Network Graph

Mapping out a network requires an understanding of the underlying *logical* topology of the network. By focusing on passive monitoring we inherently avoid performance and compatibility issues marring active monitoring systems. From the events which are useful in building a network graph (events that include information from the higher layers of the OSI Model), a subset is used for mapping the lower levels of the networks logical structure. We refer to this subset as Network Layer Events (NLE). NLEs are often produced by networking systems such as Routers, Switches, Firewalls, as well as basic network services such as DHCP, DNS, IDS Sensors, Wireless Access points, etc. Events used to populate information about an individual object's applications are referred to as Application Level Events (ALE). ALEs are produced by Operating Systems, Web Servers, FTP Servers, Mail Servers, Authentications Services, etc. By processing a set of events produced by a network firewall, our passive monitoring system can detect an internal network and the presence of hosts within that network. The sample event set used for the purposes of this paper is compromised of 1586 events produced over a time span of 4 hours and 12 minutes by a Cisco ASA firewall system. We used a relatively small event set in order to be able to produce reasonably sized figures due to space constraints.

Using the information extracted from the processed events we attempt to construct a map of the network topology by observing the connections between the communicating hosts. Figure 3 demonstrates a simple example mapping communications between one subnet of the internal network and a subset of the external networks and hosts. Subsequently, by monitoring the application level events an overall specification list for the individual systems is formed. Host specific information is then overlaid on to the network topology map to construct a comprehensive view of the network. In addition to the ALEs, the context information added to the ALEs during their processing by REAMS can also be used to create a more complete image of the system. By analyzing NLEs in our sample log file, we can detect the presence of multiple external networks. As a security feature, we resolved the external IP addresses discovered in the events and match them against a list of known malicious domains found in [13]. We discovered no correlation between the external hosts and the list of known malicious domains. However such tests could be instrumental in discovering network security breaches by network monitoring systems. Table 1 represents the 10.10.10.0/24 subnet discovered (one of the internal subnets) and displays the hosts within the subnet as well as the ports used for communication by those hosts.[2]

We resolve the port numbers against a knowledge base of port number information. Table 2 presents a mapping between the ports and the relevant information. At this point we can observe that there is no service *registered* on port 1028, although it is being used. Additionally it can be seen that in the past certain malware have used this port for communication. In a commercial setting, the NMS could raise an alert to inform the network administrators of the event's occurrence.

[2] Some ports and hosts were omitted due to size constraints

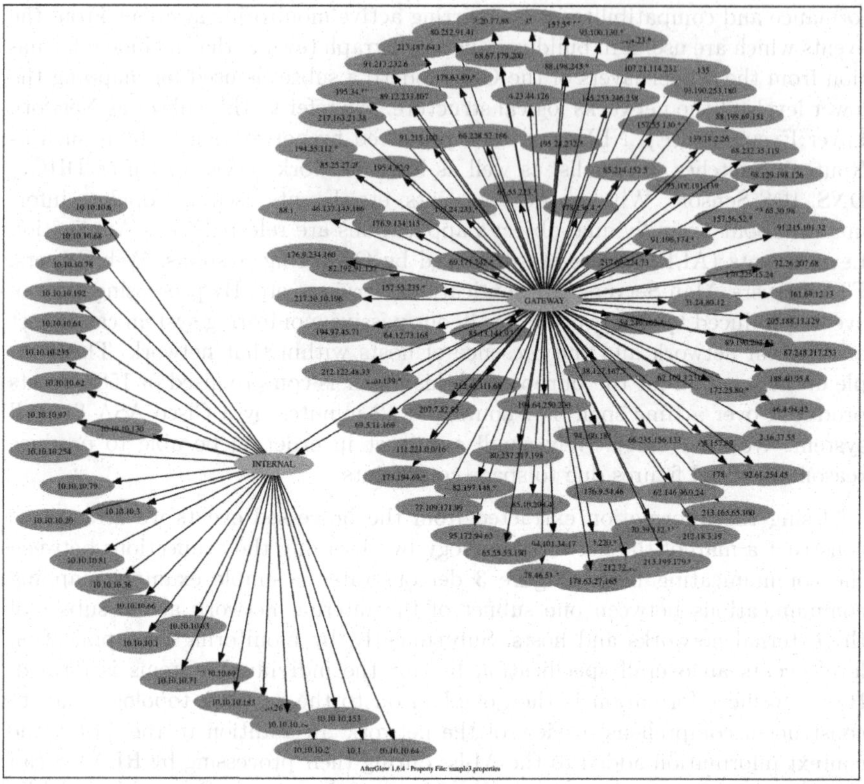

Fig. 3. A simplified view of the network topology using event driven passive network discovery.

4 Related Work

NMSs are not a novel idea. Indeed there has been considerable and promising research in network topology detection algorithms capable of mapping out complex networks. In [7], the authors describe a topology discovery algorithm (TopDisc) which using wireless neighbor discovery methods can map networks. In [6] and [3], the authors describe a central network monitoring system which maps the network with the use of network layer tools such as Ping. Although effective in mapping hosts, the design has limitations as it relies entirely on active network scanning tools. The majority of the effort directed at finding better and more accurate network mapping algorithms has been to actively scan the network using low-level technologies (e.g. Pinging — sending ICMP echo requests

---

[3] Well Known Ports: 0 through 1023.
Registered Ports: 1024 through 49151.
Dynamic/Private : 49152 through 65535.

IP Address	Roles	Listening (port #)	Connecting (port #)
10.10.10.1	DNS Server	1702, 53	53, 123
10.10.10.3	Standard Host		123
10.10.10.6	Admin Host		17, 22, 80, 123, 161, 427, 443
10.10.10.20	Standard Host		
10.10.10.26	Standard Host		**1028**
10.10.10.36	Standard Host		
10.10.10.58	Standard Host		80
10.10.10.61	Standard Host		
10.10.10.62	Standard Host		
10.10.10.63	Standard Host		80, 443
10.10.10.64	Standard Host		443
10.10.10.66	Standard Host		80, 443, **1906**
10.10.10.68	Standard Host		80, 443
10.10.10.69	Standard Host		80, 443
10.10.10.71	Standard Host		80, 443
10.10.10.74	Standard Host		80, 443, 993
10.10.10.78	Standard Host		80
10.10.10.79	Standard Host		80, 443
10.10.10.81	Standard Host		80, 443
10.10.10.97	Standard Host		80, 443
10.10.10.130	Standard Host		80, 443
10.10.10.153	Standard Host		**1027**
10.10.10.183	Standard Host		500
10.10.10.192	Standard Host		80
10.10.10.225	Standard Host		
10.10.10.235	Standard Host		80, 443
10.10.10.254	Firewall	161	

Table 1. List of hosts and services detected in the 10.10.10.0/24 subnet

— systems across the network). As a result, the overall picture of the network can be constrained and can miss information such as the applications installed on the host, user activities on the host and within the network, and more. Another issue is that most networks are protected by a gateway and firewall system. In this case, hosts beyond the reach of the systems within the Intranet are never observed. In contrast, when a network is monitored based on the events occurring on the network and the hosts within it — even if there is no access to outside systems — , more information can be extracted and a more accurate representation of the network can begin to form. For instance, such a system would be able to observe incoming and outgoing connections by processing the events produced by the firewall, gateway or the host providing a service being used over the network. Many NMSs already monitor gateway systems for mapping network connections. However they do so using active techniques (using agents). In this paper we focus solely on passive monitoring techniques.

Port #	TCP	UDP	Known Service(s)	Status
17	✗	✓	Quote of the Day	Official
22	✓	✗	Secure Shell (SSH)	Official
53	✗	✓	Domain Name System (DNS)	Official
80	✓	✗	Hypertext Transfer Protocol (HTTP)	Official
123	✗	✓	Network Time Protocol (NTP)	Official
161	✗	✓	Simple Network Management Protocol (SNMP)	Official
427	✗	✓	Service Location Protocol (SLP)	Official
443	✓	✗	Hypertext Transfer Protocol over TLS/SSL (HTTPS)	Official
500	✗	✓	Internet Security Association...(ISAKMP)	Official
993	✓	✗	IMAP over TLS/SSL (IMAPS)	Official
1027	✗	✓	Native IPv6 behind IPv4-to-IPv4 NAT	Official
1028	✗	✓	**KiLo, SubSARI**	**Unknown**
1702	✗	✓	Deskshare	Unofficial
1906	✗	✓	Unknown	Unknown

Table 2. Correlating a subset of the communication open ports with likely services.[3]

5 Conclusion

Active NMSs can impose an unacceptable performance penalty on the systems they monitor. Intrusion detection systems (security oriented monitoring) are sometimes omitted in performance conscience environments because their sensors can be an unacceptable burden on the network and the systems they monitor. As noted, passive monitoring differs from active monitoring and has some advantages (as well as disadvantages) over active monitoring. Passive monitoring allows for a much more generic solution that can work with many different systems with very little operational requirements to be met. As a result of the genericness of the approach used in our passive monitoring approach, the network landscape view can be populated by information that active monitoring methods may not have detected. Additionally, passive monitoring methods have a higher threshold for allowing dishonesty on the part of the monitored entities. Our results demonstrate how services running on different systems can be detected by processing firewall events. We used the same events to map the networks. Information within the knowledge base of the system allowed for the service type detection.

6 Future Work

As discussed, there are situations where a passive system might not yield the same results as an active NMS. Tools such as Nmap[11] are effective in detecting services running on the systems they scan. Therefore one possibility is to combine active and passive network scanning to yield better results. This may be particularly effective if the host detection is carried out passively and the service detection actively. Performance enhancements to the passive scanning mechanisms as well as information validation and information collision resolution within the network landscape would be of interest and should be considered for future enhancements to the approaches outlined in this paper.

References

1. Amir Azodi, David Jaeger, Feng Cheng, and Christoph Meinel. A New Approach to Building a Multi-Tier Direct Access Knowledge Base For IDS/SIEM Systems. In *Proceedings of the 11th IEEE International Conference on Dependable, Autonomic and Secure Computing (DASC2013)*, Chengdu, China, December 2013.
2. Amir Azodi, David Jaeger, Feng Cheng, and Christoph Meinel. Pushing the Limits in Event Normalisation to Improve Attack Detection in IDS/SIEM Systems. In *Proceedings of the First International Conference on Advanced Cloud and Big Data (CBD2013)*, Nanjing, China, December 2013.
3. Srinivas Basa and Naveen Ganji. *Enhanced NMS Tool Architecture for Discovery and Monitoring of Nodes.* PhD thesis, Master Thesis Computer Science Thesis no: MCS-2008-15 January 2008, 2008.
4. A.B. Bondi. Network management system with improved node discovery and monitoring, January 20 1998. US Patent 5,710,885.
5. Jeffery Case, Mark Fedor, Martin Schoffstall, and C Davin. A simple network management protocol (snmp), 1989.
6. Antonios G Danalis and Constantinos Dovrolis. *Anemos: An autonomous network monitoring system.* PhD thesis, University of Delaware, 2003.
7. Budhaditya Deb, Sudeept Bhatnagar, and Badri Nath. A topology discovery algorithm for sensor networks with applications to network management. 2002.
8. Nagios Enterprises. Nagios XI the industry standard in it infrastructure monitoring, 2014.
9. Rainer Gerhards. The Syslog Protocol. RFC 5424 (Proposed Standard), March 2009.
10. Hewlett-Packard. Arcsight security intelligence platform. `http://www.ndm.net/siem/main/arcsight-siem`.
11. Insecure.Org. Nmap security scanner, 2014. [Online; accessed 14-August-2014].
12. Logstash. Logstash.
13. The DNS-BH project. Malware prevention through domain blocking (black hole dns sinkhole), 2014. [Online; accessed 11-August-2014].
14. David Reid and Steve Blizzard. Standards-based secure management of networks, systems, applications and services using snmpv3 and hp openview, 2006. [Online; accessed 11-August-2014].
15. Splunk Inc. Splunk Enterprise. `http://www.splunk.com/`, 2003.
16. TORCH GmbH. Graylog2 Central Log Server. `http://www.graylog2.org/`.
17. TORCH GmbH. Graylog Extended Log Format (version 1.1). Web Site, November 2013.

Low Latency Video Transmission Device

Chanho Park, Hagyoung Kim

Cloud Computing Research Department,
Electronics and Telecommunications Research Institute (ETRI)
{jangddaeng, h0kim}@etri.re.kr

Abstract. This paper presents a low latency video transmission device which includes a protocol engine and a packet buffer for delivering data. The packet buffer is designed to transfer data packet between different clock domains. The protocol engine is designed for video transmission and especially for hardware implementation. Adopted protocol is simpler than TCP protocol to implement in hardware and has some reliability which UDP does not have. The proposed device can achieve low latency and good quality of service

Keywords: Video Transmission, Dual-clock FIFO, Reliability, Protocol

1 Introduction

Video transmission plays a key role in a cloud service, remote PC control system and real-time multimedia service. In the cloud service, many protocols such as RDP, HDX, and PCoIP are used and they use TCP or UDP protocol as the layer4 protocol. TCP protocol is useful for safe transmission, unless there is latency limit. However if latency is an important factor of service, for example real-time multimedia, TCP can degrade service quality because of repetitive retransmission. UDP protocol also has problems such as packet loss and sequence violation. It is still arguing that which one is better [1]. To solve these problems, RUDP [2] protocol was suggested and used in several systems [3], but it was not selected as an internet standard.

As the request of real-time multimedia service grows, new protocols such as RTP/RTCP [4], RSVP, RSTP are suggested and become standards in real-time multimedia service. In these protocols, it is RTP protocol which actually carries the multimedia data and it is encapsulated in UDP protocol. This is quite simple protocol, however, to realize reliability and connection control, not only RTP, but also RTCP protocol must be implemented.

To implement simple video transmission system with pure hardware, simplicity of protocol is important. Compare with UDP, TCP protocol is very complicated for hardware implementation. RTP/RTCP protocols are also quite complicated. UDP protocol is good candidate unless it has no reliability at all. In this paper, I proposed relatively simple protocol for video transmission. It is similar to RUDP protocol, but it includes short-term reliability and several parameters are changed for hardware implementation. Short-term reliability means that corrupted or lost packets are retransmitted during limited time.

K.J. Kim (ed.), *Information Science and Applications*,
Lecture Notes in Electrical Engineering 339, DOI 10.1007/978-3-662-46578-3_25

When designing a hardware system, several clocks can be used. Delivering data between different clock domains can lead to problems, for example, metastability problem. Asynchronous FIFO can be used to solve this problem [5]. This FIFO applies gray code conversion to the head and tail pointer to solve metastability problem. It is also a good solution, but in this paper, different FIFO structure which is suitable for packet processing is proposed.

A video transmission device with proposed structure can be simply implemented in hardware and has good real-time service quality. For easy management and power saving, remote PC control system is used [6]. Proposed system can be applied to here as well.

2 System Implementation

2.1 Video Transmission System

Figure 1 shows an example of video transmission system. PC (or Server) loads pass-through virtualization software. Each VM uses its own video card exclusively. The outputs of video cards are inserted into the video transmission card. The video signals are compressed, packetized, and transmitted to a network. The video receiving card in figure 1 receives the packetized video data, decompresses it, and displays it through a monitor. The proposed protocol is used between video transmission card and receiving card.

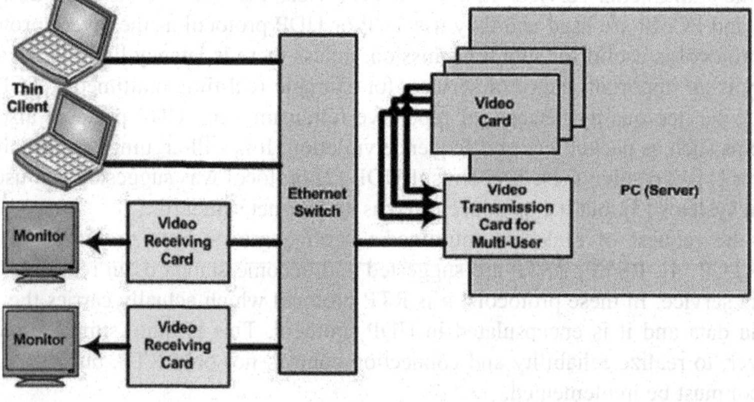

Fig. 1. An example of video transmission system

2.2 Device Structure

Figure 2 shows the block structure of video transmission & receiving cards. Video source data is supplied via a DVI or HDMI port, compressed in the compression block, and converted to network packets in the SRUDP (Short-term Reliable UDP) protocol engine. Packets are transmitted to the network via the MAC block. In the receive card, network packets are received by a MAC block and restored to the compressed image data in the SRUDP engine. The image data is converted to RGB or YUV format in the decompression block and it is displayed on a monitor.

Source data may be supplied from the CPU via the I/O bus, such as PCI express, not the DVI/HDMI port, if the image is already compressed, the compression block is not necessary. Whatever the source is, the clock used of data source may not coincide with the MAC block. In general, 125Mhz clock is provided in the FPGA for the MAC IP and the ADC chip for receiving a DVI/HDMI signal uses 75MHz and 150Mhz clock. The clock of compression block may also vary, if you do not design your own IP. Therefore, there is at least one change of clock domains in the data path. In this paper, we introduce the dual clock packet FIFO and SRUDP engine for providing data transfer over the network, and shows a video transmission device using them. In the proposed device, dual clock packet FIFO is used between MAC block and other blocks in order to take advantage of the characteristics of the packet processing, and a third-party IP is used for the compression block.

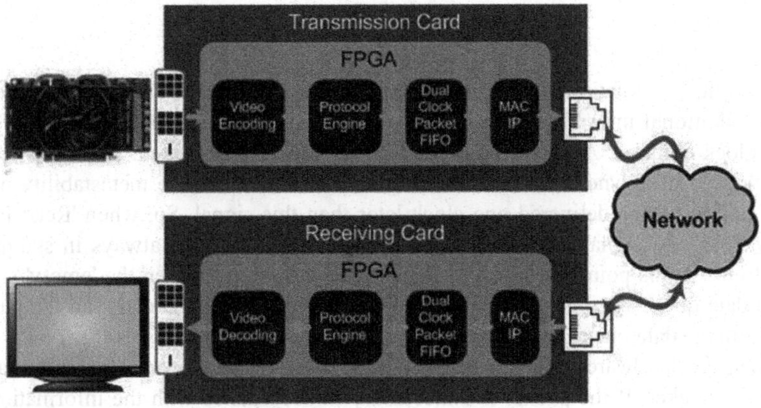

Fig. 2. An example of video transmission system

2.3 FPGA Implementation

- **Dual-clock packet FIFO.**

Figure 3 shows the structure of the dual-clock packet FIFO and the dual clock data queue which is a sub block of it. It has two FIFOs which are data queue and information queue.

When a packet is written in the data queue at the WCD (write clock domain), if the last data of packet is written, 'write_done' signal should be activated. This signal is used as the write signal of information queue. The information is tail pointer of data queue. Additionally, some information about the packet can be written together. If it is proven that the packet has a problem (for example, checksum error), all the data of the packet can be removed using the 'write_err' signal. The information queue is a common FIFO.

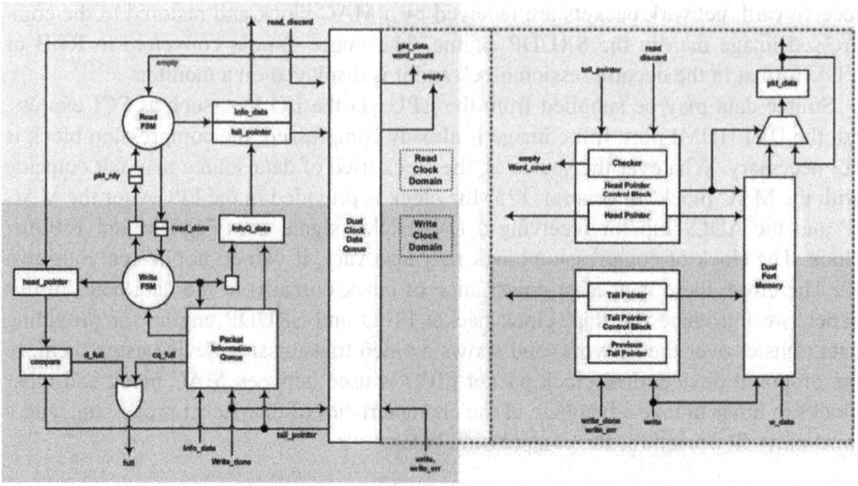

Fig. 3. Structure of Dual-clock Packet FIFO

When the information queue is not empty, 'Write FSM' read data (which are pointer and additional information) from the information queue and passes it to the RCD (read clock domain). At the same time, 'pkt_rdy' signal is activated. Since this signal is composed of a synchronizer connecting two DFF to eliminate metastability problems, tail pointer is delivered one clock later than this signal. So, when 'Read FSM' detects 'pkt_rdy' signal, tail pointer and packet information are always in stabilized condition. At this point, 'Read FSM' updates the tail pointer. Then the 'empty' signal of the data queue becomes zero, and hardware logic in the RCD can read the packet data from the data queue.

When the hardware logic in the RCD read data, it can preview the information about the packet. If the packet is unnecessary in accordance with the information, it can be discarded at one time using 'discard' signal. Or, the packet can be passed to an appropriate hardware depending on the information. When all the data of a packet are read, 'Read FSM' activates 'read_done' signal to inform 'Write FSM' of this. Then, 'Write FSM' can process with the next packet if the information queue is not empty. The relationship between the head pointer and the 'read_done' signal is same as the relationship between the tail pointer and the 'pkt_rdy' signal. The 'full' signal of the dual-clock packet FIFO becomes one, if the data queue or the information queue is full.

The data queue (right part of Fig. 3) has no data to be passed between clock domains, except a dual port memory. The tail pointer and the head pointer are not compared in the data queue, but they are transmitted through the interface between the two FSMs.

- **Network protocol for video transmission.**

Figure 4 shows the header format of suggested protocol. The header format is similar with RUDP protocol. Several distinguishing parameters are as follows.

The parameter "Expiration Timeout Value" in SYN packet represents how long a packet can be retransmitted. Each packet gets a timestamp when it is transmitted for the first time, and if the duration designated by this parameter is passed, the transmitter abandons retransmission. When a transmitter decides not to send a lost packet, it must be noticed to a receiver. The tag "UAN" is used for this. When a receiver receives a packet with "UAN" tag, acknowledge number must be replaced by the sequence number of received packet. During transmission, parameters can be updated using UPM packet (with "UPM" tag). Its format is same as SYN packet. The parameter "Time Scale" indicates unit of "Timestamp" and several "Timeout Values" in microsecond. The parameter "Max Outstanding Kbytes" indicates how many Kbytes can be sent without ACK. Because we target hardware implementation, the amount which decides the size of retransmission buffer is used instead of packet counts. The tag "IMA" orders the receiver to send an ACK packet immediately.

The parameter "Consecutive Counter" in EACK packet represents how many packets are lost from "Lost Sequence Number" in series. The parameter "Max Lost Seq" in SYN packet indicated how many lost packets can be included in EACK packet. Serial lost packets as regarded as one.

These parameters are matched to the appropriate registers in an implemented hardware [7].

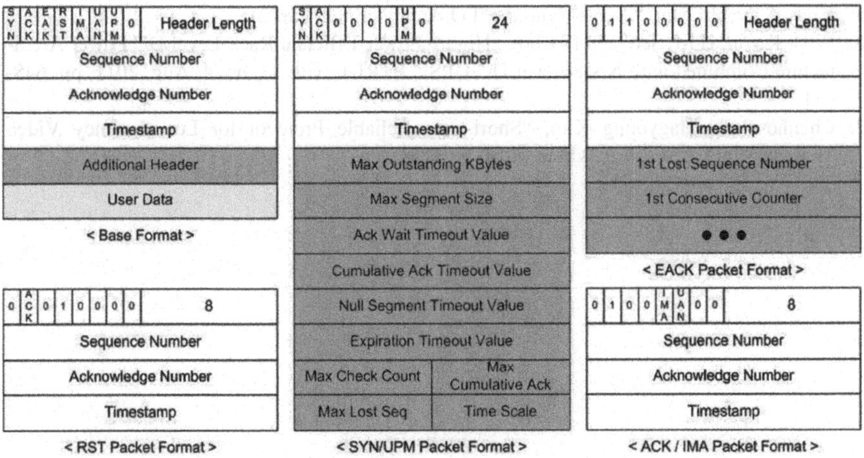

Fig. 4. Protocol Header Format

3 Conclusion and Future Works

This paper proposed a low latency video transmission device which consists of a video compression/decompression block, a transmission protocol block, a dual clock packet FIFO, and a MAC IP. Also, this paper describes the architecture of the dual clock FIFO and the structure of the protocol. The dual clock packet FIFO enables to transfer packets between different clock domains. The transport protocol provides sufficient reliability for real-time video transmission and is suitable for hardware implementation. Later, we are going to implement hardware and measure the video quality to prove our expectation

Acknowledgement

This work was supported by the ICT R&D Program of MSIP/IITP. [10041932, Development of remote control protocol and graphic acceleration technology for Full HD cloud service]

References

1. Brian Madden, "The Layer 4 protocols behind PCoIP and HDX : Which is better for VDI?", http://searchvirtualdesktop.techtarget.com, Jan. 2011.
2. David Velten, Robert Hinden, Jack Sax, "Reliable Data Protocol", RFC908, July. 1984.
3. Jiwon Yoon, Sungjin Park, Seok Lee, Hyungseok Kim, "A Reliable Data Transport Protocol for Internet-accessile Wireless Sensor Network", ITC-CSCC, MB3-1, 2007.
4. H. Schulzrinne, GMD Fokus, S. Casner, R. Frederick, V. Jacobson, "RTP : A Transport Protocol for Real-Time Applications", RFC1889, Jan. 1996.
5. Vijay A. Nebhrajani, "Asynchronous FIFO Architectures (Part 2).
6. W.O. Kwon, H.M. Seo, and P. Choi, "Highly Power-Efficient Rack-Level DC Power Architecture Combined with Node-Level DC UPS," ETRI J., vol. 33, no. 4, Aug. 2011, pp. 648-651.
7. Chanho Park, Hagyoung Kim, "Short-term Reliable Protocol for Low Latency Video Transmission", CSCI, vol 2, Mar. 2014, pp 311-312.

Part III

Comparative Investigation of Medical Contents using Desktop and Tabletop Technologies

Tahir Mustafa Madni[1], Yunus Nayan[2], Suziah Sulaiman[3], Muhammad Tahir[4]
[1,2,3] Faculty of Information Sciences, University Technology PETRONAS, Malaysia
[4] Faculty of Computing and Information Technology, King Abdulaziz University, North Jeddah Branch, Jeddah, Saudi Arabia
tahir_mustafa@comsats.edu.pk[1], yunusna@petronas.com.my[2], suziah@petronas.com.my[3], mtyousuf@kau.edu.sa[4]

Abstract. In recent years, in the field of human-computer interaction show that interactive technologies, constructive methods and appropriate interfaces have great impact in improving learning and creative education. Such interactive technologies along with suitable interfaces are beneficial for humans in various domains especially in the field of healthcare. Generally, in practice, medical monitoring and diagnosis systems are based on desktop computers and have been less explored on multi-touch tabletops. It is due to its less availability but supports collaborative activities and natural form of interaction. This paper presents an experimental investigation of technological-based comparative study, with collaborative activities using desktop technologies and tabletop technologies. A prototype was implemented based on the proposed conceptual framework. Furthermore, a quantitative investigation on the usefulness of these technologies was conducted with twenty students. The experiment showed investigation of medical elements, was more significant in tabletop technologies as compared to desktop technologies.

Keywords. medical contents, collaboration, cooperation, desktop technologies, multi-touch tabletop technologies

1 Introduction

Medical imaging is an important fragment of healthcare especially for medical students to learn human anatomy. With advancement in the research of interactive technologies along with appropriate interfaces provide an opportunity for improvement of investigation of medical elements. The investigation of medical elements is categorized as monitoring (radiological image preview), diagnosis (analyze organs abnormality), treatment planning (pre-operative surgery planning) and surgery [1]. This study only focused on human interaction for monitoring and diagnosis of medical contents for the medical students. These medical images or modalities (for example CT, MRI, PET and US) [6] along with computer technologies provide powerful opportunity for investigation of medical contents. The contents are interface elements which include high resolution medical modalities and related documentations.

Generally, students interact with desktop technologies to learn medical theories and visualize medical images. These technologies are generally available but it lacks natural form of interaction (by the use of mouse and keyboard). It also limits the interaction of one or more group members sharing same artifact. In recent years, the advancement in the research of healthcare by the use of tabletop is increased in collaborative groupware environments [1]. Usually tabletop is horizontal multi-touch

K.J. Kim (ed.), *Information Science and Applications*,
Lecture Notes in Electrical Engineering 339, DOI 10.1007/978-3-662-46578-3_26

display screen. It provides an opportunity for multiple users to collaborate with each other and make face-to-face discussions. With enriched medical modalities provide increased efficiency and lowers the risk of complications for example treatment planning [1], heart [2], lungs [3, 20], adrenal [4] and gastric [5].

The fundamental requirements for these systems are to present modalities similar to physical situation. A quantitative research was adopted to accomplish research. Varity of laboratory and non-laboratory research methods were available in the field of Human-Computer Interaction (HCI). An experimental method was used to determine usefulness of the system due to its nature of less cost and short completion time. Such investigation determines causal relationship between multiple factors.

This experimental investigation provides technological-based comparative study using desktop and tabletop technologies. For this, a conceptual framework was designed. A prototype named as Medical Investigation System (MIS) was developed based on proposed conceptual framework. The main objectives were: to design an interactive monitoring and diagnosis application, to explore the usefulness of single-mouse single display groupware and multi-touch tabletop groupware setups. It does not attempt to cover all the efforts but rather specifically focus on the perspective of human-computer interaction. In user experiments, the students experienced the system, to determine usefulness of MIS application for both technologies. The key research contributions were: to design conceptual framework called medical investigation framework, the human-computer interaction concepts of moveable and natural size zoom of medical contents and quantitative analysis about the use of system for learning medical concepts.

In the subsequent sections this paper initially describes related work. The paper then development of prototype based on conceptual framework. Later section describes the evaluation process of prototype followed by discussion and conclusion.

2 Related Work

Interactive technologies and appropriate user interfaces provide powerful opportunity to investigate human organs. Such systems are used for radiological reviews to minimize the risk of complications during medical treatments. Existing research efforts mostly focus on user interaction for image visualization. A 3D medical application was developed by Gallo [21], which supports stereoscopic displays with input of mouse or Wii remote wireless device. The mouse and keyboard represents a unnatural interaction in complex applications, for example 3D imaging [14].

Multi-touch tabletops have been applied in different domains, for example education [9], design [10], architecture [11], and scientific image visualization [12, 13]. The multi-touch tabletop systems have been less explored but present potential in healthcare [20]. In orthopedic surgery planning [1], a tabletop provides treatment planning. It intends to support real clinical situations. It was experienced by orthopedic surgeons and they gave positive feedback for its use. However in healthcare professional use, still there are fewer examples of large-size screens and tabletop applications. Some of them are used in medical tasks specifically for surgical planning and training of doctors [17] [18] [19]. The existing research work is based on theories reviewed thoroughly to achieve the design and implementation of MIS system.

3 Conceptual Framework

This study conceptualize the design space through the use of framework for surface manager [23], Noman,s theory-of-action and interaction framework [24]. The Nomans theory-of-action and interaction framework are most influenced in the field of HCI, due to its closeness of understanding the concept of user interaction.

Fig. 1 presents a conceptual medical investigation framework. It was responsible to define a set of concepts to design and build medical diagnosis and monitoring system that supports users to access medical resources in collaborative manner. The proposed conceptual medical investigation framework has four dimensions/layers that are responsible for different tasks and contribute differently to the medical application. These layers overlapping each other as layer one overlap layer two and so on.

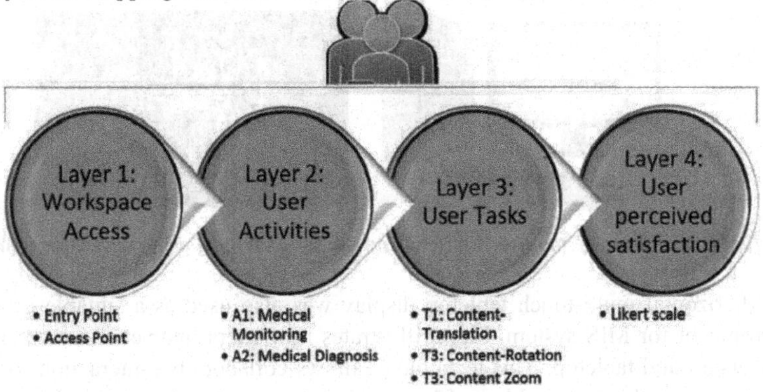

Fig. 1. Proposed conceptual medical investigation framework

These four layers involve workspace access, user activities, user tasks and user perceived satisfaction respectively. Layer 1 represents workspace and access in collaborative environment, presents entry point and access point which expresses the way to mediate interaction. The entry point facilitates users', how to enter in the working environment, understand the context of work, and start interaction. It facilitates user to distinguish computer technologies, i.e. either interact with desktop setup or tabletop setup. Whereas an access point represents characteristics that users' initiate interaction and start working as a team member in group setting. Later 2 refer user activities for monitoring and diagnosis of medical contents. The monitoring activity presents radiological image preview, whereas diagnosis presents analysis of organs abnormality. The user activities are performed through user task as presented in layer 3. The user task involves content selection, content translation and zoom-in/out medical contents. The performance of technological context and user activities are determined by user perceived satisfaction as presented in layer 4. It means how the system behave, how the people use, feel, and their perceived satisfaction about the system.

4 Medical Investigation System (MIS): Prototype

In initial stage of experimental investigation, prototypes were implemented based on proposed conceptual framework. The evaluation process was iterative in nature, it means when iteration of certain stage of prototype development was finished then evaluation process started. It helped in the improvement of next iteration of prototype development process. All iterations were tested by in-house designers and programmers, but only the final iteration of prototype was experienced by test participants.

The implementation of MIS was based on simple principle of utilizing the strength of medical modalities and collaborative working environment. It was designed and implemented for desktop and tabletop setups. As desktop technologies are commonly used due to easy availability and have been extensively tested for user experience. Fig. 2 illustrates user experience MIS by sitting/standing around desktop technology. They use application for understanding human anatomy. They can manipulate (view, zoom, rotate) medical artifacts using mouse and keyboard through interface buttons.

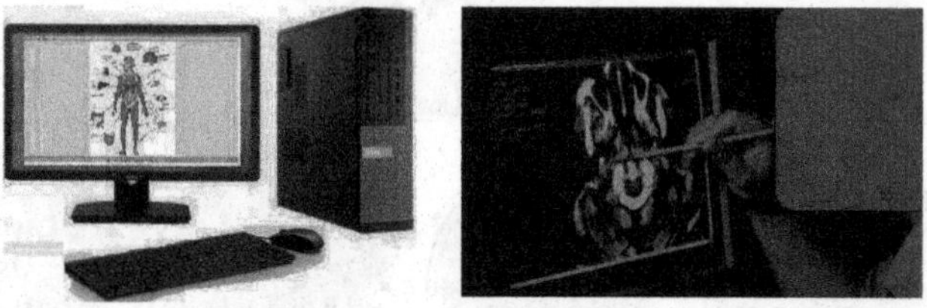

Fig. 2. **medical image monitoring (left) and diagnosis (right) using desktop technology**

A horizontal multi-touch tabletop display was also used as a suitable groupware environment for MIS system. Fig. 2 illustrates how users interact and share medical contents around tabletop. This technology allows collaborative interaction for digital contents from all sides of tabletop [7, 15, 16].

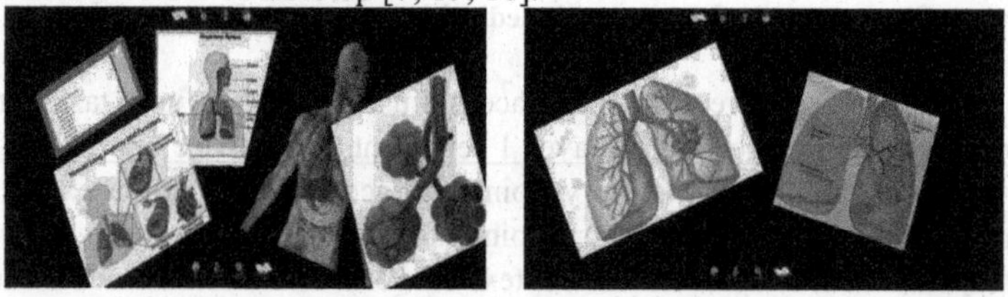

Fig. 3. Monitoring (left) and diagnosis (right) using natural size zoom of multiple finger gesture activates for calibration on region of interest using tabletop technology

MIS system provides natural form of interaction because users can manipulate (view, zoom and rotate) digital contents by using single/multiple fingers on display screen. The content selection is made through natural touch of finger(s). The content can be moved to any direction of display by placing single finger on the content and move it to the location desired (like a drag operation but more intuitively). For the natural size zoom-in/out, the calibration is made using two fingers. The zoom-out functionality is achieved by the movement of both fingers towards each other and zoom-in as movement of fingers away from each other. Furthermore, two or more fingers can be used for the rotation of content, for the best view and comprehension of user.

5 Experiment

This section explains experimental investigation of MIS application. An experiment was conducted to determine the usefulness of MIS application for both technologies. The participants have experienced groupware educational environment for understanding medical concepts. Later on, a quantitative analysis was made to represents empirical user satisfaction to determine user experience.

5.1 Participants

Twenty foundation students volunteers (12 male, 8 female) between the age of 19 and 23 (mean 22.5) participated as pairs in 10 groups. Their educational background ranged from secondary education to post-secondary education. All of the participants have a basic medical knowledge of human anatomy (for example skeletal system, respiratory system, nervous system, digestive system, etc.).

Their mean computer technological experience was 4 years and mean computer daily use was 3.1 hours. Some of the participants have an opportunity for biological investigation using desktop technology but none of them have any experience of using multi-touch tabletop technology.

5.2 Materials

There were two studies conducted in which the participants monitor and diagnose medical artifacts on display screens i.e. 1) desktop display groupware, 2) collaboration on multi-touch tabletop. The participants were able to visualize medical contents on vertical desktop screens (Fig. 1 presents session using desktop display groupware setup). In a multi-touch tabletop setup, a Samsung SUR 40 Microsoft Surface was used as a horizontal multi-touch display screen. It presents the visual display and control on same screen. The touch-sensitive area was 40” horizontal display screen. Participants were free to sit/stand around all accessible sides of tabletop and perform concurrent access of the digital contents on the screen (see Fig. 2).

5.3 User Tasks

The participants experienced abstract activities that required their visual coordination and collaboration skills. Medical monitoring and diagnosis activities were performed through user-tasks. The task consisted of manipulation of digital contents in which the participants were able to perform multiple concurrent selection and manipulation (translate, rotate and zoom) of digital contents for both desktop and tabletop setups. These tasks were experienced within-group of students, in corporative brain-storming session, in which each participant is to be exposed to multiple experimental conditions for medical image investigation of human organs.

5.4 Procedure

The experiment was conducted in University Technology PETRONAS, Malaysia. The participants were grouped together in pairs. They were provided with a basic questionnaire regarding their demographic information and their impression on interaction of technologies (single mouse and multi-touch). Initially they were explained about what they will experience during the user test. In first half of the experiment for around ten minutes, they experienced existing demo applications (Google maps, puzzle and Window photo viewer) on desktop setup and tabletop setup in a random order. After having certain level of understanding of using these technologies, they have experienced MIS application for fifteen minutes on both technologies in a random order. The interaction and collaboration of participants were observed throughout the user test. After completion of user-task a questionnaire [1] was filled by participants regarding their perceived experience of using MIS application.

6 RESULTS AND DISCUSSION

As discussed in above section, the participants experienced two groupware setups. They were able to collaborate, learn, discuss and showed positive influence towards the use of human-computer technologies. Table 2 presents the observations made during experiment of monitoring and diagnosis of medical artifacts using computer technologies.

	Desktop Technology	Tabletop Technology
User	Emphasis on 'I'	Emphasis on 'we'
Face-to face	Self-face concern	Other-face concern
Need of face-to-face	Negative	Positive
Multi-synchronous interaction	None	Unlimited interaction
Collaboration	Limited collaboration	Unlimited collaboration
Learning	Partial learning	Comprehensive learning
Verbal communication	Direct speech act	Indirect speech act
Non-verbal communication	Individual non-verbal acts, direct emotional expressions	Indirect emotional expressions, role oriented
Organizing the task	Limited	Interminable
Workspace	Small work space	Substantial space
Regulating collaboration	Minimum	Maximum
Participation awareness	Least awareness	Maximum awareness

Table 1. Observation of user study for desktop and tabletop technologies

The desktop setup provide individual working environment with unnatural interaction (using mouse and keyboard) towards less creative actions. Such environment provides less coordination among users. With the advantage of ease of use of such traditional interaction techniques (using mouse, keyboard etc.), it represents a bottleneck with naturalness of human-computer interaction.

The questionnaire was used as an instrument that covers factors as overall impression, ease of use, comfort, performance, learnability, interaction, collaboration, participation awareness, similarity with real situation and recommendation to others to use such computer technology. The response was given on 5 point likert scale range from 'strongly disagree' (1), 'disagree' (2), 'unsure' (3), 'agree' (4) to 'strongly agree' (5). Fig. 3 presents quantitative result of the experiment.

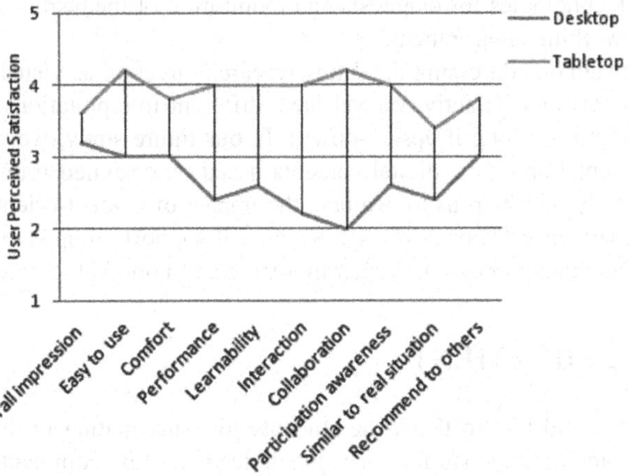

Fig. 4. The quantitative results of comparative user study. The 5 point scale range from 'strongly disagree' (1) through 'unsure' (3) to 'strongly agree' (5).

For understanding human anatomy using MIS application, the results showed subjective satisfaction of the use of tabletop technology (overall mean 3.9, SD 0.39) as compared to desktop technology (overall mean 2.64, SD 0.25). Also, paired sample t-test suggests that there is a significance difference in user perceived satisfaction within the group who used desktop and tabletop technologies ($t(9) = 7.465$, $p < 0.05$). The results highlight that interaction (desktop 2.4, tabletop 4.0) and collaboration (desktop 2.0, tabletop 4.2) plays a primary role in a groupware working environments. Regarding tabletop interaction, multi-touch and gestures were intuitive, straight forward and easy to use (desktop 3.0, tabletop 4.2).

For those participants who have never used tabletop technology provide positive feedback with an interesting potential of its use. It was also identified that the rendering of data is useful in original size as similar to real situation. But in majority of sessions a higher level of zoom was used, for instance, for close-up views. Some of the test participants were not so comfortable (desktop 3.0, tabletop 3.8) in term of interaction and viewing the content on the screen. The desktop limits the collaborative interaction whereas the tabletop confines users view while standing in opposite direction of digital content. An effective content orientation [7] technique in tabletop environment enables increased effectiveness, efficiency and user satisfaction to work in group with significance of user performance.

7 CONCLUSION AND FUTURE WORK

The use of tabletop technologies and appropriate interfaces show fundamental role of enhancing creative education and learning. The foundation of MIS prototype was designed and implemented on the bases of conceptual medical investigation framework. It have explored the space of collaborative computer technologies for medical diagnosis and monitoring system. Using theoretical concepts, observational findings and user feedback, we identified that the tabletop technologies were more effective than desktop technologies for diagnosis and monitoring of medical digital contents in collaborative working environment.

In our next step on addressing the future research direction as visual representation of digital contents and its activity level have different interpretations from different sides (for example viewing it upside-down). In our future study, we will be on controlling the orientation of the digital contents based on designed framework. An extended field study will help us to evaluate the impact of content-orientation to determine user performance for professional use. It will support groupware working environments to encourage users with higher motivation and optimal learning experience.

ACKNOWLEDGEMENT

The author would like to thank the students for participating in the experiments. Special thanks to Aiman Khan for her valuable feedback and comments on the paper. The authors also like to thank anonymous reviewers for their valuable suggestions for the paper.

REFERENCES

1. Claes Lundstr¨om, Thomas Rydell, Camilla Forsell, Anders Persson, and Anders Ynnerman, Multi-Touch Table System for medical Visualization: Application to Orthopedic Surgery Planning. IEEE. 2011
2. Packer DL, Aharon S, Johnson SB, Camp JJ, Robb RA. "Feasibility of virtual anatomy-based three-dimensional mapping in a canine model. J. Am. Coll. Cardiol., 31(2):181A-182A. 1998
3. Y. Hu and R. A. Malthaner. The feasibility of three-dimensional displays of the thorax for preoperative planning in the surgical treatment of lung cancer. European Journal of Cardio-Thoracic Surgery, 31(3):506 – 511. 2007.
4. M. Shiozawa, N. Sata, K. Endo, M. Koizumi, Y. Yasuda, H. Nagai, and H. Takakusaki. Preoperative virtual simulation of adrenal tumors. Abdominal Imaging, 34:113–120, 2009.
5. S.-W. Lee, H. Shinohara, M. Matsuki, J. Okuda, E. Nomura, H. Mabuchi, K. Nishiguchi, K. Takaori, I. Narabayashi, and N. Tanigawa. Preoperative simulation of vascular anatomy by three-dimensional computed tomography imaging in laparoscopic gastric cancer surgery. Journal of the American College of Surgeons, 197(6):927 – 936, 2003.
6. TRobb, R.A.: Three-Dimensional Visualization in Medicine and logy. Book Chapter in: Handbook of medical Imaging: Processing and Analysis, ed. Isaac N. Bankman, Academic Press, San Diego, CA, Chapter 42, pp. 685-712, 2000.

7. Tahir Mustafa Madni, Suziah Bt Sulaiman, Muhammad Tahir. Content-Orientation for Collaborative Learning using Tabletop Surfaces. IEEE, 2013.
8. Erin Walker, Nikol Rummel, Kenneth R. Koedinger. CTRL: A research framework for providing adaptive collaborative learning support. Springer Science and Business Media. 2009.
9. Alejandro Catala, Javier Jaen, Adria A. Martinez-Villaronga, Jose A. Mocholi, "AGORAS: Exploring Creative Learning on Tangible User Interfaces", 2011, 35th IEEE Annual Computer Software and Applications Conference
10. Tim Warnecke, Patrick Dohrmann, Alke Jürgens, Andreas Rausch,Niels Pinkwart, "Collaborative Learning through Cooperative Design Using a Multitouch Table", 2011, Springer-Verlag Berlin Heidelberg
11. Y. Jung, J. Keil, J. Behr, S. Webel, M. Z¨ollner, T. Engelke, H. Wuest, and M. Becker. Adapting X3D for multi-touch environments. In Proceedings of the 13th international symposium on 3D web technology, pages 27–30, 2008.
12. T. Isenberg, M. H. Everts, J. Grubert, and S. Carpendale. Interactive exploratory visualization of 2d vector fields. Computer Graphics Forum, 27(3):983–990, 2008.
13. L. Yu, P. Svetachov, P. Isenberg, M. H. Everts, and T. Isenberg. FI3D: Direct-touch interaction for the exploration of 3D scientific visualization spaces. IEEE Transactions on Visualization and Computer Graphics, 16(6):1613–1622, 2010.
14. Bruno Loureiro, Rui Rodrigues, "Multi-Touch as a Natural User Interface for Elders: A Survey", 2011, IEEE
15. Tim Warnecke, Patrick Dohrmann, Alke Jürgens, Andreas Rausch, Niels Pinkwart, "Collaborative Learning through Cooperative Design Using a Multitouch Table", 2011, Springer-Verlag Berlin Heidelberg
16. Alistair Jones, Atman Kendira, Dominique Lenne, Thierry Gidel, Claude Moulin, "The TATIN-PIC Project A Multi-modal Collaborative Work Environment for Preliminary Design" 2011, 15th International Conference on Computer Supported Cooperative
17. D. Coffey, N. Malbraaten, T. Le, I. Borazjani, F. Sotiropoulos, and D. Keefe. Slice WIM: a multi-surface, multi-touch interface for overview+ detail exploration of volume datasets in virtual reality. In Symposium on Interactive 3D Graphics and Games, pages 191–198. ACM, 2011.
18. W. Krueger and B. Froehlich. The Responsive Workbench. IEEE Computer Graphics and Applications, 14(3):12–15, May 1994.
19. C. Lin, R. Loftin, I. Kakadiaris, D. Chen, and S. Su. Interaction with medical volume data on a projection workbench. In The Proceedings of 10th International Conference on Artificial Reality and Telexistence, pages 148–152, 2000.
20. F. Volonte, J. Robert, O. Ratib, and F. Triponez. A lung segmentectomy performed with 3D reconstruction images available on the operating table with an iPad. Interactive CardioVascular and Thoracic Surgery, 2011.
21. L. Gallo, A. Minutolo, and G. D. Pietro. A user interface for VR-ready 3D medical imaging by off-the-shelf input devices. Computers in logy and Medicine, 40(3):350 – 358, 2010.
22. J. H. F. Jonathan Lazar, Harry Hochheiser, Research Methods in Human-Computer Interaction. Glasgow: John Wiley & Sons Ltd, 2010.
23. N. A.-h. Hamdan, S. Voelker, and J. O. Borchers, "Conceptual framework for surface manager on interactive tabletops," presented at the CHI '13 Extended Abstracts on Human Factors in Computing Systems, Paris, France, 2013.
24. H. S. Yvonne Rogers, Jenny Preece, *Interaction Design, Beyond Human -Computer Interaction*, Third ed. United Kingdom: John Wiley & Sons Ltd, 2011.

Dynamic-Time-Warping Analysis of Feature-Vector Reliability for Cognitive Stimulation Therapy Assessment

Tuan D. Pham

Aizu Research Cluster for Medical Engineering and Informatics
Research Center for Advanced Information Science and Technology
The University of Aizu
Aizuwakamatsu, Fukushima, 965-8580, Japan
E-mail: tdpham@u-aizu.ac.jp

Abstract. Cognitive stimulation therapy (CST) can help people with mental illness improve their health condition. In particular, CST provides an alternative treatment for people with mild to moderate dementia. Signal processing and pattern recognition methods are promising tools for automated assessment of the effectiveness of CST in treating individuals with dementia. This paper applies the dynamic time-warping for investigating the reliability of photoplethysmography-derived features extracted by the largest Lyapunov exponents and spectral distortion for CST evaluation.

Keywords: Dynamic time-warping; Largest Lyapunov exponents; Spectral distortion; Dementia; Cognitive stimulation.

1 Introduction

While traditional cognitive training interventions are delivered by humans, a recent review concluded that computer-based cognitive interventions are comparable or better than paper-and-pencil cognitive training approaches [1]. This review suggests that the utilization of computerized technology offers an effective and labor-saving method for improving and maintaining the quality of life and confidence of the individual with age-related impairment in cognitive function. In general, cognitive stimulation therapy is found to be less expensive than usual care with respect to benefits in cognition and quality of life. There is also evidence showing that cognitive stimulation therapy can be more cost-saving than dementia medication (http://www.cstdementia.com/page/cost-effectiveness). Because the evaluation of the cost-effectiveness of psychosocial interventions in dementia is becoming increasingly important, assessing the efficacy of cognitive stimulation therapy can even further contribute to the cost effectiveness.

Based on the motivation of the importance of the use of CST for mental health, photoplethysmograph (PPG) was applied for the pattern analysis of

K.J. Kim (ed.), *Information Science and Applications*,
Lecture Notes in Electrical Engineering 339, DOI 10.1007/978-3-662-46578-3_27

short-term effects on efficacy in a caregiver on the daily provision of therapeutic treatment to aging people with dementia using the feature vectors of the largest Lyapunov exponent (LLE) values and spectral distortion. PPG is an optically obtained plethysmogram as a volumetric measurement of an organ, often obtained by using a pulse oximeter which illuminates the skin and measures changes in light absorption [2]. A pulse oximeter monitors the perfusion of blood to the microvascular layer of the tissue of the skin into which infrared light is emitted [3]. PPG technology has been pervasive as a low-cost, non-invasive, and flexible tool for physiological analysis in medicine and health, such as cardiology [4, 5], paediatric intensive care [6], hypertension [7], and depression [8]. In particular, wearable devices of photoplethysmographic sensors have been progressively developed [9, 10], making the PPG technology more atractive for its applications in telemedicine and e-health.

However, given the outcome obtained from the PPG-based pattern analysis, a question of interest is: How reliable are the analysis results since they are difficult to be validated by human response? This is the motivation of the present study that presents the application of the dynamic time-warping for establishing the uncertainty of the computerized assessment of CST effectiveness.

2 Methods

2.1 Largest Lyapunov exponents

Given a time series of length N, the first step is to reconstruct the phase space of the dynamical system using the time-delay method proposed by Takens (1981). Let m and L be the embedding dimension and time delay (lag). The reconstructed phase space can be expressed in matrix form as

$$\mathbf{X} = (\mathbf{X}_1, \mathbf{X}_2, \ldots, \mathbf{X}_M)^T \tag{1}$$

where $\mathbf{X}$ is matrix of size $M \times m$, $M = N-(m-1)L$, and $\mathbf{X}_i = (x_i, \ldots, x_{i+(m-1)L})$ which is the state of the system at discrete time i.

$$d_j(0) = \min_{\mathbf{X}_{j^*}} ||\mathbf{X}_j - \mathbf{X}_{j^*}|| \tag{2}$$

where $|j - j^*| > MP$ where MP is the mean period which is the reciprocal of the mean frequency of the power spectrum.

The basic idea is that the LLE (λ_1) for a dynamical system can be defined as [13]

$$d(t) = c \; e^{\lambda_1 t} \tag{3}$$

where $d(t)$ is the average divergence of two randomly chosen initial conditions at time t, and c is a constant that normalizes the initial separation between neighboring points.

By the definition given in Eq. (3), the j pair of nearest neighbors can be assumed to diverge at a rate measured by λ_1 as follows:

$$d_j(i) \approx c_j \; e^{\lambda_1(i\Delta t)} \tag{4}$$

where $d_j(i)$ is the distance between the j pair of nearest neighbors after i discrete-time steps which is $i\Delta t$, Δt is the sampling period of the time series, and c_j is the initial separation between two neighboring points.

Taking the logarithm of both sides of Eq. (4), giving

$$\ln[d_j(i)] \approx \lambda_1(i\Delta t) + \ln(c_j) \tag{5}$$

where $d_j(i) = ||\mathbf{X}_j(i) - \mathbf{X}_{j^*}(i)||$. Eq. (5) gives a set of approximately parallel curves, one for each j ($j = 1, \ldots, M$). If these curves are approximately linear, their slopes represent the LLE (λ_1). The LLE can be computed as the slope of a straight-line fit to the average logarithmic divergence curve defined by

$$s(i) = \frac{1}{i\Delta t} < \ln[d_j(i)] >_j \tag{6}$$

where $< \cdot >_j$ denotes the average over all values of j.

2.2 Spectral distortion

Spectral distortion measures are designed to compute the dissimilarity or distance between two (power) spectra [14] (the power spectrum of a signal describes how the variance of the data is distributed over the frequency components into which the signal may be decomposed, and the most common way of generating a power spectrum is by using a discrete Fourier transform) of the two feature vectors, originally developed for comparison of speech patterns [15]. Three methods of spectral-distortion measures were used in this study, based on their popular applications in signal processing: Itakura distortion (ID), log spectral distortion (LSD), and weighted cepstral distortion (WCD) [15]. Unlike the Itakura distortion, both log spectral distortion (distance) and weighted cepstral distortion (distance) are symmetric. Only the ID was applied in this study.

Consider two signals S and S', and their two spectral representations $S(\omega)$ and $S'(\omega)$, respectively, where ω is normalized frequency ranging from $-\pi$ to π. The Itakura distortion of S and S' is a likelihood-based measure and defined as [16]

$$\begin{aligned} ID(S, S') &= \log \left[\int_{-\pi}^{\pi} \frac{|S(\omega)|^2}{|S'(\omega)|^2} \frac{d\omega}{2\pi} \right] \\ &= \log \frac{\mathbf{a}^T \mathbf{R}' \mathbf{a}}{\sigma'^2}, \end{aligned} \tag{7}$$

where σ'^2 is the prediction error of S' produced by the linear prediction coding (LPC) [15], $\mathbf{a}$ is the vector of LPC coefficients of S, $\mathbf{R}'$ the LPC autocorrelation matrix of S'. It is shown that $ID(S, S') \neq ID(S', S)$, hence to make the measure symmetrical, a natural expression of its symmetrized version, denoted as $ID_s(S, S')$, is

$$ID_s(S, S') = \frac{ID(S, S') + ID(S', S)}{2}. \quad (8)$$

2.3 Dynamic time-warping

Dynamic time-warping (DTW) measures the similarity between two sequences or vectors which may vary in time. In general, DTW is a method that calculates an optimal match between two given sequences with some constraints. The sequences are warped non-linearly in time to determine a measure of their similarity, being independent of certain non-linear variations in time. This vector alignment method, carried out via the dynamic programming procedure, is often used for time-series classification. DTW criteria that need to be determined are [18]: 1) global constraints, 2) local constraints, 3) end-point constraints, and 4) transition cost.

The global constraints define the overall stretching or compression of the pair of vectors allowed for the matching procedure. The local constraints set limits for maximum expansion or compression that successive transitions can achieve. Any local constraint must satisfy the criterion of monotonicity that means all predecessors of a node of the grid are located to its west and south orientation to ensure the matching operation following natural time evolution. The end-point constraints aim to search for the optimal complete path that starts at the first points of the pair of feature vectors (0,0) and whose first transition is to node (1,1). The Euclidean distance is commonly adopted as the cost for the transitions [18].

3 Results

The preprocessed PPG data of the care-giver and 18 selected participants with dementia, which were synchronously recorded, were used to calculate the ID and LLE-based Euclidean distance between the care-giver and each of the participants before, during and after the therapeutic session. Figure 1 shows segments of typical pre-processed PPG signals of the care-giver and a participant. The dissimilarity matrices of the PPG data between the care-giver and the participants obtained from the ID and LLE-based Euclidean distance were then used to construct the "phylogenetic" trees with the UPGMA algorithm. Figures 2 and 3 show two ID-based trees of the synchronized PPG data of two participants and the care-giver, in which the terms *Care-giver*, *Before care*, *During care*, and *After care* in the tree nodes denote the care-giver, the participated individual before, during, and after the therapeutic session, respectively. Based on the inference of the phylogenetic tree reconstruction, the evidence of influence of the care-giver over a particular participant is when the PPG patterns of the care-giver and the participant during the session belong to the same node (Figure 2). Otherwise, the CST is considered ineffective (Figure 3 is an example).

The DTW was used to evaluate the reliability of the trees obtained from the ID and LLE based measures of similarity. The global constraints adopted in

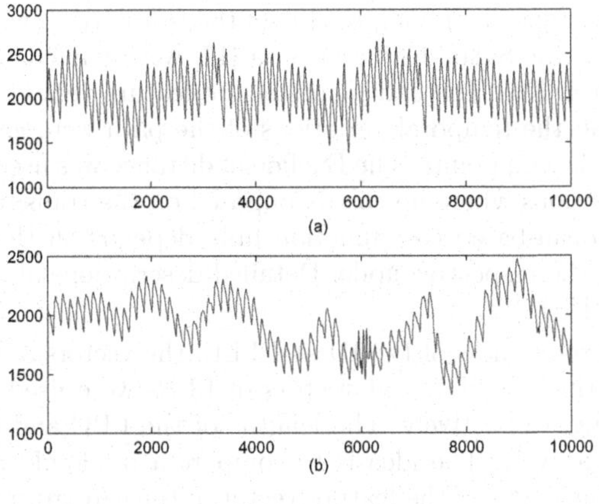

Fig. 1. First 10000 samples of synchronized PPG signals of (a) elderly participant with dementia, and (b) care-giver.

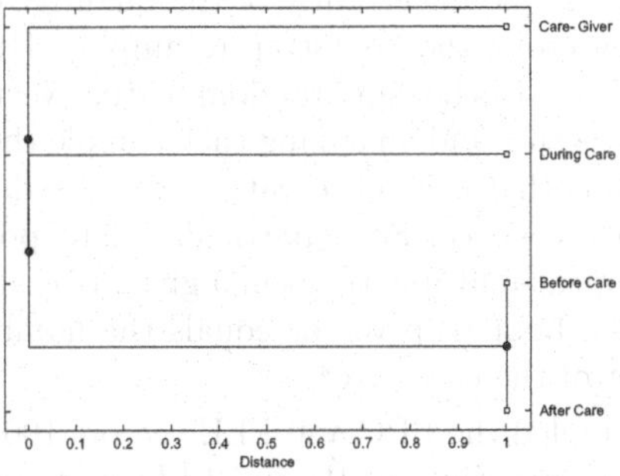

Fig. 2. Synchronized cognitive stimulation communication between care-giver and participant #1

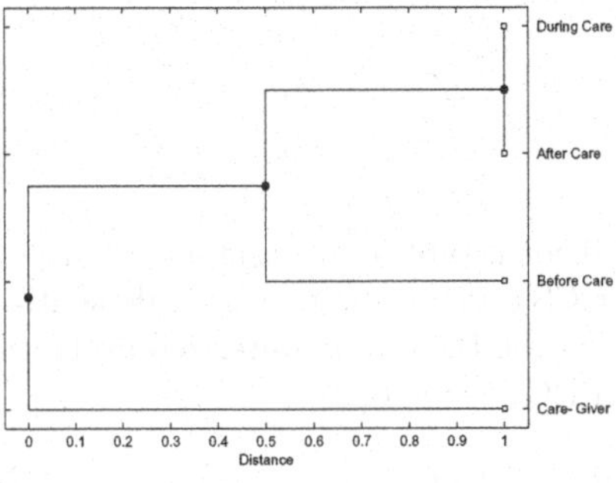

Fig. 3. Synchronized cognitive stimulation communication between care-giver and participant #2

this study are known as Itakura constraints [16] and a maximum factor of 2 were imposed for any expansion or compression of the test pattern with respect to the reference pattern. The Sakoe-Chiba criteria [17] were implemented as the local constraints in the DTW. The end points of the feature vectors were considered given a priori, and the temporal variations of the path were located within the range defined by the end points. The Euclidean distance was used to calculate the cost for the transitions where no cost is imposed on the transitions to a specific node. This Euclidean-based cost function fully depends on the feature vectors corresponding to the respective node. Detailed descriptions of these constraints can be found in [15, 18].

To assess the robustness of the ID and LLE, the vectors of LPC coefficients (used for calculating the ID), and vectors of LLEs were used for DTW-based template matching, respectively. The lengths of the LPC and LLE vectors are 16 and 103, respectively. The idea is to compare a randomly generated vector that is of the same length of the feature vector of the care-giver with the feature vectors of the care-giver and during care. Using the DTW, the calculated similarity of the feature vectors of the care-giver and during care must be smallest among other pair-wise similarity measures associated with the random vector. To establish statistics for the uncertainty of the quantity of the reliability evaluation of the feature vectors, the bootstrap resampling [19] was employed. The bootstrap procedure involves choosing random feature vectors with replacement from a sample feature vector and analyzing each sample the same way. Sampling with replacement means that each observation is selected separately at random from the original feature space. So a particular data point from the original feature space could appear multiple times in a given bootstrap vector. The number of elements in each bootstrap vector equals the number of elements in the original feature vector of the care-giver.

The boostrap resampled the LPC and LLE vectors 100 times. The reliability of using the LPC for computing the ID and LLE for calculating the Euclidean distance for the construction of the phylogenetic trees was found to be 100% and 18%, respectively. The use of the spectral distortion for CST assessment is obviously robust and preferred to the LLE-based Euclidean distance.

4 Conclusion

The DTW was applied for establishing insights into the uncertainty of the ID and LLE measures for CST for partipants with dementia. The results strongly suggest the utilization of the ID for computerized evaluation of the effectiveness of the CST using the PPG technology.

Acknowledgments: PPG and LLE data were provided by Mayumi Higa of Chaos Technology Research Lab, Shiga, Japan. Satoshi Haga assisted the author in carrying out the computer experiments.

References

1. A.M. Kueider, J.M. Parisi, A.L. Gross, G.W. Rebok, Computerized cognitive training with older adults: a systematic review, *PLoS ONE*, 7 (2012) e40588.
2. K. Shelley, S. Shelley, Pulse Oximeter Waveform: Photoelectric Plethysmography, in: *Clinical Monitoring*, C. Lake, R. Hines, C. Blitt, Eds, pp. 420-428, 2001.
3. J. Allen, Photoplethysmography and its application in clinical physiological measurement, *Physiological Measurement*, 28 (2007) R1-R39.
4. H.H. Asada, P. Shaltis, A. Reisner, S. Rhee, R.C. Hutchinson, Mobile monitoring with wearable photoplethysmographic biosensors, *IEEE Engineering in Medicine and Biology Magazine*, 22 (2003) 28-40.
5. A.T. Reisner, P.A. Shaltis, D. McCombie, H.H. Asada, Utility of the photoplethysmogram in circulatory monitoring, *Anesthesiology*, 108 (2008) 950-958.
6. B. Frey, K. Waldvogel, C. Balmer, Clinical applications of photoplethysmography in paediatric intensive care, *Intensive Care Medicine*, 34 (2008) 578-582.
7. E. Monte-Moreno, Non-invasive estimate of blood glucose and blood pressure from a photoplethysmograph by means of machine learning techniques, *Artificial Intelligence in Medicine*, 53 (2011) 127-138.
8. T.D. Pham, C.T. Truong, M. Oyama-Higa, M. Sugiyama, Mental-disorder detection using chaos and nonlinear dynamical analysis of photoplethysmographic signals, *Chaos, Solitons & Fractals*, 51 (2013) 64-74.
9. K. Sonoda, Y. Kishida, T. Tanaka, K. Kanda, T. Fujita, K. Higuchi, K. Maenaka, Wearable photoplethysmographic sensor system with PSoC microcontroller, *Int J Intelligent Computing in Medical Sciences & Image Processing*, 5 (2013) 45-55.
10. R. Yousefi, M. Nourani, S. Ostadabbas, I. Panahi, A motion-tolerant adaptive algorithm for wearable photoplethysmographic biosensors, *IEEE J Biomedical and Health Informatics*, 18 (2014) 670-681.
11. J.L. Sprott, *Chaos and Time-Series Analysis.* Oxford, New York, 2003.
12. J.B. Dingwell, Lyapunov exponents, in: Metin Akay, Ed, *Wiley Encyclopedia of Biomedical Engineering.* John Wiley & Sons, New York, 2006, 12 pages.
13. M.T. Rosenstein, J.J. Collins, C.J. DeLuca, A practical method for calculating largest Lyapunov exponents from small data sets, *Physica D: Nonlinear Phenomena*, 65 (1993) 117-134.
14. W.H. Press, B.P. Flannery, S.A. Teukolsky, W.T. Vetterling, Power Spectra Estimation Using the FFT, in: *Numerical Recipes in FORTRAN: The Art of Scientific Computing*, 2nd ed. Cambridge University Press, Cambridge, 1992, pp. 542-551.
15. L. Rabiner, B-H. Juang, *Fundamentals of Speech Recognition.* Prentice-Hall, Englewood Cliffs, 1993.
16. F. Itakura, Minimum prediction residual principle applied to speech recognition, *IEEE Trans Acoust Speech Signal Process*, 23 (1975) 67-72.
17. H. Sakoe, S. Chiba, Dynamic programming algorithm optimization for spoken word recognition, *IEEE Trans Acoustics, Speech and Signal Processing*, 26 (1978) 43-49.
18. S. Theodoridis, K. Koutroumbas, *Pattern Recognition.* Academic Press, London, 2009.
19. B. Efron, R. Tibshirani, *An Introduction to the Bootstrap.* Chapman & Hall/CRC, Boca Raton, 1993.

An Algorithm for Rock Pore Image Segmentation

Zhang Jiqun[1], Hu Chungjin[1], Liu Xin[1], He Dongmei[2], Li Hua[3]

[1] University of Science and Technology Beijing
{zjqwxs, huchungjin}@sina.com, ustb.liuxin@gmail.com
[2] Petrochina Planning and Engineering Institute
dmhe@petrochina.com.cn
[3]Research Institute of Petroleum Exploration & Development
lihua001@petrochina.com.cn

Abstract. A new algorithm used for rock pore segmentation is presented in this paper. Using the morphological erosion operator and hit-or-miss transform from mathematical morphology, the dividing lines' possible positions are derived. The dividing lines' direction can be further confirmed by dilation. The authenticity of the dividing lines can be verified using conditional dilation. Consequently, the accurate pore segmentation can be achieved. It has been proven by various practical applications that this algorithm is relatively accurate, rapidly executed, and insensitive to noise. In effect, the algorithm has excellent anti-noise ability.

Keywords: pore, throat, image segmentation, erosion, dilation, hit-or-miss transform, dividing line

1 Introduction

With the continuous development of mathematical morphology, computer has played more and more important role in the rock sample analysis in oil and gas industry. Since it provides convenient, accurate and quantitative measures for the pore structure, the image analysis gradually becomes a powerful tool for mineral identification and analysis. Pore structure analysis is based on identification of pore and throat. Pore is defined as the empty space in the rock which is not occupied by the solid minerals. The throat refers to the narrow pathway which connects the pores in different types of reservoir.

Nowadays there are state-of-the-art methods for pore segmentation using mathematical morphology, dominated by the following methods: (A) Watershed Segmentation Algorithm[1], which converts pores' binary images to grayscale images based on distance function. Afterwards the ultimate erosion and dilation will be performed. (B) Medial Axis Algorithm[2], which extracts the medial axis of the pores after erosion operation is performed based on thinning algorithm. The branch points of the medial axis are considered as the loci of the pores. The dividing line is calculated as the connecting line between two pores' loci with minimum number of erosion. (C) Grayscale

K.J. Kim (ed.), *Information Science and Applications*,
Lecture Notes in Electrical Engineering 339, DOI 10.1007/978-3-662-46578-3_28

dilation and erosion using mathematical morphology[3,4]. Erosion operation is performed first on the grayscale images. As a result, the pore central regions are obtained. The final image segmentation is achieved by applying dilation on these pore central regions.

Apparently each algorithm has its benefit and limitation. The execution time of the first method is generally acceptable, but it is only applicable to the rocks with poorly connected pores. Sometime it can lead over segmentation due to additional small seed region might be obtained by applying ultimate erosion. The second method is relatively sensitive to the image edges and is prone to over segmentation. The advantage of the latter method is easy operation, but drawback is slow execution. Furthermore, it may cause erroneous segmentation since only the information from grayscale image is consumed without utilizing the information of spatial distribution and texture.

The common feature of the above algorithms is that the pore positions are located by applying ultimate erosion first, then the dividing lines connecting pore network are explored. A different approach is presented in this paper, which finds the possible locations of dividing lines first, and then the authenticity of the pores is evaluated. It is illustrated by the following rock back scattered SEM Image (Fig.1).

2 ALGORITHM BASIC PRINCIPLE AND ITS MATHEMATICAL DESCRIPTION

2.1 Pre-Processing

Before the segmentation operation is performed, a binary process has been applied to the pore grayscale image to extract all the pores as shown Fig.2, which is based upon the iterative threshold segmentation algorithm[9].

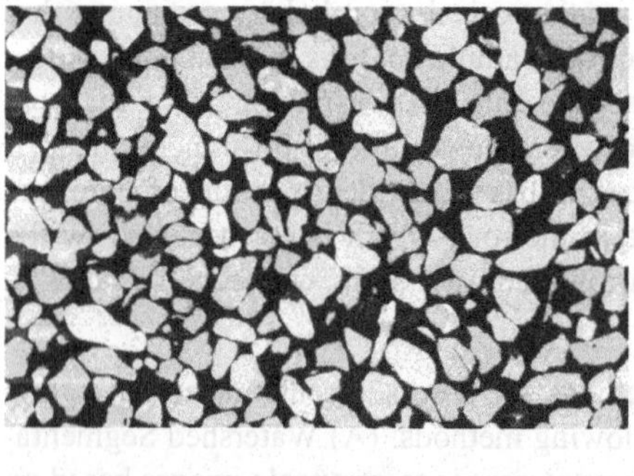

Fig. 1. Rock Back Scattered SEM Image (Dark black denotes pore, light gray denotes rock particles)

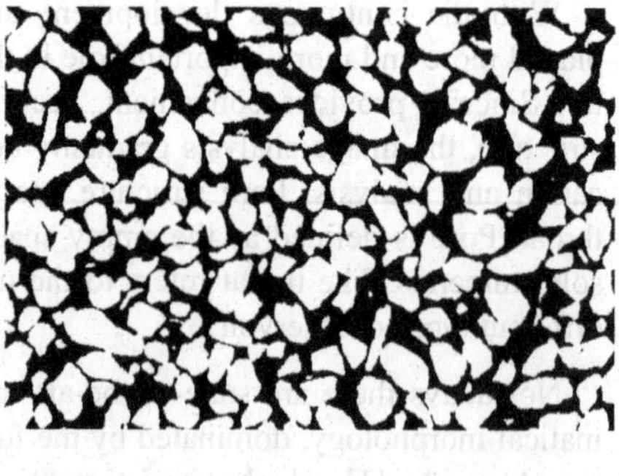

Fig. 2. Rock Pore Image after Binary Process (Black denotes pore, white denotes not pore)

Apparently various pores are connected through the throats to form an interconnected region. The purpose of the new algorithm presented here is to identify the complete throats after the connecting pores are segmented. First, a closing operation is executed before the pore segmentation in order to filter out foreground noise. Afterwards, an opening operation is executed to filter out background noise.

2.2 Basic Principle

A series of erosion is applied to the pore image after the binary process. The narrowest place connecting the pores is located after every erosion process, which is the potential dividing line between the pores. Then both ends of the dividing line are further evaluated, which also solves the noise sensitivity. As a result, a throat will be confirmed when both ends are verified as pores and they are separated at this location. Otherwise, it is not the true throat.

2.3 Algorithm mathematic descriptions and steps

The algorithm is illustrated by a simulated example shown in Fig.3. There are three pores connected by throats. The location of throats will be identified and the connected region will be split into three sub-regions.

Step 1 : Identify the possible locations of dividing lines.

Erosion is operated on the pores in Fig.4 (Formula 1)

$$A \ominus B = \{x: B + x \subset A\} \tag{1}$$

A – Input image
B – Structure element
Hit-or-Miss Transform is used (so-called Serra transform, formula 2).

$$A * B = (A \ominus E) \cap (A^C \ominus F) \tag{2}$$

A – Input image
A^C – A's complementary set
E – Studying image internal structure element
F – Studying image external structure element

During erosion, every pixel is evaluated to check whether it can split the originally connected region. If it does, it will be considered as a dividing point and Step 2 is executed, otherwise, Step 1 is executed. The process terminates if any of the following conditions is met.

Condition 1 – Erosion has been operated on all pixels of the pores and then calculation stops.

Condition 2 – For the next round of erosion, one or two pixels will be added to the existing dividing line. The calculation stops if the length of the resulting dividing line is greater than the maximum value of dividing lines from the user input.

Fig.4 shows the result after the first erosion and no dividing point is found. The second round erosion is applied and the dividing point is found with hit-or-miss transform, as the red pixel shown in Fig.5. It is the possible location of dividing line and Step 2 is executed next.

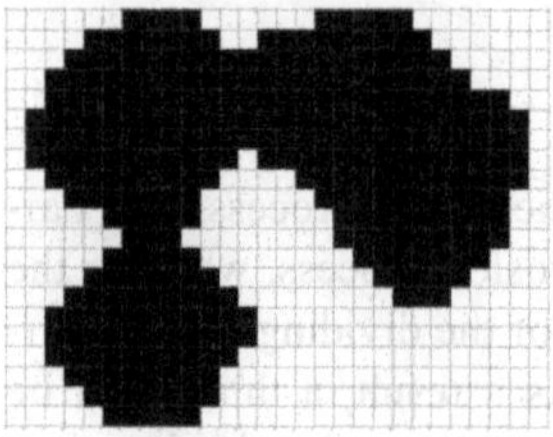

Fig. 3. Simulated image (Black pixel denotes pore)

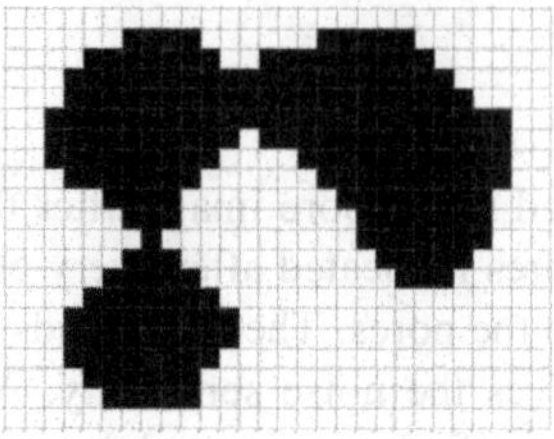

Fig. 4. Results after the first erosion

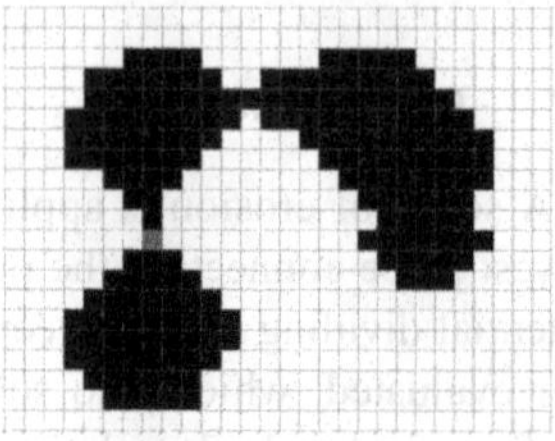

Fig. 5. A dividing point obtained during erosion

Step 2 : Derive dividing line.

After dividing point is found, dilation is operated taking this dividing point as the center (formula 3).

$$A \oplus B = [A^C \ominus (-B)]^C = \cup \{A + b : b \in B\} \quad (3)$$

A – Input image

A^C – A's complementary set

B – Structure element

-B – Rotate B 180 degree clockwise

The pixel after dilation is evaluated to check if it intercepts the previous pore edge after binary process. Dilation continues if there is no interception until the dilated pixel has two intercepted points with the pore edge. The line connecting these two intercepted points is the dividing line. As shown in Fig.6, the dilation stops after the first round since the pixel has two intercepted points. The dividing line is obtained by connecting these two points, as shown in Fig.7.

Step 3: Verify the dividing line authenticity(1).

Based on the dividing point location and dividing line direction, two neighboring points inside the pores can be identified as the "seed points", as the pink pixel shown in Fig.8.

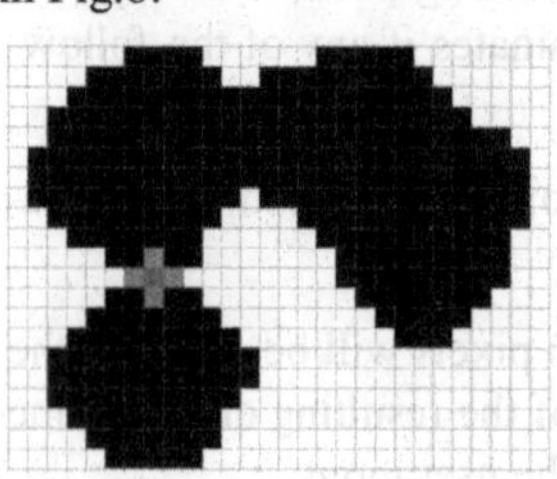

Fig. 6. Dilation of this dividing point

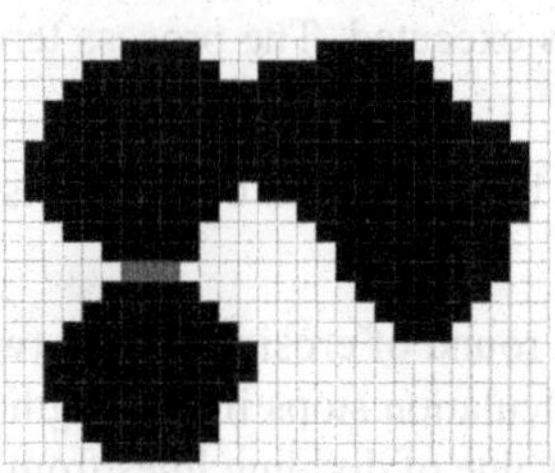

Fig. 7. The dividing line - between two pores

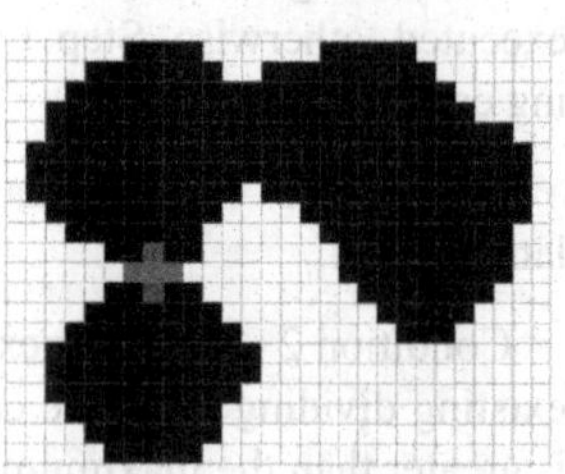

Fig. 8. Image showing the seed points in pink

From these two seed points, the conditional dilation is applied to the segmented pores (Formula 4).

$$H \oplus B: A = \cup \{(B + h) \cap A: h \in H\} \tag{4}$$

A – Input image
B – Structure element
H – Subset of A, H is the seed point before the first dilation.

During the conditional dilation, the filled area in the pores is continuously calculated, as shown in Fig.9.

If both ratios of derived diameter and dividing line length calculated from the neighboring pores are greater than the value defined by the end user, then step 4 is executed next. Otherwise, the dividing line is incorrect and the dividing point shown in Fig.5 needs to be eroded away. Step 1 is executed as the next step, continuing from this point.

Step 4: Verify the dividing line authenticity(2).

Parallel lines of the dividing line were drawn in Fig.7 along the black pixel edge points which spread on both sides of the red pixel in Fig.5. As shown in Fig.10, the length of the dividing line (D_0) is 3, the length of parallel lines above the dividing line ($D_{Up1\text{-}4}$) are 5, 7, 9, 11, and the length below ($D_{Down1\text{-}4}$) are 5, 7, 9, 11 accordingly.

The above length values were plotted as a curve in Fig.11. If the length values increase while departing from the dividing line, then authentic pores are confirmed and so is the dividing line. Step 5 is executed next. Otherwise, the dividing line is incorrect and the dividing point shown in Fig.5 needs to be eroded away. Step 1 is executed as the next step, continuing from the red point shown in Fig. 5.

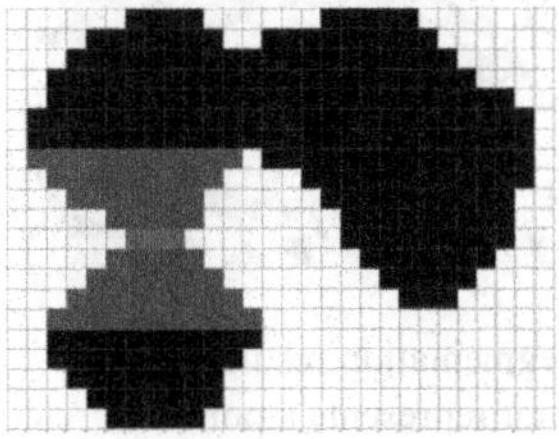

Fig. 9. Conditional dilation of pore

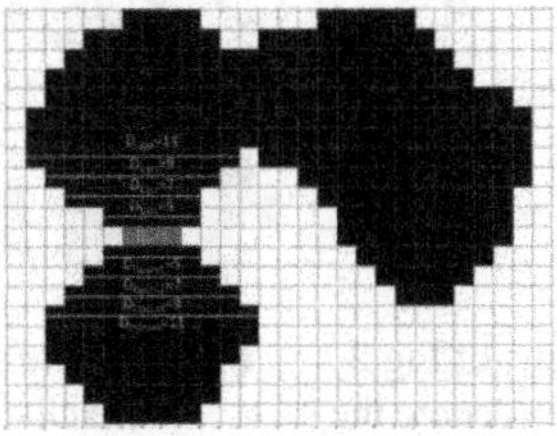

Fig. 10. Parallel lines of dividing line

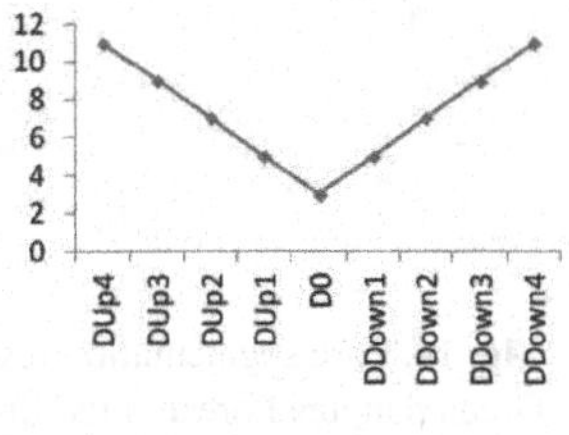

Fig. 11. Parallel lines length values

Step 5 : Indentify throat.

The ending pixels from the dividing line are saved into the dividing line array. Also, the pixel color is modified to white that indicates the non-pore pixel. As a result, the image is generated after two pores are separated, as shown in Fig.12. Loop back to Step 1 and execution continues.

Eventually the second dividing line is found after the algorithm completes. The connecting region in Fig.3 is separated as three pores by two dividing lines, which are the throats shown in Fig.13.

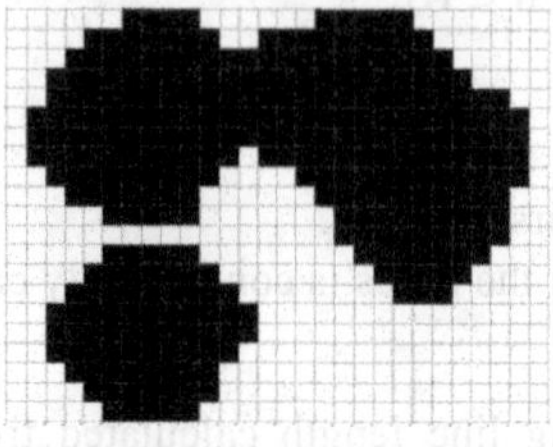

Fig. 12. The pore image after removing the dividing line

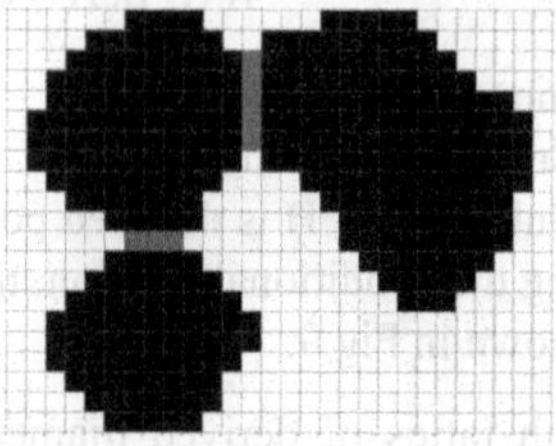

Fig. 13. Pore segmentation results of Fig.3

3 CASE EXAMPLE RESULT

3.1 Case Example 1

The pore segmentation has been performed against the pores in Fig.2 using the new algorithm and the result is shown in Fig.14. In order to clearly display the segmentation result, different colors are used for the each segmented pores, as shown in Fig.15.

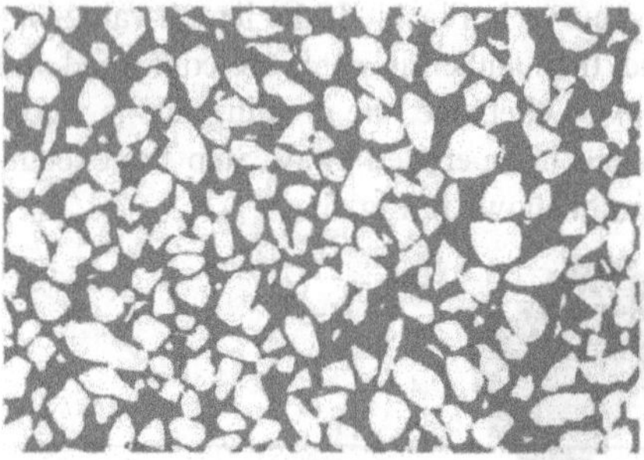

Fig. 14. Pore segmentation results of Fig.2 (Green denotes Pore and red lines denote the dividing line)

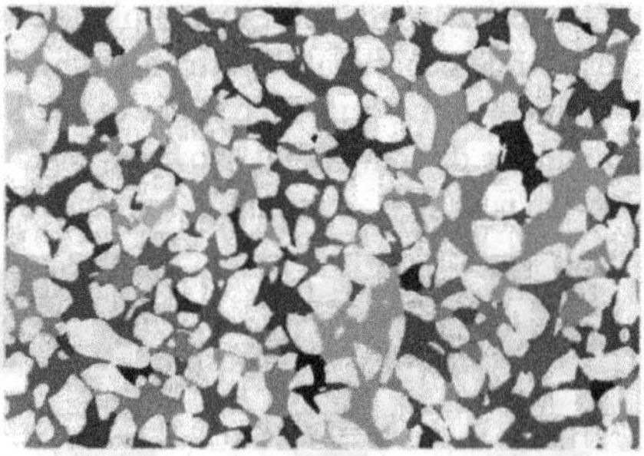

Fig. 15. Pore segmentation results of Fig.2 (Each color represents a pore)

As comparison, the pore segmentation analysis of Fig.2 (1024x768 pixels) is processed using this new algorithm versus the other three algorithms on a Lenovo laptop W500 (Processor Intel(R) Core(TM)2 Duo CPU P8600 @ 2.40GHz, 2GB memory). It takes 0.853 second to eventually obtain 241 pores and 344 dividing lines. The benchmark diagram is shown in Fig.16.

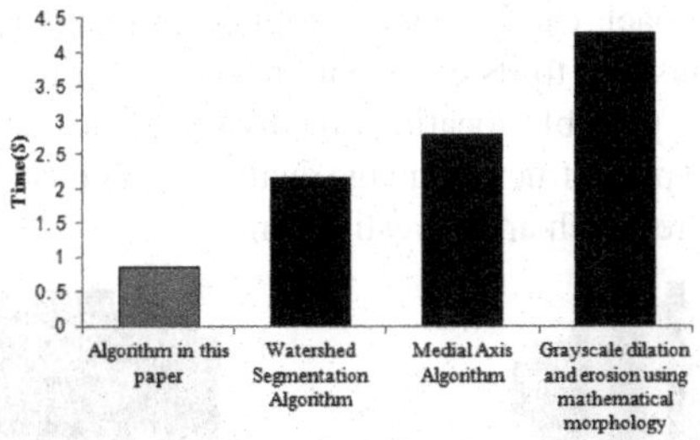

Fig. 16. The comparison of four algorithms – by processing time of Fig.2

3.2 Case Example 2 - Rock Casting Thin Section Image

Other types of images such as rock casting thin section image can also be analyzed using this method. The original rock casting thin section image was shown in Fig.17, and the pore segmentation results were shown in Fig. 18 and Fig.19.

Fig. 17. Rock casting thin section image (Blue denotes pore)

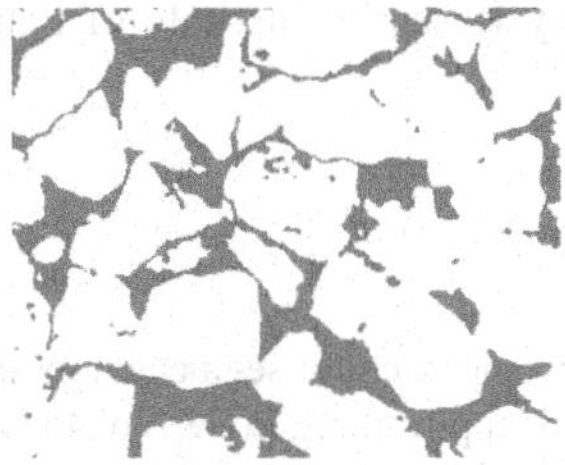

Fig. 18. Pore segmentation results of Fig.17 (Green denotes pore and red lines denote the dividing lines)

Fig. 19. Pore segmentation results of Fig.17 (Each color represents a pore)

4 CONCLUSION

A new pore segmentation algorithm is presented in this paper, which is equally as complex as image thinning algorithm. The method execution time can be significantly improved through reduction of the number of erosions by modifying the longest dividing line, which in turn, leads to the decrease of execution time of this method.

The new segmentation algorithm has excellent anti-noise ability. Step 3 and 4 can significantly reduce the effect of noise in the image. For example, step 3 can determine the pink lines in Fig.20 are not the dividing lines, and step 4 can determine the pink line in Fig.21 is not the dividing line. It will effectively lower over segmentation as compared with other algorithms.

It has already been proven by various practical applications that throats can be accurately and efficiently separated from pores by applying this new segmentation methodology. This approach can be widely applied to image analysis for segmentation of connected regions due to its excellent anti-noise ability. In addition, the algorithm of identifying the possible locations of dividing lines and then evaluating the pores can be used in 3D pore structure image analysis via constructing 3D models, but the effect needs further research and investigation.

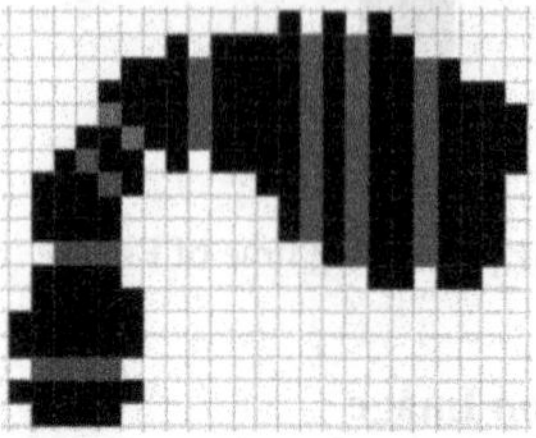

Fig. 20. Step 3 determines the pink lines are not the dividing lines

Fig. 21. Step 4 determines the pink line is not the dividing line

The software application for the pore structure image analysis based on this new algorithm has been utilized by Research Institute of Petroleum Exploration and Development in China.

References

1. YIN Yong, LI A-qiong. New approach for segmentation of overlapping blood cell images. J. Computer Engineering and Applications, 2009, Vol.45 No.35: 173-175
2. HU Dong, Mustafa Touati,Martin J. Blunt. Pore Network Modeling: Analysis of Pore Size Distribution of Arabian Core Samples. R. SPE 105156
3. LIU Li-li, WANG Zheng. Modified Watershed Algorithm for Blood Cells Image Segmentation. J. MICROEL ECTRONICS & COMPUTER. 2010, Vol.27 No.11
4. Liu Jianjun, Lin Lijun, Ji Youjun. Using Rock SEM Image to Create Pore-scale Finite Element Calculation Mesh. C. 2011 International Conference on Physics Science and Technology (ICPST 2011)
5. Dmitry B. Silin, Guodong Jin, Tad W. Patzek. Robust Determination of the Pore Space Morphology in Sedimentary Rocks. R. SPE 84296
6. WANG Xue-li, ZHANG Ji-qun. Background Correction in High Press Stratum Micromodel. J. Science Technology and Engineering. 2012, Vol.12 No.25: 6498−6502
7. V.S. Suicmez, M. Touati. Pore Network Modeling: A new Technology for SCAL predictions and interpretations. R. SPE 110961
8. M.A. Al Ibrahim, N.F. Hurley, W. Zhao, D. Acero-Allard. An Automated Petrographic Image Analysis System: Capillary Pressure Curves Using Confocal Microscopy. R. SPE 159180
9. WANG Zhi-ming. Digital Image Processing and Analysis. M. TSINGHUA UNIVERSITY PRESS. 2012
10. Chui Qi. Image Processing and Analysis - Mathematical Morphology Methods and Applications. M. Science Press. 2000
11. Milan Sonka, Vaclav Hlavac, Roger Boyle. Image Processing, Analysis, and Machine Vision, 3rd ed. M. TSINGHUA UNIVERSITY PRESS. 2011

Action Recognition by Extracting Pyramidal Motion Features from Skeleton Sequences

Guoliang LU[a,b,1], Yiqi ZHOU[a,b,2], Xueyong LI[a,b,3], and Chen LV[c,4]

[a]Key Laboratory of High-efficiency and Clean Mechanical Manufacture of MOE,
[b]School of Mechanical Engineering, Shandong University, Jinan, China
[1,2,3]{luguoliang,yqzhou,lxy88}@sdu.edu.cn
[c]Institute of Computing Technology, Chinese Academy of Sciences, China
[4]lvchen@ict.ac.cn

Abstract. Human action recognition has been a long-standing problem in computer vision. Computational efficiency is an important aspect in the design of an action-recognition based practical system. This paper presents a framework for efficient human action recognition. The novel pyramidal motion features are proposed to represent skeleton sequences via computing position offsets in 3D skeletal body joints. In the recognition phase, a Naive-Bayes-Nearest-Neighbors (NBNN) classifier is used to take into account the spatial independence of body joints. We conducted experiments to systematically test our framework on the public UCF dataset. Experimental results show that, compared with the *state-of-the-art* approaches, the presented framework is more effective and more accurate for action recognition, and meanwhile it has a high potential to be more efficient in computation.

Keywords: Action recognition; Motion skeletons; Action features

1 Introduction

Human action recognition, couple with intelligent video processing, has been received more and more attention in recent days due to its variety of potential applications in home assistant robot or system, natural human-computer interaction (HCI), video indexing/retrieval and so on.

Many studies on action recognition [1] have been already made to address this problem with RGB videos. This solution is, however, limited to some extent in practical usages due to its obtrusive un-protection for individual's privacy. Recently, by virtue of the advances in image/video capturing technique, especially the release of Microsoft Kinect, it has been feasible to capture RGB sequences couple with Depth information, sometimes called Depth maps, in data collection. The depth information enables to provide a rich and unique set of human body shapes and motion information for action discrimination, and moreover, they are relatively unobtrusive and privacy-protecting in applications [2]. Approaches on this topic are mainly divided into two groups depending on whether they are using raw depth information/maps [2, 3] or action skeletons [4]. In this

K.J. Kim (ed.), *Information Science and Applications*,
Lecture Notes in Electrical Engineering 339, DOI 10.1007/978-3-662-46578-3_29

study, we focus on the latter group due to its more received attention. On the other hand, traditional approaches for action recognition were mostly driven by improving recognition accuracy, but in some cases such as real-time HCI and online video processing, computational efficiency is strongly desired as well. Low computational efficiency will cause the system's latency in response behind the user's action performing which further results in bad usage experience for users. The latency is hence a very notable issue in the design of a system [4, 11].

In this study, we also focus on computational efficiency in action classification, which is keeping along with our previous work, e.g., [6–8]. The aim of this study is to exploit an effective and efficient framework for recognizing human actions from skeleton sequences, and thus we avoid calculating computational-complexity high or time consuming action features, e.g., optical flow based action patterns, body parts detection and tracking, features fusion. Here, one notes that the focus of our research is similar to that of [4] which is to explore the trade-off between recognition accuracy and observational latency in action recognition. That work concentrates on decreasing the observational latency by exploiting the least amount of video frames for human action recognition, while our study aims to increase the computational efficiency to decrease the system latency.

Specific contributions of this paper are summarized in three folds:

- A novel kind of pyramidal motion feature, which is informative and sufficient for action discrimination, is proposed to represent the skeleton sequences. We compute position offsets in 3D skeletal body joints with multiple time lags to produce this feature.
- In recognition, a Naive-Bayes-Nearest-Neighbors (NBNN) based framework is presented to take into account the spatial independence of body joints.
- We compared the results of our method with those of the *state-of-the-art* approaches. In two terms of recognition accuracy and computational complexity, our method is shown to be more feasible for action recognition.

The rest of the paper is organized as follows: In Section 2, we describe in details the proposed framework. Experiment results are then presented in Section 3 and followed by conclusions in Section 4.

2 Methodology

Since we base the action skeletons for action representation, we assume that the skeletons have been recovered in every frame of videos by human pose estimation, e.g., [9]. We hence do not make any discussion on action skeleton extraction.

2.1 Extracting pyramidal motion features from skeleton sequences

In our previous work [7], we have investigated the effectiveness of position offset of skeletal body joints in action recognition, which is computed as: for one given video sequence F with n frames, let us first suppose that the skeletons have been already extracted from the original RGB-D data, i.e., $F = \{f(t)\}, t \in$

$\{1, 2, ..., n\}$, where $f(t)$ corresponds to the skeleton at tth frame, and furthermore, each skeleton $f(t)$ has been described by a set of 3D positions of body joints as:

$$f(t) = \phi(t) = \{\theta^1(t), \theta^2(t), ..., \theta^m(t)\}, \tag{1}$$

where $\theta^i(t) = (x^i(t), y^i(t), z^i(t)), i = \{1, 2, ..., m\}$ is the 3D position of ith joint in $f(t)$ and m is the body joint number.

To take into account the possibly existed local fluctuations caused by random noise and/or detection error produced in the pose estimation, the position offset of ith body joint is computed with a time lag Δt as:

$$\Delta\theta^i(t) = \theta^i(t) - \theta^i(t - \Delta t). \tag{2}$$

The resulted $\Delta\theta^i(t)$ indicates the position offset of tth frame from $(t - \Delta t)$th frame. One notes that Δt balances the precision of computed offset and the capacity of robustness to noise. By collecting the offset vectors in all body joints at tth frame, $f(t)$ is then described by $\Delta\phi(t) = \{\Delta\theta^1(t), \Delta\theta^2(t), ..., \Delta\theta^m(t)\}$.

We have experimentally demonstrated the effectiveness of proposed action feature in action recognition, but this action representation still needs further improvement with considering the following practical issues:

- The value of time lag depends on observation settings in action capturing, e.g., the sampling rate of camera. As a result, we have to make an estimation to confirm the value of it prior to the use.
- The usage of one time lag brings higher computation efficiency, and also was shown to be effective for action recognition, e.g., 97.58% on UCF dataset [7]. However, a better performance is still demanded for practical usage.

In this paper, we develop a novel kind of pyramidal motion feature based on our previous work [7]. To make the computed features more informative in action representation and more feasible in practice, we employ multiple time lags to compute position offsets in this study: $\Delta t = \{2, 4, 6, 8\}$. On the basic of each time lag, we first compute the position offset $\Delta\theta^{i,l}$ that corresponds to the ih body joint at the lth time lag, by Eq.2, and then gather them as a pyramidal description: $\Delta\theta^i = \{\Delta\theta^{i,l}\}, l \in \{1, 2, 3, 4\}$, as illustrated in Fig.1.

2.2 Video expression with bag-of-words framework

As presented above, the extracted pyramidal features are local descriptors, which is low-level description of human actions due to its limitation of time scope/extent. We, therefore, employ the standard bag-of-words (BoW) framework to assemble these local descriptors to be a histogram representation for video expression, motivated by its recent successes in action recognition.

Let us assume that every skeleton sequence have been represented as our proposed pyramidal motion features, to form the code book, we first collect all the position offset vectors in body joints of every frame in all training video sequences together. Then, we construct the code words by applying a clustering

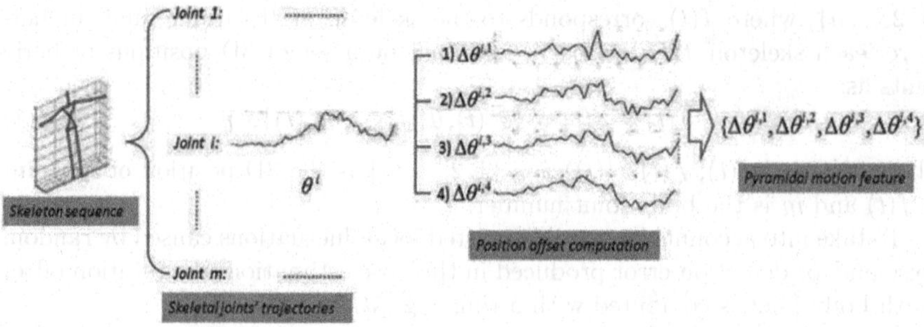

Fig. 1. The proposed pyramidal motion features.

algorithm, e.g., K-means algorithm: the code words are defined as the centres of resulted clusters. Third, the position offset vectors of each body joint at every frame in a video are assigned to be one of code words, respectively, by minimizing the Euclidean distance over all code words. The motion of each body joint is last represented by a histogram of assigned code words. In this way, the given action video is thus expressed by a set of such histograms of all joints: $F = \{h^1, h^2, ..., h^m\}$, where h^i corresponds to the histogram of ith body joint. Here, since we use pyramidal motion features for action representation, inspired by [10], we employ the pyramidal matching kernel (PMK) algorithm to speed up the computation of similarity measurement between each pair of two position offset vectors, $\Delta\theta^i$ and $\Delta\theta^j$, of compared body joints. PMK calculates interactions over pyramidal features, which takes $O(N)$ time where N is the number of features. Its computation is followed as:

$$S(\Delta\theta^i, \Delta\theta^j) = \sum_{l=1}^{4} \frac{1}{2^l}((\tau(\Delta\theta^{i,l} - \Delta\theta^{j,l})) - (\tau(\Delta\theta^{i,l-1} - \Delta\theta^{j,l-1}))). \quad (3)$$

where $S()$ is the computed similarity and $\tau()$ is the size of interaction of two compared sets.

2.3 Action recognition by Naive-Bayes-Nearest-Neighbors

For a given video F, once we have obtained its expression descriptor, i.e., a set of joint histogram, we fed it to a Naive-Bayes-Nearest-Neighbor (NBNN) classifier due to the following advantages: NBNN is (1) an un-parametrical classifier, and (2) suitable for dealing with a large number of classes and capable of avoiding the over-fitting problem. Furthermore, in this paper we extend the original concepts of NBNN for action recognition by using *joint-histograms-to-class* distance as:

$$F \to c\star, \text{where} \quad c\star = \arg\min_{c} \sum_{j=1}^{m} ||h^i - NN_c^i(h^i)||, \quad (4)$$

where $NN_c^i(h^i)$ is the nearest neighbor of h^i on ith joint in action c. Here, it should be noted that we also use PMK to speed up the computation of $NN_c^i(h^i)$.

3 Experiment

We have conducted experiments using the publicly-available datasets: UCF Kinect dataset [4] to evaluate our proposed framework, and have compared our results with the *state-of-the-art* approaches. This dataset includes 16 action classes: Balance, Punch, Duck, Run, Kick, Leap, Hop, Vault, Climb ladder, Climb up, Twist (left/right) and Step (back/front/left/right), which were performed 5 times by 16 actors, respectively. In total, we have 1280 action instances. In each frame per video, three-dimensional positions of 15 body joints are provided.

In our proposed framework, the only parameter affecting recognition performance is the size of code book K, which is used for video expression. In the experiment, we tested the values of K from 10 to 100 at an interval of 10. As previously stated, our proposed framework is extended from our previous work, and thus we compared the recognition result of the extended framework with the previous one. It is noted that since we have experimentally demonstrated that the recognition performance in our previous work [7] was obtained at the best by using $\Delta t = 8$, we only show the results for this value, here in this paper. We used two ways: leave-one-person-out cross-validation and 2) 10-fold cross-validation, to estimate the recognition performance.

3.1 Result

The recognition rates are shown in Fig.2. It can be seen that the proposed framework achieves a better performance than our previous work on both leave-one-person-out cross-validation and 10-fold cross-validation on the average, that is 98.3% (at K=30) versus 97.6% (at K=20) and 92.7% (at K=30) versus 91.0% (at K=20), respectively. In addition, the recognition rate of the proposed framework fluctuates largely over the tested values of K, which is consistent to the trends in some reports, e.g., [4]. One possible reason is, the number of training samples is not sufficient to describe actions, at least not sufficient to generate stable code words to describe actions.

Fig.3 shows the confusion matrices of action recognition corresponding to the best recognition rates by our proposed framework, i.e., 98.3% and 92.7%, respectively by leave-one-out cross-validation or by 10-fold cross-validation. It can be found that for either way, the action of **Punch** is the hardest to be recognized compared with the other actions in the dataset. Specifically, by the leave-one-out cross-validation, its recognition rate is 93% which is lower than others; the rate is even much lower by 10-fold cross-validation, that is only 57%. One possible reason is, the style of this action heavily depends on the performers, which brings large inter-class variations.

3.2 Comparison with the state-of-the-art approaches

We compared the recognition performance of our proposed framework with the *state-of-the-art* approaches. For these approaches, the recognition rates were directly taken from the corresponding references. As shown in Table 1, our frame-

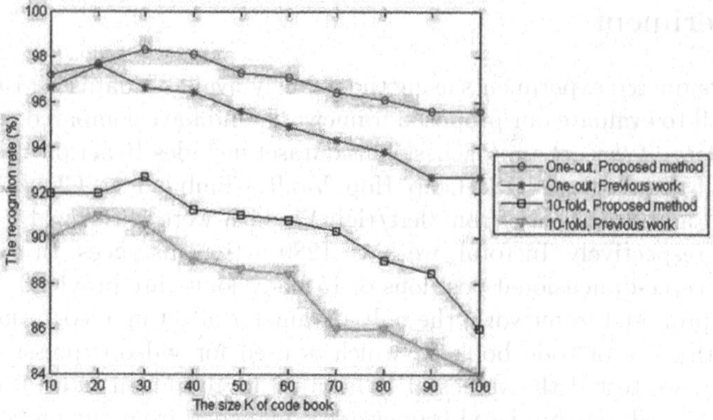

Fig. 2. Recognition accuracies in this paper and in our previous work [6]. The tested sizes K of code book are from 10 to 100 at an interval of 10.

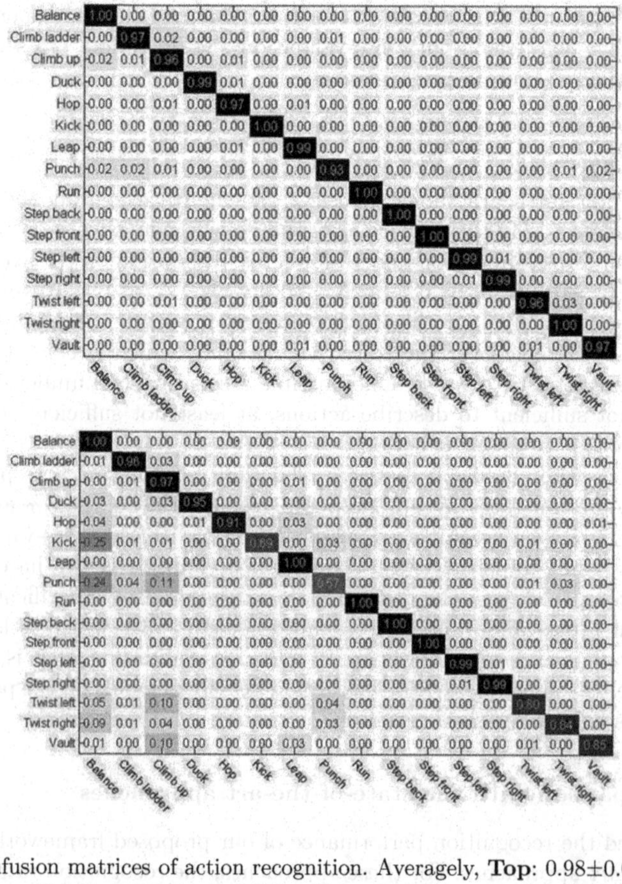

Fig. 3. Confusion matrices of action recognition. Averagely, **Top:** 0.98±0.02 by leave-one-out cross-validation; **Bottom:** 0.93±0.10 by 10-fold cross-validation.

work achieves the best performance on action recognition among the compared works including our previous work.

Table 1. Comparison of recognition rates with *state-of-the-art* methods.

Method	Year	Accuracy (%)
Masood et al.[11]	2011	92
Ellis et al.[4]	2013	95.54
Yang & Tian [5]	2013	97.1
Ohb-Bar & Trivedi [12]	2013	97.37
Our previous work [7]	2014	97.58
This study		98.27

Additionally, since our focus is on exploiting a low computationally-efficient framework for recognizing human actions, it is necessary to compare the real computation time with the compared approaches. However, it is very hard due to the hardness of faithful implementation and fair execution of those approaches under the same experimental conditions. Alternatively, we compare the proposed pyramidal motion feature used in our framework with some conventional action features in the term of computational complexity. From Table 2, we can see that our proposed feature is more efficient in computation compared with three conventional alternatives in the spatio-extent in frames. The feature also has a low computational complexity in the time extent of video, which is the same as the compared features. The merit of the proposed feature in computational complexity makes sure the fast-computation of it in practical usage.

Table 2. Computational complexities on *spatio* and *time*. m is the number of body joints in a skeleton, and n is the number of video frame.

Method	Studies	Complexity	
		Spatio	Time
Joint-difference intra-frame	[4, 5, 11, 12]	$O(m^2)$	$O(n)$
Joint-difference extra-frame	[4, 5, 11, 12]	$O(m^2)$	$O(n)$
3D skeleton histogram	[3]	$O(m^2)$	$O(n)$
Ours		$O(m)$	$O(n)$

4 Conclusion

In this study, we present a new framework for efficient action recognition. The novel pyramidal motion features are proposed to represent skeleton sequences via computing position offsets in 3D skeletal body joints. In the recognition phase, we employ a Naive-Bayes-Nearest-Neighbors (NBNN) classifier with considering the spatial independence of body joints. Experimental results conducted on the public UCF dataset show that our framework is more effective and more accurate for action recognition compared with the state-of-the-art approaches, and meanwhile it has a high potential to be more efficient in computation.

In the future work, we will explore selecting less but more discriminative features from the proposed pyramidal motion features by feature selection or feature reduction so as to further improve the performance of our framework.

Acknowledgments. The work is financially supported by National Natural Science Foundation of China (61403232), Natural Science Foundation of Shandong Province, China (ZR2014FQ025), and Fundamental Research Funds of Shandong University (2014TB004).

References

1. Poppe, Ronald: A survey on vision-based human action recognition. Image and vision computing, 28(6), 976–990 (2010)
2. Zhou, Z., Chen, X., Chung, Y., He, Z.: Activity analysis, summarization and visualization for indoor human activity monitoring. IEEE Transactions on Circuits and Systems for Video Technology, 18(11), 1489–1498 (2008)
3. Carletti, V., Foggia, P., Percannella, G., Saggese, A., Vento, M.: Recognition of Human Actions from RGB-D Videos Using a Reject Option. In New Trends in Image Analysis and Processing-IICIAP, pp.436–445 (2013)
4. Ellis, C., Masood, S. Z., Tappen, M. F., LaViola Jr, J. J., Sukthankar, R.: Exploring the trade-off between accuracy and observational latency in action recognition. International Journal of Computer Vision, 101(3), 420–436 (2013)
5. Yang, X., Tian, Y.: Effective 3d action recognition using eigenjoints. Journal of Visual Communication and Image Representation, 25(1), 2–11 (2014)
6. Lu, G., Kudo, M.: Learning action patterns in difference images for efficient action recognition. Neurocomputing, 123, 328–336 (2014)
7. Lu, G., Zhou, Y., Li, X., Kudo, M.: Efficent action recognition via local position offset of 3D skeletal body joints, Multimedia Tools and Applications. *Accepted.*
8. Lu, G., Kudo, M., Toyama, J.: Temporal segmentation and assignment of successive actions in a long-term video. Pattern Recognition Letters, 34(15), 1936–1944 (2013)
9. Zhu, Y., Dariush, B., Fujimura, K.: Kinematic self retargeting: a framework for human pose estimation. Computer vision and image understanding, 114(12), 1362–1375 (2010)
10. Lv, F., Ramakant N.: Single view human action recognition using key pose matching and viterbi path searching. in IEEE Conference on Computer Vision and Pattern Recognition, pp.1–8 (2007)
11. Masood, S. Z., Ellis, C., Nagaraja, A., Tappen, M. F., Laviola, J. J., Sukthankar, R.: (2011, November). Measuring and reducing observational latency when recognizing actions. In: ICCV Workshops, pp. 422-429 (2011)
12. Ohn-Bar, E., Trivedi, M. M.: Joint angles similarities and HOG2 for action recognition. In Computer Vision and Pattern Recognition Workshops (CVPRW), pp. 465-470 (2013)

P-PCC: Parallel Pearson Correlation Condition for Robust Cosmetic Makeup Face Recognitions

Kanokmon Rujirakul and Chakchai So-In

Department of Computer Science, Faculty of Science, Khon Kaen University
123 Mitaparb Rd., Naimuang, Muang, Khon Kaen, 40002 Thailand
kanokmon.r@glive.kku.ac.th and chakso@kku.ac.th

Abstract. The performance of face recognition has been improved over the past; however, there remain some limitations, especially with noises and defects, such as occlusion, face pose, expression, and, in particular, cosmetic makeup change. Recently, the makeup has directly impacted on face characteristics, e.g., face shape, texture, and color, perhaps leading to low classification precision. Thus, this research proposes a robust approach to enhance the recognition accuracy for the makeup using Pearson Correlation (PC) combining with the channel selection (PCC). To further optimize the complexity, the parallelism of PCC was then investigated. This technique demonstrates the practicality and proficiency by outperforming the accuracy and computational time over a traditional PCA and PC.

Keywords: Cosmetic; Face Recognition; Makeup; PCA; Pearson Correlation

1 Introduction

Face recognition has been practically used in many applications, e.g. authentication, human identification, and criminal forensics in both industry and research. However, although the accuracy of face recognition has been improved, some limitations still remain due to the effect of illumination, face pose, expression, and especially cosmetic makeup [1]. There are many changes in appearances caused by cosmetic makeup, e.g., skin tone, facial texture, and color, probably leading to face brightening, younger appearance, and reducing wrinkles, blemishes and spots. These changes have direct effects on the precision of extracted features as well as classifications for face recognition systems [2-3]. Thus, this research focuses on the investigation over robust cosmetic makeup face recognition approach for the purpose of accuracy enhancement using Pearson Correlation (PC) with color condition called PCC. To further optimize PCC, the parallelism was investigated to improve the algorimatic speed-up, called P-PCC.

2 Background of Face Recognition Systems

In general, face recognition systems consist of four key components, i.e., acquisition, pre-processing, feature extraction, and classification. To appropriately recognize the face, there are two main processes: training and testing. For training process, the face

K.J. Kim (ed.), *Information Science and Applications*,
Lecture Notes in Electrical Engineering 339, DOI 10.1007/978-3-662-46578-3_30

image which uses for training acquisition step can be acquired from various inputs and then fed those into the pre-processing step, some of which are illumination normalization, background removal, and histogram equalization. After pre-processing, the feature vectors of normalized face images will be extracted by the feature extractor, and then all of the feature vectors will be stored in a training set.

The testing process is similar to that of the training but with fewer steps. The system will feed the testing image into the pre-processing step in order to generate the normalized new face image, and then perform facial image classification by comparing those images with feature vectors stored in the training sets. There are several techniques of feature extraction and classification, e.g., PCA, ICA, and LDA [4]. One of the most well-known is PCA normally used to reduce the dimension of data. However, in this research, Pearson Correlation (PC) [5] is one of the statistical techniques typically used to find the relation of two data sets called Coefficient Correlation (CC). The value of this CC or r can be calculated given equation (1) below where $\bar{X}$ and $\bar{Y}$ are the mean of X and Y, respectively. Generally, r value will be in the range of -1.00 to +1.00.

$$r = \frac{\sum_{i=1}^{n}(X_i - \bar{X})(Y_i - \bar{Y})}{\sqrt{\sum_{i=1}^{n}(X_i - \bar{X})^2}\sqrt{\sum_{i=1}^{n}(Y_i - \bar{Y})^2}} \tag{1}$$

It should be noted that if r is close to +1.00, then both data have a direct variation relationship. In other words, if the first data are increased, then the second data will be also increased. On the contrary, if r is close to -1.00, then both data have reverse variation relationship, and in case that r is close to zero, then both data have less relationship. If r is very close to -1.00 or +1.00, then both data have more relationship.

3 Related Work

There have been several proposals of face recognition optimization for the purpose of high precision. One of the pioneering approaches is PCA including later derivatives. For example, in 2009, P. Shamna et al. and W. Zhao et al. [6-7] provided a survey of different PCA techniques given the recognition rate constraint. Recently, in 2013, A. K. Bansal and P. Chawla [8] introduced Normalized Principal Component Analysis which normalized images to remove the lightening variations and background effects.

Although the recognition accuracy has been improved, there are limitations, especially with the effect of noises and defections, one of which is cosmetic makeup noises. A few works are under focus, for instance, in 2005, S. Pamudurthy et al. [9] proposed Digital Image Skin Correlation to track the position of facial skin pores in different face expression with intensive makeup face identification suitable for high resolution image.

In addition, in 2012, A. Dantcheva et al. [2] evaluated the recognition rate among Gabor, LBP, and Verilook on YMU makeup database. In 2013, L. Wen and G. Guo [10] proposed dual-attributes for face verification with key 28 attributes used for SVM-RBF classification. In the same year, J. Hu et al. [11] applied PCA with Canonical Correlation Analysis to calculate the relationship between the face images. Recently, in 2014, G. Guo et al. [12] introduced the correlation mapping technique extracted from a local patch of (non) makeup face images using Partial Least Square. Then, PCA was applied as the feature extractor. However, most of the proposals lack of intensive investigation on either makeup databases or computational complexity of the algorithm.

The optimization of face recognition research focuses not only on the accuracy but also on the computational complexity for practicality, especially with the parallelism consideration. For example, in 2003, C. Jiang et al. [13] proposed a distributed system for face recognition by dividing trained face databases into five sub-databases and then performed an individual face recognition algorithm individually in parallel, called Parallel PCA (P-PCA). Recently, a multi-core processors architecture has been explored in a single machine; for example, in 2013, K. Rujirakul et al. [14] proposed a parallel version of fixed point PCA yielding in-significant difference of accuracy to PCA but lower computational complexity. The same authors [15-16] also introduced the faster algorithm with its deviation, i.e., (weighted) Parallel EM-PCA, utilizing expectation maximization with optimized histogram equalization to improve the traditional face and facial expression recognition while maintaining the speed-up. However, most proposals do not consider the cosmetic makeup precision.

4 Parallel Pearson Correlation Condition Face Recognition

Fig. 1 shows an overall architecture of Parallel Pearson Correlation Condition (P-PCC) consisting of three main modules: Pre-Processing, Feature Extraction, and Image Identification.

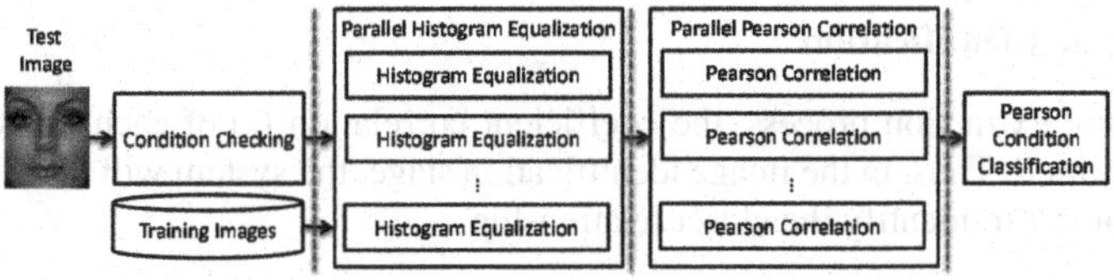

Fig. 1. Overall of Parallel Pearson Correlation Condition Face Recognition

4.1 Pre-Processing

Considering the makeup face images, generally, various cosmetic colors have direct impact on the recognition precision, and therefore, aside from the intensity metric via histogram equalizations, the selection criteria on which channel will achieve the highest accuracy were proposed in the pre-processing stage as follows:

1) Conditional Checking: at this step, the system will separately calculate the standard deviation of RGB channels as well as gray scale of testing images, each of which will be then selected as a representation using the lowest standard deviation criterion.

2) Histogram Equalization: after the first step, the system will convert the trained and testing images into the selected type (R/G/B channel or gray scale), and then calculate the histogram equalization. The details are as follows: (1) generate the counting table of each image for the entire color image in range of 0 to 255 (2) compute the cumulative color value of each table, and (3) finally apply equation (2) stating below.

$$h[v] = round\left(\frac{cd[v] - cd_{min}}{(M \times N) - cd_{min}}\right) \times (L - 1) \tag{2}$$

In this equation, cd is the cumulative of images in size of $M \times N$; v denotes as color values; and L is the entire color. It should be noted that we apply the parallelism in this step to speed-up the entire process as stated in one of our previous works [16].

4.2 Feature Extraction

Instead of a traditional PCA feature extractor for face recognition, in this research, we proposed the use of Pearson Correlation (PC). In general, a traditional PC calculates the coefficient correlation of every data pair. However, when we applied it into the face recognition, the coefficient correlation calculation among trained data is not necessary, and thus, our first optimization was to enhance this computation stating in Algorithm I.

This algorithm was derived from equation (1) to calculate the coefficient correlation of training and testing data where N and M are the number of trained images and pixels in each image. It should be noted that given this optimization, the complexity is decreased from $O(mn^2 + n^2)$ to $O(mn + n)$. To further enhance the speed-up, the parallelism was also applied, i.e., parallel matrix manipulation (discussed later in section 4.4).

ALGORITHM I: OPTIMIZED PEARSON CORRELATION

```
Input: Matrix data
Output: Matrix result
for k from 0 to M
  for j from 0 to N
    result[0,j]+= data[0,j] × data[j,0]
    sumX[0,j]+= data[j,k] × data[j,k]
  endfor
  sumY+= data[0,k] × data[0,k]
endfor
for j from 0 to N
  result[0,j]/= (√sumX[0,j] × √sumY)
endfor
return result;
```

4.3 Image Identification

After feature extraction process, the coefficient correlation (r) of each trained image will be acquired. Thus, in the image identification stage, the system will then figure out the maximum r to identify the closet relationship.

4.4 Parallel Matrix Manipulation

Based on our previous work [15], the parallel matrix manipulation was proposed to speed-up the face recognition system in five main operations, i.e. subtraction, addition, multiplication, division, and transpose. However, in this research, there are other types of matrix computation, especially used in PCC. Therefore, we introduce a novel parallel matrix manipulation in three operations, i.e., subtraction/multiplication/division each element with one dimensional matrix, sum each row, power and square root as follows:

ALGORITHM II: PARALLEL MATRIX MANIPULATION

```
Input: Matrix left, Matrix right, String operation
Output: Matrix result
parallel for i from 0 to rows
  for j from 0 to cols
//Subtraction/Multiplication/Division each element with 1D matrix
    if (operation =="Subtract")
      result[i,j] = left[i,j] − right[i];
    endif
    if (operation =="Multiply")
      result[i,j] = left[i,j] × right[i];
    endif
    if (operation =="Divide")
      result[i,j] = left[i,j] ÷ right[i];
    endif
    //Sum each row of Matrix
    if (operation =="Sum each")
      result[i]+= input[i,j];
    endif
//Matrix Power and Matrix Square Root
    if (operation =="Power")
      result[i,j] = (input[i,j])²;
    endif
    if (operation =="Root")
      result[i,j] = √input[i,j];
    endif
  endfor
endfor
return result;
```

1) Subtraction/Multiplication/Division each element with one dimensional matrix: this operation will perform the matrix operation between $M \times N$ dimensional matrix and one dimensional matrix which is calculated according to equation (3) such that "·" is an operator in a set of $\{-,\times,\div\}$. It should be noted that this operation can be performed in parallel to calculate the difference of each element at the same position.

$$c[i,j] = a[i,j] \cdot b[i] \tag{3}$$

2) Sum each row: the summation of each row in the matrix can be also performed in parallel using equation (4) since each calculation step is independent to each other.

$$c[i] = \sum_{j=0}^{cols} a[i,j] \tag{4}$$

3) Power and Square Root: the calculations of power and square root of all elements in the matrix are frequently used in PCC calculation. This operation can be also performed in parallel since each element can be independently calculated using equation (5) such that n is in a set of $\{2,\frac{1}{2}\}$.

$$c[i,j] = a[i,j]^n \tag{5}$$

It is worth noting that the parallel matrix manipulation in this research is summarized and illustrated in Algorithms II, and here, the *Parallel_for* [17] of .NET C# was applied during the actual implementation in order to achieve process concurrency.

5 Performance Evaluation

In this section, the evaluation process was performed in order to assure the performance of our mechanism (P-PCC) by comparing with a traditional PCA and PC including their parallel derivations over two public (non) makeup face image databases.

5.1 Experimental Setup

Our testbed is a standard configuration on personal computer Windows 7 Ultimate operating systems (64 bits): CPU Intel(R) Core(TM) i-3770K 8-Cores 3.50 GHz (8MB L3 Cache), 8192×2 MB DDR3-SDAM, and 500 GB 5400 RPM Disk.

The experimental testbed was implemented in .NET C# programming environment to emulate the real-world application, and especially to illustrate the computational time complexity in multi-cores architecture in single machine. For testing purposes, similar to the evaluations in other related work [14-16], there are two public face image databases: YMU (YouTube Makeup) and VMU (Virtual Makeup) [2-3]. A selection of image characteristics was over a set of color, 24-bits RGB, PNG images, ranging from 100 to 237 images (VMU) and 100 to 584 (YMU) images of 130×150 pixels [14]. Fig. 2 shows some examples of face images from both databases.

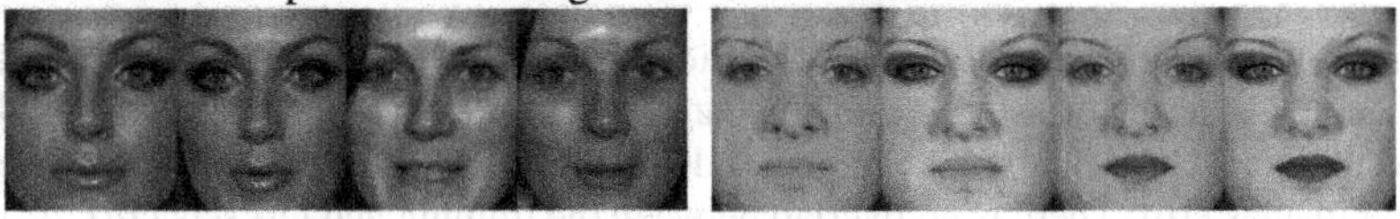

Fig. 2. Example: YMU (left) and VMU (right) face image database

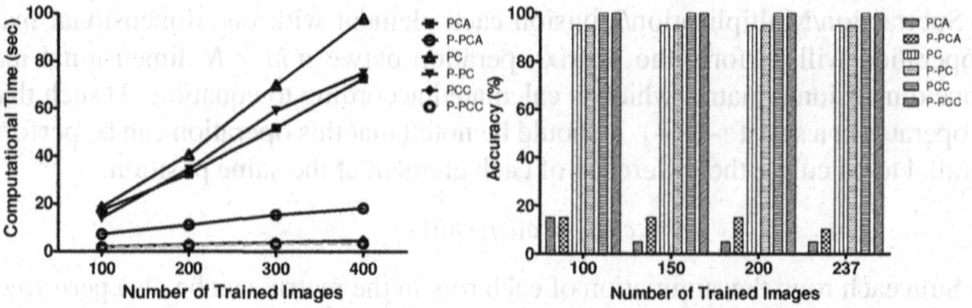

Fig. 3. Computational time and percentage of accuracy over #trained images of VMU database

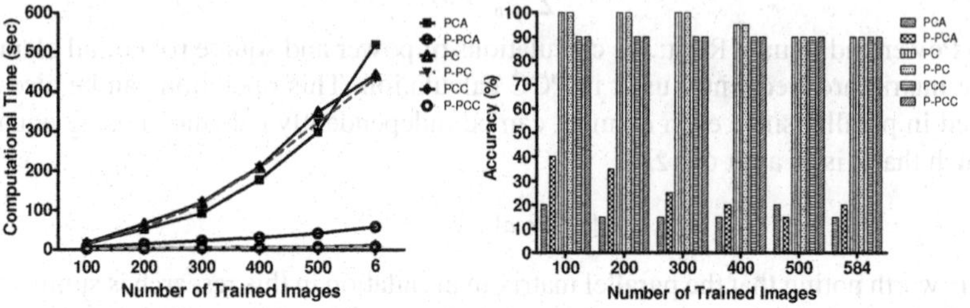

Fig. 4. Computational time and percentage of accuracy over #trained images of YMU database

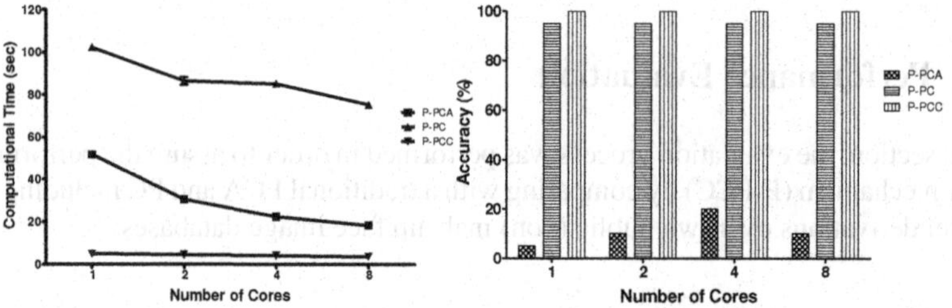

Fig. 5. Computational time and percentage of accuracy over #CPU cores of VMU database

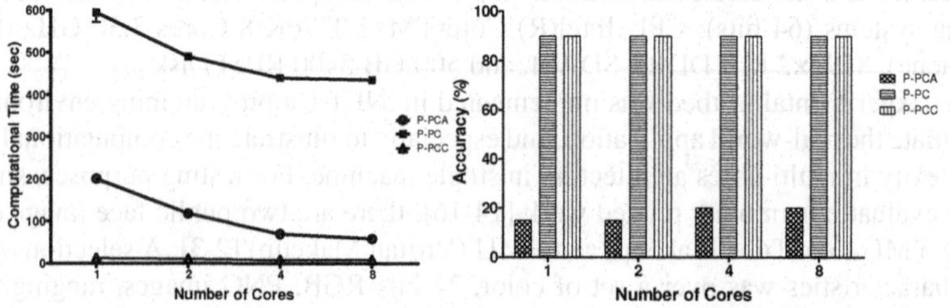

Fig. 6. Computational time and percentage of accuracy over #CPU cores of YMU database

In all of the evaluations, two main metrics were applied either computational time or accuracy in terms of average and standard deviation over seven trials. It should be noted that in each evaluation, a simple random selection was applied to select twenty testing images [16]. There is no overlap between the training and test sets [14].

There are three main scenarios as follows: firstly, to perform the comparative study on our first optimization method, i.e., PCC, comparing with traditional PC and PCA; secondly, to show the effect of parallelism, the parallel version of these three algorithms were comparatively evaluated, i.e., P-PCC, P-PC, and P-PCA [13] by dividing the training dataset into sub-datasets regarding the number of CPU cores and then performing the recognition in parallel (to achieve the best performance, we used the entire CPU cores, 8 cores, for testing purposes); and finally, to evaluate the performance of degrees of parallelization, the comparative performance of these three parallel algorithms when ranging the number of cores of 1, 2, 4, and 8 cores, respectively, testing with the maximum number of trained images (237 images for VMU and 584 images for YMU).

5.2 Experimental Results

Considering the first scenario (non-parallelism) with VMU database, as shown in Fig. 3, our proposed method, PCC, outperforms the others, i.e., only 4.76 seconds, which is a factor of 15 and 21 faster than PCA and PC, respectively. The recognition precision of PCC is also outstanding, i.e., 100%, (5% and 95% for PCA and PC). In addition, in the second scenario, our parallelism optimization, P-PCC, again, yields the best performance. For example, the computational time and accuracy of P-PCC, P-PCA, and P-PC are as follows: 3.49, 75.15, and 18.22 seconds; 100%, 10%, and 95%, respectively.

Fig. 4 shows the performance with YMU database, similar to that with VMU, the non-parallel version of PCC still maintains the best performance, i.e., 12 seconds, and that is faster than PCA and PC by a factor of 43 and 37. The accuracy of PCC outperforms the others as well, i.e., 90%, (15% and 90% for PCA and PC). With the parallelism, P-PCC is still outstanding in computational time and accuracy, i.e., 8.49, 58.29, and 434.63 seconds; 90%, 20%, and 90%, for P-PCC, P-PCA, and P-PC, respectively. Note that when a number of trained images are increasing, the computational time tends to be gradually increased in all algorithms as well as in similar trends for all datasets.

Considering the third scenario by varying the number of cores of parallel versions of PCC, PCA, and PC, Fig. 5 shows that increasing the number of cores can lower the computational time such as from 4.81 to 3.49, 102.18 to 75.15, and 51.69 to 18.22 seconds, respectively, for VMU database. The accuracy of P-PCC and P-PC has no effect on number of cores except P-PCA (higher cores higher precision). These trends are also applied for YMU database such as from 12.27 to 8.49, 593.97 to 434.63, and 199.92 to 58.29 seconds, respectively, with no accuracy gain except P-PCA.

6 Conclusions and Future Work

In this research, an optimization of face recognition using Pearson Correlation, Conditional Checking, and Histogram Equalization, called Pearson Correlation Condition (PCC), was proposed against a traditional PCA face recognition, especially for makeup cosmetic face images. To further speed-up the computational complexity, the parallelism was also investigated over PCC resulting into the outstanding approach over two main public (non) makeup databases (VMU and YMU). The proposed method outperforms a traditional PCA and its parallelism and a traditional Pearson Correlation (PC) and its parallelism by factors of 61, 7, 52, and 51, respectively, while maintaining high recognition precision, i.e., from 90% to 100%. However, although P-PCC can achieve significant speed-up with high accuracy, more investigation can be performed, e.g., heterogeneous images and further parallelism optimizations of recognition stage. Another aspect of face detection can be also investigated, and these are for future work.

7 References

1. Dantcheva A., Chen C., Ross A.: Makeup challenges automated face recognition systems. SPIE Newsroom 2013, Defense and Security, 1--4 (2013)
2. Dantcheva A., Chen C., Ross A.: Can facial cosmetics affect the matching accuracy of face recognition systems? In: IEEE Int. Conf. on Biometrics: Theory, Appl. and Syst., pp. 391--398. IEEE Press, Arlington, VA, USA (2012)
3. Chen C., Dantcheva A., Ross A.: Automatic Facial Makeup Detection with Application in Face Recognition. In: Int. Conf. on Biometrics, pp. 1--8. IEEE Press, Madrid, Spain (2013)
4. Jafri R., Arabnia H. R.: Survey of face recognition techniques. J. of Info. Process. Syst., 5(2), 41--68 (2009)
5. Xi C., Govindaraju, V.: Utilization of Matching Score Vector Similarity Measures in Biometric Systems. In: IEEE Computer Society Conf. on Computer Vision and Pattern Recognition Workshops, pp. 111-116. IEEE Press, Buffalo, NY, USA (2012)
6. Shamna P., Paul A., Tripti C.: An Exploratory Survey on Various Face Recognition Methods Using Component Analysis. Int. J. of Advanced Research in Comput. and Commun. Eng., 2(5), 2081--2086 (2013)
7. Zhao W., Chellappa R., Phillips P., Rosenfeld A.: Face recognition: a literature survey. ACM Comput. Surveys, 35, 399--458 (2003)
8. Bansal A.K., Chawla P.: Performance Evaluation of Face Recognition using PCA and N-PCA. Int. J. of Comput. Appl., 76(8), 14--20 (2013)
9. Pamudurthy S., Guan E., Mueller K., Rafailovich M.: Dynamic approach for face recognition using digital image skin correlation. In: Audio- and Video-based Biometric Person Authentication, pp. 1010--1018. Springer (2005)
10. Wen L., Guodong G.: Dual Attributes for Face Verification Robust to Facial Cosmetics. J. of Comput. Vision and Image Process., 3(1), 63--73 (2013)
11. Hu J., Ge Y., Lu J., Feng X.: Makeup-Robust Face Verification. In: IEEE Int. Conf. on Acoustics, Speech and Signal Process., pp. 2342--2346. IEEE Press, Vancouver, BC, Canada (2013)
12. Guodong G., Wen L., and Yan S.: Face Authentication with makeup changes. IEEE Trans. on Circuits and Syst. for Video Technol., 24(5), 814--825 (2014)
13. Jiang C., Su G., and Liu X.: A distributed parallel system for face recognition, In: Int. Conf. on Parallel and Distributed Comput., Appl. and Technol., pp.797--800. IEEE Press, Chengdu, China (2003)
14. Rujirakul K., So-In C., Arnonkijpanich B., Sunat K., Poolsanguan S.: PFP-PCA: Parallel Fixed Point PCA Face Recognition, In: Int. Conf. on Intell. Syst. Model. & Simul., pp. 409--414. IEEE Press, Bangkok, Thailand (2013)
15. Rujirakul K., So-In C., Arnonkijpanich B.: PEM-PCA: A Parallel Expectation-Maximization PCA Face Recognition Architecture. The Scientific World J. 2014, 1--16 (2014)
16. Rujirakul K., So-In C., Arnonkijpanich B.: Weighted Histogram Equalized PEM-PCA Face Recognition. In: Int. Conf. on Comput. Sci. and Eng. Conf., pp. 144--150. IEEE Press, Bangkok, Thailand (2014)
17. Microsoft MSDN, "Parallel.For Method," .NET Framework 4.5. (2014) Available online at http://msdn.microsoft.com/en-us/library/dd783539%28v=vs.110%29.aspx

Flight Simulator for Serious Gaming

Aruni Nisansala, Maheshya Weerasinghe, G.K.A.Dias, Damitha Sandaruwan, Chamath Keppitiyagama, Nihal Kodikara, Chamal Perera, Prabhath Samarasinghe.
University of Colombo School of Computing, Reid Avenue, Colombo 07, Sri Lanka.
{asn,amw,gkad,dsr,cik,ndk,clp,yps}@ucsc.cmb.ac.lk

Abstract. Providing entertainment is the primary concern of the gaming. Once this primary objective alters to provide learning and training materials it calls simulators or the serious gaming. Learning through experiencing or facing the actual scenario is considered as an effective learning technique. The limitations of the experiential learning and how the simulations are going to address those limitations are also reviewed in this paper. Aviation field is one of the most critical and potentially high risk areas where one has to spend lots of money and resources in training scenario. Hence the serious gaming concepts have being playing as an effective cost cutting solution in aviation training. In this paper it is intended to discuss the seriousness of a selected flight simulator and how they adopted the teaching learning concepts. How the simulator can be used in the learning curve is also discussed separately

Keywords: Serious Gaming, Flight Simulator, Flightgear, Experiential Learning

1 Introduction

Computer games are basically focused on providing entertainment and fun. When games are designed with different intensions like teaching, learning and training, then it is called serious gaming applications or simulators (Michael & Chen, 2005). Serious gaming concept copes with solving real world problems more than providing entertainment. In several fields, training is a risky, costly and demanding process. Pilot, astronaut, military, fire rescue training and medical surgery training can be pointed out as such domains. In order to reduce the potential risk and training cost serious gaming concept can also be adapted to various fields such as health, security, inland defense, communication as well as in education fields (Djaouti, Damien, Alvarez, & Pierre J, 2011).

"Serious Gaming" is a combination of the aspect of "seriousness" and the "gaming". The seriousness refers to the contents of the application which is used in teaching and learning process. Federal Aviation Administration (FAA) has proclaimed that human factors, both mental and physical, significantly affect to the aviation safety. Practically this is a common fact to all fields. Most of the time worker injuries, wasting time and accidents are caused by those factors (Aviation Maintenance Technician

K.J. Kim (ed.), *Information Science and Applications*,
Lecture Notes in Electrical Engineering 339, DOI 10.1007/978-3-662-46578-3_31

Handbook – General- Chapter 14, 2013). Training will provide a basic platform to stand in critical situations and experience it. That will improve both mental and physical fitness which will directly affect in reducing the human error factors.

As the technology grows rapidly during the past decades hardware and electronic device cost has being reduced. There is free and open source software available for physics integrations and virtual environment rendering. Some free and open source software provides the framework to implement the projects on top of that. Flightgear (Flightgear), GiPSi (GiPSi - Open Architecture Software Development Framework for Surgery Simulation), spring (Montgomery, 2002) and OpenSurgSim (OpenSurgSim) are some frameworks where developer can integrate his solution without building it from the scratch. With those available resources real experiential training has been replaced by the low cost virtual immersive environment training in numerous fields.

Albeit the serious gaming concept is widely spread, the discussion is going to be bounded to the aviation domain.

2 Experiential Learning

Experiential learning is a different paradigm other than the traditional reading writing education. This concept has the proper combination among learning, training and other life activities and extracting the knowledge itself. Acting and experimenting are identified as teach enhancers (Winett, 1972). According to the Kolb, experience creates a concrete basis of learning through experiencing (Kolb, 1984). It enhances the observation and decision making power. Lewinian Model present a common characteristics of experiential learning with four stages proceeding in spiral way.

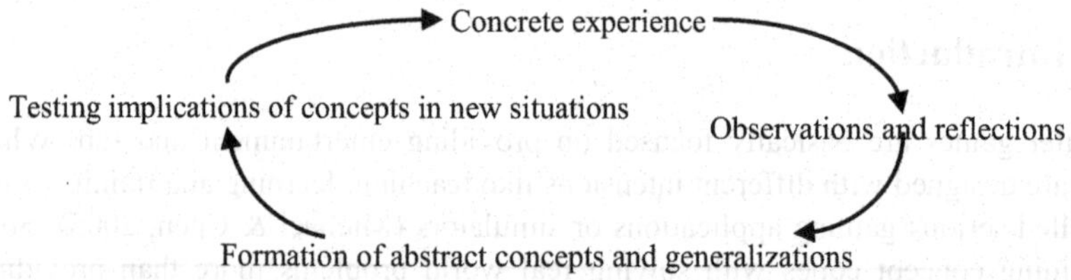

Fig. 1. The Lewinian Experiential Learning Model (Kolb, 1984)

Even though this experiential learning is significant with knowledge gaining, real life scenarios has limitation in following this process. As briefed in introduction phase cost of experiment, ethical and social reasons and the safety issues has affected in this process. Furthermore the impossibility of creating natural conditions (storm, snowing, forest fire etc.) has again condemned the concept, learning through the experience.

While concerning the importance of learning through experiencing and its limitations, Simulation environments would be the best educational environment where almost all the limitations were addressed in proper manner.

3 Development Strategy

As discussed above simulations are used in training with practically difficult scenarios. The experience gained from simulation would be impossible in real world due to the cost, risk, time and safety. For the military training of bombing scenario it is not applicable to practice in real environment. It will risk lots of properties and even the life of the pilot. Same goes with the medical surgery training process. Normally surgeons have to practice using corpse or small animals such as rats. Both scenarios don't give the proper education for a surgeon. Here the simulation plays as a great rescuer. With the prevailing technologies it is possible to integrate physics to simulate the real time pressure, bleeding and beating to replicate a real patient-doctor contact.

In pilot training there are difficult scenarios such as engine fire, stormy or snowing weather, night landing, and landing on a mother ship where it is impossible to get the real experience without a proper training. These scenarios can be successfully replicated on the virtual environment and give the real feeling to the trainee. But the training process cannot be 100% substitute from the simulation training. This simulation environment will only cover the practical section where they have to use the real equipment or flights. Even with the simulators there are impossible scenarios where it cannot be simulated to get the real feeling and apply the real time motion or physics. These activities which are within the simulation boundary should be identified in feasibility study.

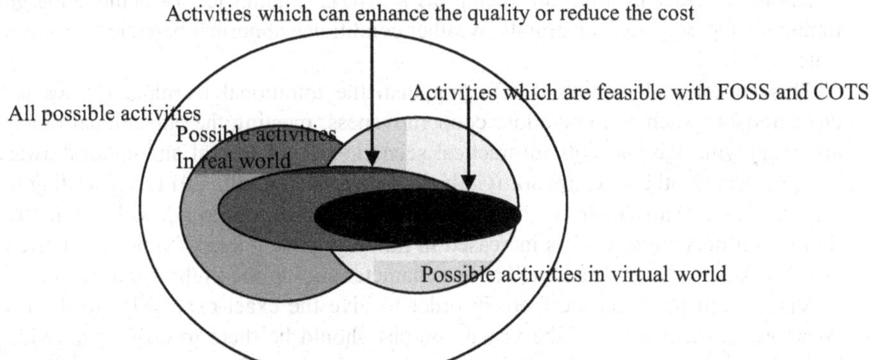

Fig. 2. Activities identified in first stage (Sandaruwan, Keppitiyagama, Dias, Kodikara, & Rosa, 2012)

Behavioral realism can be improved with higher accuracy of motion prediction and visualization (Yin, 2010). High polygon count could increase the mesh qualities with better realism. Due to the limitations of computational power there will be implementation and rendering issues with those meshes. Physical realism can be increased by interfacing the solution with real equipment. As the opening section reveals this interfacing has become a low cost procedure with the COTS hardware. In the fighter jet simulation it is used those COST hardware in interfacing process. Only the right and left consoles were interfaced with throttle. The front meter gauges and integrated warning panels were displayed in wide screen. Physical, Semi-physical and digital simulation technology is generally used in astronaut training. When considering this the fighter jet simulation can be considered as a man-in-loop emulation system where applied semi-physics and simulation method (Jian-gang, Jun-yi, & Ning, 2010). Further the fighter jet simulator can be enhanced using the motion system to increase the physical and behavioral realism.

4 Flight Simulator Based Training

As the generations are more familiar with the digital media adopting the gaming concept in learning is a successful and timely convenient. However as discussed above as the games are used in learning and training purpose other than the entertainment gaming concept should enclose some qualities to attract the student. The visual realism of the simulation should be one main point where the trainee feels the same real world mentality in the simulation environment. Hence the simulation environment should boost up the visualizing, hearing, feeling and interpreting qualities (R. Batista, 2008).

A game is a structured or semi structured context with goals that players have to obtain, overcoming challenges introduced by the context. In this process the goals and the challenges contained educational purpose then that game can be categorized as an edutainment one. When applying this with flight simulators, it should have goals such as landing to Bucharest airport, taking off a MIG-29 fighter jet. To achieve the goal trainer should be given constraints; weather conditions, abnormal behaviors or normal state.

This educational gaming is effecting than the traditional learning. Games have characteristics such as motivation, cooperativeness, meeting the educational objects; allow applying the concepts in practical scenarios, favor in oral and cultural awareness, respect to others, teamwork (Gouveia, Lopes, & Carvalh, 2011). According to a research survey carried out by Paulo David et.al in military field it has shown motivation for military lifestyles has increased in military game players (Simões & Ferreira, 2011). Likewise it can be adopted above characteristics to the flight simulations.

Visual realism should be there in order to give the exact experience to the user. Awareness, sensitivity and the visual contrast should be there in order to provide a high visualization. Simulator adaption syndrome is counted as a negative design impact (McCaffrey). In its' sever form dizziness, nausea, sweating and vomiting can be seen. This is a form of motion sickness. With the increased width FOV display system this negative impact could be reduced. In the beginning of the simulator training this was a serious challenge to the designers.

Interaction of the devices has enhanced the quality of simulator based learning. Performance of the hardware and electronic devices has increased rapidly during the past decade and cost has gone down. With the interfacing of the cockpit it can have the real world feeling. Integration of the real time physics and this commodity off the shelf hardware provide accurate scenarios.

As the interaction between younger generation and technology grows up rapidly, idea of using augmented and virtual environment in teaching and learning process have proportional advantages (Skill Evolution Report, 2008). Number of research was carried out to identify how the simulation game favors the training process. Whitehall, McDonald (Whitehal & McDonald, 1993) and Ricci (Katrina, Cannon-Bowers, & Janis, 1996)et al showed that integrating game features improves the learning. In 2007 Garris et al. presented some individual attributes that a game should have to be a good educational effective tool (Garris, Ahlers , & Driske, 2002).

"Capability" is the mental and physical skills that need to be developed in the learner by playing the game. These capabilities lie as cognitive, psychomotor and affective skills (Yusoff, Crowder, & Gilbert, 2010). "Instructional content", "Intended Learning Outcomes", "Serious Game Attributes", "Learning Activities" (challenging and goal driven), "Reflection", "Game Genre" and "Game Achievements" are the attributes which were proposed by Garris et al.

Instructional content and Learning outcome are depending on each other. Serious game attributes tied with learning activities extract the knowledge through entertainment. Reflection is measuring progress from one session to another. Game Genre is the category of the game. Open world sandboxes to strategy games or simulation. Game achievements define the training level of the learner. This can be scores, resource amount or collected assets or anything which have a numeric value. This indicates how the trainee has performed within the lesson. This attributes are compare and contrast with a selected flight simulator in a below section.

5 Comparison of Flight Simulators

Simulation is a less expensive way to have the experience of flying with relative to the real training. A lots of flight simulation applications have been introduced to the market. X Plane (X Plane 10), FSX (Microsoft_Flight_Simulator_X), Flightgear (Flightgear), YSFlight (YS Flight Simulator), Lock On (LockOn- modern air combat) and GL 117 (GL-117) are some flight simulations with training and entertainment provisions within. Mainly military and pilot training institutes are demanding the real time flying features while the others demanding the entertainment/ gaming features. Hence after analyzing basic features it can be categorized those simulators according to their intended section whether training or the entertainment. So the following briefing is done with the intention of categorizing the simulations and identifying their features. X Plane, FSX, YS Flight and Flightgear were concerned with higher priority as they are the best rating products in market. Hence FSX discontinued in 2012 (Microsoft_Flight_Simulator_X) let it apart from the comparison.

Table 1. Comparision of Existing Flight Simulators

Feature	X-Plane 10	Flightgear	YS Flight
Airports.	Facilitate both online downloading and can	Facilitate both online downloading and can create own airports.	There are default in built airport pack and bundles of community made maps.

	create own airports.		
Aircraft.	Default aircraft bundle is there. Instead of that user can build any new one or download one.	Can be populated according to the users wish. Down-loadable & inbuilt aircrafts are also there.	Can model more aircrafts except the default in built pack.
Controls /Joysticks.	Mouse, key-board and joy-stick controls are available.	Joystick, mouse and key-board inputs are allowed.	Keyboard, mouse or joy-stick assignments are available.
Processor & GPU support.	More GPU support and uses multi processors for background scenery loading.	Does not drain lots of re-sources. Multi-core proces-sor would be better for threaded tile loader.	Not much CPU power needed but GPU must be enough to populate clear visuals.
Updates.	Updates availa-ble with ongo-ing Q/A forums.	Continuing. Version 2.0 released in 2013.	Continuing with an active community.
Graphics scenery.	Lower quality.	More realistic.	Lighter visuals but less realistic.
Settings (Key assigning etc.)	Few control settings availa-ble.	Wider range of setting avail-able with weather, wind and nigh & day settings.	Multiplayer, weather changing few settings.
Commercial/ Open Source.	Commercial.	Open source.	Open source.
Simulated Emergencies.	Yes.	Yes.	No.

As the open source project Flightgear has the most realistic virtual environment. It has number of airports available either in default package or online. The ability to design runways and taxiways has encouraged the users to design the non-available airports. Although YS Flight has this extendibility realism is really low with relevant to the Flightgear. In a training realism is a significant fact in order to feel that the user is actually out there in that situation. Accuracy and the real time effect are other aspects that are significant in a training scenario and Flightgear gets more points in those two cases too. Unlike the other open source flight simulators Flightgear has a large active development community. Abundance of the resources, documentations, forums and the wiki itself boost up the user friendliness of the Flightgear compared to the other open source flight simulators. The feasibility of extending the Flightgear is also a very useful feature with relative to the other flight simulators. The real time physics can be used in the process and Flightgear also supports all three flight dynamics models. Based upon reasons it is selected the Flightgear to be discussed its seriousness or the learning characteristics against the framework that Garris et al. proposed.

6 Flightgear for Training

Among the available flight simulators Flightgear is selected for the discussion due to following concerns. First it is open source and has relatively active community than the other FOSS flight simulation communities. This will ease the customizing process. Airports, Airplanes and sceneries are available online and can be downloaded directly to the package. Editing them is simple and resources, tutorials are available. In that case Flightgear is taken as the base case for comparing the educational capabilities along with model proposed by Garris et al.

Table 2. Comparision of Flightgear with Eductional characteristics proposed by Garris et al.

Feature	Availability in Flightgear	Description
Capability	Yes	– Released under GNU license which gives the freedom to create and implement individual contributions – Adopt the sceneries or the environment – Changing the flying conditions – Experience different situations; normal flying scenarios, emergency or critical flying scenarios
Instructional content	Yes	– Include the basic instructional content with in the design. Eg: If the engine starting of the being 777 is to be taught then the steps of checking and power on process should be the instructional content.
Intended Learning Outcomes	Yes	– Ability to re-do the trained process under the relevant constraints. E.g. From engine start up training, learner should be capable to re-do the process.
Serious Game Attributes	Yes	– A vast number of setting plans: detailed and accurate model, accurate world scenery data, number of various aircrafts, multiplayer model are some features Flightgear

		– Supports multiplayer feature, voice communication between the players – Practice the formation flight or for tower simulation purposes (Flightgear Features)
Learning Activity	Yes	– Flightgear has no special missions as in Microsoft Flight Simulator X. – Predefined flying exercises such as flying between selected two destinations following a predefine route in Flightgear. – As Flightgear is free and open source it can integrate learning activities as user wants.
Reflection	Yes	– Measure user progressing from one session to the next. – Flightgear has the facility of recording and replaying the flying. – Based on that user or instructor can recognize the previous errors and re-correct them.
Game Achievements	No	– Flightgear is an open game platform. It doesn't have levels, individual accounts or achievement counter. – Flightgear loose this attributes of serious gaming application. But since the Flightgear is an open source product this features can be implemented on the framework.

As discussed above Flightgear has the attributes that need to be an educational platform. This implies that Flightgear is capable of using as a training platform. Flight simulation is used in several fields such as passenger transportation, military and helicopter flying. Flightgear can be considered as a real time serious game which serves in educational/training purpose in advanced manner.

Flightgear allows using the multiplayer mode where a group of students can practice simultaneously. This feature allows the students to cooperate with others and share their knowledge. As well as this enhance the oral expressions and their communicational skills. Flightgear facilitates applying the concepts in practical scenarios. With the simulation pilot can train in different day-time conditions, weather conditions and abnormal behaviors. Likewise Flightgear has the basic structure addressing the characteristics which are needed in effective learning process.

7 Conclusions

During the past decades simulators have proven that they are an effective and efficient solution in the teaching, learning and training process. Adopting the commodity off the shelf hardware and open source software to upgrade the virtual environments has delivered a great platform for the experiential learning. Flightgear is commonly used open source flight simulator. It has extendable modules which can be used to achieve high realism. When comparing and contrasting the educational characteristics of the Flightgear it has shown its compatibility acquiring almost all the characteristics except one according to the measurements proposed by Garris et al.

As the current generations favor digital media, simulation has enhanced consequence on learners rather than traditional read-write process. Further simulations can be used in the risky and costly training processes to overcome those barriers in training address the safety issue in well-organized manner. Experiencing the difference weather conditions, wind speeds and day-night flying are some natural conditions simulations could generate independent of the real environment. This won't be possible without the simulations in action. Thus simulators have lift up the prominence of aviation training.

Bibliography

David R. Michael and Sandra L. Chen, *Serious games: Games that educate, train, and inform*, 1592006221st ed.: Muska & Lipman/Premier-Trade, 2005.

Djaouti, Damien, Julian Alvarez, and Jean Pierre J, *Classifying serious games: The G/P/S model.*: Handbook of research on improving learning and motivation through educational games: Multidisciplinary approaches, 2011.

"Aviation Maintenance Technician Handbook – General- Chapter 14," Fedaral Aviation Adminstration, 2013.

Flightgear. [Online]. http://www.flightgear.org/

GiPSi - Open Architecture Software Development Framework for Surgery Simulation. [Online]. http://gipsi.case.edu/

Kevin Montgomery, "Spring: A general framework for collaborative, real-time surgical simulation.," in *Studies in health technology and informatics*, 2002, pp. 296-303.

OpenSurgSim. [Online]. https://www.assembla.com/spaces/OpenSurgSim/wiki

Richard A., and Robin C. Winkler Winett, "Current behavior modification in the classroom: Be still, be quiet, be docile," *Journal of Applied Behavior Analysis*, pp. 499-504, 1972.

D.A. Kolb, "Experiential learning: experience as the source of learning and development," , 1984.

Real-time computing. [Online]. http://en.wikipedia.org/wiki/Real-time_computing

Sandaruwan, D., Kodikara, N., Keppitiyagama, C., Rosa, R., Dias, K., Senadheera, R., & Manamperi, K. (2012). Low Cost Immersive VR Solutions for Serious Gaming. In A. Kumar, J. Etheredge, & A. Boudreaux (Eds.) *Algorithmic and Architectural Gaming Design: Implementation and Development* (pp. 407-429). Hershey, PA: Information Science Reference. doi:10.4018/978-1-4666-1634-9.ch017

Y., Sun, X., Zhang, X., Liu, X., Ren, H., Zhang,X., & Jin, Y. Yin, "Application of virtual reality in marine search and rescue simulator. ," *The International Journal of Virtual Reality*, pp. 19–26, 2010.

CHAO Jian-gang, SHEN Jun-yi, and HE Ning, "Space Flight Training Simulation Project Technology Research," in *2010 Second WRI World Congress on Software Engineering*, 2010.

C.V. Carvalho R. Batista, "Learning Through Role Play Games," in *38th IEEE Annual Frontiers in Education Conference*, 2008.

David Gouveia, Duarte Lopes, and Carlos Vaz de Carvalh, "Serious gaming for experiential learning," in *Frontiers in Education Conference (FIE)*, 2011, pp. T2G-1.

Paulo David da Silva Simões and Cláudio Gabriel Inácio Ferreira, "Military war games edutainment," in *Serious Games and Applications for Health (SeGAH)*, Braga, 2011, pp. 1-7.

Bernard McCaffrey, "Flight simulators and related factors," ERAU,.

Robert F. Stengel, "Aircraft Flight Dynamics ," Princeton University School of Engineering and Applied Science,.

Flight dynamics (fixed-wing aircraft). [Online]. http://en.wikipedia.org/wiki/Flight_dynamics_(fixed-wing_aircraft)

Fligh Dynamics. [Online]. http://en.wikipedia.org/wiki/Flight_dynamics

E. Bruce Jackson, "Manual for a Workstation-based Generic Flight Simulation Program(LaRCsim) ," National Aeronautics and Space Administration-Langley Research Center, Virginia , 1995.

Jon S Berndt, "JSBSim: An Open Source Flight Dynamics Model in C++," in *AIAA International Communications Satellite Systems Conference & Exhibit*, 2004.

Jon S. Berndt, "JSBSim an open source, platform-independent, flight dynamics model in C++," , 2011.

Flightgear Wiki. [Online]. http://wiki.flightgear.org/Flight_Dynamics_Model

Outerra. [Online]. http://www.outerra.com/wfeatures.html

OpenEaagles. [Online]. http://www.openeaagles.org/wiki/doku.php

YASim. [Online]. http://wiki.flightgear.org/YASim

UIUC Applied Aerodynamics Group. Aircraft Dynamics Models for Use with FlightGear. [Online]. http://aerospace.illinois.edu/m-selig/apasim/Aircraft-uiuc.html

UIUC. [Online]. http://wiki.flightgear.org/UIUC

X Plane 10. [Online]. http://www.x-plane.com/x-world/landing/

Microsoft_Flight_Simulator_X. [Online]. http://en.wikipedia.org/wiki/Microsoft_Flight_Simulator_X

YS Flight Simulator. [Online]. http://ysflight.in.coocan.jp/

LockOn- modern air combat. [Online]. http://lockon.co.uk/en/

GL-117. [Online]. http://es.wikipedia.org/wiki/GL-117

"Skill Evolution Report," UFI, Sheffield, 2008.

Whitehal and McDonald, "Improving learning persistence of military personnel by enhancing motivation in a technical training program. Simulation & Gaming," , 1993, pp. 294-313.

Ricci E. Katrina, Janis A. Cannon-Bowers, and A. Cannon-Bowers Janis, "Do computer-based games facilitate knowledge acquisition and retention? Military Psychology.," , 1996, pp. 295-307.

Rosemary Garris, Robert Ahlers , and James E. Driske, "Games, motivation, and learning: A research and practice model," , 2002, pp. 441-467.

Amri Yusoff, Richard Crowder, and Lester Gilbert, "Validation of serious games attributes using the technology acceptance model.," in *2010 Second International Conference on. IEEE*, 2010.

Flightgear Features. [Online]. http://www.flightgear.org/features.html

Potential Z-Fighting Conflict Detection System in 3D Level Design Tools

Pisal Setthawong

Assumption University,
Bang Na-Trad Km. 26, Samut Prakan, 10540, Thailand
{pisalstt@au.edu}
http://www.au.edu

Abstract. Z-Fighting is an effect that happens in 3D scenes when two co-planar surfaces share similar values in the z-buffer which leads to flicking and visual artifacts during the rendering process due to conflicting order of rendering the surface. However in 3D level design, scenes created by the tools can be complex, in which level designers can inadvertently place co-planar surfaces that would be susceptible to z-fighting. Level designers typically notice the z-fighting artifact through visual inspection through the usage of a 3D walkthrough test on the scene which is time-consuming and easy to miss. To solve the issue, a proposal of a z-fighting detection system for level design tools is proposed to streamline the process of detecting potential hotspots where z-fighting conflicts may occur from co-planar objects.

Keywords: Z-Fighting, Level Design Tools, 3D Graphics, Computer Graphics, 3D Production Pipeline

1 Introduction

Level design is an important part of creating 3D games, virtual reality applications, 3D walkthroughs, and serious games, and other forms of interactive 3D scenes[1][8]. To aid with the process, tools such as 3D level designs tools were created to help the level designer design 3D levels and scenes to use in their applications. Over the years, 3D level design tools have increased in functionality and complexity, making it possible for level designers to design more complex scenes. Though level design tools have improved over the years, there are still a number of outstanding issues that can cause visual artifacts in the final scene in which typical level design tools cannot detect in the design process.

One of the potential visual artifacts is when z-fighting artifacts occur[7]. When the level designer place objects that are co-planar, z-fighting conflict can occur causing visual artifacts such as flickering and inconsistently merging textures. Z-fighting and their resultant effects are highly visible to the end-user and is detrimental to the overall experience of the applications and should be avoided if possible.

K.J. Kim (ed.), *Information Science and Applications*,
Lecture Notes in Electrical Engineering 339, DOI 10.1007/978-3-662-46578-3_32

To deal with z-fighting artifacts, there are many approaches. One approach is to detect and fix the z-fighting artifacts is to visually inspect the 3D scene. This process is time consuming and error-prone. Another approach is to use technical approaches such as changing rendering parameters and buffers are one area in which could limit the effect of z-fighting when it occurs, but does not prevent the effect from happening. Alternatively another approach that could limit the amount of z-fighting in 3D scenes is to ask the level designer to avoid placing co-planar objects in the scene. Though many of the z-fighting artifacts could be avoided in the level design process by following best level design practices, the effect may still be visible in the end scene as it is possible that the level designer may inadvertently place co-planar surfaces/objects in the scene. To deal with the situation and avoid the time consuming and error prone process of using 3D walkthrough of the scene to visually inspect the scene[5], a tool to parse the hierarchy of the scene and detect potential z-fighting conflicts could be implemented to improve this process.

2 Z-Fighting

Z-fighting is a scenario that happens when a number of 3D objects in the scene share the same orientation and depth or considered to be co-planar[9][7][6]. In such a situation when two surfaces are co-planar or the depth difference to be very minimal, the renderer is confused by the order of which of the object to render. This dilemma causes the rendered scene to display the conflicting surface alternatively or erratically due to pixel interpolation of the texture causing flickering and other visual artifacts which is illustrated in Fig. 1. The artifacts are especially evident in large co-planar surfaces that have huge overlapping areas, and is not desirable in any 3D scenes.

Fig. 1. Overlapping co-planar objects that can lead to z-fighting conflict and the resulting visual artifact and flicker in a level design

To minimize the effect of z-fighting, changing technical parameters such as increasing the buffer resolution, using stencil buffers, post rendering techniques, and modifying clipping planes could help, but may not entirely eradicate z-fighting from occurring the in the scene. Other approaches such as avoiding the placement of co-planar objects in the scene is the most effective way to prevent z-fighting from happening, and is a guideline in which level designers are recommended to follow.

In a typical level design process, the z-fighting artifacts are usually detected during the walkthrough process in which the level designer tests the system observes areas where there are issues in the stage, and make necessary fixes. This process is time consuming and error-prone leading to the proposal of a z-fighting detection system that could help streamline the level design process by highlighting areas in which z-fighting artifacts can potentially occur in the level design tool allowing the level designer be aware of issues earlier in the design pipeline, and in turn minimizing the need to visually inspect the 3D scene to detect z-fighting artifacts.

3 Proposed System

This section describes the proposed system that is used to detect potential z-fighting conflicts. To limit the scope of the research, this paper describes the detection of planar surfaces that overlap in a co-planar fashion in which forms the majority of the cases of z-fighting and its associated visual artifacts. The system is integrated into the level design tool to allow easy access for the 3D level designer to quickly detect potential areas where z-fighting conflicts can occur. The level design tools where this system is implemented in is based on the Unity Game Engine [10].

The first step is to transverse the scene hierarchy to select objects for the test. Planar surfaces contribute most to co-planar overlapping areas which causes z-fighting conflicts and are selected.

For each object selected, the pair combination of the objects is built in which the rotation orientation of the object is compared. When a pair has the same rotation orientation, the pair is saved for later processing whereas pairs with different orientations are discarded.

For pairs that remain that have similar orientation, based on the position and scale of the planar objects, an overlapping test is conducted. The approach used is the separating axis theorem (SAT)[2]. The theorem states that if a line could be drawn to separate two convex polygons, then they do not collide and by using the SAT test on the aligned axis of the convex polygons, the system can detect if the planar objects overlap or not.

The first step of the process is to select the axis in which the SAT test would be implemented on. In typical convex 3D objects, a total of 15 axis [3] have to be calculated for OBB (oriented bounding box)[4] vs OBB tests. However in the pair that are selected in the system, all objects are coplanar objects which reduces the problem set into a 2D problem complexity requiring only 4 axis tests based on the perpendicular vector of the major parallel edges of the planar objects to calculate if the objects are overlapping via the SAT test which is displayed in Fig. 2.

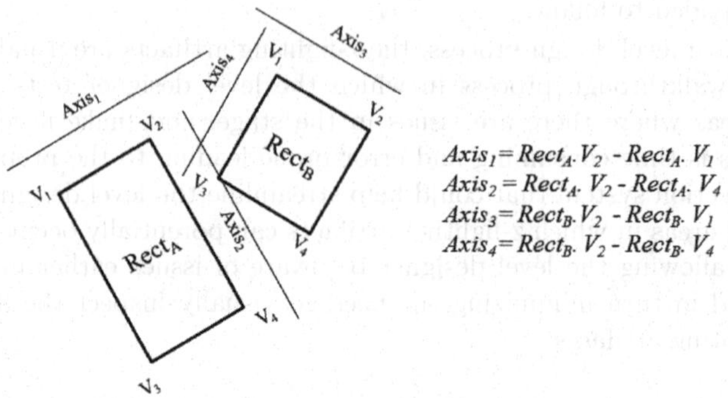

Fig. 2. Example of planar objects and the perpendicular vectors that are selected as the major axis in the SAT test to determine if planar objects overlap

The set of all the vertices from the coplanar pairs are first selected as in Equation 1.

$$V = \{Rect_A.V_{1..4}, Rect_B.V_{1..4}\} . \tag{1}$$

The vertices are then projected into the each of the axis selected for the SAT test which is illustrated in Fig. 3 and calculated as in Equation 2. Based on each of the planar object, the minimum and maximum scalar values are selected. The values are then put in an overlap test as in Equation 3. If the overlap tests fair for any of the cases, the planar objects are not overlapping, but if all the cases are overlapping, the SAT test will conclude that the planar objects are overlapped.

$$Projection_{Axis_i}^{V} = \frac{V \cdot Axis_i}{||Axis_i||^2} Axis_i . \tag{2}$$

$$(B_{Min} <= A_{Max}) \; \&\& \; (B_{Max} >= A_{Min}) . \tag{3}$$

Once the SAT test has been applied on the pairs, any overlapping co-planar objects are saved and reported. The system would report the pairs of the overlapping co-planar objects in which z-fighting conflict can occur, and mark in the level design tool for the level designer to be aware of the potential hotspot areas and fix the associated objects and reduce the potential of z-fighting conflicts.

4 Conclusions and Future Work

The proposed system was applied on tested on 57 student level design assignments in which z-fighting conflicts occur in many levels. The systems manages

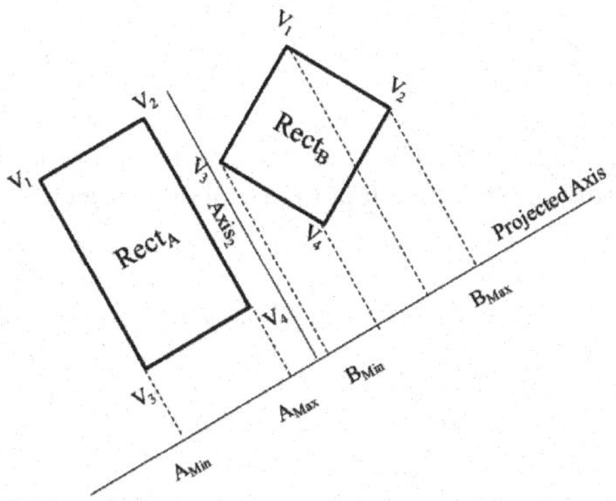

Fig. 3. Projecting the vertices of the planar objects onto the projection axis for the SAT test

to pinpoint many co-planar objects in the scenes where z-fighting conflicts may occur, though in many of the cases due to the renderer settings, the z-fighting conflict visual artifacts do not happen.

The proposed system solves an issue where potential z-fighting conflicts are detected early in the level design process for planar objects. This simplifies the process by allowing the level designer to avoid z-fighting artifacts from being caused in the scene without the need of a visual inspection, and allows fixes to be done in a more streamline fashion as co-planar objects that overlap are identified. Once overlapping co-planar objects are identified, slight changes to the objects can remove the z-fighting artifacts from the scene. The proposed system can help steam-line the review process of the level design process in removing most cases of z-fighting artifacts from potentially happening on the scene saving the level designer time and effort.

Though the proposed system works well with opaque planar objects, there are many areas in which the z-fighting detection system could be extended in the level design tools. In certain cases, perpendicular planar objects can cause z-fighting artifacts in certain viewing angles, though the artifacts happens less often when compared to co-planar objects and is illustrated in Fig. 4. Extension of the system to include more complex polygonal shapes other than planar objects could be explored. In rendering exception cases such as rendering transparent and translucent objects which is significantly more complex than opaque objects, z-fighting artifacts can occur under different circumstances. Due to depth sorting issues, detecting z-fighting conflicts for transparent and translucent objects require many exceptional cases. Both these scenarios highlighted are future work, in which the z-fighting detection system could be extended to improve the usefulness of the system in 3D level design.

Fig. 4. Z-fighting conflicts occurring at certain overlapping perpendicular planar objects which is less common than overlapping co-planar objects

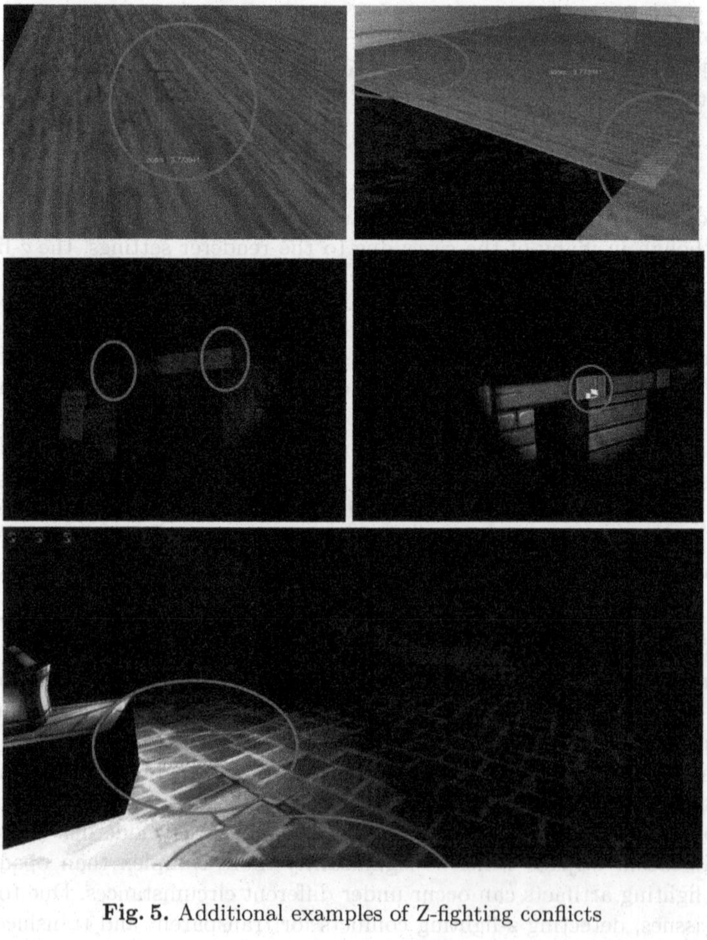

Fig. 5. Additional examples of Z-fighting conflicts

References

1. Co, P.: Level Design for Games: Creating Compelling Game Experiences. New Riders, San Francisco (2006)
2. Eberly, D.: Intersection of Convex Objects: The Method of Separating Axes, `http://geometrictools.com/Documentation/MethodOfSeparatingAxes.pdf`
3. Ericson, C.: Real-Time Collision Detection. Morgan Kaufmann, San Francisco (2005)
4. Gottschalk, S. and Lin, M. C. and Manocha, D.: OBBTree: A Hierarchical Structure for Rapid Interference Detection. In: SIGGRAPH '96, pp. 171–189. ACM, New York (1996)
5. Levy, L. and Novak, J.: Game Development Essentials: Game QA Testing. Cengage Learning, (2010)
6. Madhav, S.: Game Programming Algorithms and Techniques: A Platform-Agnostic Approach. Addison-Wesley, Boston (2013)
7. OpenGL.: The Depth Buffer, `https://www.opengl.org/archives/resources/faq/technical/depthbuffer.htm`
8. Rudolf, K.: Level Design: Concept, Theory, and Practice. A K Peters, (2009)
9. Shreiner, D. and Sellers, G. and Messenich, J. M. and Licea-Kane, B. M.: OpenGL Programming Guide: The Official Guide to Learning OpenGL, Version 4.3. Addison-Wesley, Boston (2013)
10. Unity, `http://unity3d.com/`

Efficient Motion Estimation Algorithms for HEVC/H.265 Video Coding

Edward Jaja[1], Zaid Omar[2], Ab Al-Hadi Ab Rahman[2], and Muhammad Mun'im Zabidi[2]

[1] Faculty of Electrical Engineering, Universiti Teknologi Malaysia, 81310 Johor, Malaysia
eejaja77@gmail.com
[2] Faculty of Electrical Engineering, Universiti Teknologi Malaysia, 81310 Johor, Malaysia
{zaid,hadi,raden}@fke.utm.my

Abstract. This paper presents two fast motion estimation algorithms based on the structure of the triangle and the pentagon, respectively, for HEVC/H.265 video coding. These new search patterns determine motion vectors faster than the two Tzsearch patterns - diamond and square – that are built into the motion estimation engine of the HEVC. The proposed algorithms are capable of achieving a faster run-time with negligible video quality loss and increase in bit rate. Experimental results show that, at their best, the triangle and pentagon algorithms can offer 63% and 61.9% speed-up in run-time respectively compared to the Tzsearch algorithms in HEVC reference software.

Keywords: H.265 algorithms, Fast HEVC encoding, HEVC algorithms, video coding

1 Introduction

HEVC is a new block-based video compression standard to replace the present standard, the H.264/AVC. This new standard is designed to support higher resolutions so the coding block sizes are larger. In the HEVC compression standard, each picture frame is divided into blocks of coding tree units (CTUs) which have a recursive quad-tree structure; the CTUs are the equivalents of macro blocks in the H.264/AVC compression standard. Each CTU consist of three coding tree blocks (CTBs) and syntax elements [1]. The first CTB is the luma CTB while the remaining two blocks are the corresponding chroma CTBs. Each CTU is associated to a quad-tree structure which reflects how the CTU is subdivided recursively into smaller coding units (CUs) which are at the leaf nodes of the quad-tree. Fig. 1a shows the recursive subdivision of the CTU into CUs. Fig. 1b shows the nested quad-tree structure of the CTU. The CUs are at the bases of predictions, its size ranges from 8×8 up to the entire CTU [2]. Each CU can also be divided to obtain the prediction units (PUs) that is not necessarily square in shape. Each PU is associated to a transform unit (TU); the TUs are square in shape with sizes ranging from 4x4 pixels-which is the smallest allowed in HEVC-to 32×32 pixels; this means that TUs can be formed across PU boundaries. The CU

K.J. Kim (ed.), *Information Science and Applications*,
Lecture Notes in Electrical Engineering 339, DOI 10.1007/978-3-662-46578-3_33

can be further partitioned into one, two or four PUs depending on the selected mode. All PU partition arising from a particular CU must share the same prediction-inter or intra prediction [3]. The PUs should be of such structure that can make-up boundaries of real objects in the picture frame, so it does not need to be square in shape.

The H.265/HEVC is designed to deliver in the range of 50% bit rate reduction at the same visual quality when compared to H.264 [1]. It is also designed to support higher resolutions up to the ultra-high definition UHD (8192 × 4320 pixels) often called the 8k; it supports frame rates of up to 300 frames per second to enable smooth rendering of higher resolutions. The fundamental coding tool of the HEVC is the enhanced hybrid spatial-temporal prediction model based on the recursive coding tree units (CTUs). Temporal prediction involves motion estimation and compensation to code the current frame from the data from the reference frames in the coded picture buffer. HEVC supports up to 15 past or future reference frames. Due to these enhanced features of the HEVC, motion estimation and compensation takes 40% of the encoder time [4]. In the HEVC reference software, two fast motion estimation algorithms are built into the Tzsearch pattern which is the motion estimation engine that evaluate the integer motion vector within a search window in the reference frame. These algorithms return motion vectors at comparable cost to the full search pattern but the motion estimation time still remains high and requires further improvement. In this paper two fast search motion estimation algorithms are proposed for the HEVC; they are the triangle and the inverted pentagon fast search algorithms. These algorithms are faster on the average when compared to the Tzsearch algorithms implemented in the HEVC reference software at a comparable visual quality. The rest of the paper is organized as follows. Section 2 presents overview of motion estimation; in section 3, the proposed motion estimation algorithms are explained. Experimental results and the conclusion are presented in sections 4 and 5 respectively.

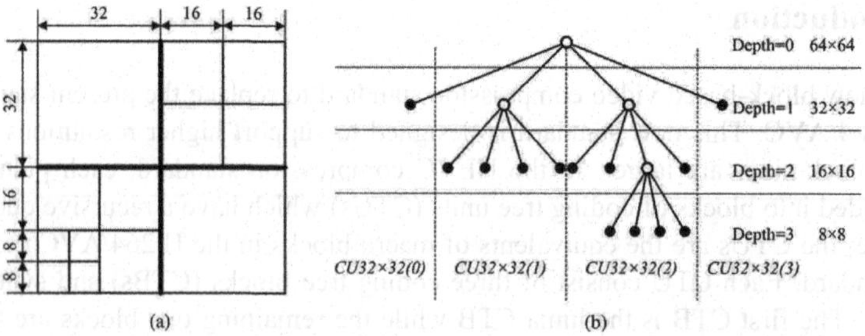

(a) Recursive division of the CTU (b) Nested quad-tree structure of the CTU

Fig. 1. The HEVC quad-tree structure [2]

2 Overview of Motion Estimation

2.1 Fast Search Motion Estimation in HEVC

In HEVC, fast search motion estimation involves determining a best match of the current block by searching the reference picture within a 64×64 pixels search window using fast search algorithms. The fast search algorithms implemented in the HEVC reference software are the square and the diamond algorithms; these algorithms return best match to the current block at comparable cost to the full search algorithm but they still remain sub-optimal in terms of motion estimation time. The following three steps are required to obtain motion vector in HEVC: start position selection, first search and refinement search [3]. The motion vector accuracy level supported in HEVC is the quarter-pel. To select the start position in the motion vector estimation process, the motion vectors of already encoded neighboring blocks, including zero motion vectors, are evaluated to determine the best start point.

The first search involves deployment of the search pattern over the search window to determine the motion vector that offers the best match. HEVC allows three deployments of the search pattern from the start point to determine the integer-pel motion vector. The matching metric used is the sum of absolute difference (SAD) at the integer level.

The refinement search process is centered on the integer-pel motion vector after up-sampling and applying interpolation filtering on the reference block samples. To obtain the quarter-pel motion vector, half-pel motion vector is first derived by determining which pixel position among 8 pixels around the integer-pel motion vector that offers the minimum cost. The 8 pixels around this half-pel motion vector are further evaluated to determine which of them offers the minimum cost; this gives the quarter-pel motion vector. The matching metric used at the half and quarter-pel levels is the sum of absolute transformed difference, SA(T)D.

Another search pattern defined within the HEVC is the raster search pattern; this only comes into operation when the derived integer-pel motion vector is more than 5 pixels away from the start position. This level is adjustable within the HM reference software. Representing the structure of these algorithms like electrons in atomic shells or grids, we have the following relations for a 64×64 search window:-

For the triangle,

$$Triangle\!: SP^0 \;\; 1sL^4 \;\; 2sL^3 \;\; 4sL^3 \;\; 8sL^3 \;\; 16sL^3 \;\; \frac{1}{2}P^8 \;\; \frac{1}{4}P^8$$

For the inverted pentagon,

$$Pentagon\!: SP^0 \;\; 1sL^5 \;\; 2sL^5 \;\; 4sL^5 \;\; 8sL^5 \;\; 16sL^5 \;\; \frac{1}{2}P^8 \;\; \frac{1}{4}P^8$$

For the square,

$$Square\!: SP^0 \;\; 1sL^8 \;\; 2sL^8 \;\; 4sL^8 \;\; 8sL^8 \;\; 16sL^8 \;\; \frac{1}{2}P^8 \;\; \frac{1}{4}P^8$$

For the diamond,

$$Diamond\!: SP^0 \;\; 1sL^4 \;\; 2sL^8 \;\; 4sL^8 \;\; 8sL^8 \;\; 16sL^8 \;\; \frac{1}{2}P^8 \;\; \frac{1}{4}P^8$$

Where SP^0 stands for the start point of the search pattern, its superscript represents the origin of the stride lengths. 1sL to 16sL represents the stride lengths of the grids from the start point; the superscripts stand for the number of search points at each of these stride lengths. $\frac{1}{2}P^8$ and $\frac{1}{4}P^8$ are the half-pel and quarter-pel search points with the superscripts representing the 8 search points at these levels. This analysis is made for a $64x64$ pixel search window; stride length $32sL^8$ is not used because its search points yields samples that over shoot the boundaries of the search window. From the way the fast search algorithm is implemented in the HM software, any motion vector returned at stride lengths greater than $4sL$ would trigger the raster search. From the analysis here, assuming the start point SP^0 is the integer motion vector, it takes 32 search points $(4 + 3 \times 4 + 2 \times 8)$ to determine the quarter-pel motion vector in the triangle. In pentagon, it takes 41 $(5 \times 5 + 2 \times 8)$ points to come-up with quarter-pel motion vector. For the square, it takes 56 (8×7) search points to evaluate the quarter-pel motion vector. For the diamond, it takes 52 points $(4 + 8 \times 6)$. The triangle and pentagon pattern use fewer search points to determine motion vectors and from the experimental results, the bit rate and the peak signal to noise ratio (PSNR) is comparable to those of the square and diamond.

2.2 Related Works

A lot of research effort has been spent to cut down motion estimation and mode decision time in HEVC. In [4], a proposal was presented to quicken the mode decision by first testing the CTU for inter-prediction before testing for early skip decision (ESD); also the Tzsearch points were also modified to horizontal diamond search pattern. These twin effects yielded an average reduction in encoding time of up to 52% at an average PSNR loss of 0.12dB with 0.5% reduction in bit rate. Early termination of Tzsearch were proposed in [5] based on the probabilities of CUs using the median predictor (MP) as the starting point of the search process as well as the centre bias characteristics of the final motion vector; 38.96% encoding time was saved through this means with acceptable PSNR value. In [6], the Tzsearch patterns were replaced with the hexagon pattern with six search points around the grid, it was reported that 15% encoding time was saved. Also the Tzsearch grids at stride lengths greater than or equal to 5 pixels were disabled and the number of search points and time were saved from those grids. This gave 50% reduction in encoding time with no effect on bit rates and PSNR values. In [7] multiple reference frame motion estimation was investigated using H.264 wherein the researchers pre-calculated cost from the centre of search window in incremental radial ring of one pixel. The closest radius to the centre of search is given a higher weight or level, and priority is given to higher levels in the motion estimation process; it was empirically determined that 83.46% time gain was achieved in the motion estimation process. This process offered a good result but is not suitable for block-based motion estimation.

From these previous works, it was discovered that the search window constitute a matching error surface with one global minimum with several minima that can also offer comparable results as the global minimum. Based on this premise, a smaller search pattern was conceived to further reduce motion estimation time in HEVC which targets higher resolutions.

3 Triangle and Inverted Pentagon Search Patterns

Based on empirical evidence from [8] that 52.76% to 98.70% of motion vectors within search windows are enclosed within a radial distance of two pixels from the center of the search window; it then follows that smaller search patterns can obtain motion vectors at comparable block matching error to the set of search patterns in use in the HM reference software. On this basis, the inverted pentagon and the triangle search patterns were designed to be used in the HM reference software. The first grid of the triangle pattern remains the four-point diamond [8]. The structure of these fast search patterns are as shown in Fig. 2.

At the start point of each of these algorithms which could be a median of neighboring motion vectors or a point-of-zero motion vector, the center of the search pattern is positioned at the chosen start point. Using this point, the matching error between the reference block centered at this point and the current block is calculated. This matching error is used to compare with matching errors at all search points on each grid. The point that generates the least matching error becomes the center of the next search step. This process is repeated two more times in HEVC to obtain the integer-pel motion vector. The flow chart for obtaining the motion vector in HEVC is as shown in Fig. 3.

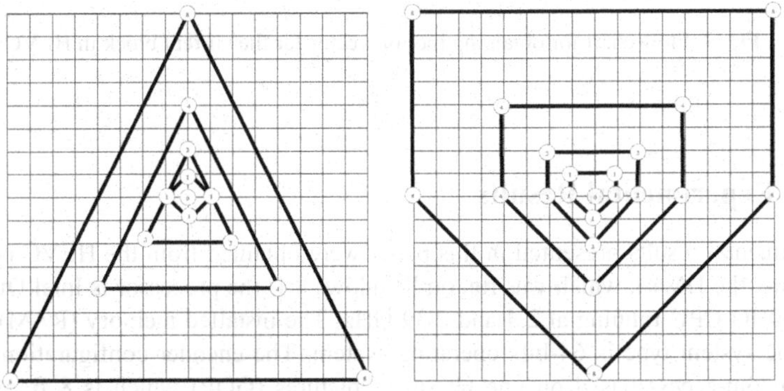

(a) Triangle search pattern (b) Inverted pentagon search pattern

Fig. 2. Triangle and inverted pentagon search pattern with stride length 8

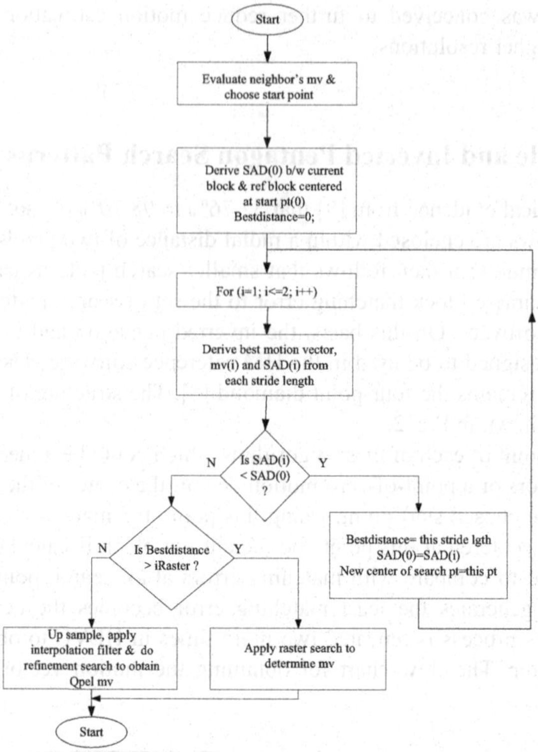

Fig. 3. Flow chart for obtaining motion vector for the current block in HEVC

4 Experimental Results

Experimental results presented in this paper were obtained from the HEVC reference software, the HM14, which was run on Windows 8.1; the processor is Intel Dual Core i7-4700HQ CPU running at 2.4 and 2.39 GHz. The installed memory (RAM) is 12.0 GB; the system type is 64-bits operating system. The encoder configuration used is the random-access based on one Group of pictures (GOP) which is 8 frames. The quantization parameter is 32 and the search range was set at 64.

Tables 1 and 2 compare the proposed inverted pentagon and the triangle fast search algorithms to the square and diamond algorithms in HM14; the bases of comparison are percentage increase in bit rate (ΔBR%), change in peak signal to noise ratio in decibel (ΔPSNR) and percentage time gain (ΔT%). These comparison metrics are evaluated as specified in [4]. From the results, pentagon had a speed-up of 53.84% and 69.9% over square and diamond respectively at a negligible loss in bit rates and PSNR values in class A. In class B, it had average of 3.3% and 28.3% speed-up respectively at a minimal loss of quality. In class C, it had 3.1% and 2.7% speed-up over square and diamond respectively with minimal quality loss. In class D, it had a similar result as in class C. In class E, pentagon ran at the same time producing the same quality as square and diamond.

For triangle, in class A it produced a speed-up of 55.34% and 70.87% over square and diamond respectively at minimal quality loss. In class B, it had 5% speed-up over square and diamond. In class D, it had 4.9% and 5.5% speed-up over square and diamond respectively at minimal bit rate gain and quality loss. In class E, the speed-up is 0.6% and 1.1% over square and diamond respectively at the same bit rate and video quality.

Table 1. Inverted pentagon compared with square and diamond algorithms in HM14

Video classification	Video Sequence	Pentagon/Square			Pentagon/Diamond		
		ΔBR(%)	ΔPSNR(dB)	ΔT(%)	ΔBR(%)	ΔPSNR(dB)	ΔT(%)
2560x1600 (class A)	PeopleOnStreet	0.416	-0.007	-53.840	0.454	0.000	-69.900
1920x1080 (class B)	Kimono1	-0.119	-0.001	-3.040	0.077	0.004	-3.900
	ParkScene	0.062	-0.004	-3.543	0.046	-0.001	-52.600
Class B average		-0.028	-0.002	-3.292	0.061	0.001	-28.250
832x480 (class C)	BQMall	0.050	-0.007	-2.800	0.163	0.001	-2.700
	PartyScene	0.171	0.004	-3.400	0.254	0.002	-2.700
Class C average		0.111	-0.001	-3.100	0.209	0.002	-2.700
416x240 (class D)	BasketballPass	0.084	-0.017	-2.020	-0.191	0.005	-2.410
	BlowingBubbles	-0.300	-0.015	-2.600	-0.410	-0.018	-3.040
	RaceHorses	-0.580	0.023	-3.940	0.580	-0.012	-4.800
Class D average		-0.265	-0.003	-2.853	-0.007	-0.009	-3.420
1280x720 (class E)	City	0.000	0.005	0.811	0.035	0.010	-0.008
	MobileCalendar	0.000	0.000	0.243	0.000	0.000	0.106
Class E average		0.000	0.003	0.527	0.018	0.005	0.049

Table 2. Proposed triangle compared with square and diamond algorithms in HM14

Video classification	Video Sequence	Triangle/Square			Triangle/Diamond		
		ΔBR(%)	ΔPSNR(dB)	ΔT(%)	ΔBR(%)	ΔPSNR(dB)	ΔT(%)
2560x1600 (class A)	PeopleOnStreet	0.490	-0.022	-55.340	0.530	-0.014	-70.870
1920x1080 (class B)	Kimono1	-0.031	-0.008	-6.000	0.164	-0.004	-6.800
	ParkScene	0.277	-0.007	-4.500	0.261	-0.003	-3.300
Class B average		0.123	-0.008	-5.250	0.213	-0.004	-5.050
832x480 (class C)	BQMall	0.266	-0.016	-3.440	0.381	-0.009	-3.400
	PartyScene	0.181	-0.006	-4.700	0.265	-0.008	-4.020
Class C average		0.224	-0.011	-4.070	0.323	-0.009	-3.710
416x240 (class D)	BasketballPass	0.560	0.015	-3.700	0.290	0.037	-4.050
	BlowingBubbles	0.260	-0.027	-4.410	0.157	-0.030	-4.900
	RaceHorses	0.025	-0.015	-6.600	0.025	-0.051	-7.400
Class D average		0.282	-0.009	-4.903	0.157	-0.015	-5.450
1280x720 (class E)	City	0.152	0.000	-0.600	0.189	0.005	-1.400
	MobileCalendar	0.000	0.000	-0.600	0.000	0.000	-0.730
Class E average		0.076	0.000	-0.600	0.094	0.002	-1.065

5 Conclusion

The proposed pentagon when compared to square across all the classes offered 12.5% encoding time gain at 0.47% increase in bit rate and at a PSNR loss of 0.002dB; also when pentagon was compared with diamond, the time gain is 20.84% at 0.147% increase in bit rate and at no loss in quality. Also comparing triangle to square and diamond offered 14.03% and 17.23% encoder time gain respectively with negligible loss in bit rate and video quality.

With an overall speed-up of 16.69% over diamond and square patterns, the inverted pentagon shows an advantage over triangle which has a corresponding speed-up figure of 15.63%. From these results, it is obvious that the presented algorithms perform better than the diamond and the square search patterns.

References

1. Sullivan, G. J., Ohm, J., Woo-Jin H., Wiegand, T.: Overview of the High Efficiency Video Coding (HEVC) Standard. In: *Circuits and Systems for Video Technology*, vol. 22, no. 12, pp. 1649-1668. *IEEE Transactions,* (2012)
2. Zhou, C., Zhou, F. and Chen, Y.: Spatio-temporal correlation-based fast coding unit depth decision for high efficiency video coding. In: Journal of Electronic Imaging, vol. 22, no. 4, pp. 043001-13, (2013)
3. Kim, L., McCann, K., Sugimoto, K., Bross, B., Han, W. and Sullivan, G.: High efficiency video coding (HEVC) test model 13 (HM 13) encoder description. In: Proceedings of the 15th JCT-VC meeting, Geneva (2013)
4. Belghith, F., Kibeya, H., Loukil, H., Ayed, M. A. B. and Masmoudi, N.: A new fast motion estimation algorithm using fast mode decision for high-efficiency video coding standard. In: Journal of Real-Time Image Processing. Springer, Berlin (2014)
5. Pan, Z., Zhang, Y., Kwong, S., Wang, X. and Xu, L.: Early termination for Tzsearch in HEVC Motion Estimation. In: *IEEE International Conference on Acoustics, Speech and Signal Processing (ICASSP),* pp. 1389-1393. IEEE Press, New York (2013)
6. Purnachand, N., Alves, L. N. and Navarro, A.: Improvements to Tzsearch motion estimation algorithm for multiview video coding. In: International Conference on Systems, Signals and Image Processing (IWSSIP), Vienna, pp. 288-391 (2012)
7. Chun-Su, P.: Level set based motion estimation algorithm for multiple reference frame motion estimation. In: Elsevier, vol. 24, pp. 1269-1275 (2013)
8. Zhu, S. and Ma, K.: A new diamond search algorithm for fast block-matching motion estimation. In: IEEE Transactions on Image Processing, vol. 9, no. 2, pp. 287-290 (2000)
9. Zhong, G., He, X., Qing, L. and Yuang Li, Y.: A fast inter-prediction algorithm for HEVC based on temporal and spatial correlations. In: Springer, New York, (2014)
10. Correa, G., Assuncao, P., Agostini, L. and Da Silva Cruz, L.A.: Complexity control of high efficiency video encoders for power-constrained devices. In: IEEE Transactions on *Consumer Electronics,* vol. 57, no. 4, pp. 1866-1874. IEEE Press, New York Nov (2011)

A Synthesis of 2D Mammographic Image Using Super-Resolution Technique: A Phantom Study

Surangkana Kantharak[1], Thanarat H. Chalidabhongse[2], and Jenjeera Prueksadee[3]

[1] Biomedical Engineering, Faculty of Engineering, Chulalongkorn University, Thailand
Surangkana.k@student.chula.ac.th
[2] Computer Engineering, Faculty of Engineering, Chulalongkorn University, Thailand
Thanarat.c@chula.ac.th
[3] Radiology, Faculty of Medicine, Chulalongkorn University, Thailand
jenjeera@hotmail.com

Abstract. The gold standard for early detection of breast cancer has been the mammogram. However, this technique still has limitation for women with dense breast. Combining mammogram with digital breast tomosynthesis overcomes the limitation but increases exposure dose approximately twice. This study focuses on reducing radiation dose by synthesizing the 2D mammographic image from multiple tomosynthesis projection images using an image Super-Resolution technique based on sparse representation. We evaluated the result images using peak signal to noise ratio (PSNR), mean structure similarity (MSSIM) and phantom passing score. We compared the synthesized 2D mammographic image from multiple projection images to the one from a single central projection image. The one from multiple images yields better result with. 27.2426 PSNR and 0.4436 MSSIM. For the phantom passing score, we obtained 5, 2, 4 for fibers, group of micocalcifications, and masses, respectively.

Keywords: Digital Breast Tomosynthesis, Super-Resolution, dose reduction, synthesized 2D mammographic image.

1 Introduction

Breast cancer is the most common type of cancer in women which can be early detected using a 2D imaging technique called Mammogram. Although, it is the gold standard, some lesions remain difficult to be detected due to the occlusion of breast tissues. This can be solved by combining mammogram with additional investigations such as a Digital Breast Tomosynthesis (DBT).

The Digital Breast Tomosynthesis can help eliminating occlusion problem of breast tissues. It is a 3D imaging technology that provides cross sectional visualization of the breast. While the breast is compressed, the x-ray tube is moved and a series of low dose images are taken at different angular positions of the x-ray tube. The acquired images are called tomosynthesis projection (TP) images which undergo a reconstruction process to reconstruct cross sectional or tomosynthsis reconstruction (TR) images that are used for diagnosis [1]. Recent studies show that DBT is better

K.J. Kim (ed.), *Information Science and Applications*,
Lecture Notes in Electrical Engineering 339, DOI 10.1007/978-3-662-46578-3_34

than mammogram in detecting masses and architecture distortions [2, 3, 4]. However, microcalcifications are better identified by mammogram. Because of long exposure time and thin reconstruction thickness of DBT system, the cluster of microcalcification is distributed over several slices [5]. These are the reason to combine both systems in diagnosis of breast disease. Not only sensitivity and specificity but also reading time is increase [5, 6]. Moreover, exposure dose increase approximately twice [7].

Few researches study the reduction of radiation dose in combining mode by synthesis a 2D mammographic image. Gur et al. [8] compared between synthesized and real 2D mammographic image, both combined with DBT. The synthesized one was generated from TR images using the primary version of the commercial software which was developed by Hologic Inc. based on Maximum Intensity Projection (MIP) [9]. The results showed that the diagnosis accuracy of the synthesized 2D mammographic image was lower than the real one when combined with DBT. However, the later version has recently been approved and can be used in clinic. Schie et al. [10] presented a novel method to generate synthesized 2D mammographic image from TR images which based on computer-aided detection (CAD) system and thin plate spline (TPS) fit. The result image had higher quality for masses detection.

In this paper, we also focus on synthesis 2D mammographic image using Super-Resolution technique.

Super-Resolution (SR) is an imaging technique to estimate a single high resolution (HR) image from one or multiple low resolution (LR) images. There are many SR algorithms have been developed and widely used in many applications such as remote sensing, surveillance, satellite imaging and, medical imaging [11, 12].

SR can be roughly divided into two main categories based on input LR image which are multiple images SR and single image SR. Multiple images SR is the process to generate HR image from a set of LR images of the same scene which are blurred, sub-sampled, and sub-pixel shifted. The performance of multiple images SR will degrade if only small numbers of LR images are available and large magnification factor are required [13, 14]. Moreover, the success of SR depends on the accuracy of motion estimation in registration step. If the motion estimation is not correct, the result will be degraded.

Single image SR can overcome the limitation of multiple images SR case, only one LR input image is used. This approach is based on learning based method which uses a co-occurrence prior between LR and HR image patches to predict a HR image. Yang et al. [13] proposed the single image SR algorithm using sparse representation which was motivated by the concept of compress sensing. This method produced superior results both in generic and more complex texture images due to adaptively choosing the most relevant patches in dictionary which was compatible with its neighbors. The improve version [15] also showed the higher performance and robust to noise. Moreover, the speed of the algorithm also increased due to learning the dictionary pairs. He and Siu [12] presented a novel method for single image SR using only original LR image and its blurred version. Each pixel in HR image was predicted by its neighbors through the Gaussian Process Regression (GPR) and learning from training set that constructed by the LR image itself. Their method can produce sharp edges reliably at large magnifications. Ning and Gao [14] combined multiple images and single image SR to generate HR image. The hierarchical iterative sub-pixel registration was introduced to generate an initial HR image. The sparse representation and special designed dictionaries were used to reconstruct the final HR image. Their dictionaries contained patches from both generic training images and interpolated input LR images. The proposed method outperforms other learning based methods, but there are still some limitations, i.e., the registration method was applicable to global translation motion and high computational time.

In this paper, we propose a feasibility study to synthesis a 2D mammographic image from multiple TP images which is different from the previous approaches proposed by Although Gur et al. [8] and Schie et al. [10]. We believe that to gain the most accuracy of the reconstruction we should capture data at source. Raw TP images are the images that are captured before further processing. Our underlying motivation to synthesis the 2D mammographic image from the multiple TP images is to help patients to reduce the X-ray dose approximately 50 percent while maintaining the same quality of examination.

The remainder of this paper is organized as follows: Section 2 briefly reviews a SR algorithm which is used in this work. The experiments are presented in section 3. The results are showed in section 4. Finally, discussion and conclusion are given in section 5.

2 Super Resolution

We applied Yang et al. [15] algorithm, a single image SR, to generate the synthesized 2D mammographic image. The brief review of this algorithm is discussed.

Yang et al. [15] algorithm is based on sparse representation. In this algorithm, two constraints are modeled. One is a reconstruction constraint and the other one is sparsity prior.

Reconstruction constraint requires that the recovered HR image $\boldsymbol{X}$ should be consistent with the input LR image $\boldsymbol{Y}$ with respect to the image observation model.

Sparsity Prior, the high resolution patches $\boldsymbol{x}$ of the HR image $\boldsymbol{X}$ can be represented in the form of sparse linear combination in a high resolution dictionary $\boldsymbol{D}_h$ trained from HR patches sampled from training images

$$\boldsymbol{x} \approx \boldsymbol{D}_h \alpha, \quad \alpha \in \mathbb{R}^K \quad \text{and} \quad \|\alpha\|_0 \ll K \tag{1}$$

The sparse representation α will be recovered by representing LR image patches $\boldsymbol{y}$ of the input LR image $\boldsymbol{Y}$, with respect to a LR dictionary $\boldsymbol{D}_l$ which is trained to have the same sparse representation with $\boldsymbol{D}_h$.

For each input low resolution patch $\boldsymbol{y}$, find a sparse representation with respect to $\boldsymbol{D}_l$. The corresponding high resolution patch bases $\boldsymbol{D}_h$ will be combine to generate the output high resolution patch $\boldsymbol{x}$. The problem of finding the sparest representation of $\boldsymbol{y}$ can formulate as follow, where F is a feature extraction operator. Then high resolution image $\boldsymbol{X}_0$ is obtained.

$$\min\|\alpha\|_0 \quad \text{s.t.} \quad \|F\boldsymbol{D}_l\alpha - F\boldsymbol{y}\|_2^2 \leq \epsilon \tag{2}$$

Then, using the result from the local sparse representation to regularize and refine the entire image X^* using reconstruction constraint. In this strategy, a local model is used to recover lost high frequency for local details. To remove artifacts and make the image more consistent and natural, the global model from the reconstruction constraint is then applied. The solution can be efficiently computed using gradient descent.

3 Experiments

Our experimental imaging system based on a Hologic Selenia Dimension DBT system (Hologic Inc., Bedford, MA, USA). Combo mode and automatic exposure control are used to acquire mammogram and DBT in the same compression. For DBT image acquisition, the x-ray tube motion covers an angular range of -7.5° to +7.5°. The TP images of the Gammex model 156 breast phantom (Gammex Inc., USA) were taken at different angle, then x-ray tube move to 0° position and mammogram is also taken. Individual TP images were extracted using the specific program provided by Hologic Inc.

The TP images are automatically registered based on feature points. Then, fuse the registered images and the reference image (i.e. central TP image). The obtained image is used as the input in single image SR step to synthesis the synthesized 2D mammographic image.

To test the performance, we evaluate both objective and subjective visual measurements. In objective measurement, the quality is defined by peak signal to noise ratio (PSNR) and mean structure similarity (MSSIM). In subjective visual measurement, the quality is evaluated by phantom passing score. We also generate the synthesized 2D mammographic image from only one TP image, central TP (0°) image. Then, the results are compared.

4 Results

The x-ray images of breast phantom, acquiring from Hologic Selenia Dimension DBT system shown in Fig. 1. In Fig. 2, the result from registration and fusion step is illustrated which is used as input image in single SR step. The synthesized 2D mammographic images generated form 15 TP images. Only central TP image are shown in Fig. 3.

4.1 Objective Measurement

The real 2D mammographic image is used as the reference HR image. We compute PSNR and MSSIM between synthesized 2D mammographic image and the reference one. PSNR and MSSIM of the results image generated from 15 TP images are 27.2426 and 0.4436, respectively. PSNR and MSSIM of the results image generated from central TP images are 27.1170 and 0.3339, respectively.

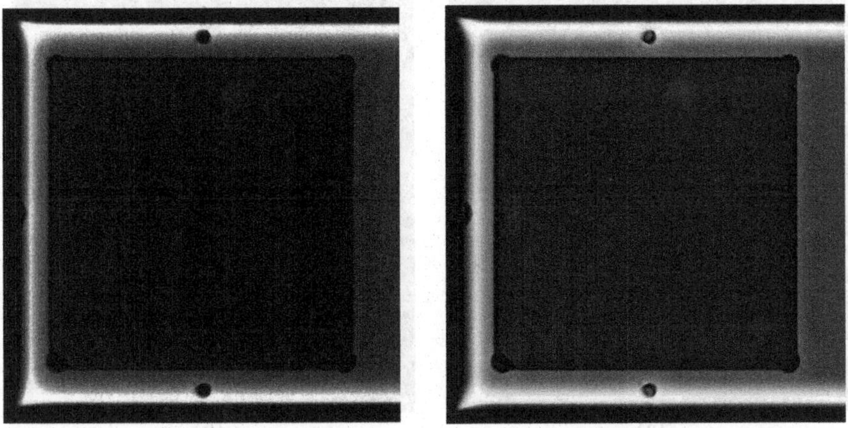

(a) (b)

Fig. 1. The x-ray images of breast phantom. (a) The sample of TP image i.e. central TP image, (b) mammographic image

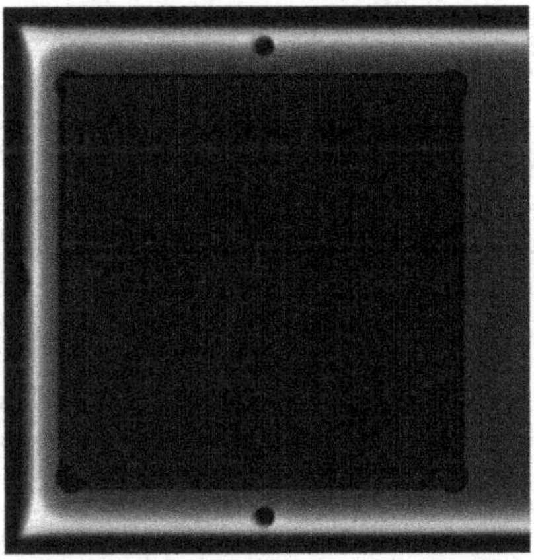

Fig. 2. The result from registration and fusion step, which is used as input LR image in single SR algorithm.

4.2 Subjective Visual Evaluation

Phantom passing score is used to evaluate the quality of the result images. By counting the number of the visible test object in the image, according to 1999 ACR Mammography Quality Control Manual [16]. Phantom passing score, the minimum needed, for 2D and 3D imaging systems are shown in Table 1. [16, 17, 18].

The results of scoring the phantom image are shown in Table 2.

(a) (b)

Fig. 3. The synthesized 2D mammographic images generated from (a) 15 tomosynthesis projection images, (b) central tomosynthesisp projection (0°) image.

Table 1. phantom passing score for 2D mammographic imaging system (mammogram) and 3D mammographic imaging system (diigital breast tomosynthesis, DBT)

Test objects	Phantom Passing Score	
	2D imaging system (Mammogram)	**2D imaging system (DBT)**
Fibers	5	4
Microcalcifications	4	3
Masses	4	3

Table 2. Comparison of phantom passing score between the synthesized 2D mammographic images which generated from 15 tomosynthesis projectio (TP) images and single central TP (0°) image

Test objects	Numbers of test object are seen in synthesized 2D mammographic image	
	generate from 15 TP images	**generated from central TP image**
Fibers	5	2
Microcalcifications	2	1
Masses	4	2

5 Discussion and Conclusion

This paper focuses on radiation dose reduction in breast imaging by generating the synthesized 2D mammographic image from 15 TP images using SR technique.

Due to DBT uses low radiation dose for maintaining total examination dose similar to or less than mammogram, the TP images are very low signal to noise ratio (SNR) compared to mammographic image, so in TP images, the details in breast phantom cannot be seen clearly only the large size of test objects are seen, as in Fig. 1(a) compared to Fig. 1(b). In Fig. 2, more details of test object are seen in the fused projection image because of fusing all the details from 15 TP images.

This study also compares the results which generated from 15 TP images and only central TP (0°) image which is similar to mammography image, except the radiation dose. The aim of comparative study is to illustrate the performance of registration and fusion step that can fuse the content in each TP image to improve the quality of input LR image before undergo SR reconstruction step, instead use only single central TP (0°) directly. So, the input LR image affects the results of SR algorithm.

In objective measurement, PSNR and MSSIM are used to evaluate the quality of the result images. The higher PSNR, the better quality. For MSSIM, contrast, luminance and structure are taken into account to determine similarity between two images. A value of MSSIM of 1 indicates perfect similarity. From the results, the synthesized 2D mammographic image which generated from multiple TP images outperform than the other one. And also in subjective visual evaluation, more numbers of test objects are seen. Although this still does not pass the standard passing score, but it is better than the synthesized 2D mammographic image generated form central TP (0°) image and the original TP image, respectively.

Although the synthesized 2D mammographic images from [8] are approved and used in clinical, the resolution of the synthesized image is still similar to TR images, while our proposed method can improve the resolution of the result image.

There are still some limitations in our approach that we are working on to improve the visualization of microcalcifications which is important in diagnosis breast disease. Our future works also include optimizing the algorithm to reduce the computational time, as well as performing the clinical trials. If the approach succeeds, this can reduce the patient dose about 50 percent

Acknowledgement

The Authors thank Andrew Smith and Baorui Ren (Hologic Inc., Bedford, MA) for providing useful background information on the Selenia Dimensions tomosynthesis system and special program for extracting individual TP images.

References

1. A Fundamental of Breast Tomosynthesis: Improving the Performance of Mammography, www.hologic.com/data/WP-00007-Tomo-0808.pdf
2. Gennaro, G., et al.: Digital Breast Tomosynthesis versus Digital Mammography: A Clinical Performance Study. European Radiology. 20(7), 1545-1553 (2010)
3. Good, W.F., et al.: Digital Breast Tomosynthesis: A Pilot Observer Study. AJR. 190(4), 865-869 (2008)
4. Noroozian, M., et al.: Digital Breast Tomosynthesis Is Comparable to Mammographic Spot Views for Mass Characterization. Radiology. 262(1), 61-68 (2012)

5. Poplack, S.P., TosTeson, T.D., Kogel, C.A., Nagy, H.M.: Digital Breast Tomosynthesis: Initial Experience in 98 Women with Abnormal Digital Screening Mammography. AJR. 189(3), 616-623 (2007)
6. Skaan, P., et al.: Comparison of Digital Mammography Alone and Digital Mammography plus Tomosynthesis in a Population-based Screening Program. Radiology. 257(1), 47-56 (2013)
7. Breast Tomosynthesis: The Use of Breast Tomosynthesis in Clinical Setting, http://www.breasttomo.com/sites/breasttomo.com/files/010-WP-00060-Rev2_June2012-TomoWhitePaper.pdf
8. Gur, D., et al.: Dose Reduction in Digital Breast Tomosynthesis (DBT) Screening Using Syntheically Reconstructed Projection Images. Acad Radiol. 19(2), 166-171 (2012)
9. Hologic Selenia Dimensions C-View Software Module, http://www.fda.gov/downloads/AdvisoryCommittees/CommitteesMeetingMaterials/MedicalDevies/MedicalDevicesAdvisoryCommittee/RadiologicalDevicesPanel/UCM325903.pdf
10. Schie, G.V., Mann, R., Imhof-Tas, M., Karssemeijer, N.: Generating Synthetic Mammograms from Reconstructed Tomosynthesis Volumes. IEEE Transaction on Medical Imaging. 32(12), 2322-2331 (2013)
11. Park, S.C., Park, M.K., Kang, M.G.: Super-resolution Image Reconstruction: A Technical Overview. In: IEEE Signal Processing Magazine. 21-36 (2003)
12. He, H., Sui, W.C.: Single Image Super-resolution Using Gaussian Process regression. In: IEEE Conference on Computer Vision and Pattern Recognition, pp.449-456. (2011)
13. Yang, J., Wright, J., Huang, T., Ma, Y.: Image Super-Resolution as Sparse Representation of Raw Image Patches. In: IEEE Conference on Computer Vision and Recognition, pp. 1-8 (2008)
14. Ning, B., Gao, X.: Multi-fram Image Super-Resolution Reconstruction Using Sparse Co-occurrence Prior and Sub-pixel Registration. Neurocomputing. 117, 128-137 (2013)
15. Yang, J., Wright, J., Huang, T., Ma, Y.: Image Super-Resolution via Sparse Represention. IEEE Transaction on Image Processing. 19(11), 2861-2873 (2010)
16. Mammography Accreditation Program Requirements, http://www.acr.org/~/media/ACR/Documents/Accreditation/Mammography/Requirements.pdf
17. MQSA Inspection Procedure 6.06, http://www.fda.gov/downloads/Radiation-EmittingProducts/MammographyQualityStandardsActandProgram/FacilityCertificationandInspection/UCM240433.pdf
18. Role of the Medical Physicist in Clinical Implementation of Breast Tomosyntheis, http://www.aapm.org/meetings/amos2/pdf/59-17210-92736-449.pdf

Using complex events to represent domain concepts in graphs

Riley T Perry, Cat Kutay, Fethi Rabhi

Computer Science and Engineering, The University of NSW

Abstract. We have developed an event based visualisation model for analysing patterns between news story data and stock prices. Visual analytics systems generally show a direct mapping from data to visualisation. We show that by inserting an intermediate step, which models an expert manipulating data, we can provide unique results that display patterns within the data being investigated and assist less expert users.

1 Introduction

As volumes of data grow the tools used to explore and analyse that data need to become more sophisticated. Some of the analysis process used by data experts can be incorporated into a visual model. Specifically we examine whether modelling financial data as events using Complex Event Processing (CEP) [2] techniques helps with data visualisation, analytics, and Exploratory Data Analysis (EDA).

An event can be defined as anything that happens, or is interpreted as happening at a particular time. Examples of events include banking transactions, financial trades and quotes, aircraft movements, updates in social media sites (e.g. Facebook), and sensor outputs [8]. EDA [1] is a term that describes techniques used to identify patterns and information in large amounts of data. Existing systems provide poor support for the grouping of data to extract conceptual patterns. This work proposes a visualisation model for presenting event data so as to incorporate expert techniques in data collation and pattern representation for event models.

The paper is structured as follows: Section 2 describes the proposed approach, introduces complex event processing concepts and contains definitions for the visualisation model. In section 3 the models and concepts are applied to financial data. Section 4 contains a brief evaluation of the visualisation. Finally, section 5 presents our conclusions and potential future work.

2 Proposed Approach

Our approach is to use a CEP model combined with an existing formal visualisation model (figure 1). Together these models allow us to visualise event occurrences on a graph, an example of which can be seen later in figure 5. The CEP and visualisation models are described next and the combined models are presented.

K.J. Kim (ed.), *Information Science and Applications*,
Lecture Notes in Electrical Engineering 339, DOI 10.1007/978-3-662-46578-3_35

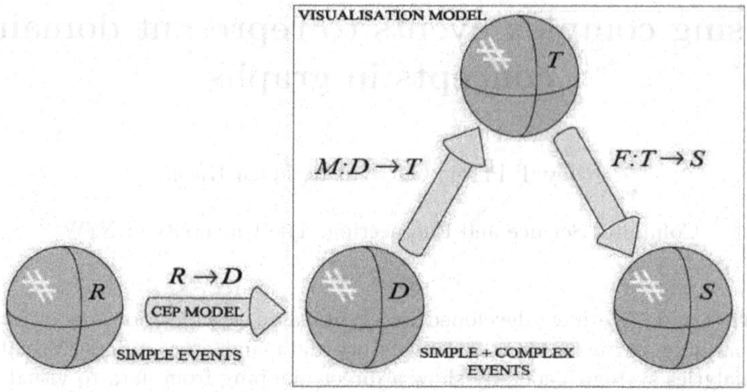

Fig. 1. The combined models

2.1 CEP Model

The role of the CEP model is to map *simple events* to *complex events* via pattern matching algorithms. Complex events are ultimately made up of simple events, which are usually *raw* events, a record of an occurrence in the real world. By utilizing a concept called CEP aggregation we can create event hierarchies to model higher level concepts.

Aggregation is a series of levels related to event complexity. Level one events are all simple events. All levels above level one are made up of complex events. The CEP model in this paper is inspired by previous work using Thomson Reuters stock market data feeds [6].

Events in a complex event are linked by relational mapping. Events can be related by *time*, *causality* and *aggregation*. E.g. events can be part of a price change and form a time sequence of changes in the price. Causality between events implies that preceding events had to happen to cause the latter event. The history of causes is called the *causal vector*. Aggregation can happen between events and they can be combined to form a more complex event. For example a news story from start to end may contain many individual news events. Formally, our events become complex when they involve aggregation.

Aspects An event is defined by the following aspects:

Form α - The main attribute store for an event.
Significance β - The type of activity the event describes within its form.
Relativity γ - A series of events that caused the event.

Formally, an event E occurred at time t and is uniquely identified by its identifier i. E is defined as a tuple: $E = (\alpha, \beta, \gamma)$. From left to right, $\alpha = (i, t, A)$. α contains the event identifier, timestamp, and n attributes where $A = \{a_1, a_2, \ldots, a_n\}$ and a_i is an attribute. β is single event type identifier which is usually the name of the event. γ is an ordered list of event identifiers: $\gamma = [i_1, i_2, \ldots, i_n]$ which are identifiers for the events that directly caused E to occur. Events in γ may be complex or simple. If $\gamma = \emptyset$ then E is a simple event, otherwise E is complex.

When modelling an event in the visualisation we wish to retain the ability to decompose back to the component events. Hence when large scale patterns are identified we can examine the separate attributes of the complex event. Hence the CEP mapping must be reversible.

Event Patterns The purpose of defining event patterns is to use them to find patterns in event data and generate new events based on these matches. The event patterns can be implemented programmatically for flexibility or reduced to a set of simple set of primitive operations as part of a pattern language.

2.2 Visualisation Model

To make meaningful statements about, and compare and contrast various visualisation techniques we are using a model of the visualisation process. The model gives us a language for talking about the differences between visualisations. The proposed model was created by Matthew Alexander Hutchins as part of his doctoral thesis [3]. Here we consider visualisation spaces as algebras and mappings between them are morphisms. His model a formal model with data (D), task (T), and scene (S) spaces. His model can be seen in figure 1 and is the part of the figure contained within the box. The model translates data to a format that can be directly mapped to display elements on the screen. Space D contains data in its original format. Space T is the mapping from D to form an algebra suitable for visual representation and S is the data display on the screen. In Hutchins' model the mapping of data to the task space ($D \rightarrow T$) automatically incorporates the data manipulation techniques used by experts in that particular domain. Hence we extend the model with a pre-mapping to incorporate CEP as representing export pattern matching.

Spaces D, T, and S are defined below. These definitions are taken from [9].

Definition 1 A space is an algebra. The space is represented by $\{\Sigma, G, K\}$ Where Σ is a collection of sets, G represents functions with domains in Σ and co domains in $\Sigma+$ and K represents constants from sets in Σ. $\Sigma = \{H_1, H_2, \ldots, H_n\}$ where $n =$ the number of sets. $G = \{G_1, G_2, \ldots, G_n\}$ where $n =$ the number of functions. $K = \{K_1, K_2, \ldots, K_n\}$ where $n =$ the number of constants.

Definition 2 A morphism between algebras $A = \{\Sigma_A, G_A, K_A\}$ and $B = \{\Sigma_B, G_B, K_B\}$ is a set of functions P with domains in ΣA and co domains in ΣB.

Definition 3 A function p is a relation from set H_1 to set H_2 if for every $x \in H_1$ there is a unique $y \in H_2$ such that $(x, y) \in p$. This can be represented by $p : H_1 \rightarrow H_2$.

An instance of the model is a tuple (d, t, s) where $t = M_p(d), s = F_p(t)$ Where d, t, and s represent the entire snapshot of values for spaces D, T, and S respectively.

2.3 Combined Model

We propose an extension to the visualisation model where data is event based and D is made up of simple and complex events. The raw data, which is made up of simple events, is contained in the space R. CEP is used between R and D to add complex events to D (also shown in figure 1). The conceptual understanding of entities and relationships in the semantic structure of the data is modelled using CEP structures.

3 Application to Financial Data Analysis

The combined models are now applied to financial data and suggested interface is presented.

3.1 Using the CEP Model: $R \to D$

R contains *Price Updates* and *News Updates*, which are simple events. The translation from R to D runs all simple events through a series of event pattern instances. Complex events will be created by these pattern instances. Simple events are then mapped directly to D and any complex events created are also mapped to D. These patterns are described next as part of an event hierarchy.

Level One Events

News Update A simple *news update* event ($\boldsymbol{NU}$) is an indivisible news related message. Each message consists of various codes including a story identifier called the *PNAC*. Simple news updates, when combined, form a partial or complete *News Story*. News stories, described next, are complex events.

Price Update A simple *Price Update* event ($\boldsymbol{PU}$) contains the trading price, volume, and trading timestamp (or period) for a particular security.

Level Two Events Events above level one are complex events. We describe here the type of complex events and how they are formed.

News Story A *News Story* (figure 2), $\boldsymbol{NS}$ with news events $\boldsymbol{NU_1}, \boldsymbol{NU_2}, \ldots, \boldsymbol{NU_n}$ can be represented by a simple tree diagram like this: $\boldsymbol{NS}$ sits in a two level *abstraction hierarchy*. $\boldsymbol{NU_1}, \boldsymbol{NU_2}, \ldots, \boldsymbol{NU_n}$ all share the same *PNAC* but may be different occurrences of same news story. Attributes of $\boldsymbol{NS}$ are drawn or derived from this pool. CEP is a realtime technology used to make decisions quickly. A set of definitive Pattern rules for $\boldsymbol{NS}$ can be hard to determine as a story can span several days with many events, or just be comprised of an instantaneous single event.

Price Jump $\boldsymbol{PJ}$ (figure 3) is a *Price Jump* which contains positive price update events $\boldsymbol{PU_1}$ and $\boldsymbol{PU_2}$. An example of a price jump for the running example is a price increase of over 2 points for a certain stock.

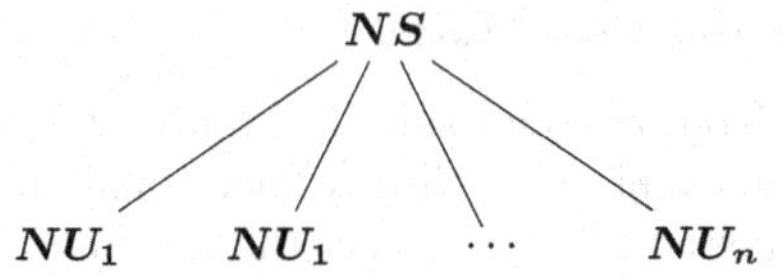

Fig. 2. A news story hierarchy

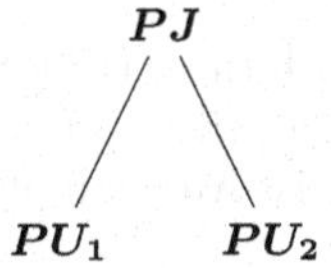

Fig. 3. A price jump hierarchy

Level Three Events We then have higher level hierarchies of aggregation, when combined event stories with pricing, that can have attributes of their own. This allows us to build higher level concepts which can in turn be metrics for analysis and further aggregation.

News Story Plus A *News Story Plus* (figure 4), $\boldsymbol{NS+}$, event is generated when there is a price jump within (and related to) the time period for a news story. Given a news story $\boldsymbol{NS}$ and a price jump $\boldsymbol{PJ}$ extra information can be gained from combining and doing calculations on the attributes of both. The concept of a news story plus (shown below) sits in three level CEP event abstraction hierarchy. Timestamps for $\boldsymbol{NU_1}, \boldsymbol{NU_2}, \ldots, \boldsymbol{NU_n}$ always

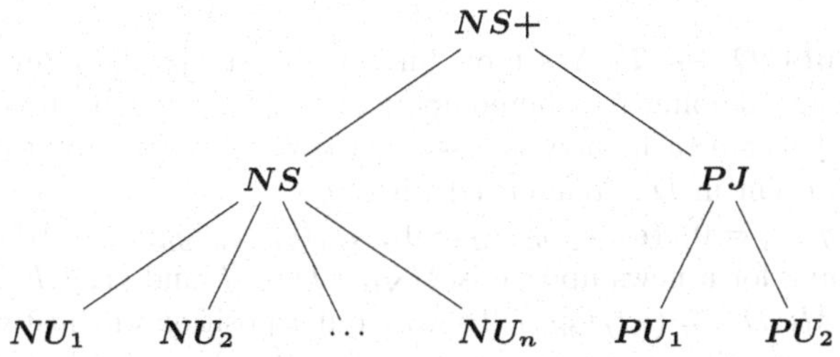

Fig. 4. A news story plus hierarchy

fall within the boundaries of $\boldsymbol{PU_1}$ and $\boldsymbol{PU_2}$, i.e.: $\boldsymbol{NU_1}^{(t)} >= \boldsymbol{PU_1}^{(t)}$ and $\boldsymbol{NU_n}^{(t)} <= \boldsymbol{PU_2}^{(t)}$. Preferably timestamps should be at regular intervals and should match exactly. i.e. $\boldsymbol{NU_1}^{(t)} = \boldsymbol{PU_1}^{(t)} \ldots \boldsymbol{NU_n}^{(t)} = \boldsymbol{PU_2}^{(t)}$.
It's useful at this point to introduce a variable, $\epsilon =$ event level-1, to represent the event level for display purposes. A news story plus for example has an ϵ of 2.

3.2 Using the Visualisation Model

The mapping of CEP (complex and simple) events $D \rightarrow$T can be described as mapping from a pool of events to a three dimensional space of discrete points. $T \rightarrow S$ is realised by filtering/zooming via the interface, which maps a subset of these points to a 2 dimensional plane. When visualizing an event on a two-dimensional plane t (time) will usually be mapped to the x axis. There are two distinct classes of significance of simple event for visualisation purposes:

2D One of which will apply to both the x (time) axis and a single form attribute for the y axis. An example is a price update.
1D One of will apply to the x axis. An example is an individual news update.

The mappings between visualisation spaces are now described in detail.

3.3 Mapping $D \rightarrow T$

The data space of CEP events D is then mapped onto the Task space T. The set of mapping functions is given as $M : D \rightarrow T$.

From Section 2.2, $D = \{\Sigma_D, G_D, K_D\}$ and $T = \{\Sigma_T, G_T, K_T\}$, where $\Sigma_D = \{d_1, d_2, \ldots, d_n\}$ and $d_1, d_2, \ldots, d_n$ are simple or complex events with the exception that relativity (γ) is separated into a set of functions $G_D = \{g_1, g_2, \ldots, g_n\}$. $M = \{m_1, m_2, \ldots, m_n\}$ where $m_1, m_2, \ldots, m_n$ are mapping functions from D to T. $K_T = \{k_1, k_2, \ldots, k_n\}$ where $n =$ the number of constants and $k_1, k_2, \ldots, k_n$ are task attributes x and y. Each event has an x, y position on a cartesian plane relative to the smallest x and y attributes as atomic units in a meta space of planes. Each ϵ (event level) in E is mapped to a different plane. This is based on the hierarchy level determined by the relativity. Events of the same type are always on the same plane.

Simple Events $D \rightarrow T$ A set of functions M maps data (events) to the task space. The mapping is a homomorphism represented by functions: $D = \{e_1, e_2, \ldots, e_n\}$ where each entry is a set of attributes for an event E. Below an instance of an event in D is referenced with D_e.
$T = \{(e_1, x_1, y_1, z_1 = 0), (e_2, x_2, y_2, z_2 = 0), \ldots, (e_n, x_n, y_n, z_n = 0)\}$

The mapping for a news update is $M_{NU} : D \rightarrow T$ and $(x, y, E_{root}) \in M_{NU}$ where $T_{e.x} = M_1(D_e) = (e.t/t_{min})$. To work out x position we must work out the minimum non zero timestamp increment t_{min} beforehand by iterating through all elements of D. i.e. the smallest non zero difference between all timestamps in D. y is always 0.

The mapping for a price update is $M_{PU} : D \rightarrow T$ and $(x, y, E_{root}) \in M_{PU}$ where $E_{root} = \boldsymbol{PU}$ (Price Update) and, $T_e.y = M_2(D_e) = (D_e.price/price_{min})$. To work out x position we must work out the minimum non zero price change $price_{min}$ beforehand by iterating through all elements of D. i.e. the smallest non zero difference between all prices in D.

If the event is the *Root Event* (E_{root}), a price update, then the event is mapped to a 2D plane, otherwise all planes are 1D and x based.

Complex Events $D \to T$ A z axis is introduced where each entry is a set of attributes for an event E.

$T = \{(e_1, x_1, y_1 = 0, z_1), (e_2, x_2, y_1 = 0, z_2), \ldots, (e_n, x_n, y_n = 0, z_n)\}$

The basic mapping for all complex events is the same. $M_{CE} : D \in T$ and $(x, z, E_o) \in M_{CE}$ where $T_{e.x} = M_{CE}(D_e) = (e.t/t_{min})$; $T_e.z = M_{CE}(D_e) = (z = \epsilon)$; where ϵ determines the z order of the plane.

Again, to work out x position we must work out the minimum non zero timestamp increment t_{min} beforehand by iterating through all elements of D. i.e. the smallest non zero difference between all timestamps in D.

UI Mapping $T \to S$ Again based on the definition from [9], for $F : T \to S$, $S = \{\Sigma_S, G_S, K_S\}$ and $F = \{f_1, f_2, \ldots, f_n\}$ where $f_1, f_2 \ldots, f_n$ are functions from T to S. $K = \{k_1, k_2, \ldots, k_n\}$ where $n =$ the number of constants and $K_S = \{s_1, s_2, \ldots, s_n)$ where $s_1, s_2, \ldots, s_n$ are scene attributes.

S represents direct visual elements on a display device. All events and 3D coordinates are stored in a 3D matrix in T (represented by x, y, and z components).

These provide the components for the mapping from the events to the visualisation space which provides the homomorphisms between the task and visual spaces.

The proposed user interface has 3 distinct sections. They are: An *Attribute Panel, Display Panel* (figure 1), and a *Filter Panel.* The display panel is the main graph and displays S_{win}, where the root event sets up the x, y plane. Events are then stacked based on their Z-order (z). There is always a selected event, E_{sel}. The attributes in the attribute panel are those of the selected event and are drawn from K_T. The filter panel allows the removal of members of T and the display panel displays all, or part of the filtered members of T depending on zoom and stacking controls which are based on relativity, i.e. G_T. Of particular interest is filtering out x, y ranges (zooming), i.e. $S_{win} = \{$ a window in $T_x, T_y\} = \{x_{start}, x_{end}, y_{start}, y_{end}\}$ and showing only events within the selected event's relativity.

With the display panel in figure 5 when we want more information on the news story plus event we would simply click on the event, in this case a large square in the top left corner. What appears then, in the panel below the arrow, is just the news story plus event and its constituent events. These parent events, and in turn, their parents, are shown here.

A news story plus occurs when a news story is generated that contains a positive price jump. You can see that the price jump (the two price updates) did indeed occur over the life of the news story, which started with news update 1 and was finalised with news update 2. Under the news story plus event there are three distinct events: news story, news update 2, and price update 2.

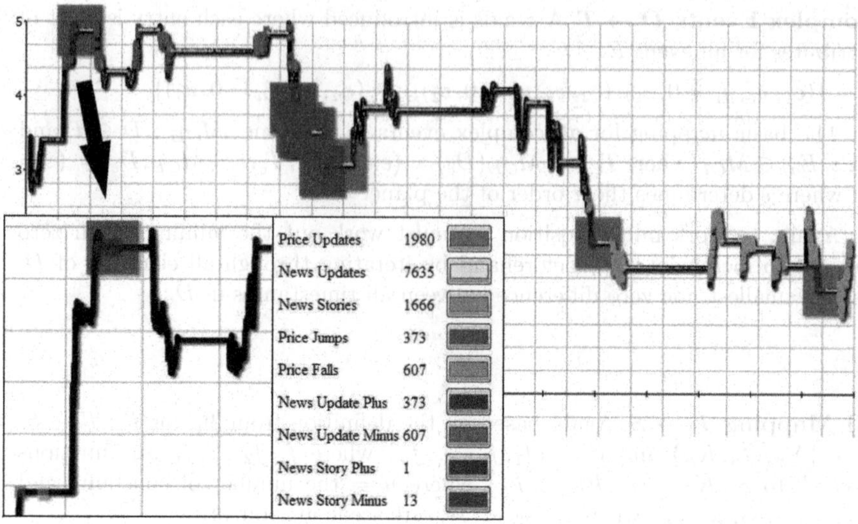

Fig. 5. The display panel

4 Visualisation Evaluation

A key method by which we may validate a process in science is the principle of falsifiability, championed by Karl Popper [4]. According to [5] a falsifiable images is a "valid pictorial representation of the truth". Falsifiability of a visual representation of data involves establishing the necessary and sufficient condition of the validity of the mapping between image and data, in the following format:

Consistent The data is consistent with the image through a homomorphism, and

Representative The image is representation of the data without distortion

The visualisation model should be such that you can detect in the image patterns that represent patterns that can then be verified in the data. Since mappings $D \rightarrow T$ are structure-preserving homomorphisms, the visualisation model mapping is falsifiable, or, the events are detectable in the final visualisation are present in D.

Other visualisation systems generally only show data representations of the R space. The event based version shows higher level entities and a hierarchy of those entities which represent the output of CEP patterns.

5 Conclusions and Future Work

By thinking of the domain conceptually and using these concepts within a visualisation framework we can enhance more simplistic visualisation techniques.

Future work would include building different CEP models, the usability of a prototype developed from this model against multiple visualisation systems, and developing an iterative model to change or add CEP patterns on the fly.

Another promising area for future research is in formal validation and visualisation falsifiability. A potential way to analyse the value of modeling tool is to verify that the visualisation is falsifiable. For this we are developing an algebraic formalism could be used to describe the CEP process from $R \to D$.

References

1. Tukey, J. A.Exploratory Data Analysis, Addison-Wesley (1977).
2. Luckham, A.D.C.: The Power of Events: An Introduction toComplex Event Processing in Distributed Enterprise Systems. Addison-Wesley, Reading (2002).
3. Hutchins, M.A. (1999) Modelling Visaualization Using Formal Algebra, A thesis for Doctor of Philosophy in The Australian National University, CSIRO ICT Centre, Canberra, Australia.
4. K. Popper, The Logic of Scientific Discovery, Routledge/Taylor & Francis e-Library, 2005.
5. Andrew J. Hanson, Putting Science First: Distinguishing Visualizations from Pretty Pictures. Computer Graphics and Applications, IEEE, 34, 4. 2014.
6. Calum S. Robertson, Fethi A. Rabhi, Maurice Peat. A (2012) Service-Oriented Approach towards Real Time Financial News Analysis University of New South Wales, Smart Services CRC, Sirca, University of Sydney.
7. Reuters NewsScope Archive v2.0 User Guide v2.4 Date of Issue: 29th May 2008.
8. W. Chen and F.A. Rabhi, An RDR-Based Approach for Event Data Analysis, In J.G. Davis, H. Demirkan and H.R. Motahari-Nezhad (eds), Service Research and Innovation, Lecture Notes in Business Information Processing Volume 177, 2014, pp 1-14.
9. T. Mala*, P. Bhargavi and T.V. Geetha, GVP model based temporal visualisation of user-centric data, International Journal of Metadata, Semantics and Ontologies Volume 3, Issue 4, 2008, pp 305-317.

Unequal Loss Protection Mechanism for Wi-Fi based Broadcasting system using a Video Quality Prediction Model

Dong Hyun Kim[1], Hyung-Yoon Seo[1], Byungjun Bae[2], Jong-Deok Kim[1*]

[1] Deppartment of Electrical and Computer Engineering, Pusan National University Busan, Republic of Korea
[2] Electronics and Telecommnications Research Institute Daejeon, Republic of Korea

Email: dhkim1106@pusan.ac.kr[1], tanyak@pusan.ac.kr[1], 1080i@etri.re.kr[2], kimjd@pusan.ac.kr[1*]

Abstract. The Wi-Fi based broadcasting system is mobile Internet Protocol Television (IPTV) technology, which transmits the multimedia content to a lot of local mobile users in real time. Unlike conventional Wi-Fi based multimedia streaming systems that use a separate unicast packets to serve each user, the Wi-Fi broadcast system uses the broadcast packets for scalability because one broadcast packet can serve many users simultaneously. However, it must use the FEC technology to recover the broadcast packet loss because an IEEE 802.11 does not support the packet loss recovery method for the broadcast packet. Considering the feature of multimedia data, equally adding FEC packet to video source is not efficient. For better efficiency Unequal Loss Protection (ULP) technology, for example, has been developed. This paper is going to modeling the characteristics of quality deterioration from effect of broadcasting packet losses in Wi-Fi based broadcasting system and suggest newly developed ULP system based on the model of quality deterioration. The system will be implemented and experimented to demonstrate its superiority.

Keywords: Wi-Fi broadcasting system, ULP, FEC, Quality prediction model, Multimedia transmission

1 Introduction

Unlike the current Wi-Fi based multimedia service using Unicast packet, the Wi-Fi based broadcasting system is Mobile-IPTV technology to serve a number of users with real-time multimedia contents. It allows one broadcasting packet to satisfy a number of users simultaneously, therefore, no matter how many users are it is able to provide sufficient resources without any problems.

However, general Wi-Fi system based on broadcasting packet has several limitations while the one based on Unicast packet does not: packet losses, Quality of Service (QoS), multi transmission rate setting, etc. One of the serious problems is packet loss. Unicast packet of Wi-Fi restores lost packet using retransmission technology. However since

*Corresponding Author

K.J. Kim (ed.), *Information Science and Applications*,
Lecture Notes in Electrical Engineering 339, DOI 10.1007/978-3-662-46578-3_36

retransmission function of MAC layer is not supported, losses are unavoidable. These losses can be restored by FEC[1].

The original data of Multimedia video to be transmitted through Wi-Fi based broadcasting system is compressed for efficient transmission by various techniques. The video compression techniques are eliminating redundancy of space and time, which divides video frames into I, P and B frames. An I-frame, intraframe, is a self-contained frame that can be decoded independently without any reference to other images. It is always the first one in a video sequence. A P-frame uses a prediction of images between frames, which decodes the frame referring to a previous I-frame and P-frame. A P-frame requires fewer bits than I-frame, but the drawback is that they are sensitive to transmission errors because of the complex dependency on earlier P and I frames. A B-frame, Bi-predictive inter-frame, is a frame that makes references to both an earlier reference frame and a future frame[2][3].

If losses incur during transmission of encoded video, frame losses incur in video frames comprised of key frames and delta frames to eliminate time redundancy and quality deteriorating waves appear from losses of other frames. If losses incur in key frames, every delta frame based on them are subject to quality deterioration. If losses incur only in certain delta frames, it deteriorates the quality of the loss delta frames and delta frames produced with them[4].

The place where loss of broadcasting packet occurs is unpredictable in the Wi-Fi based broadcasting system where standard frames and reference frames are transmitted in sequence. Lost packets are recovered by FEC techniques, which are called Equal Loss Protection (ELP). ELP recovers packet loss by applying an equal recovery rate to the whole video. Considering how encoding and decoding work, this ELP scheme is not efficient for quality of a received video because a quality of decoding video is varied according to the place of packet loss occurred even with the same packet loss rate. Therefore, several unequal loss protection (ULP) schemes have been suggested.

The existing studies about ULP have distinguished data structure for priority array into packets, macroblock, frames, etc., according to the features of network and multimedia streaming and have used macroblock and frames for priority array of determining video quality. Those studies are only considering priority array of video quality based on transmitting units. However, the place where packet losses occur is not predictable considering the features of multimedia encoding and quality of received video is subject to the place of packet loss. Therefore, the needs to study about ULP system based on a quality model for received video that is able to predict its quality after frame losses occur has arisen[5][6][7].

Allotting FEC packet only to recover important frames is not optimal in network where amount of FEC data available is limited and place of lost packet is unpredictable. Therefore, this paper uses a quality of video quality prediction model for received video as a standard to set different FEC rate to each frame depending on the importance of frames.

First of all, calculate frame loss rate using packet loss rate, multimedia data of the same numbers of packets that comprise frames and binomial to predict the quality of received video over frame loss. Then, design the video quality distortion model of received video according to video frame losses and predict the quality of received video with the model and calculated frame loss rate. The result of overall video quality will be varied depending on allocations of FEC packets.

This paper suggests the FEC rate where the quality of received video is to be the highest from the effect of packet losses based on this quality model.

2 Video Quality Prediction model based on ULP system

The video quality prediction model builds on real video quality, which calculates quality using encoding video of I-frames and P-frames and received video. It defines quality of real video as Q, quality of I-frames as $Q_I(j)$ and that of P-frames as $Q_P(m,j)$. Average quality of Group of Pictures (GOP) is defined as $Q(j)$. This paper predicts quality of received frames before transmitted by using those qualities. Expected quality of I-frames is defined as $\overline{Q_I(j)}$, P-frames as $\overline{Q_P(m,\ j)}$ and average prediction quality of GOP as $\overline{Q(j)}$. The relation between $Q(j)$ and $\overline{Q(j)}$ is shown in figure1.

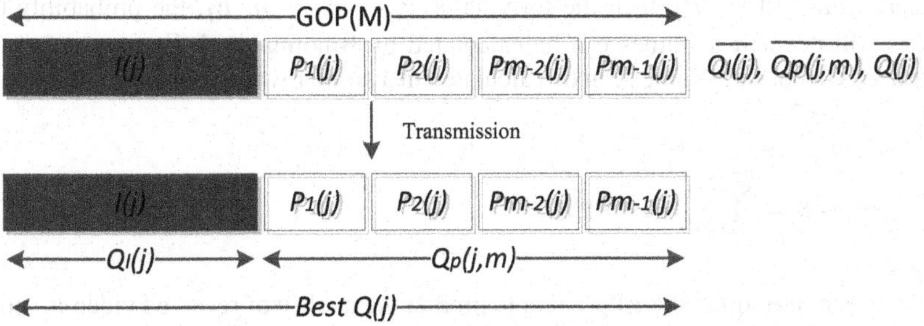

Fig1. The relation between $Q(j)$ and $\overline{Q(j)}$

Quality of received video from the effect of real packet losses was identified by using various videos to predict the quality of received video.

After considering both modeling results depending on loss occurrence in I-frame and P-frame, Formula (1) is formed. $\boldsymbol{L_i}$ represents loss rate of I-frames, $\boldsymbol{L_P}$ represents loss rate of p-frames and α and β for a quality distortion constant.

$$\overline{Q(j)} = \frac{1}{M}\sum_{h=0}^{M-1}[\beta \times L_i + (1-\alpha L_P)^h \times (1-L_i)] \tag{1}$$

This paper used Reed-Solomon (RS) code for FEC. RS(n, k) represents an additional amount of packet, n-k packets, and k represents the number of connected data packets for error correction. Once encoded data is converted into packets for transmission, one frame consists of several packets. I-frames, for example, have bigger size; therefore, several numbers of packets are needed. If any of these packets is lost, the whole frame will get lost as shown in figure2.

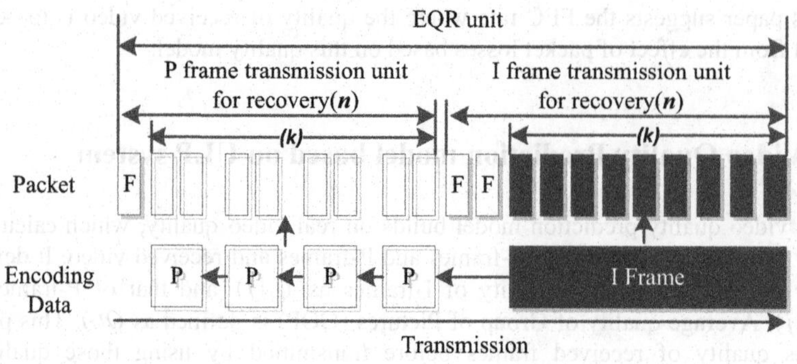

Fig2. The characteristic of FEC and encoded video stream

When packet loss is P, the number of packet for I-frames is P_I, an additional number of data from FEC is R_I then the formula is $P_I + R_I = n$. E_I, the probability that losses will occur in I-frames can be calculated by Binomial and E_P, the probability that losses occur in P-frames can be calculated in a similar way as shown in the formula (2).

$$E_I = \sum_{k=0}^{k-1} \binom{n}{k} (1-P)^k P^{n-k} \tag{2}$$

The general outline of ULP system to predict the quality of received video is shown in figure3 which has the process of calculating frame loss rate by referring to packet loss rate and allotment of FEC resources. R_I is the number of FEC packets allotted to I-frames and R_P is for P-frames.

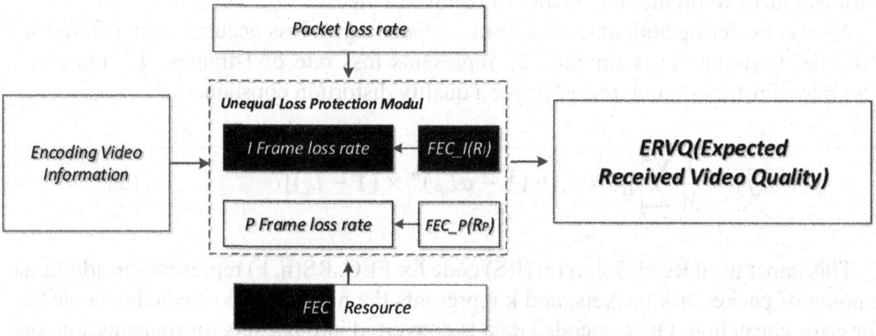

Fig.3 Suggested outlines of ULP system

3. Performance evaluation

To ensure the validity of the suggested quality prediction model, the actual experiment was carried out for received video quality from the effect of frame loss. It identified the dispersion of expected quality from suggested algorithm using Akiyo video when packet loss rates are 10, 20 and 30% and the size of FEC packets is 64kbps and 128kbps. The figure 4 shows the graphs of $\overline{Q(j)}$ and $Q_e(j)$ according to $R_I(f)$ and ρ when data size of FEC is 64kbps for Akiyo video.

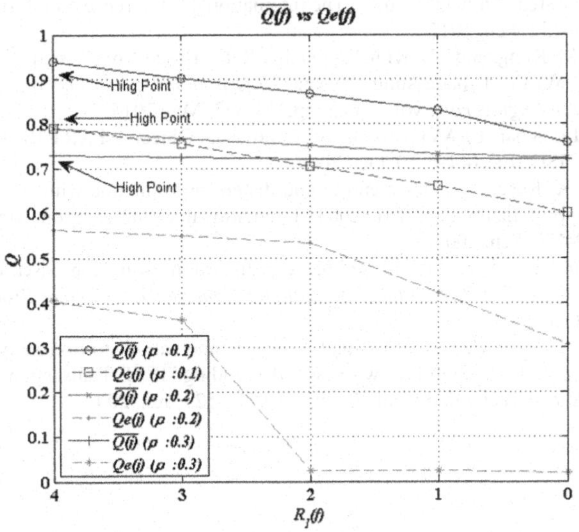

Fig.4 $\overline{Q(j)}$ and $Q_e(j)$ according to $R_I(f)$ and ρ when FEC is 64kbps

Akiyo video has frame rate of 30fps. Since the size of GOP is 15 and the size of the packet is 1024Bytes, FEC packet of 64kps can be used to recover the losses of four packets. Those four packets have five patterns for allotment of I and P frames. They were calculated for 10%, 20%, and 30%.

For each loss rate, allotting all FEC packets to I-frames was predicted to result in the highest quality of received video while lower the $R_I(f)$ of both $Q_e(j)$ and $\overline{Q(j)}$, the lower the quality of received video.

4. Conclusion

This paper suggested the video quality prediction model and frameworks to predict quality distortion of received video from effects of packet losses with the purpose of transmitting multimedia video data efficiently with Wi-Fi broadcasting system. Based on the quality prediction model ULP system has been designed and implemented. Also, the experiment was carried out to prove the validity of this model and demonstrate the superiority of the algorithm.

Acknowledgments. This work was supported by the ICT R&D program of MSIP/IITP. [14-000-02-002, Development of Service and Transmission Technology for Convergent Realistic Broadcast]

References

1. IEEE 802.11: Wireless LAN Medium Access Control(MAC) and Physical Layer(PHY) Specifications, *IEEE 802.11 Standard*, Jan. 1997.
2. D. H. Kim and J. D. Kim, "Broadcast packet collision and avoidance method in Wi-Fi based broadcasting system,"in ICOIN 2012 : The International Conference on Information Network 2012, pp. 108-113, Feb. 2012.
3. D. H. Kim, J. M. Kong, andJ. D. Kim, "A Quality Ratio-Based Novel Unequal Loss Protection Scheme in Wi-Fi Broadcasting System,"*International Journal of Computer and Communication Engineering,* vol.2, no.3, pp.313-323, May 2013.
4. ISO/IEC, MPEG-4 part 10/AVC, Coding of Audiovisual Objects – PART 10: Advanced Video Coding 2003, ISO/IEC 14496-10, Oct. 2004.
5. H. Yang and K. Rose, "Recursive end-to-end distortion estimation with model-based cross-correlation approximation,"in ICIP 2003: International Conference on Image Processing, vol.3, pp.469-472, Sep. 2003.
6. S. Ekmekci and T. Sikora, "Recursive decoder distortion estimation based on ar(1) source modeling for video," in*ICIP 2004: International Conference on Image Processing*, vol. 1, pp.187-190, Oct. 2004.
7. Z. H. He, J. F. Cai, and C. W. Chen, "Joint source channel rate-distortion analysis for adaptive mode selection and rate control in wireless video coding,"IEEE Transaction on Circuits and Systems for Video Technology, vol.12, no.6, pp.511-523, Jun. 2002.

An Approximate Matching Preprocessing for Efficient Phase-Only Correlation-Based Image Retrieval

Honghang Wang and Masayoshi Aritsugi

Computer Science and Electrical Engineering
Graduate School of Science and Technology, Kumamoto University
2-39-1 Kurokami, Chuo-ku, Kumamoto 860-8555, Japan
{king@dbms.,aritsugi@}cs.kumamoto-u.ac.jp

Abstract. In this paper, we focus on phase-only correlation-based image matching, which is often used for biometrics image retrieval. Although systems using phase-only correlation of images allow us to have image matching results of high quality, it takes long time to process phase-only correlation-based image matching especially when there are a large number of images to be processed in the matching. In this paper, we propose an approximate matching preprocessing for pruning images that do not have to be matched to a query image. An empirical study shows the superiority of our proposal.

Keywords: Phase-only correlation, biometrics image retrieval, image database

1 Introduction

It is highly convenient if a method of identifying a person can use biometrics data including face images, fingerprints, dental features [5], irises and palm prints. Such data can be useful for identifying victims of disasters such as earthquakes and tsunami, since they are not stolen or lost from a human body.

Phase-only correlation-based matching [1] is often used in biometrics image processing [6, 8] due to its ability to achieve estimation of sub-pixel image translation. For example, Ito et al. [2] proposed a phase-only correlation-based matching algorithm of dental X-ray images and showed its availability. However, it has been well known that a phase-only correlation-based matching tends to take long time to be processed because it includes the calculation of Fourier Transforms. This can be a serious problem if we need to identify victims of large-scale disaster by means of biometrics image matching.

In this paper, we focus on the fact that Fourier Transforms are used in phase-only correlation-based matching. For performance, we propose introducing an approximate matching preprocessing where we can filter candidate answers by means of a part of image data.

The rest of this paper is organized as follows. Section 2 briefly describes phase-only correlation-based matching. Section 3 presents our proposal. Section 4

K.J. Kim (ed.), *Information Science and Applications*,
Lecture Notes in Electrical Engineering 339, DOI 10.1007/978-3-662-46578-3_37

shows some experimental results and Section 5 gives our conclusion and future direction.

2 Phase-only correlation-based matching

We briefly describe phase-only correlation-based matching [4]. Figure 1 shows an overview of the matching process.

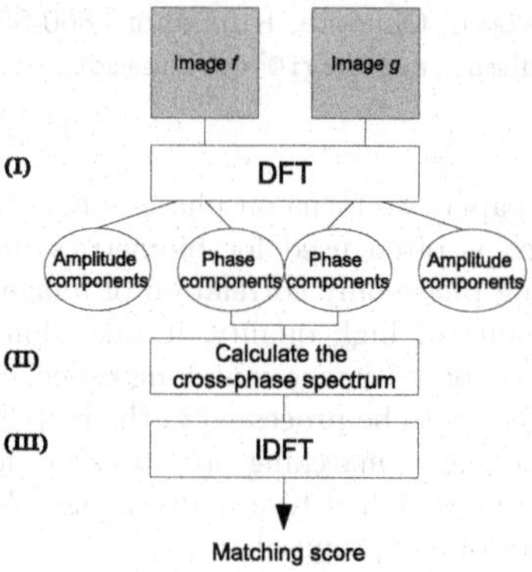

Fig. 1. Overview of phase-only correlation-based matching process

Given two images of $N_1 \times N_2$, we express them as $f(n_1, n_2)$ and $g(n_1, n_2)$, where $n_1 = -M_1, \cdots, M_1 (M_1 > 0)$ and $n_2 = -M_2, \cdots, M_2 (M_2 > 0)$, and thus $N_1 = 2M_1 + 1$ and $N_2 = 2M_2 + 1$. Let $F(k_1, k_2)$ and $G(k_1, k_2)$ be the 2D Discrete Fourier Transforms of the two images. They can be calculated as follows:

$$F(k_1, k_2) = \sum_{n_1, n_2} f(n_1, n_2) W_{N_1}^{k_1 n_1} W_{N_2}^{k_2 n_2} = A_F(k_1, k_2) e^{j\theta_F(k_1, k_2)} \quad (1)$$

$$G(k_1, k_2) = \sum_{n_1, n_2} g(n_1, n_2) W_{N_1}^{k_1 n_1} W_{N_2}^{k_2 n_2} = A_G(k_1, k_2) e^{j\theta_G(k_1, k_2)} \quad (2)$$

where $k_1 = -M_1, \cdots, M_1$, $k_2 = -M_2, \cdots, M_2$, $W_{N_1} = e^{-j\frac{2\pi}{N_1}}$, $W_{N_2} = e^{-j\frac{2\pi}{N_2}}$, $\sum_{n_1, n_2}$ expresses $\sum_{n_1=-M_1}^{M_1} \sum_{n_2=-M_2}^{M_2}$, $A_F(k_1, k_2)$ and $A_G(k_1, k_2)$ are amplitude components, and $\theta_F(k_1, k_2)$ and $\theta_G(k_1, k_2)$ are phase components of the given two images, respectively.

Then, the cross spectrum $R(k_1, k_2)$ between $F(k_1, k_2)$ and $G(k_1, k_2)$ is given as follows:

$$R(k_1, k_2) = \frac{F(k_1, k_2)\overline{G(k_1, k_2)}}{|F(k_1, k_2)\overline{G(k_1, k_2)}|} = e^{j\{\theta_F(k_1, k_2) - \theta_G(k_1, k_2)\}} \quad (3)$$

where $\overline{G(k_1,k_2)}$ denotes the complex conjugate of $G(k_1,k_2)$. The 2D Inverse Discrete Fourier Transform of $R(k_1,k_2)$, $r(n_1,n_2)$, is given by

$$r(n_1,n_2) = \frac{1}{N_1 N_2} \sum_{k_1,k_2} R(k_1,k_2) W_{N_1}^{-k_1 n_1} W_{N_2}^{-k_2 n_2} \tag{4}$$

where $\sum_{k_1,k_2}$ expresses $\sum_{k_1=-M_1}^{M_1} \sum_{k_2=-M_2}^{M_2}$. Function 4 gives a high peak if the two images are similar, and no peak otherwise. As a result, the height of the peak can be used as a similarity metric of two image, and the location of the peak indicates the amount of shift difference of them.

3 Proposal

The phase-only correlation-based matching takes longer time for processing than other matching methods including SAD (Sum of Absolute Difference) and SSD (Sum of Squared Difference) [7]. This must be a serious problem if we have a large number of images to be processed in the matching.

In this paper, we propose an approximate matching preprocessing for pruning images that do not have to be matched to a query image. Our idea is based on the fact that there is few phase difference between two images' components in a low frequency domain if an image shift is not necessary between them. Moreover it is well known that low frequency components are useful particularly when some noises are supposed to be included [4]. We thus propose a preprocessing in which low frequency components of images are extracted and stored in a database in advance and are used for pruning images that do not have to be matched. Note that the pruning process can be performed only with the low frequency components thereby performing our preprocessing very quickly. The step overview of our proposal is as follows:

1. Extract phase components with Discrete Fourier Transforms and store them in a database
2. Rank stored image data with only low frequency components of the phase components
3. Perform matching process with the phase-only correlation

3.1 Extract phase components

We can transform an image data into a spatial frequency component with the 2D Discrete Fourier Transform. Let $Re_{F(k_1,k_2)}$ and $Re_{G(k_1,k_2)}$ be real parts, and $Im_{F(k_1,k_2)}$, $Im_{G(k_1,k_2)}$ be imaginary parts of $F(k_1,k_2)$ and $G(k_1,k_2)$ in equations (1) and (2), respectively. Then, the following can be derived.

$$A_F(k_1,k_2) = \sqrt{Re_{F(k_1,k_2)}^2 + Im_{F(k_1,k_2)}^2} \tag{5}$$

$$A_G(k_1,k_2) = \sqrt{Re_{G(k_1,k_2)}^2 + Im_{G(k_1,k_2)}^2} \tag{6}$$

$$e^{j\theta_F(k_1,k_2)} = \tan^{-1}\frac{Im_{F(k_1,k_2)}}{Re_{F(k_1,k_2)}} \quad (7)$$

$$e^{j\theta_G(k_1,k_2)} = \tan^{-1}\frac{Im_{G(k_1,k_2)}}{Re_{G(k_1,k_2)}} \quad (8)$$

In this step, we store the data of (7) and (8) into a database (Fig. 2). The stored data allow us to skip step (I) in Figure 1 concerning an image of $g(n_1, n_2)$.

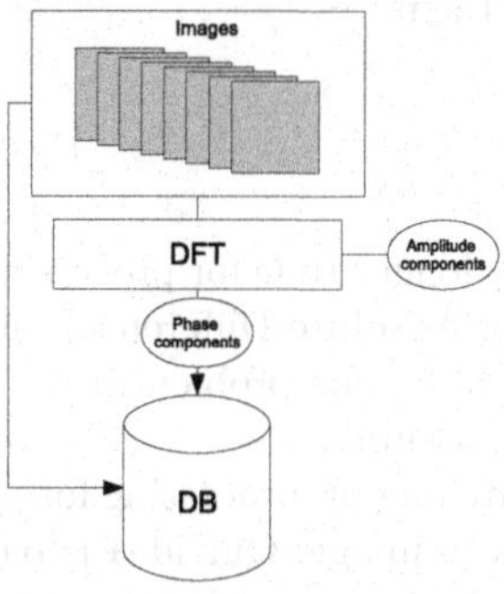

Fig. 2. Storing images with their phase components

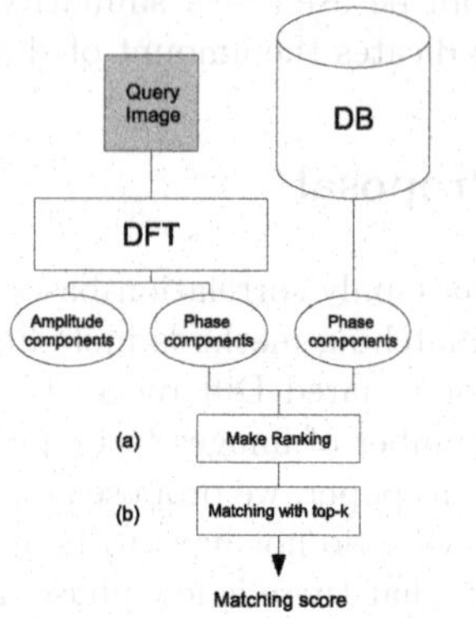

Fig. 3. Pruning process of our proposed system

3.2 Rank the stored image data

Figure 3 shows an overview of the pruning process in our proposed system. We assign a score to each image (Fig. 3 (a)) for ranking the stored images by means of stored components mentioned in the previous subsection.

Given an image, we calculate its phase component $e^{j\theta_F(k_1,k_2)}$. Getting the phase component of an image stored in a database $e^{j\theta_G(k_1,k_2)}$, let Φ_{diff} be the phase difference between the two phase components. We assume that the range of $e^{j\theta_F(k_1,k_2)}$, $e^{j\theta_G(k_1,k_2)}$, and Φ_{diff} is between $-\pi$ and π for simplicity. Φ_{diff} can be calculated as follows:

$$\Phi_{diff} = \begin{cases} e^{j\theta_F(k_1,k_2)} - e^{j\theta_G(k_1,k_2)} + 2\pi \; (condition1) \\ e^{j\theta_F(k_1,k_2)} - e^{j\theta_G(k_1,k_2)} \qquad (condition2) \\ e^{j\theta_F(k_1,k_2)} - e^{j\theta_G(k_1,k_2)} - 2\pi \; (condition3) \end{cases} \quad (9)$$

where the conditions are explained in Table 1.

If the given two images are identical, there is no phase difference between them in the spatial frequency domain. In other words, there is some phase difference between them if there is some shift difference. In fact, the larger the amount of shift difference, the larger the phase difference.

Table 1. Conditions

condition 1	$(e^{j\theta_F(k_1,k_2)} - e^{j\theta_G(k_1,k_2)} < -\pi \cup e^{j\theta_F(k_1,k_2)} - e^{j\theta_G(k_1,k_2)} > \pi) \cap e^{j\theta_F(k_1,k_2)} < e^{j\theta_G(k_1,k_2)}$
condition 2	$-\pi \leq e^{j\theta_F(k_1,k_2)} - e^{j\theta_G(k_1,k_2)} \leq \pi$
condition 3	$(e^{j\theta_F(k_1,k_2)} - e^{j\theta_G(k_1,k_2)} < -\pi \cup e^{j\theta_F(k_1,k_2)} - e^{j\theta_G(k_1,k_2)} > \pi) \cap e^{j\theta_F(k_1,k_2)} > e^{j\theta_G(k_1,k_2)}$

We define scores of stored images to the query image based on the above consideration. Let *range* be the pre-defined range of low frequency components, and τ be the absolute threshold of phase difference. Then, we calculate the score of an image as follows:

```
foreach(element in range){
  if(−τ < element < τ){
    score++;
  }
}
```

3.3 Matching processing

After getting a ranking of stored images to the query image, we select top k stored images as candidates to match the query and perform the matching process with the phase-only correlation of the whole range of the k images (Fig. 3 (b)), thereby reducing the number of images to be processed. Since we have already gotten phase components of the images, we can start step (II) of Fig. 1 for this process.

4 Experiments

We did experiments with CASIA Fingerprint Image Database Version 5.0 [1]. The database contains 20,000 fingerprint images of 500 subjects; five fingerprints of each of eight fingers other than the little fingers of the 500 subjects are stored. We used 1,000 out of the 20,000 fingerprint images and edited them as there was nothing except for the fingerprint in each image. We used additional 25 fingerprint images, which were edited similarly and also as no image shift nor rotation were necessary. The 25 images were used not only as stored images but also query images in the experiments. The resolution of the original images is 328*356, and that of the 1,025 images was 128*128. We applied the hanning window function to the 1,025 images. Note that we did not apply any other operations including distortion correction to the images, and thus maximally 1.69 and 3.51 pixels in x- and y-axis directions, respectively, in difference between two images existed.

[1] CASIA-FingerprintV5 collected by the Chinese Academy of Sciences' Institute of Automation (CASIA), `http://biometrics.idealtest.org/`

4.1 Experiments on parameters

As discussed in the previous section, *range*, τ, and k are parameters to be set in our proposal. We did experiments with varying *range* and τ. The results are shown in Table 2. A number in the table shows the worst rank of correct answer images out of the 1,025 images. For example, in the case where $range = 41 \times 41$ and $\tau = 120°$, the worst rank was 1,018th and thus we need matching process of at least 1,018 full images to obtain all the correct images.

Table 2. Results with varying *range* and τ

	τ					
range	120°	90°	60°	45°	30°	15°
41 × 41	1018	1005	1014	1023	1024	1018
31 × 31	418	440	893	888	998	956
21 × 21	107	41	159	227	781	828
15 × 15	53	116	337	438	796	600
11 × 11	502	242	265	233	619	477

According to the results, the case where $range = 21 \times 21$ and $\tau = 90°$ was the best. We therefore set the parameters as $range = 21 \times 21$, $\tau = 90°$, and $k = 50$ in the following experiments.

4.2 Performance

Next we compare three matching processes: one is our proposal, another is to perform phase-only correlation-based matching with phase components of the all images, which corresponds to steps (II) and (III) in Fig. 1 (denoted as full retrieval with phase components), and the other is to perform phase-only correlation-based matching with the all images, which corresponds to all (I), (II) and (III) steps in Fig. 1 (denoted as full retrieval with images). In fact, we compare them in terms of the times took for retrieving images matched with the 25 images. We measured five times and in the following we report the average of 2nd-4th large values in Table 3.

Table 3. Performance

	times (ms)
our proposal	10635.33
full retrieval with phase components	140303.67
full retrieval with images	250250.67

According to the results, our proposal could give more than ten times better performance than the other methods. We therefore conclude that it is valuable to pay penalty for the proposed preprocessing because the preprocessing allows us to perform phase-only correlation-based image matching efficiently.

4.3 Observations

Our proposal could retrieve all correct answers in the experiments. However, an image retrieved by the full retrieval methods was not retrieved by our proposal, although the image was an incorrect answer indeed.

The incorrect answer image was ranked as 700th in our proposal. The image has image shifts of 14.31 and -6.98 pixels in x- and y-axis directions, respectively. We shifted the data 14 and 7 pixels in x- and y-axis directions, respectively, but the shifted data was ranked as 573rd. The query, the correct answer, and the incorrect answer shifted image data are shown in Figs. 4, 5, and 6, respectively.

Fig. 4. A query image data

Fig. 5. A correct answer image data

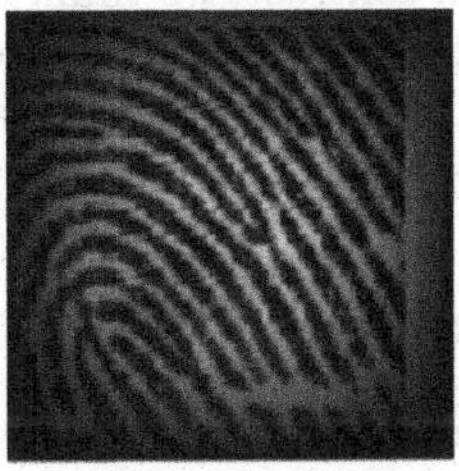

Fig. 6. The incorrect answer image data

Figures 7 and 8 visualize the differences between query and the correct answer, and query and the incorrect answer image data, respectively. Around the center of the figures corresponds to low frequency domain, and the far from the center the higher the frequency. The more difference positively, the closer to red the color, and the more difference negatively, the closer to blue the color; green points indicate that there is no phase difference.

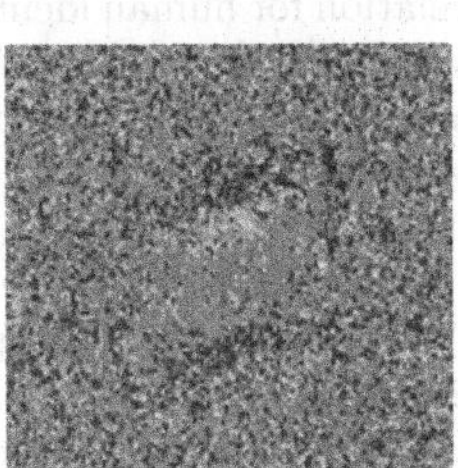

Fig. 7. Phase difference between query and the correct answer images

Fig. 8. Phase difference between query and the incorrect answer images

We can observe that a certain amount of green area exists in the center of Fig. 7, while few amount exists in the center of Fig. 8. We think that would be the main reason of this result. We may need to take account of this fact when applying our proposal to other kinds of image data.

5 Conclusions

We proposed an approximate matching preprocessing for efficient phase-only correlation-based image retrieval in this paper. The results of our experiments with fingerprint image data showed that our proposal can prune many image data and also it is valuable to use it for performance.

Several algorithms for biometrics image data retrieval based on phase-only correlations have been studied so far [3]. We think, however, they intended to devise algorithms in accordance with specific characteristics each biometrics data has. In contrast, we think that our proposal can benefit general image data. We would like to apply our proposal to other image data in future work. Also, we intend to consider image features to be indexed in databases based on the results of this work for further performance improvement.

References

1. Horner, J.L., Gianino, P.D.: Phase-only matched filtering. Appl. Opt. 23(6), 812–816 (Mar 1984), http://ao.osa.org/abstract.cfm?URI=ao-23-6-812
2. Ito, K., Aoki, T., Kosuge, E., Kawamata, R., Kashima, I.: Medical image registration using phase-only correlation for distorted dental radiographs. In: Proc. 19th International Conference on Pattern Recognition. pp. 1–4. ICPR (Dec 2008)
3. Ito, K., Morita, A., Aoki, T., Nakajima, H., Kobayashi, K., Higuchi, T.: A fingerprint recognition algorithm combining phase-based image matching and feature-based matching. In: Zhang, D., Jain, A. (eds.) Advances in Biometrics, Lecture Notes in Computer Science, vol. 3832, pp. 316–325. Springer Berlin Heidelberg (2005), http://dx.doi.org/10.1007/11608288_43
4. Ito, K., Nakajima, H., Kobayashi, K., Aoki, T., Higuchi, T.: A fingerprint matching algorithm using phase-only correlation. IEICE Trans. Fundamentals E87-A(3), 682–691 (Mar 2004)
5. Ito, K., Nikaido, A., Aoki, T., Kosuge, E., Kawamata, R., Kashima, I.: A dental radiograph recognition system using phase-only correlation for human identification. IEICE Trans. Fundamentals E91-A(1), 298–305 (Jan 2008)
6. Krichen, E., Garcia-Salicetti, S., Dorizzi, B.: A new phase-correlation-based Iris matching for degraded images. IEEE Transactions on Systems, Man, and Cybernetics, Part B: Cybernetics 39(4), 924–934 (Aug 2009)
7. Tian, Q., Huhns, M.N.: Algorithms for subpixel registration. Computer Vision, Graphics, and Image Processing 35(2), 220 – 233 (Aug 1986), http://www.sciencedirect.com/science/article/pii/0734189X86900289
8. Zhang, L., Zhang, L., Zhang, D., Zhu, H.: Ensemble of local and global information for finger-knuckle-print recognition. Pattern Recognition 44(9), 1990 – 1998 (Sep 2011), http://www.sciencedirect.com/science/article/pii/S0031320310002712

Pixel Art Color Palette Synthesis

Ming-Rong Huang and Ruen-Rone Lee

Department of Computer Science, National Tsing Hua University, Hsinchu, Taiwan
mrhuang2528@gmail.com, rrlee@cs.nthu.edu.tw

Abstract. Pixel art is created together with a color palette which greatly affects the overall visual quality. From a given image with hundreds of thousands of colors, it is difficult to pick out a limited number of colors for the color palette. We propose an automatic system, which adopts similar steps of the manual creation processes by pixel art artists, to effectively synthesize the color palette of a pixel art from a given image. Based on our approach, both artists and novices can easily derive a low-resolution image which is very close to the required final pixel art.

Keywords: Pixel Art, Color Palette, Color Quantization.

1 Introduction

Pixel art is a type of digital art where low-resolution images are created using limited number of colors in the color palette. It is commonly used in computer and console video games which have limited number of colors support and low resolution displays. Although the graphics display has been greatly improved, the unique charms of pixel art still fascinate people to keep on plugging away. Thus, it has gradually evolved into a kind of new visual design language in the modern art. Fig. 1 shows an example of pixel art (Fig. 1(a)) and its associated color palette (Fig. 1(b)). Note that the color palette is consisting of several color ramps, where a ramp is a group of colors whose hues are adjacent in the color space. In each ramp, colors are commonly sorted by their luminance values.

(a) A pixel art

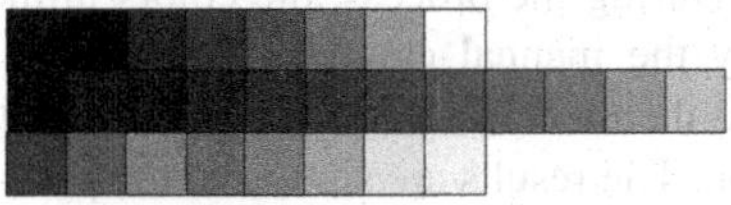

(b) Color palette

Fig. 1. An example of pixel art and its associated color palette*.

From the generalized process in creating a pixel art by artists, we found that it is a labor intensive work consisting of the processes in outlines generation, coloring, and shading, etc. Thus, it would be nice to have an automatic tool to produce a pixel art

* Image source: pixel art of Wikipe-tan from Wikipedia Commons. Created by Cpro, 2007.

K.J. Kim (ed.), *Information Science and Applications*,
Lecture Notes in Electrical Engineering 339, DOI 10.1007/978-3-662-46578-3_38

from a given image. However, it is not a trivial work, especially for the color selection from the original image to constitute the limited number of colors in color palette. Besides the outlines, the visual quality of a pixel art is greatly affected by the colors selected. It is difficult to determine the optimized color palette among the original hundreds of thousands of colors. Unlike common color quantization methods which analyze the entire image to determine the color palette with hue and luminance together; artists tend to choose some base colors in hue and then extend them in luminance to fill up the color palette with some artistic guidelines. In this paper, we will adopt similar approach of the artist to propose a system in synthesizing the color palette from a given input image.

2 Related Work

Our work can be considered as a kind of color quantization technique, which intends to reduce the number of distinct colors in an image and remain faithful to the original image as much as possible. The median cut algorithm [1] is based on dividing the color space while the *k*-means method [2] is based on cluster analysis. All these methods are with a priori fixed number of final colors. Although the results are close to the original images, there are still some noticeable color deviations or noises when examined carefully.

For pixel art related works, the methods introduced in [3, 4] focus on converting a vector line art to a pixel line art, which cover only the outlines of a pixel art. The papers in [5, 6, 7] are more related to our work. In [5, 6], they try to abstract a high-resolution image to a low-resolution image with a reduced color palette in the style of pixel art. They apply superpixels and a global optimization method to obtain the final downscaled image and color palette through iterative refinements. As for [7], it provides a content adaptive image downscaling method with the number of colors reduced by either mean shift segmentation or *k*-means clustering. However, all the mentioned previous works did not account for the artist creation flow. Instead, we synthesize the color palette by following the guidelines from manual creation of an artist.

There are many pixel art creation tutorials [8]. The common process is to fill in some essential colors first, and then add some shade to response to some virtual light sources. During the process, the colors in the color palette are derived manually. Inspired by the manual creation process, we introduce an automatic method, which similar to the manual creation flow of an artist, to help in the process of color palette derivation. The results are similar to the pixel arts that an artist would like to create.

3 System Approach

From the manual process in creating a pixel art, it is observed that the base colors determination and the ramps derivation for smooth shading are the major processes to generate the colors in the color palette. Our goal is to automatically synthesize a good color palette of pixel art from an input image to reduce the workload for artists.

Similar to the manual workflow in deriving the color palette, our approach is illustrated in Fig. 2. First, a downsampling process is required to resize the source image into a desired resolution. Then, through image analysis, N representative colors are derived as the base colors. For each base color, a ramp is derived to provide smooth color shading. Moreover, some colors among the ramps are visually similar and can be merged to further reduce the number of colors.

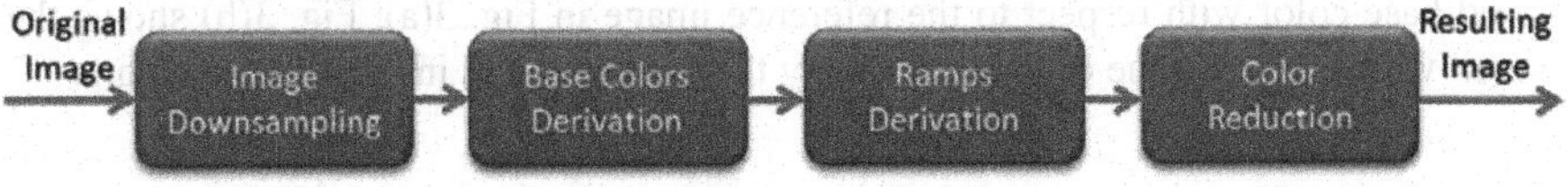

Fig. 2. System flow.

3.1 Image Downsampling

Pixel art is commonly presented in low-resolution. Starting with a smaller size reduces the amount of details you need to draw. Resizing is a well-established technique [7]. Basically, any good downscaling method can be applied. In our system, we have adopted the method, which is commonly recommended for downscaling, by simply averaging the original pixels covered by the resized pixel.

3.2 Base Colors Derivation

The downsampled image is regarded as the reference image since the size is the same as the final outcome. Even it is already downscaled, there are still a lot of colors remained as can be seen in Fig. 3(a). Artists often try to hammer out a "definitive" palette with most essential colors as the base colors to start with and then extend those base colors to come out the final color palette. So, we have to derive some representative colors from the reference image to serve as the base colors as artists will do.

(a) Reference image with 20333 colors

(b) Result of applying the16 base colors

(c) Resulting 16 base colors

Fig. 3. An example after base color derivation.

We first convert the color of each pixel from RGB to CIELAB color space. Each color is represented as a 3D point, (L, a, b), in the color space. The finding of base colors can be regarded as a three-dimensional clustering problem. We adopt *k*-means clustering algorithm [9] to derive the *k* base colors. It is hard to determine how many base colors an image should have. High computation cost makes *k*-means with larger *k* impractical. We had tested on various images, and found that the setting of *k* to be 16 would be suitable for most cases. As for the initial guess, we simply sample *k* points randomly.

Fig. 3 shows an example of base color derived as proposed. Fig. 3(c) shows the derived base color with respect to the reference image in Fig. 3(a). Fig. 3(b) shows the result which replaces the original colors by the closest colors in the 16 base colors.

3.3 Ramp Derivation

In the manual creation process, artists tend to shade a specific base color region by extending some new colors from the base color in representing levels of color gradient. The ramp for each base color is derived with respect to the levels it required.

We analyze the original colors in the regions that map to a specific base color. For each specific base color region, the minimum and maximum luminance values are detected. The number of luminance levels is then determined by $\lceil 1/\lambda \rceil$, where λ is a predefined constant for luminance level adjustment. Each pixel in the specific base color region is assigned to a closest luminance level in the ramp. Besides, if the number of pixels in a luminance level is less than a threshold value ε, then this level will be merged into the neighboring level. Finally, each color in the ramp is derived as the weighted mean of all colors in the associated level.

Fig. 4(a) shows all the ramps derived from the base colors in Fig. 3(c). For each base color region, reassign each pixel in the region by the closest color in the derived base color ramp. The result is shown in Fig. 4(b) which is very close to the reference image in Fig. 3(a). The number of colors has been reduced from 20333 to 52 colors.

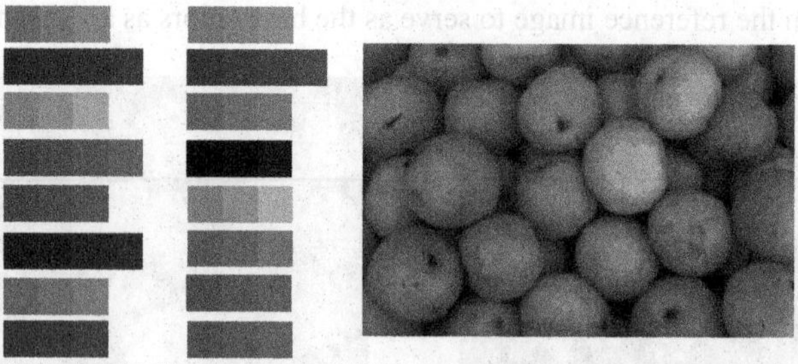

(a) Derived color ramps (b) Result of ramp derivation, 52 colors

Fig. 4. An example after ramp derivation.

3.4 Color Reduction

After all the ramps are derived from the base colors, the number of colors in the color palette is increased. It is possible to further reduce the number of colors to provide a smaller color palette for use. Some colors in different ramps resemble visually and can be merged together. The final reduced color palette is regarded as a good color palette in that the number of colors is reduced while the length of each ramp remains the same. That is, even though the colors are reduced, the levels of gradient are still retained and thus the visual quality is affected minimally.

The Euclidean distance in CIELAB color space is used for the color similarity measurement. Any two colors locating in different ramps can be merged into one color if their distance is less than a predefined threshold value δ. In our experiment, the default threshold is set to 8. The merged color is defined to be the weighted average of the two colors.

We first calculate the distance of any two colors in the palette of ramps. A binary square matrix, similarity table, with row and column indicating the colors in the palette is used to represent the table of color similarity. We label a "1" if the distance is less than the threshold δ, and a label "0" otherwise. In order to preserve the length of a ramp, or similarly the gradient levels, it is not allowed to merge any two colors from the same ramp, or merge two colors of the same ramp with a color from other ramps. According to the similarity table, we can sort the ramps by the number of colors it can be reduced (the number of "1"s in this ramp) in descending order. Choose the highest one to merge the closest colors firstly. Once merged, change the label of those colors from "1" to "0". Recalculate the number of colors each ramp can be reduced, and then sort and merge again. Repeat those steps until there is no "1" in the similarity table. Fig. 5(b) shows the final result after color reduction on the color ramps in Fig. 4(a). A new color palette is derived as shown in Fig. 5(a) (only parts of the merged ramps are shown). Each ramp is highlighted by a black outline. The red outlines indicate those colors that can be merged. The result shows that the number of colors has been reduced from 52 down to 30 while the length of each ramp remains the same.

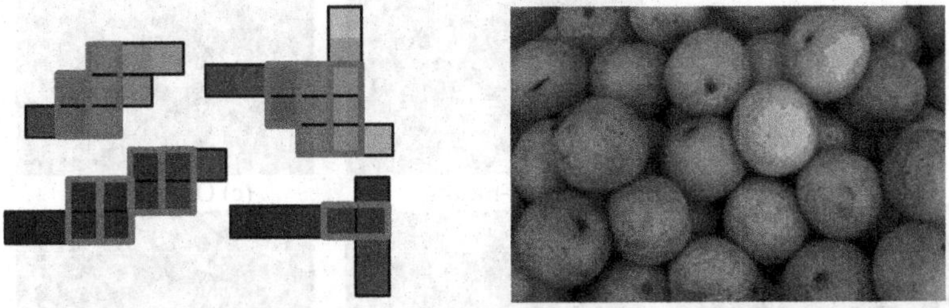

(a) Final color palette (partial) (b) Resulting image after color reduction

Fig. 5. Final result after color reduction.

4 Results

Some results are presented in Fig. 6 and the final results are similar to the original source images using only a few colors in the color palettes, respectively.

Comparison with other methods based on the same size of color palette is also given in Fig. 7. The results of optimized median cut using error diffusion (Opt-error) and nearest color (Opt-nearest) are generated by Corel PaintShop Pro which implemented the two methods. The result of Opt-error (Fig. 7(c)) results in a lot of noticeable noises. The result of Opt-nearest (Fig. 7(d)) suffers from apparent color deviation as can be seen in comparing to the input reference image in Fig. 7(a). Our result is very similar to the result of k-means (Fig. 7(b)). However, when examined carefully, our method keeps more levels of gradient and details especially on the presentation of the two clown fishes as shown in the close up look in Fig. 7(f) and Fig. 7(g).

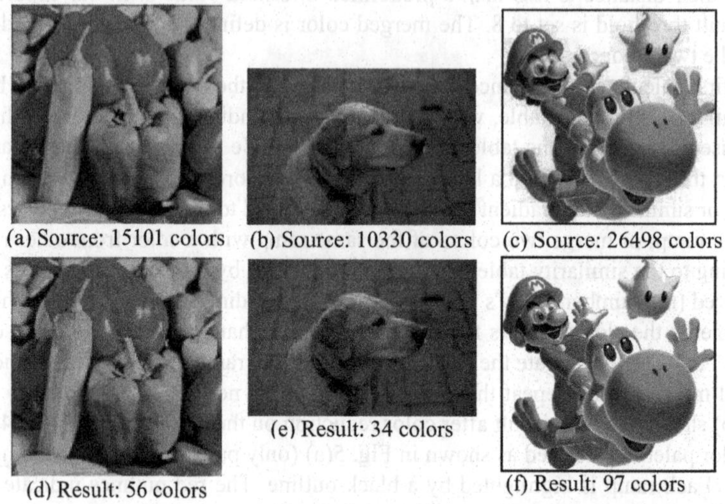

(a) Source: 15101 colors (b) Source: 10330 colors (c) Source: 26498 colors

(d) Result: 56 colors (e) Result: 34 colors (f) Result: 97 colors

Fig. 6. Some results.

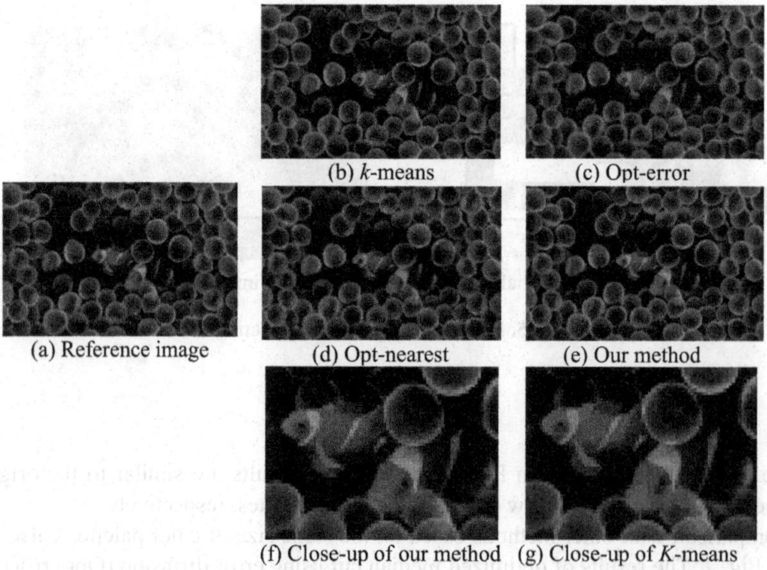

(a) Reference image (b) k-means (c) Opt-error (d) Opt-nearest (e) Our method (f) Close-up of our method (g) Close-up of K-means

Fig. 7. Comparison with other methods. (all with 47 colors in the color palette)

4.1 Quantitative Quality Assessment and User Studies

In order to evaluate the quality of our results, some quantitative analysis metrics such as mean squared error (MSE), peak signal-to-noise ratio (PSNR), and structural similarity index metric (SSIM) are applied [10]. Based on the same size of color palette, we compare our method with *k*-means, Opt-nearest and Opt-error on 20 test images of different types covering from nature, clipart, object, and scene.

From the assessment results, we found that it is very difficult to distinguish the quality solely from the numbers of quality assessments. By carefully reviewing the results of those 20 test images generated by our method and the others, we found that the quality assessment may not truthfully reveal the best one from human perception. Therefore, we try to conduct a user study to further prove that our method produces better quality than the others in most of the test cases.

The user study is designed with the reference image being placed in the center of a page. The other four images which associate with four different methods are shuffled randomly and placed in the four corners of the page. All the participants are with the background of computer graphics and image processing. We think that their background can help them to evaluate image quality more correctly and objectively. 20 test cases are prepared in a 20-page user study. Two rounds of user studies were performed by 14 and 7 participants, respectively. The 1st round was designed to ask the participants to pick out one image which they regarded as the closest one to the reference image among the four corner images without giving any hints. In the 2nd round, the participants are asked to rank the quality for the four different methods with comparators ">" (better than) and "=" (equal to) only. They were also asked to briefly write down the reasons of their judgments.

We collected all the user feedbacks and summarized them in Table 1. The numbers in the table indicate the votes of preferring our method out of the participants. From the two user studies, we found that our approach indeed produced better quality than the other methods in most of the cases.

1st round user study									
Cat	Elephant	Flower	Landscape	Fruits	Giraffe	Lena	Lion	Pears	Penguin
11/14	11/14	4/14	5/14	9/14	7/14	13/14	11/14	8/14	13/14
Pepper	Pepper2	Forest	Mario	Whale	Simba	Fish	Dog	Persons	Bird
13/14	13/14	12/14	7/14	7/14	10/14	11/14	13/14	13/14	11/14
2nd round user study									
Cat	Elephant	Flower	Landscape	Fruits	Giraffe	Lena	Lion	Pears	Penguin
6/7	7/7	1/7	6/7	6/7	5/7	7/7	6/7	4/7	7/7
Pepper	Pepper2	Forest	Mario	Whale	Simba	Fish	Dog	Persons	Bird
7/7	7/7	7/7	4/7	3/7	5/7	7/7	7/7	7/7	6/7

Table 1. Results of 1st round and 2nd round user studies

5 Conclusions and Future Works

We present a system to automatically synthesize an appropriate initial color palette of pixel art from a given image. By adopting the common manual processes in creating the color palette of a pixel art into our proposed system, we have successfully generated better quality in color presentation of a pixel art using the derived colors in color palette. It not only preserves the length of each color ramp extended from the based colors but also reduces the colors by merging colors with similar visual perception. The result is a low-resolution pixel art that is very close to a final pixel art an artist can create. We conduct both quantitative assessment and user studies to prove that our approach indeed results in better quality than the other methods generated.

From the feedbacks of user studies, we have found that people used to have an attentional preference to the foreground rather than to the background. In the future work, we will try to reserve more colors for the foreground and fewer colors for the background to create better visual experience than the one without this discrimination.

Acknowledgements. This project was supported in part by the Ministry of Science and Technology of Taiwan (MOST-103-2220-E-007-012) and the Ministry of Economic Affairs of Taiwan (MOEA-103-EC-17-A-02-S1-202).

References

1. Heckbert, P.S.: Color Image Quantization for Frame Buffer Display. SIGGRAPH Computer Graphics, vol. 16, no. 3, pp. 297-307. ACM (1982)
2. Celebi, M. E.: Improving the Performance of *k*-means for color quantization. Image and Vision Computing, vol. 29, no.4, pp. 260-271 (2011)
3. Inglis, T.C., and Craig S.K.: Pixelating vector line art. Proceedings of the Symposium on Non-Photorealistic Animation and Rendering, pp.21-28. Eurographics Association (2012)
4. Inglis, T.C., Vogel, D., and Craig S.K.: Rasterizing and antialiasing vector line art in the pixel art style. Proceedings of the Symposium on Non-Photorealistic Animation and Rendering, pp. 25-32. ACM (2013)
5. Gerstner, T., DeCarlo, D., Alexa, M., Finkelstein, A., Gingold, Y., Nealen, A.: Pixelated image abstraction. Proceedings of the Symposium on Non-Photorealistic Animation and Rendering, pp. 19-36. Eurographics Association (2012).
6. Gerstner, T., DeCarlo, D., Alexa, M., Finkelstein, A., Gingold, Y., Nealen, A.: Pixelated image abstraction with integrated user constraints. Special session on Expressive Graphics, Computer and Graphics, vol. 37, no. 5, pp.333-347 (2013)
7. Kopf, J., Shamir, A., and Peers, P.: Content-adaptive image downscaling. ACM Transactions on Graphics (TOG). Vol. 32, no.6, p. 173 (2013)
8. DeviantArt, pixel art tutorial search results `http://www.deviantart.com/browse/all/?q=pixel+art+tutorial`
9. MacQueen, J.: Some methods for classification and analysis of multivariate observations. Proceedings of the fifth Berkeley symposium on mathematical statistics and probability. vol. 1, no. 14, pp. 281-297 (1967)
10. Wang, Z., Bovik, A. C., Sheikh, H. R., and Simoncelli, E. P.: Image quality assessment: from error visibility to structural similarity. IEEE Transactions on Image Processing, vol. 13, no. 4, pp.600-612 (2004).

Introducing a New Radiation Detection Device Calibration Method and Estimating 3D Distance to Radiation Sources

Pathum Rathnayaka, Seung-Hae Baek, Soon-Yong Park

School of Computer Science & Engineering, Kyungpook National University, South-Korea
bandarapathum@yahoo.com, eardrops@vision.knu.ac.kr, sypark@knu.ac.kr

Abstract. Radiation detection devices; also known as particle detectors; are vastly used to track and identify radioactive sources within a given area. The 3D distance to such radioactive sources can be estimated using stereo radiation detection devices. In stereo vision, the devices have to be calibrated before they are used to acquire stereo images. In this work, we first introduce a new method to calibrate the stereo radiation detection devices using homography translation relationship. The radiation detection devices we have used in our approach are pinhole cameras. The calibrated pinhole cameras are then used to generate stereo images of radioactive sources using a pan/tilt device, and estimated the 3D distance using the intrinsic and extrinsic calibration data, and triangulation. Stereo vision cameras are used along with pinhole cameras to obtain coincident 2D visual information. We performed two experiments to estimate the 3D distance using different input image data sets. The inferred 3D distance results had around a 5~6% error which assures the accuracy of our proposed calibration method.

Keywords: Radiation detection devices, radioactive sources, Homography translation relationship, 3D distance, device calibration.

1 Introduction

Radiation has become one of the most widely discussed topics in around the world. Even though they are said to be having some negative corollaries, they play a significant role in many areas of science and industry including nuclear medicine, astronomy, environmental protection and nondestructive testing [1]. Many methods such as; conventional portable gamma cameras with various detectors and collimators have been introduced to acquire 2D images and information of radioactive sources in a particular area [2]. These images are processed to obtain the 3D position information along with image overlaying procedures to visualize the radioactive area in real world environments. Once the 3D distance to a radioactive source is estimated, the source activity (strength of the radiation source) can be estimated easily [2].

In 3D computer vision, device calibration is considered as the preliminary step to be followed. The necessity of device calibration is mostly encountered in many stereo-

K.J. Kim (ed.), *Information Science and Applications*,
Lecture Notes in Electrical Engineering 339, DOI 10.1007/978-3-662-46578-3_39

vision experiments. A proper calibration method always effects in higher accurate results. Many work related to camera calibration has been done throughout the past few decades [3, 4, 5, 6], [7, 8] but an accurate method to calibrate stereo radiation detection devices has not yet been introduced in computer vision society. Once the calibration relationship between radiation detection devices is known, 3D position information, 3D distance information, and homography matrices can be estimated accurately.

More accurate projector-camera calibration methods have been introduced in the recent past where the projector is considered as an inverse camera. In this inverse camera model, it is mentioned that a planar homography between the projector and the camera exists [6]. In this case, a separate calibration image; which is assumed to be the image obtained from the projector; can be generated. The generated image then satisfied with the homography translation relationship between the camera image and the projector image. A similar image acquisition method is used in this paper and the Zhang's method [3] is applied to calibrate stereo radiation detection devices.

The structure of this paper is as follows. Section 2 describes the pan/tilt scanning method used to generate radiation images. Section 3 comprehensively describes the method used to calibrate the devices along with the method used to find the homography translation relationships between radiation images and camera images. Two different experiments performed to estimate the 3D distance to radioactive sources are mentioned in section 4 which follows with the conclusion and future works.

2 Generating Radiation Images

In our approach, we have used a pan/tilt table to scan the area of the radioactive sources. The experimental setup is depicted in Fig.1. As it is mentioned in the figure, stereo pinhole cameras are mounted on the pan/tilt table. The rotation of the table in panning and tilting directions is simply controlled by a control board which is connected to a general purpose computer.

In our method, we followed two scanning procedures; a primary fast-scan to track the location of the radioactive source, and a secondary slow-precision scan to acquire the 2D information of radioactive sources within the region of interest (ROI). This procedure is depicted in Fig.2. The acquired information of radioactive sources are transmitted to the computer and radiation signal processing methods (filtering and quantizing) are applied to generate left and right stereo radiation images (Fig. 3). These generated radiation images are stored in the computer and sent for further image processing procedures.

We have used pinhole cameras as our radiation detection devices. Main reason is that pinhole cameras manage to produce radiographs and photographs of objects, which emit radiation and visible light. On the other hand, light is a form of radiation that spreads similar to gamma rays and other radiation sources. Considering it as a fact, we have used four bright LEDs (which are placed in the same plane) as our radioactive sources (Fig. 4).

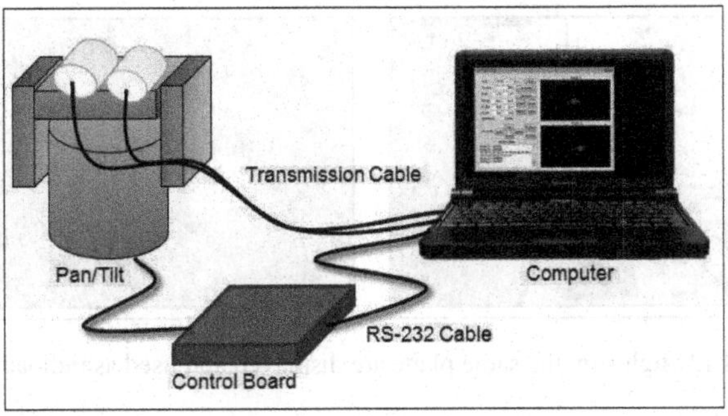

Fig. 1. The pan/tilt table with stereo pinhole cameras and the main control board connected to a pc via RS-232 cable

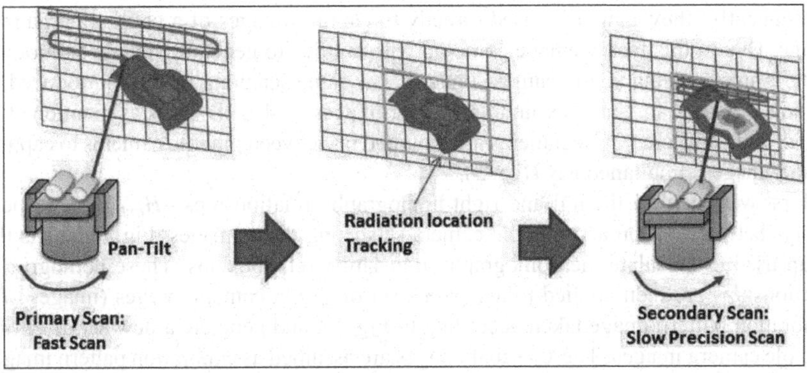

Fig. 2. Primary fast scan and secondary slow-precision scan used to track and acquire 2D information of radioactive sources. Yellow rectangle represents the ROI area

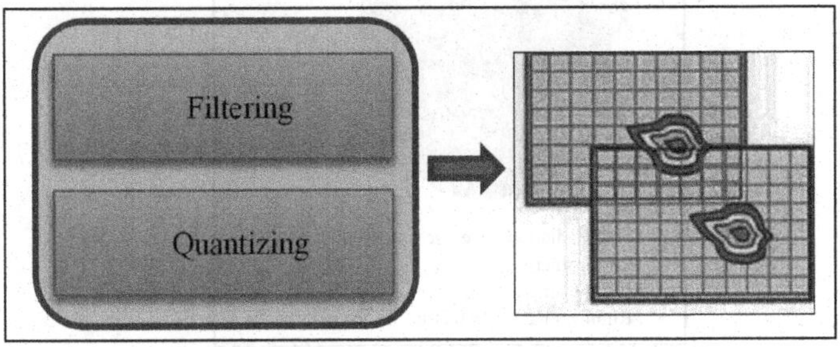

Fig. 3. Left and right radiation images are created from the processed scanned data. Filtering and Quantizing are the used signal processing methods.

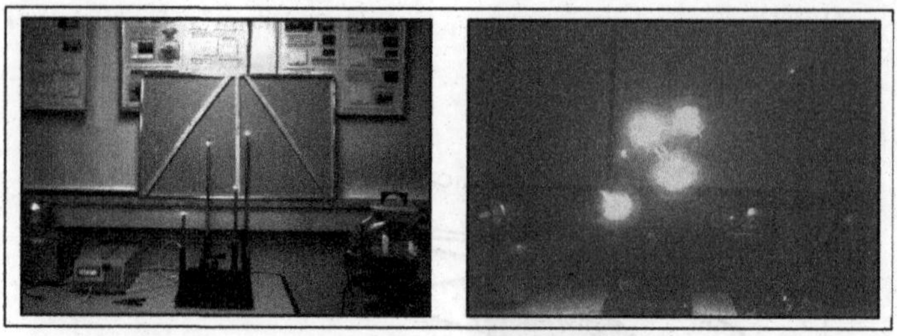

Fig. 4. Four LED lights on the same plane are displayed and used as radioactive sources

3 Calibrating Radiation Detection Devices

In device calibration, sets of left and right calibration pattern images are required. The quality of pinhole camera images compared to camera images is considerably low and consequently; they cannot be used directly to capture images of a calibration pattern image. Due to this fact, we have come-up with an idea to generate pinhole calibration pattern images using vision camera images. The same scanning method introduced in section 2 is used to generate pinhole camera images of 4 LED spots as mentioned in Fig. 4, and additional CCD cameras are mounted in-between pinhole cameras to capture vision images simultaneously (Fig. 5).

First we calculate the left and right homography relationships - H_{crl}, H_{crr} respectively- between vision and pinhole camera sets using these images. Fig. 6 depicts the setup used to calculate the homography translation relationships. These homography relationships are then applied to a separate set of vision camera images (images of a calibration pattern image taken according to Fig. 7), and generate a new set of virtual pinhole camera images. These virtual images are assumed as calibration pattern images obtained from pinhole cameras. Finally; the Zhang's camera calibration method [3] is applied on the virtual images to calibrate stereo pinhole cameras.

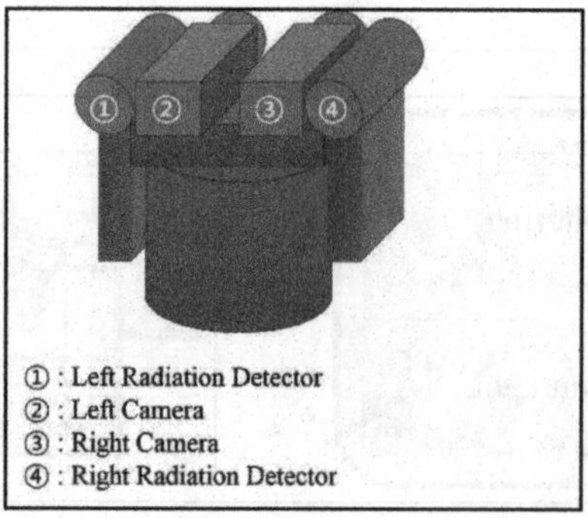

Fig. 5. Mounting additional stereo CCD cameras in-between pinhole cameras

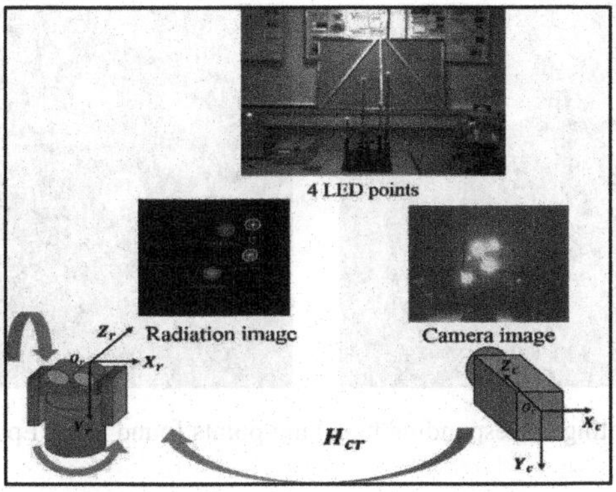

Fig. 6. Experimental setup used to calculate homography translation relationships. Pinhole cameras are mounted on the pan/tilt table.

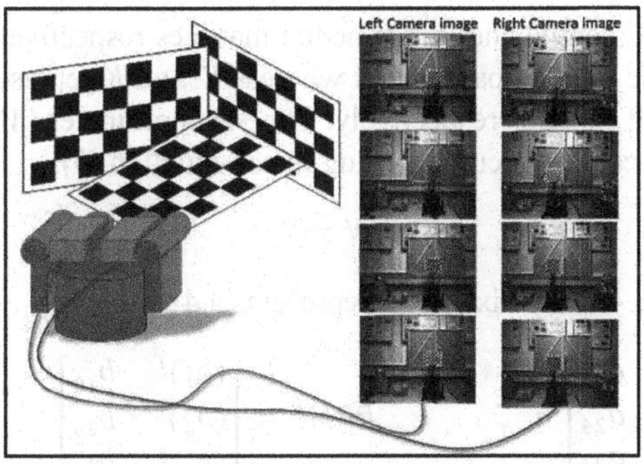

Fig. 7. Capturing images of calibration pattern image shown at a few different orientations using vision cameras

4 Experiments and Results

4.1 Estimating 3D Position

After device calibration; the intrinsic and extrinsic camera parameters can be obtained and they are used to calculate left and right perspective projection matrices (PPM). Fig. 8 shows a set of left and right stereo radiation images where image processing methods such as adaptive thresholding, contour detection and ellipse fitting are applied to find the center points $((u, v)\ and\ (i, j))$ of each radiation (LED) spot. Essential and Fundamental matrices can also be calculated using the camera parameters according to equations (1) and (2) and the Epipolar geometry is applied to identify the corresponding matching points of left and right radiation images. These correspondences are represented with identical colors in Fig. 8.

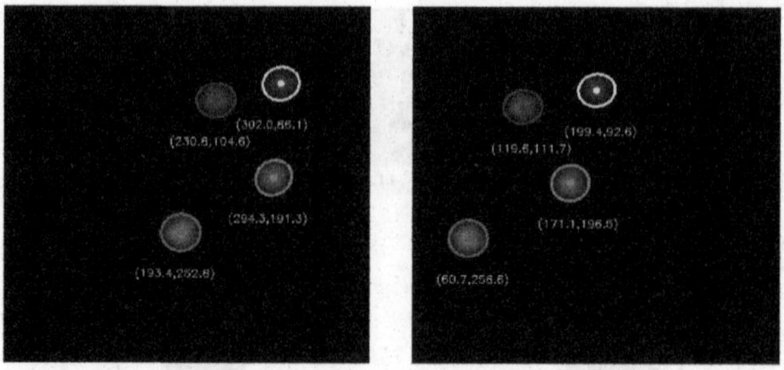

Fig. 8. Representing corresponding matching points found using Epipolar geometry

$$\mathrm{E} = \mathrm{R}\ [t]_x \tag{1}$$

$$\mathrm{F} = K'^{T^{-1}} E\, K^{-1} \tag{2}$$

E and F represent Essential and Fundamental matrices respectively. R $[t]_x$ is the cross product of extrinsic camera parameters whereas $K'$and K represent intrinsic parameters of right and left cameras respectively. The 3D coordinates ($W\ (X, Y, Z)$) for each matching point set can be calculated using linear equation (3).

$$\mathrm{AW} = \mathrm{Y} \tag{3}$$

Here A represents a 4×3 matrix, and Y represents a 4×1 matrix.

$$PPM^L = \begin{bmatrix} (a_1)^T & a_{14} \\ (a_2)^T & a_{24} \\ (a_3)^T & a_{34} \end{bmatrix} \qquad PPM^R = \begin{bmatrix} (b_1)^T & b_{14} \\ (b_2)^T & b_{24} \\ (b_3)^T & b_{34} \end{bmatrix}$$

$$A = \begin{bmatrix} (a_1 - ua_3)^T \\ (a_2 - va_3)^T \\ (b_1 - ib_3)^T \\ (b_2 - jb_3)^T \end{bmatrix} \qquad Y = \begin{bmatrix} (a_{14} - ua_{34}) \\ (a_{24} - va_{34}) \\ (b_{14} - ib_{34}) \\ (b_{24} - jb_{34}) \end{bmatrix}$$

The inverse of matrix A is calculated using singular value decomposition (SVD) method. Fig. 9 shows an example of how 3D coordinates are estimated for a one matching set. The same procedure is repeated to estimate the 3D coordinates of all the rest corresponding sets.

4.2 Distance Estimating Experiments

We did two experiments to calculate 3D distances and the results are depicted in Tables 1 and 2. In first experiment, we used 5 LED spots (similar to Fig.4) and measured the distances to each spot from a lower distance range (Table 1). The average error was around 4~6%. In second experiment, we used 4 LED spots and checked for the accuracy for a wide distance range. The average error was about 1~3% (Table 2).

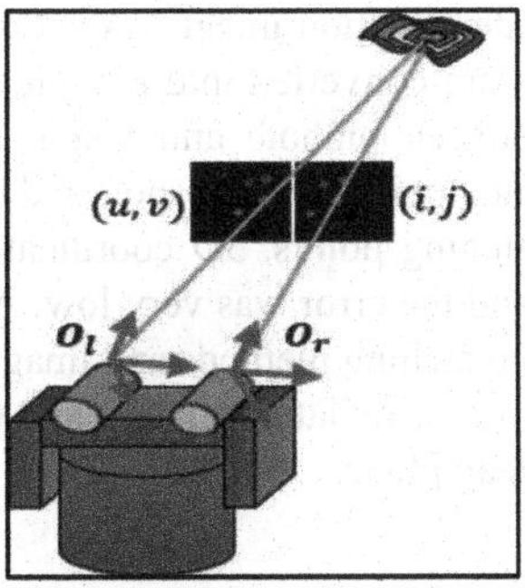

Fig. 9. Calculating 3D coordinates of a corresponding LED spot

Table 1. Estimated 3D distances of 5 points. Real distances are mentioned in *Dist* column in centimeters.

Dist (cm)	P1	P2	P3	P4	P5	Avg	Error(%)
80	76.359	79.686	72.617	74.676	77.845	76.237	4
100	91.879	95.920	91.078	92.503	94.615	93.199	6
120	113.686	114.789	112.715	113.890	113.065	113.629	5
140	132.484	134.443	132.024	129.567	131.881	132.080	5

Table 2. Estimated 3D distances of 4 points. Real distances are mentioned in *Dist* column in centimeters.

Dist (cm)	P1	P2	P3	P4	Avg	Error(%)
300	303.037	293.658	290.358	300.039	296.773	1
330	330.363	346.611	339.973	329.079	336.506	2
370	378.757	375.874	373.683	359.664	378.494	2
400	416.115	417.167	400.495	409.566	410.836	3

5 Conclusions

In this paper, we proposed a new method to calibrate radiation detection devices using homography translation relationships. Bright LED spots on a planar surface are used as radioactive sources and pinhole cameras are used as radiation detection devices. Pan/tilt scan method is used to generate radiation images. A set of calibration pattern images captured from vision cameras are converted into pinhole camera images by applying homography relationships between pinhole and vision cameras. Image processing methods are applied to find center points of LED spots and Epipolar geometry is applied to find the corresponding matching points. 3D coordinates were obtained with high accuracy using triangulation and the error was very low. As future work, we are planning to apply advanced stereo matching methods and image processing techniques and stereo matching methods to process radiation images and to detect n number of radioactive sources on a widely spread plane.

Acknowledgements.
This project is supported by civil-military technical cooperation promotion center of the defense research institute, and the Korean atomic energy research institute of South Korea.

References

1. Lee, Wonho, and David Wehe. "3D position of radiation sources using an automated gamma camera and ML algorithm with energy-dependent response functions." Nuclear Instruments and Methods in Physics Research Section A: Accelerators, Spectrometers, Detectors and Associated Equipment 531.1 (2004): 270-275.
2. Wonho Lee; Wehe, D.K., "3D position of radiation sources using an automated gamma camera and ML algorithm with energy-dependent response functions," Nuclear Science Symposium Conference Record, 2003 IEEE , vol.2, no., pp.737,741 Vol.2, 19-25 Oct. 2003
3. Z. Zhang, "A Flexible New Technique for Camera Calibration," IEEE Transactions on Pattern Analysis and Machine Intelligence. Vol. 22, No. 11, pp. 1330-1334, 1998.
4. Hongshan YU and Yaonan Wang, "An Improved Self-calibration Method for Active Stereo Camera," Proc. 6th World Congress on Intelligent Control and Automation, pp. 5186-5190, June 2006.
5. Hyukseong Kwon, Johnny Park and Avinash C. Kak, "A New Approach for Active Stereo Camera Calibration," IEEE Conf. Robotics and Automation, pp. 3180-3185, April 2007.
6. Soon-Yong Park and Go Gwang Park, "Active Calibration of Camera-Projector Systems based on Planar Homography," Conf. Pattern Recognition, pp.320-323, Aug 2010.
7. Wei, G-Q., and S. D. Ma. "A complete two-plane camera calibration method and experimental comparisons." Computer Vision, 1993. Proceedings., Fourth International Conference on. IEEE, 1993.
8. R. Hartley, A. Zisserman, Multiple View Geometry in Computer Vision, Cambridge. 2nd edition, 2003.

Integrating Feature Descriptors to Scanpath-based Decoding of Deformed Barcode Images

Poonna Yospanya[1] and Yachai Limpiyakorn[2]

[1] Department of Computer Engineering, Kasetsart University, Si Racha Campus, Chon Buri 20230, Thailand
poonna@eng.src.ku.ac.th
[2] Department of Computer Engineering, Chulalongkorn University, Bangkok 10330, Thailand
Yachai.L@chula.ac.th

Abstract. There is an extensive literature on localization and decoding of barcode images. Most traditional scanline-based methods have difficulty decoding a highly deformed barcode. This paper thus presents a scanpath-based method to manage the decoding of deformed barcode images. The notion of local orientedness based on HOG descriptors, called Modal Orientation Deviation (MOD), is introduced to help in scanpath construction. The effectiveness of the proposed method is also discussed according to the results of tests reported.

Keywords: image processing · deformed surface · scanpath · modal orientation deviation · gradient-based descriptor

1 Introduction

Current cell phones are equipped with very good cameras and powerful CPUs that enable them to be used as portable barcode scanners. These smartphones can be considered as assistive equipment for the visually impaired to improve their quality of life by allowing them to shop independently. However, most camera-based barcode reading systems [1,2,3] are designed to work with barcodes attached on flat or smooth surfaces. This seems not practical when dealing with barcodes on surfaces with high level of deformation, containing indecipherable regions in the barcode images. Creases, bends, and wrinkles create ridges and valleys on the surface on which the barcode is printed, which in turn introduce artifacts such as reflections and shadows, and produce visual distortion of various forms. These artifacts obscure parts of the barcode and make them unreadable.

The problems of localization and decoding of barcode images have been extensively studied in literature. However, little has been explored in dealing with highly deformed barcodes. Most traditional scanline-based methods [1, 2, 3, 4, 5] would have difficulty decoding such barcode. Particularly, the technique would likely fail to find a single scanline that passes through the barcode without running over some unreadable regions. We have previously explored this specific

K.J. Kim (ed.), *Information Science and Applications*,
Lecture Notes in Electrical Engineering 339, DOI 10.1007/978-3-662-46578-3_40

aspect of barcode scanning in [6], and proposed a method using a non-linear scanpath instead of a scanline. The presented approach is a departure from the traditional scanline-based methods and we have shown that it is effective on many cases of highly deformed barcodes. However, the method suffers from its reliance on binarization. Barcodes with uneven lighting sometimes cause the scanning to fail due to it missing or skipping parts of the barcodes.

Dalal and Triggs [7] first publicized the technique of Histogram of Oriented Gradients (HOG) in 2005. Histograms of Oriented Gradients are feature descriptors used in computer vision and image processing for the purpose of visual object recognition. The essential notion of HOG descriptors is that local object appearance and shape within an image can be described by the distribution of intensity gradients or edge directions.

Our current approach of the scanpath-based decoder requires a means to distinguish those parts of the barcode that contain artifacts. The computed scanpath could thus avoid them and more likely produce a correct decoding. Generally, barcodes, even when deformed, are locally oriented. If we divide a barcode into small blocks, the fragments of bars contained in a block will likely have similar orientation. Artifacts disturbing the orientedness of the blocks they reside can then be identified. In this paper, we therefore introduce the notion of local orientedness based on HOG descriptors [7], called Modal Orientation Deviation (MOD), to improve the fidelity of the scanpath constructed. This would result in successful decoding of deformed barcode images.

2 Overview of the Method

2.1 Prepare Input Image

Initially, the input image is converted to gray-scale and properly resized. We choose the width of the image to be between 300-600 pixels. If the image size does not fall within the range, then it will be resized accordingly. In actual scanning, we also employ subpixel sampling using bilinear interpolation to achieve the required sampling resolution.

2.2 Compute HOG Descriptor

To obtain the HOG descriptor, we first compute the gradient values for all the pixels by applying 1D centered derivative masks on both the horizontal and vertical directions. Specifically, we obtain the following images G_x and G_y from the intensity image I:

$$G_x = [-1, 0, 1] * I \tag{1}$$

$$G_y = [-1, 0, 1]^{\mathsf{T}} * I \tag{2}$$

We then compute the gradient magnitude and orientation as follows:

$$G = \sqrt{G_x^2 + G_y^2} \tag{3}$$

$$O = \text{atan2}(G_x, G_y) \bmod \pi \tag{4}$$

The function atan2() is the quadrant-correct version of the arctangent function. Note that the orientation is in the range $[0, \pi)$ as the sign of the gradient is disregarded, e.g., a gradient direction of $\pi/3$ and another gradient direction of $-2\pi/3$ are regarded as being of the same orientation. The floating-point modulo operator is used to remove the sign of the gradient.

The gradient image is partitioned into cells of 16×16 pixels. Within each cell, pixel gradients are casted into 9 histogram bins based on their orientation. Histogram bins are evenly spread over the range of $[0, \pi)$, and each pixel casts a weighted vote based on its magnitude. At the end of the process, each cell has its gradient orientation histogram with each bin containing the sum of magnitudes of gradients whose orientation lies within the bin range.

Cells are usually grouped and normalized together as blocks. In this work, however, we use the block size of one cell, that is, the block and the cell are equivalent. We will refer to them as blocks from this point onward.

Finally, to account for local differences in illumination and contrast, we perform the following L1-sqrt block normalization on every block in the image.

$$\text{L1-sqrt:} f = \sqrt{\frac{v}{\|v\|_1 + \epsilon}} \tag{5}$$

where v is the non-normalized histogram vector of the block, $\|v\|_1$ denotes the L1-norm of the vector, and ϵ is a small constant. The result is the HOG descriptor being used in the subsequent analysis steps. Example of the visualization of the HOG descriptor associated with the input deformed image is shown in Fig. 1.

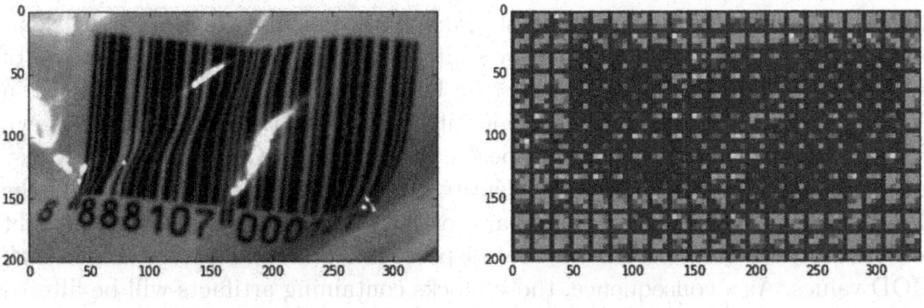

Fig. 1. Input image (left) and visualization of the HOG descriptor (right). Each block in the HOG image depicts gradient distribution as a 3×3 grid of the 9 orientation bins, with blue indicating low vote count and red showing high vote count

2.3 Determine Decodable Blocks

We observe that blocks that are parts of a barcode tend to be locally oriented. Even though the barcode is warped, blocks are usually small enough that parts of bars contained within the same block often have similar orientation. Therefore, we define a measure of orientedness, called Modal Orientation Deviation (MOD), to quantify how the gradients within a block align with the dominant orientation of the block. We consider the mode of the HOG for the block B_i as its dominant orientation and define the modal orientation deviation function $MOD(i)$ of the block as:

$$MOD(i) = \frac{\sum_{j=1}^{N} w_j^i D(i,j)}{\epsilon + \sum_{j=1}^{N} w_j^i} \tag{6}$$

where N is the number of histogram bins, w_j^i denotes the weight (the normalized sum of gradient magnitudes) of the histogram bin b_j of the block B_i, and ϵ is a small constant. $D(i,j)$ is the angular distance between the central orientations of the bin b_j and the dominant orientation bin of the block B_i. Since the orientation value can be cyclic within the range $[0, \pi)$, the smallest distance measured in either direction will be selected. $D(i,j)$ is then defined as:

$$D(i,j) = \min\left(\left|\phi_j^i - \phi_{Mo(i)}^i\right|, \pi - \left|\phi_j^i - \phi_{Mo(i)}^i\right|\right) \tag{7}$$

where ϕ_j^i is the central orientation of the bin b_j of B_i, $Mo(i)$ represents the bin number of the dominant orientation bin, defined as the mode in the orientation histogram of B_i, and $\phi_{Mo(i)}^i$ denotes the central orientation of the dominant orientation bin. Fig. 2(a) depicts the visualization of MOD corresponding to the input image shown in Fig. 1.

Blocks that are highly oriented have lower MOD values. However, a suitable threshold is required to separate highly-oriented blocks from the rest. We do so by computing the MOD histogram for the whole image. First, the MOD for each block is computed, then the range valued from the highest and the lowest MODs is divided evenly into 12 histogram bins, as shown in Fig. 2(b). Next, bimodal thresholding is performed by splitting the histogram into two halves of equal population. We find the peak (the largest bin, shown in blue and red in the Fig. 2(b)) from each half, then identify the valley (the smallest bin, shown in green) that lies between the two peaks. The central MOD value of the valley bin is used as the threshold. We identify blocks with MOD values below the threshold as potentially decodable parts of the barcode. Artifacts, such as light reflections, often create gradients that disperse in all directions, resulting in high MOD values. As a consequence, those blocks containing artifacts will be filtered out. We may also filter out isolated blocks and groups of connected blocks that contain fewer than a certain number of blocks at this stage. Example output of decodable block identification is depicted in Fig. 2(c).

2.4 Find a Scanpath

In this work, a scanpath is defined as a set of connected line segments through which the barcode is read. These line segments are arranged such that artifacts are avoided, and thus producing a clean reading of the barcode intensity levels along the scanpath.

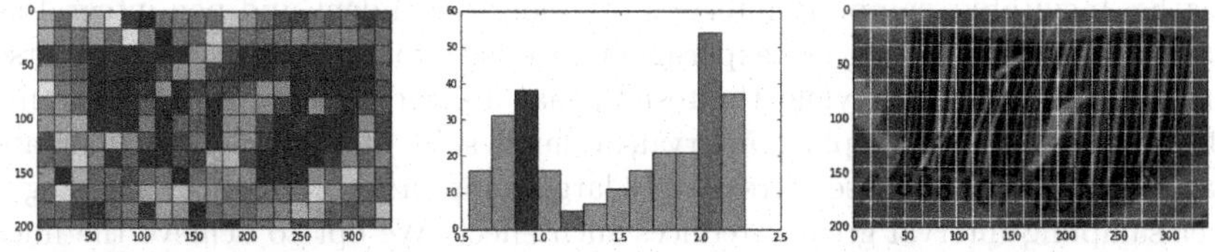

(a) (b) (c)

Fig. 2. (a) Visualization of MOD, with blue indicating a low MOD value and red showing the opposite; (b) MOD histogram to distinguish highly-oriented blocks with bimodal thresholding; (c) Result from block qualification, with disqualified blocks highlighted in red

In order to find a scanpath, a scangraph is required to be constructed from the block information obtained in the previous step. The A* pathfinding algorithm is used for searching over the barcode blocks to find the scanpath through the scangraph. Each block has a variable path cost depending on the orientation difference between the block gradients and the path segment. We plant a vertex for the scangraph at the central pixel of each qualified block. For each pair of vertices (v_i, v_j) that is within a predefined distance, we define a forward edge $e_{i,j}$ if v_i is located at a column to the left of that of v_j and there are no disqualified blocks obstructing the edge. The leftmost group of vertices is chosen as the sources and the rightmost ones as the destinations. Euclidean distance is chosen as the heuristic distance function. The actual path cost is thus defined as follows.

First, the flow of a block B_i is defined as a polar vector (W_i, Φ_i) representing the projected weight along its dominant orientation and the dominant orientation itself. Formally:

$$FLOW(i) = (W_i, \Phi_i) = \left(\sum_{j=1}^{N} w_j^i \cos D(i,j), \phi^i_{Mo(i)} \right) \tag{8}$$

A path that traverses against the flow is penalized more heavily than the one that is more in alignment with it. Also, the stronger the flow is, the larger the penalty becomes. Given path $P = (v_1, v_2, \ldots, v_n)$, with each edge $e_{i,j}$ connecting v_i and v_j, $e_{i,j}$ can be decomposed at block boundaries into multiple connected sub-edges. Let e_k be the sub-edge across the boundaries of and contained inside the block B_k, α be a constant, and Φ_{e_k} be the orientation of e_k, path cost for e_k is thus defined as:

$$f(e_k) = |e_k| \cdot \left(1 + \alpha W_k \sin^2(\Phi_{e_k} - \Phi_k)\right) \tag{9}$$

2.5 Sample the Scanpath

Scanning is conducted at subpixel resolution using bilinear interpolation. There are two reasons for this requirement. Firstly, the scanpath is not linear throughout its length, as its segments span across different blocks with differing projected lengths. Inevitably, sampling intervals are block-dependent and non-integral as a result. Secondly, during the experiments, we have found that using too coarse sampling interval would yield the results that are too sensitive to offset translation. In a coarser sampling interval, a small shift of the offset in template matching sometimes causes a relatively large jump in similarity scores. Using a finer sampling interval greatly reduces such effect. We opt to achieve the finer sampling interval through the use of subpixel sampling instead of using higher image resolution, since the latter requires more processing power at the earlier stages.

We take into account the orientation differences between path segments and their underlying blocks' flows. Assuming that there is no perspective distortion, if a path segment meets its block's flow at an angle theta, we need to sample the segment at an interval $T \sec \theta$, as opposed to T in the case of perfect alignment, to maintain a coherent projected distance between samples.

2.6 Decode with Dynamic Template Matching

The samples taken from the scanpath is decoded by matching them with a set of dynamically generated digit templates. Since the barcode is deformed, it is expected that the encoded digits are subject to position shifting and width scaling along the sample sequence L. Therefore, the objective is to find such shifting and scaling factors for each digit that would minimize matching errors. Specifically, we compute:

$$C(A) = \sum_{k=1}^{12} \min_{0 \leq n \leq 9} M(L, T_{n,k}(o_k, w_k)) \tag{10}$$

A is an arrangement $(o_1, o_2, \ldots, o_{12})$ where $o_k + w_k = o_{k+1}$ except for the case $k = 6$, in which $o_6 + w_6 < o_7 - w_g$. o_k and w_k are the position and the width of the kth digit on the sample sequence L, and w_g is the minimum width of the central guard pattern. $M(L, T_{n,k}(o_k, w_k)$ is a distance function between L and the template function $T_{n,k}$ of the number n at the digit k. We try to optimize over the global cost function $C(A)$ by finding an arrangement A that produces the minimum cost. There are positional and width constraints that limit the range of possible values for o_k and w_k to those in the zero-crossing point set that are within some margins around the ideal position and width of the kth digit. The optimization is carried out using dynamic programming over the arrangement space under such constraints.

A set of zero-crossing points is collected to aid in template matching. Starting from normalizing all samples using a small moving average window to offset local differences in contrast and brightness (Fig. 3), this process effectively 'straightens' and centers the sampled values around zero. These locally normalized samples are merely used for obtaining the zero crossing points, but cannot be reliably used for matching, as the small window size could strongly skew the samples. An example of a segment cut at normalized zero-crossing points and its matching template is shown in Fig. 3(c).

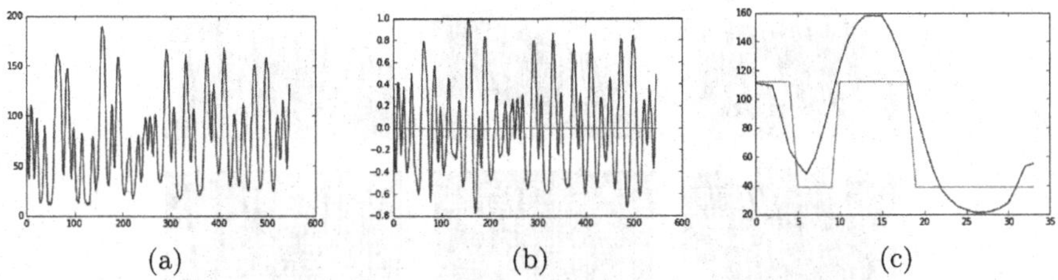

Fig. 3. (a) Sequences sampled along the scanpath; (b) Locally normalized sequences used in collecting zero-crossing points (center); (c) Digit samples (blue) and its matching template (green) showing the digit 0 with even parity

3 Experimental Result and Discussion

Some of the results are selected and shown in Fig. 4. Fig. 4(b) is a test of deformed barcode containing light reflection and merged lines caused by the camera viewpoint. Fig. 4(c) exhibits uneven brightness. Fig. 4(a)(d) and (e) contain reflections that cover most of the height of the barcodes, leaving narrow spaces that are still readable. Our method has successfully identified barcode regions that contain no lighting artifacts, and created a scanpath associated with each of them as shown underneath. In addition, the method successfully decoded all the scanpaths constructed. However, although the scanpath is obtained in Fig. 4(f), the decoding fails due to the extreme distortion in the middle part of the barcode.

The method in this research outperforms that presented in [6]. Fig. 4(a–d) and (f) are the same tests used in the previous work, the method of which fails to produce the results of decoding Fig. 4(c) and Fig. 4(f). The failure is caused by errors in binarization which is the technique our previous method relies on. Such drawback is alleviated in the current work. The distortion in Fig. 4(f) is still an issue, though.

4 Summary and Conclusions

This paper presents the scanpath-based approach for decoding deformed barcode images. The method utilizes the gradient-based HOG descriptor to provide information about the local orientation of the barcode fragments. A measure of orientedness, MOD, is also introduced and used to determine whether a block is a proper barcode region. Since MOD is orientation-invariant, we believe that it can also be used in localization of barcodes of arbitrary orientation, which is another area we have yet to explore in detail.

The method proposed in this article can be considered as the improvement of the previous work [6], which is based on a binarization technique. As this aspect of barcode reading is not yet widely studied, we have no other benchmarks to compare with. However, the current method still cannot overcome the deformation containing extreme distortion regions. Possible future directions include

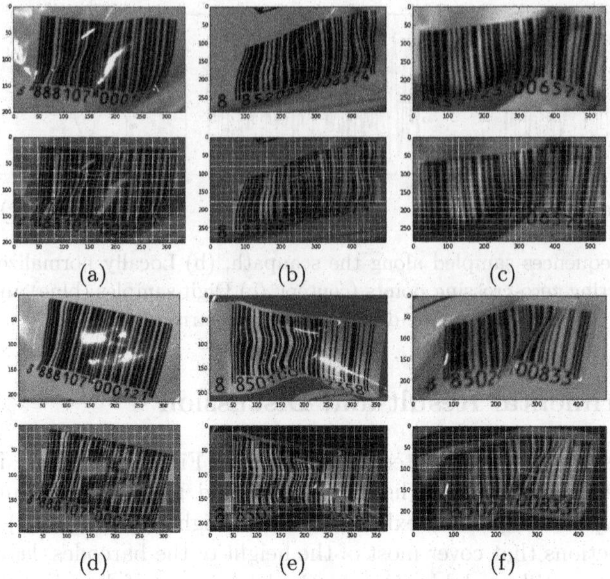

Fig. 4. Six different deformed barcodes with their block qualification results and the associated scanpaths

determining such regions either to avoid them or to devise a method that can decode them better.

References

1. Ohbuchi, E., Hanaizumi, H., Hock, L.A.: Barcode readers using the camera devices in mobile phones. Int. Conf. on Cyberworlds (2004)
2. Adelmann, R., Langheinrich, M., Flrkemeier, C.: A toolkit for bar code recognition and resolving on camera phones jump starting the internet of things. Workshop Mobile and Embedded Interactive Systems (MEIS06) at Informatik (2006)
3. Gallo, O., Manduchi, R.: Reading 1-D barcodes with mobile phones using deformable templates. In: IEEE Trans. on Pattern Analysis and Machine Intelligence, vol. 33(9), pp 1834–1843 (2010)
4. Chai, D., Hock., F.: Locating and decoding EAN-13 barcodes from images captured by digital cameras. Int. Conf. on Information, Communication and Signal Processing (2005)
5. Chen, L., Man, H., Jia, H.: On scanning linear barcodes from out-of-focus blurred images: a spatial domain dynamic template matching approach. In: IEEE Trans. on Image Processing, vol. 23, no. 6, pp. 2637–2650 (2014)
6. Yospanya, P., Limpiyakorn, Y.: Decoding deformed barcode images with scanpath walkthrough. In: International Journal of Multimedia and Ubiquitous Engineering, vol. 10, no. 1 (2015) (to appear)
7. Dalal, N., Triggs, B.: Histograms of oriented gradients for human detection. IEEE Conference on Computer Vision and Pattern Recognition (CVPR) (2005)

Detecting and extracting text in video via sparse representation based classification method

Bo Sun[1], Yang Wu[1], Feng Xu[1], Yongkang Xiao[1*], Jun He[1], Chen Chao[2]
[1] Beijing Normal University, College of Information Science and Technology, Xinjiekouwai Street No. 19, Beijing, China, 100875, Email: xiaoyk@bnu.edu.cn
[2] Naval Academy of Armament, Beijing China

Abstract – **This paper describes a new approach to detect and extract the video text more precisely and efficiently. The proposed approach combines the frame difference and the sparse representation based classification (SRC) method. The experiments demonstrated that the proposed method is effective and efficient for the both scenes, i.e. detecting and extracting the captions in a video the character in a sign.**
Keywords – **detecting and extracting text, frame difference, sparse representation based classification.**

1 INTRODUCTION

With the development of information technology, many video files are available. In many applications, we need get information from the printed text or characters in the videos by optical character recognition (OCR). While generic OCR systems cannot work very well directly with a graph-text mixed image. Thus, it is necessary to detect and extract text area in a video.

The main procedure of video text-access process is shown as Fig.1, but text areas in a video do not always appear in a particular single form. According to application, text characters can be roughly divided into two types: scene–text and artificial–text. Scene–text is contained by a shooting scene and is part of a scene, such as text on a billboard or a license plate, which are shown as Fig.2. For this type, the scale or direction of text is always random. With a moving camera, scene–text in an image maybe also moving, rotating, scaling, etc. So it is hard to locate them correctly. With some processing tools, artificial–text can be artificially added to an image such as news headlines, simultaneous voice subtitles, etc [Fig.3]. This type of text usually appears in videos and has strong contrast with background. The fonts are regular and the characters are generally arranged in a horizontal or vertical direction. Artificial–text plays an important role in frame indexing because it can reflect some specific messages in a video.

All methods can be roughly divided into four categories, namely edge features, region features, texture features and machine learning. Those are often based on two features: related features and independent features. Contrast, colors, text directions and sports positions are independent features while strokes density, strokes types and text sizes are related ones. The font, size, texture, color, background, orientation or distortion of a character is often changed in videos, so a simple algorithm can seldom provide satisfactory performance in all scenes.[1]

1 The work was supported in part by the Fundamental Research Funds for the Central Universities (2014 KJJCA15), the Research Project of China's High-resolution Earth Observation System (No.03-Y30B06-9001-13/15), the State Key Laboratory of Acoustics, Chinese Academy of Science under Grant SKLA201304 and the Fundamental Research Funds for the Central Universities under Grant (2013 NT55).

K.J. Kim (ed.), *Information Science and Applications,*
Lecture Notes in Electrical Engineering 339, DOI 10.1007/978-3-662-46578-3_41

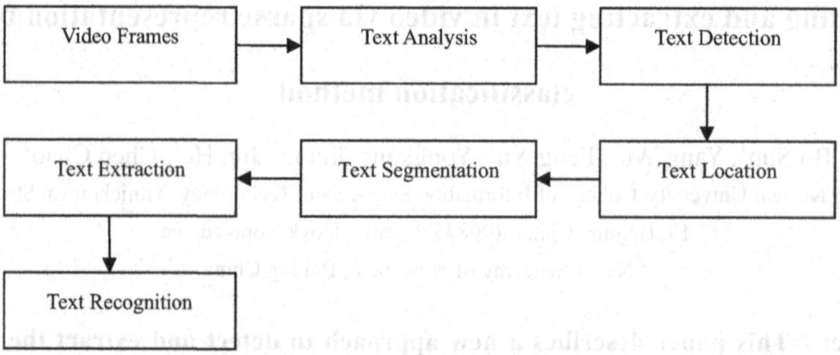

Fig.1. Procedure of video text recognition

Fig.2. Scene-text

Fig.3. Artificial-text

Early in 1994, Effelsberg and Lienhart proposed an algorithm based on color to locate text areas and then segment them by anisotropic diffusion. However, it needs to split and merge each frame of the video so it has high complexity and low efficiency [1]. In 1997, Michael et al. proposed a method to extract information from an audio or video file. They thought text generally has dense edge rectangle which depended on its scale, which has limited its application on the text with specific font size [2]. In 2000, Huiping Li and David proposed a text detection method based on wavelet transformation and neural network. They assumed the texture area was different from background area and it has poor accuracy in videos with complex background [3]. In 2002, Xiaoou Tang and Bo Luo proposed a fuzzy clustering neural network by the difference of histogram between neighboring frames which could only handle a particular video model well [4]. In 2003, Wong and Chen stated the importance and necessary of a common procedure and test data for estimating the performance of video text detection methods [5]. For this issue, Xiansheng Hua and Wenyin Liu proposed an automatic protocol in 2004, which is time-consuming [6]. In 2009, Palaiahnakote Shivakumara, Trung and Chew proposed a method used wavelet single level decomposition LH, HL and HH for computing features, which will be clustered with k means clustering method to classify the text pixels from the background of the frames, but it has not been extended to fix the bounding boxes for text lines with arbitrary direction [7]. In 2011, Palaiahnakote proposed a Max-Min clustering concept to obtain text cluster from the normalized absolute gradient feature matrix of the video text line image. It was based on the text height difference at character boundary column smaller than the other columns [8]. In 2013, Yusufu proposed a scheme to roughly identify candidate text regions with stroke model and morphological operation, which does not work well when texts are in complex non-linear motion [9].

Considering text areas with complex background, we analyze the performance of a video file and text area, and propose a framework with SRC to detect and extract text areas quickly and accurately.

The remainder of this paper is organized as follows. Section 2 gives a detailed description on our method. The experimental results are presented in Section 3 and we summarize this paper in Section 4.

2 METHODOLOGY

Obviously, a video file is composed of multiple frames. Frame difference, which is calculated as the grayscale difference between neighboring frames, may reflect some information variation. Thus, we may utilize it to detect and extract the text area quickly. Considering the regularity of print, we may apply SRC to detect and extract text areas.

2.1 FRAME DIFFERENCE

Alike with background subtraction or optical flow field method, frame difference is often used to detect the moving target in a video [10, 11]. At present, the main methods are based on two-frame-difference and three-frame-difference. As an improvement of former, three-frame-difference includes following steps: (1) select three consecutive frame images; (2) calculate the difference between neighboring frames respectively; (3) convert the difference to binary image by a suitable threshold; (4) process two binary images by AND operation to get common parts. However, three-frame-difference cannot fully detect all mutative text areas. So in this stage, we use two-frame-difference method to detect and extract them between neighboring frames. We set up the first frame as our known model. By comparing the number of statistics of white pixels with a threshold, we judge whether the previous results are applied or not. Then the results will be divided into three cases: (1) the neighboring frames are nearly same; (2) the neighboring frames are some different; (3) the neighboring frames are completely different.

For first case, we take the previous frame as unchanged and base to the after. For second case, we firstly treat the difference frame with preprocessing and processing stage, and then summary both the results of the previous frame and the difference frame, and finally merge the results as the after one. For last case, we treat the both frames separately with preprocessing and processing.

2.2 PREPROCESSING

In traditional machine learning methods, some researchers extract the desired text features directly in a whole frame without any preprocessing method. However, most of these features are only concentrated in few regions of interest (ROI). Accordingly, Using preprocessing method to recognize ROI firstly is critical and essential because it can reduce experimental time complexity. Since abundant text information concentrating on the edge of ROI, we detect and extract the edge features firstly by edge recognition operators in an original grayscale image. Prevalent operators applied in state-of-art edge detection methods include Sobel, Roberts, Canny, Prewitt, Laplace, etc [12]. Then many small gaps and isolated points occur in our operated binary image. They will hinder connectivity domain analysis which appears later. So we utilize dilation and erosion of morphological analysis to reduce strong noises and fractured strokes, which can effectively eliminate the small clearance in a binary image. After connected component analysis, many small connected blocks can be obtained which include three possibilities: text area, non-text area and partial-text area. We segment them from the original frame as all candidates in processing stage.

2.3 PROCESSING

A large number of background regions and partial text regions are recognized as text regions after preprocessing. Fortunately, both artificial-text and scene-text can be segmented out. In order to improve the accuracy, we will execute a further processing after preprocessing to distinguish the text area from other candidates correctly. Traditional methods generally base on machine learning algorithms, such as match template (MT), support vector machine (SVM) [13], artificial neural network (ANN) [14], etc. Though effectively, they still have limitation, for example, MT is sensitive for stroke shape and thickness sizes directions and have low computation efficiency, SVM is only suit for a small amount of training samples, and ANN has complex architecture with poor generality and cannot guarantee the classification results "absolutely" correct.

Considering the regularity of the print, we adopt SRC in this stage. Considering a dictionary $\mathbf{D}=[\mathbf{D}_1,\mathbf{D}_2]$, where $\mathbf{D}_1=[\mathbf{d}_{11},\mathbf{d}_{12},\ldots,\mathbf{d}_{1N_1}]$, $\mathbf{D}_2=[\mathbf{d}_{21},d_{22},\ldots,d_{2N_2}]$. Each column $\mathbf{d}_{1n}(n=1,\ldots,N_1)$ in D_1 and $\mathbf{d}_{2n}(n=1,\ldots,N_2)$ in D_2 denotes a text feature and a background feature respectively. Given a query $\mathbf{Y}=[\mathbf{y}_1,\mathbf{y}_2,\ldots,\mathbf{y}_m]$, we solve the sparse representation problem for each query feature $\mathbf{y}_i(i=1,\ldots,m)$.

$$\hat{\mathbf{x}}_i=\arg\min_{\mathbf{x}_i}\|\mathbf{x}_i\|_1,\ s.t.\ \mathbf{y}_i=\mathbf{D}\mathbf{x}_i, i=1,\ldots,m \tag{1}$$

where $\|\cdot\|_1$ denotes the l_1 norm of a vector and $\mathbf{x}_i$ is the sparse coefficient of the query feature y_i.

The identity of the query Y is determined by the following multi-task sparse representation based classification method:

$$\min_c r_c(\mathbf{Y})=\frac{1}{m}\sum_{i=1}^{m}\|\mathbf{y}_i-\mathbf{D}\delta_c(\hat{\mathbf{x}}_i)\|_2^2 \tag{2}$$

where $\|\cdot\|_2$ denotes the l_2 norm of a vector and $\delta_c(\cdot)$ is a function which selects the coefficients corresponding to the *cth* class.

Obviously, an appropriate dictionary is the key for SRC. Higher atomic differences in a dictionary contribute to a higher robustness of SRC. Thus, we will get some image blocks of different stroke to build the dictionary for detecting text area. We block a text image by a 9×9 sliding window from left to right and from top to bottom with a step of 3 pixels. Then we convert each 9×9 block into an 81-dimensional column vector as an original atom in our dictionary. Our two over-complete dictionaries (text dictionary D_1 and background dictionary D_2) are from Fig.4.

Dictionary training has a crucial influence on the ultimate classification in SRC. Michael Elad and Michal Aharon have proposed k means singular value decomposition (K-SVD) algorithm which is an iterative method that alternates between sparse coding of the examples based on the current dictionary and a process of updating the dictionary atoms to better fit the data [19]. We train our two dictionaries via K-SVD algorithm and our experimental results testify they best suit a set of given signals.

Considering different scale, we should regulate the candidates with a criterion and each standard character is 20×20 pixels. We scale our candidates with a uniform height of criterion, meanwhile equally conducted a proportional length adjustment.

After scaling processing, we execute the identical procedure on our candidates via our sliding window and then gain a set of testing samples which have same dimension with training atoms. But sparse representation optimization problem is a NP-hard problem for a redundant dictionary. Therefore, we utilize orthogonal matching pursuit (OMP) method to solve this problem [20, 21, 22]. As the improvement for MP, OMP has faster convergence rate and optimized iteration results [23, 24]. For each testing sample, we express it by two dictionaries and get two groups of sparse coefficients. As result, we determine a judgment by the coefficient sparsity, that is, the testing sample will belong to the class of atoms with sparser coefficients. Finally, we analyze the statistic results from one same candidate and finally classify it via voting with the threshold 50%.

3 EXPERIMENTAL RESULTS

Fig.4. Samples of text dictionary (left) and background dictionary (right)

In our experiments, we adopt a news video as our experimental data source. We utilize OpenCV2.3.1 to decode video into 45011 frames.

We randomly choose 3000 frames as our dictionary source and our two dictionaries are showed below [Fig.5]. In Fig.5, the left is text dictionary and the right is background. Both of them have 1024 atoms and each atom is an 81×1 vector corresponding with a 9×9 block.

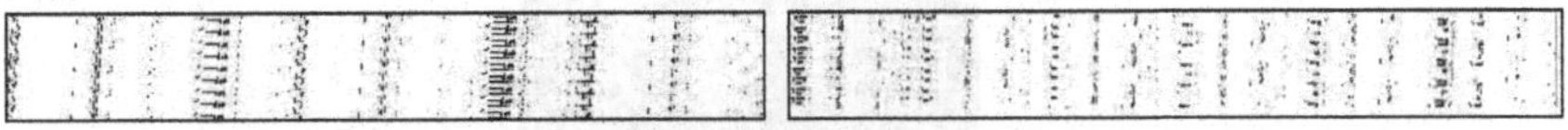

Fig.5. Text dictionary (left) and Background dictionary (right)

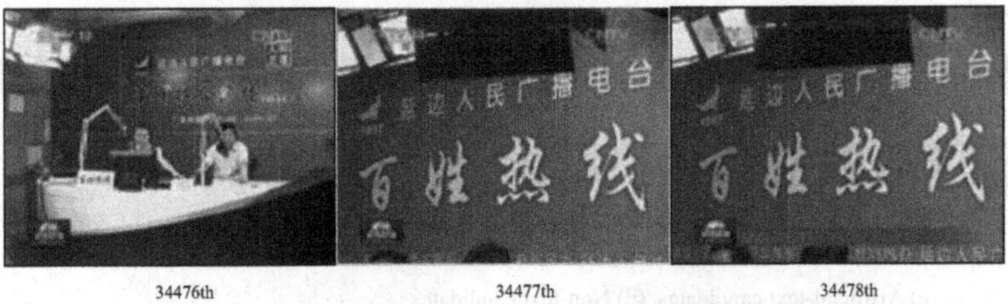

Fig.6. Neighboring frames in the test video

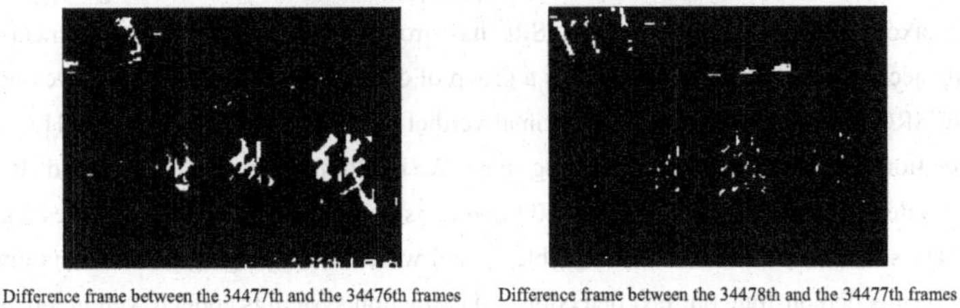

Fig.7. Difference frame between neighboring frames

Fig.6 shows the 34476th, 34477th, 34478th frame and our practical results by frame difference are showed in Fig.7. Table 1 reflects the number of white pixels in Fig.7.

Table 1. Number of white pixels in difference frame

	White Pixels	Threshold = 500	Case
34477 - 34476	653	>	3
34478 - 34477	242	<	2

So we should separately process the 34476th and 34477th frames by preprocessing and processing stage. Fig.8 shows the experimental results of 34477th in preprocessing stage.

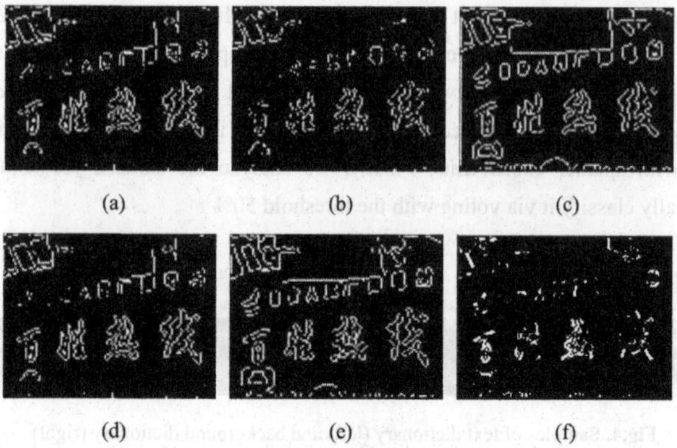

Fig.8. Preprocessing results of the 34477th frame with different operators as: (a) Sobel, (b) Roberts, (c) Canny, (d) Prewitt, (e) Laplace, (f) Dilation and Erosion

Fig.9. Segment candidates in the original frame

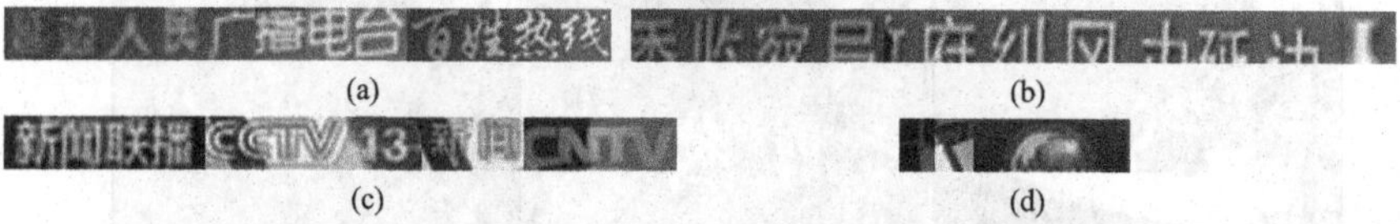

Fig.10. Classification of the segmenting results as: (a) Scene-text candidates, (b) Partial-text candidates, (c) Artificial-text candidates, (d) Non-text candidates

In order to test the interference from movement of pixels, we change our sliding step with 2 pixels, 4 pixels, 5 pixels and the results indicate that SRC has strong robustness on image movement.

While accuracy and efficiency is always a group of contradictions and it is also a vexing problem in OMP of SRC method and we apply an optimal verdict in our experiment. From Fig.11, we can find that 10 iterations are the best in our processing stage. We randomly select 500 frames and choose about 1500 candidates which conclude 500 texts, 500 non-texts and 500 partial-texts as our testing samples. The final classification is showed below [Table 2] and we consider judging partial–texts into text are right. The results demonstrate our text detection and extraction system is applied well in videos.

Table 2. Experimental correct rate

	500 texts	500 non-texts	500 partial-texts
Judging text	477	49	381
Judging background	23	451	119
Correct rate	95.4%	90.2%	76.2%

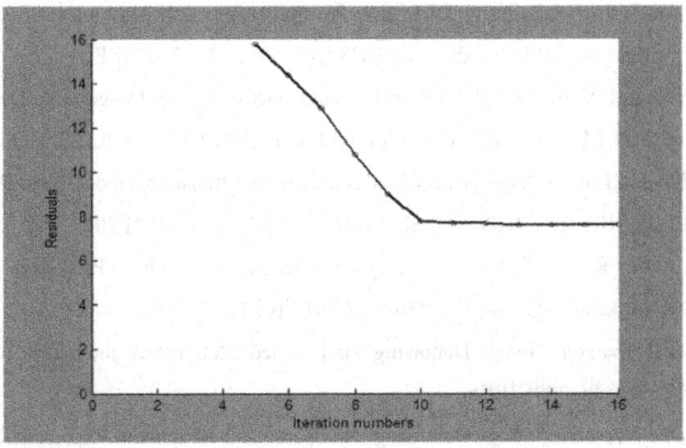

Fig.11. Accuracy and efficiency in OMP of SRC

4. CONCLUSION

In this paper, we propose a new approach to detect and extract video text more precisely and efficiently. The proposed protocol can not only be used in scene-text detection but also can extract artificial-text. We have chosen about 50 different types of web images and the results are also promising. However this method cannot be applied to various partial-texts. In future, we plan to overcome this problem and construct a compositive system with our method. With the development of OCR technology, we will also focus more attention on other scarce fonts.

Reference:

[1] Rainer Lienhart, Silvia Pfeiffer, Wolfgang Effelsberg. : The MoCA Workbench Support for Creativity in Movie Content Analysis. In: 0-8186-7436-9/96 1996 IEEE

[2] Michael A. Smith, Takeo Kanade. : Video Skimming and Characterization through the Combination of Image and Language Understanding Techniques. In: 1063-6919/97 1997 IEEE

[3] Huiping Li, David Doermann. : A Video Text Detection System Based on Automated Training. In: 0-7695-0750-6/00 2000 IEEE

[4] Xiaoou Tang, Bo Luo, Xinbo Gao, Edwige Pissalod, Hong jiang Zhang. : Video Text Extraction Using Temporal Feature Vectors. In: 0-7803-7304-9/02 C2002 IEEE

[5] E. K. Wong, M. Chen. : A New Robust Algorithm for Video Text Extraction. In: Pattern Recognit., vol. 36, no. 6, pp. 1397–1406, 2003

[6] Xiansheng Hua, Wenyin Liu, Hongjiang Zhang. : An Automatic Performance Evaluation Protocol for Video Text Detection Algorithms. In: IEEE TRANSACTIONS ON CIRCUITS AND SYSTEMS FOR VIDEO TECHNOLOGY, VOL. 14, NO. 4, APRIL 2004

[7] Palaiahnakote Shivakumara, Trung Quy Phan, Chew Lim Tan. : A Robust Wavelet Transform Based Technique for Video Text Detection. In: 978-0-7695-3725-2/09 2009 IEEE DOI 10.1109/ICDAR.2009.83

[8] Palaiahnakote Shivakumara, Souvik Bhowmick, Bolan Su, Chew Lim Tan. : A New Gradient based Character Segmentation Method for Video Text Recognition. In: 1520-5363/11 IEEE DOI 10.1109/ICDAR.2011.34

[9] Tuoerhongjiang Yusufu, Yiqing Wang, Xiangzhong Fang. : A Video Text Detection and Tracking System. In: 978-0-7695-5140-1/13 2013 IEEE DOI 10.1109/ISM.2013.106

[10] Jiwoong Bang, Daewon Kim, Hyeonsang Eom. : Motion Object and Regional Detection Method Using Block-based Background Difference Video Frames. In: 978-0-7695-4824-1/12, 2012 IEEE DOI 10.1109/RTCSA.2012.58

[11] Ralph Ewevth, Bemd Freisleben. : Frame Difference Normalization : An Approach to Reduce Error Rates of Cut Detection Algorithms for MPEG Videos. In: 0-7803-7750-8/03/ 02003 IEEE
[12] Nick Kanopoulos, Nagesh Vasanthavada, Robert L. Baker. : Design of an Image Edge Detection Filter Using the Sobel Operator. In: IEEE JOURNAL OF SOLID-STATE CIRCUITS, VOL. 23, NO. 2, APRIL 1988
[13] Chih-Wei Hsu, Chih-Jen Lin. : A Comparison of Methods for Multiclass Support Vector Machines. In: IEEE TRANSACTIONS ON NEURAL NETWORKS, VOL. 13, NO. 2, MARCH 2002
[14] Kumar Abhishek, Abhay Kumar, Rajeev Ranjan, Sarthak Kumar. : A Rainfall Prediction Model using Artificial Neural Network. In: 978-1-4673-2036-8/12 2012 IEEE
[15] Michael Elad, Michal Aharson. : Image Denoising Via Learned Dictionaries and Sparse representation. In: (CVPR'06) 0-7695-2597-0/06 2006 IEEE
[16] Michael Elad, Michal Aharson. : Image Denoising Via Sparse and Redundant Representations Over Learned Dictionaries. In: IEEE TRANSACTIONS ON IMAGE PROCESSING, VOL. 15, NO. 12, DECEMBER 2006
[17] Matan Protter, Michael Elad. : Image Sequence Denoising via Sparse and Redundant Representations. In: IEEE TRANSACTIONS ON IMAGE PROCESSING, VOL. 18, NO. 1, JANUARY 2009
[18] Yi Ma, Jianchao Yang, John Wright, Thomas S. Huang. : Image Super-Resolution via Sparse Representation. In: IEEE TRANSACTIONS ON IMAGE PROCESSING, VOL. 19, NO. 11, NOVEMBER 2010
[19] Michal Aharon, Michael Elad, Alfred Bruckstein. : K-SVD: An Algorithm for Designing Overcomplete Dictionaries for Sparse Representation. In: IEEE TRANSACTIONS ON SIGNAL PROCESSING, VOL. 54, NO. 11, NOVEMBER 2006
[20] Wan-Fung Cheung, Yuk-Hee Chan. : A Fast Two-stage OMP Algorithm for Coding Stereo Image Residuals. In: 0-7803-7750-8/03/ 82003 IEEE
[21] Hind O. Al-Misbahi, Arwa Y. Al-Aama. : The Overlay Multicast Protocol (OMP): A Proposed Solution to Improving Scalability of Multicasting in MPLS Networks. In: 0-7695-2842-2/07 IEEE
[22] Tianyun Wang, Changchang Liu, Li Ding, Hongchao Lu, Weidong Che. : Sparse Imaging Using Improved OMP Technique in FD-MIMO Radar for Target off the Grid. In: 2013 Asia-Pacific Conference on Synthetic Aperture Radar
[23] Inon Zuckerman, Ariel Felner. : The MP-MIX Algorithm: Dynamic Search Strategy Selection in Multiplayer Adversarial Search. In: IEEE TRANSACTIONS ON COMPUTATIONAL INTELLIGENCE AND AI IN GAMES, VOL. 3, NO. 4, DECEMBER 2011
[24] Siéler, Jean Pierre Dérutin, Alexis Landrault. : A MP-SoC Design Methodology for the Fast Prototyping of Embedded Image Processing System. In: 978-1-4244-8631-1/10/2010 IEEE

Part IV

A Cost Model for Client-Side CaaS

Chaturong Sriwiroj[1] and Thepparit Banditwattanawong[2]

[1,2] School of Information Technology, Sripatum University, Bangkok, Thailand,
[1] studychr8@gmail.com

Abstract. Deploying cache-as-a-service (CaaS) at the corporation level reduces network bandwidth expense and improves performance. Careful consideration must be given when choosing CaaS to achieve cost-effectiveness. This requires the service model of CaaS allowing custom-made SLA. This paper presents the flexible economic model of client-side CaaS as an early attempt in the field. The model has been evaluated to be promising based on a realistic scenario.

Keywords: Cloud computing, Cache, Service Model, Multitenancy

1 Introduction

Almost all sectors increasingly engage cloud computing [1] paradigm in various deployment models including on-premise and off-premise private clouds. The amount of personal and business data in clouds has been growing rapidly and being shared among numerous clients. This situation affects the network access latencies and probably data-transfer monetary cost of cloud services. It can be relieved by leveraging cloud cache as a service (CaaS), which is deployed on a client side [2] rather than a server side [3]. Although the client-side CaaS has been technically well established, it however lacks a service models to be turned into a real service. The service models to deliver practical client-side CaaS are two-fold: performance model and economic model. The former renders technical choices of efficiency CaaS subscribers can selectively obtain, while the latter gives both CaaS providers and consumers service pricing information. This paper focuses and proposes an economic service model for client-side CaaS.

2 Related Works

There have been a number of studies conducted to investigate the issue of cache service models. The focus of these investigations includes technological and economic dimension. With respect to CaaS models in economic dimension, Dash, Kantere and Ailamaki [4] proposed an economic model for adaptive cloud cache that supports large scientific inquiries services. The cost model considers all queries and necessary infrastructure expenditure such as bandwidth, network, disk space and CPU times. Chockler, Laden and Vigfusson [5] proposed the economic model of cloud cache service for charge balancing between customer benefit and cost of the service. Kan-

K.J. Kim (ed.), *Information Science and Applications*,
Lecture Notes in Electrical Engineering 339, DOI 10.1007/978-3-662-46578-3_42

tere et al. [6] proposed an economic model and dynamic pricing scheme designed for a cloud cache, which offers querying services. Han et al. [3] proposed an economic model of cloud cache service for value pricing between performance and operation. However, all of these studies are the models that serve server-side CaaS and differ from our work, which concentrates on a novel client-side CaaS model. Sriwiroj and Banditwattanawong [9] proposed an economic model for client-side cloud caching service and evaluated the model using a data set collected from a research organization whereas this paper uses the new data set of big-data firm.

3 Cloud Cache

Caching is a classic idea [7] that is able to improve server response times and reduce amount of the usage of network bandwidth that enhance overall system performance. Proxy caching on the client-side [8] stores data in client locality to improve the response time and reduce network bandwidth usage. The concept of client-side cloud cache is also similar to that of caching proxy except that cached data is of cloud instead of traditional WWW. The main difference of cloud data from that of WWW is average size [2]. "The cloud cache basically inherits the capability of traditional forward web caching proxy since cloud data is also delivered by using the same set of HTTP/TCP/IP protocol stack as in WWW" [2]. Fig. 1(left) shows a client-side cloud cache use case as a CaaS deployment model. The CaaS is operated by an outsourcer known as CaaS provider.

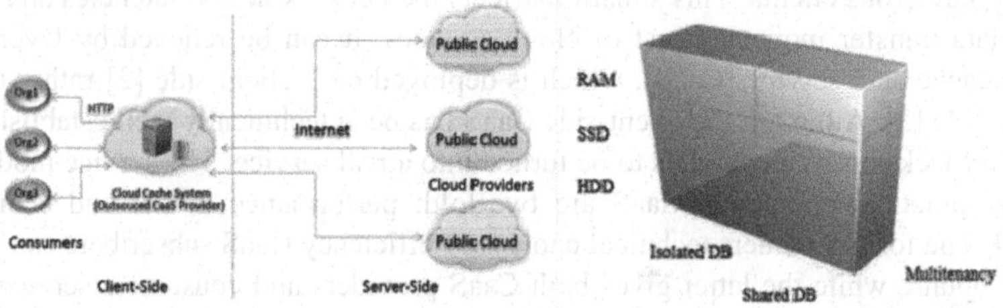

Fig. 1. Deployment of Client-Side CaaS (left) and its overall service model (right)

4 The Service Models of Cloud Cache

This section presents the overall service model in a technical perspective of client-side CaaS illustrated in Fig. 1 (right). The first dimension represents potential cache storage technologies of different access times. The second dimension represents choices of cached-data sharing either cross-consumer shared (shared DB) or consumer-specific nonshared (isolated DB). The other dimension represents the only caching logic derived from the standard characteristic of cloud computing. Based on the combination of these three dimensions, the model contains totally six types of CaaS that consumers can choose to subscribe: (1) RAM Mutitenancy Isolated DB; (2) RAM

Mutitenancy Shared DB; (3) SSD Mutitenancy Isolated DB; (4) SSD Mutitenancy Shared DB; (5) HDD Mutitenancy Isolated DB; (6) HDD Mutitenancy Shared DB. These types of CaaS offer different obtained performances and service charges to subscribers. On the other hand, they incur different performance guarantees and costs. To realize the service charges or the costs of the offerred CaaS types mandates an economic service model, which is described in the following section as our main contribution.

5 A Proposed Economic Service Model

The proposed CaaS economic model serves as service pricing guidance for providers as suggested by the three technical dimensions in Fig. 1 (right) The model basically takes into account investment costs and desired profit margin to determine what reasonable service charge is. The model involves several cost factors as depicted and defined in Table 1.

Table 1. Model parameter definitions

Notation	Description	Notation	Description
PR	Pricing model	*SO*	Set of possible data object is = $\{o_i, \ldots o_n\}_{i=1\text{->}n}$
AC	Total actual cost	o_i	Data object$_i$
PF	Profit	To_i	Cost of transfer data object$_i$ from public cloud server
CapEx Parameters		$S_c o_i(t)$	Cost of data object$_i$ using cache per space per times, $\{c{=}RAM,SSD,HDD\}$
d	Set offixed assets, {*d=infrastructures, facilities, hardware equipments, licenses, servers, networks, storages, cache medias*}	$Sg_c o_i(\Delta t)$	Cost of storing data object$_i$ from time t_{store} to t_{evict}, $\{c{=}RAM,SSD,HDD\}$
D_d	Depreciation per annum of asset	Δt	Elapsed time t_{evict} - time t_{store}
C_d	Cost of fixed assets	Oo_i	Cost of transfer-out of data object$_i$ from cache to consumers
R_d	Estimated remain value at the end of the asset	$M_p o_i$	Cost of managing CaaS process, {*p=MTIdb, MTSdb*} where MTIdb stands for Multitenancy Isolated DB and MTSdb stands for Multitenancy Shared DB
p_d	Number of periods (*year*)	$Cmiss_{c,p} o_i$	Cost of caching service of data object$_i$ in case cache-miss, $\{c{=}RAM,SSD,HDD\}$, $\{p{=}MTIdb, MTSdb\}$
CapEx	Capital expenditures per period (*month*)	$Chit_{c,p} o_i$	Cost of caching service of data object$_i$ in case cache-hit, $\{c{=}RAM,SSD,HDD\}$, $\{p{=}MTIdb, MTSdb\}$
PfCapEx	Profit generated from CapEx	*pm*	Desired profit margin in percentage
PRCapEx	Pricing model of CapEx per month	$Pfmisso_i$	Profit of caching service of data

			$object_i$ in case cache-miss
P	Projected number of consumer sites	$Pfhito_i$	Profit of caching service of data $object_i$ in case cache-hit
OpEx Parameters		c	Cache media RAM or SSD or HDD
	Parameters used in Fixed cost	p	Managing process MTIdb or MTSdb
am	Set of principal for, {*am=salaries, wages, utilities, hardware maintenance, software maintenance*}	h	Number of cache hits per month
A_{am}	Payment amount per period	m	Number of cache misses per month
P_{am}	Initial principal	j	Cache hit times, {*1,2,3,...h*}
r_{am}	Interest rate per period	l	Cache miss times, {*1,2,3,...m*}
n_{am}	Total number of payments or periods *(month)*	$PRmiss_{c,p}o_i$	Pricing of CaaS of o_i in case cache-miss
$OpExf$	OpEx fixed cost per month	$PRhit_{c,p}o_i$	Pricing of CaaS of o_i in case cache-hit
$PfOpExf$	Profit from OpExf per month	$PROpEx$	Pricing part of OpEx per month
$PROpExf$	Pricing part of OpEx fixed cost per month	$PROpExv$	Pricing part of OpEx variable cost per month
	Parameters used in Variable cost		

In order to determine an effective economic or pricing model for client-side CaaS, it is necessary to understand the cost bases in providing the service that is two-fold. First, the Capital Expenditure (CapEx) is expenses incurred for initiating business or service system to acquire or upgrade fixed long-term assets like service infrastructure (software and hardware including cache storage media), supporting facilities, constructions. The CapEx must also reflect the fact that the assets are usually depreciated in value over time or decreased in its value every period based on certain rules. Second, the Operational Expenditure (OpEx) is expenses incurred in the course of ordinary business such as general and administrative expenses that is the money spends in order to run or operate a business or service system over a short period like staff salaries, wages, utilities, maintenance. In CaaS economic model, the OpEx can be fixed cost (which is periodic constant cost) and variable cost (which is a periodic cost that varies with operated service output). The actual total cost comprises CapEx and OpEx. Based on this described rationale, we derived the following basic equation.

$$PR = AC + PF \tag{1}$$

The CaaS pricing model also conforms to *Eq. 1.* Since AC is calculated based on both CapEx and OpEx on a monthly basis, it is necessary to clarify the CapEx part and the OpEx part of CaaS, respectively, as follows. The CapEx part of CaaS must be translated into depreciation over the life of the service components. Depreciation is a method for allocating the cost of fixed asset uniformly over the life of an asset. The formula to calculate depreciation of an asset for a full period is composed cost includes the initial and any subsequent expenditure, residual value is the estimated remain value at the end of useful life of the asset, and useful life in number of periods is the estimated time period an asset is expected to be utilized. The *Eq. 2* shows a standard depreciation calculation. $D_d = (C_d - R_d) / p_d$ (2)

Hence, the depreciation of d = Set of fixed assets, {*d=infrastructures, facilities, hardware equipments, licenses, servers, networks, storages, cache medias*} per month is

$$D_d = ((C_d - R_d) / p_d) / 12 \tag{3}$$

All of D_d depreciation costs can be summed up and average by P as CapEx per month is $CapEx = \sum D_d / P$ (4)

The desired profit of CaaS of CapEx per month is

$$PfCapEx = (CapEx \times pm) / 100 \tag{5}$$

Hence, the CaaS pricing model part of CapEx per month is

$$PRCapEx = CapEx + PfCapEx \tag{6}$$

The OpEx part of CaaS is divided into fixed cost and variable cost. First, the fixed cost must be in the form of amortization, a process of accounting for an amount over a period. The basic amortization formula is described below.

$$A = (P \times r \times (1 + r)^n) / ((1 + r)^n - 1) \tag{7}$$

Hence, the amortization cost of am = Set of principal for, {*am=salaries, wages, utilities, hardware maintenance, software maintenance*} per month is

$$A_{am} = (P_{am} \times r_{am} \times (1 + r_{am})^{n}{}_{am}) / ((1 + r_{am})^{n}{}_{am} - 1) \tag{8}$$

All of A_{am} amortization costs can be summed up and average by P as total OpEx fixed cost per month is

$$OpExf = \sum A_{am} \ / \ P \tag{9}$$

The desired profit of caching service of OpExf per month is

$$PfOpExf = (OpExf \times pm) / 100 \tag{10}$$

Hence, the CaaS pricing model part of OpEx fixed-cost per month is

$$PROpExf = OpExf + PfOpExf \tag{11}$$

Second, the variable cost of the OpEx part must be calculated by considering data object$_i$. The cost of having data object$_i$ inside cache storage from time t_{store} to time t_{evict} is

$$Sg_c o_i(\Delta t) = To_i + S_c o_i(\Delta t) \tag{12}$$

Cost of managing process of object$_i$ is $M_p o_i$ and cost of transfer-out of data object$_i$ from cache to consumers is Oo_i. It is basis the cost in caching that is two-case. first, not found object in cache known as cache-miss, second, found object in cache known as cache-hit. Hence, The OpEx variable cost of caching data object$_i$ in case cache-miss is $Cmiss_{c,p} o_i = M_p o_i + Sg_c o_i(\Delta t) + Oo_i$ (13)

The desired profit of $Cmiss_{c,p} o_i$ is $Pfmisso_i = (Cmiss_{c,p} o_i \times pm) / 100$ (14)

The OpEx variable cost of caching data $object_i$ in case cache-hit is

$$Chit_{c,p}o_i = M_p o_i + Oo_i \tag{15}$$

The desired profit of $Chit_{c,p}o_i$ is $Pfhito_i = (Chit_{c,p}o_i \times pm) / 100$ (16)

The CaaS pricing model part of OpEx variable-cost of data $object_i$ in case cache-miss is

$$PRmiss_{c,p}o_i = Cmiss_{c,p}o_i + Pfmisso_i \tag{17}$$

The CaaS pricing model part of OpEx variable-cost of data $object_i$ in case cache-hit is

$$PRhit_{c,p}o_i = Chit_{c,p}o_i + Pfhito_i \tag{18}$$

Thus, the CaaS pricing model part of OpEx variable-cost, where h is number of cache hits per month, m is number of cache misses per month, j is cache hit times and l is cache miss times. For subscribed types of CaaS for caching data $object_i$ per month per consumer site by average is

$$PROpExv = \left(\sum_{l=1}^{m} PRmiss_{c,p}o_i + \sum_{j=1}^{h} PRhit_{c,p}o_i \right) / P \tag{19}$$

The precise value of $PROpExv$ of each consumer site can be figured out by simply omitting P.

Hence, the CaaS pricing model part of OpEx per month is

$$PROpEx = PROpExf + PROpExv \tag{20}$$

Finally, the pricing model of client-side CaaS per month is

$$PR = PRCapEx + PROpEx \tag{21}$$

6 Evaluation

A representative scenario used to assess the model is digital content firm where big data like high-definition videos and pictures are its products. The scenario is carefully analyzed by setting up the feasible value of variables and simulation to bring about plausible results. We used one server node as a caching server, which was equipped with a dual-core CPUs Intel Xeon 2.40 GHz, RAM 80 GB, SSD100 GB, HDD 1 TB and Red Hat Enterprise Linux 6.2 64-bits. The server-node instantiated caching server atop hypervisor Xen 4.3.0 to serve limited number of consumers. The approximated values for each cost base item in details are in Table 2:

Table 2. Input parameter values for CapEx and OpEx costing

Relevant cost item	Approximated Values	Approximated Usage	Description
Infrastructures	$20,000	15 year	Infra of power and cooling
Facilities	$10,000	25 year	Facilities construction
Equipments	$1,000	5 year	Cables, Racks
OS software licenses	$700 per annum	per annum	Red Hat Enterprise Linux
Server	$1,000	5 year	8 CPUs Intel Xeon
Network switch	$2,500	5 year	CISCO 500 users20 mbps

Storage	$500	3 year	1TB, 8X faster backups
RAM	$1,000	3 year	DDR4 clocked 3200 MHz
SSD	$800	3 year, 10,000 Mb/s	1.2TB SSD (4 x 300GB)
HDD	$650	3 year, 1TB/month	SATA interface (Gb/s)
Consumer sites	50 sites		Projected consumers sites
Systems administrators	$10,000 per annum	8 hrs/day	Salary of staff
Utilities	$2,000 per annum	24 hrs/day	Electric, telephone
Hardware maintenance	$500 per annum		
Software maintenance	$700 per annum		
$S_c o_i(t)$	HDD $0.00000000021 per GB per second		Data $object_i$ using cache space per byte unit (b) per time (t), {c=HDD}
To_i	$0.00000000002 per second		Transfer data $object_i$ from public server
$M_p o_i$	MTIdb $0.00000007 per Hour, MTSdb $0.00000015 per Hour		Managing process, {p=MTIdb, MTSdb} included cost of CPU power consumption
$O_c o_i$	HDD $0.00000000002 per second		Transfer-out data $object_i$ from cache

We show how to calculate CapEx and OpEx, respectively, for the scenario. First, CapEx costs had been calculated using *Eq. 3& 4*. Thus, CapEx per month following *Eq. 4* = ($105.56 + $30.00 + $16.50 + $58.33 + $15.00 + $37.50 + $11.11 + $68.06) / 50 = $6.84. Profit from CapEx using *Eq. 5* = $6.84 × profit margin 10% / 100 = $0.68. Hence, PRCapEx per month using *Eq. 6* = $6.84 + $0.68 = $7.52.

Second, OpEx fixed cost had been calculated using *Eq. 8& 9*. Thus, the OpEx fixed cost per month following *Eq. 9* = (867.57 + $173.51 + $43.38 + $60.73) / 50 = $22.90. Profit from OpEx fixed cost using *Eq. 10* = $22.90 × profit margin 10% / 100 = $2.29. Hence, PROpExf per month using *Eq. 11* = $22.90 + $2.29= $25.19.

To figure out OpEx variable cost, we have made further measurement with respect to caching workload. The source of workload is 50 sites subscribing our CaaS of HDD-Mutitenancy-Isolated-DB option. The approximated values are in Table 3.

Table 3. Additional input parameter values for OpEx variable costing

Digital content data object	Object size (bytes)	Avg. Δt (sec.)	No. of objects	Cache space utilized (bytes)	No. of hits	No. of misses
High-definition video files	50 GB = 53,687,091,200	86,400	55	2,952,790,016,000	16	39
Animation	20 GB = 21,474,836,480	86,400	160	3,435,973,836,800	73	87
E-Learning content	15 GB = 16,106,127,360	86,400	1,078	17,362,405,294,080	453	625
Standard-definition video files	4.7 GB = 5,046,586,572.8	86,400	3,519	17,758,938,149,683	1,556	1,963

Thus, the CaaS pricing model part of OpEx variable-cost per month per consumer site by average using *Eq. 19* with total miss per month = 2714, total hit per month = 2098 and consumer sites $P = 50$ is

$$PROpExv = \left(\sum_{i=1}^{2714} PRmiss_{HDD,MTIdb}\, o_i + \sum_{j=1}^{2098} PRhit_{HDD,MTIdb}\, o_i \right) / 50,$$

$PROpExv = (\$857.383842668 + \$0.000000091) / 50 = \$17.15.$

Hence, the averaged CaaS pricing part of OpEx per month using *Eq. 20* is $PROpEx = PROpExf + PROpExv = \$25.19 + \$17.15 = \$42.34.$

Finally, the service charge of client-side CaaS per month using *Eq. 21* is $PR = PRCapEx + PROpEx = \$7.52 + \$42.34 = \$49.86$ by average.

7 Conclusion and Discussions

This paper presents a client-side CaaS economic model, which considers both CapEx and OpEx as well as fixed costs and variable costs. The model is simple and intuitive as an early attempt in the field. The important benefit of the model is its flexibility to support the different types of CaaS. The model has been validated in a real new environment.

References

1. Peter Mell and Timothy Grance. The NIST Definition of Cloud Computing. In *National Institute of Standards and Technology Special Publication* 800-145, 2011.
2. Thepparit Banditwattanawong and Putchong Uthayopas. A Client-Side Cloud Cache Replacement Policy. In: *ECTI Transactions on Computer and Information Technology: Special section on papers selected from ECTI-CON*, vol.8, no.2, pp.113-121, 2014.
3. Hyuck Han, Young Choon Lee, Woong Shin, Hyungsoo Jung, Heon Y. Yeom and Albert Y. Zomaya. Cashing in on the Cache in the Cloud. In *IEEE Trans Parallel and Distributed Systems*, vol. 23, no. 8, pp. 1387-1399, 2012.
4. Debabrata Dash, Verena Kantere and Anastasia Ailamaki. An economic model for self-tuned cloud caching. In *Data Engineering ICDE09 IEEE 25th International Conference on IEEE*, pp. 1687-1693, 2009.
5. Gregory Chockler, Guy Laden and Ymir Vigfusson. Design and implementation of caching services in the cloud. In *IBM Journal of Research and Development*, vol. 55, no. 6, pp. 9:1–9:11, 2011.
6. Verena Kantere, Debabrata Dash, Gregory Francois, Sofia Kyriakopoulou and Anastasia Ailamaki. Optimal Service Pricing for a Cloud Cache. In *IEEE Trans on Knowledge and Data Eng*, vol. 23, no. 9, pp. 1345-1358, 2011.
7. Lin Cao, Lifu Huang, Kai Lei, Zhiming Zhang and Lian-en Huang. Hybrid caching for cloud storage to support traditional application. In *Cloud Computing Congress (APCloudCC), IEEE Asia Pacific IEEE*, pp. 11-15, 2012.
8. Greg Barish and Katia Obraczka. World Wide Web Caching: Trends and Techniques. In *IEEE Communications Magazine, Internet Technology Series*, pp. 178-185, 2000.
9. Chaturong Sriwiroj and Thepparit Banditwattanawong. An Economic Model for Client-Side Cloud Caching Service. In: *7th International Conference on Knowledge and Smart Technology (KST) on IEEE*, 2015. (accepted)

A Reliability Analysis of Cascaded TMR Systems

Hyun Joo Yi, Tae-Sun Chung, and Sungsoo Kim

Ajou University, Computer Engineering, Suwon, Republic of Korea
{sigran, tschung, sskim}@ajou.ac.kr

Abstract. Cascaded TMRs (Triple Module Redundancy) have been used in various areas like pipeline process, redundant Poly-Si TFT, and etc. Original cascaded TMR with one voter has a single point of failure problem, so that a lot of model studies of cascaded TMR have been done for making more reliable and cost-effective model than previous one. The model with one voter in previous work improves reliability compared to the original one by solving a single point of failure problem. However, the model requires strict rule to improve the reliability. The way to loosen the strict rule is to use more than one voter per stage. However, using more than one voter in the every stage is not a cost-effective solution. In this paper, we suggested a cost-effective new model of cascaded TMR. At the same time, the model also loosened the strict rule in previous works.

Keywords: Cascaded TMR, Multi-Stage TMR, Reliability, Fault-Tolerance

1 Introduction

Cascaded TMRs have been used in the fields like a pipeline process [1], or a redundant Poly-Si TFT-LCD [2]. TMR is a structure for covering Single Point of Failure (SPF). SPF means the whole system failure caused by only one failed component. Therefore, TMR is structured to mask one failure. If two of three modules are working and only one module fails, the output of TMR would correct [3].

"Cascaded TMR" is also called as "Multi-Stage TMR". It is a structure that one output of current TMR becomes three inputs of the next TMR. As a result, a long TMR chain is made.

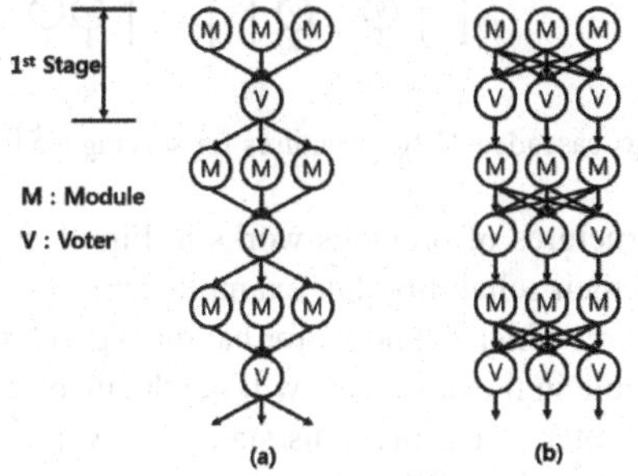

Fig. 1. Original cascaded TMRs

K.J. Kim (ed.), *Information Science and Applications*,
Lecture Notes in Electrical Engineering 339, DOI 10.1007/978-3-662-46578-3_43

Fig. 1-(a) is an original cascaded TMR with one voter. In this structure, it can cover the SPF of module, but can't cover SPF of the voter. To solve this problem, there is a cascaded TMR structure with three voters (Fig. 1-(b)). However, this structure costs much more than Fig. 1-(a) because it uses three voters per a stage. Therefore, the ways to use less voters and simultaneously get higher reliability have been studied [4,5,6,7]. However, these researches have too strict rule for making their structures more reliable than original cascaded TMR [5,6].

In this paper, we suggest a new model which has less strict rule than previous work while maintaining the reliability at the same time.

Section 2 is about the works related to our research and its problems. Section 3 is about our new model, and in Section 4, we evaluate the reliability of previous models and that of our new model. In Section 5, we summarize our research and conclude the paper.

2 Related Work

Lee et al. suggested a new structure of cascaded TMR with one voter which is applied to the shift register chains in the screen driver circuit of Poly-Si TFT LCD [5,6]. They also suggested multiple structures of cascaded TMR [6].

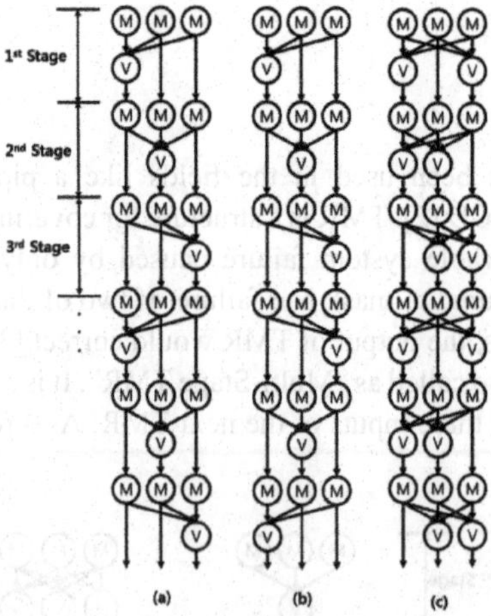

Fig. 2. Previous cascaded TMR structures for solving a SPF problem [6]

There are the core structures of previous works in Fig. 2. Fig. 2-(a) and 2-(b) have strict rule for improving their reliability. For example, let's assume the voter of the first stage in Fig. 2-(a) structure, which is known for having a good reliability value for cost, fails. Then, the first module in the next stage will get the incorrect output as an input. In this case, the failure of voter in the previous stage is covered by other two modules, which have correct inputs, so that SPF cannot occur.

However, the modules in the most left vertical line of Fig. 2-(a) will accept incorrect inputs until next three stages from the module which accepts the incorrect input first. If there is another failure among those three stages, the whole system will fail. The paper which suggested this structure supposed that there must be no more other module or voter failures among the three stages. These three stages are called as recovery window (RW). Like this way, the average RWs for voter failures in Fig. 2-(a) are 3 in all the vertical lines. Also, the RWs for module failures are (1, 3, 2), (2, 1, 3), and (3, 2, 1) in turn from the most left vertical line, so that averages of those are 2 in common.

Fig. 2-(c) has shorter RWs for module and voter failures than those of Fig. 2-(a). Its average RWs for module and voter failures are 1.33 and 1.5 respectively. However, it uses one more voter in every stage compared to Fig. 2-(a), so that it costs more than that of Fig. 2-(a).

It is a strict rule that we cannot have more failure during the long RW. In addition, long RW means low reliability. Using more than one voter in every stage for solving this problem is not a cost-effective solution. Our new model can shorten the RW compared to Fig. 2-(a), and simultaneously be more cost-effective than Fig. 2-(c).

3 Suggested New Models

There are newly suggested cascaded TMR models in the Fig. 3. In the case of Fig. 3-(a), even if a voter failure occurs at the first stage and also at the second stage, no SPF occurs unlike Fig. 2-(a). However, this kind of masking failure works only for the odd stages. RW for voter failure in the even stages is 5, so that the average RW for voter failures is 3 which is same with that of Fig. 2-(a). However the model is less reliable than Fig. 2-(a) because the longest RW for the voter failure of Fig. 2-(a) is 3, on the other hand, that of Fig. 3-(a) is 5. Lager RW requires stricter rule, and stricter rule means low reliability.

RWs for module failures in Fig. 3-(a) are (1, 1, 5, 4, 3, 2), (3, 2, 1, 1, 5, 4), and (5, 4, 3, 2, 1, 1) in turn from the most left vertical line. Averages of those are 2.7 in common, which is bigger than Fig. 2-(a). It means even though Fig. 3-(a) can mask consecutive two voter failures, it is not a good model.

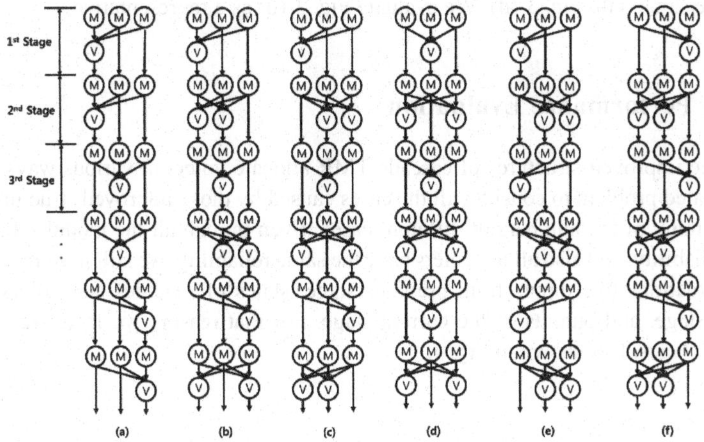

Fig. 3. New cascaded TMR structures

In the case of Fig. 3-(b), it also masks consecutive two voter failures if the voter failure is started from the odd stages. It uses two voters in the even stages unlike Fig. 3-(a). As a result, RWs for module failures are (1, 1, 4, 3, 2, 1), (2, 1, 1, 1, 4, 3), and (4, 3, 2, 1, 1, 1) in turn from the most left vertical line. The averages of those RWs are 2 in common. Even though it is same with that of Fig. 2-(a), Fig. 3-(b) is less reliable than Fig. 2-(a). It is because the longest RW for the module failure of Fig. 2-(a) is 3, on the other hand, that of Fig. 3-(b) is 4.

In the Fig. 3-(c), we can see that the consecutive voter failures can be masked if it starts from the second stage. In this case, not even but odd stages have bigger RWs. Average RW for voter failures is 2, and it is smaller than that of the Fig. 3-(a) and 3-(b). In addition, the longest RW for the voter failure of Fig. 3-(c) is 3, which is same with that of Fig. 2-(a).

RWs for module failures of Fig. 3-(c) are (1, 3, 2, 1, 2, 1), (2, 1, 1, 3, 2, 1), and (2, 1, 2, 1, 1, 3) in turn from the most left vertical line. It means averages of those are 1.6 in common. As a result, Fig. 3-(c) has the same average RW for voter failures, and smaller average RW for the module failures compared to Fig. 2-(a). However, Fig. 3-(c) has bigger average RW for the module and voter failures than that of Fig. 2-(c). In conclusion, Fig. 3-(c) can be said it is more reliable than Fig. 2-(a), but less reliable than Fig. 2-(c).

Fig. 3-(d) has 1.5 average RW for the module failures which is smaller than that of Fig. 3-(c). It is still slightly bigger than that of Fig. 2-(c), also RW of the voter failure is still bigger than that of Fig. 2-(c). However, it is the most reliable model among those using one and two voters alternatively. It is because the point of reducing RW is to place voters as close as possible to each other in each line. In detail, the way to do that is to place voters sequentially. However, whenever we try to place it sequentially, the problem like the first three stages of Fig. 3-(b) or 3-(c) happens. It means one line among three lines has to have a long RW if we place voters sequentially for the other two lines. It is the limit of using one and two voters alternatively. Using two voters per more than two stages for solving this problem is not a cost-effective.

In conclusion, we suggest Fig. 3-(d) as our final model. In addition, Fig. 3-(e) and 3-(f) also can be our final model. It is because the RW for module and voter failures of those are same with Fig. 3-(d). We evaluate Fig. 3-(d) as a representative.

4 Performance Evaluation

All the improved structures of cascade TMR allocate voters in various ways. It is a complicated problem to consider all the cases caused by those improved structures. We have to think of the numbers of different cases. Even though all the modules have the same reliability, and also all the voters have the same reliability, we still need to consider some situations like inputs from previous stage, states of modules and voters at the current stage, and outputs of the current stage. For that reason, we factorized all the cases of stages. It is shown in Fig. 4.

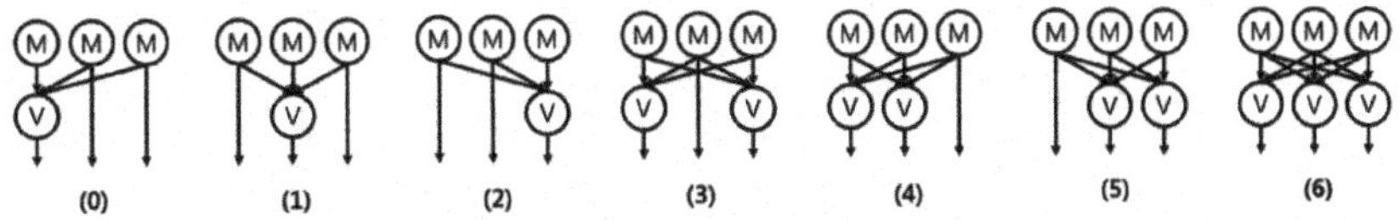

Fig. 4. The factorized stages

The first structure to be evaluated is Fig. 2-(a). It is known as the most cost-effective and also having high reliability. Fig. 2-(a) consists of the Fig. 4-(0) ➔ Fig. 4-(1) ➔ Fig. 4-(2) sequence. The second structure to be evaluated is Fig. 2-(c) which uses two voters per one stage. Fig. 2-(c) consists of the Fig. 4-(3) ➔ Fig. 4-(4) ➔ Fig. 4-(5) sequence. It will be a comparison group for the Fig. 3-(d) which is the third structure to be evaluated. The Fig. 3-(d) consists of the Fig. 4-(1) ➔ Fig. 4-(3) sequence. The last structure is Fig. 1-(b), which is known as the structure having the highest reliability. It consists only Fig. 4-(6) in every stage.

In this paper, for example, when module or voter has five nine (0.9_5) reliability, the program randomly generates a number between 1 and 100,000. If the number is equal to 100,000, then the module or voter is considered as failed module or voter. For that reason, the program requests inputs for the both reliabilities. It must be like 10, 100, 1000, and so on. Also, if there is an input for repeating, the program tests the cascaded TMR repeatedly. It means, if the input for repeating is 1,000,000, the program tests the cascaded TMR in 1,000,000 times. There are some special situations like two incorrect outputs, or two module failures. If one of these situations occurs, system will fail. Then, the program stops the current process and increases value of global variant named system-fail. After that, program begins the process from the beginning again. If the sum of the number of whole system failure and the number of whole system success are equal to 1,000,000, program stops. Finally, it returns the number of whole system failure.

The experiment was progressed in two ways below.

1. The number of stages (n), the number of repeating whole process (repeating time), and reliability of a voter (R_v) are fixed at 100, a million, and 100,000 (0.9_5) respectively. Only reliability of a module (R_m) is changed from 100 (0.9_2) to 100,000 (0.9_5).
2. The repeating time, R_v, and R_m are fixed at a million, 1,000 (0.9_3), and 100,000 (0.9_5) respectively. It means that the $(1 - R_m) / (1 - R_v)$ is equal to 100. Only the number of stage is changed from 100 to 1, 500.

Fig. 5 shows the average results from testing models 10 times by 1st way.

We divided the graph into three cases because the number of system failure increased too much rapidly compared to the previous one. Fig. 5-(a) is a graph when the module reliability is from 10,000 (0.9_4) to 100,000 (0.9_5). Fig. 5-(b) is a graph when the module reliability is from 1,000 (0.9_3) to 100,000 (0.9_5), and Fig. 5-(c) is a graph when the module reliability is from 100 (0.9_2) to 100,000 (0.9_5).

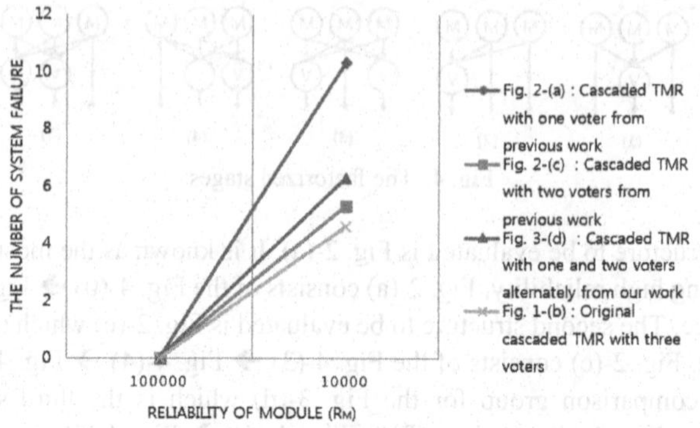

Fig. 5-(a). The result of changing module reliability from 10,000 (0.9_4) to 100,000 (0.9_5)

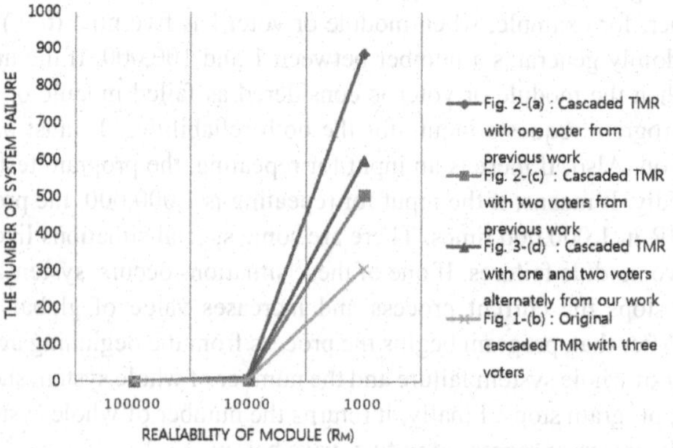

Fig. 5-(b). The result of changing module reliability from 1,000 (0.9_3) to 100,000 (0.9_5)

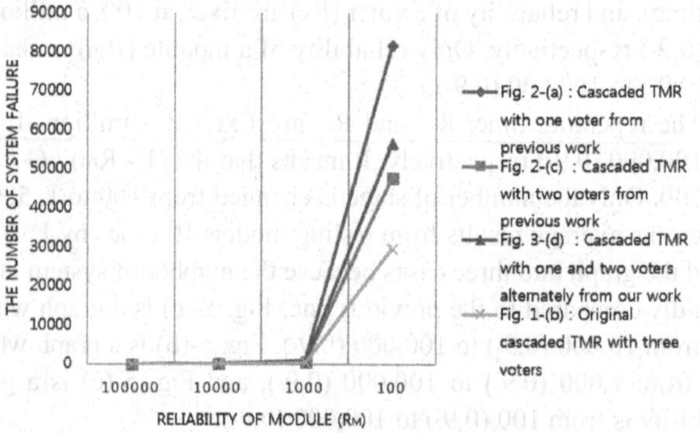

Fig. 5-(c). The result of changing module reliability from 100 (0.9_2) to 100,000 (0.9_5)

All the graphs show that the orders of the number of system failures among four models are maintained. Average number of whole system failure of the Fig. 2-(a) is much greater than other figures in all the cases. Fig. 3-(d) has a similar number of system failure with that of Fig. 2-(c). It means it has a similar reliability with Fig. 2-(c).

Fig. 6 shows the average results from testing models 10 times by 2nd way.

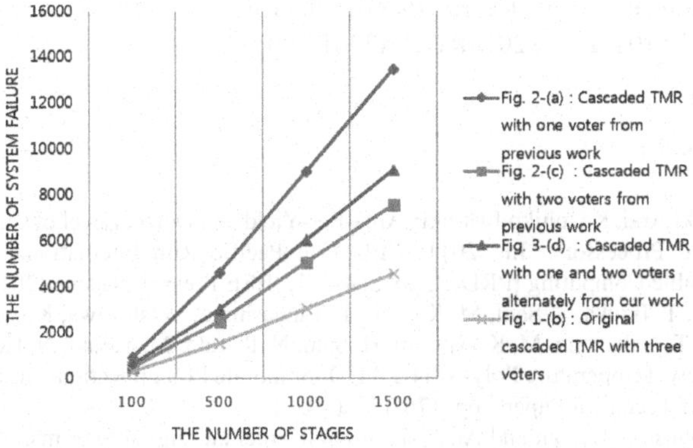

Fig. 6. The result of changing the number of stage

Regardless of the increasing number of the stage, the orders for the number of system failure among four models are maintained in all the cases. Also, it is clear that Fig. 3-(d) is always much closer to Fig. 2-(c) compared to Fig. 2-(a).

5 Conclusion

Models of cascaded TMR in the previous work were made up for higher reliability compared to the original cascaded TMRs. The original cascaded TMR with one voter per one stage has a problem that a voter failure calls whole system failure. Original cascaded TMR with three voters can solve this problem, but using three voters per one stage costs too much.

For more cost-effective and highly reliable model, many studies had been done [4,5,6,7]. They suggested two good models. One is Fig. 2-(a) and another one is Fig. 2-(c). Fig. 2-(a) is a cascaded TMR with one voter per a stage like original cascaded TMR with one voter. It is more reliable than original one because it masks SPF. However, Fig. 2-(a) has too strict rule for making their structures more reliable than original cascaded TMR. The rule does not allow to have more voter or module failure during the RW. However, it is realistically impossible. For that reason, we need to loosen the strict rule. It refers to shorten the RW. Using two voters loosen the strict rule Fig.2-(a) has, but it costs more than that.

Our newly suggested model (Fig. 3-(d)) loosens the strict rule of Fig. 2-(a) by using two voters in even stages. It means that the reliability of Fig. 3-(d) is higher than Fig. 2-(a). Also, our model improved cost efficiency compared to the Fig. 2-(c) by using one voter in the odd stages. We also maintained the reliability similar to the Fig. 2-(c).

When R_m is 1,000 (0.9_3) and R_v is 100,000 (0.9_5), 887.2 average system failures occurred in cascaded TMR of Fig. 2-(a) structure. On the other hand, our model (Fig. 3-(d)) has only 598.2 average system failures. Fig. 2-(c) has 506.6 average system failure, and it is similar to our model. In addition, our model uses less voter than Fig. 2-(c), so that it has better cost efficiency than that.

Acknowledgement

This research was supported by Basic Science Research Program through the National Research Foundation of Korea (NRF) funded by the Ministry of Education (2013R1A1A2A10012956, 2013R1A1A2061390).

References

1. Masayuki Arai, Kazuhiko Iwasaki: Area-Per-Yield and Defect Level of Cascaded TMR for Pipelined Processors. In: 2011 IEEE 17th Pacific Rim International Symposium on Dependable Computing (PRDC), pp. 264-271. IEEE Press, Pasadena (2011)
2. Y. Aoki, T. IIzuka, S. Sagi, M. Karube, T. Tsunashima, S. Ishizawa, K. Ando, H. Sakurai, T. Ejiri, T. Nakazono, M. Kobayashi, H. Sato, N. Ibaraki, M. Sasaki, N. Harada: A 10.4-in XGA Low-Temperature Poly-Si TFT-LCD for mobile PC application. In: SID Symposium Digest of Technical Papers, pp. 176-179. (1999)
3. B.W. Johnson: Design and Analysis of Fault Tolerant Digital Systems. Addison Wesley. (1989)
4. Masashi Hamamatsu Tatsuhiro Tsuchiya Tohru Kikuno: Finding the Optimal Configuration of a Cascading TMR System. In: 2008 14th IEEE Pacific Rim International Symposium on Dependable Computing (PRDC), pp. 349-350. IEEE Press, Taipei (2008)
5. Seungmin Lee, Inhwan Lee: Staggered voting for TMR shift register chains in poly-Si TFT-LCDs. In: Journal of information display, pp. 22-26. (2001)
6. Sungjae Lee, Jae-il Jung, Inhwan Lee: Voting structures for cascaded triple modular redundant modules. In: Ieice Electronic Express, pp. 657-664. (2007)
7. Masashi Hamamatsu, Tatsuhiro Tsuchiya, Tohru Kikuno: On the Reliability of Cascaded TMR Systems. In: 2010 IEEE 16th Pacific Rim International Symposium on Dependable Computing (PRDC), pp. 657-66. IEEE Press, Tokyo (2010)

Availability Analysis for A Data Center Cooling System with (n,k)-way CRACs

Sohyun Koo, Tae-Sun Chung, Sungsoo Kim
Dept. of Computer Engineering
Ajou University
Suwon, Korea
{uriinhyoung, tschung, sskim}@ajou.ac.kr

Abstract. Data center cooling systems have cooled computer equipments. Recently, these systems have been supposed to economize energy consumption and perform effectively in various environment. So, it is important to design appropriate data center cooling system infrastructure and analyze its dependability, especially availability and reliability. In this paper, we have proposed a (n,k)-way data center cooling system which composes n units of main CRAC (Computer Room Air Conditioner) and k units of spare CRAC. We have also calculated adequate amount of main CRAC units (*n*). In addition, optimal values of *n* and *k* to achieve a cost effective solution have been stated with a cost-effective equation.

Keywords: Availability, Data center cooling system, (n,k)-way CRACs

1 Introduction

Data center is a facility that contains severs, storages, network, etc. Data center energy consumption is almost 1.3% of total world energy consumption [1, 2]. Furthermore, cooling costs in data center are around 30-50% of total data center energy consumption [3]. Data center cooling system has been supposed to economize energy consumption and perform effectively in various environment. Hence, it is important to design appropriate data center cooling system infrastructure and analyze its dependability. The term of dependability encompasses the concept of reliability, availability, safety, maintainability, performability, and testability. Reliability is the probability that the system will deliver a set of services for a given period of time, and availability is the probability that a system is operating correctly at the instant of mission time *t* [4].

G. Callou et al. [5, 6] have designed data center cooling system infrastructure with duplex cooling towers and chillers which are units of data center cooling system. In this work, they have set cold-standby CRAC (Computer Room Air Conditioner) as well. They have evaluated system reliability and availability compared to spare CRAC performed as cold-standby, and hot-standby. However, they haven't considered how much CRAC units are necessary and cost-effective in each data center. D. Kim et el. [7] and B. Wei et al. [8] have developed availability model of overall data center and analyzed availability of it. But, they didn't explain details about the data center cooling system.

In this paper, we have proposed (n,k)-way data center cooling system which contains *n* units of main CRAC and *k* units of spare CRAC. The adequate amount of main CRAC (*n*) has

K.J. Kim (ed.), *Information Science and Applications*,
Lecture Notes in Electrical Engineering 339, DOI 10.1007/978-3-662-46578-3_44

been calculated by the cost-effective equation in [9]. This new cooling system infrastructure also has duplex cooling towers and chillers. These caused high availability in [5]. The purpose of this paper is to come up with an effective (n,k)-way data center cooling system and analyze its availability.

2 Related Work

G. Callou et al. [5] has represented basic data center cooling system infrastructure. Fig.1 provides an overview of their design which contains a redundant cooling tower and a redundant chiller. They have checked reliability and availability when using spare CRAC as a cold-standby and hot-standby as well. As a result, when a redundant cooling tower is added in Fig.1, availability is around 10% higher than the basic architecture one in this case. In the basic architecture case, it has 2.26 9's availability. In contrast, using a cold-standby spare CRAC caused 4.96 9's availability, and using a hot-standby spare CRAC caused 4.94 9's availability. Between the hot-standby case and cold standby case has only a small difference. Furthermore, cold-standby technique's availability is higher than hot-standby one. To consider the energy consumption and system availability, using cold-standby technique is much better.

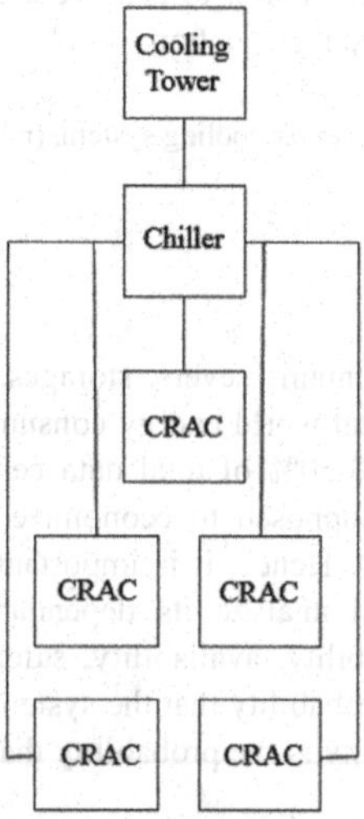

Figure 1 Basic Data Center Cooling System Infrastructure [3]

I. Cathcart [9] has calculated the proper number of CRAC units with the Equation (1) focused on the total and unit load of data center. In (1), the appropriate CRAC units are decided to the total sensible cooling load and the unit rated sensible cooling load. For example, the real data center in [9] has 535,629 BTU (British Thermal Unit)/ Hour as a total cooling load value and one CRAC unit`s cooling load value is 217,600 BTU per hour. In this case, CRAC units are calculated by 2.5, so totally 3 CRAC units should be placed in data center. He has researched about maintainability and fault-tolerance for data center cooling system. It implemented to improve availability with redundant CRAC unit.

$$\textit{CRAC Units} = \frac{\textit{Total Sensible Cooling Load}}{\textit{Unit Rated Sensible Cooling Load}} \quad (1)$$

3 (n,k)-way Data Center Cooling System Model

We have organized (n,k)-way data center cooling system to analyze how much main CRAC units and spare CRAC units are necessary for efficient data center cooling system. In (n,k)-way model, *n* units of main CRACs and *k* units of spare CRACs are placed. Its purpose is to minimize the number of CRAC hardware units. In addition, the new cooling system model is to ensure efficient performance. We have used duplex cooling towers, duplex chillers and cold standby spare CRAC technique to get high dependability by using technique in [5]. Fig. 2 is the overall view of (n,k)-way data center cooling system infrastructure. Duplex cooling towers are connected to duplex chillers, and each chiller is linked to *n* units of main CRACs and *k* units of spare CRAC. The total number of CRAC is $(n+k)$.

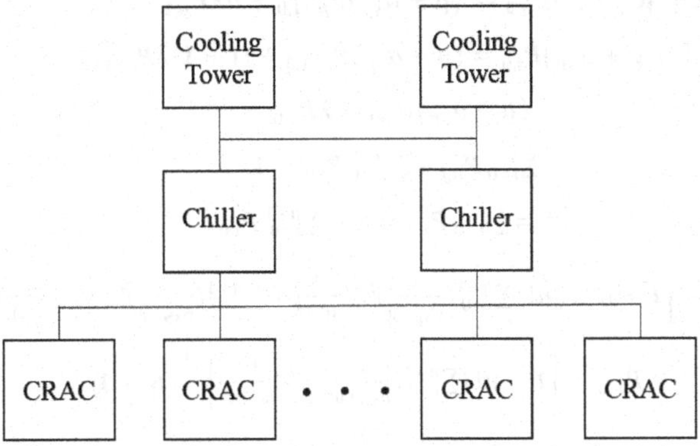

Figure 2 (n,k)-way Data Center Cooling System Infrastructure

3.1 Assumptions

- In (n,k)-way data center cooling system, *n* units of main CRACs are linked to two chillers and each CRAC performs independently.
- *k* units of spare CRAC should be less than *n* units of main CRACs (**n ≥ k**). Spare CRAC replaces main CRAC in cold-standby technique.
- Failure rate (λ) and repair rate (μ) of main CRAC and spare CRAC are constant, and they follow exponential distribution.
- Spare CRACs are activated when main CRAC fails, and failed main CRAC gets repaired by repair person. Arrival rate of repair person who repairs failed CRACs is α_{sp}.

3.2 Analysis of (n,k)-way data center cooling system availability

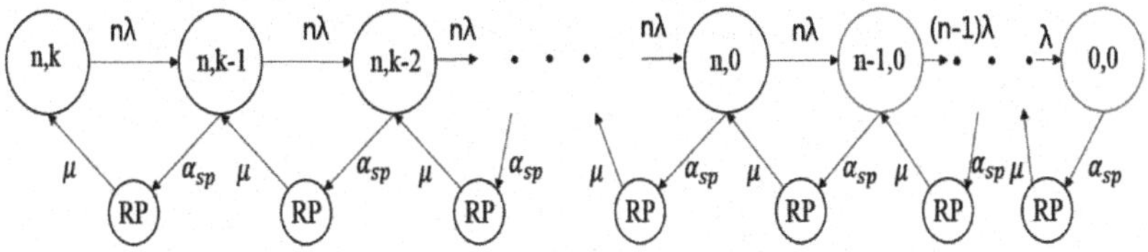

Figure 3 Failure-Repair State Diagram for (n,k)-way Data Center Cooling System

Fig. 3 is the failure-repair transition diagram about (n,k)-way data center cooling system. Active CRAC states are represented as (n,k), (n,k-1), …, (n,1), (n,0), …, (0,0); *state variables (main CRAC, spare CRAC)*. (n,k)-way data center cooling system has to calculate variable *n* which is the minimum amount of main CRAC to cover overall data center load with Equation (1). Otherwise, CRACs cannot cover system load. Normal state as an available state is defined until all spare CRACs get failed, and all main CRAC should work. State (n-1,0), (n-2,0), …, (1,0), (0,0) represent that more than one failure occurs on main CRAC. They are failure (unavailable) state, because these states could not be satisfied to overall data center load. The balance equations of each state are as follows.

$$n\lambda P_{n,k} = (\mu + \alpha_{sp})P_{n,k-1} \tag{2}$$

$$(n\lambda + \mu + \alpha_{sp})P_{n,j} = (\mu + \alpha_{sp})P_{n,j-1} + n\lambda P_{n,j+1}\ , j = 1, 2, \ldots, k-1 \tag{3}$$

$$(n\lambda + \mu + \alpha_{sp})P_{n,0} = (\mu + \alpha_{sp})P_{n-1,0} + n\lambda P_{n,1} \tag{4}$$

$$(i\lambda + \mu + \alpha_{sp})P_{i,0} = (\mu + \alpha_{sp})P_{i-1,0} + (i+1)\lambda P_{i+1,0} \tag{5}$$

$$(\mu + \alpha_{sp})P_{0,0} = \lambda P_{1,0} \tag{6}$$

$$\textstyle\sum_{j=0}^{k} P_{n,j} + \sum_{i=0}^{n-1} P_{i,0} = 1 \tag{7}$$

$$A = \textstyle\sum_{j=0}^{k} P_{n,j} = 1 - \sum_{i=0}^{n-1} P_{i,0} \tag{8}$$

$$P_{n,k} = \left[1 + n^k \times n!\,\left(\textstyle\sum_{i=0}^{k-1}\left(\frac{\lambda}{\alpha_{sp}+\mu}\right)^{k-i}\frac{1}{n^{n+i}} + \sum_{i=0}^{n-1}\left(\frac{\lambda}{\alpha_{sp}+\mu}\right)^{n+k-i}\frac{1}{(n-i)!}\right)\right]^{-1} \tag{9}$$

$$P_{n,0} = \left[1 + n!\,\left(\textstyle\sum_{i=0}^{n-1}\left(\frac{\lambda}{\alpha_{sp}+\mu}\right)^{n-i}\frac{1}{(n-i)!}\right)\right]^{-1}, n = 1, 2, \ldots \tag{10}$$

$$P_{n,j} = \left(\frac{n\lambda}{\alpha_{sp}+\mu}\right)^{k-j} P_{n,k}\ , j{=}0, 1, \ldots, k-1 \tag{11}$$

$$P_{i,0} = \left(\frac{\lambda}{\alpha_{sp}+\mu}\right)^{n-1}\frac{n!}{(n-i)!}P_{n,0}\ , i = 0, 1, \ldots, n-1 \tag{12}$$

Equation (2)-(6) are balance equations of each state in Fig. 3. (7) is the conservation equation which get 1 as a result. After solving a simultaneous equation with (2)-(7), we could get (9)-(12). So, we are able to calculate the probability of each state in (n,k)-way data center cooling system. (8) shows the availability of this cooling system model and it is settled with K. Park [10].

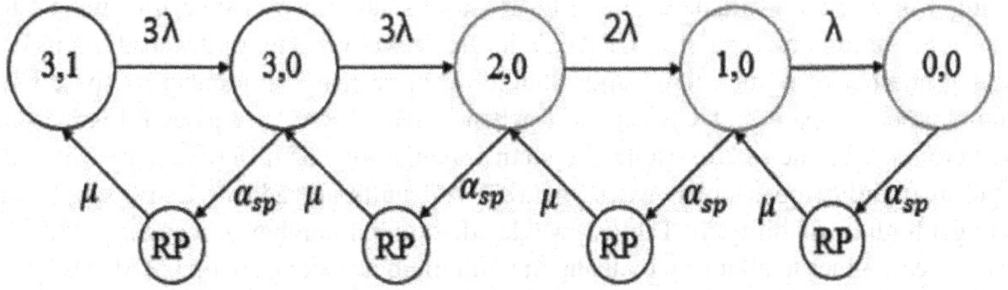

Figure 4 Failure-Repair State Transition Diagram for (3,1)-way Data Center Cooling System

Fig. 4 illustrates the failure-repair state transition diagram when the total cooling load is 535,629 BTU/Hour and the CRAC unit rated cooling load is 217,600 BTU/Hour in the real data center [9]. In this case the proper number of CRAC units are calculated as 3. We add only one CRAC as a spare CRAC for understandable example. This (3,1)-way data center cooling system model has totally 5 states. It has only 2 available states which can perform properly. Others such as (2,0), (1,0), (0,0) states are failure states.

(1) Available state: $\boldsymbol{P_{3,1}} + \boldsymbol{P_{3,0}}$
(2) Failure (unavailable) state: $\boldsymbol{P_{2,0}} + \boldsymbol{P_{1,0}} + \boldsymbol{P_{0,0}}$

An active CRAC gets failed with i* λ (i: the number of CRAC units) rate of change, and a failed CRAC repairs with $\mu + \alpha_{sp}$ rate of change.

4 Availability Evaluation

To evaluate availability of (n,k)-way data center cooling system infrastructure, we calculate different number of spare CRAC and main CRAC units. The data center cooling system parameters are illustrated in Table 1. Total sensible cooling load and CRAC unit rated sensible cooling load are from the real data center data [9]. The parameter of mean time for CRAC failure is 3,100,000 hours. The Mean Time to Repair (MTTR), (CRAC unit) and MTTR (person summoned) are 30 minutes each.

Table 1 Data center cooling system parameters [5, 7]

Parameter	Description	Value
Total load	Total sensible cooling load	535,629 BTU/Hr
Unit load	CRAC unit rated sensible cooling load	217,600 BTU/Hr
$1/\lambda$	Mean time for CRAC failure	3,100,000 hours
$1/\mu$	MTTR CRAC unit	30 minutes
$1/\alpha_{sp}$	MTTR person summoned	30 minutes

By using Table 1, we evaluate availability in various cases, we change *n* of main CRAC from 3 to 6 (the minimum number of CRAC units are 3 to cover data center load in this case). Because the number of main CRAC units should be more than the number of spare CRAC units, the number of spare unit *k* is zero to 3 when 3 main CRAC are placed. Likewise, *k* is changed zero to *n* in each case. Table 2 contains availability of different (n,k)-way CRAC units. The availability goes down when main CRAC units are added. Using spare unit is helpful to get high availability. In Table 2, we decide optimal number of *n* and *k* as (3,3) case. This guarantees 7-nines availability by using the minimum number of main CRAC units.

Table 2 Availability of Different (n,k)-way CRACs Data Center Cooling System

n \ k	0	1	2	3	4	5	6
3	0.99999928	0.999999706	0.999999786	0.999999946	.	.	.
4	0.99999808	0.999999712	0.99999994	0.99999997	0.999999971	.	.
5	0.99999032	0.999999846	0.99999997	0.999999984	0.999999988	0.999999988	.
6	0.999942403	0.999999992	0.999999992	0.999999992	0.999999993	0.999999993	0.999999993

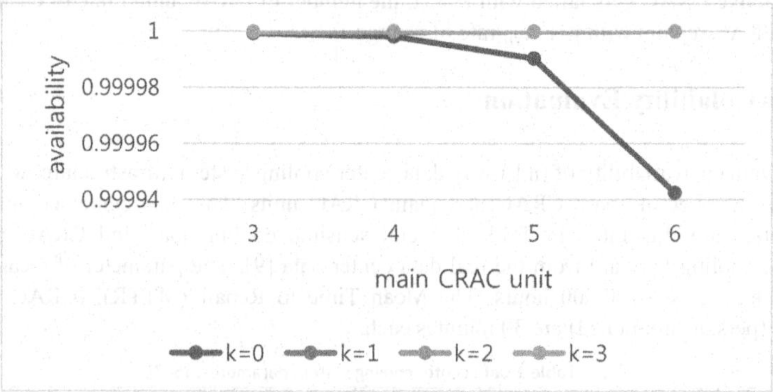

Figure 5 Availability Depending on The Number of Spare and Main CRAC Units

Additionally, Fig. 5 represents data in Table 2, and Fig. 6 is its detail. Fig. 5 illustrates availability when spare unit *k* changes as 0, 1, 2, 3. If there is no spare CRAC in data center cooling system (blue line), the availability would be decreased. Most cases are placed around availability of 1. So, we put another graph Fig. 6 to show details in specific areas. *k*=1 to *k*=6 cases are illustrated in Fig. 6. Using spare unit is helpful to get high availability. The effectiveness is prominent when many main CRACs are placed in data center cooling system.

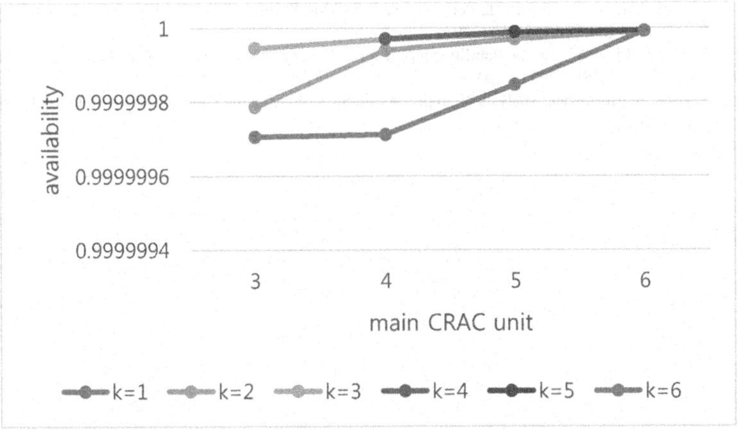

Figure 6 Detail Graph of Figure 5 with Availability between six-9 and 1

5 Conclusion

In this paper, we have suggested a (n,k)-way model with main CRAC units and spare CRAC units. In addition, we evaluate availability of (n,k)-way data center cooling system with real data center parameters. We have also determined the optimal value of *n*, *k* for a cost-effective solution. The optimal case is a data center cooling system with a cost-effective and high availability.

Acknowledgement

This research was supported by Basic Science Research Program through the National Research Foundation of Korea (NRF) funded by the Ministry of Education (2013R1A1A2A10012956, 2013R1A1A2061390).

References

1. J.G. Koomey, "Growth in Data Center Electricity Use 2005 to 2010," Analytic Press, Oakland, CA, Aug. 1, (2011)
2. S. Lee, S. Moon, J. Kim, S. Shin, Y. Seo, Y. Choi, "The Establishment Method of Green Data Center in Public Sector", Journal of KISS, Vol 27, 48-57, (2011)
3. C. Onyiorah, R. Eiland, D.Agonafer, R. Schmidt, "Effectiveness of Rack-Level Containment in Removing Data Center Hot-spots", 14th IEEE ITHERM Conference, 798-806 (2014)
4. B.W. Johnson, "Design and Analysis of Fault-Tolerant Digital Systems", Addison Wesley, (1989)
5. G. Callou, P. Maciel, D.Tutsch, J. Araujo, "Models for Dependability and Sustainability Analysis of Data Center Cooling Architectures", IEEE DSN, (2012)
6. G. Callou, P. Maciel, E. Tavares, E. Sousa, B. Silva, J. Figueiredo, C. Araujo, F.S. Magnani, F. Neves, "Sustainability and Dependability Evaluation on Data Center Architectures", IEEE SMC 2011, 398-403, (2011)
7. D.S. Kim, F. Machida, K.S. Trivedi, "Availability Modeling and Analysis of a Virtualized System", 2009 15th Pacific Rim International Symposium on Dependable Computing, 365-371 (2009)
8. B. Wei, C. Lin, X. Kong, "Dependability Modeling and Analysis for the Virtual Data Center of Cooling Computing", 2011 IEEE International Conference on High Performance Computing and Communications, 784-789 (2011)
9. I. Cathcart, "How to Build a Data Centre Cooling Budget", Bicsi conference.

10. K. Park, S. Kim, "Analysis of Available Performance Satisfying Waiting Time Deadline for (n,k)-way Systems", Journal of KISS, Vol 30, Oct, pp. 445-453 (2003)
11. K. Park, S. Kim, J. Liu, "A New Availability concept for (n,k)-way cluster systems regarding waiting time", ICCSA 2003, LNCS 2667, 998-1005 (2003)
12. K. Park, S. Kim, "Performance Analysis of Highly Available Cold Standby Cluster Systems", Journal of KISS, Vol. 28,. 173-180 (2001)

A Recovery Model for Improving Reliability of PCM WL-Reviver System

Taehoon Roh, Tae-Sun Chung, and Sungsoo Kim

Ajou University, Dept. of Computer Engineering, Suwon, South Korea
{th2v, tschung, sskim}@ajou.com

Abstract. PCM (Phase Change Memory) that takes center stage has characteristics of non-volatile memory, but as it has a limited number of write times, we wear-leveling technique is required. WL-Reviver (Wear Leveling – Reviver) system is a wear-leveling technic tailored to the PCM, but if internal and external errors occur, it gets failed. This paper proposes a recovery model based on WL-Reviver system for internal and external error situations. Compared to previous work, our newly suggested model improves the system reliability.

Keywords: PCM, Wear-Leveling, WL-Reviver, Recovery Model

1 Introduction

Nowadays IT devices are used in many kinds of systems. Especially memory, which is used in storing information, is considered as an indispensable part in IT devices. Today, DRAM (Dynamic Random Access Memory) is most widely used in various digital devices, as it is cost-effective and fast.

However, as many modern systems aim at low-power, high-efficiency and small size system. DRAM is not suitable to modern system. Thus, many researchers have addressed next generation memory. As a result, various next generation memory models were proposed and PCM (Phase Change Memory) is expected to widely use. Especially the wear-leveling system should be considered.

This paper is organized into four parts, in Section 2, characteristic and related work of the WL-Reviver (Wear Leveling - Reviver) are described, our recovery model is addressed in Section3, and in Section 4, we implement and evaluate our new model. In Section 5, we conclude the paper.

K.J. Kim (ed.), *Information Science and Applications*,
Lecture Notes in Electrical Engineering 339, DOI 10.1007/978-3-662-46578-3_45

2 Related Work

2.1 Wear-Leveling in PCM

Normally PCM has limited endurance (Average $10^7 \sim 10^8$ per one memory cell) [1]. So wear-leveling technique is necessary in the use of PCM. Wear-leveling is a technique that increases length of device's life by keeping endurance of PCM memory cell evenly. There are various wear-leveling techniques. Flash-memory also has limited endurance, but it has wear-leveling controller inside the device in contrast to PCM which highly relies on operating system for wear-leveling technique. However wear-leveling that depends on operating system is not suitable for PCM, because it is designed based on DRAM. Therefore new rule for PRAM based Wear-Leveling should be addressed.

2.2 WL-Reviver

Among PCM wear-leveling techniques, operating system based techniques still has lots of limits and fault-tolerance has difficulty in fail block remapping. In order to make up such drawbacks, WL-Reviver system was proposed [2]. WL-Reviver system minimizes the interference of the operating system and provides countermeasure for faults by using wear-leveling technique with fault-tolerant techniques. Also it is an algorithm that minimizes overheads without requesting existing DRAM-based memory system. However, the technique can't recover from sudden internal or external error.

In this paper, we suggest a system recovery model for internal system failure and external error based on architecture of the WL-Reviver System. In addition, we propose a scheme that provides higher reliability system compared to previous work.

3 WL-Reviver Recovery Model

In this chapter, WL-Reviver system will be briefly described, problems that can occur in the system is addressed, finally, recovery model for solving such problem will be described in detail.

3.1 WL-Reviver Architecture

Fig 1. represents basic structure of WL-Reviver. In the WL-Reviver system, block used in communication between operating system page and memory block is organized in 4 types (Failed Block, Shadow Block, Virtual Shadow Block, and Inverse Pointer Block). OS page can't be accessed by software. By referencing OS page, system can figure out memory block's failed block and use it in fault-tolerant techniques. In such system, it uses Start-Gap [3] technique and Security Refresh [4] technique to design wear-leveling. We can detect failed blocks applying FREE-P [5] and Zombie [6]. Zombie technique uses ECC (Error Correction Codes) or ECP (Error Correcting Pointers) algorithm to allocate failed block. And fault-tolerant technique uses the

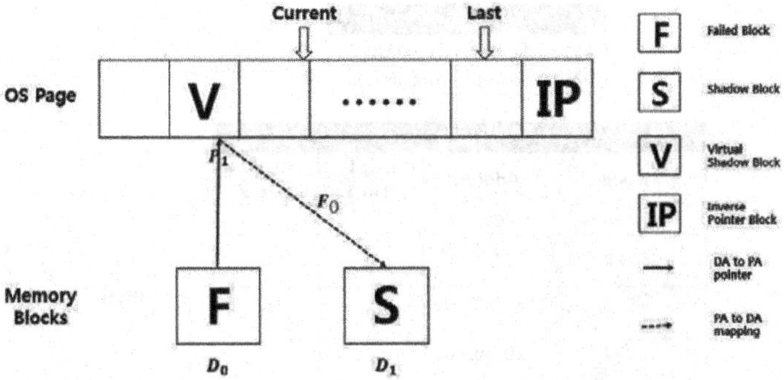

Figure 1. WL-Reviver Architecture. Failed block $\mathbf{D_0}$ points to virtual shadow block $\mathbf{P_1}$, which is mapped to shadow block $\mathbf{D_1}$ via mapping function $\mathbf{F_1}$() [2].

internal function algorithm to allocate new shadow block in memory block. This will point to virtual shadow block in the OS page. Virtual shadow block is used as block for hiding fault and is mapped to device address shadow block with a function algorithm. Shadow block is a block that creates pre-allocated memory space, which represents page with fault occurred by OS. This is not used to store data but can be linked only when failed block occurs. However, when system fails due to system errors, WL-Reviver technique doesn't give any solution.

3.2 Defining the Problem

Problem of WL-Reviver system can be defined as follows.

1) Error factors caused by inside of the system
   (1) System failure due to deadlock
   (2) Synchronization error in critical section

Beside above situations, all situation with system malfunction caused by software problems should be included.

2) Error factors caused by outside of the system
   (1) System failure due to power outage
   (2) System failure due to overheating

Beside above situations, all situation with system failure caused by hardware problem should be included.

3.3 WL-Reviver Recovery Model

In this section recovery model will be proposed to solve the WL-Reviver System problems.

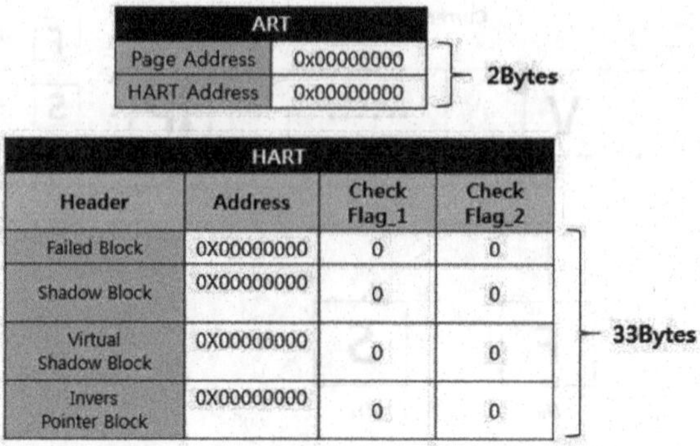

Figure 2. Recovery Model Architecture. ART contains page index address and HART index address. HART contains header address and value.

Fig 2 shows structure of the WL-Reviver recovery model. As it is shown in Fig 2, recovery model is formed with two address tables. ART(Address Recovery Table) stores index address of page, that is being used in OS and address of working HART(Header Address Index Table). ART is positioned in boot algorithm and referenced when boot loader start to reference the page and the page status before the system failure. As for HART, it is formed with address, check flag_1, and check flag_2 of each failed block, shadow block, virtual shadow block, inverse pointer that is used by WL-Reviver. The initial values of check flag_1 and check flag_2 are set to 0. It means that there are no failed blocks. When FB, SB, VSB, or other blocks occurs while WL-Reviver is running, the check flag_1's value is changed to 1 and identifies whether there is the block or not by comparing with check flag_2. The system will consider that the block exists if the check flag_1 and the check flag_2 have different value. After the end of the running process, check flag_2's value will be changed to 1, and the system will consider that the block doesn't exist. The reason for using the flags is to reduce the number of erase.

4 Implementing WL-Reviver Recovery Model

Page has two sections that is, section for virtual shadow blocks and section for inverse pointers. WL-Reviver recovery model uses one block of virtual shadow block section. Suppose practical operating system uses 4KB page, this means that it is possible to divide 64 blocks in PCM that uses 64B memory cell. Fig 3 shows the recovery model implemented in WL-Reviver system. ART with 2B size is loaded to memory data block, which can be referenced when the boot loader starts. The first 1B of ART will include address of the page. And the second 1B will store address of the page's HART. And when system restarts, the boot loader will find the value of ART and with the ART it will be possible to continue the task before the system failure. HART is allocated in block, and then it is allocated in memory data block according to result of function.

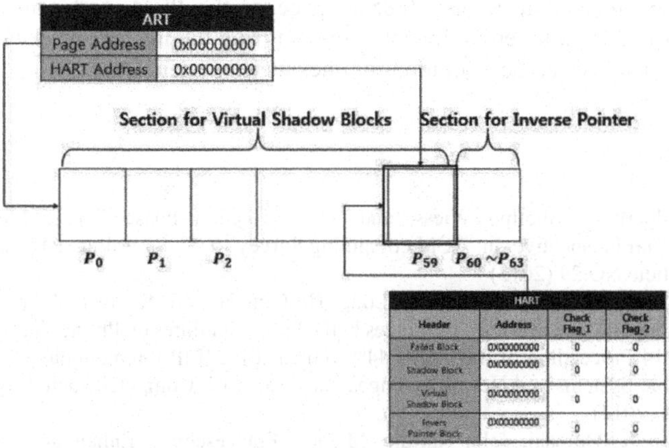

Figure 3. Using Recovery Model of WL-Reviver System.

Table 1 shows comparison between WL-Reviver system and system with recovery model. Both systems have fault-tolerant function and seem to have 60 and 59 usable blocks. This means that recovery model only takes 25B out of 4KB page. If we do the math, it only occupies 0.85% of the whole page. For example, the recovery model will be occupying only about 8~9MB if 1GB PCM is being used. Also deadlock or synchronization errors of the critical section may occur under the WL-Reviver system. In case of that, the system reboots itself. But, if we apply the recovery module, we can roll back the system to a previous error-free state.

Table 1. Comparison between WL-Reviver system and (WL-Reviver + recovery model)

	WL-Reviver	WL-Reviver + Recovery
Fault-tolerance	Possible to proceed	Possible to proceed
Available block number per Page	60/64	59/64
Internal Fault	Impossible to proceed	Possible to proceed
External Fault	Impossible to proceed	Possible to proceed

5 Conclusion

In this paper, the recovery model of WL-Reviver, which is one of PCM Wear-Leveling technique, is proposed. We implemented the recovery model for WL-Reviver system to improving the system reliability.

The proposed method uses only 0.85% space of the PCM, and can efficiently recovery the valid working environment after occurring the internal or external error. Therefore, our scheme can be a solution for unexpected system error.

References

1. Omer Zilberberg, Sholmo Wiess, and Sivan Toledo.: Phase-Change Memory: An Architectural Perspective. In: ACM computing Surveys (CSUR), Volume 45, Issue 3, June 2013, Article NO.29 (2013)
2. Jin Fan, Jiang Song, Shu Jiwu, Sung Long, Hu Oingda.: WL-Reviver: A Framework for Reviving any Wear-Leveling Techniques in the Face of Failures on Phase Change Memory. In: DSN'14 Proceeding of the 20014 44th Annual IEEE/IFIP International Conference on Dependable Systems and Networks, Pages 228-239, IEEE Computer Society Washington, DC, USA (2014)
3. Moinuddin K. Qureshi, John Karidis, Michele Franceschini.: Enhancing Lifetime and Security of PCM-Based Main Memory with Start-Gap Wear Leveling. In: MICRO 42 Proceeding of the 42nd Annual IEEE/ACM International Symposium on Microarchitecture, Pages 14-23, ACM New York, NY, USA (2009)
4. Nak Hee Seong, Dong Hyuk Woo, Hsien-Hsin S. Lee.: Security Refresh: Prevent Malicious Wear-out and Increase Durability for Chase-Change Memory with Dynamically Randomized Address Mapping. In: ISCA '10 Proceeding of the 37th annual international symposium on Computer architecture, Pages 383-394, ACM New York, NY, USA (2010)
5. Doe Hyun Yoon, Naveen Muralimanohar, Jichuan Chang.: FREE-P: Protecting Non-Volatile Memory against both Hard and Soft Errors. In: HPCA'11 Proceedings of the 2011 IEEE 17 International Symposium on High Performance Computer Architecture, Pages 466-477, IEEE Computer Society Washington, DC, USA (2011)
6. Rodolfo Azevedo, John D. Davis, Karin Strauss, Parikshit Gopalan, Mark Manasse, Sergey Yekhanin.: Zombie Memory: Extending Memory Lifetime by Reviving Dead Blocks. In: ISCA'13 Proceeding of the 40th Annual International Symposium on Computer Architecture, Pages 452-463, ACM Newyork, NY, USA (2013)
7. ERAN GAL AND SIVAN TOLEODO.: Algorithms and Data Structures for Flash Memories, In: ACM Computing Surveys (CSUR), Volume 37, Issue2, June 2005, Pages 138-163, ACM New York, NY, USA (2005)

Real-time Sitting Posture Monitoring System for Functional Scoliosis Patients

Ji-Yong Jung[1], Soo-Kyung Bok[2], Bong-Ok Kim[2], Yonggwan Won[3] and Jung-Ja Kim[4,5,*]

[1] Department of Healthcare Engineering, Chonbuk National University, College of Engineering, Jeonju, Republic of Korea
cholbun@hanmail.net

[2] Department of Rehabilitation Medicine, Chungnam National University Hospital, Daejeon, Republic of Korea
skbok@cnuh.ac.kr,bokim@cnu.ac.kr

[3] Department of Electronics and Computer Engineering, College of Engineering, Chonnam National University, Gwangju, South Korea
ykwon@jnu.ac.kr

[4] Division of Biomedical Engineering, College of Engineering, Chonbuk National University, Jeonju, Republic of Korea

[5] Research Center of Healthcare & Welfare Instrument for the Aged, Chonbuk National University, Jeonju, Republic of Korea
jungjakim@jbnu.ac.kr

Abstract. In recent years, with the increase in sitting time, real-time posture monitoring system is needed. We developed sitting posture monitoring system with accelerometer to evaluate postural balance. Unstable structure of the system was designed to assess asymmetrical balance caused by habitual sitting position. Postural patterns of normal group and patients with functional scoliosis group were analyzed. Consequently, inclination angle of scoliosis group were tilted to the posterior and left side. From these results, we concluded that real-time sitting posture monitoring system can be utilized to measure asymmetrical balance of patients and provide accurate diagnosis as well as treatment for individuals.

Keywords: Real-time monitoring system, Functional scoliosis, Postural balance, Sitting.

1 Introduction

The ability to keep balance is one of the most essential factors in activities of daily living. Human body can adjust the posture in response to continuous external perturbations for maintaining good posture. Good posture provides normal

---

* Corresponding author.

K.J. Kim (ed.), *Information Science and Applications*,
Lecture Notes in Electrical Engineering 339, DOI 10.1007/978-3-662-46578-3_46

biomechanical functions of the musculoskeletal system. On the contrary, postural changes caused by habitual bad posture with prolonged static working environment have a crucial influence on body asymmetry, muscle imbalance, and structural problem in the spine.

As the social interest in healthcare increases, many devices were developed to evaluate balance of individuals during standing and walking for preventing damage to the skeletal muscle pathologies [1,2]. Especially, in recent years, studies related to real-time measurement technology for health monitoring have been conducted in various clinical fields [3]. Accelerometer is a practical device to perform long-term monitoring of human balance and posture control in the real-life [4]. Due to advantage of its price, size, and weight, these devices are utilized to detect the motion of upper and lower extremities and trunk while static and dynamic conditions [5,6,7].

To date, there are few clinical quantitative real-time measurement systems for sitting posture. It is very important to record the movement patterns while sitting as most people spend more time sitting with change of working condition. Functional scoliosis (non-structural) patients have a temporary mild curvature caused by different leg length and life habits including lack of exercise. Real-time monitoring and analysis data of these patients can be utilized to correct bad posture as well as improve their postural balance in daily life. Accordingly, real-time sitting posture monitoring system for patients with asymmetrical balance is necessary to evaluate the effects of functional scoliosis on balance performance and to provide appropriate treatment methods for individuals.

In this paper, we suggest real-time sitting posture monitoring system with accelerometer to observe postural pattern of patients with body asymmetry caused by functional scoliosis.

2 Methods

2.1 System Configuration

The upper side of the system has soft curve shape to accommodate the hip area during sitting. Two photo sensors (SG-23FF, Kodenshi Co., Tokyo, Japan) was located to the surface of left and right side, leaving approximately 10 cm space between sensors, for the purpose of checking sitting state of users. Seat surface of the system was covered with soft material to provide comfort. The lower side was shape of a hemisphere with diameter 32 cm to observe the change of natural postural balance pattern according to prolonged sitting condition, as shown in Fig. 1. To detect inclination angle in the frontal and sagittal plane, 3-axis accelerometer (MMA 7331L, Freescale Semiconductor Inc., Austin, Texas) was attached in the middle of main board. A digital finite impulse response (FIR) low pass filter at 2 Hz was used to correct the sensor output. Acquiring data from sensors were converted to angle value in Microcontroller (MCU) and then transformed ASCII codes were transmitted to PC wirelessly by using Bluetooth communication module.

Balance evaluation program was developed to analyze sitting patterns of subjects based on accelerometer data at 100 Hz sampling rate using LabVIEW software (National Instrument Corp., Austin, Texas). The signal was presented as a two-dimensional graph in which the X-axis represents roll angle in the frontal plane and the Y-axis represents pitch angle in the sagittal plane.

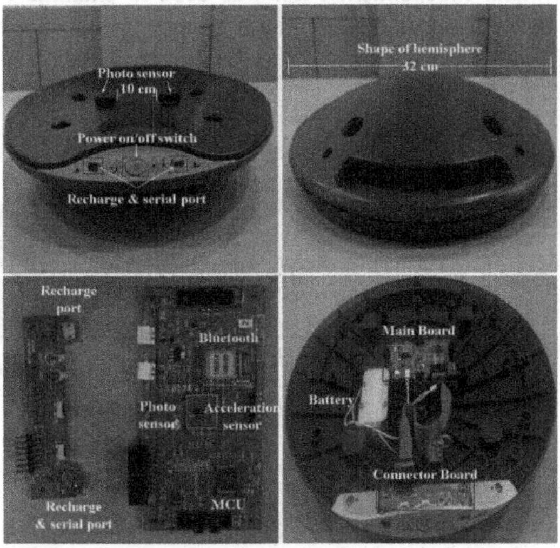

Fig. 1. Structure of sitting posture monitoring system

2.2 Experimental Procedure

10 female subjects who were diagnosed functional scoliosis using posteroanterior full spine standing X-ray were included in this study. They were recruited from the Department of Rehabilitation Medicine of Chungnam National University Hospital in Daejeon, South Korea. All participants were all informed fully about the experiment, and provided written consent prior to their participation. Characteristics of subjects are shown in Table 1.

Table 1. Characteristics of subjects

Variable	Normal Group (n=5)	Patients with scoliosis and pelvic asymmetry Group (n=5)
Age (years)	14.4±1.9	15.4±2.3
Height (cm)	155.4±7.3	165.1±11.7
Weight (kg)	45.4±8.9	57.8±17.6
Cobb angle (°)	-	13.4±4.4
Difference in height of the pelvis (mm)	-	7.9±2.6

Measurement procedure of the system was shown in Fig. 2. Real-time sitting monitoring system was located in the center of a normal wood chair without soft cushion, backrest, and armrests. All subjects were asked to sit on the calibrated system with their usual sitting position. And they kept this posture for 1 minute. Before experiment, they had enough time to adapt unstable state caused by lower part structure of the system. Independent t-test was used to examine the differences in inclination angles between normal and scoliosis group, at $p < 0.05$ level.

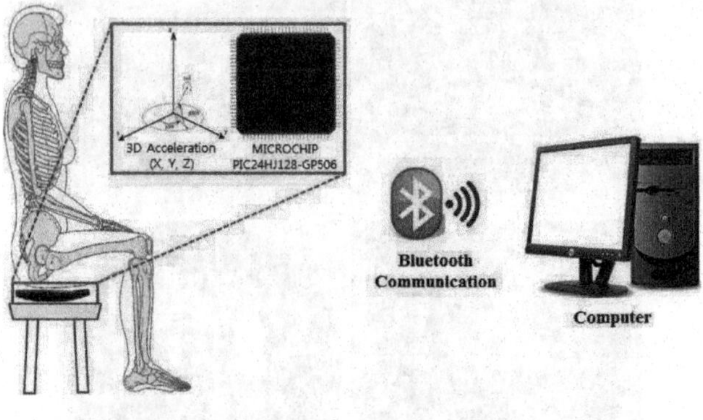

Fig. 2. Measurement procedure of the system

3 Results

Table 2 shows the results of inclination angle while sitting on the real-time posture monitoring system. As compared with inclination angles in the sagittal plane between normal and scoliosis group, tilting angle of scoliosis group more increased on the posterior than normal group. However, there was no significant difference between the groups. In contrast, the opposite tendency was observed in the frontal plane. Inclination angle of normal group increased on the right side, while lateral tilting angle of scoliosis group significantly increased on the left side.

Table 2. Results of inclination angle between normal and scoliosis groups

Inclination angle (°)	Normal Group	Patients with scoliosis and pelvic asymmetry Group	p-value
Anterior (+) / Posterior (-)	-0.88±2.03°	-2.41±4.37°	0.23
Right (+) Left (-)	0.28±0.75°	-1.06±0.69°	0.00*

Functional scoliosis patients in this study have C-shaped lumbar curve to the left side defined by cobb angle and pelvic asymmetry on the right side. Postural balance of patients was tilted to the posterior and left side during sitting. It means that pelvic position in functional scoliosis patients could be associated with spinal alignment and sitting balance due to the lumbar spine being connected to the pelvis directly [8]. Therefore, abnormal lateral curvature of the spine and length asymmetry may negatively influence balance function in compensation.

4 Conclusion

Real-time sitting posture monitoring system with accelerometer was developed to determine postural pattern of patients with functional scoliosis. The results obtained in this study indicated that spinal deformity caused by pelvic inequality can affect the postural balance in the frontal and sagittal planes during sitting. From these results, we confirmed that this system can be applied to monitor natural and comfortable posture with prolonged sitting in the office and at home for accurate diagnosis and treatment on various asymmetrical posture patterns according to the site and type of scoliosis.

Acknowledgements. This work was supported by National Research Foundation of Korea (NRF) grant funded by the Korea government (MSIP) (NRF-2013R1A2A2A04016782) and Basic Science Research Program through the National Research Foundation of Korea (NRF) funded by the Ministry of Education, Science and Technology (2012R1A1B3003952).

References

1. Gopalai, A.A., Senanayake, S.M., Kiong, L.C., Gouwanda, D.: Real-time Stability Measurement System for Postural Control. J. Bodyw. Mov. Ther. 15, 453-464 (2011)
2. Bechly, K.E., Carender, W.J., Myles, J.D., Sienko, K.H.: Determining the preferred modality for real-time biofeedback during balance training. Gait Posture. 37, 391–396 (2013)
3. Mukhopadhyay, S.C., Ihara, I.: Sensors and Technologies for Structural Health Monitoring: A Review. In: Mukhopadhyay. S.C. (eds.) New Development in Sensing Technology for Structural Health Monitoring 2011. LNEE, vol. 96, pp. 1-14. Springer, Heidelberg (2011)
4. Gupta, P., Dallas, T.: Feature Selection and Activity Recognition System using a Single Triaxial Accelerometer. IEEE Trans. Biomed. Eng. 61, 1780-1786 (2014)
5. Karantonis, D.M., Narayanan, M.R.: Implementation of a real-time human movement classifier using a triaxial accelerometer for ambulatory monitoring. IEEE Trans. Int. Tech. Biomed. 10, 156-167 (2006)
6. Curone, D., Bertolotti, G.M., Cristiani, A., Secco, E.L., Magenes, G.: A real-time and self-calibrating algorithm based on triaxial accelerometer signals for the detection of human posture and activity. IEEE Trans. Inf. Tech. Biomed. 14, 1098-1105 (2010)
7. Bliley, K.E., Schwab, D.J., Holmes, D.R., Kane, P.H., Levine, J.A., Daniel, E.S., Gilbert, B.K.: Design of a compact system using a MEMS accelerometer to measure body posture and ambulation. In: 19th IEEE International Symposium on Computer-Based Medical Systems, pp. 335-340. IEEE Press, Utah (2006)
8. Wang, Z.W., Wang, W.J., Sun, M.H., Liu, Z., Zhu, Z.Z., Zhu, F., Qiu, X.S., Qian, B.P., Wang, S.F., Qiu, Y.: Characteristics of the pelvic axial rotation in adolescent idiopathic

scoliosis: a comparison between major thoracic curve and major thoracolumbar/lumbar curve. Spine J. 14, 1873-1878 (2014)

Part V

Decreasing Size of Parameter for Computing Greatest Common Divisor to Speed up New Factorization Algorithm Based on Pollard Rho

Kritsanapong Somsuk

Department of Electronics Engineering, Faculty of Technology, Udon Thani Rajabhat University, UDRU, Udon Thani, Thailand
kritsanapong@udru.ac.th

Abstract. Pollard Rho is one of integer factorization algorithms for factoring the modulus in order to recover the private key which is the one of two keys of RSA and is kept secret. However, this algorithm cannot finish all values of the modulus. Later, New Factorization algorithm (NF) which is based on Pollard Rho was proposed to solve the problem of Pollard Rho that cannot finish all value of the modulus. Nevertheless, both of Pollard Rho and NF have to take time – consuming to find two large prime factors of the modulus, because they must compute the greatest common divisor for all iterations of the computation. In this paper, the method to speed up NF is presented by reducing the size of the parameter which is used to be one of two parameters to compute the greatest common divisor. The reason is, if the size of one of two parameters is reduced, the computation time for computing the greatest common divisor is also decreased. The experimental results show that the computation time of this method is decreased for all values of the modulus. Moreover, the average computation time of the proposed method for factoring the modulus is faster than NF about 6 percentages.

Keywords: RSA, Pollard Rho, Computation time, Factorization Algorithm, Prime Number

1 Introduction

Integer Factorization has become an interesting problem since RSA was introduced. RSA [1] is a public key cryptosystem using a pair of keys for encryption and decryption. This algorithm can be applied with a lot of techniques such as data encryption and digital signature. For data encryption, the public key, which is disclosed for everyone, is used to encrypt the message and the private key, which is kept secret, is used to decrypt the encrypted message. On the other hand, for digital signature, the private key is used for the encryption process and the public key is used for the decryption process. However, if the modulus, n, is factored as the two large prime numbers by using some of factorization algorithms i.e. [2, 3, 4, 5], the private key will be recovered and then RSA will be broken.

K.J. Kim (ed.), *Information Science and Applications*,
Lecture Notes in Electrical Engineering 339, DOI 10.1007/978-3-662-46578-3_47

Pollard Rho, which was discovered by J.M. Pollard in 1975 [6], is one of integer factorization algorithms. This algorithm can factor n into the two large prime numbers very fast whenever the size of n is little. In fact, Pollard Rho is very efficient when the size of n is not larger than 10^{15}. Nevertheless, Pollard Rho cannot finish factoring for some values of n.

In 2011, B.R. Ambedkar and S.S Bedi [7] proposed a new modified integer factorization algorithm which is improved from Pollard Rho, is called New Factorization (NF). This algorithm can finish all values of n. The key of this algorithm is to find the greatest common divisor between s^2-1 and n (GCD(s^2-1,n)) where the initial value of s is $\lceil \sqrt{n} \rceil$ and is increased by 1 whenever the result of GCD(s^2-1,n) is equal to 1 or n. Assume n = p*q, where p and q are two large prime factors of n, one out of two these prime factors is found when the result of GCD(s^2-1,n)) is not equal to 1 or n. However, for every iteration, NF has to take time – consuming to compute the greater common divisor.

In this paper, a method to reduce computation time of NF for factoring n is proposed. The key of this method is that, the size of $s^2 - 1$ will be reduced before using the reduced parameter as one out of two parameters instead of $s^2 - 1$ to compute the greatest common divisor while the accuracy of the result remains the same with the traditional NF, the accuracy is 100% but the computation time of the proposed method is decreased. The experimental results show that the computation time of applying the proposed method with NF is decreased when compares with the traditional NF.

2 Related Work

2.1 Pollard Rho

This is a probabilistic algorithm for factoring composite integers. This algorithm can factor n very fast when the size of n is small (less than 10^{15}). The algorithm of Pollard Rho is as follows:

Algorithm: Pollard Rho
Input: n

1. $x_0 = 2$, i = 1
2. p = 1
3. While(p is equal to 1 or n)
4. $\quad x_i = (x_{i-1}^2 + 1) \% n$
5. $\quad$ If(i % 2 is equal to 0)
6. $\qquad p = GCD(|x_i - x_{i/2}|, n)$
7. $\quad$ EndIf
8. $\quad i = i + 1$
9. EndWhile
10. q = n/p

Output: p and q

However, Pollard Rho cannot factor some values of n. Example 1 shows the case of n which cannot be implemented by using Pollard Rho.

Example 1: Let n = 21

1. $x_0 = 2$
2. $x_1 = (x_0^2 + 1)\%\ n = (2^2 + 1)\ \%\ 21 = 5$
3. $x_2 = (x_1^2 + 1)\%\ n = (5^2 + 1)\ \%\ 21 = 5$
4. $x_i = (x_{i-1}^2 + 1)\%\ n = 5$

From Example 1, we can see that the prime factors of n cannot be found, because the result of x_i is always equal to 5 that the result of GCD($|x_i - x_{i/2}|$,n) is always equal to 0.

2.2 New Factorization (NF)

NF is an improved probabilistic algorithm which is modified from Pollard Rho. The advantage of this method is that it can finish factoring all values of n while Pollard Rho cannot finish factoring some values of n. The algorithm is as follows:

Algorithm: NF
Input: n

1. $s = \lceil \sqrt{n} \rceil$
2. p = 1
3. While(p is equal to 1 or n)
4. p = GCD(s^2 – 1, n)
5. s = s + 1
6. EndWhile
7. q = n/p

Output: p and q

From this algorithm, we can see that, the value of s is always increased by 1 whenever the value of p is equal to 1 or n. Therefore, NF becomes to take time – consuming to compute p = GCD(s^2 – 1, n) especially s^2 is very large.

3 The Proposed Method

The aim of this paper is to reduce the size of the parameter s^2 - 1 in order to increase computation speed before computing the greatest common divisor between the reduced size of s^2 – 1 and n. In addition, if the result of GCD(s^2 – 1, n), which is more than 1 and less than n, is found, it implies that:

$$s^2 - 1 = r*a \text{ and } n = r*b \quad (1)$$

Where r is the result of $GCD(s^2 - 1, n)$ and a, b are any positive integers and a is more than b.

Let, k is the maximum integer that $u - 1 = r*a - k*r*b$ is still less than n and more than 1 but not equal to b, where $GCD(a - k*b, b)$ is 1. Then, the result of $GCD(u - 1, n)$ is also equal to r as follows:

$$GCD(u - 1, n) = GCD(r*a - k*r*b, r*b)$$
$$= GCD(r*(a - k*b), r*b)$$

GCD(a - k*b, b) is 1, $= r$

That means $$(u - 1) \equiv (s^2 - 1) \bmod n \quad (2)$$

Notation: because of b which is represented as the factor of n, therefore b is always a prime number. Moreover, if the result of $a - k*b$ is more than 1 and not equal to b, it implies that the result of $GCD(a - k*b, b)$ must be equal to 1.

Although the total iterations of the computation of the proposed method is always equal to the total iterations of the computation by using NF, one of parameters of the proposed method to compute the greatest common divisor is less than the parameter of NF, while the other parameter of the both methods are same. Therefore, the computation time of the proposed method is decreased when compared with NF.

Furthermore, from theory of mathematical, the distance of sequence perfect square is always increased by 2. For example, the distance between 1^2 and 2^2 is 3, 2^2 and 3^2 is 5, 3^2 and 4^2 is 7. So, if the distance between x^2 and $(x+1)^2$ is $2x + 1$, where x is any integer, then the distance between $(x+1)^2$ and $(x+2)^2$ is $2x+1+2 = 2x +3$. In addition, this technique will be also applied with the proposed method. Therefore, the speed up algorithm of NF is as follows:

Algorithm: The Proposed method
Input: n

1. $s = \lceil \sqrt{n} \rceil$
2. $u = s^2$ % n;
3. dis = 2*s + 1 // dis is an initial distance between of s^2 and $(s+1)^2$
4. While(p is equal to 1 or n)
5. p = GCD(u – 1, n)
6. u = u + dis;
7. dis= dis + 2
8. EndWhile
9. q = n/p

Output: p and q

This algorithm implies that the size of u is always less than the size of s^2 because the value of u is from: $u = s^2$ % n, s^2 is always more than n. Therefore, the computation time of the proposed method to find two large prime factors of n is always less than NF, because the proposed method has to compute GCD(u – 1, n) for every iteration of the computation while NF has to compute GCD($s^2 - 1$, n) for every iteration of computation.

Example 2: Let n = 60491, Find p and q by using NF and the proposed method

NF: (The previous study)

1. $s = \lceil \sqrt{n} \rceil = \lceil \sqrt{60491} \rceil = 246$
2. Find p = GCD($s^2 - 1$, n), 1 < p < n

   Round 1: (s = 246),
   GCD($s^2 - 1$, n) = GCD(**<u>60515</u>**, 60491) = 1

   Round 2: (s = 247),
   GCD($s^2 - 1$, n) = GCD(**<u>61008</u>**, 60491) =1

   Round 3: (s = 248),
   GCD($s^2 - 1$, n) = GCD(**<u>61503</u>**, 60491) =1

   Round 4: (s = 249),
   GCD($s^2 - 1$, n) = GCD(**<u>62000</u>**, 60491) =1

   Round 5: (s = 250),
   GCD($s^2 - 1$, n) = GCD(**<u>62499</u>**, 60491) = 251

   So, p = 251, q = n/p = 60491/251 = 241

The proposed method:

1. s = 246
2. u = s^2 % n = 246^2 % 60491 = 25
3. dis = 2*s + 1 = 2 * 246 + 1 = 493
4. Find p = GCD(u – 1, n), 1 < p < n

   Round 1: (u = 25, dis = 493),
   GCD(u - 1, n) = GCD(**<u>24</u>**, 60491) = 1
   u = u + dis = 25 + 493 = 518
   dis = dis + 2 = 495

   Round 2: (u = 518, dis = 495),
   GCD(u – 1, n) = GCD(**<u>517</u>**, 60491) =1
   u = u + dis = 518 + 495 = 1013
   dis = dis + 2 = 497

   Round 3: (u = 1013, dis = 497),
   GCD(u – 1, n) = GCD(**<u>1012</u>**, 60491) =1
   u = u + dis = 1013 + 497 = 1510
   dis = dis + 2 = 499

Round 4: (u = 1510, dis = 499),

GCD(u – 1, n) = GCD(**1509**, 60491) =1

u = u + dis = 1510 + 499 = 2009

dis = dis + 2 = 501

Round 5: (u = 2009, dis = 501),

GCD(u – 1, n) = GCD(**2008**, 60491) =251

So, p = 251, q = n/p = 60491/251 = 241

From the example above, although, the number of iterations of the computation of the both algorithms are same, the size of u – 1 which is used to be the one out of two parameters of the proposed method to compute the result of the greatest common divisor, is very small when compared with the size of $s^2 - 1$ which is used to be the one out of two parameters of NF. Therefore, the computation time of the proposed method is less than NF.

4 Experimental Results

The experimental is the comparison about computation time for factoring n between NF and the proposed method. The size of n which is used in the experimental is from 32 to 64 bits. Both of two algorithms are implemented by using the same environment, java programming language which is run on 2.53 GHz an Intel® Core i3 with 4 GB memory.

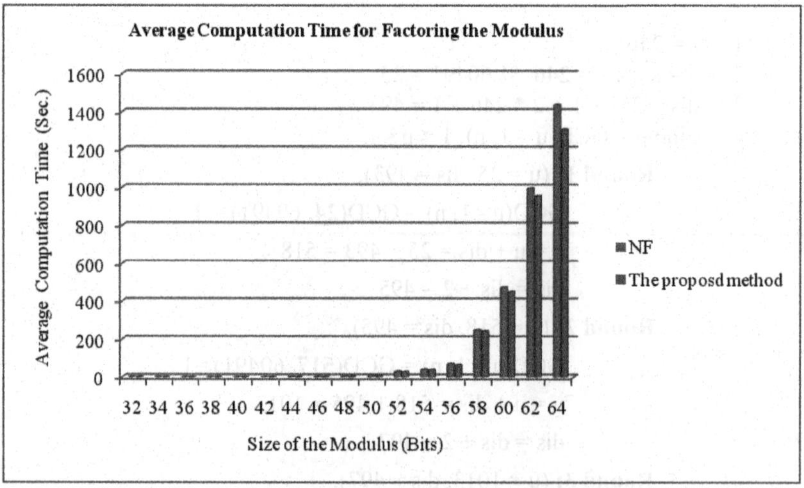

Fig. 1. The average computation time for factoring n when the size of n is between 32 – 64 bits

The information in Fig. 1 shows the average computation time of two algorithms, NF and the proposed method, for factoring n when the size of n is between 32 to 64 bits. Each bit size of n is run about 10 times to find the average computation time. The experimental results show that, although the graphs of the both algorithms are very nearest to each other, the graph of the proposed method is always under the graph of NF for all values of n. That means, the computation time of the proposed method for factoring n is always less than the computation time of NF. Moreover, the average computation time of the proposed method is faster than NF about 6 percentages.

5 Conclusion

In this paper, a new method is proposed to speed up NF which is an improved integer factorization algorithm, modified from pollard Rho and can finish all value of n. The key is that the size of parameter which is used to compute the greatest common divisor to find the prime factor of n is decreased while the accuracy of the result is still the same with the traditional NF. The experimental results show that the computation time for factoring n by using the proposed method is decreased in comparison to the traditional NF.

References

1. Rivest, R.L., Shamir, A., Adleman, L.: A method for obtaining digital signatures and public key cryptosystems. In: Communications of ACM, vol. 21, pp. 120 – 126 (1978)
2. Ambedkar, B. R., Gupta, A., Gautam, P., Bedi,S.S.: An Efficient Method to Factorize the RSA Public Key Encryption . In: International Conference on Communication Systems and Network Technologies, pp. 108 – 111, (2011)
3. Huang, Q., Li, Z.T., Zhang, Y., Lu, C.: A Modified Non-Sieving Quadratic Sieve For Factoring Simple Blur Integers. In: International Conference on Multimedia and Ubiquitous Engineering, pp. 729 – 732, (2007)
4. Bishop, D.: Introduction to Cryptography with java Applets. In: Jones and Bartlett Publisher, (2003)
5. Sharma, P., Gupta, A.K., Vijay, A.: Modified Integer Factorization Algorithm using V-Factor Method. In: Advanced Computing & Communication Technologies, pp.423 – 425, (2012)
6. Pollard, J.: Monte Carlo methods for index computation (mod p). In: Math. Comp., vol. 32, pp.918-924, (1978)
7. Ambedkar, B. R., Bedi,S.S.: A New Factorization Method to Factorize RSA Public Key Encryption. In: International Journal of Computer Science Issues., vol. 8, pp.242-247, (2011)

Network Security Situation Assessment: A Review and Discussion

Yu-Beng Leau* Selvakumar Manickam and Yung-Wey Chong

National Advanced IPv6 Centre (NAv6), Universiti Sains Malaysia,
11800, Bayan Lepas, Penang, Malaysia.
{beng,selva,chong}@nav6.usm.my
http://www.springer.com/lncs

Abstract. The number of network intrusion attempts have reached an alarming level. Questions have been raised about the efficiency of deploying intrusion detection and prevention system which are more concern on single device instead of overall network security situation. Researchers have shown an increased interest in designing network security situation awareness which consists of event detection, situation assessment and situation prediction. Generally, Network Security Situation Assessment is a process to evaluate the entire network security situation in particular time frame and use the result to predict the incoming situation. In this paper, we review existing network security situation assessment methods from three major categories in the aspect of its strengths and limitations. A list of consideration criteria has been summarized for future situation assessment model design.

Keywords: Network Security Situation Assessment, Statistical Approach, Relationship Analysis, Artificial Intelligence

1 Introduction

The rapid development of Internet, which offers convenient services and information sharing, has opened up an opportunity for it to become a breeding ground for malware and cyber criminals [1, 2]. In a report revealed by G Data Software, there were 1,509,934 new malware found in the first half of 2013. This means that an average of 8,342 new malware program types is produced every day. [3]. In 2012, Symantec also encountered a 58% increase in new mobile malware compared to previous year [4]. In Malaysia, the number of incidents is also rising progressively which hit 10636 cases with different types of attacks such as denial of service, intrusion attempt, malicious codes, spamming and etc [5]. These increasing numbers brings serious challenges and problems to network security. In addition, most prevention system responds directly to these attacks without any assessment on the alert, creating a lot of false positive and false negative notification [6]. Therefore, instead of concerning single asset in the network, security

* National Advanced IPv6 Centre (NAv6), Universiti Sains Malaysia, 11800, Bayan Lepas, Penang, Malaysia.

K.J. Kim (ed.), *Information Science and Applications*,
Lecture Notes in Electrical Engineering 339, DOI 10.1007/978-3-662-46578-3_48

administrator nowadays are more interested on assessing security situation in overall network in order to provide useful information to Intrusion Prevention System for predicting the incoming network security situation and be ready with proper action taken.

2 Network Security Situation Assessment in Situation Awareness Framework

Network Security Situation Assessment is the second level in Network Security Situation Awareness framework. The concept of Security Awareness (SA) was first introduced by Endsley [7] in the aviation and aerospace realms throughout the research on human factors. Its objective is to ensure the necessary information is readily accessible and understood by various levels of decision makers and analysts by providing them in an abstract visual format. To firmly expand upon their perspective of SA, three hierarchical phases which begins with Perception, followed by Comprehension and the highest level is Projection were introduced [8, 9]. Perception classifies information about the status, attributes and dynamics of relevant elements within the environment into understood representation. Comprehension of the situation includes how people integrate multiple pieces of information, interpret them in terms of their relevance to an individual's underlying goals and able to infer conclusions about the goals. Based on the knowledge from previous levels, Projection represents a forecasting of the elements of the situation into the near future [10]. The range of awareness levels is progressively increasing from basic perception of important data to interpretation and combination of data into knowledge, then prediction of future situation and their implementation.

With concerning environment of cyberspace, in 1999, Tim Bass proposed the concept of SA into Network Security field called Network Security Situation Awareness (NSSA). NSSA can be divided into three stages which are event detection, current situation assessment and future situation prediction [11]. Event Detection identifies the abnormal and malicious activity in the network and translates them into logical format. Current Situation Assessment is a process to evaluate the security situation of the entire network by using the information from previous stage. Last stage is Future Situation Prediction is aimed to forecast the future network security tendency according to the current and historical network security situation status.

Basically, Network Security Situation Assessment is the core of SA where it extracts the situation elements, analyses the association of security events among them and infers the degree of threats in each layer of the network in order to reflect the security situation in the whole network. It provides an all-in-one and reliable frame of reference to administrators in order to make the right and timely decisions. In this paper, we only focus on reviewing existing network security situation assessment mechanisms.

3 Existing Network Security Situation Assessment Mechanisms

Today, many assessment mechanisms have been proposed to attain the goal and there can be categorized into three oriented-bases which are logical relationship, artificial intelligence and mathematical model.

3.1 Based on Relationship Analysis

Assessment method based on relationship analysis is a means to examine interrelationship among the alerts which might symbolize an attack scenario in a network. Some subjective criterions such as expert's knowledge, experience and historical data are required in this method. By correlating detected alerts on the basic of prerequisites and consequences of attacks, the attack scenarios can be constructed to represent security situation of the network such as the frameworks in [12, 13]. The intuitive conviction, a prerequisite of an attack, is the necessary condition for an attack to be successful, whereas the consequence of an attack is the possible outcome of the attack. These methods are not only heavily depending on the completeness of predefined reference templates but also involve high labour cost and time.

In 2010, Zhaowen et al proposed a real-time prerequisite and consequence (RIAC) intrusion alert correlation model [14]. It employs distributed agents to collect alert information online and uses them to produce hyper-alerts. Sets of prerequisite and expended consequence have been derived by using information in knowledge base. By identifying the "prepare-for" relationships among hyper-alerts, attack scenario and intent intrusion behind the alerts were discovered.This model can be applied in large scale environment but it highly depends on expert knowledge to completely prepare alert information knowledge base. To overcome this constraint, Anbarestani et al [15] proposed a Bayesian Network-based alert correlation which discovers attack strategies without the need for expert knowledge. The approach extracted attack scenarios using actions sequence classification. It leverages upon historical data from log sources and classify them based on observed intrusion objective as class variables. The possible attack scenarios constructed from hyper alerts sequences are examined and the most plausible strategies for constructing a cooperative attack are extracted. The model eliminates the redundant relationships but a set of adequate and reliable historical data from log sources is required. In the same year, Zali et al [16] applied graph model called Causal Relations Graph to represent attack pattern in knowledge base. Some trees related to alerts probable correlations are constructed offline while the correlations of each received alert in real time with previously received alerts will be identified by performing a search only in the corresponding tree. Although the model is able to identify the relationship between the alerts in a short time, the construction of complete tress related to alerts probable correlations is laborious and challenging.

In general, situation assessment method based on relationship analysis is easy to understand and efficient if a comprehensive relation template which includes prerequisites and consequences of various types of attacks is in place. Nevertheless, this method has some limitations. The reference model with prerequisite and consequence of various attacks is difficult to construct and it is hard to identify the logical relationship among the alerts which come from multiple sources. For this reason, this method is unable to illustrate the uncertainty exist in the whole network system which is required by security managers as their guidance to configure network security mechanism.

3.2 Based on Artificial Intelligence

Assessment method based on artificial intelligence, particularly artificial immune system, uses biological immune theoretic as references to search and design relevant models and algorithms to solve the various problems. With the help of biological technology such as self-tolerance, self-learning and evolution mechanisms, the intrusion detection system utilizes the biological immunity theory as the base of security awareness to detect known and unknown intrusions. In 2009 and 2011, Liu Nian et al.[17] and Zhang Ruirui et al.[18] have applied the concept of artificial immune technology in their proposed NSSA structure respectively . They suggested that by determining the corresponding relationship of the change of antibody concentration of human immune system and pathogens intrusion rate with attack power, researchers can use this information to calculate the security situation of hosts. Network security situation can be quantitatively determined in real time and dynamic manner. The proposed model is claimed to be able to realize overall evaluation of the attacks in each level of host computer, sub-network and whole network, observe the current system risk which reflects the present security situation in network timely and accurately. Unfortunately, there are some limitations in this model particularly in scalability and coverage which lead to low efficiency and high false negative. Besides that, a huge number of detectors are needed and intolerable time is required to improve the coverage rate.

In order to overcome these deficiencies, a network security situation assessment model based on cooperative artificial immune system has been suggested [19]. In this model, the memory detectors in different computers which spread in the whole system will share the differences of their detected packet. The function of the cooperative module is to send and receive the information of cooperative detector. It will also decide which collaboration relationship should be used and when it has to quit. Although this model is to share the efficient detectors in a certain range with which shorten the training time of individual immune system, yet it is not an easy task to train the detector to be act intelligently in sharing the information.

3.3 Based on Statistical Approach

Assessment method based on statistical approach is aimed to build an evaluation function by considering a range of elements which influence the security situation of network. The function maps the relation of all situational elements, R into a situational space, S which can represents as

$$S = (r_1, r_2, r_3, \ldots, r_n), \qquad r_1 \in R(1 \leq i \leq n) \tag{1}$$

Many researchers had proposed weighted average method to assess the network security situation [20–34] and most of them are hierarchical approach applied. They divided their proposed framework into three layers from bottom to top which covers service layer, host layer and network layer [20–22, 26–29, 31] . In service layer assessment, the focus is more on the vulnerabilities of services and its times to be utilized. The threat degree of each situational element in the layer is taken into account to product the weight of the element in order to calculate the situation value of the layer. The sum of situation value from each layer represents the situation of element in higher layer. The same evaluation process has been done at host and network layer and the overall security situation of the network is the total of situation value in host layer.

Undoubtedly, weighted average method is simple and easy to be used in any assessment process which a weight value will be assigned to each situational element or resource according to its importance in the network. However, the main limitation of this method is lack of solid and standard guidelines to determine the significance of each element in the network and it causes the weight assignment subjectively. Due to this, there are several alternatives have been used to obtain the weight value such as through the expert's perception and organization policy [20, 22, 28, 30, 32–34], comparison between the current and historical observed value at the layers [24, 25], calculation from Grey Relational Analysis [22] as well as Analytic Hierarchy Process [29].

4 Conclusion

In this paper, we identified the existing problems of network security situation and explained the role of situation assessment in a NSSA framework. We also categorized and presented a comprehensive review on existing network security situation assessment methods based on their approaches. Each of them has own strengths and weaknesses in terms of the completeness of chosen criteria in assessment process. Table 1 stated the consideration criteria in some recent existing security situation assessment methods.

From the table, we noticed that most of the related works focus on statistical approaches which quantitatively determine the asset importance, attack severity and its likelihood of occurrence. Based on our best knowledge, there is no single study exists which adequately considers the cost factor in the assessment. In this context, it referred to damage and response costs which usually used in selecting appropriate responses after the situation evaluation [35, 36]. While damage cost characterizes the amount of damage to a target resource by an attack when intrusion detection is unavailable or ineffective, response cost is the cost of acting upon an alarm or log entry that indicates a potential intrusion [37].

Table 1. Criteria has been considered in Existing Network Security Situation Assessment Methods

Category	Existing Assessment Method	Severity of Attack	Frequency of Attack	Importance of Service	Likelihood of Occurrence	Damage Degree	Types of Attacks
Relationship Analysis	Ning, Cui et al.[12]		/				
	Xu and Ning [13]	/					/
	Zhaowen, Li et al. [14]						/
	Anbarestani, Akbari et al. [15]	/			/		
	Zali, Hashemi et al. [16]						/
Artificial Intelligence	Nian, Diangang et al.[17]						/
	Ruirui, Tao et al.[18]			/		/	/
	Qiao and Xu [19]				/	/	
Statistical Approach	Hu, Li et al. [20]		/	/		/	
	Yong, Xiaobin et al.[21]			/	/	/	/
	Zhang, Wang et al. [22]		/	/	/		
	Wang, Zhang et al. [23]	/		/			/
	Liqun and Xingyuan [24]			/			/
	Song and Zhang [25]	/	/	/	/		
	Zhang, Huang et al. [26]	/		/		/	
	Cheng and Lang [27]		/	/			
	Xiaorong, Su et al. [28]	/		/			
	Bian, Wang et al. [29]	/	/	/	/		/
	Xiangdong Cai, Yang Jingyi et al. [30]	/		/	/		
	Xiaoli and Hui [31]	/		/			
	Szwed and Skrzynski [32]			/			
	Zheng, Wei et al. [33]	/	/	/	/	/	
	Zhang, Chen et al. [34]			/	/		

As a conclusion, all the criteria including the damage and response costs should take into consideration in designing an efficient and cost-sensitive network security assessment mechanism in order to provide a more reliable and accurate current network security situation.

Acknowledgement The authors would like to thank the National Advanced IPv6 Centre (NAv6), Universiti Sains Malaysia for supporting this research project.

References

1. Yu Beng, L., et al., A Survey of Intrusion Alert Correlation and Its Design Considerations. IETE Technical Review. 31(3), pp. 233-240. (2014).
2. Beng, L.Y., S. Manickam, and T.S. Fun, A Framework for Analytic Hierarchy Process-Entropy Network Security Situation Assessment and Adaptive Grey Verhulst-Kalman Prediction in Intrusion Prevention System. Australian Journal of Basic & Applied Sciences, 8(14), pp.34-39.(2014).
3. G Data PC Malware Report in Half-yearly Report (January - June 2013). G Data SecurityLabs: Germany. pp. 1-12. (2013).
4. Internet Security Threat Report 2013. Symantec Corporation: United States. pp. 1-58. (2013).
5. MyCERT Incident Statistics Year 2013-2014, http://www.mycert.org.my
6. Jawdekar, A., V. Richariya, and V. Richariya, Minimization of False Alarm Prediction in IDS Based On Frequent Pattern Mining. International Journal of Emerging Technology and Advanced Engineering, 2(4), pp. 511-514. (2012).
7. Endsley, M.R. Situation awareness global assessment technique (SAGAT). In: National Aerospace and Electronics Conference, pp. 789-795. IEEE, (1988).
8. Endsley, M.R., Toward a theory of situation awareness in dynamic systems. The Journal of the Human Factors and Ergonomics Society, 37(1), pp. 32-64. (1995).
9. Endsley, M.R., et al., Situation awareness information requirements for commercial airline pilots. International Center for Air Transportation. pp. 1-7. (1998).
10. Jajodia, S., et al., Cyber situational awareness. 14, pp. 3-14. Springer, Heidelberg (2010).
11. Bass, T., Multisensor data fusion for next generation distributed intrusion detection systems. In: 1999 IRIS National Symposium on Sensor and Data Fusion pp. 24-27. (1999).
12. Ning, P., et al., Techniques and tools for analyzing intrusion alerts. ACM Transactions on Information and System Security (TISSEC), 7(2), pp. 274-318.(2004).
13. Xu, D. and P. Ning. Alert correlation through triggering events and common resources. In: 20th Annual Computer Security Applications Conference, pp. 360-369. IEEE, (2004).
14. Lin, Z., et al., Real-Time Intrusion Alert Correlation System Based on Prerequisites and Consequence. In: 6th International Conference on Wireless Communications Networking and Mobile Computing (WiCOM), pp. 1-5, IEEE, (2010).
15. Anbarestani, R., et al., An iterative alert correlation method for extracting network intrusion scenarios. In: 20th Iranian Conference on Electrical Engineering (ICEE), pp. 684-689 , IEEE, (2012).
16. Zali, Z., et al., Real-time attack scenario detection via intrusion detection alert correlation. In: 9th International ISC Conference on Information Security and Cryptology (ISCISC), pp. 95-102, IEEE, (2012).
17. Nian, L., et al. Research on network security situation awareness technology based on artificial immunity system. In: International Forum on Information Technology and Applications, pp. 472-475. IEEE, (2009).
18. Ruirui, Z., et al., A Network Security Situation Awareness Model Based on Artificial Immunity System and Cloud Model. In: Computing and Intelligent Systems, pp. 212-218. Springer, Heidelberg (2011).
19. Qiao, Y. and J. Xu. A network security situation awareness model based on cooperative artificial immune system. In: International Conference on Computer Science and Service System (CSSS), pp. 1945-1947. IEEE, (2011).

20. Hu, W., J. Li, and J. Shi. A novel approach to cyberspace security situation based on the vulnerabilities analysis. In: The Sixth World Congress on Intelligent Control and Automation, pp. 4747-4751. IEEE, (2006).
21. Yong, Z., T. Xiaobin, and X. Hongsheng. A novel approach to network security situation awareness based on multi-perspective analysis. In: International Conference on Computational Intelligence and Security, pp. 768-772. IEEE, (2007).
22. Zhang, F., J. Wang, and Z. Qin. Using gray model for the evaluation index and forecast of network security situation. In: International Conference on Communications, Circuits and Systems, pp. 309-313. IEEE, (2009).
23. Wang, J., et al. Alert analysis and threat evaluation in Network Situation Awareness. In: International Conference on Communications, Circuits and Systems, pp. 278-281. IEEE, (2010).
24. Liqun, T. and Z. Xingyuan. A method of service-oriented network security situational assessment in transport layer. In: International Conference on Multimedia Technology, pp. 4759-4763. IEEE, (2011).
25. Song, S. and Y. Zhang. A Novel Extended Algorithm for Network Security Situation Awareness. In: International Conference on Computer and Management, pp. 1-3. IEEE, (2011).
26. Zhang, Y., et al., Multi-sensor Data Fusion for Cyber Security Situation Awareness. Procedia Environmental Sciences. 10, pp. 1029-1034. (2011).
27. Cheng, X. and S. Lang. Research on network security situation assessment and prediction. In: Fourth International Conference on Computational and Information Sciences (ICCIS), pp. 864-867. IEEE, (2012).
28. Xiaorong, C., L. Su, and L. Mingxuan, Research of Network Security Situational Assessment Quantization Based on Mobile Agent. Physics Procedia. 25, pp. 1701-1707. (2012).
29. Bian, N., X. Wang, and L. Mao. Network security situational assessment model based on improved AHP_FCE. In: Sixth International Conference on Advanced Computational Intelligence, pp. 200-205. IEEE, (2013).
30. Xiangdong Cai, X.C., Y.J. Yang Jingyi, and H.Z. Huanyu Zhang, Network Security Threats Situation Assessment and Analysis Technology Study. International Journal of Security and Its Applications. 7(5), pp. 217-224.(2013).
31. Xiaoli, G. and W. Hui, Research on the Network Security Situation Awareness Model for the Electric Power Industry Internal and Boundary Network. Journal of Applied Sciences. 13(16), pp. 3285-3289. (2013).
32. Szwed, P. and P. Skrzynski, A new lightweight method for security risk assessment based on fuzzy cognitive maps. International Journal of Applied Mathematics and Computer Science. 24(1), pp. 213-225. (2014).
33. Zheng, R., et al., Network Security Situation Evaluation Strategy Based on Cloud Gravity Center Judgment. Journal of Networks. 9(2), pp. 283-290. (2014).
34. Zhang, B., et al., Network security situation assessment based on stochastic game model, in Advanced Intelligent Computing, pp. 517-525. Springer, Heidelberg (2012).
35. Jumaat, N.B.A., Incident Prioritisation for Intrusion Response Systems, in School of Computing and Mathematics, pp. 25-37, Plymouth University: United Kingdom, (2012).
36. Stakhanova, N., et al., A Cost-Sensitive Model for Preemptive Intrusion Response Systems. In: AINA, 7, pp. 428-435, (2007).
37. Lee, W., et al., Toward cost-sensitive modeling for intrusion detection and response. Journal of Computer Security. 10(1), pp. 5-22. (2002).

ProtectingBinary Files from Stack-Based Buffer Overflow

Sahel Alouneh[1],HebaBsoul[2], Mazen Kharbutli[2]
[1]German Jordanian University, [2]Jordan University of Science and Technology
{sahel.alouneh@gju.edu.jo, habsoul09@cit.just.edu.jo,kharbutli@just.edu.jo}

Corresponding author: Sahel Alouneh, Ph.D

Abstract

Vulnerabilities that exist in many software systems can be exploited by attackers to cause serious damages to the users. One of such attacks that have become widely spread in the last decade is the buffer overflow attack. The attacker can, if successful, execute an arbitrary code with the same access privileges as the attacked process. Thus, if the attacked process is a root process, the attackers can execute any kind of code they want and therefore causing a security breach in the system. In this paper, we propose a new solution to the buffer overflow attacks that can protect return addresses from being overwritten. Our solution works with string library functions, such as *strcpy()* by preventing access to memory locations beyond the frame pointer of a function, and thus preventing overwriting the return address. Unlike other approaches that have been used to solve the buffer overflow attack, our solution can detect and fix buffer overflow vulnerabilities in executable (i.e.,the *.exe* or binary files). In other words, our solution does not require the availability of the program source code, which may not be available for many applications, and does not require any hardware modifications, which can be expensive. Therefore, we developed a tool that can be used to convert a vulnerable program to a safe version that is protected against buffer overflow attacks.

Keywords- Buffer overflow, Security, Executable file, Stack, Return address

1. INTRODUCTION

Nowadays, computer users are very worried about computer attacks that come in the form of computer viruses, spyware, malware, etc. Computer attackers exploit the vulnerabilities that exist in the software applications to hijack the program control and to take illegal root privilege. Buffer overflow attack is one of the most common forms of attack that violate the security of software applications. This paper presents a new method for protecting software against this kind of attacks. The buffer overflow attack relies on changing the control flow of a program by overwriting the return address of a routine by the address of an arbitrary code, leading to executing that arbitrary code that may lead to a security breach in the system.

The attackers exploit the stack more than other memory regions to hijack the control flow of a program [17]. The stack region is chosen because it is the easiest to be targeted by this kind of attacks. Additionally, the stack is an executable region in which any injected code will be executed if the control flow is transferred to it. Thus, to complete the attack, the attacker needs to modify the control flow of the execution to refer to that previously injected code. Figure 1.1 is an example of buffer overflow attack.

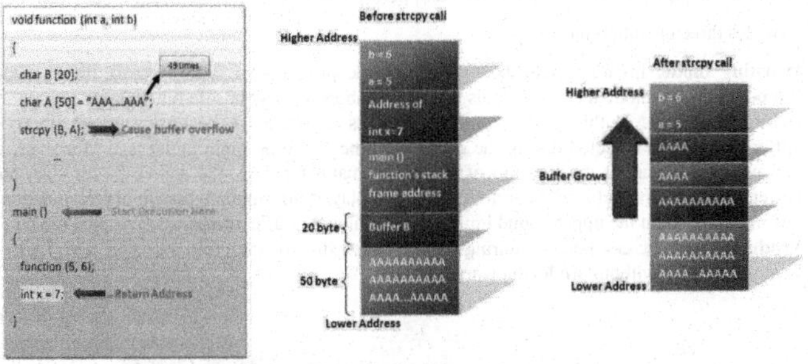

Figure 1.1Example on buffer overflow in C

In this paper, we propose, implement, and test a new comprehensive solution and tool that provides protection against buffer overflow attacks. Our proposed solution detects and prevents the attacker from overwriting the return address and the previous frame pointer regions. Unlike most of the available solutions, our solution tool does not require the availability of the original source code, only the executable (i.e. the .exe or binary) files are needed. Moreover, it does not require any architectural modifications while available hardware solutions doe, and it does not require operating system's modifications.

K.J. Kim (ed.), *Information Science and Applications*,
Lecture Notes in Electrical Engineering 339, DOI 10.1007/978-3-662-46578-3_49

This paper is organized as follows: Section 2 presents the literature of related proposed solutions to buffer overflowattacks. In Section 3, our proposedsolution and tool are discussed. After that, the proposed tool is evaluated in Section 4 with discussionson the obtained results. Finally, we conclude the paper in Section 5.

2. RELATED WORK

Many solutions were proposed to solve the buffer overflow problem. However, none solves the problem fully. Moreover, these solutions suffer from many shortcomings. These solutions are either hardware solutions or software solutions. The hardware solutions [10, 14, 16, 25, and 26] require architecture's changes which are very expensive and thus making this kind of solutions undesirable. The software solutions [1, 2, 3, 4, 6, 7, 8, 9, 12, 13, 15, 18, 19, 20, 21, 22, 23, 24, and 27] on the other hand may require compiler's modifications which are not easy to implement.Creating a solution that is compatible for most applications is a difficult task. Most of proposed the solutions require a source code to solve the problem [1, 2, 6, 7, 8, 12, 13, 15, 19, 23, and 24]; unfortunately, the source code sometimes may not exist especially for legacy systems.

Several approaches have been proposed to reduce the severity of buffer overflow attacks. The approaches in [1], [6], [7], [8], [14] (compiler-based version), [2], [12], [13], [14], [15], [19], [23] and [24] require source code which may not be available such as in legacy systems. The approaches in [1], [9], [14], [7], [8], [4], [18], [19], [23], [17] and [16] proposed to save a copy in return address repository (return stack) have a problem that the allocated storage may not be enough to store all data pushed in it.

3. PROPOSED SOLUTION

In this section we present the details of our solution that provides protection against buffer overflow attacks in the absence of the source code of the program that is under testing. Our tool detects and prevents the attack from overwriting the return address and the previous frame pointer regions. The fundamental idea of our tool is to intercept every call to unsafe library functions and then redirect them to the corresponding safe version implemented by our tool. The safe version overcomes the vulnerability that exists in C programming language by checking the bounds for the buffers of the C library functions.

Most importantly, the input to our tool is an executable (binary) file. Indeed, this is a very unique feature in our solution as we assume that the source code is not available. This assumption is actually valid as in many cases the source code is not made open to public, or the software is a legacy one where the source code cannot be found.The output of the tool is a modified safe assembly code of the original input binary file. This output assembly code is assembled and compiled using the gcc compiler to produce a new safe executable file.The following steps (Figure 3.1) summarize the main procedures performed by ourtool to produce a secure binary file:

1. Disassemble the input binary file using "objdump" disassemble tool that converts the input file into an assembly file.
2. Convert the resulted assembly file into a compatible assembly x86 file that is ready for compilation.
3. Add the appropriate instructions and functions to make the assembly x86 file invulnerable to buffer overflow attacks.
4. Compile the produced file using the gcc compiler; this step results in the final secure executable (binary) file.

Our tool provides three operation modes:

a) **Warning mode**: In this mode, the tool alarms the programmer if any unsafe library function is detected. Moreover, in this mode the tool outputs the line numbers of these unsafe functions.
b) **Debugging mode**: In this mode, the tool generates an alarm whenever any unsafe library function causing buffer overflow is detected during the execution time of the program under test. Moreover, the tool prints the line number indicating the location of that problematic function, the source buffer size, and the destination buffer safe upper bound size. In addition to the displayed information, unsafe library functions are corrected in this mode based on the upper bound limit of the destination buffer size.
c) **Production mode**: can be used during the execution time of the program under test. This mode fixes unsafe library functions without producing warnings.

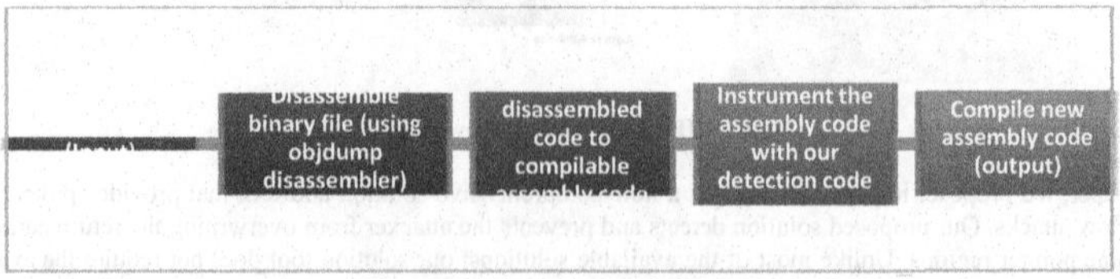

Figure 3.1 Procedures of our tool

As mentioned previously, the input for our tool is the "objdump" output file produced from the original unsafe executable file. The output file from "objdump" disassembler contains, in addition to the assembly instructions of the program under test, other information such as labels and addresses of each instruction. This format, therefore, is not suitable to be used for compilation; rather, the output files from "objdump" tool need to be converted to a valid x86 assembly files that are suitable for compilation. The information to be extracted from the "objdump" output file are labels, globally-initialized variables, globally-uninitialized variables, functions, and jump instructions. This conversion process is the first step that our tool does, and it is done in several stages.An "objdump" file uses the addresses of the variables and labels instead of their names, whereas the x86 assembly file uses the names of the variables or labels; thus, the first step in the conversion process is to convert these addresses to their corresponding names. We have three types of addresses: the addresses of global, initialized variables that are located in .data segment, the addresses of global, uninitialized variables that are located in the symbol table (*.bss* segment), and the addresses of the strings that are located in *.rodata* segment. The names of strings (labels) are created by our tool, and the names of variables are extracted from the symbol table.

We first convert the addresses of the strings in .rodata section to the appropriate labels. Each label should have a unique name, so we declare a counter that starts from zero and increment it by one for every label, and then we concatenate the counter value with the string "LC". For example, the first label name is LC0, the second label name is LC1, and so on.

Figure 3.2 shows the segments of .rodata section. These segments are the addresses, the hexadecimal values, and the strings themselves. Every address in a line differs from the address in the next line by sixteen byte (i.e. 0x8048548 – 0x8048538 = 0x10). The addresses are represented in hexadecimal representation.Every line in the .rodata segment also contains thirty two hexadecimal digits and sixteen characters. The address at the beginning of every line refers to the address of the first character in that line, and it also refers to the address of the first two digits in the hexadecimal values. Every string ends with a null character that appears as two consecutive zeros in the hexadecimal values segment.

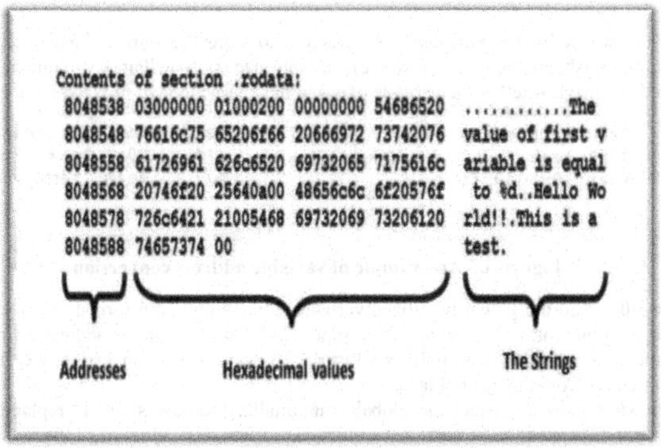

Figure 3.2 .rodata Segment in the Output File of Objdump Disassemble Tool

For example, to calculate the address of the third label (This is a test), we use the start address of the second string (Hello world!!), which is 0x8048570 (as shown in Table 3-1).Then we add the number of characters in the second string (0xd) to its address (0x8048570) plus one (0x1) for the null character, so that the address of the third string will be:
0x8048570 + 0xd + 0x1 = 0x804857e , note that the first string in the first line of the .rodata section starts after twelve characters. Thus, the address of the first string in our example is:0x8048538 + 0xc = 0x8048544

Table 3-1Extracting information from .rodata section

Characters	Addresses	Hexadecimal Values	Characters	Addresses	Hexadecimal Values
	8048568	20	H	8048570	48
T	8048569	74	E	8048571	65
O	804856a	6f	L	8048572	6c
	804856b	20	L	8048573	6c
%	804856c	25	O	8048574	6f
D	804856d	64		8048575	20
.	804856e	0a	W	8048576	57
.	804856f	00	O	8048577	6f

In our implementation, we ignore the first twelve characters in the first line, and we ignore their equivalent hexadecimal values.Every label in x86 assembly starts with ".string", and is enclosed between two double quotation marks.Therefore, we should perform the required modification on the labels that exist in the "objdump" file of the program under test.We create a structure called **Labels** (shown in Figure 3.3) to store all information related to .rodata data section. These information are *name, startAdrs, lastAdrs*, and the *labelContent.* The *name* attribute stores the unique name to label (LC0, LC1, ...). The *startAdrs* attribute stores the address of the first character in the specific string. The *lastAdrs* attribute stores the address of the null character at the end of the specific label. The *labelContent* attribute stores the content (charchters or string) of the specific string.

```
typedef struct Labels labelNode;
struct Labels
{
   char *name;              //label name
   char startAdrs[8];       //start address of the label
   char lastAdrs[8];        //last address of the label
   char *labelContent;      //label content
   labelNode *next;
};
```

```
typedef struct VariableNode varNode;
struct VariableNode
{
   int size;
   char *name;
   char address[8]; //start address
   char lastAdrs[8]; //last address
   char *value;
   char flag;  //to determine if the variable is exist in the code
   varNode *next;
};
```

Figure 3.3 Labels StructureFigure 3.4 VariableNode Structure

In our implementation, we defined a *VariableNode* structure to store the start address (*address* attribute), the last address (*lastAdrs* attribute), variable name (*name* attribute), variable size (*size* attribute), variable hexadecimal value (*value* attribute) and *flag* attribute to check whether the variable exists in the instructions or not (shown in Figure 3.4 and 5).

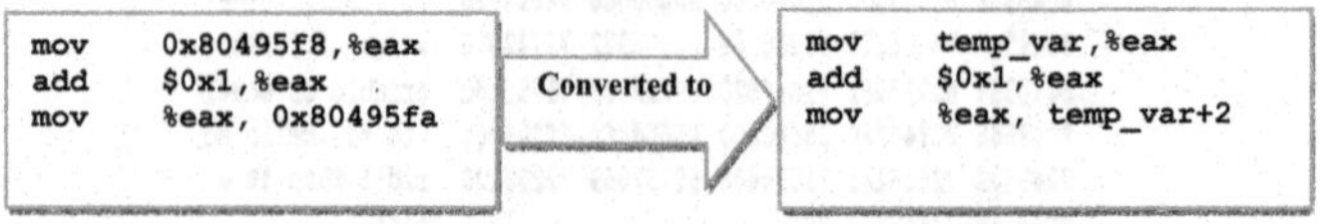

Figure 3.5 An example of variable address conversion

After converting the labels and the global initialized variables to assembly x86 format, we need to convert the global uninitialized variables to the same format. These variables appear in the instructions as addresses (in the "objdump" output file), while they must appear as names in assembly x86 format. To perform the conversion, we extract the names, sizes, and addresses of these variables from the symbol table.

We execute the following steps to extract the globally uninitialized variables and to replace their addresses in the assembly instructions with their names:

1. *Search for an uninitialized global variable in the symbol table (.bss).*
2. *When a variable is found, create an instance of VariableNode type.*
3. *Save the address of the variable in the address attribute, the variable size in the size attribute, and the variable name in the name attribute.*
4. *Calculate the last address by adding the variable address to its size – 1 and store the resulted address in the lastAdrs attribute.*
5. *If bss_head, the head of the bss linked-list, is NULL, then make the bss_head points to the created instance. Otherwise, add the newly created instance to the tail of the bss linked-list.*
6. *Repeat the steps from 1 to 5 for all uninitialized global variable in the symbol table.*
7. *Replace the addresses of the uninitialized global variable in the assembly instructions with the variable name (or variable name + offset).*
8. *Add .comm section at the end of the x86 assembly file for every instance. The .comm section has the name of the variable followed by its size (in decimal).*

After the conversion of all instructions in a program under test to the x86 assembly format from an "objdump" file, .comm entries are inserted in the x86 assembly file to account for global, unintialized variables. In our previous example, the following line is inserted:

```
.comm          uninitialized_variable, 4
```

The next step in the conversion process of the "objdump" output file into assembly x86 file is to build a structure data type and to copy the important information from "objdump" output file to it. This structure is shown in Figure 3.6. The section that starts with **Disassemble of section .text** string in the "objdump" output file contains all user-defined functions. These functions are located between the **<frame_dummy>** and the **<__libc_csu_fini>** functions. The entire lines between these two functions are copied into a temporary buffer called *textLines*.Figure 3.6 shows the buffer *textLines* with three user-defined functions, *Fun1*, *Fun2*, and *Fun3*, stored in it. Figure 3.7 gives an example to illustrate this conversion stage.

There are string library functions that are vulnerable to buffer overflow attacks in C programming language. This vulnerability comes from the fact that C programming language does not check buffers bounds. Solving this problem is not an easy task with the absence of the source code for many applications.In this paper, a new comprehensive toolthat provides protection against buffer overflow attacks is proposed, implemented, and tested. The proposed tool provides protection for string library functions that are vulnerable to buffer overflow attacks without the need for the source code of the vulnerable application. This is an important property for the applicability of the tool when the vulnerable application source code is not available.The proposed tool overcomes the vulnerability to buffer overflow attacks that exists in C programming language by checking the bounds for the buffers of the C library functions. To carry out this bound checking on a certain buffer, the number of memory locations that can be written on that buffer without causing a buffer overflow problem is calculated. These memory locations are called *safe upper bound limit of the destination buffer*. In this paper, our solution handles six vulnerable library functions, *strcpy, strncpy, strcat, memcpy, memmove* and *bcopy*,. For each one of them, a safe version is created in which buffers' bounds are checked. Moreover, the original function code is executed with considering only the safe upper bounds of the destination buffers. Figure 3.13 illustrates the main procedures for handling the library functions.

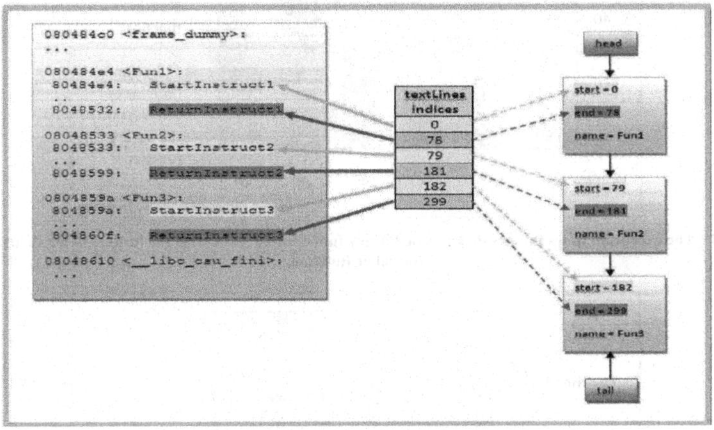

Figure 3.6 textLines Buffer and funNode linked-list

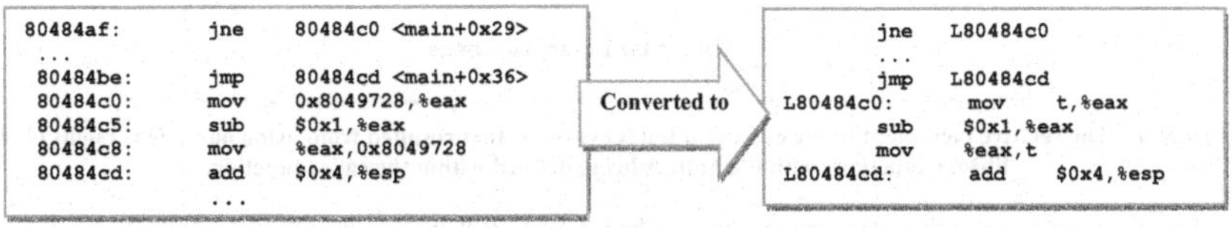

Figure 3.7 Jump Conversion Stage

4.EVALUATION

The evaluation of any new tool or algorithm is essential to judge its success. Here, we create programs to evaluate the execution time overhead caused by using the proposed tool due to the execution of our safe versions of the library functions rather than the execution of the unsafe library functions. More importantly, we create micro-benchmarks to validate the functionality of our tool and its ability to detect the buffer overflow attacks. This section provides details about thetime overhead performance. We use a machine with Intel® Core™ i5-2450M CPU @2.5 GHz (4CPU), 4096MB of RAM, and GUN/Linux Fedora 9 OS for all evaluation tests. GCC compiler is used to re-compile the newly generated code (assembly x86 code) and to run it.

We evaluate the time overhead due to using our tool by creating two sets of programs. In the first set of programs, the buffers to be used as parameters for the library functions are defined within the scope of the caller functions. In the first set of programs, in which the buffers (vulnerable buffers) are defined, we evaluate the time overhead for every unsafe library function we have handled separately by calling every one of them 10^9 times. The time overhead is calculated by finding the difference in the execution times for the unsafe versions of the library functions and our safe versions. Figure 4.1 shows a comparison between the two execution times for the library functions. Figure 4.2 shows the relative increment in the execution times that resulted from using our safe versions of the library functions and we define this relative increment as the overhead. We calculate the overhead of every function by applying the following equation:

Overhead = execution time for our safe versions of the library function / execution time of the original library functions *(Equation 4.1).*

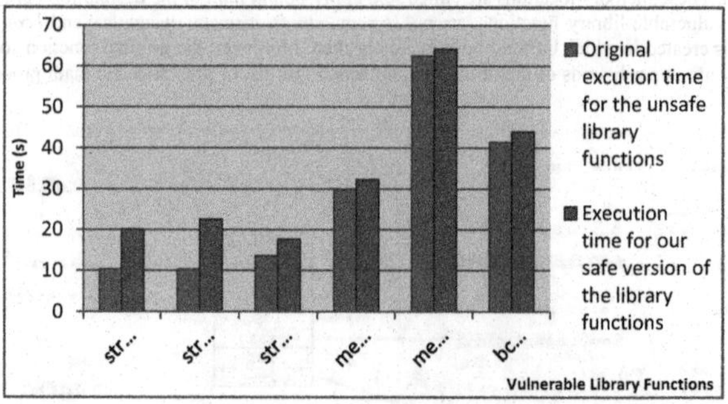

Figure 4.1 The execution times for unsafe and safe library functions 10^9 time with the buffers being defined within the caller function.

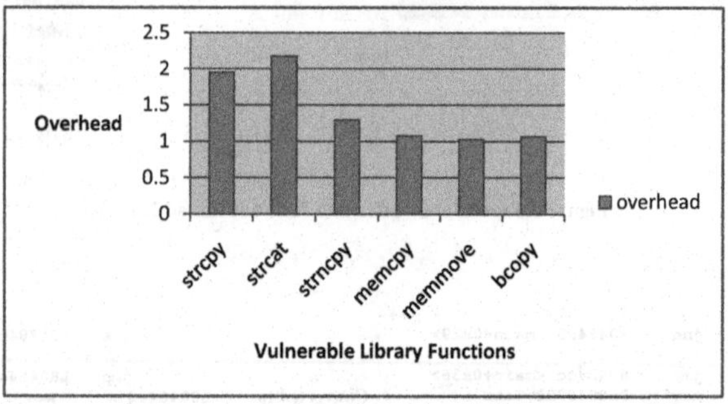

Figure 4.2 The relative increment in the execution times overhead that resulted from using our safe versions of the library functions with the buffers being defined within the caller function.

For the second set of testing programs, in which the buffers (vulnerable buffers) are defined outside the scope of the caller functions, we evaluated the execution times and the overhead in similar manner we used for the first set of programs. Figure 4.3 shows a comparison between the two execution times of calling the library functions 10^9 times for the original unsafe library functions and our safe versions.

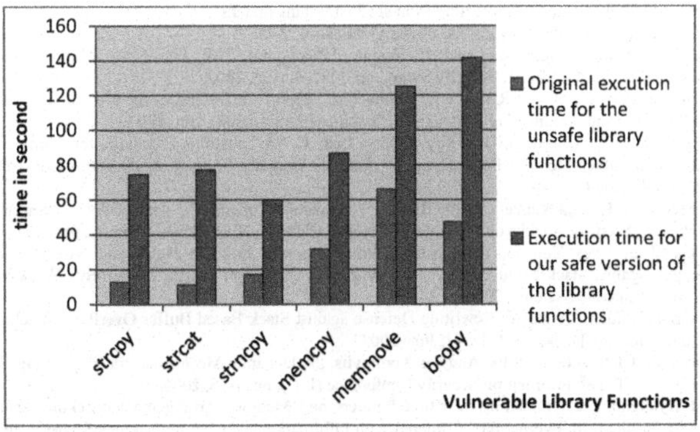

Figure 4.3The execution times for unsafe and safe library functions of 10^9 time with the bufferes being defined outside the caller function.

5.CONCLUSION

The buffer overflow attack is a serious problem that exploits vulnerabilities in available software systems. In this attack, the return address of a routine in the affected process is hijacked to redirect the control flow in the process to an injected code; this can give the attacker the ability to manipulate the whole system. In this paper, we proposed, implemented and tested a novel tool to protect the return addresses from being overwritten. Our tool takes as an executable program as input, disassembles it back to assembly code, analyzes the assembly code, and modifies it by substituting vulnerable code segments with safer code. The developed tool supports three modes of operations; debugging mode, production mode, and warning mode. The debugging mode alarms the user when there is a buffer overflow attack and corrects the unsafe library functions that cause this alarm. The production mode corrects unsafe library functions to prevent the attack. The warning mode alarms the user with every library function that exists in the code.Unlike other approaches that have been used to solve the buffer overflow attack, our solution does not require the availability of the program source code, and does not require any hardware modifications, which makes our solution more desirable than other solutions.

REFERENCES

[1] T. Chiueh and F. Hsu, "RAD: A Compile-Time Solution to Buffer Overflow Attacks," Proc. 21st Int'l Conf. Distributed Computing Systems (ICDCS '01), pp. 409-417, Apr. 2001.

[2] C. Cowan, C. Pu, D. Maier, H. Hinton, P. Bakke, S. Beattie, A. Grier, P. Wagle, and Q. Zhang, "StackGuard: Automatic Adaptive Detection and Prevention of Buffer- Over flow Attacks, " Proc. Seventh USENIX Security Conf., pp. 63-78, Jan. 1998.

[3] A. Baratloo, T.K. Tsai, and N. Singh, "Libsafe: Protecting Critical Elements o f Stacks," technical report, Bell Labs, Lucent Technologies, Murray Hill, N.J., Dec. 1999, http://www.bell-labs.com/ org/11356/libsafe.html.

[4] Steven Van Acker, Nick Nikiforakis, Pieter Philippaerts, Yves Younan, and Frank Piessens. "ValueGuard: Protection of native applications against data-only buffer overflows", in Proc. ICISS, pp. 156-170, 2010.

[5] Bindu Padmanabhuni, Hee Beng Kuan Tan. "Techniques for Defending from Buffer Overflow Vulnerability Security Exploits", IEEE Internet Computing, Sep. 2011.

[6] J. Wilander and M. Kamkar, "A Comparison of Publicly Available Tool s f or Static Intrusion Prevention", Proc. Seventh Nordic Workshop Secure IT Systems, pp. 68-84, Nov. 2002.

[7] Vendicator, "StackShield: A 'Stack Smashing' Technique Protection Tool for Linux," http://www.angelfire.com/sk/stackshield/ download.html, Jan. 2001.

[8] Bulba and Kil3r, "Bypassing StackGuard and StackShield", Phrack Magazine, vol. 10, no. 56, May 2000, http://www.phrack.org/ show.php?p=56&a=5.

[9] A. Baratloo, N. Singh, and T. Tsai , "Transparent Run-Time Defense against Stack Smashing Attacks", Proc. USENIX Ann. Technical Conf., pp. 251-262, June 2000 .

[10] Aurelien Francillon, Deniele Perito, and Clude Castelluccia. "Defending embedded systems against control flow attacks", in Proc. of the first ACM workshop on secure execution of untrusted code, 2009.

[11] C. Cowan, P. Pu, S. Beattie, and J. Walpole, "Buffer overflows: Attacks and defenses for vulnerability of the decade", In Proc. of the DARPA Information Survivability Conference and Expo (DISCEX), pp. 119—129, 2000.

[12] Hiroaki Etoh and Kunikazu Yoda, "Protecting from stack-smashing attacks", http://www.research.ibm.com/trl/projects/security/ssp/main.html, 2000.

[13] C. Cowan, S. Beattie, J. Johansen, and P. Wagle, "Pointguard TM: Protecting Pointers from Buffer Overflow Vulnerabilities," Proc. 12th USENIX Security Symp., pp. 91-104, Aug. 2003.

[14] J. Xu, Z. Kalbarczyk, S. Patel, and R.K. Iyer, "Architecture Support for Defending against Buffer Overflow Attacks," Proc. Workshop Evaluating and Architecting System Dependability (EASY-2002), Oct. 2002.

[15] Rinard, M.; Cadar, C.; Dumitran, D.; Roy, D.M.; Leu, T.; "A Dynamic Technique for Eliminating Buffer Overflow Vulnerabilities (and Other Memory Errors)", in 20th Annual Computer Security Applications Conference, IEEE Computer Society, pp. 82 – 90, 2004.

[16] Hilmi Özdoganoglu, T. Vijaykumar , Carla Brodley , Benjamin Kuperman , Ankit Jalote. "SmashGuard: A Hardware Solution to Prevent Security Attacks on the Function Return Address", IEEE Trans. Comput., pp. 1271 – 1285, 2006.

[17] Francesco Gadaleta, Yves Younan, Bart Jacobs, Wouter Joosen, Erik De Neve, and Nils Beosier, "Instruction-level countermeasures against stack-based buffer overflow attacks", in Proc. of the 1st EuroSys Workshop on Virtualization Technology of Dependable Systems, 2009.

[18] M. Prasad and T. Chiueh, "A Binary Rewriting Defense against Stack Based Buffer Overflow Attacks," Proc. Usenix Ann. Technical Conf., General Track, pp. 211-224, June 2003.

[19] Stelios Sidiroglou, Giannis Giovanidis, Angelos Keromytis, "A Dynamic Mechanism for Recovering from Buffer Overflow Attacks", In Proc. of the 8th Information Security Conference (ISC), pp. 1-15, 2005.

[20] C. Pyo and Gyungho Lee, "Encoding Function Pointers and Memory Arrangement checking against Buffer Overflow Attack", in Proc. of the forth International Conference on Information and Communications Security, pp. 25-36, 2002.

[21] Chun-Chung Chen, Shih-Hao Hung, Chen-Pang Lee. "Protecting against Buffer Overflow Attacks via Dynamic Binary Translation", in Proc. Reliable and Autonomous Computational Science, pp. 305-324, 2010.

[22] Zhenkai Liang, R. Sekar, and Daniel C. DuVarney, "Automatic Synthesis of Filters to Discard Buffer Overflow Attacks: A Step towards Realizing Self Healing Systems", In Proc. of the 14th USENIX Annual Technical Conference, pp. 375-378, 2005.

[23] André Zúquete, "StackFences: a run-time approach for detecting stack overflows", the 1st International Conference on E-business and Telecommunication Networks, 2011.

[24] kyn Kyung-suk Lhee and Steve J. Chapin, "Type-Assisted Dynamic Buffer Overflow Detection", in Proc. of the 11th USEBIX Security Symposium, 2002, pp. 81-88.

[25] M. Frantzen and M. Shuey, "StackGhost: Hardware Facilitated Stack Protection," Proc. 10th USENIX Security Symp, pp. 55-66, Aug. 2001.

[26] K. Piromsopa, and R Embody, "Dependable and Secure Computing", IEEE Transactions on (Volume:3 , Issue: 4), IEEE Computer Society, 365-376, 2006

[27] A. Slowinska, T. Stancescu, and H. Bos,"Body armor for binaries: preventing buffer overflows without recompilation". In Proceedings of the 2012 USENIX conference on Annual Technical Conference (USENIX ATC'12). USENIX Association, Berkeley, CA, USA, 2012.

Meet-in-the-middle Attack with Splice-and-Cut Technique on the 19-round Variant of Block Cipher HIGHT

Yasutaka Igarashi[1], Ryutaro Sueyoshi[1], Toshinobu Kaneko[2], Takayasu Fuchida[1]

[1] Kagoshima University, 1-21-40 Korimoto, Kagoshima, 890-0065 Japan
igarashi@eee.kagoshima-u.ac.jp, fuchida@ibe.kagoshima-u.ac.jp
[2] Tokyo University of Science, 2641 Yamazaki, Noda, Chiba, 278-8510 Japan
kaneko@ee.noda.tus.ac.jp

Abstract. We show a meet-in-the-middle (MITM) attack with Splice-and-Cut technique (SCT) on the 19-round variant of the block cipher HIGHT. The original HIGHT having 32-round iteration was proposed by Hong et al. in 2006, which applies the 8-branch Type-2 generalized Feistel network (GFN) with 64-bit data block and 128-bit secret key. MITM attack was proposed by Diffie and Hellman in 1977 as a generic method to analyze symmetric-key cryptographic algorithms. SCT was proposed by Aoki and Sasaki to improve MITM attack in 2009. In this paper we show that 19-round HIGHT can be attacked with 2^8 bytes of memory, 2^8+2 pairs of chosen plain and cipher texts, and $2^{120.7}$ times of the encryption operation by using MITM attack with SCT.

Keywords: block cipher HIGHT, meet-in-the-middle attack, Splice-and-Cut technique

1 Introduction

We show a meet-in-the-middle (MITM) attack with Splice-and-Cut technique (SCT) on the 19-round variant of the block cipher HIGHT. The original HIGHT having 32-round iteration was proposed by Hong et al. in 2006, which applies the 8-branch Type-2 generalized Feistel network (GFN) with 64-bit data block and 128-bit secret key [1]. The designers said that HIGHT does not only consist of simple operations to be ultra-light but also has enough security as a good encryption algorithm.

Table 1 shows the complexity of attack on HIGHT. Sasaki et al. studied integral attack on 22-round HIGHT with data complexity 2^{62} and time complexity $2^{102.35}$ [2]. Chen et al. studied impossible differential attack on 27-round HIGHT with data complexity 2^{58} and time complexity $2^{126.6}$ [3]. Wen et al. studied zero-correlation attack on 27-round HIGHT with data complexity $2^{62.79}$ and time complexity $2^{120.78}$ [4]. Özen et al. studied related-key impossible differential attack on 31-round HIGHT with data complexity 2^{63} and time complexity $2^{127.28}$ [5]. Koo et al. studied related-key rectangle attack on full-round HIGHT with data complexity $2^{57.84}$ and time complexity $2^{125.833}$ [6]. Song et al. studied biclique attack on full-round HIGHT with

K.J. Kim (ed.), *Information Science and Applications*,
Lecture Notes in Electrical Engineering 339, DOI 10.1007/978-3-662-46578-3_50

data complexity 2^{48} and time complexity $2^{125.93}$ [7]. Previously security of HIGHT against MITM attack was not investigated.

MITM attack was proposed by Diffie and Hellman in 1977 as a generic method to analyze symmetric-key cryptographic algorithms [8]. Its basic idea is that if a target algorithm can be decomposed into two small consecutive segments and the computation of each segment only involves portions of a master key, then we can check the consistence of the intermediate data of each segment. Because separately analyzing two small segments does not require much effort, the overall time complexity to analyze the whole algorithm could decrease significantly compared to a brute force attack. Recently MITM attack has developed into multidimensional MITM attack [9], [10].

SCT was proposed by Aoki and Sasaki to improve MITM attack in 2009 [11]. In SCT an attacker chooses an arbitrary intermediate state of cipher by supposing a chosen plain text scenario. The data complexity of SCT would increase as compared with simple MITM attack, because the complexity is depend on key bits that we need to partially decrypt or encrypt an intermediate state to obtain a plain text or a cipher text. However the time complexity may decrease, because we have the freedom to choose the intermediate state.

In this article we decompose 19-round HIGHT into a 9.5-round forward segment and a 9.5-round backward segment, and show that HIGHT can be attacked with 2^8 bytes of memory, 2^8+2 pairs of chosen plain and cipher texts, and $2^{120.7}$ times of an encryption operation by MITM attack with SCT.

Table 1. Complexity of attack on HIGHT. Imp. diff. and zero-c. denote impossible differential and zero-correlation, respectively. Data complexity is represented by the number of pairs of plain and cipher text. Time complexity is represented by the number of encryption operations.

Attack	MITM	Integral	Imp. diff.	Zero-c.	Related-key imp. diff.	Related-key rectangle	Biclique
Round	19	22	27	27	31	32	32
Data	2^8+2	2^{62}	2^{58}	$2^{62.79}$	2^{63}	$2^{57.84}$	2^{48}
Time	$2^{120.7}$	$2^{102.35}$	$2^{126.6}$	$2^{120.78}$	$2^{127.28}$	$2^{125.833}$	$2^{125.93}$
Reference	Sect. 3	[2]	[3]	[4]	[5]	[6]	[7]

2 Overview of data mixing part of 19-round HIGHT

We describe the brief overview of data mixing part of 19-round HIGHT to understand this manuscript. Refer to the original proposal [1] for more details.

Figure 1 shows the data mixing part of 19-round HIGHT, which consists of XOR (⊕), arithmetic addition modulo 16 (⊞), linear functions F_0 and F_1. F_0 and F_1 are given by

$$F_0(x) = x^{<<<1} \oplus x^{<<<2} \oplus x^{<<<7}, \quad F_1(x) = x^{<<<3} \oplus x^{<<<4} \oplus x^{<<<6} \quad (1)$$

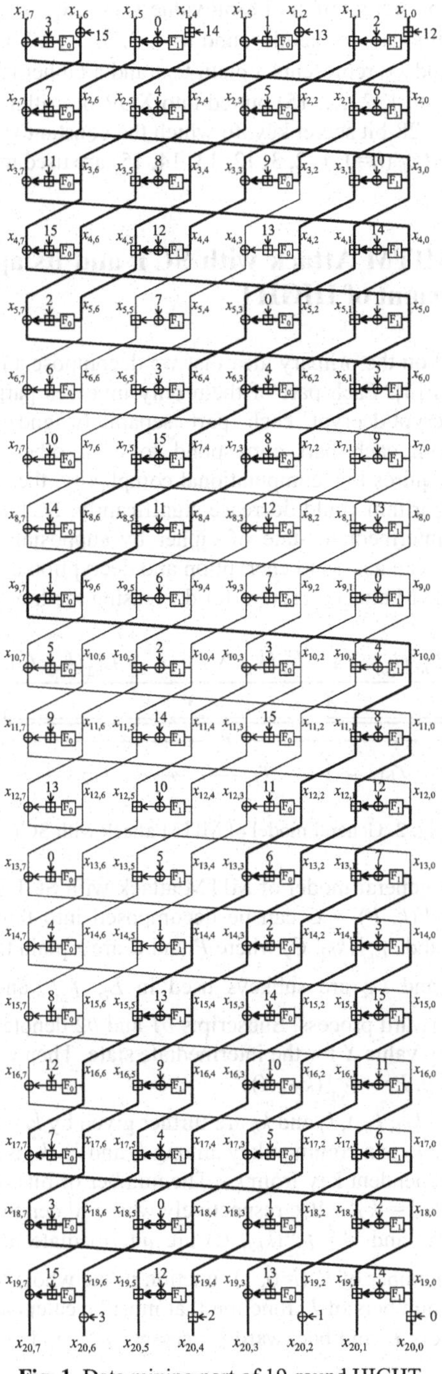

Fig. 1. Data mixing part of 19-round HIGHT.

where $x^{<<<i}$ denotes i-bit left rotation of 8-bit value x. $(x_{i,7}, x_{i,6}, x_{i,5}, x_{i,4}, x_{i,3}, x_{i,2}, x_{i,1}, x_{i,0}) = x_i$ denotes 64-bit input data to the ith round (i=1, 2, 3, …, 20) where $x_{i,j}$ is 8-bit data (j=0, 1, 2, …, 7). x_0 and x_{20} represents a plain text and a cipher text, respectively. The numerical symbol h (=0, 1, 2, …, 15) putted into XOR or arithmetic addition denotes 8-bit segment MK_h of 128-bit secret key, to which the constant value [1] is added. For example, 8 pieces of MK_h (h=0, 1, 2, 3, 12, 13, 14, 15) are used in the rounds 1 and 19.

3 Outline of MITM Attack with SCT and its application to the 19-round variant of HIGHT

MITM attack is based on the primary idea that we decompose a cipher algorithm into two consecutive parts [8]. Each part of them only involves partial information of a secret key. We encrypt/decrypt each part separately and check whether the intermediate data from each part correspond to each other. Because separately analyzing each part requires low computational complexity, the overall complexity to analyze the whole algorithm could decrease significantly. SCT allows an attacker to choose an arbitrary intermediate state of cipher by supposing a chosen plain text scenario as long as we can access to encryption and decryption oracles. SCT increases data complexity of MITM attack in return for decreasing the time complexity.

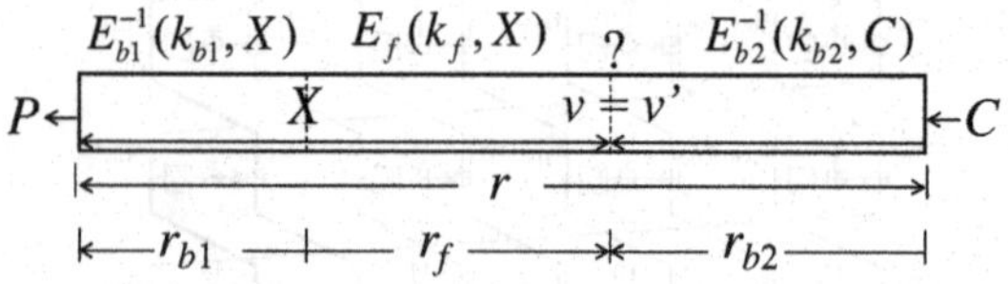

Fig. 2. General model of MITM attack with SCT.

Figure 2 shows the general model of MITM attack with SCT [10]. We suppose an encryption algorithm $E(k, P) = C$ can be decomposed into three consecutive parts $E_{b1}^{-1}(k_{b1}, X)$, $E_f(k_f, X)$, and $E_{b2}^{-1}(k_{b2}, C)$ where P and C are a plain text and a cipher text, respectively. k_f, k_{b1}, and k_{b2} are subkeys used in E_f, E_{b1}^{-1}, and E_{b2}^{-1}, respectively. Subscript f denotes forward process. Subscripts $b1$ and $b2$ denote back processes. We can choose an arbitrary value X for the intermediate state. Then we partially decrypt X to obtain a plain text i.e. $P = E_{b1}^{-1}(k_{b1}, X)$.

Supposing that $k_b = (k_{b1}, k_{b2})$, k_f and k_b are further given by $k_f = (k'_f, k_c)$ and $k_b = (k'_b, k_c)$, respectively where k_c is a common key among k_f and k_b. k'_f is the independent key from k'_b. k'_b is the independent key from k'_f. The number of bits of k_f and k_b are given by $|k_f| = |k'_f| + |k_c|$ and $|k_b| = |k'_b| + |k_c|$, respectively where $|x|$ denotes the number of bits of data x. $v = E_f(k_f, X)$ and $v' = E_{b2}^{-1}(k_{b2}, C)$ are intermediate data of an encryption process. r is the total number of F_i (i=0, 1) function in the whole algorithm. r_f, r_{b1}, and r_{b2} represent the total numbers of F_i function that must be calculated to derive v and v' through forward process E_f and backward processes E_{b1}^{-1} and E_{b2}^{-1}, respectively. We can usually derive $r \neq r_f + r_{b1} + r_{b2}$, because some of F_i functions in the whole algorithm are not calculated for the attack. The detailed steps of MITM attack with SCT are as follows [9], [10]:

1. Choose a fixed value for X.
2. For each guess of the common key k_c,
   (A) Encrypt X and obtain all possible value of v through $E_f(k_f, X)$ for all possible key k'_f. And then collect all k'_f in a set V indexed by v.
   (B) For each guess of the subkey k'_b,
      (a) Decrypt X and obtain P through $E_{b1}^{-1}(k_{b1}, X)$.
      (b) Obtain the corresponding cipher text C.
      (c) Decrypt C and obtain v' through $E_{b2}^{-1}(k_{b2}, C)$.
      (d) Check whether $v' \in V$. If so output the corresponding key triangle (k_c, k'_f, k'_b) as a possible key.

The memory complexity of this attack is given by the size of V in step 2(A), which is $2^{|k'_f|}$. The time complexity for the step 2(A) in terms of complete encryption is given by $2^{|k_f|} \times r_f/r$. In other words, the time complexity is the number of times of encryption or decryption operation of a target cipher. Similarly the time complexity of the step 2(B) is given by $2^{|k_b|} \times (r_{b1} + r_{b2})/r$. Therefore the time complexity T for these steps is given by

$$T = 2^{|k_f|} \times r_f / r + 2^{|k_b|} \times (r_{b1} + r_{b2}) / r. \tag{2}$$

Because we check for a match as $v = v'$, the total number of possible keys is reduced to $2^{|k'_f|+|k'_b|+|k_c|-|v|}$ when we perform these steps, where $|v|$ is the number of bits of v.

We next apply this MITM attack with SCT to 19-round HIGHT. Bold data lines in Fig. 1 are necessary for MITM attack with SCT. We set the intermediate state X as the 64-bit state given by

$$X = (x_{1,6}, x_{1,5}, x_{1,4}, x_{1,3}, x_{1,2}, x_{1,1}, x_{1,0}, x_{2,0}). \tag{3}$$

We set v and v' as the 8-bit state given by

$$v = v' = x_{10,0}. \tag{4}$$

From (3) and (4), the keys k'_f, k'_b, and k_c are given by 8-bit segment, 8-bit segment, and 112-bit segment of 128-bit secret key as

$$k'_f = MK_8, \quad k'_b = MK_3, \tag{5}$$

$$\begin{aligned} k_c = (&MK_0, MK_1, MK_2, MK_4, MK_5, MK_6, MK_7, \\ &MK_9, MK_{10}, MK_{11}, MK_{12}, MK_{13}, MK_{14}, MK_{15}). \end{aligned} \tag{6}$$

The numbers of F_i functions r, r_f, r_{b1}, and r_{b2} are also derived from (3) and (4) with Fig. 1 as

$$r = 4 \times 19, \quad r_f = 23, \quad r_{b1} = 1, \quad r_{b2} = 24. \tag{7}$$

In other words, r_f, r_{b1}, and r_{b2} are the numbers of F_i functions on the bold line in forward process and backward processes, respectively. From (5)-(7), (2) can be rewritten as

$$T = 2^{8+112} \times 23/76 + 2^{8+112} \times (1+24)/76 \approx 2^{119.3}. \tag{8}$$

When we perform the steps 1 and 2, the number of possible keys is reduced to $2^{8+8+112-8} = 2^{120}$. These 2^{120} pieces of possible key are furthermore reduced to $2^{120-64} = 2^{56}$ when we check these possible keys by exhaustive search with one independent pair of plain text and cipher text, because the block size of HIGHT is 64 bits. Similarly, these 2^{56} pieces of possible key are furthermore reduced to $2^{56-64} = 2^{-8}$ when we check these keys by exhaustive search with another independent pair of plain text and cipher text. In this way we can identify a true key because the true key definitely survives although the number of possible keys is less than 1. Therefore overall time complexity T_a for this attack is given by T and 2 times of exhaustive key search as

$$T_a = T + 2^{120} + 2^{56} \approx 2^{120.7}. \tag{9}$$

Memory complexity is given by $2^{|k'_f|} = 2^8$ bytes. Because $x_{1,7}$ has 2^8 varieties depending on MK_3 and MK_{15} on the step 2(B)-(a), a plain text P has 2^8 varieties. Therefore data complexity D of this attack is given by

$$D = 2^8 + 2. \tag{10}$$

In other words, the data complexity is the number of pairs of pain text and cipher text required for the attack. The constant 2 on the right side of (10) is derived from 2 times of exhaustive key search. We believe that an experimental proof is not required because this theoretical result is not a hypothesis.

4 Conclusions

We have shown MITM attack with SCT on the 19-round variant of the block cipher HIGHT, which have not been studied so far. We decomposed 19-round HIGHT into a 9.5-round forward segment and a 9.5-round backward segment, and showed that HIGHT can be attacked with 2^8 bytes of memory, 2^8+2 pairs of chosen plain and cipher texts, and $2^{120.7}$ times of an encryption operation by MITM attack with SCT. Future work is the application of multidimensional MITM attack to HIGHT.

References

1. Hong, D., Sung, J., Hong, S., et al.: HIGHT: A New Block Cipher Suitable for Low-Resource Device. CHES 2006, Lecture Notes in Computer Science, vol. 4249, pp 46-59, Springer (2006)
2. Sasaki, Y., Wang, L.: Meet-in-the-Middle Technique for Integral Attacks against Feistel ciphers. SAC 2012, Lecture Notes in Computer Science, vol. 7707, pp. 234-251, Springer (2013)

3. Chen, J., Wang, M., Preneel, B.: Impossible Differential Cryptanalysis of the Lightweight Block Ciphers TEA, XTEA and HIGHT, AFRICACRYPT 2012, Lecture Notes in Computer Science, vol. 7374, pp. 117-137, Springer (2012)
4. Wen, L., Wang, M., Bogdanov, A., Chen, H.: Multidimensional Zero-correlation Attacks on Lightweight Block Cipher HIGHT: Improved Cryptanalysis of an ISO Standard, Information Processing Letters, vol. 114, issue 6, pp. 322-330, ELSEVIER (2014)
5. Özen, O., Varıcı, K., Tezcan, C., Kocair, Ç.: Lightweight Block Ciphers Revisited: Cryptanalysis of Reduced Round PRESENT and HIGHT, Information Security and Privacy, Lecture Notes in Computer Science, vol. 5594, pp. 90-107, Springer (2009)
6. Koo, B., Hong, D., Kwon, D.: Related-Key Attack on the Full HIGHT, ICISC 2010, Lecture Notes in Computer Science, vol. 6829, pp. 49-67, Springer (2011)
7. Song, J., Lee, K., Lee, H.: Biclique Cryptanalysis on Lightweight Block Cipher: HIGHT and Piccolo, International Journal of Computer Mathematics, vol. 90, issue 12, pp. 2564-2580, Taylor & Francis (2013)
8. Diffie, M.E., Hellman, W.: Special Feature Exhaustive Cryptanalysis of the NBS Data Encryption Standard. Computer, vol. 10, issue 6, pp. 74-84, IEEE (1977)
9. Zhu, B., Gong, G.: Multidimensional Meet-in-the-Middle Attack and Its Applications to KATAN32/48/64. Cryptology ePrint Archive: Report 2011/619.
10. Boztaş, Ö., Karakoç, F., Çoban, M.: Multidimensional Meet-in-the-Middle Attacks on Reduced-Round TWINE-128. Lecture Notes in Computer Science, vol. 8162, pp. 55-67, Springer (2013)
11. Aoki, K., Sasaki, Y.: Meet-in-the-Middle Attack against Reduced SHA-0 and SHA-1. CRYPTO 2009, Lecture Notes in Computer Science, vol. 5677, pp 70-89, Springer (2009)

Nonintrusive SSL/TLS Proxy with JSON-Based Policy

Suhairi Mohd Jawi [1], Fakariah Hani Mohd Ali [2], Nurul Huda Nik Zulkipli [2]
[1] Cybersecurity Malaysia, {suhairi}@cybersecurity.my
[2] Universiti Teknologi MARA, {fakariah, nhuda}@tmsk.uitm.edu.my

Abstract. The placement of an interception proxy in between a client and web server has its own implications. Therefore, it is more practical to take a "middle" approach that can moderate the ongoing and future SSL/TLS sessions while not compromising the user privacy. A policy rule in JSON schema and data is proposed in handling SSL/TLS connection delegated by a non-intrusive, pass-through proxy.

Keywords. SSL/TLS - proxy - policy - JSON

1.0 Introduction

SSL/TLS proxy is used for decrypting and re-encrypting HTTPS traffic from browser to SSL-enabled web server, applying security policy and caching [1]. Additional security features such as HTTPS traffic monitoring, control, acceleration, content filtering for phishing, spyware and virus can be applied as well.

This paper presents an overview of an adaptive policy for monitoring, analyzing and responding towards HTTPS requests for security purposes using a custom-made Perl-based HTTPS proxy controlled via web-based interface. The proxy does not decrypt and re-encrypt HTTPS traffic at the first place likes its original purpose which is to access to the plain-text session contents while transferring between the two encrypted sessions [2].

The main objective is to analyze the attributes obtained from SSL/TLS-enabled web server based on requests by browsers connected to the proxy. The attributes from SSL handshakes and server certificates are checked against a set of policies written in JSON, based on predefined JSON Schema.

2.0 Background

There exist several issues regarding placement of proxy in between SSL/TLS connection as well as inherent problem of X.509 certificates.

2.1 Proxy Implementation Issues

Acting as a real target server, interception proxy must generate a new certificate (signed by CA) and key pair for each session [2]. To sign a certificate under a CA that a client will trust, we need either use private CA or public SubCA. One of the analysis done on forged SSL certificates by measuring over 3 million SSL connections [3], results show that 0.2% SSL connections are in fact tampered with forged certificates mostly by antivirus software and interception proxies. Using private CA will pose a logistical challenge to install. As a workaround, we can convince a trusted public root CA to issue a SubCA certificate for interception proxy.

Proxy may negotiate less secure cipher suite that both endpoints support [2]. This happens when cipher suites are not commonly preferred in practice that support forward secrecy or also know as Perfect Forward Secrecy (PFS).

Vulnerabilities on interception proxies pose an increased risk to all clients [2]. Encrypted sessions at interception proxies can be viewed in plain text, inspected and potentially modified by attacker. Exposure of public SubCAs signing key exposure are even greater risk because it has trust of the issuing root.

2.2 Legal Issues

There is also a legal issue regarding the proxy [2]. Organization face increased legal exposure and risks between employee and employer for inspecting communications intended to be encrypted and private. Organizations should have a policy for handling of encrypted communications for its employees. A survey on 1,261 users regarding SSL/TLS proxy showed most of them comfortable with benevolent purposes [4]. However, they were wary of attackers and government intrusion and need for better transparency and user opt-in.

K.J. Kim (ed.), *Information Science and Applications*,
Lecture Notes in Electrical Engineering 339, DOI 10.1007/978-3-662-46578-3_51

2.4 Certificate Issues

Browser relies on digitally signed certificates by validating a chain of trust from the site's certificate back to one of a set of trusted root certificates bundled with the operating system or browser [4].

Self-signed, expired and revoked certficates

Several issues arise with X.509 certificates being used for SSL/TLS [2]. A proxy needs to manage trusted roots on the interception proxy as it is on the client endpoint. It must be able to recognize self-signed certificates, expiry date stamps against current time for expired certificates and revoked certificates.

SubCA certificate

Another certificate issue is on the SubCA itself. A proxy must check that intermediate certificate's basic constraints field is not "false" to ensure it is authorized to act as an intermediate CA [2]. There were past flaws regarding the failure to check this field by software such as IE6, OpenSSL and iOS operating system.

Null prefix injection

Null-terminated subject name of a certificate is not properly handle by validation routine in the programming language such as C by appending a null character (0x00) on the end [2]. Therefore, a validation routine will interpret the certificate subjectwww.victim.com0x00.example.com is www.victim.com.

3.0 SSL Surveys

There are several surveys were done to examine the SSL connections that gave insights to the state of SSL/TLS in the wild.

In 2012, there was a client-side Shockwave (SWF) applet developed by [3] and hosted at Facebook that tried to detect forged certificate. The applet imitated browser's SSL handshake by opening a socket connection to the Facebook site. The applet records the SSL handshake and report the observed certificate chain to logging servers.

Between Nov 2009 and Apr 2011, an analysis of the x.509 PKI using active and passive measurements methodology was done [5]. Active scan on Alexa Top 1 Million Hosts list were using nmap and OpenSSL for probing full certificate chain and TLS/SSL connection properties. Passive scan used two-step processing using Bro as TLS/SSL processing tool in a port-independent way.

Results from above SSL/TLS surveys for the abnormal conditions of a certificate can be summarized as below. These results will be used as a guide for writing the policy rules.

- Certificates size is small in size that is less than a kilobyte.
- Certificate chains have a depth of one without any intermediate certificates. If more than one chain, it is usually broken chain, no issuing root CA certificate, certificate in the chain has expired or their signatures in the chain cannot be verified.
- Certificate subjects with empty issuer organization and use unrelated domains such as IP addresses, wildcard domain. The value may belong to device name, proxy software, antivirus, firewalls, parental control software, adware, and malware. It may has fraudulent value such as "Facebook" or default common name such as 'plesk', 'localhost', 'localhost.localdomain' for a self-signed certificate using the OpenSSL library.

4.0 Proposed Nonintrusive Method for SSL/TLS Proxy

This research will focus on engaging a nonintrusive, forward proxy with adaptive security features for SSL/TLS connections. The methodology used does not require clients to install binary executable file or a custom browser extension in order to examine SSL certificates and connections. The only requirement on the user side is to set proxy setting to point to the proxy. It is scalable to a large number of normal users with little technical knowledge.

Common method for SSL proxy-ing is called SSL tunneling or HTTP CONNECT method as specified by an Internet-Drafts [6]. However, CONNECT method is a purely pass-through connection. The proxy only transparently redirects the data between client and server. It only knows the source and destination addresses; and does not interfere with the connection. Therefore, CONNECT method that employs blind passing of SSL/TLS transaction can be enhanced to include adaptive security features to solve some SSL/TLS problems.

Fig. 1. Placement of proxy for SSL/TLS connection

The proxy runs as service and is listening on 0.0.0.0:8080 which means the service is listening on all the configured network interfaces. Inside the proxy there are three components for monitoring, analysis and response for each connection. The analysis component is coupled with a security policy. It feeds information of each SSL/TLS connection for static and dynamic testing before proceeding further to response component.

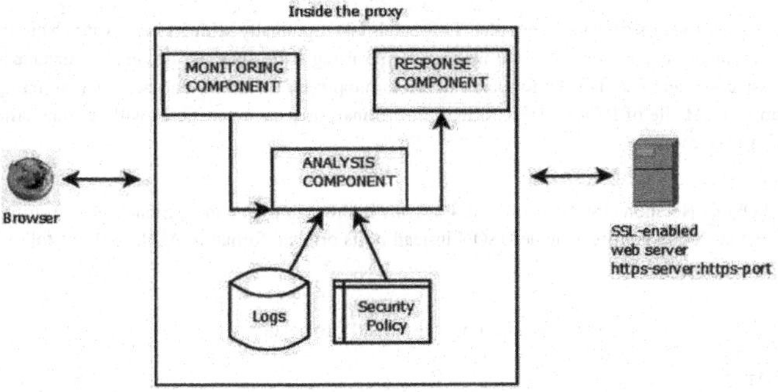

Fig. 2: Nonitrusive proxy components.

4.1 SSL/TLS Security Policy

Security policy is the foundation for the analysis component and is created under the characteristics of SSL/TLS attributes for static and dynamic testing. Static attributes consists of the list of versions for SSL/TLS, ciphersuites, root certificates, key exchanges, ciphers and hashes. These attributes are stored in a database for the analysis component.

4.1.1 Static Attributes

1) Cipher suites

Collection of cipher suites are required in order to diagnose supported cipher suite on a remote host using SSL/TLS. There are 357 cipher suites had been gathered so far for SSL v2.0 up to TLS v1.2 from RFCs and OpenSSL documentation. These cipher suites are categorized according to their SSL/TLS version (SSL v2.0, TLS v1.0, etc), key exchanges, encryption algorithms and signature algorithms as well as their equivalent names in OpenSSL.

For example, cipher suite TLS_RSA_WITH_RC4_128_MD5 is also known as RC4-MD5 in OpenSSL. However, not all cipher suites have OpenSSL equivalences. Further information such as 2-byte encoding is also captured. Example for SSL, the first byte is 0x00 and the second byte is the hexadecimal representation of the cipher suite number [6].

2) SSL/TLS versions, key exchanges, ciphers and hashes

Each cipher suite is categorized according to its SSL/TLS version, key exchange, cipher and hash. For example TLS_RSA_WITH_RC4_128_MD5 belongs to TLS v1.0 and uses RSA for key exchange, RC4 for encryption and MD5 for its signature algorithm. Each element in the categories is given a weightage in order to determine a weak cipher suite. For example, cipher suites belong to SSL v2.0 should no longer be used because it is an older implementation and has a number of security flaws.

3) Root certificates

Preliminary model of this proxy is developed on top of Ubuntu Linux 12.04 for 32-bit machine. We made used the files for root certificates that are distributed with Mozilla software products which can be found at "/usr/share/ca-certificates/mozilla" folder in the operating system. A Perl script was used to extract the information from each root certificate using Crypt::OpenSSL::X509 and Crypt::X509 perl modules. All important fields returned by this script for each certificate are stored in a database for future references when verifying a certificate chain up to the root certificate.

4.1.2 Dynamic Attributes

1) SSL certificates

SSL certificate can be considered as a dynamic attribute due to its constant state of change. A site certificate should be the same. However, website spoofing or man-in-the-middle attack may result in two different certificates for a website.

A server will provide its own certificate during SSL/TLS handshake optionally with its certificate chain up until its root certificate. Certificates can be extracted from the network dump using tools such as 'ssldump' or obtained directly from server using 'openssl' command line. For the former, a network dump need be run concurrently. The resulting certificate(s) can be in the form of PEM file of BER or DER format. In the analysis part, a certificate will be transformed into JSON formatting for policy evocation.

4.2 JSON Schema

We use JavaScript Object Notation (JSON) with draft 4 schema in order to define the attributes and security language. For example, to represent an X.509 certificate under JSON instead of its original format in ASN.1 [7], the following schema is used.

```
{
   "$schema": "http://json-schema.org/draft-04/schema#",
   "type": "object",
   "properties":
   {
      "version"              : { "type": "string" },
      "serialNumber"         : { "type": "string" }, . . .
      "validity":
      {
        "type": "object",
         "properties":
         {"notBefore"      : { "type": "string", "format": "date-time" },. . .}
      }, . . .,
      "extensions":{. . .}
   },
   "required":
   ["version", "serialNumber", "signatureAlgorithm", "issuer", "validity"]
}
```

[A portion of certificate in JSON Draft 4 schema]

The above schema is validated using a JSON Schema validation implementation in pure Java that support latest JSON Schema draft version 4. We make use the validator program 'json-schema-validator' version 2.2.5 obtained from its project site hosted at GitHub [8].

X.509 certificate as a digital object in ASN.1 of BER, DER or PEM format must be decoded to JSON format. This research uses Crypt::OpenSSL::X509 and Crypt::X509 Perl modules in order to do this. However, the conversion is not straight forward to follow the order in the schema for consistency and readability although order can be interchangeable. Both modules are used to complement each other since problems arise when decoding the certificate extensions because each module does not handle them thoroughly for every object ID (OID) of each extension. More of this will be discussed in Section 5. The following is a snippet of a certificate based on JSON schema defined above.

```
{
        "version" : "3",
        "serialNumber": "2",
        "signatureAlgorithm": "sha512WithRSAEncryption",
        "issuer": "C=US, O=Manufacturing Business, CN=Customer CA",
        "validity": {
        "notBefore": "2014-07-01T13:50:05Z",
        "notAfter": "2015-07-01T13:50:05Z"
        }
        . . . .
}
```

[Snippet of a certificate according to the schema]

Note that the 'date-time' type format is defined by RFC 3339, section 5.6 [9]. Although JSON gives simple formatting decision and more direct mapping for application data, it has exception or the absence of date/time literal [10].

4.3 Policy Rule in JSON

In order to show how the schema rule evolves from case to case basis, we start with the expired certificate detection since in JSON the date object is a string. Comparing the expiry date requires support program that can convert JSON date into format that can be compared with current date. The rule in JSON can be defined as followed:

```
{
  "name": "cert-expired",
  "description": "SSL certificate has expired.",
  "type" : "certificate",
  "test":
   {
       "field": "notAfter",
       "datatype":"date-time",
       "engine": "perl",
       "script": "jsondate.pl",
       "op":"lt",
       "level":"medium",
       "action":"alert",
       "message":"The server %servername%'s SSL certificate has expired."
   }
}
```

[Rule for expired certificate]

The rule starts with the name, its brief description, type of the test and test configuration. Since JSON does not have comment directive, the description in the rule becomes the data itself under 'description' field. Followed the description is the type of test which is on a certificate as the rule above.

The 'test' field point to the test configuration on a certificate. The field to be tested on the certificate is on the "notAfter" which has "date-time" data type. Since value of "date-time" cannot be compared directly from current time, a Perl script "jsondate.pl" will be called to compare if the "notAfter" value is less than the current time and this is marked by "lt" in the "op" for the operation field. The last three fields are for the security actions to be done if the comparison returns true. The "%servername%" is the replacement marker to allow name of the server to be inserted the message.

The execution of external script likes for comparing current date will become the standard norm for the rules. The same scripts can be reused on other rules and can embed external program such as OpenSSL for SSL/TLS analysis. The rule above can be represented by the following schema.

```
{
   "$schema": "http://json-schema.org/draft-04/schema#",
   "type": "object",
   "properties":
```

```
  {
     "name": { "type": "string" }, "description"  : { "type": "string" },
     "type": { "type": "string" },
     "test":
     {
        "type": "object",
        "properties":
        {
           "field"    : { "type": "string" }. . . .
        }
     }
  }
}
```

[JSON schema for policy rule]

4.4 Schema as a Rule

JSON schema has 'pattern' and Pattern Properties keywords that use regular expressions to express constraints [11]. It is possible to turn a schema into a rule for detecting irregularities in certificates. A certificate that conforms to predefined certificate schema can be passed to another schema for policy rule checking.

Supposed a self-signed certificate that contains a value "localhost.localdomain' on its subject and its representation in x.509 schema as follows:

```
{
      "version" : "3",
      . . .
      "subject": {
      "C": "--",
      . . .
      "CN": "localhost.localdomain/Email=root@localhost.localdomain"
      }
      . . .
}
```

[Snippet of a certificate with anomalous subject]

The above certificate can be matched against the following schema as policy rule checking:

```
{
   "$schema": "http://json-schema.org/draft-04/schema#",
   "type": "object",
   "properties":
   {
      "subject":
      {
         "type": "object",
         "properties":
         {
            "CN" : { "type": "string", "pattern": "^localhost.localdomain" }
         },
         "required" : [ "CN" ]
      }
   },
   "required": [ "subject"]
}
```

[Snippet of a schema for anomalous subject]

If any certificate matches the above schema, the validator will raise no error message that means the certificate has dubious subject. A typical certificate will encounter error by schema checking but that indicates the certificate is normal in the analysis component in reverse. Here is how the error returned by the validator when rule (schema) is mismatched.

```
[error] /subject/CN : ECMA 262 regex "^localhost.localdomain" does not match input
string "www.test.com/Email=root@test.com"
```

[Error when rule (schema) is unmatched]

5. Discussions and Future Work

5.1 Rationale for Opting JSON Format

ASN.1, XML and JSON are the popular data interchange formats with many other formats have been derived from them especially for XML. For example, XHTML, RSS, Atom, KML and YAML are some extensions of XML. Due to ASN.1 long existence, it is widely used in data encapsulation mainly over network link. Basically ASN.1 itself is incomparable with XML and JSON since ASN.1 is a schema language with sets of encoding rules (BER, DER) and the latter are data format. However, ASN.1 and JSON Schema at the very least they are comparable.

ASN.1 encoded data, in this case X.509 certificate, can be decoded and values can be arranged in either XML or JSON format for the ease of data manipulation. We choose JSON over XML for its recentness, simplicity, extensibility, interoperability and openness as described in [12]. However, the interpretation for extensions of X.509 using Perl modules based on ASN.1 specification is error prone, even the ubiquitous OpenSSL sometime mishandles them as shown on its vulnerabilities list [13].

Another issue with certificate extensions is the value derived from each Object ID. Due to Perl module limitation, some values when represented by a string are decoded into a complex object instead of its real value. Storing an improper string value inside a JSON format is undesirable. Therefore, this value shall be represented by other format other than decoded text such as Hex string or Base64 format.

In addition, the policy and its schema will always be in constant changes. Choosing JSON schema and data as policy or rule configuration gives flexibility for extending detection of abnormal behavior of SSL/TLS connection. Defined policies that get handed on to analysis component can be incorporated with supporting program likes OpenSSL or custom script such as in 4.3. Examples for JSON schemas and data given in the previous sections are hard coded in a file. It is also possible to create the JSON schema on-the-fly based on input data by a schema generator [14].

5.2 Auxiliary SSL Tools

There are several tools available that provide results in JSON format. *sslprobe* [15] decodes SSL/TLS protocol and cipher suite scanning results into JSON output. *sslprobe* strips down certificate outputs. Nevertheless, it provides more details on each SSL/TLS protocols from SSL v2.0 to TLS v1.2 which can be piped into the policy for detecting weak key exchanges, ciphers and hashes collected earlier for static attributes in 4.1.1.

5.3 Proxy Applications

For end users, they always consider their web connections in HTTPS as a universally capable security solution. Non-standardized error messages sometimes do appear across browsers. Besides stuffing the policy with all rules, more meaningful error messages can also be included in the policy. The proxy can do HTTP rewriting for CONNECT request with error messages or advices to be displayed on user browsers.

Another relevant application from the proxy is for conducting future SSL survey. HTTPS requests to all possible IPv4 address space can be created using custom scripts instead of web browsers. With the modification of the schemas and rules, detection of vulnerabilities such as POODLE, Heartbleed or BEAST are possible.

6. Conclusions

On the technical site, JSON data and schema provide flexibility for future policy modification for SSL/TLS connection and data-interchange with other security components such as security appliances and software. The next stage of this proxy development is in the response component and JSON will be used again for data interchange with analysis component.

Another future consideration is for storing JSON data whether to keep it as a flat file or inside a database. The availability of document-oriented database such as MongoDB and Apache CouchDB opens an opportunity to store JSON data format directly without worrying about the structure of table and rows as in traditional RDMS. Furthermore, MongoDB is a Big Data ready solution. Although data captured by EFF SSL Observatory is about 12 GB of MySQL database file may not be considered as big, the adoption of IPV6 and 'Internet of Things' may see the analysis of SSL/TLS connection require Big Data solution in the future.

7. Acknowledgment

This research is supported by the Research Management Institute, UniversitiTeknologi MARA and registered under the Research Acculturation Grant Scheme (RAGS) #600-RMI/RAGS 5/3 (204/2013)

8. References

1. Van der Linden, Maura A. Testing code security. CRC Press, (2007)

2. Jarmoc, Jeff. "SSL/TLS Interception Proxies and Transitive Trust." Transitive Trust. Dell SecureWorks, (2012) http://www.secureworks.com/cyber-threat-intelligence/threats/transitive-trust/.

3. L.-S. Huang, A. Rice, E. Ellingsen, and C. Jackson. Analyzing forged ssl certificates in the wild. In To appear, IEEE Symposium on Security and Privacy, (2014)

4. O'Neill, Mark, et al. "TLS Proxies: Friend or Foe?." arXiv preprint arXiv:1407.7146 (2014)

5. Holz, Ralph, et al. "The SSL landscape: a thorough analysis of the x. 509 PKI using active and passive measurements." Proceedings of the 2011 ACM SIGCOMM conference on Internet measurement conference ACM, (2011)

6. Rolf Oppliger, "SSL and TLS: Theory and Practice", Artech House / Horizon, (2009)

7. Hoffman, P. and J. Schaad, "New ASN.1 Modules for the Public Key Infrastructure Using X.509 (PKIX)", RFC 5912, (2010)

8. fge/json-schema-validator, https://github.com/fge/json-schema-validator

9. Newman, Chris, Graham Klyne. "Date and Time on the Internet: Timestamps", RFC 3339, (2002)

10. "An Introduction to JavaScript Object Notation (JSON) in JavaScript and .NET." An Introduction to JavaScript Object Notation (JSON) in JavaScript and .NET. Microsoft, (2007). http://msdn.microsoft.com/en-us/library/bb299886.aspx

11. M. Droettboom et al, "Understanding JSON Schema Release 1.0", Space Telescope Science Institute, (2014)

12. JSON: The Fat-Free Alternative to XML, http://www.json.org/xml.html

13. OpenSSL vulnerabilities, https://www.openssl.org/news/vulnerabilities.html

14. "JSON schema generator", http://www.jsonschema.net

15. sslprobe, "SSL/TLS protocol and cipher suite scanner with JSON output", https://github.com/noahwilliamsson/sslprobe

Secure Problem Solving by Encrypted Computing

Hiroshi Yamaguchi, Phillip C.-Y. Sheu,and Shigeo Tsujii

Chuo University, Tokyo, Japan, Research and Development, Tokyo, Japan, Department of Electrical Engineering and Computer Science, University of California Irvine, USA

yamaguchivc@cap.ocn.ne.jp

Abstract. The advancement of encrypted computing technologies has led to increase the demands for problem solving system which allows customers and problem solving providers to guarantee the privacy and confidentiality in the cloud environment. Meanwhile, along with the diversity of the customer levels, an improvement of easiness in describing the requirements (problem) has become a serious challenge issues. In this paper, we propose the structured natural language based query scheme improving descriptiveness of problem and efficient private information retrieve scheme for enhancing the security level of encrypted computing.

1. Introduction

Cloud computing has been introduced to optimize the general usage of IT infrastructures. According to NIST, "cloud computing is a model for enabling ubiquitous, convenient, on-demand network access to a shared pool of cofigurable computing resources (e.g. networks, servers, storage, applications, and services) that can be rapidly provisioned and released with minimal management effort or service provider interaction". Meanwhile, new demand has been arising for secure and practical computing technologies which address the challenge to safely outsource data processing onto remote computing resources by protecting programs and data even during processing in the cloud system. This allows users to confidently outsource computation over sensitive information from the security level of the remote delegate.Consider a customer that makequery (problem) to a problem solving system. For example, an investor (customer) that queries problem solving system for the value of a stock may wish to keep private the identity of the stock he is interested in. In this scheme, problem solving system have to give the solution to the investor without knowing no information about query he received from investor. These schemes will be called as "Secure problem solving scheme.For the sake of resolving these new demand, encrypted computing such as problem solving scheme, Privacy Preserving Data Mining (PPDM) scheme, encrypted search scheme, applications in bio-informatics, and hybrid (partially encrypted) applications have been researched so far. Together with these privacy, confidential issues, the new demand on "Easiness of requiring the processing to the computer system as easily as possible" has created considerable interest.In other words, underlying assumption is that people (customers have to be willing to submit processing requirement in natural language if possible. Explosion progress in cloud computing will lead these demands in the near future. Cloud computing environment is constructed by various computing and networking technologies. Virtual machine can support individual processes depending on the abstraction level, while an important challenge of distributed computing is location transparency. These technologies of virtualization and transparency offer benefits on cost-effectiveness and scalability, while the extensive use of virtualization in implementing cloud infrastructure brings unique security concerns for customers and

K.J. Kim (ed.), *Information Science and Applications*,
Lecture Notes in Electrical Engineering 339, DOI 10.1007/978-3-662-46578-3_52

service providers. A malicious administrator can secretly attack a virtual machines and distributed computing environment in a way that no one can notice using his higher privileged access.

We describe the required properties in Section II, explain basic idea in Section 3, and argue the our contribution in Section 4 and workflow of problem solving scheme in section 5, and building blocks in Section 6, Example is presented in Section 7, present performance and security analysis in Section in Section 8, and concluding remarks in Section 9.

2. Required Properties

We have shown a strong set of required properties in which the user desires on security issues for a problem solving services in the cloud depicted inFigure 1.

2.1. Privacy of user have to be preserved:

Users who submit a problem to problem solving system usually like to protect the privacy "sensitive date such as knowhow, private information, experience, data" involved in his problem.

2.2. Confidentiality of problem solving provider have to be preserved:

Problem solving provider usually like to protect the "know how, experience, technology, information" of a solution if possible against any third party.

2.3. Easiness of descriving the problem:

User who submit a problem to problem solving system usually like to describe a problem in the form that is almost a natural language. Problem written by keywords is usually insufficient for problem solving system.

2.4. Feasible system response time:

In cloud based system, the number of users is generally unspecified. Therefore, the feasible response time is important service issue.

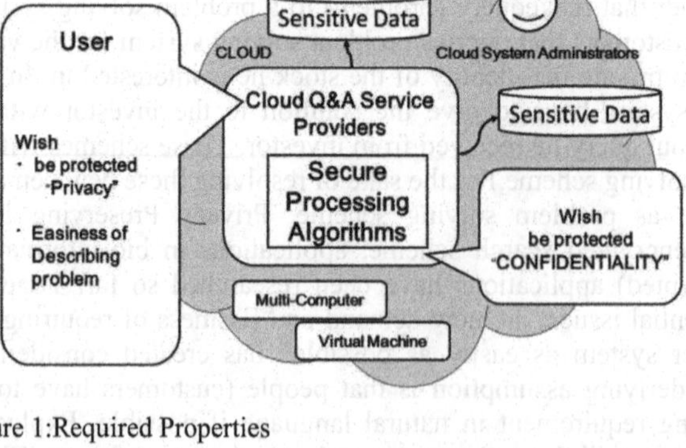

Figure 1:Required Properties

3. Basic Ideas

The basic ideas of this paper are summarized as follows:

3.1. Easiness of descriving the problem:

Structural Natural Language (SNL) is developed for describing a problem (query). Complex but restricted feature compared to natural language. We quote this idea for our research due to being sufficient for a lot of usage except for some artistic or literature use.

3.2. Privacy, Confidentiality protecting approach:

We share the problem described by SNL into two kinds of parts;

- Procedure describing part constructed by procedure, logic, data (not sensitive)
- Sensitive data describing part is constructed by sensitive data

Each part is processed separately for protecting privacy and confidentiality.

This approach is depicted in Fig. 2.

3.3. Advance privacy protecting approach:

Efficient block-based Private Information Retrieval (Block PIR) is presented and work together with the conventional encrypted computation algorithms comprehensively. This approach enhances the range of privacy protection compared to conventional encrypted computation algorithms.

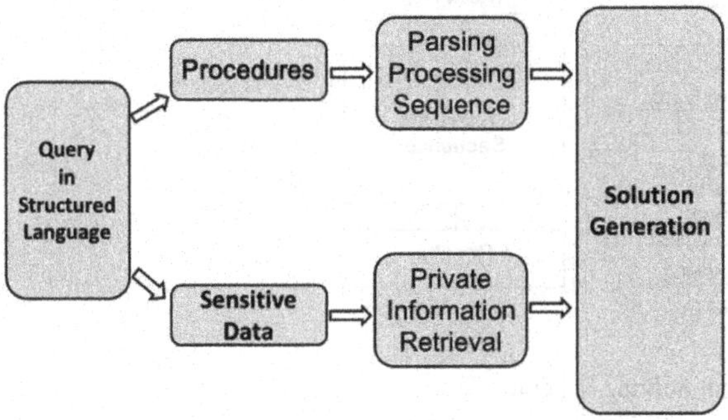

Figure 2: Shared Operation

4. Our Contributions

4.1. To expand user layer

Users have to be willing to describe his problem and consequently can benefit from the solutions.

4.2. To enhance a availability of privacy policy

Combined use of keyword-based private information retrieval scheme makes privacy policy more flexible, due to the feature that the keywords and pertained data are not leaked to the administrator.

4.3. To enhance a cloud based service system

Efficient block private information retrieval scheme plays an important role implement practical problem solving system.

5. Workflow of Problem Solving Scheme

Fig. 3 depicts the workflow of problem solving schemein which the participants are user, privacy preserving algorithm and processing algorithm.

(i) User

Users describe the problem in structured natural language.(SNL)

(ii) Privacy preserving algorithm:

Problem written in SNL is shared into the processing part and sensitive data part. Identification (ID) of sensitive data is retrieved is sent to keyword-based private information retrieval (PIR) scheme as the keyword. Result of PIR is sent to processing algorithm which is constructed form encrypted data. On the other hand, processing sequence part is parsed and converted to process flow chart depicted in Fig. 6 and finally to programming language.

(iii) Processing algorithm inputs the encrypted data retrieved and processes along with the processing sequence generated by parser and generate the result. The result is post-processed by privacy preserving algorithm and sent to user as a solution.

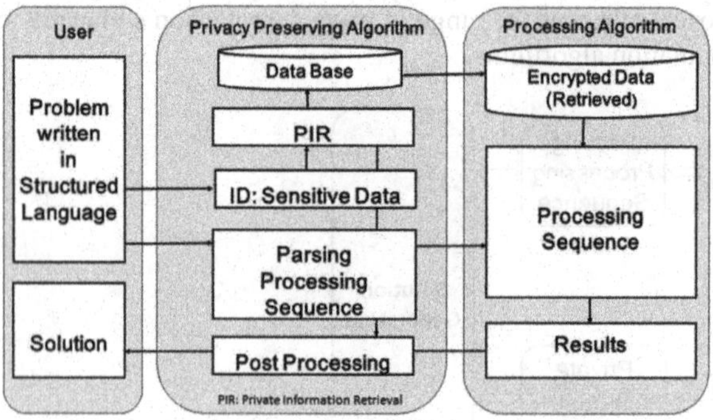

Figure 3: Work Flow of Problem Solving Scheme

6. Bulding Block

6.1. Structured Natural Language

Structured Natural Language (SNL)allows user to compose their query. SNL is a subset of natural language.SNL is restricted feature compared to natural language. This idea for our research due to being sufficient for a lot of usage except for some artistic or literature use. Syntax of SNL is similar to SQL-99. SNL adds to SQL-99 the concepts of sets such as set variable, permutation variable and interactive variable so that queries involving combinatorial problems can be expressed easily. The basic syntax of SNL is shown as follows.

SNL adds to SQL-99 the concepts of sets such as set variable, permutation variable and interactive variable so that queries involving combinatorial problems can be expressed easily.

6.2. Improved Block PIR Protocol

Basic techniques are consist from single bit Private Information retrieval (PIR) protocol, block PIR protocol, keyword PIR protocol, and encrypted computing scheme.

In cryptography a private information retrieval (PIR) protocol allows a user to retrieve an item from a server in possession of a data without revealing which item is retrieved. PIR is a weaker version of 1-out-of-n oblivious transfer where it is also required that the user should not get information about other database items. One trivial, but very inefficient way to achieve PIR is for the server to send an entire copy of the database to the user. In fact, this is the only possible protocolthat gives the user information theoretic privacy for their query in a single-server setting. There are two ways to address this problem: one is to make the server computationally bounded and the other is to assume that there are multiple non-cooperating servers, each having a copy of the database. The problem was introduced in 1995 by Chor, Goldreich, Kushilevitz and Sudan [CGKS95], [CG97]in the information-theoretic setting and in 1997 by Kushilevitz and Ostrovsky [KO97] in the computational setting. Since then, very efficient solutions have been discovered. Single database (computationally private) PIR can be achieved with constant (amortized) communication depicted in Figure 2. Before going any further let us make the problem more concrete. We view the database as a binary string $x = x_1, \dots x_n$ of length n. Identical copies of this string are stored by $k = 2$ servers. The user has some index i, and he is interested in obtaining the value of the bit xi. To achieve this goal, the user queries each of the servers and gets replies from which the desired bit xi can be computed. The query to each server is distributed independently of i and therefore each server gains no information about i. The user has some index i and he is interested in obtaining the value of contents R_i from the database server, while each database server gains no information about i. The user sends queries S and $S \oplus i$ in the form of sequences of subsets $S \subset \{1, \dots, n\}$ and each server replies with a corresponding sequence of blocks, $\oplus_{j \in S} R_j$ and $\oplus_{j \in s \oplus i} R_j$. The user exclusive-or $\oplus_{j \in S} R_j$, $\oplus_{j \in s \oplus i} R_j$ and obtains the value of R_i.depicted in Figure 4. In a more realistic model of private information retrieval, the data is organized in records (or blocks) rather than single bits. Clearly, a block may be retrieved by retrieving each of its bits. The scheme combines PIR solutions together with data structures that support search operations inorder to retrieve information privately in the keyword model was presented in [CGN97]. The main idea in constructions is the following: the databases insert $s_1, \dots, s_n$ into a data structure, which supports search operations on strings. The user U conducts an oblivious walk on the structure until either the word w is found, or U is assumed of the fact that w is not one of $s_1, \dots, s_n$. A typical search in the data structure involves a sequence of operations, where each operation, which depends on the keyword and the fetched contents, and either determining a new address based on the computation, or terminating the search. This sequence of operations can be viewed as a walk on the data structure. For simplicity, each block contains the same number of bits, l. Clearly $PIR(l,n)$ can be solved by l invocations of $PIR(n)$ (Namely, l interactions of a solution to $PIR(n)$. We are interested in more efficient solutions. The user has some index i and he is interested in obtaining the value of contents R_i from the database server, while each database server gains no information abou i.

The user sends queries S and $S \oplus i$ in the form of sequences of subsets $S \subset \{1,\ldots,n\}$ and each server replies with a corresponding sequence of blocks, $\oplus_{j\in S} R_j$ and $\oplus_{j\in S\oplus i} R_j$. The user exclusive-or $\oplus_{j\in S} R_j$, $\oplus_{j\in S\oplus i} R_j$ and obtains t he value of R_i.

S. Tsujii, et. al. presented this scheme in [TYM13], [TYM12] depicted in Fig. 4.

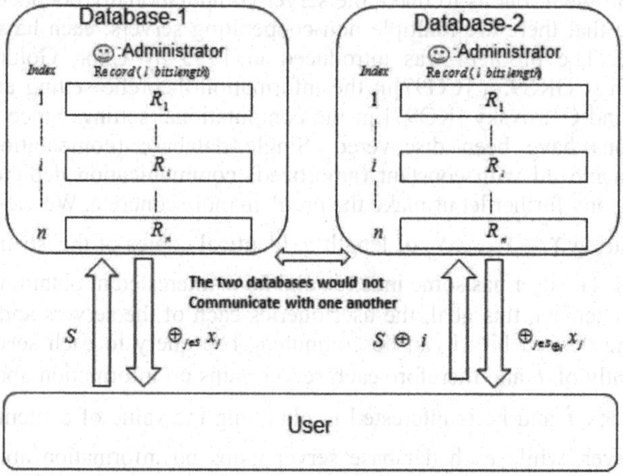

Figure 4: Improved Block PIR

PIR by keywords is presented by (CGN98) and depicted in Figure 5.

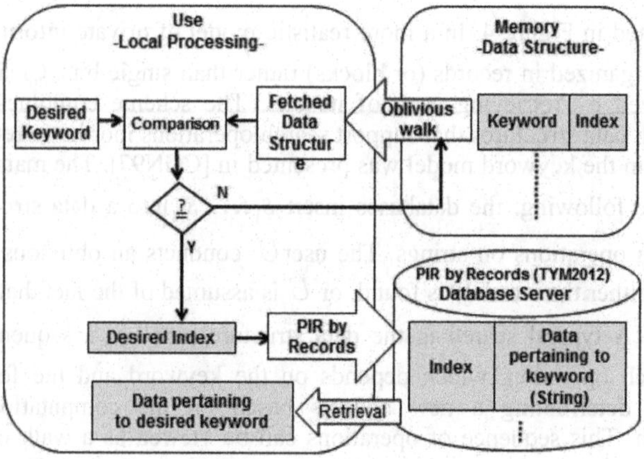

Figure 5: Improved Keyword PIR Scheme

6.3. Encrypted Computing Scheme

The demand for encrypted computing technologies preserving the privacy and confidentiality of users are getting critical issue in cloud environment. Many papers have challenged to safely outsource data processing onto remote computing resources by protecting programs and data even during processing. This allows users to confidently outsource computation over confidential information independently from the trustworthiness or the security level of the remote delegate. Secret sharing (also called secret splitting) refers to methods for distributing a *secret* amongst a group of participants, each of whom is allocated a *share* of the secret. The secret can be reconstructed only when a sufficient number, of possibly different types, of shares are combined together; individual shares are of no use on their own.In one type of secret sharingscheme there is one *dealer* and *nplayers*. The dealer gives ashare of the secret to the players, but only when specific conditions are fulfilled will the players be able to reconstruct the secret from their shares. The dealer accomplishes this by giving each player a share in such a way that any group of *t* (for *threshold*) or pplayers can together reconstruct the secret but no group of fewer than *t* players can. Such a system is called a *(t, n)*-threshold scheme (sometimes it is written as an *(n, t)*-threshold scheme).Secret sharing was invented independently by Adi Shamir and George Blakley in 1979.Secure multi-party computation (also known as secure computation or multi-party computation (MPC)) is a subfield of cryptography. The goal of this field is to create methods that enable parties to jointly compute a function over their inputs, while at the same time keeping these inputs private. For example, two millionaires can compute which one is richer, but without revealing their net worth. Titled "Protocols for secure computations" was formally introduced in 1982 by A. Yao [Yao822].Meanwhile,another type of encrypted computing scheme, "homomorphic encryptionand full homomorphic encryption" are a form of encryption which allows specific types of computations to be carried out on ciphertext and generate an encrypted result which, when decrypted, matches the result of operations performed on the plaintext. Full homomorphic encryption scheme is expected to be a useful tool, but currently many theoretical problems remaining. In many approaches mentioned above, they exists a computational and communication costs for evaluating the correctness of processing problems to be resolved so far. The main reason in these problem is due to the exponential computation which requires cost consuming computations.

The encrypted computing scheme has to be applied in consideration of the required property of system.

7. Example

Since the development of chain termination of a DNA sequencing method by Sanger and his colleagues in 1977, the subsequent development of computational methods for data retrieval and analysis, bioinformatics have been rapidly developed that include systematic analysis of gene expression profiles. Many computational biomedical applications have been developed such as sequence alignment, gene finding, genome assembly, analysis of differential expression. In the meanwhile, with the rapid advancement of Web services, the challenge to safely outsource data processing onto remote computing resources by protecting programs and data even during processing in biomedical analysis services in the Web. This allows users to confidently outsource computation over confidential information independently from the trustworthiness or the security level of the remote delegate. We show some examples for privacy preserving biomedical data analyses [WHHHNCST08], [SR09] below:

The ultimate objective of the BioSemantic System is to provide an integrated framework for prospective users to facilitate their works, such as biological and biomedical knowledge retrieval, management, discovery, capture, sharing, delivery and presentation.

Notations and definitions are as follows:

The notations used in the examples are listed below:

- A DNA sequence is a string of nucleotide bases q_i,

  Where $q_i \in NA = \{A,T,C,G\}, i = 1,\ldots,n; n \in Z^{+;}$
- A protein sequence is a string of nucleotide bases a_i.
- The predicate blast (A, B) is true if nucleotide sequence A blasts nucleotide sequence B;
- λ^{NA} designates the set of all possible DNA sequences;
- ξ^{AA} designates the set of all possible protein structures;
- The function (s_u, s_v) Similar( ) calculates the structural and/or sequence similarity to compare it with a predefined threshold t. Primary sequence analyses of genes and proteins represent a fundamental class of applications that are routinely performed. These analyses depend solely on the underlying nucleic acid sequences for genes, and the amino acid sequence for proteins. These analyses often cover the BLAST, alignment, and prediction of protein demonstrates the wide applicability of SCDL and Bio-Semantic System to primary sequence, structural analyses and alignment problems.

Example

- **SNL;**

  "Find nucleotide sequences from a database that are similar to a given sequence"
- **SNL Syntax :**

  SELECT N

  FROM λ^{NB} *(input)* s, λ^{NBA} *(input)* s'

  WHERE (s, s')

  Similar$(\,) \leq t$

Where s and s' are input protein that is a number of the protein structured. The values of s and s' are sent to keyword Private Information Retrieval (PIR) scheme as keyword, and data pertaining to the keywords are retrieved without revealing the value of keywords to PIR scheme. Therefore, privacy of users is preserved. λ^{NB} designates the set of all possible DNA sequences, where s designates input protein that is a number of the protein structured database. λ^{NBA} designates the set of all possible DNA where s' designates input protein that is a number of the protein structured database. (s, s') is calculations of the structural sequence similarity to compare it with a predefined threshold t. The values of s and s' are sent to keyword Private Information Retrieval (PIR) scheme as keyword, and data pertaining to the keywords are retrieved without revealing the value of keywords to PIR scheme. Therefore, privacy of users is preserved. Above SNL procedure is sent to arithmetic operation algorithm and computed followed by the instructions and outputted to user.

Figure 5 depicts the processing sequence described above. This flow chart is conv erted into programming language such as C^{++}, JAVA etc. In this figure, ◎ desig nate input, output for keyword PIR.

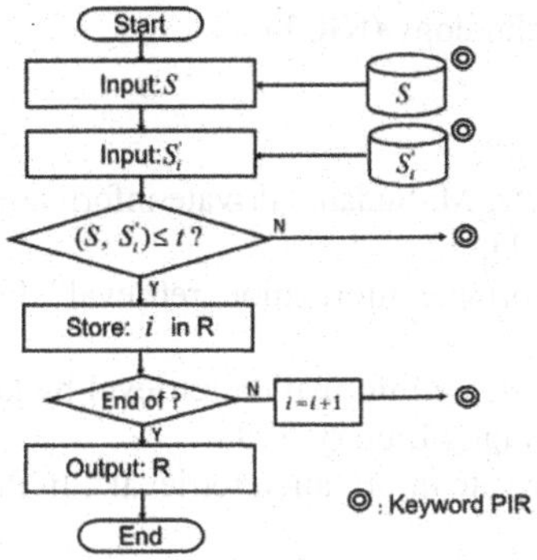

Figure 6: Flow Chart of Processing Sequence

8. Performance Evaluationpl and Secureity Analysis

[CGKS95] are based on multiple databases and toward information-theoretical secu rity applied.However, the assumption that the multiple databases would not communic ate with one another is somewhat strong for practical implementation. Later in [CG97], [KO97], and [CMS99], private information retrieval schemes with single database wer e proposed. The computational costs of these solutions are very large due to their bit-b y-bit processing manner. In a more realistic model of private information retrieval, the data is organized in records (or blocks) rather than single bits. Clearly, a block may be r etrieved by l invocat*ion. We introduced efficient private block retrieval in* [TYM12], [TYM13] where a block (length of block: m bits) can be retrieved by one invocation a nd computational cost is $O(m)$. This scheme is applied in keyword PIR scheme in Fig ure 4. In Three-party secure function evaluation protocol in [ICT10] and block private i nformation scheme, information-theoretic security is guaranteed.

9. Concluding Remarks

In this paper, we considered a practical model of secure problem solving scheme in computational cost and security level. This paper addresses the challenge to safely outsource data processing onto remote computing resources by protecting programs and data even during processing. This allows users to confidently outsource computation over confidential information independently from the trustworthiness or the security level of the remote delegate. The technologies and techniques discussed in this paper are key to extend the range of applications that can be securely and practically outsourced. As a conclusion, all issues described in 2 Required Properties are satisfied. The required item "Privacy of user have to be preserved"is satisfied because the three-party secure evaluation protocol can obtain the result of function without disclosing input data. The required item "Confidentiality of problem solving provider" is satisfied because the PIR by keywords protocol obtains the keyword pertaining contents without revealing the keywords (Problem). The required item "Optimal computational and communicational complexity" is also satisfied because the PIR by keywords protocol and three-party secure evaluation protocol are effective.

Acknowledgement

This study is supported by the Project entitled "Development of Public-key Cryptosys tem for confidential communication among Organizations (17201)" of the National In stitute of Information and Communications Technology (NICT).

References

[CGKS95] B. Chor, O. Goldreich, E. Kushilevitz, M. Sudan, "Private information retr ieval", Proc. of 36th FOCS, 1995, pp. 41-50, 1995.

[CG97] B. Chor, N. Gilboa, "Computationally private information retrieval", Proc. of 29th STC, 1997, pp. 304-313, 1997.

[CGN98] B. Chor, N. Gilboa, and M. Naor, "Private information retrieval by keywor d," Technical Report 98-03, Theory of Cryptography Library, 1998.

[Amb97] A. Ambainis, "An upper bound for private information retrieval", In Proc. of 24th ICALP, to appear, 1997.

[WHHHNCST08] S. Wang, R.-M. Hu, H. C. W. Hsiao, D. A. Hecht, K.-L.Ng, R.-M. Chen, P. C.-Y. Sheu and J. J. P. Tsai, "Using SCDL for integrating tools and data for c omplex biomedical applications", IJSC Vol. 2, No. 2, June 2008, pp, 291-308.

[SR09] P. C.-Y. Sheu and C. V. Ramamoorthy, "Problems, Solutions, and Semantic C omputing", IJSC Vol. 3, No. 3, September 2009, pp. 383-394.

[HGHS11] K. Hao, Z. Gong, C. Huo and P. C.-Y. Sheu, "Semantic computing and co mputer science(Tehnical Survey)", IJSC Vol. 5, No. 1, March 2011, pp. 95-120.

[WS10] Q. Wang and P. C.-Y. Sheu, "Synthesis of relational Web services", IJSC Vol. 4, No. 3, September 2010, pp. 385-417.

[KOYS13] J. Kim, D. A. Ostrowski, H. Yamaguchi and P. C.-Y. Sheu, "Semantic com puting and business intelligence", IJSC Vol. 7, No. 1, March 2013, pp. 87-117.

[KO97] E. Kushilevita and R. Ostrovsky, "Single-database Computational private info rmation retrieval, Proc. of 38th FOCS, 1997.

[TYM13] S. Tsujii, H. Yamaguchi and T. Morizumi, "Proposal on concept of encrypti on theory based on logic -Toward realization of confidentiality preserving retrieval an d creation of answer by natural language-", SCIS 2013,

[TYM12] S. Tsujii, H. Yamaguchi and T. Morizumi, "Proposal on concept of encrypte d state processing at semantic layer -Toward realization of confidentiality preserving retrieval and creation of answer by natural language-",ISEC 2012 Technical Report, 2 012.

[SR09] P, C.-Y.Sheu and C. V. Ramamoorthy, "Problems, solutions, and semantic com puting", IJSC Vol. 3, No. 3 pp. 383-394, 2009.

[CIT10] K. Chida, D. Igarashi, and K. Takahashi, "Efficient 3-party secure function e valuation and its application," CSEC 48, 4th, March, 2010.

[SR09] P.C.Y. Sheu, and C.V. Ramamoorthy, "Problems, Solutions, and Semantic Co mputation," INTERNATIONAL JOURNAL OF SEMANTIC COMPUTING, Vol. 3, No. 3, September 2009.

Anomaly Detection from Log Files Using Data Mining Techniques

Jakub Breier[1,2] and Jana Branišová[3]

[1]Physical Analysis and Cryptographic Engineering, Temasek Laboratories@NTU
[2]School of Physical and Mathematical Sciences, Division of Mathematical Sciences, Nanyang Technological University, Singapore
jbreier@ntu.edu.sg
[3]Faculty of Informatics and Information Technologies, Slovak University of Technology, Bratislava, Slovakia
branisovaj@gmail.com

Abstract. Log files are created by devices or systems in order to provide information about processes or actions that were performed. Detailed inspection of security logs can reveal potential security breaches and it can show us system weaknesses.
In our work we propose a novel anomaly-based detection approach based on data mining techniques for log analysis. Our approach uses Apache Hadoop technique to enable processing of large data sets in a parallel way. Dynamic rule creation enables us to detect new types of breaches without further human intervention. Overall error rates of our method are below 10%.

Keywords: Log Analysis, Anomaly Detection, Data Mining, Apache Hadoop, MapReduce

1 Introduction

A system or a device can provide information about its state in the form of log files. These files contain information if it works properly or which actions or services have been executed. By analyzing such information we can detect anomalies which can reveal potential security breaches. Since complex systems provide huge amounts of log records, it is not feasible to analyze them manually, therefore it is necessary to use automatized methods for this purpose. Outcomes of this process can help with a correct setup of network devices, it is possible to reveal non-privileged access to the system or even to find a person who performed the breach [8].

There are various types of logging. *Security logging* encompasses obtaining information from security systems and can be used to reveal potential breaches, malicious programs, information thefts and to check the state of security controls. These logs also include *access logs*, which contain data about user authentication. *Operational logging* reveals information about system errors and malfunctions, it is useful for a system administrator that needs to know about current system

K.J. Kim (ed.), *Information Science and Applications*,
Lecture Notes in Electrical Engineering 339, DOI 10.1007/978-3-662-46578-3_53

state. *Compliance logging* provides information about compliance with security requirements and it is sometimes similar to security logging. There are two subtypes of these logs: logging of security of information systems with respect to data transfer and storage, e.g. PCI DSS or HIPAA standard compliance, and logging of system settings.

In our work we focus on anomaly detection in log files with the help of data mining techniques. By comparing different approaches of anomaly and breach detection we decided to use a method based on dynamic rule creation with the data mining support. The main advantage of this method is its ability to reveal new types of breaches and it minimizes the need of manual intervention. We have also investigated possibilities for more effective log analysis of large data volumes because of the size and amount of log files, which has increasing tendency. By implementing the Apache Hadoop technology we created a single node cluster for parallel log processing using the *MapReduce* method. Time needed for log analysis has greatly increased and the algorithm implemented in Hadoop was able to process data faster than the standard algorithm using tree-based structure. Also, by adding more computing nodes it is easy to improve processing time, if necessary.

The rest of the paper is organized as follows. Section 2 provides an overview of related work in this area. Section 3 describes methods that could be used for anomaly detection. In section 4 we present the design of our solution, and finally, section 6 concludes this paper.

2 Related Work

There are several works proposing usage of data mining methods in log file analysis process or in detecting security threats in general.

Schultz et al. [6] proposed a method for detecting malicious executables using data mining algorithms. They have used several standard data mining techniques in order to detect previously undetectable malicious executables. They found out that some methods, like Multi-Label Naive Bayes Classification can achieve very good results in comparison to standard signature-based methods.

Grace, Maheswari and Nagamalai [4] used data mining methods for analyzing web log files, for the purposes of getting more information about users. In their work they described log file formats, types and contents and provided an overview of web usage mining processes.

Frei and Rennhard [2] used a different approach to search for anomalies in log files. They created the Histogram Matrix, a log file visualization technique that helps security administrators to spot anomalies. Their approach works on every textual log file. It is based on a fact that human brain is efficient in detecting patterns when inspecting images, so the log file is visualized in a form that it is possible to observe deviations from normal behavior.

Fu et al. [3] proposed a technique for anomaly detection in unstructured system logs that does not require any application specific knowledge. They also included a method to extract log keys from free text messages. Their false positive rate using Hadoop was around 13% and using SILK around 24%.

Makanju, Zincir-Heywood and Milios [5] proposed a hybrid log alert detection scheme, using both anomaly and signature-based detection methods.

3 Detection Methods

According to Siddiqui [7], there are three main detection methods that are used for monitoring malicious activities: scanning, activity monitoring and integrity check. *Scanning* is the most widely used detection method, based on searching for pre-defined strings in files. Advanced version of scanning includes heuristic scanning which searches for unusual commands or instructions in a program. *Activity monitoring* simply monitors a file execution and observes its behavior. Usually, APIs, system calls and hybrid data sources are monitored. Finally, *integrity checking* creates a cryptographic checksum for chosen files and periodically checks for integrity changes.

Data mining is relatively new approach for detecting malicious actions in the system. It uses statistical and machine learning algorithms on a set of features derived from standard and non-standard behavior. It consists of two phases: data collection and application of detection method on collected data. These two phases sometimes overlap in a way that selection of particular detection method can affect a data collection.

Data can be analyzed either statically or dynamically. Dynamic analysis observes a program or a system during the execution, therefore it is precise but time consuming. Static analysis uses reverse-engineering techniques. It determines behavior by observing program structure, functionality or types of operation. This technique is faster and it does not need as much computational power as dynamic analysis, but we get only approximation of reality. We can also use hybrid analysis - first, a static analysis is used and if it does not achieve correct results, dynamic analysis is applied as well.

3.1 Parallel Processing

A huge amount of log data is generated every day and it is presumed that this amount will grow over time. Therefore, it is necessary to improve the process of the log analysis and make it more effective. We have chosen Apache Hadoop[1] technology for this purpose with the *MapReduce* [1] programming model. *MapReduce* is used for processing large data sets by using two functions. *Map* function processes the data and generates a list in the key-value form. *Reduce* function can be then used by the user for joining all the values with the same key. Hadoop architecture is based on distributing the data on every node in the system. The resulting model is simple, because MapReduce handles the parallelism, so the user does not have to take care about load balancing, network performance or

[1] http://hadoop.apache.org/

fault tolerance. Hadoop Distributed File System can then effectively store and use the data on multiple nodes.

4 Design

The main idea for the design of our solution is to minimize false positives and false negatives and to make the anomaly identification process faster.

The steps of the algorithm are following. First, a testing phase is performed and rules are made from the testing data set. The outcome of this phase is an anomaly profile that will be used to detect anomalies in network devices log files. For creating rules, log file is divided into blocks instead of rows. A block is identified by the starting time, session duration and type of a service. We will use a term 'transaction' for particular block. This approach allows us to create a rule based on several log files from different devices or systems, so that one transaction can contain information from various sources. For creating uniform data sets, which can be processed by different algorithms, each transaction is transformed in a binary string form. For a spatial recognition of a log record in transaction, each record in the original log file will be given a new attribute - transaction ID.

For the detection program it is necessary to be able to process various log file formats, therefore we decided to use configuration files which will help to determine each attribute position.

4.1 Data Transformation

Data transformation includes creation of a new data set that contains the binary data only. The advantage of such data representation is ability to process it with various algorithms for association rules creation. Example of such a transformation is depicted in Table 1.

Table 1. Binary Transformation Example.

Session Time	Type of Service
00:00:02	telnet
00:00:04	http
00:00:05	telnet

ST1	ST2	ST3	ToS1	ToS2
1	0	0	1	0
0	1	0	0	1
0	0	1	1	0

Session Time	ST
Type of Service	ToS

ST1	00:00:02
ST2	00:00:04
ST3	00:00:05

ToS1	telnet
ToS2	http

To avoid a problem with large dimension number by using binary representation of log records, we propose a data reduction. This reduction is achieved by inserting values into categories and using an interval representative instead of a scalar or time value. Binary string contains a numerical value of 1 for values which are present in the record and a numerical value of 0 otherwise. However, some of the values are unable to reduce, such as IP addresses or ports.

4.2 Transaction and Rule Creation

Transaction and rule creation algorithm works as follows. It loads each record from log files line by line and stores them in the same block if they were created in the same time division, within the same session and if the IP addresses and ports are identical. If they can be identified as related, a transaction is created. Then it is decided whether this transaction fulfills conditions to be included in the anomaly profile. If yes, a new rule is created, if no, this transaction is ignored for the further rule creation process.

A rule contains attributes in a binary form that are defined in configuration file. It always contains some basic attributes related to time and session parameters and also a device ID, from which particular log record originates.

The anomaly finding algorithm first loads a set of previously created rules from the database. Then it sequentially processes the log files intended for analysis and creates transactions from these files. This transaction is then compared with the set of rules and if it is identified as an anomaly, it is stored for further observations.

4.3 Processing of Large Data Sets

As stated in Section 3.1, we decided to use Hadoop technology with MapReduce programming model to process large data sets. Hadoop enables us to easily add processing nodes of independent device types. After program starts, *JobTracker* and *TaskTracker* processes are started, which are responsible for *Job* coordination and for execution of *Map* and *Reduce* methods. *JobTracker* is a coordinator process, there exists only one instance of this process and it manages *TaskTracker* processes which are instantiated on every node. MapReduce model is depicted in Fig. 1, however in our case, only one *Reduce* method instance is used. First, a file is loaded from Hadoop Distributed File System (HDFS) and it is divided into several parts. Each *Map* method accepts data from particular part line by line. It then processes the line and stores them until a rule is created (if all the conditions are met). The rule is then further processed by the *Reduce* method, which identifies redundant rules and if the rule is unique, it is written into the HDFS.

To allow nodes to access same files in the same time, but without loading them onto the each node separately, a Hadoop library for distributed cache is used.

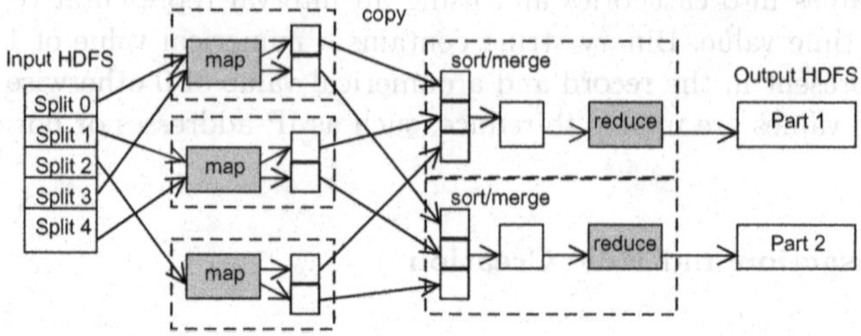

Fig. 1. MapReduce Algorithm.

5 Testing

Our anomaly detection method was implemented in Java programming language. For testing, we used Intel i7-4500U CPU with 8 GB of RAM, running Ubuntu 12.04 operating system. For testing purposes, two data sets were used: 1998 DARPA Intrusion Detection Evaluation Set[2], that was created by monitoring a system for two weeks, and Snort logs, created by analyzing the DARPA data set[3]. Snort logs contain information, if the attack was performed, or not. Based on that, we were able to determine if our anomaly detection method was able to successfully identify an intrusion, or not.

Testing was performed on a log records set of a size of 442 181 records. This set was made by merging DARPA and Snort data sets. We have split this data set into ten subsets for cross-validation purposes. In a cross-validation, a set is divided into subsets with similar sizes and each subset is used as many times as is the number of subsets. In each testing, one subset is used as a test set and the other subsets are used as training sets. For our validation, we split the main data set in a way that each subset contained log records from every day when monitoring was performed.

5.1 Data Transformation

After data sets merging, it was necessary to determine how many unique values are present in each table column. These values are stated in Table 2.

As we have already stated, high-dimensional data increases memory requirements of anomaly detection algorithm. It is possible to reduce some of the attribute values so that it can still be able to detect anomalies on a reduced set. We can analyze the 'Session Time' attribute and a process of reducing its values into intervals. These intervals are stated in Table 3.

Since the majority of records has a session time value 00:00:01, it was decided to take this value as a standalone interval. The same holds for value 00:00:02.

[2] http://www.ll.mit.edu/mission/communications/cyber/CSTcorpora/ideval/data/
[3] https://www.snort.org

Table 2. Occurrence of Unique Values in Merged Data Set.

Attribute	Unique Values	Min	Max
Date	10	07/20/1998	07/31/1998
Time	63 299	00:00:00	23:59:59
Session Time	884	00:00:01	17:50:36
Service	4664	n/a	n/a
Source Port	38 637	-	65 404
Destination Port	7887	-	33 442
Source IP	1 565	000.000.000.000	209.154.098.104
Destination IP	2 640	012.005.231.198	212.053.065.245
Attack Occurred	2	0	1
Attack Type	47	n/a	n/a
Alert	61	n/a	n/a

Table 3. Intervals with Highest Number of Occurences.

Session Time	00:00:01	00:00:02	(00:00:02,01:00:00>	(01:00:00,18:00:00)
Occurrences	795 421	13 873	7 987	759

Two other intervals cover longer time sessions, but since there are not many values present in each of these intervals, it was possible to make the reduction. Therefore, after reduction it was possible to change the range of values from 884 to 4 in this case, which enables significantly faster data processing.

5.2 Error Rate

Overall accuracy of the algorithm can be determined by Equation 1, where FP = false positives, FN = false negatives, TP = true positives, TN = true negatives.

$$Error\ Rate = \frac{FP + FN}{TP + TN + FP + FN} \tag{1}$$

The anomaly detection algorithm was implemented both in Java and in Hadoop. Table 4 shows values for both implementations. The table shows us that Hadoop implementation has around 1% lower error rate than Java implementation.

Table 4. Error Rate for Each Subset Using Java and Hadoop Implementations.

Error Rate	Set 1	Set 2	Set 3	Set 4	Set 5	Set 6	Set 7	Set 8	Set 9	Set 10
Java	0.095	Set 0.087	0.144	0.091	0.093	0.090	0.090	0.086	0.087	0.123
Overall	**0.098576**									
Hadoop	0.087	Set 0.077	0.142	0.083	0.083	0.083	0.080	0.077	0.077	0.121
Overall	**0.091465**									

5.3 Processing Speed

Important factor in anomaly detection is both speed of rules generation and speed of data processing. We compared our algorithm with two other anomaly detection algorithms, Apriori and FP-Growth. Apriori algorithm serves as a base for several rule-creation methods. Its disadvantage is that it needs to process the data set several times. FP-Growth algorithm uses tree-based storages for storing intermediate values. We used Weka libraries[4] for these algorithms implementations. Testing results are stated in Table 5. We can see that Hadoop implementation was the fastest among the tested algorithms. Therefore we can conclude that parallelization can bring very good results in terms of speed into the rule generation process.

Table 5. Comparison of Rule Generation Speed.

Algorithm	Java Implementation	Hadoop Implementation	Apriori	FP-Growth
Time (s)	163.1	15.6	226	93

Speed of data set processing for anomaly detection is stated in Table 6. We were comparing standard implementation in Java and implementation in Hadoop. Tests were performed on three data sets of sizes 10, 50, and 500 GB. As we can see, Hadoop can speed up this process more than ten times, even by using a single node. The Hadoop configuration was set to pseudo-distributed operation, which allowed it to run on a single-node. It is, of course, possible to add more nodes in order to improve throughput. We have tested a 10 GB data set on a three-node cluster, one node was configured as a master+slave, the other two nodes were configured as slaves only. Running time was 2040s, which gives us approximately 1.55 times better throughput than using a single-node.

Table 6. Comparison of Anomaly Detection Speed.

Data Size	10 GB	50 GB	500 GB
Number of Records (in millions)	84	423	851
Java Implementation Time (s)	32 622	164 050	330 152
Hadoop Implementation Time (s)	3 164	13 531	29 042

6 Conclusion

In our work we have proposed a way for anomaly-based breach detection from log files. We have chosen an approach based on dynamic rule creation using data

[4] http://www.cs.waikato.ac.nz/ml/weka/

mining techniques. The main advantage of such an approach is minimization of tasks requiring human interference and it is possible to detect new types of breaches.

The second goal was to reduce the time required for the log analysis, since log files are becoming larger and their number grows. We have implemented the application using Hadoop technology. We have created a single-node cluster for parallel processing, using *MapReduce* technology. This allowed us to make analysis more than ten times faster compared to using standard Java implementation. Also, it is easy to add more nodes for improving the analysis speed.

Log records were aggregated into transactions, identified by the time, session time and service type, so the manipulation with the data became more convenient. After that these transactions were transformed into binary format for faster processing.

Acknowledgement This work was supported by VEGA 1/0722/12 grant entitled "Security in distributed computer systems and mobile computer networks."

References

1. J. Dean and S. Ghemawat. MapReduce: Simplified Data Processing on Large Clusters. *Commun. ACM*, 51(1):107–113, January 2008.
2. A Frei and M. Rennhard. Histogram matrix: Log file visualization for anomaly detection. In *Availability, Reliability and Security, 2008. ARES 08. Third International Conference on*, pages 610–617, March 2008.
3. Q. Fu, J.-G. Lou, Y. Wang, and J. Li. Execution anomaly detection in distributed systems through unstructured log analysis. In *Proceedings of the 2009 Ninth IEEE International Conference on Data Mining*, ICDM '09, pages 149–158, Washington, DC, USA, 2009. IEEE Computer Society.
4. L.K.J. Grace, V. Maheswari, and D. Nagamalai. Web log data analysis and mining. In Natarajan Meghanathan, BrajeshKumar Kaushik, and Dhinaharan Nagamalai, editors, *Advanced Computing*, volume 133 of *Communications in Computer and Information Science*, pages 459–469. Springer Berlin Heidelberg, 2011.
5. A Makanju, A.N. Zincir-Heywood, and E.E. Milios. Investigating event log analysis with minimum apriori information. In *Integrated Network Management (IM 2013), 2013 IFIP/IEEE International Symposium on*, pages 962–968, May 2013.
6. M.G. Schultz, E. Eskin, E. Zadok, and S.J. Stolfo. Data mining methods for detection of new malicious executables. In *Security and Privacy, 2001. S P 2001. Proceedings. 2001 IEEE Symposium on*, pages 38–49, 2001.
7. M. A. Siddiqui. *Data mining methods for malware detection.* ProQuest, 2011.
8. R. Winding, T. Wright, and M. Chapple. System Anomaly Detection: Mining Firewall Logs. In *Securecomm and Workshops, 2006*, pages 1–5, Aug 2006.

Improved Remote User Trust Evaluation Scheme to apply to Social Network Services *

Youngwoong Kim, Younsung Choi, and Dongho Won+

College of Information and Communication Engineering
Sungkyunkwan University, KOREA
{ykim, yschoi, dhwon}@security.re.kr

Abstract. Social networking services are interactive services that allow users to build friendships or social relations with others on the Internet. They have played a central role in human interchange networks based on openness. Recently, however, groundless rumors have been spread by making bad use of their openness and a lot of users have been adversely affected by various attacks violating private information. Therefore, information on social networks should be able to be delivered to trustworthy persons or be provided from them. Accordingly, this paper proposes a remote trust evaluation scheme that improves the existing schemes, and provides a method for improving security by applying the proposed scheme to access control on social networks.

Keywords: Social network services, Remote trust evaluation scheme, Trust estimation, Relation and public trust

1 Introduction

The recent advent of social networks has led to great changes in our lives. Due to the emergence of social networks, however, users can produce and share information by themselves beyond one-way communication; this trend has become a driving force for the ability to interactively communicate. As a result of using the 2012 Korea Media Panel survey results to analyze the current status of SNS use, it was found that the average daily usages of SNS, phone, text message and chatting/messenger users are 73.2, 59.3, 58.8 and 39.3 minutes, respectively; SNSs represent the highest average daily usage per user [15]. Considering that the number of smartphone users is on the rise, in the future, it is expected that social networks' influence will be even greater. The interactivity of social networks has provided open and various information markets to many people, but on the other hand, a lot of unreliable information has also

* This research was supported by Basic Science Research Program through the National Research Foundation of Korea (NRF) funded by the Ministry of Science, ICT & Future Planning (2014R1A1A2002775).

+ Corresponding author.

K.J. Kim (ed.), *Information Science and Applications*,
Lecture Notes in Electrical Engineering 339, DOI 10.1007/978-3-662-46578-3_54

been produced due to the flood of indiscriminate information. Consequently, due to this information, normal users have gotten into situations where they experience confusion due to distorted information spread by malicious users when using social networks. In addition, the risk of violating privacy is even greater than before. These attacks, abusing the openness of social networks, have become a serious problem; they cause damage due to their easy and fast spreading characteristics. Furthermore, it is considerably difficult for users to understand the source of information propagated in the deluge of information. Recently, an inter-node dynamic user trust estimation scheme was proposed to be applied to SNSs [1]. In addition, studies are also in progress on methods to quantify trust based on the various components of social networks. However, most of these schemes are restrictive for the trust calculation between several nodes. Two people connected directly can be measured accurately, but it is difficult to specifically measure the trust estimation for persons across more than two connections due to various restrictions, thus it is difficult to carry out specific trust calculations. This paper would like to compensate for the defects of the existing papers. This paper is organized as follows. Chapter 2 covers related studies and analyzes the characteristics and limits of the existing trust studies; it describes a six-step separation theory and background on basic access control for applying the theory to social networks. Chapter 3 proposes an improved trust scheme model and a method for applying it to access control based on the improved trust scheme. Chapter 4 analyzes whether the proposed scheme is effective or not through different experiments. Chapter 5 presents conclusions drawn from this study.

2 Related studies

2.1 Characteristics and Limits of the Existing Trust Studies

Recently, many studies have been carried out for evaluating the specific trust of SNSs and a variety of schemes have been suggested to make a trust model that could apply to any social network. In the case of the research in [1], the trust could be measured for an interval, but it was limited in that it could not measure trust between users at several intervals. Golbeck proposed the tidal trust algorithm [3]. Tidal trust is an algorithm in which an original user asks for the trust levels of an evaluation target from primary neighbors who directly know the evaluation target in order to understand them; the trust level is predicted by a method that takes a weighted average of the trust levels. At that time, the weight is represented by a value for the degree of trust of the reliable path leading to the evaluation target's primary neighbor from the original user, and then the weight is determined as the minimum value of the trust link. However, the paths used to predict trust are limited to only the shortest distance that reaches the evaluation target from the original user. The paper in [2] used transition and combination for trust propagation to consider the calculation of distances over two or three nodes in order to better improve the content argued by Golbeck. In addition, it proposed a method for using the principle of min-max and weighted averages to calculate a distance over two nodes. However, because this paper's method focused only on calculating distances between several nodes, the distance calculation on a node is less accurate than the paper in [1]. In addition, it used the min-max method, and it is unclear in this method whether or not the exact numerical value should also be reflected if the trust deviation is large. Therefore, the scheme in [4] was proposed to improve upon the disadvantages of [1] and [2]. The scheme in [4] overcame the limits in [4] in which only the shortest path is considered because it could measure the trustworthiness of users at several intervals and consider the trust of various paths. In [4], however, the trust was represented as a relative value when calculating between 0 and 100%; it therefore had a problem when the trust exceeds 100%. To reflect the combination property in [2], this scheme greatly considered a variable element called the number of paths into the trust; therefore it is unreliable for representing the trust within 100%. The trust represents a numerical value for how much a user is able to trust others; therefore it needs an additional study for models which could accurately quantify when the trust is represented as 100%.

2.2 Six-step Separation Theory

The six-step separation theory states that all people are connected to each other when they respectively cross six persons; this has become known to the public through the 'The Small World Problem', a paper presented in Psychology Today by Stanley Milgram, a professor in the Psychology department at Harvard University in 1967 [8]. In general, his experimental results showed that there were 5.2 persons between any two persons; this result was therefore reflected in what was named the six-step separation theory. In 2010, Sysomos had studied this problem based on the relations of 5.2 billion Twitter users [6]. As a result, Sysomos had arrived at a conclusion that the average distance between people was 4.67 steps on Twitter, and up to 83% and 96% of people could reach almost everyone around the world by crossing 5 and 6 steps, respectively [6][7]. It was discovered that the average distance decreased more and more, coming down to 4.67 in 2011 and 2012.

3 Proposed Model

3.1 Basic Terms

3.1.1 Transition

If all people build reliable relationships with each other, not only could persons with a close relationship be trusted but also those with a distant relationship. This is possible because trust could be propagated endlessly along the trust path. For example, when A trusts B, B trusts C and C trusts D; the condition in which A could trust D is called a transition.

3.1.2 Combination

Combination means that information for an evaluation target judged through several paths is more reliable than judging the evaluation target through a person. For example, suppose A want to know person B. There are two options for getting to know B: on one is through a person and the other is through several paths. At this time, judging with information acquired through several paths is more reliable than judging only with one person's account. Nevertheless, the trust does not always increase when the number of combination paths is greater. A hop represents an interval between nodes.

3.2 Proposed Model

In [4], an expression was proposed to measure the trust beyond two steps. This paper assumed that T was between 0 and 100% when the trust of a node was T. The problem with this is that it is difficult to represent a sufficient trust value in certain situations with a numerical value for this expression. For example, if the trust value of all possible full paths is 100% (1 in decimal form), a problem exists for being able to have a value exceed 1. Therefore, we would like to improve the expression. By first looking into the background of the suggested expression in [4], the improved expression is as follows. In the recent USA subgraph, Facebook graph and Italy, Swiss graph model experiments, the research findings suggested that almost all people could be known if respective social networks have only about four hops [6]. Therefore, this paper does not handle combinations if the target to be evaluated is above five hops. Therefore, it assumed that an expression is organized with trust paths containing two, three and four hops. Furthermore, this paper considers the number of paths in the trust coefficient to strengthen the characteristics of combination. However, according to the experimental results in [2], it was found that the unconditionally growing number of paths does not guarantee an increase in trust. To compensate for the above problems, this paper improves the expression as follows.

$$T_r = aT_2+bT_3+cT_4$$

where T_2 indicates an arithmetic average of the trust obtained from the total number of paths in the case of two hops. T_3 and T_4 indicate arithmetic averages of the trust obtained from the total number of paths with three and four hops, respectively. The variables a, b and c are weights reflecting the network's influences from respective absorption and the trust coefficients of respective hops, which become $a+b+c = 1$. The values for a, b and c are determined within the range of ($a>b>c$) by considering this to be suitable for the network's situation. According to [2], the trust to be calculated tends to decrease as the number of hops increases. For example, suppose there are three paths between A and B, which have two, three and four hops, respectively. In addition, suppose the trust between nodes is 0.5. Then, A's trusts for B becomes 0.25, 0.125 and 0.0625 in the case of two-, three- and four-hop paths, respectively. By comparing these numerical values, it is seen that the trust decreases as the number of hops increases. Therefore, this paper considers such a characteristic for determining the expression's coefficient condition within the range of a>b>c.

Table 1. Parameters of scheme.

Sign	Content
a	Factor of 2 hops
b	Factor of 3 hops
c	Factor of 4 hops
T_2	Average trust value for 2 hops
T_3	Average trust value for 3 hops
T_4	Average trust value for 4 hops
T_r	Total remote trust value

3.3 Total Trust

This paper also uses public, relationship and total trust based on the definition in [1]. The total trust is evaluated by reconciling objective and subjective evaluations well. At this time, the total trust between nodes across them is represented as follows.

$$\mathrm{T} = \sqrt{(m_p T_p)^2 + (n_r T_r)^2},\ (m_p)^2 + (n_r)^2 = 1, m_p = \sin\ \theta,\ n_r = \cos\theta$$

where T_r and T_p represent the relationship and open trust, respectively. The values for m_p and n_r could be changed appropriately to adjust the relative importance of the public and relationship trust. On the other hand, a great number of nodes and links are connected in social networks. Therefore, considerable time and energy is consumed if it would like to understand all these nodes and links. Furthermore, nodes in social networks have the dynamic tendency of being newly created or removed. Therefore, depending on the situation, considering all paths and nodes may be inefficient. Thus, to prepare for the above limitations, this paper considers a concept called the centrality of networks [9] for using the path finding algorithm in [11] which is optimized for social network environments [10][12].

4 Experiments and Analysis

4.1 Experimental Environment Configuration

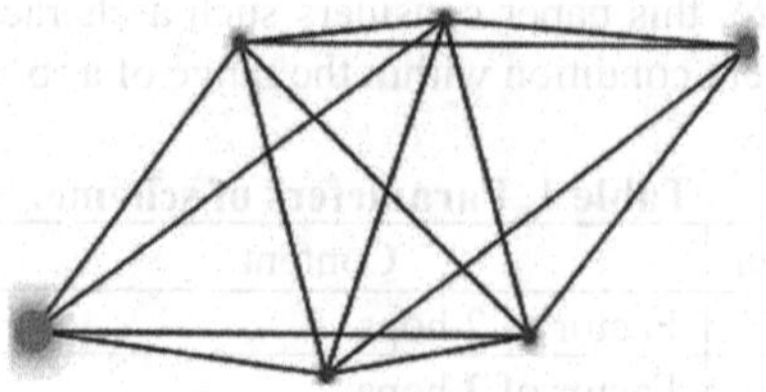

Figure 1. Six nodes in Facebook.

To confirm the applicability to the remote trust evaluation scheme proposed in this paper, the environment was configured to represent relationships in SNSs. To do this, six nodes were used as an experimental sample (Figure 1), and these nodes targeted people who actually make friends on Facebook. Of the six nodes, four intermediate nodes became friends with the evaluation target and evaluator, and all four nodes were connected to each other by friend requests. This relationship is represented by a node excel as shown Figure 1 [13, 14]. In this figure, the largest left and upper right nodes represent the evaluation target and the evaluator who will carry out the evaluation, respectively. The blue users in the middle represent the intermediate nodes and a line represents a friend relationship. The measurement of trust was calculated with two, three and four hops. For the trust between the respective number of hops, a random function in the node excel was used to obtain objective and arbitrary numerical values. To do this, the possible range of the trust was set to [1, 100]. A total of 20 experiments were carried out within the range of [1, 100]. For convenience, the trust between [1, 100] was represented in decimal form as [0, 1]. The improved scheme was compared with the min-max method used in [2], the scheme of in [4] and the open trust. This experiment set $\theta=60^{\circ}$ to raise the importance of the open trust values. In addition, weights were set as a=0.7, b=0.2 and c=0.1 to reflect that the trust decreases as the distance increases. There were 4, 12 and 24 paths for 2, 3 and 4 hops, respectively, and the total number of paths was 40. Based on these paths, the values obtained from the Excel random function were substituted for the paths to apply to the respective schemes.

4.2 Analysis Results

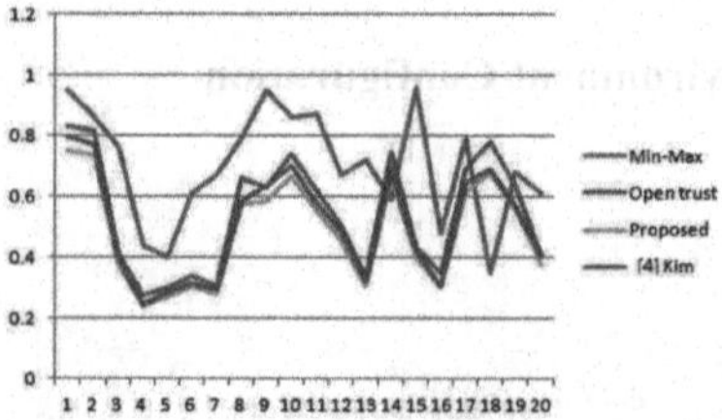

Figure 2. Trust graphs of schemes.

As shown in Figure 2, the min-max exhibited higher trust than other methods in general. Allowing for large deviations in data, min-max hardly seems to uniformly consider the trust of many people. Finally, the min-max method may not experience problems when using it in situations with small deviations, but if there are high trust values on a path when the deviation is large, a problem occurs when one of the values is used as the trust. In addition, the min-max method had large deviations when compared with the open trust. The open trust is an average trust of intimate people who know the evaluation target well; it can be recognized that its trust is high because it is more objective than the trust of typical nodes. When the min-max method is compared with the open trust, there are many cases with large deviations, as shown in Figure 2. In this respect, it hardly appears as if the min-max method reflects the exact trust. By comparing the values in [4] with the experiment values in the improved paper, it can be observed that the trust values are consistent in the sense that they are close to the open trust values. On the other hand, one of the advantageous important characteristics of the two schemes is that users can adjust the variables of the schemes to suit the network's situation. By doing this, the trust scheme should be able to clearly determine whether or not to sufficiently trust the other party. To clearly make a trust determination for the evaluation target, it should clearly represent the difference in the trust result values by the angle control. Figure 3 compares the result values in [4] and the improved paper by their angles. In Figure 3, the angles are 60 degrees for [4] and 45 degrees for the improved paper. As a result, the improved paper exhibits larger differences from the angle than the existing papers. This means that is easier to evaluate trust with the improved scheme than the existing papers when viewed from the evaluator's perspective. Therefore, evaluators can determine the evaluation target more clearly with the improved scheme.

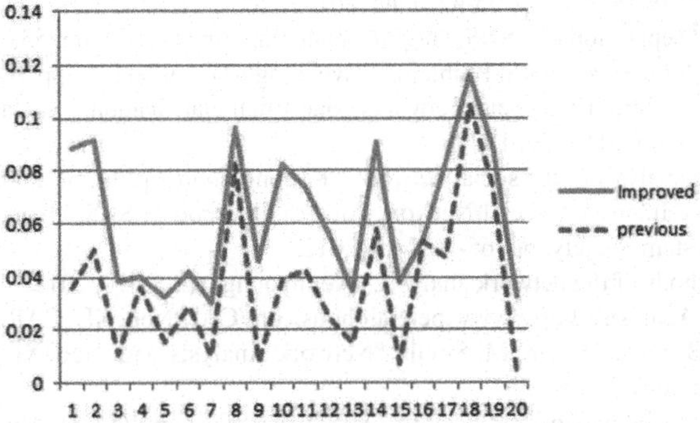

Figure 3. Differences between improved and previous schemes.

5 Conclusion

This paper would like to improve a problem existing with the scheme in [4]. The scheme in [4] has the limitation of not being able to represent a trust range within 100% for a very special case. However, by using the proposed scheme, trust between 0 and 100% can be represented for any situation; this solves the problem in [4]. Furthermore, by selecting a path algorithm that is suitable for social networks, it is possible to efficiently evaluate trust in social network environments. Besides, because weights can be set in the scheme, that is, the weights of the open and relation trusts from the evaluator's tendency and the network characteristics viewed by the evaluator, this can be used as a user-friendly scheme.

References

1. C. Lee, D. Won, "Dynamic User Reliability Evaluation Scheme for Social Network Service," Korea Institute of Information Security, 23(2), pp. 157-168, April 2013.
2. S. Song, "A Study on Transitivity and Composability of Trust in Social Network," Journal of Information Technology Applications & Management, 18(4), pp. 41-53, Dec 2011.
3. J. Golbeck, "Personalizing Applications through Integration of Inferred Trust Values in Semantic Web-Based Social Networks," ACM Transactions on Internet Technology (TOIT) Volume 6 Issue 4, pp. 497-529, November. 2006.
4. Y. Kim, D. Won, "Dynamic Remote User Trust Evaluation Scheme for Social Network Service", Korea Institute of Information Security and Cryptology, pp. 373-384, April 2014
5. J. Leskovec, E. Horvitz, "Planetary-Scale Views on a Large Instant-Messaging Network," Proceedings of the 17th international conference on World Wide, pp. 915-924, Apr. 2008.
6. L.Backstrom, P. Boldi, "Four degrees of separation," Proceedings of the 3rd Annual ACM Web Science Conference, pp. 33-42, June. 2012.
7. Six degree of separation and SNS, "http://socialcomputing.tistory.com/53,"
8. S. Milgram, "The Small world Problem," Psychology Today, vol. 1, pp. 61-67, May. 1967
9. L. Freeman, "Centrality in social networks conceptual clarification," Social Network, Volume 1, Issue 3, pp. 215–239, 1979
10. D. Son, "The analysis of the social network." Kyeongmoon, pp. 1-254, 2002
11. J. Son, S. Jo, "Improved Social Network Analysis Method in SNS," Korea Intelligent Information System Society, pp. 65-70, May 2012.
12. S. Lee, "Methods of the network analysis." Nonhyoung, pp. 5-370, 2012.
13. NodeXL for Analysis, http://www.peteraldhous.com/CAR/NodeXL_CAR2012.pdf
14. D. Hansen, B. Shneiderman, M. Smith, "Network Analysis with NodeXL tutorial," Draft, pp. 1-28, Feb 2009
15. KISDI, "The social network service use," KISDI STAT Report (13-4), Apr
16. K. Kwon, D. Won, "Relationship-based Dynamic Access Control Model with Choosable Encryption for Social Network Service", Korea Institute of Information Security and Cryptology, pp. 59-74, Feb 2014
17. Y. Kim, D. Won, "Remote User Trust Evaluation Scheme by combining geometric and truncated average", Korea Conference on Software Engineering, pp. 146-149, Jan 2014
18. K. Son, D. Won, "Design for Zombie PCs and APT Attack Detection based on traffic analysis", Journal of the Korea Institute of Information Security and Cryptology, pp. 491-498, June 2014

Threat assessment model for mobile malware

Thanh v Do[1] Fredrik B. Lyche[2] Jørgen H. Lytskjold[2] and van Thuan Do[3]

[1] Telenor & NTNU, Snaroyveien 30, 1331 Fornebu, Norway – thanh-van.do@telenor.com
[2] Norwegian University of Science and Technology, Department of Telematics, O.S. Bragstads Plass 2E, NO-7491 Trondheim, Norway – {fredrik}{jorgen}@item.ntnu.no
[3] Linus AS, Martin Linges vei 15, 1364 Fornebu, Norway – t.do@linus.no

Abstract. Today the smartphone is definitely the most popular and used device, surpassing by far the laptop. Unfortunately, its popularity makes it also the most targeted goal for malicious attacks. On the other hand mobile security is still in its infancy and improvements are needed to provide adequate protection to the users. This paper contributes with threat assessment model which enables a systematic analysis and evaluation of mobile malware. The same model can also be used to evaluate the effectiveness of security protection measures.

Keywords: threat assessment model, threat analysis, mobile device security, mobile phone security, smartphone security, security mechanism.

1 Introduction

The smartphone has in the last decade become the quintessential technical device of our time. Ever since Apple introduced its iPhone in 2007 and defined what one could expect of a modern smartphone, its popularity has soared. A smartphone has the capabilities of a computer and phone combined in one, making it a device suitable for both work and private use. Unfortunately, with a wide array of functions and applications, comes an equally wide array of vulnerabilities. Indeed, a smartphone is much more exposed compared to a basic low-cost phone [1] with only telephony and short message service or a feature phone [2]. Figures provided by Juniper Networks [3] show malware findings in smartphones has increased almost exponentially, growing with 155 % in 2011 to 614 % from March 2012 to March 2013. Sophos reveals the same trends in its mobile security threats report issued in 2014 [4]. The biggest problem is the lack of awareness. Indeed, both the users, enterprises and service providers were not aware of the threats that the smartphones are exposed to. Furthermore, smartphone security is still in its infancy and improvements have to be done to provide adequate protection.

The goal of this paper is to contribute to the strengthening of the smartphone security by proposing a threat assessment model which enables a systematic analysis and evaluation of mobile malware. It starts with a brief review of the state-of-the-art of smartphone security and a summary of existing security protection measures for smartphones. The central part of the paper is the description of the proposed Threat

K.J. Kim (ed.), *Information Science and Applications*,
Lecture Notes in Electrical Engineering 339, DOI 10.1007/978-3-662-46578-3_55

Assessment Model. To illustrate the usage and usefulness of the model two threat assessment case studies, namely Trojans and Advertisement are carried out.

2 State-of-the-art Mobile Security

Most of the related works were aiming to assess the vulnerabilities and threats of smartphones such as [13][15][16] and the fewer ones [14][17] that consider mobile malware only came up with an evaluation at generic type level and do not provide an assessment framework for individual malware. Although still insufficient smartphones do have today a number of security protection measures that will be successively introduced in this section.

2.1 Operating System measures

Smartphone's operating systems such as Android [5] provides different protection measures as follows:

A. Kernel measures:

- *User based permission access:* Applications are not allowed to access resource and capabilities without permission.
- *Isolated processes:* Each application is executed separately and not access resources of another application.
- *Secure inter-process communication:* To provide secure communication between processes OS like Android offers four methods: binder, services, intents and content providers [6].
- *Sandbox:* The kernel enforces a security barrier between the application and the system at a process level.

B. Application framework measures:

Some smartphones OS such as Android has an Application Framework Layer which provides a variety of functionalities and basic tools to build applications e.g. Activity Manager managing application life cycle, Content Provider handling data sharing between applications, Telephony Manager managing all voice calls and Resource Manager managing various resources used in applications.

To get access to restricted functionalities provided by this layer an application has to declare the permission in its manifest file (AndroidManifest.xml). An example of permission is to grant Internet access to an application internet access. At installation these permissions are presented to the user, who has then the option to accept or decline the installation.

C. Memory Security

Some OS like Android provide a few additional safety features to prevent memory exploitation such as by setting heaps and stacks of memory to nonexecutable [7], or randomization of address space layout.

2.2 Application vetting

The process of verifying that a software application meets the requirements imposed by an organisation is known as vetting [8]. Vetting can be applied to either binary code or source code and the code can be examined as static artifacts or the binary can be examined while running dynamically. An example of application vetting is the Google's Bouncer that helps prevent malware spreading through the Google Play Store by vetting all the application submitted to the Play Store.

2.3 Anti-virus

Antivirus or anti-virus software (often abbreviated as AV), sometimes known as anti-malware software, is computer software used to prevent, detect and remove malicious software. An AV program differs from the other security measures featured above in that it is optional. It usually includes the following:

- *Malware detection and protection:* The traditional signature-based detection in which signatures are used to compare and validate software, is the obligatory feature that all AV solutions employ.
- *Theft protection:* It does not protect you from thieves, but rather mitigates the impact should the smartphone be stolen or lost. Typical components are remote wipe, remote lock and locate device.
- *Safe web surfing:* It protects the user whilst surfing the Internet, primarily against phishing attacks and threats. It is often accompanied by a parental control mode, where it is possible to define a blacklist or whitelist of specific websites.

3 The proposed Threat assessment model for mobile malware

Let us now start with the elaboration of the Threat assessment model malware on smartphone.

Every malicious attack has usually as objective to access and misuse certain assets in the smartphone as organized follows:

1. **Telephonic features**
   - Read/send SMS
   - Call function
2. **User data**
   - Camera
   - Global Positioning System (GPS)
   - Microphone
   - Contacts
   - Email
   - Calendar
   - Phone number
   - SD card

- o International Mobile Equipment Identity (IMEI) number
- o Gyro-/accelerometer

3. Internet communication

- o Remotely impose instructions (Push option)
- o Remotely retrieve information (Pull option)

4. Computing Power

- o Central Processing Unit (CPU)

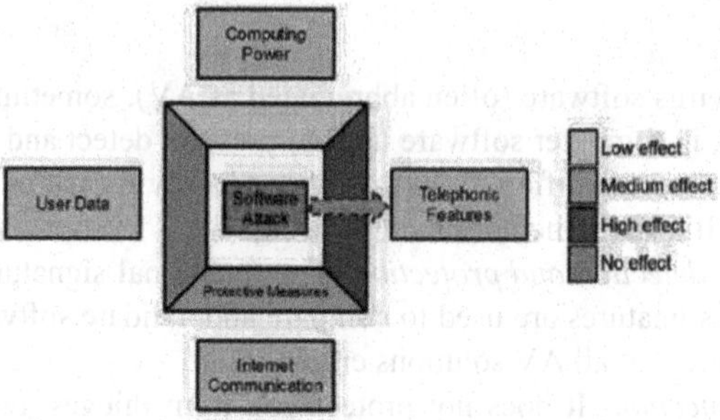

Figure 1 The Threat assessment model

The assets are classified into 4 classes with their underlying items. It is worth emphasizing that only the most targeted assets are listed and the other assets with low risk are omitted. For example, many attackers would probably like to access other applications' internal storage and resources, but the existing security solution renders this option so unlikely that the risk is negligible, and thus this asset has not been included in the list above.

The Threat assessment model in Figure 1 can be explained as follows:

- A malware type is represented by a red square in the center of the model the colour red indicates its malicious nature.
- The blue boxes around serve as the four assets: Computing Power, User Data, Internet Communication and Telephonic features
- The red arrow represents a specific attack from the given malware class towards an asset group (which in the picture Telephonic Features).
- The four edges that constitute a frame, symbolize protective measures restricting the attack's access to the smartphone's assets. Each edge is colored according to the protective measure's level of effectiveness in preventing the attack. The effectiveness can be Low effect, Medium effect, high effect and no effect.

It is worth noting that the proposed Threat Assessment Model cannot only be used to assess a malware but also in the evaluation of the effectiveness of the countermeasures against a certain types of malware.

Figure 2 shows the Threat Assessment Model with the current security protection measures and the effectiveness of the Anti-virus program is concluded to be a medium effect against a certain software attack.

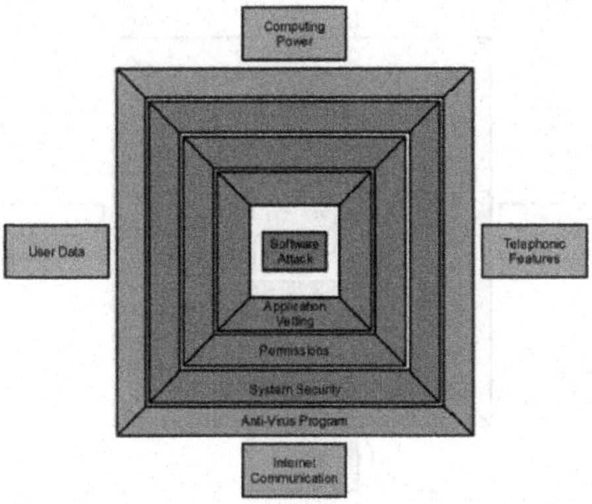

Figure 2 The Threat Assessment Model with current security protection measures

3 Malware assessment case studies

Let us now demonstrate the usability of the proposed Threat Assessment Model by assessing a few malware types.

3.1 Trojans

Trojans are the most prevailing attacks by far. They are programs or codes that disguise as something useful and then deliberately perform harmful actions once installed. They do not exploit technical weaknesses in the OS or device itself in order to be deployed, but make use of people's credulity to get a foothold for further/future misconducts. As such, new security mechanisms have not been able to assuage fears of malign Trojans, as the concern is not primarily technical. Human errors are very hard to eliminate, whereas technical vulnerabilities are constantly being rectified by manufacturers and developers. Together, these elements have contributed to the widespread success of Trojan attack vectors.

Trojans can be further classified into sub-groups as follows:

- *Fake installers:* aims at making profit by sending premium SMS [10].
- *Spyware:* collects information from the user's device without the user's consent [11].
- *Backdoor:* enables the attacker to either steal information stored on the phone, or take advantage of its hardware resources

- *Botnet:* Botnets are capable of executing computationally demanding tasks in feasible time, but may also be used as a multi-headed beast in a distributed denial-of-service attack.

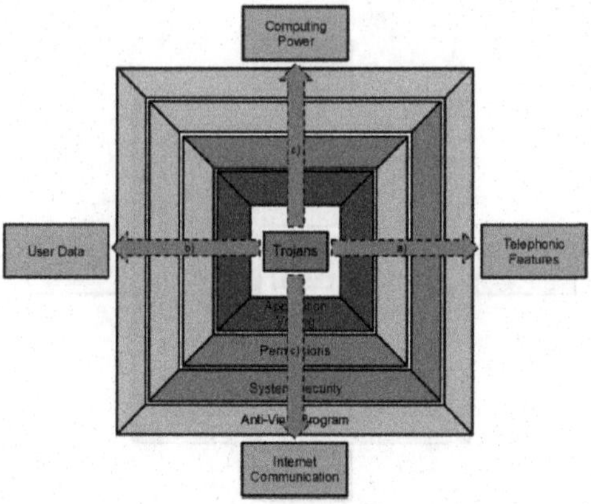

Figure 3 The Threat Assessment Model for Trojans

Since Trojans exploits the user's credulity the application vetting is classified as highly effective against all forms of Trojan attacks.

In Figure 3 Permissions is colored light green i.e. medium effect for fake installer indicated the arrow pointing right a) and also spyware, indicated by the arrow pointing left b) because some users may refrain from installing applications requesting intrusive permissions. Most users ignore that warning, though.

Permissions do not thwart attacks that aim to hijack device resources, as using computational power requires no particular approval from the user.

System security, however, does make sure no application uses the entire hardware capacity. It also prevents a malicious app from accessing other application's internal data.

The effectiveness of Anti-Virus programs relies heavily on the issuing company's database. If a malware is known, the AV program may warn the user before any harmful actions are done, but it cannot ease the impact of the attack further.

3.2 Advertisement

A very popular way for developers to get paid for their work is to include advertisements in their applications. By incorporating an advertisement library, they can release the application for free and still generate revenue [9] The result is some sort of a banner, e.g. at the bottom of the screen, or an advertisement covering the whole screen for some seconds. At runtime, the advertisement library connects with the advertisement server and receives advertisements to be displayed. The advertisement network pays the developers based on the exposure of the advertisement, either per time it is displayed, per click or some other measures.

Advertisement can bring the following threats:

- *Permission abuse:* The Advertisement Library activity and the advertisement that is loaded into the application will have the same permissions as the host application. Malicious advertisements included in applications with a lot of permissions can have uncomfortable effects for the user.
- *Acquire user data:* An advertisement can for instance gather user location information
- *Send spam:* This can be done either from their servers or through the use of email or SMS from the user's phone.
- *Fetching and loading dynamic code*: Some advertisement networks are capable of fetching and downloading code from the Internet and pose a great threat for two reasons. [12] First, the dynamically loaded code cannot be analyzed, thus bypassing any static analysis efforts. Secondly, the downloaded code can change at any time, making it very hard to predict and defend against its behavior.

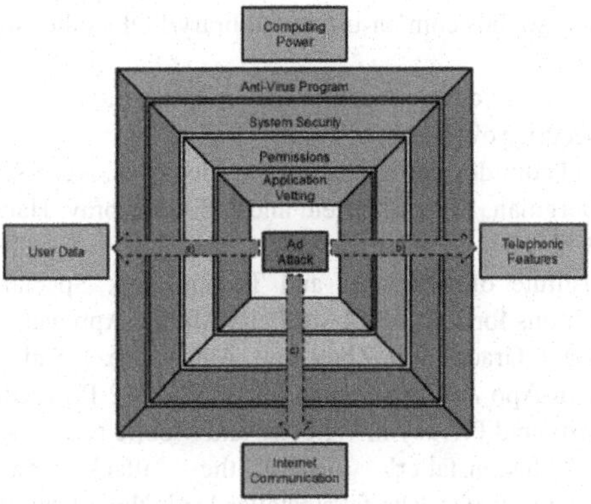

Figure 4 The Threat Assessment Model for Advertisement

As shown in Figure 4 advertisement attacks target three different assets: User Data, Telephonic Functions and Internet Communication. To reach these assets the threats has to pass through the different security "walls" in the model:

- *Application Vetting*: Advertisement attacks are hard to detect using application vetting, because they are delayed in striking and often the malicious code is not part of the host application. Malicious Advertisement Libraries and threats that that fetches dynamic code might be detected with Application vetting, because the malevolent code is present in the application from publication.
- *Permissions*: Malicious advertisements often attach themselves to legitimate applications and inherit the host applications permissions. However, Telephonic Features require more permissions then User Data and Internet Communication, thus it has a darker grade.

4 Conclusion

This paper proposes a Threat Assessment Model which provides a systematic and qualitative analysis and evaluation of the threats that a mobile malware could incur the user. Furthermore, this model could also be used to assess the effectiveness of a given protection measure such as application vetting, permissions, anti-virus, etc. As further work it would be really useful to make the model quantitative by introducing metrics allowing to quantify the level of threat of mobile malware and the level of effectiveness of the protection measures.

References

1. Low cost phone.
2. Feature phone.
3. Juniper Networks. Juniper Networks Third Annual Threats Report, 2013.
4. SophosLabs Vanja Svajcer. Sophos mobile security threat report. Technical report, 2014. http://www.sophos.com/en-us/medialibrary/PDFs/other/sophos-mobile-security-threat-report.pdf
5. Android developers – http://developer.android.com/index.html
6. Android. Android security overview. https://source.android.com/devices/tech/security/index.html. Accessed May 6, 2014.
7. Jason Rouse Neil Bergman, Mike Stanfield and Joel Scambray. Hacking Exposed Mobile. McGraw Hill, 2013.
8. NIST (National Institute of Standards and Technology). Special Publication 800-163 Technical Considerations for Vetting 24 3rd Party Mobile Applications (Draft) Aug 2014
9. Xuxian Jiang Michael Grace, Wu Zhou and Ahmad-Reza Sadeghi. Unsafe Exposure Analysis of Mobile In-App Advertisements. In WISEC '12 Proceedings of the fifth ACM conference on Security and Privacy in Wireless and Mobile Networks, 2012.
10. Fernando Ruiz. 'fakeinstaller' leads the attack on android phones. http://blogs.mcafee.com/mcafee-labs/fakeinstaller-leads-the-attack-on-android-phones.Accessed November 12, 2013.
11. Lookout Mobile Security. Lookout Mobile Threat Report, 2011.
12. Xuxian Jiang Michael Grace, Wu Zhou and Ahmad-Reza Sadeghi. Unsafe Exposure Analysis of Mobile In-App Advertisements. In WISEC '12 Proceedings of the fifth ACM conference on Security and Privacy in Wireless and Mobile Networks, 2012.
13. ENISA: Hogben, G., Dekker, M.: Smartphones: Information security risks, opportunities and recommendations for users. Technical Report, 2010
14. Jeon, W., Kim, J., Lee, Y., Won, D.: A Practical Analysis of Smartphone Security: HCII 2011, Smith, M.J., Salvendy, G. (eds.), Part I. LNCS. Springer, Heidelberg (2011)
15. Mylonas, A., Dritsas, S., Tsoumas, B., Gritzalis, D.: Smartphone Security Evaluation: The Malware Attack Case: International Conference on Security and Cryptography (SECRYPT 2011)
16. NIST: Souppaya M. and Scarfone, K. (2013), Guidelines for Managing the Security of Mobile Devices in the Enterprise, Special Publication 800, June 2013
17. D. Damopoulos, G. Kambourakis, S. Gritzalis, S. O. Park (2014), Exposing mobile malware from the inside (or what is your mobile app really doing?), Peer-to-Peer Networking and Applications, Vol. 7, No. 4, 2014, Springer.

Performance Evaluation of System Resources Utilization with Sandboxing Applications

Tarek Helmy* Ismail Keshta Abdallah Rashed

Department of Information and Computer Science,
College of Computer Science and Engineering,
King Fahd University of Petroleum and Minerals,
Dhahran 31261, Mail Box 413, Kingdom of Saudi Arabia,
* *On leave from Faculty of Engineering, Department of Computers & Automatic Control Engineering, Tanta University, Egypt*
[helmy, ismailk, abdrashed]@kfupm.edu.sa

Abstract. Sandboxing is a popular technique that is used for safely executing untested code or testing un-trusted programs inside a secure environment. It can be employed at the operating system level or at the application level. In addition, it limits the level of access requested by the untested programs in the operating system by running them inside a secure environment. Therefore, any malicious or improperly coded programs that are aiming to damage hardware or software recourses will be prevented by the sandboxing. In this paper, we want to assess the effect of sandboxing on the system's recourses utilization. We will examine and evaluate the operating system performance with Sandboxie, Bufferzone and Returnil sandboxes applications. Different performance parameters are considered, such as the execution time by the CPU for each sandbox and the read/write speed for various input output devices like memory and disks. Moreover, it is important to highlight that we have evaluated the mentioned sandboxing applications under the effect of having a virus that is attacking the operating system. We defined our own performance metrics that contain the most important parameters used in evaluating the related research work.

Keywords: Sandboxing, Computer Security, Operating System.

1 Introduction

One of the most important challenges that face the Operating System (OS) is detection and reaction upon Malware. Malware is malicious software that is designed to damage a computer system, including the hardware and software recourses, and we know that the OS has the full responsibility to manage these recourses. Not only that but also, it has the ability to make the OS in unavailable or unusable state as described in [1, 2]. Many antivirus vendors use signature-based method to detect this kind of software. However, the update rates of the virus signature database cannot find out the creation rate of the new Malware variants. There are many Malwares such as Adware, Trojan, Worm, and Botnets. Such programs want to access computer systems that include the data and recourses in order to damage them or make them unavailable or in unusable state [12]. One of the main goals of the OS is resource management.

K.J. Kim (ed.), *Information Science and Applications*,
Lecture Notes in Electrical Engineering 339, DOI 10.1007/978-3-662-46578-3_56

Thus, the OS should protect the access to system resources from untested programs. Sandboxing is a new mechanism that is used for safely executing untested code or testing un-trusted programs. It has the ability to run programs inside itself without changing them but their access to the system resources is limited. It can be applied at the application level or the OS level. In addition, it has the ability for monitoring and controlling Malwares. It can limit the level of access requested by these programs in the OS. Thus, this will obviously prevent any damage that might happen by these programs, because they are either Malware software or not coded in a proper way [1, 3]. The most important feature provided by sandboxing is the security mechanism for separating running programs. In other words, it provides a minimum level of isolation between the running programs. In addition, it ensures some level of privacy for the users. It is important to highlight that by using sandboxing, an additional protection layer will be added to the system, which means extra overhead. Therefore, the performance of a running program will decade as proofed by [2, 4]. This paper is organized as follows. Section 2 provides a literature survey that reviews some related work. Also, under the same section, we describe different types of sandboxes that are commonly used and give examples of each type. In Section 3, the main contribution of this paper is highlighted. Section 4 demonstrates the experimental setup and discusses the obtained results. Finally, Section 5 concludes the paper and highlights the future research directions.

2 Sandboxing Overview

The authors of [5] studied the sandboxes application based on the system call context. They proposed a sandboxing model to capture the system calls that are generated from the sandboxes application. Their model shows low overhead in terms of time in addition to correct capturing for system calls. They did an experimental work over a certain number of sandboxing applications in order to see the performance of the OS with and without them over different scenarios. They used for example the execution time and the usage of physical memory as performance metrics in the evaluation. In addition, the authors of [6] presented a sandbox created for analyzing Android cloud service applications. They proposed an Android Application Sandbox (AASandbox) and used it for evaluating some of the Android cloud service applications. One of their experiments was about counting the number of system calls throughout the application runtime. About 150 applications were used by them to analyze and test the system. They run a self-written fork on the OS with and without sandboxing; they found that without sandboxing application the OS is not able to respond for a certain time. Sandboxing is classified into different classes based on their method of implementation. These classes are Jails, Virtual Machines, Rule-based execution and Stand-alone applications [2]. As shown in Figure 1. Applets sandboxing is one of the important sandboxing applications. In general, it is a very small application that performs one specific task that runs inside a large program. Java applets are the famous type of the Applet sandboxing. They are usually written in Java programming language as in Web sites [2, 12]. Another type of the sandboxing is Jail. In this type, the kernel will limit the program accessibility to the recourses such as network, file system, disk space limit and etc. Therefore, it can be defined as a set of resources that limit imposed on programs by the OS kernel [2, 12]. Jail sandboxing is commonly used in virtual hosting. A virtual machine (VM) [2, 12] is a completely isolated environment. Therefore, it has the ability to emulate a complete host computer. In other words, the guest OS will not have access to the resources of the system because it runs inside a sandbox environment. Such guest OS could only access host resources through the emulator. As in [12], rule-based execution represents another category of sandboxing techniques. It allows the user to have full control over what processes are started. It also provides a special environment to make viruses and Trojans having fewer opportunities to infect a computer. Two famous rule based execution implementations for Linux are the SELinux and Apparmor security frameworks. In this paper, we will focus on stand-alone sandboxing applications that have similar aims of providing safe environment or protecting the OS for running un-trusted software as discussed in [2] and we will extend the work being done by using more metrics and experiments. It is important to point out that in our work we will study the sandboxes which run on Windows platform such as Sanboxie, Bufferzone and Returnil. One of the sandboxing tools that are mentioned in sandboxing applications is Sanboxie. It is a software tool that has the ability to run other software's or programs in an isolated environment. Thus, these software's or programs will not be able to affect the underlining OS. The purpose for these tools is to help the users who are surfing the Web browsing activities or downloading un-trustable software by mistake. Running such un-trustable programs in an isolated environment will help in protecting the OS from malware or malicious software that could harm both the OS and the computer system recourses.

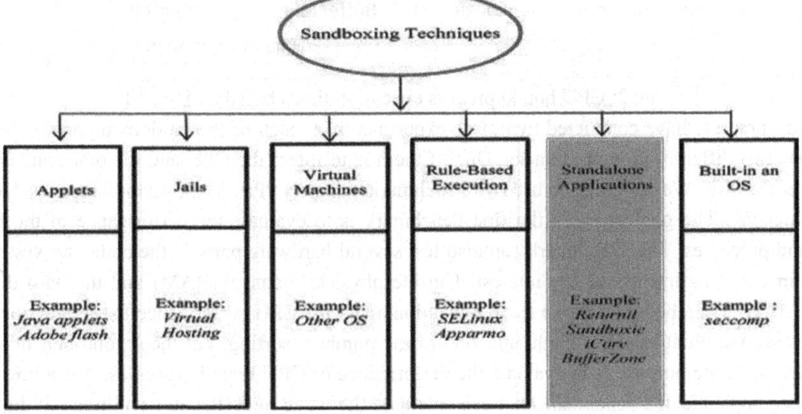

Fig. 1: Sandboxing classification

Sandboxie will achieve this by running a program or software in a virtual space. As soon as this tool is terminated, this space is released. In addition, it will erase everything that has been stored in it. Another tool of sandbox application that is widely used is Bufferzone [9]. It creates a virtual zone/container on user's device. Upon termination of any program executed in the virtual zone then all the changes of the computer system including recourses and data will be deleted. It is important to highlight that virtual zone is created on a special directory and stored in a special location in the C drive on Windows OS [2, 9]. Returnil is another sandboxing application that is categorized under the stand-alone applications category. This tool has the capability to copy the system drive into a virtual disk. Therefore, it will install anything in virtual disk instead of installing it on the system drive. After rebooting a process, everything will be deleted or cleared. A system partition which is a part of the system drive is created by this tool and stored in the RAM or in the hard disk. In additional, Returnil tool is able to create virtual partitions to store/retrieve the files even after rebooting process [11].

3 Proposed Work

In this paper, a stand-alone sandboxing application class will be the main focus. Sandboxie, Bufferzone and Returnil will be selected as examples of sandboxing applications. The performance of the three sandboxes will be examined and evaluated in terms of CPU-bound processes and I/O-bound processes in order to point out their benefits and limitations. It is important to point out that all experiments have been conducted on Windows 7 OS with and without the sandboxing applications to see how the OS will perform.

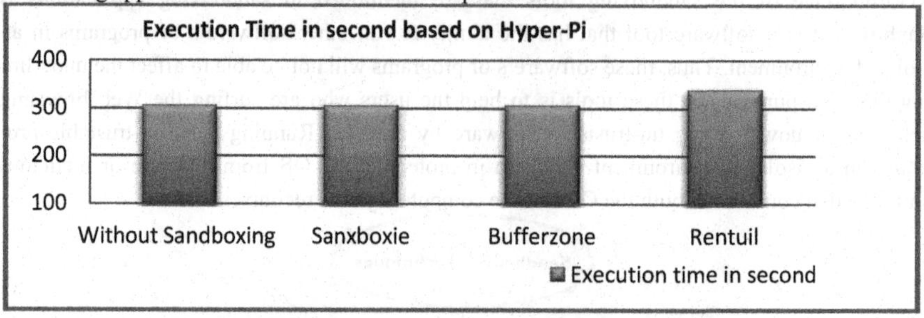

Fig. 2: CPU bound process execution times based on Hyper PI

In addition, we have conducted two more experiments on each of the sandboxing applications under two different program attacks. One of them is to infect the CPU and the other one is to infect the disk. We are going to use two benchmarks, namely "PassMark performance test" and "Hyper-Pi". The goal of using the first benchmark is to evaluate the performance of the I/O bound processes. This benchmark can also test several hardware parts of the computer system, but in our experiments we are interested in the physical memory (RAM) and the hard disk. Moreover, it can be used to give us an indication about the CPU performance for certain operations such as floating point math, and find prime numbers sorting. For the second benchmark "Hyper-Pi", the purpose is to evaluate the performance of CPU bound processes. It is a powerful and sophisticated benchmark that can calculate the value of Π (PI) up to millions of digits. In addition, we will provide and make a comparison between them. After the analysis and comparison of the obtained results, we will point out the most important parameters that highlight the effect of the tools and showing where we have big differences between them.

4 Results and Discussion

Different experiments are conducted in order to evaluate the performance of the three sandboxing applications on the CPU and I/O bound processes. The following sections will present the results of the experiments followed by detailed discussion and analyze for each one.

4.1 CPU-Bound Processes Evaluation

For evaluating the performance of the three sandboxing applications on the CPU, we conducted two experiments. The first one was by using Hyper PI benchmark while the second one was by using PassMark performance test. It is important to point out that Hyper PI is a very sophisticated benchmark and it demonstrates the performance of the CPU based on CPU bound process better than the other benchmarks. This is because it gives the user more opportunity and flexibility in evaluating the performance of the CPU. Therefore, we were able to configure Hyper PI to calculate Π (PI) value up to 8 million digits. Figure 2 shows the result of the average time taken by the CPU to calculate the value of PI without and with sandboxing applications. Figure 3 shows that Sandboxie and BufferZone sandboxes do not affect the CPU performance too much. In other words, the difference between the average execution time with BufferZone Sandboxie, and without sandboxing is almost the same. However the average execution time is quit high for Returnil virtual system. In other words, evaluating the value of PI up to 8 million digits inside Returnil will consume more time from CPU than evaluating it without sandboxing. Thus, comparing to the other tools, we can conclude that Returnil virtual system will significantly affect the CPU's performance. This is might be because it consumes more CPU resources from the system.

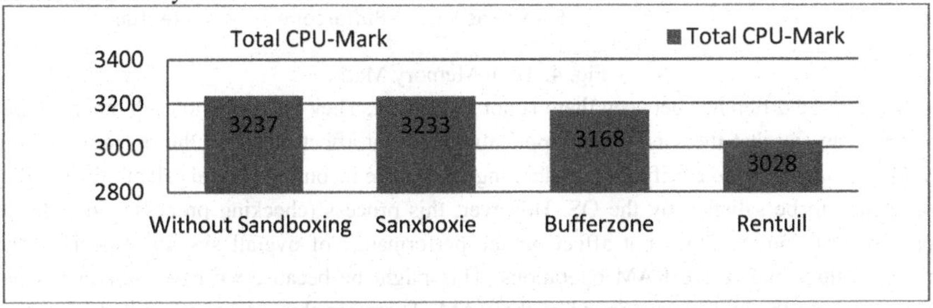

Fig. 3: CPU bound process based on PassMark

In order to evaluate the performance of the sandboxes applications on the CPU-bound, we used the PassMark performance test. This benchmark allows us to test the performance of the CPU according to different CPU test parameters. The CPU test includes various parameters such as CPU-Integer math, CPU-floating point math, CPU-compression, CPU-encryption and CPU sorting. Figure 3 shows the result of this test under PassMark with and without sandboxing applications. As we can see from Figure 3, the results of CPU Mark by Sandboxie and BufferZone were close to the result of CPU Mark without sandboxing. In other words, the difference between them is insignificant. Based on this, we can point out that Sandboxie and BufferZone do not affect the performance of the CPU bound processes. However, CPU Mark by Returnil virtual system was low when compared to other application sandboxes. This observation clearly supports what we concluded when we used Hyper PI benchmark previously.

4.2 I/O Bound Processes Evaluation

We conducted mainly two tests in order to evaluate the performance of the sandboxes applications on the I/O-bound processes. Both of these tests are conducted by using the PassMark performance test benchmark.

4.2.1 Memory Test

Memory testing includes many tests such as memory-read cached, memory–read un-cached and memory-write. All of these tests are used to calculate Memory-Mark. The results of this test are shown in Figure 4. It shows the difference between running the Passmark benchmark without and with sandboxing applications in terms of the Memory-Mark composite average.

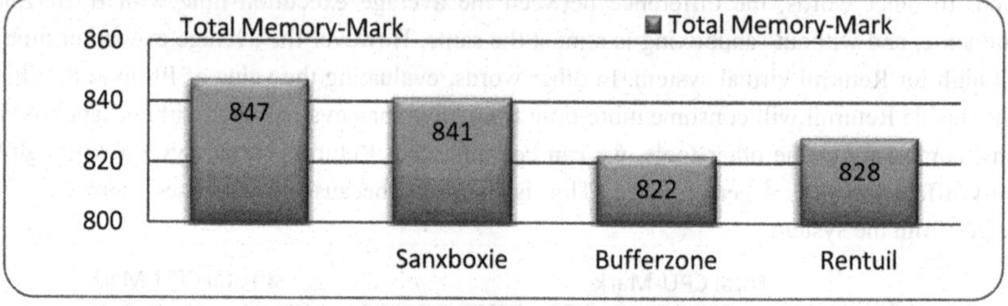

Fig. 4: Total Memory Mark

We can see the difference between them is not significant. They are closer to each other. Therefore, we can say that the sandboxing applications do not affect much on the operation of the RAM. It is important to clarify that sandboxing checks the incoming system calls to the OS that are needed to be satisfied by the OS. However, this process (checking processes in order to apply security policy) does not affect on the performance of overall system, especially the processes that involve the RAM operations. This might be because we have four fast speed processors and the high access speed of the RAM. Therefore this might reflect that, checking process does not have any effect on the overall performance of the system.

4.2.2 Disk Test

Disk test includes Disk-Sequential read, Disk-Sequential write and Disk-Random Seek read/write. All of them are used to calculate the total Disk-Mark. Figure 5 shows the results of the Disk test without and with sandboxing. We can say that the performance difference with Sandboxie, BufferZone sandboxes and without the sandboxing is small. However, Returnil Virtual System is significantly affecting the disk performance of the system. Moreover, out of Figure 5, we can clearly find that the Disk-Sequential read has the major role in the difference between Returnil virtual system and the others. We can read sequentially from disk with a rate of 120.8 Mbytes/sec without Sandboxie. It is 119.4 Mbytes/sec and 119.8 Mbytes/sec for Sandboxie and BufferZone application, respectively. However, we can read sequentially from disk with a rate 77.6 Mbytes/sec with Returnil virtual system. Our interpretation for this result is that Returnil uses reading from a virtual space rather than from the normal space. We know that reading from virtual space on the hard disk can be slower than the reading from normal locations on the hard disk. Also, we may point out that hard disk space chosen by Returnil virtual system is fragmented. Thus, if we read using Returnil from disk, it will read the files as fragments, fragment by fragment, where each fragment is checked according to the security policy which causes the delay for the other parameters.

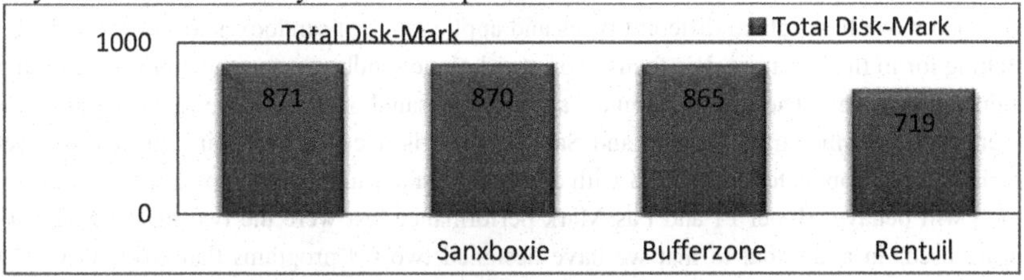

Fig. 5: Total Disk Mark

4.3 CPU&IO-Bound Processes Evaluation under the effect of Attacking Codes

In this section, we implemented two attacking codes in C# programming language. The first one was a code for making the CPU to get busy by making it going on an infinite looping and do useless work, i.e. just doing dummy arithmetic operations. In general, if a real virus attacked the system while we are running one of the sandboxing tools, the effect of this virus will be limited, and as soon as we quit from the sandboxing tool, all corrupted sections or effect of the virus will be vanished, this is the idea of using the sandboxing tools from the beginning. As shown in Figure 6, Returnil virtual system has the best performance in preventing the attacks on CPU. In order to evaluate the tools under the effect of having a virus attacking the OS by infecting the disk, we implemented the second code for infecting the disk and making it to keep asking the CPU to perform disk operation attacks. This was implemented by making CPU going on an infinite looping for writing nothing and does useless work, i.e. just doing dummy writing operations to a file on a disk. In this test, we focused on disk tests include Disk-Sequential read, Disk-Sequential write and Disk- Random seek read/write. All of them are used to calculate the total Disk-Mark. The results point out that the Bufferzone has the best performance in detecting such attacking operation and harming operation as shown in Figure7.

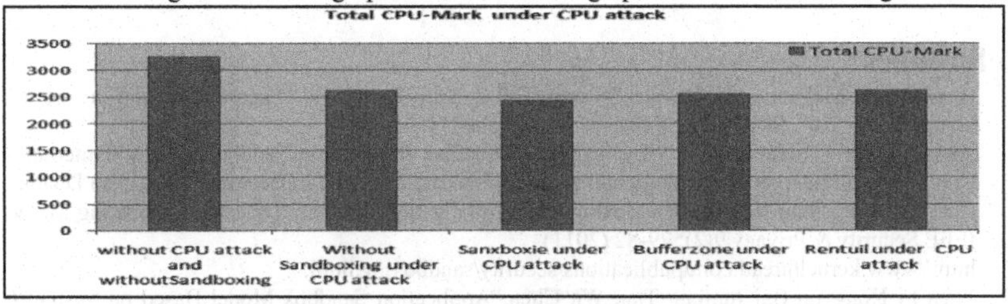

Fig. 6: Total CPU Mark under CPU attack

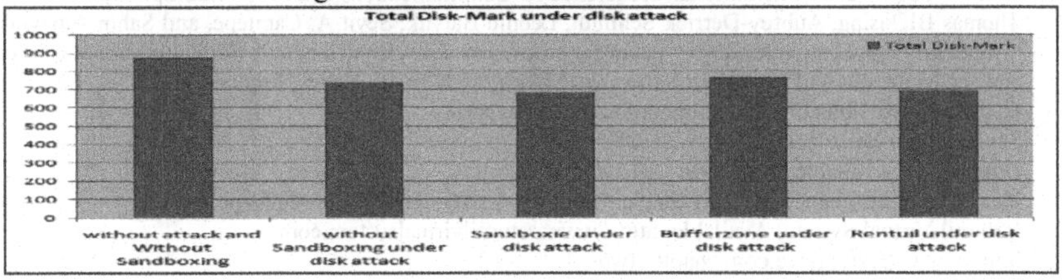

Fig. 7: Total Disk Mark under disk attack

5 Conclusion and Future Work

In this paper, we highlighted different types and applications of sandboxes that we found while searching for in the literature. We focused on stand-alone sandboxing applications which run on Windows platform. The three popular application sandboxes that we have chosen are, BufferZone, Returnil virtual system and Sandboxie. It is necessary to point out that different experiments are conducted on the OS with and without the sandboxing applications to see how the OS will behave. Hyper PI and PassMark performance test were the two main benchmarks that are used. In additional to that we have designed two C# programs that work as virus to attack the OS and keep it busy by doing useless arithmetic and disk operations. In the experiments, where we did not run virus program, the CPU bound processes are not affected by running sandboxing except in the case of Returnil virtual system as it consumes more CPU resources. In addition, Returnil virtual system is affecting the disk performance of the system while Sandboxie and BufferZone are not. Also, checking processes for applying security policy in these three sandboxing applications does not affect on the performance of the overall system, especially the processes that involve the RAM operations. We evaluate each one of the three sandboxing under the effect of having a virus attacking the OS. The experimental results for this part prove that even you are adding extra overhead cost which affects the speed and performance but you will be secured and in the safe environment from the malicious software program. Our future goal is to use more performance metrics to validate its ability to support secure and safe environment.

Acknowledgment

We would like to acknowledge the support from both King Fahd University of Petroleum & Minerals and King Abdulaziz City for Science and Technology (KACST) through project No.10-INF1381-04 for supporting this research work. Thanks extended to the anonyms reviewers whose valuable comments enhance the paper presentation.

References

1. Hrushikesha Mohanty, M Venkata Swamy, Thilak, Srini Ramaswa, "Secured Networking by Sandboxing Linux 2.6", SMC 2009, Texas, USA, October 11-14.
2. Ameiri, Faisal; Salah, Khaled, "Evaluation of Popular Application Sandboxing", 6th International Conference on Internet Technology and Secured Transactions, 11-14 December 2011, Abu Dhabi.
3. Chris Greamo, Anup Ghosh: Sandboxing and Virtualization: Modern Tools for Combating Malware. IEEE Security & Privacy 9(2): 79-82 (2011).
4. http://www.kernelthread.com/publications/security/sandboxing.html,
5. Zhen Li Hongyun Cai Junfeng Tian Wu Chen,"Application Sandbox Model Based on System Call Context", IEEE 2010 International Conference on Communications and Mobile Computing.
6. Thomas Bl□asing, Aubrey-Derrick Schmidt, Leonid Batyuk, Seyit A. Camtepe, and Sahin Albayrak, "An android application sandbox system for suspicious software detection", 5th International Conference on Malicious and Unwanted Software (MALWARE'2010), Nancy, France.
7. Hyper PI, Available: http://virgilioborges.com.br/hyperpi
8. Sandboxie, Available: http://www.sandboxie.com
9. BufferZone, Available: http://www.trustware.com/
10. Pass Mark Performance Test, Available: http://www.passmark.com/products/pt.htm
11. Returnil Virtual System, Available: http://www.returnilvirtualsystem.com.
12. http://www.securingjava.com/chapter-two/

Fault Attacks by using Voltage and Temperature Variations: An Investigation and Analysis of Experimental Environment

Young Sil Lee[1], Non Thiranant[1], HyeongRag Kim[2], JungBok Jo[3], HoonJae Lee[3]

[1]Department of Ubiquitous IT, Graduate School of Dongseo University, Sasang-Gu, Busan 617-716, Korea
[2]Department of IT & Electronics, Pohang University
[3]Div. of Information and Communication Engineering, Dongseo Univeristy
{youngsil.lee0113, thiranant.non}@gmail.com
hrkim@pohan.ac.kr, {jobok, hjlee}@dongseo.ac.kr

Abstract. Physical attacks are a powerful tools to exploit implemented weaknesses of embedded devices even if that using robust cryptography algorithms. Various physical attack techniques have been researched, both to make practical several theoretical or physical fault/error models proposed in open literature and to outline new kinds of vulnerabilities. In this paper, we investigated and summarized the classification of physical attacks; especially we focused on fault attack by using voltage and temperature variations which is one of the types of glitch attacks in non-invasive attacks. Also we investigated and compared several case of experimental environment for fault attacks by using temperature and voltage variations.

Keywords: Physical attack; Glitch attack; Voltage attack; Temperature attack; Side-Channel Attack;

1 Introduction

Traditionally the term 'physical security' has been used to describe of material assets from fire, water damage, theft, or similar perils. However, ongoing concerns in computer security have caused physical security to take on a new meaning – Technologies used to safeguard information against physical attacks [1]. The goal of all these attacks is to reveal secret keys of cryptographic devices, secure microcontrollers, and smartcards.

Physical attacks are also several ways of categorizing in [2-3]: the first criterion is whether an attack is passive or active and second is an invasive, semi-invasive and non-invasive attack.

- Invasive attacks: is the strongest type of attack that can be mounted on a cryptographic device. In such an attack, there are essentially no limits to what is done with the cryptographic device in order to reveal its secret key. Typi-

K.J. Kim (ed.), *Information Science and Applications*,
Lecture Notes in Electrical Engineering 339, DOI 10.1007/978-3-662-46578-3_57

cally starts with the depackaging of the device. Subsequently, different components of the device are accessed directly using a probing station.

- Semi-invasive attacks: the cryptographic device is also depackaged. However, in contrast to invasive attacks, no direct electrical contact to a chip surface is made – the passivation layer stays intact. The goal of passive semi-invasive attacks is typically to read out the content of memory cells without using or probing the normal read-out circuits and the goal of active semi-invasive attacks is to induce faults in the device.
- Non-invasive attacks: the cryptographic device is essentially attacked as it is, i.e. only directly accessible interfaces are exploited. The device is not permanently altered and therefor no evidence of an attack is left behind. Most of attacks can be conducted with relatively inexpensive equipment, and hence, these attacks pose a serious practical threat to the security of cryptographic devices.

Below the Fig.1 presents the classification of physical attack and kind of known attacks. From among these, Toothpick attack is special case which needs two different techniques between invasive and semi-invasive attack. Reverse Engineering also require the technique that include the three regions.

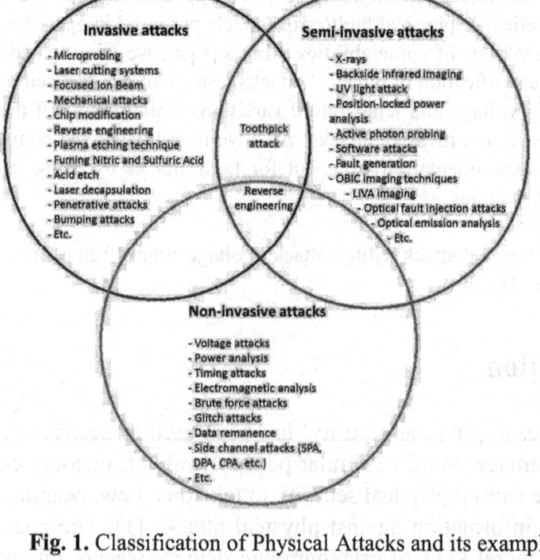

Fig. 1. Classification of Physical Attacks and its examples

2 Definitions of fault attack by using voltage and temperature variations

2.1 Fault Attack by using High/Low Voltage Variations

Secure devices, such as smartcards and cryptographic devices, have a built-in voltage regulator. Its main function is converting the external power source level, defined by the standard, into a power source level the internal technology operates at. A side effect of having a voltage regulator is that it can filter some of the noise present at its input (V_{cc}) and still provide a clean and steady power source at its output (V_{dd}) [4]. A non-invasive attack technique designed to disturb the computation of a cryptographic primitive by feeding voltage glitches on the supply voltage, in order to leak secret information inferring it from the erroneous results. By spiking CMOS RAM with short duration, high-voltage pulses, it may be possible to imprint the contents in a manner similar to radiation imprinting [1]. In addition, by changing V_{cc} to abnormally high or low value, erratic behavior may be included in many circuits. The erratic behavior may include the processor misinterpreting instructions, erase or overwrite circuitry failing, or memory retaining data when not desired.

2.2 Fault Attack by using High/Low Temperature Variations

An attack actively tampers the environmental temperature of a device, either by cooling or heating. It is possible to freeze the data and recover the content from the memory, or heat to leak memory errors after hours of extensive heating. In the 1980s, it was realized that low temperatures can increase the data retention time of SRAM too many seconds or even minutes. With the devices available at that time, it was found that increased data retention started about – 20℃ and increased as temperature fell further. Some devices are therefore designed with temperature sensors; any drop below – 20℃ is treated as a tempering event and results in immediate memory zeroisation [5].

3 An Investigation and Analysis of Experiment Environment

In this chapter an experiment environment analysis of fault attack is discussed, including the comparison table and explanation of each parameter. We mainly focus on fault attack by using high/low voltage variation (we called 'Voltage Attack') and high/low temperature variation (we called 'Temperature Attack') on various embedded devices. Recent case studies are explained with the general implementation approach. However, there is not an ideal method that is able to break all types of embedded devices, because each one is different from others. There are several factors that contribute to success in breaking embedded devices such as temperature, voltage supply, electric current, duration, time and cycle, etc.

3.1 High/Low Voltage Attacks

The parameters necessary in high/low voltage attacks include: target chip/ algorithm, modulus size, voltage supply, duration, time/ cycle. These parameters are needed because each embedded device comprises of different components, thus voltage and time consumption varied. The table below shows target chip/ algorithm and parameters of high/ low voltage attacks. The description of each parameter is explained below:

Table 1. Parameters high/low voltage attacks

Parameters	Description
Target chip/algorithm	Chip or algorithm selected as the aim of the attack
Key size	Size or length of the key measured in bits, used in cryptographic algorithms
Voltage	Electric potential difference between two points, measured in volts
Duration	Time consumption of the experiment
Time/Cycle	The number of times or cycles of the operations performed

From below the table 2, different experiments are performed according to the type of target embedded devices and algorithms. The environment to perform the experiment for each case has almost no difference from others. The target chips are MC68HC05B6 microcontroller, PIC16F84 microcontroller, FPGAs, EEPROMs, CMOS digital circuits. The main target algorithms are AES and RSA of different key size. Most of the researches make use of low voltage to induce faults from microcontrollers. The duration varies depending upon the target, the minimum time taken from the table above is 10ns and the maximum time is 100 hours. However, the time unit is typically measured in nanosecond, because there are only few cases that are time-consuming up to minutes or hours. Low voltage attack is usually performed with the normal supply voltage and gradually decreasing, on the other hand, high voltage attack starts with the normal supply voltage and gradually increasing until faults occur.

Table 2. An experimental environment of high/low voltage attacks

Case	Target chip/ algorithm	Key size	Voltage	Duration	Time/ Cycle
S.P. Skorobogatov [6]	MC68HC05B6 microcontroller	-	Low voltage, reduced to 50 -70%	period of time that the "AND $0100" instruction is executed	-
	PIC16F84 micro-controller	-	10V	-	-
A. Barenghi et al. [7]	CRT based version of RSA	512, 1024, 2048 bits	905.5 - 912.0 mV 1.5mV (1single fault)	-	10000 runs
A. Pellegrini et al.[8]	RSA	1024-bit	1.23 - 1.30V	100 hours	-

G. Canivet et al.[9]	FPGA	-	between 5 and 80 volts	ranging from 10ns to 100ns by steps of 10 ns	49 to 60 cycles
G.Y. Asier et al.[10]	EEPROMs	-	2V - 17V	10ns - 500ns	-
G. Kamil et al.[11]	CMOS digital circuits	-	1.1 - 1.22V	-	-
A. Barenghi et al.[12]	AES	128, 192, 256-bit key	-	execution time 1, 2, 2.21 (128, 192, 256-bit key)	10, 12 and 14 rounds
H. Bar-El et al.[13]	EEPROM	RSA, DES	--	-	-
K.M. Zick et al.[14]	FPGA-based system	-	> 688mV	10ns	-
A.G. Yanci et al.[15]	Glitch detectors proposed	All secure devices	-	detection time 611 - 24755ns	-

3.2 High/Low Temperature Attacks

The parameters necessary in high/low temperature attacks include: target chip/ algorithm, source of heat, temperature/electric current, and duration. The parameters are similar to voltage attack, because they have the same objective which is to induce faults. Temperature and time taken are different for each case depending on the target chips. The table below shows target chip/ algorithm and parameters of high/ low temperature attacks. The description of each parameter is explained below:

Table 3. Parameters for temperature attacks

Parameters	Description
Target chip/algorithm	Chip or algorithm selected as the aim of the attack
Source of heat	The heat source, usually from the supply electric current and heater
Temperature/Current	Temperature or electric current needed in the experiment to induce faults
Duration	Time consumption of the experiment

From the table 4, the parameters are similar to voltage attacks and experiments are performed differently according to the type of target devices and algorithms. The environment is slightly different from one another, depending on the source of heat. Most of the temperature attack experiments use electrical current to increase the heat in FPGAs. The target chips are ATmega162, FPGAs, and SRAM chips. Temperature attack mainly targets hardware devices, unlike voltage attack where algorithms are also targeted. The target devices are usually de-capsulated first before performing temperature attack. The table above shows different cases which need different parameter values. For low temperature attack, the minimum is -55C, the experiment usually starts with the room temperature and gradually decreasing. The maximum is 300C for high temperature attack, the process is similar to low temperature attack, but gradually increasing from the room temperature. Duration of each experiment is different, normally less than 1 second but it can take up to 18 hours in case 3. In case 2, SRAM chips data retention needs both low and high temperature between -55~150℃ with duration 12~13100ms and the source of heat is from LH135H temperature sensor. In case 3, a program is designed for purposely provoking errors, used to adjust CPU speed, and heat CPU for attacking FPGA. The experiment takes up to 18 hours and temperature varies according to the fan speed. Most of the cases typically target FPGAs, but the temperature and source of heat, and duration are different from one another. Both low and high temperatures are used for the experiment to induce faults. In case 8, power supplied to the heater is used with the temperature of 20~300℃ on FPGA for thermos luminescence measurement. Case 10 uses supply voltage for low temperature attack purpose with 3~4℃.

Table 4. An experimental environment of high/low temperature attacks

Case	Target chip/ algorithm	Source of heat	Temperature/ Current	Duration
M. Hutter et al.[16]	ATmega162	Electrical current	152-158℃	70 minutes
S.P. Skorobogatov[6]	SRAM chips data retention	LM135H temperature sensor	-55 - 150℃	12 - 13100ms
J. Brouchier et al.[17]	FPGA	Programs for purposely provoking errors e.g., CPUBurn-in (used to adjust CPU speed, heat CPU)	varies according to fan speed	up to 18 hours
W. Zhao et al.[18]	FPGA logic circuit	electrical current	-	20-30ns
X. Xhang et al.[19]	FPGA-based system	electrical current	15 - 85℃	5 - 75 seconds
S. Mondal et al.[20]	FPGAs	electrical current	17 - 46℃	-

P.H. Jones et al.[21]	FPGAs	electrical current	24.5 - 35℃	up to 1000s
N. Bharathi et al.[22]	FPGA for thermo luminescence measurement	power supplied to the heater	20 - 300℃	-
P.H. Jones et al.[23]	FPGA	electrical current	10 - 90℃	8100 - 8700ns
S. Velusamy et al.[24]	FPGA	supply voltages	3 - 4℃	-

4 Conclusions

The objective of this paper is to explain the accuracy analysis of physical attacks on cryptographic devices, especially microcontrollers. The general introduction of physical attacks and the categories are given. The physical attacks and techniques are classified into 3 main categories, which are non-invasive, semi-invasive, and invasive attacks. They are classified according to the way of approaching, such as dealing with temperature, voltage, electromagnetic emission, etc. The comparison is made between each setup experiment and approach to attack the target chip.

The paper highlights the 2 main physical attacks, including the parameters necessary to conduct the experiment. The comparison table shows that each target needs to undergo different operations and approaches to let faults occur. In addition, there is no absolute ideal method that is able to break all types cryptographic device. Tools, resources, and time consumption differ according to the target and expertise that perform the experiment. With the rapid development of the technology, manufacturers literally need to come up with better protection of hardware devices, and appropriate hardware security evaluation methods are necessary.

Acknowledgements. This work was supported by Basic Science Research Program through the National Research Foundation of Korea (NRF) funded by the Ministry of Education, Science and Technology (grant number: NRF-2014). And it also supported by BB21 project of Busan Metropolitan City.

References

1. Steve H. Weingart, "Physical Security Devices for Computer Subsystems: A Survey of Attacks and Defenses 2008 (updated from the CHES 2000 version)," ©atsec information security corporation, March 2008.
2. Sergei P. Skorobogatov, "Semi-Invasive Attacks (Definition)," http://www.cl.cam.ac.uk/~sps32/semi-inv_def.html
3. Stefan Mangard, Elisabeth Oswald and Thomas Popp, "Power Analysis Attacks: Revealing the Secrets of Smart Cards," Book, Springer-Verlag new York, Inc. Secaucus, NJ, USA, 2007.
4. Asier GOIKOETXEA YANCI, Stephen PICKLES and Tughrul ARSLAN, "Characterizaton of Voltage Glitch Attack Detector for Secure Devices," Symposium on Bio-inspried Learning and Intelligent Systems for Security, pp.91-96, Aug. 2009.

5. Sergei Skorobogatov, "Low temperature data remanence in static RAM," Technical Report, University of Cambridge, Computer Laboratory, June 2002.
6. S.P. Skorobogatov, "Semi-invasive attacks – A new approach to hardware security analysis," Technical report, University of Cambridge, Computer Laboratory, 2005.
7. A. Barenghi, G. Bertoni, E. Parrinello, G. Pelosi, "Low voltage fault attacks on the RSA cryptosystem," 6th International Workshop on Fault Diagnosis and Tolerance in Cryptography (FDTC), pp.21-31, Sep. 2009.
8. A. Pellegrini, V. Bertacco, T. Austin, "Fault-based attack of RSA authentication," Design, Automation & Test in Europe Conference & Exhibition (DATE), pp.855-860, Mar. 2010.
9. G. Canivet, P. Maistri, R. Leveugle, J. Clédiére, F. Valette, M. Renaudin, "Glitch and laser fault attacks onto a secure AES implementation on a SRAM-based FPGA," Journal of cryptology, vol.24, No.2, pp.247-268, Apr. 2011.
10. Asier GOIKOETXEA YANCI, Stephen PICKLES, Tughrul ARSLAN, "Characterization of a Voltage Glitch Attack Detector for Secure Devices," Symposium on Bio-inspired Learning and Intelligent Systems for Security, pp.91-96, 20-21 Aug. 2009.
11. Kamil Gomina, Philippe Gendrier, Jean-Baptiste Rigaud, Assia Tria, "Detecting positive voltage attacks on CMOS circuits," Proceedings of the First Workshop on Cryptography and Security in Computing Systems(CS2'14), pp.1-6, Jan. 2014.
12. Barenghi, A., Bertoni, G. M., Breveglieri, L., Pellicioli, M., Pelosi, G., "Low voltage fault attacks to AES," In Hardware-Oriented Security and Trust(HOST), pp. 7-12, June, 2010.
13. Bar-El, H., Choukri, H., Naccache, D., Tunstall, M., Whelan, C., "The sorcerer's apprentice guide to fault attacks," Proceedings of the IEEE, Vol.94, Issue.2, pp. 370-382, Feb, 2006.
14. Zick, K. M., Srivastav, M., Zhang, W., French, M., "Sensing nanosecond-scale voltage attacks and natural transients in FPGAs," In Proceedings of the ACM/SIGDA international symposium on Field programmable gate arrays, pp. 101-104, Feb, 2013.
15. Yanci, A. G., Pickles, S., Arslan, T., "Detecting Voltage Glitch Attacks on Secure Devices," In Bio-inspired Learning and Intelligent Systems for Security, pp. 75-80, Aug, 2008.
16. M. Hutter, J. M. Schmidt, "The Temperature side Channel and Heating Fault Attacks," IACR Cryptology ePrint Archive, 2014.
17. Brouchier, J., Kean, T., Marsh, C., Naccache, D., "Temperature Attacks," Security&Privacy(IEEE), Vol.7, Issue.2, pp.79-82, Apr, 2009.
18. Whisheng Zhao, Eric BELHAIRE, Bernard DIENY, Guilaume PRENAT, Claude CHAPPERT, "TAS-MRAM based Non-volatile FPGA logic circuit," International Conference on Field-Programmable Technology, pp.153-160, Dec, 2007.
19. Xun Zhang, Wassim Jouini, Pierre Leray, Jacques Palicot, "Temperature-Power Consumption Relationship and Hot-spot Migration for FPGA-based System", Green Computing and Communications & International Conference on Cyber, Physical and Social Computing, pp.392-397, Dec, 2010.
20. Mondal, S., Mukherjee, R., Memik, S. O., "Fine-Grain Thermal Profiling and Sensor Insertion for FPGAs," International Symposium on Circuits and Systems, pp.4387-4390, May, 2006.
21. Phillip H. Jones, Young H. Cho, John W. Lockwood, "An Adaptive Frequency Control Method using Thermal Feedback for Reconfigurable Hardware Applications," International Conference on Field Programmable Technology, pp.229-236, Dec, 2006.
22. N. Bharathi, P. Neelamegam, "FPGA Based Linear Heating System for Measurement of Thermoluminescence," Measurement Science Review, Vol.11, Issue.6, pp.207-209, Jan, 2011.
23. Jones, P. H., Moscola, J., Cho, Y. H., Lockwood, J. W., "Adaptive Thermoregulation for Applications on Reconfigurable Devices," International Conference on Field Programmable Logic and Applications, pp.246-253, Aug, 2007.
24. Velusamy, S., Wei Huang, Lach, J. Stan, M., "Monitoring temperature in FPGA based SoCs," International Conference on Computer Design: VLSI in Computers and Processors, pp.634-637, Oct, 2005.

Pseudoinverse Matrix over Finite Field and Its Applications*

Dang Hai Van, Nguyen Dinh Thuc

Faculty of Information Technology
University of Science, VNU-HCM, Vietnam
Email:{dhvan,ndthuc}@fit.hcmus.edu.vn

Abstract. With the development of smart devices like smart phones or tablets, etc., the challenging point for these devices is risen in the security problems required limited computational capacity, which attracts researchers in both academia and industrial societies. In this paper, we tackle two interesting security problems: changing a shared key between two users and privacy-preserving auditing for cloud storage. Our solution is based on pseudoinverse matrix, a generalization concept of inverse matrix. In addition, this can be applied for devices with limited computational capacity due to its fast computation.

Keywords: Pseudoinverse, Data Audit, Shared Key

1 Introduction

Apart from developments of modern technology, smart phones and smart devices such as home appliances and wearable devices have affected a lot on users' behaviour. More and more users in over the world often use their smart phones and tablets to access internet, listen to music, exchange email,... Furthermore, they require their personal privacy that is guaranteed and confidential. This leads to requirements of security solutions for devices with limited capacities. There are many approaches which can be applied for such devices. However, we realize that matrices are feasible due to its simple and flexible operations.

In 2013, T.D.Nguyen [6] refers to quasi-inverse monoid and presents its application in cryptography. A monoid $\mathcal{M}$ is regular, or quasi-inverse if for every element a of $\mathcal{M}$ there exists an element b of $\mathcal{M}$ such that $a \cdot b \cdot a = a$ and $b \cdot a \cdot b = b$. An instance of quasi-inverse monoid is pseudoinverse matrix. Generally speaking, pseudoinverse is an extension concept of inverse for non-square or singular matrix.

In this paper, we will propose an approach to construct a pseudoinverse. After that, we will suggest two others applications of pseudoinverse.

The first application is to change a pre-shared key, which is an extension to key exchange protocol. After a period of time since exchanging a key, users may

* This work was supported by The Department of Science and Technology (DOST) in Ho Chi Minh city, Vietnam

K.J. Kim (ed.), *Information Science and Applications*,
Lecture Notes in Electrical Engineering 339, DOI 10.1007/978-3-662-46578-3_58

need to change the shared key. One solution is that the users repeat key exchange process again. However, we propose another simpler way based on pseudoinverse to support changing key. The condition for our proposal is that the pre-shared key is a matrix. Therefore, this solution can be combined with the key exchange protocol in [6] to support both exchanging key and changing shared key.

Next, the second application is to audit data in cloud. In database as a service, data owner sends their data to store in a cloud server. Additionally, the server is often assumed to be semi-trusted. It is believed not to use or reveal users data without permission. However, when data loss or corruption happens, the server may not inform to users to protect its image. As well the server may possibly decide to delete some data sections that have been rarely accessed by users. In order to avoid such risks, data owner can hire a third-party auditor to audit its data without knowing them, meaning privacy-preserving auditing. Previous works of auditing methods by third party auditors consist of private verification [3, 5, 8] and public verification [1, 8–10]

2 Preliminaries

At first, we review the concept of pseudoinverse and its properties without proofs. For the proofs, we refer the reader to [2, 7].

Definition 1. *For every matrix (square or rectangular) A of real or complex elements, there is a unique matrix $A^\dagger$, called pseudoinverse of A, satisfying all of the four equations:*

$$AA^\dagger A = A \tag{1}$$

$$A^\dagger AA^\dagger = A^\dagger \tag{2}$$

$$(AA^\dagger)^T = AA^\dagger \tag{3}$$

$$(A^\dagger A)^T = A^\dagger A \tag{4}$$

where ^{T} is the transpose operation of a matrix.

Proposition 1. *The pseudoinverse of a matrix is unique.*

Proposition 2. *Over the field $\mathbb{R}$ of real numbers,*

1. *If the columns of A are linearly independent, then A^TA is invertible, and*

$$A^\dagger = (A^TA)^{-1}A^T \tag{5}$$

2. *If the rows of A are linearly independent, then AA^T is invertible, and*

$$A^\dagger = A^T(AA^T)^{-1} \tag{6}$$

Next we present the concept of row-self-orthogonal matrix and a way to construct it from J.L.Massey [4].

Definition 2. *A non-square matrix C is row-self-orthogonal if $CC^T = \mathbb{O}$ where $\mathbb{O}$ denotes a zero matrix of appropriate dimension.*

Proposition 3. *1. Given a matrix Q of size $k \times m$ over a field of characteristic 2. The matrix $P = [Q : Q]$ is row-self-orthogonal.*

2. *Given a matrix Q of size $k \times m$ over a field of characteristic p such that there exists α in $GF(p)$ for which $\alpha^2 = -1$. The matrix $P = [Q : \alpha Q]$ is row-self-orthogonal.*
3. *Given a matrix Q of size $k \times m$ over a field of characteristic p. The matrix $P = [\alpha Q : \beta Q : \gamma Q : \delta Q]$ with $\alpha^2 + \beta^2 + \gamma^2 + \delta^2 = 0$ [1] is row-self-orthogonal.*
4. *Given a matrix Q of size $k \times k$, a row-self-orthogonal matrix C of size $k \times m$ and an orthogonal matrix A of size $m \times m$. The matrix $P = QCA$ is row-self-orthogonal.*

3 Pseudoinverse matrix over finite field $\mathbb{Z}_p$

The pseudoinverse presented and discussed in [2, 7] are of real or complex numbers only. Applying real matrices leads to difficulties about accuracy and computation. Consequently, we concern about pseudoinverse over a finite field $\mathbb{Z}_p$ where p is prime. In this section, we will review properties of pseudoinverse over the field $\mathbb{Z}_p$ in [6], then propose an approach to construct pseudoinverse matrix.

Proposition 4. *Given a matrix A in $\mathbb{Z}_p^{n\times m}$, where p is a prime. If a pseudoinverse of A exists, it is unique.*

Proof. Assume that B and C are two pseudoinverses of A, then

$$AB = (AB)^T \text{ (due to Equation 3)} = B^T A^T = B^T (ACA)^T \text{ (due to Equation 1)}$$
$$= B^T(A^T C^T A^T) = ABAC = AC$$
and similarly, $BA = CA$

Thus, we have:

$$B = BAB = BAC = CAC = C.$$

□

Proposition 5. *Over the finite field $\mathbb{Z}_p$, a matrix A has size of $m \times n$*

1. *If $m > n$ **and A^TA is non-singular** then*

$$A^\dagger = (A^TA)^{-1}A^T \tag{7}$$

2. *If $m < n$ **and AA^T is non-singular** then*

$$A^\dagger = A^T(AA^T)^{-1} \tag{8}$$

[1] $\alpha, \beta, \gamma, \delta$ always exist due to the Lagrange's Four Square Theorem

Compared to the proposition 2, in the proposition 5, we need the condition $A^T A$ is non-singular (or AA^T is non-singular) directly. The reason is that in the field $\mathbb{Z}_p$, a matrix A with linearly independent columns (or linearly independent rows) does not ensure that $A^T A$ is non-singular (or AA^T is non-singular).

Proposition 6. *Given a matrix $A \in \mathbb{Z}_p^{m\times n} (m < n)$ such that AA^T is non-singular. We have:*

(i) $A^\dagger = A^T \left(AA^T\right)^{-1}$
(ii) $AA^\dagger = \mathbb{I}$ *where* $\mathbb{I}$ *is the identity matrix.*
(iii) $A^\dagger A \in \mathbb{Z}_{n\times n}$ *is a projection matrix.*

Proof. (i) Because AA^T is non-singular, according to the proposition 5, $A^\dagger = A^T \left(AA^T\right)^{-1}$.

(ii) $A^\dagger = A^T \left(AA^T\right)^{-1} \Rightarrow AA^\dagger = AA^T \left(AA^T\right)^{-1} = \mathbb{I}$

(iii) $\left(A^\dagger A\right)^2 = \left(A^T \left(AA^T\right)^{-1} A\right) \left(A^T \left(AA^T\right)^{-1}\right) = A^T \left(AA^T\right)^{-1} = A^\dagger A \Rightarrow$ $A^\dagger A$ is a projection matrix.

□

According to proposition 5, if we can construct a matrix A so that AA^T is non-singular, we can compute pseudoinverse of A using Equation 8. Consequently, we will suggest a way to construct such a matrix A. Our proposal is based on non-singular matrix and row-self-orthogonal matrix.

Proposition 7. *Given a non-singular matrix $Z \in \mathbb{Z}_p^{m\times m}$ and a row-self-orthogonal matrix $V \in \mathbb{Z}_p^{m\times(n-m)} (m < n)$. The matrix A created by combining by column the two matrices Z and V, $A = \left[Z\ V\right]$, then the product of the matrix A and its transpose, AA^T, is non-singular.*

Proof.

$$A = \left[Z\ V\right] \Leftrightarrow A^T = \begin{bmatrix} Z^T \\ V^T \end{bmatrix}$$

$$\Rightarrow AA^T = ZZ^T + VV^T = ZZ^T + \mathbb{O} \textit{ (because } V \textit{ is row-self-orthogonal)} = ZZ^T$$

$$\Rightarrow det(AA^T) = det(ZZ^T) = det(Z)det(Z^T) \neq 0 \textit{ (because } Z \textit{ is non-singular)}$$

It deduces AA^T is non-singular.

4 Application of pseudoinverse matrix

Rely on the ability to construct a matrix and its pseudoinverse over a finite field $\mathbb{Z}_p$ (p is prime), we can apply pseudoinverse in real applications. In this section, we present two applications of pseudoinverse matrix: changing a shared key and privacy-preserving auditing data in cloud storage.

4.1 Application to change the shared key between two users

Assuming that two users shared a key already and the key is a matrix K. After a period of time, they want to change the shared key. We propose a method to change the shared key using pseudoinverse matrix.

Table 1. Steps to change the shared key between two users

	Alice(A)	Bob(B)
	Shared key K	Shared key K
(1)	Random a matrix K_1 and find K_2 such that $K_1 \cdot K_2 = K$	
(2)	Generate a non-singular matrix Q, compute $PK = K \cdot K^{\dagger} \cdot K_1 \cdot Q$, $SK = Q^{-1} \cdot K_2$, and send PK	
(3)		Receive PK. Generate a matrix M, compute $C = M \cdot PK$ and send C
(4)	Receive C and compute $K' = C \cdot SK$	Compute $K' = M \cdot K$

Correctness The protocol is correct if the two keys K' of Alice and Bob are the same. Indeed,

$$\begin{aligned} C \cdot SK &= M \cdot PK \cdot SK = M \cdot (K \cdot K^{\dagger} \cdot K_1 \cdot Q) \cdot Q^{-1} \cdot K_2 \\ &= M \cdot K \cdot K^{\dagger} \cdot K_1 \cdot K_2 = M \cdot K \cdot K^{\dagger} \cdot K \\ &= M \cdot K \end{aligned}$$

Security We consider two following attack scenarios.

Scenario 1: Given $PK = K \cdot K^{\dagger} \cdot K_1 \cdot Q$, the attacker finds $SK = Q^{-1} \cdot K_2$
Scenario 2: Given $C = M \cdot PK$ and PK, the attacker finds M.

In the scenario 1, the attacker needs to factorize $PK = K \cdot K^{\dagger} \cdot K_1 \cdot Q$ to find Q, K and K_1.

In the scenario 2, the attacker can solve the system of linear equations $C = M \cdot PK$ given PK to find M (Notice that even if the attacker can find M, he cannot deduce the value of the new shared key $K' = M \cdot K$ without knowing K). Assuming that size of K is $m \times n$ ($n < m$), size of K_1 and K_2 are $m \times k$ and $k \times n$ ($k < n < m$), size of M is $h \times m$. It deduces size of PK is $m \times k$. All matrices are over the field $\mathbb{Z}_p$. Hence in order to prevent Brute Force attack to find M, we need to ensure $p^{(h \times m - h \times k)} > 2^{80}$ ($h \times m$ is size of M, $h \times k$ is the number of linear equations). It is easy to choose non-large values p, h, m, k that satisfy the condition.

4.2 Application for privacy-preserving auditing data in cloud

Finally, we present a protocol for a third-party auditor (TPA) to audit data in a cloud server (CS) without knowing the content of data.

In this application, the model consists of three factors:

- Cloud server (CS): server that provides cloud database-as-a-service (DaS).
- Data owner (DO): user who use DaS to keep their data. User wants to detect if any data has been lost or deleted without its knowing, then it hires a third-party to audit data. However, the user does not want its data to be revealed to the third-party.
- Third-party auditor (TPA): a third-party is hired by data owner to audit its data without knowing them. When TPA wants to audit data in cloud server, it sends a query to cloud server. Cloud server then returns corresponding metadata back so that TPA can rely on to audit if data has been kept unchanged or not.

In our proposal, we assume that

- A function f to generate a non-singular matrix is given. Its inputs consist of two values as seed.
- A key is pre-shared between the data owner and the auditor, or it is shared every time before the auditor audits data. The pre-shared key is a matrix X over finite field $\mathbb{Z}_p$ (p is prime) such that there exists its pseudoinverse.
- Data consists of multiple data records. Each record is a matrix.
- Every time the auditor audits at least 2 data records.

Table 2. Steps to privacy-preserving audit data

	Data owner (DO)	Cloud server (CS)	Auditor (TPA)
	Shared key X		Shared key X
The data owner sends data to store in the cloud			
(1)	For every data record M_i with index i, generate a non-singular matrix $Q_i = f(X, i)$ using the generator f where the shared key and the index are inputs. Compute $T_i = Q_i \cdot M_i \cdot X \cdot X^{\dagger}$. Send (i, M_i, T_i) to the server.		
(2)		Store data $\{(i, M_i, T_i)\}$	
The auditor audits data in the cloud			
(1)		Store data $\{(i, M_i, T_i)\}$	Random a list of indexes $\{i_j\}, j = 1 \dots \ell, \ell \geq 2$. Generate a non-singular matrix R and $Q_{i_j} = f(X, i_j), j = 1 \dots \ell$. Send $\{(i_j, R \cdot Q_{i_j})\}$ to the server.
(2)		Compute $A = \Sigma_{j=1}^{\ell} \left(R \cdot Q_{i_j} \cdot M_{i_j}\right)$ and $B = \Sigma_{j=1}^{\ell} T_{i_j}$, and send to the auditor.	
(3)			Return 1 if $A \cdot X = R \cdot B \cdot X$, return 0 otherwise (1 means that data is intact and 0 means that data has been modified).

Correctness The protocol is correct if $A \cdot X = R \cdot B \cdot X$ in case the server sends intact data and $A \cdot X \neq R \cdot B \cdot X$ in case the server sends non-intact data.

- In case the server sends intact data:
  $A \cdot X = \Sigma_{j=1}^{\ell} \left(R \cdot Q_{i_j} \cdot M_{i_j}\right) \cdot X = R \cdot \Sigma_{j=1}^{\ell} \left(Q_{i_j} \cdot M_{i_j}\right) \cdot X \cdot X^{\dagger} \cdot X = R \cdot \Sigma_{j=1}^{\ell} \left(Q_{i_j} \cdot M_{i_j} \cdot X \cdot X^{\dagger}\right) \cdot X = R \cdot \left(\Sigma_{j=1}^{\ell} T_{i_j}\right) \cdot X = R \cdot B \cdot X$
- In case the server modified data. Without loss of generality, assuming that the server modified M_{i_1} into M'_{i_1}.
  $A \cdot X = R \cdot Q_{i_1} \cdot M'_{i_1} \cdot X + \Sigma_{j=2}^{\ell} \left(R \cdot Q_{i_j} \cdot M_{i_j}\right) \cdot X \neq \Sigma_{j=1}^{\ell} \left(R \cdot Q_{i_j} \cdot M_{i_j}\right) \cdot X = R \cdot \Sigma_{j=1}^{\ell} \left(Q_{i_j} \cdot M_{i_j}\right) \cdot X \cdot X^{\dagger} \cdot X = R \cdot \Sigma_{j=1}^{\ell} \left(Q_{i_j} \cdot M_{i_j} \cdot X \cdot X^{\dagger}\right) \cdot X = R \cdot \left(\Sigma_{j=1}^{\ell} T_{i_j}\right) \cdot X = R \cdot B \cdot X \Rightarrow A \cdot X \neq R \cdot B \cdot X$

Security We consider two following attack scenarios.

Scenario 1: Given $T_i = Q_i \cdot M_i \cdot X \cdot X^{\dagger}$ and M_i, the cloud server finds Q_i and $X \cdot X^{\dagger}$.

Scenario 2: Given $A = \Sigma_{j=1}^{\ell} \left(R \cdot Q_{i_j} \cdot M_{i_j}\right)$ and $B = \Sigma_{j=1}^{\ell} T_{i_j} = \Sigma_{j=1}^{\ell} \left(Q_{i_j} \cdot M_{i_j} \cdot X \cdot X^{\dagger}\right)$ and $R, \{Q_{i_j}\}, X$, the auditor finds $\{M_{i_j}\}$.

In the scenario 1, the attacker needs to factorize $T_i = Q_i \cdot M_i \cdot X \cdot X^{\dagger}$ given M_i to find Q_i and $X \cdot X^{\dagger}$.

In the scenario 2, in order to prevent the auditor to solve the systems of linear equations, we can modify the protocol a little as below

$A = \Sigma_{j=1}^{\ell} \left(c_{i_j} \times R \cdot Q_{i_j} \cdot M_{i_j}\right)$ and $B = \Sigma_{j=1}^{\ell} \left(c_{i_j} \times T_{i_j}\right)$

where $\{c_{i_j}\}$ are random numbers generated by the cloud server.

Without knowing $\{c_{i_j}\}$, the auditor cannot solve $\{M_{i_j}\}$ from A, B.

5 Conclusion

In this paper, we propose an approach to generate a matrix and its pseudoinverse over finite field $\mathbb{Z}_p$ (p is prime) based on row-self-orthogonal matrix and non-singular matrix. After that, we describe two applications of pseudoinverse. The first application supports to change the shared key between two users. The second application supports a third-party auditor to audit data stored in a cloud server such that the auditor learns nothing about data content. This is an indication of potential applications of pseudoinverse over finite field in cryptography and security.

References

1. G. Ateniese, R. Burns, R. Curtmola, J. Herring, O. Khan, L. Kissner, Z. Peterson, and D. Song. Remote data checking using provable data possession. *ACM Transactions on Information and System Security (TISSEC)*, 14(1):12, 2011.
2. A. Ben-Israel and T. N. Greville. *Generalized inverses.* Springer, 2003.
3. A. Juels and B. S. Kaliski Jr. Pors: Proofs of retrievability for large files. In *Proceedings of the 14th ACM conference on Computer and communications security*, pages 584–597. ACM, 2007.
4. J. L. Massey. Orthogonal, antiorthogonal and self-orthogonal matrices and their codes. *Communications and coding*, 2:3, 1998.
5. M. Naor and G. N. Rothblum. The complexity of online memory checking. In *Foundations of Computer Science, 2005. FOCS 2005. 46th Annual IEEE Symposium on*, pages 573–582. IEEE, 2005.
6. T. D. Nguyen and V. H. Dang. Quasi-inverse based cryptography. In *Computational Science and Its Applications–ICCSA 2013*, pages 629–642. Springer, 2013.
7. R. Penrose. A generalized inverse for matrices. In *Proc. Cambridge Philos. Soc*, volume 51, pages 406–413. Cambridge Univ Press, 1955.
8. H. Shacham and B. Waters. Compact proofs of retrievability. In *Advances in Cryptology-ASIACRYPT 2008*, pages 90–107. Springer, 2008.
9. C. Wang, S. S. Chow, Q. Wang, K. Ren, and W. Lou. Privacy-preserving public auditing for secure cloud storage. *Computers, IEEE Transactions on*, 62(2):362–375, 2013.
10. C. Wang, Q. Wang, K. Ren, and W. Lou. Privacy-preserving public auditing for data storage security in cloud computing. In *INFOCOM, 2010 Proceedings IEEE*, pages 1–9. Ieee, 2010.

Multi-Agent based Framework for Time-correlated Alert Detection of Volume Attacks

Abimbola Olabelurin[1], Georgios Kallos[2], Suresh Veluru[3], Muttukrishnan Rajarajan[1]

[1]School of Mathematics, Computer Science and Engineering, City University London, UK

[2]Security Futures Practice, Research & Innovation, British Telecom, Adastral Park, Ipswich, UK

[3]United Technologies Research Centre Ireland, Ltd, Ireland *

Abstract

Recent and emerging cyber-threats have justified the need to keep improving the network security technologies such as Intrusion Detection Systems (IDSs) to keep it abreast with the rapidly evolving technologies subsequently creating diverse security challenges. A post-processing filter is required to reduce false positives and large number of alerts generated by network-based IDSs for the timely detection of intrusions. This paper investigates statistical-based detection approach for volume anomaly such as Distributed Denial-of-Service (DDoS) attacks, through the use of multi-agent framework that hunt for time-correlated abnormalities in different behaviours of network event. Employing statistical process-behaviour charts of Exponentially Weighted Moving Average (EWMA) one-step-ahead forecasting technique, the framework correlates undesirable deviations in order to identify abnormal patterns and raise alarm. This paper provides the architecture and mathematical foundation of the proposed framework prototype, describing the specific implementation and testing of the approach based on a network log generated from a 2012 cyber range simulation experiment as well as the DARPA 2000 datasets. Its effectiveness in detecting time-correlated anomaly alerts, reducing the number of alerts and false positive alarms from the IDS output is evaluated in this paper.

Keywords: DDoS detection, EWMA, Intrusion Detection Systems (IDS), time-series analysis, volume attacks.

1. Introduction

Intrusion Detection Systems (IDSs) are one of the key components for securing computer infrastructures, using several techniques to detect security violations, securing them against being a source of network attacks as well as being a victim of attacks. However, several limitations still exist [11, 14] as modern cyber-attack activities justified.

Anomalies are patterns in the data that do not conform to a well-defined notion of normal behaviour. Such patterns are also known as outliers. A significant problem when diagnosing anomalies is that their forms and causes can vary considerably. Volume anomaly in a network refers to a sudden change (positive or negative) in an Origin-Destination (OD) flow's traffic [10], and typical examples include Denial-of-Service (DoS) attacks, flash crowds and alpha events, and outages. *Volume attacks originate outside the network, propagate from the origin nodes to the destination nodes and visible on each of the link.* The key challenge steps in detecting volume anomaly are grouped into identification, quantification and detection.

Distributed Denial-of-Service (DDoS) attacks was the leading attack technique in 2013 and according to recent report from VeriSign, it has been increasingly growing in volume in 2014 with 83 percent rise in the average DDoS attack size [6]. It remains a serious and permanent threat to users, organisations and infrastructures on the Internet, as the scale and cost of business cyber security breaches doubled compared to previous year.

Because of the characteristics and properties of volume attacks, this study redefines *an anomaly as an unexpected 'state change' in time across multiple behaviours in a network system.* Hence, this study proposes a multi-agent network-based framework known as Statistical model for Correlation and Detection (SCoDe), that looks for time-correlated anomalies by leveraging the statistical properties of a large network, monitoring the rate of events occurrence based on their intensity.

SCoDe architecture is proposed to be a plug 'n' play module that utilizes time-series analysis and Exponential Weighted Moving Average (EWMA) one-step-ahead forecasting technique to find event behaviours and its anomalies. Because of its impending implementation in BT SATURN [4], the approach was made generic so that it can easily be modified to fit particular types of problem, with a defined attribute. The statistical technique used made the model to be robust and suitable for real-time analysis. The ability of EWMA to monitor the rate of event occurrence, its sensitivity to small variation in process mean, its ability to customize the detection of small shift and large shift in the process and ease of parameter modification makes it useful for anomaly detection in event intensity.

The framework was evaluated using network log files generated from a 2012 cyber-range simulation experiment [25] of an industrial partner as well as the DARPA (Defense Advance Research Project Agency) dataset from the 2000 simulation programme at the MIT Lincoln Labs, with prime results showing high detection rate of time-correlated events corresponding to network intrusions and reduction in huge alarm generated by IDS.

*This work was done prior to joining UTRC Ireland

K.J. Kim (ed.), *Information Science and Applications*,
Lecture Notes in Electrical Engineering 339, DOI 10.1007/978-3-662-46578-3_59

The idea to have time-correlated anomaly detection system was initially discussed in our previous work [15]. However, the full architecture of the proposed framework, as well as new experimental results using DARPA 2000 dataset is detailed in this report. Furthermore, this paper also investigates the model's assumption that aggregated network traffic follows a Gaussian distribution. The paper begins with literature review of related works in Section 2, detailed the proposed SCoDe framework and it architecture in Section 3, with experimental scenarios and results evaluation in section 4. Finally, this paper concludes in Section 5.

2. Related Works

Many researchers have discussed the module of hybrid detection technique for IDS to gain both advantages of anomaly detection and misuse detection techniques [2, 12]. The combination of signature detection technique and anomaly detection technique that formed *Hybrid Intrusion Detection Systems (HIDS)* can detect unknown attacks with the high detection rate of anomaly detection and the high accuracy of misuse detection. However, the interoperation of techniques, the complexity and high processing time are major challenges [17].

Some other studies have also explore the possibilities of having IDS layered in hierarchy where the alert output of the lower stage is processed by a second IDS [3, 22] in a post IDS analysis. This is especially useful when a large number of alerts are produced and currently culminating into Security Information and Event Management Systems (SIEM). This often used for correlation [7]. It can also be used to generate statistical reports, trend analysis, grouping of alert and detect outlier.

A range of techniques have been investigated for the above analyses. However, Farshchi [8] further conversed that the best IDS implementation can be a combination of a Signature based IDS and a statistical anomaly based IDS, since 95% of the attacks are detected by signature based IDSs but one successful zero day attack can change the whole nice looking secure implementation. Statistical analysis can easily fit into a real-time solution, however the assumption of its effectiveness highly depends on whether the assumptions made for the statistical model hold true for the given data.

Though previous similar studies have employed different statistical process-behaviour charts to detect network intrusions, they either spot abnormalities by treating IDS alerts as a single flow [24], use network traffic data [21], or/and evaluate results using only DARPA 1999/2000 dataset [19, 26] that can be regarded as obsolete data compared to current network security technologies among other criticisms.

This study builds on the principle of Syzygy model [16] that appropriately uses the aggregate behaviour of a community to decide whether to raise an alarm for the community and not individual clients, making strong guarantees and better efficient anomaly detector even with a noisy model. Hence, the main contribution of our novel approach is to use multi-agents structure to monitor *n* number of network behaviours through post-IDS analysis, employing statistical process-behaviour charts (such as EWMA, CUSUM, and regression control charts) to analyse and find deviation from each "immediate" client's behaviour in real time, and raise an alarm only when there is significant time-correlated deviation in the network behaviours.

SCoDe is specifically looking for time-correlated activities, as might be expected from a propagating worm or a coordinated attack [16]. Using IDS alert logs from DARPA dataset and modern cyber-range experiment, this study investigates the approach above and evaluate the proposed model in detecting volume attacks and reducing false positive alerts.

3. Proposed Framework

The SCoDe architecture as shown in Fig. 1 is proposed to be a "plug n play" module with specialised function of monitoring IDS alerts, in order to detect time-correlated anomaly events that could result from volume anomaly attacks using statistical based techniques. Other researchers have discussed about the idea of detectors having "pluggable" modules, with a specialized functions and capabilities [9].

In order to achieve its objectives, SCoDe consists of two major stages, which include the data pre-processing stage and the data analytical stage. The breakdown of these stages showing its architecture and relationship among its components are shown in Fig. 1.

3.1 Stage 1: Data Pre-processing Stage

An *event* is defined as a low level observable behaviour that is analysed on a network that may imply harm or a potential compliance violation as detected by the IDS within the network environment. An *alert* is generated by the IDS to log and notify parties of interesting events. This data pre-processing stage involves transforming and normalising unstructured or semi-structured alerts into suitable, uniform objects, as input to the proposed framework are IDS alerts from heterogeneous sources with wide varieties of technologies. Security alerts from firewall systems can also be integrated. As described in Fig. 1, *raw alerts* from IDSs are collected and stored in a *database* for pre-processing and transformation purpose, in order to extract a set of relevant features.

Each alert have at least five to six major features which include *Time, Type, Description, Priority level, Protocol, Source IP and Port, Destination IP and Port, Sensor Information.* In the *feature selection + extraction phase*, a raw alert is taken and fragmented into several available features. Merging and elimination of duplicate alert in order to filter irrelevant alerts are processed at this stage.

Alert prioritization is performed to assess and classify the relative importance of alerts generated by the IDS sensors. The IDS alerts is of three different priority levels labelled as priority 1, 2 or 3, with priority 1 as the most severe alert. This implies that all network behaviours can be categorised into three.

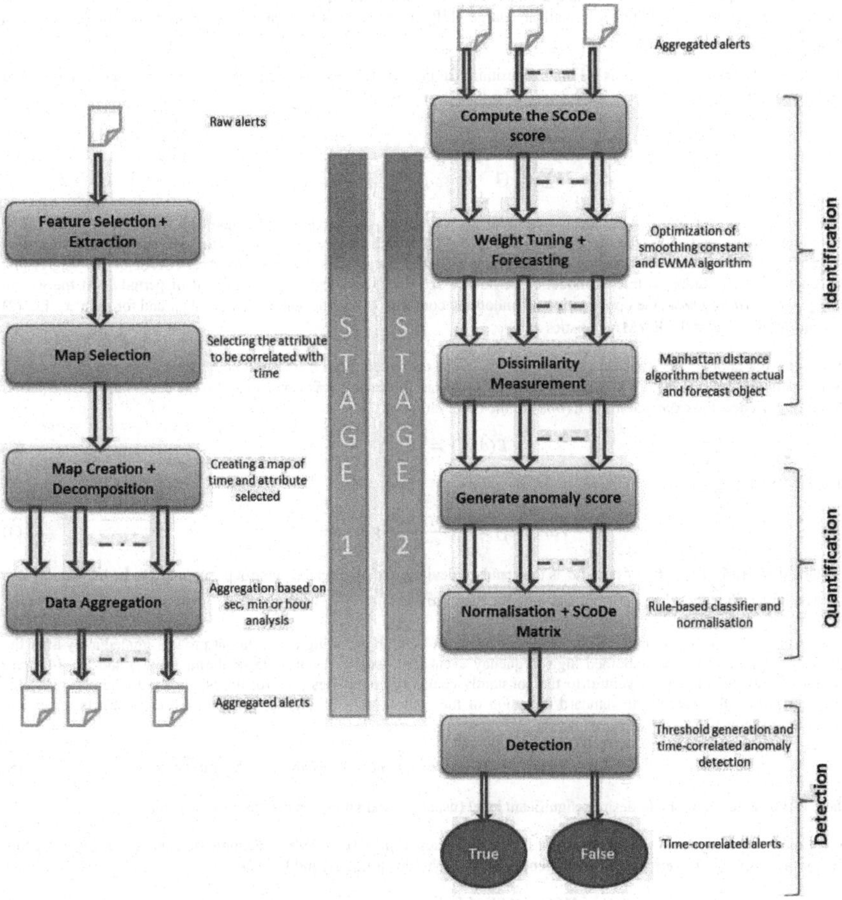

Fig. 1: *SCoDe* architecture of time-correlated alert anomaly detection

In order to spot and distinguish different type of events in the proposed model, SCoDe uses the combination of different *priority levels* and *time as* main features extracted to define *SCoDe agent* and these are mapped together at the *map selection phase.* The priority levels are then decomposed and map with corresponding alert timestamp at the *map creation + decomposition phase.* The idea behind the *data aggregation phase* is to group together alerts that belong to the same level (behaviour) within a particular time window. In order to perform different type of analysis, SCoDe allows this time window to be varied between second, minute or hour time window by the SA. The output of this stage is an *aggregated alert* that is more structured for further analysis.

3.2 Stage 2: Data analytical stage

Having decomposed and grouped related alerts into a number of aggregated alert per time window, the second stage is to perform statistical analysis in order to detected anomaly in each aggregated alert and label time-correlated alerts into true positive or false positive.

3.2.1 Identification process

Control charts are specialised time series plots. However, EWMA control chart overtakes other control charts such as the Shewhart chart because it pools present and past data in such a way that a small shift in the time-series can be detected more easily and quickly [20]. This technique was adopted because of its sensitivity to small variation in process mean, its ability to customize the detection of small and large shift in the process and ease of parameter modification to fit a particular type of problem.

EWMA one-step ahead technique is a simple algorithm that predicts the next value in a given time series sequential data. It is simply defined as

$$z_i = \lambda x_i + (1 - \lambda) z_{i-1} \tag{1}$$

$$z_i = \lambda \sum_{j=0}^{i-1} (1 - \lambda)^j x_{i-j} + (1 - \lambda) z_{i-j} \tag{2}$$

where x_i is the observation; z_i is the current EWMA value; z_{i-1} is the previous smoothed value; $(1 - \lambda)$ is called smoothing factor; and λ is the decay factor $(0 \leq \lambda \leq 1)$ which decides the importance of current and historical observations. During the *identification process*, each *SCoDe agent* tracks and updates the network of the aggregated frequency of each features extracted modelled as *SCoDe score* x_i , at a regular time interval of period i. At the *weight tuning + forecasting phase*, the optimization of smoothing constant, λ is performed to achieve λ_{opt} and for each x_i, Eq. (2) is employed to forecast the EWMA statistics z_i.

The framework adopts the initial value of EWMA z_o as the estimated mean of the first aggregated subgroup of n samples taken at time instant i. If the EWMA is a suitable 1-step-ahead forecast, it was suggested [20] the center line for the period i to be z_{i-1}. Hence, the expected mean $E(obs)$ of the observation is

$$E(obs_i) = z_{i-1} \tag{3}$$

Also, the variance given as

$$Var(obs_i) = \frac{\lambda \left[1 - (1-\lambda)^{2i}\right]}{2 - \lambda} Var(x) \tag{4}$$

is obtained where $Var(x) = \sigma^2 / n$ and σ is the standard deviation of aggregated subgroup, supposed to be known a priori. As i gets larger, the term $(1 - \lambda)^{2i}$ approaches unity, then $Var(obs_i) = \frac{\lambda}{2-\lambda} Var(x)$.

In order to significantly improve the performances of the SCoDe, in detecting an off target process immediately after the EWMA is started and after smoothed high frequency event that usually result in false alarm, Eq. (2) and Eq. (4) were substantially exploited. This is central to the constantly changing boundaries (control limits) for the framework which is usually based on the asymptotic standard deviation of the control statistic. Hence, the framework constructs upper and lower control limit UCL_i and LCL_i set as

$$UCL_i = E(obs_i) + L\sqrt{Var(obs_i)};\ LCL_i = E(obs_i) - L\sqrt{Var(obs_i)} \tag{5}$$

where L is a value set to obtain desire s-significant level (usually 3 and known as the "three-sigma rule").

To optimize the decay factor, this framework iterates λ values from 0.01 to 0.99. Obtaining the sum of the squared errors (SSE_λ), the mean squared error (MSE) was generated for each λ from Eq. (6) and Eq. (7).

$$MSE_\lambda = SSE_\lambda \,/\, (n - 1) \tag{6}$$

where

$$SSE_\lambda = \sum_{i=1}^{n} err_i^{\,2} \tag{7}$$

and n is the total number of events (aggregated alerts). The λ value that resulted in least MSE_λ is the λ_{opt} chosen for the model. Although this process increases the computational cost of the system, it significantly improves the sensitivity and accuracy of the model. This approach was tested and fully investigated in early studies [5, 26].

The dissimilarity of the actual *SCoDe score* x_i obtained by the agents and forecasted z_i calculated by the framework is measured at the *dissimilarity measurement phase* using the Minkowski distance algorithm defined as

$$d(x,z) = \sqrt[h]{|x_{i1} - z_{i1}|^h + |x_{i2} - z_{i2}|^h + \cdots + |x_{ip} - z_{ip}|^h} \tag{8}$$

where h is a real number such that $h \geq 1$. The Minkowski is a generalisation of the Euclidean distance (when $h = 2$) and Manhattan distance (when $h = 1$) measurement, that both satisfy the following mathematical properties: *Non-negativity, Identity of indiscernible, symmetry and triangle inequality.*

Having formed an estimate of the alert rate, SCoDe now proceeds to estimate the number that constitutes the anomaly in the *quantification process*. The framework assumes that the frequency of the alert and its error term follows a stochastic model. This assumption can be justified by assuming that the error term is actually the sum of a large number of independent error terms; even if the individual error terms are not normally distributed, by the central limit theorem (CLT), their sum can be assumed to be normally distributed [1].

3.2.2 Quantification process

For each x_i and z_i, an anomaly score s_i is generated from the output of the dissimilarity measurement and normalised by a rule-based classifier described below. For each of the priority levels and corresponding *SCoDe* agents, the model defines a t-dimensional feature vector $A_m = \langle s_1, s_2, s_3, \ldots, s_t \rangle$ to represent an agent m update for its feature and each A_m can be plotted as a point in a dimensional feature space. Using Eq. (5) above, the instantaneous value s_i is set to 1 if the $x_i > z_i + \delta$ where δ is the defined feature threshold, otherwise x_i is set to 0. The larger the value of δ, the better, though any positive value will be sufficient.

In order to minimize the classification error and loss of information that could result from the binary vector (0 and 1), *SCoDe* applied a five levels scoring method to its anomaly score as described below

$$s_1 = \begin{Bmatrix} 0 & if\ \delta = 0 \\ 0.25 & 0 < \delta < \sigma \\ 0.50 & \sigma < \delta < 2\sigma \\ 0.75 & 2\sigma < \delta < 3\sigma \\ 1.0 & \delta > 3\sigma \end{Bmatrix} \tag{9}$$

This approach enables *SCoDe* to be very sensitive to slight parameter variations, classify and detect any small deviation from the smoothed profile in order to achieve anomaly detection.

3.2.3 Detection process

Given a detection window, *SCoDe* computes the average score among all attributes anomaly signal to generate the *network score C* that represents the state of the client. If C > V (network threshold), then the model reports an anomaly.

Consider a network of n attributes and let $s_i \sim X$ where X is a random variable with finite mean and finite positive variance. By Central Limit Theorem [18], as n → ∞, the network scores are distributed normally with mean μ_X and variance $\sigma_z{}^2$/n:

$$\text{C} = \text{average}_i\ (s_i) = \frac{1}{n}\sum_i(\text{X}) \sim Norm\ (\mu_x, \sigma_x{}^2/n) \tag{10}$$

When $(E(|X|^3) = \rho < \infty$ where $E()$ denotes the expected value, convergence happens at a rate on the order of $1/\sqrt{n}$ (Berry-Esseen theorem). Concretely, let C' = C - μ_x and let F_n be the cumulative distribution function *(cdf)* of C'$\frac{\sqrt{n}}{\sigma_x}$ and ϕ the standard normal cdf. Then, there exist a constant B > 0 such that $\forall x, n, |F_n(x) - \phi(x)| \leq \frac{B\rho}{\sigma_x{}^3\sqrt{n}}$. Now consider when some numbers of the attributes d ≤ n of the network have been exploited. The anomaly score, as n, d → ∞ will be

$$\text{C} = \frac{1}{n}\left(\sum_{i=1}^{n-d}(\text{X}) + \sum_{i=1}^{d}(\text{Y})\right) \sim Norm\left(\frac{(n-d)\mu_X + d\mu_Y}{n}, \frac{(n-d)\sigma_X{}^2 + d\sigma_Y{}^2}{n^2}\right) \tag{11}$$

The rate of convergence guarantee that we get this asymptotic behaviour at relatively small values of n and d, and even when $d \ll n$.

According to Oliner et al [16], we can choose any positive V between $\sigma_X{}^2/n$ and $\sigma_X{}^2/n + \delta$ and guarantee that there exist n and d that give arbitrarily high probability of perfect detection (FP = FN =0). Without knowing δ however, the best strategy is to pick the lowest value of V such that the false positive is acceptable.

And in line with Oliner et al [16], V threshold is generated by

$$\text{V} = \mu_H + 2\sigma_H \tag{12}$$

where *H is the distribution of anomaly score for network with attribute n.*

4. Experimental Scenarios and Results Evaluation

4.1. Data 1-DARPA 2000 Intrusion Detection dataset

The first dataset evaluated using the proposed framework is the DARPA (Defense Advance Research Project Agency) LLDOS 1.0 from the simulation programme at MIT Lincoln Lab in 2000. The experiment are of two attack scenarios (LLDOS 1.0 and LLDOS 2.0.2) which includes 5 attack phases leading to a DDoS attack run by a novice attacker and more stealthy attacker respectively. The datasets from this experiment remain one of the most well-known and determined the IDS assessment to date. Although these datasets appear to be the most preferred evaluation data sets used in IDS research, it has received in-depth criticisms on how the data was collected, the question of the degree to which the stimulated background traffic represent the real traffic, and the how the data were obsolete compared to current network traffic and security technologies [13,23].

LLDOS 1.0 was examined through the SCoDe model as shown in Fig. 2.1 to Fig. 2.4. The hour analysis revealed that alerts were collected between 15:21:41 and 18:21:41 (GMT) on March 7, 2000. The total alerts were 27,145 over the 3 hour intervals. During this period, 5 grouped attack phases were performed, of which the attacker probes the network, breaks in to a host by exploiting the Solaris sadmind vulnerability, installs Trojan mstream DDoS software and launches a DDoS attack at an off site server from the compromised host.

Using Fig. 2.4, SCoDe was able to detect intrusion as early as the first and second phase of the attack which involves ICMP Ping and RPC sadmind UDP Ping. The generated network threshold for this dataset was 0.41 and the number of time-correlated intrusive alerts detected was 11,995 alerts representing a 55.81% reduction rate. The detail of the time-correlated anomaly events detected is described in Table 1 below.

Table 1: Detail analysis of time-correlated events detected by SCoDe framework

Attack Type Detected	No of Host Exposed
ICMP Ping	20
RPC portman sadmind request UDP	11
RPC sanmind UDP ping	3

Though there was decline in all the priority levels of alerts towards the end of the experiment including the last phase of the attack which was the main DDoS attack, SCoDe network score indicates a rise in time-correlated events. This is because the framework employs both changing upper and lower control limit UCL_i and LCL_i as the acceptable boundaries for the model.

4.2 Data 2-Cyber-Range Experiment (CRE) dataset

The second dataset is from the cyber-range experiment performed by an industrial partner who simulated a medium-scale organisations network operation for over 140 hours between March 1, 2012 and March 7, 2012 [25]. The case study is taken from the perspective of two attackers who aims to attack an organisation with the injected malicious traffic such as casual port scanning, intrusive port scanning, nexpose vulnerability assessment, nexpose exhaustive assessment, brute force attacks, leading to full scale DDoS attacks. The overview of the network architecture and full event description can be found in another report.

Two different IDSs (DMZ IDS and INT IDS) and one firewall were configured for the organisation with a total of over 1.5 million alerts logged on the system during the experiment. The DMZ implementation runs behind the firewall, protecting the LAN from the unsecure public network (Internet), while the INT implementation defends the LAN from intrusion within the organisation.

In this study, the DMZ IDS log files were analysed. Fig. 3.1 to Fig. 3.4 shows the time-series hour analysis of CRE events, with the generated network threshold of 0.52. The early stages of the attack with attack type such as casual port scanning, intrusive port scanning were detected, as well as some alerts from the main DDoS attack stage. Overall, using the available experiment data sheet to evaluate the result, the SCoDe model detection rate was 90.9% and the time-correlated anomaly alerts was found to be 63.8% of the alerts, with a low false positive of 9.09%.

Table 2: Model comparison with similar EWMA (λ, L) techniques

Techniques	% Alert Detection Rate	% False Positive Rate
EWMA (0.2, 3) - Ye et al Approach	81.8	9.09
EWMA (0.3, 3) - Ye et al Approach	72.2	0
SCoDe (Hr) – Proposed Approach	90.9	9.09

Using the 2012 CRE dataset, the model was compared with similar EWMA techniques from previous similar studies [26], where λ values of 0.2 and 0.3 was experimented with *s-significant levels L* of 0.3 (99.7% confidence interval). The results in Table 2 shows that the alert detection rate stands between 70% and 80% for the Ye et al [26] approaches compared with 90.9% achieved with SCoDe. However, relative 0% false positives but much reduced detection rate was achieved when EWMA (0.3, 3) was used.

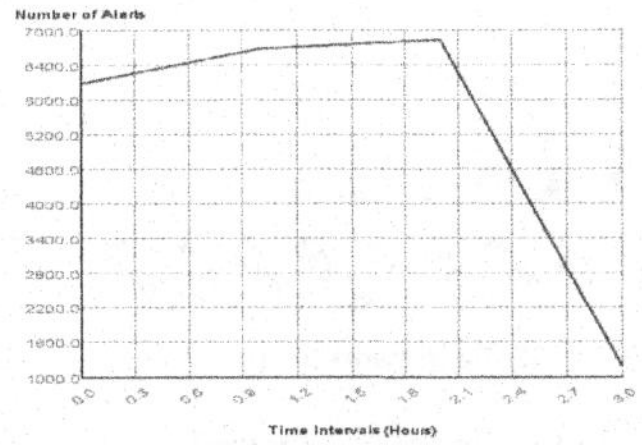

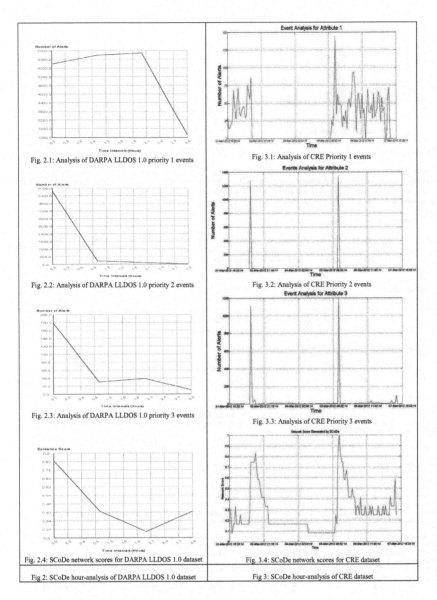

Fig. 2.1: Analysis of DARPA LLDOS 1.0 priority 1 events

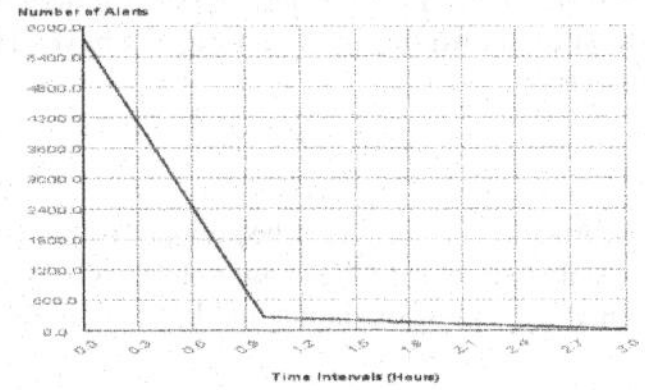

Fig. 2.2: Analysis of DARPA LLDOS 1.0 priority 2 events

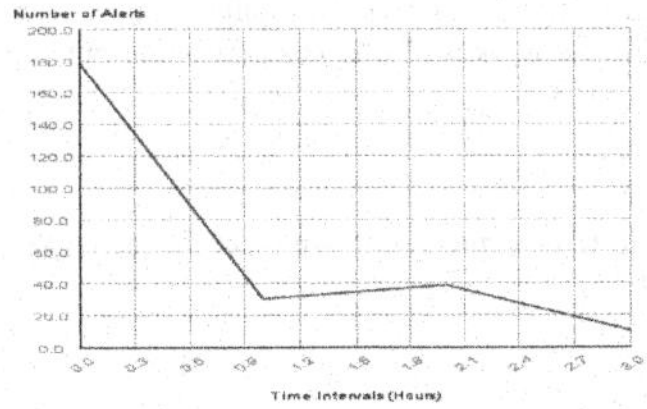

Fig. 2.3: Analysis of DARPA LLDOS 1.0 priority 3 events

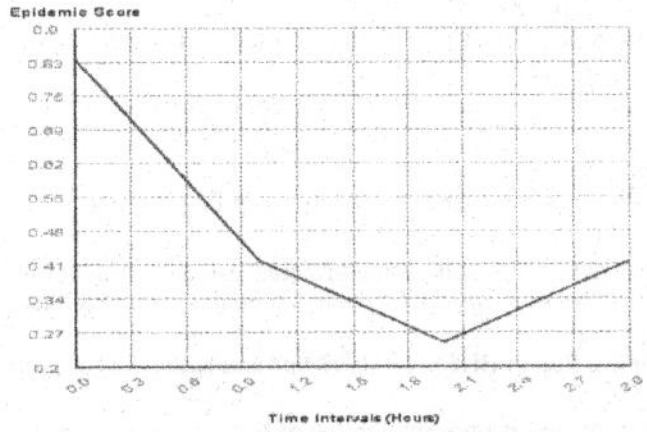

Fig. 2.4: SCoDe network scores for DARPA LLDOS 1.0 dataset

Fig 2: SCoDe hour-analysis of DARPA LLDOS 1.0 dataset

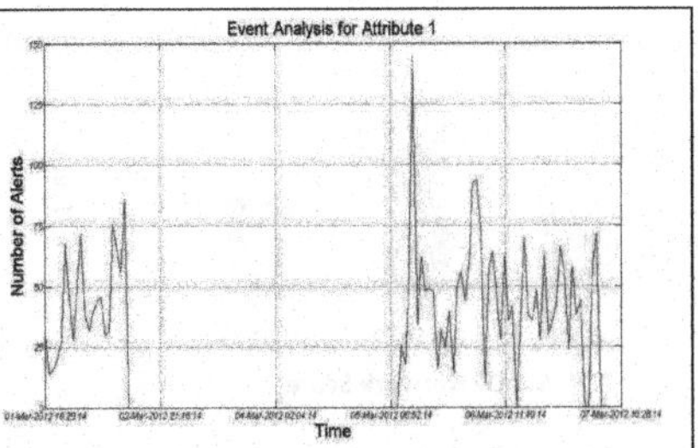

Fig. 3.1: Analysis of CRE Priority 1 events

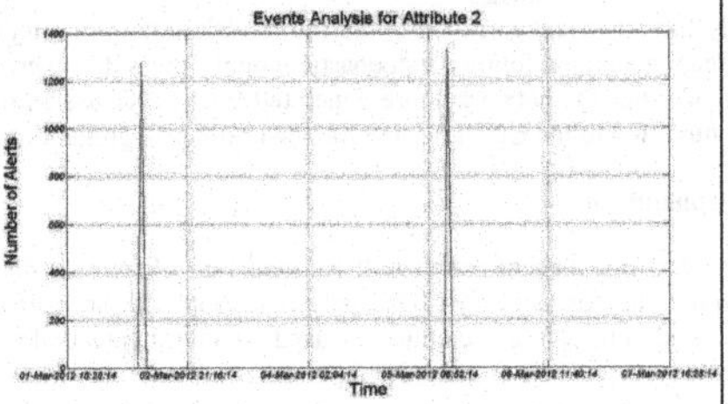

Fig. 3.2: Analysis of CRE Priority 2 events

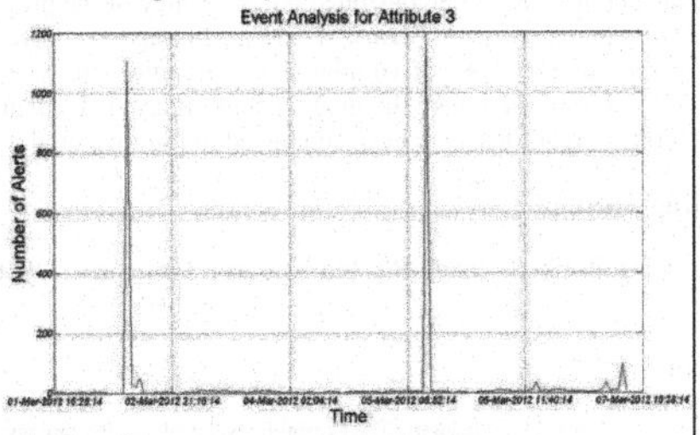

Fig. 3.3: Analysis of CRE Priority 3 events

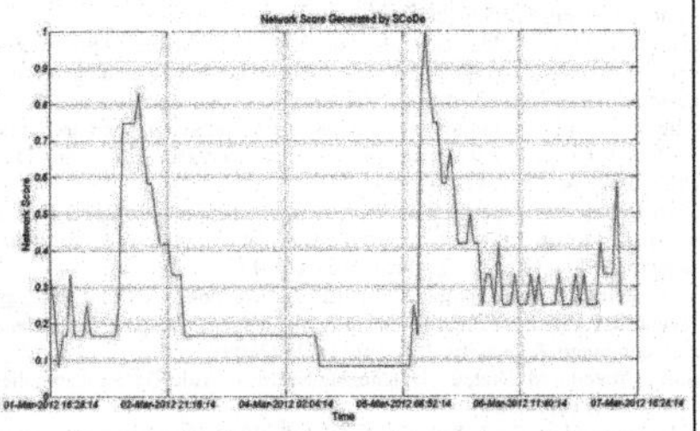

Fig. 3.4: SCoDe network scores for CRE dataset

Fig 3: SCoDe hour-analysis of CRE dataset

4.3 Distribution of the network scores generated

Many statistical inferential procedures such as hypothesis testing and the estimation of confidence intervals are based on the assumption that the distribution of a sample statistic is normal. This does not always hold true. However, the central limit theorem (CLT) which is a fundamental and profound concept of statistics justifies the above assumption and explains why many distributions tend to be close to normal distribution [1]. CLT states that "*given certain conditions, the mean of a sufficiently large number of iterates of independent random variables, each with a well-defined mean and well-defined variance will be approximately normally distributed*".

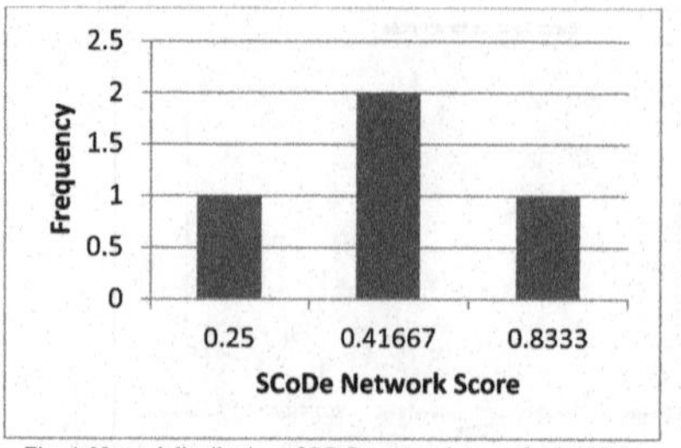

Fig. 4: Normal distribution of SCoDe network score for DARPA datasets

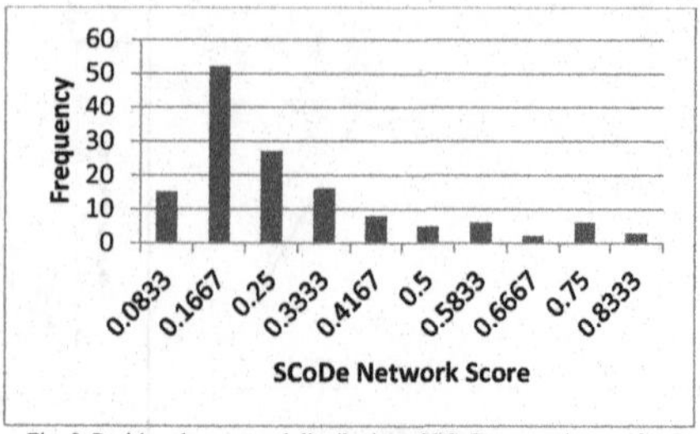

Fig. 5: Positive skew normal distribution of SCoDe network score for CRE datasets

This study further investigates about the framework assumption that the frequency of the alert behaviour and the SCoDe network score generated follows a stochastic model. It was discovered that the alert behavioural pattern does not follow any definite model but the network score generated tracks a Gaussian distribution as shown in Fig. 4 and Fig. 5 above. This could be attributed to the CLT after the framework data aggregations and analysis.

5. Conclusion

In this paper, a multi-agent framework for detection of time-correlated abnormalities in different behaviours of network event, through the means of post-IDS analysis is proposed. Specifically, the statistical process-behaviour charts of EWMA one-step-ahead forecasting technique is used to investigate undesirable time-correlated deviation of different network behaviours in order to identify abnormal patterns.

The experimental tests have demonstrated the efficiency of the proposed SCoDe model in terms of reducing false positives alerts and the number of alerts that a security administrator have to investigate. Because of its approach, the model is suitable for real-time analysis of volume attack detection and works well with existing IDSs in a plug 'n' play manner. Though the framework is robust, the model is not highly scalable and might not work well with categorical attribute of the network. Further experiments are been conducted to improve on this model.

Acknowledgement

The authors would like to thank BT and Northrop Grumman UK, the industrial partners of this project for their valuable contributions.

References

1. Anderson, C.-J.: Central limit theorem. The Corsini Encyclopedia of Psychology, John Wiley & Sons, Inc. (2010)
2. Aydın, M.-A., Zaim, A.-H., Ceylan, K.-G.: A hybrid intrusion detection system design for computer network security. Computers & Electrical Engineering, 35(3), pp. 517-526 (2009)
3. Beng, L.-Y., Ramadass, S., Manickam, S., Fun, T.-S.: A Comparative Study of Alert Correlations for Intrusion Detection. In: International Conference on Advanced Computer Science Applications and Technologies (ACSAT), IEEE, pp. 85-88 (2013)
4. Rowlingson, R., Healing, A., Shittu, R., Matthews, S.-G., Ghanea-Hercock, R.: Visual Analytics in the Cyber Security Operations Centre. Proceedings of the Information Systems Technology Panel Symposium on Visual Analytics (2013)
5. Cisar, P. Maravic Cisar, S.: *EWMA statistic in adaptive threshold algorithm*, In: 11th International Conference on Intelligent Engineering Systems, 2007 (INES 2007), pp. 51 –54 (2007)
6. eWeek Report: http://www.eweek.com/small-business/ddos-attacks-target-online-gaming-sites-enterprises.html (2014)
7. Elshoush, H.-T., Osman, I.-M.: Intrusion Alert Correlation Framework- An Innovative Approach. In IAENG Transactions on Engineering Technologies, pp. 405-420, Springer Netherlands (2013)
8. Farshchi, J.: Statistical based approach to Intrusion Detection, SANS Institute, http://www.sans.org/security-resources/idfaq/statistic_ids.php (2003)
9. Garcia-Teodoro, P., Diaz-Verdejo, J., Maciá-Fernández, G., Vázquez, E.: *Anomaly-based network intrusion detection: Techniques, systems and challenges*, computers & security, 28(1), pp. 18-28 (2009)
10. Lakhina, A., Crovella, M. Diot, C.: Diagnosing Network-Wide Traffic Anomalies. Technical Report BUCS-TR-*2004*-008 and RR04-ATL-022666, Boston University, Feb. (2004)
11. Liao, H.-J., Richard Lin, C.-H., Lin, Y.-C., Tung, K.-Y.: Intrusion detection system: A comprehensive review. Journal of Network and Computer Applications, 36(1), pp. 16-24 (2013)
12. Kim, G., Lee, S., Kim, S.: A novel hybrid intrusion detection method integrating anomaly detection with misuse detection. Expert Systems with Applications, 41(4), pp. 1690-1700 (2014)
13. Mahoney, M.-V., Chan, P.-K.: *An analysis of the 1999 DARPA/Lincoln laboratory evaluation data for network anomaly detection*, Florida tech. report CS-2003-02 (2003)
14. Modi, C., Patel, D., Borisaniya, B., Patel, H., Patel, A., Rajarajan, M.: A survey of intrusion detection techniques in cloud. Journal of Network and Computer Applications, 36(1), pp. 42-57 (2013)
15. Olabelurin, A., Kallos, G., Xiang. Y., Bloomfield, R., Veluru, S., Rajarajan, M.: Time correlated anomaly detection based on inferences. In: Proceedings of the 12th European conf. on Information Warfare and Security, ECIWS (Finland), Academic Conference and Publishing International Limited, pp. 351-360 (2013)
16. Oliner, A.-J., Kulkarni, A.-V., Aiken A.: Community epidemic detection using time-correlated anomalies. In: Proceedings of the 13th international conf. on Recent advances in intrusion detection, RAID'10, (Berlin, Heidelberg), Springer-Verlag, pp. 360 – 381 (2010)
17. Patcha, A., Park, J.-M.: An overview of anomaly detection techniques: Existing solutions and latest technological trends. Computer Networks, 51(12), pp. 3448-3470 (2007)
18. Papoulis, A.: *Probability, Random Variables, and Stochastic Processes*. Mc-Graw Hill (1984)

19. Park, Y., Baek, S.-H., Kim, S.-H., Tsui, K.-L.: Statistical Process Control-Based Intrusion Detection and Monitoring. Quality and Reliability Engineering International, 30(2), pp. 257-273 (2014)
20. Raza, H., Prasad, G., Li, Y.: EWMA model based shift-detection methods for detecting covariate shifts in non-stationary environments. *Pattern Recognition* (2014)
21. Salem, O., Vaton, S., Gravey, A.: A scalable, efficient and informative approach for anomaly-based intrusion detection systems-theory and practice. International Journal of Network Management, 20(5), pp. 271-293 (2010)
22. Smith, R., Japkowicz, N., Dondo, M., Mason, P.: Using unsupervised learning for network alert correlation. In: Advances in Artificial Intelligence pp. 308-319, Springer Berlin Heidelberg (2008)
23. Tjhai, G.-C.: *Anomaly-based correlation of IDS Alarms*. PhD Thesis, The University of Plymouth, UK (2011)
24. Viinikka, J., Debar, H., Mé, L., Lehikoinen, A., Tarvainen, M.: Processing intrusion detection alert aggregates with time series modeling. Information Fusion, 10(4), pp. 312-324 (2009)
25. Winter, H.: *System security assessment using a cyber-range.* 7th IET International Conference on System Safety, incorporating the Cyber Security Conference, pp. 41 (2012)
26. Ye, N., Chen, Q., Borror, C. EWMA forecast of normal system activity for computer intrusion detection. IEEE Transactions on Reliability, 53, pp. 557 – 566 (2004)

Side Channel Attacks on Cryptographic Module: EM and PA Attacks Accuracy Analysis

HyunHo Kim[1], Ndibanje Bruce[1], Hoon-Jae Lee[2], YongJe Choi[3], Dooho Choi[3]

[1]Department of Ubiquitous IT, Graduate School of Dongseo University,
Sasang-Gu, Busan 617-716, Korea
feei_@naver.com
bruce.dongseo.korea@gmail.com
[2]Division of Computer and Engineering Dongseo University
Sasang-Gu, Busan 617-716, Korea
hjlee@dongseo.ac.kr
[3]Electronics and Telecommunication Research Institute (ETRI)
choiyj@etri.re.kr
dhchoi@etri.re.kr

Abstract. Extensive research on modern cryptography ensures significant mathematical immunity to conventional cryptographic attacks. However, different side channel techniques such as power analysis and electromagnetic attacks are such a powerful tool to extract the secret key from cryptographic devices. These techniques bring serious threat on hardware implementations of cryptographic algorithms. In this paper an extensive analysis of side channel analysis on cryptographic device is presented where we study on the EM and PA attacks methods as sideways attacks on the hardware implementation of the crypto-module. Finally we establish a comparison table among different attacks tools methods for the accuracy analysis.

Keywords:

1 Introduction

In order to ensure the secrecy of the transmitted message a cryptosystem should be in a way where the adversary cannot recover the original message. In this regards, protecting a message involves cryptographic algorithms such as stream ciphers. The contemporary study of cryptography can be separated into different areas, together with symmetric-key cryptography, public-key cryptography, etc. For various applications, in the technically advanced society, from military, electronic commerce to computer security and wireless communication, stream ciphers are well known and accepted for providing data confidentiality by the usage of their secret key cryptosystem, faster speed and high efficiency. During the encryption operation, a pseudorandom cipher digit stream called keystream is combined with the plaintext to give a digit of cyphertext stream. From the cryptographers' analysis, ciphers are secure in view of the traditional cryptanalysis to the security ciphers. On the straightforward example of tradi-

K.J. Kim (ed.), *Information Science and Applications,*
Lecture Notes in Electrical Engineering 339, DOI 10.1007/978-3-662-46578-3_60

tional cryptanalysis is the brute force attack. To reveal to secret key of the cipher, the cryptanalyst tries all possible combinations key until the correct one is found. Therefore, if the number of combinations is large, an exhaustive research becomes unfeasible. Nevertheless, Side Channels Attacks (SCA) is the most well known attacks on the hardware implementation of ciphers through a cryptographic device. Meanwhile, the hardware implementation of a cryptographic algorithm can be particularly difficult to control and often result in the leakage of side-channel information. Concerning the SCA attacks like Simple Power Analysis (SPA), Differential Power Analysis (DPA), Simple EM-Analysis (SEMA), Differential EM-Analysis (DEMA), Timing Attacks, Fault Injection Attacks, the literature on stream cipher is not so much as the other cryptographic algorithm[1-3].

In this paper we study on the Power Analysis (PA) attack which is a form of side-channel attacks based on analyzing the power consumption of a cryptographic device (such as a smart card, tamper-resistant "black box", or integrated circuit) while it performs the encryption operation. The attacker can non-invasively extract secret keys and other information from the device by combining with other cryptanalysis techniques. Power analysis attacks generally consist of two catalogues, namely Simple Power Analysis and Differential Power Analysis. They were first introduced by Kocher et al. in [4]. Electromagnetic (EM) analysis using the EM field generated by a cryptographic module is also presented. One of the characteristics of EM analysis is that EM waveforms can be obtained by noncontact probing. Conventional attacks usually require direct physical access to the target module in order to acquire side-channel information.

2 Analysis of EM and PA attack on Cryptographic Devices

2.1 Electromagnetic Attack on Cryptographic Devices

When a cryptographic module performs encryption or decryption, information on secret parameters that correlate to the intermediate data being processed can be leaked as side-channel information, via operation timing, voltage/ current fluctuation, or electromagnetic (EM) radiation. EM analysis (EMA) measuring the electromagnetic field generated by a cryptographic module has also been presented as an extension of the power analysis [5-6].In this study of EM attack on the crypto-module, we first analysis the experiment in [7] and we state the evaluation result. Figure 1 shows the scheme of the experiment setup.

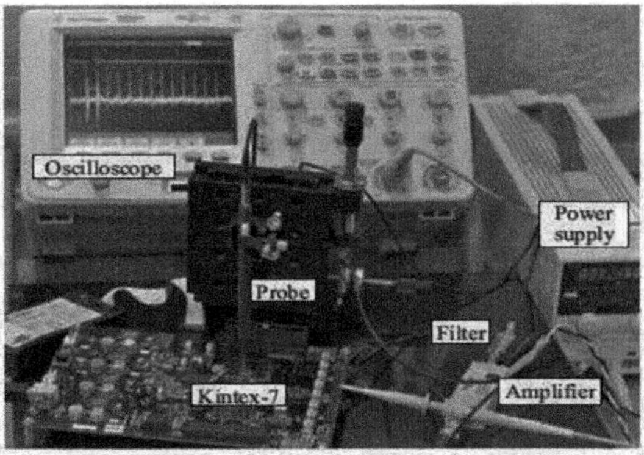

Fig. 1. EM environmental Experiment

The main components of this experiments are as follow: a Langer LF-B 3 EM probe, a Miteq AU-3A-0150 amplifier (50 dB, 0.3–600 MHz), a fifth-order Bessel low-pass filter, and an Agilent DSO6104A oscilloscope. The EM probe was placed at a position where the peak voltage of the measured radiation waveform was maximal. To test the feasibility of SCA against state-of-the-art large-scale integration (LSI) technology, we have developed a new experimental environment, the Side-channel Attack Standard Evaluation Board (SASEBO)-GIII [8], equipped with the latest 28-nm Kintex-7 field-programmable gate array (FPGA) device. Table 1 gives the functional components of SASEBO-GIII.

Table 1. SASEBO-GIII componets functions

Component name	Description
Board Size	250×200 mm^2, 8 layers
Cryptographic Device	Kintex-7 325T, 28nm, 1.0v
Control Device	Spartan-6
Communication Interface	USB 2.0 and two FMCs (LPC)
Memory	1-Gbit DDR3-DRAM
FPGA config. Interface	BPI, JTAG and Select Map
Monitoring Point	V_{core} line of Kintex-7

From this experiment, Figure 2 shows example waveforms from the emission of EM radiation for SASEBO-GIII. The sampling interval here was 500 ns, and so a single waveform contains 10,000 values. The amplitude of SASEBO-GIII's waveform is one fifth. A lower noise environment and more expensive instruments with higher precision are hence required to measure the side-channel information of state-of-the-art devices when evaluating physical vulnerability. In the experiment, 50,000 waveforms were acquired for each of two cryptographic keys: key1 $\{00\ 01\ 02\ 03\ 04\ 05\ 06\ 07\ 08\ 09\ 0A\ 0B\ 0C\ 0D\ 0E\ 0F\}_{16}$ and key2 $\{2B\ 7E\ 15\ 16\ 28\ AE\ D2\ A6\ AB\ F7\ 15\ 88\ 09\ CF\ 4F\ 3C\}_{16}$.

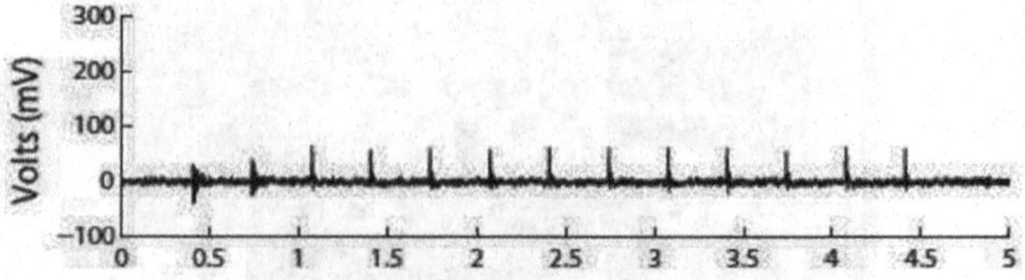

Fig. 2. EM radiation waveform

2.2 Power Analysis on Cryptographic Devices

This subsection introduces the most known side channel attack introduced by Kocher et al [8]. We investigate on the experiment in [9] for the DPA attack on AES algorithm implementation on SASEBO board. Figure 3 gives an overview of the procedure for construction simulation model from the case studied. The aim of the experiment is to measure the power consumption of entire AES processing sequence and record the power traces. In this experiment, the total experimental setup consists of SASEBO-W, Host PC, Experimental smartcard and Digital Oscilloscope. The experimental smartcard is an ATMega 163 microcontroller with 8-bit architecture and 1KB of data and instruction memory. This operates at 3.57MHz and supports AES function. Power consumption waveforms were collected using any 2 cryptographic keys and the collected waveforms are analysed using Correlation Power Analysis. The power consumption H_{ij} of CPA was set to the Hamming weight of SubBytes output.

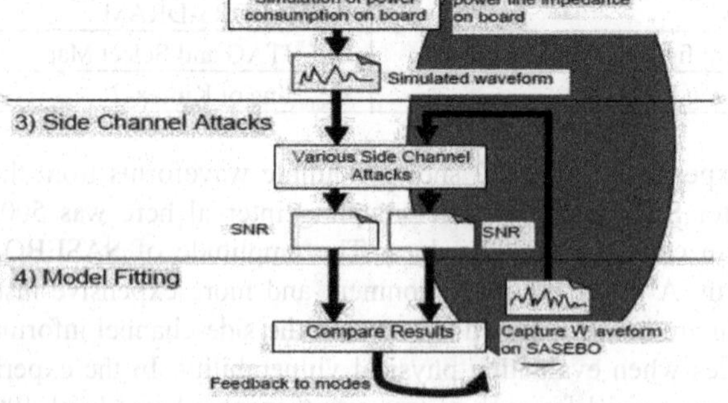

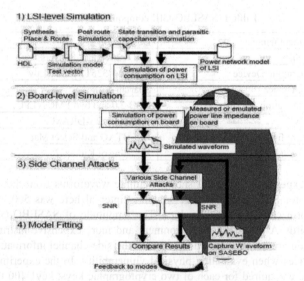

Fig. 3. Framework of Simulation Model

The power waveforms are captured clearly from SASEBO-GII, even though their peak-to-peak amplitude is less than 20 mV, for each capacitance setup, respectively. In order to evaluate the SNR, a 128-bit data register is divided into sixteen 8-bit registers, and Hamming distances (range of 0-8) for all 8-bit registers are calculated by changing the input data (secret key is fixed). Then, the 10,000 power traces are divided into nine groups depending on their Hamming distances, and the average waveforms are calculated. The variance in the averaged waveforms and in the 10,000 power waveforms are defined as side-channel information and noise, respectively. The SNR is calculated as the ratio between the information and noise. Table 2 gives the conditions measurements of the experiments

Table 2. Measuremnt parameters

Cryptographic circuit	AES with Composite _eld S-box
Logic usage	2221 LUTs and 525 FFs
Operation frequency	4 MHz
Probing point	1.0-Ωresistor on VCORE line (1.0 V)
Capacitance	No caps., Seven 0.1uF, 10 uF
Instrument	Agilent DSO 8104A oscilloscope
Probe	ProbeMaster 1024A and SMA cable
Sampling rate	4GSa/sec, 14,000 Sa / trace
Filter	20 MHz BWL

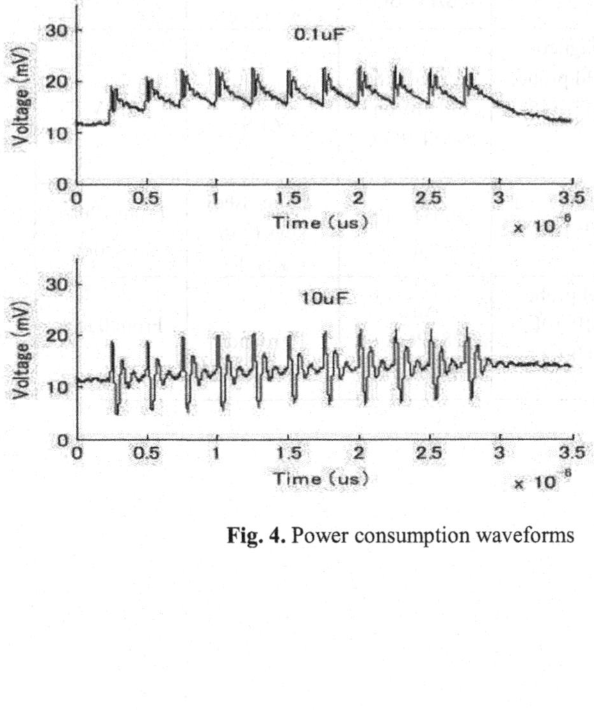

Fig. 4. Power consumption waveforms

3 Accuracy Analysis : SCA Attacks Setup Environment Indicators

This section introduces the study on the parameters using to perform the side channel attack. In this analysis, we investigate broadly on the setup environment parameters in order to recommend the best platform to the researchers. The table 3 gives an overview about the EM attack accuracy analysis where we investigated on the device type, attack target chip or algorithm, attack function, area/distance accuracy, time accuracy, energy(or voltage) and oscilloscope type. Table 4 summarizes the PA accuracy analysis with the following indicators: case studied, attack type (DPA/SAP), attack target chip or algorithm, attack function, oscilloscope (model and sampling rate), energy (voltage) and traces.

Table 3. EM Accuracy Analysis

Case	EM Device(ex : probe)	Attack Target Chip/Algorithm	Attack Function	Energy or Voltage
Yohei et al., 2012[10]	Langer LF-B 3 EM probe	FPGA on SASEBo-III AES	S-Box	1.0V
Hayashi et al., 2013 [11]	Fisher F-2000	AES FPGA on SASEBO	one encryption operation in eleven cycles	1Vor 120 dB·μV
Hayashi et al., 2013 [12]	magnetic field probe (MT-545)	AES-128 FPGA on SASEBO-G	All rounds of AES	3.3 V,
Nakai et al., 2014 [13]	EM radation: HC020	ROM AES chip	all guessed keys (256 keys)	From 100mV to 300mV
Hayachi et al., 2013 [14]	EM probe: (MP-10L, NEC)	AES-128FPGA on SASEBO	10 rounds	From 0 to 1 V

Table 4. PA Accuray Analsysi

Case	Attack Type(ex: DPA/SPA)	Attack Target Chip/Algorithm	Attack Function	Trace
T. Katashita et. al., 2010[15]	DPA	FPGA on SASEBO-GII/ AES	S-box	14,000 Sa / trace
Seo et. al., 2005[16]	DPA	ARM 32bit CPU core/ ARIA	S-Box	5,000
Han et. al., 2006[17]	DPA	8bit Microprocessor/ARIA, SEED	S-Box	1,000
Yoo et. al., 2006[18]	DPA	AVR 8bit Microprocessor/Masking applied ARIA	S-Box	1,000

4 Conclusion

In this paper, we have presented the study on the side channel attack using sideways such as Electromagnetic and Power Analysis attack on cryptographic devices. We have explained previous experiments done in this field and presented the configuration environment mostly used by SASEBO boards. In addition we stated a wide table where we investigated on cases studied in order to evaluated each research environment and then establish the SCA parameters for the accuracy analysis. From this study, we recommend to the cryptographic researchers groups to the using of the SASEBO boards because it has many advantages in the side channel attack field research.

References

1. D. Strobel, "Side Channel Analysis Attacks on Stream Ciphers", Masterarbeit Ruhr-Universität Bochum, March 3, 2009
2. P. Kocher, J. Jaffe, B. Jun, "Differential Power Analysis", In proc. *19th Annual International Advances in Cryptography Conference, CRYPTO' 99*, pp 388-397 (1999).
3. F. Koeune, J. J. Quisquater, "A timing attack against Rijndael", Technical Report CG-1999/1, Université Catholique de Louvain, June 1999

4. P. C. Kocher, J. Jaffe, and B. Jun, "Differential Power Analysis," in CRYPTO '99: *Proceedings of the 19th Annual International Cryptology Conference on Advances in Cryptology*. London, UK: Springer-Verlag, 1999, pp. 388-397.
5. K. Gandolfi, C. Mourtel, and F. Olivier, "Electromagnetic analysis: Concrete results," *CHES 2001, Lecture Notes in Computer Science*, vol. 2162, pp. 251–261, May 2001.
6. T. Plos, M. Hutter, and M. Feldhofer, "On comparing side-channel preprocessing techniques for attacking rfid devices," *WISA 2009, Lecture Notes in Computer Science*, vol. 5932, pp. 163–177, Aug.2009.
7. Yohei. H, Toshihiro. K, Akihiko.S, and Akashi S. Electromagnetic Side-channel Attack against 28-nm FPGA Device.
8. Akashi, S., Toshihiro, K., Sakane, H.: Secure implementation of cryptographic modules-development of a standard evaluation environment for side channel attacks. Synthesiology 3(1) (2010) 56–65
9. Katashi. T, Satoh.A, Katsuya. K, Nakagawa. H, Aoyagi. M Evaluation of DPA Characteristics of SASEBO for Board Level Simulations. COSADE 2010 - First International Workshop on Constructive Side-Channel Analysis and Secure Design. Pp.36-39.
10. Yohei Hori, Toshihiro Katashita, Akihiko Sasaki, Akashi Satoh, "Electromagnetic Side-channel Attack against 28-nm FPGA Device", The 13th International Workshop on Information Security Applications(WISA), 16-18 Aug. 2012.
11. Y-I. Hayashi, N. Homma, T. Mizuki, T Aoki, H. Sone, L. Sauvage and J-L. Danger "Analysis of Electromagnetic Information Leakage from Cryptographic Devices with Different Physical Structures". IEEE Transaction on Electronics Compatibility, Vol. 55, NO. 3, pp. 571-580, June 2013.
12. Y-I. Hayashi, N. Homma, T. Mizuki, T Aoki, H. Sone, L. Sauvage and J-L. Danger "Efficient Evaluation of EM Radiation Associated With Information Leakage From Cryptographic Devices". IEEE Transaction on Electronics Comptability, Vol. 55, NO. 3, pp. 555-563, June 2013.
13. T. Nakai, M. Shibatani, M. Shiozaki, T. Kubota, T. Fujino "Side-Channel Attack Resistant AES Cryptographic Circuits with ROM reducing Address-Dependent EM Leaks" In the Proceedings of the International Symposium on the Circuit Systems (ISCAS), pp. 2547-2550, 2014.
14. Y-I. Hayashi, N. Homma, T. Aoki, Y. Okugawa and Y. Akiyama "Transient Analysis of EM Radiation Associated with Information Leakage from Cryptographic ICs" In the Proc of the 9th International Workshop on Electromagnetic Compatibility of Integrated Circuits(EMC Compo), pp. 78-82, 15-18December, 2013.
15. T. Katashita, A. Satoh, K. Kikuchi, H. Nakagawa, M. Aoyagi "Evaluation of DPA Characteristics of SASEBO for Board Level Simulations" In the Proc of the 1st International Workshop on Constructive Side-Channel Analysis and Secure Design (COSADE), pp.36-39, 2010
16. JungKab Seo, ChangKyun Kim, JaeCheol Ha, SangJae Moon, IlHwan Park, "Differential Power Analysis Attack of Block Cipher ARIA", Journal of The Korea Institute of Information Security & Cryptology, Vol.15 No.1, pp. 99-106, 2005.
17. DongHo Han, JeaHoon Park, JaeCheol Ha, SungJae Lee, SangJae Moon, "Development of Side Channel Attack Analysis Tool on Smart Card", Journal of The Korea Institute of Information Security & Cryptology, Vol.16 No.4, pp. 59-68, 2006.
18. HyungSo Yoo, JaeCheol Ha, ChangKyun Kim, IlHwan Park, SangJae Moon, "A Secure Masking-based ARIA Countermeasure for Low Memory Environment Resistant to Differential Power Attack", Journal of The Korea Institute of Information Security & Cryptology, Vol.16 No.3, pp.143-155, 2006.

Exploring Bad Behaviors from Email Logs

Supachai Kanchanapokin, Sirapat Boonkrong
Faculty of Information Technology,
King Mongkut's University of Technology North Bangkok, Thailand

Abstract

Human is the weakest factor in organization security and is always the target of attackers. Attackers find vulnerability of human cause by his/her bad behaviors and exploit it. Organization shall proactively detect those bad behaviors and prevent it on the first hand. This paper studied employees with bad habit of sending of corporate file(s) to public email addresses that are not allowed by the organization. By using email log as a source to generate social network graph and collect 2 types of out-degree of each node, regular out-degree and out-degree to "not allowed" public email addresses. We analyze correlation of these 2 numbers and found that people with bad habit tend to send mail to public email addresses when they have to send file outside the company. The result of this paper gave us parameters to calculate risk score for further study using the integration of Social Network-Attack Graph (SN-AG) analysis approach.

1. Introduction

Information security has been known since 1960s and it mostly involves human, the weakest factor in security of an organization. Attackers today are now using social engineering as a major attack to an organization by exploiting bad behaviors of employees in the organization. If we know who cause or able to be exploited in social engineering attacks, we can find the way to prevent them on the first hand. We can say that an organization must improve their protection technology together with eliminating risk behaviors of its employees. This paper proposes the way to detect bad behaviors from existing information of an organization. Email system is one source of information in any organizations because employees use email in everyday activities and email is one of the most frequent channels in corporate information leakage. For example, users may send corporate file(s) to public email via corporate email without company consent. Even though most company has policy for data loss prevention but some employee might not realize. Sometimes the organization itself does not realize it as well. Thanks to most corporate email systems products which provide event logs in almost activities done by emails. This is the source of information we will use to explore bad habits from users. However, there will be a lot of user behaviors that can be explored from email log and it would be too much to be covered in one paper. In this

K.J. Kim (ed.), *Information Science and Applications*,
Lecture Notes in Electrical Engineering 339, DOI 10.1007/978-3-662-46578-3_61

paper, we will be focusing on a behavior of sent mails with attachment to public email service which is the worst habit that result in an organization information leaks.

In order to study the behavior, we know that people use email to communicate with others, and this forms the social network of its kind. With this fact, we are able to generate a network from email information and apply social network analysis (SNA) as well as graph theory in the study. However, we use 2 sets of email data in this study. The first set of data contains email logs from the organization we are studying. The other set are Enron Corpus email data which have been openly shared and used for several purposes [9].

We provide approach to extract required information from the logs and analyze it using statistical technique (data correlation) to determine the users we should aware of. We also raise some discussion points from what we have found in the study, and where we are going to study further using the results from this paper.

2. Related Works

Graph Theory and Networks

The Graph Theory has been used to study much kind of networks for years. Leonhard Euler used it to explain a "Bridges of Königsberg" in 1736. Equations and formulas from his works were further developed and improved to model more broaden mathematical problems such as the "Four Color Problem" by Francis Guthrie and De Morgan in 1952, the study of trees by A. Cayley in 1857, the "External Graph Theory" by Turan in 1941, and the "Random Graph" by Erdös and Rényi in 1959.

In Graph Theory, analysts use picture of nodes (vertices) and links (edges) to generate mathematical model for specific problem. Developing graph requires information and data collection in certain structure such as matrices and linked lists. Model of the graph depicts relationship between vertices and edges in several types of graph such as undirected graph, directed graph, weighted graph. These types of graph provide different graph measures. Analysts must carefully choose what type of the graph they are studying.

Social Network Analysis (SNA)

SNA also utilizes graph theory to generate mathematical model of the networks. SNA is the analysis of relationships between members in a social network. As Newman has put email studies into one of the social networks studies[1], it can be studied using graph theory as well. Ebel *et al.*[4] studied email communication of students at Kiel University and have found that the email network is scale-free and exhibits Milgram's small-world effect. Newman *et al.*[5] have found that email viruses spread in organization network differently from human viruses because it is directed, unlike human network which is undirected. He also suggests that controlling email viruses problem in an organization shall use network structure to help identifying risk node in the network. Newman studied several social networks and found that social graphs that were generated for different social network types provide different properties measured as well [1]. For example, network of email provides properties of directed graph while network of telephone provides undirected graph. However, his studies are on a few behaviors and properties of networks but there are more to study on the behaviors and functions of the network that will help analysts understand more of the real world networks.

3. How we studied.

In order to explore bad behaviors from email information, we have to be sure that our organization has its email traffic recorded, and those logs contain all information we need for investigation. For example, message tracking log of Microsoft Exchange 2007 has basic information that provides senders and receivers of each message including date, time, message ID and subject of the message. However, if we would like to know that those message has attachments or not, we would have to turn on message log in our mail server system.

Once we have sufficient information on message tracking log we will extract information we need from the log and transform it to sparse matrix format. We can prepare social network graph from these information. Let a node represent each email address and links represent direction of mail transmission. By counting the number of sent mails with attached file(s), we will have out-degree of each node that can be processed to explore some behaviors.

Once the above process has been carried out, we should have enough information to begin our study. However, before we start our exploration, we need some parameters to compare our records with. We accumulate number of attachments, and group to individual senders. By sorting the table by number of attachments, we can see who the top email senders with attachments are.

Before we can tell what email messages have been sent with inappropriate habit, we must know whether the destinations of each message sent are allowed or not. At this point, the list of Allowed Host names (AH) for receivers' email addresses shall be created. This is the list of host names that our organization is working with as a business partners, co-worker, or the organizations that we usually communication with.

Allowed Host Names (AH)

- @mycompany.co.th
- @mypartners.com
- @mypartners.co.th
- @myvendors.com
- @myvendors.co.th
- @government.go.th

Figure 1 Allowed Host Name list

With the list of AH, we can then filter out email messages that were sent to the AH and accumulate number of attachments for only messages sent to the host names not in the AH. By grouping and sorting the number of attachments by individuals, we will know who might have a bad habit.

However, some people may violate company policy sometimes with no intention of creating bad behavior. For example, Mr.A wanted to take document file to his home to continue his work. Unfortunately, his thumb drive was full and he did not have a mobile computing device. Therefore, he sent his working document via email to his public email address so that he could download the file and continue his work at home. This kind of situation is not the habit but it sure did not follow the company's policy. On the contrary, Mr.B happened to send his file to public email addresses every possible time during the day. This is the bad habit that we would like to detect.

The technique we will use to detect the above habit is to find correlation between the number of sent mails with attached file(s) and the number of sent mails with attached file(s) to public email addresses that are not in the AH list. We assume that if an employee regularly sends emails to public email addresses that are not in AH list, the two numbers shall be correlated. This means that employees have a bad habit.

For each node, let *x* and *y* represent the out–degree of sent mails with attached file(s) and the number of sent emails with attached file(s) to public email addresses not in AH list respectively. By collecting *x* and *y* for period of time *i* to *n,* we can tell that *x* and *y* are correlated if data of both variables form a linear function as below.

$$y_i = ax_i + b$$

Where *a* and *b* are constants and can be calculated as below.

$$a = \frac{(\sum x_i)(\sum y_i) - n(\sum x_i y_i)}{(\sum x_i)^2 - n(\sum x_i^2)}$$

$$b = \frac{(\sum x_i)(\sum x_i y_i) - (\sum y_i)(\sum x_i^2)}{(\sum x_i)^2 - n(\sum x_i^2)}$$

By using scattered diagram, the graph of x_i and y_i will form linear pattern as below.

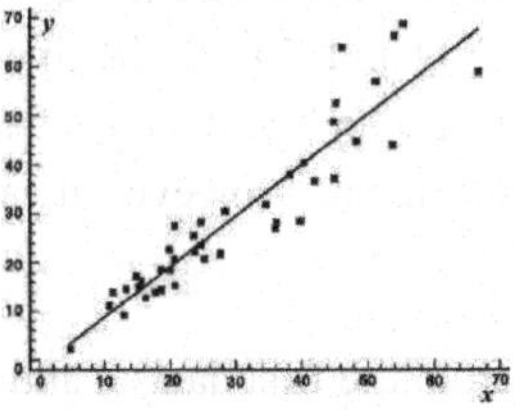

Figure 2 Correlation between x and y

In order to determine correlation between 2 variables, correlation coefficient, r, will be calculated. The r tells us how strong of the relation between 2 variables. If r of variable x and y are getting close to 1 or -1, it means that x and y are strong correlated. The r can be calculated by the below equation.

$$r = \frac{n\sum x_i y_i - \sum x_i \sum y_i}{\sqrt{n\sum x_i^2 - (\sum x_i)^2}\sqrt{n\sum y_i^2 - (\sum y_i)^2}}$$

For our study, we processed 2 datasets to see whether their patterns are similar or not. The first dataset was 10 weeks of weekly email log from one of organization email system (Microsoft Exchange 2007). The other dataset was monthly Enron Corpus email data which is a plain-text and we have to convert that plain text to graph-related data before using. We select 7 nodes from both sets with the out-degree "x" and out-degree to public email "y" and calculate r from all parameters above. The results are shown in Figure 3. Please note that we have changed name of employees to secure their privacy. Employees "A" to "G" represent the dataset from the studied organization, and employees "Z" to "T" represent the dataset from Enron email data.

Employee -->	A		B		C		D		E		F		G	
Week# (i)	x_A	y_A	x_B	y_B	x_C	y_C	x_D	y_D	x_E	y_E	x_F	y_F	x_G	y_G
1	141	10	101	34	80	20	73	8	40	5	56	30	153	25
2	135	15	119	40	115	9	74	3	38	5	47	24	143	7
3	102	16	97	33	118	8	62	9	61	14	71	35	128	30
4	142	5	66	25	93	3	51	7	58	30	82	39	141	36
5	122	8	73	27	99	18	50	7	38	12	78	37	121	37
6	106	18	82	29	87	20	70	5	53	27	70	32	121	21
7	142	27	77	20	103	10	50	10	42	9	55	27	153	31
8	107	6	90	28	109	17	81	15	33	4	49	24	148	32
9	116	6	68	26	118	25	71	13	56	15	63	33	147	33
10	101	20	87	22	120	4	68	5	50	12	81	42	150	23
r (Correlation Coeff.)	-0.0445		0.8075		-0.2805		0.1850		0.7369		0.9665		-0.0962	

Employee -->	Z		Y		X		W		V		U		T	
Month#	x_Z	y_Z	x_Y	y_Y	x_X	y_X	x_W	y_W	x_V	y_V	x_U	y_U	x_T	y_T
1	11	0	11	1	16	15	8	1	13	1	8	0	36	0
2	4	2	3	0	19	11	17	3	11	0	6	5	4	2
3	31	10	18	0	5	5	2	1	2	1	8	0	5	0
4	123	20	38	1	10	4	20	2	12	0	4	1	3	0
5	23	2	6	0	10	2	1	0	11	0	2	2	5	1
6	614	8	8	0	10	5	20	4	7	0	1	1	12	0
7	792	18	8	0	4	3	10	3	4	0	8	0	6	0
8	1444	66	31	0	12	7	26	12	5	0	2	0	11	0
9	486	41	15	0	23	4	22	8	4	0	1	0	7	0
10	335	19	12	1	8	2	25	3	31	0	1	0	2	0
11	106	6	0	0	29	19	26	14	15	0	1	1	7	0
12	100	26	28	3	3	3	33	6	30	1	12	1	5	2
r (Correlation Coeff.)	0.7917		0.4763		0.7399		0.6807		0.0186		0.0417		-0.2190	

Figure 3 Correlation Coefficient of x and y for selected employees of both datasets

4. What we found and discussion-

As the result in Figure 3 show, we have found is that some employees send file(s) to public email addresses not in AH but their *r* are not quite high. This means that it is not the habit. For example, A, C, D and G from the first dataset and V, U and T from Enron dataset have their | *r* | lower than 0.3. It shows that sometimes they send file(s) to public email address which deem bad. However, it cannot be claimed that this is a bad behavior from them.

On the other hand, F, B and E from the first dataset and Z, X and W from Enron dataset have been sending to public email addresses not in AH regularly because their *r* is quite high. The organization should have them warned or take some action to further investigate their email.

The scattered graph of *x* and *y* of all selected employee nodes are presented in Figure 4 below. The graphs illustrate as we expected from the data in Figure 3. A, C, D, G, V, U and T have their graphs scattered while B, F, E, Z, X, W have their graphs in mostly linear shape. Both graphs also show that, in an organization, there will be some employees who have bad habit of sending file to public email address outside the organization.

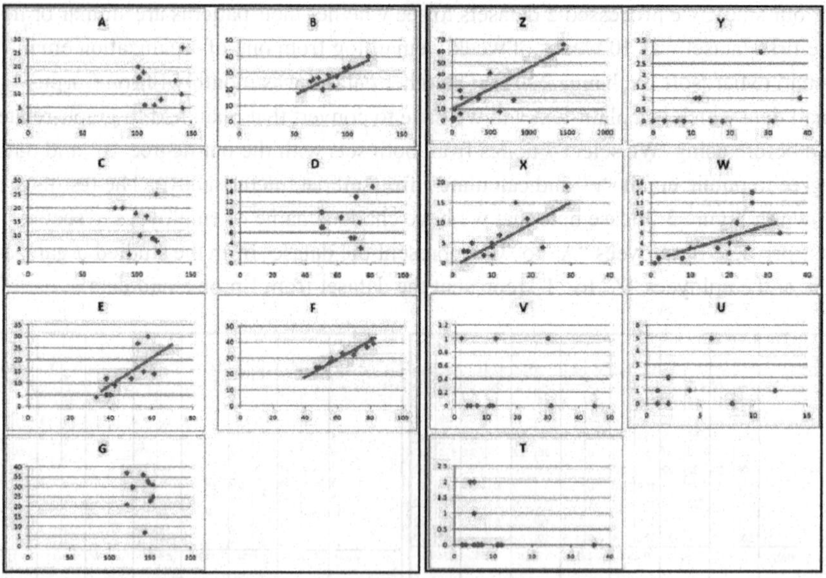

Figure 4 – Scattered Diagram of *x* and *y* of selected employees of both datasets

However, there are some issues that we need further discussion and exploration.

- The lower *r* value of and employee does not always mean that they are doing the right thing. Their email usages are still unconformity so the organization should consider taking some action regarding this.
- There should be other parameters that can be explored from email log of an organization. For example, the cluster of employee nodes and non-employee nodes will help us find the people outside the organization that they have been contacting.
- More parameters are better for more accurate analysis on employees' behaviors.
- Collecting information for the study should be planned carefully so that unnecessary works could be limited. In this study, the information from the studied organization was well recorded in the format that helped analysts work on their study. On the contrary, Enron Corpus mail data was in a pure plain-text format that needed to be extracted, cleansed, and reformatted before studying. Its mail headers were also from very old version of email standard before MIME. This means that we could not accurately identify which emails were with attached file(s). Analysts who plan to use Enron email data should be aware and prepared for data cleansing.

5. Conclusion

The study in this paper shows that we can find bad behavior by exploring existing information of the organization using graph theory and social network analysis. We have found that there are more behaviors that can be explored via mathematical models using graph theory and social networks. The results of this study help analysts to find parameters of the relationship model for organization graph so we can determine risk score of the organization, and evaluate overall risk of the organization. It is the important part to prove that we can integrate social network analysis with attack graph analysis, and determine risk of organization from user aspect. We come up with the proposed methodology, SN-AG, for further studying.

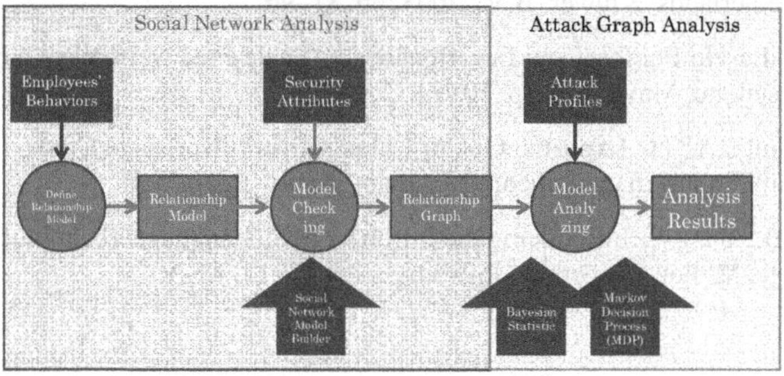

Figure 5 - The SN-AG approach

From the proposed approach, we create the model and the graph using social network analysis approach instead. Second, we add employee's behavior and security attributes from available information in organization email environment such as hub-like user node, virus/malware ratio, and spam ratio. The rest of the methodology stays the same.

Information from email service system can be used for generating a social network because it provides necessary information that social network analysis require such as relationship of sender and receiver, properties of the individual nodes, properties of each relationship. The social network generated from email information can also include risk properties based on information from virus protection system. Computer asset system also provides linkage of personal computer assigned to individual employee.

6. References

[1] M. E. J. Newman, "The structure and function of complex networks," ArXiv-cond-Mat0303516, Mar. 2003.

[2] B. B. Madan, K. Go\vseva-Popstojanova, K. Vaidyanathan, and K. S. Trivedi, "A Method for Modeling and Quantifying the Security Attributes of Intrusion Tolerant Systems," Perform Eval, vol. 56, no. 1–4, pp. 167–186, Mar. 2004.

[3] J. O. Aagedal, F. den Braber, T. Dimitrakos, B. A. Gran, D. Raptis, and K. Stolen, "Model-based risk assessment to improve enterprise security," in Enterprise Distributed Object Computing Conference, 2002. EDOC '02. Proceedings. Sixth International, 2002, pp. 51–62.

[4] H. Ebel, L.-I. Mielsch, and S. Bornholdt, "Scale-free topology of e-mail networks," ArXivcond-Mat0201476, Jan. 2002.

[5] M. E. J. Newman, S. Forrest, and J. Balthrop, "Email networks and the spread of computer viruses," Phys. Rev. E, vol. 66, no. 3, p. 035101, Sep. 2002.

[6] J. R. Tyler, D. M. Wilkinson, and B. A. Huberman, "Communities and Technologies," M. Huysman, E. Wenger, and V. Wulf, Eds. Deventer, The Netherlands, The Netherlands: Kluwer, B.V., 2003, pp. 81–96.

[7] Albert-Laszlo Barabasi and Eric Bonabeau, "Scale-Free Networks," Scientific American, no. May 2003, pp. 50–59

[8] Mandiant's APT1: Exposing One of China's Cyber Espionage Units. http://intelreport.mandiant.com/Mandiant_APT1_Report.pdf

[9] Klimt B., Yang Y. Introducing the Enron Corpus. Language Technology Institute, Carnegie Mellon University, Pittsburgh, PA 15213, USA.

Using PKI to Provide Credential Delegation in non Web-based Federations

Daniel Kouřil, Marcel Poul, Michal Procházka

CESNET z.s.p.o.,
Zikova 4, 160 00 Praha 6, Czech Republic
first.last@cesnet.cz

Abstract. Authentication is basic functionality required by most services that provide access to protected resources or personalized content. In order to authenticate to services users maintain sets of credentials that they use to prove their identity. Credential delegation allows users to seamlessly access multiple services across the network. The concept manifested their utility in the scope of single domain authentication mechanisms. Therefore, emerging identity federations are expected to provide similar functions, too. Recently, various non web-based federation models have emerged, unfortunately they do not cover properly delegation of credentials. In this paper we introduce a mechanism utilizing digital certificates and PKI, which provides support for credential delegation in non web-based federations. The viability of the concept is demonstrated on integration of the mechanism with the Moonshot federation framework. However, the solution forms an independent middleware layer that can be used by several federation models.

1 Introduction

Authentication and access control are two crucial security functions that are needed by virtually all services on the contemporary Internet. Both of them have attracted a lot of attention and plenty of solutions exists that provide support for them. Nevertheless, there are still areas requiring further notice, especially with regards to new technologies that are emerging nowadays. In this paper we focus on delegation of credentials in the scope of identity federations, aiming specifically at non web architectures.

Credential delegation is a mechanism to transfer authentication credentials of a user so that that the transferred credentials can be used to authenticate their owners as if they were the original credentials. A delegated credential can be further delegated to facilitate secure access to services at multiple locations. It is common to restrict the applicability of delegated credentials to limit the impact of abusing credentials that were compromised.

In this paper we focus on the domain of identity federations, especially on non web-oriented architectures that have been emerging. Provisioning of credential delegation is insufficient in federated mechanisms, which either neglect

K.J. Kim (ed.), *Information Science and Applications*,
Lecture Notes in Electrical Engineering 339, DOI 10.1007/978-3-662-46578-3_62

the problem or support just a generic framework lacking a particular delegation mechanism that could be profiled for real deployment.

There are two basic ways how support of credential delegation can be enabled in existing federation mechanisms. A straightforward solution is to provide an ad-hoc support for particular mechanism. It is relatively easy to accomplish, however, such adaptations would have to be done independently for every single mechanism. Another option is to provide a more generic solution that can be easily integrated with existing federated mechanisms.

In this paper we address the lack of delegation support in federated mechanisms by providing a general middleware layer. We decouple the functions for credential delegation from actual authentication mechanisms, which makes them possible to be used independently. The solution forms itself a middleware layer providing delegation, which can be reused with multiple federation frameworks. The viability of the solution is demonstrated on integration with the Moonshot federated protocol suite.

2 Delegation in non web-based federations

This paper aims primarily at identity federations and expects their common arrangement, with service providers providing access to services and identity providers acting as authorities that perform user's authentication and issue assertions claiming user's characteristics. The concept is quite common nowadays and several middleware suites exist nowadays to implement identity federations.

There are a number of scenarios demonstrating the need for credential delegation and SSO in a federated environment. The typical example comes from the area of high-performance computing where user submit batch jobs that need to access large datasets locates at remote storages. In order to retrieve the data the computing jobs needs to authenticate using its owner's credential. Ideally the credentials are delegated during the submission process.

2.1 Delegation tokens

A common mechanism to provide delegation of credentials is to introduce a notion of *delegation tokens* that are used instead of the main credentials. The token must prove the authentication of its owner in the same way as the original credentials but its usage is limited to limit damage caused by its potential abuse. Tokens are used for delegation to another services or users. A token can be created as a result of previous delegation and can be delegated further.

Delegation tokens must provide a few basic functions to be suitable for delegation of credentials. In terms of identity proving, they must provide the same level of assurance as the original credential and should be issued by the same identity provider. Especially tokens cannot be forged without access to the credentials. Delegation tokens should be limited in time and/or scope for which they are permitted to be used. Tokens are created on user's request and another token can be derived from an existing token, providing additional delegation levels to another service. Tokens can be used without user's explicit intervention.

2.2 Delegation in existing mechanisms

Delegation tokens are often provided as part of a particular authentication schema, for instance Kerberos, X.509 proxy certificates, SAML2 and OAuth2, however all of them have some limitations.

The most recognized one from the non-web oriented domain is Kerberos [9], which addresses area of a single administrative domain. Kerberos supports delegation, but the functionality is limited to a single realm that is managed by the authentication instance. A well known example is Microsoft's Active Directory (AD) which is based on Kerberos additionally supporting restricted delegation. Deploying Kerberos in a large scale is practically impossible due to the centralized authentication component (KDC).

In the grid community X.509 proxy certificates [12] are used for delegation, and they are much more suited for a federated environment. Nevertheless, X.509 proxy certificates are not suitable for end users due to complicated usage and limited support in client applications.

Current web-based federations using protocols SAML2 [3] or OAuth2 [6] usually employ time-limited HTTP cookies as authentication tokens, but delegation is supported only by few of them. OAuth2 is currently one of the most often used protocol for delegation, however the delegation can be done only for the first service, subsequent delegation is not supported. In OAuth2 after the successful authentication with the user's identity provider user gets delegation token which is transferred to the end service. Additionally the user can specify what kind of information will be available for the service which receives the delegated token. SAML2 protocol which is widely used in academia has build-in support for delegation, but it is not commonly used due to non trivial deployment and configuration. SAML2 *Enhanced Client Proxy* (ECP) defined in the SAML2 profile is designed for non-web usage of SAML2 together with delegation support, but missing support on most client applications and libraries is the main reason why it is not widely used. ECP version 2.0 has been recently standardized [4] which strengthens security and part of the specification is available in latest versions of middlewares[1]. Both SAML2 and OAuth2 requires initial authentication to be done using web-based tools, therefore support for non-web applications is very limited.

None of them was convenient for the delegation support in the non web domain or they were too complex to integrate. Fortunately, all of the mentioned authentication schemes support transport of additional information from the identity provider to the service, which can be used to integrate delegation support.

In order to address the lack of delegation support we have designed a solution providing generic delegation tokens. The X.509 [5] format was chosen as a suitable mechanism to support delegation. The design of the method is described in the following section.

[1] `https://shibboleth.net/`

3 A PKI-based middleware layer for delegation

To introduce support for X.509 it is necessary to provide a public key infrastructure to manage certificates and their life-cycles. The designed PKI maintains short-lived X.509 certificates that are used for subsequent authentications of clients with their identity providers. Users use the certificates and corresponding keys instead of their long-lived credentials that do not leave their desktops. The creation of a new certificate is governed by the IdP that provides the delegation certificate to the service during standard authentication of the user. Resulting certificates serve as a replacement of the standard, long-lived credentials of users.

X.509 certificates along with the appropriate private keys form delegation tokens fulfilling the requirements. Their management is completely transparent for the users who do not need to be aware of the credentials at all and are not exposed to usual certificate and key management issues. Private keys are not encrypted and can therefore be used transparently by applications.

It is a good security precaution that delegated credentials are restricted in order to prevent from damage caused by stolen or misused delegated credentials. The certificates produced by the PKI are therefore only short-lived, limited to couple of hours of lifetime. Beside the short lifetime, it is also possible to add additional restrictions specifying where the credential is allowed to be used. The user can describe the allowed usage and bound this limitation to the credentials. In our PKI we use X.509v3 critical extensions to convey the limitations, similarly to the concept of X.509 proxy certificates [13]. There are several languages to express the policy, such as ODRL [8] or XACML [1].

To issue X.509 certificates in our concept we use an on-line CA that signs certificates on-demand and is available all the time. Even though the concept breaks some conservative notions of PKI, it is well established in the domain of grid computing and brings several advantages.

In our concept the on-line CA is installed along with the identity provider and provides a mechanism to express the assertion claimed by the IdP in the X.509 format. If requested in the authentication request, upon successful authentication of a user the CA issues certificates to service that the user is authenticating to.

The IdP in turn accepts authentication using these certificates, if presented by the clients as if they were standard, long-lived credentials that are owned by the user. In this way it is possible for a user to delegate their credential to the services.

The design requires that certificate requests and final certificates are transported between the IdP and service. Support for such a transport must be provided as part of the actual protocol that is being extended.

Compared to general-purpose PKIs, our PKI is simpler, which streamlines its deployment and does not increase operations cost. For instance, because of the short lifetimes, the CA do not run any revocation service. Relying party is only the IdP, which also eases operations.

Even though the CA certifies the users in its constituency, it actually does not constitute a new trusted service. From the operations standpoint the CA is a module of the IdP that transforms assertions to the X.509 format. That fact is important to comprehend since the introduction of the CA does not change the main trust relationships in the federated arrangement as the certificate are only consumed by the IdP itself.

X.509 is widely supported in various middleware and applications, which makes it easier to enable its support in existing application and protocols.

An integration with a particular federation middleware is shown in the next section.

4 Credential delegation for Moonshot

The main goal of the Moonshot project [7,10] is to develop a generic interface for access to end services, using the identity federation approach. The primary focus of Moonshot is aimed at non-web based environments, however its results can be utilized in classic web-based federations, too. The main goal of the project is to design a mechanism that combines the EAP [2], RADIUS [11] protocols and SAML. Although authentication of users is well covered by Moonshot, a suitable delegation mechanism is not supported.

4.1 Moonshot architecture

Moonshot is built upon well known and proven mechanisms for user authentication and authorization. It leverages from well-tested approaches and systems that have been operated for a long time. The first pilots relied on the Eduroam authentication fabric, which provided a scalable hierarchy of authentication servers.

In Moonshot, the EAP protocol is used for the user authentication. From non-web identity federations, Moonshot adopted the user's attribute distribution mechanism based on the SAML. The language is used to carry the assertions about the user made by the IdP to the federated service. Authorization based on the SAML assertions is hence available in Moonshot and application can make use the same attributes that are distributed by web identity providers.

Moonshot benefits from the usage of EAP since the user's credentials can be safely delivered to the authentication server inside the RADIUS protocol. EAP is a widely used protocol that is independent on the underlying transport protocol and supports many authentication mechanisms such as TLS and various password-based methods. Moonshot therefore supports as many authentication mechanisms as EAP does.

The RADIUS protocol serves for the authentication, authorization and accounting. In Moonshot, federated services adopt the role of the RADIUS clients. The home RADIUS server (user's Identity Provider), is chosen upon the user's realm information in RADIUS attributes in the message. Servers are connected in a hierarchical manner so that each of them knows its parent and child. RADIUS protocol can carry different types of messages and when needed it can be extended to convey additional attributes, encoded as *Attribute Value Pairs* (AVPs).

The RADIUS infrastructure and protocol provides the actual federation framework connecting identity providers of multiple institutions. It is used for the credentials delivery from the service to the authentication server as well as the authentication result and user's attribute delivery back to the service.

In the rest of the section we describe how the Moonshot framework was extended to support delegation using PKI.

4.2 Changes to the protocol and profile

We have applied several enhancements on the level of the RADIUS protocol. The integration did not require any significant changes to the core protocol and all adaptations are easily achieved by using custom AVP attributes and appropriate fields of conveyed in SAML messages.

To allow the client to request credential delegation we have introduced two AVP attributes, `Request-Delegation` and `Delegation-Policy`. The former is meant to bear a binary indicator whether delegation should be performed or not, and the latter contains optional delegation policy restricting the usage of the delegated credentials.

In our pilot implementation we have chosen a simple language to enumerate IP addresses. By specifying a list of allowed IP addresses the user can determine the set of machines from which it is allowed to use the delegated credentials. Authentication attempts originating from other machines would be immediately rejected by the RADIUS server. Instead of IP addresses it is also possible to use other kinds of identifications of the servers, namely the authentication service names.

In order to provide a means how a certificate request can be carried from the service to the RADIUS server, we added another AVP attribute called `Certificate-Request`. The attribute is supposed to convey a certificate request in the standard PKCS #10 format.

Final certificates issued by the IdP CA are sent as part of the standard SAML attributes that the RADIUS server encloses to the final reply to the service. The certificate is accommodated in a dedicate SAML field of the response and thus is available as a normal attribute.

Moonshot benefits from the use of the EAP and its authentication mechanisms, which includes TLS support. There were no additional changes needed to the authentication protocol with respect to the usage of the delegated certificate for authentication, simple client and server configuration was sufficient.

4.3 Changes to the Identity Provider

To integrate the PKI with the RADIUS server two crucial steps had to be accomplished. It was necessary to provide the CA support and also make sure that certificates issued by the CA can be used for authentication of clients, subject to possible delegation restrictions.

To provide the function of the CA we provided a set of tools using the OpenSSL command suite that implements manipulations with certificate requests and certificate signing. Using these tools we implemented a module for the FreeRADIUS server that is used to produce X.509 certificates during the authentication exchange. The module is invoked after successful authentication of a client and is passed authenticated username of the client. After it has been invoked, the module constructs an X.509 certificate structure, optionally inserts policy restrictions into the certificate extensions and signs the certificate using the CA signing key. If the client authentication was done using an already delegated certificate, the expiration time of the new certificate is set equally. In other cases, a lifetime of ten hours is used.

While support of authentication using X.509 was trivial, additional steps were required to check the restrictions imposed on the delegation token. Time-related restrictions of the certificate (expiration time) are checked in the standard authentication process but checking the delegation policy had to be done separately. We provided another module that is invoked by FreeRADIUS server during the credential verification step. The module evaluates the policy rules embedded in the certificate and matches them with the IP address of the server that was made available as a standard RADIUS attribute. If the IP address is not allowed in the list, the whole RADIUS request is denied.

4.4 Changes to the client and server sides

To finish the integration we extended the Moonshot implementation to enable client's utilization of the new features.

The server side of Moonshot API routines was extended to expose the delegated credentials to the calling application so that the credentials could be handled properly. We provide an implementation of routines that store the credential on the local filesystem, similarly to how Kerberos tickets or X.509 proxy certificates are stored.

On the client side we added support for utilization of the X.509 credentials if they are found on the local filesystem. Similarly to e.g. Kerberos an environment variable has been introduced (`MOONSHOT_X509_CRED`), which refers to a file keeping the credentials.

5 Conclusion

Both delegation of authentication credentials and support of the Single Sign-On principle are two very crucial functions that users and providers and services require. However, these functions are not provided by all mechanisms. Their lack is especially visible in the field of identity federations, especially non web-based.

In this paper we have introduced a general way how support for restricted delegation can easily be added to a range of existing mechanisms, without changing the way how they do authentication of users and services.

The designed mechanism is based on a public-key schema utilizing the X.509 format. Even though the approach introduces a PKI to transport the delegated information, it actually does not require any changes to the existing trust relationships among the peers.

The viability of the solution was demonstrated by its integration with the Moonshot middleware.

Acknowledgment

This work was supported by program "Projects of Large Infrastructure for Research, Development, and Innovations" (LM2010005) funded by the Ministry of Education, Youth and Sports of the Czech Republic. Support of the EU GN3+ project is also appreciated.

References

1. eXtensible Access Control Markup Language (XACML) Version 3.0. (2013), http://docs.oasis-open.org/xacml/3.0/xacml-3.0-core-spec-os-en.html
2. Aboba, B., Blunk, L., Vollbrecht, J., Carlson, J., Levkowetz, H.: Extensible Authentication Protocol (EAP). RFC 3748 (2004), http://www.ietf.org/rfc/rfc3748.txt
3. Cantor, S.: Assertions and Protocols for the OASIS Security Assertion Markup Language (SAML) V2.0 (2005), http://docs.oasis-open.org/security/saml/v2.0/saml-conformance-2.0-os.pdf, document ID saml-conformance-2.0-os
4. Cantor, S.: SAML V2.0 Enhanced Client or Proxy Profile Version 2.0 (Aug 2013), http://docs.oasis-open.org/security/saml/Post2.0/saml-ecp/v2.0/cs01/saml-ecp-v2.0-cs01.html
5. Cooper, D., Santesson, S., Farrell, S., Boeyen, S., Housley, R., Polk, W.: Internet X.509 Public Key Infrastructure Certificate and Certificate Revocation List (CRL) Profile. RFC 5280 (May 2008)
6. Hardt, D.: The OAuth 2.0 authorization framework. RFC 6749 (Oct 2012)
7. Howlett, J., Hartman, S.: Project Moonshot. Briefing paper for IETF 77, Anaheim (2010), http://www.painless-security.com/wp/wp-content/uploads/2010/03/moonshot-ietf-77-briefing-paper.pdf
8. Iannella, R., Guth, S., Pahler, D., Kasten, A.: ODRL V2.0 Core Model (2012), http://www.w3.org/community/odrl/two/model/
9. Neuman, C., Yu, T., Hartman, S., Raeburn, K.: The Kerberos Network Authentication Service (V5). RFC 4120 (Jul 2005)
10. Painless Security: Project Moonshot: Feasibility Analysis (2010), http://www.painless-security.com/wp/wp-content/uploads/2010/02/moonshot-feasibility-analysis.pdf
11. Rigney, C., Rubens, A., Simpson, W., Willens, S.: Remote Authentication Dial In User Service (RADIUS). RFC 2865 (2000), http://www.ietf.org/rfc/rfc2865.txt
12. Tuecke, S., Welch, V., Engert, D., Pearlman, L., Thompson, M.: Internet X.509 Public Key Infrastructure (PKI) Proxy Certificate Profile. RFC 3820 (Jun 2004)
13. Welch, V., Foster, I., Kesselman, C., Mulmo, O., Pearlman, L., Tuecke, S., Gawor, J., Meder, S., Siebenlist, F.: X.509 Proxy Certificates for Dynamic Delegation. In: Proceedings of the 3rd Annual PKI R&D Workshop (2004)

Modeling and Simulation of Identification Protocol for Wireless Medical Devices in Healthcare System Communication

Hyun Ho Kim[1], Ndibanje Bruce[1], SuHyun Park[2], JungBok Jo[2], Hoon-Jae Lee[2]

[1]Department of Ubiquitous IT, Graduate School of Dongseo University,
Sasang-Gu, Busan 617-716, Korea
[2]Division of Computer and Engineering, Dongseo University,
Sasang-Gu, Busan 617-716, Korea
feei_@naver.com
bruce.dongseo.korea@gmail.com
{subak,jobok,hjlee}@dongseo.ac.kr

Abstract. The digital era is changing the nature of health care delivery system with the integration of Information Technology's potential to improve the quality, safety, and efficiency of health care. Medical devices may be connected on wireless and wired networks and communicate with nearby receivers that are connected to landline networks, cellular systems or broadband facilities that provide more ubiquitous coverage of connectivity, allowing uninterrupted monitoring of patients in transit by accessing the Internet. This promise of universal connectivity allow data availability service from the high-end systems such as routers, gateways, firewalls, and web servers to the low-end systems such as smart phone, tablet, etc…Hence, security has become an essential part of today's computing world regarding the ubiquitous nature of the entities and wireless technology evolution. This paper presents a framework for Healthcare System Communication where wireless medical devices are challenged to an identification protocol procedure which enables negotiation between wireless medical devices and specify authorization requirements that must be met before accessing the network and patients 'data. To illustrate the feasibility of the work to real world protocols, we simulate the scheme using Scyther tool by applying the IEEE-802.16-2004 PKM protocols for WiMAX protocol standards.

Keywords: identification protocol, security, privacy, scyther

1 Introduction

We are witnessing an explosive growth in medical devices that use wireless technologies, some implanted and some worn on the body, to control bodily functions and to measure an array of physiological parameters. Implanted devices can control heart rhythms, monitor hypertension, provide functional electrical stimulation of nerves, operate as glaucoma sensors, and monitor bladder and cranial pressure. Patients no

K.J. Kim (ed.), *Information Science and Applications*,
Lecture Notes in Electrical Engineering 339, DOI 10.1007/978-3-662-46578-3_63

longer need to be tethered to one spot by a tangle of cables, creating a safer workplace for medical professionals and a more comfortable environment for the patient, with a reduced risk of infection. Wireless monitoring permits patients to thrive outside medical environments, reducing health care costs and enabling physicians to obtain vital information on a real-time basis without the need for office visits or hospital admissions. With this regards, security is primary concern to ensure the whole communication between the devices passing through networks systems. Following the problem statement, many novel challenges ore offered by the growth of the application's wireless healthcare offers, like, reliable data transmission, node mobility support and fast event detection, timely delivery of data, power management, node computation and middleware [1-2] .This paper presents a framework for Healthcare System Communication where wireless medical devices are challenged to an identification protocol procedure which enables negotiation between wireless medical devices. To illustrate the feasibility of the work to real world protocols, we simulate the scheme using Scyther tool by applying the IEEE-802.16-2004 PKM protocols for WiMAX protocol standards. The remainder of this paper is organized as follows: Section 2 presents the proposed solution. The analysis is done in Section 3 before concluding in Section 4.

2 Proposed Solution

In this experiment, the IEEE-802.16-2004 PKM standard [3] is used in order to satisfy the authentication and key establishment protocols. To achieve the step of mutual authentication protocol we have used the PKM version 2 RSA protocol [3]. The aim of this protocol is to establish a shared secret preliminary primary authentication key so called pre-PAK in WiMAW technology between the *WD* and *GW* node. The main components of the proposed protocol are described as following:

- Wireless Device (WD): We consider a wireless device such as smart phone, tablet, laptop… which can communicate using Wi-Fi technology.
- Gateway (GW): In this work we take in consideration a gateway, as a router or bridge to give access right to the wireless device to enjoy the network services

For convenience, Table 1 provides a list of some notations and symbols will be used throughout the rest of paper.

Table 1. Notation and Description

Notation	Description
MACid	Mobile Authentication Code id
WDid	Wireless Device id
GWid	Gateway id
SAID	Security Association Id
WDRand	Wireless Device Secret Number
GWRand	Gateway Secret Number
pPAK	Pre Primary Authentication Key
Sk	Secret Key
Pk	Public Key

2.1 Mathematical Concepts

The pPAK is utilized to obtain the authentication key from which the keys for hashed message authentication codes and symmetric encryptions are derived. The structure of the mutual authentication protocol is defined by the following exchanged messaged between the two entities: *Request, Reply* or *Reject* and *Acknowledgement*

Before going into the flow chart of the protocol, some mathematical concepts are described using the equations as follow:

$$WDid = \{wd, Pk(wd)\}_{Sk(MACid)} \quad (1)$$

$$GWid = \{gw, Pk(gw)\}_{Sk(MACid)} \quad (2)$$

The wireless device id is given in equation *(1)* where it is encrypted with the *MACid* using his secret key. In the same way the gateway id is encrypted using his identification parameters in equation *(2)*. As aforementioned, the mutual authentication is processed by the request message *(RqMssg)* as the first message sent by the GW to ask the right of network access; the mathematical expression is as follow:

$$RqstMssg = WDid, \{WDRnd, SAID, WDid\}_{Sk(wd)} \quad (3)$$

After receiving the request message from the WD, there are two possibilities from the GW; whether it replies (4) or reject (5) the request, from the sender. Those events are mathematically uttered as follow:

$$RpMssg = GWid, \left\{WDRnd, GWRnd, \{pPAK, wd\}_{Pk(wd)}, GWid\right\}_{Sk(gw)} \quad (4)$$

$$RjctMssg = WDid, \{WDRnd, GWRnd, GWid\}_{Sk(gw)} \quad (5)$$

In the last message the wireless device has to acknowledge to the gateway if or not he got a notification from the gateway. The equation (6) describes the mathematical relation of the message.

$$AckMssg = \{GWRnd\}_{Sk(wd)} \quad (6)$$

2.2 Exchanged Message Concept: Mutual Identification Process

The negotiation of security capabilities is done during this protocol between wireless device and gateway. Both devices can agree on unilateral or mutual authentication or no authentication at all, and on a key management protocols assortment. In this research we consider the mutual authentication protocol as well. In order to keep the traffic encryption keys updated, the key management protocols are periodically repeated to the entire authentication chain on a less frequent basis. Once the traffic encryption keys are established, user data protocols start. To avoid service interruptions, traffic encryption keys have overlapping lifetimes. The Figure 1 describes the messages handled by the protocol as expressed in the mathematical concepts. The details of the exchanged messages are given in subsection 3.1.

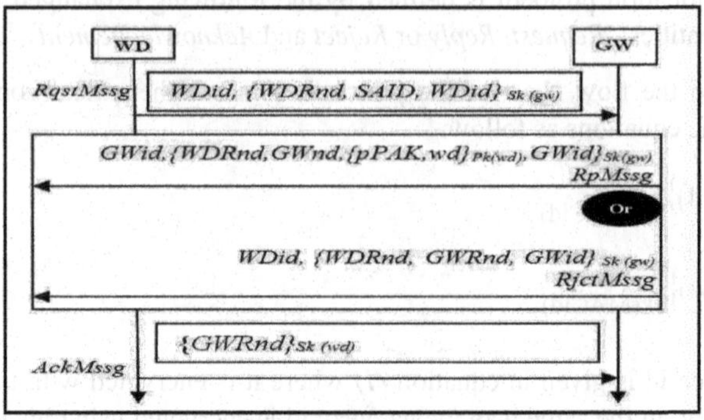

Fig. 1. Proposed Mutual Identification Process

2.3 Simulation Experiment with Scyther

This subsection presents the details of the simulation process by scyther tool. Scyther's input language is loosely based on C/Java-like syntax [4]. The main purpose of the language is to describe protocols, which are defined by a set of roles. Roles, in turn, are defined by a sequence of events, most of which send or receive terms. The technical requirements to achieve the simulation experiment are defined as follow:

- Hardware equipment: In our experiment we use a PC with Intel® Core™2 Duo CPU E8500 @ 3015GHZ, 3017GHz, and 4GB of RAM.
- Software tool: In this simulation we have used Microsoft Windows XP, Professional SP3. Scyther tool, the GraphViz library, Python, wxPython libraries

Step I. In order to execute the proposed protocol, the Python language v3.3.2 has been used. The IEEE-802.16-2004 PKM standard for WIMAX technology is followed in this research and then PKMv2 RSA is applied because of it is the mutual authentication protocol according to the standard of protocols. Figure 2 is the Scyther main window of the protocol, saved as "*Proposed_Check.sdl*". All the claims of the proposed protocol (*claim_muti3, claim_muti4, claim_muti5*) are correct for unbounded number of runs.

Step II. From the main menu "Verify" of Scyther tool, we verify the protocol and a report is generated in Figure 3 named *"Scyther results: verify"* where Scyther reports that is found at least one attack on those claims. In the proposed protocol, the attacks occurred in the GW device for *claim_mutr3* and *claim_mutr4* which are *Niagree* and *Nisynch* respectively.

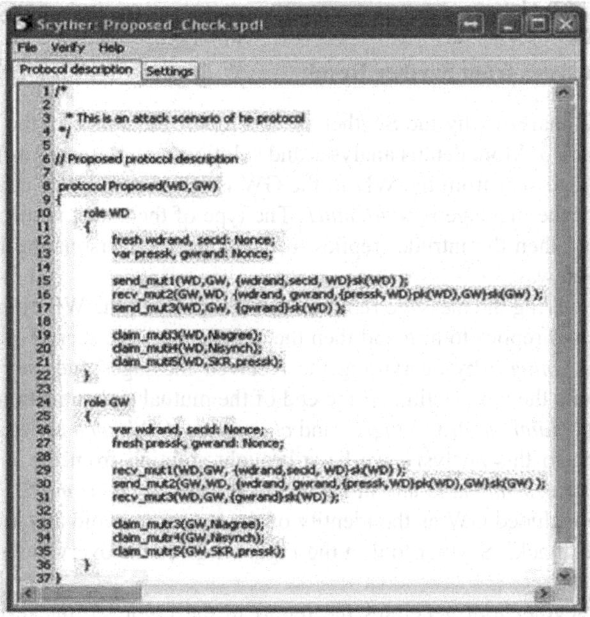

Fig. 2. Proposed Protocol Scyther Window with Python scripture

Step III. From the report in Figure 3, the tool gives us a possibility to check where the attack has been done. From the *Patterns column*, we can see that the proposed protocol suffer from two attacks in the GW device. For more details of the attacks we check the *Niagree* and *Nisynch* where the secrecy of the generated and received nonce is attacked, thus the non-injective agreement and non-injective synchronization are not verified, finally the *Nonce* has been compromised.

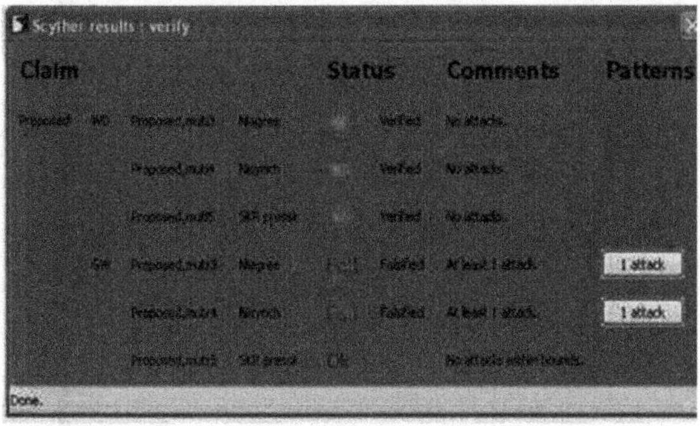

Fig. 3. .Attacked Scyther Result: Verification of the Proposed Protocol

3 Security Analysis

3.1 Formal Analysis from Scyther Result

In this Section we analyze why the Scyther system report the attack in the proposed protocol from Figure 3. More details analyses and solutions are presented as follow:

- The message sent from the WD to the GW is interrupted by an attacker; the content of the message is *send_mut1*. The type of the attack is man-middle–attack, and then the intruder replies to the WD by impersonating him to be the real gateway.
- After receiving the message replied from the attacker, the WD computes the message and replies to him and then the attacker gets the secrecy parameters *(secid and others)* by decrypting the received messages and encrypts it to impersonate the new victim. At the end of the mutual authentication, Scyther found that *claim_ mutr3: Niagree* and *claim_mutr4: Nisynch* are attacked.
- According to this analysis, we have brought solution from the WD in the sent message: *send_mut3* and in the received message: *recv_mut3.*
- We have included GW as the identity of the sender to avoid any kind of impersonate attack. Scyther tool in the Figure 4 gives the overview of the corrected code.
- Finally, Scyther tool generates the report in the Figure 5, the result shows that all claims now are verified.

Fig. 4. Verified Python Scripture's Protocol

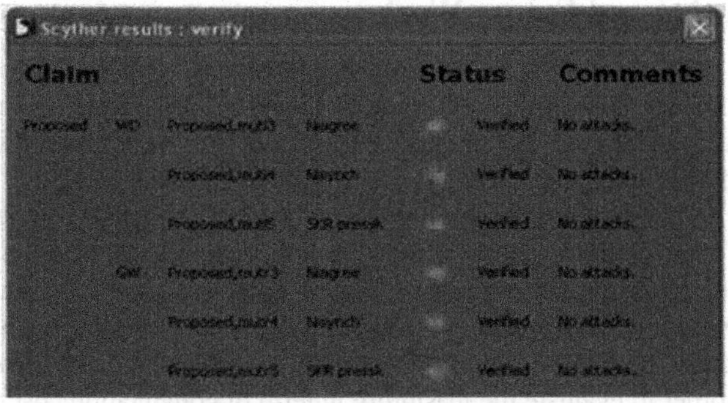

Fig. 5. Final result of the Verified Protocol

3.2 Formal Analysis with Security Properties Specifics Standards

Pseudonymity, information confidentiality, no theft of service possible [5], secrecy and uniqueness of the session keys are selected for our security analysis. The goals of the authentication protocol should be:

- *Pseudonymity:* This claim is fulfilled if an outsider, who keeps track of the communication, cannot relate the traffic to a specific WD. In order to fulfill pseudonymity the MAC of the WD which identifies it must remain secret. The formal definition of pseudonymity is given below.

  Property 1: *{claim (WD, Secret, MACid)}*

- *Information confidentiality:* This claim is fulfilled if the WD ensures that all exchanged messages are secret. The exchanged messages between the WD and the GW are designated by *RqstMssg*. Each information (σ) in *RqstMssg* should remain secret. The formalization of information confidentiality is given below.

  Property 2: $\forall \sigma \in$ *RqstMssg (claim (WD, Secret, σ))*

- *No theft of Service possible*: This claim is fulfilled if the GW has the guarantee that neither an unauthenticated user should gain access to the services provided, nor should an unauthenticated user be able to impersonate another user. A service should always be bound to an authenticated user. This claim is similar to information confidentiality. Its formal definition is given as follows:

  Property 3: $\forall \sigma \in$ *{RpMssg* $\vee$ *RjctMssg} (claim (GW, Secret, σ))*

- *Secrecy and uniqueness of the session keys:* This claim is fulfilled if the GW and the WD have the guarantee that all exchanged keys (described as key) are secret and unique. The formal of this security evaluation is described as follows:

  Property 4*:* $\forall$*key (claim (GW|WD, Secret, key))*

4 Conclusion and Future Work

This paper discussed an identification protocol for data and network security where wireless and gateway devices are mutual authenticated in a healthcare system. We have performed a simulation test using Scyther tool to verify the feasibility and security level of the protocol. From the results we saw that known attacks can be launched to the system. Thus, patients 'data and network are in danger. Within enhancement, the performance analysis has been done in view of those attacks and the finally result reveals that the protocol is efficient and resilient. In the future work, we plan to implement the proposed protocol using a wireless device following the WiMAX standard protocol in order to evaluate the feasibility in term of computation cost and power consumption.

Acknowledgment. This research was supported by Basic Science Research Program through the National Research Foundation of Korea (NRF 2014) funded by the Ministry of Education, Science and Technology. And it also supported by the BB21 project of Busan Metropolitan City.

References

1. Chung, W.Y.; Yan, C.; Shin, K. A Cell Phone Based Health Monitoring System with Self Analysis Processing Using Wireless Sensor Network Technology. In *Proceedings of 29th Annual International Conference on the IEEE EMBS*, Lyon, France, 23–26 August 2007.
2. Gravina R.; Guerrieri A.; Fortino G.; Bellifemine F. Giannantonio R.; Sgroi M. Development of Body Sensor Network Application Using SPINE. In *Proceedings of IEEE International Conference on Systems, Man and Cybernetics (SMC 2008)*, Singapore, 12–15 October 2008.
3. Sameni K., Nasser Y., Ali P., "Analysis of Attacks in Authentication Protocol of IEEE 802.16e." *International Journal of Computing and Network Technology* 1, No. 1, 33-44, @ 2013 UOB SPC, University of Bahrain.
4. Cremers. C.J.F., Scyther: Automatic verification of security protocols. http://www.win.tue.nl/~ccremers/scyther/.
5. E. Kaasenbrood, "WiMAX Security - A Formal and Informal Analysis," Master's thesis, Eindhoven University of Technology, Department of Mathematics and Computer Science, Groningen, Netherlands, August 2006.

A Method for Web Security Context Patterns Development from User Interface Guidelines based on Structural and Textual Analysis

Pattariya Singpant, Nakornthip Prompoon

Software Engineering Lab, Center of Excellence in Software Engineering
Department of Computer Engineering, Chulalongkorn University, Bangkok, Thailand
Pattariya.S@student.chula.ac.th, Nakornthip.S@chula.ac.th

Abstract. Currently, only a small number of user agents present information on the web security context to the user in an easy way for understandability. W3C has created WSC-UI documents as a security suggestion standard for web security context. The application in designing user agents to be secure requires human resources in identifying specifications, which takes much time and expense, and may also result in incompleteness. Security patterns have been used to collect solutions to recurring problems. Therefore, this research proposes a method for creating web security context patterns, based on WSC-UI documents, and identifying the relationship structure of the patterns. The proposed patterns are validated and refined according to the initial validation list. The developers can specify the security requirements based on the proposed patterns according to the specified application approach, for the benefits in designing a user agent to be aware of the web security context.

Keywords: Web Security Context Patterns; Web User Agents; Structure and Textural Analysis

1 INTRODUCTION

Web security context is very important to the user's trust in the system while using a web user agent, which is an intermediate way for accessing web content. Most display information relating to the web security context to the user; however, only a small part is understood by the user. Also, it may cause confusion if the user decides to trust a forged website, and provide sensitive information to the forged website. When the data winds up in the possession of thieves, it may cause damage to the user information. Therefore, it is highly important to consider the web security context displayed via the web user agent before the system design. In order to support security in the web context, organizations related to web development have created documents [1, 2] collecting best practices for developers to use as a standard in system design and development. The Open Web Application Security Project (OWASP) [2] proposed technical practices to support development using the J2EE and .NET frameworks, while the World Wide Web Consortium (W3C) [1] proposed recommendations for user interface design with consideration of security contexts, from the study of best practices in us-

K.J. Kim (ed.), *Information Science and Applications*,
Lecture Notes in Electrical Engineering 339, DOI 10.1007/978-3-662-46578-3_64

age. For appropriateness in applying them to specify requirements, recommendation documents should not provide limits on programming languages used for development

The recommendations in the documents were collected in a natural language format. Their usage requires analysts and system designers to analyze the documents to extract requirements appropriate for the system being developed. Such methods require time and resources, resulting in high expenses, and the risk of incompleteness. Researches [3, 4] use the recommendation documents for identifying system requirements. However, those formats do not provide enough detail for system analysis and design, making it difficult for them to be used directly, while the security patterns play a role in collecting security solutions [5]. Therefore, applying security patterns in collecting best practices from recommendation documents is challenging.

This research presents a method for creating web security context patterns by analyzing the content and structure relation from the WSC-UI recommendation documents, along with approaches in applying the proposed patterns.

The content of the paper is as follows: Section 2 explains background related to the research. Section 3 explains the concepts and research method, starting by specifying the elements for researching the content of the pattern and analyzing the relationship structure within and between patterns, as well as pattern validation. Section 4 explains the application of the patterns. Finally, Section 5 summarizes the results and approaches for further development.

2 Background and Related Work

2.1 Security Patterns

Security patterns [5] have the purpose of collecting solutions related to requirements collected from practices and methods relating to security. The patterns creation supports the concept of reuse. The pattern creation activity [6] begins by specifying the Knowledge Sources, Knowledge Synthesis and Pattern Identification methods, Pattern Representation, Pattern Refinement, and lastly, Pattern Application. The activity depends on the context of the problem, whereas the pattern representation step has important characteristics that most research works on security pattern modification according to the context of the research. The popular characteristics are those presented in [5]. The key pattern elements that reveal the characteristics of a security pattern are represented by the pattern template as follows: Name, Also Known As, Example, Context, Problem, Solution, Structure, Dynamics, Implementation, Example Resolved, Variants, Known Uses, Consequences, and See Also.

To facilitate understanding the overall context of the patterns, appropriately organizing the security patterns will help the user apply the patterns, understand their functions and relationships between patterns. The simplest method for organizing the patterns is patterns classification based on domain concepts [7], for use in categorizing the patterns, in the context of Web Core Security Services [2]. A popular categorization is CI4A, Confidentiality, Integrity, Authentication, Authorization, Availability, and Accountability. Categorization requires considering overlap between groups, as well as the cooperation and relationships between patterns [8].

Applying patterns from the initial step of software development, during the specification of software requirements for design and development helps reduce the errors and gaps in the developed system through best practices. However, most specification of requirements from patterns and reuse is still in a cut-and-paste process [9]. Therefore, it is a challenge to create a tool that assists in specifying requirements from patterns [10] to allow patterns to be applied effectively.

This research proposes patterns as a collection of practices and categories the patterns according to the core security services classes, considering the application of the patterns in specifying system requirements.

2.2 Web Security Context: User Interface Guidelines

The source document for knowledge used in developing patterns for this research is Web Security Context: User Interface Guidelines (WSC-UI) [1] proposed by W3C, which considers the security context from the user interface design, by collecting best practices from the observation of user behavior, in order to control and maintain security and trustworthiness in using the web. The approach does not have limits on development languages, so it is appropriate for use in specifying requirements. There are four sections as follows:

- Section 5: Applying TLS to the Web, referring to the use of Transport Layer Security (TLS) in specifying the security level for communicating data between the client and the server through HTTP transactions.
- Section 6: Indicators and Interactions, the specification of the data properties of the identity signal, indicator, and interactions with the users through various alerts.
- Section 7: Robustness Best Practices, to avoid system actions that resemble attacks.
- Section 8: Security Considerations, recommendations on security to prevent unwanted events that may occur to the system.

The Product Classes referred to in the WSC-UI document consist of Web User Agents, Extensions which call operating system functionality, and Assistive Technologies, which may be installed on a desktop, mobile phone, or multimedia player.

The WSC-UI document has a Markup Language structure, with Hyper to documents to accompany analysis, such as scope defining document: Web Security Experience, Indicators and Trust: Scope and Use Cases (WSC-USECASES) [11] identifies use cases that require considering the web security context. This research will use the aforementioned document in explaining example problem situations.

3 Proposed Method

The research process for creating web security context patterns from the WSC-UI documents can be separated into 4 steps, starting with analyzing the content of WSC-UI documents , identifying the relationships of internal pattern elements, categorizing the topics of the content, and lastly, verifying the results with experts, as shown in **Fig. 1**, the overall context of the research.

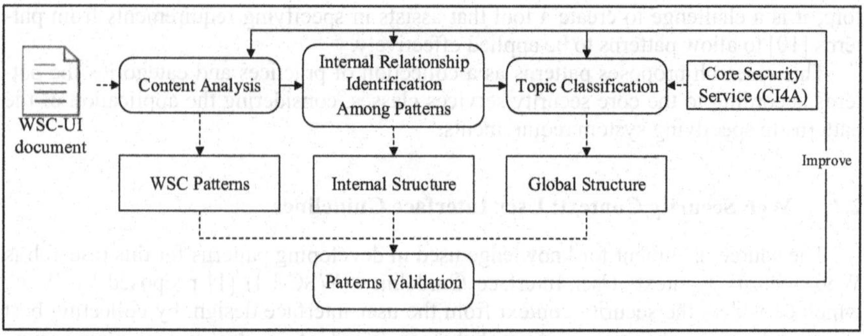

Fig. 1. The proposed method overview

3.1 Content Analysis

Creating a web security context pattern requires defining a structure to obtain a format for collecting solutions from analyzing the content of the WSC-UI documents. Pattern elements cam be categorized, based on the sources of the content, into two groups; 1) Elements extracted directly from the content of the WSC-UI documents, and 2) Elements required consideration from other documents.

The elements extracted directly from the contents of the WSC-UI content are as follows: Section, ID, Name, Description, Context, Solution, See Also, and the elements requiring consideration, from the WSC-USECASES documents are as follows: Example, Problem, Example Resolved, Known Uses, Variants, Consequences.

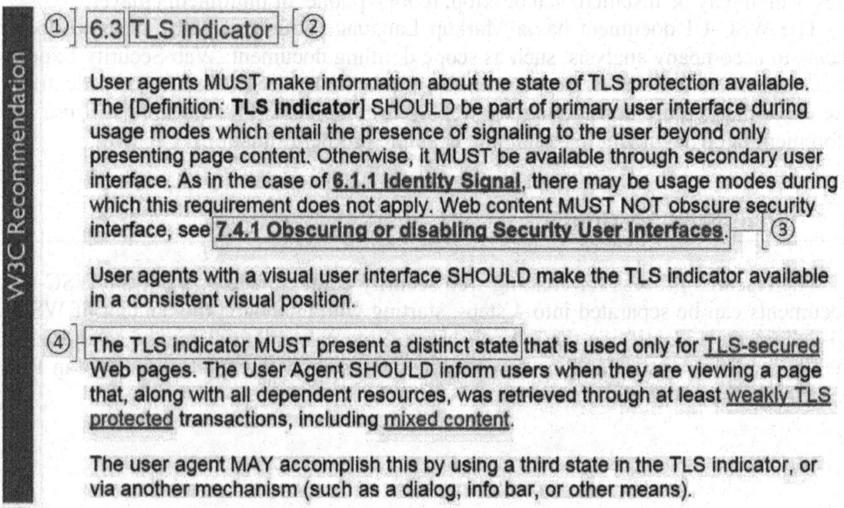
W3C Recommendation
① 6.3 TLS indicator ②
User agents MUST make information about the state of TLS protection available. The [Definition: **TLS indicator**] SHOULD be part of primary user interface during usage modes which entail the presence of signaling to the user beyond only presenting page content. Otherwise, it MUST be available through secondary user interface. As in the case of 6.1.1 Identity Signal, there may be usage modes during which this requirement does not apply. Web content MUST NOT obscure security interface, see 7.4.1 Obscuring or disabling Security User Interfaces. ③
User agents with a visual user interface SHOULD make the TLS indicator available in a consistent visual position.
④ The TLS indicator MUST present a distinct state that is used only for TLS-secured Web pages. The User Agent SHOULD inform users when they are viewing a page that, along with all dependent resources, was retrieved through at least weakly TLS protected transactions, including mixed content.
The user agent MAY accomplish this by using a third state in the TLS indicator, or via another mechanism (such as a dialog, info bar, or other means).

Fig. 2. Content on defining the TLS indicator [1]

The WSC-UI content in the topic of 6.3 TLS indicator as shown in Fig. 2 is the case study for the proposed methods.

Table 1. Elements and acquisition of web security context pattern content

Element	Description
Class	Identify core security service through topic classification.
Section	Category of the content used to create pattern, identified from the section topicof the WSC-UI document.
ID	Referencing the number of the topic. The letters "WSCP" (Web Security Context Pattern) in front of the numbers. For example, In Fig. 2 ①, when converted to a reference number, becomes WSCP63.
Name	The text following the reference number. For example, in Fig. 2 ②, TLS indicator, will be used as the heading name.
Description	Brief details on what the content of the pattern is relevant to.
Example	An example situation for the problem, from the WSC-USECASES document, in this case corresponding with CASE7.
Context	Extracted from descriptions for the state of the system on which regulations will be enforced.
Problem	Identify the problem, or the question, analyzed from the WSC-USECASES document.
Solution	Text identifying the level of obligation to follow, using these words: MUST, SHOULD, MAY, MUST NOT, SHOULD NOT. Extract each sentence where such words appear, such as "The TLS indicator MUST present a distinct state, as shown in Fig. 2 ④.
Structure	Identify entities and relations obtained from the analysis.
Example Resolved	Discussion on the situation when the problem has been solved using the pattern.
Variants	Explain other attributes of relationships between entities for implementation.
Consequences	Benefits from applying the patterns. Can be studied from WSC-USECASES documents.
See Also	Other contents appearing or referenced within the regulations, as words or groups of words with Hyperlinks, such as, in Fig. 2 ③, the text "See 7.4.1 Obscuring or disabling Security User Interfaces."

The method for extracting the content and the structure of the recommendation documents into the requirement pattern format, in order to display the elements, their original and applications of each element, has criteria according to Table 1, summarizing the acquisition and application of pattern elements.

Content analysis can be split as follows, starting with elements that can be extracted directly from WSC-UI documents. The elements that require analysis of the structure and the relationships between entities will then be analyzed subsequently. In order to clarify, the WSC-UI content relating to defining the TLS indicator as shown in Fig. 2 will be used as an example of input to the process of extraction and pattern element categorization according to Table 1, obtaining results shown in Table 2. The content from defining the TLS indicator has had its content extracted into pattern elements. The results obtained will have their structure and relationships analyze subsequently, for use in validation and pattern application.

Table 2. Proposed TLS indicator Pattern

<table>
<tr><th>Class</th><td>Availability</td><th>Section</th><td>Indicators and Interactions</td></tr>
<tr><th>Name</th><td>TLS indicator</td><th>ID</th><td>WSCP-63</td></tr>
<tr><th colspan="4">Description</th></tr>
<tr><td colspan="4">This pattern describes how to define state and mechanism of TLS indicator for the usage mode of web user agent by TLS protection.</td></tr>
<tr><th colspan="4">Example</th></tr>
<tr><td colspan="4">Example Inc. has a popular online service that processes many credit card transactions a day. Betty occasionally uses the service and trusts it with her credit card information. Malcolm is a thief with an idea. He creates an imitation of the Example web site and begins directing users to it. He's also given his imposter site a domain name that is just a typo away from Example's authentic web site, so some victims will arrive by accident. Betty is about to enter her credit card information into a site that looks just like Example's.</td></tr>
<tr><th colspan="4">Context</th></tr>
<tr><td colspan="4">User agents with a visual user interface during usage modes, there may be usage modes during which this requirement does not apply.</td></tr>
<tr><th colspan="4">Problem</th></tr>
<tr><td colspan="4">How does the user know if it's the authentic site, or the imposter?</td></tr>
<tr><th colspan="4">Solution</th></tr>
<tr><td colspan="4">• User agents MUST make information about the state of TLS protection available.
• The TLS indicator SHOULD be part of primary user interface during usage modes which entail the presence of signaling to the user beyond only presenting page content. Otherwise, it MUST be available through secondary user interface.
• User agents with a visual user interface SHOULD make the TLS indicator available in a consistent visual position.
• The TLS indicator MUST present a distinct state that is used only for TLS-secured Web pages. The User Agent SHOULD inform users when they are viewing a page that, along with all dependent resources, was retrieved through at least weakly TLS protected transactions, including mixed content</td></tr>
<tr><th colspan="4">Example Resolved</th></tr>
<tr><td colspan="4">User agent display TLS-indicator that is used only for Example's authentic web site to inform Betty that the resource was retrieve through TLS-protected transaction. Betty should enter her credit card information into the authentic site that TLS indicator was shown only. In contrast, the disappearance of TLS indicator warns Betty that she must not provide any data to the imposter site.</td></tr>
<tr><th colspan="4">Variants</th></tr>
<tr><td colspan="4">The user agent MAY accomplish this by using a third state in the TLS indicator, or via another mechanism (such as a dialog, info bar, or other means).</td></tr>
<tr><th colspan="4">Consequences</th></tr>
<tr><td colspan="4">Benefits are the user better understands the web security context through the TLS indicator that informs the user about the state of TLS-secured web page that user interaction with and support them to make a safe decision to the TLS-secured web page only.</td></tr>
<tr><th colspan="4">See Also</th></tr>
<tr><td colspan="4">• As in the case of 6.1.1 Identity Signal, there may be usage modes during which this requirement does not apply.
• Web content MUST NOT obscure security interface, see 7.4.1 Obscuring or disabling Security User Interfaces.</td></tr>
</table>

3.2 Internal Elements Relationship Identification Among Patterns

The structure and relationship between entities among patterns will be analyzed to extract data for the elements of pattern application, that is, Structure, Dynamics, and Implementation. The results are also used in verifying the Internal Structure, to obtain the problem solved achievement and the purpose of pattern application context.

Analysis of the structure and relationships within the content is performed using a class diagram as shown in Fig. 3, as entity classes and their relationships. This figure shows both internal (WSCP-63 Package) and external (WSCP-61 Package) relationships. Internal relationships show the entity classes and properties of the TLS-indicator, and the external relationships reference the Usage Mode identified in the WSCP-61 pattern. Here, only the portions of the pattern related to the WSCP-63 pattern will be displayed. The structure and relationship analysis will be used to verify the correctness of the pattern in terms of objective achievement.

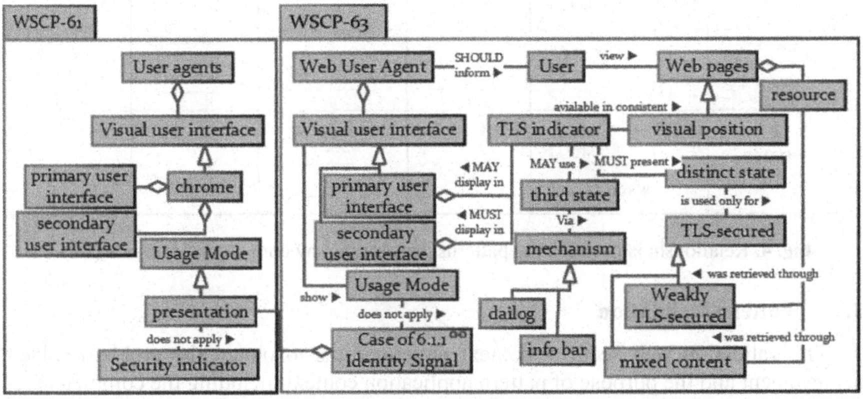

Fig. 3. Class diagram showing structure and relationships between patterns

3.3 Topic Classification

To show the overall global structure of the content, the WSC-UI headings are categorized according to Core Security Services CI4A. In the context of web user agents, confidentiality is consideration in parallel with policies on personal data transmitted or stored within the system, including system assets. Integrity is the trustworthiness of data, to ensure that the data being used is free of alterations by non-privileged persons. Authentication refers to verifying a person, and confirming that received messages actually come from that person. Authorization focuses on access rights to system services and data. Availability requires system assets to be accessible when requested by those with privileges. Accountability allows the system user to check their own actions, including Non-Repudiation to prevent the sender and receiver of data from denying that they have sent or received electronic data. From the aforementioned Core Security Services, the contents of WSC-UI can be categorized into classes as the global structure of the patterns as shown in Fig. 4. These relationships are obtained from considering the correspondence to Core Security Services of entities obtained from analysis of internal patterns elements. For example, from the internal structure of pattern

WSCP-63 in figure Fig. 3, there exists a property of requiring access for the TLS indicator, therefore it is categorized under the Availability class according to the Core Security Services, and its context is validated by experts. The aforementioned relationship is used in verifying relationships between the patterns, and applying the patterns in identifying the requirements.

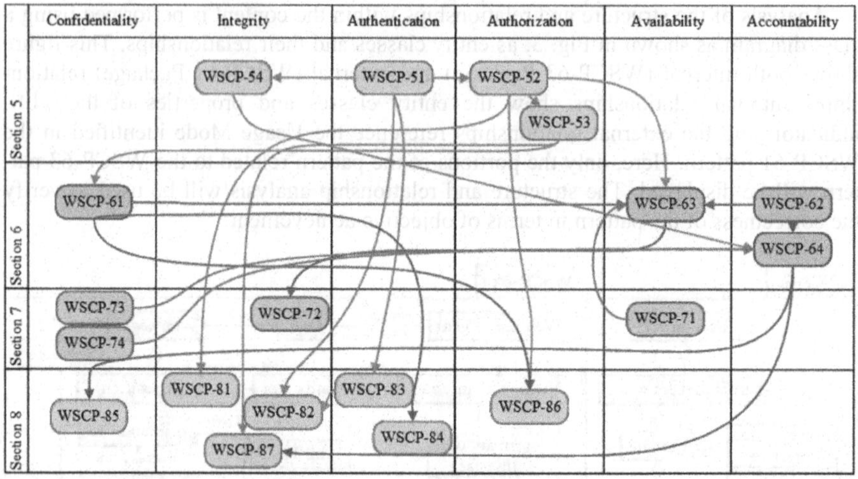

Fig. 4. Relationships between the patterns, categorized by core security service class

3.4 Pattern Validation

The validation web security context patterns aims to obtain the problem solved achievement and the purpose of pattern application context including the completeness, correctness, and consistency of the content appearing in the pattern. The security experts with intense knowledge background in web security perform validation on four levels: 1) WSC Patterns validation, 2) Internal Structure validation, 3) Global Structure validation, and 4) Patten Application Context, using a list of guideline questions as shown in Table 3. The results from validation will lead to pattern review and refinement.

From the process of partial creation of patterns and validation, the initial results from the four levels of validation will show that whether the content in the proposed patterns conform to the WSC-UI documents. For example, one of the validation result of Internal Structure validation may be a situation that uses different words means the same thing such as "user agent" and" web user agent" in different patterns. The patterns were refined by adding the "also" to identify the entity with the same meaning. At the Global Structure validation, if it is found that the pattern does not have an entity corresponding to Core Security Services, refinement of the Global Structure must further be considered the purpose of the pattern. At last, feedback from expert on scenario application for each pattern in both application achievement and ease of application criteria are reviewed for patterns improvement.

Table 3. Web security context pattern validation list

No.	Question
WSC Patterns Validation	
1	Is the pattern content complete, when compared to the WSC-UI documents?
2	Is the analyzed content consistent with the WSC-UI documents?
Internal Structure Validation	
3	Is the identified Internal Structure complete, when compared to the WSC-UI docs?
4	Is the identified Internal Structure consistent with the WSC-UI documents?
5	Is the identified Internal Structure consistent with the WSC Patterns?
Global Structure Validation	
6	Has the pattern been used to create the complete Global Structure?
7	Does the Global Structure display complete relationships between patterns?
8	Does the Internal Structure of the pattern correspond to the identified CI4A Class?
Patterns Application Context	
9	What is the degree level of each pattern application in the real situation?
10	For a specific situation of web user agent security point of view, what is the level of proposed patterns coverage?
11	Will you use the proposed patterns?

4 Patterns Application

To show the context of proposed patterns application, an example is shown in Fig. 5. The pattern shows recommendation for designing web security context-aware user agents. The developer can define the security requirements for the user agent from the proposed pattern. Thus, without proper display of the Web security context from the web user agent, the user may become a victim of system attackers, through the transmission of personal data to forged websites. To close the loophole, from the pattern usage context in Table 2, the problem is solved by defining the list Web Agent Requirements to require the Web agent to have a TLS indicator, to inform the user that the website being accessed has been authenticated. To provide convenience for developers, applications of patterns should have a support tool for identifying requirements from the pattern.

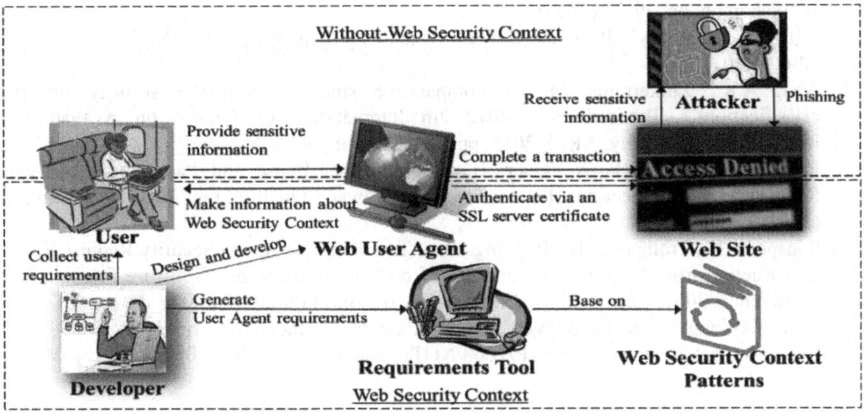

Fig. 5. Web Security Context Architecture

5 Conclusion and Future Work

This research proposes a method for creating a web security context patterns through analysis of WSC-UI and WSC-USECASES documents, identifying the internal elements pattern structure and external relationships among patterns, validating and performing initial refinement in order to meet the predefined evaluation criteria. The created patterns are beneficial to developers for designing web security context aware user agents.

In order to fulfill the purpose of creating patterns to be applied for defining requirements, for usage in designing the web security context of web user agents, the created patterns will be validated by experts concerning the completeness, correctness and usage coverage against WSC-UI and WSC-USECASES documents and ease of application, and feedback will be analyzed and input to improve the pattern content. In addition, a tool will be developed based on an analysis of pattern elements and construct an appropriate grammar using Extended-BNF as a requirements generator in order to help specify web security context requirements.

References

1. W3C: Web Security Context: User Interface Guidelines, http://www.w3.org/TR/2010/REC-wsc-ui-20100812/
2. Lebanidze, E.: Securing enterprise web applications at the source: an application security perspective. OWASP-The Open Web Application Security Project (2006)
3. Bolchini, D., Colazzo, S., Paolini, P.: Requirements for Aural Web Sites. Proceedings of the Eighth IEEE International Symposium on Web Site Evolution, pp. 75-82. IEEE Computer Society (2006)
4. Dias, A.L., Fortes, R.P.d.M., Masiero, P.C.: Increasing the Quality of Web Systems: By Inserting Requirements of Accessibility and Usability. Proceedings of the 2012 Eighth International Conference on the Quality of Information and Communications Technology, pp. 224-229. IEEE Computer Society (2012)
5. Schumacher, M., Fernandez-Buglioni, E., Hybertson, D., Buschmann, F., Sommerlad, P.: Security Patterns: Integrating Security and Systems Engineering. Wiley (2013)
6. Riaz, M., Williams, L.: Security requirements patterns: understanding the science behind the art of pattern writing. Requirements Patterns (RePa), 2012 IEEE Second International Workshop on, pp. 29-34 (2012)
7. Hafiz, M., Adamczyk, P., Johnson, R.E.: Organizing security patterns. IEEE Software 24, 52-60 (2007)
8. Alvi, A.K., Zulkernine, M.: A comparative study of software security pattern classifications. Proceedings - 2012 7th International Conference on Availability, Reliability and Security, ARES 2012, pp. 582-589 (2012)
9. Palomares, C., Franch, X., Quer, C.: Requirements Reuse and Patterns: A Survey. Requirements Engineering: Foundation for Software Quality, pp. 301-308. Springer (2014)
10. Supaporn, K., Prompoon, N., Rojkangsadan, T.: Enterprise Assets Security Requirements Construction from ESRMG Grammar based on Security Patterns. Software Engineering Conference, 2007. APSEC 2007. 14th Asia-Pacific, pp. 112-119 (2007)
11. W3C Working Group Note: Web Security Experience, Indicators and Trust: Scope and Use Cases, http://www.w3.org/TR/2008/NOTE-wsc-usecases-20080306/

A Comparative Study of Combination with Different LSB Techniques in MP3 Steganography

Mohammed Salem Atoum

Faculty of Science, Information Technology and Nursing
Irbid National University

Moh_atoom1979@yahoo.com

Abstract. Steganography hides the existence of the data inside any cover file. There are different file formats used in steganography like text, image, audio and video. Out of these file formats audio steganography is followed in this paper .One of the major objective of hiding data using audio steganography is to hide the data in an audio file, so that the changes in the intensity of the bits of host must not be detect by human auditory system. The focus of this paper is on time domain technique i.e. LSB technique of audio steganography. Method used in the paper hides the data in combination of LSBs instead of hiding the data in least significant bit. Results are compared by using parameters PSNR, BER and correlation.

KeYwords: LSB, PSNR, BER

1 Introduction

The increasing Internet usage stems from the growing availability of the global communication technology that has led to electronically induced information gathering and distribution. However, the challenge it presents in terms of information security is enormous. Every Internet user interest lies in having a secure transaction, communication and information across the transmission link, but in reality, much communication are infiltrated, jabbed and altered. Information confidentiality was enacted by the CIA as one of the key principles of a secure communication and if abused attracts penalty. However, many communications still fall short of achieving a secured information transmission across the global network (the Internet). The need to secure information within the global network is of paramount importance so that user information is preserved until it reaches its destination undisclosed.

A lot of sensitive information goes through the Internet on frequent basis. This information could be military codes, government dealings, and personal data, the route, sender/ receiver, the content of such information requires that they are protected against hacking and infiltration. Therefore, providing a secure framework that con-

K.J. Kim (ed.), *Information Science and Applications*,
Lecture Notes in Electrical Engineering 339, DOI 10.1007/978-3-662-46578-3_65

ceals information content and sender/receiver identity should be an urgent matter of interest. There are two known approaches to information confidentiality; they are cryptography and steganography [1]. Cryptography has long existed as the method for securing data; it works with set of rules that transforms information into unrecognizable format. The rules are used to serve for authentication purposes, because only the one who knows the rules can decipher the encrypted information [2]. The advent of steganography provides more security features since the information is disguised in the sense that the information does not give away its content and identity of sender and receiver within the communication link. Cryptography and steganography techniques both make use of data encryption approach but Cryptography encrypts plainly its secret message thereby making the content and the user's details vulnerable to exploitation. Steganography technique protects both information content and identity of a person's transmitting the information, whereas only information is concealed with cryptography [3].

Steganography operates by embedding a secret message which might be a copyright mark, or a covert communication, or a serial number in a cover such as a video film, an audio recording, or computer code in such a way that it cannot be accessed by unauthorised person during data exchange. A cover containing a secret data is known as a Stego-object [3]. After data exchange; it is advisable for both parties (sender and receiver) to destroy the cover in order to avoid accidental reuse. The basic model of a steganography system is shown in the Figure1 [4]. The model contains two inputs and two processes, the inputs are a cover medium and secret message both can be any image, audio, video and so on. Two processes contain embedding and extracting processes.

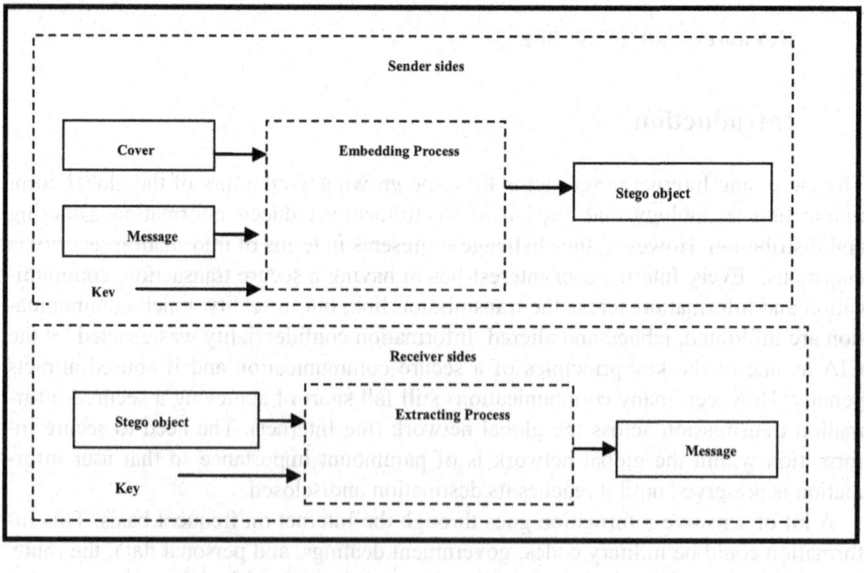

Fig. 1. Basic Model of Steganography

The embedding process is used to hide secret messages inside a cover; the embedding process is protected by a key if using secret key steganography and public key Steganography types, or without a key if using pure Steganography type. When using key, only those who possess the secret key can access the hidden secret message, while the extracting process is applied to a possibly modified carrier and returns the hidden secret message. Until recently, steganography utilized image files for embedding information across the Internet network. However recently, its use has been extended to audio steganography. The usage of audio signal as an embedding platform for information hiding is due to the fact that it has sophisticated features that allow information hiding, though difficult its robustness counts. In audio steganography, various signal processing techniques can be utilized to hide information in an audio file in such a way that it cannot be visually interpreted [5]. This approach has brought about the growing research interest in the use of digital audio signal for embedding information. The sensitiveness of audio files to delay presents more challenges to the design objective of steganography. There are three fundamental properties to the design of steganography. They are: 1) imperceptibility, 2) robustness and 3) capacity. However, there are other properties such as computational time that must be considered when dealing with different types of applications (information) such as broadcast monitoring applications in a global network. In most cases it requires real time processing and thereby cannot tolerate any form of delay [6].

The rest of this paper is organized as follows: A detailed introduction of audio steganography, the existing methods for MP3 steganography, then experimental results and conclusion.

2 AUDIO STEGANOGRAPHY

The techniques of Steganography were originally developed and used for images. Researchers in the field then started studying on how the techniques can be used on audio media. Hence, the introduction and development of the known algorithms for audio steganography was founded. As the known steganography techniques are mostly used for images there are not many methods for audio steganalysis. Thus, audio Steganography provides considerably better security [5].

In audio Steganography, many types of file can be used as a cover of steganography such as Waveform Audio File Format (WAVE, or more commonly known as WAV due to its filename extension) or MPEG-1 or MPEG-2 Audio Layer III (MP3). Similarly, secret messages that are embedded can be of secured types such as text or speech. MP3 is the most popular compression format for digital audio. In steganography, which uses MP3 as a cover, secret message can be embedded during compression and after compression [7-9].

3 RELATED WORKS

Little research's using MP3 as a cover in steganography [9]. Most researchers work in wave steganography. The following methods explain and discuss the current MP3 steganography methods:

3.1 Using Audio data of MP3 file

3.1.1 M4M method

Bhattacharyya and et al [10] proposed a new method for imperceptible audio data hiding for an audio file of wav or mp3 format. This approach based on the Mod 16 Method (M16M) [11] designed for image named Mod 4 Method (M4M) along with a Number Sequence Generator Algorithm to avoid embedding data in the consecutive indexes of the audio, which will eventually help in avoiding distortion in the audio quality. The input messages can be in any digital form, and are often treated as a bit stream. Embedding positions are selected based on some mathematical function which de-ends on the data value of the digital audio stream. Data embedding is performed by mapping each two bit of the secret message in each of the seed position, based on the remainder of the intensity value when divided by 4. Extraction process starts by selecting those seed positions required during embed-ding. At the receiver side other different reverse operation has been carried out to get back the original information.

3.1.2 M16M method

Bhattacharyya and et al [12] proposed a new method for imperceptible audio data hiding for an audio file of wav or mp3 format. This approach based on the Mod 16 Method (M16M) [11] designed for image named Mod 16 Method for audio (M16MA) along with a Number Sequence Generator Algorithm to avoid embedding data in the consecutive indexes of the audio, which will eventually help in avoiding distortion in the audio quality. The input messages can be in any digital form, and are often treated as a bit stream. Embedding positions are selected based on some mathematical function which de-ends on the data value of the digital audio stream. Data embedding is performed by mapping each four bit of the secret message in each of the seed position, based on the remainder of the intensity value when divided by 16. Extraction process starts by selecting those seed positions required during embedding. At the receiver side other different reverse operation has been carried out to get back the original information.

3.1.3 New Secure Scheme in Audio Steganography (SSAS)

Atoum et al [16] proposed model uses Mathematical equation to partition secret message into blocks, and applies permutation equation to re-order the blocks position, which will change and create new secret message addressing. So, if attackers detect the Stego object they can't know the secret message contains. In addition, the proposed model applied a mathematical model for embedding, and also used map for extracting secret message. All these techniques increase the security of Stego object and also reduce the probability to attack the secret message or exchange it.

3.2 Using Header of MP3 File

3.2.1 Unused Header Bit Stuffing (UHBS):

The MP3 frame headers are made up of fields such as the private bit, original bit, copyright bit, and emphasis bit but its usage are mostly omitted in some MP3 players. These fields are the important aspect of the frame that aids the interpretation of information concealed within the audio signal. They can be properly used to embed undisclosed massage by replacing the bit stream of undisclosed massage through the bits in the field. However, if in the process of replacing the bit stream with the bit in the field fails, the actual content of the secret message received within the frame will be lost and that will make the signal recovery more challenging [13]. The work by [13] highlighted on the possibility that audio steganography can achieve good capacity and robustness through the use of 4 bits in each header frame of the audio signal to embed secret messages.

3.2.2 Padding Byte Stuffing (PBS)

Padding byte stuffing was recently established as one of the techniques for steganography. Its approach is relatively straightforward in terms of implementation. It represents fine regular storage capability and has the ability to program 1 byte of information for each frame as long as padding bytes are accessible. The MP3 file is a given example of the material medium that can well utilize the padding byte stuffing method because it can allow for hundreds of frames in one secret message, especially when the filling bytes cannot take any more audio information [13].

3.2.3 Embedding in Before All Frames (BAF)

BAF was developed by [14]. Their approach embeds text file to MP3 file. The text file is encrypted by using RSA algorithm to increase the security of undisclosed secret message. The first frame will be filled with encrypted information. This process is repeated sequentially until the frame headers are filled. The capacity of about 15 KB is utilized when encryption algorithm is used otherwise it takes about 30 KB for the MP3 file. Even though there are chances of the secret message being sniffed, for this approach, its advantages are enormous, for instance, the method of padding and the unused bit even after the frames must have been filled, provides more encoding capability.

3.2.4 Embedding in Between Frames (BF)

[15] Developed steganography technique that embedded between frames (BF). It also embeds text file to MP3 file like the BAF and encrypts information in bits format by using RSA algorithm in order to increase the protection of concealed secret messages. The BF differs from the BAF in the way the text files are inserted into the frames. It does not start with the first frame it sees but selects the frame of its choice. On the other hand, the capacity of the BF in comparison to the BAF utilizes the capacity of about 40 MB with encryption algorithm but requires 80 MB on original format. BF likewise provides good capacity for embedding text file in more capacity but it is still prone to attack. We draw inferences based on the literatures accessed that the method of embedding information after compression is a challenging task since the embedding process is done after compression and the text file are located in the unused bit location and not in the audio data. This technique provides a platform that is prone to attack because the content of the secret message sent can be easily deciphered by a third party sniffing through the communication link. It also provides only limited capacity for secret message hiding. However, if the LSB technique is used to insert speech in MP3 file with the use of 2, 3 and 4 bit exchange in audio data (8-bit for sample), the problem of capacity can be resolve. In addressing the problem of security, the use of key as the lock for concealed secret message is a foreseeable approach that can achieve maximum security for concealed secret messages.

Literature review shows that the method of embedding information after compression. There are several methods for embedding secret message after compression namely: M4M, M16M, UHBS, PBS, BF, BAF and SSAS. The strengths and weaknesses are shown in Table1.

Table 1. shows the strength and weakness for MP3 methods

Techniques		Strength	Weakness
Using Audio Data	M4M	➢ Good capacity and security but can be improved.	➢ Low imperceptibility and robustness.
	M16M	➢ Good capacity and security but can be improved.	➢ Low imperceptibility and robustness.
	SSAS	➢ High capacity and security	➢ Don't have message integrity checking.
Using Header Frames	UHS	➢ Simple	➢ Not all MP3 contains padding stuffing, just constant bit rate. ➢ Low security and capacity.
	PBS	➢ Low capacity.	➢ Low security. ➢ Using cryptography.
	BF	➢ High capacity. ➢ Simple	➢ Low security. ➢ Using cryptography.
	BAF	➢ Simple	➢ Low security and capacity.

4 EXPERIMENTAL RESULTS

The dataset will be used for analysis shown in the Table 2. The dataset uses different size for cover audio file and using 100KB for secret message size to evaluate all methods [17-18].

Table 2. Data Set

Name of genre	Time (Minute)	Size (MB)
Pop	4:00	2.75
rock	4:33	3.13
Blues	4:41	3.22
Hip-hop	5:27	3.74
Dance	6:12	4.26
Metal	6:28	4.40

According to the Table.3 below the results for SSAS methods is better than other methods results, this because the method using scrambling techniques to prepare the secret message before embedding and using randomly first byte chosen to begin embedding secret message. In addition, using different jump equations to jump from previous byte was used to embed into next byte should be used to embed. However, SSAS method is applying message integrity to check the secret message was altered after extraction. Thus, other methods avoid this checking. UHBS, PBS and BAF are stopped to embed because the capacity for the secret message is 100KB.Figure 3 and 4 shows the PSNR and BER results for Table.3

Table 3. Experimental results

Genre name	Methods	M4M	M16M	BF	SSAS
Pop	PSNR	63.035	53.569	48.212	68.117
	Corr	0.9949	0.9949	0.9199	1.000
	BER	0.0132	0.0098	0.0212	0.0032
Rock	PSNR	63.601	50.184	48.650	68.794
	Corr	0.9949	0.9949	0.9199	1.000
	BER	0.0112	0.0091	0.0198	0.0029
Blues	PSNR	63.708	0.232	49.806	68.830
	Corr	0.9969	0.9969	0.9398	1.000
	BER	0.0092	0.0081	0.0175	0.0027
Hip-hop	PSNR	64.390	51.019	49.3759	69.425
	Corr	0.9979	0.9979	0.9398	1.000
	BER	0.0091	0.0074	0.0169	0.0017
Dance	PSNR	64.986	51.523	49.651	70.050
	Corr	0.9997	0.9997	0.9449	1.000
	BER	0.0087	0.0070	0.0145	0.0015
Metal	PSNR	65.117	1.667	49.748	70.142
	Corr	0.9997	0.9997	0.9454	1.000
	BER	0.0081	0.0067	0.0142	0.0011

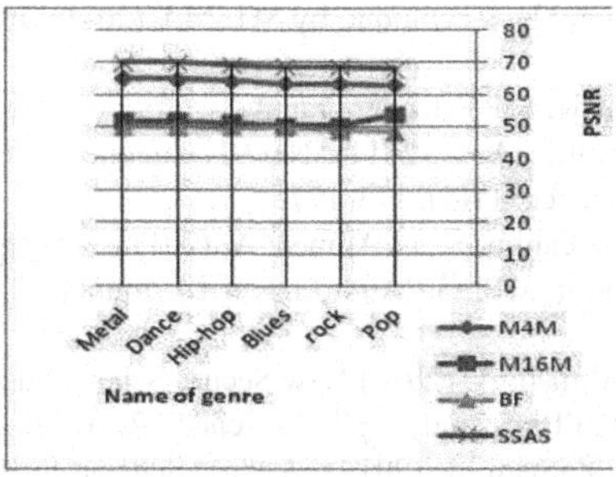

Fig. 2. PSNR Results

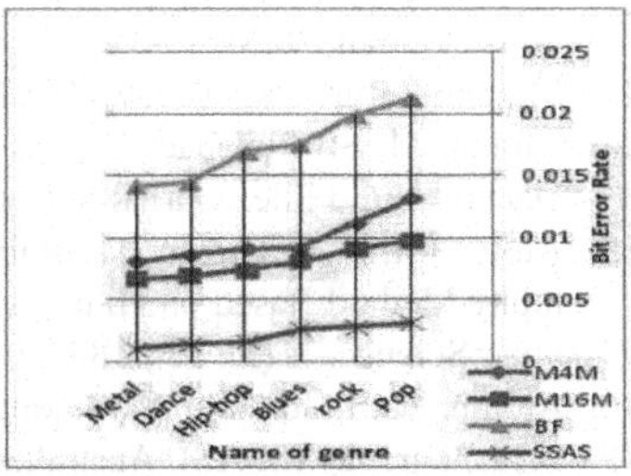

Fig. 3. Bit Error Rate results

5 Conclusion

This paper concluded the comparison between different methods which that using MP3 steganography techniques. The experimental results show that the SSAS method result is better than other methods results. A test is made on various MP3 files which include different size and time. Different measurement methods are used to evaluate the methods. PSNR, BER and correlation are three critical factors to evaluate the methods. In the future, we can also extend SSAS method to add authentication code to prove the secret message is sent by the right person.

6 REFERENCES

1. Lentij J. (2000), Steganographic Methods, Department Of Control Engineering And Information Technology, Budapest University. Periodica Poltechnica Ser. El. Eng. Vol.44, No. 3–4, P. 249–258.
2. Katzenbeisser S., Peticotas F. (2000), Information Hiding Techniques For Steganography And Digital Watermarking, Artech House Inc.
3. Petitcolas F.A, Anderson R.J., Kuhn M.G. (1999), Information Hiding – A Survey, IEEE, Special Issue On Protection Of Multimedia Content: 1062-1078.
4. Cacciaguerra S., Ferretti S.(2002), Data Hiding: Steganography And Copyright Marking, Department Of Computer Science, University Of Bologna, Italy.
5. Nedeljko C. (2004). Algorithms for Audio Watermarking and Steganography. Acta Universitatis Ouluensis. Series C.
6. Andres G. (2002). Measuring and Evaluating Digital Watermarks in Audio files. Washington Dc.
7. Chan, P. (2011). Secret Sharing in Audio Steganography. Industrial Research.
8. Deng, K., Tian, Y., Yu, X., Niu, X., Yang, Y., & Technology, S. (2010). Steganalysis of the MP3 Steganographic Algorithm Based on Huffman Coding. Test, (1), 79-82.
9. Atoum, M. S., Ibrahim, S., Sulong,G. & Ahmed, A. (2012). MP3 Steganography: Review. Journal of Computer Science issues, 9(6).
10. Bhattacharyya, S., Kundu, A., Chakraborty, K., & Sanyal, G. (2011). Audio Steganography Using Mod 4 Method. Computing, 3(8), 30-38.
11. Arko Kundu, Kaushik Chakraborty and Souvik Bhattacharyya (2011),Data Hiding in Images Using Mod 16 Method, In the Proceedings of ETECE.

12. Bhattacharyya, S, A Novel Audio Steganography Technique by M16MA (2011). International Journal, 30(8), 26-34.
13. Mikhail Zaturenskiy (2009), Behind the Music: MP3 steganography.
14. Atoum, M. S., Rababah, O. A. A.-, & Al-attili, A. I. (2011). New Technique for Hiding Data in Audio File. Journal of Computer Science, 11(4), 173-177.
15. Atoum, M. S., Suleiman, M., Rababaa, A., Ibrahim, S., & Ahmed, A. (2011). A Steganography Method Based on Hiding secrete data in MPEG / Audio Layer III. Journal of Computer Science, 11(5), 184-188.
16. Atoum, M. S., Ibrahim, S., Sulong, G., & Ahmed, A. (2013).New Secure Scheme in Audio Steganography (SSAS). Australian Journal of Basic and Applied Sciences, 7(6), 250–256.
17. Atoum, M. S., Ibrahim, S., Sulong, G., & Zamani.M. (2014). A New Scheme for Audio Steganography Using Message Integrity, Journal of Convergence Information Technology, 8(14), 35-44.
18. Atoum, M. S., Ibrahim, S., Sulong, G., Zeki, A and Abubakar, A. (2013). Exploring the Challenges of MP3 Audio Steganography. Proceding IEEE from 2nd International Conference on Advanced Computer Science Applications and Technologies (ACSAT) ,Sarawak, Malaysia.

Review of Digital Forensic Investigation Frameworks

Ritu Agarwal[1], Suvarna Kothari[2]

[1] Delhi Technological University, New Delhi, India.
ritu.jeea@gmail.com
[2] Delhi Technological University, New Delhi, India
suvarnakothari91@gmail.com

Abstract. Digital Forensic Investigation has seen a tremendous change in the past 25 years. From the age of early computers to the current day mobile devices and storage devices, the crime rate has also followed growth. With the diversity in crimes, frameworks have also been modified over time to cope-up with the pace of crimes being committed. The paper amalgamates all major approaches and models presented that have helped in shaping the digital forensic process. Each discussed model is followed by its advantages and shortcomings.

Keywords: Digital Forensics, models, review.

1 Introduction

Digital Forensics is "the use of scientifically derived and proven methods towards the preservation, collection, validation, identification, analysis, interpretation and presentation of digital evidence derived from digital sources for the purpose of facilitating or furthering the reconstruction of events found to be criminal or helping to facilitate the unauthorized actions shown to be disruptive to planned actions" [5]. With the advent of time, significant changes have been observed in the digital forensic process.

The statistical analysis based on trends from 2004 till present, of the papers from Elsevier journals, IEEE and magazine articles shows the frequency of articles published under Framework and Architecture are the least as compared to Challenges and Opportunity, Security and Privacy Issues and Cloud Forensic Investigation [25].This leads to much scope of future research being done on building a consistent and standardized framework for conducting digital investigation.

A concise survey on digital forensic models is being presented that may help researchers explore new ideas and provide new solution to challenges in the field. The literature review is divided into three phases: Phase1 consolidates papers from 1995 to 2003; Phase 2 combines papers from 2004 to 2007 and Phase 3 from 2008 to present. The paper tries to include major publications that have helped in shaping the digital forensic process.

K.J. Kim (ed.), *Information Science and Applications*,
Lecture Notes in Electrical Engineering 339, DOI 10.1007/978-3-662-46578-3_66

2 Literature Review

2.1 Phase 1:1995-2003

One of the earliest papers that clearly mapped the forensic process was given by Mark M. Pollitt [1] where he proposed four distinct steps "Acquisition, Identification, Evaluation and Admission as Evidence" so that evidence could be documented in the court of law. The result of these phases or methods is "media (physical context), data (logical context), information (legal context) and evidence". But except for this paper, people created guidelines that were focused on the details of the technology and a generalized process was not considered.

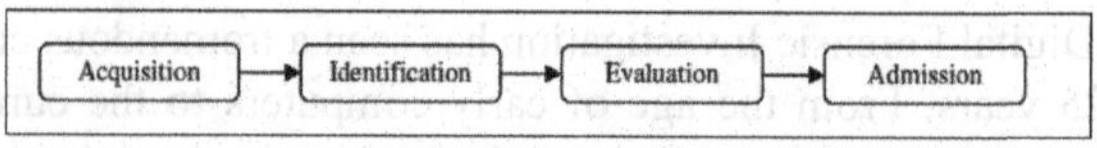

Figure 1: Computer Forensic Investigation Process

Farmer and Venema [2] gave steps as "secure and isolate, record the scene, conduct a systematic search for evidence, collect and package evidence, and maintain a chain of custody" which formed the foundation for further research but it was aimed at UNIX forensic procedures.

Mandia and Prosise [3] proposed a methodology which had step as "pre-incident preparation, detection of incidents, initial response, response strategy formulation, duplication, investigation security measure implementation, network monitoring, recovery, reporting, and follow up". This was advancement over the previous approach but was targeted for explicit platforms such as UNIX, Windows NT/2000 and Cisco Routers. The drawback is that other digital devices like mobile phones, personal digital assistants etc. are not addressed by this approach.

This was succeeded by the abstract model given by the U.S. Department of Justice [4] whose process included "collection, examination, analysis, and reporting". This is helpful as it attempts to shape a comprehensive process that will be valid for most electronic devices but the drawback is that analysis phase of this model is improperly defined and is ambiguous.

The Digital Forensic Research Workshop [5] was the first big consortium headed by the academic community rather than law enforcement. It worked towards developing framework that contains steps such as "identification, preservation, collection, examination, analysis, presentation and decision." In this framework, elements refer to individual tasks and classes of tasks are called processes. This framework lays foundation for future work.Working on this framework, many more models were proposed.

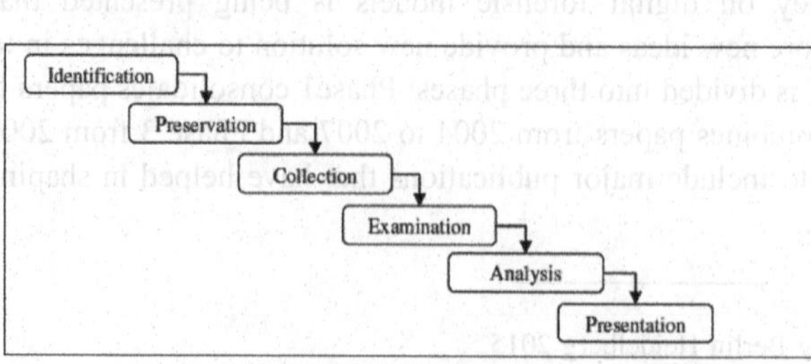

Figure 2: DFRDWS Investigative Model

The Abstract Digital Forensics Model [6] was one of them. It standardized the digital forensics process into nine components "identification, preparation, approach strategy, preservation, collection, examination, analysis, presentation, returning evidence." Categorizing of incidents can be done very well using this framework. This broad method has many advantages as proposed by the authors such as the same framework being applicable to forthcoming digital technologies. As we can see, the second step is almost the same as the third step.

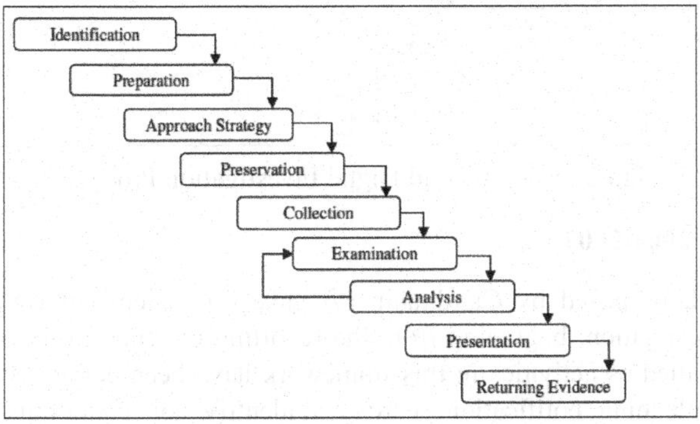

Figure 3: Abstract Digital Forensic Model

The Integrated Digital Investigation Model [7] proposed another model that consists of total of 17 phases generalized into five groups. It has "Readiness phase, Deployment phase, Physical Crime Scene Investigation Phase, Digital Crime Scene Investigation Phase and Review Phase". Physical crime scene was analysed using high level phases. High-level phases are used in this framework for the analysis of both the digital crime scene as well as the physical crime scene. This model covers all the cyber terrorism capabilities and the incidents that led to the events are also reconstructed. However, there are some shortcomings as well. It does not clearly differentiate amongst investigations at the suspect's scene and the victim's scene and moreover it seems impossible to make out whether a digital crime was committed or not unless some prior examination has been made.

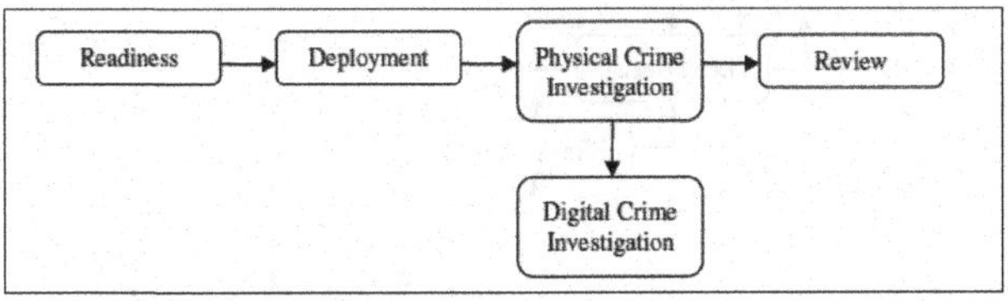

Figure 4: Integrated Digital Investigation Process

A Comprehensive Approach to Digital Incident Investigation [8] given by Stephenson sights class as a process of the DFRWS framework [5] elements of the class is called an action. The investigative process is divided into six classes. He then prolonged the processes into nine steps which formed the End-to- End digital Investigation Process (EEDI). The investigator performs these nine steps in order to "preserve, collect, examine and analyse digital evidence". Critical activities in the collection process were defined by him so as to "collect the images of effected computers, to collect logs of intermediate devices especially those on the internet, to collect logs of effected computers and to collect logs and data from intrusion detection systems, firewalls, etc". Digital Investigation Process Language (DIPL) and Coloured Petri-net Modelling was then developed by him working on these steps. The principle focus of the framework was on analysis process and integrating events from different locations.

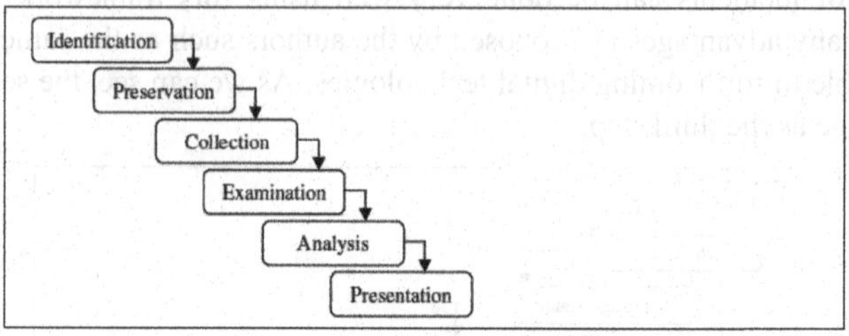

Figure 5: End-to-End Digtal Investigation Process

2.2 Phase 2: 2004-2007

The framework proposed by Ciardhuain [9] gave crisp steps for carrying out the process of investigation, beginning from the reporting of crime to the closure of the case. Phases called as activities in this framework have been defined as "awareness, authorization, planning, notification, search and identify, collection, transport, storage, examination, hypotheses, presentation, proof, defence and dissemination". A basis for the development of techniques and tools to assist in the work of investigators was provided by this framework. Therefore, this is the most complete framework till date.

Baryamueeba and Tushabe [10] made some additions to the Integrated Digital Investigation Model [7] and removed one of its disadvantages by showing clear difference between primary and secondary crime scene by adding two supplementary phases "Trace back phase and Dynamite phase". The aim was to recreate the two crime scenes simultaneously to avoid discrepancies. The primary and secondary crime scenes were separated by the framework while the phases were depicted as iterative instead of linear.

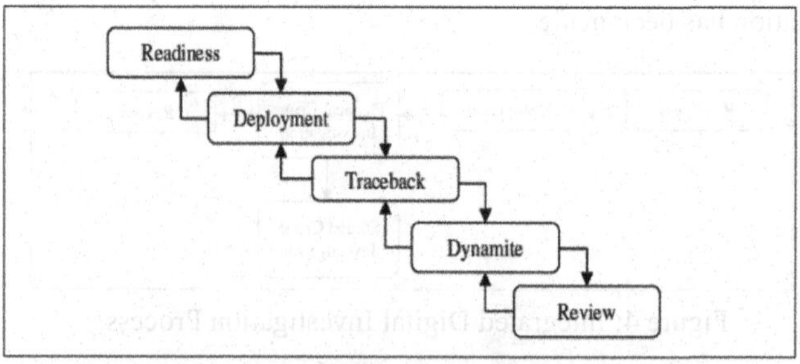

Figure 6:Enhanced Digital Investigation Process

In the Hierarchical Objectives based Framework for the Digital Investigations Process [11] by Beebe and Clark a multi tired model is proposed as opposed to the single tier approach being followed till now. It also introduces the objectives based task concept where analysis tasks are selected by investigative goals. Survey, extract and examine approach is suggested by the author to propose subtasks for analysis of data. The first tier comprises of phases "preparation, incident response, data collection, data analysis, presentation and incident closure". The second tier consists of "survey phase, extract phase and examine phase" Concept of objective-based tasks is used for analysing tasks in this framework. As stated by the authors, exclusive advantages in the field of realism and specificity are offered by this framework.

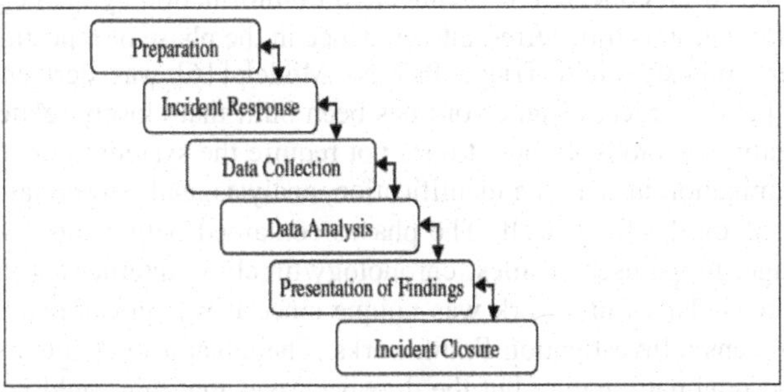

Figure 7: Hierarchical Objectives based Framework

In their 2004 paper, Carrier and Spafford [12] added events and event reconstruction to the digital forensic framework. Reconstruction is done using evidence so that hypothesis can be developed and tested. The framework comprises of three phases "Preservation, Search and Reconstruction Phase" and is based on sources and consequence of events. However completeness of each phase in not mentioned and it cannot be proven that this framework is satisfactory enough for investigation.

Rubin, Yun and Gaertner[13] carried on the work of Carrier[12][7] and Beebe [11] an introduced the concepts of seek knowledge, knowledge reuse and case-relevance. Seek knowledge refers to the investigative clues by which the analysis of data is driven. Case Relevance is "The property of piece of information, which is used to measure its ability to answer the investigative "who, what, where, when, why and how" questions in a criminal investigation" [13].The various levels of Case Relevance are "Absolutely irrelevant, Probably Irrelevant, Possibly irrelevant, Possibly Case-Relevant, Probably Case Relevant".

A paper on network forensics by Erbacher, Christensen and Sunderberg[14] brought up a number of grave matters as visualization of data in intrusion and network forensic situations. They suggested different aspects require different visualization techniques of examination but they also have to be combined.

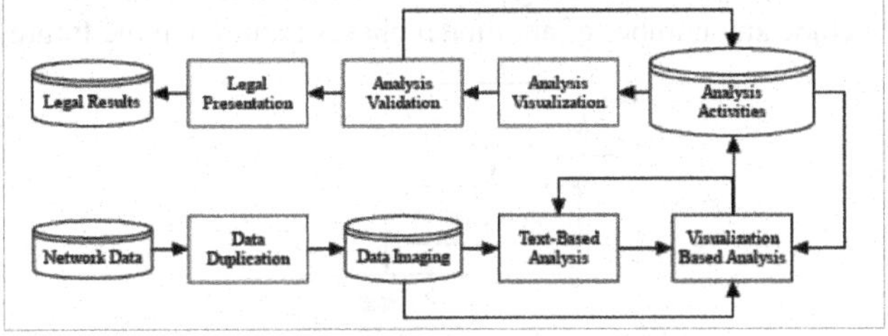

Figure 8: Visual Forensic Techniques and Processes

Kent, Chevalier, Grance and Dang[15] published a guide to Integrated Forensics into Incident Response where they have summarized the forensic process in four basic steps "Collection, Examination, Analysis and Reporting". This is very similar to [1].

Media is transformed into evidence by the forensic process in accordance with this framework either for an organization's inside usage or law enforcement. First, the data gathered from the media is transformed into a format that is readable by forensic tools. After the data has been collected, it is converted to information by the help of analysis and finally information is transferred into evidence in the phase of reporting.
The Computer Forensic Field Triage Process Model [16] was derived from IDIP Framework [7] and a process framework has been built that closely relates to the real world investigative methods. Hence it does not require the system to be taken back to the lab for examination instead the identification, analysis, and interpretation of digital evidence is done on the field itself. The phases contained within this framework are "planning, triage, usage/user profiles, chronology/timeline, internet activity and case specific evidence". This framework was unique since it was developed in reverse to most Digital Forensic Investigation Frameworks. The advantage of this model was its practicality and pragmatic nature but the drawback was that this could not be applied to all situations.

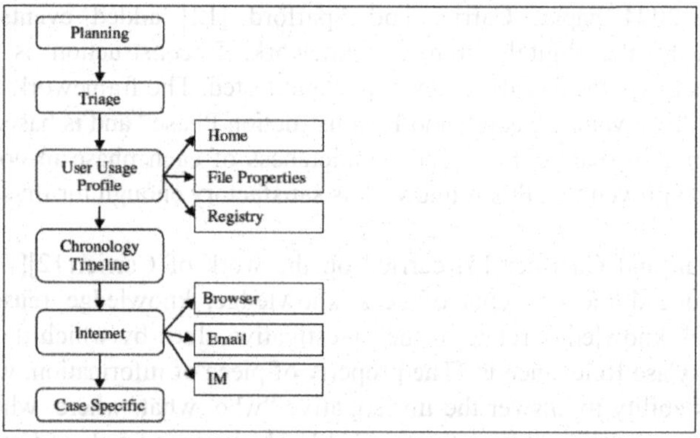

Figure 9: Computer Forensics Field Triage Process Model

In Framework for a Digital Forensic Investigation by Kohn, Eloff and Oliver [17] the aim was to merge the existing frameworks already proposed earlier [10][7][9][6] as it was discovered that a many steps or phases coincided with each another and the differed primarily in the terminology used. So similar tasks were grouped together and three stages were formed "preparation, investigation and presentation". Here the point to be noted is that knowledge of the relevant legal base was essential prior to setting up of the framework. The advantage of this framework is that it can be easily expanded to include any number of additional phases required in the future.

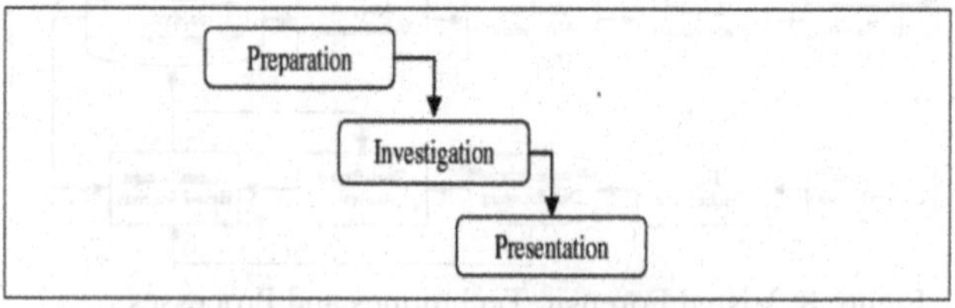

Figure 10: Framework for a Digital Forensic Investigation

The Common Process Model for Incident and Computer Forensics[18] proposed by Freiling and Schwittay has introduced a new framework in overall process of investigation is improved by combining the two conceptions of Incident Response and Computer Forensics. This model fixated significantly on analysis and it comprises of "Pre- Incident Preparation, Pre-Analysis, Analysis and Post- Analysis". All phases and actions that are completed before the actual analysis starts are combined in the Pre-Analysis Phase and Post-Analysis Phase deals with the documentation of the all actions undertaken during the course of an investigation. Computer Forensics can be applied during the analysis phase. Thus a proper technique to conduct incident response and integrating forensic analysis into Incident Response is suggested by this framework.

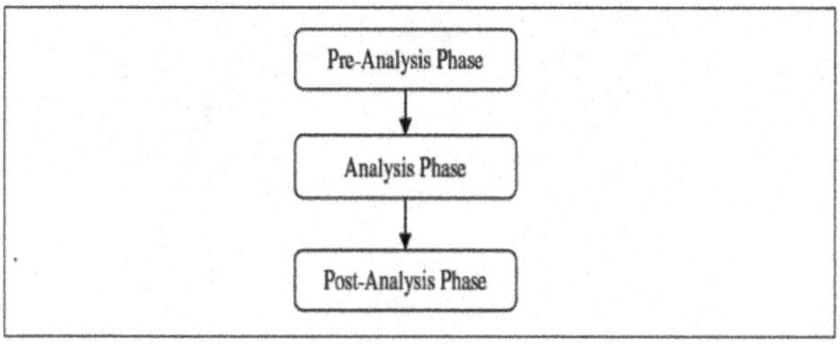

Figure 11:Common Process Model for Incident and Computer Forensics

2.3 Phase 3: 2008-2014

Perumal [19] proposed a model based on Malaysian Investigation Process in which more emphasis was given on "live data acquisition and static data acquisition" to focus on fragile evidence. It included steps of "Planning, Identification, Reconnaissance, Transport and Storage, Analysis, Proof and Defence and Archive Storage."

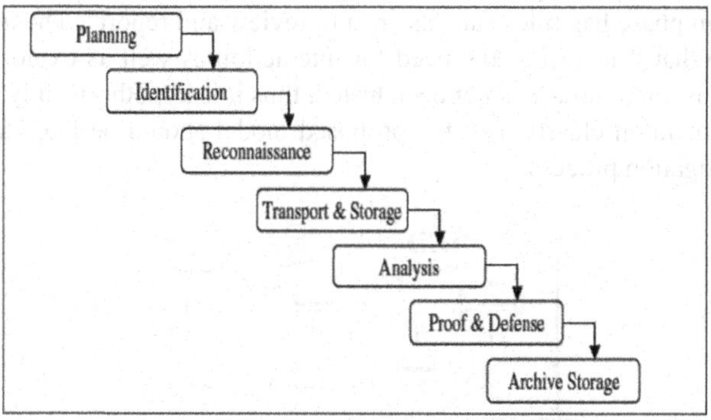

Figure 12: Digital Forensic Model based on Malaysian Investigation Process

The Digital Forensic Process Model proposed by Cohen [20] consists of seven listed processes or phases as "Identification, Collection, Transportation, Storage, Examination and Traces, Presentation and Destruction." Thus as we can see the focus of given model is on the examination of digital evidence.There is no need to include page numbers or running heads; this will be done at our end. If your paper title is too long to serve as a running head, it will be shortened. Your suggestion as to how to shorten it would be most welcome.

Agawal [21] established a systematic model for assisting forensic practitioners and organizations in making suitable strategies and processes .The proposed model suggests eleven stages and the diverse methods involved in the investigation of cyber fraud and cyber-crime -"Preparation, Securing the scene, Survey and Recognition, Documenting the scene, Communication Shielding, Evidence Collection, Preservation, Examination, Analysis, Presentation, Result and Review". The model emphasizes on study cases of cyber-crimes and computer frauds. The drawback of the model is that application of the model is limited to computer frauds and cyber-crimes only.

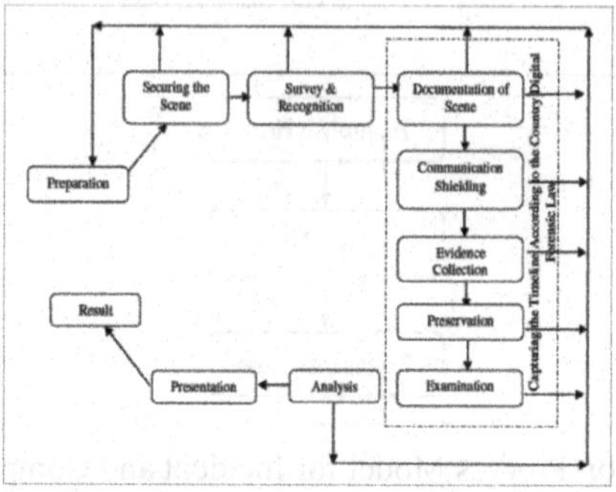

Figure 13: Systematic Digital Forensic Investigation Model

A new approach [22] was proposed by Ademu, Chris and David in which they the digital forensic investigation process was generalized into 4 tier iterative approach. The first tier will have 4 rules for digital forensic investigation which involves "preparation, identification, authorization and communication". The second tier has rules such as "collection, preservation and documentation", the third tier has rules consisting "examination, exploratory testing, and analysis" and the 4th tier which is the presentation phase has rules such as "result, review and report". The advantages of this model are that it identifies the need for interaction as well as exploratory testing but this model is ambiguous has not been tested, thus it is hypothesis only at present. It also does not mention clearly how the proposed model should be integrated with the forensic investigation process.

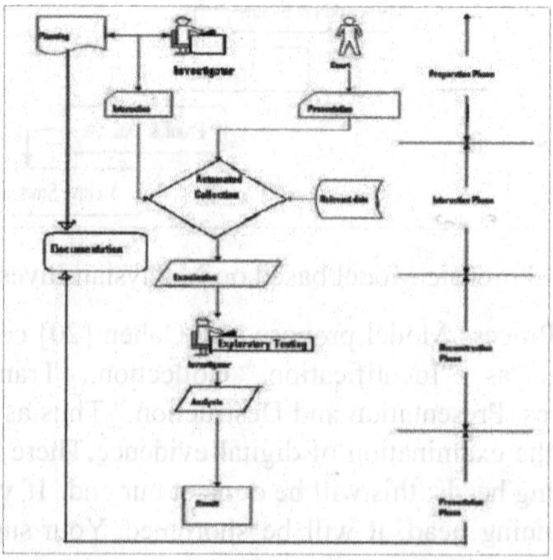

Figure 14:A New Approach of Digital Forensic Model for Digital Forensic Investigation

Valjarevic and Venter [23] defined a digital forensic investigation process model intended at harmonizing existing models. The model is quite similar to other models propsed by different authors as it is inclusive, but it differs from the others as it offers different placement of the phases and presents a new method for executing some of digital forensic principles through actionable items calls "parallel actions". The proposed model comprises the following twelve phases: "incident detection, first response, planning, preparation, incident scene documentation, potential evidence identification, potential evidence collection, potential evidence transportation, potential evidence storage, potential evidence analysis, presentation and conclusion". They propose a multi-tiered model which was built by accumulating a set of sub-phases. The drawback is that this model is yet to be verified for its accuracy and efficiency.

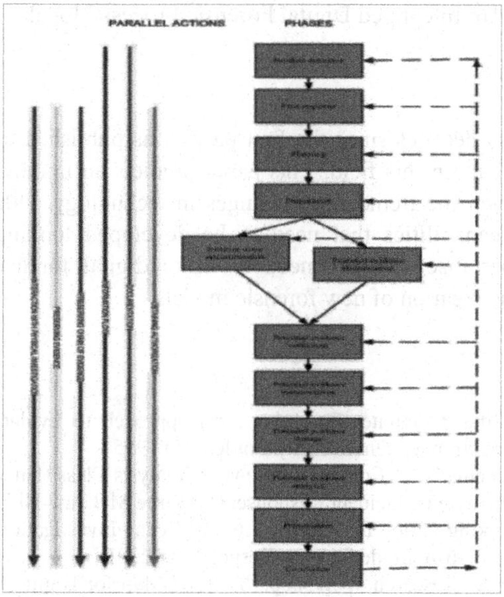

Figure 15:Harmonized Digital Forensic Investigation Process Model

The Integrated Digital Forensic Process Model [24] consists of the following processes: "preparation, incident, incident response, physical investigation, digital forensic investigation and presentation". Numerous complications were recognized in the present models, such the same processes or steps being written by dissimilar names, or altered explanations of a phase. "Therefore, the IDFPM is not just a merging of existing DFPMs, but an integration of the discussed DFPMs and a purification of the terminology used, resulting in an all-encompassing standardized IDFPM" [24]. The disadvantage is that this model is not applicable everywhere as it was made by considering only a few of the forensic models.

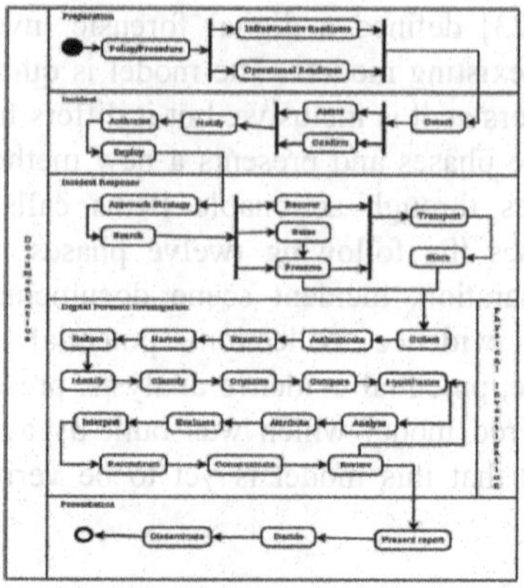

Figure 16: Integrated Digital Forensic Process Model

3 Conclusion

It has been over two decades since the first paper was published but we can see that much needs to be done in this field. This paper predicts an imminent predicament in digital forensics given the tremendous changes in technology. Other papers propose precise calculated capabilities that need to be developed looking at the future of forensics, this paper discusses the need to make digital forensics research more effective through the creation of new forensic models.

References

[1] Mark M. Pollitt. "Computer Forensics : An approach to Evidence in Cyberspace". National Information System Security Conference.1995

[2] Farmer D.,Venema W. :" Computer Forensics Analysis Class Handouts."1999

[3] Mandia K., Prosisse C. Incident Response. Osbourne/McGraw-Hill.2001

[4] Technical Working Group for Electrical Crime Scene Investigation. "Electronic Crime Scene Investigation:A Guide for First Responders."2001

[5] Digital Forensics Research Workshop. "A Road Map for Digital Forensics Research" 2001.

[6] Reith,M.,Carr,C., Gunsch,G., "An Examination of Digital Forensic Models". International Journal of Digital Evidence, 2002.

[7] Carrier,B., Spafford,E. "Getting Physical with the Investigative Process". International journal of Digital Evidence, 2003.

[8] Stephenson P. "A Comprehesive Approach to Digital Incident Investigation". Elsevier Information Security Technical Report.2003.

[9] Ciardhuain,SO. "An Extended Model of Cybercrime Investigations", International Journal of Digital Evidence.2004

[10] Baryamureeba V., Tushabe F. "The Enhanced Digital investigation Process Model". DFRWS 2004.

[11] Beebe N., Clark J. " A Hierarchical Objectives Based Framework for the Digital Investigations Process" DFRWS 2004.

[12] Carrier,B. , Spafford, E. "An Event based Digital Forensic Investigation Framework". DFRWS 2004.

[13] Rubin,G.,Yun C., Gaertner,M. "Case-Relevance Information Investigation : Binding Computer Intelligence to the Current Computer Forensic Framework" International Journal of Digital Evidence. 2005.

[14] Erbacher Robert F., Christensen Kim, Sunderberg Amanda."Visual Forensic Techniques and Processes"2006.
[15] Kohn M., Eloff JHP., Olivier MS., "Framework for a Digital Forensic Investigation". Proceedings of Information Security South Africa(ISSA) 2006.
[16] Kent K., Chevalier S.,Grance T., Dang H. "Guide to Integrating Forensics into Incident Response"NIST Special Publication 800-86.2006.
[17] K.Rogers M., Goldman J., Mislan R., Wedge T. and Debrota S. "Computer Forensics Field Triage Process Model" Conference on Digital Forensics Security and Law.2006.
[18] Freiling F., Schwittay B. "A Common Process Model for Incident Response and Computer Forensics".Conference on IT Incident Management and IT Forensics.2007
[19] Perumal S. "Digital Forensic Model based on Malaysian Investigative Process" International Journal of Computer Science and Network Security .2009
[20] Cohen F. "Towards a science of Digital Forensic Evidence Examination".Advances in Digital Forensics VI,IFIP Advances in Information and Communication Technology,Springer.2010
[21] Agarwal A., Gupta M., Gupta S., Gupta C. "Systematic Digital Forensic Investigation Model"International Journal of Computer Science and Security.2011
[22] Ademu O., Chris O., David S. " A New Approach of Digital Forensic Model for Digital Forensic Investigation." International Journal of Advanced Computer Science and Application.2011
[23] Valjarevic A., Venter H. "Harmonized Digital Forensic Investigation Process Model.IEEE.2012
[24] Kohn M., Eloff M., Eloff JHP. "Integrated Digital Forensic Process Model" International Journal of Computer and Security 2013
[25] Daryabar F., Dehghantanha A., Nur Izura Udzir, Nor Fazlida binti Mohd Sani, Solahuddin bin Shamsuddin, Farhood Norouzizadeh F.,"A Survey about Impact of Cloud Computing on Digital Forensics" International Journal of Cyber-Security and Digital Forensics 2013.

Part VI

Knowledge Discovery in Dynamic Data Using Neural Networks

Eva Volna, Martin Kotyrba, Michal Janosek

University of Ostrava, 30. dubna 22, 70103 Ostrava, Czech Republic
{eva.volna,martin.kotyrba,michal.janosek}@osu.cz

Abstract. This article aims at knowledge discovery in dynamic data via classification based on neural networks. In our experimental study we have used three different types of neural networks based on Hebb, Adaline and backpropagation training rules. Our goal was to discover important market (Forex) patterns which repeatedly appear in the market history. Developed classifiers based upon neural networks should effectively look for the key characteristics of the patterns in dynamic data. We focus on reliability of recognition made by the described algorithms with optimized training patterns based on the reduction of the calculation costs. To interpret the data from the analysis we created a basic trading system and trade all recommendations provided by the neural network.

Keywords: Pattern recognition, knowledge discovery, neural networks, automated trading system.

1 Introduction

This article aims at knowledge discovery in dynamic data via classification based on neural networks. The develop pattern recognition algorithm is based on artificial neural networks in order to recognize the market pattern in real time with maximum reliability. These patterns do not cover every time point in the series, but are optimized to be suitable candidates in experimental tasks so that the developed classifiers would be able to learn key characteristics of these patterns and accurately recognize them. Such optimized inputs for neural network reduce the calculation costs.

Currently, there are mainly two kinds of pattern recognition algorithms: an algorithm based on rule-matching [1] and an algorithm based on template-matching [8]. Nonetheless, both of these two categories have to design a specific rule or template for each pattern. Both types of algorithms require participation of domain experts, but for last few decades, neural networks have shown to be a good candidate for solving problems with the pattern analysis. More than sixty important technical patterns are detailed in [2].

Before we start it is crucial to choose particular system on which we are able to conduct desired experiments. We need a system, where we can very easily observe certain patterns and where we can obtain observed data detail and without delay. For

K.J. Kim (ed.), *Information Science and Applications*,
Lecture Notes in Electrical Engineering 339, DOI 10.1007/978-3-662-46578-3_67

that reason we use Forex market. Forex, or foreign exchanges, also known as the international foreign exchange market is a world foreign exchange trading [5, 9]. There are several reasons why to choose Forex market. There is a lot of data freely available in real time and in very good quality. Almost every broker company offers a demo platform and free online data for study. There also exist many described patterns whose emerge in the market as well.

2 Neural network based tool

2.1 Training set optimization

We applied data from X-Trade Brokers [3] that shows the development of market values of EUR/USD, which reflect the exchange rate between EUR and USD. The testing time scale was eight months from Mar 2012 to October 2012 on a 5 - minute chart. It means that every 5 minutes a new record is created in the table. For our experiments, we use 49946 records totally.

It is necessary to remark that determination of training patterns is one of the key tasks that need our attention. Improperly chosen patterns can lead to confusion of neural networks. A neural network "adapted" on incorrect patterns can give meaningless responses. During experiments, we used 21 patterns found by expert, representing classes with denotations "sell" and "buy". We have marked both classes of patterns with corresponding symbols "sell" and "buy". The training set contained 12 patterns "sell" and 9 patterns "buy".

We choose the flip pattern for our experiments. This pattern occurs in markets with both decreasing and increasing tendency. We can split this pattern to two subpatterns; a pattern anticipating price decrease and a pattern anticipating price increase. In the following Fig.1 – Fig.2, vertical lines on the left and on the right side of the each chart represent marks of the beginning and the end of the pattern as it was marked by the user. Following figures represent some particular patterns in our dataset. We have defined the ideal patterns anticipating uptrend (Fig.4) and the ideal pattern anticipating downtrend (Fig.3) to compare pattern recognition process on real pattern vs. user defined ideal patterns.

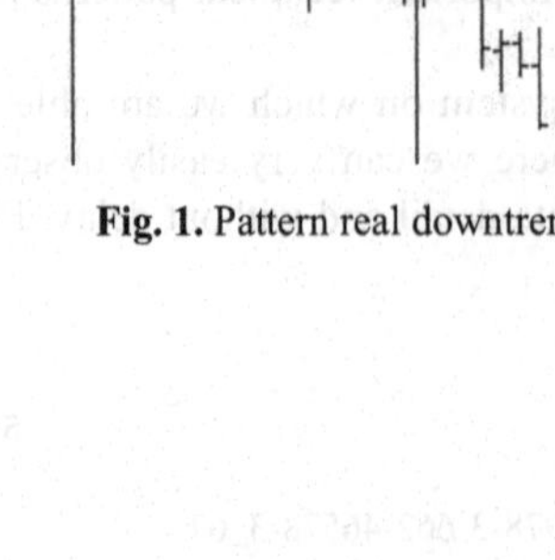

Fig. 1. Pattern real downtrend

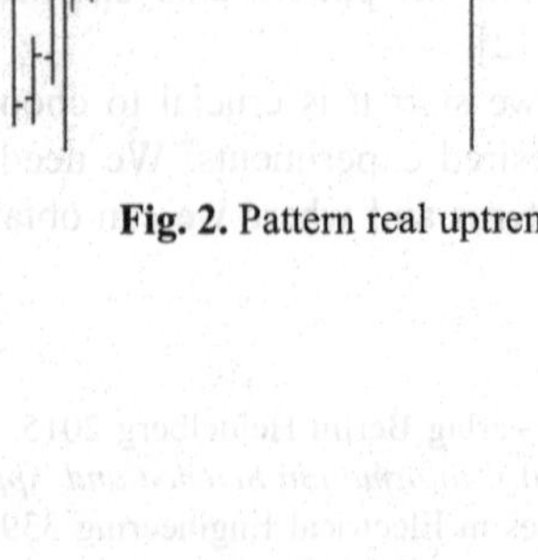

Fig. 2. Pattern real uptrend

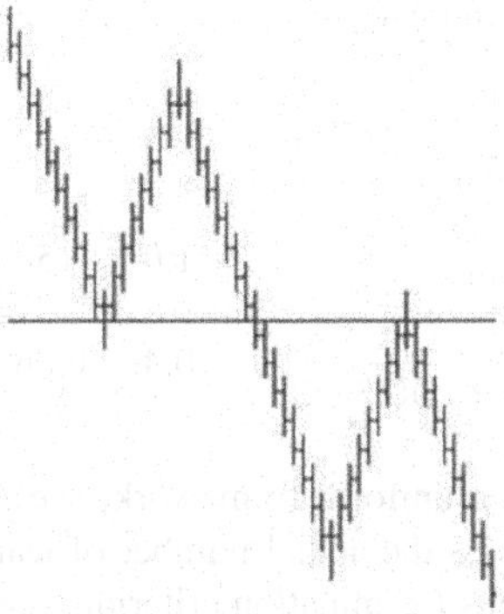

Fig. 3. Pattern ideal downtrend

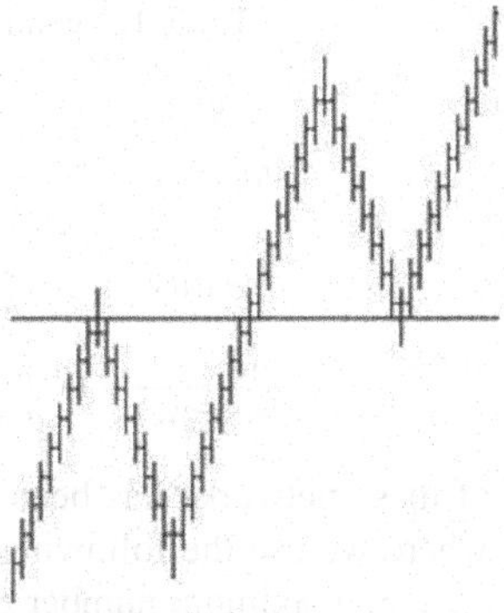

Fig. 4. Pattern ideal uptrend

Based on our experimental study we chose methods AREA (area under the graph) and LINE (points connected by lines) as appropriate binarization [6]. The reason was the fastest backpropagation adaptation with all patterns AREA and the ability of Hebb network to work only with patterns LINE and AREA. Based on our experimental study, we proposed a method of evaluating the relevance of the input vector components [7], see Algorithm 1.

```
1.Mark all items as irrelevant.
2.Load input vector of the first pattern and remember
  values of its items.
3.Repeat with all successive patterns:
  3.1 Load input vector.
  3.2 Mark every irrelevant item as relevant in case
   that its actual value differs from that in the first
   pattern.
4.End.
```

Algorithm 1: Searching for irrelevant items

2.2 Training algorithms optimization

In our experimental study we have used three different types of neural networks. Hebb network, Adaline and Back Propagation network. All classifiers work with the same set of inputs. Details about initial configurations of the used networks are shown in Table 1, where we use the following nomenclature: **Chyba! Nenalezen zdroj odkazů.**All the neural networks used the winner-takes-all strategy for output neurons (Y1,.....,Yn) when worked in the active mode, i.e. the Y_i is considered the winner if and only if $\forall j, i \neq j: y_j < y_i \vee (y_j = y_i \wedge i < j)$ i.e. the winner is the neuron with the highest output value y_i [7]. In the case that more neurons have the same output value, the winner is considered the first one in the order. It is important to note that the confidence parameter of such winner is set to zero.

Table 1. Neural networks initial configuration

Type	φ	Δw	α	C_l	C_c
Modified Adaline	*Identity*	$\alpha x\left(t - {y_o}/{\lambda}\right)$	0.3	20	0
Hebb	*Identity*	αxt	1.0	150	0
Back propagation	$\frac{2}{1+\exp(-y_{in})} - 1$	$\alpha x(t - y_o)\cdot\frac{1}{2}(1+y_o)(1-y_o)$	0.4	0.86	0

Each of these networks has been embedded into a uniform framework, see Algorithm 2, where we use the following nomenclature: e - the actual number of learning epoch; e_{max} - the maximum number of learning epochs (termination criterion); c_l - the confidence parameter; r - network is ready flag.

```
1.Initialize f = false
2.Repeat
  2.1 Do one learning epoch (process all training pat-
      terns and modify weights according to the network
      adaptation rule).
  2.2 Switch the network to an active mode, process all
      the training patterns and remember the confidence
      parameter of the worst learned pattern c_min.
3.Until e > e_max ∨ c_min < c_l
```

Algorithm 2: Learning of a neural network

We introduced the confidence concept in our previous work [6], which defined the confidence parameter in the following way: As Hebb algorithm tends to find almost all the introduced patterns familiar, its modification has been designed, tested and used for the active mode of Hebb rule. In a common operation, the output value y_{out} of each output neuron was derived from its input y_{in} value in a very simple way (1)

$$y_{out} = \begin{cases} -1, & y_{in} < 0 \\ 1, & y_{in} \geq 0 \end{cases} \tag{1}$$

The problem is that y_{out} takes the same output no matter how big its y_{in} value is (apart from sign rules). In other words, we do not have any information about the certainty of the network result. Our network output function was replaced with a simple equivalence: $y_{out} = y_{in}$. Then we can use parameter which means minimum $|y_{out}|$ value required to accept the neural network result. We call the parameter *confidence* in case, that $|y_{out}| < confidence$ for some output, the neural network result is ignored. Using the *confidence* parameter, we can easily and accurately regulate count (and a minimum of quality) of found patterns. As a matter of course it is not easy to find an optimal value for the *confidence* parameter. One of the reasons is the fact that the optimal value vary with patterns' size, patterns amount and analysis results requirements.

Unfortunately, thus defined parameter confidence works only for the Hebb network. In this experimental study, we have slightly changed the determination of the confidence parameter, which is specified separately for both learning and classification phases of neural network activity. A separate setting of both confidence parameters allows to control the accuracy of the learning phase and the benevolence of the network during the classification phase independently. Provided that $y_w \geq Y_s \wedge \forall i \neq w, i \neq s: y_s \geq y_i$ C is the confidence parameter of the classification phase, which is defined as follows: $C = y_w - y_s$, where y_w represents the output value of the winner neuron and y_s represents the output of the second-in-order neuron.

2.3 Data analysis - pattern recognition

The data analysis is the main task of the experiment. Data files were presented to the learned networks during the analysis. The aim of the analysis was to obtain a list of occurrences of learned patterns in data. Data analysis was conducted as a simulation of real operation. Entries were submitted to the networks successively, so as they accumulate in time. The software tried to use neural networks to find a familiar pattern in the data section ranging from the past to the present.

We used a stretching-window approach. Searching always proceeded within the designated permissible minimum and maximum length (number of records) of pattern. Minimal length *pMin*=15rec and maximal length *pMax*=55rec were used during our experiment. These values were received experimentally from minimal and maximal lengths of the training patterns. Searching began at the length of *pMin*, which gradually grew larger until it reached the length of *pMax*. If more than one pattern was found (the most cases), then only one with the greatest degree of learned pattern fitness (e.g. the highest confidence value) was chosen. Then its occurrence was at the actual (the last) record position, see Fig.5.

Patterns usually tend to form clusters, which we call bunches, around positions of their occurrences. As the working window passes over the actual pattern, the system recognizes patterns at several positions.

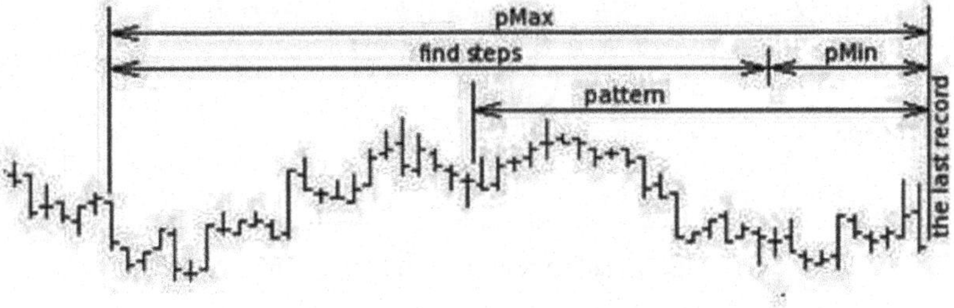

Fig. 5. The occurrence of pattern at the actual record position

3 Knowledge discovery via data analysis

To interpret the data from the analysis it is necessary to create a basic trading system and trade all the recommendations provided by the neural network. Our goal is to select at least two kinds of similar patterns: one for short positions - *sell signal* and the second one for long positions - *buy signal*. Then, an adaptation of our system ensues based on neural networks to recognize these patterns with desired response. The neural network should present a recommendation if "buy" or "sell" at the moment. Based on the neural network recommendation our automatic trading system will enter short (sell) or long (buy) position.

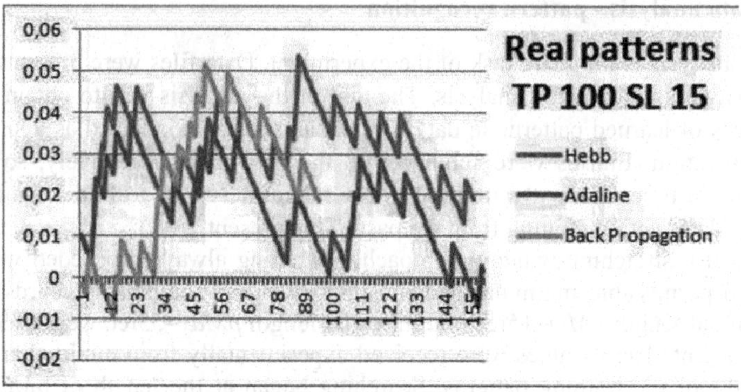

Fig. 6. Trading results with real patterns

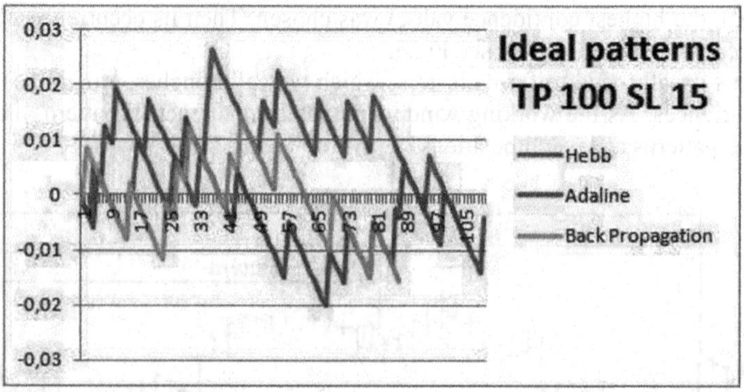

Fig. 7. Trading results with ideal patterns

We present results of trading in the months March - October 2012 based on recommendations using our Hebb, Adaline and Back Propagation neural networks implementation. Fig.6 shows a graph with trading results based on the recommendations using outputs from our Hebb, Adaline and Back Propagation network implementation with real patterns. As we can see in the graph, the result of trading settled just between -0.01 and 0.05. Fig.7 shows a graph with trading results based on the recommendations using outputs from our Hebb, Adaline and Back Propagation network implementation with ideal patterns. As we can see in the graph, the result of trading settled just between -0.02 and 0.03.

Even though each neural network produced the same number of pattern marks, so some of these marks were too close together and therefore were skipped during one trade operation. Because of that some networks produced less number of trades than the others. If we would like to compare obtained results, a Hebb network seems to deliver the best trading results independent on the training pattern type (real or ideal). Our goal was to keep our account balance around zero and that was achieved. But the trading results are one of the possible criteria only.

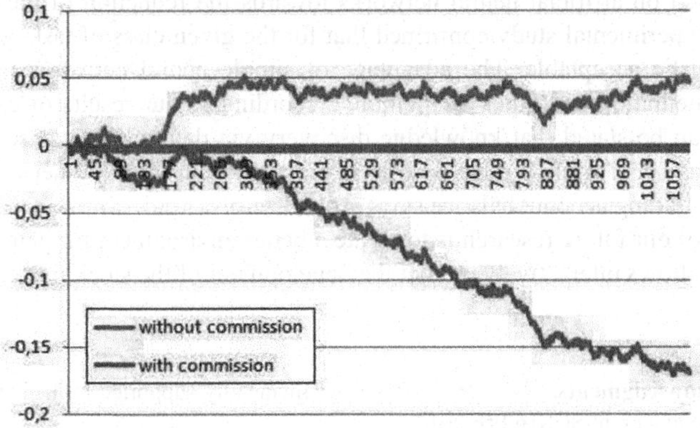

Fig. 8. Hebb algorithm - managed trading

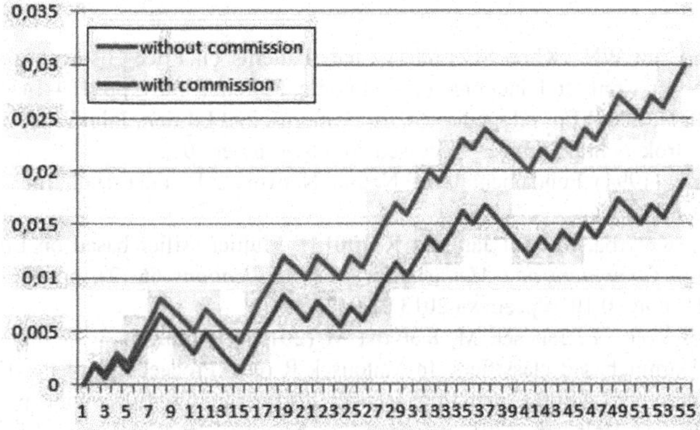

Fig. 9. Human managed trading

We present results of trading in the months April - June 2011 based on both recommendations: using our Hebb neural network implementation, Fig.8 and human expert,Fig.9. Results of the trading without commissions and with commissions are displayed in each graph. In the case of trading with commissions, value 0.0002 was subtracted from each transaction. The x axis indicates the number of trades during the given period, the y axis indicates the profit. As for the profit, the absolute value of the trading account is not important. More important is the trading equity itself.

4 Conclusion and future work

In this paper we have showed a short introduction into the field of at knowledge discovery in dynamic data via classification based on neural networks. We focus on reliability of recognition made by the described algorithms with optimized training patterns based on artificial neural networks towards the reduction of the calculation costs. Our experimental study confirmed that for the given class of tasks simple classifiers could be acceptable. The advantage of simple neural networks is their very easy implementation and quick adaptation. According to our results of experimental studies, it can be stated that knowledge discovery via data analysis based on neural networks was successful. We have fulfilled our goal with all three networks as well because our trading account balance stays around zero.

The aim of our future research is to create a better environment for pattern recognition, which allows interactively predict the development of the stock market.

Acknowledgments. The research has been financially supported by University of Ostrava grant SGS16/PřF/2014.

References

1. Anand S, Chin WN, Khoo SC (2001) Chart Patterns on Price History. Proc. of ACM SIGPLAN Int. Conf. on Functional Programming, Florence, Italy, pp. 134-145.
2. Bulkowski N (2005) Encyclopedia of Chart Patterns, 2nd Edition. John Wiley and Sons.
3. X-Trader Brokers http://xtb.cz. Accessed 20th November 2012.
4. Fausett LV (1994) Fundamentals of Neural Networks. Prentice-Hall, Inc., Englewood Cliffs, New Jersey.
5. Volna, E., Kotyrba, M. and Jarusek, R. (2013). Multiclassifier based on Elliott wave's recognition. *Computers and Mathematics with Applications* **66** (2), pp. 213-225. ISSN: 0898-1221. doi: 10.1016/j.camwa.2013.01.012.
6. Kocian V, Volna E, Janosek M, Kotyrba M (2011) Optimizatinon of training sets for Hebbian-learningbased classifiers. In Matoušek R (ed.) Proceedings of the 17th International Conference on Soft Computing, Mendel 2011, Brno, Czech Republic, pp. 185-190.
7. Kocian, V. and Volná, E. (2012) Ensembles of neural-networks-based classifiers. In R. Matoušek (ed.): Proceedings of the 18th International Conference on Soft Computing, Mendel 2012, Brno, Czech Republic, pp. 256-261.
8. Leigh W, Modani N, Hightower R (2004) A Computational Implementation of Stock Charting: Abrupt Volume Increase As Signal for Movement in New York Stock Exchange Composite Index. Decision Support Systems, 37 (4) 515-530.
9. Moore M, Roche M (2002) Less of a Puzzle: A New Look at the Forward Forex Market. Journal of International Economics 58, 387-411.

Data Mining for Industrial System Identification: A Turning Process

Karin Kandananond

Valaya Alongkorn Rajabhat University, Thailand
kandananond@hotmail.com

Abstract. The modeling of an industrial process is always a challenging issue and has a significant effect on the performance of the industry. In this study, one of the most important industrial processes, a turning process, is considered as a black box system. Since it is also a dynamic system, i.e., its characteristics changing over time, the system identification method has been applied on the measurement data in order to obtain an empirical model for explaining a system output, surface roughness. The inputs of the system are feed rate, cutting speed and tool nose radius. According to the study, three non-parametric models, Box-Jenkins, autoregressive moving average with exogenous inputs (ARMAX) and output error (OE), are recommended to be used to construct mathematical models based on data mining available from the manufacturing process. These system identification models are appropriate to model the dynamic turning process since they have the capability to construct both dynamic and noise parameters separately.

Keywords: Surface roughness, System identification, Turning process.

1 Introduction

Due to the manufacturing process of metal workpieces, surface finish is an important characteristic of the workpieces achieved from machine operations. To achieve the best surface roughness, machining process is considered as a black box so the experimental data from the operation is collected and utilized to construct an empirical model. For turning processes, input factors, e.g., depth of cut, cutting speed, feed rate, rake angle, cutting edge angle and tool nose radius, are the outputs of the system while the common output is the surface roughness. As a result, different data mining methods, artificial neural network (ANN) and support vector machine (SVM), have been widely applied in order to model the surface roughness. However, if the behavior of the system is dynamic, another approach, system identification method, might be an alternative option for modeling the turning process.

K.J. Kim (ed.), *Information Science and Applications*,
Lecture Notes in Electrical Engineering 339, DOI 10.1007/978-3-662-46578-3_68

2 Literature Review

According to the literature, ANN method was utilized by Pontes et al. [1] to construct a model for predicting the surface roughness of workpieces operated from a turning process. Natarajan, Muthu and Karuppuswamy [2] also applied ANN to determine the relationship between cutting speed, feed rate, depth of cut and surface roughness in order to increase the efficiency of a turning process. The comparison of ANN and multiple linear regression was done by Asilturk and Cunkas [3] and Asilturk [4]. The performance of another method, SVM, was compared with the one of ANN by Caydas and Ekici [5] in the turning process of stainless steel. Gupta [6] also compared the modeling capability of ANN, SVM and a design of experiment approach, response surface method (RSM). Garg, Bhalerao and Tai [7] pointed out that empirical modelling techniques such as artificial neural networks, regression analysis, fuzzy logic and support vector machines were used for predicting the performance of the process. According to the literature, a number of data mining methods were utilized to characterize and optimize the relationship between the surface roughness and other machining parameters.

On the other hand, another data mining method to obtain an empirical model from the experimental data is done through system identification method. System identification is popular in the modeling of a plant system. However, the black box system must be the dynamic system which evolves with time. Therefore, if the turning process is continuous and changes over time, system identification method should be another interesting option for modeling the surface roughness.

3 Method

System identification method is the utilization of different statistical methods to construct a mathematical model of a dynamic system by focusing on inputs and outputs. The general equation of dynamic model is the discrete linear time-invariant or the composition polynomial models where input (y(t)) and output (u(t)) are shown as follows:

$$y(t) = G(q)u(t) + H(q)e(t) \tag{1}$$

, where G(q) is the relationship between inputs and output and it is represented in the form of the following transfer function or the parameter of dynamic system. The pole of dynamic system is determined in F(q).

$$G(q) = B(q)/F(q) \tag{2}$$

On the contrary, H(q) is the relationship between noise and output or parameters of noise system as shown in (3). D(q) determines the noise of system.

$$H(q) = C(q)/D(q) \tag{3}$$

, where B(q), F(q), C(q) and D(q) are polynomials with

$$B(q) = b_1q^{-1} + \ldots + b_nq^{-n} \tag{4}$$

$$F(q) = 1+f_1q^{-1} + \ldots + f_nq^{-n} \tag{5}$$

$$C(q) = 1+c_1q^{-1} + \ldots + c_nq^{-n} \tag{6}$$

$$D(q) = 1+d_1q^{-1} + \ldots + d_nq^{-n} \tag{7}$$

, where q^{-n} is the shift operator so $q^{-n}u(t)$ is the output signal at time $t = t-n$ or $u(t-n)$. According to (1)-(7), (8) is called the Box-Jenkins (BJ) model.

$$y(t) = [B(q)/F(q)]u(t) + [C(q)/D(q)]e(t). \tag{8}$$

BJ model has the capability to estimate both dynamic and noise parameters. Therefore, the construction of a model is based on the measurement noise, not the input noise. As a result, the structure of BJ model provides high flexibility for modeling noise. Moreover, the above model can be extended into different subclasses of how to model dynamic and noise parameters. The dynamic model can be extended into these following models:

3.1 ARX (Autoregressive Model with Exogenous Input)

ARX is the extended model of BJ but ARX has coupled the parameter estimation of noise and dynamic independently as shown in (9) and (10). A(q) corresponds to poles that are common for the dynamic model and the noise model.

$$A(q)y(t) = B(q)u(t) + e(t) \tag{9}$$

$$y(t) = (B(q)/A(q))u(t) + (1/A(q))e(t) \tag{10}$$

3.2 ARMAX (Autoregressive Moving Average Model with Exogenous Input)

For ARMAX model, the ARX model is modified to estimate noise by using the C(q) parameters (the moving average of random noise). As depicted in (11) and (12), ARMAX is utilized when there are disturbances at the input. The advantage of ARMAX model is that the deterministic and stochastic parts are calculated separately.

$$A(q)y(t) = B(q)u(t) + C(q)e(t) \tag{11}$$

$$y(t) = (B(q)/A(q))u(t) + (C(q)/A(q))e(t) \tag{12}$$

3.3 Output error (OE)

OE model focuses on the estimation of dynamic model only while a noise model is equal to 1 or the noise is completely random as shown in (13).

$$y(t) = [B(q)/F(q)]u(t) + e(t) \tag{13}$$

4 Process

The application of system identification method is illustrated through one of the material processing techniques, turning process, and the material used for the experiment is bronze. The workpieces produced are bronze bushing as shown in Fig. 1. The cutting fluid is also used in the process to reduce the heat generated while the workpieces are produced. The surface finish of workpieces at the outside surface is measured in the form of surface roughness (R_a: unit, microinch) or the output signal (y_1) of the system. The system has three factors or inputs, namely, u_1 or speed (ft/min), u_2 or feed rate (inch per round) and u_3 or nose radius (inch). The range of each input is represented in Table 1.

Table 1. Range of input signals.

Input	Range
Speed	900-1150 ft/min
Feed rate	0.008-0.0095 inch per round
Nose radius	0.0156-0.0625 inch

The inputs and output signal are shown in Fig. 2 and each workpiece is produced every minute which is the same number as the sampling interval. Therefore, the total number of samples are 320 pieces. Since the characteristic of turning process has changed over time or it can be considered as a dynamic system.

Fig.1. Workpiece.

5 Results

The system identification is facilitated and carried out by using a toolbox provided by MATLAB software, system identification or ident. The available methods to estimate the output are linear parametric, process, non-linear, spectral and correlation models. All methods are run to find the optimal model to fit the observations. The results show that parametric identification method is the best model. The top performing models are selected by determining how accurate each model can fit the measured data. Among these models are autoregressive moving average exogenous (ARMAX), Box-Jenkins (BJ) and output error (OE).

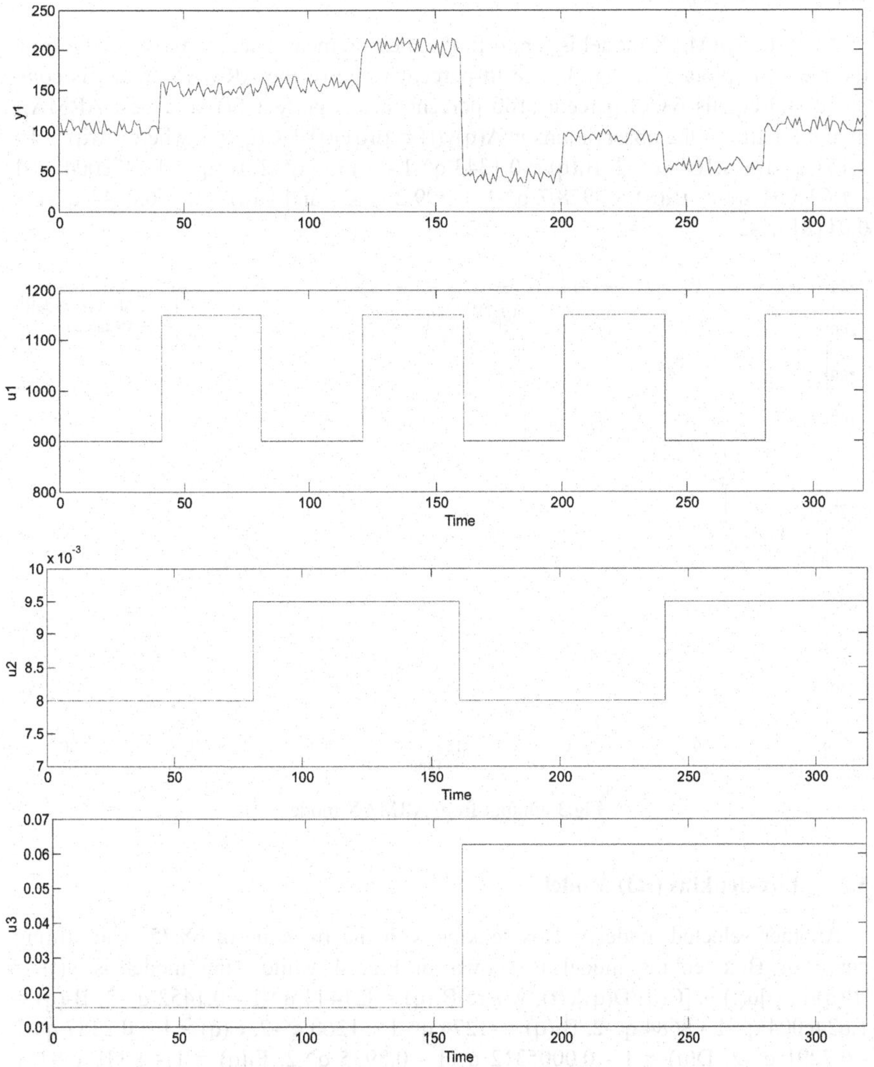

Fig.2. Output and input signals.

5.1 ARMAX Model

After the ARMAX model is run to fit the measurement data, the fitting curve from the model is plotted in Fig. 3. The fit percent value of the ARMAX model is computed and equals 68.73 percent (100 percent means perfect fit). The best ARMAX model for fitting the observations is A(q)y(t) = B(q)u(t) + C(q)e(t), where A(q) = 1 - 1.151 q^-1 + 0.1684 q^-2, B_1(q) = 0.1343 q^-1 - 0.1336 q^-2, B_2(q) = 1.647e004 q^-1 - 1.62e004 q^-2, B_3(q) = -928.7 q^-1 + 899.2 q^-2 and C(q) = 1 - 0.8144 q^-1 + 0.01181 q^-2.

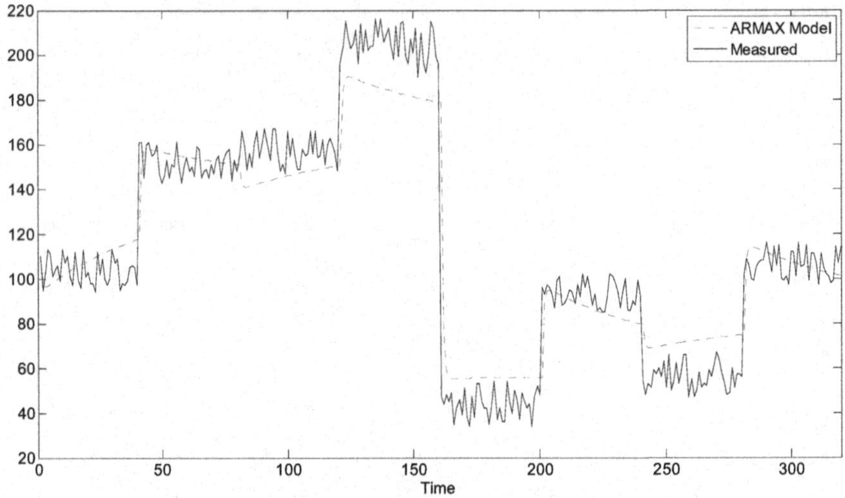

Fig.3. Output from ARMAX model.

5.2 Box-Jenkins (BJ) Model

Another selected model is Box-Jenkins with the fit value of 68.75. The fitting curve of Box-Jenkins model is shown in Fig. 4 while the model is y(t) = [B(q)/F(q)]u(t) + [C(q)/D(q)]e(t), where B_1(q) = 0.1444 q^-1 - 0.1452 q^-2, B_2(q) = 1.624e004 q^-1 - 8684 q^-2, B_3(q) = -1274 q^-1 + 1260 q^-2, C(q) = 1 + 0.2387 q^-1 - 0.7291 q^-2, D(q) = 1 - 0.0005312 q^-1 - 0.8935 q^-2, F_1(q) = 1 - 1.117 q^-1 + 0.1258 q^-2, F_2(q) = 1 - 0.6679 q^-1 + 0.02203 q^-2 and F_3(q) = 1 - 0.8562 q^-1 - 0.1263 q^-2.

5.3 Output Error (OE) Model

Another parametric method is output error (OE) model and the result of the fitting by this model is illustrated in Fig.5. Its fitting value is 74.69 and the optimal OE model is y(t) = [B(q)/F(q)]u(t) + e(t), where B_1(q) = 0.2114 q^-1 - 0.212 q^-2, B_2(q) = 2.641e004 q^-1 - 2.549e004 q^-2, B_3(q) = -1663 q^-1 + 1635 q^-2, F_1(q) = 1 - 0.8202 q^-1 - 0.1706 q^-2, $F_2(q)$ = 1 - 0.3369 q^-1 - 0.619 q^-2 and $F_3(q)$ = 1 - 1.064 q^-1 + 0.08577 q^-2

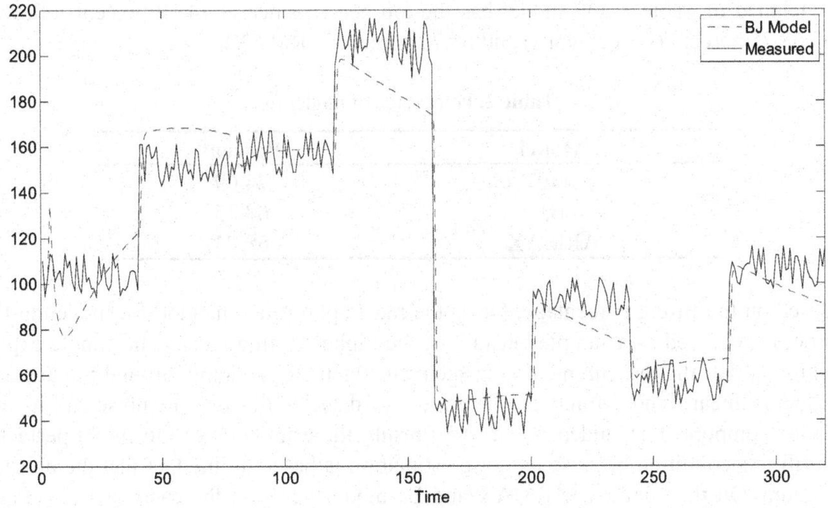

Fig.4. Output from BJ model.

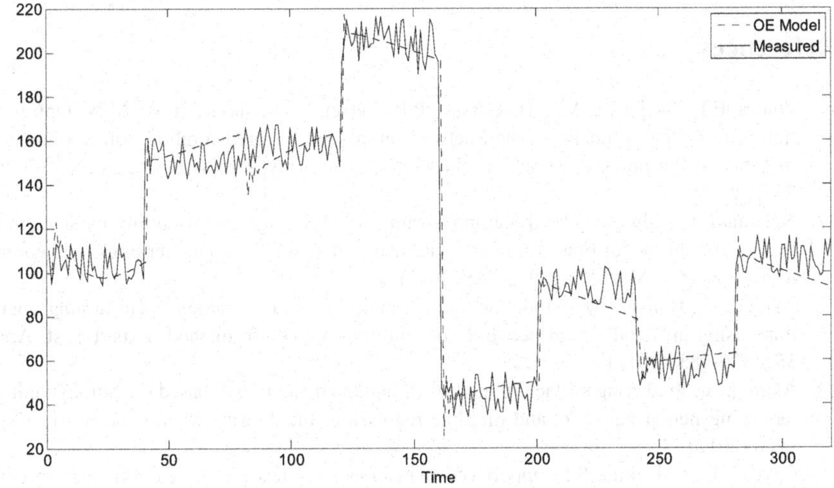

Fig.5. Output from OE model.

6 Conclusions and Discussions

According to the results, the percentage of model fit from the above three model is concluded in Table 2. OE model has the top performance at 74.69 percent followed by 68.75 percent (Box-Jenkins) and 68.73 percent (ARMAX).

Table 2. Percentage of model fit.

Model	Percent
OE	74.69
BJ	68.75
ARMAX	68.73

Based on the fitting percentage, OE is the most appropriate method for modeling the process deployed as a sample because of the highest fitting value. The simple explanation is that the measurement data seems to fluctuate randomly around the average value or linear trend. Moreover, the pattern of data implies that the noise and its dynamic component are independent. As a result, the estimation of OE might perfectly fulfill this condition because its noise estimation is based on the fact that the noise is random. On the contrary, ARMAX and BJ model work on the basis that noise and dynamic parameters are not independent. Therefore, OE is preferred to ARMAX or BJ.

References

1. Pontes, E.F., De Paiva, A.P., Balestrassi, P.P., Ferreira, J.R. and Da Silva, M.B.: Optimization of radial basis function neural network employed for prediction of surface roughness in hard turning process using Taguchi's orthogonal array. Expert Syst. Appl. 39, 7776–7787(2012)
2. Natarajan, C., Muthu, S. and Karuppuswamy, P.: Investigation of cutting parameters of surface roughness for brass using artificial neural networks in computer numerical control turning. Aust. J. Mech. Eng. 9, 35-45 (2012)
3. Asilturk, I., Cunkas, M.: Modeling and prediction of surface roughness in turning operations using artificial neural network and multiple regression method. Expert Syst. Appl. 38, 5826-5832(2011)
4. Asilturk, I.: Predicting surface roughness of hardened AISI 1040 based on cutting parameters using neural networks and multiple regression. Int. J. Adv. Manuf. Tech. 61, 1263-1268 (2012)
5. Caydas, U. and Ekici, S.: Support vector machines models for surface roughness prediction in CNC turning of AISI 304 austenitic stainless steel. J. Intell. Manuf. 23(3), 639-650 (2012)
6. Gupta, A.K.: Predictive modelling of turning operations using response surface methodology, artificial neural networks and support vector regression. Int. J. Prod. Res. 48(3), 763-778 (2010)
7. Garg, A., Bhalerao, Y. and Tai, K.: Review of empirical modelling techniques for modelling of turning process. Int. J. Model Ident. Contr. 20, 121-129 (2013)

Investigating the Effectiveness of E-mail Spam Image Data for Phone Spam Image Detection Using Scale Invariant Feature Transform Image Descriptor

So Yeon Kim[1], Yenewondim Biadgie[1] and Kyung-Ah Sohn[1*]

[1]Department of Information and Computer Engineering, Ajou University, Suwon, S. Korea
{jebi1771, kasohn}@ajou.ac.kr

Abstract. The increased number of spam images in mobile phones has become a big trouble by annoying users steadily. One big issue in developing an effective phone spam image detection system using machine learning and data mining techniques is unavailability of sufficient phone spam image data. In this study, we demonstrate that the utilization of similar email spam image data obtained by chi-square similarity distance is an effective solution to develop phone spam image classifier. We compared the performance of our approach with the one using randomly selected email spam image data and showed that this approach works better than the one using randomly selected images. Our analysis further illustrates that a more sophisticated clustering algorithm is expected to improve the performance.

Keywords: Spam detection · Image spam · similarity measure · Image classification · dense SIFT feature · k-means clustering

1 Introduction

The growing number of spam messages in a smart phone has become a big issue in these days. Due to the impact of recent massive personal information leaks from credit card companies, the amount of fraudulent spam messages is increasing. As a result of this existing spam filtering systems cannot detect them. Especially, unlike in an email, those phone spams are annoying because it contains unsolicited commercial advertisements and also links for users to automatically charge a fee. Recent spam filtering systems can detect spam phone numbers from a user-supplied database. However, it is a temporary solution in that spammers can hide themselves by changing their sending numbers.

Although there have been many previous approaches to filter spams, they typically focused on text spams, but not on image spams. In recent months, however, the number of image spams in mobile phones has steadily increased. Such fraudulent image spams are going to increase not to be caught by an advanced filtering system.

Image spams are usually appeared in emails and there have been many studies for detecting email spams. But those were not specifically designed to detect recently

* Corresponding author

K.J. Kim (ed.), *Information Science and Applications*,
Lecture Notes in Electrical Engineering 339, DOI 10.1007/978-3-662-46578-3_69

increased phone spam images. In comparison with email spams, the size of phone image spam data is smaller and image data among them is too small to train a model.

This paper carefully investigates the contribution of using large email spam data to detect phone spam images. As a feature of email spam image is quite different from the one from phone spam images, we collect email spam images that show high similarity to phone spam images. Consequently, by training with both email and phone spam images, we propose a predictive model for phone spam image classification. We show that the use of email spam images which are similar to phone spam is effective on filtering phone spam images by comparing its performance with randomly selected email spams.

2 Methodology

We proposed two main steps to classify phone spam images. Thus, the system is divided into two main steps, 'Data acquisition' and 'Spam classification'.

2.1 Data Acquisition

In the first stage (Fig. 1), we gained similar e-mail spam images as phone ones to keep spam data big enough to train the model. We obtained the similarity matrix between e-mail spam images and phone spam images by measuring their color histogram similarity using chi-square method. The similarity matrix is partitioned into K mutually exclusive clusters by K-means clustering algorithm. The cluster which has most similar e-mail spam images is added with phone spams as a whole dataset.

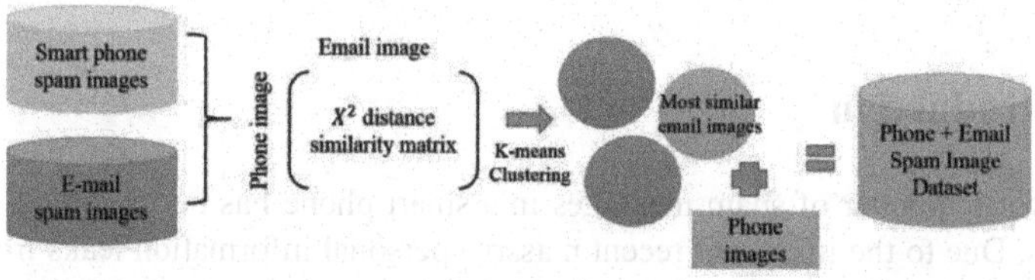

Fig. 1. Procedure to select email spam images which have high similarity with Phone Spam Images.

Collecting phone spam image data is one big challenge in that there is not any public database and furthermore, there are not a variety of spam images in phone yet. In this study, we manually gathered embedded images in phone messages which are 66 spams and 405 non-spams. Such phone spam images are too small to train the predictive model. To solve this issue, we gained some similar email spam images from a huge public spam data based on similarity measure.

Image Spam Hunter [2] collected 929 email spam images. As shown in Fig. 3, email spam images have visually different features from phone spam images. Most of email spam images are like left one in Fig. 3(b) which include sentences and use a few colors just to highlight some sentences or the title. In contrast with the email, phone spam images (in Fig. 3(a)) contain various colors and images. They emphasize key words rather than sentences. However, we figured out some email images have such phone-like features as a second image in Fig. 3(b). In this respect, it is positively necessary to extract some email images which have similar features to phone images.

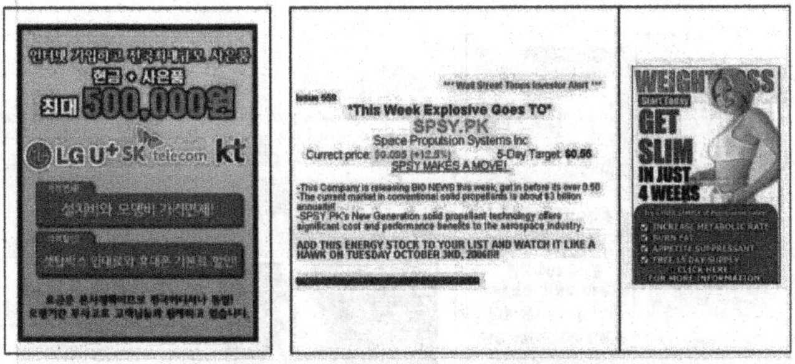

(a) Phone spam image (b) Email spam image

Fig. 2. A sample image from smart phone and e-mail respectively.

Similarity matrix of histogram distances. In order to find the similarity between email and phone spam images, we compute the chi-square distance [3] between the color histograms of each pair of email and phone spam images. To compute the distance between two images, we first extract two set of histograms $E_{email} = [E_1, E_2, \ldots, E_n, \ldots, E_N]$, $P_{phone} = [P_1, P_2, \ldots, P_m, \ldots, P_M]$ with N email images and M phone images respectively. Chi-square distance between histograms of email $E_n = \{e_1, e_2, \ldots, e_b, \ldots, e_B\}$ and those of phone $P_m = \{p_1, p_2, \ldots, p_b, \ldots, p_B\}$ with B bins is defined as

$$D = d_{\chi^2}(E_n, P_m) = \sum_{b=1}^{B} \frac{(e_b - m_b)^2}{m_b}, \quad (1)$$

where $m_b = \frac{e_b + p_b}{2}$ [3].

Each image is represented RGB color histograms which have 4 bins per red, green and blue. Totally, a single image has B = 64 bins. We used N = 929 email spam images from Image Spam Hunter and M = 66 manually collected phone spam images. Given every pair of email and phone spam image histogram, similarity matrix of chi-square distance D is obtained which is N-by-M matrix. Each element of this matrix represents how similar email and phone spam image are, that is, the lower the distance is, the more similar they are.

***k*-Means clustering.** For every email spam image, it has similarity value to each phone spam image in the similarity matrix D. To find the most similar group of email spam images, we partitioned distance matrix D into k mutually exclusive clusters by k-means clustering. The centroid of a cluster is the representative chi square distance of it. The cluster with the smallest centroid is the most similar one. We performed k-means clustering iteratively when k = 2, 3, 4, 5. We determined optimal k = 3 heuristically because 3 clusters are best separated with respect to the number of shared features with phone spam images as shown in Fig. 4. As a result, we added 353 email spam images from the cluster which has the smallest centroid. To sum up, 419 spam images and 405 non-spam images are used as described in Table 1.

(a) Phone spam images (b) Email spam images

Fig. 3. Sample phone spam images (a) and email spam images among the cluster which shows high similarity to phone spams (b).

Table 1. Dataset used with email spam images by k-means clustering with similarity measure.

	Spam	*Non-spam*
Phone	66	405
Email	353	-
Total	419	405

2.2 Image Representation

In the second stage (Fig. 2), each input image is represented by PHOW (a Pyramid Histogram Of visual Words) descriptor [1]. First, multiple scale invariant feature transform (SIFT) descriptors are computed in each single image. By K-means clustering, the extracted descriptors are vector quantized into K visual words. The spatial bag (histogram) of visual words are obtained in p × q regions. After all, using p × q spatial histograms together are represented for PHOW descriptor of each image. Training with SVM (Support Vector Machine), this descriptor finally goes to classify the image as spam or non-spam.

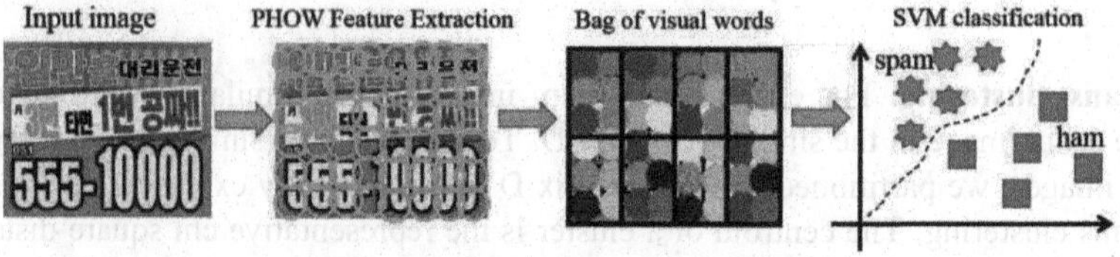

Fig. 4. Feature extraction of an input image and spam image classification.

For image representation and classification, we followed implementation of image categorization in publicly available VLFeat open computer vision library [4].

Feature extraction. To extract features of an image, we used dense scale SIFT features (PHOW features [1]). PHOW feature is state-of-the-art algorithm for extracting features in high-dimensional space. To construct visual words, SIFT descriptors are extracted at each point on a regular grid with 3 pixels spacing at four scales using concentric circles with radius 4, 6, 8, 10 pixels respectively.

Bag of visual words. The descriptors of images are clustered into visual words by k-means clustering [5]. To learn visual word, 500 visual words from training images are obtained to construct the visual vocabulary dictionary. KD-tree [6] is built to quantize descriptors of a test image into visual words from visual dictionary.

Spatial Histograms. To get PHOW descriptors of an image, first the image is subdivided into 2x4 sub-regions. Second, the SIFT descriptor of each sub-region is computed. The image is subdivided into sub-regions to consider the spatial co-occurrence of the histograms of each sub-regions of an image. Finally spatial histograms [7] of each region of an image is obtained, which are histograms of 500 visual words. The result feature vector of an image is the concatenation of 8 spatial histograms as an image descriptor.

2.3 Spam Image Classification

To achieve our final goal, we trained our model with SVM (Support Vector Machine) from the obtained image feature vector. For better classification, χ^2-kernel SVM is used in VLFeat library, which trains the model with SVM using homogeneous kernel map [8]. The homogeneous kernel map transforms the large bag-of-words image data into the linear representation and computes non-linear χ^2-kernel SVM. A. Vedaldi and B. Fulkerson [8] proved that χ^2-kernel SVM performs better than one with other kernels. Therefore, we performed SVM when the soft margin of which is 10 on each training image data and classified test images into spam or non-spam. Finally, we evaluated our result with respect to sensitivity and specificity.

2.4 Cross validation and Evaluation

To prevent overfitting to the training data, we performed 5-fold cross validation and evaluated the performance. Note that we would classify phone spam images by training with phone and email spam images. We used 20% of phone images as a validation set, and 80% of phone and email images as a training set. At each run, accuracy, sensitivity, specificity and F-measure of the model are obtained as an evaluation measure.

Additionally, in order to demonstrate that applying email spams by k-means clustering with similarity measure is effective for phone spam image classification, we compared our model with the one using randomly selected email spam images. In this manner, 10, 20, ⋯, 80, 90% of email spam images are randomly chosen. Also we experimented with 38% of those (353 images) which is the same size as the most similar cluster by k-means clustering algorithm.

3 Results

Performance comparison with 5-fold cross validation. We examined the performance of our model using email spam images by similarity measure as compared to it using randomly selected ones. We focused on True Positive rate (Sensitivity) and True negative rate (specificity) for evaluating the performance. The higher sensitivity is, the more accurately the model can predict actual spam image as spam. The higher specificity is, the less it is likely to filter legitimate image as spam.

In Fig. 5, we compared the model when we use email spam images by k-means clustering with similarity measure, and randomly selected ones respectively. The x-axis represents the percentage of random email spam data used in training set.

Overall, the best performance is shown when similarity between email and phone image is measured. This shows that email spam image data make a positive contribution to phone spam image classification. Also, we find that the bigger the size of random email spam data, the more different features from phone spams are trained, therefore it has high sensitivity (low false negative rate) and low specificity (high false positive rate). In other words, it is more likely to filter non-spam image as spam. It implies that if the email data size is increased, more accurate clustering algorithm is needed because there are many different types of features from phone spams in a large-scale email spam data.

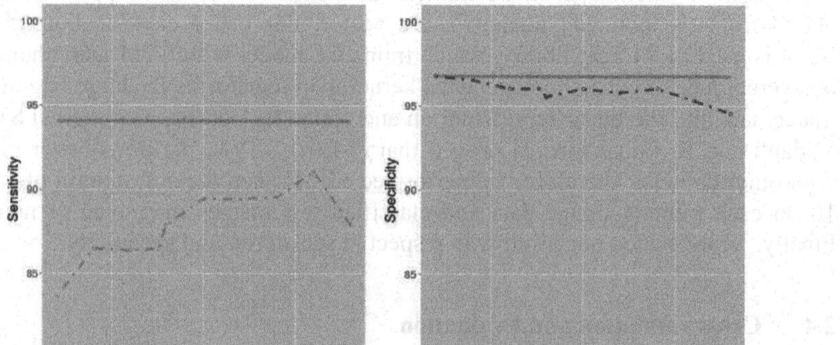

Fig. 5. Average sensitivity and specificity using 5-fold cross validation on training data using email spam data by k-means clustering with similarity measure(solid line) and randomly selected one across the data size(dash-dotted line).

Misclassified images on validation set. The following images in Fig. 6 are misclassified samples on validation set when we train with similar email spam image by k-means clustering. Images in Fig. 6(a) are actual legitimate image predicted as spam. Some of them are coupons with barcode which user may have agreed to get or the news or alert message from mobile service provider which can be annoying for someone but needed for others. Although those images are hard to classify accurately, only a few images are misclassified. In Fig. 6(b), images are classified as legitimate even though it has visually similar features as phone spam images. But, the size of false negatives is quite small as compared to other validation set which is trained with random email spams.

(a) False positives (b) False negatives

Fig. 6. Misclassified sample images on validation set (phone spam images) when it is trained with similar email spam image by k-means clustering.

Average performance. The average accuracy, sensitivity, specificity and F-measure on best performed validation set (has the highest F-measure) are shown in Table 2. When we trained the model with random images, the best performance is obtained when 50% of email spam images (465 random images) are used in training set. Although the specificity of k-means method is slightly better than randomized method, it outperforms randomized method with respect to sensitivity. In general, we show that the performance of the model using email spam data with similarity measure is better than that with random images.

Table 2. Average performance comparison between k-means clustering with similarity measure and randomized method on validation set.

	k-means	**random (50%)**
Accuracy	**96.39%**	95.12%
Sensitivity	**94.07%**	89.45%
Specificity	**96.79%**	96.05%
F-measure	**87.94%**	83.80%

4 Conclusions

In this paper, we investigated the effectiveness of using email spam image data for phone spam image classification. We stated the problem of data acquisition for phone spam image classification and brought the idea of using public email spam image dataset. In addition, we computed similarity distance between email and phone spam images and obtained the most similar email spam image group. Our results show that using email spam image data gained by similarity measure is quite effective for phone spam image classification. It is interesting in that we demonstrate the effective use of dataset from other source with similarity measure.

In our results, we also find that if email data size becomes larger, it has many kinds of feature group. In future works, a more precise clustering algorithm such like topic modeling or similarity distance could be used for a larger scale of spam image data.

Acknowledgements This research was supported by the Basic Science Research Program through the National Research Foundation (NRF) of Korea funded by the Ministry of Science, ICT, and Future Planning (MSIP) (NRF-2014R1A1A3051169), and by the MSIP under the Global IT Talent support program (NIPA-2014- H0904-14-1004) supervised by the NIPA(National IT Industry Promotion Agency).

References

1. Bosch, A., A. Zisserman, and X. Munoz, *Image classification using random forests and ferns.* 2007.
2. Gao, Y., et al. *Image spam hunter.* in *Acoustics, Speech and Signal Processing, 2008. ICASSP 2008. IEEE International Conference on.* 2008. IEEE.
3. Rubner, Y., C. Tomasi, and L.J. Guibas, *The earth mover's distance as a metric for image retrieval.* International Journal of Computer Vision, 2000. **40**(2): p. 99-121.
4. Vedaldi, A. and B. Fulkerson. *VLFeat: An open and portable library of computer vision algorithms.* in *Proceedings of the international conference on Multimedia.* 2010. ACM.
5. Elkan, C. *Using the triangle inequality to accelerate k-means.* in *ICML.* 2003.
6. Muja, M. and D.G. Lowe. *Fast Approximate Nearest Neighbors with Automatic Algorithm Configuration.* in *VISAPP (1).* 2009.
7. Lazebnik, S., C. Schmid, and J. Ponce. *Beyond bags of features: Spatial pyramid matching for recognizing natural scene categories.* in *Computer Vision and Pattern Recognition, 2006 IEEE Computer Society Conference on.* 2006. IEEE.
8. Vedaldi, A. and A. Zisserman, *Efficient additive kernels via explicit feature maps.* Pattern Analysis and Machine Intelligence, IEEE Transactions on, 2012. **34**(3): p. 480-492.

Features extraction for classification of focal and non-focal EEG signals

Khushnandan Rai, Varun Bajaj, and Anil Kumar

Discipline of Electronics and Communication Engineering.
PDPM Indian Institute of Information Technology, Design and Manufacturing
Jabalpur, 482005 Jabalpur, India.
Email: bajajvarun056@yahoo.co.in

Abstract. The electroencephalogram (EEG) is most often used signal to detect epileptic seizures. For a successful epilepsy surgery, it is very important to localize epileptogenic area. In this paper, a new method is proposed to classify focal and non-focal EEG signals. EEG signal is decomposed by empirical mode decomposition (EMD). The average Renyi entropy and the average negentropy of IMFs for EEG signals have been computed as features. The class discrimination ability of these features are quantified using Kruskal–Wallis statistical test. These features are set to input in neural network classifier for classification of focal and non-focal EEG signals. The experimental results are presented to show the effectiveness of the proposed method for classification of focal and non-focal EEG signals.

Keywords: Focal and non-focal EEG signals, Empirical mode decomposition, Average Renyi entropy, Average negentropy, Neural network classifier.

1 Introduction

Epilepsy seizure is a group of long term neurological disorders which causes abnormalities in brain and it is not controllable by medicines therefore patients turn to surgery in case of drug resistance epilepsy. Now detection of the brain epileptogenic area, is an important task. The signals recorded from this area are called focal EEG signals and other signals are called as non-focal EEG signals [1]. Focal seizure (also called partial seizure or localized seizures) occurs in a portion of the brain, the area is identified by observing the affected body part of the patient such as movement, sensation or behavior. Around 20% of the patients become for surgery, have primary epilepsy and around 60% have focal epilepsy [2]. Classification of the focal and non-focal signals is an automatic detection of focal seizure which identifies the portion needed for surgery.

Focal EEG signals are less random, more nonlinear as compare to non-focal EEG signals, it has been shown [2] with help of combined surrogate analysis method. Other pre-surgical techniques like, magnetic resonance imaging (MRI) [3], positron emission tomography (PET) [4] and ictal single photon emission

K.J. Kim (ed.), *Information Science and Applications*,
Lecture Notes in Electrical Engineering 339, DOI 10.1007/978-3-662-46578-3_70

computed tomography (SPECT) [5] are able to focus the epileptic area with less accuracy about 50-80% [6-8]. Recently another method, based on delay permutation entropy (DPE) [9] provides 84% classification accuracy. One another method by extracting features average sample entropy (ASE) and average variance of instantaneous frequencies (AVIF) from separated EEG signal through EMD [10] gives 85% accuracy for classification of focal and non-focal EEG signals. In this paper, new features namely average Renyi entropy and average negentropy of IMFs for classification of focal and non-focal EEG signals is proposed.

2 Methodology

2.1 Dataset

The EEG dataset used for simulation is available online at www.dtic.upf.edu/~ralph/sc/ in [1]. It is intracranial EEG recording of five epilepsy patients (candidates for surgery due to longstanding drug resistant temporal lobe epilepsy). Sampling frequency is 512 or 1024 Hz as according to number of channels being more or less than 64 respectively. The simulation has been done using 750 focal and 750 non-focal EEG signals. Each signal contains a pair of two EEG signals namely x and y which are recorded from adjacent channels. The pairs x and y of focal and non-focal EEG signals are shown in Fig. 1.

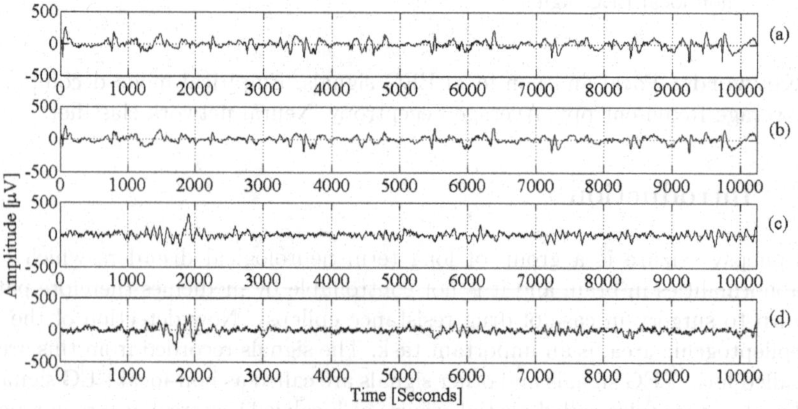

Fig. 1. A sample of focal and non-focal EEG signals: (a) focal x signal, (b) focal y signal, (c) non-focal x signal, (d) non-focal y signal.

2.2 Empirical Mode Decomposition (EMD)

EMD is an adaptive method to decompose nonlinear and non-stationary signals into N intrinsic mode functions (IMFs) and a residue. EMD algorithm is based on a sifting process which ends when residual $r_N(t)$ remains either a constant, a monotonic slope, or a function with only one extreme [11]. EEG pairs x and y are separately decomposed by EMD in its IMFs $c_k(t)$ and a residue $r_N(t)$. A focal x EEG signal, its first 10 IMFs and its residue are shown in Fig. 2.

$$x(t) = \sum_{k=1}^{N} c_k(t) + r_N(t) \tag{1}$$

2.3 Features Extraction

The features extracted from IMFs obtained by EMD of EEG signals are as follows:

Renyi Entropy Probability distribution P($p_1, p_2 ...., p_n$) is obtained for each IMF $c_k(t)$ and Renyi entropy of the IMF is calculated as:

$$H_\alpha(P) = \frac{1}{1-\alpha}\log(\sum_{i=1}^{n} p_i^\alpha) \tag{2}$$

Where α is entropic index. $\alpha > 0$ and $\alpha \neq 1$ [12]. In this study $\alpha = 2$ taken for computational efficiency.

Negentropy Negentropy is difference between differential entropy $H(P_G)$ of Gaussian density with the same mean and variance as P and differential entropy *H(P)* of P [13] calculated as:

$$H(P) = -\sum_{i=1}^{n} p_i \log(p_i) \tag{3}$$

$$H(P_G) = \frac{1}{2}\log((2\pi e)^n \sigma^2) \tag{4}$$

Where $p_i = [p_1, p_2 ...., p_n]$ is probability distribution obtained for each IMF $c_i(t)$, P_G is Gaussian random vector of the same mean and variance σ^2 as P and n is dimension of P. Negentropy is calculated as:

$$J(P) = H(P_G) - H(P) \tag{5}$$

These features are calculated for both x and y EEG signal. The average Renyi entropy and the average negentropy have been computed as follow:

$$AH_\alpha(P) = \frac{[H_\alpha(P_x) + H_\alpha(P_y)]}{2} \tag{6}$$

$$AJ(P) = \frac{[J(P_x) + J(P_y)]}{2} \tag{7}$$

Both features average Renyi entropy and average negentropy have been given as inputs in neural network classifier to perform classification of focal and non-focal EEG signals.

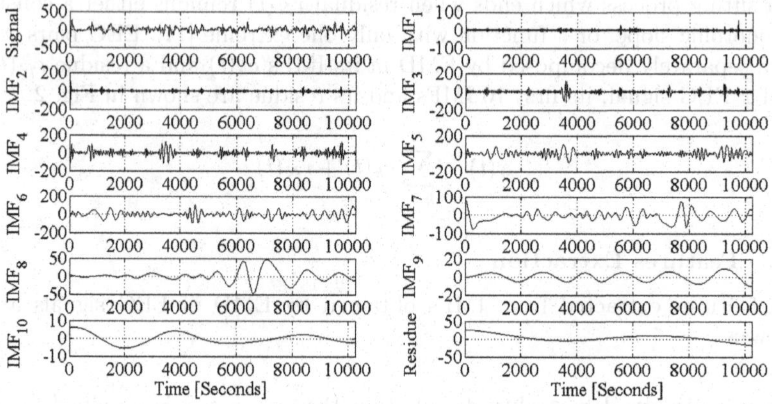

Fig. 2. Empirical mode decomposition of the focal x EEG signal.

2.4 Neural Network Classifier

The classification has been done using the neural network toolbox of MATLAB. The design of the neural network classifier contains a network of multiple processing elements called artificial neurons or neurons, these neurons are adjusted by the function of training process on their optimal values according to target matrix [14, 15]. The structure consists of three layers: input layer, hidden layer and output layer. The input-hidden layer transfer functions and hidden-output layer transfer functions are hyperbolic tangent sigmoid and linear transfer functions respectively. The weights and bias values of neurons are updated corresponding to target matrix, while the incoming data is processed by neurons repeatedly to obtain the final output. The target matrix contains two values 0 and 1 corresponding to focal and non-focal EEG signals respectively. More number of hidden layers contain more neurons with different weights and biases that increase complexity, it has been noticed that the classification accuracy remains same when adjusting the number of neurons from 10, 15, 20. Therefore we have selected 10 neurons for classification of focal and non-focal EEG signals. Classification has been done ten different times for more accurate results and the classification accuracy was calculated by averaging of all ten sets using confusion matrix of testing data.

3 Results and Discussion

In the proposed method, the features average Renyi entropy and average negentropy have been calculated for every IMF obtained from EMD based decomposition of EEG signal. The Kruskal-Wallis statistical test is applied for both features to find the p-value that enumerated distinction performance, which is shown in Fig. 3 and Fig. 4 respectively. It was found that for Renyi entropy

IMF_7, IMF_8, IMF_9 and IMF_{10} and for negentropy IMF_7 and IMF_8 have very less p-values, that show better distinction performance of the features for these IMFs. The features was given as input in neural network classifier to classify the focal and non-focal EEG signals. Accuracy has been computed ten times for each IMF and the final accuracy was calculated by averaging of all ten sets. The p-values and classification accuracy is shown in Table 1.

The proposed method has been compared with other existing methods [9] and [10] in which same dataset is used and classification accuracy is achieved as 84% and 85% respectively. Comparisons with these two methods are shown in Table 2. The classification accuracy is achieved by proposed method is 98.33% which is evidently better than existing methods [9] and [10].

Table 1. The p-values obtained from Kruskal-Wallis statistical test for both features and The classification accuracy (%) with Neural Network classifier

IMFs	p-values for average Renyi entropy	p-values for average negentropy	Accuracy (%)
IMF1	3.78×10^{-54}	5.96×10^{-2}	69.73
IMF2	1.88×10^{-12}	1.02×10^{-13}	65.74
IMF3	1.23×10^{-32}	7.52×10^{-25}	69.06
IMF4	3.38×10^{-28}	6.30×10^{-1}	60.20
IMF5	3.74×10^{-1}	1.13×10^{-17}	56.48
IMF6	2.46×10^{-2}	1.13×10^{-17}	60.33
IMF7	1.49×10^{-238}	8.49×10^{-245}	97.74
IMF8	1.99×10^{-244}	6.30×10^{-180}	96.61
IMF9	4.04×10^{-246}	6.96×10^{-5}	97.54
IMF10	2.16×10^{-246}	1.61×10^{-26}	98.33

4 Conclusion

EMD is a most suitable technique to decompose the EEG signals into set of narrow-band IMFs. In this paper, the parameters average Renyi entropy and average negentropy extracted from IMFs of EEG signal have been found useful to classify focal and non-focal EEG signals. The simulation results indicated that neural network classifier with these features had provided 98.33% accuracy in classification of focal and non-focal EEG signals. Future direction of research may include to find out more accurate classifier for classification of focal and non-focal EEG signals. Localization of the epileptogenic area is essential process for

Table 2. Comparison of the proposed method with the existing method.

Authors (year)	Method	Accuracy (%)
Zhu et al. (2013) [9]	Delay permutation entropy (DPE) methodology and SVM	84.00
Sharma et al. (2014) [10]	ASE, AVIF and LS-SVM	85.00
proposed work	Renyi entropy, negentropy and neural network classifier	98.33

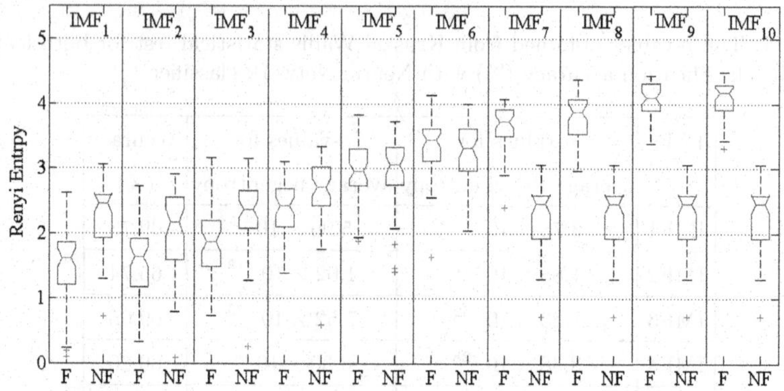

Fig. 3. Comparison of Average Renyi Entropy for focal (F) versus non-focal (NF) EEG signals for different IMFs.

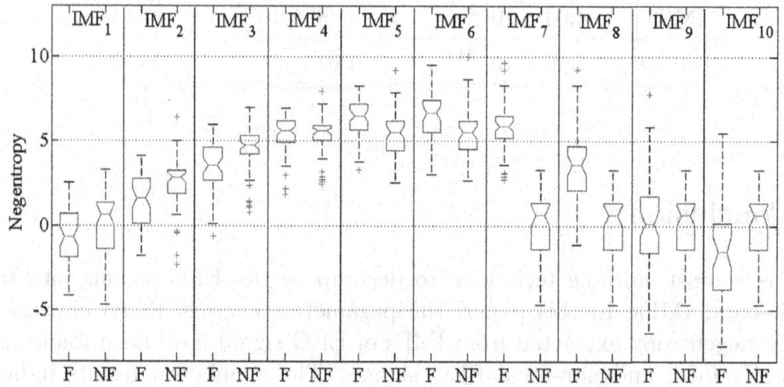

Fig. 4. Comparison of Average Negentropy for focal (F) versus non-focal (NF) EEG signals for different IMFs.

a success surgery, the proposed method can be conducive for surgery specialist to localize the epileptogenic area.

References

[1] Andrzejak, R.G., Schindler, K., and Rummel, C.: Nonrandomness, nonlinear dependence, and nonstationarity of electroencephalographic recordings from epilepsy patients. Physical Review E 86, 046206 (2012).

[2] Pati, S., and Alexopoulos, A.V.: Pharmacoresistant epilepsy: From pathogenesis to current and emerging therapies. Cleveland Clinic Journal of Medicine 77, 457-467 (2010).

[3] Seeck, M., Lazeyras, F., Michel, C.M., Blanke, O., Gericke, C.A., Ives, J., Delavelle, J., Golay, X., Haenggeli, C.A., Tribolet, N.D., and Landis, T.: Non-invasive epileptic focus localization using EEG-triggered functional MRI and electromagnetic tomography. Electroencephalography and Clinical Neurophysiology 106, 508-512 (1998).

[4] Savic, I., Thorell, J.O., and Roland, P.: Flumazenil positron emission tomography visualizes frontal epileptogenic regions. Epilepsia 36, 1225-1232 (1995).

[5] Newton, M.R., Berkovic, S.F., Austin, M.C., Rowe, C.C., McKay, W.J., and Bladin, P.F.: SPECT in the localisation of extratemporal and temporal seizure foci. Journal of Neurology, Neurosurgery & Psychiatry 59, 26-30 (1995).

[6] Spanaki, M.V., Spencer, S.S., Corsi, M., MacMullan, J., Seibyl, J., and Zubal, I.G.: Determining surgical candidacy in temporal lobe epilepsy. Journal of nuclear medicine: official publication, Society of Nuclear Medicine 40, 730-736 (1999).

[7] Henry,T.R., and Heertum, R.L.V.: Positron emission tomography and single photon emission computed tomography in epilepsy care. Seminars in Nuclear Medicine 33, 88-104 (2003).

[8] Knowlton, R.C., Elgavish, R.A., Bartolucci, A., Ojha, B., Limdi, N., Blount, J., Burneo, J.G., Hoef, L.V., Paige, L., Faught, E., Kankirawatana, P., Riley, K., and Kuzniecky, R.: Functional imaging: II. prediction of epilepsy surgery outcome. Annals of Neurology 64, 35-41 (2008).

[9] Zhu, G., Li, Y., Wen, P.P., Wang, S, and Min Xi,: Epileptogenic focus detection in intracranial EEG based on delay permutation entropy. International Symposium on Computational Models For Life Sciences 1559, 31-36 (2013)

[10] Sharma, R., Pachori, R.B., and Gautam, S.: Empirical mode decomposition based classification of focal and non-focal EEG signals. International Conference on Medical Biometrics, 135-140 (2014)

[11] Flandrin, P., Rilling, G., and Goncalves, P.: Empirical mode decomposition as a filter bank. IEEE Signal Processing Letters 11, 112-114 (2004)

[12] Renyi, A.: On measures of entropy and information. Proceedings of the fourth Berkeley Symposium on Mathematics, Statistics and Probability 1, 547-561 (1961)

[13] Leon, B.: Negentropy principle of information. Journal of Applied Physics 24, 1152-1163 (1953)

[14] Yegnanarayana, B.: Artificial Neural Networks, Prentice-Hall of India, New Delhi, (1999)

[15] MacKay, D.J.C.: Bayesian interpolation, Neural Computation 4, 415-447 (1991)

Relations on Intuitionistic Fuzzy Soft Multi Sets

Anjan Mukherjee[1] and Ajoy Kanti Das[2]

[1]Department of Mathematics,
Tripura University, Agartala-799022
Tripura, INDIA,anjan2002_m@yahoo.co.in

[2]Department of Mathematics,
Iswar Chandra Vidyasagar College,
Tripura, INDIA, ajoykantidas@gmail.com

Abstract. The theory of soft set and multiset are vital mathematical tools used in handling uncertainties about vague concepts. In this paper we define a new operation product of two intuitionistic fuzzy soft multi setsand present the concept of relations in intuitionistic fuzzy soft multi set. The notions of null relation and absolute relation are to be defined. Also we study their properties and discuss reflexive, symmetric and transitiveintuitionistic fuzzy soft multi relations. Some new results along with illustrating examples have been put forward in our work.

Keywords: Intuitionistic fuzzy set, Intuitionistic fuzzy soft set, Intuitionistic fuzzy soft multi set, Intuitionistic fuzzy soft multi relations.

1 Introduction

Mostof the problems in engineering, computer science, medical science, economics, environments etc. have various uncertainties. In 1999, Molodstov [11] initiated the concept of soft set theory as a mathematical tool for dealing with uncertainties. Research works on soft set theory are progressing rapidly. Maji et al.[9] defined several operations on soft set theory. Based on the analysis of several operations on soft sets introduced in [11], Ali et al.[2] presented some new algebraic operations for soft sets. Combining soft sets [11] with fuzzy sets [14] and intuitionistic fuzzy sets [5], Maji et al. [[8], [10]] defined fuzzy soft sets and intuitionistic fuzzy soft sets, which are rich potential for solving decision making problems. Alkhazaleh and others [[1], [3], [6], [7],[13]] as a generalization of Molodtsov's soft set, presented the definition of a soft multi set and its basic operations such as complement, union, and intersection etc. In 2012, Alkhazaleh and Salleh [4] introduced the concept of fuzzy soft multi set theory and studied the application of these sets and recently, Mukherjee and Das [12] studied the concepts of intuitionistic fuzzy soft multi topological spaces and constructed the fundamental theory on intuitionistic fuzzy soft multi topological spaces.

In fact all these concepts having a good application in other disciplines and real life problems are now catching momentum. But, it is seen that all these theories have their own difficulties that is why in this paper we are going to study the concept of rela-

K.J. Kim (ed.), *Information Science and Applications*,
Lecture Notes in Electrical Engineering 339, DOI 10.1007/978-3-662-46578-3_71

tions in intuitionisticfuzzy soft multi sets, which is another one new mathematical tool for dealing with uncertainties.
The concept of fuzzy relation on a set was defined by Zadeh[15] and several authors have considered it further. In this paper we define a new operation product of two intuitionistic fuzzy soft multi setsand present the concept of relations in intuitionistic fuzzy soft multi set. The notions of null relation and absolute relation are to be defined. Also we study their properties and discuss reflexive, symmetric and transitiveintuitionistic fuzzy soft multi relations. Some new results along with illustrating examples have been put forward in our work.

2 Preliminary Notes

In this section, we recall some basic notions in soft set theory, soft multi set theory, fuzzy soft multi set theory and intuitionistic fuzzy soft multi set theory. Molodstov defined soft set in the following way. Let U be an initial universe and E be a set of parameters. Let P(U) denotes the power set of U and A⊆E.

Definition 2.1[11]

A pair (F, A) is called a soft set over U, where F is a mapping given by F: A→P(U). In other words, soft set over U is a parameterized family of subsets of the universe U.

Definition 2.2[8]

Let $\{U_i : i \in I\}$ be a collection of universes such that $\bigcap_{i\in I} U_i = \phi$ and let $\{E_{U_i} : i \in I\}$ be a collection of sets of parameters. Let $U = \prod_{i\in I} FS(U_i)$ where $FS(U_i)$ denotes the set of all fuzzy subsets of U_i, $E = \prod_{i\in I} E_{U_i}$ and $A \subseteq E$. A pair *(F, A)* is called a fuzzy soft multiset over U, where F is a mapping given by $F: A \to U$.

Definition 2.3[12]

Let $\{U_i : i \in I\}$ be a collection of universes such that $\bigcap_{i\in I} U_i = \phi$ and let $\{E_i : i \in I\}$ be a collection of sets of parameters. Let $U = \prod_{i\in I} IFS(U_i)$ where $IFS(U_i)$ denotes the set of all intersection fuzzy subsets of U_i, $E = \prod_{i\in I} E_{U_i}$ and $A \subseteq E$. Apair (F, A) is called an intuitionistic fuzzy soft multiset over U, where F is a mapping given by $F: A \to U$, i.e. $\forall e \in A$,

$$F(e) = \left(\left\{ \frac{u}{(\mu_{F(e)}(u), \nu_{F(e)}(u))} : u \in U_i \right\} : i \in I \right).$$

Definition 2.4[12]

The complement of an intuitionistic fuzzy soft multiset*(F, A)* over U is denoted by $(F,A)^c$ and is defined by $(F,A)^c=(F^c,A)$, where

$$F^c(e)=\left(\left\{\frac{u}{\left(\nu_{F(e)}(u),\mu_{F(e)}(u)\right)}:u\in U_i\right\}:i\in I\right),\ \forall e\in A.$$

3 Relations on Intuitionistic fuzzy Soft Multi Sets

In this section, we define a new operation product of two intuitionistic fuzzy soft multi setsand present the concept of relations in intuitionistic fuzzy soft multi set. The notions of null relation and absolute relation are to be defined.

Definition 3.1

The product $(F,A)\times(G,B)$ of two intuitionistic fuzzy soft multi sets (F, A) and (G, B) over U is an intuitionistic fuzzy soft multi set (H, $A\times B$) where H is mapping given by H: A×B→U and $\forall(a,b)\in A\times B,\ H(a,b)=\cap(F(a),G(b))$

$$=\left(\left\{\frac{u}{\left(\min\{\mu_{F(a)}(u),\mu_{G(b)}(u)\},\max\{\nu_{F(a)}(u),\nu_{G(b)}(u)\}\right)}:u\in U_i\right\}:i\in I\right).$$

Example 3.2

Let us consider two universes $U_1=\{h_1,h_2,h_3\}$, $U_2=\{c_1,c_2\}$. Let $\{E_{U_1},E_{U_2}\}$ be a collection of sets of decision parameters related to the above universes, where $E_{U_1}=\{e_{U_1,1},e_{U_1,2}\}$, $E_{U_1}=\{e_{U_2,1},e_{U_2,2}\}$. Let $U=\prod_{i=1}^{2}FS(U_i)$, $E=\prod_{i=1}^{2}E_{U_i}$ and $A,B\subseteq E$, such that $A=\{a_1=(e_{U_1,1},e_{U_2,1}),a_2=(e_{U_1,1},e_{U_2,2})\}$ and $B=\{b_1=(e_{U_1,2},e_{U_2,1}),b_2=(e_{U_1,2},e_{U_2,2})\}$.

Table 1. The tabular representation of the fuzzy soft multi-set *(F, A)* be:

	a_1	a_2
h_1	(0.7,0.2)	(0.5,0.4)
h_2	(0.3,0.6)	(0.1,0.8)
h_3	(0.7,0.2)	(0.3,0.6)
c_1	(0.7,0.2)	(0.3,0.6)
c_2	(0.4,0.5)	(0.4,0.5)

Table 2. The tabular representation of the fuzzy soft multi-set (G, B) is:

	b_1	b_2
h_1	(0.5,0.4)	(0.3,0.6)
h_2	(0.7,0.2)	(0.5,0.4)
h_3	(0.3,0.6)	(0.5,0.4)
c_1	(0.4,0.5)	(0.7,0.2)
c_2	(0.7,0.2)	(0.4,0.5)

Table 3.The product set (H, $A \times B$):

	(a_1, b_1)	(a_1, b_2)	(a_2, b_1)	(a_2, b_2)
h_1	(0.5,0.4)	(0.3,0.6)	(0.5,0.4)	(0.3,0.6)
h_2	(0.3,0.6)	(0.3,0.6)	(0.1,0.8)	(0.1,0.8)
h_3	(0.3,0.6)	(0.5,0.4)	(0.3,0.6)	(0.3,0.6)
c_1	(0.4,0.5)	(0.7,0.2)	(0.3,0.6)	(0.3,0.6)
c_2	(0.4,0.5)	(0.4,0.5)	(0.4,0.5)	(0.4,0.5)

Definition 3.3

Let (F, A) be an intuitionistic fuzzy soft multi set over U. Then an intuitionistic fuzzy soft multi relation R on (F, A) is an intuitionistic fuzzy soft multi subset of the product set $(F,A)\times(F,A)$ and is defined as a pair $(R, A\times A)$, where R is mapping given by R: A×A→U.
The collection of all intuitionistic fuzzy soft multi relations R on (F,A) over U is denoted by $IFSMR_U(F,A)$. We denote an intuitionistic fuzzy soft multi relation $(R, A\times A)\in IFSMR_U(F,A)$ as simply R.

Example 3.4

Consider the intuitionistic fuzzy soft multi set (F, A) given in example 3.2, then a relation $(R, A\times A)$ on (F, A) is given by

Table 4.R:

	(a_1, b_1)	(a_1, b_2)	(a_2, b_1)	(a_2, b_2)
h_1	(0.5,0.4)	(0.4,0.5)	(0.3,0.6)	(0.2,0.7)
h_2	(0.3,0.6)	(0.1,0.8)	(0.1,0.8)	(0.1,0.8)
h_3	(0.3,0.6)	(0.3,0.6)	(0.3,0.6)	(0.3,0.6)
c_1	(0.4,0.5)	(0.1,0.8)	(0.1,0.8)	(0.3,0.6)
c_2	(0.4,0.5)	(0.4,0.5)	(0.4,0.5)	(0.4,0.5)

is an intuitionistic fuzzy soft multi relation.

Definition 3.5

Let $R_1, R_2 \in IFSMR_U(F, A)$, then we define for (a,b)∈ A×A,

(i) $R_1 \le R_2$ iff R_1 (a, b) $\subseteq R_2$ (a, b), for (a, b)∈ A×A.

(ii) $R_1 \vee R_2$ as($R_1 \vee R_2$)(a,b)= R_1 (a,b)∪ R_2 (a,b), where ∪ denotes the intuitionistic fuzzy union.

(iii) $R_1 \wedge R_2$ as($R_1 \wedge R_2$)(a,b)= R_1 (a,b)∩ R_2 (a,b), where ∩ denotes the intuitionistic fuzzy intersection.

(iv) R_1^C as R_1^C (a, b)=C[R_1 (a,b)], where C denotes the intuitionistic fuzzy complement.

Result 3.6

Let $R_1, R_2, R_3 \in IFSMR_U(F, A)$. Then the following properties hold:

(a) $(R_1 \vee R_2)^C = R_1^C \wedge R_2^C$.
(b) $(R_1 \wedge R_2)^C = R_1^C \vee R_2^C$.
(c) $R_1 \vee (R_2 \vee R_3) = (R_1 \vee R_2) \vee R_3$
(d) $R_1 \wedge (R_2 \wedge R3_1) = (R_1 \wedge R_2) \wedge R_3$
(e) $R_1 \wedge (R_2 \vee R_3) = (R_1 \wedge R_2) \vee (R_1 \wedge R_3)$
(f) $R_1 \vee (R_2 \wedge R_3) = (R_1 \vee R_2) \wedge (R_1 \vee R_3)$

Definition 3.7

A null relation $R_\Phi \in IFSMR_U(F, A)$ is defined as $R_\Phi = (R_\Phi,\ A \times A)_\Phi$ and an absolute relation $R_U \in IFSMR_U(F, A)$ is defined as $R_U = (R_U,\ A \times A)_U$.

Remark 3.8

For any $R \in IFSMR_U(F, A)$, we have

(i) $R \vee R_\Phi = R$

(ii) $R \wedge R_\Phi = R_\Phi$

(iii) $R \vee R_U = R_U$

(iv) $R \wedge R_U = R$

4 Various types of intuitionistic fuzzy soft multi relations

Definition 4.1

Anintuitionistic fuzzy soft multirelation $R \in IFSMR_U(F, A)$ is said to be reflexive if $\mu_{R(a,a)}(u)=1$ and $\nu_{R(a,a)}(u)=0$, $\forall u \in U_i$, $\forall i \in I$ and $\forall a \in A$.

Example 4.2

Let us consider two universes $U_1 = \{h_1, h_2, h_3\}$, $U_2 = \{c_1, c_2\}$. Let $\{E_{U_1}, E_{U_2}\}$ be a collection of sets of decision parameters related to the above universes, where $E_{U_1} = \{e_{U_1,1}, e_{U_1,2}, e_{U_1,3}\}$, $E_{U_1} = \{e_{U_2,1}, e_{U_2,2}\}$. Let $U = \Pi_{i=1}^{2} FS(U_i)$, $E = \Pi_{i=1}^{2} E_{U_i}$ and $A \subseteq E$, such that $A = \{a = (e_{U_1,1}, e_{U_2,1}), b = (e_{U_1,1}, e_{U_2,2})\}$.

Table 10. The tabular representation of the intuitionistic fuzzy soft multi-set (F, A)

	a	b
h_1	(1,0)	(1,0)
h_2	(1,0)	(1,0)
h_3	(1,0)	(1,0)
c_1	(1,0)	(1,0)
c_2	(1,0)	(1,0)

Then a relation $(R,\ A \times A)$ on (F, A) is given by

Table 11.R= $(R,\ A \times A)$:

	(a, a)	(a, b)	(b, a)	(b, b)
h_1	(1,0)	(0.4,0.2)	(0.3,0.6)	(1,0)
h_2	(1,0)	(0.1,0.7)	(0.1,0.5)	(1,0)
h_3	(1,0)	(0.7,0.2)	(0.4,0.6)	(1,0)
c_1	(1,0)	(0.1,0.5)	(0.5,0.1)	(1,0)
c_2	(1,0)	(0.4,0.5)	(0.5,0.4)	(1,0)

is a reflexive intuitionistic fuzzy soft multi relation on (F, A).

Definition 4.3

Anintuitionistic fuzzy soft multirelation $R \in IFSMR_U(F, A)$ is said to be symmetric if

$\mu_{R(a,b)}(u) = \mu_{R(b,a)}(u)$ and $\nu_{R(a,b)}(u) = \nu_{R(b,a)}(u)$, $\forall u \in U_i$, $\forall i \in I$ *and* $\forall (a,b) \in A \times A$

Example 4.4

Consider the intuitionistic fuzzy soft multi set (*F, A*) given in 4.2. Then the relation $R \in FSMR_U(F,A)$ be defined as follows:

Table 12.R:

R:	(a, a)	(a, b)	(b, a)	(b, b)
h_1	(0.5.0.3)	(0.4,0.3)	(0.4,0.3)	(0.2,0.7)
h_2	(0.3,0.6)	(0.1,0.7)	(0.1,0.7)	(0.1,0.5)
h_3	(0.3,0.4)	(0.3,0.5)	(0.3,0.5)	(0.3,0.7)
c_1	(0.4,0.5)	(0.1,0.7)	(0.1,0.7)	(0.3,0.1)
c_2	(0.4,0.2)	(0.4,0.1)	(0.4,0.1)	(0.4,0.4)

is a symmetric intuitionistic fuzzy soft multi relation on (*F, A*).

Definition 4.5

Let $R_1, R_2 \in IFSMR_U(F,A)$ be two intuitionistic fuzzy soft multi relation on (*F, A*). Then the composition of R_1 and R_2, denoted by $R_1 O R_2$, is defined by $R_1 O R_2 = (R_1 O R_2, A \times A)$ where $R_1 O R_2$: A×A→U is defined as $\forall u \in U_i$, $\forall i \in I$ and $\forall a,b,c \in A$

$$\mu_{R_1 OR_2(a,b)}(u) = \max_c \left\{ \min\left(\mu_{R_1(a,c)}(u), \mu_{R_2(c,b)}(u) \right) \right\}$$

and

$$\nu_{R_1 OR_2(a,b)}(u) = \min_c \left\{ \max\left(\nu_{R_1(a,c)}(u), \nu_{R_2(c,b)}(u) \right) \right\}$$

Definition 4.6

Anintuitionistic fuzzy soft multi relation $R \in IFSMR_U(F,A)$ is said to be a transitive if $RoR \subseteq R$.

Example 4.7

Consider the intuitionistic fuzzy soft multi set (*F, A*) given in 4.2. Then the relation

Table 13.R:

	(a, a)	(a, b)	(b, a)	(b, b)
h_1	(0.5,0.4)	(0.4,0.5)	(0.3,0.6)	(0.4,0.5)
h_2	(0.3,0.6)	(0.1,0.8)	(0.1,0.8)	(0.1,0.8)
h_3	(0.3,0.6)	(0.3,0.6)	(0.3,0.6)	(0.3,0.6)
c_1	(0.4,0.5)	(0.1,0.8)	(0.1,0.8)	(0.3,0.6)
c_2	(0.4,0.5)	(0.4,0.5)	(0.4,0.5)	(0.4,0.5)

is a transitive intuitionistic fuzzy soft multi relation on (F, A).

5 Conclusion

In this paper the theoretical point of view of intuitionistic fuzzy soft multi set is discussed. We extend the concept of relations in intuitionistic fuzzy soft multi set theory context. These are supporting structure for research and development of set theory.

6 References

1. Alhazaymeh, K., Hassan, N.: Vague Soft Multiset Theory. Int. J. Pure and Applied Math. 93, 511-523(2014).
2. Ali, M.I., Feng, F., Liu, X., W.K. Minc, W.K., Shabir, M.: On some new operations in soft set theory.Comp. Math. Appl. 57,1547-1553(2009).
3. Alkhazaleh, S.,Salleh, A.R., Hassan, N.: Soft Multisets Theory. Applied Mathematical Sciences, vol. 5, pp. 3561-3573, 2011.
4. Alkhazaleh, S., Salleh, A.R.: Fuzzy Soft Multi sets Theory.Hindawi Publishing Corporation, Abstract and Applied Analysis, Vol. 2012, Article ID 350603, 20 pages, doi: 10.1155/2012/350603.
5. Atanassov, K.: Intuitionistic fuzzy sets. Fuzzy Sets and Systems. 20, 87-96(1986).
6. Babitha, K.V., John, S.J.: On Soft Multi sets. Ann. Fuzzy Math. Inform. 5, 35-44(2013).
7. Balami, H.M., A. M. Ibrahim, A.M.: Soft Multisetand its Applicationin Information System.International Journal of scientific research and management. 1, 471-482 (2013).
8. Maji, P.K., Roy, A.R., Biswas, R.: Fuzzy soft sets. J. Fuzzy Math. 9, 589-602(2001).
9. Maji, P.K., Biswas, R., Roy, A.R.: Soft set theory.Comp. Math. Appl. 45,555-562(2003).
10. Maji, P.K., Biswas, R., Roy, A.R.: Intuitionistic fuzzy soft sets. J. Fuzzy Math. 12, 669-683(2004).
11. Molodtsov, D.: Soft set theory-first results.Comp. Math. Appl. 37, 19-31(1999).
12. Mukherjee, A., Das, A.K.: Parameterized Topological Space Induced by an Intuitionistic Fuzzy Soft Multi Topological Space. Ann. Pure and Applied Math. 7, 7-12(2014).
13. Tokat, D., Osmanoglu, I.: Soft multi set and soft multi topology.NevsehirUniversitesi Fen BilimleriEnstitusuDergisiCilt. 2, 109-118(2011).
14. Zadeh, L.A.: Fuzzy sets. Inform. Control. 8, 338-353(1965).
15. Zadeh, L.A.:Similarity relations and fuzzy orderings. Inform. Sci. 3, 177 – 200(1971).

Towards Automatically Retrieving Discoveries and Generating Ontologies

Kenneth Cosh

Department of Computer Engineering, Chiang Mai University, Thailand

Abstract. For the web to become intelligent, machines needs to be able to extract the nature and semantics of various concepts and the relationships between them. Most approaches focus on methods involving manually teaching the machine about different entities, their properties manually constructing an ontology. This paper discusses an approach where the necessary metadata is extracted automatically from Wikipedia, the online encyclopedia. This metadata is then used to compare documents allowing them to be clustered together so that similar documents can be identified allowing alternative knowledge to be discovered. The results show that an ontology indicating the relationships between types of documents can be automatically identified and also alternative knowledge can be discovered.

Keywords: Information Retrieval, Knowledge Discovery, Natural Language Processing, Social Media.

1 Introduction

The development and evolution of the Internet is continuing to impact virtually every field, by making information and knowledge accessible to an unprecedented audience. The web is continuing to evolve, with the community today taking responsibility for the creation, organization and critique of new content [1][2]. This era known as Web 2.0, or the social web, is expected to continue to involve towards an intelligent web, where the machine can use metadata coupled with the web to intelligently perform tasks for users [3]. For example, the machine could assist with planning a vacation by intelligently suggesting alternative destinations, identifying suitable activities to help the user to discover more possibilities. Currently the machine can assist us with locating information if we know what to search for, but it doesn't understand the semantics of a 'destination' or 'flight' or 'hotel', or the properties of such entities, such as 'climate', 'activities' or 'geography'. Nor can the machine comprehend the complex relationships between these classes.

The idea of a web which can understand these semantics is not a new one, but it is still yet to be realized. Much of the existing work has focused on developing standards and languages for describing information and representing relationships, such as with RDF (Resource Description Framework) and OWL (Web Ontology Language) [4]. Large challenges remain though, particularly given the enormous complexity of the web, vague, uncertain and inconsistent concepts used within it, and

K.J. Kim (ed.), *Information Science and Applications*,
Lecture Notes in Electrical Engineering 339, DOI 10.1007/978-3-662-46578-3_72

the rate at which it is constantly growing [5]. One of these challenges is that manual effort is needed to create the ontology, essentially requiring web masters to create a machine readable version of their content as well as the human readable one. Given how social media is being constantly created, this seems to be a non-scalable approach. This research investigates the application of Natural Language Processing techniques to automatically extract the semantics from Wikipedia articles, to automatically retrieve discoveries, and generate an ontology.

2 Wikipedia

Wikipedia is a crowd sourced encyclopedia which has grown to over 31 million articles in 285 languages. Considering only the English language version there are more than 4 million articles containing more than 2.5 billion words covering a wide variety of subjects. Unlike much of the content online, the content of Wikipedia is available for fair use, so it affords an interesting source for data mining, information retrieval and knowledge discovery.

Opinion is divided about the reliability of Wikipedia, with many restricting its use for research or student work [6] and there have been several statistical approaches to assess the reliability [7] and the credibility [8] of Wikipedia content. Other studies have investigated the coverage of the articles, with the finding that there is significantly greater coverage of natural sciences over other disciplines [9]. As Wikipedia is co-written by volunteers of different backgrounds, experience and ability, it is by its nature open to abuse or errors, whether intentional or unintentional. However, as mistakes can be and are easily corrected for this study it is considered a valuable resource for knowledge discovery. In this study a subset of the full English Wikipedia was used, known as version 0.8, which contains around 47,000 articles, which were selected based on their importance and quality.

While Wikipedia is a large knowledge base, automatically retrieving information from it is not a straightforward task, due at least in part to a lack of consistent structure. While it is written in the style of an online encyclopedia, the styles and language used by multiple diverse authors creates challenges, for example, in one country a region may be known as a state, while in others it is a province, a county, commune or division. Within the architecture of Wikipedia there are some opportunities to structure data, using templates for example. Templates allow an info box to appear on each page with standard information included, for example the area, population and climate of a region. Unfortunately different templates have been used by different authors, and in many cases important information is missing [10].

Other projects have attempted to formalize the ways in which concepts are interrelated and hence add structural knowledge, and the Semantic Wikipedia project allows contributors to add extra syntax to each page with links and attributes [11][12]. There are questions concerning the scalability and reliability of this approach as it increases the workload for the voluntary contributors requiring additional off page content to be created for a machine on top of the on page content for the human reader. This research automatically extracts enough semantics from each page in order to classify it, and identify relationships between different classes of content, with the intention of reducing the need for manual efforts allowing for the machine to identify the semantics.

3 Natural Language Processing

Natural Language Processing (NLP) refers to techniques to enable computer programs to extract information from a document. There are two key approaches to NLP, rule based or statistically based. This research focuses on a statistical or probabilistic approach and a couple of key methods are used to extract information. This section introduces the log likelihood calculation and the RV coefficient, both of which depend on having a corpus of words and their respective frequencies.

The first step in this research was to create a corpus of work frequencies from Wikipedia to establish the standard usage of English across a wide variety of pages written in an online encyclopedia. Around 47,000 articles were previously identified by a group of volunteers and taken from a variety of subject areas, based on the quality and importance of each article. These pages were extracted and Kiwix was used to host this portion of Wikipedia. This created a corpus of over 181 million words, with over 500,000 different words. 'The' was the commonest word, occurring 11.1 million times, followed by 'of' (6.1), 'and' (4.5), 'in' (4) and 'a' (3.1). These words are often considered as a skip list and excluded from NLP studies, however in this research, all words were included as it shall be shown that even words such as 'he' or 'at' contain important semantics about the kind of information they describe. 'He' may suggest that the article is about a person, while 'at' may suggest it is about a place.

3.1 Log Likelihood Calculation

Having established a corpus of the standard frequencies of the words in the pages, a log likelihood comparison is then used to compare each individual article with the standard [13]. For any article words like 'the', 'of' and 'and' have the highest frequency, but the interesting part is finding the significantly overused words. First the expected frequency, E, is calculated for each word that occurs in the page based on the corpus (1), and then this is compared with the observed frequency, O, as in (2). This is used to highlight the significantly overused (and underused) words within a page. The page about Thailand is more likely to overuse words like 'bangkok', 'temple' or 'beach' than it is to use words like 'ferret' or 'gravity'.

$$E_i = \frac{N_i \sum_i O_i}{\sum_i N_i} \quad (1)$$

$$-2\ln\lambda = 2\sum_i O_i \ln\left(\frac{O_i}{E_i}\right) \quad (2)$$

This process is repeated for every word in every page in the collection, so that each page is represented as a profile of the words used in it, their frequencies and the log likelihood of their frequencies. Table 1 shows the 10 most significantly used words in the article about Thailand, showing that whilst 'constitution' only appears 28 times, this is more significant than 'and' which occurs 322 times.

Table 1. Significantly overused words on the Thailand page.

Word	Frequency	Log Likelihood
Thailand	227	2617.9
Thai	158	1711.9
Bangkok	43	452.5
The	790	312.0
Muay	18	229.6
Nakhon	15	197.6
Malay	19	159.9
Asia	31	148.1
Constitution	28	144.3
Thaksin	14	143.5

The same information can be shown in a content cloud [14], which offers a better visualization of the profile of the page, as in Fig 1, which shows the top 50 words with size indicating a higher log likelihood score.

Fig. 1. Content Cloud of the page about Thailand.

3.2 RV Coefficient

Having created a profile for each page, the next step is to compare profiles to identify articles that have similar profiles. The expectation is that articles covering similar topics would have similar profiles. To do this, a multivariate correlation is performed using the RV coefficient (3). This measures the closeness of two matrices, in this case the words and their log-likelihood scores [15].

$$\varphi = r_c = \frac{\sum_{i,j} x_{i,j} y_{i,j}}{\sqrt{\left(\sum_{i,j} x_{i,j}^2\right)\left(\sum_{i,j} y_{i,j}^2\right)}} . \tag{3}$$

The results of the RV coefficient can help with discovering pages which are similar to each other. Experiments have shown that the approach can be used to identify suitable alternative destinations, with the approach suggesting Bangkok, Luang Prabang or Phuket as alternatives to Chiang Mai [16]. Table 2 shows a selection of articles along with their RV Coefficient when compared with the article titled "Thailand". Notably other places score much higher than articles about people, bands or disorders. The places highlighted include cities in Thailand, as well as other countries. While these results demonstrate some interesting results that can be used to automatically discover an alternative, to compare a page with every other page is not a scalable solution - even with 47,000 articles it is not efficient. This research investigates a way to reduce this by initially placing each page into a subset, for example first placing Thailand into a place class, and then a country class before making comparisons.

Table 2. RV Coefficient comparisons the 'Thailand' page and a selection of other pages

Page	RV Coefficient
Bangkok	0.3190
Laos	0.1070
Pattaya	0.1053
Singapore	0.0441
England	0.0322
Cardiac cycle	0.0175
Faces (Band)	0.0055
Discrete cosine transform	0.0040
Donald Trump	0.0027
Bipolar disorder	0.0021

4 Classifying Pages

The main purpose of this research is to establish whether the approaches discussed in the previous section could be used to automatically classify a page, clustering them into a particular type. The main objective is to be able to take a page, decide which categories, or sub-categories, the page belongs in, so that alternative pages could be discovered. A further objective is to investigate the relationships between categories and investigate the construction of an ontology by analyzing the categories that pages are placed in.

It should be noted, that pages don't necessarily belong in a single particular category, they may have degrees of membership for multiple categories. Traditionally the objective of librarianship has been to place information in a particularly category

– the physical properties of books require that they are placed on a shelf, and so they can be found they need to be placed in a logical category [2]. Moving into an electronic space, the physical constraints are lifted, and information can be placed in multiple categories, with different memberships for each category.

The process began by using a training set of random articles, and manually classifying them into one or more broad categories. For example the article 'Bangkok' would be placed into the 'Place' and the category 'City', while the article 'Bob Dylan' would be placed into the categories 'Music', 'Person', 'Singer', 'Musician' and 'Song-writer'. Other pages were placed initially in single categories, such as 'Iodine' was initially only placed in 'Chemical'. Attempts were made to generalize categories using 'Region' to combine 'State', 'County', 'Area' and other synonyms.

As pages were placed in categories, logically some categories grew in size more quickly than others. After 1500 pages had been initially categorized, the main two categories that emerged were 'Person' and 'Place'. At this point any category which had more than 25 members was selected to predict the membership of future articles. Table 3 shows the categories with more than 25 members.

Table 3. Number of members of larger categories

Category	Member Count	Category	Member Count
Person	344	Place	247
Music	92	City	90
Region	86	Politician	49
Ruler	48	Sportsperson	46
Chemical	44	Plane	42
Animal	42	Vehicle	40
Weapon	38	Business	36
Date	35	Musician	34
Singer	33	Football Team	32
Medical Condition	30	Band	29
Movie	27	Footballer	26

The next step was to create further corpora for each of these popular categories. Each of the articles that had been placed in each of these categories were re-examined and combined to create a new frequency list for each word that occurred in the articles. Again, the frequency of the words is not as important as the log likelihood of that frequency, so each corpus was compared with the original complete corpus. The result is a matrix of words and log likelihood scores for each category. Examining the largest category 'Person', the most significantly overused word is 'his', while 'her' is the seventh most significant. In the 'Place' category significant words include 'city', 'area', 'population', 'sea', 'town' and 'region'. For the 'Music' category the key words include 'album', 'band', 'music', 'rock' and 'song'.

Each of these corpora create a new standard to which future pages can be compared, using the RV coefficient, i.e. if an article scores a high RV coefficient with the 'Place' corpus, it is likely to belong in the 'Place' category. Having created the training data, further pages were chosen at random to test if their categories could be accurately predicted. In each case, if the page did belong in an existing cluster, the RV coefficient was highest for that cluster and if the page belonged in multiple clusters, it scored highest for all. Table 4 shows some randomly selected articles and their highest RV scores.

Table 4. RV coefficient comparison between page and categories

Page	Category	RV Score	Category	RV Score
Hai Phong	City	0.056	Place	0.034
Mitsubishi Heavy Industries	Business	0.030	Plane	0.026
Monty Python Life of Brian	Movie	0.165	Place	0.128
Iain Duncan Smith	Politician	0.106	Person	0.071
Cuba	Place	0.055	Region	0.048
Dalarna	Place	0.116	Region	0.112
Scarborough, Ontario	Place	0.059	City	0.058
Raja Ravi Varma	Person	0.038	Ruler	0.021
Oskar Lafontaine	Politician	0.090	Person	0.048
Chamonix	Region	0.058	Place	0.056
Clover	Animal	0.007	Business	0.002

In each of these cases the top suggestion matches where the page would be manually placed, except for 'Clover'. Clover should be placed in the 'plant' category, but as the training set didn't include enough 'plants' there was no valid category to place it in. It is worth noting that its closest match was with Animal, and that was a very low RV score. In 8 of the examples given above, the 2nd closest category also matched a category in which the page should be placed. It is worth noting that the cities in the sample (Hai Phong and Scarborough, Ontario), were matched more closely with 'city' than 'region', while the areas (Dalarna, Chamonix and Cuba) we more closely matched with 'region' even though all were matched with 'place'.

5 Conclusions and Future Work

One of the challenges in creating an intelligent web is enabling the machine to understand the semantics of the concepts used in it. Efforts have focused on teaching the machine by manually creating an ontology, while this research proposes that there is enough information contained within a pages content to automatically create semantic links. First, subject matter can be automatically placed in a correct categories, even with a relative small initial corpus (built from only around 25 articles. Once articles are placed in categories, those categories can be mined to identify other closely related articles, allowing knowledge discovery. Given the problem of wanting to organize a vacation to a somewhere like Bangkok – a discovery type question, rather than a search type question. The machine is first able to place Bangkok in the place category, and then identify other places with similar profiles to Bangkok.

Beyond this discovery, further aspects of creating an ontology are possible. There is a relationship between the 'musician', or 'singer' category and the 'person' category, with musician being a subclass of person. However, musicians also belong in the music category, along with bands, styles of music, instruments and other music themed entities. Further work is needed to investigate the types of relationships that can be identified and further kinds of discoveries that can be found. This research demonstrates the extraction of metadata from an article and how that meta data can be used to relate similar classes of entity.

References

1. O'Reilly, T.: What is Web 2.0: Design Patterns and Business Models for the next generation of Software, Communications and Strategies, no.1, p.17, first quarter, (2007)
2. Wesch, M.: Information R/Evolution, Kansas State University, www.youtube.com/watch?v=-4CV05HyAbM
3. Berners-Lee, T., Hendler, J., Lassila, O.: The Semantic Web, in Scientific American, May 2001.
4. Allemang, D, Hendler, J.: RDF – the Basis of the Semantic Web, in Semantic Web for the Working Ontologist, 2nd Ed. 2011.
5. Gardenfors, P.: How to make the Semantic Web more Semantic, in Formal Ontology in Information Systems, FOIS 2004.
6. Waters, N.: Why you can't Cite Wikipedia in my Class, Communications of the ACM, Vol. 50, No. 9, (2007)
7. Lih, A.: Wikipedia as Participatory Journalism: Reliable Sources? Metrics for Evaluating Collaborative Media as a News Resource, 5th International Symposium on Online Journalism, Austin Texas (2004)
8. Chesney, T.: An Empirical Examination of wikipedia's Credibility, First Monday, Vol.11, No. 11 (2006)
9. Halavais, A., Lackoff, D.: An Analysis of Topic Coverage of Wikipedia, Journal of Computer-Mediated Communication, Vol. 13, Issue 2, (2008), 429-440
10. Auer, S., Lehman, J.: What have Innsbruck and Liepzig in common? Extracting Semantics from Wiki Content, The Semantic Web Research and Applications, Lecture Notes in Computer Science, Volume 4519, (2007) 503-517
11. Völkel, M., Krötzsch, M. Vrandecic, D. Haller, H, Studer, R.: Semantic Wikipedia, WWW '06 Proceedings of the 15th International Conference on World Wide Web, (2006) 585-594.
12. Krötzsch, M. Vrandecic, D. Völkel, M.: Wikipedia and the Semantic Web, the Missing Links", Proceedings of Wikimania 2005, Frankfurt, Germany, (2005)
13. Rayson, P, Garside, R.: Comparing Corpora using Frequency Profiling, in Procceedings of the Workshop on Comparing Corpora held in conjunction with the 38th annual meeting of the Association for Computational Linguistics, ACL 2000, Hong Kong, (2000).
14. Cosh, K., Burns, R., Daniel, T.: Content Clouds, Classifying Content in Web 2.0, Library Review, Vol. 57, Issue 9, (2008) 722-729.
15. Robert, P., Escoufier, Y.: A Unifying Tool for Linear Multivariate Statistical Methods: the RV-Coefficient, Journal of the Royal Statistical Society, Vol. 25, No. 3 (1976)
16. Cosh, K.: On Automatically Extracting Discoveries from User Generated Content, In Proceedings of CISIS 2014, the 8th International Conference on Complex, Intelligent, and Software Intensive Systems (2014)

Improving process models discovery using AXOR clustering algorithm

Hanane Ariouat, Kamel Barkaoui, and Jacky Akoka

Lab. Cedric, Cnam
Paris, France,
{hanane,barkaoui,jacky}@cnam.fr

Abstract. The goal of process mining is to discover process models from event logs. Real-life processes tend to be less structured and more flexible. Classical process mining algorithms face to unstructured processes, generate spaghetti-like process models which are hard to comprehend. One way to cope with these models consists to divide the log into clusters in order to analyze reduced sets of cases. In this paper, we propose a new clustering approach where cases are restricted to activity profiles. We evaluate the quality of the formed clusters using established fitness and comprehensibility metrics on the basis distance using logical XOR operator. throwing a significant real-life case study, we illustrate our approach, and we show its interest especially for flexible environments.

Keywords: Process mining, process discovery, clustering, fitness

1 Introduction

Data clustering is one of the most important fields in data mining and a lot of solutions have been proposed in the literature [1]. There is an increasing interest in process mining and many case studies have been performed to show the applicability of process mining [2]. The field of process mining [4] is a new area of research, whose purpose is to develop techniques to gain insight into business processes based on the behavior recorded in event logs. An event log corresponds to a set of process instances following a particular business process. A process instance is manifested as a trace (a trace is defined as an ordered list of activities invoked by a process instance from the beginning of its execution to the end).

For processes with a many cases and high diversity of behavior, the models generated tend to be very confusing and difficult to understand. These models are usually called *spaghetti models*. This is due to by the inherent complexity of processes (i.e.all possible behaviors are shown in a single diagram). An approach to surmount this is to divide the set of cases into more homogeneous subsets that can each be adequately represented by a process model, what is called *Trace Clustering*.

K.J. Kim (ed.), *Information Science and Applications*,
Lecture Notes in Electrical Engineering 339, DOI 10.1007/978-3-662-46578-3_73

Trace clustering is one possible method to resolve the problem that currently available process discovery algorithms are unable to discover accurate and comprehensible process models out of event logs stemming from highly flexible environments. The significance of trace clustering to process mining has been studied in [6], [7]. Greco et al. [6] used trace clustering to discover expressive process models, this technique allows for a multitude of so-called profiles to determine the vector associated with each process instance. As such, they define activity, transition, performance, case attribute profiles. Furthermore, the implementation of the technique presents a full range of distance metrics and clustering techniques. Song et al. [5] have proposed the idea of clustering traces by considering a combination of different perspectives of the traces (such as activities, transitions, data, performance). They elaborate on the idea of constructing a vector space model for traces in an event log. Context-aware trace clustering described by Jagadeesh Chandra Bose and van der Aalst [8], [9]. They extend contemporary approaches by improving the way in which control-flow context information is taken into account.
Inspired by the work of Song et al [5], they proposed a new distance measure between cases. We rely on the traces profiles by considering only the activities in a case. We propose a new distance measure by using the Logical operator XOR between cases. This provide homogeneous subsets, and for each subset a better process model is created. Our technique presented in the next section is similar to [5], notwithstanding that the clustering strategy is different. We implemented our approach in ProM [1], an extensible framework for process mining that already includes many techniques to address challenges in this area.

The remaining of this article is organized as follows: in section 2, we introduce the activity profiles which are used to characterize cases. In Section 3, we present our clustering approach. Section 4 discussion and Section 5 concludes the article.

2 Activity Profiles

For every clustering application, it is crucial to appropriately design a way to determine the similarity of points to be clustered. In this paper, points correspond to cases, i.e., process instances that left a trace in the log. In our trace clustering approach, each case is characterized by a defined set of activities which can be extracted from the corresponding trace. Activities for comparing traces are organized in trace profiles. To run an example, we took the same event log which is used by song et al [5]. Each row refers to a single case and is represented as a sequence of events. Events are represented by the case identifier (denoted by the row, e.g., 1), activity identifier (first element, e.g., A), and originator (second element, e.g., John).

[1] ProM is an extensible framework that provides a comprehensive set of tools/plugins for the discovery and analysis of process models from event logs. See www. processmining.org for more information and to download ProM.

Case ID	log events
1	(A,John),(B,Mike),(D,Sue),(E,Pete),(F,Mike),(G,Jane),(I,Sue)
2	(A,John),(B,Fred),(C,John),(D,Clare),(E,Robert),(G,Mona),(I,Clare)
3	(A,John),(B,Pete),(D,Sue),(E,Mike),(F,Pete),(G,Jane),(I,Sue)
4	(A,John),(C,John),(B,Fred),(D,Clare),(H,Clare),(I,Clare)
5	(A,John),(C,John),(B,Robert),(D,Clare),(E,Fred),(G,Robert),(I,Clare)
6	(A,John),(B,Mike),(D,Sue),(H,Sue),(I,Sue)

Table 1: Example process logs (A: Receive an item and repair request, B: Check the item, C: Check the warranty, D: Notify the customer, E: Repair the item, F: Test the repaired product, G: Issue payment, H: Send the cancellation letter, I: Return the item)

In our approach, traces are characterized by profiles, where a profile is a vector $Profile\_Vector$ composed of n items (hich support binary values) as follow :

$$Profile\_Vector_i = (a_1, a_2, ..., a_n).$$

Where $a_{j=1..n}$ are binary values representing if an activity a_j is present in the trace or not. 1 means the activity occur at least once, and 0 means that the activity does not occur in the trace. These resulting vectors will be used to calculate the distance between two cases.

Table 2 shows the result of profiling the example log from table 1. Each row of the table corresponds to the profile vector of one trace in the log. The Trace profile defines one item per type of activity (i.e., event name) found in the log. Measuring an activity item is performed by simply counting if the activity has at least one occurrence in the trace.

Case ID	Trace Profile								
	A	B	C	D	E	F	G	H	I
1	1	1	0	1	1	1	1	0	1
2	1	1	1	1	1	0	1	0	1
3	1	1	0	1	1	1	1	0	1
4	1	1	1	1	0	0	0	1	1
5	1	1	1	1	1	0	1	1	0
6	1	1	0	1	0	0	0	1	1

Table 2: traces profiles for the example log from 1

2.1 Clustering Algorithm

The application of process mining techniques to traces from such clusters should generate clusters such that the process models mined from the clustered traces show a high degree of fitness and comprehensibility. Fitness quantifies how much of the observed behavior is captured in the model. Using the activity profiles presented in the previous section, we briefly explain our clustering algorithms:

To calculate the distance between two cases, we use the XOR operator as follow : let consider two traces a and b : a = (0001111) and b = (1101011) We

Algorithm 1 AXOR Algorithm

```
Require: Given an event log L consisting of N traces,
Ensure: Partition the N traces into M-clusters,
int MinDistance[i][j] = 0; //The minimum distance from trace i is to trace j.
list Clusters;
for i:= trace 1 to trace N-1 do
    for j:= trace i + 1 to trace N do
        distance(i,j)= trace i xor trace j;
        if (MinDistance[i][j] > distance(i,j) and distance(i,j)<3) then
            MinDistance[i][j] = distance(i,j);
        end if
    end for
end for
int ClusterNumber = 0;
for i:= 1 to N do
    if NotClustered(trace i) then
        ClusterNumber++;
        Clusters_ClusterNumber = CollectClusterTraces(trace i);
        //CollectClusterTraces(): a method to collect all the traces j which have
        their (MinDistance[i][j]) to trace i
    end if
end for
```

have a XOR b = (1100100). To calculate the distance between a and b, we use this measure (D) = the sum of the resulted vector item : $D = 1 + 1 + 0 + 0 + 1 + 0 + 0 = 3$. The value 3 can be seen as the degree of difference between the trace a and b.

3 Experimentation

The presented trace clustering technique achieves its goal to improve currently available trace clustering techniques when evaluation is considered from a process discovery perspective.

In this paper, we propose two hypotheses to evaluate the goodness of clusters from a process mining point of view. A good cluster tends to cluster traces such that: *(1) the discovered process model has a high fitness* and (2) *the process model mined is less complex.* To validate the approach discussed in this paper, we have implemented the trace clustering plug-in in ProM. It allows us to cluster cases in a log and further apply other process mining techniques to each cluster. The case study uses a process log from the Dutch academic hospital in the Netherlands [3]. In the log, patients correspond to cases, thus there are 1143 cases and 150291 events.

We use the heuristic mining algorithm to derive the process model. Figure 1 shows the process model for all cases obtained using the Heuristics Miner. The generated model is spaghetti-like and too complex to understand easily, since it consists of a lot of activities and many links exist between them.

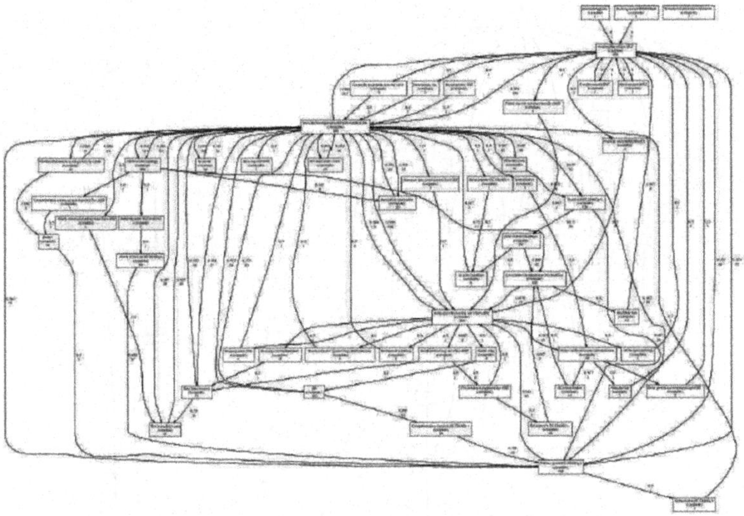

Fig. 1: Process model based on the entire log

By using our approach, we obtained several clusters of reasonable size. In this paper, we show only the results for two clusters that contain 116 cases and 15 cases respectively. Figure 2 shows the two heuristic nets derived from these two clusters.

We see that the two models are less complex than the first schematic, and we obtained a fitness that is 0.91 and 0. 96, respectively.

As future prospects, we plan to apply the techniques of game theory for the creation of clusters based on the " cooperative game". This might provide us with some more insights toward defining a measure of inter-cluster dissimilarity, a feature that is currently lacking from a cluster evaluation perspective.

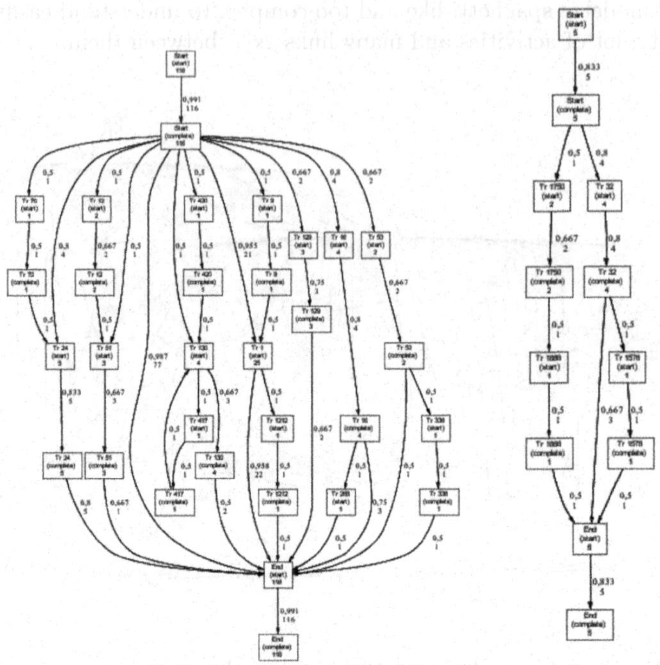

Cluster 1 Cluster 2

Fig. 2: Process models for the two clusters

4 Conclusion

In the context of process mining, clustering (or cluster analysis) aims to organize a collection of cases into clusters, such that cases within a cluster are more "similar" to each other than they are to items in the other clusters. In this paper, to measure of similarity between cases we have proposed a new metric based on logical operator (XOR). this concept of distance has proved relevant to discover to discover accurate and comprehensible process models out of event logs stemming from highly flexible environments. We have developed and presented in this paper a solution for such environments. The approach has been implemented as a plug-in for the ProM framework and has been applied in a real-world case-study.

References

1. A. K.Jain, and R.C.Dubes: *Algorithms for Clustering Data.* Prentice-Hall Inc. (1988).
2. W.M.P. van der Aalst, H.A. Reijers, A.J.M.M. Weijters, B.F.van Dongen, A.K.A. de Medeiros, M. Song, and H.M.W. Verbeek: *Business Process Mining: An Industrial Application.* Info. Sys. 32(5) 713-732 (2007).
3. R.S. Mans, M.H. Schonenberg, M. Song, W.M.P. van der Aalst, and P.J.M. Bakker, *Process Mining in Health Care.* In L. Azevedo and A.R. Londral, editors, International Conference on Health Informatics (HEALTHINF'08), pages 118-125. IEEE Computer Society, (2008).
4. W.M.P. van der Aalst, A.J.M.M. Weijters, *Process Mining: A Research Agenda*, Computers in Industry, 53(3):231-244, (2004).
5. M. Song, and C.W. Gunther, and W.M.P. van der Aalst: Trace Clustering in Process Mining, BPM Workshops. (2008).
6. G. Greco, and A. Guzzo, and L. Pontieri, and D. Sacca: *Discovering Expressive Process Models by Clustering Log Traces.* IEEE Trans. Knowl. Data Eng. 1010-1027 (2006).
7. A.K.A. de Medeiros, and A. Guzzo, and G. Greco, and W.M.P. van der Aalst, and A.J.M.M. Weijters, and B.F. van Dongen, and D. Sacca: *Process Mining Based on Clustering: A Quest for Precision.* BPM Workshops 17-29 (2007).
8. R.P. Jagadeesh Chandra Bose and W.M.P. van der Aalst, *Context Aware Trace Clustering: Towards Improving Process Mining Results*, Proc. SIAM Int'l Conf. Data Mining (SDM), pp. 401-412, (2009).
9. R.P. Jagadeesh ChandraBose and W.M.P. van der Aalst, *Trace Clustering Based on Conserved Patterns: Towards Achieving Better Process Models*, Proc. Int'l Business Process Management Workshops, pp. 170-181, (2009).

Simple approaches of sentiment analysis via ensemble learning

Tawunrat Chalothorn and Jeremy Ellman

Department of Computer Science and Digital Technologies
University of Northumbria at Newcastle, United Kingdom
{tawunrat.chalothorn,jeremy.ellman}@northumbria.ac.uk

Abstract. Twitter has become a popular microblogging tool where users are increasing every minute. It allows its users to post messages of up to 140 characters each time; known as 'Tweets'. Tweets have become extremely attractive to the marketing sector, since the user can either indicate customer success or presage public relations disasters far more quickly than web pages or traditional media. Moreover, the content of Tweets has become a current active research topic on sentiment polarity as positive or negative. Our experiment of sentiment analysis of contexts of tweets show that the accuracy performance can improve and be better achieved using ensemble learning, which is formed by the majority voting of the Support Vector Machine, Naive Bayes, SentiStrength and Stacking.

Keywords: Twitter, Tweet, sentiment, analysis, natural language processing, ensemble learning.

1 Introduction

Natural language processing (NLP) is a research area composed of various tasks; one of which is sentiment analysis. The main goal of sentiment analysis is to identify the polarity of natural language text [1]. Sentiment analysis can be referred to as opinion mining; studying opinions, appraisals and emotions towards entities, events and their attributes[2]. Sentiment analysis is a popular research area in NLP, which aims to identify opinions or attitudes in terms of polarity. Consequently, various researchers are interested in classifying Tweets using sentiment analysis. Many studies focus on using a single classifier, such as Naïve Bayes and Support Vector Machine (SVM), to analyze sentiment. However, this paper demonstrates that the use of multiple classifiers in ensemble leaning can improve the performance accuracy of sentiment classification. Moreover, we investigate the used of sentiment lexicons that could affect the classification.

The main contribution can be broken down as follows: (i) the ensemble classifiers have been formed using supervised and semi-supervised learning; (ii) sentiment lexicons and bag-of-words (BOW) have been combined for the comparison and clearly shown; (iii) the combinations of lexicons and BOW for use in supervised, semi-supervised and ensemble learning are explained and discussed. The remainder of this

K.J. Kim (ed.), *Information Science and Applications*,
Lecture Notes in Electrical Engineering 339, DOI 10.1007/978-3-662-46578-3_74

paper is constructed as follows: related work is discussed in section 2; the methodology, experiment and results are presented in sections 3 and 4, respectively. Finally, a conclusion and recommendations for future work are provided in section 5.

2 Related works

Twitter is a popular social networking and microblogging site that allows users to post messages of up to 140 characters; known as 'Tweets'. Tweets are extremely attractive to the marketing sector, since they can be searched in real-time. The word 'emoticon' is a neologistic contraction of 'emotional icon'. Specifically, it refers to the combination of punctuation characters to indicate sentiment in a text. Well-known emoticons include :) to represent a happy face, and :(a sad one. Emoticons allow writers to augment the impact of limited texts (such as in SMS messages or tweets) using fewer characters. [3] used supervision to classify sentiment of Twitter. Emoticons were used as noisy labels in training data; thereby facilitating the performance of supervised learning (positive and negative) at a distance. Three classifiers were used: Naïve Bayes, Maximum Entropy and SVM. Respectively, these classifiers were able to obtain more than 81.30%, 80.50% and 82.20% accuracy on their unigram testing data.

Moreover, [4] used stacking and majority voting to analyse sentiment of the dataset obtained from IBM's Predictive Modelling Group. The datasets are concerned with posts related to the 2008 U.S. presidential election. The datasets were labelled as positive, neutral and negative by the service of Amazon Mechanical Turk. However, only positive, neutral and negative labels were used. Three features were used in the experiment: social network features, sentiment analysis features and unigram BOW features. The use of each feature was separated into four sections: social network features used with Logistic Regression, named SNA; sentiment analysis feature used with NBM, named SA; NBM used with unigram BOW features, named as BOW; and NBM used with all features and named as ALL. Next, two ensemble learning called majority voting and stacking were used with the first three sections. The results showed that they achieved F-scores of 36.30%, 44.63%, 48.41% and 47.71% for SNA, SA, BOW and ALL, respectively. Conversely, stacking and majority voting achieved F-scores of 44.33% and 46.68%, respectively. In the comparison, stacking and majority voting achieved lower F-scores than BOW and ALL.

3 Methodologies

3.1 Classifier

Two machine learning, one sentiment resource and two ensemble learning are used in this research and are detailed below.

Naïve Bayes (NB) algorithm [5] was used from NLTK. NB is a classification algorithm based on Bayes' theorem that underlies the naïve assumption that attributes within the same case are independent given the class label [6]. This is also known as the state-of-art of Bayes rules [7]. NB [5] constructs the model by adjusting the distribution of the number for each feature. For example, in the text classification, NB [5] regards the documents as a BOW and from which it extracts features [8,9]. NB [5] model follows the assumption that attributes within the same case are independent given the class label [10]. Tang et al. [11] considered that Naïve Bayes assigns a context X_i(represented by a vector X_i^*) to the class C_j that maximizes $P(C_j|X_i^*)$ by applying Bayes's rule, as in (1).

$$P(C_j|X_i^*) = \frac{P(C_j)P(X_i^*|C_j)}{P(X_i^*)} \tag{1}$$

where $P(X_i^*)$ is a randomly selected context X. The representation of vector is X_j^*. $P(C)$ is the random select context that is assigned to class C.

To classify the term $P(X_i^*|C_j)$, features in X_i^* were assumed as f_j from $j = 1\ to\ m$ as in (2).

$$P(C_j|X_i^*) = \frac{P(C_j)\prod_{j=1}^{m} P(f_j|C_j)}{P(X_i^*)} \tag{2}$$

Support Vector Machine (SVM) [12] was used from SVMLight. SVM is a binary linear classification model with the learning algorithm for classifying and regression analysing the data and recognising the pattern. The purpose of SVM is to separate datasets into classes and discover the decision boundary (hyper-plane). To find the hyper-plane, the maximum distance between classes (margin) will be used with the closest data points on the margin (support vector). The equation of SVM can present as:

$$\vec{w} = \sum_j \alpha_j c_j \vec{d_j}, \qquad \alpha_j \geq 0 \tag{3}$$

where vector $\vec{w}$ represented as hyperplane. c_j is a polarity (negative and positive) of the data d_j which $c_j \in \{-1, 1\}$. $\alpha_j$are obtained by solving he dual optimisation problem. Those $\vec{d_j}$ such that α_j is greater than zero are called, support vectors, since they are the only document vectors contributing to $\vec{w}$. Classification of test instances consists simple of determining which side of $\vec{w}$ hyperplane they fall on.

SentiStrength (SS) [13] is also available to use free of charge and has been adopted by some researchers. SentiStrength is the analysis methodology used to judge whether a sentence has a positive or negative sentiment. The methodology was developed by [14], using nearly 4,000 comments on MySpace. They used three annotators and Krippendorf's alpha [15] to measure their agreement. The data have been separated into two groups: trail data and testing data. Trail data was used to identify algorithms for judgement and suitable scales. Algorithms were identified, ranging from 1 to 5, and used alongside testing data for final judgement. These will be SentiStrength's lexicon.

Majority voting [16] or called, majority rules are basic and simple algorithm in ensemble learning that uses the combination of various classifiers. The decisions of the voting are depended on agreement among more than half of the classifiers otherwise the input is rejected. The equation of majority voting [16] can present as:

$$\sum_{i=1}^{L} d_{i,k} = \max_{j=1,\dots,c} \sum_{i=1}^{L} d_{i,j} \tag{4}$$

where it is assumed that the label outputs of the classifiers are given as c dimensional binary vectors (for majority rules only two classes, i.e. $[d_{i,1}, d_{i,2}]^T \in \{0,1\}^c, i = 1, \dots, L$), and where $d_{i,j} = 1$ if D_i labels x in w_j and 0 otherwise.

Stacking (ST) [17] was used from WEKA. ST is ensemble learning technique that uses the prediction of the base learning algorithms as a training data to produce the final prediction, whereby the ST techniques can be represented by any learning algorithm. The idea of stacking is that, when given a dataset $L = \{(y_n, x_n), n = 1, \dots, N\}$, where y_n is the class value and x_n is a vector representing the attribute values of n instance, randomly split the data into J set. Define L_j and $L^{(-j)} = L - L_j$ to be the testing and training set for j fold of J-fold cross validation. Given K learning algorithms, which are called, $level - 0\ generalizers$ produce k algorithm on the data in training set $L^{(-j)}$ to induce a model $M_k^{(-j)}$, for $k = 1, \dots, K$ which are called, $level - 0\ models$. For each instance x_n in L_j, the test set for J-fold cross validation, let z_{kn} mean the prediction of model $M_k^{(-j)}$ on x_n. At the end of the entire cross-validation process, the data set assembled from the output of K models is $L_{cv} = \{(y_n, z_{1n}, \dots, z_{Kn}), n = 1, \dots, N\}$. These are the $level - 1\ data$.

$level - 1\ generalizer$ is a learning algorithm that derive from these data of model $\widetilde{M}$ for y as a function of $(z_1, \dots, z_K)$. These are called, $level - 1\ model$. For completing the training process, the final level-0 models $M_k, k = 1, \dots K$ are derived using all the data in L. For considering of the classification process, which used the model $M_k, k = 1, \dots K$, in conjunction with $\widetilde{M}$. Given a new instance, model M_k produce a vector $(z_1, \dots, z_k)$. This vector is input to the level-1 model $\widetilde{M}$, whose the output is the final prediction results of the instance.

3.2 Pre-processing

The datasets used in our experiment are from SemEval 2013 Task 2A [18]. The datasets are composed of Tweets and SMS. Tweets were used as major dataset while SMS was used for evaluating the system. For data pre-processing, emoticons were labelled by matching those collected manually from the dataset against a well-known collection of emoticons. Subsequently, negative contractions were expanded in place and converted to full form (e.g. don't -> do not).

Moreover, the features of Twitter were also removed or replaced by words, such as twitter usernames, URLs and hashtags. A Twitter username is a unique name displayed in the user's profile and may be used for both authentication and identification. This is shown by prefacing the username with an @ symbol. When a Tweet is directed at an individual or particular entity, this can be shown in the tweet by including @username. For example, a Tweet directed at 'som' would include the text @som. Before URLs are posted to Twitter, they are shortened automatically to use the t.co domain, whose modified URLs are a maximum of 22 characters. However, both features have been removed from the datasets. Hashtags are used to represent keywords and topics in Twitter by using # followed by words or phrases, such as #newcastleuk. This feature has been replaced with the following word after the # symbol. For example, #newcastleuk was replaced by newcastleuk.

Frequently repeated letters are used to convey emphasis in Tweets. These were reduced and replaced using a simple regular expression by two of the same character. For example, happpppppy will be replaced with happy, and coolllll will be replaced by cool. Next, special characters were removed, such as [,{,?,and !. Slang and contracted words were converted to their full form; for example, 'fyi' became 'for your information'. Finally, NLTK [19] stopwords were removed from the dataset, such as 'a', 'the', etc..

Moreover, three sentiment lexicons were used in this experiment. Bing Liu Lexicon (HL) [20], MPQA Subjective Lexicon (MPQA) [21] and AFINN Lexicon (AFINN) [22].

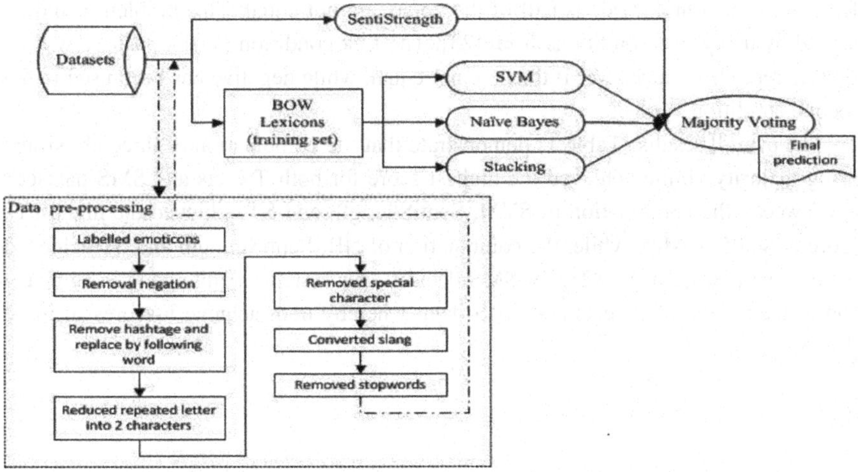

Figure 1. Flowchart of our approach in the experiment

4 Experiments and results

The experiments were tested using SentiStrength, NB model in NLTK and SVMLight for individual classification. Both datasets used the same method of pre-processing. For ensemble leaning, ST was used based on WEKA, and majority voting was implemented using Python. SS has been used as a server and accessed by my application in Python for passing the testing data directly to the SentiStrength website to calculate the testing data scores.

The contents of BOW are from the training datasets of Tweets. The use of BOW has been tested against the combination of BOW and sentiment lexicons. These were merged by removing words that duplicate, overlap and contradict in sentiment. They were tested using SVM and NB on two datasets: Tweets and SMS. The results showed that the F-score accuracy improved after being combined with BOW; reaching 83.55% for Tweets and 87.85% for SMS dataset, as illustrated in Table 1. For the processing of ensemble learning, known as stacking, the combination of SVM and NB as level 0 classifier and bagging as level 1 classifier. The training data used in ST was chosen from the combination of BOW and sentiment lexicons that obtained that highest F-score. For Tweet datasets, the training dataset was used from the combination of BOW, HL, MPQA and AFINN, which obtained the highest F-score of 83.55% from using the SVM classifier.

Conversely, for SMS datasets, the training dataset was used from the combination of BOW and MPQA, which obtained the highest F-score of 87.85% from using the NB classifier. After testing both datasets in ST, he results demonstrated their ability to obtain F-scores of 84.05% and 85.57% for Twitter and SMS datasets, respectively. Next, majority voting was used for the combination of all classifiers. The combination used in majority voting was separated into two, three and four voters. There are problems in the first and third, as half of the voters are not equal. This problem could be solved by using two conditions from [23]. The first condition (V01), positive will be used to represent the answer if they are not equal, while negative has been used in the second condition (V02).

The overall results (Table 1) demonstrate that the combination of three classifiers using majority voting achieved the highest score for both Tweets and SMS datasets. For Tweets, the combination of SVM, SentiStrength and ST achieved the highest F-score at 86.05%. Meanwhile, the combination of NB, SentiStrength and ST achieved the highest F-score at 88.82% for SMS dataset. Our system is quite good in comparison to the results of Tweets and SMS data; whereby both achieve F-scores of more than 85%.

Table 1. All results of Tweets and SMS dataset

Methods of Tweet dataset	Avg. F-score (%)	**Methods of SMS dataset**	Avg. F-score (%)
NB - BOW	81.94	NB - Bow	85.49
SVM - BOW	83.55	SVM - BOW	85.05
SS	78.37	SS	79.83
NB - BOW + HL	80.84	NB - BOW + HL	84.51
NB - BOW + HL + MPQA	81.26	NB - BOW + HL + MPQA	84.56
NB - BOW + HL + MPQA + AFINN	81.94	NB - BOW + HL + MPQA + AFINN	85.03
NB - BOW + HL + AFINN	81.74	NB - BOW + HL + AFINN	84.98
NB - BOW + MPQA	82.57	NB - BOW + MPQA	**87.85**
NB - BOW + MPQA + AFINN	81.73	NB - BOW + MPQA + AFINN	84.84
NB - BOW + AFINN	82.91	NB - BOW + AFINN	87.25
SVM - BOW + HL	82.47	SVM - BOW + HL	85.54
SVM - BOW + HL + MPQA	82.81	SVM - BOW + HL + MPQA	85.45
SVM - BOW + HL + MPQA + AFINN	**83.55**	SVM - BOW + HL + MPQA + AFINN	85.78
SVM - BOW + HL + AFINN	83.32	SVM - BOW + HL + AFINN	85.96
SVM - BOW + MPQA	81.99	SVM - BOW + MPQA	85.63
SVM - BOW + MPQA + AFINN	83.20	SVM - BOW + MPQA + AFINN	86.05
SVM - BOW + AFINN	83.00	SVM - BOW + AFINN	84.95
ST	84.05	ST	85.57
ENS (SVM + NB) (V01)	83.82	ENS (SVM + NB) (V01)	86.68
ENS (SVM + NB) (V02)	81.65	ENS (SVM + NB) (V02)	86.74
ENS (SVM + SS) (V01)	84.44	ENS (SVM + SS) (V01)	84.87
ENS (SVM + SS) (V02)	77.33	ENS (SVM + SS) (V02)	80.30
ENS (SVM + ST) (V01)	84.02	ENS (SVM + ST) (V01)	85.51
ENS (SVM + ST) (V02)	83.57	ENS (SVM + ST) (V02)	85.68
ENS (NB + SS) (V01)	83.30	ENS (NB + SS) (V01)	86.40
ENS (NB + SS) (V02)	76.82	ENS (NB + SS) (V02)	81.14
ENS (NB + ST) (V01)	83.46	ENS (NB + ST) (V01)	86.63
ENS (NB + ST) (V02)	82.39	ENS (NB + ST) (V02)	86.72
ENS (SS + ST) (V01)	82.33	ENS (SS + ST) (V01)	85.74
ENS (SS + ST) (V02)	79.66	ENS (SS + ST) (V02)	79.28
ENS (SVM + NB + SS)	84.09	ENS (SVM + NB + SS)	87.90
ENS (SVM + NB + ST)	84.28	ENS (SVM + NB + ST)	86.19
ENS (SVM + SS + ST)	**86.05**	ENS (SVM + SS + ST)	86.54
ENS (NB + SS + ST)	85.91	ENS (NB + SS + ST)	**88.82**
ENS (SVM + NB + SS + ST) (V01)	84.54	ENS (SVM + NB + SS + ST) (V01)	87.13
ENS (SVM + NB + SS + ST) (V02)	83.87	ENS (SVM + NB + SS + ST) (V02)	87.58

5 Conclusion and future work

In this research, the demonstration of using machine and ensemble learning formed by different components can provide state-of-the-art results for this particular domain. Moreover, we compared the use of BOW with the combination of lexicon and BOW. The results showed that the F-score of the combination of BOW and sentiment lexicons achieved greater accuracy than using only BOW. Furthermore, the results demonstrate that the size of training data does not always affect performance accuracy, provided they did not have sufficient information related to the test data. Our results show that the combination of three classifiers was able to achieve higher F-scores than the combination of two and four classifiers. Although our system was tested by using the contexts of Tweets and SMS, we believe that our system could be used with the contexts of other datasets. In future work, we are going to study other methods of ensemble learning, which we believe could be used in combination with our system for improving performance.

References

1. Shaikh MA, Prendinger H, Mitsuru I (2007) Assessing Sentiment of Text by Semantic Dependency and Contextual Valence Analysis. Paper presented at the Proceedings of the 2nd international conference on Affective Computing and Intelligent Interaction, Lisbon, Portugal,

2. Pang B, Lee L (2008) Opinion Mining and Sentiment Analysis. Foundations and Trends in Information Retrieval 2 (1-2):1-135

3. Go A, Bhayani R, Huang L (2009) Twitter sentiment classification using distant supervision. CS224N Natural Language Processing, Project Report, Stanford:1-12

4. Gryc W, Moilanen K (2014) Leveraging Textual Sentiment Analysis with Social Network Modelling. From Text to Political Positions: Text analysis across disciplines 55:47

5. Tan S, Cheng X, Wang Y, Xu H (2009) Adapting naive bayes to domain adaptation for sentiment analysis. In: Advances in Information Retrieval. Springer, pp 337-349

6. Elangovan M, Ramachandran KI, Sugumaran V (2010) Studies on Bayes classifier for condition monitoring of single point carbide tipped tool based on statistical and histogram features. Expert Systems with Applications 37 (3):2059-2065. doi:10.1016/j.eswa.2009.06.103

7. Cufoglu A, Lohi M, Madani K (2008) Classification accuracy performance of Naive Bayesian (NB), Bayesian Networks (BN), Lazy Learning of Bayesian Rules (LBR) and Instance-Based Learner (IB1) - comparative study. Paper presented at the International Conference on Computer Engineering & Systems (ICCES),

8. Liu B (2007) Web data mining: exploring hyperlinks, contents, and usage data. Springer,

9. Liu B (2012) Sentiment Analysis and Opinion Mining. Morgan & Claypool,

10. Hope LR, Korb KB (2004) A bayesian metric for evaluating machine learning algorithms. Paper presented at the Proceedings of the 17th Australian joint conference on Advances in Artificial Intelligence, Cairns, Australia,
11. Tang H, Tan S, Cheng X (2009) A survey on sentiment detection of reviews. Expert Systems with Applications 36 (7):10760-10773
12. Kecman V (2005) Support Vector Machines – An Introduction. In: Wang L (ed) Support Vector Machines: theory and applications, vol 177. Springer Science & Business Media, pp 1-48
13. Thelwall M, Buckley K, Paltoglou G, Cai D, Kappas A (2010) SentiStrength. University of Wolverhampton,
14. Thelwall M, Buckley K, Paltoglou G, Cai D (2010) Sentiment strength detection in short informal text. Journal of the American Society for Information Science and Technology 61 (12):2544-2558
15. Krippendorff KH (1980) Content analysis: an introduction to its methodology. Calif Sage Publications, Beverly Hills, Califonia
16. Polikar R (2012) Ensemble learning. In: Ensemble Machine Learning. Springer, pp 1-34
17. Wolpert DH (1992) Stacked generalization. Neural networks 5 (2):241-259
18. Wilson T, Kozareva Z, Nakov P, Ritter A, Rosenthal S, Stoyanov V SemEval-2013 Task 2: Sentiment Analysis in Twitter. In: Proceedings of the 7th International Workshop on Semantic Evaluation (SemEval), 2013. Association for Computational Linguistics,
19. Bird S, Klein E, Loper E (2009) Accessing Text Corpora and Lexical Resources. In: Natural Language Processing with Python. O'Reilly Media, p 60
20. Hu M, Liu B (2004) Mining and summarizing customer reviews. Paper presented at the Proceedings of the tenth ACM Special Interest Group on Knowledge Discovery and Data Mining (SIGKDD) international conference on Knowledge discovery and data mining, Seattle, WA, USA,
21. Wilson T, Hoffmann P, Somasundaran S, Kessler J, Wiebe J, Choi Y, Cardie C, Riloff E, Patwardhan S (2005) OpinionFinder: a system for subjectivity analysis. Paper presented at the Proceedings of HLT/EMNLP on Interactive Demonstrations, Vancouver, British Columbia, Canada,
22. Nielsen FÅ (2011) A new ANEW: Evaluation of a word list for sentiment analysis in microblogs. arXiv preprint arXiv:11032903
23. Martin-Valdivia M-T, Martinez-Camara E, Perea-Ortega J-M, Urena-Lopez LA (2013) Sentiment polarity detection in Spanish reviews combining supervised and unsupervised approaches. Expert Systems with Applications 40 (10):3934-3942

Effective Trajectory Similarity Measure for Moving Objects in Real-world Scene

Moonsoo Ra, Chiawei Lim, Yong Ho Song, Jechang Jung, and Whoi-Yul Kim

Hanyang University, Electronics and Computer Engineering,
222 Wangsimni-ro, Seongdong-gu, Seoul, 133-791, Korea
{msna,chiawei}@vision.hanyang.ac.kr,{yhsong,jjeong,wykim}@hanyang.ac.kr
http://hanyang.ac.kr

Abstract. Trajectories of moving objects provide fruitful information for analyzing activities of the moving objects; therefore, numerous researches have tried to obtain semantic information from the trajectories by using clustering algorithms. In order to cluster the trajectories, similarity measure of the trajectories should be defined first. Most of existing methods have utilized dynamic programming (DP) based similarity measures to cope with different lengths of trajectories. However, DP based similarity measures do not have enough discriminative power to properly cluster trajectories from the real-world environment. In this paper, an effective trajectory similarity measure is proposed, and the proposed measure is based on the geographic and semantic similarities which have a same scale. Therefore, importance of the geographic and semantic information can be easily controlled by a weighted sum of the two similarities. Through experiments on a challenging real-world dataset, the the proposed measure was proved to have a better discriminative power than the existing method.

Keywords: Video surveillance, trajectory clustering, moving objects

1 Introduction

Trajectories of moving objects are frequently used metadata to analyze behaviors of the moving objects and there are several trajectory clustering methods [1–4] have been introduced in recent years. Most of trajectory clustering methods utilize dynamic programming (DP) based similarity measures to cope with different lengths of the trajectories. However, in order to properly cluster trajectories, using only DP based similarity measures is not desirable, since trajectories acquired from real-world scene have a large shape variation, and may have missing data or noises from inaccurate measurement. Therefore, Liu and Schneider [2] developed a trajectory similarity which combines a geographic similarity with a semantic similarity. As the geographic similarity, [2] used a center of mass for the trajectory and a displacement vector which represents an approximated direction of the trajectory. In addition, as a semantic similarity, Longest Common Subsequence (LCSS) algorithm [5] is used to penalize a similarity between trajectories which have different shapes.

K.J. Kim (ed.), *Information Science and Applications*,
Lecture Notes in Electrical Engineering 339, DOI 10.1007/978-3-662-46578-3_75

However, the similarity measure introduced in [2] is not well applied on the trajectories acquired from a real world scene because, LCSS algorithm has a limited ability to capture semantic relationship between the trajectories. Furthermore, geographic and semantic similarities cannot equally contribute to a total similarity, since the total similarity is defined as a ratio of geographic and semantic similarities. In order to overcome the limitations, this paper proposed an effective trajectory similarity measure. The proposed measure is similar to the existing measure, in terms of using the concept of geographic and semantic similarities; however, it utilizes the starting point and angle difference of the displacement vector to capture the geographic relationship. Furthermore, Hausdorff distance [6] is applied on the normalized trajectory to capture the semantic relationship. One big advantage of the proposed measure is that both geographic and semantic similarities have a same scale; thus, both similarities can equally contribute to the total similarity. Through challenging experiments, the proposed measure is proved to have improved performance on the moving object trajectories acquired from the real-world scene.

The rest of the paper is organized as follows. In Section 2, trajectory obtaining process is explained in detail. The proposed trajectory similarity measure is presented in Section 3. An improved performance of the proposed measure is evaluated using the real world trajectories in Section 4. Finally, this paper is concluded in Section 5.

2 Trajectory Obtaining Process

Trajectory obtaining process consists of three stages: moving object detection, moving object association, and pruning. In the moving object detection stage, Gaussian Mixture Model (GMM) is used to separate foreground and background from an input surveillance video; then labeling algorithm is utilized to obtain blobs of the moving objects. As an implementation of GMM, algorithm introduced in [7] is used and as a labeling algorithm, simple grassfire algorithm [8] is utilized.

In order to associate the moving objects, similarity measure between moving objects should be defined first. Assume that i^{th} moving object in the previous frame is denoted as O^i_{prev} and j^{th} moving objects in the current frame are denoted as O^j_{curr}. Then, similarity S_{ij} between O^i_{prev} and O^j_{curr} are defined through following equations:

$$S_{ij} = s^{HSV}_{ij} \times exp(-dist(\mathbf{p}^i_{prev}, \mathbf{p}^j_{curr})/\lambda), \quad (1)$$

$$s^{HSV}_{ij} = \sum \min(H^i_{prev}, H^j_{curr}) / \sum \max(H^i_{prev}, H^j_{curr}), \quad (2)$$

where $\mathbf{p}^i_{prev}$ and $\mathbf{p}^j_{curr}$ are center of masses for O^i_{prev} and O^j_{curr}; then, H^i_{prev} and H^j_{curr} are HSV color histograms of O^i_{prev} and O^j_{curr}, respectively. Using S_{ij} for all i and j, bipartite graph B is constructed to associate moving objects from previous and current frames. In order to solve the bipartite graph association

Fig. 1. Example of the second pruning condition. Assume that t_1 and t_2 are obtained from the trajectory obtaining process. However, dotted trajectory, t_2, is not started or ended at boundaries of the image (not shaded region). Trajectories like t_2 are removed from the set of trajectories during the pruning process.

problem, traditional Hungarian algorithm [9] is applied on B. As a result of moving object association stage, set of moving object trajectories $T = \{t_1, t_2, ..., t_N\}$ are obtained.

Finally, inappropriately segmented or associated moving objects are pruned to improve quality of T. In order to determine pruned moving objects, following two conditions are used. First condition is that moving objects should not be disappeared at least 20 frames. Second condition is that moving objects should be started and ended at boundaries of the image. Through the pruning stage, subset of moving object trajectories $T_{sub} = \{t_1, t_2, ..., t_K\}$ is obtained. In order to give the readers better understanding of the pruning conditions, example of the second pruning condition is depicted in Fig. 1.

3 Trajectory Similarity Measure

In this paper, the trajectory $t_i \in T_{sub}$ of a moving object is defined as a sequence of 2D points, $t_i = \{(x_1^i, y_1^i), (x_2^i, y_2^i), ..., (x_L^i, y_L^i)\}$, where L is a length of the trajectory. The proposed similarity measure is defined on the two trajectories, t_i and t_j, and the proposed measure consists of two distinctive similarities: geographic similarity and semantic similarity. Geographic similarity captures spatial adjacency of the trajectories and semantic similarity captures shape difference of the trajectories.

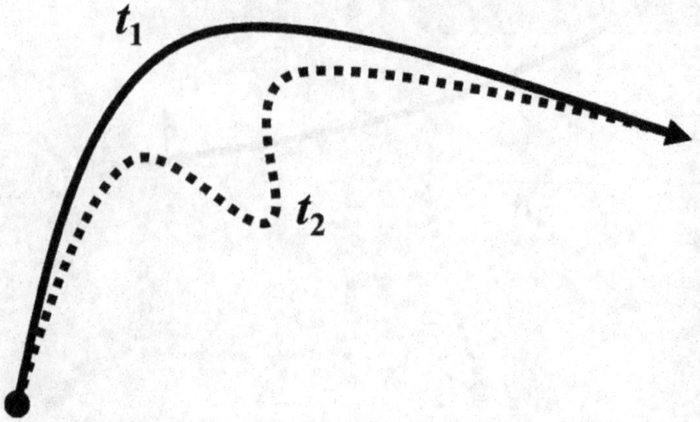

Fig. 2. Example of the trajectories which cannot distinguish by the geographic similarity, $s_{geo}(t_1, t_2)$, since start and end points of solid trajectory and dotted trajectory are same. In order to overcome such problem, semantic similarity will be introduced.

3.1 Geographic Similarity

In this section, geographic similarity is defined to satisfy following two properties: the similarity has a higher value when start points between the trajectories are spatially adjacent and approximated directions of the trajectories are similar. Spatial similarity of the start points is calculated as a traditional Euclidean distance, $d(\mathbf{s}_i, \mathbf{s}_j) = \|\mathbf{s}_i - \mathbf{s}_j\|$, where $\mathbf{s}_i$ and $\mathbf{s}_j$ are start points of t_i and t_j, respectively. The approximated direction is defined as a displacement vector, $\mathbf{d} = \mathbf{e} - \mathbf{s}$, where $\mathbf{e}$ is an end point of the trajectory; then similarity $s_{disp}(\mathbf{d}_i, \mathbf{d}_j)$ between $\mathbf{d}_i$ and $\mathbf{d}_j$ is calculated as an angle difference of $\mathbf{d}_i$ and $\mathbf{d}_j$. Using $d(\mathbf{s}_i, \mathbf{s}_j)$ and $s_{disp}(\mathbf{d}_i, \mathbf{d}_j)$, the proposed geographic similarity s_{geo} is defined as:

$$s_{geo}(t_i, t_j) = d(\mathbf{s}_i, \mathbf{s}_j) + s_{disp}(\mathbf{d}_i, \mathbf{d}_j), \quad (3)$$

$$s_{disp}(\mathbf{d}_i, \mathbf{d}_j) = \max(\|\mathbf{d}_i\| + \|\mathbf{d}_j\|) \exp(\lambda|\theta_i - \theta_j|/\pi), \quad (4)$$

where θ_i and θ_j are angles of $\mathbf{d}_i$ and $\mathbf{d}_j$, respectively, and λ is a parameter for controlling a decreasing rate of the exponential function, and $\|\mathbf{d}\|$ indicates a magnitude of the displacement vector $\mathbf{d}$.

As denoted in equation (3) and (4), geographic similarity does not consider shape of the trajectories; therefore, it cannot distinguish trajectories which have same start and end points but different trajectory shapes as illustrated in Fig. 2.

3.2 Semantic Similarity

The proposed semantic similarity is designed to have a higher value when shapes of the trajectories are similar. As a semantic similarity $s_{sem}(t_i, t_j)$, Hausdorff

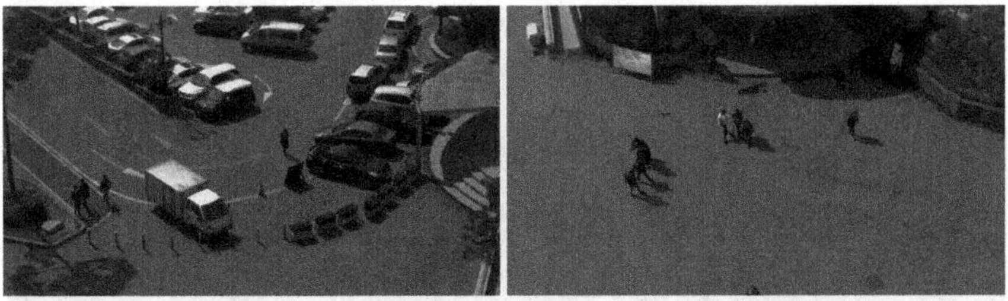

Fig. 3. Sample images from the challenging real-world dataset. Left image is a sample image from video A, and right image is a sample image from video B. Camera view of the left image is including a parking lot; therefore, vehicles are appearing frequently. While, camera view of the right image is targeted on the entrance of a subway station; therefore, a lot of people can be observed even in short duration of time.

distance is adopted, since it is known as having a good performance in comparing shapes of objects [6, 10, 11]. Hausdorff distance of two trajectories, $d_H(t_i, t_j)$, is defined as

$$s_{sem}(t_i, t_j) = d_H(t_i, t_j) = \max(h(t_i, t_j), h(t_j, t_i)), \quad (5)$$

$$h(t_i, t_j) = \max_{\mathbf{u} \in t_i} \min_{\mathbf{v} \in t_j} \|a - b\|, \quad (6)$$

where $\mathbf{u}$ and $\mathbf{v}$ are 2D points belong to t_i and t_j, respectively.

Since Hausdorff distance is calculated based on the Euclidean distance as denoted in equation (6), s_{sem} and s_{geo} have a same scale; thus, contributions of each similarity can be easily controlled by weighted sum of $s_{sem}(t_i, t_j)$ and $s_{geo}(t_i, t_j)$.

3.3 Proposed Similarity Measure

The proposed similarity measure s_{total} is defined as a weighted sum of $s_{geo}(t_i, t_j)$ and $s_{sem}(t_i, t_j)$ as following:

$$s_{total}(t_i, t_j) = \alpha s_{geo}(t_i, t_j) + (1 - \alpha) s_{sem}(t_i, t_j), \quad (7)$$

where α is a parameter for controlling an importance between geographic and semantic similarities. In this paper, α is empirically set as 0.4.

In order to cluster moving object trajectories by using the proposed measure, Affinity propagation algorithm [12] is utilized. Advantage of using Affinity propagation is that it automatically selects a number of clusters; therefore, parameter optimization process is not necessary throughout the experiments.

4 Experimental Result

In this section, performance of the proposed similarity measure was evaluated by using a challenging real-world dataset. The dataset consists of video A and

Fig. 4. Result of the clustering algorithm with the existing measure [2]. Video A is used for the source of the trajectories and total six clusters were acquired. As you can see in the top-left or bottom-left images, obviously different trajectories are considered as a single cluster.

B with 1280×720 resolution, and sample images from the dataset are depicted in Fig. 3. Video A and B were captured in Seoul campus of Hanyang university, and had 30 and 60 minutes duration, respectively. The evaluation process was conducted on Intel i5-2500 3.3 GHz computer with 4 GB memories.

Experiments were carried out for comparing performance of the proposed measure with existing measure in [2]. In detail, numbers of moving object trajectories obtained from two videos were 85 and 243, respectively; then, obtained trajectories were clustered by Affinity propagation with different similarity measures. When the proposed measure was utilized for the clustering, 11 and 22 clusters were obtained for video A and B, respectively. On the other hand, for the existing measure, 6 and 14 clusters were obtained. Details of clustering results for the video A is illustrated in Fig. 4 and Fig. 5; while, only subset of the clustering results for video B is depicted in Fig. 6, since number of pages for the paper is limited. In the figures, all trajectories are colored in rainbow, and the color has its own meaning. Points colored in green are closer to the start point; while, points colored in red are closer to the end point.

As shown in Fig. 4 and Fig. 5, there were a lot of people walking through a road which is located on the left side of the video. By using the existing measure, all the detail movements (some of the people are oriented to the left) of the people were grouped to a single cluster; while, clustering algorithm with the proposed measure could discriminate the detail movements.

In Fig. 6, difference of discriminative powers for the similarity measures could be observed more obviously. A cluster grouped by using the proposed measure only contains trajectories started from top-left to bottom-right of the image; however, trajectories clustered by using the existing measure have two distinctive shapes.

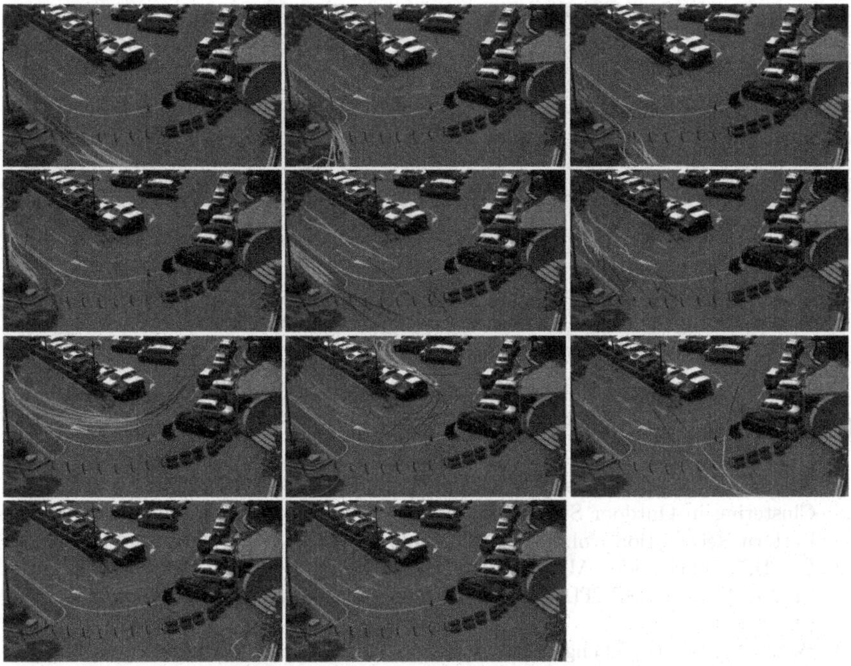

Fig. 5. Result of the clustering algorithm with the proposed measure. Video A is used for the source of the trajectories and total 11 clusters were obtained. Different from the clustering result of the existing measure, the proposed has a higher discriminate power than the existing measure.

5 Conclusion

In this paper, the similarity measure based on geographic and semantic similarities which have a same scale is proposed. Through the experiments on the challenging real-world dataset, the proposed measure is proved to have a better discriminative power than the existing method. Furthermore, a balance between the geographic and semantic similarities can be easily controlled, since they are combined by a form of weighted sum. However, the proposed method has a lack of ability to discriminate unusual trajectories, so that future research direction will be detecting unusual trajectories from the dataset. In addition, performance measure for the trajectory clustering quality is going to be researched to numerically analyze the clustering results.

Acknowledgments "This research was supported by the MSIP (Ministry of Science, ICT & Future Planning), Korea, under the ITRC (Information Technology Research Center) support program supervised by the NIPA (National IT Industry Promotion Agency)" (NIPA-2014-H0301-14-1018)

Fig. 6. Subset of the clustering results for video B. Left image is one of the results obtained by using the existing measure, and right image is one of the results acquired by using the proposed measure.

References

1. Zhang, Z., Huang, K., Tan, T.: Comparison of Similarity Measures for Trajectory Clustering in Outdoor Surveillance Scenes. In: 18th International Conference on Pattern Recognition, vol. 3, pp. 1135–1138 (2006)
2. Liu, H. and Schneider, M.: Similarity Measurement of Moving Object Trajectories. In: 3rd ACM SIGSPATIAL International Workshop on GeoStreaming, pp. 19–22 (2012)
3. Wang, H., Su, H., Zheng, Kai., Sadiq, S., Zhou, X.: An Effective Study on Trajectory Similarity Measures. In: 24th Australasian Database Conference, vol. 137, pp. 13–22 (2013)
4. Chen, P., Gu, J., Zhu, D., Shao, F.: A Dynamic Time Warping based Algorithm for Trajectory Matching in LBS. In: International Journal of Database Theory and Application, vol. 6, no. 3, pp. 39–48 (2013)
5. Bergroth, L., Hakonen, H., Raita, T.: A Survey of Longest Common Subsequence Algorithm. In: 7th International Symposium on String Processing and Information Retrieval, pp. 39–48 (2000)
6. Huttenlocher, D. P., Klanderman, G. A., Rucklidge, W. J.: Comparing Images Using the Hausdorff Distance. In: IEEE Transaction on Pattern Recognition and Machine Intelligence, vol. 15, no. 9, pp. 850–863 (1993)
7. Zivkovic, Z.: Improved Adpative Gaussian Mixture model for Background Subtraction. In: 17th International Conference on Pattern Recognition, vol. 2, pp. 28–31 (2004)
8. Gonzalez, R. C., Woods, R. E.: Digital Image Processing. Pearson Education, New Jersey (2010)
9. Munkres, J.: Algorithms for the Assignment and Transportation Problems. In: Journal of the Society for Industrial and Applied Mathematics, vol. 5, pp. 32–38 (1957)
10. Dubuisson, M.-P., Jain, A. K.: A Modified Hausdorff Distance for Object Matching. In: International Conference on Pattern Recognition, pp. 566–568 (1994)
11. Jesorsky, O., Kirchberg, K. J., Frischholz, R. W.: Robust Face Detection Using the Hausdorff Distance. In: Third International Conference on Audio- and Video-based Biometric Person Authentication, pp. 90–95 (2001)
12. Frey, B. J., Dueck, D.: Clustering by Passing Messages Between Data points. In: Science, vol. 315, 972–976 (2007)

A Prediction of Engineering Students Performance from Core Engineering Course using Classification

Nachirat Rachburee[1], Wattana Punlumjeak[2], Sitti Rugtanom[3], Deachrut Jaithavil[4], and Manoch Pracha[5]

Department of Computer Engineering
Faculty of Engineering, Rajamangala University of Technology Thanyaburi, Pathum Thani, Thailand
nachirat.r@en.rmutt.ac.th[1] wattana.p@en.rmutt.ac.th[2]
sitti.r@en.rmutt.ac.th[3] deachrut.j@en.rmutt.ac.th[4]
manoch.p@en.rmutt.ac.th[5]

Abstract. All Engineering students in Thailand must complete four core engineering courses which consist of mechanic, material, drawing and computer programming. These four core courses are essential basis and fundamental. Prediction of student academic performance helps instructors develop good understanding of how well or poor students perform. Thus, instructors can take proper proactive evaluation to improve student learning. Students can predict themselves to gain higher performance in the future. This paper focuses on developing a predictive model to predict student academic performance in core engineering courses. A total of 6,884 records has been collected from year 2004 to 2010. Five classification models are developed using decision tree, naïve bayes, k-nearest neighbors, support vector machine, and neural network, respectively. The results show that the neural network model generates the best prediction with 89.29% accuracy.

Keywords: data mining, prediction, classification, core engineering course

1 Introduction

Nowadays, many information technology researches related to educational area have been developed to support students with decision making information for academic course selection, chance for completion, and so on. These researches in educational field which data can be used for improving student performance are called educational data mining researches.

In Thailand, many universities have engineering course which is one of the major faculties for country development. Engineering student data are being kept in a large database. Engineering core course is the course that every engineering student is required to take and must be passed. Core courses consist of mechanic, material, drawing

K.J. Kim (ed.), *Information Science and Applications*,
Lecture Notes in Electrical Engineering 339, DOI 10.1007/978-3-662-46578-3_76

and computer programming. This group of courses is one of the most difficult and important courses. Prediction of student performance from core engineering course can play very important roles by helping not only the instructors develop an effective course curriculum and assist low performance student, but also the student improve and take more attention to achieve high Cumulative Grade Point Average (CGPA) in the future.

Main objective of this research is to use data mining techniques to classify engineering students performance using decision tree, naïve bayes, k-nearest neighbor (k-NN), support vector machine (SVM) and neural network to predict students education scores. The research will also construct the most confidence model for future engineering student performance prediction.

2 Background and Related Work

Knowledge Discovery in Databases (KDD) is an automatic, exploratory analysis and modeling of large data repositories. Data Mining is the core of the KDD process, involving the inferring of algorithms that explore the data, develop the model and discover previously unknown patterns. Data mining techniques are used to build a model according to which the unknown data will try to identify the new information or new relationship [1]. Data mining techniques are divided into two basic groups: unsupervised algorithms and supervised algorithms.

Classification techniques are supervised learning techniques that classify data items into predefined class label. The generated model will be able to predict a class for given data depending on previous learned information from historical data [2].There are five algorithms for data classification: decision tree, naïve bayes, k-nearest neighbor (k-NN), support vector machine (SVM) and neural network.

2.1 Decision Tree

Decision Tree is a classifier technique that consists of nodes. Root node is a node that has no edge. Each edge represents output of the test. The number of edges of an internal node is equal to the number of possible values of the corresponding input attribute. The terminal node represent the conclusion of the result.

We have to select attributes that have a correlation with class or label. Root node will be selected from an attribute that has the highest information gain (IG). Then, we will look for relative attributes and recursively split the data into smaller subsets by testing for a given attribute at each node [3].

$$\text{Entropy}(c1) = -p(c1) \log p(c1) \tag{1}$$

$$\text{IG}(\text{parent},\text{child}) = \text{Entropy}(\text{parent}) - [p(c1) \times \text{Entropy}(c1) + p(c2) \times \text{Entropy}(c2) + \ldots] \tag{2}$$

The high IG has a high potential to improve the classification process [1].

Decision tree model (ID3, C4.5) is used to predict student final grades. The research also applied the decision tree model to predict student grades using data from year 2005 C++ course grade of Yarmouk University, Jordan. The result indicated that decision tree model had more accuracy than other models [4]. Another research shows result from comparative of predicting using ID3, C4.5 and CART. Data set of 48 students from Computer Applications Department of MCA (Master of Computer Applications) VBS Purvanchal University, session 2008 to 2011 was obtained using sampling method and used 10-cross validation to evaluate results of each approach. The best result was produced by CART [5].

2.2 Naïve Bayes

Naïve Bayes classifier is one of classification techniques that is used to predict a target class. It depends in its calculations on probabilities, namely Bayesian theorem.

Let A and B be events in a sample space. Then the conditional probability P(A |B) is defined as [6]

$$P(A \mid B) = \frac{P(A \cap B)}{P(B)} \tag{3}$$

Naïve Bayes is one of the classifier methods to measure the performance of student academic success. The research found that result of Naïve Bayes classification algorithm had the highest success percentage [7]. Bayesian classification method is also used in research to predict students data which gathered from different degree colleges and institutions affiliated with Dr. R. M. L. Awadh University, Faizabad, India. The study shows that academic performances of the students are not always depending on their own effort, and shows that other factors have got significant influence over student performance [8].

2.3 k-nearest neighbor (k-NN)

k-nearest neighbor (k-NN) classification finds a group of k objects in the training set that are the closest to the test object. The method used to determine the class of the target object based on the classes and distances of the k nearest neighbors.

$$d(x,y) = \sqrt{\sum_{k=1}^{n}(x_k - y_k)^2} \tag{4}$$

k-NN can involve assigning an object the class of its nearest neighbor or of the majority of its nearest neighbors. k-NN were used in research that compared five classification algorithms for predicting students grade particularly for engineering students. The data set consisted of 1,000 instances and 11 attributes. The overall result of all five algorithms was good but the results of individual classes for naïve bayes and NB tree was not sufficient enough for the individual class prediction [9].

2.4 Support Vector Machine

The Support Vector Machine (SVM) classifier is one of the most important tasks for different applications such as text categorization, image classification, data Classification. Several recent studies have reported that the SVM generally are capable of delivering high performance in terms of classification accuracy.

SVM is applied on different data which have two or multi class. Therefore, SVM is a powerful machine method developed from statistical learning and has made significant achievement in some field. SVM method does not suffer the limitations of data dimensionality and limited samples [10]. SVM classifier based on Rough set (RS) reducts is researched in order to enhance the predicting performance. Experiment results explain the validity and feasibility of proposed algorithm [11].

2.5 Neural Network

The multi-layer feedforward neural networks, also called multi-layer perceptrons (MLP). MLP is a network consisting of a number of highly interconnected simple computing units called neurons, nodes, or cells, which are organized in layers. Each neuron performs simple task of information processing by converting received inputs into processed outputs.

The architecture of a three-layer feedforward neural network that consists of input layer, hidden layer, and output layer. Neurons in the input layer receive the data patterns and then pass them into the neurons into the next layer. Neurons in the hidden layer are connected to both input and output neurons and are key to learning the pattern in the data and mapping the relationship from input variables to the output variable [1]. Five hundred Computer and Communication Engineering students' records had been gathered covering a period of almost seven years. The results confirmed that students, who are weak performs particularly in this pool of courses prediction results demonstrated a high level of accuracy and offered efficient analysis and information pertinent to the management of engineering schools [12]. Radial Basis Function can create prediction model. Its aim is to predict marks obtained by students in a subject that is related to subjects taken during previous semesters. There are more than 10,000 student records and over 250 subjects. After predicting the performance, students can be classified under four categories Excellent, Good, Average, and Poor [13].

3 Methodology

This paper intends to establish a classification model consisting of three steps: data preparation, modeling and evaluation. Four attributes have been selected including mechanic, material, drawing, and computer programming. These four attributes are core engineering courses in Thailand that students must pass within the first academic year. RapidMiner Studio version 6.1 is used in this research.

3.1 Data Preparation

463,956 records of data set used in this study have been obtained from Faculty of Engineering, Rajamagala University of Technology Thanyaburi, Pathum Thani, Thailand from year 2004 to 2010. In this step, data are filtered only for these 4 interested core engineering subjects. Later, this data set is grouped into 6,884 records by Student ID. The collected data consist of student id, student cumulative GPA (numerical values 0.0-4.0), and scores (letter grades A, B+, B, C+, C, D+, D, or F).

3.2 Data Cleaning

Data cleaning tasks are to identify outliers, smooth out noisy data, and correct inconsistent data. For Example, Data such as grade F has been removed. Student's grade from regrade records kept in the form of 'B$' means this student has grade 'B' after regrade. Therefore, '$' shall be removed from this attribute. Records with blank letter grade shall be removed.

3.3 Data Transformation

We have discretized the numerical values of the student cumulative GPA into a categorical classes. A class is divided into 3 classes consisting of High, Medium, and Low as shown below:

Table 1. Class discretization

Class	Possible Value (GPA Range)
High	3.00 - 4.00
Medium	2.00 - 2.99
Low	1.00 - 1.99

3.4 Implementation of Mining Model

Decision tree. Decision tree is generated using the C4.5 algorithm. The model indicates that the Mechanic attribute is the most effective attribute.

```
mech = C
| com = A
| | mat = A: high {high=12, medium=1, low=0}
| | mat = B
| | | draw = A: high {high=5, medium=2, low=0}
| | | draw = B: high {high=9, medium=1, low=0}
| | | draw = C+: high {high=21, medium=2, low=0}
```

Fig. 1. Part of the rules generated by C4.5.

Naïve bayes. We calculate an example of a prior probability and posterior probability as shown below. Amount of students in medium class are 4,957 records and amount of total students are 6,884 records. From formula (3)

$$P(com= A \mid GPA = medium) = P(com = A \cap GPA = medium) / P(GPA = medium)$$
$$= ((353 / 4957) * (4957 / 6884)) / 0.72$$
$$= 0.071 \text{ and so on.}$$

K-nearest neighbors (k-NN). This paper defines k as 10 and then calculates a distance to find a nearest neighbors as following:

Data test example was Mech = 2, Mat = 3,Com = 2.5 and Draw = 2. From formula (4)

$$\text{Row 1}: \ d(x,y) = \sqrt{(1-2)^2 + (1-3)^2 + (2-2.5)^2 + (1.5-2)^2} = 2.345$$

Table 2. Example student data set

Row no.	Mech	Mat	Com	Draw	Class
1	1	1	2	1.5	low
2	2	1.5	3	1.5	medium
3	2.5	3	3	2.5	high

Then, repeat the calculation for all distances of data set. Select the k lowest distances. From these k lowest distances, most distances fall in to which class. Data test will be then determined as belong to that class.

Support vector machine (SVM). SVM in this paper uses linear model to classify classes into 3 classes. SVM classification is performed by LIBSVM with kernel type is linear, and C parameter is 100.

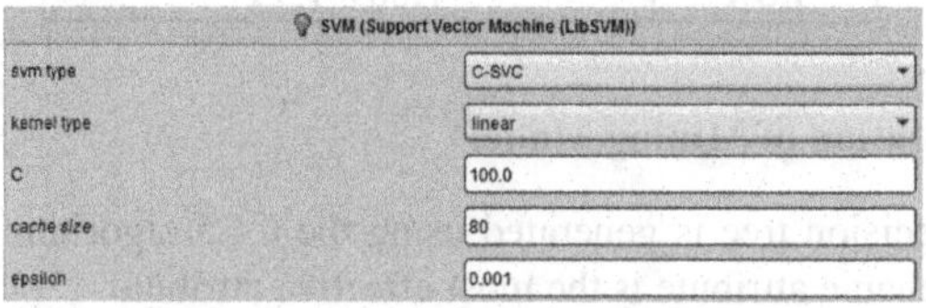

Fig. 2. Parameter setting of SVM

Neural network. Neural network in this paper generates 3 layers as described in the picture below

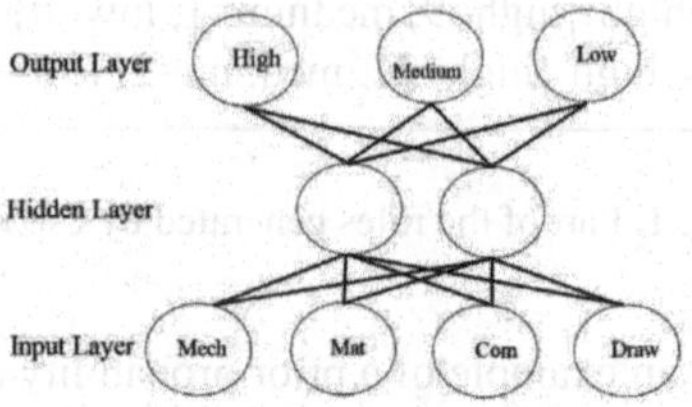

Fig. 3. Neural network

4 Result and Discussion

The experiment uses five classification algorithms in order to try to obtain the highest classification accuracy. The algorithms used are decision tree, naïve bayes, k-nearest neighbors, support vector machine, and neural network.

Accuracy calculation uses 10 cross-validation method. Cross-validation method is a statistical algorithm by dividing data into two segments: 70% for model learning or training, and 30% for model validation.

Table 3. Classification Accuracy

	Correct	Incorrect
Decision Tree	86.77%	13.23%
Naïve Bayes	88.37%	11.63%
K-NN	89.08%	10.92%
SVM	89.25%	10.75%
Neural Network	89.29%	10.71%

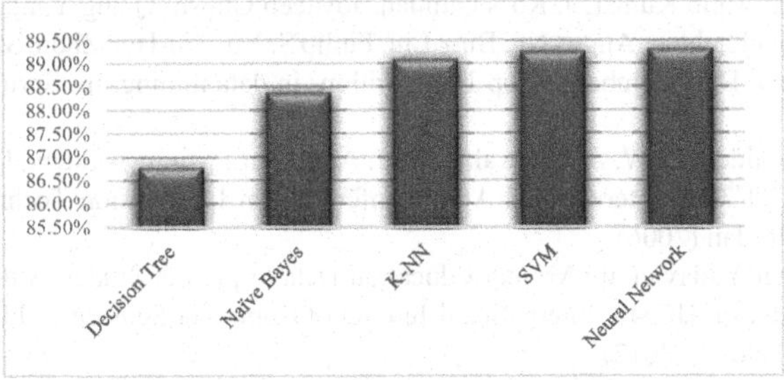

Fig. 4. Result of five Algorithms

From the result, neural network and support vector machine give the best percent correctness of classified instances of 89.29% and 89.25% respectively. Decision tree gives the worst percent correctness of classified instances of 86.77%.

5 Conclusion

This research aims to predict engineering student performance from core engineering courses. Before implementing the systems, the experiment is executed to define the best classification model.

This paper proposes five algorithms to classify the student performance. From the experiment, the best classification method is neural network with the percent accuracy of 89.29%. We believe this model is considerable and can be a great advantage for both students and instructors. Instructors can be able to assist low performance students properly. Students; whereby, will stay highly motivated to successfully graduate with high academic achievement.

We plan to use more features by selecting other significant features from bio data such as age, sex, hometown, income, and etc. We hope the selected features will be able to provide higher accuracy of prediction model.

Acknowledgement

The authors cordially thank Office of Academic Promotion and Registration, Rajamangala University of Technology Thanyaburi for their student data support.

Reference

1. Oded Maimon, Lior Rokach.: Data Mining and Knowledge Discovery Handbook. Second Edition Springer Science+Business Media, LLC, New York p6, p133 (2010)
2. Galit Shmueli, Nitin R. Patel, Peter C.Bruce.: Data mining for business intelligence: concepts, techniques, and applications in Microsoft Office Excel with XLMiner. 2nd ed. john Wiley & Sons,Inc, New Jersey (2010)
3. Xindong Wu, Vipin Kumar, J. Ross Quinlan, Joydeep Ghosh, Qiang Yang, Hiroshi Motoda, Geoffrey J. McLachlan, Angus Ng, Bing Liu, Philip S. Yu, Zhi-Hua Zhou, Michael Steinbach, David J. Hand, Dan Steinberg.: Top 10 algorithms in data mining. In: Springer-Verlag, London (2007)
4. Q. A. AI-Radaideh, E. W. AI-Shawakfa, M. I. AI-Najjar.: Mining student data using decision trees. In: ACIT'2006 International Arab Conference on Information Technology. Yarmouk University, Jordan (2006)
5. Surjeet Kumar Yadav et al.: Mining Education Data to predict Student's Retention: A comparative Study. In: IJCSIS International Journal of Computer Science and Information Security, Vol. 10, No. 2, (2012)
6. Daniel T. Larose. : Data mining methods and models. John Wiley & Sons, Inc., New Jersey (2006)
7. Hanife GÖKER, Halil Ibrahim BÜLBÜL, Erdal IRMAK.: The Estimation of Students' Academic Success by Data Mining Methods. In: 12th International Conference on Machine Learning and Applications. (2013)
8. Brijesh Kumar Bhardwaj, Saurabh Pal.: Data Mining: A prediction for performance improvement using classification. In: IJCSIS International Journal of Computer Science and Information Security, Vol. 9, No. 4, April (2011)
9. S.Taruna, Mrinal Pandey.: An Empirical Analysis of Classification Techniques for Predicting Academic Performance. In: IACC2014 IEEE International Advance Computing Conference , (2014)
10. Durgesh K. Srivastava, Lekha Bhambhu.: Data Classification using Support Vector Machine. In: JATIT Journal of Theoretical and Applied Information Technology© (2005 - 2009)
11. Guojun Zhang, Jixiong Chen. : A SVM Classifier Research Based on RS Reducts. In: 2009 International Conference on Information Management, Innovation Management and Industrial Engineering. (2009)
12. Chady El Moucary: Data Mining for Engineering Schools Predicting Students' Performance and Enrollment in Masters Programs. In: IJACSA International Journal of Advanced Computer Science and Applications, Vol. 2, No. 10, (2011)
13. Yojna Arora, Abhishek Singhal, Abhay Bansal.: PREDICTION & WARNING: A Method to Improve Student's Performance. In: ACM SIGSOFT Software Engineering Notes, January Volume 39 Number 1, New York (2014)

Evolutionary Circular-ELM for the reduced-reference assessment of perceived image quality

Sarutte Atsawaraungsuk[1], Punyaphol Horata[2],

[1]Department of Computer Education, Rajabhat Udonthani University, Thailand,
[2]Department of Computer Science, Khonkaen University, Thailand,

sarutte4@gmail.com, punhor1@kku.ac.th

Abstract. At present, the quality of the image is very important. The audience needs to get the undistorted image like the original image. Cause of the loss of image quality such as storage, transmission, compression and rendering. The mechanisms rely on systems that can assess the visual quality with human perception are required. Computational Intelligence (CI) paradigms represent a suitable technology to solve this challenging problem. In this paper present, the Evolutionary Extreme Learning Machine (EC-ELM) is derived into Circular-ELM (C-ELM) that is an extended Extreme Learning Machine (ELM) and the Differential Evolution (DE) to select appropriate weights and hidden biases, which can proves performance in addressing the visual quality assessment problem by embedded in the proposed framework. The experimental results, the EC-ELM can map the visual signals into quality score values that close to the real quality score than ELM, Evolutionary Extreme Learning (E-ELM) and the original C-ELM and also stable as well. Its can confirms that the EC-ELM is proved on recognized benchmarks and for four different types of distortions.

1 Introduction

Computational Intelligence (CI) paradigms that handle the quality assessment task from a different perspective, since they aim at mimicking quality perception instead of designing an explicit model of the human visual system. Quality assessment based on CI paradigms follows 2 steps (Fig.1.). In the first step, a meaningful feature-based representation of the visual signal is defined which is reduced from the dimensionality of the original image. In the second step, a learning machine handles the actual mapping of the feature vector into quality scores. Since such an approach can rely on the ability of CI tools to deal with complex, nonlinear problems, relatively simple metrics can be designed. Most of the computational power is spent in the off-line training phase; indeed, a trained system can support real-time quality assessment with a minimal overhead to the feature extraction computational cost. Several studies proved the effectiveness of methodologies that exploit CI tools to address image quality [1–2].

K.J. Kim (ed.), *Information Science and Applications*,
Lecture Notes in Electrical Engineering 339, DOI 10.1007/978-3-662-46578-3_77

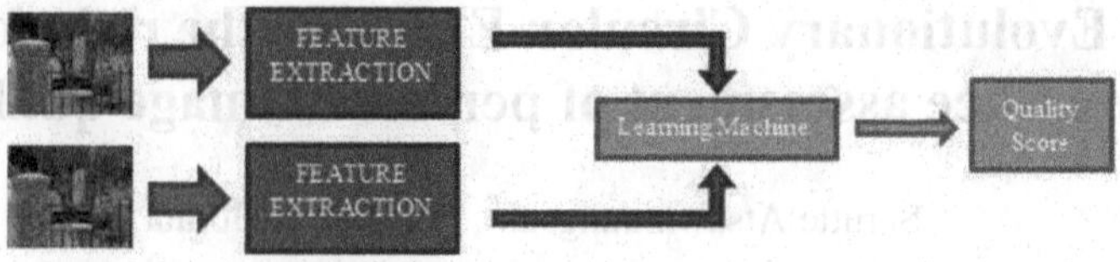

Fig.1. A CI-based reduced-reference schema for image quality assessment

C-ELM [3] is an extended ELM [4-6]. It works on ELM structure, single hidden layer feed forward network (SLFNs) and Circular Back Propagation (CBP) [7-8] architecture. In C-ELM, the input weights (linking the input layer to the hidden layer) and hidden biases are randomly chosen. However, C-ELM may be affected from randomly unacceptable input weights and biases. It may cause a poor accuracy rate and also uses a large number of hidden nodes. To solve this problem, DE [9-10] is a problem optimizer by iteratively and most commonly used to improve generalization of some classifies and also selecting appropriate weight and hidden biases which is a key idea to increase the correctly data classification. Furthermore, G.-B.Huang et al. applied DE to E-ELM [13]. From this concept, a new hybrid algorithm between DE and C-ELM by applying DE to take appropriate input weights and biases for C-ELM and also calls it as EC-ELM.

Experiments comparing the performance of ELM, E-ELM, C-ELM and EC-ELM by embed all of these networks into a Reduced-Reference objective quality metric. Reduced Reference (RR) methods [11–12] combine the information on the received signal with a (limited) number of features extracted from the original signal.

Section 2 is an overview of the Reduced-reference framework for objective quality assessment. **Section 3** introduces the algorithms for comparison and parameters **Section 4** the experimental set up and experimental results and the last is conclusions in **Section 5**.

2 Reduced-reference framework for objective quality assessment

A CI-based approach to objective quality assessment allows to reduce the overall complexity of the perception prediction problem. We can summarize feature extraction in reduced-reference framework in Fig.2.

Feature extraction
Input: a picture $\overline{I}$ ($\overline{I}$ is the input image), a descriptive feature $f_u \in \Phi$, and a value for distance k
Steps: 1. Block level feature extraction a. Split $\overline{I}$ into N_b non-overlapping square blocks, and obtain the set: B = $\{b_m; m=1, \ldots, N_b\}$

b. For each block $b_m \in$ B: compute the associate color correlogram $Z^k_{b_m}$

c. For each matrix $Z^k_{b_m}$: compute the value $x_{u,m}$ of feature f_u, and obtain the set : $X_u = \{x_{u,m};\ m = 1,\ldots, N_b\}$

2. Global level numerical representation

a. Compute a percentile-based description of X_u; let p_α be the α-th percentile: $\varphi_{\alpha,u} = p_\alpha(X_u)$

b. Assemble the objective descriptor vector, x_u, for the feature f_u on the image $\bar{I}$: $x_u = \{\varphi_{\alpha,u}\ ;\ \alpha \in [0,100]\}$ $u = 1, \ldots, N_u$,

3. Output: A global descriptor x_u for image $\bar{I}$ and use it that is input of machine.

Fig.2. The pseudo-code of feature extraction in reduced-reference framework

3 Algorithms for comparison and parameters

ELM: Extreme Learning Machine [4] is a new SLFNs used in classification and regression proposed by G.B.Huang et al. This algorithm consists in k hidden nodes, randomly assigned input weight, hidden layer bias, and activation function. Applied to the learning data for N arbitrary distinct samples (x_i,t_i), i=1, 2, …, N where $x_i = (x_{i,1}, x_{i,2}, \ldots, x_{i,D}, \|x_i^2\|)^T \in R^D$ is a vector having D-dimensional data, which is appended with the square of norm and $t_i = [t_1, t_2, \ldots, t_D]^T \in R^M$ is a vector of target values. ELM can be written in a compact form as

$$\mathbf{H}\beta = \mathbf{T} \tag{1}$$

$$\mathbf{H} = \begin{bmatrix} g(w_1x_1+b_1) & \ldots & g(w_jx_1+b_j) \\ \ldots & \ldots & \\ g(w_1x_N+b_1) & \ldots & g(w_jx_N+b_j) \end{bmatrix} \tag{2}$$

$$\beta = \mathbf{H}^\dagger\mathbf{T} \tag{3}$$

where $\mathbf{H}^\dagger$ is the MP generalized inverse of $\mathbf{H}$.

C-ELM: Sergio Decherchi et al. [3] proposed C-ELM is an extended ELM. It works on ELM structure and CBP architecture. But **H**'s C-ELM is different form ELM as

$$\mathbf{H} = \begin{bmatrix} g(z_1 \bullet \|x_1-c_1\|^2-b_1) & \ldots & g(z_j \bullet \|x_1-c_j\|^2-b_j) \\ \ldots & \ldots & \ldots \\ g(z_1 \bullet \|x_N-c_1\|^2-b_1) & \ldots & g(z_j \bullet \|x_i-c_j\|^2-b_j) \end{bmatrix} \tag{4}$$

Where

$$z_j = w_{j,K+1} \tag{5}$$

$$c_j = \left[-\frac{w_{j,1}}{2w_{j,K+1}}, \ldots, -\frac{w_{j,K}}{2w_{j,K+1}} \right] \quad (6)$$

$$b_j = \frac{1}{w_{j,K+1}} \left(\sum_{k=1}^{K} \frac{w_{j,K}^2}{4w_{j,K+1}} - br_j \right) \quad (7)$$

E-ELM: Qin-Yu Zhu et al. [13] applies DE with ELM. They provided that DE can find the suitable weights and biases for ELM to obtain more accuracy rate.

EC-ELM: [14] is a new hybrid algorithm between DE and C-ELM by applying DE to take appropriate input weights and biases for C-ELM

Table 1. Parameters used in the Extreme Learning Machines

Algorithm	Parameters
ELM, E-ELM	Activation function = 'Sigmoid'
C-ELM, EC-ELM	Activation function = 'Gaussian'
DE in E-ELM and EC-ELM	F = 0.9
	$\mathcal{E} = 0.005$
	The number of cycles is 20 rounds.

4 Experimental results

In the experiment was trained on the four distorted datasets in the second release of the LIVE database [15]: the JP2K and JPEG were divided into 2 sets, the White noise and Gaussian Blur. The 2 JP2K datasets included 82 and 87 patterns, respectively; the White noise and Gaussian Blur dataset each contained 145 patterns; the last is the number of JPEG(1) and JPEG(2) are 87 and 88 images, respectively.

For each images have a Difference Mean Opinion Score (DMOS) was a target which provided in the LIVE database, indicating the perceived difference in quality between the original and processed image, and for the reason in the computational, remapped into ranging between [-1, 1] form [1,100].

For calculating the features, the image was split in blocks of 32x32 pixels. For each block the features described in Table 2. Are calculated from the color correlogram that is computed for distance k = 1 using the L1 norm. The objective vectors $y_{u,RR}^{(n,r)}$ (the luminance layer in the YC_bC_r color space) and $h_{u,RR}^{(n,r)}$ (the hue layer in the HSV color space) were constructed by assembling six percentiles (i.e.,{0,20,40,60,80,100}) of the distributions y_u, h_u over all blocks.

Table 2. Features used for the C-ELM predictors

	JP2K	WN	GB	JPEG
Luminance	Entropy	Entropy	Entropy	Entropy
	Homogeneity	Contrast	Homogeneity	Homogeneity
Hue	Homogeneity	Contrast	Entropy	Diagonal Energy
	Contrast	Energy Ratio	Homogeneity	Entropy

To prepare the datasets, the datasets were divided each dataset into 5 groups of the *k*-fold test strategy (the whole experimental session adopts the same composition of the 5 folds used in [11] each containing 29 original images) because the overall performance evaluation of the CI-based quality estimator proves effective to obtain reliable results when few data are available [16].

The inputs of the network were $y_u^{(n,r)}$ and $h_u^{(n,r)}$ both sized 12 and work on an ensemble. To find the optimal number of hidden node was selected form 1-100 nodes

In the experiments, the meta-metrics evaluation (the greatest results are highlighted by the bold letters on all the models) [17-18] is used to measure the performance of the ELM, E-ELM, C-ELM and EC-ELM which is divided into 5 criteria;

1) Pearson's correlation coefficient between target values and predicted values.
2) Root Mean Square Error (RMSE).
3) The number of used hidden node for the highest Pearson's correlation.
4) Standard deviation (SD) of Pearson's correlation.
5) SD of RMSE.

Table 3. Pearson's correlation (the higher is better)

Dataset	ELM	C-ELM	E-ELM	EC-ELM
JPEG(1)	0.9480	0.9427	0.9417	**0.9512**
JPEG(2)	0.9349	0.9264	0.9294	**0.9358**
JP2K(1)	0.8215	0.8373	0.8465	**0.8565**
JP2K(2)	0.8852	0.8704	0.8851	**0.8927**
White Noise	0.9587	0.9494	0.9605	**0.9651**
Gaussian Blur	0.8244	0.8247	0.8455	**0.8476**

From Table 3., EC-ELM has the winning score (the highest accuracy rates) in all of datasets.

Table 4. RMSE (the lower is better)

Dataset	ELM	C-ELM	E-ELM	EC-ELM
JPEG(1)	0.2004	0.2158	0.2204	**0.1923**
JPEG(2)	0.2397	0.2576	0.2437	**0.2285**
JP2K(1)	0.3932	0.4066	0.3774	**0.3690**
JP2K(2)	0.3409	0.3666	0.3309	**0.3115**
White Noise	0.1423	0.1508	0.1336	**0.1261**
Gaussian Blur	0.2601	0.2862	0.2655	**0.2514**

From Table 4., the winning scores of EC-ELM is 6 which it can provide the lowest RMSE in all of datasets.

Table 5. The number of used hidden node for the highest Pearson's correlation (the lower is better)

Dataset	ELM	C-ELM	E-ELM	EC-ELM
JPEG(1)	31	22	**21**	25
JPEG(2)	18	25	**15**	20
JP2K(1)	31	**20**	32	29
JP2K(2)	**28**	30	37	34
White Noise	24	**21**	26	35
Gaussian Blur	26	**12**	29	13

From Table 5., EC-ELM doesn't have the winning scores of usage the lowest hidden nodes.

Table 6. SD of Pearson's correlation (the lower is better)

Dataset	ELM	C-ELM	E-ELM	EC-ELM
JPEG(1)	0.0091	0.0121	0.0189	**0.0078**
JPEG(2)	0.0235	0.0329	**0.0215**	0.0316
JP2K(1)	0.1034	0.1400	0.2428	**0.0823**
JP2K(2)	0.0998	0.0388	**0.0381**	0.0915
White Noise	0.0265	0.0217	0.0158	**0.0121**
Gaussian Blur	0.1006	**0.0577**	0.0599	0.1676

From Table 6., it shows that EC-ELM has the lowest SD of Pearson's correlation 3 datasets from 6 datasets: JPEG(1), JP2K(1) and White noise.

Table 7. SD of RMSE (the lower is better)

Dataset	ELM	C-ELM	E-ELM	EC-ELM
JPEG(1)	0.0321	0.0352	0.0300	**0.0217**
JPEG(2)	0.0549	0.0774	**0.0275**	0.0762
JP2K(1)	0.1039	0.0940	0.1409	**0.0851**
JP2K(2)	0.1488	**0.0422**	0.1114	0.0879

White Noise	0.0445	0.0458	0.0307	**0.0271**
Gaussian Blur	0.0832	0.0702	**0.0578**	0.0953

From Table 7., it shows that the winning scores in the part of SD of RMSE. EC-ELM is 3 which is the highest scores in 6 datasets: JPEG(1), JP2K(1) and White noise.

From the experimental results, we can summarize from the best rates of each table (If an algorithm has the best of each of their values then it's a winning score is increased by 1, all other algorithms not counted as a winning score), as in Table 8.

Table 8. Total of the winner score (the higher is better)

Result	ELM	C-ELM	E-ELM	EC-ELM
Pearson's correlation	0	0	0	**6**
RMSE	0	0	0	**6**
#Node	1	**3**	2	0
Total SD	0 (0+0)	2 (1+1)	4 (2+2)	**6 (3+3)**
Total	1	5	6	**18**

From Table 8., the total winning scores of EC-ELM is 18 which is higher than the score of ELM, E-ELM and C-ELM in all criteria (except winner score in the part of number of hidden nodes).

5 Conclusions

This research presents EC-ELM to deal with addressing the visual quality assessment problem. From the experimental results, the meta-metrics were used to measure the performances of comparative methods based on ELM, there results show that EC-ELM has a total winning score higher than of ELM, E-ELM and C-ELM, which it can prove the performance of EC-ELM. Furthermore, EC-ELM is suitable to solve this challenging problem.

References

1. Eskicioglu, Ahmet M., and Paul S. Fisher. Image quality measures and their performance. Communications, IEEE Transactions on 43.12 (1995): 2959-2965.
2. Gastaldo, Paolo, and Rodolfo Zunino. Neural networks for the no-reference assessment of perceived quality. Journal of Electronic Imaging 14.3 (2005): 033004-033004.Huang, G. B., Wang, D. H., & Lan, Y. (2011).
3. Decherchi, Sergio, et al. Circular-ELM for the reduced-reference assessment of perceived image quality. Neurocomputing 102 (2013): 78-89.

4. Huang, Guang-Bin, Qin-Yu Zhu, and Chee-Kheong Siew. Extreme learning machine: a new learning scheme of feedforward neural networks. Neural Networks, 2004. Proceedings. 2004 IEEE International Joint Conference on. Vol. 2. IEEE, 2004.
5. Huang, Guang-Bin, Qin-Yu Zhu, and Chee-Kheong Siew. Extreme learning machine: theory and applications. Neurocomputing 70.1 (2006): 489-501.
6. Huang, Guang-Bin, Dian Hui Wang, and Yuan Lan. Extreme learning machines: a survey. International Journal of Machine Learning and Cybernetics 2.2 (2011): 107-122.
7. Ridella, Sandro, Stefano Rovetta, and Rodolfo Zunino. Circular backpropagation networks for classification. Neural Networks, IEEE Transactions on 8.1 (1997): 84-97.
8. Gastaldo, Paolo, et al. Circular back-propagation networks for measuring displayed image quality. Artificial Neural Networks—ICANN 2002. Springer Berlin Heidelberg, 2002. 1219-1224.
9. Storn, Rainer, and Kenneth Price. Differential evolution–a simple and efficient heuristic for global optimization over continuous spaces. Journal of global optimization 11.4 (1997): 341-359.
10. Coello Coello, Carlos A. Theoretical and numerical constraint-handling techniques used with evolutionary algorithms: a survey of the state of the art.Computer methods in applied mechanics and engineering 191.11 (2002): 1245-1287.
11. Redi, Judith A., et al. Color distribution information for the reduced-reference assessment of perceived image quality. Circuits and Systems for Video Technology, IEEE Transactions on 20.12 (2010): 1757-1769.
12. Wang, Zhou, and Eero P. Simoncelli. Reduced-reference image quality assessment using a wavelet-domain natural image statistic model. Electronic Imaging 2005. International Society for Optics and Photonics, 2005.
13. Zhu, Qin-Yu, et al. Evolutionary extreme learning machine. Pattern recognition 38.10 (2005): 1759-1763.
14. Atsawaraungsuk, Sarutte, et al. Evolutionary Circular Extreme Learning Machine. Computer Science and Engineering Conference (ICSEC), 2013 International. IEEE, 2013.
15. H.R. Sheikh, Z. Wang, L. Cormack, A.C. Bovik: LIVE Image Quality Assessment Database at (http://live.ece.utexas.edu/research/quality).
16. Bartlett, Peter L., Stéphane Boucheron, and Gábor Lugosi. Model selection and error estimation. Machine Learning 48.1-3 (2002): 85-113.
17. Stefani, Antonia, and Michalis Xenos. Meta-metric evaluation of e-commerce-related metrics. Electronic Notes in Theoretical Computer Science 233 (2009): 59-72.
18. Horata, Punyaphol, Sirapat Chiewchanwattana, and Khamron Sunat. Robust extreme learning machine. Neurocomputing 102 (2013): 31-44.

Enhanced Web page Cleaning for Constructing Social Media Text Corpora

Melanie Neunerdt[1], Eva Reimer[2], Michael Reyer[1], and Rudolf Mathar[1]

Institute for Theoretical Information Technology[1],
Textlinguistics/Technical Communications[2],
RWTH Aachen University, Germany

Abstract. Web page cleaning is one of the most essential tasks in Web corpus construction. The intention is to separate the main content from navigational elements, templates, and advertisements, often referred to as *boilerplate*. In this paper, we particularly enhance Web page cleaning applied to pages containing comments and introduce a new training corpus for that purpose. Beside extending an existing boilerplate detection algorithm by means of a comment classifier, we train and test different classifiers on extended feature sets solving a two-class problem (content vs. boilerplate) on our and an existing benchmark corpus. Results show that the proposed approach outperforms existing methods, particularly on comment pages from different domains. Finally, we point out that our trained classifiers are domain independent and with small adjustments only transferable to other languages.

Keywords: information retrieval, Web page cleaning, content extraction, boilerplate detection, social media corpora

1 Introduction

The social media aspect of the World Wide Web is constantly growing, which leads to a high amount of user generated content, such as comments. These data provide a convenient source for different Natural Language Processing tasks like, e.g., opinion mining or trend analysis. An essential task of automatically compiling such social media text corpora is to separate the main Web page content including comments from, e.g., navigation bars, advertisements or related articles, which are out of interest for the corpus. Several approaches have been introduced to solve the problem of Web page (main/relevant) content extraction, Web page cleaning or boilerplate detection. However, existing approaches do rather study the results achieved on pages containing comments. This is particularly reasoned by the fact that existing training corpora suffer from low numbers of Web pages containing comments, e.g., the *L3S-GN1 corpus* introduced in [5], which contains only 1% comments.

In this paper, we introduce a new Web page training corpus particularly designed to train and test boilerplate classifiers on Web pages with comments.

K.J. Kim (ed.), *Information Science and Applications*,
Lecture Notes in Electrical Engineering 339, DOI 10.1007/978-3-662-46578-3_78

We first explore the performance of a state-of-the art boilerplate detection algorithm on that corpus. Furthermore, we enhance that algorithm by means of a comment classifier leading to significantly improved results. Based on our corpus, we train different classifiers on an extended feature set solving the two-class problem (content vs. boilerplate) and show substantial improvements compared to state-of-the art algorithms. Finally, we benchmark the proposed classifiers on an existing corpus and point out, that our trained classifiers are domain independent and with small adjustments transferable to other languages.

The outline of this paper is as follows. We shortly summarize related work in Section 2. In Section 3, we propose different adapted classification approaches to solve the two-class problem for content vs. boilerplate classification on Web pages containing comments. In Section 4 and 5, we introduce a new annotated training corpus and discuss results for our and a benchmark corpus. Section 6 concludes our work.

2 Related Work

Automatically extracting the main content from Web pages has been well studied and a wide range of methods has been proposed. Common alternative terms for Web page cleaning are content extraction or boilerplate detection. Although Web page cleaning is a very crucial step in the construction of Web corpora, only relatively little literature can be found in this area. The *CleanEval* shared task and competition, [1], is one of few works, which particularly aims at cleaning arbitrary Web pages with the goal of preparing Web data for the use as a corpus. The *shared task* competitors basically apply different classification algorithms based on a variety of classification features [1]. For instance, NCleaner, [2], extracts content by deleting boilerplate with regular expressions and uses n-gram language models to separate content segments from non-content segments. Sousa et al., [7], employ a Conditional Random Field (CRF) based on multiple features and thereby treat the problem of boilerplate detection as a sequence labeling task.

Kohlschütter et al., [5], analyse a representative set of features used by the approaches proposed by the *CleanEval* competitors. In addition to a boilerplate detection algorithm working with a small set of shallow text features, they introduce a new Web page training corpus called *L3S-GN1 corpus*. Furthermore, they extend their approach with handcrafted rules for comment detection (11 indicator strings like, e.g., User comments:), but achieve lower precision. Both corpora, *CleanEval* and *L3S-GN1*, contain only a little amount of comment pages, which are annotated as part of the main content. Kohlschütter et al., [5], even explicitly annotate comments as separate class. Nevertheless, their publication reveals few insights about the efficiency of extracting comments as main content.

3 Boilerplate Detection

Our Web cleaning methods are built up on Kohlschütter's work presented in [5]. We take results of Kohlschütters boilerpipe detection tool and have a particular look at the boilerplate classified parts. For those parts, we apply a comment classifier in order to identify not detected content. In an alternative approach, we propose a new training corpus consisting of Web pages with immanent comments and train different classifiers on that corpus. Moreover, we extend the feature set representing the Web page segments.

In the following, we first mathematically describe the boilerplate detection problem. A Web page p is segmented into a sequence of N_p text blocks (segments), where each block is represented by an n-dimensional feature vector

$$(\mathbf{b}_n^p)^T \in \mathcal{X} = \mathcal{B}_1 \times \mathcal{B}_2 \times \ldots \mathcal{B}_n.$$

The aim is to predict the to $\mathbf{B}^p = \left(\mathbf{b}_1^p, \ldots, \mathbf{b}_{N_p}^p\right)$ associated class sequence

$$\mathbf{c}^p = \left(c_1^p, \ldots, c_{N_p}^p\right),$$

with $c_n^p \in \mathcal{C}$ for $n = 1, \ldots, N_p$. In the task of Web page cleaning, the set of text classes $\mathcal{C}$ comprises *CONTENT* and *BOILERPLATE*.

The feature vector we use is built up on Kohlschütter's [5] approach combining 67 features. However, in contrast to his approach we introduce token-based and POS-based features. The language in comments is particularly characterized by a communicational style or particular tokens, e.g., emoticons or interjections. Since this is reflected in the usage of particular tokens and their corresponding part-of-speech (POS), we believe that classification results can be improved. A detailed description of the feature set can be found in [8].

For Web page segmentation, we use the segmentation from Kohlschütters boilerpipe tool. Web pages p are segmented into atomic *text blocks (TB)*, represented by $\mathbf{b}_n^p$, by a simple split at each HTML tag, except for the $<a>$ tag. The result is a very fine-grained segmentation, which might result in splitted articles or comments. To counteract this problem, we apply the same block fusion algorithm delivered with Kohlschütters *Article Extractor* and achieve a sequence of fused text blocks represented by

$$\mathbf{S}^p = \left(\mathbf{s}_1^p, \ldots, \mathbf{s}_{M_p}^p\right),$$

with $M_p \leq N_p$, where feature vectors $\mathbf{s}_m^p$ are calculated analogously to $\mathbf{b}_1^p$. The fused text blocks are later referred to *Article Segments (AS)*. In general, this problem is a sequence labeling task and is solved by the optimization problem

$$\hat{\mathbf{c}}^p = \arg\max_{\mathbf{c}^p} \left\{ g\left(\mathbf{S}^p, \mathbf{c}^p\right) \right\}, \tag{1}$$

where g represents any decision function. In the following paragraphs, we describe our different suggestions for g.

In the first approach, we propose an extension to the *Article Extractor* algorithm provided in Kohlschütters boilerpipe tool. The *Article Extractor* is tuned towards articles and works like a sensitive minimum content extractor with particularly high CONTENT precision rates. The basic idea is to improve the web cleaning accuracy by detecting comments in such areas, which are classified as BOILERPLATE by the *ArticleExtractor* and classify those as CONTENT.

The concept of combining the Article Extractor with a comment classifier is described in Algorithm 1. Algorithm 1 splits the Web page p into text blocks. Then, text blocks are fused and classified into CONTENT and BOILERPLATE in ArticleExtractor$(\mathbf{b}_1^p, \ldots, \mathbf{b}_{N_p}^p)$. All resulting article segments classified as BOILERPLATE are further investigated. For all fused text blocks of such an article segment our comment classifier is applied. If it detects at least one comment it classifies the whole article segment as CONTENT. The comment

Algorithm 1:

ADAPTED ARTICLE EXTRACTOR(p)

Input: Web page p
Output: Predicted class sequence $\tilde{c}_1^p, \ldots, \tilde{c}_{M_p}^p$
$\mathbf{b}_1^p, \ldots, \mathbf{b}_{N_p}^p \leftarrow$ SEGMENTWEBPAGE(p)
$(\mathbf{s}_1^p, \ldots, \mathbf{s}_{M_p}^p, \tilde{c}_1^p, \ldots, \tilde{c}_{M_p}^p) \leftarrow$
ARTICLEEXTRACTOR($\mathbf{b}_1^p, \ldots, \mathbf{b}_{N_p}^p$)

for each position $m = 1 \ldots M_p$ **do**
  if IsBOILERPLATE($\tilde{c}_m^p$) **then**
    if EXISTCOMMENT($\mathbf{s}_m^p$) **then**
      $\tilde{c}_m^p \leftarrow$CONTENT
    end if
  end if
end for
return $\tilde{c}_1^p, \ldots, \tilde{c}_{M_p}^p$

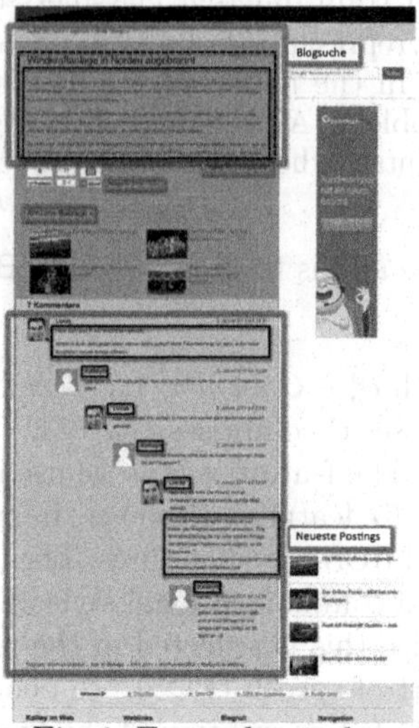

Fig. 1: Example results.

classifier solves a two class problem, where comments are separated from the remaining content of a Web page. A detailed description about different comment classifiers, the classification features, and accuracies achieved with different approaches can be found in [8].

In the second approach, we train different classifiers on a self annotated representative training corpus based on an extensive feature set. In order to describe the different methods, we build on the optimization problem given by the sequence labeling problem from (1). This is a huge optimization problem, which is simplified differently by our two approaches.

First, we make very strong independence assumptions: (1) The classes $\hat{c}_m^p$ are predicted independently from predictions $\hat{c}_j^p$ for $j \neq m$. (2) We assume that the class for a given text segment $\mathbf{s}_m^p$ at position m only depends on some - here k - preceeding and succeeding text segments. Hence, the optimization problem is reformulated as

$$\hat{c}_m^p = \arg\max_{c_i^p} \left\{ g\left(\mathbf{s}_{m-k}^p, \ldots, \mathbf{s}_m^p, \ldots, \mathbf{s}_{m+k}^p, c_m^p\right) \right\}.$$

In this approach, we apply different classifiers, which operate on a single feature vector as input. Therefore, we represent our segment $\mathbf{s}_m^p$ by a feature vector

$$\mathbf{s}_{mk}^p = \left((\mathbf{s}_{m-k}^p)^T, \ldots, (\mathbf{s}_m^p)^T, \ldots, (\mathbf{s}_{m+k}^p)^T \right)^T$$

Hence, the decision functions is reformulated as $g\left(\mathbf{s}_{mk}^p, c_m^p\right)$. As decision function we apply three different approaches: a K-Nearest Neighbor (KNN) algorithm, a decision tree (J48), and a Support Vector Machine (SVM). Note that we apply the proposed classifiers to both segmentation results AS and TB. The optimization problem for the TB segmentation is formulated and solved in the same way. For our experiments we use the WEKA tool, [4].

Additionally, we apply a Conditional Random Field (CRF) to the sequence labeling task. CRFs, first introduced in [6], are probabilistic models, which make only few independency assumptions, where the sequence of text classes $(c_1^p, \ldots, c_{M_p}^p)$ is predicted jointly, dependent on the whole sequence of observed text segments $(\mathbf{s}_1^p, \ldots, \mathbf{s}_{M_p}^p)$. The probabilistic model has exponential form depending on some feature functions. A detailed description would go beyond the scope of this paper, but more details can be found in [6]. For our experiments the free software CRFSuite, [9], is used, which implements a first-order CRF, as described in [6]. CRFSuite only supports binary observation feature functions, which check if the corresponding feature equals a particular value or not. Since this would lead to a huge amount of feature functions and an overfitted model, when applied to continuous valued features, we apply Fayyads multi-interval feature discretization, [3].

4 Experimental Results

We perform our evaluations on three different corpora, two self annotated corpora for training and testing, the German Web page corpus *GWebTrain* and the English Web page corpus *EWebTrain*, and the English benchmark corpus *CleanEval* for validation. The *GWebTrain* corpus consists of 200 manually assessed Web pages all from different domains. Web pages contain forums, blogs and different news sites dealing with different topics. We manually annotate the corpus by means of a self-implemented annotation tool, which supports visual annotation in a Web browser. The annotators mark the CONTENT of the Web page depicted in the Web browser, areas, which are not marked, are automatically assigned to the BOILERPLATE class. Since we are particularly interested in the accuracy achieved on Web page parts, where comments are posted, we additionally annotate such areas as COMMENT_AREA. Furthermore, the annotators mark the main text of each comment in order to produce adequate training data for the comment classifier. Figure 1 shows an exemplary annotated Web page, with the CONTENT depicted in green, the COMMENT_AREA depicted in yellow, and non-highlighted BOILERPLATE. The English Web page corpus *EWebTrain* consists of 50 Web pages and is annotated in the same way as the *GWebTrain* corpus. It is significantly smaller, but particularly serves to show the portability considering language and domain of our methods. Distributions of the classes CONTENT, BOILERPLATE, and COMMENT_AREA at token, text block (TB), and fused segment (AS) level for *GWebTrain* are depicted in Table 1. Note that the standard deviation is depicted by $\pm$. Additionally, we use the *CleanEval* corpus particularly for benchmarking our methods on English Web pages. The manually annotated corpus includes two division for English, a development/training set, and an evaluation set. For our experiments, we only consider the evaluation set.

Table 1: Class distributions in the *GWebTrain* Corpus with 200 Web pages.

Class	# TB	Mean TB	# AS	Mean AS	# Tokens	Mean Tokens
Total	41,796	222 ± 120	31,305	167 ± 98	312,282	$1,661 \pm 1,171$
BOILERPLATE	24,694	131 ± 126	23,179	123 ± 121	79,521	423 ± 522
CONTENT	17,102	91 ± 109	8,126	43 ± 56	232,761	$1,238 \pm 1,587$
NON-COMMENT_AREA	4,193	22 ± 22	1,935	10 ± 13	83,893	446 ± 507
COMMENT_AREA	12,909	69 ± 87	6,191	33 ± 43	148,868	$792 \pm 1,081$

Table 2: *GWebTrain* 10-fold cross validation results for different methods.

Method	P_{CONT}	R_{CONT}	P_{BOIL}	R_{BOIL}	$\overline{F_1}$*-Score*	R_{COM}
ARTICLE Extractor + SVM COMMENT classifier based on TB segmentation						
	0.83	0.61	0.77	0.91	0.77	0.62
BOILERPLATE classifier based on TB segmentation						
KNN	**0.94**	0.76	0.85	0.97	0.87	0.80
J48	0.87	0.83	0.89	0.91	0.88	0.88
SVM	0.92	0.88	0.92	0.95	0.92	0.94
CRF	0.91	0.91	0.94	0.94	**0.93**	**0.96**
BOILERPLATE classifier based on AS segmentation						
KNN	**0.94**	0.64	0.79	0.97	0.82	0.71
J48	0.88	0.84	0.89	0.92	0.89	0.88
SVM	0.92	0.87	0.92	0.95	**0.92**	**0.94**
CRF	0.90	0.84	0.90	0.94	0.90	0.94
Extractors provided by Kohlschütter						
DEFAULT Extractor	0.85	0.50	0.73	0.94	0.73	0.34
ARTICLE Extractor	0.92	0.25	0.65	0.99	0.59	0.18
LARGEST CONTENT Extractor	0.94	0.13	0.63	0.99	0.50	0.07

4.1 Training and Testing on *GWebTrain*

In order to evaluate our methods, we apply a 10-fold cross validation to the *GWebTrain* corpus and solve the two-class problem (CONTENT vs. BOILERPLATE). We split the corpus by randomly selecting Web pages. Note that the comment classifier used in the extended ArticleExtractor approach is a trained SVM based on the same 10-fold cross validation. Classification accuracy is measured by CONTENT/BOILERPLATE precision (P_{CONT}, P_{BOIL}), recall (R_{CONT}, R_{BOIL}), mean F_1-scores ($\overline{F_1}$*-Score*) based on TB segments. In addition, we calculate the recall achieved on COMMENT_AREA (R_{COM}). Overall, cross validation results for the different classifiers and extractors provided by the Kohlschütter's tool are depicted in Table 2. The first row depicts results achieved with the extended Kohlschütter algorithm. Generally, results can be improved compared to the base line algorithm, however, the recall achieved for the COMMENT_AREA is not satisfying. Particularly good results are achieved by applying a CRF on TB or a SVM on AS segmentation, with F_1-Scores up to 0.93. More importantly, the CRF on TB achieves a good balance with both precision and recall above 0.91. Furthermore, it leads to the highest F_1-Score and recall on the COMMENT_AREA. This can easily be explained by the fact that the classes of the preceding and succeeding segments are good indicators for the class of the current segment. This dependency is particularly modeled in a CRF. The KNN classifiers attract attention by their high CONTENT precisions, which is particularly interesting for Web corpus construction. Comparing our results to those achieved by Kohlschütter depicted in the lower part of Table 2 shows that F_1-Scores can be increased by 20 percentage points.

Figure 1 illustrates the result for an example Web page with three different approaches, (1) the SVM based on AS segmentation (red), (2) the DefaultExtractor (black), (3) Article Extractor (dotted black). Detected CONTENT parts

Table 3: Evaluation of different methods applied to the *CleanEval* corpus.

Micro-average	TB CRF	AS SVM	LCExtractor	AExtractor	NCleaner
Mean F_1-*Score*	0.84	0.90	0.69	0.81	0.90
Mean Precison	0.95	0.93	0.97	0.96	0.90
Mean Recall	0.75	0.86	0.54	0.70	0.90

of the Web page are depicted by the particular box. The two Kohlschütter's extractors do not even detect any or only small parts of the comment area as CONTENT. Our classifier successfully detects the complete COMMENT_AREA except for the headline. However, some keywords like *Blogsuche (Blogsearch)* or *Neuste Postings (recent posts)* are wrongly detected as CONTENT. This can be explained by the fact that these keywords are strongly related to the comment area and in other Web pages are part of the content.

4.2 Application to *CleanEval*

It is important to us to show that the methods are portable to English Web pages and to test the domain-independence of the classifiers. Therefore, we train the same classifiers on our manually annotated English Web page corpus *EWebTrain*. We use the same set of features, but adapt the calculation of language-dependent features. In particular, calculation of token- and POS-based features are adapted by using English word lists and integrating a POS tagger for English.

We evaluate the two-class problem (CONTENT vs. BOILERPLATE) on the *CleanEval* test set for two of our proposed classifiers that have been trained for the *EWebTrain* collection and for the **A**RTICLE and **L**ARGEST **C**ONTENT extractor proposed by Kohlschütter. Exemplarily, we choose the two classifiers, BOILERPLATE CRF based on TB segmentation (TB CRF) and BOILERPLATE SVM based on AS segmentation (AS SVM). Note that the *CleanEval* corpus only contains few comment Web pages and no specific annotation of the comments. Therefore, no explicit evaluation for such Web page types is possible. Because the CleanEval corpus only provides content annotation on a text-level, we cannot directly use the evaluation setup applied to the *GWebTrain* corpus. In order to produce comparable results to existing approaches, we use the python script *cleaneval.py* proposed by [2] and calculate mean precision, recall, and F_1-Score based on token level. Note that the accuracy measure is based on a weighted Levenshtein distance at token-level proposed by the *CleanEval* initiative and, hence, is not directly comparable to the results depicted in Table 2.

Results are depicted in Table 3 in the same form than published in [2]. For a direct comparison, the published results of *NCleaner*, [2], trained on the *CleanEval* training set are depicted in the right column of the table. Similarly to the cross-validation results, the CRF based on TB and the SVM based on AS outperform the results achieved with Kohlschütter's extractors. However, in contrast to the previous results, the SVM based on AS significantly outperforms the CRF based on TB and hence seems to be more robust against the data basis. Note that the *EWebTrain* corpus is significantly smaller than the German *GWebTrain* corpus. Experiments have shown that enlarging the amount of training Web pages from 50 to 180 leads to a F_1-Score increase of 1.5 and 4.2 percentage points for SVM AS and CRF TB. A final comparison of our classifiers to *NCleaner* shows that we can still compete with the results achieved by the *NCleaner*, which has specifically been trained for *CleanEval*. Overall, this shows that the classifiers trained on the proposed Web page corpora can be applied to Web page corpora from other domains without significant performance loss. Furthermore, the methods are portable to English Web pages with little effort.

5 Conclusion

In this paper, we have proposed an effective approach for boilerplate detection applied to Web pages containing comments. Enhancements are particularly achieved by providing a representative training corpus and the usage of an extended feature set with token- and POS-based features. The proposed methods achieve mean F_1-Scores up to 0.93 applied to our corpus, which significantly outperforms existing approaches. Applying a Conditional Random Field yields the highest recall of 0.96 on the COMMENT_AREA, which is of special interest in our task. Furthermore, we analyse our boilerplate detection methods on the English benchmark corpus *CleanEval*. Results show that the train classifiers are domain-independent and can be transferred to other languages with little effort. Satisfying results are achieved on the *CleanEval* evaluation corpus, already for small training corpora.

References

1. M. Baroni, F. Chantree, A. Kilgarriff, and S. Sharoff. Cleaneval: A Competition for Cleaning Web Pages. In *LREC*, 2008.
2. S. Evert. A Lightweight and Efficient Tool for Cleaning Web Pages. In *Proceedings of the Sixth International Conference on Language Resources and Evaluation*, Marrakech, Morocco, May 2008. European Language Resources Association (ELRA).
3. U. M. Fayyad and K. B. Irani. Multi-interval discretization of continuous-valued attributes for classification learning. In *International Joint Conference on Artificial Intelligence*, pages 1022–1029, 1993.
4. M. Hall, E. Frank, G. Holmes, B. Pfahringer, P. Reutemann, and I. H. Witten. The WEKA data mining software: An update. *SIGKDD Explor. Newsl.*, 11(1):10–18, November 2009.
5. C. Kohlschütter, P. Fankhauser, and W. Nejdl. Boilerplate Detection Using Shallow Text Features. In *Proceedings of the Third ACM International Conference on Web Search and Data Mining*, pages 441–450, New York, NY, USA, 2010. ACM.
6. J. D. Lafferty, A. McCallum, and F. C. N. Pereira. Conditional random fields: Probabilistic models for segmenting and labeling sequence data. In *Proceedings of the Eighteenth International Conference on Machine Learning*, ICML '01, pages 282–289, San Francisco, CA, USA, 2001. Morgan Kaufmann Publishers Inc.
7. M. Marek, P. Pecina, and M. Spousta. Web page cleaning with conditional random fields. In *Building and Exploring Web Corpora (WAC3 - 2007), Proceedings of the 3rd Web as Corpus Workshop, Incorporating Cleaneval*, 2007.
8. M. Neunerdt, M. Reyer, and R. Mathar. Automatic Genre Classification in Web Pages Applied to Web Comments. In *12th Conference on Natural Language Processing*, pages 145–151, Hildesheim, Germany, October 2014.
9. N. Okazaki. CRFsuite: A fast implementation of Conditional Random Fields (CRFs), 2007.

Finding Knee Solutions in Multi-objective Optimization using Extended Angle Dominance Approach

Sufian Sudeng and Naruemon Wattanapongsakorn

Department of Computer Engineering, King Mongkut's University of Technology Thonburi, Bangkok, 10140, Thailand.

Abstract. The aim of this paper is to develop a knee-based multi-objective evolutionary algorithm (MOEA) which is a method to find optimal trade-off solutions among all available solutions. In multi-objective optimization, the knee regions represent implicitly preferred parts to the decision maker (DM). The proposed approach uses the extended angle dominance concept to guide the solution process towards knee regions. The extent of the obtained solutions can be controlled by the means of user-supplied density controller parameter. The approach is verified by two and three objective knee-based test problems. The results have shown that our approach is competitive to well-known knee-based MOEAs in convergence view point.

Keywords: Multi-objective optimization; knee region; knee-based MOEAs; knee solutions; pareto-optimal solutions

1 Introduction

Real-world problems commonly require the simultaneous consideration of multiple performance measures. Most often, the multiple objective problems are in conflict and compete with each other. It can be solved by mathematical modeling-based approaches that are also effective in finding trade-off solutions. However, these approaches consume huge computing resources and do not generate multiple solutions in single simulation run. Multi-Objective Evolutionary Algorithms (MOEAs) overcome this limitation. These algorithms imitate the genetic evolution in nature and apply it to optimization, and they work with a population of solutions. MOEAs are also suitable for solving multi-objective problems in a reasonable time. The main goal of MOEAs is to assist the DM in selecting the final alternative which satisfies most or all of his/her preferences. Instead of having a single solution, there is a set of trade-off solutions called non-dominated solutions (i.e., no solution dominates or is better than the other solutions in the set). All non-dominated solutions are also called Pareto-optimal solution. All solutions that are Pareto-optimal constitute the Pareto set. The objective values of the Pareto set in the objective space constitute the Pareto front.

To simplify the decision making task, the DM can incorporate his/her preferences into the search process. These preferences are used to guide the search towards the preferred parts of the Pareto front [1], called the region of interest (ROI), the preferred portion of the Pareto front from the DM's perspective. The decision maker (DM) has to decide on an individual solution based on certain preferences and objectives' priorities. In the absence of explicit DM's preference information in multi-objective optimization, there exist special parts of available solutions that represent implicitly preferred parts for the DM called "knee regions". Knee regions are potential parts presenting the maximal trade-offs between objectives. Solutions residing in knee regions are characterized by a small improvement in either objective can cause a large deterioration in at least another one which makes moving in either direction not attractive. Fig. 1(a) exemplifies the Pareto front with a knee region, and Fig. 1(b) illustrates the Pareto-optimal solutions with four knees. In this paper, we present an approach to find promising knee solutions in multi-objective optimization problems using an extended angle dominance concept.

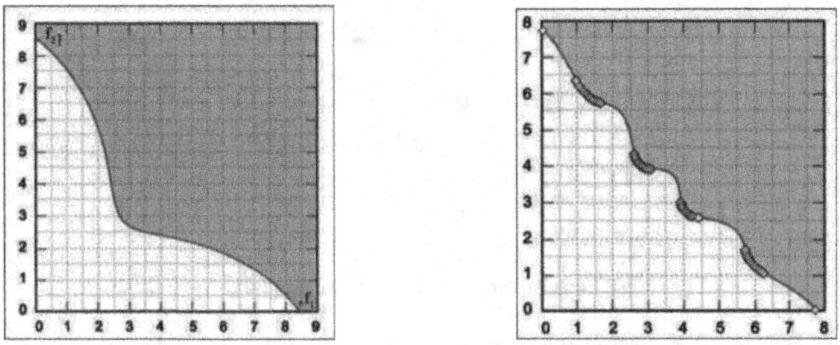

(a) Simple Pareto-optimal solutions with a knee region (b) A Pareto-optimal solutions with four Knee regions

Fig.1. Example knee region(s).

An erratum of the original chapter can be found under DOI 10.1007/978-3-662-46578-3_133

K.J. Kim (ed.), *Information Science and Applications*,
Lecture Notes in Electrical Engineering 339, DOI 10.1007/978-3-662-46578-3_79

In the next section, we review the concept of multi-objective optimization. In Section 3, we review existing knee-based MOEAs. In Section 4, we present our knee-based searching approach. In Section 5, we show some example problems. Section 6 provides simulation results in convergence view point. Lastly, section 6 concludes our research work and contribution.

2 Multi-objective optimization

In its general form, a multi-objective optimization problem can be formulated as follows.

$$\begin{array}{lll} \text{Minimize/maximize} & f_m(x), & m = 1,2,\ldots,M; \\ \text{Subject to} & g_j(x) \le 0, & j = 1,2,\ldots,J; \\ & h_k(x) = 0, & k = 1,2,\ldots,K; \\ & x_i^{(L)} \le x_i \le x_i^{(U)}, & i = 1,2,\ldots,n. \end{array}$$

where,

- x is a vector of n decision variables: $x = (x_1, x_2,\ldots, x_n)^T$.
- $f(x) = (f_1(x), f_2(x), \ldots, f_M(x))^T$ is objective functions.
- $g_j(x)$ is inequality constraint.
- $h_k(x)$ is equality constraint.
- Decision space is constituted by variable bounds that restrict each variable x_i to take a value within a lower bound $x_i^{(L)}$ and an upper bound $x_i^{(U)}$.
- A solution x that satisfies all constraints and variable bounds is a **feasible solution.** Otherwise it is called **infeasible solution**.
- **Feasible space** is a set of all feasible solutions.
- Objective function $f(x) = (f_1(x), f_2(x), \ldots, f_M(x))^T$ constitutes a multi-dimensional **objective space.**
- For each repetition x in the decision space, there exists a point in the objective space.

$$f(x) = z = (z_1, z_2, \ldots, z_M)^T$$

In multi-objective problems, the concept of "dominance" is used to determine if one solution is better than others. A solution 'x' is said to dominate a solution 'y' if the following two conditions are true [2]: (1) 'x' is no worse than 'y' in all objectives and (2) 'x' is better than 'y' in at least one objective. In this case 'y' is said to be "dominated" by 'x', or alternatively, 'x' said to be "non-dominated" by 'y'. The concept of dominance is exemplified in a two-objective minimization example shown in Fig. 2.

Since both functions are to be minimized, the following dominance relationships can be observed: solution 2 dominates solutions 1, 3 and 5; solution 3 only dominates solution 5 and solution 4 only dominates solution 5. Conversely, solutions 2 and 4 are non-dominated solutions because there is no solution that dominates them.

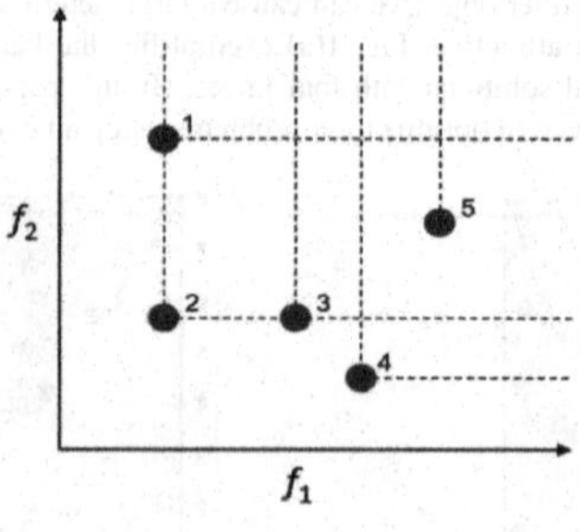

Fig.2. Regular Pareto dominance example [1-2].

Note that even if solution 2 is equal in one objective to solutions 1 and 3, it still dominates them, given the concept of dominance. The non-dominance relationship determines the concept of Pareto optimality. A solution is said to be Pareto optimal if it is non-dominated by any other solution. In other words, a Pareto optimal solution cannot be improved in one objective without losing in another one. In this case, solutions 2 and 4 are Pareto-optimal solutions. All solutions that are Pareto optimal constitute the Pareto set. The objective values of the Pareto set in the objective space constitute the Pareto frontier.

3 Existing knee-based MOEAs

This section reviews the well-known knee-based MOEAs that have been developed for a decade. Das [3] proposed a method based on the Normal-Boundary Intersection (NBI) to locate the knee of the Pareto front. Das characterized the knee solution in terms of the Convex Hull of Individual Minima (CHIM). The knee regions can be characterized by the farthest solution from the CHIM. Branke et al. [4] proposed two modification strategies to NSGA-II algorithm to make it concentrating on knee regions. In the first strategy, the algorithm utilizes a geometrical property of knee regions. This approach has been shown to be effective for bi-objective case but not for higher dimensional cases. In the second strategy, the author suggested a Marginal Utility function Approach (MUA) to approximate angle-based measure in more than three objective cases. Rachmawati and Srinivasan [5] developed a MOEA focusing on knees. The proposed MOEA computes a transformation of original objectives based on weighted sum approach called WSNA (Weighted Sum Niching Approach). Rachmawati and Srinivasan [6] proposed a fitness scheme that applies preference-based selection pressure in a MOEA to obtain solutions concentrated on knee regions. The proposed method is a two-step algorithm. In the first step, the MOEA seeks a rough approximation of the Pareto optimal solutions, and in the second step, the linear weighted sums of the original objective functions are optimized to guide the solutions towards the knee regions. Bechikh et al. [7] proposed a knee-based MOEA called KR-NSGA-II. KR-NSGA-II uses mobile reference points. They called mobile reference point because a reference point is updated automatically in each generation of the MOEA. Bechikh et al. [8] also proposed an enhanced version of their KR-NSGA-II called TKR-NSGA-II (Trade-off-based KR-NSGA-II). In KR-NSGA-II, the knee region characterization was modified. Instead of using the distance to the extreme line, they used a trade-off worth metric designed by Rachmawati and Srinivasan[6].

In summary, the existing knee-based MOEAs can be divided in two groups:

1. Weighted-sum based approaches.
2. Reference-point based approaches.

In particular, most knee-based MOEAs are evaluated based on convergence view point [3-8]. The Generational Distance (GD) performance metric is considered as the standard convergence metric in the multi-objective optimization research community. The GD indicator was introduced for measuring how far the elements in the set of non-dominated solutions are from those in the Pareto front. In this paper, we are interested in developing our knee-based approach based on extended angle dominance technique.

4 Proposed Approach

This section explains our approach in details. Our approach is called *k*-ASA-NSGA-II algorithm. The process can be divided into two steps inside one optimization algorithm. Initially, the NSGA-II algorithm seeks a rough approximation of the Pareto front, and *k*-ASA is applied in the second step.

4.1 *k*-ASA algorithm (Knee- Angle based with Specific bias parameter Algorithm)

The angle between two non-dominated solutions is calculated by using Eq. 1. The geometric angle is denoted by θ_n where n is the nth objective. For the minimizing objective context, θ_n is given by

$$\theta_n = \tan^{-1}\left[\frac{\sqrt{\sum_{m=1,m\neq n}^{N}\left(\Delta f_m\right)^2}}{\Delta f_n}\right] \tag{1}$$

where, N denotes the number of objective functions. n denotes the nth objective functions. Δf_n denotes the difference between the nth objective values of two non-dominated solutions. For example, considering multi-objective optimization with two-objective functions f_1 and f_2, the angle of extended dominated area for solution A is shown in Fig. 1(a).

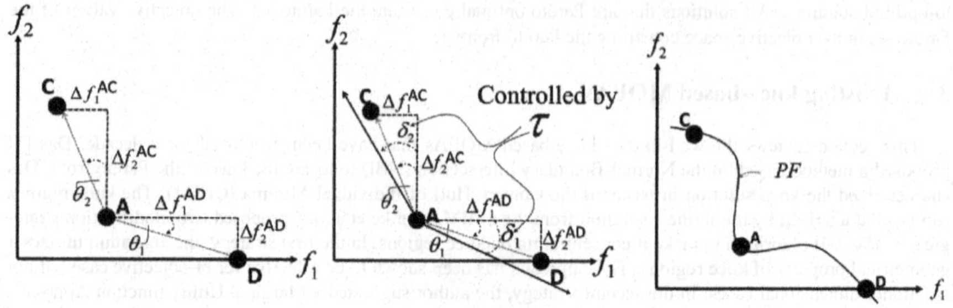

(a) The angle calculation of two-objective problem (b) The effect of τ in controlling threshold angle size (c) Knee solution

Fig. 3. *k*-ASA approach

We define the dominance condition as follows.

$$i \prec j \Leftrightarrow \sum_{n=1}^{N}\left(\left[f_n(i) \le f_n(j)\right] \wedge \left[|\theta_n(i,j)| \le |\delta_n|\right]\right) > 0$$

Threshold angle (δ) is a reference angle, uses to compare to an angle between pair of solutions (θ). For any pair of solutions (*i*, *j*), if an angle between a pair solutions is smaller than threshold angle, solution *j* will be discarded, and the algorithm keeps only solution *i*. The process to determine threshold angle can be elaborated as follow: (1) Each non-dominated solution is sorted in ascending order for each objective. (2) Inter-quartile range of sorted data of each objective is calculated, denoted by IQS_n. (3) The inter-quartile range of average distance of nth objective value between two consecutive non-dominated solutions is calculated, denoted by IQ_n.

The threshold angle (δ_n) of each objective value can be calculated as follows:

$$\delta_n = \left[\tan^{-1}\left(\frac{IQS_n}{IQ_n}\right)\right]^{\tau} \quad (2)$$

where *n* is objective number; τ is bias intensity of each objective, ranging from 0.0-1.0.

Consider Fig.3(b), solutions A, C and D have an opportunity to be indexed as *i* or *j* solution to form a pair of solutions. A pair of solutions could be solutions C and A, solutions A and D, and solutions C and D. Assuming that solutions A and C are formed as pair of solutions, solution A is indexed as *i* , and solution C is indexed as *j*. In fact, the process of indexing pair of solutions is allocated randomly by the algorithm. We see that the threshold angle (δ_2) is greater than the angle between solutions A and C (θ_2), in this case, solution C will be discarded. The algorithm keeps solution A for pairing to another solution. Next, solutions A and D are selected for next pair of solutions. We see that δ_1 is greater than θ_1.Thus solution D will be discarded, so that the algorithm keeps solution A. Consider Fig. 3(c), supposedly, we have a problem that contains a Pareto front with a knee region, and then we put the solutions A, C and D from Fig.3(b) at the same location to Fig. 3(c). In Fig. 3(c), only solution A is kept as final Pareto optimal solution, and solutions C and D are discarded in angle based view point. We see that solution A is located at knee region of Pareto front. Fig. 3(b)-(c) explain clearly how our *k*-ASA algorithm acquires the knee solutions. *k*-ASA is then incorporated to a baseline algorithm. We choose NSGA-II algorithm [x] in order to make a fair comparison with existing knee-based MOEAs since most well-known knee-based MOEAs use NSGA-II as a baseline algorithm. Our algorithm is presented as follows.

k-ASA-NSGA-II Algorithm :

Set:
R_t = total population.
P_t = preserved population.
Q_t = recombined population at t generation.
Front number = Fi

***Pseudo code of k-ASA-NSGA-II Algorithm*:**

1. Combine P_t and Q_t to R_t after the first generation.
2. Fitness assignment.
3. Assign R_t to the front.
4. Calculate crowding distance of each front
5. Sort R_t in ascending order and crowding distance in descending order.
6. Select the first half of R_t and assign to P_{t+1}
7. Recombine P_{t+1} and assign Q_{t+1}
8. If number of evaluation <= 90% of maximum number of evaluation
   Repeat steps 1-8, Else Go to step 9
9. Incorporate *k*-ASA
10. Update P_{t+1} and Q_{t+1}
11. Repeat steps 1-7 (except step 4) until the maximum iteration is met

5 Example Problems

This section demonstrates the simulation of selected problems including a two and a three objective knee-based test problems using *k*-ASA-NSGA-II. The detail description for knee-based test problems can be found in [6] and [8]. All experiments are made with JAVA software under Eclipse IDE environment. Once the results from our *k*-ASA-NSGA-II algorithm are obtained, we plot the graph with MATLAB software to observe the result visually. We considered DEB2DK with K=4 for two-objective knee-based problem. For three-objective problem, DEB3DK with K=1 is considered and visualized.

Experiment setting

A set of 10 simulation runs was conducted for each test problem. For all MOEAs, the termination criterion is set to 25000 FEs (Function Evaluations) for DEB2DK. 30000 FEs for DEB3DK problem. The used genetic operators are the simulated binary crossover (SBX) and the polynomial mutation [9] with crossover probability of 0.9 and mutation probability of $1/n$ (where n is the number of decision variables). The population size is settled to 100 for all problem instances. Note that marginal utility approach (MUA) does not have specific parameters.

Table 1 MOEA specific parameter setting

Algorithm	MOEA specific parameter values
WSNA [5]	Q = 100, P=20
PLWSO [6]	Q=100, δ' = 0.1, Sp_2 = 80%
KR-NSGA-II [7]	ζ = 0.05 for two-objective problems and
TKR-NSGA-II[8]	0.08 for three-objective problems.
	e = 0.001, KN = K
k-ASA-NSGA-II	τ = 0.95for two-objective problems
	τ = 0.90 for three-objective problems

Fig. 4(a) shows the results for DEB2DK test problem with K=4 (K= number of knees to discover). τ is used to control the density of each knee and knee number. For example, with τ=0.90, the search found four knees with high density and the solutions in the extreme regions are also discovered. When τ=0.95, the search found four knees with lower density than τ=0.90, and the extreme regions are omitted. When τ=0.98, the search loses two knees and the density of remaining knees are quite low. Fig. 4(b) shows the result for DEB3DK problem with K=1, with the density controller parameter τ=0.90. It's noted that our approach does not need to specify the number of knees to discover while the other Knee-based MOEAs do. This is one of the advantages of our approach comparing to others.

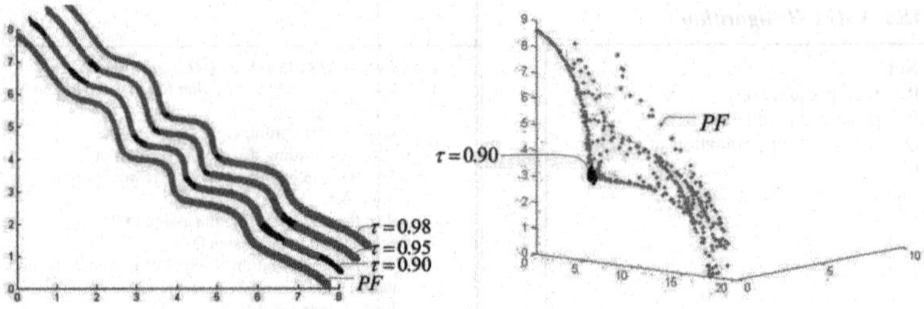

(a) (b)

Note: PF= Pareto Front

Fig. 4. (a) DEB2DK Problem (K=4). (b) DEB3DK Problem (K=1),τ =0.90

6 Simulation Result

We compare the k-ASA-NSGA-II with most representative works in this research area from convergence view point. k-ASA-NSGA-II is confronted to the Marginal Utility Approach (MUA) [4], the Weighted sum Niching Approach (WSNA) [5], the Sp2 version of Parallel Local Weighted Sum Optimization approach [6], KR-NSGA-II [7] and TKR-NSGA-II [8]. We set the environment of the experimental design similarly to the study of Rachmawati and Srinivasan [6] and Bechick et al. [8] in order to make fair comparisons.

Result Discussion

Table 2 shows the *GD* values for the six algorithms under comparison. We exploit the results published in [8]. We remark, the reference point-based algorithms (i.e., KR-NSGA-II and TKR-NSGA-II) present better convergence than the three other weighted sum-based algorithms (MUA, WSNA and PLWSO). The results in Table 1 also show that k-ASA-NSGA-II is competitive to all previous Knee-based MOEAs. The superiority of k-ASA-NSGA-II over KR-NSGA-II and TKR-NSGA-II on DEB2DK and DEB3DK test problems may be explained by the output provided by k-ASA-NSGA-II depends on the comparison between each pair of solutions in angle-based view point which eliminates the solutions that are too far from the Pareto front. Thus, the final solutions can be considered as close as possible to the Pareto front while KR-NSGA-II and TKR-NSGA-II are incorporated to MOEA on the early stage of optimization.

Table 2. : GD Values (Small value is better)

Problem	K	MUA		WSNA		PLWSO		KR-NSGA-II		TKR-NSGA-II		ASA-NSGA-II	
Year		2004		2006		2009		2010		2011		2014	
		Mean	S.D.	Mean	S.D.	Mean	S.D.	Mean	S.D.	Mean	S.D.	Mean	S.D.
DEB2DK	1	0.0406	0.00708	0.4352	0.39414	0.0162	0.00125	0.14971	0.00549	0.01232	0.00271	**0.00732**	0.0023
	2	0.02497	0.00628	0.00396	0.00043	0.0029	0.00066	0.00303	0.00052	0.00238	0.00061	**0.002095**	0.0045
	3	0.02511	0.0036	0.00463	0.00038	0.0036	0.00029	0.00501	0.00074	**0.00294**	0.00033	0.00563	0.0033
	4	0.01368	0.00139	0.31743	0.47184	0.0067	0.00114	0.01883	0.00211	0.00831	0.00319	**0.00775**	0.0024
DEB3DK	1	0.17511	0.11718	0.13382	0.02361	0.1136	0.01958	0.10322	0.02355	0.09461	0.01992	**0.01458**	0.0292
	2	0.7632	0.25049	0.29348	0.07652	0.3293	0.1569	0.33782	0.14088	0.29917	0.12063	**0.01587**	0.1359

7 Conclusion

In this paper, we have contributed the search to find special points of the Pareto front that correspond to DM's implicit preferences. The proposed algorithm used the new concept of extended dominance in angle based view point. We have proposed a Knee-based algorithm called k-ASA-NSGA-II algorithm, where we modified the crowding estimator technique in NSGA-II algorithm. We have tested and evaluated our Knee-based algorithm using well-known test problems. The results have shown that our approach is competitive to previous weighted sum based approaches (i.e., MUA, WSNA and PWLSO) and reference point based approaches (i.e., KR-NSGA-II and TKR-NSGA-II) for the selected test problems.

References

1. S. Sudeng, and N. Wattanapongsakorn, Post Pareto-optimal Pruning Algorithm for Multiple objective Optimization using Specific Extended Angle Dominance, J. Eng. Appl. of Artificial Intelligent. – *In Press*.

2. S. Sudeng, and N. Wattanapongsakorn, "Finding robust Pareto-optimal solutions using geometric angle-based pruning algorithm," Intelligent System for Science and Information- Sudies in Computational Intelligence, vol. 542, pp. 277-295, 2014.

3. Das, "On characterizing the knee of the Pareto curve based on normal-boundary intersection," Structural Optimization, vol. 18, no. 2), pp.107-105, 1999.

4. J. Branke, K. Deb, H. Dierolf, and M. Osswald, "Finding knees in multiobjective optimization" In: Proceeding of 8th international conference on Parallel Problem Solving from Nature (PPSN), pp 722-731, 2004.

5. L. Rachmawati and D. Srinivasan, "A multi-objective evolutionary algorithm with weighted sum niching for convergence on knee regions" In: Proceeding of the 8th Genetic and Evolutionary Computation Conference (GECCO'06), pp 749-750, 2006.

6. L. Rachmwati D. Srinivasan, "Multiobjective evolutionary algorithm with controllable focus on the knees of the Pareto front" IEEE Transaction on Evolutionary Computation, vol. 13, no. 4, pp. 810-824, 2010.

7. S. Bechikh, L. Ben Said, and K. Ghedira, "Searching for knee regions in multi-objective optimization using mobile reference points," In: Proceedings of the 25th ACM symposium on Applied Computing, pp 1118-1125, 2010.

8. S. Bechikh, L. Ben Said, K. Ghedira, "Searching for knee regions of the Pareto front using mobile reference points," Soft computing- A Fution of Foundations, Methodologies and Applications, vo.15, no.9, pp. 1807-1823, 2011.

9. K. Dep, A. Pratab, S. Agarwal, and T. Meyarivan, "A fast and elitist multi-objective genetic algorithm: NSGA-II," IEEE Transaction of Evolutionary Computation, vol. 6, no. 2, pp. 182-197, April 2002.

Affective Learning Analysis of Children in a Card Sorting Game

Marini Othman, Abdul Wahab, Abdul Qayoom, Mohd Syarqawy Hamzah and Muhamad Sadry Abu Seman

Kulliyyah of Information and Communication Technology, International Islamic University Malaysia, Kuala Lumpur, Malaysia

{omarini,abdulwahab,syarqawy,msadri}@iium.edu.my, qauiam@gmail.com

Abstract. The purpose of this paper is to provide an affective learning analysis on children while they were playing a card sorting game. The electroencephalogram (EEG) signals of 8 preschoolers aged between 4 to 6 years were collected (a) while they were playing a card sorting game; and (b) observing affective faces. The features from EEG signals were extracted using Kernel Density Estimation (KDE). The Multi-Layer Perceptron (MLP) was used to classify and generate the affective maps of the EEG signals while the children were playing the game. The initial results show that the children's affective states are unique and there might be different affects that drive a child's performance. This analysis shows the potential of using the affective learning analysis approach in assessing educational tools such as computer games.

Keywords. Emotion, computer games, preschoolers, affective computing

1 Introduction

Many researchers have reported the negative effects of playing computer games such as aggressiveness due to the violent nature of computer games [1]. However, recent experimental studies have suggested that playing computer games might improve cognitive abilities of the players [2]. Thus, the biggest challenge may actually lies in the identification of suitable games genre for educational purpose. A possible solution is setting clear learning outcomes for computer games and how they can be aligned with the cognitive skills that can be obtained by the children [3].

From another perspective, the over-emphasis on cognitive abilities might be at the price of relative neglect on affect [4]. The capability for affect regulation is known to be partly responsible for children academic performance [5] that leads toward confidence, creative thinking and problem solving [6]. Thus, the contribution of an affective learning analysis simply cannot be denied.

In this paper, the affective states of children with the aged of 4 to 6 years old were investigated by monitoring their EEG signals while they were playing a card sorting game, called the Dimensional Change Card Sorting [7]. Through observation of the EEG affective maps and EEG dynamic affect satisfied the affective component

K.J. Kim (ed.), *Information Science and Applications*,
Lecture Notes in Electrical Engineering 339, DOI 10.1007/978-3-662-46578-3_80

of our analysis. These works might lead toward the establishment of intelligent machines that can replace a skilled tutor in teaching, learning and assessment in the future.

1.1 Computer Games and Learning

Currently there are thousands of computer games in the market. A frequently asked question by educators is how to adapt the games for teaching and learning. A study has proposed a few videogames genres as follows [8]:

1. Imagination: *consists of fantasy role-playing, action-adventure, strategy and sims.*
2. Traditional: *consists of classic board games, arcade, card dice, quiz-trivia and puzzles.*
3. Physical enactment: *consists of fighters, shooters, sports and racing speed.*

To date, a few of the widely investigated games genre in relation to education are action based and puzzle games [8]. However, it is acknowledged that many of these action based games can be violent in nature. Thus, puzzle based games is chosen since it is easily adaptable for a classrom setting.

1.2 Blooms Taxonomy

For the past few decades, the most renowned assessment model for cognitive skills in education is Bloom's taxonomy [9]. Traditionally, the Bloom's taxonomy is described as 5 orders of assessment (i.e. knowledge, comprehension, application, analysis, synthesis and evaluation). It is thought that the experiential environment in computer games might present a better opportunity for developing higher order cognitive processes such as analysis, synthesis and evaluation (i.e. up to the fifth order of Bloom's) [10].

1.3 Affective Space Model

In the past, psychologists have attempted to explain affect using a two-dimensional structure namely *pleasant-unpleasantness* and *relaxation-attention* [11]. The problem with this approach is that certain affective states such as fear and anger are placed in almost similar position in the two dimensional structure, but significantly differs in terms of physiological manifestation.

The affective space model was later introduced with the belief that affect can be placed in varying degree on the circumplex of *valence* and *activation* (Fig. 1 below) [12]. The circumplex model placed feeling-related concepts in a circular order, departing from the bipolar dimensions [11]. Based on the model, it is possible to quantify an affect based on the level of valence and arousal.

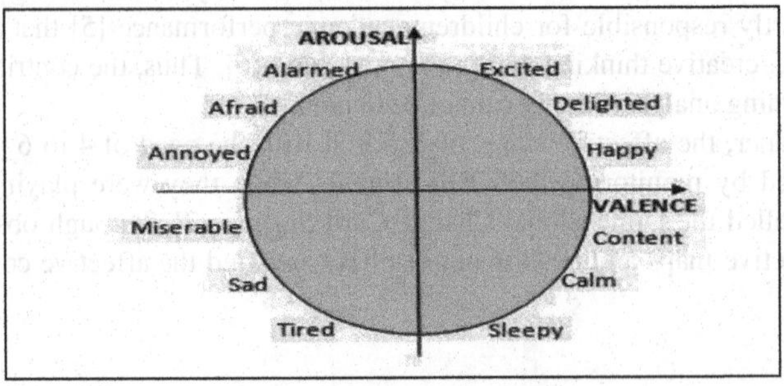

Fig. 1. The circumplex model of affect (adapted from [12])

2 The Affective Learning Analysis

Fig. 2 below shows the method for our affective learning analysis. It consists of 3 stages of analysis (i.e. data collection and pre-processing, feature extraction and classification).

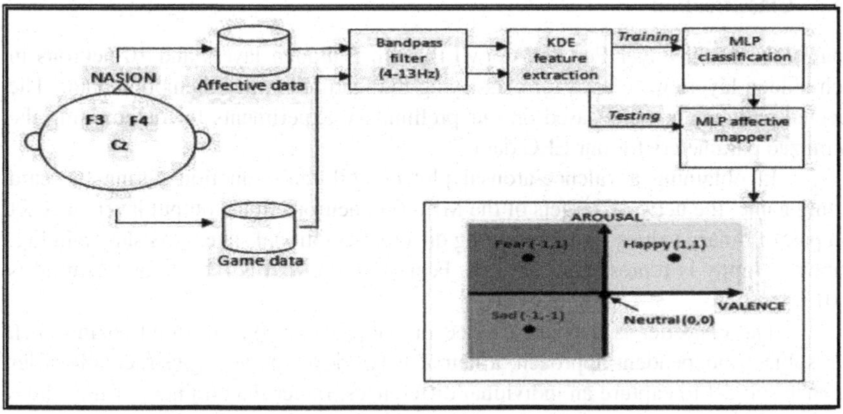

Fig. 2. The proposed method for the affective learning analysis

2.1 Data Collection and Pre-processing

In the proposed method, children's EEG signals were collected at 250Hz sampling frequency. The electrode locations are F3 and F4, with Cz as reference. The captured bio-signals were then band-pass filtered using elliptic filter at 4-13Hz to retain theta and alpha waves that correlates with affective experiences [13].

2.2 Feature Extraction

The feature extraction method used in this work is the Kernel Density Estimation (KDE). KDE provides a probability density estimation of the EEG data regardless of the data distribution (i.e. normal or skewed). For a given input in sample points (x_1, x_2,..x_n), the estimate of the kernel density $f(x)$ is given by:

$$f_h(x) = \frac{1}{n}\sum_{i=1}^{n} K_h(x - x_i) = \frac{1}{nh}\sum_{i=1}^{n} K\left(\frac{x-x_i}{h}\right) \quad (1)$$

Where,

- K Represents the kernel function, which is taken as normal distribution for our experiment.
- n Represents the number of samples per frame. In our case, n = 10.
- H Represents the window width, which is 64 in our experiments.

Thus, at 10 equally spaced feature points that covers the range of data in each channel (F3 & F4) and frequency band that includes theta and alpha bands, a final 40-feature matrix was obtained for this method. Similar filter design and feature extraction method were employed for brainwaves from the affective tasks and game tasks.

2.3 Classification

Finally, the Multi-Layer Perceptron (MLP) with 2 hidden layers and 10 neurons in each hidden layers were used for classifying the feature extracted emotion data. The selected parameters were based on our preliminary experiments for determining the optimized parameters for our EEG data.

In obtaining a valence-arousal plot of children's emotion during the card sorting game, the network targets of the MLP (i.e. neurons in the output layer) was set to a pair of binary values for representing different emotional states. As shown in Fig. 2 above, Happy is represented by (1,1), Fear is (-1,1), Sad is (-1, -1) and Neutral is (0,0).

The classifier is evaluated based on subject-independent blind testing [14]. In a subject-independent approach, a neural network learns and constructs classifier for each subject to capture an individual differences, rather than trained for the whole data in subject-dependent network training. On the other hand, a blind testing is a situation where the training data (affective data) are different from the testing data (game data). The result of the MLP network training is a model that serves as an affective mapper for the game data.

3 Experiments

The participants were 8 right-handed preschoolers (male: 4; female: 4) aged from 4 to 6 years old. In this work, boys were labeled with ID of 1 until 4 while girls were labeled with ID of 5 until 8. An informed consent was obtained from their respective parents. All children received a souvenir for their participation in the study

There are 3 main stages for the EEG experimental, including the baseline recording, the card sorting game and the affective faces task. During baseline recording, the EEG signals were captured while participants were closing their eyes for a minute and opening their eyes for another minute.

In the card sorting game, children watched a game demonstration and followed by a set of 10 practice cards. In the first round of the game known as "pre-switch', children were presented with two target cards (i.e. a blue car and a red flower) and asked to sort the card using the "shape" dimension. Afterwards, the children were asked to sort the cards using the "color" dimension, hence the second round is known as "post-switch". Pre-switch and post switch are standard protocol in DCCS, where children are required to sort cards according to a changing dimension. In both phases, their EEG signals were captured while they were sorting as many test cards that they could in the duration of a minute.

Finally, the children were required to watch selected affective faces from the Radboud Faces Database (RafD) for the affective state of happy, sad, neutral, and fear [15].

4 Results

Fig. 3 below shows inter-subject classification results as measured by the mean squared error (mse) of the MLP model. The highest accuracy as measured was recorded for participant 5 (mse: 0.152) followed by participant 3 (mse: 0.153). The worst performance was for participant 6 (mse: 0.206).

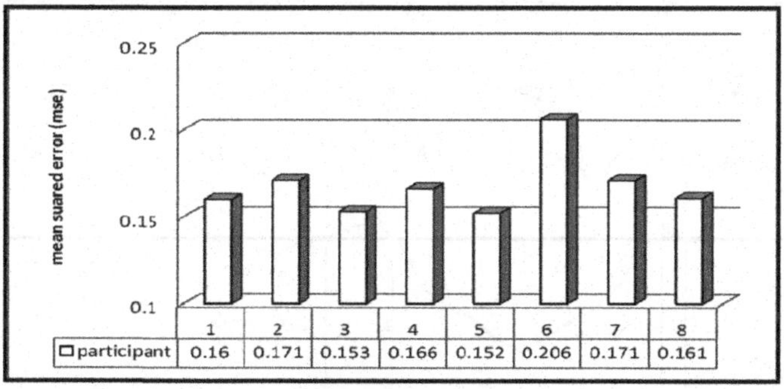

Fig. 3. Accuracy of affective recognition based on the mean squared error of the MLP model

The overall affective state distribution of children is mapped based on the KDE/MLP model (Appendix 1).

5 Conclusion

From Fig. 3 we can conclude that the affective states of children cannot be generalized and there are different affective states that drive a child's performance.

Therefore, we decided to select one of the children (ID of 3) for a closer investigation at the dynamics of human affect (Fig. 4 below). Based on Fig. 3, it shows that the boy predominantly experienced negative emotions (fear and sad) in the pre-switch phase although there was also a brief moment of "happy" toward the end of pre-switch. Similar affective states of "happy" can be found mostly at the end of "post-switch". This reflects a sense of relief for almost completing the assigned game.

In this paper, we have analyzed the affective responses of children aged 4 to 6 years old, which are mapped based on affective space model [12]. Results obtained have provided evidences that it is not possible to generalize human affect hence a subject independent blind testing was appropriate for capturing their affective differences. Furthermore, by measuring several of assessment defined in Bloom's taxonomy through the card sorting game, this study can be used for assessment of various educational tools.

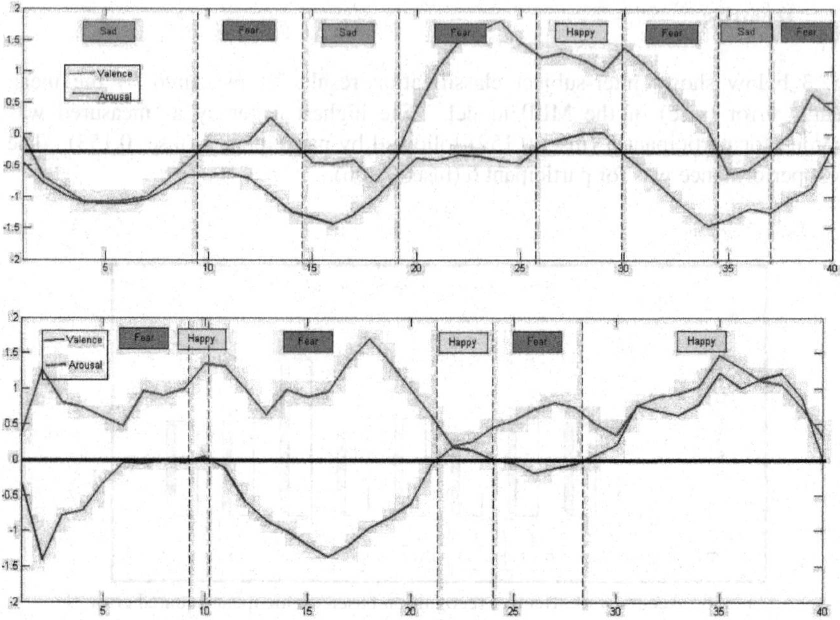

Fig. 4. Dynamic EEG Affect (ID of 3)

References

1. Hasan, Y., L. Bègue, et al. (2013). The more you play, the more aggressive you become: A long-term experimental study of cumulative violent video game effects on hostile expectations and aggressive behavior. Journal of Experimental Social Psychology 49(2): 224-227.
2. Spence, I., & Feng, J. (2010). Video Games and Spatial Cognition. *Review of General Psychology, 14*(2), 92-104.
3. Sherry, J., & Pacheco, A. (2006). Matching Computer Game Genres to Educational Outcome. *The Electronic Journal of Communication, 16*(1).
4. Picard, Rosalind W., Seymour Papert, Walter Bender, Bruce Blumberg, Cynthia Breazeal, David Cavallo, Tod Machover, Mitchel Resnick, Deb Roy, and Carol Strohecker. Affective learning—a manifesto. *BT Technology Journal* 22, no. 4 (2004): 253-269.
5. J. S. Matthews, C.C. Ponitz, F.J. Morrison. Early gender differences in self-regulation and academic achievement. *Journal of Educational Psychology*, vol 101(3), pp. 689-704, 2009.
6. B. Hoffman. I think I can, but I'm afraid to try: The role of self-efficacy beliefs and mathematics anxiety in mathematics problem-solving efficiency. *Learning and Individual Differences*, vol. 20(3), pp. 276-283, 2010.

7. Zelazo, P.D. (2006). The Dimensional Change Card Sort (DCCS): A method for assessing executive function in children. *Nature Protocols,* vol. 1 (2), pp. 297-301.
8. Lucas, K. & Sherry, J.L. (2004). Sex differences in video game play: A communication-based explanation. *Communication Research*, 31(5), 499-523.
9. Bloom, B. S., Engelhart, M. D., Furst, E. J., Hill, W. H., Krathwohl, D. R. (1956). Taxonomy of educational objectives: The classification of educational goals. *Handbook I: Cognitive domain*. New York: David McKay Company.
10. Sherry, J. L. (2013). Formative research for STEM educational games: Lessons from the Children's Television Workshop. *Zeitschrift für Psychologie*, 221(2), 90.
11. Schlosberg, H. (1952). The description of facial expressions in terms of two dimensions. *Journal of experimental psychology*, vol. *44*(4), pp. 229.
12. Russell, J.A. (1980). A circumplex model of affect. *Journal of Personality and Social Psychology*, vol. 39, pp. 1161-1178.
13. C. M. Krause, V. Viemerö, A. Rosenqvist, L. Sillanmäki, and T. Åström. (2000). Relative electroencephalographic desynchronization and synchronization in humans to emotional film content: an analysis of the 4–6, 6–8, 8–10 and 10–12 Hz frequency bands. *Neurosci. Lett.*, vol. 286, no. 1, pp. 9–12.
14. Yaacob, H., Abdul Wahab & Kamaruddin, N. (Oct, 2014). Subject-dependent and subject-independent emotional classification of CMAC-based features using EFuNN. Paper presented at the *ISCA 27th International Conference on Computer Applications in Industry and Engineering (CAINE 2014)*, Oct 13-15, New Orleans, Lousiana.
15. Langner, O., Dotsch, R., Bijlstra, G., Wigboldus,D.H.J., Hawk,S.T.,& van Knippenberg, A. (2010).Presentation and validation of the Radboud Faces Database, *Cognition and Emotion*, vol. 24(8), pp. 1377—1388.

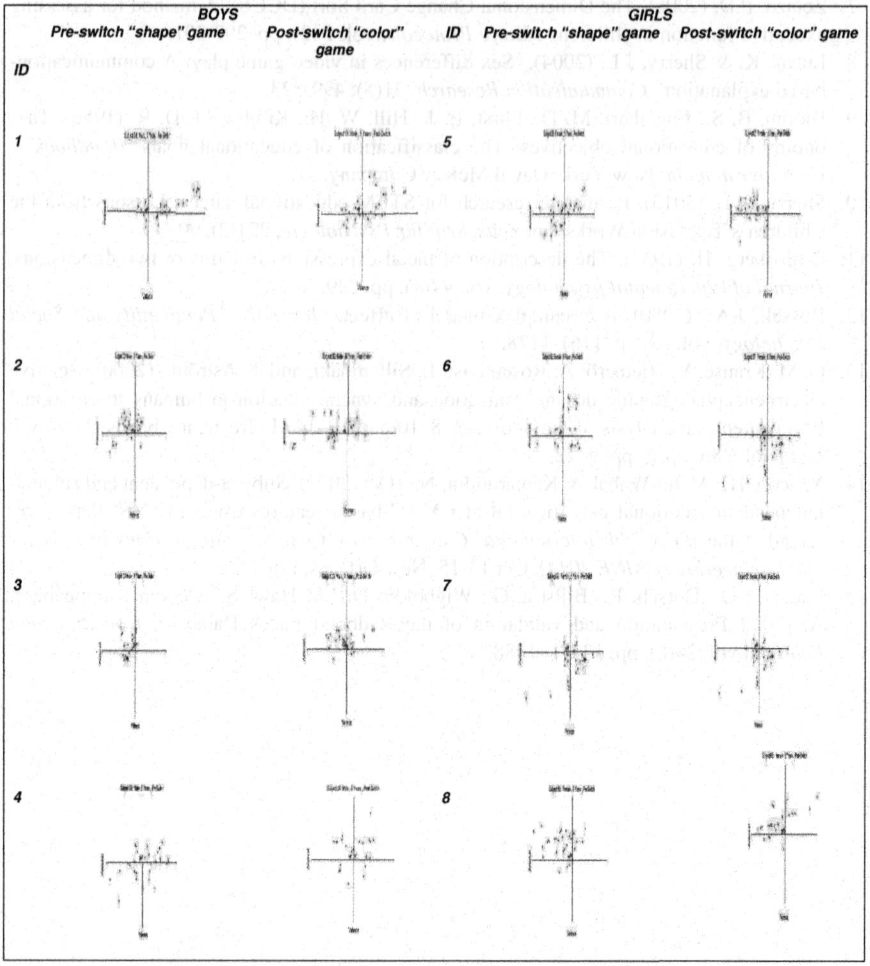

Appendix 1. Affective state distribution of boys and girls during pre-switch "shape" and post-switch "color" games

Application of Mean and Median Frequency Methods for Identification of Human Joint Angles Using EMG Signal

Sirinee Thongpanja[1,*], Angkoon Phinyomark[2], Chusak Limsakul[1], and Pornchai Phukpattaranont[1]

[1] Department of Electrical Engineering, Prince of Songkla University, Thailand
sirinee.th@gmail.com, chusak.l@psu.ac.th, pornchai.p@psu.ac.th
[2] Faculty of Kinesiology, University of Calgary, AB, Canada
angkoon.p@hotmail.com

Abstract. The analysis of surface electromyography (EMG) signals is generally based on three major issues, i.e., the detection of muscle force, muscle geometry, and muscle fatigue. Recently, there are no any techniques that can analyse all the issues. Mean frequency (MNF) and median frequency (MDF) have been successfully applied to be used as muscle force and fatigue indices in previous studies. However, there is the lack of consensus upon the effect of muscle geometry on the basis of varying joint angles. In this paper, the modification of MNF and MDF using a min-max normalization technique was proposed to provide a consistent relationship between feature value and joint angle across subjects. The results show that MNF and MDF extracted from normalized EMG showed a stronger linear relationship with elbow joint angle compared to traditional MNF and MDF methods. Modified MNF and MDF features increased with increasing elbow angle during isometric flexion. As a result of the proposed technique, modified MNF and MDF features could be used as a universal index to determine all the issues involving muscle fatigue, muscle force, and also muscle geometry.

Keywords: Feature extraction · Frequency analysis · Muscle fatigue · Spectral analysis · Surface electromyography signal

1 Introduction

Surface electromyography (EMG) signal is one of the useful electrophysiological signals, which is applied in many medical and engineering applications. The analysis of EMG signals is generally based on three issues, i.e., the detection of muscle force, muscle geometry, and muscle fatigue [1]. Many classical and advanced feature extraction techniques based on time and/or frequency domain have been used, such as root mean square, fractal dimension, wavelet transform as well as mean and median frequencies (MNF and MDF) [2]. Unfortunately, there are no any techniques that can be used to analyse all the issues as a universal index. Moreover, when one of the three issues was analysed, other issues could affect the results of the analysis, for example, the assessment of muscle fatigue during dynamic contraction [1].

K.J. Kim (ed.), *Information Science and Applications*,
Lecture Notes in Electrical Engineering 339, DOI 10.1007/978-3-662-46578-3_81

Generally, MNF and MDF are held as the gold standard for the assessment of muscle fatigue [3]. During the fatigue of a muscle, the EMG power spectrum is shifted toward lower frequencies, so MNF/MDF decreases when muscle fatigues. However, daily activities involve dynamic contraction where the changes of muscle force and/or geometry produce non-stationary EMG signal. The effects of muscle force and geometry are also still not clear [1], [4]. More details about contradictory findings of muscle force and geometry effects on MNF and MDF can be found in [5].

To develop a technique that can detect muscle force and muscle fatigue during dynamic contraction, in our previous studies [6,7,8] the modification of MNF and MDF is proposed based on the basis of consecutive fast Fourier transforms (FFTs). As a result of this proposed technique, the modified MNF and MDF features can be used to identify both muscle fatigue and muscle force. Muscle fatigue can be detected by a decrease of MNF/MDF during one dynamic contraction [3], and by a slope of the regression line that fits maximum MNF/MDF during several cyclic dynamic contractions [9]. For the muscle force detection, a mean value of the effective range of the consecutive MNFs/MDFs can provide a strong linear relationship between feature value and muscle contraction level for upper-limb muscles, such as biceps brachii and flexor pollicis longus [8,9,10]. When muscle contraction levels increased, the modified MNF and MDF features decrease [6]. This relationship is consistent across all the studied subjects but it is not consistent across the subjects using traditional MNF and MDF methods.

Generally, the effect of muscle geometry including electrode configuration, fibre diameter, and subcutaneous tissue thickness, has been evaluated by the resulting from changes in joint angles [1], [11]. Two relationships have been found for the joint angle effect on MNF/MDF [5]. First, MNF and MDF are unaffected by changes in joint angle [12]. Gerdle et al. [13] reported that no significant change in the EMG power spectrum measured from biceps brachii under constant load while joint angle varied. Second, MNF/MDF increases as a joint angle increased (or a muscle length decreased) [1], [14,15]. This relationship has been found in most of the previous studies. Although the second case has been found frequently in literature compared to the first case, in our previous study [5] both cases are found (i.e., subject-dependent). In this paper, to develop the universal MNF and MDF indices which can analyse all the issues this problem was revisited by proposing the modification of MNF and MDF using a min-max normalization technique for the characterization of EMG signals at different angles of elbow joint.

2 Materials and Methods

2.1 EMG data collection

Surface EMG signals were acquired at five different isometric contraction levels: 1, 2, 3, 4, and 5 kg, and five different elbow joint angles: 30°, 60°, 90°, 120°, and 150° of flexion. Nineteen normal subjects, Subj:#1-#19 (10 males and 9 females, 21.10±0.81 years) participated in this study. The subject was asked to lift the required load at the specific elbow joint angle. EMG data were recorded for 5 s in each trial after the subject's elbow joint angle was stable. Five trials of each of the twenty-five (5 loads × 5 angles) possible combinations were performed per day for four separate days. A sequence of the 25 combinations was randomized. EMG data were collected from the biceps brachii muscle using bipolar Ag/AgCl electrodes (H124SG, Kendall ARBO) with a common ground reference on wrist. The signals were amplified at a gain of 19x with a bandwidth of 20-500 Hz and were sampled at a rate of 1024 Hz with an analog-to-digital resolution of 24 bits using an EMG measuring system (Mobi-6b, TMS International BV, Netherlands).

2.2 Normalization Technique

There are many techniques that can use to normalize EMG signals, such as the maximal voluntary isometric contraction method, the peak and mean activation-levels methods, the peak-to-peak amplitude of the maximum M-wave method, and the z-score normalization method [16]. Usually, these techniques have been used to reduce the variation of EMG signals between subjects, between days within a subject, and/or within a day in a subject if the electrode positions are shifted [16]. However, in this study the normalization technique is used based on the observation of asymmetric and symmetric between the maximum positive and the minimum negative EMG amplitudes at different angles of elbow joint in time domain. This can be confirmed by a skewness value, which is a measure of the asymmetry of the probability distribution of data.

The min-max normalization technique is used which performs a linear transformation on raw EMG data by setting the highest (positive) value to 1 and the lowest (negative) value to -1. The interval maximum and minimum of raw EMG data x is transformed into a new interval, which can be expressed as

$$y = \frac{(x - x_{\min})(y_{\max} - y_{\min})}{x_{\max} - x_{\min}} + y_{\min}, \tag{1}$$

where $x_{\max}$ and $x_{\min}$ represent the maximum and the minimum values of x respectively, $y_{\max}$ and $y_{\min}$ are the new range in which the normalized data will fall, and y is the normalized EMG data within the interval $y_{\min}$ to $y_{\max}$. In this paper, $y_{\min}$ and $y_{\max}$ are -1 and 1, respectively.

2.3 Mean Frequency (MNF) and Median Frequency (MDF)

MNF is an average frequency which is defined as a sum of the product of EMG power spectrum and frequency divided by a total sum of power spectrum. It can be expressed as

$$\mathrm{MNF} = \sum_{j=1}^{M} f_j P_j \Big/ \sum_{j=1}^{M} P_j, \tag{2}$$

where P_j is the EMG power spectrum at a frequency bin j, f_j is the frequency of the spectrum at a frequency bin j, and M is the total number of frequency bins.

MDF is a frequency at which the EMG power spectrum is divided into two regions with an equal integrated power. It can be expressed as

$$\sum_{j=1}^{M} P_j = \sum_{j=\mathrm{MDF}}^{M} P_j = \frac{1}{2}\sum_{j=1}^{M} P_j . \quad (3)$$

The proposed MNF and MDF features are extracted from the normalized EMG data y instead of the raw EMG data x. After the modified MNF and MDF values were calculated, the relationship between MNF/MDF and elbow joint angle was evaluated for all the different loads. Correlation analysis was also used to measure the strength of linear association between feature value and elbow joint angle.

3 Experimental Results and Discussion

First, the relationship between traditional MNF/MDF feature and elbow joint angle was re-investigated. Three different relationships were found in our experiments for the effect of elbow joint angle on MNF and MDF: (1) MNF/MDF is unaffected by changes in elbow joint angle, as shown in Figs. 1(a) and 1(b). This relationship was found for 8 subjects. (2) MNF/MDF increases as elbow flexion angle increases or muscle length decreases, as shown in Figs. 1(c) and 1(d). This relationship was found for 10 subjects. (3) MNF/MDF decreases as elbow flexion angle increases or muscle length decreases, as shown in Figs. 1(e) and 1(f), for a subject.

There are several possible reasons for the lack of consensus upon the effect of muscle geometry found in the previous and current studies. Differences in the experimental conditions between studies may be the main reasons for the conflicting results. For instance, Doheny et al. [19] suggested that the difference of muscle type and electrode location over the muscle is one of the reasons based on the experiments on three muscles: biceps brachii, triceps brachii, and brachioradialis. However, the three conflicting cases were found in the present investigation on the same muscle, i.e., the biceps brachii. Cechetto et al. [1] suggested that an inter-electrode distance is another possible reason for the conflicting results. Several inter-electrode distances have been used in the literature such as 10, 30, and 40 mm. In the present investigation, an inter-electrode distance of 20 mm was used and the three conflicting cases were still found. Further, it is important to note that due to a fixed load used in the experiments in literature at the different angles, the change of MNF and MDF is not due to the change of only muscle length but also due to the change of muscle contraction levels.

Second, to reduce the variation of intrinsic and extrinsic factors, raw EMG data was normalized in a comparison of activity. Figures 2(a) and 2(b) show the raw EMG data in time domain at a constant load with two different elbow angles, i.e., 30° and 150° of flexion. The results show that at a small elbow flexion angle, the distribution of the positive and negative EMG amplitudes was asymmetry. On the other hand, the distribution of the positive and negative EMG amplitude was symmetry at a large elbow flexion angle. Their power spectrums are shown in Figs. 2(c) and 2(d).

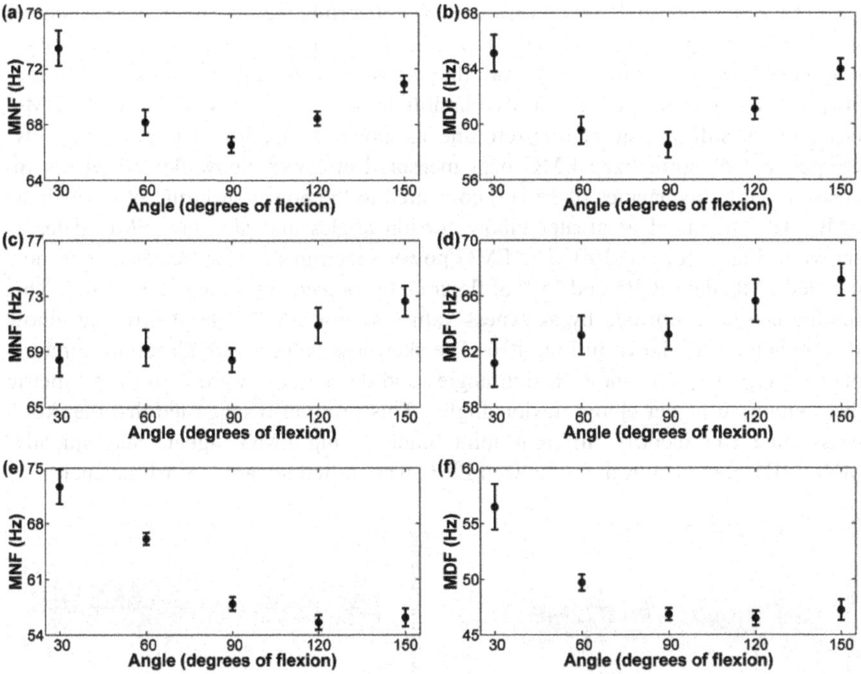

Fig. 1. (a, c, e) MNF and (b, d, f) MDF of raw EMG data at a constant load (4 kg) as a function of elbow angles (30°-150° of flexion): (a-b) the first case from Subj: #11, (c-d) the second case from Subj: #13, (e-f) the third case from Subj: #18. The error bars shown are given by the standard deviation of the mean value over 20 trials.

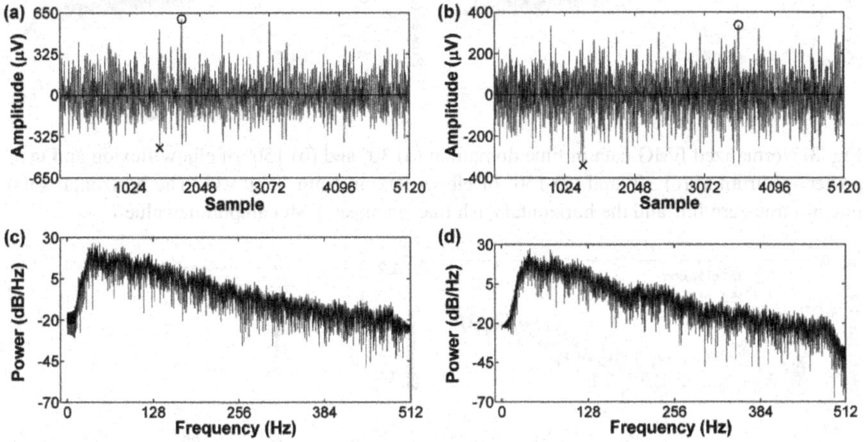

Fig. 2. Raw EMG data in time domain at (a) 30° and (b) 150° of elbow flexion and their power spectrum at (c) 30° and (d) 150° of elbow flexion at a constant load of 3 kg from Subj: #11. The maximum peak (positive value) is marked with a circle and the minimum peak (negative value) is marked with a cross.

If raw EMG amplitude was normalized by converting the maximum peak value (○) to 1 and the minimum peak value (×) to -1 for the asymmetric signal, the EMG baseline (dash line) was shifted away from the true zero line (solid line), as shown in Fig. 3(a). On the other hand, if raw EMG amplitude had a symmetry property, the EMG baseline was still been at the true zero line, as shown in Fig. 3(b). Therefore, the power spectrum of normalized EMG data measured at lower elbow flexion angles increased at low frequencies (0-15 Hz) compared to the power spectrum of normalized EMG data measured at greater elbow flexion angles and also raw EMG data, as shown in Figs. 3(c) and 3(d). The EMG power spectrums at low frequencies of normalized EMG data at 30° and 150° of flexion are zoomed and shown in Fig. 4(a). This finding can be confirmed by skewness values of raw EMG data at different elbow flexion angles, as shown in Fig. 4(b). The skewness value was high (an asymmetric distribution) at lower elbow flexion angles and decreased toward zero (a symmetric distribution) at higher elbow flexion angles. This result also suggested that the skewness value can identify different joint angles using EMG signal. Subsequently, MNF/MDF extracted from normalized EMG was shifted toward lower frequencies.

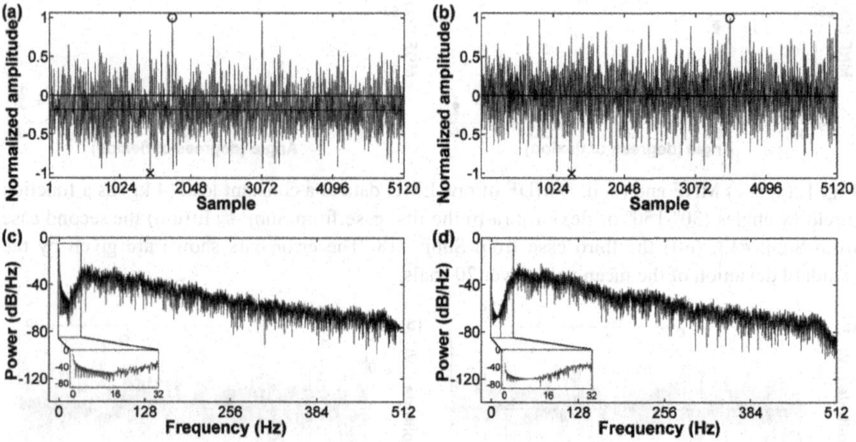

Fig. 3. Normalized EMG data in time domain at (a) 30° and (b) 150° of elbow flexion and their power spectrum at (c) 30° and (d) 150° of elbow flexion from Subj: #11. The horizontal solid line is a true zero line and the horizontal dash line is a mean EMG amplitude value.

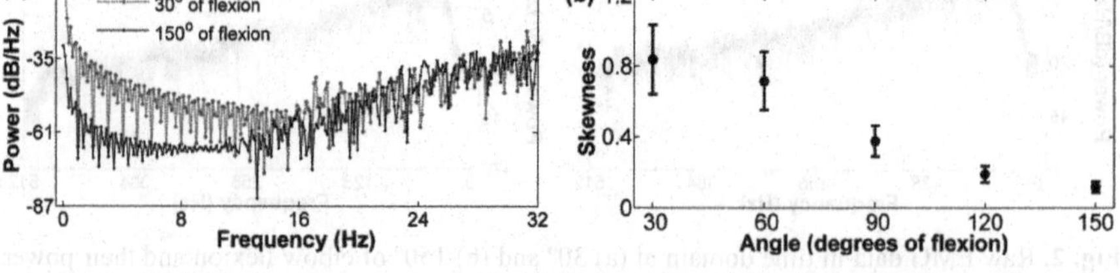

Fig. 4. (a) EMG power spectrum at low frequencies (0-32 Hz) of normalized EMG data at (dash line) 30° and (solid line) 150° of elbow flexion. (b) Skewness values of raw EMG data at a constant load (4 kg) as a function of elbow angles (30°-150° of flexion) from Subj: #18.

A certain relationship was found across all the studied subjects and loads (i.e., subject- and muscle force-independent), as shown in Fig. 5. Specifically, modified MNF and MDF features increased with increasing elbow flexion angle or decreasing muscle length (i.e., the second relationship case for traditional MNF and MDF features). To measure the increasing strength of linear relationship between modified MNF (and MDF) feature and elbow joint angle, the correlation analysis was performed. On average, normalized EMG data exhibited greater correlation coefficients compared to raw EMG data for both MNF (0.89>0.78) and MDF (0.86>0.80). This result confirmed a stronger linear relationship of the modified MNF/MDF feature and elbow joint angle.

In conclusion, a concept of using the normalization technique to provide the MNF and MDF features to determine the muscle geometry based on changing elbow joint angle was proposed in this paper. As a result of the proposed normalization technique, the modified MNF and MDF features can be used to analyse all three issues consisting of muscle force, muscle geometry, and muscle fatigue. Additional normalization procedure does not require any additional hardware.

Acknowledgements. This work is jointly funded by Prince of Songkla University and Thailand Research Fund through the Royal Golden Jubilee Ph.D. Program (Grant No. PHD/0155/2554). It is partially supported by NECTEC-PSU Center of Excellence for Rehabilitation Engineering, Faculty of Engineering, Prince of Songkla University.

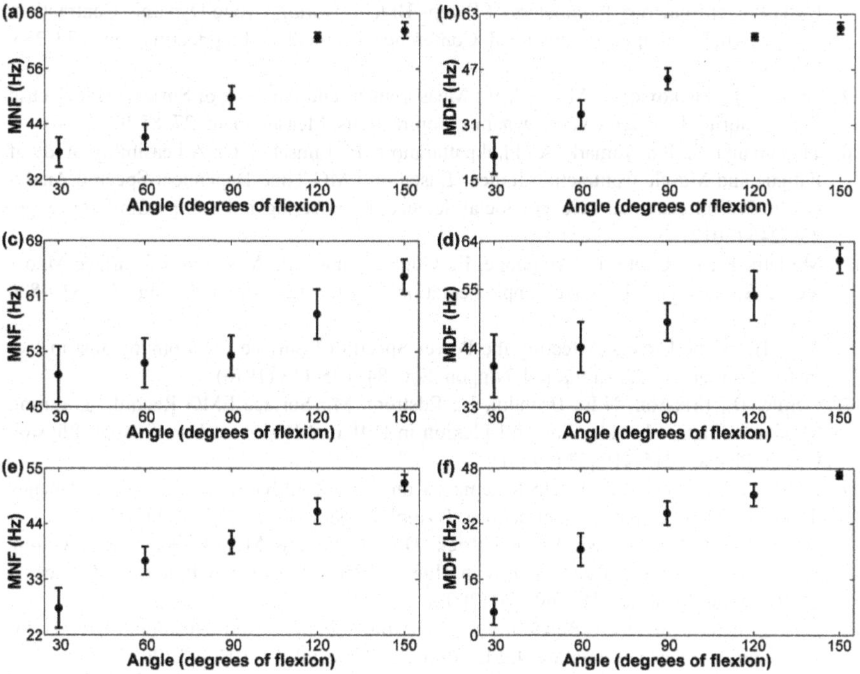

Fig. 5. (a, c, e) MNF and (b, d, f) MDF of normalized EMG data at a constant load (4 kg) as a function of elbow angles (30°-150°): (a-b) Subj: #11, (c-d) Subj: #13, and (e-f) Subj: #18.

References

1. Cechetto, A.D., Parker, P.A., Scott, R.N.: The Effects of Four Time-Varying Factors on the Mean Frequency of a Myoelectric Signal. J. Electromyogr. Kinesiol. 11, 347–354 (2001)
2. Phinyomark, A., Phukpattaranont, P., Limsakul, C.: Feature Reduction and Selection for EMG Signal Classification. Expert Syst. Appl. 39, 7420–7231 (2012)
3. Cifrek, M., Medved, V., Tonković, S., Ostojić, S.: Surface EMG Based Muscle Fatigue Evaluation in Biomechanics. Clin. Biomech. 24, 327–40 (2009)
4. Kaplanis, P.A., Pattichis, C.S., Hadjileontiadis, L.J., Roberts, V.C.: Surface EMG Analysis on Normal Subjects Based on Isometric Voluntary Contraction. J. Electromyogr. Kinesiol. 19, 157–71 (2009)
5. Phinyomark, A., Thongpanja, S., Hu, H., Phukpattaranont, P., Limsakul, C.: The Usefulness of Mean and Median Frequencies in Electromyography Analysis. In: Naik G.R. (ed.). Computational Intelligence in Electromyography Analysis: A Perspective on Current Applications and Future Challenges. pp. 195–220. InTech (2012)
6. Thongpanja, S., Phinyomark, A., Phukpattaranont, P., Limsakul, C.: Mean and Median Frequency of EMG Signal to Determine Muscle Force Based on Time-Dependent Power Spectrum. Elektron. Elektrotech. 19, 51-56 (2013)
7. Thongpanja, S., Phinyomark, A., Phukpattaranont, P., Limsakul, C.: Time-Dependent EMG Power Spectrum Features of Biceps Brachii During Isotonic Exercise. J. Sports. Sci. Technol. 10, 314–318 (2010)
8. Thongpanja, S., Phinyomark, A., Phukpattaranont, P., Limsakul, C.: Time-Dependent EMG Power Spectrum Parameters of Biceps Brachii During Cyclic Dynamic Contraction. In: 5th Kuala Lumpur International Conference Biomedical Engineering, pp. 233-236. (2011)
9. Cifrek, M., Tonković, S., Medved, V.: Measurement and Analysis of Surface Myoelectric Signals during Fatigued Cyclic Dynamic Contractions. Measurement. 27, 85-92 (2000)
10. Thongpanja, S., Phinyomark, A., Phukpattaranont, P., Limsakul, C.: A Feasibility Study of Fatigue and Muscle Contraction Indices Based on EMG Time-Dependent Spectral Analysis. In: 3rd International Science, Social Science, Engineering and Energy Conference, pp. 239-245 (2012)
11. Merletti, R., Lo Conte, L., Avignone, E., Guglielminotti, P.: Modeling of Surface Myoelectric Signals–Part I: Model implementation. IEEE Tran. Biomed. Eng. 46, 810-820 (1999)
12. Sato, H.: Some Factors Affecting the Power Spectra of Surface Electromyograms in Isometric Contractions. J. Anthropol. Nippon. Soc. 84, 105-113 (1976)
13. Gerdle, B., Eriksson, N.E., Brundin, L., Edstrom, M.: Surface EMG Recordings during Maximum Static Shoulder Forward Flexion in Different Positions. Eur. J. Appl. Physiol. Occup. Physiol. 415-419, 57 (1988)
14. Potvin, J.R.: Effects of Muscle Kinematics on Surface EMG Amplitude and Frequency during Fatiguing Dynamic Contractions. J. Appl. Physiol. 82, 144-151 (1997)
15. Doheny, E.P., Lowery, M.M., FitzPatrick, D.P., O'Malley, M.J.: Effect of Elbow Joint Angle on Force-EMG Relationships in Human Elbow Flexor and Extensor Muscles. J. Electromyogr. Kinesiol. 18, 760-770 (2008)
16. Bolgla, L.A., Uhl, T.L.: Reliability of Electromyographic Normalization Methods for Evaluating the Hip Musculature. J. Electromyogr. Kinesiol. 17, 102-111 (2007)

A study of Big Data solution using Hadoop to process connected Vehicle's Diagnostics data

Lionel Nkenyereye[1], Jong-Wook Jang[1]

[1] Department Computer Engineering, Dong-Eui University,176 Eomgwangro, Busanjin-Gu,Busan,614-714, Korea

lionelnk82@gmail.com,jwjang@deu.ac.kr

Abstract. Recently, we have witnessed a period where things are connected to the Internet; the vehicle is not leaved back because connected car field is currently explored. It is without doubt that connected vehicles will generate a huge of vehicle's diagnostics data which will be sent to remotely servers or to vehicle's cloud providers. As the amount of vehicle's diagnostics data increases, the actors in automotive ecosystem will encounter difficulties to perform a real time analysis in order to simulate or to design further services according to the data gathered from the connected vehicle. In this paper, Apache Hadoop framework and its ecosystems particularly Hive, Sqoop have been deployed to process vehicle diagnostics data and delivered useful outcomes that may be used by actors in automotive ecosystem to deliver new services to car owners. A study of big data solution to process vehicle diagnostics data from connected vehicles using Hadoop is proposed.

Keywords: Connected car, Hadoop project, HIVE open source, big data Problem, on-board diagnostics, Map Reduce

1 Introduction

The information technology authorizes information sharing from vehicles, roads infrastructure's sensors, vehicle–to vehicle, vehicle-to infrastructure, road condition [1]. Based on data received from Intelligent Transportation System (ITS), the third party interested in automobile ecosystem such as car manufacturers, repair shops, road and transportations authorities will continuously support a multitude of applications as for instance, monitoring performance of vehicles sold in the market by leveraging reporting of Diagnostic Troubles codes (DTC), status diagnosis information, emergency services management to locate and help victims of accidents or injured persons, traffic transportation authorities to increase safety, accident prevention, car repairs to analyze vehicle diagnosis in real time while the vehicle breaks down.

In fact, the data from connected vehicle will serve as a source to the actors in automotive ecosystem to afford value added services to the car owners. It seems that as long as data from connected vehicles increases, a big solution is expected to be implemented by car manufacturers or transportation authorities in order to process it and make available a reliable database of outcomes to all interested in connected car field. Apache Hadoop is nowadays suitable to process a huge amount of data. In this paper, we present a study of big data solution using Hadoop to process connected Vehicle's Diagnostics data.

Hadoop is an open source framework for writing and running distributed applications [2] that process large amounts of data. Data is broken up and distributed across the cluster and as much as possible, computation on a piece of data taken place on the same machine where the piece of data resides. The Hadoop projects that are used in this paper to keeping processing vehicle diagnostics data are: HDFS a distributed File System that runs on large clusters of commodity clusters, HIVE a distributed data warehouse that manages data stored in HDFS and provides a query language based on SQL, Sqoop a tool for efficiently moving data between databases and HDFS [3]. Effectively analysis of data is considered as a new way to increase productivity and also accelerate services as in automotive industry in now day where we witness Connected Car.

The rest of the paper is organized as fallows. Section I presents an introduction. The section II provides a background and existing work. The section III describes the proposed system design. At the last of this paper, we explain the implementation and its results. The results from the implementation are followed by a conclusion and future works.

2 Related Studies

The evolution of Internet of Things (IOT) has brought one line many little nodes plainly exists today [4].Google has considered the vehicle as one of the big node comprise by a lot of little node. Google 's WAZE division gathers traffic

K.J. Kim (ed.), *Information Science and Applications*,
Lecture Notes in Electrical Engineering 339, DOI 10.1007/978-3-662-46578-3_82

information from all enabled WAZE geographical application based on smartphone out and distribute that information to all car's owner who has opted to use WAZE.

Like other Global Positioning System (GPS) software, it considers from user's driving times to provide routing and real-time traffic updates. People can report accidents, traffic jams, speed and police traps, and from the online map editor, they can update roads, landmarks, house numbers, etc. [5].

The Google's WAZE GPS application involves only car owners or drivers and requires uploading their current reports accidents in order to allow them to make their optimal navigation routing decisions. This appear to be a weakness of the Connected Car vehicle platform that enables sharing vehicle sensors data and allows others services providers interest to automotive industry to design and propose their respectively services to the car owners. The proposed solution in this paper learns that a cloud based storage, real time filtered , sorted vehicle information by Hadoop platform is available and can be accessed by actors in automotive ecosystem via internet through their respectively web portal . Once processing outcomes performed using Hadoop are available, actors in automotive ecosystem can provide new services to the car owners.

3 System Design

3.1 Design of the solution for handling vehicle diagnostics data

The proposed solution is related to use Hadoop framework. After vehicle Diagnostic software based on android performs uploading vehicle diagnostics data to the datacenters based storage and processing, Hadoop framework will process them and the outcomes will be stored on the web server on which third party can access them. Fig. 1 shows the design of solution proposed based on Hadoop.

3.2 Data acquisition of vehicle diagnostics data

The acquisition of data describes in this paper consists of OBD scan tool, an android smartphone and a remote storage database (MySQL database) known to us as a Relational Database Management System. OBD scan tool interacts with the Engine Control Unit (ECU) via OBD-II connector, the android device is used as the client to handle data acquisition and data transmission, the remote database realizes reception of data and real time replication on the processing system based on Hadoop.

The HttpClient protocol in this paper is used to complete the communication between android and web server. The auto scanner tool used to connect with OBD-II interface start reading on board diagnostics PIDs codes since the profile of the driver like car owner name, car owner password, and valid vehicle identification based on Bluetooth Machine Access Address (MAC) of the OBD scan tool have been authenticated on the database, and then sent back to the car owner. The failure of authentication leads to a new car owner registration. The acquisition of vehicle diagnostics data are stored into a RDBMS know to us as MySQL database which is considered as a reliable source of data to the Hadoop platform during processing steps and also a reliable data outcomes after processing by Hadoop framework. Fig. 2 summarizes the design of the acquisition of vehicle diagnostics data to the remote data centers.

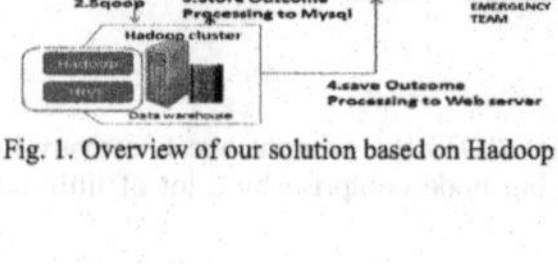

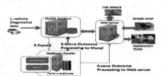

Fig. 1. Overview of our solution based on Hadoop

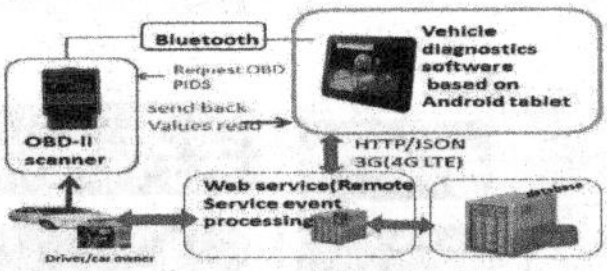

Fig. 2. Overview of the acquisition vehicle diagnostics data to the remote datacenter.

3.3 Processing vehicle diagnostics data using Hadoop framework

The processing of the vehicle diagnostics data on the remote database is divided in four phases: import data from MySQL to Hadoop clusters, loading data from HDFS to Hadoop data warehouse platform (HIVE), analysis using Map Reduce framework, upload outcome files in CVS format from HDFS to the virtualized web server.

1) import data from MySQL to Hadoop data file system(HDFS)

In this phase, the process consists of importing data from MySQL into Sqoop. Sqoop is a relational database import and export system which was created by Cloudera [5].

2) loading data from HDFS to Hadoop data warehouse platform(importing to HIVE)

After Sqoop keeping parallelizing import across multiple mappers, the next step consists of loading data into a Hive table using Sqoop. The import data into hive relies on sake of efficiency that has a post processing step where HIVE table is created and loaded [6]. When the data is loaded into HIVE from HDFS directory, HIVE moves the Sqoop replication table which is viewed as a directory into its warehouse rather than copying data.

3) analysis through Map Reduce framework

Hadoop acts as a distributed computing system that holds a distributed file system across Hadoop cluster. It relies to a distributed computational, Map Reduce, which coordinates and runs jobs in the cluster to operate on a part of the overall processing task in parallel. Some of this analysis needs to be performed by using customs map and reduce scripts. In conjunction with HIVESQL which is a SQL like query language provided by HIVE to create a variety of summaries outcomes, the developers can write SQL scripts to perform historical analysis over data loading in HIVE warehouse.

4) Save the outcomes to the web server

In this paper, when the final result from processing is available, HIVE is able to copy the outcome to the web server. Sqoop can at it turn exports the result back to the MySQL database.

4 Implementation and its results

The purpose of the big data solution is to process vehicle diagnostics data, save the outcomes into a MySQL database and on the web server on which data is remained available for the third party (actors in automotive ecosystem) for their further needs. Therefore, we set up a plan of required useful outcome information. For each useful information, we develop and execute a Map Reduce jobs on single node. Fig. 3 describes in summary the result expected for each job according to the vehicle diagnostics data and on which event the outcomes may be used to fulfill car owners' needs or services.

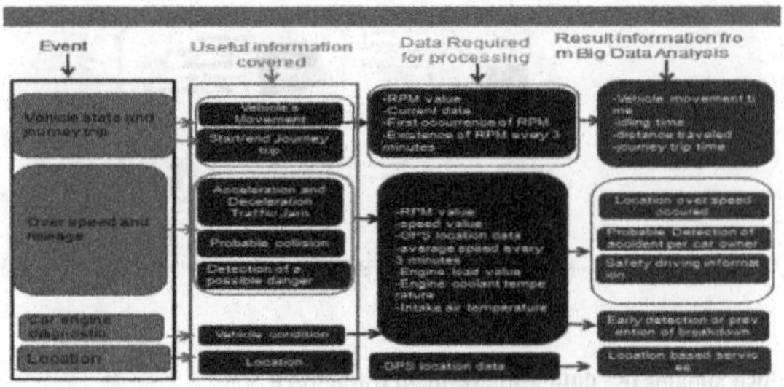

Fig. 3. Summary of processing vehicle diagnostics data using Hadoop.

4.1 Data acquisition and storage

In this paper, we develop a Remote Vehicle Diagnostics application based on android. It will be used by the vehicle owner via his mobile device based android. The car owner starts by connecting the Bluetooth OBDII scan tool adapter into OBD-II car's connector. Once the connection with the android based smartphones is established, he starts request data and the action is performed in background services. Fig. 4 and Fig. 5 show login before requesting engine performance and values uploaded into database.

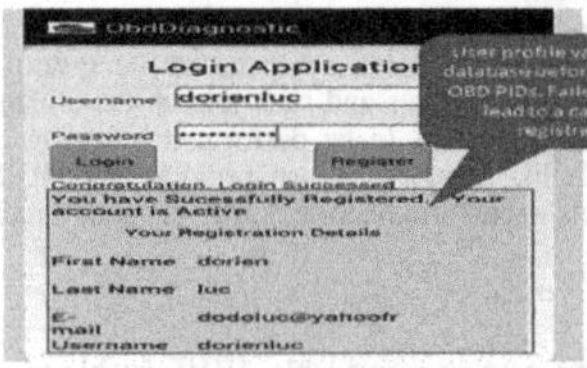

Fig. 4. Validation car user profile on cloud before requesting OBD-PIDs

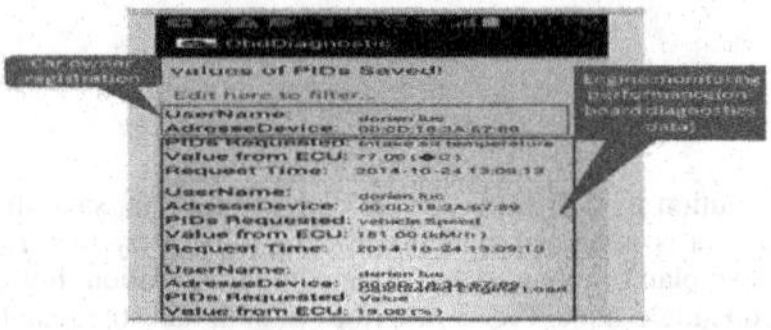

Fig. 5. Request of on-board diagnostics data saved on the remote server by the car owner

4.2 Processing using Hadoop platform

Acquisition of on-board diagnostics is carried out by the software based in android that features sending OBD-PIDs[7] to the ECU engine via OBD scanner tool equipped with Bluetooth capability, a device on the CAN-bus(Controller Area

Network) recognizes the PIDs as on it is responsible for and reports the value for that PID to the bus. The OBD scanner tool reads the response and displays it and at the same time uploads it to the remote database. Apache Hadoop open source and its projects are used to process the on-board diagnostics data and make available useful information to the third party as described in the Fig. 3. When on-board diagnostics data are uploaded to the database, Apache Sqoop performs a replication import of data required to run Map Reduce jobs. Before the import can be initiated, Sqoop uses JDBC [6] to examine the table to import. HIVE open source has an important role especially for storing vehicle diagnostics data unto a relational database. Sqoop generates a Hive table based on it originally definition and at the same time stores data on HDFS. Data are imported into HIVE data warehouse and stored on HDFS as well. Fig. 6 shows importation data to HIVE warehouse.

One of the most key of Map Reduce jobs implementation is to take the vehicle on-board diagnostics data replicated from MySQL database and stored into HIVE warehouse; process them according to the useful information we design and then store it to the MySQL database or generate it into CVS, JSON format that can be readable by users . HIVESQL is used to execute Map Reduce jobs. Fig. 7 and Fig. 8 show our Map Reduce functions definition structure using HIVESQL query. In this paper, we consider that the value of the RPM for example helps to compute vehicle's movement by taking in consideration current data and value of RPM after every 3 minutes. Fig. 9 and Fig. 10 show how Hadoop converts HIVESQL queries (Fig. 7) into a set of Map Reduce jobs.

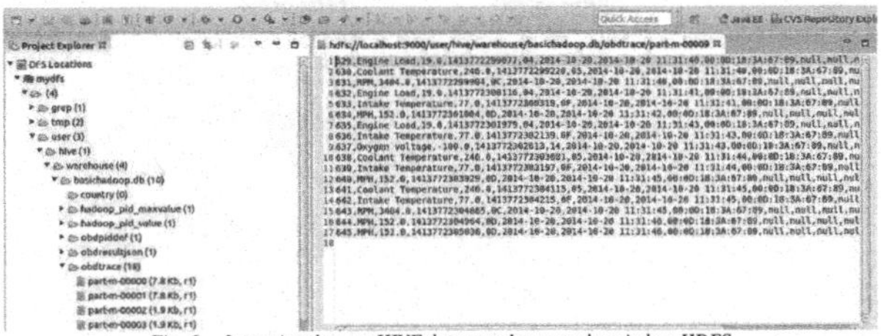

Fig. 6. Importing data on HIVE data warehouse and copied on HDFS

```
INSERT OVERWRITE TABLE basichadoop.obdresultjson
select concat(b.firstname,"_",b.lastname) as fullname,c.description_pid as DTC_description,max(a.pid_value) as measure_units
from basichadoop.obdtrace a JOIN basichadoop.userregistration b ON a.obddeviceaddress =b.addressdevice
right outer JOIN basichadoop.obdpiddef c ON    a.pid = c.pid_code
where a.pid_desc is not null and a.obddeviceaddress is not null
and c.pid_code in ('0C','0D','04','05','0F','0B','10','14','RV')
group by concat(b.firstname,"_",b.lastname),c.description_pid;
INSERT OVERWRITE LOCAL DIRECTORY '/home/mysmart/workspacejee/studyBasicOBD/ApplyLicenceproject/hadoop/output.cvs'
select * from    basichadoop.obdresultjson;
```

Fig. 7. Map Reduce function definition structure using HIVESQL query

```
import org.apache.hadoop.hive.ql.exec.Description;
import org.apache.hadoop.hive.ql.exec.UDF;
import org.apache.hadoop.io.Text;
@Description(
    name="DateFormatRequest",
    value="returns 'date x,y', where x is whatever stored datetime you give it (TIMESTAMP), y is the format you like to have",
    extended="SELECT DateFormatRequest(obd_daterequest,Y) from basichadoop.obdtrace limit 1;"
    )class DateFormatRequest extends UDF {
  }

  }
```

Fig. 8. UDF function to retrieve the current date and time

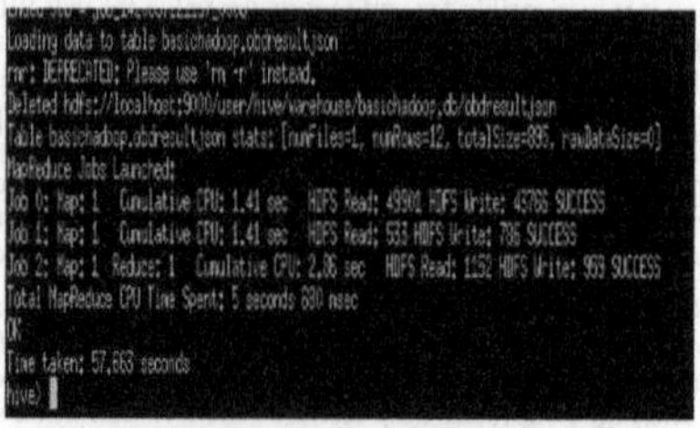

Fig . 9. MapReduce function in execution on single cluster node.

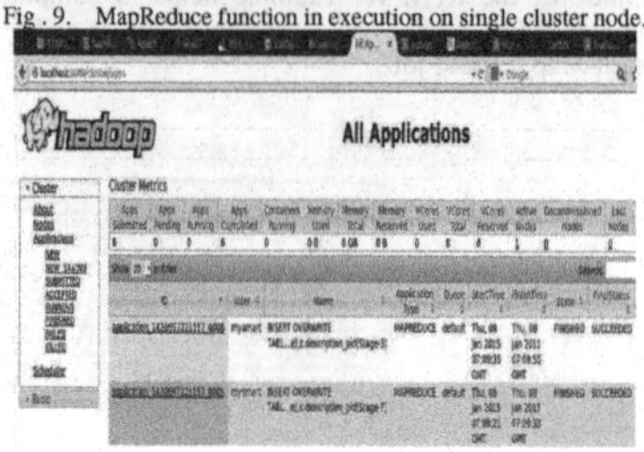

Fig. 10. HIVESQL query runned on Hadoop converted into a set of Map Reduce jobs

4.3 Car user front end web portal for remote monitoring

The main idea of user-friendly front end web portal is to track down: monitor on board diagnostics data that are already uploaded periodically from android based smartphone from different car owners to the MySQL database. Car owner can decide to share on board diagnostics data with services providers that offer up services. The vehicle owner can promptly through android based smartphone retrieves and saves its own vehicle diagnosis data at any time and shares them with the vehicle repair house while vehicle has malfunction indicator lamp also known as check engine light [7]. The front web page allows also vehicle owners to have an idea of where they have been by analyzing Google maps which displays the path they took via various markers that correspond to the various locations provided by the global Positioning System (GPS). Fig. 11 describes the information available on the web front web portal accessible by the car user meanwhile he is not driving his car for instance at office or at home or any time he wants or when a car repair shops contacts him for more details about value uploaded on the remote database.

4.4 Car manufacturer web portal

Fig. 12 shows a screenshot of the car manufacturer dashboard that is used to display and monitoring on-board diagnostics for a particular mark of vehicle sold on the market. The dashboard has several options for displaying the following information:

1) Car owner information;
2) Engine performance;
3) Current location;
4) Speed Analysis;
5) Tracking OBD-PIDs using charts.

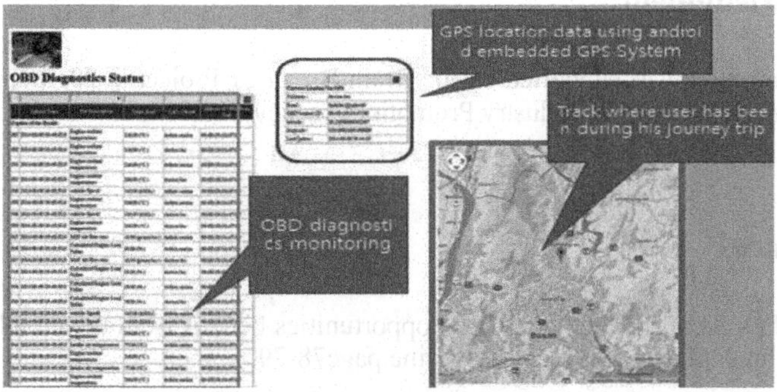

Fig. 11. Car User-friendly front end web portal

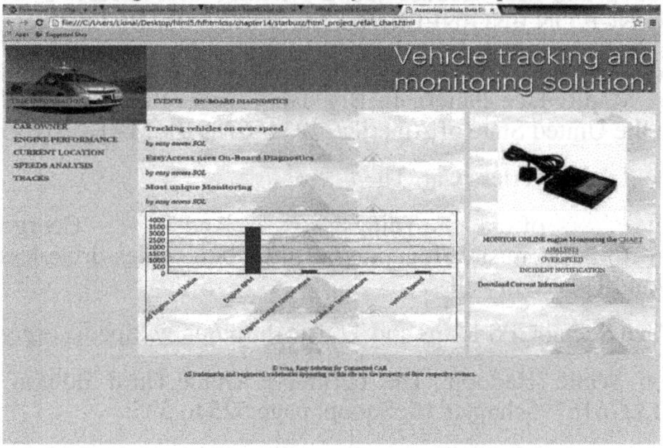

Fig. 12. Screenshot of the car manufacturer dashboard.

5 Conclusion and Future works

With the development of Hadoop platform project, now is possible to build big data solution using open source projects integrated with Hadoop. In this paper, a study of big data solution for processing data from vehicles using Hadoop and making final results available, allows accessing useful information via web services to the third party such as car manufacturer, transportation and road operators, car dealers, police, emergency services has been conducted on a single node cluster. The outcome obtained from various Map Reduce functions managed after executing HIVESQL query indicate favorable results in term of time taken. Future work will center on implementing and evaluating performance on cloud platforms for instance Amazon Elastic compute Cloud (EC2).

Acknowledgment

This research was supported by the Brain Busan 21 Project in 2014 and Nurimaru R&BD Project (Busan IT Industry Promotion Agency) in 2014.

References

[1]Jukka Ahola."Vehicle services opportunities benefit from the cloud",Applying cloud technologies for business Magazine,page78-79.

[2]Aditya B.Patel,Manashvi Birla,Ushma Nair,"Addressing Big Data Problem using Hadoop and Map Reduce",2012 NIRMA UNIVERSITY INTERNATIONAL CONFERENCE ON ENGINEERING,NuiCONE-2012,06-08December,2012

[3] Alex Holmes, "Hadoop in Practice, Including 85 techniques", chapter 4, "Applying MapReduce Pattern to Big data"' page 139-173, ISBN:9781617290237, Printed in the United State of America, 12345678910- MAL-171615141312

[4]Google's Driverless Car, The I

nternet of Things, And Georges Orwell, http://www.forbes.com/sites/rogerkay/2014/09/08/googles-driverless-car-the-internet-of-things-and-george-orwell/.

[5] Waze Social GPS Maps and Traffic, http://en.wikipedia.org/wiki/Waze

[6]Tom White,"Hadoop: The definitive Guide,Third Edition", ISBN:978-449-31152-0,1327616795chapter 15, Sqoop, page 525 to 535.

[7]Wikipedia, OBDIIPIDs, http://en.wikipedia.org/wiki/OBD-II_PIDs.

Wiki SaGa: an Interactive Timeline to Visualize Historical Documents

Daniel Yong Wen Tan[1], Bali Ranaivo-Malançon[2], Narayanan Kulathuramaiyer[2]

[1]University College of Technology Sarawak, Sibu Sarawak Malaysia
danieltan@ucts.edu.my
[2]Universiti Malaysia Sarawak, Kota Samarahan Sarawak Malaysia
{mbranaivo, nara}@fit.unimas.my

Abstract. Searching for information inside a repository of digitised historical documents is a very common task. A timeline interface that represents the historical content which can perform the same search function will reveal better results to researchers. This paper presents the integration of *SIMILE Timeline* within a wiki, named Wiki SaGa, containing digitised version of *Sarawak Gazette*. The proposed approach allows display of events and relevant information search compared to traditional list of documents.

Keywords: historical documents, timeline, Wiki SaGa, Sarawak Gazette.

1 Introduction

Timeline is a common method used to display events with respect to time. The earliest timeline system was introduced by Joseph Priestley, in 1769, when he published "A New Chart of History". Timeline was presented on physical paper in the past. Today, timeline is also available as a computer-based tool. According to Allen [1], utilizing a timeline interface to visualize events provide several advantages for a user. First, a timeline can inform as it shows basis information such as the title of an event and when the event itself occurs. Second, a timeline shows the context of events, which allows the comparison of these events across time. Third, a timeline encapsulates as "an event may be part of a bigger event". And fourth, a timeline can show links illustrating the attributes shared by events "or a hypothesized causal relationship."

The domain that utilized timeline the most is historical domain. History consists of events and their respective date. Timeline could, for example, display these events in a chronological way based on their date. The hidden relationship between events can start to emerge [2]. Thus, a timeline as a graphical representation provides a more interesting insight of historical events. This can be true for historians working on *Sarawak Gazette*, called henceforth SaGa, if this is displayed in a timeline. SaGa is the oldest newspaper published in Sarawak. The first publication was on August 26, 1870, during the reign of the first White Rajah, James Brooke. The gazette was published monthly and was edited by the Rajah's Civil Service. The published articles reflected the official thinking

K.J. Kim (ed.), *Information Science and Applications*,
Lecture Notes in Electrical Engineering 339, DOI 10.1007/978-3-662-46578-3_83

of the Rajah on major issues. The copies of SaGa used for this research are scanned images in PDF files.

This paper presents the integration of a timeline, called *SIMILE Timeline*, into the SaGa repository hosted in a wiki, called *JSPWiki*. Thus, the name of Wiki SaGa. The integrated timeline is interactive, that is, the timeline is not static. Users can scroll through the timeline and zoom in or out to view in details the contents of the timeline.

The rest of the paper is organized as follows. Section 2 reviews the main features of five current available timeline systems. Section 3 explains the integration of *SIMILE Timeline* and J*SPWiki*, the components of Wiki SaGa, an interactive timeline. Section 4 concludes the paper and provides some future works.

2 Timeline Systems Review

A basic timeline consists of the titles of events and their respective date of occurrence or reported date. Nowadays, the domain of journalism, for example *Le Monde*, *Time Magazine* and *Al Jazeera,* tends to make use of timeline to displays events. Besides, there are now many easily available timeline interface systems. A study has been conducted on few timeline interface systems available commercially and by open-source. The study focuses on the main features of these systems and how they generate timeline event.

2.1 myHistro

myHistro[1] is a free geolocated timeline interface that displays events alongside maps. The web-based application allows timeline creation with only the need of a signup. The main feature of *myHistro* is that it combines timeline and map to present historical events with spatial temporal element. The system is also available in app store. Timeline can be generated or viewed using Apple smartphone or tablet easily. The creation of an event is done through a page by filling in event information, which will be shown on Google Map. Descriptions and images are displayed when an individual event is selected.

2.2 Timeglider

Timeglider[2] is a commercial web-based timeline system. It displays events under their respective category. Using this feature, multiple dimension of events can be displayed accordingly. Another feature that *Timeglider* offers is a search function. Using this search function, specific information in the timeline can be searched by using keyword. This feature allows users to quickly access specific content of the timeline.

Event creation using this system is through a page. Each event is given an importance rating, from one to 100. Events that are rated with lower score, meaning least important,

1 myHistro, http://www.myhistro.com/

2 Timeglider, http://timeglider.com/how_it_works

will be omitted if the timeline is zoomed out, allowing the more important events to stand out. The timeline created can be shared by embedding it in a webpage.

2.3 Timeline JS

Timeline JS[3] is an open-source tool that is used by several organizations to display their timelines. Among them are *Le Monde*, *Time Magazine* and *Al Jazeera*. Creating events involves filling in a Google spreadsheet provided by *Timeline JS*. The file is then uploaded to *Timeline JS* generator and a link for embedding the timeline into HTML will be generated. The system also allows files to be read from database through a JSON file input. Contrary to most timeline interfaces, which display the description (if any) of an event through a popup box, *Timeline JS* shows the content of an event in a frame above the timeline.

2.4 Simile Timeline Widget

SIMILE Timeline Widget[4] system is an open-source tool that is available as Application Programming Interface (API). Being an API, *SIMILE Timeline* is more customizable in its implementation because users can easily configure what and how they want to present in the timeline. This system implements a search function that locates specific information within the timeline. The system also provides a highlight function that allows events that contain the keyword searched to be highlighted. This system used event sources to generate its timeline. The accepted file types are XML, JSON from MYSQL and SPARQL. However, since this system comes as an API, implementing it required some knowledge of programming regarding scripting and HTML.

2.5 Timeline

Timeline[5] is an open-source and desktop-based timeline system. The system offers features such as different representations based on zoom level, events organized in hierarchical categories and duplicate events. Although it also offers a search function, it can only search for keyword of the title of an event only. Being a desktop-based system, timelines created using *Timeline* cannot be shared through internet and intranet.

2.6 Summary

All the reviewed systems share the following features: event descriptions, hyperlink, timeline zoom and user interaction of scrolling through time. Table 1 shows other features that are not necessary shared.

---

[3] Knight Lab, http://timeline.knightlab.com/

[4] Massachusetts Institute of Technology, "Timeline," http://www.simile- widgets.org/timeline/

[5] R. Lindberg, "Welcome to timeline," http://thetimelineproj.sourceforge.net/#getting-started

Table 1. Comparison of timeline system

	myHistro	*Timeglider*	*TimelineJS*	*SIMILE*	*Timeline*
Licensing	Free	Commercial, free for students with limited function	Free	Free	Free
Platform	Web-based	Web-based	Web-based	Web-based and desktop-based	Desktop-based
Keyword search	No	Yes	No	Yes	Yes, title only
Event source	Input through UI	Input through UI	Input through source file	Input through source file	Input through UI
Out of box use	Yes	Yes	Yes	Configuration is needed	Yes
Display multimedia	Yes	Yes	Yes	Yes	No
System integration	Embedded in HTML	Embedded in HTML	Embedded in HTML	API	No

3 Integrating SIMILE Timeline within Wiki SaGa

Before the actual integration of *JSPWiki* and *SIMILE Timeline*, SaGa documents need to go through some preprocessing stages to allow Wiki SaGa to display the articles in its page and for *SIMILE Timeline* to generate the timeline. The integration of *SIMILE Timeline* into *JSPWiki* is shown in Fig. 1.

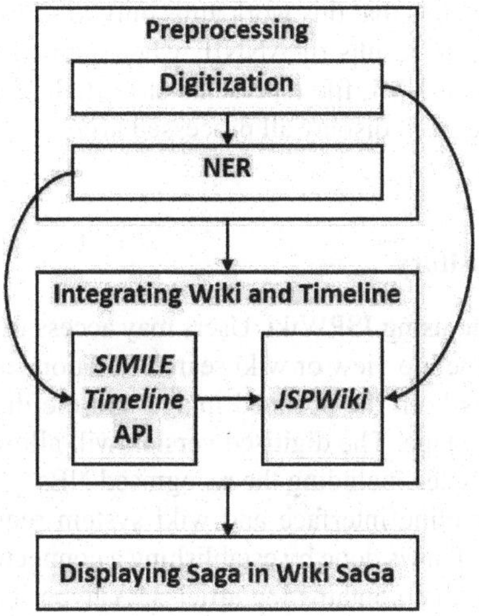

Fig. 1. Integrating timeline and wiki

3.1 Preprocessing

The preprocessing stage involves the digitisation of SaGa documents and the recognition of named entities.

Digitisation of SaGa Documents. Since all current SaGa documents are in the form of either printed or scanned documents, they need to be converted into plain texts (TXT format) to serve as input files for both the wiki page and the Named Entity Recognition (NER) system. This process is done using optical character recognition system *ABBYY Fine Reader* and manual cleanup of the output.

NER Applied to SaGa Documents. The term "named entity" (NE) was first coined in 1996 at the Six Message Understanding Conference (MUC-6) to refer to "unique identifiers of entities". However, the definition of the term itself is not clear. Marrero [3] came to the conclusion that the definition of NE is "according to the purpose and domain of application". There are many kinds of NEs but the basic NEs used to describe an event are Date, Person, and Location. Thus, this work is focusing only on the identification and extraction of these basic NEs to describe an event in the timeline interface. NER is "a task in Information Extraction consisting in identifying and classifying just some types of information elements, called Named Entities (NE). As such it, serves as the basis for many other crucial areas in Information Management, such as Semantic Annotation, Question Answering, Ontology Population and Opinion Mining." [3]. There are many available NER systems [4] [5] but ANNIE [6], an open-source NER available in GATE, was selected to perform the recognition of the basic NEs in the SaGa documents. By default, *ANNIE* can recognise NEs in any given English texts. This by-default setting is used in this work without any modifications or improvements.

The reason of such decision is that this work aims only to show the integration of NER into a timeline system. The results of ANNIE when applied to SaGa documents are saved into XML files. Each XML file represents one article extracted from one edition of SaGa. Therefore, in order to display all processed articles, all the XML files need to be consolidated.

3.2 Wiki SaGa Repository

Wiki Saga is implemented using JSPWiki. Users may access to SaGa processed articles through two methods: timeline view or wiki search function, as shown in Fig. 2. The article page contains both the scanned image and the digitised version obtained during the preprocessing stage. The digitised version will allow users to search for any word contained in the articles including the recognised NEs.

The integration of timeline interface and wiki system requires the content of the timeline to be structured. This is done by establishing a connection between the timeline and the actual wiki page that contains the articles. A hyperlink is inserted in a node-article so that it can refer to the wiki page.

The next part of the integration is to determine what to display within the timeline interface. For this project, NEs, titles of articles, and hyperlinks are selected to be displayed.

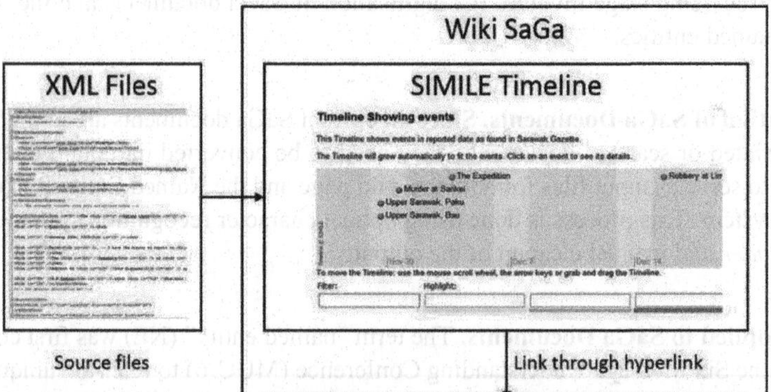

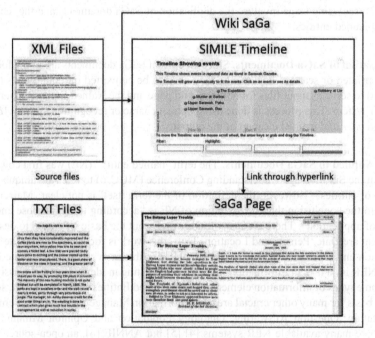

Fig. 2. Components of Wiki SaGa

3.3 Using SIMILE Timeline Functions

This section discusses the use of search function and highlight function to produce the desired results. Both functions filter the content but display their result in different view. Depending on how the results are shown, patterns will emerge.

Search function. The search function allows users to search for specific information that is within each node of the timeline. The search result will only display the nodes containing the keyword searched by the users.

Highlight function. Highlight function acts the same way as the search function. The difference between the two functions is that highlight function foregrounds the nodes that contain the keyword searched without omitting the other nodes that do not contain the keyword. This feature allows users to highlight the other nodes with different searched keywords in different color, as shown in Fig. 3. Therefore, highlight function allows users to detect patterns in the timeline that involves different entities that are not reported in the same article.

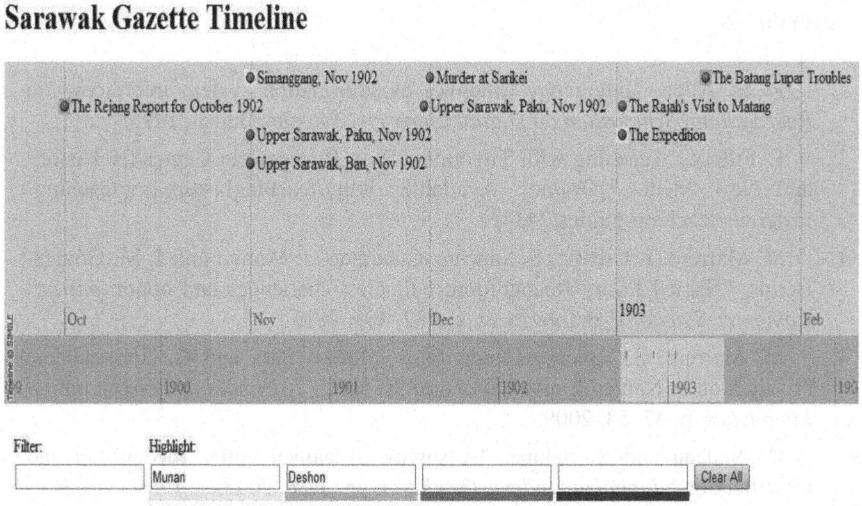

Fig. 3. Filter and highlight function of *SIMILE Timeline*

4 Conclusions

From our knowledge, Wiki SaGa is the first work on generating automatically a timeline to visualize the contents of historical documents like SaGa. Currently, due to copyright issue with Sarawak State Library, Wiki SaGa is only available for internal research and reference.

The framework is easily replicable as it does not constrain the tools used in the framework's components. Tools that are more suitable based on specific requirements or newer tools can be replaced anytime.

The proposed framework is feasible but difficult to be evaluated. The system itself is not difficult to evaluate based on performance and user friendliness. However, the automatic generated content is difficult to be evaluated due to the lack of human experts in SaGa domain. Nonetheless a small scale survey was conducted, revealing positive feedback from three historians specialised in SaGa. For future work, a method to evaluate the automatic generated timeline content will be explored.

Acknowledgements. The authors would like to thank the Universiti Malaysia Sarawak, mainly the Sarawak Language Technology (SaLT) research group, for providing the resources used in the conduct of this study. The first author thank University College of Technology Sarawak for the use of their Conference Grant to support the publication of this paper. The authors would like also to thank the anonymous reviewers for the critical reading of the paper.

References

1. R. B. Allen, "Interactive timelines as information system interfaces," in *International Symposium on Digital Libraries*, Tsukuba, Japan, 1995.
2. E. Fillpot, "Teaching with Timelines," Roy Rosenzweig Center for History and New Media, [Online]. Available: http://teachinghistory.org/teaching-materials/teaching-guides/24347.
3. M. Marrero, J. Urbano, S. Sánchez-Cuadrado, J. Morato and J. M. Gómez-Berbís, "Named Entity Recognition: Fallacies, challenges and opportunities," *Computer Standards & Interfaces,* p. 482–489, 2013.
4. M. Marrero, S. Sánchez-Cuadrado, J. Morato Lara and G. Andreadakis, "Evaluation of Named Entity Extraction Systems," *Advances in Computational Linguistics,* p. 47–58, 2009.
5. D. Nadeau and S. Sekine, "A survey of named entity recognition and classification," *Linguisticae Investigationes,* pp. 3-26, 2007.
6. H. Cunningham, "Chapter 6: ANNIE: a Nearly-New Information Extraction System," in *Developing Language Processing Components with GATE Version 8 (a User Guide)*, The University of Sheffield, Department of Computer Science, 2014, p. 117.

Discovering genomic associations on cancer datasets by applying sparse regression methods

Reddy Rani Vangimalla, Kyung-Ah Sohn*

[1]Department of Information and Computer Engineering, Ajou University, Suwon, S. Korea
{jebi1771, kasohn}@ajou.ac.kr

Abstract. Association analysis of gene expression traits with genomic features is crucial to identify the molecular mechanisms underlying cancer. In this study, we employ sparse regression methods of Lasso and GFLasso to discover genomic associations. Lasso penalizes a least squares regression by the sum of the absolute values of the coefficients, which in turn leads to sparse solutions. GFLasso, an extension of Lasso, fuses regression coefficients across correlated outcome variables, which is especially suitable for the analysis of gene expression traits having inherent network structure as output traits. Our study is about considering combined benefits of these computational methods and investigating the identified genomic associations. Real genomic datasets from breast cancer and ovarian cancer patients are analyzed by the proposed approach. We show that the combined effect of both the methods has a significant impact in identifying the crucial cancer causing genomic features with both weaker and stronger associations.

Keywords: Lasso · GFLasso · gene expression · Breast cancer · Ovarian cancer

1 Introduction

Cancer is a result of uncontrollable growth of cells. Unlike regular cells, cancer cells do not experience programmatic death and instead continue to grow and divide. Breast and Ovarian cancers are the most predominant malignancy in women. The estimated new cases and expected mortality rate is rapidly rising [1]. The ongoing study of gene expression with respect to multi layered genomic features is highly useful to overcome the poor prognosis of cancer.

The Cancer Genome Atlas (TCGA) [2] provided a platform and exceptional opportunity for biomedical researchers and practitioners to explore disease mechanisms and to identify clinically important biomarkers by data mining. The International Cancer Genome Consortium (ICGC) [3] is another platform with comprehensive description of genomic, transcriptomic and epigenomic changes with 50 different tumor types and their subtypes. ICGC is also widely used for discovering genomic associations.

Genome-wide association study (GWAS) is a well-known study, which uncovers genetic variants associated with complex traits [4]. The identified genetic association information can be used by the researchers for better decease prognosis and also in finding genetic variations that contribute to common, complex diseases, such as asth-

* Corresponding Author.

K.J. Kim (ed.), *Information Science and Applications*,
Lecture Notes in Electrical Engineering 339, DOI 10.1007/978-3-662-46578-3_84

ma, cancer, diabetes, heart disease and mental illnesses [5 – 7]. In our study we employed a multivariate regression techniques typically used in GWAS studies for identifying genomic associations on ovarian and breast cancer datasets. Studies revealed that, a possible genetic contribution to both breast and ovarian cancer risks is highly based on hereditary factors [8]. A person with breast cancer or ovarian cancer has a parallel risk of developing both cancers. The increased risk of developing either of these cancers is identified as inherited mutations of two particular genes BRCA1 and BRCA2 [9, 10].

In this study, we employ and compare two sparse regression techniques to identify genomic associations observed in cancer patients' data. Lasso (least absolute shrinkage and selection operator) [11] is first considered as a baseline, which produces sparse regression coefficients in a high-dimensional setting. As Lasso deals with each phenotype independently and it doesn't use any structural information of genomic features and expression traits, the second method we use in our study is GFLasso (Graph-Fused Lasso) [12] that utilizes the structural information about correlated output variables or traits. This is especially suitable for our study that considers gene expression traits as output variables because gene expression traits have been shown to be under natural network structure. We consider combined benefits of these computational methods and investigate the identified genomic associations in real genomic datasets from breast cancer and ovarian cancer patients.

2 Materials and Methods

2.1 Data & Preprocessing

From TCGA, gene expression data and methylation data were collected for both ovarian cancer and breast invasive carcinoma (BIC). Expression data is acquired from UNC-Agilent-G4502A-07 platform for BIC with 17,814 genes and from level 3 data of TCGA for ovarian cancer with 12,042 genes. Methylation data is from JHU-USC-Human-Methylation-27 platform for BIC with 23,094 methylation probes and from the beta-values of Infinium methylation 27 BeadChip for ovarian cancer with 27,578 types of methylation probes. The total sample size of breast and ovarian cancer data is 105 and 381 respectively [14, 15].

The preprocessing is applied to each individual type of dataset following the steps typically done in previous studies [14,15]. The methylation probes were first mapped into gene features, filtered by removing all non-zero values and even further filtered by variance such that features with lower 25% variance were removed. The final dataset for ovarian cancer is with 6,913 DNA methylation features, and 12,042 expression traits. Breast cancer dataset is compared with four other cancer datasets (GBM, LSCC, KRCCC and COAD) [14], to experiment with more essential methylation features and gene expression traits, the common methylation genes and expression genes of all the 5 cancer types (including BIC) were collected. This resultant final BIC dataset is with 597 methylation features and 10299 expression traits. We further filtered the expression traits with respect to cancer related genes that are collected from Cosmic website [13], by which the size of gene expression traits is reduced to 385. This type of filtration facilitates in identifying strong influencing predicators of cancer. The table below refers the final datasets of all cancer types used in this experiment. To focus on highly influencing cancer genomic associations, methylation data of ovarian cancer dataset also filtered as BIC, but for analysis of such filtration behavior we included both ovarian cancer dataset and ovarian-filtered dataset in our study.

Table 1. Dataset details before and after preprocessing

Cancer Type	Samples	Methylation Features		Gene Expression Traits	
		Before	After	Before	After
Breast	105	23,094	597	17,814	385
Ovarian	381	27,578	6,913	12,042	413
Ovarian-Filtered		27,578	467		

The feature values of all the datasets are finally standardized such that each feature has a zero mean and standard deviation of one, which in turn results in representing different genomic features on expression traits properly and without any bias.

2.2 Least absolute shrinkage and selection operator (Lasso)

Lasso is a sparse regression framework. This method is used to identify genes whose expressions are associated with DNA methylation features. The impact of J possible features $X_{1i},...,X_{Ji}$ to a gene expression trait value Y_i is modeled as a multivariate linear regression as follows, where i is the index of different samples:

$$Y_i = \beta_0 + \beta_1 X_{1i} + \beta_2 X_{2i} + \cdots \beta_J X_{Ji} + \epsilon_i, \epsilon_i \sim N(0, 6^2) \quad (1)$$

The linear model in (1) is for multiple independent phenotypes. The L_1 penalized regression function lasso is used for optimizing and finding relatively small number of effective covariates affecting the trait

$$\boldsymbol{Min}\ \sum_i \left(Y_i - \left(\beta_0 + \beta_1 X_{1i} + \beta_2 X_{2i} + \cdots . \beta_J X_{Ji}\right)\right)^2 + \lambda \sum_j |\beta_j| \quad (2)$$

The second term of equation (2) induces a sparse solution by reducing the number of non-zero coefficients in β. The value of λ was identified by cross validation. Finally the solution derived by lasso is a set of a few independent features which are in association with given traits. The association strength of each effective feature j is given by β_j [15]. This is implemented in R using *glmnet* package.

2.3 Graph Guided Fused Lasso (GFLasso)

Along with lasso penalty, 'fusion penalty' is applied in GFLasso, this fusses regression coefficients across correlated phenotypes, using weighted connectivity [12]. The method deals with multiple correlated phenotypes, instead of multiple independent phenotypes (Lasso). An additional penalty term that fuses two regression coefficients β_{jm} and β_{jl} for each marker j if traits m and l are connected with an edge in the graph is added. In equation (3) λ is Lasso regularization parameter and γ is a GFLasso regularization parameter

$$\widehat{\boldsymbol{B}}^{GC} = \boldsymbol{argmin} \sum_k (y_k - X\beta_k)^T . (y_k - X\beta_k) + \boldsymbol{\lambda} \sum_k \sum_j |\beta_{jk}| + \boldsymbol{\gamma} \sum_{(m,l)\epsilon E} \sum_j |\beta_{jm} - sign(r_{ml})\beta_{jl}| \quad (3)$$

After considering the edge weights in graph G, in addition to the graph topology, the equation (3) becomes.

$$\hat{B}^{GW} = \boldsymbol{argmin} \sum_k (y_k - X\beta_k)^T \cdot (y_k - X\beta_k) + \boldsymbol{\lambda} \sum_k \sum_j |\beta_{jk}| + \boldsymbol{\gamma} \sum_{(m,l)\in E} f(\mathbf{r}_{ml}) \sum_j |\beta_{jm} - sign(r_{ml})\beta_{jl}|, \quad (4)$$

Where $f(r_{ml})$ is the correlation between the two phenotypes that are being fused. If the two phenotypes m and l are highly correlated in graph G with a relatively large edge weight, the regression coefficients β_{jm} and β_{jl} is penalized more than for other pairs of weaker correlation. The correlation weight can be $f_1(r) = |r|$ (absolute value) or $f_2(r) = r^2$ (Squared value). $f_2(r)$ is used in this work, as both mean squared error and non-zero beta values density is less for $f_2(r)$ compared to $f_1(r)$. This is implemented in matlab with the help of the code available at *http://www.sailing.cs.cmu.edu/main/?page_id=462*

We choose the tuning parameters of λ and γ using the following steps. Initially, median of non-zero beta coefficient is chosen as λ_0, multiplied it with total count of gene expression features. Initial Gamma is fixed as 1. The 2/3rd of dataset is used as training data and rest of 1/3rd as test data and verified the mean squared error (MSE), non-zero beta coefficients density and time to execute the dataset. The observations is carried out on different λ and γ values, for example fixing γ at γ_0 and applying on different values of λ as $\lambda_0/2$, λ_0, 2 λ_0, then fixing λ_0 and changing γ to $\gamma_0/2$, γ_0, $2\gamma_0$. After iterations, λ and γ values are fixed as $\lambda = 12$ and $\gamma = 1$. The correlation threshold $f(r_{ml})$ was fixed as 0.7 for all the datasets throughout the experiments, considering only very highly correlated gene expression features.

3 Results

Identifying the GFLasso and Lasso performance in terms of MSE, density and execution time We first compare the behavior of both the methods. Table 2 shows the mean squared error (MSE) on two types of cancer data. As the smaller MSE implies the better performance, GFLasso consistently outperformed Lasso, even for the high dimension of predicate datasets (ovarian cancer methylation dataset (6,913) which is almost 11.6 times larger than other datasets).

Table 2. MSE of different types of cancer datasets

Cancer Type	Mean Squared Error (MSE)	
	GFLasso	Lasso
Breast	1.113988	1.12009
Ovarian	0.339665	0.360145
Ovarian-Filtered	0.35867	0.369149

Figure 1A displays the density of the regression coefficient matrix. We can conclude that, due to Lasso's regular behavior of shrinkage of coefficients to zero the least number of non-zero betas are obtained with Lasso, whereas due to the additional fusion penalty of GFLasso, it has the larger densities than Lasso. Figure 1B compares the execution time. Except for the very high dimensional dataset (Ovarian) GFLasso's computational time is much smaller than Lasso, but Lasso executes faster for high dimensional datasets (ovarian dataset). Though GFLasso has larger density, it facilities in identifying weaker signals along with stronger signals, as GFLasso considered highly correlated phenotypes (0.7 is the correlation threshold). The integrated results of both the methods are used in this study, to identify influential predicators of cancer.

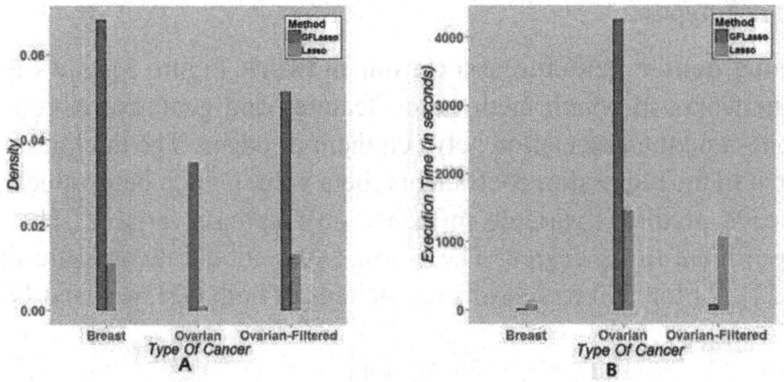

Fig. 1. Comparison of GFLasso and Lasso based on **A.** Regression coefficients density, **B.** Computation time.

Discovering common genomic features of both the methods As a further study, we tried to identify the common predicators that were retrieved using both the regression methods. Figure 2 A, B and C are the Venn diagrams of breast, ovarian-filtered and ovarian cancer types respectively. As the expression traits we use are recognized cancer census genes, we focused on the genomic features those are identified using both the methods (as they are the strongest predicators of the expression traits). Though, the combined results may increase signal to noise ratio, it certainly helps in identifying the stronger as well as weaker signals. Even though the non-zero beta densities are larger for GFLasso, the final identified genomic features are lesser than Lasso, therefore it fairly discarded unwanted predicators, and the same can be observed in Figure 2A and 2B.

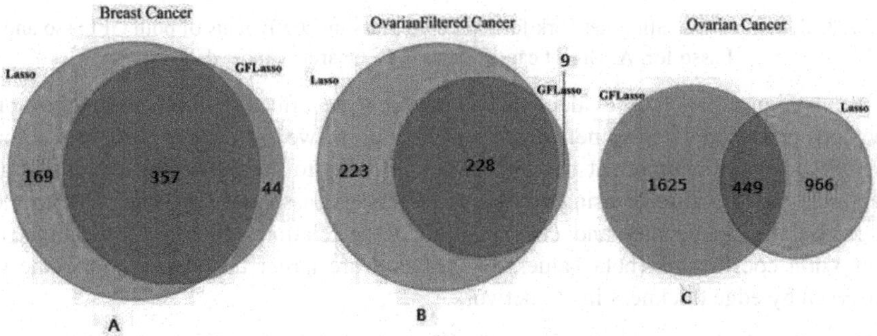

Fig. **2.** Venn diagram of all nonzero beta methylation features. A. Common methylation feature pairs of breast cancer are 357/597. B. For ovarian-filtered dataset 228/467 are identified as common features. C. For a high dimensional dataset, ovarian it is 449/6913.

The larger number of predictors identified by GFLasso in Figure 2C is due to the higher dimension of genomic feature dataset (almost 11.6 times larger than other datasets) and also due to the additional fusion penalty. The identified common (by both GFLasso and Lasso) genomic feature and expression trait pairs, that are associated to each cancer type is 141, 53 and 135 for breast, ovarian-filtered and ovarian cancer types respectively. These are the strongest predicator and response variable couples identified using both the methods, they are in turn a true highly influential pairs for respective cancer types.

Heterogeneous denser genomic association network Figure 3 shows the genomic association networks in which methylation features and gene expressions are represented as nodes and the association between them as edges. The thickness of the edge is proportional to the regression coefficients (beta value). Each beta value signifies the strength of each predictor variable influence on response variable. The size of the node is proportional to its degree. The below association networks are drawn using Cytoscape [22], for top 500 regression coefficients of both GFLasso and Lasso.

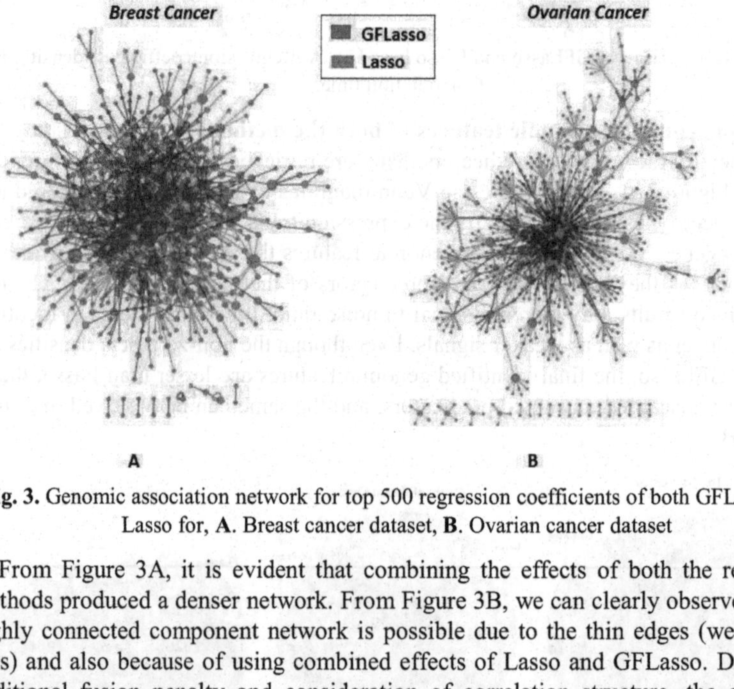

Fig. 3. Genomic association network for top 500 regression coefficients of both GFLasso and Lasso for, **A**. Breast cancer dataset, **B**. Ovarian cancer dataset

From Figure 3A, it is evident that combining the effects of both the regression methods produced a denser network. From Figure 3B, we can clearly observe that the highly connected component network is possible due to the thin edges (weaker signals) and also because of using combined effects of Lasso and GFLasso. Due to the additional fusion penalty and consideration of correlation structure, the estimated regression coefficients (beta values) of GFLasso are larger than Lasso, the same is observed by edge thickness in the network.

Functional characterization of the affected genes using the tool DAVID The functional annotation test was executed for gene-enrichment analysis with respect to GO Biological Process (BP) to the set of feature genes that are common in both Lasso and GFLasso i.e. 357 methylation feature genes of Breast cancer (Figure 2.A) and 449 of Ovarian cancer (Figure 2.C). Studies revealed that GO: 0042127 regulation of cell proliferation and Tyrosine protein kinase are overexpressed in high percentages (more than 70%) of human breast cancers [19, 20], cancer pathway genes were also recognized. Similarly for ovarian cancer identification of plasma membrane proteins from SKOV3 cells is an important preliminary step for identifying the cancer bio markers [21].

Table 3. Significantely enriched GO (top 5 terms) for common methylation features of both GFLasso and Lasso (Breast cancer - 357, Ovarian Cancer – 449 genomic features)

Cancer	Category	Most Significant Term	N	p - value	FDR
Breast	GOTERM_BP_FAT	GO:0042127 regulation of cell proliferation	96	8.65E-40	1.55E-36
	INTERPRO	IPR001245 Tyrosine protein kinase	35	1.06E-29	1.61E-26
	INTERPRO	IPR008266 Tyrosine protein kinase, active site	31	2.36E-27	3.59E-24
	KEGG_PATHWAY	hsa05200:Pathways in cancer	62	9.53E-26	1.10E-22
	SP_PIR_KEYWORDS	tyrosine-protein kinase	30	1.33E-25	1.87E-22
Ovarian	SP_PIR_KEYWORDS	signal - GO:0005576 Name extracellular region	133	4.50E-12	6.39E-09
	UP_SEQ_FEATURE	signal peptide	133	7.19E-12	1.18E-08
	GOTERM_CC_FAT	GO:0044459 plasma membrane part	105	3.71E-10	5.06E-07
	UP_SEQ_FEATURE	sequence variant	337	4.65E-10	7.64E-07
	SP_PIR_KEYWORDS	disulfide bond	116	1.37E-09	1.95E-06

4 Discussion & Conclusion

GFLasso utilizes complete information of correlation structure in phenotypes available as a graph, where the subgroup information is embedded implicitly within the graph as densely connected sub graph. In this study we used this graph information and also the effects of Lasso. This facilitated in identifying the strongest possible signal and as well as weaker signals, but with some accepted false positive rate. Along with each of the method's advantages, the limitations also influenced the results. To guarantee the strong active predicators applying strong rule, that is combing the screening methods with Karush-Kuhn-Tucker (KK) will effectively discard the inactive predicators and will produce promising results and reduced the signal to noise ratio [18].

Group regression approaches use clustering algorithms to detect pleiotropic effect by learning subgroups of traits and searching for genetic variations that perturb the subgroup [16, 17]. In our future study, we plan to explore different statistical techniques to utilize such information on input or output structure, or both as in [16,17].

Acknowledgements This research was supported by the Basic Science Research Program through the National Research Foundation (NRF) of Korea funded by the Ministry of Education (2012R1A1A2042792), and by the MSIP under the Global IT Talent support program (NIPA-2014-H0904-14-1004) supervised by the NIPA(National IT Industry Promotion Agency).

References

1. SEER Stat Fact Sheets: Breast, Ovary National Cancer Institute. http://seer.cancer.gov/statfacts/html/breast.html
2. The Cancer Genome Atlas (TCGA). http://www.cancergenome.nih.gov/.
3. International Cancer Genome Consortium (ICGC). https://icgc.org/icgc
4. National Human Genome Research Institute. http://www.genome.gov/20019523
5. Guillaume Lettre and JohnD.Rioux, Autoimmune diseases: insights from genome-wide association studies. Human Molecular Genetics, 2008 , R116–R121.
6. Dirkje S. Postma, and Gerard H. Koppelman Genetics of Asthma, Proceedings of the American Thoracic Society, Vol. 6, No. 3 (2009), pp. 283-287.
7. McPhersonR, PertsemlidisA, KavaslarN, StewartA, RobertsR, CoxDR, HindsDA, Pennacchio LA, Tybjaerg-Hansen A, Folsom AR, Boerwinkle E, Hobbs HH,Cohen JC (5830). A common allele on chromosome 9 associated with coronary heart disease, 2007 May 3.
8. National cancer Institute. http://www.cancer.gov/cancertopics/pdq/genetics/breast-and-ovarian/HealthProfessional/page1#Reference1.3
9. What is the Link between Breast Cancer and Ovarian Cancer? http://www.wndu.com/16buddycheck/headlines/28313989.html
10. TCGA,http://cancergenome.nih.gov/newsevents/multimedialibrary/videos/BreastOvarianMartignetti2014
11. Robert Tibshirani, Regression Shrinkage and Selection via the Lasso, J.R. Statistics, 1996, pp.(267 – 288)
12. Seyoung Kim, Kyung-Ah Sohn, Eric P. Xing. A multivariate regression approach to association analysis of a quantitative trait network, ISMB 2009, pages i204–i212.
13. Catalogue of somatic mutations in cancer. http://cancer.sanger.ac.uk.
14. Bo Wang, Aziz M Mezlini, Feyyaz Demir, Marc Fiume, Zhuowen Tu, Michael Brudno,Benjamin Haibe-Kains & Anna Goldenberg.Similarity network fusion for aggregating data types on a genomic scale, Published online 26 January 2014 Nature Methods 11, 333–337.
15. Kyung-Ah Sohn, Dokyoon Kim, Jaehyun Lim and Ju Han Kim. Relative impact of multi-layered genomic data on gene expression phenotypes in serous ovarian tumors, BMC Systems Biology 2013.
16. Seunghak Lee and Eric P. Xing, Leveraging input and output structures for joint mapping of epistatic and marginal eQTLs , ISMB 2012, pages i137–i146.
17. Noah Simon, Jerome Friedman, Trevor Hastie & Robert Tibshirani A Sparse-Group Lasso, Journal of Computational and Graphical Statistics, 30 May 2013.
18. Robert Tibshirani, Jacob Bien, Jerome Friedman, Trevor Hastie, Noah Simon, Jonathan , Taylor, Ryan Tibshirani , Strong Rules for Discarding Predictors in Lasso-type Problems Genes-to-Systems Breast Cancer (G2SBC) Database, Departments of Statistics and Health Research and Policy, November 11, 2010.
19. Genes-to-Systems Breast Cancer (G2SBC) Database http://www.itb.cnr.it/breastcancer/php/GOTree.php?idGO=GO:0042127
20. Jacqueline S Biscardi, Rumey C Ishizawar, Corinne M Silva, and Sarah J Parsons, Tyrosine kinase signalling in breast cancer: Epidermal growth factor receptor and c-Src interactions in breast cancer, Breast Cancer Research, Published online Mar 7, 2000 (203 -210).
21. P.J. Adam, R. Boyd, K.L. Tyson, G.C. Fletcher, A. Stamps, L. Hudson, H.R. Poyser, N. Redpath, M. Griffiths, G. Steers, A.L. Harris, S. Patel, J. Berry, J.A. Loader, R.R. Townsend, L.Daviet, P. Legrain, R. Parekh and J.A. Terrett, Comprehensive proteomic analysis of breast cancer cell membranes reveals unique proteins with potential roles in clinical cancer, The Journal of Biological Chemistry, published online December 10, 2002, 6482–6489.
22. Cytoscape. http://www.cytoscape.org/cy3.html

The Adaptive Dynamic Clustering Neuro-Fuzzy System for Classification

Phichit Napook[1], Narissara Eiamkanitchat [1,2]

[1] Department of Computer Engineering, Faculty of Engineering,
[2] Social Research Institute,
Chiang Mai University, Thailand
phichit_n@cmu.ac.th, narisara@eng.cmu.ac.th

Abstract. This paper proposes a method of neuro-fuzzy for classification using adaptive dynamic clustering. The method has three parts, the first part is to find the proper number of membership functions by using adaptive dynamic clustering and transform to binary value in a second step. The final step is classification part using neural network. Furthermore the weights from the learning process of the neural network are used as feature eliminates to perform the rule extraction. The experiments used dataset form UCI to verify the proposed methodology. The result shows the high performance of the proposed method.

Keywords. Neuro-Fuzzy System, Adaptive Dynamic Clustering, Classification.

1 Introduction

The neural network is a popular model used for classification. It has been widely used since the strength of robustness for unknown data. Furthermore, the reason of it wildly used is the merit of using the back propagation for learning process. However, the neural network has a weak point of interpreting ability or usually called the black box [1-3]. The fuzzy system has been widely used in classification problems the same as the neural network. The fuzzy system can handle the uncertainty and proper for creating the linguistic value. The fuzzy rule base extraction is easy to understand and important for many applications. The weakness of the fuzzy system is no adaptive learning methodology [2]. In order to improve the classification accuracy, many researchers have proposed the combination of the neural network and fuzzy system methodology called neuro-fuzzy system. The strength of the two algorithms is selected to use in the neuro-fuzzy system, which is the learning algorithm and the linguistic for rule extraction that is easy to understand [5-6]. The experimental in [4-7] shown that the neuro-fuzzy system has a good performance for classification. In [5-6] has proposed the neuro-fuzzy model for improving the performance of classification. The method is used three fuzzy membership function for converting the original input to the linguistic value then fed into the neural network. The method applied back propagation to update neural network and fuzzy membership function. The results from training process are used to generate rule extraction for classification and can use as

K.J. Kim (ed.), *Information Science and Applications*,
Lecture Notes in Electrical Engineering 339, DOI 10.1007/978-3-662-46578-3_85

feature selection. The classification results compare with K-mean SVM, DE-SVM and DE-SVM showed a high performance. The method proposed Enhance Neuro-Fuzzy (ENF) in [7] is the development of [6], the algorithm to find number of memberships functions using dynamic clustering is proposed. The result of classification in the method shown that the better performance. However, the high number of memberships from the dynamic clustering method cause the higher complexity.

This research focuses on the development of Neuro-fuzzy proposed in [6-7]. The concept proposed to reduce the number of membership functions. In addition, the proposed method also improves performance of classification and reduce the complexity. The rule extraction method from [6] also improve by using the Golden Section Search method (GSS). The GSS, with less complexity, is applied to find the proper number of linguistic to use in the classification rule.

2 Structure of Adaptive Dynamic Clustering of Neuro-Fuzzy

This paper proposes the new methodology for reducing the number of clusters from ENF by using fuzzy union operation. The new methodology can be divided into three parts. Firstly, using dynamic clustering to find the number of clusters for each feature. The fuzzification process using the number of clusters from first step to generate membership value using Gaussian membership function. The adaptive dynamic clustering is processed to find the optimal number of membership functions. The fuzzification process is regenerated to a binary value. Finally, using neural network for classification. The structure of the proposed method shown in figure 1

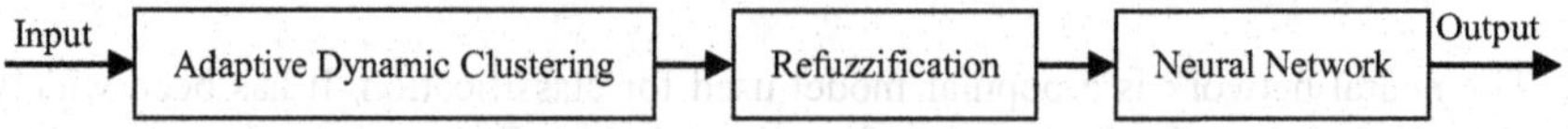

Fig. 1. Structure of Adaptive Dynamic Clustering of Neuro-Fuzzy System.

2.1 Adaptive Dynamic Clustering process

The Adaptive Dynamic Clustering (ADC) begins by sorting each feature of the original data order. The first original data is assigned to cluster 1. The following data is considered as if original class is same as previous data then assign to previous cluster, otherwise assign to new cluster. The process is recursive until all original data are considered. The second step, calculate the average number of member in each to define the threshold of each feature to use in agglomerative step.

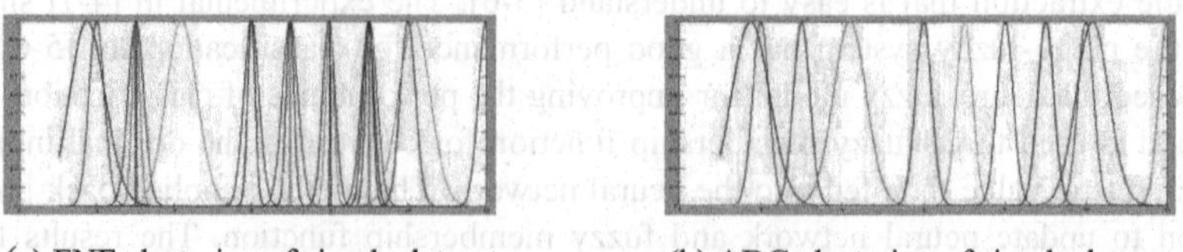

(a) ENF clustering of simple dataset (b) ADC clustering of simple dataset

Fig. 2. Example results of membership functions from each method

The cluster which has member less than threshold is combined to the nearest neighboring. The process is recursive until all clusters are considered. The ENF algorithm uses these clusters to generate the Gaussian membership function as displayed in figure 2. From the experimental results there are many membership functions generated. The high number of clusters the high complexity of the classification model. Clearly see figure 2(a) that the Gaussian membership graphs which have the same mean are overlap.

In order to reduce the number of membership functions, the parameter of Gaussian membership function is considered. There are 2 parameters in this function that is mean (μ) and standard deviation (σ). The fuzzy union operation is applied by considering the mean of each graph. The union operation is defined as

$$u_{\widetilde{A}_1 \cup \widetilde{A}_2 \cup \ldots \widetilde{A}_n} = \max(u_{\widetilde{A}_1}(x), u_{\widetilde{A}_2}(x), \ldots u_{\widetilde{A}_n}(x)) \quad (1),$$

where $\widetilde{A}_1, \widetilde{A}_2, \ldots, \widetilde{A}_n$ are the fuzzy set of $A_1, A_2, \ldots, A_n$ those have the same mean value. The combined results from the fuzzy union operation is demonstrated in figure 2(b). Consider figure 2(a) and figure 2(b), both have the same number mean points of membership graph, but number of membership from ADC is dramatically reduced. Clearly see the proposed method is reduced complexity from ENF algorithm.

2.2 Refuzzification process

The optimal number of membership function from previous process are used in the refuzzification and transform to binary value as displayed in figure 3.

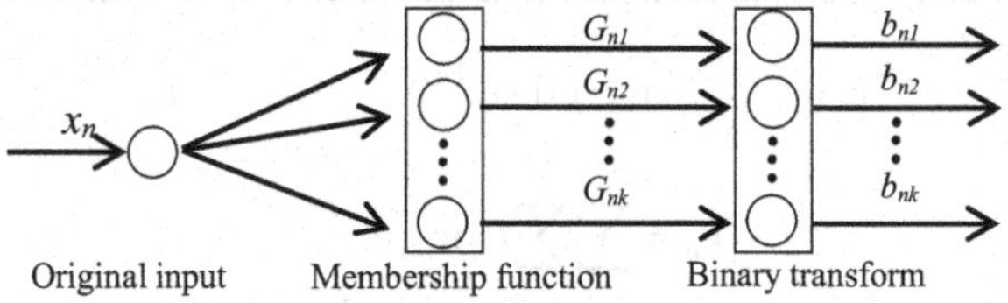

Fig. 3. The structure of refuzzification process.

The original inputs are fed to Gaussian membership function and calculate the membership value of original input as

$$G_{nk} = \begin{cases} 0, \text{ if } \sigma_{nk} = 0 \text{ and } x_n \neq \mu_{nk} \\ e^{(-(x_n - \mu_{nk})^2 / 2\sigma_{nk}^2)}, \text{ if } \sigma \neq 0 \\ 1, \text{ if } \sigma_{nk} = 0 \text{ and } x_n = \mu_{nk} \end{cases} \quad (2).$$

The G_{nk} is the membership value, n represent the feature and k represent the cluster of original dataset. The x_n is the original input, μ_{nk} is the mean value and σ_{nk} is the standard deviation value respectively. The results from membership function are fed to binary transform. The binary value are defined as

$$b_{nk} = \begin{cases} 1, & \text{if } G_{nk} \text{ is maximum} \\ 0, & \text{otherwise} \end{cases} \quad (3),$$

where b_{nk} is binary value in cluster k of feature n.

2.3 Neural network process

The structure of neural network process of ADCNF is shown in figure 4. The input binary values from previous process are used for classification by neural network.

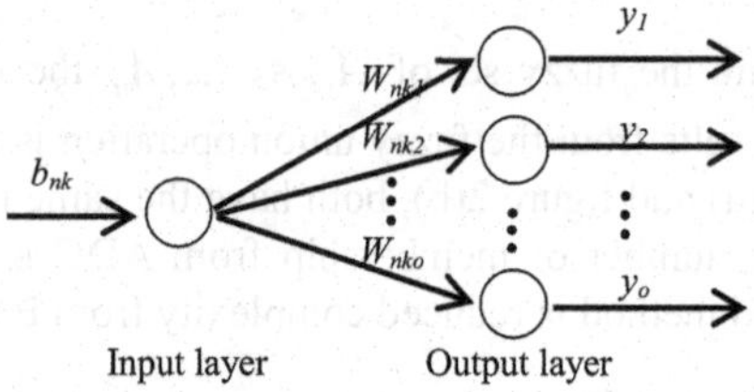

Fig. 4. The structure of neural network process

The numbers of input nodes are equal to the number of cluster from ADC. The number of output nodes are equal to number of class in dataset. The activation function is used Sigmoid in all output nodes. The calculate output value from equations as

$$y_o = 1/(1+e^{-V_o}) \quad (4),$$

where

$$V_o = \sum_{i=1}^{n}\sum_{j=1}^{k} b_{ij} w_{ijo} \quad (5).$$

The b_{ij} is the binary value, i represent feature and j represent the considered node. The w_{ijo} is weight of output node. The error of output node o is defined as

$$e_o(l) = d_o(l) - y_o(l) \quad (6),$$

where $e_o(l)$ is error of output node, $d_o(l)$ is desired output and $y_o(l)$ is output value. The delta value of output layer is defined as

$$\delta_o(l) = e_o(l)\varphi'(V_o(l)) \quad (7).$$

The weights are update by

$$w_{ijo}(l+1) = w_{ijo}(l) - \eta\delta_o(l)b_{ij}(l) \quad (8).$$

3 Complexity Evaluation

The result of the complexity evaluation between ADCNF and ENF computed from equation (9) of the UCI machine learning repository datasets [11] shown in table 1.

Table 1. The comparison of the complexity of ENF and ADCNF method in all datasets.

Dataset	Complexity		% of reduce from ENF method
	ENF	ADCNF	
Wine	1,254	**1,209**	3.59%
Breast Canner	488	**218**	55.33%
Iris	132	**123**	6.82%
Vote	1,162	**122**	89.50%
Zoo	1,960	**252**	87.14%
Ionosphere	2,374	**2,074**	12.64%
Seeds	660	**648**	1.82%
Pima	1,860	**1,280**	31.18%
Hayes	471	**210**	55.41%
Glass	3,094	**2,926**	5.43%
Liver Disorders	654	**500**	23.55%
Forest Fire	1,656	**1,316**	20.53%
Yeast	24,480	**10,010**	59.11%

The complexity evaluation between ADCNF and ENF for classification problem defined by the following conditions. The structures are used the number of clusters to generate membership function and transform to the binary value to use as input in neural networks. Considering the training process in neural networks. The equation use to compare the complexity is defined as

$$K_{dm} = \sum_{i=1}^{n} C_{im} * O_d \qquad (9).$$

In equation (9) d is represented dataset, m is represent method and i is represent the feature. The C_{im} is number of clusters and O_d is number of output nodes in dataset. It is clear to see from table 1 that the proposed ADCNF can reduce the complexity from ENF method. The reduction percentage depend on the number of the function that overlapped be mean value.

4 Feature Selection and Rule Extraction

The results from the learning process of ADCNF can be applied as feature selection by sorting weights in ascending order. All weights that have a value less than 0 are eliminated, because the input value is binary (0, 1) the eliminated weights are unimportant. All features that weights are not eliminated are selected. The selected features can use by the ADCNF algorithm for classification problem with desirable performance as displayed in table 2. The experimental observation is that the selected features are the feature that has high numbers of clusters

Algorithm in [6] proposed method to select weights for creating the classification rules, but not have algorithm to define the appropriate linguistic number. In [9] proposed the GSS to search for minimum or maximum value of continue function. The application of GSS in [10-11] showed that the computation time and the search speed are reduced. The concept of GSS is applied for select the number of weight in our experiments since it has less complexity than other GA algorithms. First define the boundary of the answer as $[L,U]$, that is the summation of 0 with all weights. Then the next point calculates by

$$d = 0.618 * (U - L) \tag{10}$$

$$x_1 = L + d \tag{11}$$

$$x_2 = U - d \tag{12}$$

The next step, considering our binary input at a hidden level of neural network. The weight at position $[x_1, x_2]$ those have been rounded up are considering. All weights at each position those have binary input equal to 1 are summed, define as following

$$f(x) = \sum_{j=1}^{m} A_j, A = \begin{cases} 1, & \text{if } b_{nk} = 1 \\ 0, & \text{otherwise} \end{cases} \tag{13}$$

The previous processes in equation (10) - (13) are iterative until $f(x_1) \neq f(x_2)$. Otherwise calculate the new position by consider if $f(x_1) \geq f(x_2)$, then set $U = x_1, L = L$ else if $f(x_1) < f(x_2)$, then set $U = U, L = x_2$. The final value of $[x_1, x_2]$ is the boundary of answer. The number of weights the values x_1 is use as number of linguistic in rule extraction process.

5 Experimental Methods and Results

5.1 Classification results from direct calculation

The experiments has done on 13 standard datasets from UCI to compare the performance of the ADCNF, ADCNF with feature selection and the ENF [7]. The 10 fold cross validation is applied to verify the performance, the result shown in table 2. Clearly see in the table 2 that both ADCNF have better than ENF. The results of classification of direct calculation show that 11 datasets classify by ADCNF are better than ENF. The classification results of the ADCNF feature selection show that 10 datasets are better than ENF, where 5 datasets have the highest among 3 methodologies.

Table 2. Comparison results of the proposed ADCNF and ENF by direct calculation.

Dataset	Average accuracy of 10 fold cross validation		
	ENF[7]	ADCNF	ADCNF feature selectioı
Wine	97.19%	**99.44%**	**100.00%**
Breast Canner	97.13%	**97.14%**	**97.14%**
Iris	96.67%	96.67%	96.67%
Vote	95.18%	**96.78%**	**96.32%**
Zoo	**94.27%**	**97.03%**	94.06%
Ionosphere	92.04%	**93.73%**	**94.02%**
Seeds	88.57%	**93.81%**	**92.86%**
Pima	**86.85%**	**75.26%**	72.14%
Hayes	74.23%	**86.36%**	**76.52%**
Glass	67.25%	**75.23%**	**74.77%**
Liver Disorders	65.54%	**69.86%**	**66.96%**
Forest Fire	55.50%	**59.97%**	**60.93%**
Yeast	51.08%	**51.89%**	**51.95%**

5.2 Classification results of rule extraction

The experiments of rule based classification results by ADCNF compare with the NF method [6] are displayed in table 3. The average accuracy of 10 fold cross validation show that the ADCNF rule extraction of 9 datasets is higher than NF method.

Table 3. Comparison results of the proposed ADCNF and NF by classification rule.

Dataset	Average accuracy of 10 fold cross validation	
	NF[6]	ADCNF
Wine	95.49%	**98.88%**
Breast Canner	**96.99%**	95.71%
Iris	96.00%	**96.67%**
Vote	95.18%	**96.32%**
Zoo	**94.18%**	93.07%
Ionosphere	**93.17%**	90.88%
Seeds	88.57%	**93.33%**
Pima	**85.16%**	74.48%
Hayes	61.26%	**68.94%**
Glass	56.67%	**61.68%**
Liver Disorders	67.91%	**69.86%**
Forest Fire	58.23%	**60.46%**
Yeast	19.20%	**37.33%**

6 Conclusion

The new propose algorithm an adaptive dynamic clustering of neuro-fuzzy system shows the high classification performance. The total average accuracy rate is higher than other high performance Neuro-fuzzy methods [5-7], those have been proposed earlier. The ADCNF shown the reduction of complexity and high performance than [7]. The GSS algorithm is applied to select the proper number of linguistic for rule base classification. The performance of rule base classification of ADCNF is better than the method proposed in [6].

References

1. Haykin, S.: Neural Networks and Learning Machines. Prentice Hall, New York (2008).
2. Badiru, A.D., Cheung, J.Y.: Fuzzy Engineering Expert System with Neural Network Applications. John Wiley, New York (2002).
3. Kriesel, D.: A Brief Introduction to Neural Networks. Retrieved August, 15, 2011, pp. 37-124 (2007)
4. Chakraborty, D., Pal, N.R.,: A Neuro-Fuzzy Scheme for Simultaneous Feature Selection and Fuzzy Rule-Based Classification. In IEEE Transactions on Neural Network, Vol., 15, NO. 1, January 2004, pp 110-123, (2004).
5. Eiamkanitchat, N., Theera-Umpon, N., Auephanwiriyakul, S.: A Novel Neuro-Fuzzy Method for Linguistic Feature Selection and Rule-Based Classification. In: The 2nd International Conference on Computer and Automation Engineering (ICCAE), pp. 247-252. IEEE Press (2010).
6. Eiamkanitchat, N., Theera-Umpon, N., Auephanwiriyakul, S.: Colon Tumor Microarray Classification Using Neural Network with Feature Selection and Rule-Based Classification. In: Zeng Z., Wang J. (eds.). LNEE, Vol. 67, pp. 363-372. Springer, Heidelberg (2010).
7. Wongchomphu, P. Eiamkanitchat, N.: Enhance Neuro-Fuzzy System for Classification Using Dynamic Clustering. In: The 4th Joint International Conference on Information and Communication Technology, Electronic and Electrical Engineering (JICTEE), pp. 1-6. IEE Press (2014).
8. J. Kiefer.: Sequential minimax search for a maximum. Proceedings of the American Mathematical Society, vol. 4, pp. 502–506, (1953).
9. Zuo, Q., Yin, X., Zhou, J. ,Kwak, BJ., Chung, K.: Implementation of Golden Section Search Method in SAGE Algorithm. In: Proceedings of the 5th European Conference on Antennas and Propagation (EUCAP), pp 2028-2032, (2011).
10. Yeam, DH., Park, J.b., Joo, Y.H.: Selection of coefficient for Equalizer in Optical Disc Drive by Golden Section Search. IEEE Transactions on Consumer Electronics, Vol. 56, No. 2, May (2010).
11. Bache, K, Lichman, M, (2013). UCI Machine Learning Repository [http://archive.ics.uci.edu/ml]. Irvine, CA: University of California, School of Information and Computer Science.

A Hybrid Approach of Neural Network and Level-2 Fuzzy set

Jirawat Teyakome [1], Narissara Eiamkanitchat [1, 2]

[1] Department of Computer Engineering, Faculty of Engineering,
[2] Social Research Institute,
Chiang Mai University, Thailand
jirawat_te@cmu.ac.th, narisara@eng.cmu.ac.th

Abstract. This paper presents a new high performance algorithm for the classification problems. The structure of A Hybrid Approach of Neural Network and Level-2 Fuzzy set, including two main processes. The first process of this structure is the learning algorithm. This step applied the combination of the multilayer perceptron neural network and the level-2 fuzzy set for learning. The outputs from learning process are fed to the classification process by using the K-nearest neighbor. The classification results on standard datasets show better accuracy than other high performance Neuro-Fuzzy methods.

Keywords: Neural Network, Level-2 Fuzzy set, Neuro-Fuzzy

1 Introduction

The development of the better classification algorithm has continued interest by many researchers. Classification algorithms are applied in numerous areas of work. There are many popular classification algorithms such as neural network, decision tree and fuzzy logic. Each algorithm has different advantage and disadvantage in data classification. This paper proposes the new methodology that combines 3 algorithms including neural network, Fuzzy logic and k-nearest neighbor. The Hybrid Approach of Neural Network and Level-2 Fuzzy set result the better accuracy of classification.

The neural network is one of the best performance classification tools [1-3]. The structure is derived from the human brain. An algorithm of learning process can use the error to update the structure for better performance. Although the neural network is robust for classifying the unknown input and has good classification results. But the flaw is the structure cannot explain in natural language for human.

The fuzzy system is a technique that can deal with the uncertainty information by using the fuzzy set. However the results of fuzzy membership value also have some uncertainty information. The level-2 fuzzy set is the technique to dealing with this problem [4-8].

The evolution of neuro-fuzzy show that this algorithm can result the better performance in classification problem than many other popular algorithms. The algorithm proposed in [9-10] using fuzzification method with original information

K.J. Kim (ed.), *Information Science and Applications*,
Lecture Notes in Electrical Engineering 339, DOI 10.1007/978-3-662-46578-3_86

and divide into 3 membership functions. The results from fuzzification process that used for classification are better than using the original value. The method of dividing the membership function using incremental clustering has developed and presented in [11]. Original feature is regrouped to obtain the number of membership functions. The results are improved and confirm that the proper number of fuzzy membership functions is one thing that makes classification more effective. The algorithm proposed in [12] show another alternative neuro-fuzzy algorithm for classification. This algorithm learning the information with neural network. The error of output from neural network are retrained using mandani fuzzy method. Finally, apply the K-nearest neighbor to classification the fuzzy output. The neuro-fuzzy algorithm modified from [12] used for recognition and classification of infant cry in [13].

2 Structure of Hybrid Approach of Neural Network and Level-2 Fuzzy set

In this paper presents the new development neuro-fuzzy algorithm for classification. In Fig.1 shows overall structure of the Hybrid Approach of Neural Network and Level-2 Fuzzy set (HANN-L2F) algorithm.

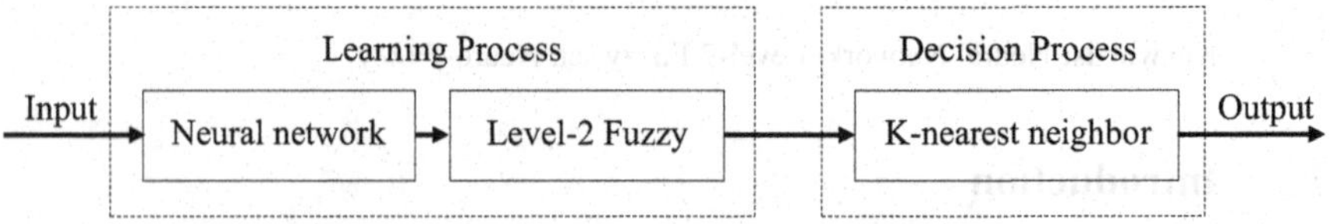

Fig. 1. Structure of Hybrid Approach of Neural Network and Level-2 Fuzzy set (HANN-L2F)

The structure including two parts that is learning process and decision process. The learning process is a combination of 2 algorithms, neural network and fuzzy system. The neural network gets input into the process for learning. After that the result will feed forward to create the level-2 fuzzy set. In the decision process of structuring, KNN is used to define the classification result.

2.1 Learning Process

The first step in this section use 3 layer neural network for learning. The outputs from neural network are preprocess by using Min-Max normalization method, before fed to fuzzification in the next step. The level-2 fuzzy set is used to handle the uncertainty of neural network results. An incremental clustering with class applied from [11] is used to find the reasonable number to create membership function. Membership values output from level-2 fuzzy set are fed to decision processes in the next step.

Neural network part. The structure from Fig.2 have three parts include input layer, hidden layer and output layer. The number of nodes in the input layer is equal to attribute size of the dataset. The number of nodes in the hidden layer is assigned to 3 nodes. Finally, output nodes in the output layer are equal to class size from data set. All weights and bias parameter are randomly initialized. The back propagation algorithm is applied in the structure.

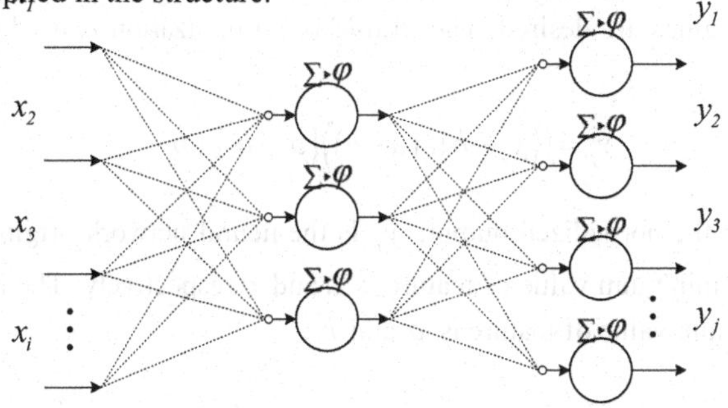

Fig. 2. Structure of Neural Network in Learning Process

Every nodes use log sigmoid as activation function. The output from each node calculate by

$$y_j = 1/\left(1+e^{-v_j}\right) \tag{1}$$

where
$$v_j = \sum_i w_{ij} x_i + b_i \tag{2}$$

In equation (1) and (2), j is represents the node in each layer and i is represents the number of input feature relevant to the considered node. In (2) w_{ij} is all weight parameters in the considered layer. The back propagation algorithm is used to update the structure parameters from learning process by using the value of undesired output error. This error will be used to find the gradient for update weights and bias can be defined as

$$e_j = d_j - y_j \tag{3}$$

In the equation (3), d is the desired output. The gradients used for update weights and biases for output layer and the hidden layer are displayed in equation (4) and (5) respectively. These equations are as follows

$$\delta_j(n) = e_j(n)\varphi_j'(v_j) \tag{4}$$

$$\delta_j(n) = \varphi_j'(v_j)\sum_i \delta_i w_{ij} \tag{5}$$

The weights are updated by

$$w_{ij}(n+1) = w_{ij}(n) + \Delta w_{ij}(n) \tag{6}$$

where $$\Delta w_{ij}(n) = \eta \delta_j(n) y_j(n) \tag{7}$$

Before the outputs are sent to create fuzzy membership value, the data are normalized. Although classify by neural network results a good performance, but does not mean the values are desired. The Min-Max normalization is used for preprocess output as

$$y_j' = \left(\left(y_j - b\right)/\left(a - b\right)\right)\left(a' - b'\right) + b' \tag{8}$$

In (8) y_j' is the normalized output, y_j is the neural network original output. The maximum and minimum value of feature is a and b respectively. The new maximum and new minimum value of feature is a' and b'.

Level-2 Fuzzy set part. This step uses the level-2 fuzzy set to transfer normalized data from the previous section to fuzzy value. The structure of this part is displayed in Fig. 3.

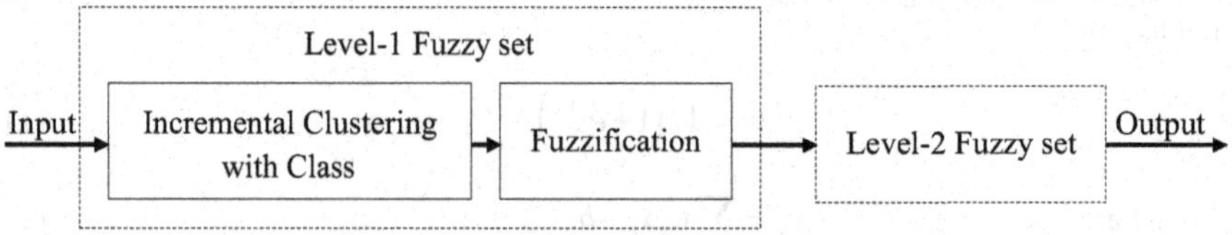

Fig. 3. Structure of Level-2 Fuzzy set process

In order to consider the uncertainty of classification results from neural network, the gaussian membership function is applied. The incremental clustering use class is applied to determine the level-1 membership function number. The incremental clustering use class algorithm is as follows.

1. **Assume** [x_i , c_i , c_i'] where x_i is input , c_i is original class and c_i' is new class.
2. Sort rows by x_i
3. Assign $class \leftarrow 1$, Let c_1' equal to $class$.
4. **For** $i \leftarrow 2$ to n
5.   **If** c_i is equal c_{i-1}
6. $c_i' \leftarrow class$
7.   **Else**
8. $class \leftarrow class + 1$
9. $c_i' \leftarrow class$
10.   **End if**
11. **End for**

The fuzzification process of level-1 fuzzy set used y'' as input of the gaussian membership function. The membership value of the level-1 fuzzy set is as follows

$$\alpha_k = \left(u_{\alpha(k)}\left(y_k''\right) / y_k'' \right) \tag{9}$$

where
$$u_\alpha\left(y''\right) = e^{\left(\left(y''-c\right)^2/2\sigma^2\right)} \tag{10}$$

Where y'' is the input, c is mean and σ is standard deviation. From (10) is the fuzzy set equation. Assign α_k is membership fuzzy set. The example of the class scatter plot and membership function graphs of Seeds dataset from level-1 fuzzy process is displayed in Fig.4.

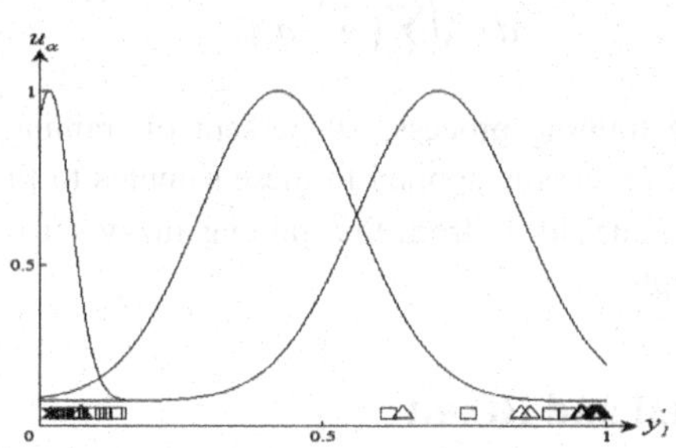

Fig. 4. Example of the scatter plot and membership function graphs of class in Seeds dataset

The seeds dataset in Fig. 4 has 3 classes the different symbols represent to each class. Between each class has interrupted with class. The uncertainty is clearly observing since between each class has interrupted with other classes. The level-2 fuzzy set is applied to handle this uncertainty problem. The results from level-1 Fuzzy set are re-fuzzification using gaussian membership function. The example of the level-2 fuzzy set is demonstrated in Fig 5.

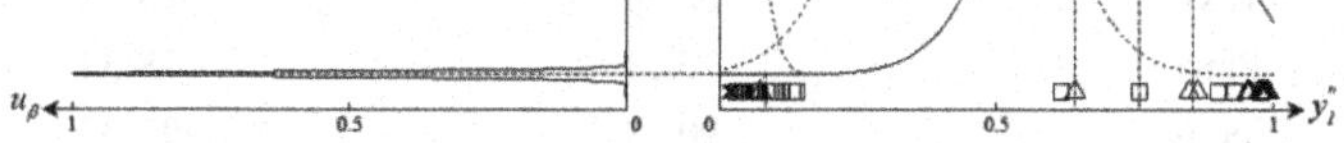

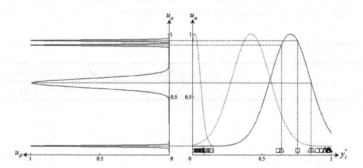

Fig. 5. Example clustering results and membership graph of the Level-2 fuzzy set.

The membership value of level-2 fuzzy set is defined by

$$\beta_k = \left(u_{\beta(k)}\left(u_{\alpha(k)}\right) / u_{\alpha(k)} \right) \tag{11}$$.

From equation (11) β is a finally training set from learning process. This value is used to classification in decision process. The $u_{\alpha(k)}$ is the membership value of level-1 fuzzy, which use to calculate the value of the level-2 fuzzy set.

2.2 Decision Process

The k-nearest neighbor is applied with the membership value from level-2 fuzzy to identify the class of classification process. The euclidean distance is used as distance function as displayed

$$d=\sqrt{\sum(p_i-q_i)^2} \quad (12)$$

After finishing the training process, 10 percent of training sample is randomly selected. The decision process is applied to these samples to find the best k of each data set. The best k equal to k from the training fuzzy set is used in the decision process of testing sample.

3 Experimental and Result

The performance measure results for algorithm in this paper. Using 10 datasets from UCI [14]. Each dataset has different in feature and details. The propose HANN-L2F is verified the performance by comparing the classification result with the popular neural network and the 2 high accuracy algorithms of neuro-fuzzy [10-11].

3.1 Experimental result

The experimental results displayed in table 1 compare the results from HANN-L2F with the popular classification algorithm neural network.

Table 1. The result from the propose HANN-L2F compare with Neural network

Dataset	Average Accuracy of 10-fold cross validation	
	Neural Network	HANN-L2F
Breast Cancer	96.77	**97.23**
Glass	61.38	**68.78**
Pima	76.23	**78.31**
Seeds	90.48	**93.33**
Zoo	96.09	**98.09**
Average	84.19	**87.15**

The neural network that used in the experiment has the same structure as the one used in our proposed neural network. The experimental results displayed in table 2 compare with NF[10] and ENF[11], those have verified their high performance with other algorithms. Accuracy displayed in both tables are the average 10-fold cross validation. Table 1, result from all 5 datasets show the higher accuracy of HANN-L2F than neural network those have the same structure.

Table 2. The result from HANN-L2F compare with Neuro-fuzzy[10] and ENF[11]

Dataset	Average Accuracy of 10-fold cross validation		
	NF[10]	ENF[11]	HANN-L2F
Hayes	61.26	74.23	**78.13**
Iris	96.00	96.67	**98.67**
Liver	67.91	65.54	**74.86**
Vote	95.18	95.18	**97.27**
Wine	95.49	97.19	**97.78**
Average	83.17	85.76	**89.34**

The algorithm NF[10] and ENF[11] verify their high performance comparing with the other popular classification algorithm. The classification results in table 2 show that our propose algorithm HANN-L2F has better accuracy than NF and ENF in all datasets. The accuracy of Hayes and Liver are drastic improve by our propose classification algorithm.

4 Conclusion

The Hybrid Approach of Neural Network and Level-2 Fuzzy set show higher classification accuracy. This algorithm used key feature from neural network and level-2 fuzzy system for learning and then used k-nearest neighbor for classification. The experimental results show that Hybrid Approach of Neural Network and Level-2 Fuzzy set can deal with uncertainty output from neural network to improve the accuracy.

Reference

1. Haykin, S. S.: Neural Networks and Learning Machines (3rd Ed.). Prentice Hall, New York (2008).
2. Kriesel, D.: A Brief Introduction to Neural Networks. Retrieved August, 15, 2011, pp. 37-124 (2007).
3. Han, J., Kamber, M., & Pei, J.: Data mining: Concepts and Techniques. pp. 325-370. Morgan Kaufmann, SanFancisco (2006).
4. Dubios , D., & Prade , H.: The Three semantics of fuzzy sets, Fuzzy sets and systems, 90(2), pp. 141-150 (1997).
5. Dubios , D., & Prade , H.: Fuzzy sets and systems theory and applications, Academic Press.

6. Gandhi, S.K., & Nimse, S.B.: A Comparative study of Level II Fuzzy sets and Type II Fuzzy sets. In: International Journal of Advanced Research in Computer Engineering & Technology (IJARCET), pp. 215-219. IEEE Press (2012).
7. Gandhi, S.K.: Level 2 Fuzzy Relations in Database Modeling. In: IBMRD's Journal Management and Research, pp. 129-135.
8. Gandhi, S.K.: Emotional Intelligence in Robots Using Level 2 Fuzzy Sets. In: International Journal of Recent Trends in Electrical & Electronics Engg. pp. 89-92.
9. Eiamkanitchat, N., Theera-Umpon, N., & Auephanwiriyakul S.: Colon Tumor Microarray Classification Using Neural Network with Feature Selection and Rule-Based Classification. In: Zeng Z., Wang J. (eds.). LNEE, Vol. 67, pp. 363-372. Springer, Heidelberg (2010).
10. Eiamkanitchat, N., Theera-Umpon, N., & Auephanwiriyakul S.: A Novel Neuro-Fuzzy Method for Linguistic Feature Selection and Rule-Based Classification. In: The 2nd International Conference on Computer and Automation Engineering (ICCAE), pp. 247-252. IEEE Press (2010).
11. Wongchomphu, P. Eiamkanitchat, N.: Enhance Neuro-Fuzzy System for Classification Using Dynamic Clustering. In: The 4th Joint International Conference on Information and Communication Technology, Electronic and Electrical Engineering (JICTEE), pp. 1-6. IEE Press (2014).
12. Saetern, K., Eiamkanitchat, N.: An Ensemble K-Nearest Neighbor with Neuro-Fuzzy Method for Classification. In: The 10th International Conference on Computing and Information Technology (ICCIT), pp. 43-51. (2014).
13. Srijiranon., K., Eiamkanitchat, N.: Application of Neuro-fuzzy approaches to recognition and classification of infant cry. In: IEEE TENCON. (2014) [Accepted to appear].
14. Bache, K. & Lichman, M. (2013). UCI Machine Learning Repository [`http://archive.ics.uci.edu/ml`]. Irvine, CA: University of California, School of Information and Computer Science.

Developing Term Weighting Scheme based on Term Occurrence Ratio for Sentiment Analysis

Nivet Chirawichitchai [1]

[1] Faculty of Information Technology, Sripatum University, Thailand.
nivet99@hotmail.com

Abstract. Term weighting is an important task for sentiment classification. Inverse document frequency (IDF) is one of the most popular methods for this task; however, in some situations, such as supervised learning for sentiment classification, it doesn't weight terms properly, because it neglects the category information and assumes that a term that occurs in smaller set of documents should get a higher weight. In this paper, I purpose sentiment classification framework focusing on the comparison of various term weighting schemes, including Boolean, TF, TFIDF and a novel term weighting (TOW). I have evaluated these methods on Internet Movie Database corpus with four supervised learning classifiers. I found TOW weighting most effective in our experiments with SVM NB and NN algorithms. Based on our experiments, using TOW weighting with SVM algorithm yielded the best performance with the accuracy equaling 93.45%.

Keywords: Sentiment classification, Term weighting, Term Occurrence

1 INTRODUCTION

In recent years, we have seen an exponential growth in the volume of text documents available on the Internet. While more and more textual information is available online, effective retrieval is difficult without organization and summarization of document content. Text classification is one solution to this problem and very important and can be used in many applications, such as text filtering, documents organization, opinion mining and sentiment classification. It consists in assigning labels to documents based on a set of labeled documents called training set. For classification, the documents are firstly represented as vectors in a high-dimension vector space[1-3]. Each dimension corresponds to the value of a feature, and here, a term. Then classifiers such as SVM DT and NB will work on these vectors, by examining the similarities to classify them to the right categories. Studies have shown that these three classifiers significantly outperform others for sentiment classification tasks. How to decide the importance of every term, represented as term weighting, is critical to the performance of the sentiment classification result. In this research, I try to explore more information from the term distribution among different categories. I propose a novel term weighting scheme (TOW) that not only considers the frequency of documents that contains term , but also considers the number of times that *t* occurs in different documents of different categories. The experiments show

K.J. Kim (ed.), *Information Science and Applications,*
Lecture Notes in Electrical Engineering 339, DOI 10.1007/978-3-662-46578-3_87

that our method really performs well. The rest of the paper is organized as follows. Section 2 describes the feature extraction methods. Section 3 describes the dimensionality reduction. Section 4 describes the classification algorithms for empirical validation. Section 5 presents sentiment experiments and results. Finally, section 6 conclusions.

2 FEATURE EXTRACTION

The first step in sentiment classification is to transform documents, which typically are strings of characters, into a representation suitable for the learning algorithm and the classification task.

2.1 Vector Space Model

The role of text classification is to assign a document into one of a number of pre-defined categories. Normally we wish this process to be automated -- a computer program inspects the document and decides on the category without any manual involvement. Furthermore, in supervised machine learning approaches, I ask the computer program automatically to learn the classification function from a number of labeled example documents. To do this, a representation of the document in a form understandable by a computer is necessary. This process of translating a document from text into a feature vector representation is called Document Indexing [3-5].

2.2 Document Indexing

Creating a feature vector or other representation of a document is a process that is known in the IR community as indexing. There are a variety of ways to represent textual data in feature vector form, however most are based on word co-occurrence patterns. In these approaches, a vocabulary of words is defined for the representation, which are all possible words that might be important to classification. This is usually done by extracting all words occurring above a certain number of times, and defining your feature space so that each dimension corresponds to one of these words. When representing a given textual instance, the value of each dimension is assigned based on whether the word corresponding to that dimension occurs in the given textual instance. If the document consists of only one word, then only that corresponding dimension will have a value, and every other dimension will be zero. This is known as the ``bag of words" approach [3-5].

2.3 Term weighting

In the vector space model, documents are represented by vectors of words. Term weighting method aims to indicate the significant of a term in a document. In sentiment classification, Boolean is simplest approach to let the weight be 1 if the word occurs in the document and 0 otherwise. TF and TF-IDF are widely applied to count the weight of a term. TF represents the number of times a term occurs in a document, and TF-IDF is the combining of TF and IDF weights. IDF indicates the general importance of a term in overall documents.[3] If a term's score of TF-IDF is high, it means this term occurs frequently and only appears in the part of overall documents. IDF and TF-IDF can be calculated as equations (1) and (2) [3-6]

$$idf = \frac{\text{the number of total documents}}{\text{the number of documents include a term}} \tag{1}$$

$$TFIDF = tf * idf \tag{2}$$

2.4 Term Occurrence Weighting Scheme

The section 2.3 summaries three traditional weighting scheme with term frequency components which were used in my experiments. The traditional weighing representation scheme might lose its ability to discriminate these positive documents from the negative ones. The normal raw term frequency has been widely used. The logarithm operation is used to scale the effect of unfavorably high term frequency in one document. Inspired by the inverse document frequency, which represent multipliers of a inverse collection frequency factor (idf). The traditional idf component which was thought to improve the term's discriminating power by pulling out the relevant documents from the irrelevant documents was borrowed from the information retrieval domain. I will validate whether this has more discriminating power than idf factor in the later experiments. The weighting representation is probability factor based on the analysis of term's distribution. I call a new formula is "Term Occurrence Weighting" (TOW) to improve efficiency the Sentiment Analysis classification. TOW is based on probabilistic theory ideas from quantitative analysis of large sets of data and investigate term occurrence ratio from the dataset between group, an than I assigned the constant value 2 in the formula because to improve of discriminating power of class. In the TOW, the weight for word i in document k is given by:

$$TOW_{ik} = \log(2 \times tf_{ik}) \times \log\left(2 + \frac{\begin{array}{c}\textit{the number of documents}\\ \textit{include a term positive}\end{array}}{\begin{array}{c}\textit{the number of documents}\\ \textit{include a term negative}\end{array}}\right) \tag{3}$$

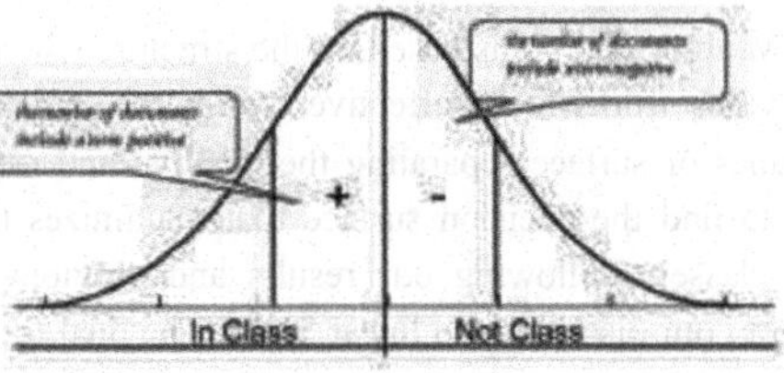

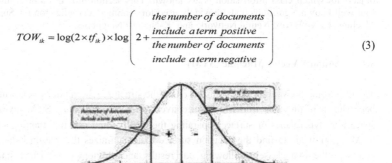

Fig. 1. Distribution of Term Occurrence Weighting

3 DIMENSIONALITY REDUCTION

With increasing of the textual data in cyberspace, how to extract significant information from a huge amount of data have been become a serious problem. The objective of feature selection is to extract the important terms in the documents, and achieve the goal of dimension reduction. Emotion classification can be considered as a classification approach in machine learning in which features are words or terms extracted from a given text corpus. A problem in statistical sentiment classification is the high dimensionality of the feature space. There exists one dimension for each unique word found in the collection of documents, typically ten thousand. Hence, there is a need for a reduction of the original feature set, which is commonly known as dimensionality reduction in the pattern recognition literature. Most of the dimensionality reduction approaches can be classified into feature selection. Therefore, we applied feature selection technique by chi-square method. [7] The x^2 statistic measures the lack of independence between word w and class c_j. It is given by:

$$x^2(w,c_j) = \frac{N(ad-cb)^2}{(a+c)(b+d)(a+b)(c+d)} \tag{4}$$

Here a is the number of documents from class c_j that contains word w, b is the number of documents that contains w but does not belong to class c_j, c is the number of documents from class c_j that does not contain word w, and d is the number of documents that belongs to class c_j nor contains word w. N is still the total number of documents. Two different measures can be computed based on the chi-square statistic. [8]

4 CLASSIFICATION ALGORITHMS

The goal of classification is to build a set of models that can correctly predict the class of the different objects. Once such a predictive model is built, it can be used to predict the class of the objects for which class information is not known. This section provides a brief introduction to four well-known algorithms that are widely used for sentiment classification i.e. Support Vector Machine, Naive Bayes, Neural Network and K-Nearest Neighbors.

4.1 Support Vector Machine (SVM)

The idea of the SVM algorithm [8] is based on the structure risk minimization principle .It has been shown in previous works to be effective for text classification. SVM divides the term space into hyperplanes or surface separating the positive and negative training samples. The SVM algorithm is to find the decision surface that maximizes the margin between the data points of the two classes. Following our results and previously published studies in text classification, we limit our discussion to linear SVM. The dual form of the linear SVM optimisation problem is to maximize :

$$\alpha^* = maximise_\alpha \sum_{i=1}^{l} \alpha_i - \frac{1}{2}\sum_{i=1}^{l}\sum_{j=1}^{l} y_i y_j \alpha_i \alpha_j (x_i x_j), \tag{5}$$

$$\text{Subject to } \sum_{i=1}^{l} y_i \alpha_i = 0, 0 \le \alpha_i \le C, i = 1...l$$

with α_i the weight of the examples and C the relative importance of the complexity of the model and the error.

4.2 Naive Bayes (NB)

The idea of the NB classier [9] is to use a probabilistic model of text. To make the estimation of the parameters of the model possible, rather strong assumptions are incorporated. In the following, word-based unigram models of text will be used, i.e. words are assumed to occur independently of the other words in the document. Let $P(y_i)$ be the prior probability of the class y_i and $P(a'_j \mid y_i)$ be the conditional probability to observe attribute value a'_j given the class y_i. Then, a naive Bayes classifier assign to a data point x' with attributes $(a'_1....a'_d)$ the class $\hat{\Phi}(x')$ maximizing :

$$\hat{\Phi}(x') = argmax_{y_i \in c} P(y_i) \prod_{j=1}^{d} P(a'_j \mid y_i) \tag{6}$$

4.3 Neural Network (NN)

The idea of the Neural network classifiers [10] are most important for the classification. The neural networks have the advantages of self adaptive method. It means adjusting the weight themselves to the data without any specification and also it should be have the arbitrary accuracy. There are several types of neural networks are used for the classification task. There are feed forward multilayer networks and multilayer perceptrons are mostly used for neural network classifiers. Multilayer perceptrons have been applied successfully to solve many problems using the algorithm called Error back propagation algorithm. It has two passes. There are forward and backward pass. Feed forward back propagation neural network have the signal flow through the forward direction. Neural networks are very competitive to traditional classifiers for solving the classification problems. The basic unit of the neural network layer is neuron unit. Each unit receives a set of inputs called X_i and associated set of weights W, corresponding to the term frequencies in the i document. P_i is the prediction function. The simplest form of neural network is perceptron. It is the linearly separable for classification in the neural network. The linear function is denoted as

$$P_i = W.X_i \tag{7}$$

4.4 K-Nearest Neighbors (KNN)

The idea of the KNN algorithm [11] is a method for classifying objects based on closest training examples in the feature space. KNN classifier is a case-based learning algorithm that is based on a distance or similarity function for pairs of observations, such as the Euclidean distance or Cosine similarity measure's. This method is try for many application Because of its effectiveness, non-parametric and easy to implementation properties, however the classification time is long and difficult to find optimal value of k .The KNN algorithm is amongst the simplest of all machine learning algorithms: an object is classified by a majority vote of its neighbors, with the object being assigned to the class most common amongst its k nearest neighbors (k is a positive integer, typically small). If k = 1, then the object is simply assigned to the class of its nearest neighbor. The best choice of k depends upon the data; generally, larger values of k reduce the effect of noise on the classification, but make boundaries between classes less distinct.

5 SENTIMENT CLASSIFICATION FRAMEWORK AND RESULTS

The input comes from Internet Movie Database (IMDB) pre-classified into a set of categories. The IMDB are first pre-processed by the text processing module, I must first apply a word segmentation to tokenize text strings into series of terms, in this Bag-Of-Words approach. Once a set of extracted words are obtained from the training corpus, the removal of HTML tags, stop words, and words stemming from the dictionary begins. The output from this step will be used in the various term weighting schemes to assign the feature values as described in Section II. We reduce the number of word features by applying the chi-square feature selection technique as described in Section III, by using the statistics ranking metric for feature selection, the top p features per category were selected from the training sets. In my experiments, I set p = {50, 100, 200,300 , 400,500 ,600 ,700 ,800,900 , 1000, 1100, 1187 (Full feature)} respectively. Figure 2 illustrates the my research framework. I performed experiments using a data set of classified movie reviews prepared by Pang and Lee [1-2]. This data set contains 2000 movie reviews: 1000 positive and 1000 negative. The reviews were originally collected from the Internet Movie Database (IMDB). Their classification as positive or negative is automatically extracted from the ratings, as specified by the original reviewer. Only reviews where the author indicated the movie's rating with either stars or some numerical system were included. No single author could have more than 20 reviews in the original data set. I used WEKA an open-source machine learning tool, to perform the experiments. Then, I used the default settings for all algorithms, classification effectiveness is usually measured using accuracy. I tested all algorithms using the 10-fold cross validation. The results in terms of accuracy are the averaged values calculated across all 10-fold cross validation experiments. The experimental results of these four term weighting scheme with respect to accuracy on Internet Movie Database corpus in combination with three learning algorithms are reported from Figure 2 to Figure 5.

Figure 2, summarizes the Sentiment Classification on the chi-square feature selection method results using SVM algorithms on Internet Movie Database corpus after feature weighting via Boolean, TF, TFIDF, and TOW weighting, respectively. Six observations from the classification results follows. First, performance of the different term weighting schemes with a small feature size can not be summarized in one sentence but the trends are distinctive that the accuracy points of different term weighting schemes increase as the number of the features grows. Second, TOW weighting is more effective than another weighting with SVM. Third, TOW term weighting schemes reached a maximum of accuracy point at the full feature. Fourth, the best accuracy points on TOW weighting with SVM were 93.45 % at a feature size of 1187. Fifth, the TOW weighting scheme shown significantly higher performance than other techniques when the number of features was larger than 200. Finally, the TOW weighting scheme shown significantly higher performance than other techniques.

Figure 3, summarizes the Sentiment Classification on the chi-square feature selection method results using NB algorithms on Internet Movie Database corpus after feature weighting via Boolean, TF, TFIDF, and TOW weighting, respectively. Six observations from the classification results follows. First, performance of the different term weighting schemes with a small feature size can not be summarized in one sentence but the trends are distinctive that the accuracy points of different term weighting schemes increase as the number of the features grows. Second, TOW weighting is more effective than another weighting with NB. Third, TOW term weighting schemes reached a maximum of accuracy point at the full feature. Fourth, the best accuracy points on TOW weighting with NB were 92.25% at a feature size of 1100. Fifth, the TOW weighting scheme shown significantly higher performance than other techniques when the number of features was larger than 200. Finally, the TOW weighting scheme shown significantly higher performance than other techniques.

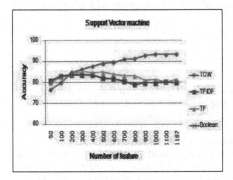

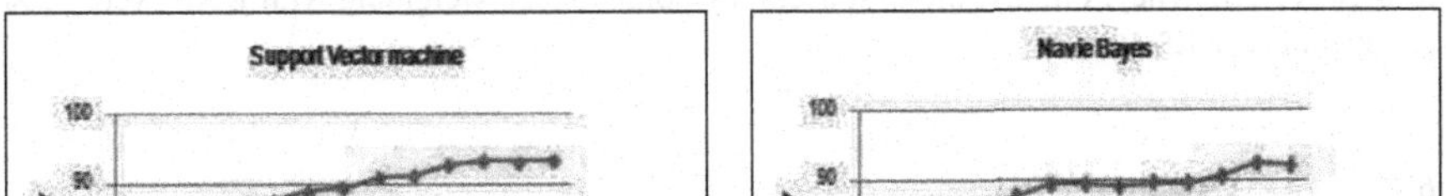

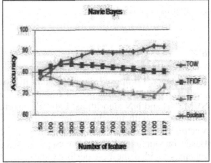

Fig. 2-3. Results of different weighting on SVM and NB algorithm.

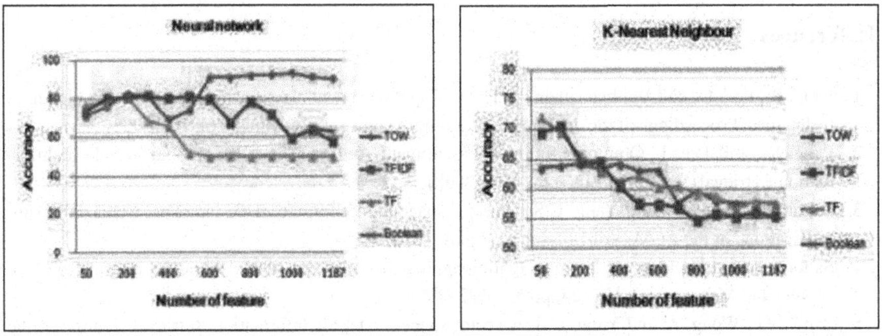

Fig. 4-5. Results of different weighting on NN and KNN algorithm.

Figure 4, summarizes the Sentiment Classification on the chi-square feature selection method results using NN algorithms on Internet Movie Database corpus after feature weighting via Boolean, TF, TFIDF, and TOW weighting, respectively. Five observations from the classification results follows. First, performance of the different term weighting schemes with a small feature size can not be summarized in one sentence but the trends are distinctive that the accuracy points of TOW term weighting schemes increase as the number of the features grows. Second, TOW weighting is more effective than another weighting with NN. Third, TOW term weighting schemes reached a maximum of accuracy point at the full feature. Fourth, the best accuracy points on TOW weighting with NN were 92.5% at a feature size of 1000. Finally, the

TOW weighting scheme shown significantly higher performance than other techniques when the number of features was larger than 600.

Figure 5, summarizes the Sentiment Classification on the chi-square feature selection method results using KNN algorithms on Internet Movie Database corpus after feature weighting via Boolean, TF, TFIDF, and TOW weighting, respectively. Three observations from the classification results follows. First, TOW weighting is not more effective than another weighting with KNN, but all term weighting have a similar performance. Second, the all term weighting scheme the reduction of the feature size does not affect the accuracy but another contribution to improve efficiency. Finally, the best accuracy points on TF weighting with KNN were 72% at a feature size of 50.

6 CONCLUSIONS

In this paper, I proposed sentiment classification framework focusing on the comparison of various term weighting schemes. I found TOW weighting most effective in our experiments with SVM, NB and NN algorithms. I also discovered that the TOW weighting schemes is suitable for combination with the chi-square feature selection method. The TOW weighting with Support Vector Machine algorithm yielded the best performance with the accuracy over all algorithms. Based on our experiments, the Support Vector Machine algorithm with the TOW weighting schemes yielded the best performance with the accuracy of 93.45%. Our experimental results also reveal that feature weighting methods have a positive effect on sentiment classification.

References

1. Pang, B., Lee, L., and Vaithyanathan, S.:Thumbs up?: sentiment classification using machine learning techniques. Proceedings of the conference on natural language processing, Vol. 10. (2002)
2. Pang, B., and Lee, L.:Opinion Mining and Sentiment Analysis. Foundations and Trends in Information Retrieval, Vol.2, No.1–2, pp.1-135 (2008)
3. Salton, G., Wong, A.,and Yang, C.S.:A vector space model for automatic indexing. Magazine Communications of the ACM, Vol.18 Issue 11, pp.613-620 (1975)
4. Salton, G., and Buckley, C.: Term weighting approaches in automatic text retrieval. Information Processing and Management, Vol. 24, pp.513-523 (1988)
5. Salton, G., Wong, A.,and Yang, C.S.:A vector space model for Information Retrieval. Journal of the American Society for Information Science, Vol.18, pp.613-620 (1975)
6. Polettini, N.:The Vector Space Model in Information Retrieval Term Weighting Problem. Department of Information and Communication Technology. University of Trento (2004)
7. Nicholls, C., and Song, F.:Comparison of Feature Selection Methods for Sentiment Analysis. Advances in Artificial Intelligence, Lecture Notes in Computer Science, Vol. 6085, pp. 286-289 (2010)
8. Yang, Y., Xu, C., and Ren, G.:Sentiment Analysis of Text Using SVM. Electrical, Information Engineering and Mechatronics Lecture Notes in Electrical Engineering, Vol. 138, pp.1133-1139 (2012)
9. Lewis, D.:Naive bayes at forty: The independence assumption in information retrieval. Proceeding of the European Conference on Machine Learning. (1998)
10. R.Moraes , J.F.Valiati and Wilson P.:Document-level sentiment classification: An empirical comparison between SVM and ANN. Expert Systems With Applications, Vol. 40 (2013)
11. K. Jędrzejewski and M. Zamorski.:Performance of K-Nearest Neighbors Algorithm in Opinion Classification. Foundations of Computing and Decision Sciences, Vol. 38 (2013)

A Comparison of Artificial Neural Network and Regression Model for Predicting the Rice Production in Lower Northern Thailand

Anamai Na-udom[1,*] and Jaratsri Rungrattanaubol[2]

[1]Department of Mathematics, Faculty of Science, Naresuan University, Phitsanulok, Thailand.
anamain@nu.ac.th
[2]Department of Computer Science and Information Technology, Faculty of Science, Naresuan University, Phitsanulok, Thailand.
jaratsrir@nu.ac.th

Abstract. Lower Northern Thailand is one of the main regions which can produce the highest rice yield. If the emphasis is on producing the rice yield in order to meet the standard yield, then the key factors, such as characteristics of rice farm, rice seed types, cultivation period, quantity of fertilizer usage, number of seeds, must be clearly studied and understood. This paper studies factors influencing the rice products and develops a model to predict rice yield per rai that can support farmers to plan their rice farming in Lower Northern Thailand. The aim of this paper is to compare the prediction accuracy between two popular predictive techniques for modelling rice yield namely, artificial neural network (ANN) and Regression. Root mean square of error (RMSE) and mean absolute error (MAE) values are used to compare prediction accuracy of the predictive models. The result shows that ANN is superior over regression model in terms of prediction accuracy and it is flexible to develop.

Keywords: Predictive Model; Artificial Neural Network; Regression model; Rice product in Lower Northern Thailand.

1 Introduction

Over the past three decades, Thailand has been recognized as the largest rice exporter in the world. In 2013, the department of agricultural extension reported that over 65 million hectares of land have been used to grow the rice field. There were 3.1 million households with farmers cultivating rice crops. According to the export records, Thailand earned more than a billion baht from rice export. Hence it is very clear that rice productivity is the major source of income of the agricultural sector and the industrial sector which contributes the employment for several million households.

Normally Thai farmers cultivate rice twice a year, in rainy season and in summer time, respectively. Hence rice production in Thailand can be classified into 2 groups according to the season of cultivation. The first group is called major rice which cultivated during June to December and another group is called second rice which normally grown during summer period. According to the empirical studies, it has been observed that rice yield has been affected by two aspects: the environment and the

K.J. Kim (ed.), *Information Science and Applications*,
Lecture Notes in Electrical Engineering 339, DOI 10.1007/978-3-662-46578-3_88

farmer's practice [1, 2]. Environmental factors include characteristics of cultivation area, amount of water, climate etc. The farmer's practice consists of selection of rice grain, selection of the appropriate method of rice cultivation area, and choice of fertilizer.

The lower northern Thailand consists of 8 provinces and it is suitable for rice cultivation as the landscape is very rich and moisture. Approximately 16.8 of cultivation are situated in the irrigation areas. It was reported that lower northern Thailand can produce the most major rice comparing to other parts in Thailand. Though the farmers in this part can cultivate rice through the year but it has been observed that the current yield of rice per rai is well below the standard yield (698 kg per rai). Hence the challenging is to investigate the factors influencing the rice yield so a suitable plan can be made prior to the cultivation time. This will benefit the farmers to increase their yields and income.

The development of accurate prediction models of the rice yields is important for the government organization to maximize the value to the farmer income [1, 3]. In the past, mathematical model such as linear regression model was used to predict to crop yield. However, the weakness of this method is that it relies on linear relationship assumption. Hence, non-linear approaches such as artificial neural network (ANN) and Bayesian classification are used to overcome the complex situation [1, 4]. Various modelling methods have been used to find an accurate predictive model. For instance Ji et al. [5] compared the performance of ANN and Regression models for rice yield prediction in mountainous regions and the results showed that ANN is superior over Regression. Shabri et al. [6] used time series forecasting technique to predict rice yield in Malaysia and compared the prediction accuracy with ANN and the results showed ANN performs better than forecasting technique. Paswan and Begum [7] discussed the performance of ANN and regression models in predicting the crop production. Raorane et al. [4] claimed that reliable and accurate forecasting techniques are required for decision making in the government office prior to pre-harvest crop. Uno et al. [8] did a comparison between ANN and stepwise multiple linear regression models in predicting corn yield and the results revealed that there was no clear difference between the two methods in terms of prediction accuracy.

Hence the aim of this paper is to compare the prediction accuracy between the two popular modelling methods including Regression and artificial neural network models. The selection of input factors will be presented and then the selected factors will be taken to the predictive model. The prediction accuracy of each model is implemented by using root mean square error (RMSE). In the next section, we present the research method including details of statistical models used in this study. The results based on prediction accuracy will be given in section 3 and the conclusion is summarized in section 4 respectively.

2 Research Method

In order to compare the prediction accuracy of Regression model and ANN, a data set of rice production in Lower Northern Thailand collected during 2007 to 2010 is used [9]. We first screen the important factors that influence the rice product using correlation analysis when nonparametric correlation is used. Then the key factors are applied to fit Regression and ANN models. The total number of 9,206 records is split into training set and test set with 80:20 proportions; hence, 7,380 records for training set and 1,826 for test set. The training set is used to construct a predictive model and the test set is applied as an unseen data for testing. The prediction accuracy is validated through RMSE and MAE values by using the training and test set. In this section, we present the details of statistical models.

2.1 Regression Model

Regression analysis is one of the most effective methods that have been successfully used in the context of yield prediction since it is simple to construct and provides information on input variables sensitivity [10]. This method is based on the assumption of random error arising from a large number of insignificant input factors. Given an output response, y, and input variables $= (x_1, \ldots, x_d)$, the relationship between y and x can be mathematically written as

$$y = f(x) + \varepsilon \quad (1)$$

where ε is a random error which is assumed to be normally distributed with mean zero and variance σ^2. Since the true response surface function $f(x)$ is unknown, a response surface $g(x)$ is created to approximate $f(x)$. Therefore the predicted values are obtained by using $\hat{y} = g(x)$, which $g(x)$ can be treated as a polynomial function of $(X_1, X_2, \ldots, X_d)$. The observed data set can be expressed in the matrix form using the data matrix X as

$$y_0 = X\beta + \varepsilon \quad (2)$$

where $y_0 = (y_1, y_2, \ldots, y_n)^T$, x is a $n \times \alpha$ design matrix, β is a $(\alpha \times 1)$ vector of the regression coefficients, and ε is a $(n \times 1)$ vector of random error. The number of unknown parameters in equation (2) is determined by α, where $\alpha = 2d + \binom{d}{2} + 1$. The vector of least squares estimators, $\hat{\beta}$, can be determined subject to the minimization of

$$L = \sum_{i=1}^{n} \varepsilon_i^2 = (y_0 - X\beta)^T\,(y_0 - X\beta) \quad (3)$$

Minimization of equation (3) yields

$$X^T X\hat{\beta} = X^T y_0 \quad (4)$$

Hence, the least squares estimator of β is

$$\hat{\beta} = (X^T X)^{-1} X^T y_0 \quad (5)$$

, provided that $(X^T X)$ is invertible.

Once β is estimated, equation (5) can be used to predict the rice yield value at any untried settings of input variables.

2.2 Artificial Neural Network Model

Artificial neural network (ANN) is commonly used in complex decision making problems [11]. Unlike a usual statistical approximation model, ANN does not require any assumptions of the model, making it easy to use in many applications such as science, engineering and health science [12]. The inspiration for neural networks was the recognition that complex learning systems in animal brains consisted of closely interconnected sets of neurons. A particular neural may be relatively simple in structure but dense networks of interconnected neurons could perform complex learning tasks such as pattern recognitions and approximation models. ANN consists of input (*p*), a data set, which is combined through a combination function such as summation ($\sum$) then pass such information into an activation function (*f*) to produce an output response (*y*) and *b* is a bias as shown in Fig. 1.

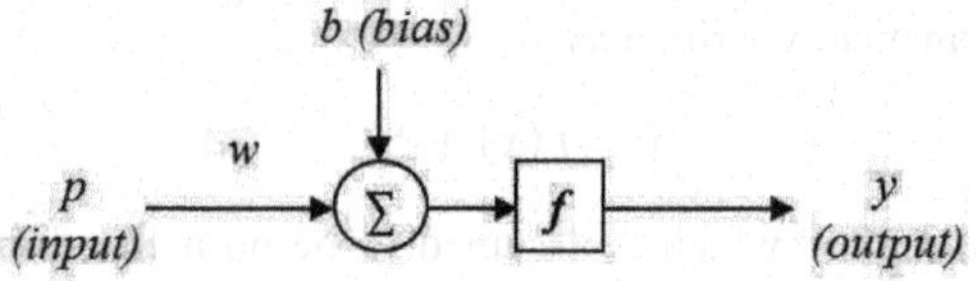

Fig. 1. A basic layout of ANN

The summary of ANN process can be rewritten as

$$y = f(wp + b) \quad (6)$$

, where w is the weight of each input variable.

Typically ANN is formed by multiple nodes (and probably multiple layers) as depicted in Fig. 2. Each node is symbolized by $\sum$. An activation function (*f*) can be the same or different. Examples of activation function are a linear, sigmoid and symmetrical hard limit. In this research, a sigmoid function is used. The weights between each node are adjusted by back propagation method. Fig. 2 displays ANN structure, which contains *m* input variables, one hidden *n* node layer and one output.

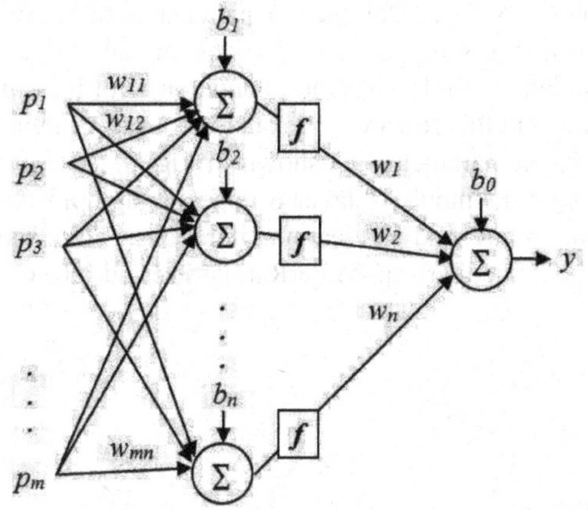

Fig. 2. ANN with multiple nodes

The entire process can be rewritten as,

$$y = \left[\sum_{i=1}^{n} w_i \times (f(\sum_{i=1}^{n}\sum_{j=1}^{n} w_{ij}p_i + b_i))\right] + b_0 \quad (7)$$

, where n is the number of nodes and m is the number of inputs.

2.3 Statistical Data Analysis

The data set used in this study is secondary data of rice yield collected from 2007 to 2010 in 8 provinces in the lower northern part of Thailand. The data set consist of 12 variables, in order to conduct a reliable model, only important variables would be included in the model. In this paper the spearman rank correlation coefficient (r) was used as a criterion to select the variable for the model. The input variables that are statistically related to the rice yield at the significance level of 0.05 are presented in Table 1. There are only 6 input variables included in the development of the prediction models, plus one output, which is a rice yield measured in terms of kilogram per rai.

Table 1. The selected significant input variables and output.

Variable name	Meaning	Range	r	P-value
RiceType	Rice seed type	{1,2,3,4,5,6}	0.243	<0.001
Method	Cultivating method	{1,2,3}	0.258	<0.001
SeedPerRai	Seed quantity, kg per rai	Min=7, Max=160	0.194	<0.001
PuiPerRai	Fertilizer used, kg per rai	Min=2.5, Max=116.7	0.287	<0.001
Period	Period of cultivation	Min=3, Max=8	-0.255	<0.001
InChon	Irrigation area	{0,1}	0.128	<0.001
ProdPerRai	**Rice yield, kg per rai**	**Min=50, Max=1900, Avg=562.22**		

To build a predictive ANN model, the key parameters to be considered are number of inputs, a number of layers and number of nodes for each layer, activation function, learning rate, momentum rate for weight calculation with back propagation method and learning iterations. In this research, Weka [13] is used as a tool to create the ANN model, by varying those parameters as shown in Fig. 3. The most optimized ANN model is obtained with 13 inputs, 1 layer 6 nodes, a sigmoid function, 0.1 learning rate and 0.1 momentum rate, and 500 learning iterations. The 13 inputs are formed by 6 RiceType, 3 Method, SeedPerRai, PuiPerRai, Period and InChon.

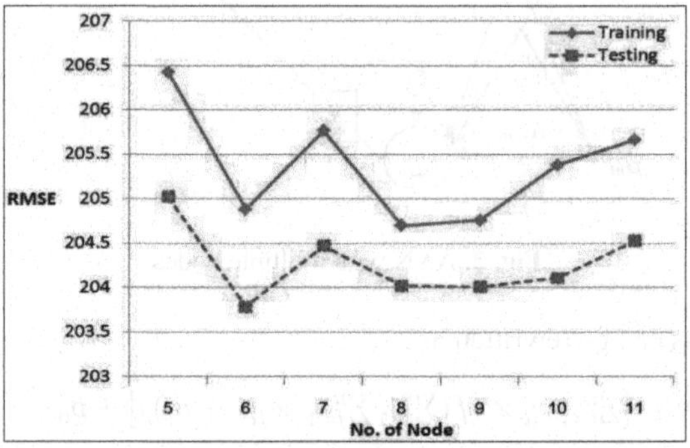

Fig. 3. RMSE values of ANN models with different number of node.

The stepwise multiple regression models have been fitted through the least squares method as described in section 2.1. The model that minimizes the RMSE is considered as the best searched model for predicting rice production.

3 Results

In this section the regression and ANN models are compared on the basis of RMSE and MAE values. The performance of each model for both of training set and test set are calculated using RMSE and MAE, defined as

$$RMSE = \sqrt{\frac{\sum_{i=1}^{k}(y_i-\hat{y}_i)^2}{k}} \quad (8)$$

$$MAE = \frac{1}{k}\sum_{i=1}^{k}\left|\frac{y_i-\hat{y}_i}{y_i}\right| \quad (9)$$

, where k is the number of test points, y_i is the actual response of the ith test point and $\hat{y}_i$ is the predicted response from statistical models for the ith test point. Lower values for RMSE and MAE imply a more accurate prediction model.

In order to calculate the prediction accuracy of ANN and regression models, the dataset is randomly split into training set and test set with a proportion of 80:20. After the best model of each method is found, the prediction accuracy is validated and the result is presented in Table 2.

Table 2. The RMSE and MAE for both methods on training and test data.

Methods	Training data		Test data	
	RMSE	MAE	RMSE	MAE
Regression	207.05172	0.450052	204.2636	0.443842
ANN	204.88462	0.293767	203.7862	0.297208

It can be clearly seen from Table 2 that ANN performs better than regression as the RMSE and MAE values obtained from both of training and test set are lower than that of regression model. As the structure this data set is quite complex and most of the input factors are qualitative data, hence the assumption free approach like ANN is more suitable comparing to regression model.

4 Conclusions

This paper presents the performance of the two popular predictive models, regression and ANN for predicting rice production in the lower northern Thailand. The results on training and predicting the rice yield reveal that ANN performs better than regression. The ANN model is also flexible in order to set all related parameters. Furthermore ANN seems to be robust to different structure of complex data. Hence ANN model would be recommended to use for modelling rice yield especially when the information of factors is not known.

References

1. Khairunniza-Bejo, S., Mustaffha, S. and Ismail, W. I. W.: Application of Artificial Neural Network in Predicting Crop Yield: A Review. Journal of Food Science and Engineering 4: 1-9 (2014).
2. Zhange, G. et al.: Predicting with artificial neural network. International Journal of Predicting, 14, 35-62 (1998)
3. Kaul, M., Hill, R. L., Walthall, C.: Artificial neural network for corn and soybean prediction. Agricultural System, 85, 1-18 (2005)
4. Raorane, A. A. and Kulkarni, R. V.: Review-Role of Data Mining in Agriculture. International Journal of Computer Science and Information Technology. 4(2), 270-272 (2013)
5. Ji, B., Sun, Y., Yang, S., and Wan, J.: Artificial neural networks for rice yield prediction in mountainous regions. The Journal of Agricultural Science 145(03), 249-261 (2007).
6. Shabri, A., R. Samsudin, et al.: Forecasting of the rice yields time series forecasting using artificial neural network and statistical model. Journal of Applied Sciences 9(23): 4168-4173 (2009).

7. Paswan, R. P. and S. A. Begum.: Regression and Neural Networks Models for Prediction of Crop Production. International Journal of Scientific & Engineering Research 4(9), 98-108 (2013).
8. Uno, Y., Prasher, R., Laeroix, R., Goel, P.K., Karimi, A., and Viau, et al.: Artificial neural network to predict corn yield from compact airborne spectrographic imager data. Computers and Electronics in Agriculture, 47, 149-161 (2005)
9. Office of Agricultural Economics, http://www.oae.go.th
10. Montgomery, D. C., Peck, E. A. and Vining, G. G. Introduction to Linear Regression Analysis. Fifth Edition, John Wiley & Sons, New Jersey (2012).
11. Bozdogan, H. : Statistical Data Mining and Knowledge discovery, Chapman & Hall/CRC, New York (2003).
12. Ripley, B.D., Statistical aspects of neural networks. In: Barndoff-Nielsen, O.E., Jensen, J.L., Kendall, W.S., editors. Networks and chaos-statistical and probabilistic aspects, Chapman & Hall, New York, 1993, 40-123.
13. Hall, M., Frank, E., Holmes, G., Pfahringer, B., Reutemann, P., Witten, I.H., (2009); The WEKA Data Mining Software: An Update; SIGKDD Explorations, 11(1), (2009)

Study on Local Path Planning for Collision Avoidance in Vehicle-to-Vehicle Communication Environment

Gyoungeun Kim[1], Byeongwoo Kim[2]

[1]Graduate School of Electrical Engineering, University of Ulsan, 93 Daehak-ro, Ulsan, Republic of Korea
gyg509@gmail.com
[2]School of Electrical Engineering, University of Ulsan, 93 Daehak-ro, Ulsan, Republic of Korea
bywokim@ulsan.ac.kr

Abstract. Local path planning is a path planning method that utilizes the dynamics and probabilities of a vehicle as criteria for determining the possibility of the corresponding vehicle to drive on a candidate path. The proposed system in this study utilizes the environmental data through a vehicle-to-vehicle (V2V) communication environment to create the collision risk index. The collision risk index takes into account the time to collision (TTC) as a criteria for determining path planning. The results of analyzing the performance of the proposed algorithm obtained through simulation verified that the proposed algorithm performed route planning in a more efficient manner when compared with the algorithm that does not take into account the TTC value. Therefore, the proposed collision avoidance algorithm that considers the TTC value is expected to contribute to the reduction of car accidents, as well as provide a more efficient route when applied to the actual system.

Keywords: Collision avoidance · Vehicle-to-vehicle · Path Planning · TTC(Time to Collision)

1 Introduction

Development of future vehicle technology focuses on intelligent technology to maximize driver safety and convenience, reduce traffic accidents, increase road traffic efficiency, and develop environmentally friendly technologies with reduced fuel consumption. The ultimate goal of intelligent technology lies in unmanned self-driving technology that allows driverless driving, and prestigious universities and automakers worldwide are currently collaborating on the study of unmanned vehicles [1].

The core technology in unmanned self-driving is to sense and determine a self-driving vehicle's environment and generate a driving path. In preceding studies, because the environment was sensed only by the sensors installed on the vehicle, the sensing range was limited, thus creating a blind zone where the sensors failed to sense vehicles outside such range [2]. Therefore, such unmanned self-driving vehicles require technology that can identify the blind zone undetected by the driver and the

K.J. Kim (ed.), *Information Science and Applications*,
Lecture Notes in Electrical Engineering 339, DOI 10.1007/978-3-662-46578-3_89

vehicle's internal sensors, allow the vehicle to sense its environment, identify the risks, and plan a driving path [3].

Thus, in this study, we propose a path planning algorithm combined with Vehicle-to-Vehicle (V2V) communication technology—vehicle safety communication, to overcome the limits of vehicle-installed sensors, i.e., the failure to detect the blind zone. The proposed algorithm generates candidate paths by applying cubic spline, and selects the optimal path using the path candidate index in consideration of Time To Collision (TTC), which is the collision risk index. In addition, using the proposed algorithm, we present the usability of collision avoidance path planning by finding the optimal path considering TTC in a real-life V2V environment.

The remainder of the paper is organized as follows: in Section 2 we introduce the structure of the collision avoidance system. Then, path generation and path selection method are described in detail in Section 2. Section 3 presents the simulation results. Followed by concluding remarks and future work in Section 4.

2 Collision Avoidance System

The proposed path planning algorithm for collision avoidance is presented in Fig. 1.

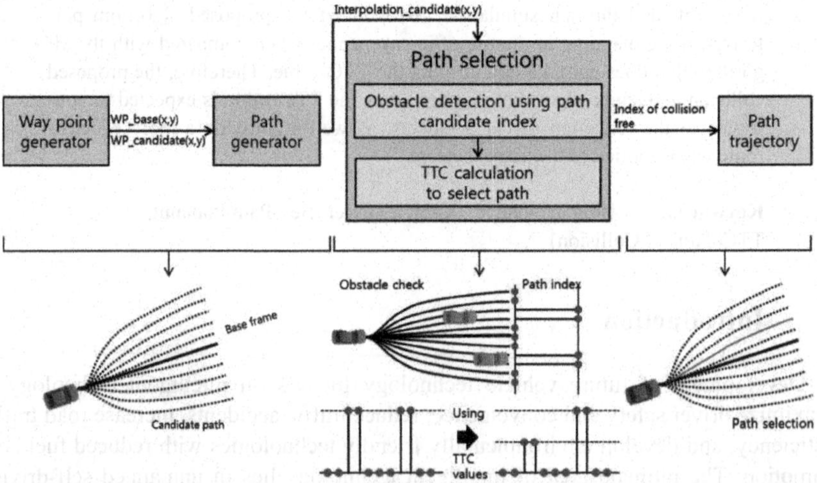

Fig. 1. Collision avoidance system

The proposed algorithm is largely divided into the generation of waypoints, selection of the optimal candidate path, and following of the selected optimal candidate path. A waypoint generator provides a global path plan that is generated based on the current point and target point of the vehicle, and a local path plan that provides candidate paths considering the nearby environment [4]. The waypoints provided by the waypoint generator are interpolated with a cubic spline, which generates paths and recognizes each candidate path as an index [5]. Each index shows the degree of risk to avoid collision; the higher the index, the greater is the risk. TTC is applied as an index to determine the degree of risk [6]. The optimal collision avoidance path, selected through TTC, is followed to present the usability and stability of collision avoidance path planning.

2.1 Path Generation

The generation of candidate paths is performed with the cubic spline interpolation method for linking waypoints smoothly. Provided that i number of waypoints are assumed to be (x_i, y_i), a cubic spline interpolation spline can be expressed with the following equation.

$$S(x) = y(x) = a_i(x - x_i)^3 + b_i(x - x_i)^2 + c_i(x - x_i) + d_i \quad i = 1, 2, 3, \ldots, n \quad (1)$$

2.2 Path Selection

After candidate paths are generated, the optimal candidate path needs to be selected based on the nearby environment. The optimal candidate path is determined by turning each candidate path to an index, and the index value is determined by the collision risk index. We proposed a method for selecting the optimal candidate path based on the location of an obstacle and the collision risk index, or TTC.

The collision risk index is determined by the location of an obstacle, and has a value of either one or zero, depending on whether there is an obstacle in the candidate path. The closer the collision risk index is to one, the higher is the degree of risk. When an obstacle is detected, the index value is one, and when no obstacle is detected, the index value is zero. The collision risk index has a value of one because the detected obstacle is adjusted by TTC. TTC, which is the collision risk cost between the ego vehicle and the preceding vehicle, is calculated by a relational equation similar to Equation (2). TTC is based on the time when an automatic braking force is applied by an automatic emergency braking (AEB) system that satisfies the required performance standard defined by the AEB Group, and it calculates the collision risk index of each candidate path. Table 1 lists the collision risk cost groups with the TTC when the AEB system applied automatic braking force.

$$\text{TTC [s]} = \frac{\textit{Relative Distance}\ [m]}{\textit{Relative Speed}\ [m/s]} \quad (2)$$

Table 1. Index cost using TTC

TTC [sec]	Collision risk cost
≤ 2.0	0.3
≤ 1.6	0.5
≤ 0.7	1.0

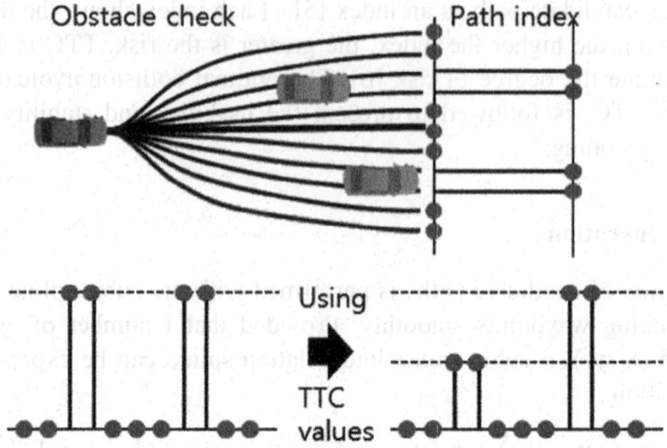

Fig. 2. Collision risk index by candidate path index with TTC

3 Simulation and Results

In this study, we used PreScan to allow exchanges of information on nearby vehicles (obstacles) through V2V communication. Moreover, the proposed algorithm was verified with a simulation in which PreScan and MATLAB/Simulink were used.

3.1 Simulation scenario

To verify the V2V-based path planning algorithm, the driving environment and the scenario were defined as shown in Fig. 3. The driving environment was set to a situation in which detecting a preceding obstacle was difficult because of a preceding vehicle on a normal straight road. For the Ego Vehicle, path planning for obstacle avoidance considering the preceding vehicle was allowed with the path planning algorithm. Table 2 presents the simulation scenario in detail.

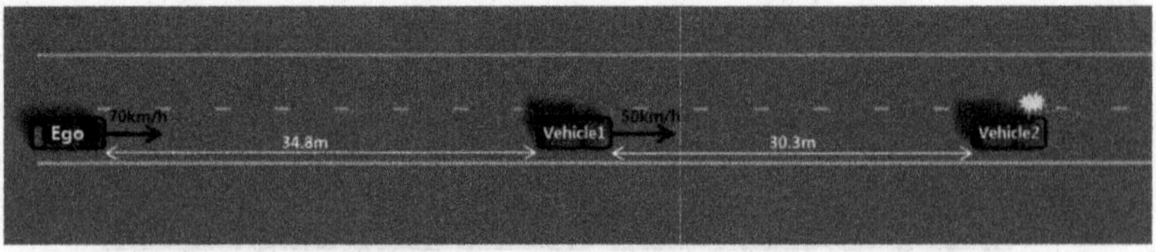

Fig. 3. Simulation viewer

Table 2. Simulation scenario

Vehicle	Initial speed	Final speed	Note
Ego	70 km/h	70 km/h	Adapting algorithm

Vehicle 1	50 km/h	30 km/h	Non-adapting algorithm
Vehicle 2	0 km/h	0 km/h	Vehicle fault

3.2 Results

Based on the scenario defined above, a simulation was run both with and without the algorithm; the simulation results are presented in Fig. 4. Which the relative distance and TTC are zero indicates the time elapsed to collision between the ego vehicle and the preceding vehicle. When the avoidance algorithm is not applied, the seconds elapsed to collision between the Ego Vehicle and Vehicle 1 and between the Ego Vehicle and Vehicle 2 are approximately 3.00 [sec] and 3.20 [sec], respectively. On the other hand, when the avoidance algorithm is applied, the seconds elapsed to collision between the Ego Vehicle and Vehicle 1 and between the Ego Vehicle and Vehicle 2 are approximately 3.23 [sec] and 3.25 [sec], respectively.

Fig. 5 presents the results of running collision avoidance simulations by applying the collision risk indexes by TTC to the index values of the candidate paths, where the Ego Vehicle avoids Vehicles 1 and 2. In Fig. 5(a), transverse collision avoidance control begins at 60 [m] from the x-axis direction, whereas it starts at 65 [m] in Fig. 5(b).

Therefore, Fig. 5(b) presents the results of the avoidance algorithm where the collision risk indexes of the candidate paths considering TTC, and the following result presents the usability of the collision avoidance algorithm: the Ego Vehicle can safely avoid the preceding vehicles, and the transverse collision avoidance control simultaneously considers a straight road, thus increasing the efficiency of the path.

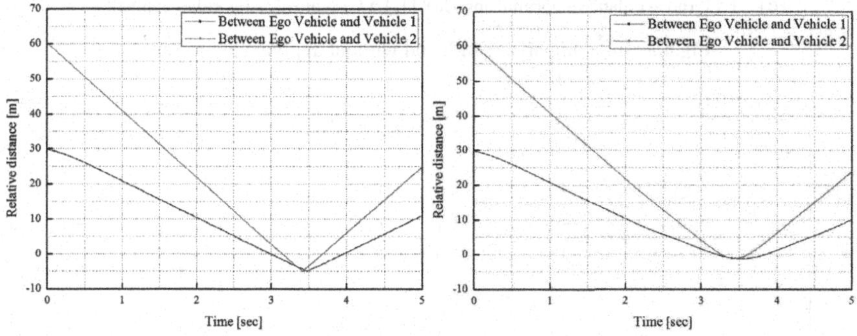

(a) Relative distance (no avoidance algorithm applied) (b) Relative distance (avoidance algorithm applied)

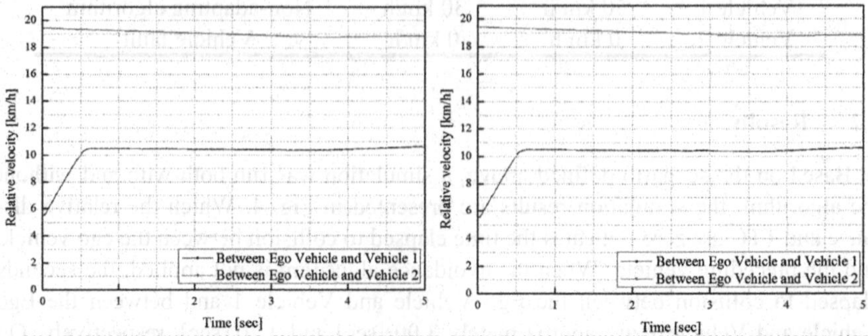

(c) Relative velocity(no avoidance algorithm applied) (d) Relative velocity(avoidance algorithm applied)

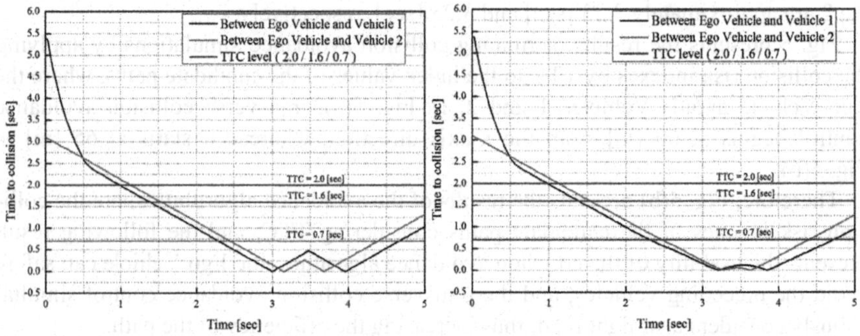

(e) TTC(no avoidance algorithm applied)(f) TTC(avoidance algorithm applied)

Fig.4. Result comparison with and without applying algorithm

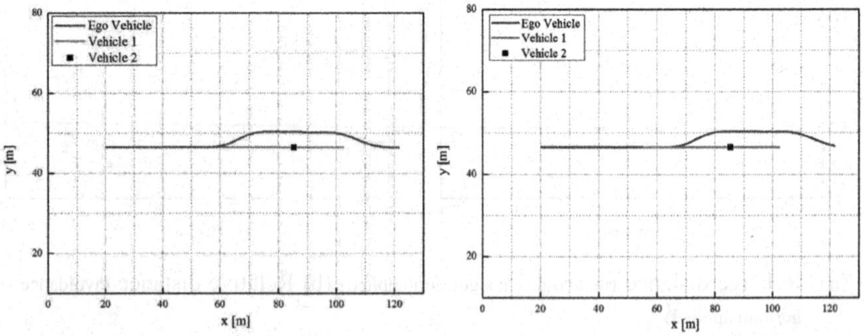

(a) No TTC considered (b) TTC considered

Fig.5. Avoidance paths by algorithm with and without considering TTC

4 Conclusion

In this study, with collision avoidance indexes considering TTC, we were able to suggest a method for selecting an optimal candidate path to avoid collisions.

For overcoming the limit where the vehicle and its internal sensors fail to detect blind zones, we structured an analytical model that considered a V2V communication environment, which is vehicle safety communication. As a result of the simulations, when the proposed algorithm was not applied, transverse collision avoidance control was initiated early, thus decreasing the efficiency of the avoidance path. However, when the proposed algorithm was applied, transverse collision avoidance control was initiated later, thus increasing the efficiency of the avoidance path. A goal of this study was to show the efficiency of the collision avoidance path planning method by drawing a TTC-considered optimal path in a real-life V2V environment using the proposed algorithm. From the results of the simulations, we were able to present the usability of the collision avoidance algorithm, using the V2V communication in comparison to vehicle-installed sensors, by applying the collision risk indexes of TTC to the indexes of the candidate paths. In a future study, we plan to test the proposed algorithm in a V2V communication environment in which real-life vehicles are used, and will perform a systematic analysis in various scenarios and driving environments.

Acknowledgments. This research was supported by the MSIP (Ministry of Science, ICT & Future Planning), Korea, under the C-ITRC (Convergence Information Technology Research Center) support program (NIPA-2013-H0401-13-1008) supervised by the NIPA (National IT Industry Promotion Agency).

Reference

1. Behringer, R., Sundareswaran, S., Gregory, B., Elsley, R., Addison, Bob., Guthmiller,W.: The DARPA Grand Challenge - Development of an Autonomous Vehicle. In: IEEE Intelligent Vehicles Symposium, pp. 226–231. IEEE Press, Parma (2004)
2. Gietelink, O.J., Verburg, D.J., Labibes, K., Oostendorp, A.F.: Pre-Crash System Validation with PRESCAN and VEHIL. In: IEEE Intelligent Vehicles Symposium, pp. 913–918. IEEE Press (2004)
3. Hong, C., Gyoung-Eun, K., Byeong-Woo, K.: Usability Analysis of Collision Avoidance System in Vehicle-to-Vehicle Communication Environment. Journal of Applied Mathematics, vol. 2014 (2014)
4. Xiaohui, L., Zhenping, S., Daxue L., Qi, Z., Zhenhua H.: Combining Local Trajectory Planning and Tracking Control for Autonomous Ground Vehicles Navigating along a Reference Path. In: 17th IEEE International Conference on Intelligent Transportation Systems (ITSC), pp. 725–731. IEEE Press, Qingdao (2014)
5. Keonyup, C., Minchae, L., Myoungho, S.: Local Path Planning for Off-Road Autonomous Driving With Avoidance of Static Obstacles. In: IEEE Transaction on Intelligent Transportation Systems, vol. 13, issue. 4, pp. 1599–1616. IEEE Press (2012)

6. Donghwi, L.,Kwangjin, H., Kunsoo, H.: Collision detection system design using a multi-layer laser scanner for collision mitigation. In: International Journal of Autonomous Technology, pp. 341–346. Korean Society Automotive Engineers-KSAE (2014)

Vehicle Position Estimation using Tire Model

Jaewoo Yoon[1], Byeongwoo Kim[2]

[1]Graduate School of Electrical Engineering, University of Ulsan, 93 Daehak-ro, Ulsan, Republic of Korea
jewos0127@gmail.com
[2]School of Electrical Engineering, University of Ulsan, 93 Daehak-ro, Ulsan, Republic of Korea
bywokim@ulsan.ac.kr

Abstract. GPS is being widely used in the location estimation technology, which is essential for stable driving of autonomous vehicle. However, GPS has problems such as reduction in location accuracy during abrupt vehicle behavior at high speed, and limitations such as signal interruption in tunnels and downtown areas. To overcome this problem, an algorithm that combines various sensor information and longitudinal/lateral slip is required. This paper proposes a three-degree of freedom (3-DoF) vehicle dynamics model, in which Dugoff's tire model is applied, and an algorithm, which combines various sensor information inside the vehicle by using extended Kalman filter. The performance of proposed location estimation algorithm was analyzed and evaluated through simulations. As a result, it is confirmed that the location estimation result of proposed algorithm is more accurate than that of method using GPS even during abrupt changes in motion.

Keywords: Autonomous vehicle · Dead Reckoning · Extended Kalman filter · Dugoff's tire model · Localization

1 Introduction

Countries with an advanced automobile industry, including the United States, European countries, and Japan, have recently developed autonomous vehicle technologies at a national level [1]. An autonomous vehicle is a vehicle that senses its environment, generates a path, and autonomously drives based on the position of the vehicle. In order for the autonomous vehicle to drive stably, accurate and high-reliable information on position is necessary.

Today, most autonomous vehicles receive position information from the GPS (Global Positioning System). However, a full dependence on GPS could substantially lower the reliability of position information because of failures, such as signal drops in tunnels and downtown, multipath errors, and electromagnetic interference. To overcome such problems, the dead reckoning technology is under development, with the INS (Inertial Navigation System) being installed in addition to GPS [2 - 5]. How-

K.J. Kim (ed.), *Information Science and Applications*,
Lecture Notes in Electrical Engineering 339, DOI 10.1007/978-3-662-46578-3_90

ever, the disadvantages of dead reckoning are the need for additional high-priced sensors, and reduced accuracy by vehicle slips caused by abrupt changes in vehicle behavior at high speeds [6].

Therefore, to overcome such problems, we propose a vehicle position estimation algorithm that considers the longitudinal/lateral slips of each vehicle tire. In addition, we used the ABS (Anti-lock Breaking System)'s wheel speed sensor, steering angle sensor, and ESC (Electronic Stability Control)'s acceleration sensor already installed on vehicles as the sensors for position estimation. The EKF (Extended Kalman Filter) algorithm is used because it can be used easily even for strongly nonlinear models, and can integrate information from a variety of sensors.

2 Vehicle Position Estimation

EKF that is based on the Bayesian filter is used the nonlinear model and integrated information from variety of sensors [7]. The applied EKF is divided mainly into Time Update and Measurement Update, as expressed in Fig. 1.

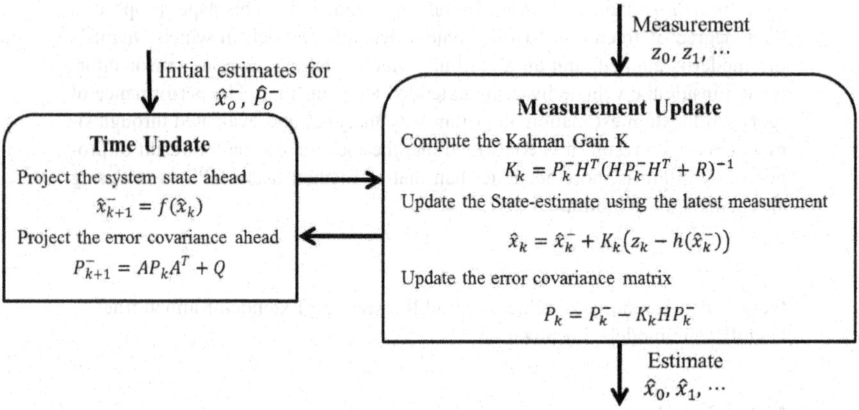

Fig. 1. Extended Kalman Filter algorithm

In the Time Update step, the system state ahead $\hat{x}_{k+1}^-$ and the error covariance ahead P_{k+1}^- are projected. The state is represented as

$$x_k = \left[x,\ v_x,\ y,\ v_y,\ \varphi,\ \dot{\varphi}\right]^T \tag{1}$$

Where (x , y) represent the vehicle position, (v_x , v_y) represents the vehicle speed, (φ) represents the yaw, and ($\dot{\varphi}$) represents the yaw rate.

In the Measurement Update, the Kalman gain K_k is computed, and the state-estimate using the latest measurement $\hat{x}_k$ and the error covariance P_k is updated. The measurement vector can be represented as

$$z_k = \left[X,\ Y,\ a_x,\ a_y,\ \dot{\varphi}\right]^T \tag{2}$$

Where (X , Y) represent the measuring position, (a_x , a_y) represent the measuring accelerations, and ($\dot{\varphi}$) represent the measuring yaw rate.

3 Dynamic Vehicle Model

3.1 Dynamic Vehicle Model

In this paper, a three-degree of freedom vehicle model in which the longitudinal, lateral, and yaw directions are considered is used for the EKF estimation variables, as shown in Fig. 2 [8]. It is nonlinear vehicle model that considers each tire/road contact point force. The dynamic motion equation is expressed in Equation (3).

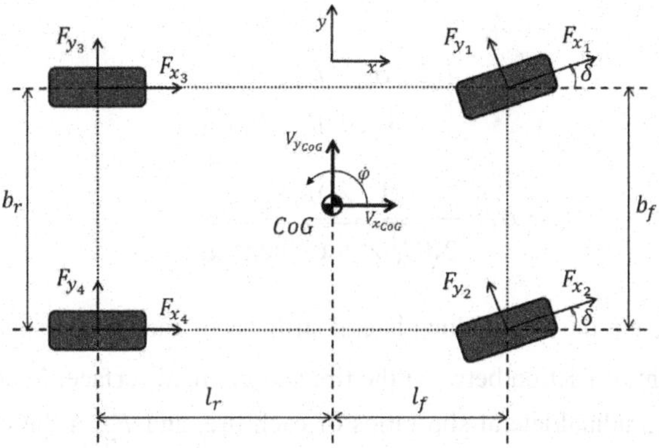

Fig. 2. Three-degree of freedom vehicle model

$$\begin{cases} \dot{v}_x = \dfrac{1}{m}\Sigma F_{x_i} + \dot{\varphi} v_y \\ \dot{v}_y = \dfrac{1}{m}\Sigma F_{y_i} - \dot{\varphi} v_x \\ \ddot{\varphi} = \dfrac{1}{I_z}\Sigma M_{z_i} \end{cases} \tag{3},$$

where m is the mass at the center of the vehicle; I_z is the moment of inertia from the vehicle's center of mass; v_x, v_y, and $\dot{\varphi}$ are the speeds in the longitudinal/lateral/yaw directions; and $\ddot{\varphi}$ is the acceleration in the yaw direction from the vehicle's center of mass. In addition, M_{z_i} is the moment at the center of mass of the vehicle, and F_{x_i} and F_{y_i} are the longitudinal/lateral forces of each tire, respectively. ($i = 1,2,3,4$ denote the four vehicle tires.)

3.2 Tire Model

In the model (3), a tire/road contact point force model is required. In this paper, the Dugoff tire model [9] is selected. Because it requires fewer parameters, and the model equations are simple.

$$\begin{cases} F_{x_i} = C_{xx} \dfrac{\lambda_i}{1-\lambda_i} k_i \\ F_{y_i} = C_{yy} \dfrac{\tan \alpha_i}{1-\lambda_i} k_i \end{cases} \tag{4}$$

where,

$$k_i = \begin{cases} (2-\sigma_i) & if \quad \sigma_i \prec 1 \\ 1 & if \quad \sigma_i \geq 1 \end{cases} \tag{5}$$

$$\sigma_i = \frac{(1-\lambda_i)\mu_i F_{z_i}}{2\sqrt{C_{xx}^2 \lambda_i^2 + C_{yy}^2 \tan^2 \alpha_i}} \tag{6}$$

C_{xx} and C_{yy} are the longitudinal/lateral stiffness of each tire, respectively, and μ_i is the coefficient of friction between the tire and the road surface. In addition, λ_i and α_i are the longitudinal/lateral slip ratios of each tire, and F_{z_i} is the normal force of each tire. The longitudinal/lateral slip ratios of each tire can be expressed by the following equations:

$$\begin{cases} \lambda_i = \dfrac{r_{eff} w_i - v_{x_i}}{\max(r_{eff} w_i, v_{x_i})} \\ \alpha_i = \delta - A\tan 2(v_{y_i}, v_{x_i}) \end{cases} \tag{7},$$

where r_{eff} is the radius of the tire, and v_{x_i} and v_{y_i} are the velocities at the tire/road contact point.

4 Simulation

4.1 Simulation Construction

The proposed position estimation algorithm was analyzed by simulation studies and used by the commercial vehicle dynamic simulator, dSPACE's ASM (Automotive Simulation Models). Because the ASM is composed of multi-body models with 24-degrees of freedom, it can provide the highly accurate vehicle dynamic information. And accordingly, the vehicle and sensor parameters for the simulation were obtained as show in Table 1, Table 2. The details of the simulation scenario to make a highly side slip are as follows: the longitudinal acceleration was set to a maximum 5 m/s^2 and the lateral acceleration to a maximum 10 m/s^2 on a 2,500-meter Hockenheimring Track.

Table 1. Vehicle parameters

Vehicle parameter	Value	Unit
Vehicle mass, m	1418	kg
Moment of inertia, I_z	1850	Kgm^2
Distance from C.G. to front wheel, l_f	1.064	m
Distance from C.G. to rear wheel, l_r	1.596	m
Longitudinal stiffness of each tire, C_{xx}	63000	N
Lateral stiffness of each tire, C_{yy}	50000	N/rad

Table 2. Sensor parameters

Sensor	Variance	Unit
Wheel speed sensor	1.0	rad/s
Acceleration sensor	1.0	m/s^2
Yaw rate sensor	1.0	rad/s
Position sensor	10	m

4.2 Simulation Results

The analyses on the simulation results of the proposed position estimation algorithm are presented in Fig. 3 – Fig. 5. Fig. 3 shows the estimation result of the GPS positions, estimated positions, and reference positions. The proposed position estimation algorithm followed the reference positions more accurately than the GPS sensor. Fig. 4 and Fig. 5 present the position errors of the X[m] and the Y[m], respectively: each error has a maximum of 2.3 m. This result is similar to that with a highly accurate sensor [5] and has improved [6].

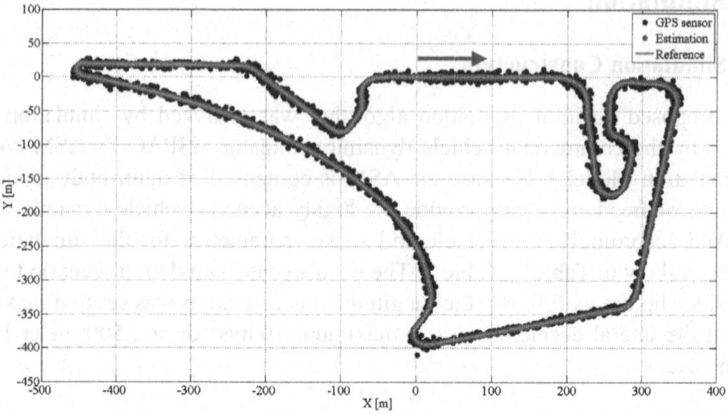

Fig. 3. Vehicle position estimation

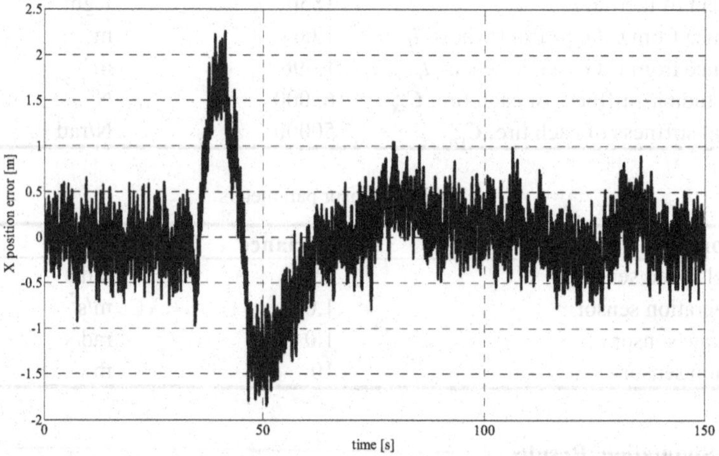

Fig. 4. Position error (X [m])

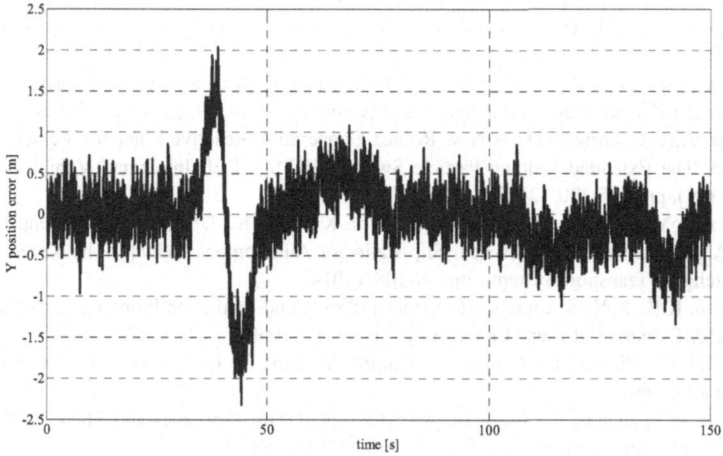

Fig. 5. Position error (Y [m])

5 Conclusion

In this paper, we proposed a position estimation algorithm for an autonomous vehicle for stable driving. To implement the position estimation algorithm, we used various sensors installed on a vehicle and a low-priced GPS sensor, and applied EKF based on the dynamic vehicle model and Dugoff's tire model. In addition, the performance of the proposed algorithm was verified with a commercial simulator—the ASM. As a result, the X and Y positions showed an error of maximum of 2.3 m, which verified the usability of the proposed position estimation algorithm.

Acknowledgments

This research was supported by the MSIP (Ministry of Science, ICT & Future Planning), Korea, under the C-ITRC (Convergence Information Technology Research Center) support program (NIPA-2013-H0401-13-1008) supervised by the NIPA (National IT Industry Promotion Agency).

References

1. An, K.H., Lee, S.W, Han, W.Y, Son, J.C.: Technology Trends of Self-Driving Vehicles, Electronics and Telecommunications Trends, vol. 24, pp. 35-44 (2013)
2. Qi, H., Moore, J.B.: Direct Kalman Filtering Approach for GPS/INS Integration. IEEE Transactions on Aerospace and Electronic Systems, vol. 38, pp. 687-693 (2002)

3. Rezaei, S., Sengupta, R.: Kalman Filter-based Integration of DGPS and Vehicle Sensors for Localization, IEEE Transactions on Control Systems Technology. vol. 15, pp. 1080-1088 (2007)
4. Cui, Y., Ge, S.S.: Autonomous Vehicle Positioning With GPS in Urban Canyon Environments. IEEE Transactions on Robotics and Automation, vol. 19, pp. 15-25 (2003)
5. Bacha, A.R.A., Gruyer, D.: A New Robust Cooperative-Reactive Filter for Vehicle Localization: The Extended Kalman Particle Swarm 'EKPS': IEEE Intelligent Vehicles Symposium (IV), pp. 195-200. (2013)
6. Kwon, J., Yoo, W. Lee, H., Shin, D. R., Park, K., Park, K.: Development of Dead Reckoning Algorithm Considering Wheel Slip Ratio for Autonomous Vehicle: The Korea Institue of Inteligent Transport Systems, pp. 99-108. (2014)
7. Kalman, R.E.: A New Approach to Linear Filtering and Prediction Problems, Transaction of the ASME-Journal of Basic Engineering, pp. 35-45 (1960)
8. Kiencke, U., Nielsen, L.: Automotive Control Systems for Engine, Driveline, and Vehicle. Springer (2000)
9. Dugoff, H., Fancher, P., Segel, L.: An Analysis of Tire Properties and Their Influence on Vehicle Dynamic Performance. SAE Technical Paper 700377 (1970)

I know when you are happy - Emotion Detection

Abhishek Singh Kilak and Namita Mittal

2012rcp9516@mnit.ac.in and nmittal.cse@mnit.ac.in

Malaviya National Institute of Technology, Jaipur

Abstract. Affective computing is an interdisciplinary research field. Various approaches have been proposed for the detection of emotions. One of the approaches is emotion detection from facial expressions. People react to situations in real life and they have little or limited control over their facial expressions. These facial expressions can be examined to know about their emotional state. In this paper we propose an approach based on Laplacian of Gaussian filters for frontal facial images. This technique removes the disadvantages of Laplcian filters susceptibility to noise. Also the simplicity of these filters makes them eligible for timely computations. For emotion recognition all possible two class emotion combinations are used. Experiments are conducted on expression dataset available for research purpose.

Keywords: Emotion detection, Face expression recognition, Laplacian of Gaussian filters, support vector machine

1 INTRODUCTION

Emotions play an important part in most of the areas of our day to day life. The general attitude of a person depends on the emotional state that he or she is in. The point of concern is that how can we identify the mental state of a person. In real life we may describe the emotional state of a person by observing his traits like the way he is sitting, walking, talking and also by observing the expressions. But can the computers detect the emotional state of people? One possible way is by examining the facial expressions of the person involved. This is true in general as the mental state of a person generally affects the facial expressions. The facial expressions give an insight as to what the particular person is going through. There are Six basic emotions namely Anger, Disgust, Fear, joy, Sadness and Surprise as defined by Ekman [19]. In addition Contempt is also considered as a basic emotion. Several techniques have been proposed and used for recognizing emotional state of people. These techniques include:

1. Brain Mapping: Brain mapping involves planting electrodes on the head of the human subject. The surges generated in various areas of the brain are studied to detect the response. This is promising but the use of electrodes,

K.J. Kim (ed.), *Information Science and Applications*,
Lecture Notes in Electrical Engineering 339, DOI 10.1007/978-3-662-46578-3_91

the subject's limited mobility and constant knowledge of the experiment being conducted on him hampers its applicability.

2. Voice Sampling: Voice is also an indicator of a person's emotional state. This can be seen in our day to day examples where a person sounds different in states when he is glad, astonished or angry. The drawbacks of using this approach are that the person might not respond to the stimulus. This is possible as he might not speak, and if this happens, there is no possibility of using this approach for detecting his mental state.

3. Facial Expression Detection [1][3][16]:People react to situations in real life and they have little or limited control over their facial expressions. Even if they have, they don't exercise it all the time. These facial expressions can be examined to know about the emotional condition of those people. Therefore this technique is explored further in our paper.

The application areas are numerous and some of them are stated here. Computer-Human interaction as to the computer or robot will interact to the human according to the emotional state that he is in. They can be applied in age old homes, neo-natal and infant care systems, recommender systems and driver assistance. Last but not the least they can be applied for diagnosis of affective disorders like anxiety and depression.

2 RELATED WORK

Expression recognition is an ongoing research field. The general steps involved are registering face followed by feature extraction, feature reduction and finally classification. Various approaches have been proposed and experimented upon.

Facial images were cropped into lower and upper part [1] and then fiducial points were applied to the two portions. Afterwards gabor filter was applied to extract features. Multi layer perception neural network was used for classification. They have reported accuracy of 81.6% and 86.6% accuracy for lower and upper face respectively. The use of neural network ensemble was done for classification of images by Fengjun Chen [6]. They have reported average recognition rate of 83.7. As an extended approach Median ternary pattern was proposed and used by Farhan Bashar [7] for extraction of features from CK [8] dataset. They have reported recognition rate of 89.1 for a grid of 3*3 using Support Vector Machine for classification.

Expression recognition is also affected by aging and this was explored by Guodong Guo [3]. They worked upon lifespan [2] dataset. The fiducial points were manually labeled and Gabor filters [4] were applied to extract facial features.SVM was used for classifier learning. Test was performed using 10-fold cross validation and average accuracy of 69.32% was reported for cross age group. The same setup when repeated for FACES [5] dataset by Guodong Guo [3] yielded average accuracy of 64.04%.

Some approaches propose personalized classifiers as they argue that there may be huge difference in Train pattern as compared to an unseen person. The unseen person may have heavy eyebrows, wrinkled face and such difference in pattern may lead to erroneous results. But such approach is unrealistic. Selective Transfer machine has been proposed by Wen Sheng [9] for a generic classifier that removes person specific biases. This approach is somewhat between personal and generic classifier. This approach uses unsupervised learning and hence labels are not required.

Use of Viola and Jones approach [10] for detection of face and afterwards mouth, nose and lips were detected using SDAM [11] by Yubo Wang [12]. They use a classifier based on SVM to classify on JAFFE dataset [13]. The same experiment was repeated on images from web and images captured by them in their laboratory. They have reported accuracy of 92.4%.

Actual or spontaneous expressions may vary from posed expressions. Difference between posed and natural expression was taken in account by Bihan Jiang [14]. They considered apex level of emotion and used Gentle Boost followed by SVM for MMI dataset [15].The difference between posed and spontaneous expressions was explored in depth by Marko Tkalčič [16]. They used posed images from Cohn Kanade dataset. The spontaneous expressions were captured by them using content images taken from International Affective Picture system [17]. Dataset acquisition was done on a set of 52 users. For LDOS-PerAff-1[18] dataset the images from all frames were averaged to yield neutral frame. They [16] filtered the images using Gabor filters and then calculated standard deviation and means for the filtered images. Feature vector of 240 elements was reduced to 80 using PCA for LDOS-PerAff-1 and 72 features for CK dataset describing 95% variance. KNN was used for emotion detection and 62% accuracy was reported by them for the spontaneous expression set.

3 PROPOSED APPROACH

The steps involved in emotion detection include

(i) Pre-Processing; (ii) Filtering; (iii) Feature Extraction; (iv) Classification

This is formalized as $I \rightarrow I_p \rightarrow I_f \rightarrow \breve{E} \rightarrow \bar{E}$

3.1 PREROCESSING

In the first phase we detect the face part from the images using Viola Jones [10]. The first and last images I were taken from the CK+ dataset as they correspond to neutral and extreme expression respectively. Once the face is detected, then the process involves taking out the region of interest and leaving the rest portion of the image. Therefore the image is cropped to contain the face part only. Sample of first and last images I_p of S52, S55 and S74 are shown in figure 1 and figure2.

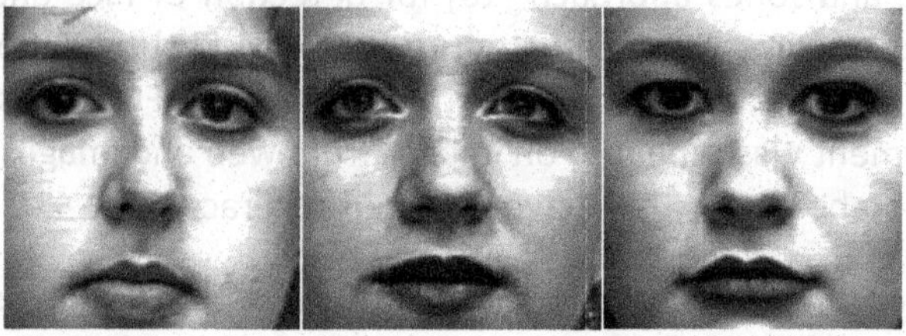

Fig. 1. Cropped first frame I_p of S52, S55 and S74 respectively

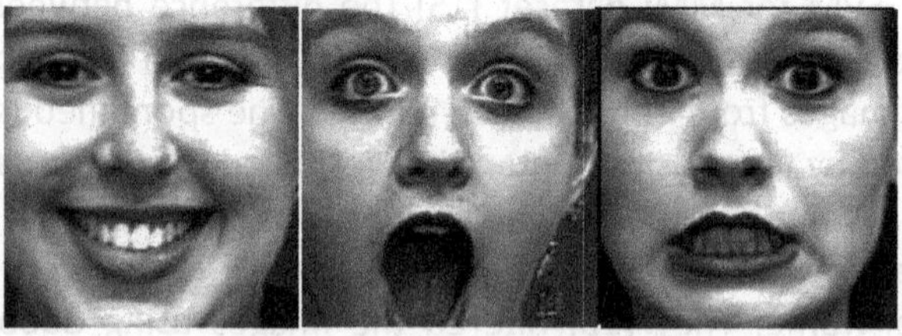

Fig. 2. Cropped last frame I_p of S52, S55 and S74 respectively

3.2 Filtering:

In the next step the face part of the images were subjected to Laplacian of Gaussian filters. This technique removes the disadvantages of Laplacian filters susceptibility to noise. The labeled images of CK+ dataset which had valid emotions were taken and filtered. The equations used for Laplacian of Gaussian filter are stated below.

$$f_g(x,y) = e^{-\frac{x^2+y^2}{2\sigma^2}} \qquad (1)$$

$$f(x,y) = \frac{(x^2+y^2-2\sigma^2) f_g(x,y)}{2\pi\sigma^2 \sum_x \sum_y f_g} \qquad (2)$$

For filtering ten incremental values of sigma were taken. This resulted in ten filtered images I_f corresponding to each I_p. The filtered images I_f for these subjects are shown in figure 3 and 4 for filter size of 5*5 and sigma value of 0.25.

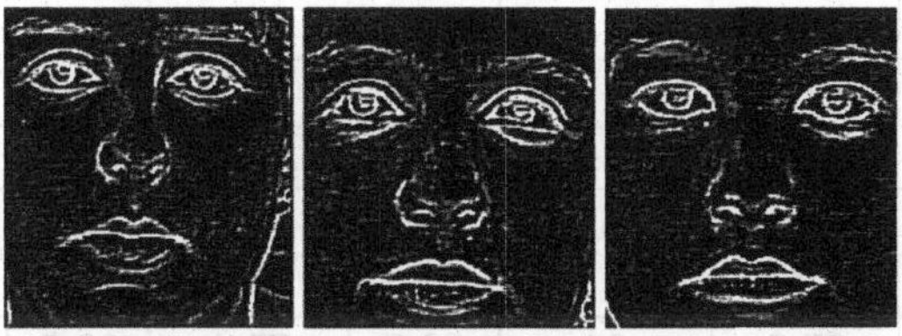

Fig. 3. Filtered cropped first frame I_f of S52, S55 and S74 respectively

Fig. 4. Filtered cropped last frame I_f of S52, S55 and S74 respectively

3.3 Feature extraction:

For each LOG filtered image the first two statistical moments (mean and standard deviation) were calculated. This yielded μ_f, σ_f for the first frame and μ_l, σ_l for the last frame respectively. Other than these independent features, six more dependent features were calculated. These were $\mu_f - \mu_l$, $\sigma_f - \sigma_l$, μ_f/μ_l , σ_f/σ_l , $(\mu_f - \mu_l)/\mu_f$ and $(\sigma_f - \sigma_l)/\sigma_f$. This resulted in a feature vector Ě having 100 elements .

3.4 Classification

For emotion recognition all possible two class emotion combinations were used. The pairing with Contempt is left as few images are present for this emotion label. Binary class support vector machine was used for the purpose of classification.

4 RESULTS

The results at various split ratios of train and test set are shown in table 1, 2 and 3. The accuracy obtained at 10 fold cross validation is shown in table 4.

Table 1. Accuracy in percentage for 70-30 split of train and test set

	Anger	Disgust	Fear	Happy	Sad	Surprise
Anger	-	82.1	58.8	78.4	75	85.3
Disgust	82.1	-	73.9	96.7	82.6	60
Fear	58.8	73.9	-	72	83.3	89.7
Happy	78.4	96.7	72	-	92	72.1
Sad	75	82.6	83.3	92	-	82.8
Surprise	85.3	60	89.7	72.1	82.8	-

Table 2. Accuracy in percentage for 80-20 split of train and test set

	Anger	Disgust	Fear	Happy	Sad	Surprise
Anger	-	84.2	81.8	75	72.7	82.6
Disgust	84.2	-	60	95	93.3	55.6
Fear	81.8	60	-	76.5	75	89.5
Happy	75	95	76.5	-	94.1	82.1
Sad	72.7	93.3	75	94.1	-	73.7
Surprise	82.6	55.6	89.5	82.1	73.7	-

Table 3. Accuracy in percentage for 85-15 split of train and test set

	Anger	Disgust	Fear	Happy	Sad	Surprise
Anger	-	85.7	87.5	72.2	75	82.4
Disgust	85.7	-	58.3	93.3	90.9	70
Fear	87.5	58.3	-	76.9	50	86.7
Happy	72.2	93.3	76.9	-	84.6	76.2
Sad	75	90.9	50	84.6	-	85.7
Surprise	82.4	70	86.7	76.2	85.7	-

Table 4. Accuracy in percentage at 10 fold cross validation

	Anger	Disgust	Fear	Happy	Sad	Surprise
Anger	-	74.2	69.6	74.6	77.8	83.2
Disgust	74.2	-	80.5	87.1	75	65.7
Fear	69.6	80.5	-	74.1	79.5	79.5
Happy	74.6	87.1	74.1	-	85.7	81.7
Sad	77.8	75	79.5	85.7	-	80.2
Surprise	83.2	65.7	79.5	81.7	80.2	-

The accuracy obtained is fair for the binary classification. There is high degree of accuracy for the Happy-Disgust class across all splits and also for 10 folds cross validation. This pair is important for recommender system applications where it can be applied to know user reaction. The system would then recommend items based on previous responses. The Sad-Happy classes also provide good accuracy. This has a potential application in finding affective disorders by recording sudden transitions between happy and sad mood. The Surprise-Fear classes can be used for application in elderly care homes.

5 CONCLUSION AND FUTURE WORK

In this paper we have proposed the use of LOG filter for emotion detection. Experiments show the effectiveness of proposed method. The simplicity of LOG filters makes them computationally efficient. This approach can be extended for use in other real time applications. Also this approach will be extended to cover all the classes in one single classification and yet not make it computationally intensive.

6 REFERENCES

1. Juliano J. Bazzo and Marcus V. Lamar, "Recognizing Facial Actions Using Gabor Wavelets with Neutral Face Average Difference", Sixth IEEE International Conference on Automatic Face and Gesture Recognition, 2004.
2. M. Minear and D.C. Park, "A Lifespan Database of Adult Facial Stimuli", Behavior Research Methods, Instruments & Computers, vol. 36, pp. 630-633, 2004.

3. Guodong Guo, Rui Guo and Xin Li, "Facial Expression Recognition Influenced by Human Aging", IEEE TRANSACTIONS ON AFFECTIVE COMPUTING, VOL. 4, NO. 3, JULY-SEPTEMBER 2013.
4. J. Daugman, "Uncertainty Relation for Resolution in Space, Spatial Frequency and Orientation Optimized by Two-Dimensional Visual Cortical Filters", Journal of the Optical Society of America A, vol. 2, pp. 1160-1169, 1985.
5. N. Ebner, M. Riediger, and U. Lindenberger, "Faces—A Database of Facial Expressions in Young, Middle-Aged, and Older Women and Men: Development and Validation", Behavior Research Methods, vol. 42, no. 1, pp. 351-362, 2010.
6. Fengjun Chen, Zhiliang Wang, Zhengguang Xu, Donglin Wang, "Research on a Method of Facial Expression Recognition", The Ninth International Conference on Electronic Measurement & Instruments (ICEMI), 2009.
7. Farhan Bashar, Asif Khan, Faisal Ahmed and Md. Hasanul Kabir, "Robust Facial Expression Recognition Based on Median Ternary Pattern (MTP)", International Conference on Electrical Information and Communication Technology (EICT), 2013.
8. P. Lucey, J.F. Cohn, T. Kanade, J. Saragih, Z. Ambadar and I. Matthews, "The Extended Cohn-Kanade Dataset (CK+): A complete dataset for action unit and emotion-specified expression", Proceedings of IEEE workshop on CVPR for Human Communicative Behavior Analysis, San Francisco, USA, 2010.
9. Wen-Sheng Chu, Fernando De la Torre, Jeffery F. Cohn, "Selective Transfer Machine for Personalized Facial Action Unit Detection", IEEE Conference on Computer Vision and Pattern Recognition, 2013.
10. P. Viola and M. Jones, "Rapid Object Detection using a Boosted Cascade of Simple Features", in IEEE Proc. International Conference on CVPR, Hawaii, USA, 2001, pp. 511-518.
11. Tong WANG, Haizhou AI, Gaofeng HUANG, "A Two-Stage Approach to Automatic Face Alignment", in Proc. SPIE International Symposium on Multispectral Image Processing and Pattern Recognition, Beijing, China, 2003, pp. 558-563.
12. Yubo WANG, Haizhou AI, Bo WU, Chang HUANG, "Real Time Facial Expression Recognition with Adaboost", Seventeenth International Conference on Pattern Recognition (ICPR'04), 2004.
13. The Japanese Female Facial Expression (JAFFE) http://www.mis.atr.co.jp/~mlyons/jaffe.html
14. Bihan Jiang, Michel Valstar, Brais Martinez, and Maja Pantic, "A Dynamic Appearance Descriptor Approach to Facial Actions Temporal Modeling", IEEE TRANSACTIONS ON CYBERNETICS, VOL. 44, NO. 2, FEBRUARY 2014.
15. M. Pantic, M. F. Valstar, R. Rademaker, L. Maat, "Web-based database for facial expression analysis", IEEE International Conference on Multimedia and Expo (ICME'05). Amsterdam, The Netherlands, pp. 317 - 321, July 2005.
16. Marko Tkalčič, AnteOdić, Andrej Košir, and Jurij Tasič, "Affective Labeling in a Content-Based Recommender System for Images", IEEE TRANSACTIONS ON MULTIMEDIA, VOL. 15, NO. 2, FEBRUARY 2013.
17. P. J. Lang, M. M. Bradley, and B. N. Cuthbert, "International Affective Picture System (IAPS): Affective Ratings of Pictures and Instruction Manual", University of Florida, Tech. Rep. A-8, 2005.
18. M. Tkalčič, J. Tasič, and A. Košir, "The LDOS-PerAff-1 corpus of face video clips with affective and personality metadata", Proc.Multimodal Corpora: Advances in Capturing, Coding and Analyzing Multimodality (Malta, 2010), LREC, 2009, p. 111.
19. P. Ekman, "Basic emotions", Handbook of Cognition and Emotion. New York: Wiley, 1999, pp. 45–60 Rafael A. Calvo and Sidney D' Mello.

Interconnection Learning between Economic Indicators in Indonesia Optimized by Genetic Algorithm

Saadah S[1], Wulandari G.S[1]

[1]*Telkom School of Computing, Telkom University, Bandung, Indonesia.*

{sitisaadah, giaseptiana}@telkomuniversity.ac.id

Abstract— Economic is important issue in a country since it is conducted by many sectors. Knowing stability of economic condition can be looked at by predicting the economic indicator. Unfortunately, the prediction that had been done to each of economic indicator did not concern about interconnection of them while predicting it. Because of that, learning about dependability of economic indicator still need further research about it. In other side, economic as knowledge that complex and chaos need differential dynamic to face the problem inside. Based on those reasons, this research not only observed about interconnection between indicators economic while predicting, but also needed differential dynamic which had been optimized by genetic algorithm. System got 20% until 80% for the accuracy system. The reason of why accuracy 80% was gotten because of using the same characteristics of economic indicators, ex. when system observed GDP and GNI together. Using similar data trend influenced the fitness function in GA able to optimized differential dynamic while doing prediction of economic indicators. Whereas, the decrease accuracy around 20% until 40% was came by using different characteristic of economic indicator. It can be found when learning dependable of GDP and Inflation while predict times series for GDP. Based on this research, it can be concluded that GA is able in optimizing learning the dependability of economic indicator's Indonesia. Moreover, it can be said that using the same characteristic indicator economic give better result for GA to learning the dependability of economic indicator than not. It can be said indirectly that government should concern about value of indicators economic that have the same characteristics when monitoring economic condition.

Keywords— Economic Indicator's Indonesia, Interconnection Learning, Differential Dynamic, Genetic Algorithm.

1. Introduction

Economic is a field that show either the uncertainty or chaos and complex [11, 12]. The uncertainty caused of dynamic issue in economic, while chaos and complex affected from many aspects. Dynamic aspects come from economic indicators. In Indonesia, economic indicators lay in data per year about inflation, GDP, GNI and export-import. In other word, it can be said these indicators represent the condition of economic in Indonesia indirectly. Because of that the condition of economic stability in Indonesia can be gotten from learning the dependability of them while predicting one of it [2, 4, 8, and 14]. The concern about it is important since Indonesia ever done with the crisis economic before. Unfortunately, until this time the research about it is rarely to find. So, the research about area concern in economic and informatics engineering-data will be conducted here.

Learning concept will use differential dynamic that will be optimized by genetic algorithm (GA) [2, 3, 7, 10, and 11]. GA as an optimization algorithm adopt from nature genetic, will be set to find the best individual data using fitness function. Parameter GA will be tested in differential dynamic to learn dependability of economic indicator. Representation of the best individual indicates how big influence of learning connectivity each of economic indicator. So, in this research will concern about GA in learning dependability each of economic

K.J. Kim (ed.), *Information Science and Applications*,
Lecture Notes in Electrical Engineering 339, DOI 10.1007/978-3-662-46578-3_92

indicator with assumption economic is stable using differential dynamic. And the result of it can be used by Government while monitoring Indonesian's economic.

2. Dynamic Model is Embedded In Genetic Algorithm

2.1 Genetic Algorithm (GA)

GA is iterative procedure which works with set of chromosome that is called by population. Population as a candidate solution with constant number then develop from one to other generation through operator genetic. Every single iterate, which is called by generation, in population will be evaluated. Then it will be selected by other generation in next population while it finds out best candidate solution in every population. Each of candidates will have fitness value which indicates excess of one solution than others. More high the value fitness is, bigger opportunity that individual to survive and give offspring. Recombination genetic in GA is simulated by operator genetic, like cross over and mutation.

2.2 Dynamic Model is Embedded in GA

GA manipulates population become potential solution to solved optimization problem. The problem that will be optimized here is learning about dependability of economic indicators. Application GA in learning attribute of stabilize economic condition of Indonesia can be illustrated by the figure below.

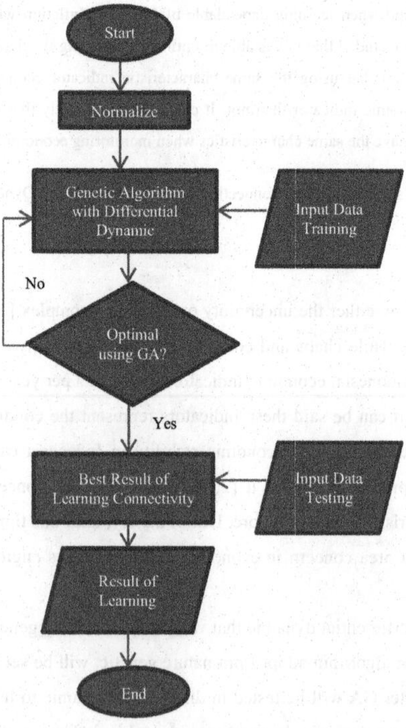

Fig.1 GA Application in Learning Interconnection Economic Indicators

Figure 2 shows dependability in dynamic system between two economic indicators. GDP increased had been followed by GNP increased, it can be vice versa. Stability of economic indicates economic situation that did not trigger to crisis since it is signed by value of economic indicator lay on in certain. This statement emphasize that this research did not concern to make early warning system of economic crisis. For example, value of inflation in Indonesia is in 1%-5%. Then, if value of inflation starts to increase or even increase sharply, then economic condition head for instable. Resume of instability one economic indicator need to observe with other indicator. Since statement unstable cannot be measured by one indicator only.

Fig.2 Dynamic Illustrations

So that, GA will be used to learn the connectivity between economic indicators refer to economic stabilization. Connected to the theory aforementioned, dependability of economic indicator will be writing down in the equation below [12].

$$\frac{dx(t)}{dt} = f(x(t),u(t)) + g(x(t)) \tag{1}$$

Where, $x(t)$ is vector state in economic system, $u(t)$ is variable input control, $f(x(t),u(t))$ become vector function between vector state and vector control. $x(t) = (x_1,...,x_h)^T$ where, h vector state temp and $u(t) = (u_1,u_2,...,u_p)^T$ where p is input vector control. For system non-linier, equation (1) will be created below.

$$\frac{dx(t)}{dt} = (Ax(t) + Bx(t)) - (Cx(t) + Dx(t)) \tag{2}$$

Where, A and B is matrix vector for one indicator, $A \in R^{hxh}$ and B^{hxp}. C and D is matrix vector for one indicator, $C \in R^{hxh}$ and D^{hxp}. In economic system, characteristic of state vector become $x, x \in R^h$ and control vector output become $u, u \in R^p$. With p and h gotten from equation below.

$$x = \{x_i(k)\}, k = 1,2,...,n; i = 1,2,...,h;$$

$$u = \{u_s(k)\}, k = 1,2,...,n; s = 1,2,...,p.$$

Using data time series of economic indicator equation (2) is structured below.

$$x(t) = (Ax(t) + Bx(t)) - (Cx(t) + Dx(t)) \tag{3}$$

2.3 Genetic Algorithm Application In Economic

Operator genetics about representation individual, population initialization, recombination, survivor selection and individual evaluation will be explained here. To represent individual become chromosome used real representation. Process representation of chromosome using element matrix $[A \mid B]$, where, $A = (a_{ij})_{hxh}$; $B = (b_{ij})_{hxp}$. One chromosome represents gain chain. The structure of chromosome is $hxh + hxp$

Population initialization is to arouse population that fill amount of chromosome, where every chromosome consists of number of gen input that needed is size of population and sum of gen. generally, population initialization figure out below.

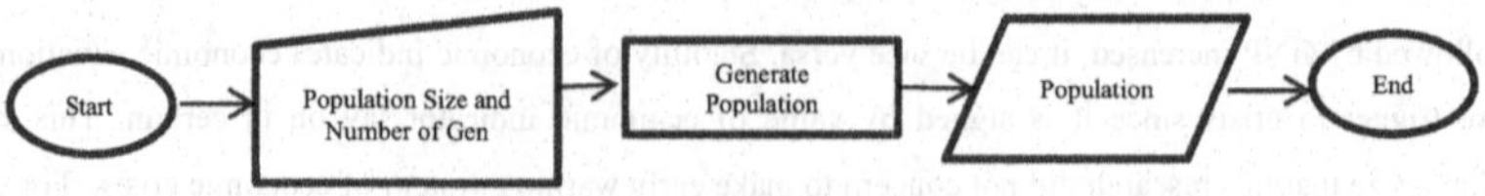

Fig.3 Population Initialization Process

After that, will be doing process individual evaluation to count fitness value and MAPE. Then parent selection will be passed. It to choose individual that will be parents. GA generate new individual with involved two individual based on roulette wheel concept. Where, every chromosome can be chosen with the same probability without concern with fitness value. Next is recombination. Recombination creates to the best solution. It comes from crossover process. Crossover that is used here is one-cross point, Pc. Besides GA use mutation to product best individual, Pmut. Selection survivor that had been used is generational replacement. The purpose of GA in learning is to minimalized error. And optimum solution is modelled to create error nol. Because of that, equation that will be used is.

$$\mathbf{f} = \frac{1}{(\mathbf{K+b})} \tag{4}$$

b is very small number to avoided divided by zero.

K is average of absolute representation error.

Besides, the equation of K is.

$$K = \frac{1}{N}\sum_{i=1}^{N}\left|\frac{z-z^*}{z}\right| \tag{5}$$

N is number all of data prediction. Z is data result of learning. And z* is data actual.

3. Presentation, Analysis, and Interpretation of Data

Here, will be looked result of GA application in learning interconnection between two economic indicators.

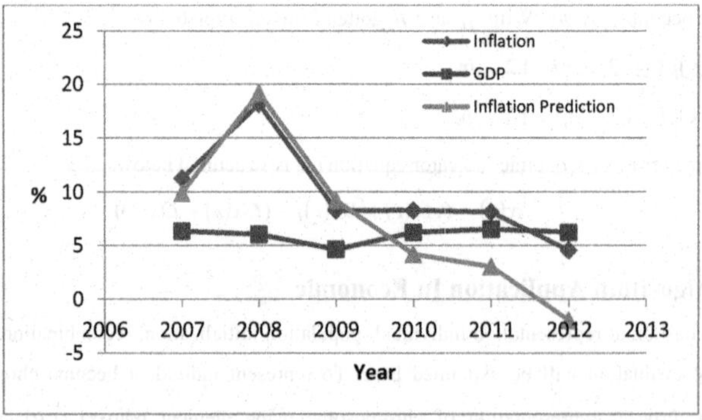

Fig. 4 Inflation Learning from 2007-2012

It can be seen that value of connectivity between inflation and GDP while predicting inflation indicate accuracy around 70%. It means interconnectivity while predicting inflation influence enough from GDP. And it was contributed also by using few data (around 5 years, from 2007-2012) in learning. The result came from GA component border up (Ra) 5 and border down (Rb) -5 with the size of population is 100, Pc 0.9, and Pmutation 0.2.

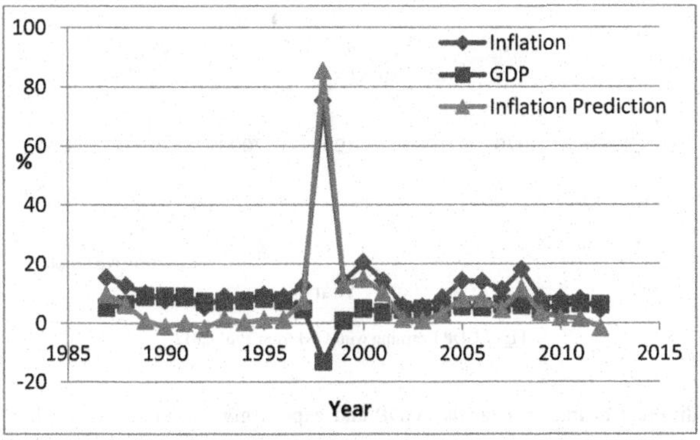

Fig. 5 Inflation Learning from 1980-2012

The result of observation in figure 5, using more data from 1980 until 2013 decreased the accuracy. It caused from two conditions, first, it caused by influenced by chaos and complex in economic indicator itself. Second, by influence of GDP that more is tight to the inflation. The accuracy is 33%.

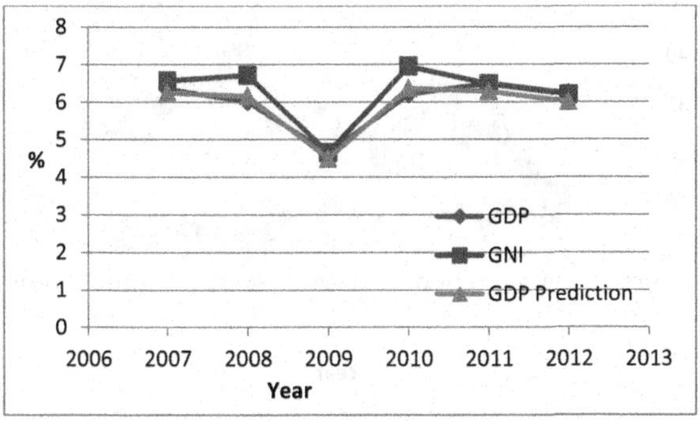

Fig. 6 GDP Learning from 2007-2012

Learning interconnectivity from GDP and GNI showed the strong one. Since measure of the accuracy raise up to 90%. It can be inferred from the figure 6 and 7 that the same characteristic both from GDP and GNI make influencing one another. Or it can be said that predicting GDP should be good enough using another economic indicator that have the same pattern data. The length of data that had been used was not influence much for the accuracy result.

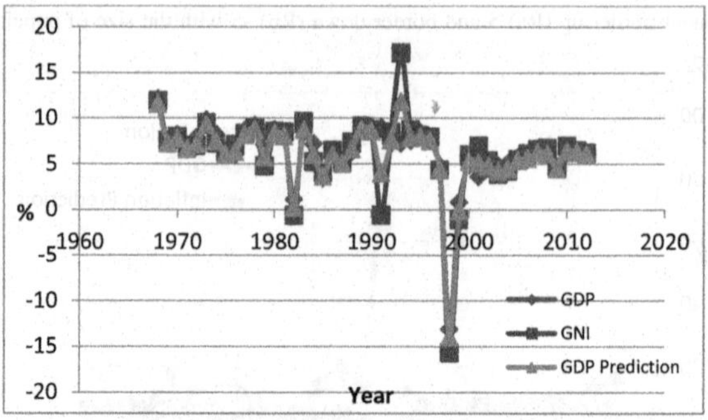

Fig. 7 GDP Learning with GNI from 1961-2012

Yet, result of different character data, like GDP and export, give accuracy 1,4% while using differential dynamic y = (A+B) – (C+D), with Pmut 0.9, Pc 0.2, Ra 5 and Rb -5. This different character caused fitness function cannot find best individual of GDP even if Pc and Pmut had been adjusted randomly. By this accuracy, it can be said that influence of GDP to export is small enough to distract stabilized of economic condition in Indonesia while using the long data (from 1967-2012). The result can be seen both in figure 8.

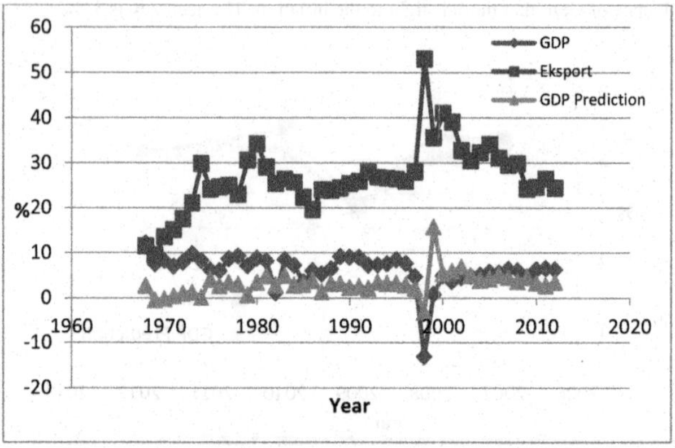

Fig. 8 GDP Learning with Export from 1970-2012

Result from figure 9 shown another analysis that using few data make the accuracy raise significantly. The range of the accuracy is 60% - 80%. It was contributed from the equation dynamic also. Besides, it was indicated by using number of data that affected to the interconnection of economic indicator; especially when the character of data is differ.

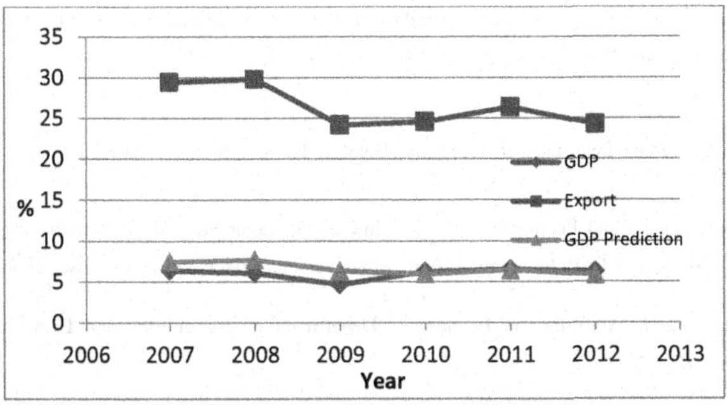

Fig. 9 GDP Learning with Export from 2007-2012

4. Conclusion

The conclusion is testing using two different indicators not only gives result low accuracy's but also dependability one of other indicator became bias. It had been showed by amount of accuracy 20% - 60%. It proved that GA did not able to follow pattern of data actual. It was caused by fitness function in GA gave best individual that failed to reach data to be predicted. Furthermore, by this result had been analysed that different two data character made learning of dependability economic indicators became chaos and complex.

In other side, the results differ while using data that have same characteristic. Since the data show similar trend, like GDP and GNI, give rise effect of accuracy at around 20%. Based on that reason, it can be said that connection between economic indicator seen clearly. Moreover, it proved that GA able to follow data historical optimally while using differential dynamic. Using various parameter of GA did not give much side-affect to the range of accuracy.

To conclude, GA application can be said succeed enough in learning interconnectivity the same characteristic of economic indicator using differential dynamic when economic condition in Indonesia is stable or not. Using another differential dynamic maybe can give more accurate result. And government should concern about two economic indicators that have the same characteristics when concern about economic condition.

References

1. Alquier, P., Wintenberger O.: Fast rates in learning with dependent observations. JMLR: Workshop and Conference Proceedings vol 1–15 (2012)
2. Anufriev, M., etc.: Learning-To-Forecast with Genetic algorithm. Amsterdam, 15 May (2012)
3. Arifovic, J.: Learning by Genetic Algorithms in Economic Environments. University of Chicago. SFI Working Paper: 1990—001. Oktober (1989)
4. Bonde, G., Khaled, R.: Stock Price Prediction using Genetic Algorithms and Evolution Strategies. Institute of Artificial Intelligence, University of Georgia. Athens. (2010)
5. Bruno, G.: Forecasting Using Functional Coefficients Autoregressive Models. ISAE, Institute for Studies and Economic Analyses, P.zza dell'Indipendenza, Rome, Italy. June (2008)
6. Davis, E.P., Karim, D.: Could Early Warning System Had Helped To Predict The Sub Prime Crisis. Brunel university and NIESR London. August (2007)

7. Drake, E.A.: Genetic Algorithms in Economics and Finance: Forecasting Stock Market Prices and Foreign Exchange. University of Stuttgart and University of New South Wales. Sydney.

8. Dietterich, T.G.: Machine Learning for Sequential Data: A Review. Oregon State University, Corvallis, Oregon, USA.

9. Inggrid.: Sector Financial and Economic Growth in Indonesia: Using Method of Causality in Multivariate Vector Error Correction Model (VECM). Economic Faculty of Universitas Kristen Petra Surabaya. (2006)

10. Lucas R.J. The Genetic Algorithm in Economics. Department of Agricultural and Economics, Montana State University. 4 May (2007)

11. Mahfoud, S., Mani, G.: Financial Forecasting using Genetic Algorithms. Applied AI, 10:543-565. (1996)

12. Shuhong, Z.: The Self Adaptable Modelling Of Socio-Economic Complex System Based On Genetic Algorithm. Journal of International for General Systems Studies. Hubei University of Economics, Wuhan. China. December 12, (2007)

13. Visco, I.: The Financial Crisis and Economic's Forecast. Originally published in BIS review 49/2009. Estudos Avancados 23-66 (2009)

14. Van, D.B.J., Candelon, B., Urbain J-P.: A Cautious Note on The Use of Panel Model to Predict Financial Crises. Universiteit Maastricht, The Netherlands. June (2008)

Classification Prediction of the Foot Disease Pattern Using Decision Tree Model

Jung-Kyu Choi[1], Yonggwan Won[2], and Jung-Ja Kim[3,4,*]

[1] Department of Healthcare Engineering, College of Engineering, Chonbuk National University, Jeonju, Republic of Korea
chaosjoozak@nate.com

[2] School of Electronics and Computer Engineering, College of Engineering, Chonnam National University, Jeonju, Republic of Korea
ykwon@jnu.ac.kr

[3] Division of Biomedical Engineering, College of Engineering, Chonbuk National University, Jeonju, Republic of Korea

[4] Research Center of Healthcare & Welfare Instrument for the Aged, Chonbuk National University, Jeonju, Republic of Korea
jungjakim@jbnu.ac.kr

Abstract. Datamining is used to find out desired important and meaningful knowledge in large scale data. The decision tree in classification algorithms has been applied to categorical attributes and numeric attributes in different domains. The purpose of study was to acquire significant information between singular disease groups and biomechanical parameters related with symptoms by developing prediction model. Sample data of 90 patient's records diagnosed with a singular disease was selected for analysis, in total 2418 data. A dependent variable was composed of 9 singular disease groups. 18 of 32 independent variables closely related to disease were selected and optimized. After object data was divided into training data and test data, C5.0 algorithm was applied for analysis. In conclusion, 10 diagnosis rules were created and major symptom information was verified. On the basis of the study, additional analysis with utilizing other datamining methods will be performed to improve accuracy from now on.

Keywords: Datamining, Decision tree, Foot, Disease, Clinical data.

1 Introduction

The foot is the most important part of body to basic activity and movement motion of human on daily living [1]. It has a highly complex anatomical- and biomechanical structure of 28 bones and 55 small joints worked organically [2]. There is about 15t

* Corresponding author.

K.J. Kim (ed.), *Information Science and Applications*,
Lecture Notes in Electrical Engineering 339, DOI 10.1007/978-3-662-46578-3_93

weight-bearing on the foot, only 5% of the total surface area of body, while a human being walks for 1 km [3]. The push-off exercise with this pressure causes stress or soft tissue strain, and it also deform the foot shape such as pes planus and pes cavus [4]. These abnormal foot shapes have a bad influence on the balance of the spine, pelvis, knee and ankle, and the unstable body cause unnatural and excessive movement in the joint [5]. In this condition, there are close connections between form of the abnormal foot and the various disease in lower limbs.

In modern society, the amount of data on medical field has significantly increased by development of information technology [6]. The result of analysis for data are possible to be used for information or index extracted on the basis of clinical experience. Datamining, the extraction of hidden predictive information, is used to deal with large amounts of data which are stored in the database, to find out desired important and meaningful knowledge [7]. The decision tree, which is one of representative technique in the datamining, in classification algorithms has been applied to categorical attributes and numeric attributes in different domains by observing relationship and modeling regulation in data [8]. This method is useful to discover meaningful information, and has advantage to understand analysis flow easily in comparison with other datamining method [9]. According to previous studies, the decision tree was adopted to analyze postoperative status of ovarian endometriosis patients. Experimental results showed new interesting knowledge about recurrent ovarian endometriomas under different conditions [10]. In addition, age, associated disease, pathology scale, course of hospitalization, respiratory failure and congestive heart failure were came out to be danger factors on death of pneumonia by using the decision-tree model for analysis of death factor on pneumonia patients [11].

More intelligent and delicate analysis is necessary because symptoms of the foot disease are not obvious, basically. However, most previous studies about the foot disease just figured out simple correlation by quantitative analysis, not integration analysis based on datamining technique. Analysis to understand various interconnection of several parameters with reference to the foot disease is important. Accordingly, the purpose of study was to acquire significant information between singular disease groups and biomechanical parameters related with symptoms by developing prediction model based on clinical data of the foot clinic.

2 Study Procedure

2.1 Subjects

The first clinical data of total 2418 patients in the foot clinic of Inje University Ilsan Paik Hospital and Chungnam National University Hospital was utilized for study. The data was made up 37 attributes diagnosed by podiatrists. 420 patient's records with missing value were excepted. Sample data of 90 patient's records diagnosed with a singular disease was selected for analysis.

2.2 Variables

A dependent variable was composed of 9 singular disease groups such as (A) Left Pes cavus, (B) Left Pes planus, (C) Right Pes cavus, (D) Right Pes planus, (E) Gastro-Soleaus muscle tightness, (F) Intoe, (G) Pes cavus, (H) Pes planus and (I) Scoliosis.

Independent variables were preprocessed through statistical validity and importance analysis. Therefore, 18 of 32 independent variables closely related to disease were selected and optimized, as shown in Table 1.

Table 1. Independent variable.

Variable	Type	Description
Sex	Nominal	Male, Female
Age	Nominal	Adolescent, Adult, Child, Infants
(L) STJ Inversion	Numeric	Inversion angle of the left subtalar joint
(R) STJ Inversion	Numeric	Inversion angle of the right subtalar joint
(L) STJ Eversion	Numeric	Eversion angle of the left subtalar joint
(R) STJ Eversion	Numeric	Eversion angle of the right subtalar joint
(L) STJ ROM	Numeric	Range of Motion of the left subtalar joint
(R) STJ ROM	Numeric	Range of Motion of the right subtalar joint
(L) FF to RF	Numeric	Angle of the left forefoot to rearfoot
(R) FF to RF	Numeric	Angle of the right forefoot to rearfoot
(L) RCSP	Numeric	Angle of the left Resting Calcaneal Stance Position
(R) RCSP	Numeric	Angle of the right Resting Calcaneal Stance Position
Pelvis Tilting	Nominal	0 : Not, 1 : left PSIS tilting, 2 : right PSIS tilting
Pelvis Rotation	Nominal	0 : Not, 1 : left lateral rotation, 2 : right lateral rotation
(L) Pelvis Trendelen	Numeric	Angle of the left Trendelenburg position
(R) Pelvis Trendelen	Numeric	Angle of the right Trendelenburg position
(L) Pelvis Elevation	Numeric	Angle of ASIS
(R) Pelvis Elevation	Numeric	Angle of ASIS

2.3 Study Process

In the study, compound of independent variables explained 9 singular disease groups effectively. Therefore, 18 independent variables were inserted to analyze in the decision tree algorithm at the same time. Data analysis was performed by IBM SPSS statistics 18 (SPSS Inc., Chicago, IL, USA) and IBM SPSS Modeler 14.2 (SPSS Inc., Chicago, IL, USA). For an ideal model, it was efficient to make a number of predictive model and conduct comparison analysis [10]. The data, therefore, was

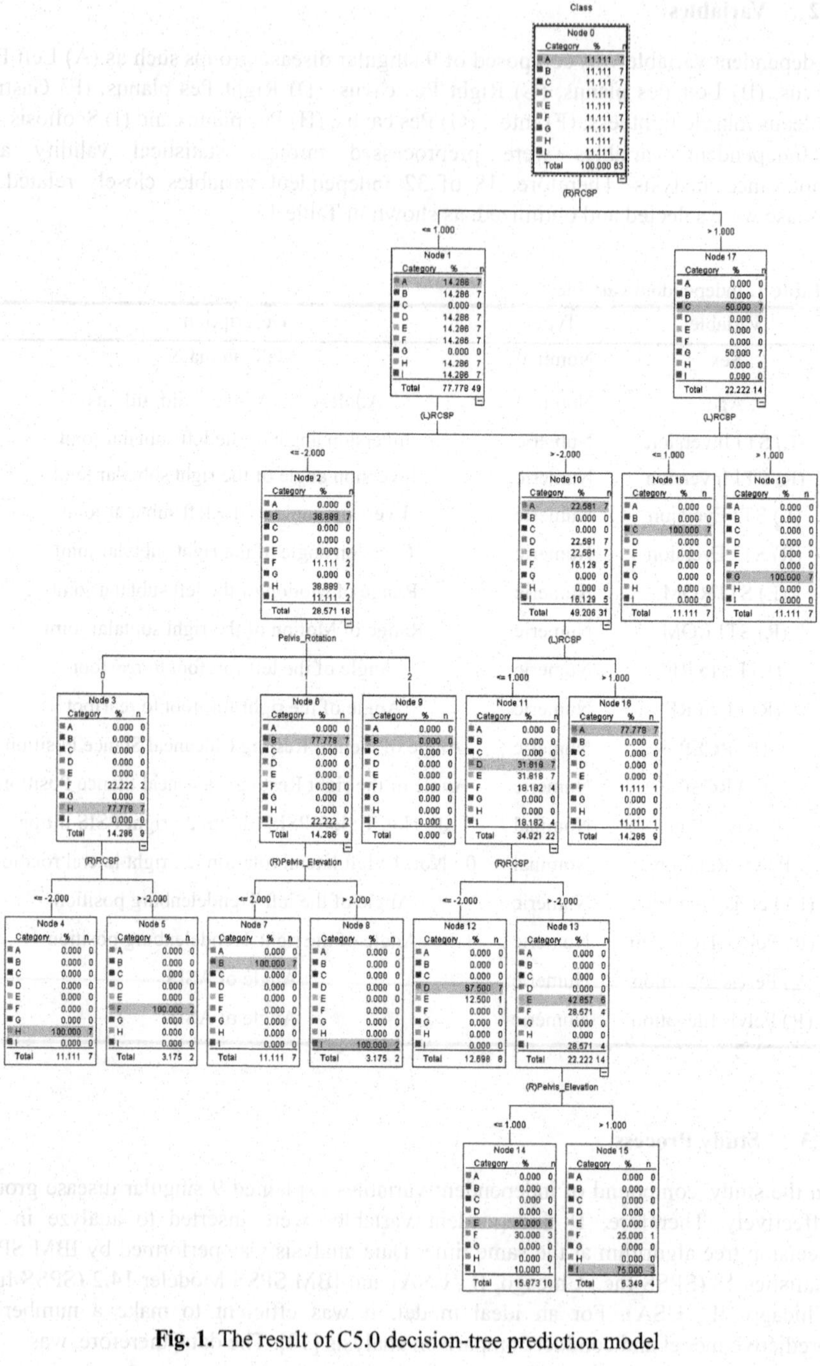

Fig. 1. The result of C5.0 decision-tree prediction model

separated into training data (70%) and test data (30%) by the partition node. In the study, C5.0 algorithm was utilized to develop model, and other methods will be applied for additional model and comparison analysis in further study. The prediction rate was verified by the analysis node after creation of model. Tree-structured decision tree model, which is knowledge discovery technique for using purpose of classification, is comprised of organization as 'If A, then B. Else B2' [12].

3 Result

The first clinical data of 90 patients (Male: 37, Female: 53) diagnosed with a singular disease were used for study, in the whole 2418 data. The prediction model was generated for analysis of disease category by applying the C5.0 algorithm. The measured prediction rate was Correct: 87.3 % and Wrong: 12.7 % in the training data, and Correct: 70.37 % and Wrong: 29.63 % in the training data.

As a result of analysis on 9 singular disease groups by using the C5.0 algorithm, the significant predictor importance was shown in the order: (L) RCSP (0.49), (R) RCSP (0.40), Pelvis_Rotation (0.09), (R) Pelvis_Elevation (0.02). In addition, total 10 rules were verified : (1) '(R) RCSP' was below 1° and '(L) RCSP' was above 1°, then 'A', (2) '(R) RCSP' was below 1°, '(L) RCSP' was below -2°, 'Pelvis_Rotation' was '1' and '(R) Pelvis_Elevation' was below 2°, then 'B', (3) '(R) RCSP' was above 1° and '(L) RCSP' was below 1°, then 'C', (4) '(R) RCSP' was below -2° and '(L) RCSP' was above -2° ~ below 1°, then 'D', (5) '(R) RCSP' was above -2° ~ below 1°, '(L) RCSP' was above -2° ~ below 1° and '(R) Pelvis_Elevation' was below 1°, then 'E', (6) '(R) RCSP' was above -2° ~ below 1°, '(L) RCSP' was below -2 and 'Pelvis_Rotation' was '0', then 'F', (7) '(R) RCSP' was above 1° and '(L) RCSP' was above 1°, then 'G', (8) '(R) RCSP' was above -2°, '(L) RCSP' was below -2 and 'Pelvis_Rotation' was '0', then 'F', (9) '(R) RCSP' was below 1°, '(L) RCSP' was below -2°, 'Pelvis_Rotation' was '1' and '(R) Pelvis_Elevation' was above 2°, then 'I', (10) '(R) RCSP' was above -2° ~ below 1°, '(L) RCSP' was above -2° ~ below 1° and '(R) Pelvis_Elevation' was above 1°, then 'I', as shown in Fig. 1.

4 Discussion and Conclusion

The object of this study was to classify category of the foot singular disease and examine diagnosis rule based on clinical data by developing prediction model of decision tree algorithm. The sample data of 90 patient's first clinical data in the foot clinic was used for analysis. Dependent variable was composed of 9 groups, and 18 attributes were selected for independent variable. The data was separated into training data and test data for ideal model, and the prediction rate was confirmed after creation of model by C5.0 algorithm. As the result, we were able to confirm that variable of each node was a key diagnosis factor to identify the foot disease such as RCSP, Pelvis rotation and right Pelvis elevation, and 10 rules were created. As follow rules, we could find out major symptom information : (B) Left Pes planus had right RCSP below 1°, left RCSP below -2°, left pelvis lateral rotation and right pelvis elevation below 2°, (C) Right Pes cavus had right RCSP above 1° and left RCSP below 1°, (D) Right Pes planus had right RCSP below -2° and left RCSP above -2° ~ below 1°, (F) Intoe had right RCSP above -2°, left RCSP below -2 and not pelvis rotation. (G) Pes cavus had right RCSP above 1° and left RCSP above 1°, (F) Pes planus had right RCSP above -2°, left RCSP below -2 and not pelvis rotation, (I) Scoliosis had right RCSP below 1°, left RCSP below -2°, left pelvis lateral rotation and right pelvis elevation above 2°.

In conclusion, the result was similar to that of previous studies which classified the shape of foot by RCSP [13, 14]. In addition, we could know that there was any relationship between the foot disease and the pelvis rotation & elevation. In the study, the error rate of prediction rate was relatively high because small sample data was used, and symptom of the foot disease was complicated, not obvious, basically. Therefore, additional analysis with utilizing other datamining methods will be performed to improve accuracy, and comparison analysis will be also carried out for producing an ideal model from now on. Ultimately, valuable knowledge by datamining would be help to raise the quality of diagnosis and treatment for patients.

Acknowledgements. This research was supported by Basic Science Research Program through the National Research Foundation of Korea (NRF) funded by the Ministry of Education, Science and Technology (2012R1A1B3003952) and National Research Foundation of Korea (NRF) grant funded by the Korea government (MSIP) (NRF-2013R1A2A2A04016782).

References

1. Ko, Y.J., Kim, H.W.: Diagnosis and Conservative Treatment of Common Foot Diseases. Journal of the Korean Medical Association. 47. 3. 247-257 (2004)
2. Kim, H.C.: Management of Foot and Ankle Disorders. Journal of the Korean Medical Association. 48. 663-671 (2005)
3. Choi, S.B., Lee, W.J.: Influence of Shoe Shape and Gait Characteristics on feet Discomforts according to Women's Foot Type. The Costume Culture Association. 10. 306-317 (2002)
4. Lott, D.J., Hastings, M.K., Commean, P.K., Smith, K.E., Mulle, M.J.: Effect of footwear and orthotic devices on stress reduction and soft tissue strain of the neuropathic foot. Clinical Biomechanics. 22. 352–359 (2007)
5. Benedetti, M.G., Catani, F., Ceccarelli, F., Simoncini, L., Giannini, S., Leardini, A.: Gait analysis in pes cavus. Gait & Posture. 5. 2. 169 (1997)
6. Hyun, Y., Jung, H.I., Chung, K.Y.: Development of Pain Prescription Decision Systems for Nursing Intervention. In: Kim, K.J., Ahn, S.J. (eds.) Proceedings of the International Conference on IT Convergence and Security 2011. LNEE, vol. 120, pp. 435-444. Springer, Heidelberg (2011)

7. ISLAM, A.R., CHUNG, T.S.: An Improved Frequent Pattern Tree Based Association Rule Mining Technique. In: Information Science and Applications (ICISA). 2011 International Conference on. IEEE. pp. 1-8 (2011)
8. Chang, M.Y., Shih, C.C., Chiang, D.A., Chen, C.C.: Mining a Small Medical Data Set by Integrating the Decision Tree and t-test. Journal of Software. 6. 12. 2515-2520 (2011)
9. Song, J.Y., Kim, H.K.: A Study of Decision Tree in Detecting Intrusions. Journal of the Korean Data Analysis Society. 12. 983-996 (2010)
10. Kim, Y. M.: A study on analysis of factors on in-hospital mortality for community-acquired pneumonia. Journal of the Korean Data Information Science Society. 22. 389-400 (2011)
11. Park, M., Choi, S., Shin, A.M., Chul, H.: Analysis of the Characteristics of the Older Adults with Depression Using Data Mining Decision Tree Analysis. J Korean Acad Nurs. 43. 1-10 (2013)
12. Huh, M.H., Lee, Y.G.: Data mining modeling and case 2nd ed. Hannarae, Seoul (2008)
13. Dahle, L.K., Muller, M., Delitto, A.: Visual assessment of foot type and relationship of foot type to lower extremity injury. J. Ortho. Sports Phys. Ther. 14. 70-74 (1991)
14. Root, M.L., Orien, W.P., Weed, J.H.: Normal and abnormal function of the foot. Clincal Biomechanics Corp., Huddersfield (1977)

Non-Preference Based Pruning Algorithm for Multi-Objective Redundancy Allocation Problem

Tipwimol Sooktip[1], Naruemon Wattanapongsakorn[2] and Sanan Srakaew[3]

Department of Computer Engineering
King Mongkut's University of Technology Thonburi
Bangkok, Thailand
[1]s.tipwimol@gmail.com, [2]naruemon@cpe.kmutt.ac.th,
[3]sanan@cpe.kmutt.ac.th

Abstract. A non-preference based pruning algorithm is proposed to rank the Pareto-optimal solutions according to the cost and reliability trade-off for solving multi-objective redundancy allocation problem. The proposed method demonstrates on multi-objective redundancy allocation problem with mixing of non-identical component types in each subsystem. The objectives of system design are to maximize system reliability and minimize system cost simultaneously while satisfying system requirement constraints. Non-dominated sorting genetic algorithm-II (NSGA-II) finds an approximation of Pareto-optimal solutions. After obtaining the approximation of Pareto-optimal solutions by NSGA-II, K-means clustering is used to cluster the approximation of Pareto-optimal solutions in to some trade-off regions. Thereafter, the Pareto-optimal solutions are ranked based on the cost and reliability trade-off compare to the centroid solution of each cluster. The results show that the proposed method is able to identify the most-compromised solution.

Keywords: multi-objective optimization · redundancy allocation problem · pruning algorithm · non-preference based

1 Introduction

Generally, a multi-objective optimization problem has a large set of trade-off solutions. The set of non-dominated solutions or Pareto-optimal solutions have trade-off between the objective functions in which a gain in one objective causes sacrifices in the other objective. Therefore, the most-compromised solution is difficult to identify. This can be challenging for selecting one Pareto-optimal solution that can be practically implemented and compromised between the objectives as the system is designed.

The redundancy allocation problem (RAP) is NP-hard problem. The RAPs have been researched for finding the approximation of Pareto-optimal solutions by using NSGA-II [2], which is a well-known algorithm and efficient in searching the Pareto-optimal solutions [3 and 4] for multi-objective optimization problem.

K.J. Kim (ed.), *Information Science and Applications*,
Lecture Notes in Electrical Engineering 339, DOI 10.1007/978-3-662-46578-3_94

K. Deb et al [5] proposed the reference pointed based non-dominated sorting genetic algorithm-II (R-NSGA-II). The decision maker (DM) specifies the reference points of all objective function. After that, NSGA-II ranks the non-dominated solutions and search for the optimal solutions that close to the reference points in objective space. J. Branke et al. [6] presented a method that modified the definition of dominance. The DM needs to specify the minimum and maximum acceptable trade-offs for each pair of objectives, which represented by slope of straight line. The dominated areas are expanding while compare to traditional definition of dominance. Therefore, some non-dominated solutions are pruned. As the number of objectives increases, specify minimum and maximum trade-offs is need to specify the trade-off values for all pair of objectives. Tilahun and Ong [7] proposed the fuzzy preferences incorporate with genetic algorithm (GA) for multiple DMs. This method collected preferences as fuzzy conditional trade-offs then formulated the acceptability of preference membership functions. GA generates weight values for objective functions according to the DM's trade-off values. This method provided flexible trade-off however it is difficult to specify trade-off for every alternative solution.

This paper proposed a non-preference based pruning algorithm for ranking the optimal system design of RAP according to cost and reliability trade-off. In this research, the RAP considers series-parallel system with mixing of non-identical component types. NSGA-II is applied to find the approximation of Pareto-optimal solutions. K-means clustering is used to cluster the approximation of Pareto-optimal solutions in to some trade-off regions. After clustering, the proposed method ranks the approximation of Pareto-optimal solutions according to cost and reliability trade-off. Therefore, the DM is able to select the final system design from a large size of the solutions.

The remaining of the paper is organized as follows. In Section II, multi-objective RAP is described. In Section III, the non-preference based pruning algorithm is presented. In Section IV, the experimental results and discussion are provided. Finally, the conclusion is in Section V.

2 Multi-Objective Redundancy Allocation Problem

The RAP is to determine the optimal design configuration from the redundant alternatives. The subsystems are connected in series, while the redundant components connected in parallel in each subsystem. The RAP with a series-parallel structure is shown in Fig. 1. The redundant components improve system reliability, while system cost and weight is increasing. Due to mixing of non-identical component type is allowed, the problem can be very complex and the search space is extended to large size. Therefore, it is difficult to find the Pareto-optimal solutions and identify the selected solutions.

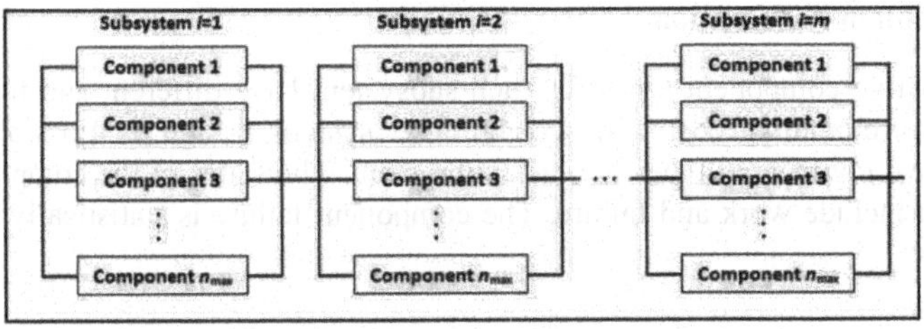

Fig. 1. General series-parallel redundancy system [8]

2.1 Problem Formulation

The model of RAPs has been proposed by previous researches [1, 8, 9 and 10]. The system consists of m subsystems that connected in series. The number of component type j allocated in subsystem i is x_{ij}, which represented in vector by $\mathbf{x}$ or $\mathbf{x}_i$. The reliability in subsystem i is R_i. The optimal configuration of system design with mixing of non-identical components has to be determined from t_i different component types in subsystem i^{th}. In this research, the objective functions are to maximize system reliability, R_{sys} and minimize system cost, C_{sys} simultaneously while satisfying system weight constraint, W_{sys_con}. The mathematical model formulation is presented as follows:

$$\max R_{sys}(\mathbf{x}) = \prod_{i=1}^{m} R_i(\mathbf{x}_i) \tag{1}$$

$$\min C_{sys}(\mathbf{x}) = \sum_{i=1}^{m}\sum_{j=1}^{t_i} c_{ij}x_{ij} \tag{2}$$

$$\text{s.t.} \quad \sum_{i=1}^{m}\sum_{j=1}^{t_i} w_{ij}x_{ij} \le W_{sys\_con}$$

$$1 \le \sum_{j=1}^{t_i} x_{ij} \le n_{max}$$

$$R_i(\mathbf{x}) = 1 - \prod_{j=1}^{t_i}\left(1 - R_{ij}(\mathbf{x})\right)^{x_{ij}}$$

where $x_{ij} \in \{0, 1, 2, \ldots, n_{\max}\}$, $i = 1, 2, \ldots, m$ and $j = 1, 2, \ldots, t_i$. The jth component in subsystem i has reliability (r_{ij}), cost (c_{ij}) and weight (w_{ij}). The maximum number of components in subsystem i is n_{max}.

2.2 Problem Assumption

The alternative component types for each subsystem have different and independent component reliability, cost and weight. The system design is possible to mix non-identical component types for each subsystem. The states of the components and the system include work and failure. The component failure is statistically independent.

3 The Non-Preference Based Pruning Algorithm

The non-preference based pruning algorithm aim to obtain the ranking of Pareto-optimal solutions for multi-objective RAP. The steps of this method are following:

1. NSGA-II searches the approximation of Pareto-optimal solutions.
2. The DM specifies the number of clusters, *k*.
3. K-means clustering finds a centroid of each cluster. The Pareto-optimal solution that is closest to each centroid is obtained and called centroid solution.
4. The cost and reliability trade-off between the non-dominated solution and its centroid solution in each cluster is calculated using the following equation.

$$\text{Trade}-\text{off}_{ij} = \frac{\left| f^{cost}_{centroid_j} - f^{cost}_{ij} \right|}{\left| f^{reliability}_{centroid_j} - f^{reliability}_{ij} \right|} \tag{3}$$

   where $f^{cost}_{centroid_j}$ and $f^{reliability}_{centroid_j}$ are the system cost and system reliability of centroid solution in cluster *j*, respectively. f^{cost}_{ij} and $f^{reliability}_{ij}$ are the system cost and system reliability of solution *i* in cluster *j* where $j = 1, 2, \ldots, k$. The trade-off$_{ij}$ is a ratio of the absolute difference between the system cost of the centroid solution in cluster *j* and the system cost of solution *i* in cluster *j* to the absolute difference between the system reliability of the centroid solution in cluster *j* and the system cost of solution *i* in cluster *j*. The trade-off$_{ij}$ represents sacrificing units in the objective to order to gain one unit in the other objective.
5. In each cluster, the non-dominated solutions are sorted according to the cost and reliability trade-off value in ascending order. The alternative that has the smallest cost and reliability trade-off value is rank 1 which is the least amount to sacrifice in one objective when compare to its centroid solution.
6. The ranking of Pareto-optimal solutions with the cost and reliability trade-off values are presented to the DM. The cost and reliability trade-off value indicates the cost-effective solutions. The solution with low trade-off value is preferred to the other solutions.

4 Experimental Results and Discussion

In our experiment, the system configuration of RAP [6] consists of 7 subsystems, with different component types, as presented in Table 1. The objectives are to maximize system reliability and minimize system cost subject to a weight constraint. NSGA-II is used as the searching algorithm. In order to achieve the optimal solutions, NSGA-II requires parameter tunings including a population size, a mutation probability, a crossover probability and a max generation. After significant trial and error experiments, we obtain the optimal parameter settings of NSGA-II as shown in Table 2. The binary tournament selection, simulated binary crossover (SBX) [11] and polynomial mutation operators [12] are used in NSGA-II.

Table 1. Component input data

Note: Sub = subsystem, Comp = component, The symbol "-" means that design alternative is not available.

Sub *i*	Design Alternative *j*											
	Comp Type 1			*Comp Type 2*			*Comp Type 3*			*Comp Type 4*		
	r_{ij}	c_{ij}	w_{ij}	r_{ij}	c_{ij}	w_{ij}	r_{ij}	c_{ij}	w_{ij}	r_{ij}	c_{ij}	w_{ij}
1	0.90	1	3	0.93	1	4	0.91	2	2	0.95	4	5
2	0.95	4	8	0.94	2	10	0.93	1	9	-	-	-
3	0.85	2	7	0.90	3	5	0.87	1	6	0.92	4	4
4	0.83	3	5	0.87	4	6	0.85	5	4	-	-	-
5	0.94	2	4	0.93	2	3	0.95	5	5	0.94	2	4
6	0.99	6	5	0.98	4	4	0.97	2	5	0.96	2	4
7	0.91	4	7	0.92	4	8	0.94	5	9	-	-	-

Table 2. Parameter setting for NSGA-II

Parameter	Value
Population size	100
Mutation probability	0.07
Crossover probability	0.9
Max generation	1000

We consider k-means clustering $k = 3$, so that 3 clusters are obtained representing groups of solutions with low system reliability, medium system reliability and high system reliability, respectively. Two test cases are considered as follows.

Case 1: Two objectives without system weight constraint are considered. The maximum number of components is 8 for each subsystem. The approximation of Pareto-optimal solutions is shown in Fig. 2.

Case 2: Two objectives with system weight constraint, 100 are considered. The maximum number of components is 4 in each subsystem. The approximation of Pareto-optimal solutions is shown in Fig. 3.

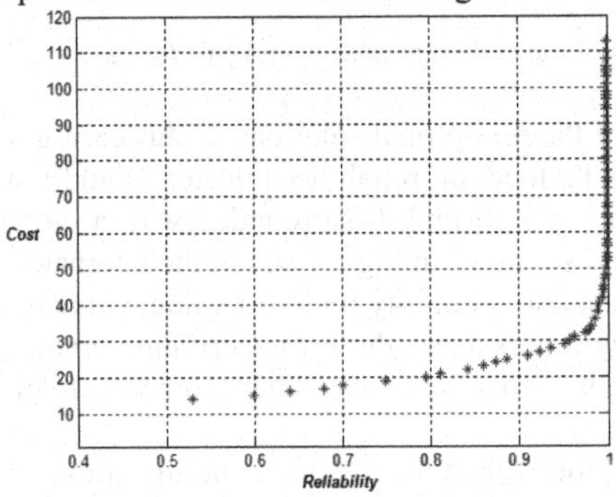

Fig. 2. The solutions of the system reliability and cost for case 1

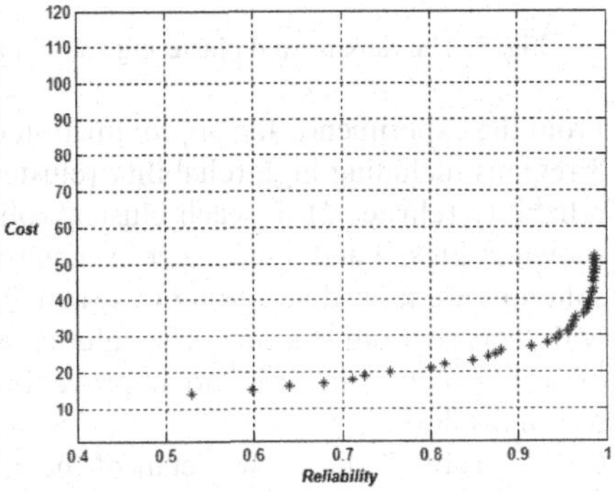

Fig. 3. The solutions of the system reliability and cost for case 2

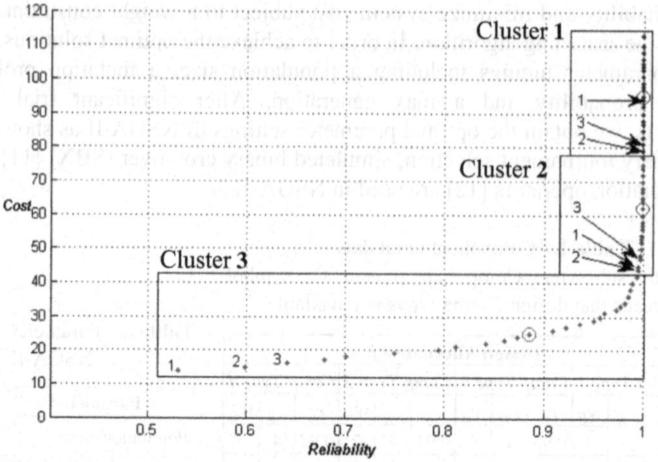

Fig. 4. The ranking and clustering solutions of the system reliability and cost for case 1

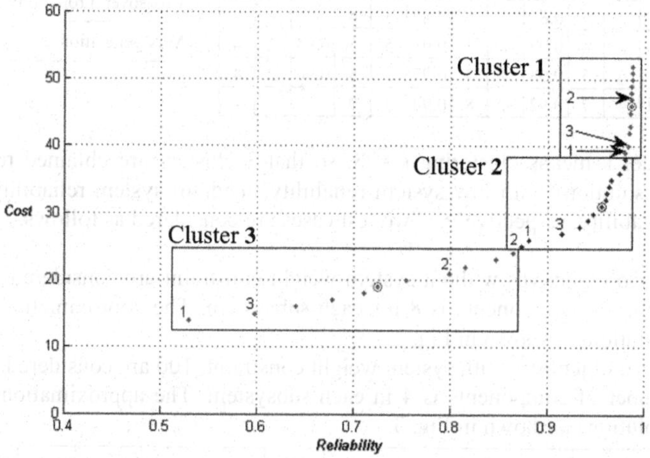

Fig. 5. The ranking and clustering solutions of the system reliability and cost for case 2

From the experiments, the approximation of Pareto-optimal solutions is clustered into 3 regions including high reliability (cluster 1), medium reliability (cluster 2) and low reliability (cluster 3). In each cluster, solution with rank 0 surrounded with a circle shown in Figs. 4 and 5 represent the centroid solution. In Figs. 4 and 5, the alternative solutions are ranked according to system cost and reliability trade-off value compared with their centroid solution. The solution with rank 1 has the least sacrificing amount of one objective in order to improve in the other objective when compare to its centroid solution.

Case 1: Table 3 shows the detail of the solutions including component allocation. In detail at solution rank 0 (cluster 1), subsystem 1 has 2 components of type 1 and 4 components of type 2, while subsystem 2 has 5 components of type 2 and so on. From the table, we can see that component mixing is obtained for all the solutions. In cluster 1, the system reliability of solution rank 1 is less than the system reliability of the centroid solution, while the system cost is lower.

Case 2: The component allocation obtained from case 2 is presented in Table 4. In cluster 1, the system reliability of solution rank 1 is less than the system reliability of the centroid solutions, while the system cost is lower.

The traditional k-means method identifies only the centroid solutions. On the other hand, our method suggests the most-compromised solution according to the system cost and reliability trade-off value besides the centroid solution. Our method is suitetable for multi-objective optimization problems where the DM has no preference in any objectives.

Table 3. Component allocation for case 1
Note: Sub = subsystem, *R* = reliability, *C* = cost, *W* = weight

Rank	R	C	W	Cluster	Sub 1				Sub 2			Sub 3				Sub 4			Sub 5			Sub 6				Sub 7		
					1	2	3	4	1	2	3	1	2	3	4	1	2	3	1	2	3	1	2	3	4	1	2	3
0	0.999995692	95	237	1	2	4	0	0	0	5	0	1	1	5	0	3	3	1	0	2	3	0	3	1	0	0	6	0
1	0.999993263	94	235	1	2	4	0	0	0	5	0	1	1	5	0	3	3	1	1	1	4	0	3	1	0	0	5	0
0	0.999763233	62	173	2	0	4	0	0	0	4	0	0	0	5	0	3	2	0	1	2	1	0	1	1	1	0	4	0
1	0.995507911	45	134	2	0	3	0	0	0	3	0	0	0	5	0	0	3	0	1	2	0	0	0	2	0	0	3	0
0	0.885793781	25	77	3	0	2	0	0	0	2	0	0	0	2	0	2	0	0	2	0	0	0	0	2	0	0	0	1
1	0.529536088	14	42	3	0	1	0	0	0	1	0	0	0	1	0	1	0	0	1	0	0	0	0	1	0	0	1	0

Table 4. Component allocation for case 2.
Note: Sub = subsystem, *R* = reliability, *C* = cost, *W* = weight

Rank	R	C	W	Cluster	Sub 1				Sub 2			Sub 3				Sub 4			Sub 5			Sub 6				Sub 7		
					1	2	3	4	1	2	3	1	2	3	4	1	2	3	1	2	3	1	2	3	4	1	2	3
0	0.98599356	46	98	1	1	2	0	0	0	2	0	0	0	1	2	1	1	1	2	0	0	0	0	0	2	2	1	0
1	0.979962339	39	96	1	1	2	0	0	0	2	0	0	1	2	0	1	1	1	2	0	0	1	0	1	0	1	1	0
0	0.955865976	31	91	2	0	2	0	0	0	2	0	1	0	2	0	1	1	0	2	0	0	0	0	2	0	1	1	0
1	0.88067852	26	75	2	0	2	0	0	0	2	0	0	0	2	0	1	1	0	1	0	0	1	0	0	0	1	1	0
0	0.726052908	19	63	3	0	2	0	0	0	2	0	0	0	2	0	0	1	0	1	0	0	1	0	0	0	0	1	0
1	0.529536088	14	42	3	0	1	0	0	0	1	0	0	0	1	0	1	0	0	1	0	0	0	0	1	0	0	1	0

5 Conclusion

The proposed method aims to solve the multi-objective RAPs and rank the most-compromised solutions among the large set of the optimal solutions. The final pruned solutions are presented to determine choices of system design for the RAP. After the approximation of Pareto-optimal solutions is obtained by NSGA-II, K-means clustering is used to cluster the Pareto-optimal solutions in to some trade-off regions. Then, this algorithm ranks the Pareto-optimal solutions that emphasizes on the cost and reliability trade-off. The alternatives are ranked from the lowest to the highest trade-off value. The cost and reliability trade-off value indicates the least amount of one objective to sacrifice in order to improve in the other objective when compare to its centroid solution. The alternative that has the smallest cost and reliability trade-off value is preferred. This is a simple method that can provide most cost-effective solutions to the decision maker.

6 Acknowledgment

Financial support from the National Research University Project of Thailand's Office of the Higher Education Commission, Thailand Research Fund through the Royal Golden Jubilee Ph.D. Program and King Mongkut's University of Technology Thonburi are acknowledged.

7 References

1. Fyffe, D.E., Hines, W.W., Lee, N. K.: System Reliability Allocation and a Computational Algorithm. IEEE Trans. Reliab. R-17, 64-69 (1968)
2. Deb, K., Pratab, A., Agrawal, S., Meyarivan, T.: A Fast and Elitist Nondominated Sorting Genetic Algorithm for Multi-objective Optimization : NSGA II. IEEE Trans. Evol. Comput., vol. 6, 182 -197 (2002)
3. Murugan, P., Kannan, S., Baskar, S..: NSGA-II Algorithm for Multi-objective Generation Expansion Planning Problem. Electric Power Systems Research, 79 (4), 622-628 (2009)
4. Nebro, A.J., Durillo, J.J. , Coello Coello, C.A., Luna, F., Alba, E.: A Study of Convergence Speed in Multi-objective Metaheuristics. Lecture Notes in Computer Science (including subseries Lecture Notes in Artificial Intelligence and Lecture Notes in Bioinformatics), 5199 LNCS, 763-772 (2008)
5. Deb, K., Sundar, J., Udaya Bhaskara Rao, N., Chaudhuri, S.: Reference Point Based Multi-Objective Optimization Using Evolutionary Algorithms. International Journal of Computational Intelligence Research, 2(3), 273–286 (2006)
6. Branke, J., Kaussler, T., Schmeck, H.: Guidance in Evolutionary Multi-objective Optimization. Advances in Engineering Software, 32:499–507 (2001)
7. Tilahun, S.L., Ong, H.C.: Fuzzy Preference of Multiple Decision-makers in Solving Multiobjective Optimisation Problems Using Genetic Algorithm. Maejo International Journal of Science and Technology, 6 (2), 224-237 (2012)
8. Sooktip, T., Wattanapongsakorn, N.: Pruning Algorithm for Multi-Objective Optimization with Decision Maker's Preferences of the System Redundancy Allocation Problem. In: 4th International Conference on IT Convergence and Security: ICITCS2014, October 28-30, Beijing, China (2014)
9. Onishi, J., Kimura, S., James, R.J.W., Nakagawa, Y.: Solving The Redundancy Allocation Problem with a Mix of Components Using the Improved Surrogate Constraint Method. IEEE Trans. Reliab., vol.56, no.1, 94-101, March (2007)
10. Sooktip, T., Wattanapongsakorn, N., Coit, D.W.: System Reliability Optimization with k-out-of-n Subsystems and Changing k. In: 9th International Conference on Reliability, Maintainability and Safety, art. no. 5979487, pp. 1382-1387 (2011)
11. Deb, K., Agrawal, R.: Simulated Binary Crossover for Continuous Search Space. Complex Systems, 9, 115-148 (1995)
12. Deb, K. Goyal, M.: A Combined Genetic Adaptive Search (Geneas) for Engineering Design. Computer Science and Informatics, Vol. 26, No. 4, 30-45 (1996)

A Combined AdaBoost and NEWFM Technique for Medical Data Classification

Khaled A. Abuhasel[1], Abdullah M. Iliyasu[1,2,*], Chastine Fatichah[3]

[1]College of Engineering, Salman Bin Abdulaziz University,
P.O. Box 173, Al-Kharj 11942, Kingdom of Saudi Arabia
[2] Department of Computational Intelligence and Systems Science,
Tokyo Institute of Technology, Japan
[3]Department of Informatics, Institut Teknologi Sepuluh Nopember, Surabaya, Indonesia
*a.iliyasu@sau.edu.sa

Abstract. A hybrid technique combining the AdaBoost ensemble method with the neural network with fuzzy membership function (NEWFM) method is proposed for medical data classification and disease diagnosis. Combining the Adaboost, a general method used to improve the performance of learning methods, with the 'standard' NEWFM, which uses as base classifiers, ensures better accuracy in medical data classification tasks and diagnosis of diseases. To validate the proposal, four medical datasets related to epileptic seizure detection, Parkinson, cardiovascular (heart), and hepatitis disease diagnoses were used. The results show an average classification accuracy of 95.8% (made up of best accuracy of 99.5% for epileptic seizure, 87.9% for Parkinson, 97.4% for cardiovascular (heart) disease, and 98.7% for Hepatitis dataset classifications), which suggests that the proposed technique is capable of efficient medical data classification and potential applications in disease diagnosis and treatment.

Keywords: biomedical engineering, adaboost ensemble method, medical data classification, disease diagnosis, fuzzy membership, neural network

1 Introduction

The growth of information technology (IT) has accelerated the development of research in medical data classification and disease diagnosis. The process has become an interesting pursuit with numerous challenges of its own for researchers. Data for diagnostic procedures are usually data types including text, signals, images, voice, etc. [1-11] that are sourced from a variety of sources within the health care records, such as the doctor's notes, the laboratory results, radiological results, pathological results, and many other sources. These health records of patients are collected into a dataset that is used to diagnose new patients based on the dataset. The use of medical dataset for diagnosis, called medical data classification tasks, is usually used to infer the nature of ailment based on data training/examples from health records of patients.

K.J. Kim (ed.), *Information Science and Applications*,
Lecture Notes in Electrical Engineering 339, DOI 10.1007/978-3-662-46578-3_95

Electroencephalogram (EEG) and electrocardiogram (ECG) signals are usually used in diagnosing the diseases such as epileptic seizure, schizophrenia, asthma, and arrhythmia [2-7]. Such uses in [2-6] focus on epileptic seizure detection, schizophrenia detection and Alzheimer based on EEG signals. While the work in [8-11] focus on medical data classification tasks for ailments such as diabetes, heart disease, hepatitis, Parkinson, liver, and cancer.

Most of the available literature utilise Neural Network methods as the common classifiers to classify medical data [2, 3, 4, 6], which together with fuzzy theory have been proven as classification methods that can effectively predict patterns required to generate predictive rules. One of such approaches, the neural network with fuzzy membership functions (NEWFM), maintains important information without decreasing the classification capability by updating weighted fuzzy membership functions, and it has been applied in [12] in feature selection, and the antibody deficiency syndrome prediction, with high accuracy. The real-time premature ventricular contraction detection using NEWFM was proposed in [13]. Forecasting KOSPI based on NEWFM was also presented in [14]. A hybrid Principle Component Analysis based NEWFM using EEG signals was recently proposed in [6] for epileptic seizure detection. To further improve the performance of classification methods, some hybrid approaches combining different optimization algorithms with other classification methods have also been proposed [15, 16].

In this study, we propose a hybrid method that uses NEWFM as base classifiers integrated into the 'standard' AdaBoost methods. By combining AdaBoost and NEWFM, the proposed method ensures improved accuracy in diagnosing diseases that require medical data classification tasks. To validate the proposal, we applied in a medical classification task for the diagnosis of epileptic seizures, Parkinson, cardiovascular (heart), and hepatitis ailments.

Section 2 presents a brief description of the AdaBoost ensemble method and the NEWFM method. The proposed AdaBoost-based NEWFM technique is presented in Section 3. The experimental results validating the use of the proposed method in medical data classification tasks are presented and discussed in Section 4.

2 Review of the Adaboost Ensemble and NEWFM Methods

The AdaBoost ensemble method [17], short for Adaptive Boosting, is a widely used Boosting algorithm for combining multiple base classifiers to produce a form of committee whose performance can be significantly better than that of any the base classifier [18].

The principal of boosting is that the base classifiers are trained in sequence, and each base classifier is trained using a weighted form of the data set where the weighting coeffecient associated with each data point depends on the performance of the previous classifiers. In particular, points that are missclasiffied by one of the base classifiers are given greater weight when used to train the next classifier in the sequence [17]. Once all the classifiers have been trained, their predicitons are then combined through a weighted majority voting scheme [18].

At each stage of the AdaBoost algorithm a new classifier is trained using a dataset in which the weighting coefficients are adjusted according to the performance of the previously trained classifier, thereby giving more weight to the misclassified data points. Finally, when the required numbers of base classifiers have been trained, they are combined to form a committee using coefficients that assign weights to different base classifiers. Detailed discussion on the AdaBoost method can be found in [17].

Fuzzy neural network methods with various architectures for learning, adaption, and rule extraction have been studied in many literatures as an adaptive decision support. A Neural Network with Weighted Fuzzy Membership Function or simply NEWFM is a method that uses weighted membership functions [12, 13, 14] considers the weights for the membership functions, a property that makes it different from other fuzzy neural network approaches [12]. The structure of the NEWFM consists of three layers, namely: the input, the hyper box layer, and the class layer, as discussed in [6].

The input layer contains n input nodes and the hyper box layer is composed of m hyper box nodes (B_l). Each hyper box nodes is connected to a class node and it consists of three weighted fuzzy membership functions (WFM). There are p-class nodes that are connected to one or more hyper box in the output layer. The input pattern is represented as $A_h = (a_1, a_2, ..., a_n)$, while the connection between a hyper box node B_l and a class node C_i is represented by a weight w_{li}. The initialisation of w_{li} parameters are 0 but w_{li} will be set to 1 when there is a connection from B_l to C_i. The parameter C_i has one or more connections from hyper box nodes; whereas B_l is restricted to have one connection to C_i [12]. There are n fuzzy sets on the hyper box B_l and each fuzzy set of B_l has three WFM (μ_j for j=1, 2, 3) that are adjusted by the learning algorithm.

The learning algorithm for NEWFM is tailored towards adjusting the positions of vertices and weights on the fuzzy membership function, and to connect the hyper box nodes to class nodes. The centre vertices of fuzzy membership functions are denoted as v_1, v_2, and v_3 representing small, medium, and large respectively. The new vertices and new weights are updated equations (1) and (2) [13].

$$\text{new}\ (v_j) = v_j \pm \alpha E_j \mu(a_i) W_j \tag{1}$$

$$\text{new}\ (W_j) = W_j \pm \beta(\mu(a_i)\text{-}W_j) \tag{2}$$

The parameters α and β are the learning rate and the E_j is the difference value between v_j and a_i. If E_j value is bigger than the adjacent of E_j (E_{j-1} and E_{j+1}), the smaller one is selected. The output of the B_l is calculated using equation (3).

$$Output(B_l) = \frac{1}{n} \sum_{i=1}^{n} \sum_{j=1}^{3} B_l^i\left(\mu(a_i)\right) W_j \tag{3}$$

The fuzzy if-then rules that are obtained from the result of the NEWFM learning process are used to classify the input data. The weighted membership functions (WFM) from the learning results are placed in the hyper box layer [12]. Therefore, the rules can be generated from the WFM directly that it will be used to predict classes.

There are two learning rate parameters for the NEWFM method: the alpha (α) parameter, used to update new vertices, and the beta (β) parameter β, which used to update new weights. The learning rate parameters of NEWFM determine the speed (where optimal vertices or weights selection is crucial) of the algorithm.

3 Proposed Adaboost-based NEWFM Technique

This section discusses our proposed method that combines the AdaBoost ensemble method with the NEWFM classification method for efficient medical data classification tasks. Accuracy is often paramount in medical classification tasks, especially in those required for efficient disease diagnosis. In addition, this diagnosis needs to be performed in real time. Since the proposed technique integrates the utility inherent to the two methods separately, we envisage that the proposed method will vastly improve the classification accuracy required for efficient disease diagnosis.

1. **Input:** a set of training data with labels
2. **Initialize**: the weight of training data: $w_n^{(m)} = 1/N$ for $n = 1, 2, \ldots, N$.
3. **Do For** $m = 1, \ldots, M$

(a) Train NEWFM with respect to distribution w_n and obtain hypothesis y_m

(b) Calculate the training error of y_m:

$$\varepsilon_m = \frac{\sum_{n=1}^{N} w_n^{(m)} I(y_m(x_n) \neq t_n)}{\sum_{n=1}^{N} w_n^{(m)}} \tag{4}$$

(c) Set weight of base classifier y_m:

$$\alpha_m = \ln\left\{\frac{1-\varepsilon_m}{\varepsilon_m}\right\} \tag{5}$$

(d) Update the weight of training data:

$$w_n^{m+1} = w_n^m \exp\{\alpha_m I(y_m(x_n) \neq t_n)\} \tag{6}$$

3. **Output:**

$$Y_M(x) = sign\left(\sum_{m=1}^{M} \alpha_m y_m(x)\right) \tag{7}$$

Table 1. Proposed AdaBoost based NEWFM Method

As seen from the algorithm in Table 1, in the first step of the proposed algorithm the weight $w_n^{(1)}$ of training data with the same value are initialised. This is then iterated from 1 until *M* of base classifiers with NEWFM. The first iteration, $y_1(x)$, is trained using NEWFM classifiers and the result is used to compute the error with weight coefficient $w_n^{(1)}$. In equation (4), an update weight coefficient w_n^m that increases weight coefficient value for misclassified the training data and decrease the weight coefficient value for correctly classified the training data. Finally, when desired number of base classifiers has been trained, they are combined to form a committee using coefficients that give different weights to different base classifiers.

4 Experimental Validation of Proposed Hybrid Method

In this section we assess the extent to which the proposed hybrid Adaboost-based NEWFM technique outlined in the previous section improves medical data classification tasks.

We start with a description of the dataset to be used together with an overview of the scenario in which the experiments were carried out.

4.1 Description of dataset

Our experimental set up consists of a dataset for from four different ailments: epileptic seizure detection in epilepsy, Parkinson disease, cardiovascular diseases and hepatitis disease.

The first dataset, sourced online via [19], consists of EEG signals for epileptic seizure classification classified as normal or infected (or seizure) datasets. This dataset contains of 2 sets of data labelled as Z and S. Each set contains single-channel scalp EEG segments with duration 23.6s and each sampled at 173.61Hz [6].

The remaining three datasets used for our classification tasks are sourced from the UCI Machine learning Repository [18] comprising of data used in Parkinson, cardiovascular and hepatitis disease classification and diagnosis. The Parkinson dataset contains 22 attributes from 195 data types, of which 46 indicate (abnormal) presence of Parkinson and 149 as normal patients. The heart disease dataset contains 13 attributes from 270 data types out of which 120 indicate presence and 150 indicate absence of heart disease. The hepatitis dataset consists of 18 attributes taken from 80 data types, of which 45 indicate inactive and 35 indicate active Hepatitis causing tissue.

These four medical data types classification is employed to evaluate the prediction performance of the proposed technique (see columns 2 and 3 in Table 2).

4.2 Evaluation methods and Experimental Results

As used in other literature, the evaluation parameters used in classification problems, like the one proposed in this study, are sensitivity, specificity and classification accuracy [4].

Table 3 also presents the experimental results on four medical data classification tasks based on the specified parameter values (number of hyperboxes and the optimal learning rate parameters (α and β)).

These results show that the best accuracy for the epileptic seizure detection task is 99.5% with hyperbox = 10, $\alpha = 0.1$ and $\beta = 0.1$. For the Parkinson disease classification task, the best accuracy is 87.69% with hyperbox = 15, $\alpha = 0.1$, and $\beta = 0.2$, while for the heart disease classification task, the best accuracy is 97.40% with hyperbox = 15, $\alpha = 0.1$ and $\beta = 0.4$. Finally, the best accuracy obtained for the hepatitis disease classification task is 98.75% with hyperbox = 10, $\alpha = 0.1$ and $\beta = 0.2$.

The comparison of results on medical data classification tasks between the separate 'standard' NEWFM and Adaboost ensemble methods and the proposed (AdaBoost+NEWFM) technique is presented in Table 3.

1 Classification Task	2 Category	3 Data size	4 Number of Hyper boxes	5 Optimal learning rates		6 Accuracy (%)
				α	β	
Epileptic Seizure	Seizure	100	10	0.1	0.1	99.50
	Normal	100	15	0.1	0.2	98.50
Parkinson	Abnormal	46	10	0.2	0.6	87.17
	Normal	149	15	0.1	0.2	87.69
Heart disease	Presence	120	10	0.1	0.2	95.92
	Absence	150	15	0.1	0.4	97.40
Hepatitis	Active	45	10	0.1	0.2	98.75
	Inactive	35	15	0.1	0.1	97.50

Table 2. The experimental results on medical data classification tasks for proposed (AdaBoost+NEWFM) technique with parameter value variations

Classification Task	Number of Hyper boxes	Optimal learning rates		Accuracy (%)		
		α	β	Adaboost	NEWFM	AdaBoost + NEWFM
Epileptic Seizure	10	0.1	0.1	98.50	99.50	99.50
Parkinson	15	0.1	0.2	82.05	77.95	87.69
Heart disease	15	0.1	0.4	85.18	92.59	97.40
Hepatitis	10	0.1	0.2	87.50	92.50	98.75

Table 3. Comparison of results on medical data classification tasks between standard methods and the proposed (AdaBoost+NEWFM) technique

In all tasks considered the accuracy obtained from our proposed hybrid (AdaBoost+NEWFM) method outperforms those from both the 'standard' NEWFM and Adaboost methods that are presently used in most medical data classification tasks.

Although the computational time required for the training process in the proposed hybrid method is thrice higher than that in the 'standard' NEWFM method, it is within the same range as the 'standard' Adaboost ensemble method.

Notwithstanding this, the improved classification accuracy obtained via the proposed method suggests that it has potential uses as an alternative medical data classification technique.

In addition, as seen from Table 4, the proposed method has an average accuracy of 95.84% with excellent results for specificity and sensitivity across all classification tasks. For example, specificity is as high as 100% for the epileptic seizure detection classification task, which means that all EEGs are correctly predicted as epileptic seizure in the test EEG data while those classified as normal reach up to 99%. For Parkinson disease classification tasks, the proposed method obtains 93.48% sensitivity and 85.90% specificity, while the proposed technique achieved 93.48% sensitivity and 85.90% specificity in heart disease diagnosis. For hepatitis disease classification, the proposed (AdaBoost + NEWFM) technique achieved 99% sensitivity and 100% specificity.

Classification Task	Sensitivity (%)	Specificity (%)	Accuracy (%)
Epileptic Seizure	100	99.00	99.50
Parkinson	93.48	85.90	87.69
Heart disease	98.33	96.67	97.40
Hepatitis	99.00	100	98.75
Average accuracy rate			**95.84%**

Table 4. Experimental results on medical data classification

5 Conclusions

To improve the performance and accuracy of classification methods, several approaches, such as robust feature enhancement and hybrid techniques that combine different optimization algorithms with other classification methods, are employed. Our study proposes a technique that combines the AdaBoost ensemble method, a general method for improving the performance of almost any learning algorithm, with the NEWFM method to enhance the efficiency of medical data classification tasks. The use of NEWFM makes a better decision function than traditional base classifiers, which ensures improved classification accuracy. In this manner, the proposed method guarantees accuracy in disease diagnosis and treatment. Employing four medical datasets for classification of epilepsy, Parkinson, cardiovascular, and hepatitis diseases we validated the accuracy of the proposed hybrid method. Based on the experiments, we see that the best accuracy of epileptic seizure dataset is 99.5% with the optimal parameters: hyperbox = 10, $\alpha = 0.1$, and $\beta = 0.1$. For Parkinson dataset, the best accuracy is 87.90% with the optimal parameters: hyperbox = 15, $\alpha = 0.1$, and $\beta = 0.2$. While for the heart disease dataset, the best accuracy is 97.40% with the optimal parameters: hyperbox = 15, $\alpha = 0.1$, and $\beta = 0.4$. Finally, the best accuracy of hepatitis dataset is 98.75% with the optimal parameters: hyperbox = 10, $\alpha = 0.1$, and $\beta = 0.2$. These results indicate that the proposed hybrid AdaBoost-based NEWFM technique offers better accuracy than the 'standard' NEWFM and Adaboost methods in most of medical data classification tasks, which suggests that the proposed method could be deployed to a wider range of classification tasks, medical disease diagnosis and treatment.

6 Acknowledgements

This work was sponsored by the Salman Bin Abdulaziz University via the Deanship for Scientific Research.

7 References

1. S. Al-Muhaideb, et al., "Hybrid Metaheuristics for Medical Data Classification," Hybrid Metaheuristics, Studies in Computational Intelligence, Vol. 434, 2013, pp 187-217
2. V. A. Golovko, et al., "Application of neural networks to the electroencephalogram analysis for epilepsy detection," Proc. of Int. Joint Conf. on Neural Networks, Orlando, Florida, USA, August 12-17, 2007
3. S. Ghosh-Dastidar, H. Adeli, and N. Dadmehr, "Principal component analysis-enhanced cosine radial basis function neural network for robust epilepsy and seizure detection," IEEE Trans. on Biomedical Eng., Vol. 55, No. 2, February 2008, pp. 512-518
4. L. Guo, D. Rivero, J. Dorado, J. R. Rabu˜nal, and A. Pazos," Automatic epileptic seizure detection in EEGs based on line length feature and artificial neural networks," Journal of Neuroscience Methods, Vol. 191, 2010, pp. 101–109
5. M. Sabetia, S.D. Katebi, R. Boostani, G.W. Price, "A new approach for EEG signal classification of schizophrenic and control participants," Expert Systems with Applications, Elsevier, Vol. 38, Issue 3, 2011, p.2063–2071
6. C. Fatichah, A. M. Iliyasu, K. A. Abuhasel, et al., "Principal Component Analysis-based Neural Network with Fuzzy Membership Function For Epileptic Seizure Detection," Proc. of 10th Int. Conf. on Natural Comput., Xiamen, China, 19-21 August 2014, pp. 186-191
7. J. P. Betancourt, et al., "Similarity-based fuzzy classification of ECG and capnogram signals", J. of Adv. Comp. Intelligence and Intelligent Informatics, Vol. 17, No. 2, 2013, pp. 302-310
8. M. C. Tu, et al., "Comparative Study of Medical Data Classification Methods Based on Decision Tree and Bagging Algorithms," Proc. of IEEE Int. Conf. on Dependable, Autonomic and Secure Computing, 12-14 Dec. 2009, Chengdu, China, pp. 183 - 187
9. Oh S, et al., "Ensemble learning with active example selection for imbalanced biomedical data classification," Trans. Comput. Biol. Bioinform., Vol. 8, No. 2, 2011, pp. 316-25
10. M. Seera, C. P. Lim, "A hybrid intelligent system for medical data classification," Expert Systems with Applications, Vol. 41, No. 5, 2014, pp. 2239–2249
11. C. Fatichah, M. L. Tangel, et al., "Interest-Based Ordering for Fuzzy Morphology on White Blood Cell Image Segmentation", J. of Adv. Comp. Intelligence and Intelligent Informatics, Vol. 16, No. 1, 2012, pp. 76-86
12. J. S. Lim, et al., "Feature Selection for Specific Antibody Deficiency Syndrome by Neural Network with Weighted Fuzzy Membership Functions", LNAI 3614, 2005, pp. 811 – 820
13. J. S. Lim, "Finding features for real-time premature ventricular contraction detection using a fuzzy neural network system," Trans. on Neural Networks, Vol. 20, No. 3, pp. 522-527
14. S. H Lee, J. S. Lim, "Forecasting KOSPI based on a neural network with weighted fuzzy membership functions," Expert Systems with Applications, Vol. 38, 2011, pp. 4259–4263
15. H. Schwenk, "Using Boosting To Improve A Hybrid HMM/Neural Network Speech Recognizer," Proc. of IEEE Int. Conf. on Acoustics, Speech, and Signal Proc., Vol. 2, pp. 1009 - 1012, Phoenix, Arizona, 15-19 March, 1999
16. M. R. Widyanto, C. Fatichah, "Boosting with Kernel Base Classifiers for Human Object Detection", Asian J. of Inf. Tech., Vol. 7, No. 5, 2008, pp. 183-190.

17. Y. Freud, R.E. Scaphire, “A Short Introduction To Boosting“, J. of Japanese Society for Artificial Intell., Vol. 14, No. 5, 1999, pp. 771-780.
18. C. M. Bishop, “Pattern Recognition and Machine Learning”, Springer, 2006.
19. R. G. Andrzejak, et al., “Indications of nonlinear deterministic and finite-dimensional structures in time series of brain electrical activity: dependence on recording region and brain state,” Physical Review E 64(6):061907, 2001, pp. 1–8.

On sentence length distribution as an authorship attribute

Miro Lehtonen

Faculty of Science at Si Racha
Kasetsart University
sfscimrl@src.ku.ac.th

Abstract. Understanding what makes written texts sound like they are written by their author has been an unsolved problem for hundreds of years. The attributes of authorship are often clumped together as an attempt to solve the case of an unknown author while the practice of investigating a single attribute by eliminating the effect of all others has been paid little attention. One of the debated attributes is the size of the text segments which authors use to group words together. Texts consist of these segments — sentences — which are of different lengths, the values being distributed in ways that are assumed to be characteristic of the author. Comparing the statistics of paired text samples, we can show that differences in the statistics in fact indicate difference in the authorship of the texts. However, certain choices of metrics and units easily lead to random and meaningless results.

1 Introduction

Author identification and verification are some of the key areas where authorship attributes play an important role. They are both closely related to plagiarism detection which in turn requires a combination of technologies from multiple disciplines such as information retrieval, computational linguistics, and even artificial intelligence. The common research question in the field concerns the author of a text when there is great uncertainty about the origin. Answering questions like "What does the text tell about its author?" is an inexact science, and outside the general scope of the research, but we can get a good start if we can identify which textual features best describe the author.

Instead of trying to solve the bigger problem of finding the optimal feature set with optimal feature weights that attributes authorship, we choose to focus on *sentence length*, a single feature which has been debated in the past without any conclusive results [16]. We will investigate the following research questions:

- Is sentence length a relevant authorship attribute?
- What is the exact relation between different authors and different sentence length statistics?
- If the difference in the sentence lengths of two texts is statistically significant, can we assume that the texts are written by two different authors?

K.J. Kim (ed.), *Information Science and Applications*,
Lecture Notes in Electrical Engineering 339, DOI 10.1007/978-3-662-46578-3_96

- Exactly which statistics and statistical tests are relevant to sentence length as an authorship attribute?

Many previous studies about authorship attribution have studied questions such as "Is a text written by single author?" or, given some known sample texts, is an unknown sample written by the same author as the known samples. However, the ultimate answers to these questions can easily be disputed in future research, in particular in areas like biblical studies where the true answers either never existed or were lost thousands of years ago [12]. Unlike previous analyses of texts where the author is unknown or disputable, we look into texts with known authors and compare the statistical variance in the texts written by a single author to that in texts written by different authors, but not crossing the border of the language or genre.

One part of the results clearly shows the connection between authors and their writing styles as reflected in the statistics while the other part shows randomness, thus giving us clues about methods to avoid.

2 Feature characteristics

Selecting features for authorship analysis is not only about which features best fit the model of an author but we also want to maximize the utility of the feature set, i.e. we want it to be easily applicable to as many languages or text genres as possible. Learning algorithms which require some kind of training data in order to function properly are therefore low on priority list, as well as lexicographic features, e.g. character frequencies, where the dominating choice is that of a language more than it is of a writing style.

Sentence length — when understood as the length of an utterance — is one of the universal characteristics of human languages. It is more specific to the physical and mental representation of the speaker in the context of the discourse than it is to their natural language competence. The consequent qualities much appreciated by computer scientists include that the relevant methods are non-parametric and straightforwardly applicable to any natural language.

At first sight, it may seem that there are big differences in the encoding of sentences in different languages and writing systems. But as we consider the definition and semantics of sentence borders, we learn that *punctuation* is defined as "the use of spacing, conventional signs, and certain typographical devices as aids to the understanding and correct reading, both silently and aloud, of handwritten and printed texts" [2] and that the punctuation marks are meant to "encode and facilitate purely discourse relational links between text units in text sentences" [1]. Although the punctuation marks differ across languages, we can safely assume that sentence length is directly proportional to the distance between punctuation marks in all natural languages.

3 Text analysis

The test corpus used in the analysis comes from the PAN 2014[1] benchmark evaluation for author identification and author attribution algorithms which was published in October 2014. The corpus consist of texts in four different genres and four different languages as described in Table 1. Each test case consists of 1–5 texts written by one known author and one text written by an unknown author which can be the same as the known author.

Code	Language	Genre	Test cases
DE	Dutch	Essays	96
DR	Dutch	Reviews	100
EE	English	Essays	200
EN	English	Novels	200
GR	Greek	Articles	100
SP	Spanish	Articles	100

Table 1. Test problems detailed.

We conduct the same procedures on all the texts, regardless of the language or genre. The first step is to split the texts into sentences using punctutation characters included in the POSIX US-ASCII character class `punct`. This operation is safe despite being somewhat language-specific as long as we only compare texts written in the same language because punctuation practices vary across different languages. The second step is tokenization by the Java `StringTokenizer` class, after which we use the token count as the length of the sentence. For the sake of comparison, we also measure the length of the sentence as the number of characters. From arrays of sentence lengths, we move on to probability distributions of the sentence being a certain length. The final step results in a cumulative probability distribution which consists of probabilities of sentences being at most a certain length.

The Kolmogorov-Smirnov test on the equality of two continuous distributions is based on the maximum deviation of the two empirical distributions from each other [11]. It evaluates the null hypothesis that the samples come from the same distribution, which would be the case when the texts for both samples are written by the same author. We use the two-sample KS test which is defined as

$D_{n,n'} = \sup_x |F_{1,n}(x) - F_{2,n'}(x)|$,

where $F_{1,n}$ and $F_{2,n'}$ are the cumulative probability distributions of the sentence lengths measured in each pair of a known text and the corresponding unknown text. If the value of $D_{n,n'}$ is greater than the critical value, we reject the null hypothesis and conclude that the texts were written by different authors. Given the common confidence level of 95% in significance testing in the field of information retrieval (p value of 0.05), we estimate the critical value as

[1] http://pan.webis.de/

$$c(\alpha)\sqrt{\frac{n+n'}{nn'}}$$

where $c(0.05) = 1.36$ [9]. Following the guidelines of PAN submissions and facilitating further analysis, the results are scaled to the range [0,1] so that the critical value will be always have a score of 0.5.

4 Results

We first apply the KS test on all sentence length distributions derived from the test data and see if the positive test cases have different results from the negative test cases. Then we conduct a similar analysis by measuring mean sentence lengths in Section 4.2. Finally, we focus on the part of the test data that gets the lowest scores on the KS test.

4.1 Scoring based on the KS test

Table 2 shows the average scores grouped by the language-genre code for both positive and negative test cases. The scaled scores reflect the distance to the case-specific critical value of the KS test so that scores near 1 denote certainty about the case having a single author whereas scores near 0 indicate big differences in the distributions and a high likelyhood of the samples having different authors.

Unit	Words		Characters	
Code	Same	Diff	Same	Diff
DE	0.526	0.516	0.975	0.981
DR	0.730	0.783	0.991	0.998
EE	0.453	0.432	0.902	0.934
EN	0.472	0.456	0.970	0.932
GR	0.526	0.571	0.994	0.977
SP	0.545	0.571	0.968	0.980
T-test	0.379		0.975	

Table 2. Average scoring of positive and negative test cases based on the Kolmogorov-Smirnov test.

We notice that using the raw character count as the length of the sentences leads to seemingly random results. The texts being quite short, the frequencies are so low that they do not deviate much between distributions, which is why the KS test determines 97.6% of the cases to have a single author. The true figure is 50% (398 out of 796). In order to get past this issue, we divide the sentence lengths into buckets, each bucket holding five lengths[2]. For example, the first bucket holds sentence lengths from 0 to 4, whereas the tenth bucket holds lengths 45–49.

[2] The choice of 5 is arbitrary but justified as the average word length in the English language is 5 characters.

Unit	5-length buckets	
Code	Same	Diff
DE	0.559	0.569
DR	0.794	0.764
EE	0.496	0.507
EN	0.560	0.579
GR	0.665	0.740
SP	0.623	0.656
T-test	0.219	

Table 3. Average scoring as in Table 2 but with similar sentence lengths grouped into buckets.

Using buckets of sentence lengths instead of raw character counts results in fewer lengths, higher probabilities, and higher KS scores which after scaling and conversion into the PAN submission format show as lower scores in Table 3. The results are also more evenly distributed: 367 cases have a score below 0.5 suggesting different authors while 429 cases are assessed having a single author. The difference between the values is still not significant at any common confidence level. The result of a two-tailed paired t-test indicates a 21.9% probability of chance between the positive and negative test cases.

4.2 Mean sentence length

Apart from distributions of sentence lengths, we also want to consider the more basic statistical figures in order to better comprehend the relation between the sentence length feature and the authorship of written texts. Measuring the mean sentence length of a text is straightforward but when calculating the difference between two means, we have some choices. Using the absolute difference in the mean sentence lengths, we compare the mean difference between mean sentence lengths of the test cases in Table 4.

Unit	Words		Characters	
Code	Same	Diff	Same	Diff
DE	1.094	1.776	8.011	11.566
DR	2.607	2.940	16.447	19.180
EE	1.712	2.415	10.238	13.442
EN	1.450	1.660	9.020	10.490
GR	1.212	2.090	8.752	16.037
SP	1.392	2.072	8.572	13.464
T-test	0.0025		0.0054	

Table 4. Mean difference in mean sentence length grouped by text type.

There is a strong correlation between the difference in mean sentence length and the difference in the authorship of the compared texts. According to the t-test, this result holds at a 99% confidence level, regardless of the unit with which sentence length is measured. Despite the positive results, there is no trivial way to convert the numbers into a score between 0 and 1 or any other measure of uncertainty of authorship. Therefore, we settle with concluding that the absolute difference in mean sentence lengths should be taken into account but how exactly it is incorporated into a scoring algorithm will be left for future research.

4.3 The most different distributions

Understanding that two different authors may share stylometric features in their texts including statistics related to sentence length, we want to be cautious about drawing too many conclusions from two texts having similar sentence length distributions. At the same time, we assume that one author only has one writing style within a language and genre, therefore justifying the connection between different distributions and different authors. The lowest scoring test cases are summarized in Table 5.

N	Words		Characters		Buckets	
5	2	.40	4	.80	4	.80
10	5	.50	7	.70	8	.80
15	7	.47	9	.60	11	.73
20	9	.45	10	.50	12	.60
25	13	.52	14	.56	15	.60
30	17	.57	16	.53	17	.57

Table 5. Proportion of test cases with different authors in the lowest scoring test cases.

N is the number of test cases from the bottom of the sorted list of KS test scores. We expect the lowest scores to have the greatest certainty of two texts having different authors. However, when sentence length is measured in number of words, the lowest KS test scores seem unreliable having only 2/5 and 5/10 negative cases among them. Counting characters instead gives us more convincing results: At the bottom of the KS scores, 80% of the test cases have different authors. On the contrary to the observations in Section 4.1, we now understand that combining the KS test with character count based sentence length is useful if we disregard the actual scores and only consider the ranking of the top scores. We are pleased with these results because counting characters instead of words also has another advantage: it is easier to apply to arbitrary languages, including ones where words are not separated by spaces. Whether the results also generalise to other data sets, including other genres and other languages will be under discussion in the future.

5 Related work

A lot recent work in the area of author identification and verification has been conducted as part of the PAN evaluation during the past 12 years of its existence [14]. The algorithms developed by the participants use large feature sets that consist of up to 70,000 automatically generated features per genre [6]. Sentence length is included in many approaches in one form or another, e.g. average sentence length [3] and sentence-length profiles built from raw differences of frequencies [10].

According to Stamatatos, sentence length counts are one of the oldest features attributed to authorship [13] as demonstrated in the 1887 publication by Mendenhall [7] who looked into the average length and in the 1888 publication by Smith [12] who did not count the words in each sentence but measured the average number of sentences per page which is proportional to sentence length, given that the page size is a constant. They both claimed that sentence length is an indicator of authorship, a claim that was quickly refuted by Parker [8].

Later noteworthy publications include Yule in 1944 [16] which supports the claim and Williams [15] who warned against it, and Holmes [5] who also argued against it. For a quite exhaustive list of publications excluding the past decade where sentence length is considered, the author refers to the M.A. thesis of Grieve [4] who notes that most of the recent research is not in favor of using sentence length as an authorship attribute but this is because of a lack of good techniques, not because it does not attribute authorship.

6 Conclusion

Based on the analysis of the PAN 2014 test corpus on author identification, we conclude that sentence length is a relevant authorship attribute, which answers our first, rather high-level research question. The second question about the exact relation between different authors and different statistics turns out to be highly multidimensional and cannot be exhaustively answered in one article. However, we are convinced that, under certain conditions, major differences in sentence length statistics are reliable evidence of texts being written by different authors. For example, we find that character count based units for sentence length are more reliable and versatile than those based on word count. As for the statistical tests, the two-sample Kolmogorov-Smirnov test is highly relevant when we test whether two continuous probability distributions differ, but in the case of author identification, we have to carefully consider how the distributions are constructed so that we can avoid too sparsely distributed values resulting from relatively short document samples.

References

1. Briscoe, T.: The syntax and semantics of punctuation and its use in interpretation. In: Proceedings of the Association for Computational Linguistics Workshop on Punctuation. pp. 1–7 (1996)

2. Encyclopaedia Britannica. Encyclopaedia Britannica, Inc. (1768–2014), https://www.britannica.com
3. Ghaeini, M.: Intrinsic author identification using modified weighted knn. In: Notebook for PAN at CLEF 2013 (2013)
4. Grieve, J.W.: Quantitative authorship attribution: a history and an evaluation of techniques. Master's thesis, Simon Fraser University, British Columbia, Canada (2005)
5. Holmes, D.: The analysis of literary style — a review. Statistical Society A 148, 328–341 (1985)
6. Khonji, M., Iraqi, Y.: A slightly-modified gi-based author-verifier with lots of features (asgalf). In: Notebook for PAN at CLEF 2014 (2014)
7. Mendenhall, T.C.: The characteristic curves of composition. Science 11, 237–249 (1887)
8. Parker, H.A.: Curves of literary style. Science 13(321), 245 (1890)
9. Pearson, E.S., Hartley, H.O.: Biometrika tables for statisticians. vol. 2. University Press, Cambridge (1972), http://opac.inria.fr/record=b1080107
10. Rygl, J.: Automatic adaptation of authors stylometric features to document types. In: Proceedings of 17th International Conference, TSD 2014: Text, Speech and Dialogue. pp. 53–61. Springer (2014)
11. Simard, R., L'Ecuyer, P.: Computing the two-sided Kolmogorov-Smirnov distribution. Journal of Statistical Software 39(11), 1–18 (3 2011), http://www.jstatsoft.org/v39/i11
12. Smith, W.B.: Curves of pauline and pseudo-pauline style i-ii. Unitarian Review 30, 452–460, 539–546 (1888)
13. Stamatatos, E.: A survey of modern authorship attribution methods. Journal of the American Society for Information Science and Technology 60(3), 538–556 (2009), http://dx.doi.org/10.1002/asi.21001
14. Stamatatos, E., Daelemans, W., Verhoeven, B., Potthast, M., Stein, B., Juola, P., Sanchez-Perez, M.A., Barrón-Cedeño, A.: Overview of the Author Identification Task at PAN 2014. Analysis 13, 31 (2014)
15. Williams, C.B.: A note on the statistical analysis of sentence-length as a criterion of literary style. Biometrika 31, 363–390 (1940)
16. Yule, G.U.: On sentence length as a statistical characteristic of style in prose: With application to two cases of disputed authorship. Biometrika 30(3-4), 363–390 (1939), http://biomet.oxfordjournals.org/content/30/3-4/363.short

Learning models for activity recognition in smart homes

*Labiba Gillani Fahad, Arshad Ali, *Muttukrishnan Rajarajan

*School of Mathematics, Computer Science and Engineering, City University London, Northampton Square, London EC1V 0HB, United Kingdom
School of Electrical Engineering and Computer Science, National University of Sciences and Technology, Islamabad, Pakistan
labiba.gillani.2@city.ac.uk, arshad.ali@seecs.nust.edu.pk
r.muttukrishnan@city.ac.uk

Abstract. Automated recognition of activities in a smart home is useful in independent living of elderly and remote monitoring of patients. Learning methods are applied to recognize activities by utilizing the information obtained from the sensors installed in a smart home. In this paper, we present a comparative study using five learning models applied to activity recognition, highlighting their strengths and weaknesses under different challenging conditions. The challenges include high intra-class, low inter-class variations, unreliable sensor data and imbalance number of activity instances per class. The same sets of features are given as input to the learning approaches. Evaluation is performed using four publicly available smart home datasets. Analysis of the results shows that Support Vector Machine (SVM) and Evidence-Theoretic K-nearest Neighbors (ET-KNN) in comparison to the learning methods Probabilistic Neural Network (PNN), K-Nearest Neighbor (KNN) and Naive Bayes (NB) performed better in correctly recognizing the smart home activities.

1 Introduction

The concept of ambient assisted living in smart homes equipped with sensors can enable people with chronic physical and cognitive impairments to live independently at home under a continuous remote monitoring [1]. Activity recognition is a fundamental task in assisted living through which the functional ability of a smart home resident can be identified and tracked [2] [3]. The resident should be able to complete the basic activities [3], for example: meal preparation, eating and sleeping etc. It can facilitate doctors to timely and constantly monitor and identify changes in patients' daily routine and ensures immediate medical aid if needed. Activity recognition using non-intrusive sensors is a preferable choice. Sensors deployed in a smart home capture events about the occupants and their interaction with multiple objects within the environment. The obtained data takes the form of sequences of events ordered in time, where each sequence is related with an activity class [4]. Activity recognition is a challenging problem; challenges arise due to sensor errors, variations in the execution of activities due to user preferences, overlapping among similar activities due to being performed at the same place [5], limited training data and imbalanced activity in-

K.J. Kim (ed.), *Information Science and Applications,*
Lecture Notes in Electrical Engineering 339, DOI 10.1007/978-3-662-46578-3_97

stances when some activities occur more frequently than others. Similar activity classes results in less inter-class variations, while variations in execution of same activity by multiple residents may result in high intra-class variations [2, 3]. These challenges affect the performance of learning classifiers in the recognition of activities [6].

Several learning based activity recognition approaches exist [2, 3, 5, 7, 8, 9]. NB exploits the information of user interaction with multiple objects in a home for activity recognition [1, 7] such that the activity class with the highest probability is assigned to the new instance; however performance of NB can be affected by limited training data. An unsupervised approach mines the discontinuous frequent patterns and groups similar patterns into clusters, while Hidden Markov Model (HMM) is used to recognize the activities [2]. HMM is compared with Conditional Random Fields to recognize the daily activities [10]. Similar activities can be clustered and learning method such as ET-KNN can be applied within each cluster for classification [9]. In contrast, frequent item set mining can be used to find the activity patterns and then density based clustering is applied to recognize activities [5]. PNN and K-means clustering are applied to monitor the daily routine of a smart home occupant [6]. The confidence of assignments obtained through a learning model is estimated by exploiting the distribution of activities through sub-clustering [3]. In [8] , SVM is used for classification of activities, while significant features are selected by PCA. The temporal information of domain knowledge (start time and activity duration) is incorporated in the Dempster Shafer Theory of evidence to improve the recognition accuracy [11].

In this paper, we apply five learning models and compare their performance for activity recognition in smart homes. The features extracted from annotated activity instances are used to train a recognition model. The trained models are used to classify new activity instance. We analyze the performance of each learning model taking into account the challenges involved in activity recognition. The models are evaluated using four publicly available single and multiple resident smart home datasets. The performances of classifiers in the presence of different challenges are analyzed to identify the most robust classifier.

2 Learning Classifiers for Activity recognition

Let $\mathbf{A} = \{A_1, ..., A_k, ..., A_K\}$ be a set of K activity classes and $\mathbf{I}_k = \{I_{1k}, ..., I_{jk}, ..., I_{Jk}\}$ be a set of J activity instances of A_k in the training data observed by R binary sensors installed at different locations in a smart home. Fig. 1 shows the analysis framework for activity recognition. Each activity instance I_{jk} is represented by a set $\mathbf{F}_{jk} = \{f_{jk}^r\}_{r=1}^R$ of R features. Each feature f_{jk}^r represents the number of times a sensor is activated during the activity.

2.1 Naive Bayes (NB)

NB is a probabilistic classifier that estimates the parameters of a distribution based on the independence assumption, such that each feature is assumed to be independent from others within a class [7]. I_{jk} is assigned to the class A_k for which it has maximum posterior probability $P(A_k|I_{jk}) > P(A_m|I_{jk}),\ \forall m\ s.t. 1 \geq m \leq K, k \neq j$, in accordance with the Bayes Theorem. The classifier resulting from the feature independence assumption is known as the Naive Bayes classifier

$$P(I_{jk}|A_k) = \prod_{r=1}^{R} P(f_{jk}^r|A_k), \quad (1)$$

where $P(I_{jk}|A_k)$ is the product of the values of features $\{f_{jk}^r\}_{r=1}^R$ of an activity instance I_{jk} for a given class A_k.

2.2 Probabilistic Neural Network (PNN)

PNN is a feed forward neural network, derived from Bayesian and Fisher discriminant analysis [12]. It has four layers: input, pattern, summation and decision layer. Each pattern neuron receives the feature vector F_{jk} from the input layer. The number of neurons $x_{j'k'}$ in pattern layer is equal to the training samples.
The exponential activation function $\phi_{j'k'}(F_{jk})$ to compute the output is given as

$$\phi_{j'k'}(F_{jk}) = \frac{1}{(2\pi)^{R/2}\sigma^R} \exp\left[-\frac{(F_{jk}-x_{j'k'})^T(F_{jk}-x_{j'k'})}{2\sigma^2}\right], \quad (2)$$

where R denotes the dimension of the pattern vector F_{jk}, σ is the smoothing parameter to control the width of activation function and $x_{j'k'}$ is the neuron vector. The output corresponding to a particular class from the pattern layer is summed in the summation layer given as

$$p_k(F_{jk}) = \frac{1}{(2\pi)^{R/2}\sigma^R} \frac{1}{J} \sum_{j=1}^{J} exp\left[-\frac{(F_{jk}-x_{j'k'})^T(F_{jk}-x_{j'k'})}{2\sigma^2}\right], \quad (3)$$

where J denotes the total number of activity instances in A_k. In the decision layer, F_{jk} is assigned to the class in accordance with the Bayes decision rule assuming an equal a priori probability and equal losses associated with every incorrect decision

$$C^{F_{jk}} = \arg\max_k\{p_k(F_{jk})\}, k = 1, 2, \cdots, K, \quad (4)$$

where $C^{F_{jk}}$ is the estimated class of F_{jk} and K is the total number of activity classes.

2.3 Support Vector Machine (SVM)

SVM finds the most optimal hyper plane to discriminate the data points of two classes with maximum margin [13]. Consider the training sample set $(\mathbf{\Omega}_i, T_i)$, where $\mathbf{\Omega}_i$ is R dimensional feature vector and T_i is the target. A hyper plane is defined as

$$T_i(\mathbf{w}_i.\mathbf{\Omega}_i + b) > 1 - \xi_i, \quad (5)$$

where $\mathbf{w}_i$ is the weight vector, $\mathbf{\Omega}_i$ is the data point on the hyper plane, b is the bias and ξ_i is the slack variable added for the non-linear support vector machine, such that a penalty term $C\sum_{i=1}^{J\times K^2} \xi_i$ is added. The optimization problem is then given by:

$$min_i \left[\frac{1}{2}\|\mathbf{w}_i\|^2 + C\sum_{i=1}^{J\times K^2} \xi_i\right], \quad (6)$$

where C is a positive regularization constant used for the trade-off between a large margin and misclassification error, while ξ_i controls the distance of $\mathbf{\Omega}_i$ from the decision boundary. The final decision function for the non-linear problems is obtained as:

$$f(\mathbf{\Omega}) = sign\ (\sum_{i=1}^{J\times K} \alpha_i T_i Ker(\mathbf{\Omega}.\mathbf{\Omega}_i) + b), \quad (7)$$

where α_i represents the Lagrange multiplier and the kernel function $Ker(\Omega.\Omega_i)$ can be Linear, Radial Basis or Polynomial satisfying the Mercer condition [13].

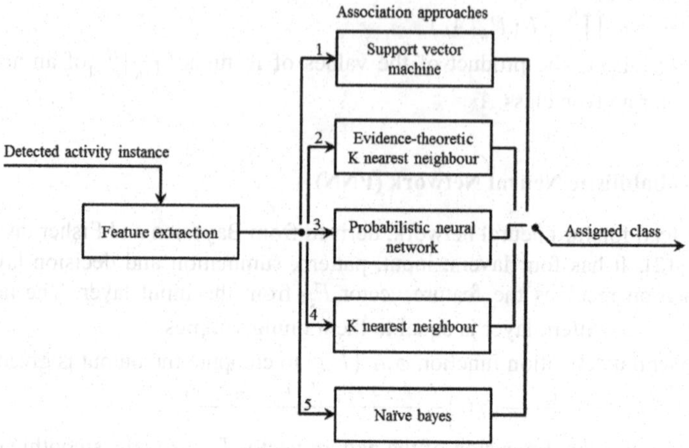

Fig. 1. Framework of learning models for activity recognition.

2.4 K-nearest neighbor (KNN)

KNN [9] classifies activity instances based on their nearest training samples in the feature space, a new activity instance X in the feature space is assigned to the activity class A_k, if it has the minimum distance $D(.,.)$ from A_k.

$$D_{min} = min\{D(X, P_l)\}, l \in K, \tag{8}$$

where P_l are the reference patterns of each activity class A_k.

2.5 Evidence theoretic K nearest neighbor (ET-KNN)

ET-KNN [14] extends KNN to address the issue of uncertainty in data. ET-KNN is based on DST [11], where each neighboring pattern is considered as evidence supporting the hypothesis of the class membership. Consider a classification problem, where a pattern (new activity instance) X has to be assigned to one of the K activity classes. We compare X with K reference patterns: $P_1, P_2, ..., P_K$. Each pattern P_l is supposed to have a degree of membership μ_k^l to each class A_k with $\sum_{l=1}^{K} \mu_k^l = 1$. We perform the Basic Belief Assignment for each pattern using the degree of membership and aggregate using Dempster's rule of combination. The label of the activity class with the maximum belief of membership is assigned to X.

3 Evaluation and Discussion

We evaluate the performance of five learning approaches multi-class SVM, ET-KNN, PNN, KNN and NB on four smart home datasets Kasteren, Aruba, Kyoto2 and Tulum1. Four performance evaluation measures for the comparisons are Precision, Recall, F1score and Accuracy, using True Positives, False Positives, True Negatives and False Negatives [3].

Table 1. Summary of datasets used in the evaluation.

Datasets	No. of resident	Activity classes	Activity instance	Name of activities
Kasteren	1	7	245	Breakfast, Dinner, Drink, Leave home, Shower, Sleep and Toileting
Aruba	1	11	6477	Bed to Toilet, Eating, Enter Home, House Keeping, Leave Home, Bed to Toilet, Eating, Enter Home, House Keeping, Leave Home
Kyoto2	20	5	100	Clean, Cook, Eat, Phone Call, Wash Hand
Tulum1	2	10	1513	Cook Breakfast, Cook Lunch, Enter Home, Group Meeting, Leave Home, R1 Eat Breakfast, R1 Snack, R2 Eat Breakfast, Wash Dish, Watch TV

3.1 Datasets

Table 1 shows the summary of publicly available smart home datasets: Kasteren [10] and Aruba, Kyoto2 and Tulum1 [1] used in the evaluation. Binary sensors are used such as absent/present, open/close, contact switch sensors and motion sensors. Kasteren and Aruba datasets have one, while Tulum1 has 2 residents. In Kyoto2, activities are performed by 20 individuals, while some erroneous features are added such as leaving the stove burner on, leaving the water turned on or not bringing the medicine container to dining room, etc. Activities performed at the same location such as, 'Breakfast', 'Lunch', 'Dinner', 'Enter Home' or 'Leave Home', share same sensors, which results in less discriminative information, highlight less inter-class variations. In multi-residents activities, high intra-class variations are also observed. The results are obtained using fivefold cross validation.

3.2 Analysis and discussion

Table 2 shows the performance comparison of learning approaches: SVM, ET-KNN, PNN, KNN and NB on four datasets. In Kasteren, ET-KNN shows an overall better performance than other classifiers with the precision and recall values of 95.30% and 93.64%, respectively, while NB remains least accurate in the recognition of activities. In Aruba, SVM shows the highest performance with precision and recall rates of 77.16% and 81%, respectively. The accuracy remains comparable to ET-KNN and KNN, and higher than PNN and NB. For Kyoto2 dataset, despite the availability of only 100 instances, activities are correctly recognized due to high inter-class discriminations. SVM obtains the highest F1score (0.92) and accuracy (96.80%). PNN, KNN and ET-KNN remain comparable in performance to SVM. NB performed poor due to fewer amounts of training data and achieve an F1score of 0.57. Also in Tulum1 with high intra-class discriminations, SVM attains a high performance and remains comparable to ET-KNN and KNN. SVM shows an overall better performance in activity recognition in the presence of both less inter and high intra-class variations.

Table 2. Performance comparison of learning approaches on four datasets.

Datasets	Approach	Precision	Recall	F1score	Accuracy
Kasteren	SVM	91.95	91.91	0.91	98.72
	ET-KNN	**95.30**	**93.64**	**0.94**	**98.83**
	PNN	81.70	58.34	0.64	93.47
	KNN	88.25	83.66	0.84	97.20
	NB	4.19	17.14	0.06	75.04
Aruba	**SVM**	**77.16**	**81.00**	**0.78**	**98.83**
	ET-KNN	76.30	77.52	0.74	98.37
	PNN	63.78	37.54	0.41	93.99
	KNN	74.73	76.29	0.72	98.14
	NB	25.22	27.79	0.18	85.99
Kyoto2	**SVM**	**93.73**	**92.00**	**0.92**	**96.80**
	ET-KNN	92.47	92.00	0.91	96.80
	PNN	91.95	90.00	0.90	96.00
	KNN	93.29	91.00	0.91	96.40
	NB	66.26	61.00	0.57	84.40
Tulum1	**SVM**	**64.77**	**61.82**	**0.62**	**96.15**
	ET-KNN	62.69	62.49	0.61	95.64
	PNN	48.33	35.40	0.37	91.28
	KNN	59.18	57.12	0.57	95.19
	NB	33.37	43.45	0.30	86.15

We perform activity based analysis by selecting the best learning methods on two challenging datasets through confusion matrices. Table 3 shows the confusion matrix of ET-KNN on Kasteren dataset. Almost all activities are identified with high recognition accuracy. The activities of 'Dinner' and 'Breakfast' share errors among themselves due to being performed at the same location and the use of similar features.

Table 3. Confusion matrix of Kasteren dataset for ET-KNN. Rows represent the actual activities and columns represent the predicted activities

Activities	Leave	Toilet	Shower	Sleep	Breakfast	Dinner	Drink
Leave	97.06	0	0	2.94	0	0	0
Toilet	0	97.37	2.63	0	0	0	0
Shower	0	0	100.00	0	0	0	0
Sleep	0	8.33	0	91.67	0	0	0
Breakfast	0	0	0	0	95.00	5.00	0
Dinner	0	0	0	0	20.00	80.00	0
Drink	0	5.00	0	0	0	0	95.00

Table 4 shows the confusion matrix of SVM on Tulum1 dataset. Most of the activities are highly correlated with less inter-class and high intra-class variations, which results in decreased performance of all the learning classifiers, while SVM comparatively obtains better performance. Meal activities are transferring errors among each other such as, 'Cook Breakfast' (CBF), 'Cook Lunch' (CL), 'R1 Eat Breakfast' (R1BF),

'R1Snack' (R1Snk) and 'R2 Eat Breakfast' (R2BF) are sharing their errors. CL for example is correctly recognized for 25% of instances, while it sends 11% of errors to both CBF and R1BF activities, 22% to R1Snk and 28% to R2BF activities. Another example is of 'Enter Home' (EH) and 'Leave Home' (LH), since both are recognized by the single door sensor. The EH activity transfer its 10% errors to LH, while LH send its 40% errors to EH activity class.

Table 4. Confusion matrix of Tulum1 for multi-class SVM. Rows represent the actual activities and columns represent the predicted activities. Key: Cook Breakfast – CBF, Cook Lunch – CL, Enter Home – EH, Group Meeting – GM, Leave Home – LH, R1 Eat Breakfast – R1BF, R1 Snack – R1Snk, R2 Eat Breakfast - R2BF, Wash Dishes – WD and Watch Tv – WT.

Activities	CBF	CL	EH	GM	LH	R1BF	R1Snk	R2BF	WD	WT
CBF	78.75	7.50	2.50	0	5	3.75	1.25	1.25	0	0
CL	11.27	25.35	0	0	0	11.27	22.53	28.17	0	1.41
EH	1.37	0	69.86	0	10.96	0	5.48	0	0	12.33
GM	9.09	0	0	63.64	0	0	9.09	0	0	18.18
LH	1.33	0	40	0	49.34	0	1.33	0	0	8
R1BF	6.06	18.18	1.51	0	0	25.77	46.97	0	0	1.51
R1Snk	2.04	1.63	0.20	0	0.20	1.83	90.23	0.61	1.22	2.04
R2BF	6.38	40.42	0	0	0	0	17.02	31.93	4.25	0
WD	1.41	0	0	0	0	1.41	12.68	1.41	83.09	0
WT	0	0	1.14	0.38	0.19	0	1.32	0	0	96.97

4 Conclusions

We performed the comparison of five learning methods on four smart home datasets. SVM showed an overall better performance on all datasets, while ET-KNN and KNN remained comparable to each other after the SVM. PNN shows better performance in the case of well discriminative activities; however, its performance is affected in case of less inter-class and high intra-class variations such as in Aruba and Tulum1. NB obtains the least performance in all datasets due to the challenges involved. Less amount of training data, imbalance activity instances results in declining the performance of all the learning approaches, which highlights the requirement of additional pre-processing steps such as, features ranking to select the discriminative enough features, data balancing and by defining a degree of confidence to ensure the reliability of recognitions.

Bibliography

[1] D. J. Cook, "Learning Setting-Generalized Activity Models for Smart Spaces," *IEEE Intelligent Systems,* vol. 27, pp. 32-38, 2012.

[2] P. Rashidi, D. J. Cook, L. B. Holder and M. Schmitter-Edgecombe,

"Discovering Activities to Recognize and Track in a Smart Environment," *IEEE Trans. Knowledge and Data Engg.*, vol. 23, pp. 527-539, 2011.

[3] L. G. Fahad, A. Khan and M. Rajarajan, "Activity recognition in smart homes with self verification of assignments," *Neurocomputing,* Vol. 149, Part C, pp. 1286-1298, 2015.

[4] B. Chikhaoui, S. Wang and H. Pigot, "Activity discovery and recognition by combining sequential patterns and latent Dirichlet allocation," *Pervasive and Mobile Computing,* vol. 8, pp. 845-862, 2012.

[5] E. Hoque and J. Stankovic, "AALO: Activity recognition in smart homes using Active Learning in the presence of Overlapped activities," in *Proc. of IEEE Int. Conf. on Pervasive Computing Technologies for Healthcare*, San Diego, CA, 2012.

[6] L. G. Fahad, A. Ali and M. Rajarajan, "Long term analysis of daily activities in smart home," in *Proc. of the European Symp. on Artificial Neural Networks, Computational Intelligence and Machine Learning*, Bruges, Belgium, 2013.

[7] E. M. Tapia, S. S. Intille and K. Larson, "Activity Recognition in the Home using Simple and Ubiquitous Sensors," in *Pervasive Computing*, Vienna, Austria, 2004.

[8] A. Fleury, M. Vacher and N. Noury, "SVM-based multimodal classification of activities of daily living in health smart homes: sensors, algorithms, and first experimental results," *IEEE Trans. Inf. Technol. in Biomed.,* vol. 14, pp. 274-283, 274-283.

[9] L. G. Fahad, S. F. Tahir and M. Rajarajan, "Activity Recognition in Smart Homes using Clustering based Classification," in *Proc. of IEEE Int. Conf. on Pattern Recognition*, Stockholm, Sweden, 2014.

[10] T. V. Kasteren, A. Noulas, G. Englebienne and B. Krose, "Accurate activity recognition in a home setting," in *Proc. of Int. conf. on Ubiquitous computing*, Seoul, Korea, 2008.

[11] S. Mckeever, J. Ye, L. Coyle, C. Bleakley and S. Dobson, "Activity recognition using temporal evidence theory," *J Ambient Intelligence & Smart Environment,* vol. 2, pp. 253-269, 2010.

[12] D. F. Specht, "Probabilistic neural networks," *Neural Networks,* vol. 3, pp. 109-118, 1990.

[13] V. N. Vapnik, "Statistical learning theory," New York, USA, John Wiley and Sons, p. 1998.

[14] L. M. Zouhal and T. Denoeux, "An evidence-theoretic k-NN rule with parameter optimization," *IEEE Trans. on Systems, Man, and Cybernetics, Part C,* vol. 28, pp. 263-271, 1998.

Tackling Class Imbalance Problem in Binary Classification using Augmented Neighborhood Cleaning Algorithm

Nadyah Obaid Al Abdouli, Zeyar Aung*, Wei Lee Woon, and Davor Svetinovic

Institute Center for Smart and Sustainable Systems (iSmart),
Department of Electrical Engineering and Computer Science,
Masdar Institute of Science and Technology, Abu Dhabi, UAE.
{nalabdouli,zaung,wwoon}@masdar.ac.ae,ds@davors.com

Abstract. Many natural processes generate some observations more frequently than others. These processes result in an imbalanced distributions which cause classifiers to bias toward the majority class because most classifiers assume a normal distribution. In order to address the problem of class imbalance, a number of data preprocessing techniques, which can be generally categorized into over-sampling and under-sampling methods, have been proposed throughout the years. The Neighborhood cleaning rule (NCL) method proposed by Laurikkala is among the most popular under-sampling methods. In this paper, we augment the original NCL algorithm by cleaning the unwanted samples using CHC evolutionary algorithm instead of a simple nearest neighbor-based cleaning as in NCL. We name our augmented algorithm as NCL+. The performance of NCL+ is compared to that of NCL on 9 imbalanced datasets using 11 different classifiers. Experimental results show noticeable accuracy improvements by NCL+ over NCL. Moreover, NCL+ is also compared to another popular over-sampling method called Synthetic minority over-sampling technique (SMOTE), and is found to offer better results as well.

Keywords: Data Preprocessing, Class Imbalance, Under-Sampling, Neighborhood Cleaning, Evolutionary Algorithm.

1 Introduction

Natural processes often generate some observations more frequently than others. Therefore, they produce samples that may not have a normal class distribution. In many cases the class distribution could be highly imbalanced (a.k.a. skewed). This phenomenon exists in many real-world datasets in different real-world applications such as text classification [19], fraud detection [21], intrusion detection [10], customer behavior prediction [17], and environmental event monitoring [4]. The ratio between the numbers of samples of the majority class and those of the minority class is called the imbalance ratio (IR).

* Corresponding author.

K.J. Kim (ed.), *Information Science and Applications*,
Lecture Notes in Electrical Engineering 339, DOI 10.1007/978-3-662-46578-3_98

In general, most classification algorithms assume normal class distribution. However, in the case of an imbalanced dataset, it is the majority class that creates a bias in the classifiers' decision. The classifiers tend to focus on the majority class and ignore the minority class. There are many cases in which the minority class represents the most important class of interest. For example, in diseased tree detection [4], it is extremely important to correctly classify the minority class. Generally, a normal classifier would misclassify many samples of the minority class that represents diseased trees. The imbalance distribution poses many challenges to widely used-classifiers such as decision tree, induction models, and multilayer perceptron neural networks [15].

In order to address the challenge of class imbalance, a number of solutions have been proposed throughout the years. These can be categorized into method-level (internal) and data-level (external a.k.a. preprocessing) approaches. Data-level processing approaches can be further categorized into over-sampling and under-sampling methods. Among a number of under-sampling methods [16, 12, 25, 24, 6], the Neighborhood cleaning rule (NCL) algorithm [16] is a simple and effective one.

In this paper, our objective is to further improve the accuracy performance of NCL by employing CHC evolutionary algorithm [9] in removing the unwanted samples. We name our proposed method as NCL+. The performance of NCL+ is compared to that of NCL on 9 UCI benchmark datasets [1] in conjunction with 11 commonly-used classifiers. Experimental results show noticeable accuracy improvements by NCL+ over NCL. Furthermore, NCL+ is also compared to another popular over-sampling method called Synthetic minority over-sampling technique (SMOTE), and is found to offer better results as well.

This paper is an extended version of a portion of research work presented in the master's thesis [3] of the first author.

2 Related Work

2.1 Method-level (Internal) Approaches

Many normal classification algorithms can be modified to take imbalanced data in account. Specifically, the modification may focus on adjusting the cost function, changing the probability estimation, or adapting recognition-based learning [12]. The algorithms that work at method level could be efficient. However, in many cases these algorithms are application specific. Thus, there needs special knowledge about both the classifier itself and the application domain in order to use them effectively [12, 13].

2.2 Data-level (Preprocessing) Approaches

The purpose of preprocessing is to balance and normalize the class distribution to certain extent before passing the dataset to the classifier. Sampling is the most used approach for overcoming misclassification problems due to imbalanced datasets [18]. There are two broad categories in sampling: over-sampling and under-sampling.

Over-Sampling: The purpose is to increase the size of the minority class by adding new synthetic samples. Some of the well-known over-sampling methods include Synthetic minority over-sampling technique (SMOTE) [8], SMOTE + Tomek's links (SMOTE-TL) [6], Selective preprocessing of imbalanced data 2 (SPIDER2) [20], Random over-sampling (ROS) [12], and Adaptive synthetic sampling (ADASYN) [14].

Under-Sampling: The aim is to reduce the size of the majority class set by removing some of its samples. Some of the well-known over-sampling methods include Neighborhood cleaning rule (NCL) [16], Tomek links (TL) [6], Condensed nearest neighbor rule (CNN) + Tomek's links (CNN-TL) [12], Under-sampling based on clustering (SBC) [24], and Class purity maximization (CPM) [25].

Here, we will elaborate the Neighborhood cleaning rule (NCL) [16] method because our proposed algorithm in this paper is based and improved upon it. NCL employs Wilson's edited nearest neighbor rule (ENN) [23]. NCL model maintains all the samples of the class of interest C and reduce those from the rest of the data O. This process is accomplished in two phases. In the first phase, ENN is used to find the noisy data A_1 in O. Specifically, 3-ENN is used to remove the samples with different class than the majority class of the three nearest neighbors. Subsequently, the neighborhoods are processed again and the set A_2 is initially created. Then, the three nearest neighbor samples that belong to O and lead to C samples misclassification are iteratively inserted in the set A_2. Finally, the data is reduced by eliminating the samples that belong to either sets A_1 or A_2 (i.e., $A_1 \cup A_2$). Figure 1 describes the NCL algorithm.

Algorithm NCL
1. Split data T into the class of interest C and the rest of data O.
2. Identify noisy data A_1 in O with edited nearest neighbor rule.
3. For each class C_i in O
   if ($x \in C_i$ in 3-nearest neighbors of misclassified $y \in C$) and ($|C_i| \geq 0.5 \cdot |C|$)
   then $A_2 = \{x\} \cup A_2$
4. Reduced data $S = T - (A_1 \cup A_2)$

Fig. 1. NCL algorithm (reproduced from [16]).

3 Proposed NCL+ Algorithm

The proposed augmented algorithm, named NCL+, follows a similar architecture of the original neighborhood cleaning rule (NCL) algorithm described above in Section 2.2 and Figure 1. The data reduction process is carried out in two phases. In the first phase, the edited nearest neighbor (ENN) algorithm [23] is used to create a subset named A_1. Unlike the original NCL algorithm which uses simple 3 nearest neighbors, the second phase is augmented by utilizing an evolutionary instance selection algorithm called CHC [9] to create a more carefully scrutinized subset, named A_2, of samples to be removed. Finally, $A_1 \cup A_2$ will be removed from the original dataset. The algorithm is shown in Figure 2.

Algorithm NCL+
1. Split data T into the class of interest C and the rest of data O.
2. For the subset O do the following:
   2.1. Using ENN to identify the noisy data and insert them to subset A_1.
   2.2. Run CHC with the following fitness function:
   $Fitness(S) = \alpha \cdot$ `clas_rate` $+ (1 - \alpha) \cdot$ `pred_red`, where:
   S: a subset of the data set.
   α: Alfa equilibrate factor.
   `class_rate`: classification rate (% of correctly classified samples in T using only S to find nearest neighbors).
   `perc_red`: percentage of reduction (calculated as: $100 \cdot (|T| - |S|)/|T|)$.
   2.3. Insert the result into subset A_2.
   2.4. Reduce the dataset $S = T - (A_1 \cup A_2)$.

Fig. 2. Proposed NCL+ algorithm.

In many cases, there could be too many samples in the majority class. However, generally, not all the samples are equally informative during the training phase. Thus, instance selection algorithms such as CHC are useful to interpret the data independently of their location in the search space. CHC chooses the most representative samples. Consequently, it provides a high reduction rate while maintaining the accuracy. Cano et al. [7] has showed that CHC offered the best ranking in data reduction rates.

The CHC algorithm relies on reducing the data by means of evolutionary algorithm (EA) for instance selection. The EAs are adaptive models that rely on the principle of natural evolution. In CHC, an EA is used as instance selector to select the data to be removed. The C4.5 decision-tree induction algorithm [22] is used to build a decision tree using the selected instances. Then the new examples are classified using the resultant tree.

During each generation, CHC follows some basic steps that can be summarized as follows. First, it generates an intermediate population of size N using the parent population of size N. Then, it randomly pairs them and use them to produce N potential offspring. Then, a survival competition is held in order to select the next generation population. The best N from the parent population and offspring are selected to form the next generation [7].

4 Experimental Results

In this section, we will present performance companions of the proposed NCL+ algorithm with its predecessor NCL [16] as well as a widely-used over-sampling algorithm, SMOTE [8].

4.1 Experimental Setup

We compare NCL+ to NCL and SMOTE on 9 imbalanced UCI datasets [1] using 11 commonly-used classifiers. The experimental setup is described below.

- **Software tool:** KEEL [5].
- **Classifiers:** 11 algorithms (Names are as given in KEEL.)
  - Cost-sensitive: `C_SVMCS-I`, `C4.5CS-I`, `NNCS-I`.
  - Ensemble: `AdaBoost-I`, `AdaBoostM2-I`, `AdaC2-I`, `Bagging-I`, `OverBagging2-I`, `MSMOTEBagging-I`, `UnderBagging2-I`, `UnderOverBagging-I`.
- **Datasets:** 9 UCI datasets [1] (Downloaded from KEEL dataset website [2]. The imbalance ratio (IR) for each dataset is given in Table 1.)
  - High IR (> 9): Abalone19, Yeast6, Glass6, Glass5, New-thyroid2.
  - Low IR (1.5 to 9): Vowel0, Vehicle1, Wisconsin, Ecoli-0_vs_1.
- **Preprocessing algorithms compared:** NCL+ (our proposed method), NCL [16] (`NCL-I` in KEEL), SMOTE [8] (`SMOTE-I` in KEEL).
- **Experimental mode:** 5-fold cross validation.
- **Evaluation criteria:** Average area under receiver operating characteristic curve (AUC) [11] for both classes.

4.2 NCL vs. NCL+

We can observe a recognizable improvement in the accuracy performance of NCL+ over the original NCL in terms of AUC. We can observe significant improvements in average AUC by 29.411% in the case of Abalone19 dataset (which has a very high IR of 128.87) and 11.231% in Vehicle dataset (with a relatively low IR of 3.23). Moderate improvements in average AUC between 0.809% and 5.192% are also observed for 4 datasets (Wisconsin, New-thyroid2, Vowel0, and Yeast6). However, for 2 datasets, NCL+ exhibits slight declines (less than 1%) in average AUC (−0.268% for Glass6 and −0.723% for Glass5). Table 1(a) presents the detailed comparison of accuracy results between NCL and NCL+ for each classifier.

Looking at the data distribution of processed data of both models and given the dataset size in each case, it is noticeable the CHC does not operate on quantity. It removes relatively fewer samples than NCL. However, the removed samples are the most defining ones whose absence can noticeably contribute to the improvement in classification accuracy. These results suggest that, in addition to CHC, other adaptive learning and evolutionary training sample selection algorithms, such as generational genetic algorithm (GGA) and population-based incremental learning (PBI), could also be used to further improve the quality of nearest neighbor-based under-sampling.

Table 1. Detailed comparisons of NCL and SMOTE vs. NCL+ on 9 different UCI datasets using 11 different classification algorithms in terms of average AUC values.

Dataset	Abalone19		Yeast6		Glass6		Glass5		New-thyroid2		Vowel0		Vehicle1		Wisconsin		Ecoli-0_vs_1	
IR	128.87		32.78		22.81		15.47		10.1		5.14		3.23		1.86		1.86	
(a) NCL vs. NCL+.																		
Classifier	NCL	NCL+	NCL	NCL+	NCL	NCL+	NCL	NCL+	NCL	NCL+	NCL	NCL+	NCL	NCL+	NCL	NCL+	NCL	NCL+
C_SVMCS-I	0.760	0.798	0.874	0.880	0.912	0.912	0.954	0.973	0.997	0.997	0.963	0.972	0.815	0.818	0.971	0.979	0.976	0.980
C4.5CS-I	0.548	0.800	0.846	0.907	0.923	0.890	0.988	0.943	0.949	0.989	0.942	0.983	0.755	0.864	0.957	0.971	0.959	0.987
NNCS-I	0.507	0.508	0.718	0.621	0.894	0.902	0.854	0.880	0.933	0.852	0.664	0.709	0.649	0.602	0.964	0.970	0.969	0.980
AdaBoost-I	0.515	0.799	0.793	0.896	0.923	0.885	0.938	0.948	0.946	0.997	0.970	0.988	0.761	0.909	0.970	0.981	0.973	0.980
AdaBoostM2-I	0.516	0.799	0.793	0.896	0.886	0.875	0.938	0.948	0.949	0.997	0.970	0.988	0.754	0.911	0.970	0.979	0.973	0.983
AdaC2-I	0.554	0.800	0.793	0.886	0.867	0.900	0.926	0.973	0.952	0.986	0.958	0.987	0.784	0.888	0.979	0.979	0.969	0.980
Bagging-I	0.500	0.500	0.782	0.798	0.872	0.830	0.940	0.840	0.940	0.949	0.930	0.988	0.761	0.884	0.969	0.974	0.983	0.987
OverBagging2-I	0.530	0.792	0.814	0.889	0.915	0.920	0.890	0.888	0.926	0.963	0.964	0.996	0.764	0.874	0.970	0.980	0.976	0.987
MSMOTEBagging-I	0.579	0.795	0.852	0.888	0.918	0.936	0.978	0.888	0.943	0.983	0.959	0.993	0.732	0.804	0.959	0.971	no res.	0.987
UnderBagging2-I	0.713	0.733	0.867	0.888	0.909	0.926	0.949	0.949	0.918	0.958	0.947	0.977	0.766	0.854	0.962	0.973	0.980	0.983
UnderOverBagging-I	0.547	0.792	0.836	0.884	0.909	0.926	0.940	0.990	0.941	0.969	0.961	0.990	0.780	0.845	0.970	0.973	0.976	0.983
Average	0.570	0.738	0.815	0.858	0.902	0.900	0.936	0.929	0.945	0.967	0.930	0.961	0.756	0.841	0.968	0.975	0.973	0.983
Improvement (%)		29.411		5.192		-0.268		-0.723		2.363		3.352		11.231		0.809		1.004
(b) SMOTE (denoted as SM) vs. NCL+.																		
Classifier	SM	NCL+	SM	NCL+	SM	NCL+	SM	NCL+	SM	NCL+	SM	NCL+	SM	NCL+	SM	NCL+	SM	NCL+
C_SVMCS-I	0.765	0.798	0.887	0.880	0.926	0.912	no res.	0.973	0.978	0.997	0.967	0.972	0.807	0.818	0.971	0.979	0.980	0.980
C4.5CS-I	0.596	0.800	0.834	0.907	0.892	0.890	0.954	0.943	0.944	0.989	0.972	0.983	0.684	0.864	0.958	0.971	0.983	0.987
NNCS-I	0.790	0.508	0.819	0.621	0.858	0.902	0.879	0.880	0.983	0.852	0.895	0.709	0.590	0.602	0.948	0.970	0.973	0.980
AdaBoost-I	0.539	0.799	0.816	0.896	0.855	0.885	0.880	0.948	0.955	0.997	0.976	0.988	0.733	0.909	0.964	0.981	0.973	0.980
AdaBoostM2-I	0.539	0.799	0.815	0.896	0.855	0.875	0.880	0.948	0.963	0.997	0.975	0.988	0.720	0.911	0.964	0.979	0.973	0.983
AdaC2-I	0.537	0.800	0.816	0.886	0.855	0.900	0.880	0.973	0.955	0.986	0.976	0.987	0.724	0.888	0.966	0.979	0.973	0.980
Bagging-I	0.527	0.500	0.834	0.798	0.898	0.830	0.966	0.840	0.958	0.949	0.981	0.988	0.742	0.884	0.962	0.974	0.980	0.987
OverBagging2-I	0.527	0.792	0.834	0.889	0.898	0.920	0.966	0.888	0.958	0.963	0.981	0.996	0.742	0.874	0.962	0.980	0.980	0.987
MSMOTEBagging-I	no res.	0.795	0.840	0.888	0.920	0.936	0.932	0.888	0.969	0.983	0.954	0.993	no res.	0.804	0.963	0.971	no res.	0.987
UnderBagging2-I	0.527	0.733	0.834	0.888	0.898	0.926	0.966	0.949	0.958	0.958	0.981	0.977	0.742	0.854	0.962	0.973	0.980	0.983
UnderOverBagging-I	0.550	0.792	0.829	0.884	0.918	0.926	0.968	0.990	0.966	0.969	0.983	0.990	0.750	0.845	0.971	0.973	0.969	0.983
Average	0.590	0.738	0.833	0.858	0.888	0.900	0.927	0.929	0.963	0.967	0.967	0.961	0.723	0.841	0.963	0.975	0.976	0.983
Improvement (%)		25.111		3.001		1.317		0.200		0.495		-0.634		16.292		1.301		0.715

4.3 SMOTE vs. NCL+

When compared to SMOTE, NCL+ also offers better results in terms of average AUC in 8 out of 9 datasets with improvements ranging from 0.200% to 25.111%. However, for one dataset (Vowel0), its performance is slightly inferior with −0.634% decline in average AUC. Table 1(b) shows the detailed comparison of accuracy results between SMOTE and NCL+ for each classifier.

5 Conclusions and Future Work

Class imbalance is a serious problem for classification applications when the minority class is important. In this research, an improvement to the well-known neighborhood cleaning rule (NCL) under-sampling algorithm by Laurikkala [16] was suggested by employing an effective CHC-based instance selection approach. The new NCL+ method was compared to both the original NCL method as well as another popular preprocessing method named SMOTE [8]. Experimental results on 9 imbalanced datasets with 11 commonly-used classifiers showed that the proposed NCL+ generally provided significantly better results than both NCL and SMOTE did.

This research was focused on imbalanced binary datasets. The proposed NCL+ in its current form cannot be directly applied to multi-class classification because of the substantial difference between binary and multi-class classifications. Thus, for future work, we have a plan to extend NCL+ to adapt it to the multi-class scenario.

References

1. `http://archive.ics.uci.edu` (2014)
2. `http://sci2s.ugr.es/keel/datasets.php` (2014)
3. Al Abdouli, N.O.: Handling the Class Imbalance Problem in Binary Classification. Master's thesis, Masdar Institute of Science and Technology, Abu Dhabi, UAE (2014)
4. Alan, J.B., Ryutaro, T., Hoan, N.: A hybrid pansharpening approach and multiscale object-based image analysis for mapping diseased pine and oak trees. International Journal of Remote Sensing 34, 6969–6982 (2013)
5. Alcalá-Fdez, J., Fernandez, A., Luengo, J., Derrac, J., Garcia, S., Sanchez, L., Herrera, F.: KEEL data-mining software tool: Data set repository. Journal of Multiple-Valued Logic and Soft Computing 17, 255–287 (2011)
6. Batista, G.E., Prati, R.C., Monard, M.C.: A study of the behavior of several methods for balancing machine learning training data. SIGKDD Explorations 6, 20–29 (2004)
7. Cano, J., Herrera, F., Lozano, M.: Using evolutionary algorithms as instance selection for data reduction in KDD: An experimental study. IEEE Transactions on Evolutionary Computing 7, 561–575 (2003)
8. Chawla, N.V., Bowyer, K.W., Hall, L.O., Kegelmeyer, W.P.: SMOTE: Synthetic minority over-sampling technique. Journal of Artificial Intelligence Research 16, 321–357 (2002)

9. Eshelman, L.J.: The CHC adaptive search algorithm: How to have safe search when engaging in nontraditional genetic recombination. In: Proc. 1st Workshop on Foundations of Genetic Algorithms. pp. 265–283 (1990)
10. Faisal, M.A., Aung, Z., Williams, J., Sanchez, A.: Data-stream-based intrusion detection system for advanced metering infrastructure in smart grid: A feasibility study. IEEE Systems Journal (2014), in press
11. Fawcett, T.: An introduction to ROC analysis. Pattern Recognition Letters 27, 861–874 (2006)
12. Fernándeza, A., Garcíaa, S., Jesusb, M., Herreraa, F.: A study of the behaviour of linguistic fuzzy rule based classification systems in the framework of imbalanced data-sets. Fuzzy Sets and Systems 159, 2378–2398 (2008)
13. Galar, M., Fernandez, A., Barrenechea, E., Bustince, H., Herrera, F.: A review on ensembles for the class imbalance problem: Bagging-, boosting-, and hybrid-based approaches. IEEE Transactions on Systems, Man, and Cybernetics Part C 42, 463–484 (2011)
14. He, H., Bai, Y., Garcia, E., Li, S.: ADASYN: Adaptive synthetic sampling approach for imbalanced learning. In: Proc. 2008 International Joint Conference on Neural Networks. pp. 1322–1328 (2008)
15. Jo, T., Japkowicz, N.: A multiple resampling method for learning from imbalanced data sets. Computational Intelligence 20, 18–36 (2004)
16. Laurikkala, J.: Improving identification of difficult small classes by balancing class distribution. In: Proc. 8th Conference on AI in Medicine in Europe. pp. 63–66 (2001)
17. Liu, N., Woon, W.L., Aung, Z., Afshari, A.: Handling class imbalance in customer behavior prediction. In: Proc. 2014 IEEE International Conference on Collaboration Technologies and Systems. pp. 100–103 (2014)
18. Lokanayaki, K., Malathi, A.: Data preprocessing for liver dataset using SMOTE. International Journal of Advanced Research in Computer Science and Software Engineering 3, 559–562 (2013)
19. Mladenii, D., Grobelnik, M.: Feature selection for unbalanced class distribution and naive Bayes. In: Proc. 16th International Conference on Machine Learning. pp. 258–267 (1999)
20. Napieralla, K., Stefanowski, J., Wilk, S.: Learning from imbalanced data in presence of noisy and borderline examples. In: Proc. 7th International Conference on Rough Sets and Current Trends in Computing. pp. 158–167 (2010)
21. Perera, K.S., Neupane, B., Faisal, M.A., Aung, Z., Woon, W.L.: A novel ensemble learning-based approach for click fraud detection in mobile advertising. In: Proc. 2013 International Conference on Mining Intelligence and Knowledge Exploration. Lecture Notes in Computer Science, vol. 8284, pp. 370–382 (2013)
22. Quinlan, J.R.: C4.5: Programs for Machine Learning. Morgan Kaufmann Publishers (1993)
23. Wilson, D.R., Martinez, T.R.: Reduction techniques for instance-based learning algorithms. Machine Learning 38, 257–286 (2000)
24. Yen, S.J., Lee, Y.S.: Under-sampling approaches for improving prediction of the minority class in an imbalanced dataset. In: Proc. 2006 International Conference on Intelligent Computing. pp. 731–740 (2006)
25. Yoon, K., Kwek, S.: An unsupervised learning approach to resolving the data imbalanced issue in supervised learning problems in functional genomics. In: Proc. 5th International Conference on Hybrid Intelligent Systems. pp. 303–308 (2005)

Part VII

Software Effective Risk Management: An Evaluation of Risk Management Process Models and Standards

[1]Jafreezal Jaafar, [1]Uzair Iqbal Janjua, [2] Fong Woon Lai

jafreez@petronas.com.my, uzair_iqbal@comsats.edu.pk, laifongwoon@petronas.com.my

[1]Department of Computer & Information Science

[2]Department of Management & Humanities

UNIVERSITI TEKNOLOGI PETRONAS, MALAYSIA

Abstract. Different software risk management process models, professional standards and specific techniques have been presented in literature by researchers and practitioners in the software industry to make the development of software projects more likely to succeed. In this study different software risk management process models and Professional standards have been evaluated against the most effective risk management techniques and processes proposed by the different researchers in the last 13 years to highlight the strengths and weaknesses of different risk management process models. The results show that, there is no model which can be called the de facto effective risk management process model.

Keywords. Risks, Risk Management (RM), Effective Risk Management, Risk Management Process Model, Risk Management Standards.

1 Introduction

Software risk management deals with risks that suffer the success of the under development software project. Since, late 1980s many researchers and practitioners proposed the risk management guidelines in the form of process, models and professional standards to increase the success rate of software project development. However, still according to different survey reports, the success rate of software development is very low. Survey results [1] conducted in 2008 shows that out of 100 software development projects, only 32 software development projects succeeded. This leads researchers and practitioners to address the need of effective risk management for software development based on effective techniques and processes to increase the success rate of software project development [2].

Researchers and practitioners, on the basis of their theoretical knowledge and practical experiences, have suggested different techniques to make current practices of risk management more effective. Two authors of this study, have already conducted an exploratory literature review [3] of IEEE and SCOPUS electronic databases from the year 1996 to 2012 with the key stream "software effective risk management" to identify the effective techniques to make current practices of software project risk management

K.J. Kim (ed.), *Information Science and Applications*,
Lecture Notes in Electrical Engineering 339, DOI 10.1007/978-3-662-46578-3_99

more effective. The purpose of this research work is to evaluate different risk management process and professional standards against a few of identified effective techniques to find a de facto risk management process model for software development project managers. So the objective of the research work was:

a) *To evaluate the different risk management process models and professional standards against identified effective risk management techniques in order to find an effective risk management model.*

The paper is structured as follows: Section 2 explain briefly the different risk management process models. Section 3 discusses the evaluation and findings of different models. Section 4 describes the conclusion.

2 Software Risk Management

Software development projects constitute of many small and large programs with many connections and dependencies. Although the development processes of different projects are similar, but it involves a creation of new endeavors. Because of a theses features of software projects, many software development projects often faced many risks like, schedule overrun, cost overrun ,usability and quality of issues [4]. To handle these issues in literature several risk management approaches has been presented in the form of process models, steps and professional standards.

2.1 Professional Risk Management Standards

Professional standards are the general and simple guidelines prepared by the professionals. Usually, professional standards are not industry or company oriented so can be used in any industry or company [5]. For this study, following two professional standards have been selected.

- IEEE 1540-2001's Risk Management Standard [6].
- AusAID' Risk Management Standard [7].

First selected [6] standard was stereotypically designed for the software industry. Whereas, second selected standard [7] was not specifically designed for the software industry. However, the purpose of this research work was to evaluate software industry based risk management process models and professional standards. So, authors first confirmed from the documentation of standard, whether can it be used for the software industry or not. In the documentation, it was found that it can be used for the software industry.

2.2 General Software Development Risk Management Processes

Industry-specific risk management process models are specifically designed to handle risks of a specific industry [5]. This type of process models are general and can be applied to all types of projects related to a specific industry. As the objective of this research is to evaluate only risk management process models which were proposed for the software development, so following process models have been selected.

- BARRY W. BOEHM Risk Management Steps [8].
- Riskit Risk Management Process Model [9].
- NPG 7120.5A NASA's Risk Management guidebook [10, 11].
- George Holt Risk Management Program [12].

2.3 Specific Process Oriented Risk Management Process Model

In Kerzner [5] risk management plan classification, specific process-oriented risk management plan was not discussed. However, authors of this study found a few risk management process plans in the field of software development risk management proposed by researchers with an intention to add a new process/ processes to make current practices of risk management more effective. Following are the risk management process models which have been selected for the evaluation.

- Mitigation Phase Oriented Software Development Risk Management [13].
- Goal-driven software development risk management model [14].
- Developer Oriented Software Development Risk Management Model [15].
- Software Development Impact Statement Process [16].
- Software Project Risk Assessment Model [17].

2.4 Effective Risk Management Techniques

Two of authors [3] of this paper had explored the Scopus and IEEE online databases from the year 1996 to 2012 with the key stream "software effective risk management". The aim of that research work was to explore the literature to develop the understanding of phenomena, "Software Project Effective Risk Management". While analyzing the data of secondary study authors have found 68% of the papers in the secondary study recommend techniques that need to be incorporated in current software risk management activities in order to make them more effective. Some of techniques and all basic processes required to make software project risk management more effective identified from selected papers [3] are used to evaluate a few risk management process models and standards to find a de facto effective risk management for software development.

3 Evaluation, Findings and Discussion of Risk Management Process Models and Professional Standards

During the evaluation, 2 professional standards, 4 general risk management process model /steps and 5 specific-process oriented risk management models have been studied. Evaluation of professional standards and models have been presented in table 1 and table 2. In table 1 and table 2, 'Y" represent that in this particular article author have conferred specific effective strategy or process. "N' represents that author remain silent about effective strategy or process. "P" mean in article effective strategy/ process was discussed partially. "U" symbol in table 2 was used to represent that in particular article's author used the term but didn't discuss.

Following the literature survey that the authors undertook, and after the evaluation of different risk management process models and professional standards following are the findings of this research work:

- There are two types of risk management professional standards; one which are not specific to the single industry like [7] and the second type of professional standards which are industry specific [17]. However, main processes are common in both types of standards. Moreover, risk management professional standards discussed the different processes of risk management in more detail as compared to other industry-specific risk management processes models.
- Risk management plan, risk identification, risk assessment/analysis, risk mitigation/treatment/controlling/resolution and risk monitoring are the most common processes in all risk management process models and professional standards. In some model, processes even with similar functionality have been named differently. For example, risk analysis and risk assessment process have the same functionality. Similarly, "Risk Management Mandate Definition" process in [9] and "Establishing the Context and Objectives" [7] have the same function as risk management plan process of many other risk management process models.

Fig. 1. Most Common and Minimum Required Processes for Effective Risk Management for Software Development

- 60% of all models, which have risk documentation process have also a lesson learned process.
- Only 45% of risk management process models/standards have a project identification process. However, all models/standards have a project identification process also have a stakeholder identification process.
- Although 45% of risk management process models/standards have stakeholder identification process, but only 20% of these risk management process models/standards also identify the goal of different stakeholders for risk management. Moreover, all risk management process models/standards which have a stakeholder's identification process use open and continuous communication among the different stakeholders except then [16].

- Generally, main processes of software risk management are sequenced in nature. For example, risk assessment cannot be started before the risk identification. Most of the models and standards follow the same sequence. However, in a few models some activities of processes are sequenced differently.

Table 1. Evaluation of Risk Management Model and Professional Standards Against Effective Processes

		Risk Management Plan	Risk Identification	Risk Assessment	Risk Mitigation	Risk Monitoring	Risk Documentation	Project Type Identification for RM	Use of Historical Data for RM Activites	Stakeholder Identification For RM	Lesson Learned	Stakeholder Goal Identification for RM
M Professional Standards	[6]	Y	Y	Y	Y	Y	Y	N	P	Y	Y	N
	[7]	Y	Y	Y	Y	Y	Y	N	N	Y	N	N
General RM Process Model	[8]	Y	Y	Y	Y	Y	N	N	N	N	N	N
	[9]	Y	Y	Y	Y	Y	N	Y	P	Y	Y	Y
	[11]	Y	Y	Y	Y	Y	Y	N	N	N	N	N
	[12]	Y	Y	Y	Y	Y	N	N	N	P	Y	P
Specific Process Oriented RM Model	[13]	Y	Y	Y	Y	P	N	N	Y	N	N	N
	[14]	P	Y	Y	Y	N	N	P	N	N	N	Y
	[15]	Y	Y	Y	Y	Y	N	N	P	P	Y	N
	[16]	Y	Y	Y	Y	N	Y	Y	N	Y	P	P
	[17]	Y	Y	Y	Y	Y	Y	N	N	N	N	N

- 60% of all general and specific process oriented models and standards, recommend continuous risk management. Except then the [8] almost all general models and professional standards recommended the continuous risk management processes. However, risk management proposed in [8] was at its initial stage of software development at that time. Moreover, author's intention at that time was to propose the general steps, it can be estimated that at that time author didn't recognize the significance of continuous risk management.
- Although the involvement of client/customer/user in risk management to make it more effective has been found very significant [3]. However, only 50% of all models and professional standards authors considered the involvement of a customer for risk management. Furthermore, 86% of all risk management process models/standards are proactive in nature. During the evaluation of models/standards it has also been observed that all models/standards which support the involvement of customers in risk management activities also promote proactive risk management.
- Literature review of software effective risk management [3] recommended the involvement of all team members with clear role and responsibilities in the risk management activities. Moreover, a few authors like [18], even proposed team oriented

risk management approaches when they realized the significance of the team member's role in risk management activities. However during the evaluation of different risk management models and professional standards it has been found that only 36% models and professional standards authors recommended the involvement of an entire team in risk management activities.

Table 2. Evaluation of Risk Management Process Models and Professional Standards Against Software Development Effective Risk Management Techniques

		Open and Continuous Communication among all Stakeholders	Entire Project Team Involvement with Clear Roles and Responsibilites	Support of Top Management for RM	User/Customer Involvement in RM	Proactive Risk Management	Circulation of Established Guidelines of RM among all Stakeholders	Skilled and Experienced Project Manager for RM	Designated Risk Manager	Continous Risk Management
Professional Standards	[6]	Y	N	P	P	Y	P	N	N	Y
	[7]	Y	P	P	P	Y	P	Y	U	Y
General RM Process Model	[8]	N	N	N	Y	Y	N	Y	N	N
	[9]	P	Y	N	N	Y	P	N	Y	Y
	[11]	Y	P	N	N	Y	N	N	N	Y
	[12]	Y	Y	N	P	P	N	N	N	P
Specific Process Oriented RM Model	[13]	N	N	N	P	Y	N	N	U	N
	[14]	N	N	N	Y	Y	N	N	N	N
	[15]	P	N	N	P	Y	N	N	N	Y
	[16]	N	Y	P	Y	Y	N	N	P	N
	[17]	N	N	N	N	N	N	N	N	Y

- Use of "risk manager" other than the project manager for risk management activities is well-known in different fields, such as banking, civil engineering, enterprise risk management, etc. [19].However, it has been found that despite the significance of a designated risk manager other than project manager only Riskit model [9] emphasis on the role of a designated risk manager. During evaluation, it has also been realized that some authors of the risk management models considered risk manager just a title for the person who perform risk management activities [11].
- Involvement of top management can make the risk management more effective. However, most of the authors of evaluated risk management models/standards remained quiet about it. Even, in risk management professional standards, involvement of top management is not discussed in detailed.

Different general risk management process models have been presented in the literature. More or less all these models have many common features. However, even with common features different researchers and practitioners used the different keywords and phrases while representing the risk management process models. For example, in [8] a leading expert on software risk management used the phrase "SOFTWARE RISK MANAGEMENT STEPS" for the graphical representation of the risk management activities. However, in [9] while citing [8], Jyrki Kontio, used, "BOEHM'S RISK MANAGEMENT MODEL" phrase instead of "Boehm's SOFTWARE RISK MANAGEMENT STEPS". However, Jyrki Kontio himself represented his work with phrase "Risk Management Process". In [6] risk management processes were represented as "Risk Management Process Model" and in [7], risk management processes are graphically represented as "Risk Management Processes diagram". In [10], graphical representations of risk management activities are represented as "The Risk Management Process". However, after the evaluation of different general/specific risk management processes models and professional standards, it can be concluded despite the use of some loose term, purposes of these processes are to provide the guidelines of software risk management.

4 Conclusion

The objective to conduct this research work was to investigate different risk management process models with a particular focus on effective risk management processes and techniques, with the aim of gaining a better understanding of their valuable features and limitations. For the purpose, different risk management process models were assessed against effective techniques found in the literature.

It has been found that all risk management models and professional standards have their own limitations. Hence, there is no model which can be called the de facto software project effective risk management process model. There is a need to incorporate the suggestions addressed in the literature by different researchers to make the models more effective to increase the success rate of software development.

References

[1] The Standish Group.CHAOS Summary 2009.

[2] Y. Hu, X. Zhang, E. Ngai, R. Cai, and M. Liu, "Software project risk analysis using Bayesian networks with causality constraints," *Decision Support Systems,* vol. 56, pp. 439-449, 2012.

[3] U. Janjua, A. Oxley, and J. Jaffer, "Effective Risk Management of Software Projects (ERM): An Exploratory Literature Review of IEEE and Scopus Online Databases," in *Proceedings of the First International Conference on Advanced Data and Information Engineering (DaEng-2013).* vol. 285, T. Herawan, M. M. Deris, and J. Abawajy, Eds., ed: Springer Singapore, 2014, pp. 445-452.

[4] Y. Kwak and J. Stoddard, "Project risk management: lessons learned from software," *Personnel,* vol. 124, p. 125.

[5] H. R. Kerzner, *Project management: a systems approach to planning, scheduling, and controlling*: John Wiley & Sons, 2013.

[6] IEEE Standard for Software Life Cycle Processes - Risk Management, *IEEE Std 1540-2001*, pp. i-24, 2001.

[7] G. Purdy, "ISO 31000: 2009—setting a new standard for risk management," *Risk analysis*, vol. 30, pp. 881-886, 2010.

[8] B. W. Boehm, "Software risk management: principles and practices," *Software, IEEE*, vol. 8, pp. 32-41, 1991.

[9] J. Kontio, *Software engineering risk management: a method, improvement framework, and empirical evaluation*: Helsinki University of Technology, 2001.

[10] NASA, "NASA Procedures and Guidelines NPG 7120.5 B," in *Risk Management*, ed, 2002, p. 122.

[11] Linda H. Rosenberg, Al Gallo, Ted Hammer, and F. Parolek, "Continuing Risk Management at NASA," *CrossTalk, The Journal of Defense Software Engineering*, February, 2000.

[12] G. Holt, "Risk Management Fundamentals in Software Development," *Crosstalk: Journal of Defense Software Engineering*, August 2000.

[13] A. S. Khatavakhotan and S. H. Ow, "Rethinking the Mitigation Phase in Software Risk Management Process: A Case Study," in *Computational Intelligence, Modelling and Simulation (CIMSiM), 2012 Fourth International Conference on*, 2012, pp. 381-386.

[14] S. Islam, "Software development risk management model: a goal driven approach," in *Proceedings of the doctoral symposium for ESEC/FSE on Doctoral symposium*, 2009, pp. 5-8.

[15] P. K. Dey, J. Kinch, and S. O. Ogunlana, "Managing risk in software development projects: a case study," *Industrial Management & Data Systems*, vol. 107, pp. 284-303, 2007.

[16] D. Gotterbarn and S. Rogerson, "Responsible risk assessment with software development: creating the software development impact statement," *Communications of the Association for Information Systems*, vol. 15, p. 40, 2005.

[17] A.-G. Tang and R.-l. Wang, "Software project risk assessment model based on fuzzy theory," in *Computer and Communication Technologies in Agriculture Engineering (CCTAE), 2010 International Conference On*, 2010, pp. 328-330.

[18] R. P. Higuera and Y. Y. Haimes, "Software Risk Management," DTIC Document1996.

[19] U. I. Janjua, A. Oxley, and J. B. Jaafar, "Classification of software project risk managers: Established on roles and responsibilities," in *Computer and Information Sciences (ICCOINS), 2014 International Conference on*, 2014, pp. 1-6.

Validating Reusability of Software Projects using Object-Oriented Design Metrics

[1] Zhamri Che Ani, [2] Shuib Basri, and [2] Aliza Sarlan

[1] School of Computing, Universiti Utara Malaysia, 06010 UUM Sintok, Kedah, Malaysia.
zhamri@uum.edu.my

[2] Department of Computer and Information Sciences, Universiti Teknologi PETRONAS, 31750 Tronoh, Perak, Malaysia
{shuib_basri, aliza_sarlan}@petronas.com.my

Abstract. Reusability is the highest priority that needs to be considered before developing good software projects. Without reusability, software projects would be hard to maintain or extend. In order to ensure the minimum reusability levels of quality attribute is achieved in designing software project, two versions of jUnit which were developed by software professionals have been assessed by using Quality Model for Object-oriented Design (QMOOD). Even though the obtained result showed that the design of jUnit projects achieved the standard, the result was still not validated by other object-oriented design metrics framework. Therefore, this paper will validate the design reusability of jUnit using CK Metrics. The result showed that jUnit has a good quality design and the reusability level can be used as a benchmark for designing other software projects.

Keywords. reusability, software metrics, object-oriented design, jUnit

1 Introduction

All software engineers should be able to produce high quality software projects. With the new object-oriented paradigm, a strong emphasis on object-oriented design (OOD) is necessary because it is a faster development process, contains high reusable features, increases design quality and so on. However, it is hard to determine what is a good software design because the concept of good design is so subjective. Without quantitative measurements, it is difficult for software engineers to manage the quality of software projects.

Among all software design quality attributes, reusability is the highest priority that needs to be considered before developing good software projects. Without reusability, software projects would be hard to maintain or extend [1- 4]. In order to ensure the minimum reusability levels of quality attribute to be achieved in designing software project, two versions of jUnit which were developed by software professionals have been assessed by using Quality Model for Object-oriented Design (QMOOD) [5]. Even though the obtained result showed that the design of jUnit projects has achieved the standard, the result was still not validated by other object-oriented design metrics framework. Therefore, this paper will validate the design reusability of jUnit using CK Metrics. To maintain the consistency of the result, Metrics-1.3.6 [6] will be also used for this study. The results will be summarized based on the original comments by Chidamber and Kemerer [7], and the latest case study done by Kulkarni, Kalshetty, and Arde [8].

The rest of the paper is structured as follows. Section 2 provides some basic concept of CK metrics. Section 3 describes the case study, and experimentation results and discussions are presented in section 4. Section 5 includes conclusion and suggestion for future work.

K.J. Kim (ed.), *Information Science and Applications*,
Lecture Notes in Electrical Engineering 339, DOI 10.1007/978-3-662-46578-3_100

2 Chidamber And Kemerer (Ck) Metrics Framework

Object-oriented metrics can be used as a basis for comparing different versions of software programs [9]. One of the most popular object-oriented metrics and the most thoroughly investigated is CK metrics [10]. It was developed by Chidamber and Kemerer [7], and can be used as early quality indicators [11]. Even though it was proposed long time ago, the usefulness in analyzing open-source software are still significant [12 -15], including examining the reusability of software projects [16].

The original CK metrics has been a subject of discussion for many years and the authors themselves and other researchers has continued to improve the metrics. However, the next discussion will focus on original comments by Chidamber and Kemerer [7], and the latest case study done by Kulkarni, Kalshetty, and Arde [8]. Kulkarni used six java based open source projects: three Charting and Reporting Tools, and three Chat Servers Tools. All of them were developed in Java.

Chidamber and Kemerer proposed six metrics described in Table I. These metrics are widely accepted stardard and theoretically validated that they are useful to be used as early quality indicators [11], [15], [17], [18], [19], [20]. By using several of CK metrics collectively will help software engineers to make better design decision for any object-oriented language.

Table 1. Six Object-Oriented Design Metrics [7]

No.	Metric	Description
1.	Weighted Methods per Class (WMC)	WMC is a weighted sum (complexity) of all the methods defined in a class. High value is suggested that the class is too large and should be split.
2.	Response For a Class (RFC)	RFC is the count of the methods that can be potentially invoked in response to a message received by an object of a particular class. Complexity increases as the RFC increases, as well as the need for increased testing.
3.	Lack of Cohesion in Methods (LCOM)	LCOM is a count of the number of method pairs of similarity, that is zero, minus the count of method pairs whose similarity is not zero. Complexity and design difficulty increase as LCOM increases.
4.	Coupling Between Objects (CBO)	CBO is a count of the number of other classes whose methods or attributes are used by it. Large values imply more complexity, reduced maintainability and reduce reusability.
5.	Depth of Inheritance Tree (DIT)	DIT is the length of the longest path from a given class to the root class in the inheritance hierarchy. Large values imply more design complexity, but also more reuse.
6.	Number of Children (NOC)	NOC is a count of the number of immediate child classes that have inherited from a given class. Large values imply that abstraction of the parent class has been diluted, the increased need for testing, and more reuse. Redesign should be considered.

3 A Case Study

In order to validate the reusability of software design, two versions of jUnit software projects, jUnit-3.8.1 and jUnit-4.11 have been chosen as a case study. jUnit is an open source multi-version software project to help developers in testing every single unit in their programs. The main reason why an open source case study was used is due to the easiness of data and results distribution. Thus, later on it will help in supporting the industrial acceptability of software metrics.

Those two versions of jUnit were developed by software professionals using Java programming language and were selected after impressive results obtained using Quality Model for Object-oriented Design (QMOOD) [5]. In summary, jUnit-3.8.1 is the highest version of version-three which has 5267 lines of source code, with 12 packages and 99 classes. Meanwhile, so far jUnit-4.11 is the highest version of version-four which has 8822 lines of source code, with 31 packages and 174 classes.

Theoretically, when calculating software metrics using different tools, the result will also differ due to different technology used in metrics tools. Metrics-1.3.6 was used in our previous study and to ensure the consistency in data analysis, therefore again Metrics-1.3.6 was chosen as a tool to support CK metrics. Metrics-1.3.6 [6], an Eclipse IDE's plug-in, both versions of jUnit software project were required to be imported into eclipse IDE. In our case, eclipse IDE version 4.2 (Juno) was used to compile the source codes and both versions of jUnit must be compiled successfully before they can be analyzed by Metrics-1.3.6.

4 Experiment Results And Discussions

As mentioned earlier, Metrics-1.3.6 was used to measure both versions of jUnit software projects. This tool accurately measured all the metrics except RPC [21]. Therefore, only five metrics will be discussed in details. The next discussion will focus on original comments by Chidamber and Kemerer [7], and the latest case study done by Kulkarni, Kalshetty, and Arde [8].

4.1 Weighted Method per Class (WMC)

The number of methods and the complexity of methods involved indicate how much time and effort is required to develop and maintain a class. The larger the number of methods in a class, the greater the potential impact on children, since children will inherit all the methods defined in the class. Classes with large numbers of methods are likely to be more application specific, limiting the possibility of reuse. It was suggested that most classes should have a small number of methods, maximum up to 10 methods in a class. If WMC value is one, it is suggested that to merge the class in some order classes within the same package without affecting the LCOM value, that is without affecting the abstraction and encapsulation of the classes. If WMC value is zero, the possibility of redesign is high. If WMC over than 20, the class should be refactored to reduce complexity of the software project.

Table 2. Descriptive statistics for WMC

Software Product	Total	Mean	Std.Dev	Max	Min
jUnit-4.11	1736	9.977	12.148	87	1
jUnit-3.8.1	1012	10.222	15.552	105	2

Based on the WMC statistics shown in Table 2, we can see that the distribution of the WMC metric values in both projects is nearly the same. The mean values for both projects are less than 11. This means both projects are well designed. In addition, there is also no WMC value equal to zero for both projects. This means both projects have less possibilities of necessity of redesign. However, jUnit-4.11 contains classes with WMC value one. The involved classes can be improved by merging the classes in some other classes with the same package without affecting the LCOM value, which is without affecting the abstraction and encapsulation of the classes. It also found that, there are also few classes have WMC value more than 20 for both projects. Thus, the involved classes should be refactored to reduce the complexity of the software projects.

4.2 Lack of Cohesion in Methods (LCOM)

Cohesiveness of methods within a class is desirable since it promotes encapsulation and decreases the complexity of the objects. Low cohesion increases complexity, thereby increasing the likelihood of errors during the development process. Therefore, a high LCOM value is a serious issue in all the software projects. High LCOM value indicates poor encapsulation and abstraction at class level. It may introduce more possibilities of errors in the software development.

Table 3. Descriptive statistics for LCOM

Software Product	Total	Mean	Std.Dev	Max	Min
jUnit-4.11	-	0.122	0.246	1	0
jUnit-3.8.1	-	0.117	0.26	0.914	0

Based on the LCOM statistics shown in Table 3, the mean LCOM values for both projects are less than 0.2. These low values indicate that both projects are less complex. This result shows that WMC and LCOM are strongly co-related which is consistent with the studies done by Yadav [22].

4.3 Coupling Between Objects (CBO)

Ideally, objects should be loosely coupled, which means there should be little as dependence between objects as possible. The larger the number of couples, the higher the sensitivity to changes in other parts of the design, and therefore maintenance is more difficult. In addition, the more independent a class is, the easier it is to reuse it in another application. A measure of coupling is useful to determine how complex the testing of various parts of a design are likely to be. The low CBO values show that most of the classes refer to few other classes. Therefore, it increases understandability, efficiency and reusability of the design. In contrast, high CBO metrics may indicate poor design. A class which is tightly coupled will cause a large ripple effect when an object is changed and increasing the risk of regressions. Testing all the changes will be time consuming and difficult to verify. Classes with a CBO more than 19 should be analyzed to reduce coupling.

Table 4. Descriptive statistics for CBO

Software Product	Total	Mean	Std.Dev	Max	Min
jUnit-4.11	-	4.806	4.138	18	1
jUnit-3.8.1	-	5	4.262	15	0

Based on the CBO statistics shown in Table 4, the mean CBO values for both projects are less than 6. None of the CBO values is greater than 19. Therefore, we can conclude that the possibilities of understandability, efficiency and reusability of the design is very high.

4.4 Depth of Inheritance Tree (DIT)

The deeper a class is in the hierarchy, the greater the number of methods it is likely to inherit, the greater the potential of reuse. However, it constitutes greater design complexity. Classes with lots of child classes need to be very carefully modified to avoid regressions in those children. One possible solution is that designers should keep the number of levels of abstraction to a manageable number. It was suggested that, the maximum value of DIT should be less than 10. If DIT value is zero or one, it indicates poor reusability of the class.

Table 5. Descriptive statistics for DIT

Software Product	Total	Mean	Std.Dev	Max	Min
jUnit-4.11	-	1.914	1.005	5	1
jUnit-3.8.1	-	2.657	1.415	6	2

Based on the DIT statistics shown in Table 5, the mean DIT values for both projects are more than one and less than 10. Only few classes of jUnit-4.11 contain WMC value equal to one. This indicates that overall design of the projects is high reusability. Only few classes of jUnit-4.11 need to be improved.

4.5 Number of Children (NOC)

Inheritance is a form of reuse. So, the greater the number of children for a class, the greater of reuse. However, most probably it may introduce improper abstraction of the parent class or misuse of sub-classing. Moreover, more testing is required for methods in that class. It is better to have depth than breadth in the inheritance hierarchy. In other word, high DIT and low NOC is the ideal combination for a software project. Low values of DIT and NOC are strongly indicated that reuse through inheritance may not be fully adopted.

Table 6. Descriptive statistics for NOC

Software Product	Total	Mean	Std.Dev	Max	Min
jUnit-4.11	118	0.678	2.065	16	0
jUnit-3.8.1	80	0.808	5.235	52	0

Based on the NOC statistics shown in Table 6, the mean NOC values for both projects are less than the average value of DIT. These indicate that overall projects have achieved minimum criteria of reusability. This study also shows that both projects have few classes with zero children. This means that some of the classes need to be redesigned.

In summary, based on WMC, CBO, DIT and NOC obtained by Metrics-1.3.6, it can be concluded that both projects were well designed with high reusability standard. Low values of LCOM indicate that both projects are also less complex. The result is consistent with our previous study using QMOOD and this is a good indicator where jUnit is strongly recommended to be used as a design quality benchmark particularly in terms of reusability.

The standard deviation (Std.Dev) measures the amount of variation or dispersion from the mean. It is commonly used to measure confidence in statistical conclusions. Based on the statistics described earlier, we found that all Std.Dev values of jUnit-4.11 for each metric are closer to the mean compared to jUnit-3.8.1. This shows that the distribution of values for jUnit-4.11 is more significant compared to jUnit-3.8.1. Therefore, it indicates that most probably the new version of jUnit has better quality design compared to the previous version.

5 Conclusion And Future Work

This study presents a reusability validation of design for two software projects by using object-oriented design metrics. Two versions of jUnit were analyzed using CK metrics to find out whether jUnit has achieved certain level of reusability standard. The results show that jUnit has a good quality design and the reusability level can be used as a benchmark for designing other software projects.

Since Metrics-1.3.6 did not has the capability to analyze the RFC in CK metrics, we are planning to further investigate on this topic by using other measuring tools and make comprehensive comparison against these results.

References

1. M. Grand, Java enterprise design patterns. Wiley, 2002.
2. C. Alexander, S. Ishikawa, and M. Silverstein, *A pattern language: towns, buildings, construction.* Oxford University Press, USA, 1977, vol. 2.
3. P. Kuchana, Software architecture design patterns in Java. CRC Press, 2004.
4. C. Lasater, *Design patterns.* Jones & Bartlett Learning, 2006.
5. Z. Ani, S. Basri, and A. Sarlan, "Assessing quality attributes of object-oriented software design: A case study," in *2nd International Conference on Computer and Information Sciences,* Kuala Lumpur, 2014.
6. "Metrics 1.3.6," January 2014. [Online]. Available: http://metrics.sourceforge.net/

7. S. Chidamber and C. Kemerer, "A metrics suite for object oriented design," *IEEE Transactions on Software Engineering,* vol. 20, pp. 476–493, 1994.
8. U. Kulkarni, Y. Kalshetty, and V. Arde, "Validation of ck metrics for object oriented design measurement," in *3rd International Conference on Emerging Trends in Engineering and Technology (ICETET).* IEEE, 2010, pp. 646–651.
9. A. Kaur, S. Singh, D. K. Kahlon, and P. S. Sandhu, "Empirical analysis of ck & mood metrics suite," *Int. Journal of Innovation, Management and Technology,* vol. 1, no. 5, pp. 447–452, 2010.
10. M. Genero, M. Piattini, and C. Calero, "A survey of metrics for uml class diagrams," *Journal of Object Technology,* vol. 4, no. 9, pp. 59–92, 2005.
11. V. R. Basili, L. C. Briand, and W. L. Melo, "A validation of object-oriented design metrics as quality indicators," *IEEE Transactions on Software Engineering,* vol. 22, no. 10, pp. 751–761, 1996.
12. K.JohariandA.Kaur,"Validation of object-oriented metrics using open source software system: an empirical study," *ACM SIGSOFT Software Engineering Notes,* vol. 37, no. 1, pp. 1–4, 2012.
13. T. Honglei, S. Wei, and Z. Yanan, "The research on software metrics and software complexity metrics," in *International Forum on Computer Science-Technology and Applications, IFCSTA'09.,* vol. 1. IEEE, 2009, pp. 131–136.
14. S. Srivastava and R. Kumar, "Indirect method to measure software quality using ck-oo suite," in *International Conference on Intelligent Systems and Signal Processing (ISSP).* IEEE, 2013, pp. 47–51.
15. R.SubramanyamandM.Krishnan,"Empirical analysis of ck metrics for object-oriented design complexity: Implications for software defects," *IEEE Transactions on Software Engineering,* vol. 29, no. 4, pp. 297–310, 2003.
16. D.P.DarcyandC.F.Kemerer,"Oo metrics in practice,"Software,IEEE, vol. 22, no. 6, pp. 17–19, 2005.
17. M. Hitz and B. Montazeri, "Chidamber and kemerer's metrics suite: a measurement theory perspective," *Transactions on Software Engineering,* vol. 22, no. 4, pp. 267–271, 1996.
18. N. I. Churcher, M. J. Shepperd, S. Chidamber, and C. Kemerer, "Comments on" a metrics suite for object oriented design," *IEEE Transactions on Software Engineering,* vol. 21, no. 3, pp. 263–265, 1995.
19. G. Succi, W. Pedrycz, S. Djokic, P. Zuliani, and B. Russo, "An empirical exploration of the distributions of the chidamber and kemerer object-oriented metrics suite," *Empirical Software Engineering,* vol. 10, no. 1, pp. 81–104, 2005.
20. L. Cheikhi, R. E. Al-Qutaish, A. Idri, and A. Sellami, "Chidamber and kemerer object-oriented measures: Analysis of their design from the metrology perspective," *International Journal of Software Engineering & Its Applications,* vol. 8, no. 2, 2014.
21. R. Lincke, J. Lundberg, and W. Löwe, "Comparing software metrics tools," in *International Symposium on Software Testing and Analysis.* ACM, 2008, pp. 131–142.
22. V. Yadav and R. Singh, "Validating object oriented design quality using software metrics," in *International Conference on Advances in Electronics, Electrical and Computer Science Engineering,* 2012.

Plasticine Scrum: An Alternative Solution for Simulating Scrum Software Development

Sakgasit Ramingwong* and Lachana Ramingwong

Department of Computer Engineering, Faculty of Engineering,
Chiang Mai University, Thailand
sakgasit@eng.cmu.ac.th, lachana@gmail.com

Abstract. Scrum is an efficient software development process. However, in order to master its concept, practitioners need to have a considerable level of software development skills and go through cycles of actual activities. This could be expensive and time consuming. "Scrum simulation with LEGO bricks" is a very popular solution to tackle such limitations. Unfortunately, due to its high costs, implementing this workshop in a large software engineering class could be financially challenging. This paper proposes an alternative low cost solution for simulating scrum software development by using low cost plasticine as main material. It also reports results of two pilot studies on this alternative simulation.

Keywords: Software development. Scrum. Simulation. Education.

1 Introduction

Software engineering is a discipline which strives to develop reliable and efficient software [1]. It traditionally involves essential activities from requirement engineering to designing, construction, testing, deployment and maintenance [2]. In these modern days, a number of enhancive techniques are further implemented to increase success and improve stakeholder satisfaction.

Software development models are sets of continuous development processes, which are usually followed by software engineers. They could be briefly classified into two categories e.g. traditional (heavyweight) and agile (lightweight) development [3, 4]. Traditional development models involve extensive planning, design and other forms of documentation. Their ideal scenario is when all of the requirements are frozen. Traditional models are usually criticized for their lack of ability to cope with changes and excessive unnecessary documents. Examples of traditional software development models include the Waterfall model, the Spiral model and the incremental model [5-7]. On the other hand, as in its manifesto, agile developments highlight interactions, working product, customer collaboration and changes over plans and documents [8]. One cycle of agile development iteration is also much shorter than the

K.J. Kim (ed.), *Information Science and Applications*,
Lecture Notes in Electrical Engineering 339, DOI 10.1007/978-3-662-46578-3_101

traditional ones. Examples of agile processes involve Scrum, Extreme Programming and feature-driven development [9-11].

Variations of both traditional and agile models are widely implemented in software industry [12]. In university level, it is necessary for the students to learn as many development models as possible. Unfortunately, there are several critical limitations in teaching of software engineering courses. Time management is obviously one of these challenges. For example, in Thailand, general three credits software engineering course usually involves only 45 hours of sessions. Each session lasts 90 minutes. All necessary software development elements as well as other relevant topics such as quality and ethical issues need to be fit within this time period. As a result, the students are usually taught the theories by traditional lectures. Software engineering models, both traditional and agile, are mostly taught in conceptual level altogether within several hours. In addition, high number of students in a class limits degree of their participation. One undergraduate class could involve more than one hundred students. This makes activity-based sessions complicate and strenuous.

Indeed, to maximize the learning outcome of software engineering course, the students should have experiential trainings on, at least, a few software development models. However, this would also require a considerable amount of time and close supervisory to be effective under such limited time and large number of participants.

A number of game-based activities and workshops have been introduced to software engineering classes in order to overcome such constraints [13]. These involve both physical and computerized actions. Some of them focus on certain software engineering phases while others simulate the entire development cycle.

"Scrum simulation with LEGO bricks" is an innovative and motivational workshop which attempts to replicate Scrum software development [14]. Instead of actual software processes, the participants are to build a city with LEGO, the popular construction set. During the workshop, the participants gradually gain knowledge on essential Scrum artifacts. This activity requires only 100-120 minutes and is suitable even for people with low technical knowledge. Therefore, it is applicable for every level of software stakeholders. The major problem of this simulation is the cost of the construction set, especially when being implemented in a large group of participants.

This paper proposes a variation of workshops based on "Scrum simulation with LEGO bricks". Instead of using the expensive construction set, it suggests the use of low cost material, i.e. plasticine, as the main development component. It also discusses potential settings, application, and feedbacks of pilot sessions of the Plasticine Scrum. The second section of this paper discusses essential elements of Scrum and details on "Scrum simulation with LEGO bricks" and similar simulations. Then, the third section introduces the Plasticine Scrum and its settings. Results of a pilot study on this alternative workshop are discussed in section four. Finally, the fifth section concludes this paper.

2 Scrum and Scrum Simulations

Scrum is an agile software development methodology which highlights increased communication between stakeholders, reduced time to market and embracing of changes [9]. Daily standup meeting, cross-functional team, servant leadership, and enhancive retrospection are some of the essential features of this engineering process.

Scrum has become increasingly popular in modern software industry [15]. Implementation of this software development methodology requires certain efforts. Basically, the practitioners need to recognize the principles of agile manifesto as well as essential roles and artifacts [8]. They also need a certain level of software development skills, which not only includes programming but also designing, testing and other techniques relevant to the entire software development life cycle. Ideally, Scrum can be gradually learnt through cycles of recurring practice. Yet, in order to master all of its concepts, a considerable amount of time and effort are needed.

"Scrum simulation with LEGO bricks" is a game which aims to educate software practitioners on the concepts of Scrum [14]. In this workshop, the participants form teams of Scrum developers while the moderator portrays the roles of product owner and scrum master. However, instead of actual programming, their task is to build a city with the LEGO bricks. The simulation includes three sprints of fifteen minutes. In each sprint, the participants select their product backlogs, attend a daily standup meeting, build and release their product, and review their process. As a result, within these three development cycles, the participants gradually gain essential knowledge on the aforementioned Scrum elements. Approximately 100-120 minutes is needed to complete this simulation. This includes the introduction and the final debriefing. The "Scrum simulation with LEGO bricks" is widely implemented in software industry and has been translated into fifteen languages [14].

The "Scrum simulation with LEGO bricks" is mostly suitable for 8-18 people. One box of 650-piece LEGO is needed for every 4-6 participants. In Thailand, if a standard construction set is used, this costs approximately $40 per group for a session [16]. For a class of 80 students, the cost of implementing this simulation could reach $640. This is approximately equal to a monthly salary of a newly graduate Thai software engineers [17]. Obviously, the more players participate in the workshop, the higher the cost of material increases. Although the bricks are completely reusable, the initial cost could be intolerable.

Other variations of this simulation have been implemented in software engineering classes. "Agile Hour" is a game which groups of players follow 7 phases of Extreme Programming to build a product from LEGO bricks within 70 minutes [18]. The winner is the team whose product is accepted. The characteristics of the game are similar to the "Scrum simulation with LEGO bricks". Unfortunately, since this simulation is a role-based game, players who act as customer do not have opportunity to actually play the game itself, and therefore do not fully experience the process like players who act as developer.

Another group of researchers implement a LEGO brick building workshop to teach and practice Agile project management principles [19]. LEGO bricks were used as a common programming language to reduce gaps in background knowledge and experience between learners. Players need to build a LEGO city in analogy to the software system. The process was run in iterations of 10 minutes each. Due to time constraints, only basic concepts of Agile such as retrospectives, iterations, burn-down charts and backlogs are covered.

3 Plasticine Scrum

Although LEGO could be the ideal material of educational Scrum simulation, the cost of these bricks could be unaffordable in some cases. In order to tackle this problem, a use of an alternative material, Plasticine, is proposed.

Obviously, plasticine is vastly cheaper than LEGO bricks. One kilogram of medium-grade plasticine costs only $2 in Thailand [20]. For one round of Scrum simulation, two kilograms or $4 worth of plasticine are more than sufficient as the main material for a group of 4-6 practitioners. This means the cost of this substitute material is ten times cheaper than the LEGO bricks.

Compared to LEGO, plasticine is obviously softer and is less likely to be an ideal material to build rigid structures. In other words, it is more compatible to more curvy and high detailed objects. Thus, in Plasticine Scrum, the participants are instructed to build a zoo instead of a city. One product backlog is simulated by a habitat of animals. Examples of these backlogs include elephants, lions, monkeys, snakes, kangaroos and bears. Similar to the original workshop, the participants are allowed to use additional stationeries to complete the requirements. Several minor adjustments based on pilot studies discussed in the next section are made in order to support the characteristics of this alternative material.

3.1 Overviews and Objectives

In Plasticine Scrum, the players form a team of 4-6 developer who are assigned to build a zoo. Each team is considered as an individual organization. However, since there are no monetary elements involved in this simulation, they do not compete with others. The main objective of the game is to complete as many backlogs as possible while gradually learning the Scrum development concepts. This workshop simulates one Scrum release cycle which include three series of sprints.

3.2 Instructions

Each team is given two kilograms of assorted-color plasticine and printed product backlogs. The team should to define a standard scale of animal size in order to standardize their products.

One release cycle of three sprints is simulated in the workshop. Each sprint lasts 20 minutes. Including the 15 minutes of introduction at the beginning and another 15 minutes debriefing and presentations at the end, the entire Plasticine Scrum takes approximately 90 minutes.

During the introduction, the moderator explains all workshop elements, the sequence of activities, as well as the objectives. Release backlogs are selected from the product backlogs at the end of the introduction by the product owner. The participants are allowed to prepare the plasticine during this stage.

In each sprint, the teams are given the first three minutes to perform estimation, sprint planning, breaking down tasks, standup meeting and preparation of the burndown chart. Then, the next twelve minutes are spent for the sprint execution. At the ninth minute, which is at the middle of the execution, the players perform a sprint backlog refinement and renegotiate with the product owner if needed. The last five minutes are used for sprint review, updating the burn-down chart and retrospective. These are illustrated in Figure 1.

A representative of each Scrum team is given an opportunity to make a quick summary of the lesson learnt at the end of the workshop. The moderator then summarizes other missing key learning items. Finally, they are encouraged to dissemble their products and restore the plasticine to their original state as much as possible.

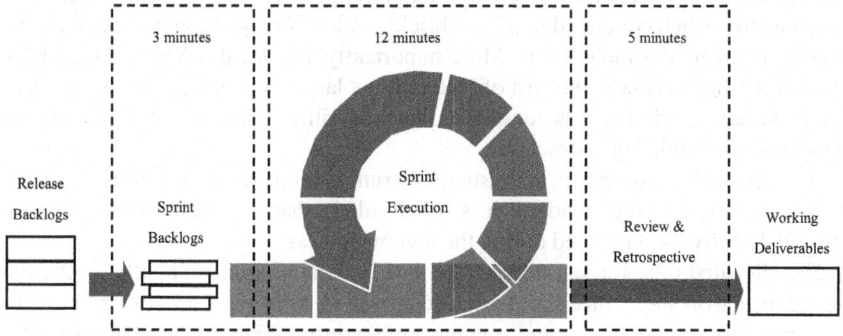

Fig. 1. A sprint in Plasticine Scrum.

3.3 Roles

There are three roles in Plasticine Scrum. Firstly, all participants play as the Scrum team. Most of all essential Scrum activities are performed by them. The second role is the product owner. The major responsibilities of the product owner include careful selection on release backlogs, approving sprint backlogs, reviewing the products and make all customer-related decisions. Generally, the product owner strives for excellence. Certain aspects such as the creature sizes and their general appearances need to be strictly inspected. Finally, the Scrum master acts as a moderator who motivates the team as well as facilitates the development. Both product owner and Scrum master can be portrayed by the lecturer.

3.4 Components

Three main components are allocated to each Scrum team. Firstly, plasticine is used as the main material. The second component, backlogs, are provided to the participants. During the game, only parts of the backlogs are selected and developed as the sprint backlogs. Elephants, lions, monkeys, snakes and other well-known animals are basic example of backlogs in this workshop. The moderator could even add mythical creatures, such as dragons, gryphons and hydras, in order to add fun factors to the game. Certain simple structures such as the entrance and park could also be included in the backlogs. Moreover, special backlogs, such as a safari zone which includes a peaceful community of various mammals in a larger habitat, can also be introduced if the moderator wants to highlight reusability aspects. Burn-down chart is the third main component in Plasticine Scrum. It is prepared and updated by the Scrum team.

3.5 Major variations from the original "Scrum simulation with LEGO bricks"

Major differences of Plasticine Scrum and the original workshop involve the main material. Since plasticine is soft and flexible, it is more suitable to build free shapes than rigid structures as in the original. Arguably, plasticine allows the players to be more creative when compared to plastic bricks. Additionally, plasticine can be mixed in order to create desirable color. More importantly, it is vastly cheaper than LEGO. This substantially reduces the cost of material in a large workshop. The downsides of this material include the less reusability, less mobility, much less eco-friendly and unsuitability to build solid structures.

One potential major defect in Plasticine Scrum is the relative size of the creatures. Without a proper sizing standard, it is highly likely that the misfit products will be deemed defective and rejected during the review process.

The allocated time for sprint execution in Plasticine Scrum is slightly longer than the original workshop since the participants need slightly more time to prepare and develop the products. Five more minutes are therefore given to the sprint execution thus this process takes 12 minutes instead of 7 minutes as in the original. However, the sprint planning and reviewing remain 3 and 5 minutes respectively. As a result, no significant emphasizing of Scrum activities is changed.

In Plasticine Scrum, special backlogs are encouraged. These backlogs highlight the importance of standards and reusability of the software components by simulating situations which the players need to reuse their finished products. Examples of the special backlogs include a safari zone, an elephant village and a circus. These special backlogs, if being introduced in the later sprints, force the players to reuse and reposition their creatures as well as expand their existing habitats. This imitates the actual development of software where standards are important and changes are commonly expected.

4 Pilot Studies of Plasticine Scrum

Two pilot batches of Plasticine Scrum workshop were implemented in software engineering related classes in Chiang Mai University, Thailand. The first simulation was participated by 20 senior undergraduate students. The students formed four groups of five members. In this pilot, the only changes from the original "Scrum simulation with LEGO bricks" were the main material and the backlogs. There were no change of timing and no special backlogs. Cartoonish backlogs of animals and their habitats from random zoo management online games were given to the students. This strictly defined the specification of the products such as the number of the animals, their characteristics and details of their habitats. At the end of the workshop, the students gained knowledge of necessary Scrum artifacts. Although their first sprints ended up with unfinished tasks, all participants completed their simulation without any remain backlogs in the final sprint. The main feedback from the participants was the development time was too short.

Based on lessons learnt from the first simulation, five more minutes were added to the sprint execution in the second batch of the pilot Plasticine Scrum. Ten graduate students formed two groups of Scrum team and participated in this workshop. A few adjustments were also made. This included change of backlog style and introduction of the special backlogs. The requirements were changed from illustrated to text backlogs. Although this caused more work to the product owner, the players became more imaginative. Also, the less details made negotiation between the product owner and the Scrum team became more important. The second adjustment was the increased execution time. The five extended minutes in the sprint execution significantly improved the overall quality and quantity of the finished products. The final adjustment of Plasticine Scrum was the introduction of the special backlogs in order to emphasize the importance of reusability and standardization. This forced several players to revise their products in order to be able to satisfy these new requirements.

Similar to the first simulation, the students could not finish all of their commitment in the first and second round. However, even with the introduction of the special backlogs, both teams successfully cleared their backlogs in the final round. The number of the finished products increased in the first two rounds of development. The students were able to build 3-4 habitats in each sprint execution.

Based on the presentation of the group representatives at the end of both pilot sessions, the feedbacks from both batches of participants were largely positive. The student were satisfied with the workshop and insisted that they learnt essential concepts of Scrum.

5 Conclusion

Plasticine Scrum is an alternative workshop based on "Scrum simulation with LEGO bricks". Its main objective is to teach essential concepts of Scrum software development. Plasticine Scrum provides an economic and efficient solution to classes with a large number of students. The cost of the workshop is approximately ten times less than the use of LEGO. Yet, it is important to note that, in a very long term, the cost of both workshops could be similar.

The details on this alternative workshop as well as the required changes are proposed in the paper. Several major adaptations such as the backlogs, the special backlogs and the extended development time are also highlighted and discussed. These variations also interestingly introduce more perspectives to the simulations.

The two pilot simulations of Plasticine Scrum suggest that this alternative workshop is promising. Feedbacks from the participants suggested that Plasticine Scrum was an efficient and informative workshop.

Further in-depth comparative experiments are to be conducted in order to empirically investigate the degrees of differences on the effectiveness of the original and alternative workshops.

Acknowledgement

This research is supported by the *"New Researcher Scholarship"*, provided by the Research Administration Center, Chiang Mai University under grant T5761.

References

1. Association for Computing Machinery.: Computing Degrees & Careers. ACM Press (2006)
2. Laplante, P.A.: What Every Engineer Should Know about Software Engineering. CRC Press, Florida (2007)
3. Glass, R.L.: Agile Versus Traditional: Make Love, Not War!. Cutter IT Journal, 14, 12-18 (2001)
4. Rudnick, B.: Agile Versus Traditional - A Tale of Two Methodologies. In Ground System Architecture Workshop, Los Angeles (2013)
5. Royce, W. W.: Managing the Development of Large Software Systems: Concepts and Techniques. In IEEE WESTCON, Los Angeles, CA (1970)
6. Boehm, B.: A Spiral Model of Software Development and Enhancement. ACM SIGSOFT Software Engineering Notes, 11, 14-24 (1986)
7. Pressman, R.S.: Software Engineering: A Practitioner's Approach. 7th edition ed.: McGraw-Hill (2010)
8. Beck, K., Beedle, M., et al.: Manifesto for Agile Software Development. Agile Alliance (2001)
9. Schwaber, K.: Agile Project Management with Scrum: Microsoft Press (2004)
10. Beck, K.: Extreme Programming Explained: Embrace Change. 2nd ed.: Addison-Wesley, (2004)
11. Palmer, S.R. and Felsing, J.M.: A Practical Guide to Feature-Driven Development. Prentice Hall (2002)
12. Yau, A. and Murphy, C.: Is a Rigorous Agile Methodology the Best Development Strategy for Small Scale Tech Startups?. University of Pensylvania (2013)
13. Caulfield, C., Xia, J., et al.: A Systematic Survey of Games Used for Software Engineering Education. Modern Applied Science, 5, 28-43 (2011)
14. Krivitsky, A.: Scrum Simulation with LEGO Bricks (2011)
15. Rubin, K.S., Essential Scrum: A Practical Guide to the Most Popular Agile Process. Addison-Wesley Professional (2012)
16. Lazada Thailand.: LEGO Basic Brick Deluxe #6177. (2013)
17. Adecco Thailand.: Thailand Salary Guide 2014. Adecco Thailand (2014)
18. Lübke, D. and Schneider, K.: Agile Hour: Teaching XP Skills to Students and IT Professionals. Lecture Notes in Computer Sciences, 3547, 517-529 (2005)
19. Velić, M., Padavić, I., and Dobrović, Ž.: Metamodel of Agile Project Management and the Process of Building with LEGO® Bricks. In 23rd Central European Conference on Information and Intelligent Systems (CECIIS 2012), 481-493, Varaždin, Croatia (2012)
20. OfficeMate Thailand.: GOLDHAND Plasticine 1 kg Mixed Color. (2014)

Understanding of Project Manager Competency in Agile Software Development Project: The Taxonomy

KamalrufadillahSutling[1],Zulkefli Mansor[2], Setyawan Widyarto[3], Sukumar Lecthmunan[4], Noor Habibah Arshad[5]

[123]Faculty of Computer Science and Information Technology, Universiti Selangor, 45600 Bestari Jaya, Selangor DarulEhsan, Malaysia.

[4]School of Computer Sciences, UniversitiSains Malaysia, 11800 Minden,Pulau Pinang, Malaysia.

[5]Faculty of Computer and Mathematical Science, UniversitiTeknologi Mara, 40450 Shah Alam, SelangorDarulEhsan, Malaysia.

[1]kamaldillah@unisel.edu.my,[2]kefflee@unisel.edu.my,[3]swidyarto@unisel.edu.my,[4]sukumar@cs.usm.my,[5]habibah@tmsk.uitm.edu.my

Abstract. The current growth of agile software development project (ASDP) continues to be more significant in the software industry. Project managers have important role to play in ensuring success of project. The success of a project depends on the competency of the project manager. Realizing on the lack of research on the project manager competency in ASDP such skill, knowledge, personal attribute and behavior. This research had taken initiatives in introducing; skill, knowledge, personal attribute and behavior that is needed by a project manager in ASDP. This paper contributes the relevant theory by developing taxonomy of the agile project manager's competency. Practitioners can use this taxonomy as a sensitizing device that will help project manager.

Keywords: Skill • Knowledge • Personal Attribute • Behaviour • agile project manager

1 Introduction

Many studies have evaluated the successfulness of agile software development project (ASDP). One of the factors contributing to the project success is the agile project manager competency[1]. The success of a project depends on the skills, knowledge, and experience [1], [2].

However, there is limited information about understanding ofthe project manager competency in ASDP particularly in Malaysia.The objective of this study is to analyze what are the bases of skills, knowledge, personal attributes and behavior required project manager in ASDP. Hence, this finding hopes to make a significant contribution to the project manager in determining the success of ASDP.

This research was funded by Fundamental Research Grant Scheme (FRGS/2/2013/ICT01/UNISEL/03/2 - Agile Project Managers Competency Model) the Ministry of Education Malaysia.

K.J. Kim (ed.), *Information Science and Applications*,
Lecture Notes in Electrical Engineering 339, DOI 10.1007/978-3-662-46578-3_102

2 Literature Review

2.1 Project Manager Competency

Competency of the project manager is very important in ASDP [1]. The competency of the project manager provides the basis for overall project performance. Therefore, project manager must understanding of four bases theproject manager competency of ASDP such as bellow:

2.2 Skill

As project manager in ASDP, he or she has to work together in order to achieve the project goals. There are three types of skills of project manager need to understood and important to the basic ofskills a project manager in ASDP such:

1. **Communication skill**- Communication is an important factor in software development [3].To improve these communication skills, project manager needs to actively listen, need to build relationships based on trust relationships and must understand the differences of personalities among the team members to improve the work processes, reduce conflict and promote understanding in strengthening cooperation [8], [4].Therefore, project manager need asking open questions, asking information, asking analysis, asking opinions and views of the team members and project participants agile. In addition, project manager needs to interact in each meeting of iteration planning, iteration retrospective and daily stand-ups to build relationships between team members [10]. Moreover, project manager must resolve the disputes and encourage the team to maintain focus.

2. **Team building skills-** project manager should build a strong bonding,with other team members and he or she needs to know how these interactions will contribute to their software development tasks. Measures such as these are important since face-to-face communication is so vital in ASDP[5].However, project manager have to help the team to move through the four stages of team development such as forming, storming, norming and performing[6].

3. **Problem-solving skills-** this refersto the ability of the agile project manager to visualize, and solve complex problems by making decisions that aresensible based on the available information. However, project manager need to make decisions which is direct and indirect to have an impact on the overall software cost, quality, and productivity[7].Furthermore, project manager needalso to be creative and address the relevant agile practices that will help yield the best results to solve their problems. Moreover, project manager need to Daily Scrum meetings involve the project team and the client, these meetings allow for the problems to be raised, and addressed in the early stages of the project [8].

2.3 Knowledge

In asset of the organization, the knowledge of the agile project manager is essential to be productive, to deliver competitive products and services [9]. Furthermore, knowledge of a project manager facilitating the increase on the production of software development, and allowing agile project manager to make efficient decisions, control complexity, and improve productivity.Therefore, the three basic knowledge of apro-ject manager that is required in ASDP success are:

1. **Strategic agility**-Strategic agility is important to pay attention to the strategic direction that leads to the big goal, and make decisions accordingly. However, knowledge in strategic agility can assist organizational leaders to assess their company's level of Strategic Agility [10].Therefore, project manager needs to know howto increase strategic agility. The following are techniques to increase strategic agility of project manager required in the past: interaction strategies[11],Transformational strategy [12] and Coordination strategy[13].

2. **Planning-**Project manager must determine what the project will accomplish, when it will be completed, and how it will be implemented or monitored. Project manager needs to be responsible for creating the project plans and defining the goals, objectives, activities and resources needed. The project plans will be the map to guide the project team and management. Moreover, project manager is also responsible for updating any new changes and plans thus communicating about it with the stakeholder[14][1].In the XP practice, project manager should know of the release planning and iteration planning which allows the customer to define the business value of desired features [6].In addition, scrum also has a set of procedures associated with it [15]. Project manager needs to know the sprint planning meeting is between the customer and the team. Hence, project manager need perform fifteen minute session of reviewing the work that is done regarding on development in Daily Scrum meeting.

3. **Coaching-** Coaching is about teamwork, motivation, communication skills and strategies. The main role of the project manager is to train staff and help the team into a cohesive unit, to facilitate interaction, optimize skills and build motivation towards a common goal. Hence the project manager should act in developing and implementing the tactics and strategies in much the same way as sports coaches [16]. Usually, the team will face more challenges and need the assistance of a project manager to enable them to stick with the practice in the core. In addition, project manager also need to make sure the team is on the right path to reinforce each other, so that the team does not to discard a practice [17]. Otherwise, project manager need to modify or replace with something equivalent. Moreover, project manager's need to remind the team with the basic principles and need to assist and guide the team to adapt their practices.

2.4 Personal Attribute

In this research, five Personal attributes acquired by aproject manager in ASDP is depicted. Usually, a project manager quickly becomes well-known in a very short period of time; clients identify those project managers who are good and those who cannot perform well. The following arethe basis personal attributes of a good project manager:

- **Common Sense:**As project manager in ASDP, project manager must use his common sense in most of the project situation for project success [18]. Project managers usually use these principles in responding to pressures the project through reason and intuition.
- **Good Listener:** Project manager in ASDP needs to listen to what the customers need and to understand the need to provide feedback on the technical aspects of how this problem might be solved or cannot be solved. Therefore, the communication between the customer and project manager can handle in the Planning Game [6]. In addition, project manager needs to provide a forum for open discussion and listening to what people want to say. The success of project management in ASDP depends on good listener and open communication from the a project manager [2], [19].
- **Good Communicator:**Project managers in ASDPwho have good communication are able to clearly outline what each member of their team should be doing. Agile methodology places a strong emphasis on communication between team members, especially face-to-face interaction. [20][5].
- **Motivator:** The project manager in ASDP must have a high level of self-motivation. As project managers must ensure that their teams produce quality work and ensure that the team members make decisions and complete tasks in a timely manner [21] and [22]. Therefore, project manager should have a confident and positive attitude in the group about software development projects are carried out. This will create a better working environment. In addition, project manager should give each team members the freedom to control their profession. This keeps them motivated for a long time. Moreover, project manager must ensure that no member of the team suffers form stress on the job and project manager needs to work with the team to ensure woks unbiased delegation. Lastly, project manager must recognize the team members for their hard work in software development projects. When the team reached a milestone, project manager can recognize their hard work such as lunch, trophy,bonuses and certificates of appreciation or mention in team meetings [23],[24].
- **Courageous:** Project manager must be courageous to develop confidence in the leadership. A leader with great challenges and risks in ASDPrequires greater courageand confidence. Therefore, project managers can build more confidence [25]. For example, courage enables project manager to feel comfortable with refactoring their code when necessary[19].

2.5 Behaviour

This paper has identifies seven behaviorsrequired by theproject manager to determine the success of anASDP; Leadership, Openness, results Orientation, Ethics, Communication, Strategic and Creative and Innovative. This behaviour will contribute to factor of the high competence and expertise on team members, increase to good customer relationship, increase to managers knowledgeable in agile process, increase to self-organizing teamwork and increase the motivation of team members[26].

1. **Leadership:** [27] the different leadership styles are more likely to lead to a successful outcome on different types of project. Furthermore, leadership style an adopted includes patterns of behaviour such as communication, conflict resolution, criticism, teamwork, decision making and delegation. However, the leadership is primarily accomplished through communication [28]. It involves many of behavior such as oral and written communication. Furthermore, the leadership requires good communication skills. Therefore, the leaders communicate a lot with personnel will contributes to the factor of employee's experience of communicating efficiency.

2. **Creative and Innovative:** Creativity and innovation will enhancecreativity and innovation of project manager behaviourin agile software development. The project manager must be creative in communication through effective use of colors, charts, and pictures to communicate concepts visually [29]. In addition, communication in the team has to be open using problems, tips and options shared freely between particular people[30].Moreover, project manager need to provide expertise or training or encourages travel and foster collaboration to ensure the team member is not depressed in completing the development project [17], [31][29]. Lastly, project manager must be creative in meeting [32].

3. **Openness:**The behaviour of Openness an project manager involving to ideas, collaboration and communication[33]. When using the agile approach, manager is needed to do collaboration with client within a constant stakeholder discussion. The particular agile manifesto places the main client relationship [30]. Furthermore, the behavior of feedback and transparency can improve venture performance in addition to productivity and facilitates open communication and the early discovery of problems [34][35].

4. **Communication:** Communication behaviour is an important to project manager in agile software development project [36]. The effective of communication behaviour a project manager in agile software development project must have Feedback Face to face and frequent communication among developers and between developers and customers[37],[5],[38]. Listening to what the customers need to do and understand these needs well enough to give the customer feedback about the technical aspects of how the problem might be solved, or cannot be solved [6]. In addition, the effective of communication behaviour a project manager in agile software development project must have osmotic communication for small agile teams [39]. Osmotic communication behavior makes the cost associated with communications low along with the feedback rate high, and so that errors are

corrected extremely easily as well as knowledge can be disseminated quickly [40].

5. **Result Orientation:** A project manager required ensure project results satisfy ones stakeholder relevant and to help focus current teams and also attention on key objectives to obtain orientation optimum outcome [41].Therefore, as project manager should work with the customer toward a shared definition of done for the requires the further trusting relationship and more flexible contract equipment [42]. However, the trust between client and the team lets the parties avoid waste connected in addition to effort [43]. In addition, credibility is the single most important quality every project manager must possess. Credibility is a combination of being seen to be trustworthy, convincing, and reliable [44]. The behavior of results-oriented leader is usually to be able to broaden section members' learning along with capabilities, and also that creates credibility. Furthermore, respect the stakeholders very important in aspect behaviourbecause project manager will benefit being realistic for having project's interests at heart [6], [45].

6. **Strategic:** Strategic is especially important for knowledge throughout project manager behaviour in ASDP .Strategic is to inspire and guide team members[46]. Strategy is usually important to cover attention to the strategic direction this leads towards big goal, and make decisions accordingly. Strategic can contributes to organizational leaders to assess their company's level of Strategic Agility [10]. Behavior that should be required is a project manager of customer interaction strategies, need to change the role of the training plan to team members to make them more comfortable with the new responsibilities [12] and coordination strategy [47].

7. **Ethics:** Ethical behaviour leads to better projectsuch honesty, respect and be fair[48],[49], [50]. Therefore, Honesty is important in order to be an ethical and also effective project manager in agile software development. One of the most important issues in any line of work is the honesty with which project manager deal with other people [49]. In addition, Project manager need to respect the Stakeholders, Project manager need to remember in mind will be this is a professional relationship as well as demands to be expressed respect at all times [50],[51]. Moreover, project manager need to be fair in dealings with everyone in the agile software development project. If project manager do this then project manager are sure to build good relationships and gain the kind of reputation for ethical behaviour[49].

3 Research Method

The research method used is a systematic literature analysis to collect data and information in order to achieve the objectives of this research. Systematic literature analysis serves to present new perspectives about project manager competency in ASDP. Therefore, this research always begins with problems along with questions during the literature review process as:

Q1: What are the skill, knowledge, personal attribute and behavior needs of a project manager to work inASDP?
Q2: What is the skill, knowledge, personal attribute and behavior a project manager that contributes to success of ASDP?

Therefore, in the process of systematic literature review, the researcher may read from one text to another to answer the research questions, as well as to identify the agile project manager competency. Furthermore, research questions serve as guidance to this study and influence the selection and collection of data. Table 1 shows 71 articles identified regarding agile project manager competency such as skill, knowledge, personal attribute and behavior. However, the data used is composed of written materials such as books, articles, journals, website, conference and thesis by using the ACM Digital Library, IEEE Xplore, ScienceDirect, Elsevier, SpringerLink, Wiley Inter Science Journal Finder and etc.

Table 1. Analysis of Literature Review

Materials such	Authour	No. of Relevant Articles
IEEE	[31],[28],[51],	3
Elsevier	[3],[5],[12],[20],[26],[27],[47]	7
ACM Digital Library	[2],	1
JCSE	[17]	1
IPMA	[41],	1
Microsoft Press	[45]	1
IJARCSSE	[15],	1
IJACA	[9],	1
IJIPM	[7]	1
Website	[4],[21],[22], [23],[24], [25],[33], [34], [44],[46],[49],[50]	12
Book	[6],[30],[36],[40],[42],	5
Others	[1],[8],[10],[11],[13], [14], [16], [18],[19],[29], [32], [35],[37],[38],[39], [43],[48],	17
Total		**51**

4 Discussion

Figure 1 shows the taxonomy regarding agile project manager competency required. The result of reading articles in systematic literature review would help in identifying the competency of an agile project manager discussed in this paper. All the research contents contribute with the sub contents with realistic and based on the Literature Review.

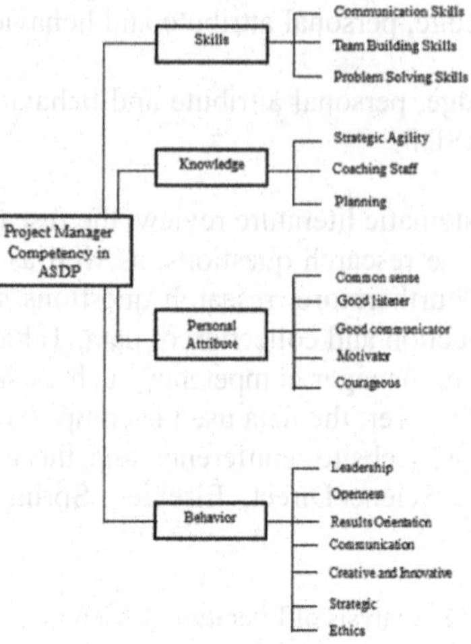

Fig.1. The Taxonomy of Project Manager Competency in ASDP

5 Conclusion

In conclusion, three basis of skill, knowledge and five basis of personal attribute and seven behavior required by agile project manager have been analyze and will be adopted in the process of software development project. However, this research contributes to the relevant theory by providing taxonomy of agile project manager in the area of software development project. Furthermore, practitioners can use this taxonomy as a sensitizing device to ensure the agile project manager considers the competency that can promote success of a project. This competency has significant importance to ensure the process of software development will be going smooth as planned by agile project manager, and increase productivity and profitability in business strategy of software development projects.

References

1. Hazimah N, Redzuan B, Mansor Z (2013) Competency Model of Agile Project Manager in Software Development Projects : A Systematic Literature Review. UJSET - UNISEL J Sci Eng Technol 1–7.
2. Begel A (2008) Pair Programming : What 's in it for Me? ACM
3. Moe NB, Dingsoyr T, Dyba T (2010) A teamwork model for understanding an agile team: A case study of a Scrum project. Inf Softw Technol 52:480–491. doi: 10.1016/j.infsof.2009.11.004
4. Jo, Ann S (2010) Top Five Communication Skills for Project Managers. http://www.projectsmart.co.uk/top-five-communication-skills-for-project-managers.php.
5. Hazzan O, Hadar I (2008) Why and how can human-related measures support software development processes? J Syst Softw 81:1248–1252. doi: 10.1016/j.jss.2008.01.037
6. Doug D (2004) eXtreme Project Managment. Jossey-Bass

7. Ahmed F, Bouktif LF, CapretzSalah, Campbell P (2013) Soft Skills and Software Development: A Reflection from Software Industry. Int J Inf Process Manag 4:171–191. doi: 10.4156/ijipm.vol4.issue3.17
8. Deemer BP, Benefield G (2007) The An Introduction to Agile Project Management. 1–16.
9. Bari M a, Ahamad S (2011) Managing Knowledge in Development of Agile Software. Int J Adv Comput Sci Appl 2:72–76.
10. Adler N (2012) The strategically agile organization: development of a measurement instrument. 2012.
11. Serena (2007) An introduction to agile software development. Danube Technol.
12. Kissi J, Dainty A, Tuuli M (2013) Examining the role of transformational leadership of portfolio managers in project performance. Int J Proj Manag 31:485–497. doi: 10.1016/j.ijproman.2012.09.004
13. Strode DE, Huff SL (2012) A Taxonomy of Dependencies in Agile Software Development. Australas. Conf. Inf. Syst. pp 1–10
14. PM4DEV (2011) Project Management for Development Organizations. Ski. a Proj. Manag.
15. Pathak K, Anju S (2013) Review of Agile Software Development Methodologies. Int J Adv Res Comput Sci Softw Eng 3:270–276.
16. Fraser S, Eckstein J, Kerievsky J, et al. (2003) Xtreme Programming and Agile Coaching. OOPSLA'03, Oct 26-30, 2003, Anaheim, California, USA ACM 265–267.
17. Shrivastava SV, Date H (2010) Distributed Agile Software Development. J Comput Sci Eng 1:10–17.
18. Lajos M (2013) Why Agile Isn ' t Working: Bringing Common Sense to Agile Principles. http://www.cio.com/article/2385322/agile-development/why-agile-isn-t-working--bringing-common-sense-to-agile-principles.html.
19. Kendall JE, Kendall KE (2005) Agile Modeling. Syst. Anal. Des. 7/e. Prentice Hall, pp 185–196
20. Wang X, Conboy K, Cawley O (2012) "Leagile" software development: An experience report analysis of the application of lean approaches in agile software development. J Syst Softw 85:1287–1299. doi: 10.1016/j.jss.2012.01.061
21. Reich, Gemino, Sauer (2013) Igniting the Passion - What Motivates Project. In: PMP Perspect. http://www.pmperspectives.org/article.php?aid=58&view=full&sid=a63080f3458f0a051dac435241c8f333.
22. Amiryar H (2012) Scrum Definition. http://www.pmdocuments.com/2012/09/26/scrum-definition.
23. Liz S (2009) Agile Coaching. http://www.agilecoach.co.uk/Articles/Motivation.html.
24. Young ML (2012) The Importance of Motivation in Project Management. In: Pm hut. http://www.pmhut.com/the-importance-of-motivation-in-project-management.
25. Anais N (2012) It Takes Courage to Develop Confidence in Leadership. http://www.tcnorth.com/courage-and-confidence-in-leadership-how-to-develop-them/.
26. Chow T, Cao D-B (2008) A survey study of critical success factors in agile software projects. J Syst Softw 81:961–971. doi: 10.1016/j.jss.2007.08.020
27. Trivellas P, Drimoussis C (2013) Investigating Leadership Styles, Behavioural and Managerial Competency Profiles of Successful Project Managers in Greece. Procedia - Soc Behav Sci 73:692–700. doi: 10.1016/j.sbspro.2013.02.107
28. Skovolt K (2009) Leadership Communication in a Virtual Team. 1–12.
29. Warner, Paul D (2012) Creativity and Innovation in Project Management. 1–10.
30. Jim Highsmith (2009) Agile Project Management. Pearson Education Inc
31. Yi L (2011) Manager as Scrum Master. IEEE Trans. Softw. Eng. Ieee, pp 151–153
32. Kieran C, Wang X, Fitzgerald B (2005) Creativity in Agile Systems Development: A literature Review.

33. Goran K (2013) Agile as a humane way of software development – the road from cooperation to collaboration. http://www.operatingdev.com/2013/03/agile-as-a-humane-way-of-software-development/.
34. Althea T (2013) Types of Communication Behavior. http://www.ehow.com/info_8075513_types-communication-behavior.html.
35. Andersen ES (2008) Rethinking Project Management: An Organisational Perspective. pearson
36. Thomas (2012) The Concept of Transparency in Agile Project Manageament. http://p-a-m.org/2012/03/the-concept-of-transparency-in-agile-project-management/.
37. Turk D, France R, Bernhard R (2004) Assumptions Underlying Agile Software Development Processes. J. database Manag.
38. Eykelhoff M (2007) Communication in global software development : A pilot study. Twente student Conf. IT, Univ. Twente,Faculty Electr. Eng. Math. Comput. Sci.
39. Chandana (2012) Osmotic Communication Agile : Agile Certification Training.
40. Cockburn A (2004) Osmotic Communication (The Crystal Clear Book). Pearson Education Inc
41. Gerrit K (2006) Contextual competences Behavioural competences Technical competences The Eye of Competence competences. IPMA Competence Baseline Version 3.0. pp 83–122
42. Mike G (2012) Chapter 2 : Agile Project Management Framework. 1–11.
43. MyMG T (2012) Agile PM – Building Trustful Relationships Between Customer And Developer.
44. Lynda B (2013) Credibility. https://stakeholdermanagement.wordpress.com/2013/04/27/733/.
45. Ken S (2004) Agile Project Management with Scrum. In: Microsoft Press.
46. Steven J S (2013) The 4 Characteristics of Strategically Agile Leaders. http://www.cmoe.com/blog/strategically-agile-leaders.htm.
47. Strode DE, Huff SL, Hope B, Link S (2012) Coordination in co-located agile software development projects. J Syst Softw 85:1222–1238. doi: 10.1016/j.jss.2012.02.017
48. John M (2011) Project management – a question of ethical and moral responsibility. McManus.indd 188–189.
49. Ben F (2012) 5 Ethical Codes of Conduct for Project Managers Email Updates. http://cobaltpm.com/5-ethical-codes-of-conduct-for-project-managers/.
50. Scott , W. A (2013) Active Stakeholder Participation : An Agile Best Practice. http://agilemodeling.com/essays/activeStakeholderParticipation.htm.
51. Power K (2010) Stakeholder identification in agile software product development organizations: A model for understanding who and what really counts. Proc. - 2010 Agil. Conf. Agil. 2010. IEEE, pp 87–94

The Design and Implementation of Around-View System using Android-based Image Processing

Gyu-Hyun Kim, Jong-Wook Jang

Computer Science and Engineering, Busan, Republic of Korea, kim33276@naver.com
Computer Science and Engineering, Busan, Republic of Korea, jwjang@deu.ac.kr

Abstract. Nowadays, image processing products such as car black boxes and CCTV are sold in the market and offer convenient to users. Especially the black boxes helps determine the cause of the accidents while driving. However they can check only the front and the rear of vehicles so they can't check for the driver's blind spot and scenes of angle of view. To solve these problems, the Black box system was improved further and AVM(Around-View Monitoring) system was developed. AVM system gets images which has a bird's-eye view of a sight on the top so it can obtain all directions, namely 360-degree images of vehicles. To obtain images, there is a condition attached to this system. Desktop should be installed in vehicles.

AVM system that this article suggests can remedy the disadvantages such as installing PC in vehicles. this system was designed, using tablet equipment.

Keywords: Black box System, Around View, Image Processing, Wireless Camera

1 Introduction

AVM system is a system to show image on the street level like it is being viewed directly from above. The system utilizes four cameras on a car, strategically placed on the front, at the back and two more on each side. This is a very convenient system in which the system is able to show the image of what is in front, at the back and on the sides of the car on demand on the screen; therefore, you will be able to see what's around your car at all times. This is very useful especially when parking the vehicle.[1].

AVM was invented to complement the front and rear black box. Because people can't find the exact cause of driving accidents which happen from various angles through this. The images received from the cameras are recorded which is very useful to determine the cause of accident. The function of black box of AVM system is that secure a clear view drivers as well as recording accidents. Prominently female drivers cause car accidents when they reverse into a parking space or the driver's carelessness. Because they can secure a clear view. Therefore there are lots of fender benders. To

K.J. Kim (ed.), *Information Science and Applications*,
Lecture Notes in Electrical Engineering 339, DOI 10.1007/978-3-662-46578-3_103

reduce these accident late, image processing technique is combined to 4 channels images. To reduce these accident rate, AVM system was developed to see a image around vehicle by combining image processing technique to 4 channels images and transforming into integrated, bird's-eye view image.. AVM system is not a normal technique that just record each image but a distinctive technique that makes person to feel their own car around 360 degrees by receiving 4 images, so it is not easy to develop technology. First, each image was connected by in three dimensions, not in two dimensions. For these reasons, image processing techniques are needed. It is expensive to install because desktop has to be installed and the system needs 4 cameras and wiring of desktop. Thus it is urgent to develop the technology simplifying wiring and excluding installation of desktop.

In this paper, it is planned to develop new system using the widely spread tablet and wireless camera instead of installing desktop. Image process technology on tablet is fairly difficult. But through R&D it will provide convenience for driving, traffic reduction and developments in the auto industry to drivers and automobile industry.

.

2 Related research

This article suggests that an Android OS-based tablet and 4 wireless cameras are required for developing Android OS-based AVM(Around-View Monitoring) system. In order to develop this system, It is planned that the images of 4 channels wireless camera are transferred into a tablet and then they incorporate the images into OpenCV(free library for processing image). In the past, OpenCV libray just worked in Window OS in the past. However Intel Corporation built up a library even works in various platforms through R&D This system is designed with Android OS equipment instead of IOS equipment, it will provide familiar circumstances. [2]

Before grafting image processing of camera, the image has to appear on Android tablet. But it can't use because the tablet does not have 4 USB Ports that install a cable camera. So wireless camera is necessary but it supports RTP / RTSP protocol, Html protocol just for picture and video.[3]

There are various ways to send and receive data with wireless camera. Typically, the first method is to load images through VLC server after encoding. The second is to load images by doing encoding works directly. In this paper, the main object of this system is simplification of installation of equipment. Thus it doesn't use VLC server which has to be installed on desktop. For this reason, the 2 method are needed. First, obtaining image using Rtsp protocol on tablet. Second, grafting OpenCV library using loading image. Generally there is a method loading images using Html protocol, but Html protocol does not fit in speed to load images in real time. If the program which helps additionally while driving can't receive images in real time it can lead to a disaster, Even it can happen due to delays. Therefore it is hoped to reduce delay time for loading image using RTSP(Real Time Streaming Protocol).[4]

Figure 1. Camera video display using VLC

Figure 2. A direct Camera video display into device

More specifically, if a video is loaded in tablet by simply using RTSP protocol, the video can't be loaded because H.264 codec and compatibility among devices make problems. All devices support international standards such as Mpeg-4 or H.264. But images can't appear if encoding has not been done. Thus synchronization with image of camera is performed by hand[5].

The object of this study is that loading 4 channels images using RTSP protocol on Android equipment after encoding progress and image processing based on these images

3 SYSTEM ARCHITECTURE AND DESIGN

3.1 System Architecture

In order to implement AVM(Around-View Monitoring)system, 4 cameras, 4 IP camera modules, a router, an Android OS-based tablet are required. Network environment should be created in vehicles because wireless cameras are used to minimize wiring. In the circumstances, the IP cameras from every direction in vehicles and an Android tablet send or receive data through network. Since there is no additional server in the vehicle, encoding process is needed to conjure up RTSP protocol image in the transmission/reception process respectively. With 4 cameras working smoothly, it shows 4 images on the android tablet. Image processing is implemented at 4 channels on each using OpenCV library for android. The whole process provides users with the AVM(Around-View Monitoring) system.

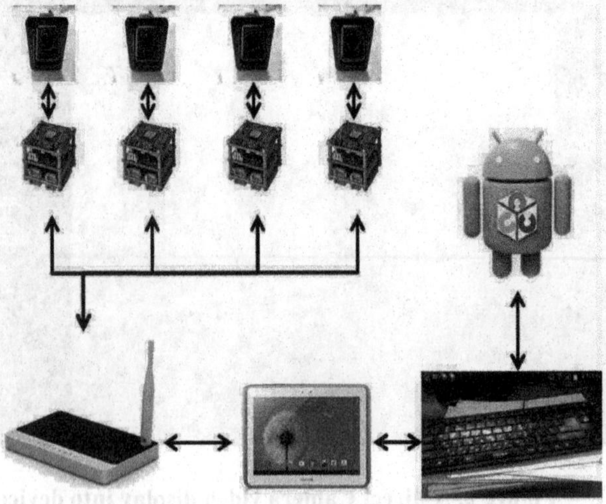

Figure 3. System Structure

3.2 System Design

Management of this system begins with the successful connection of a tablet and a network. Once they are successful in connecting, they attempt access with 4 cameras through manipulating the tablet. This connection might be failed because of wireless camera, not cable cameras. So it must be avoided to create the surroundings affecting the network connection. If they are successful in connecting with 4 cameras, they transmit images on the tablet in real time. Though transmission is achieved, the proper images can't be seen on the tablet. Therefore during sending and receiving process between tablet and wireless camera, H.264 CODEC goes through encoding process. If it fails, it repeats infinitely until encoding process has been completed into 4 cameras. 4 images are printed out into tablet after the repetitive works are completed. The part up to this point is 4 channels black box system on the market today. Tablet carry OpenCV library for android and it applies 4 screens on each. There are various techniques in image processing. Of these, 3 techniques are used in this system. homography technique, image distortion compensation, image registration. The edges of image curves when the images are far away from a focal point. This is called image distortion and correcting this situation flatways.

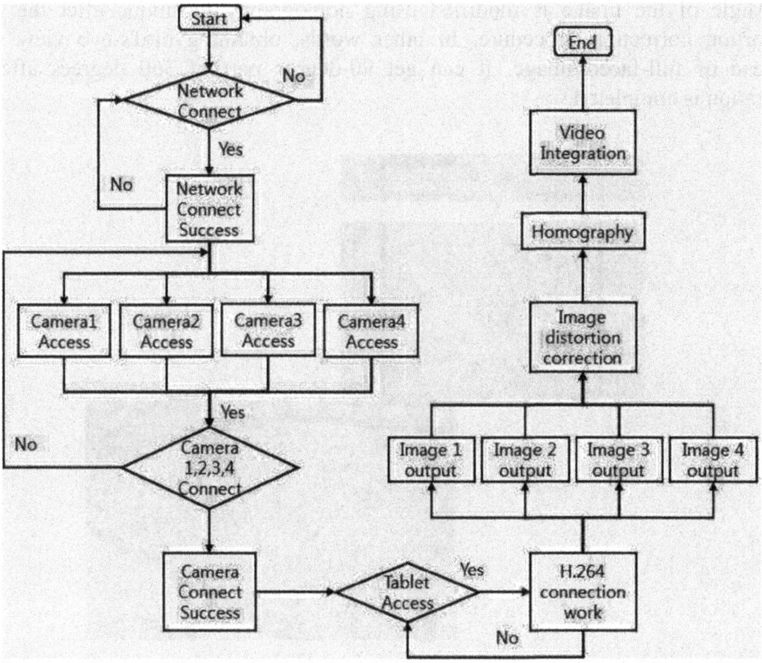

Figure 4. System Design

is called image distortion compensation(image distortion correction)

Figure 5. Picture of image distortion

Angle of the image is modified using homography technique after the image distortion correction procedure. In other words, obtaining bird's-eye view image instead of full-faced image. It can get 90-degree part of 360 degrees after this operation is completed.

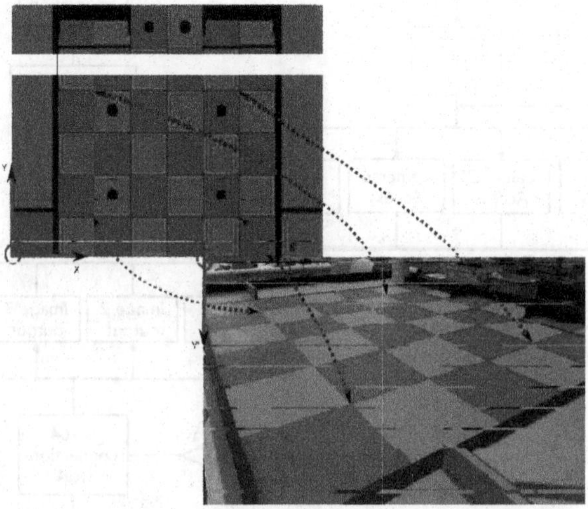

Figure 6. Pictures using homography technique

If image distortion compensation and homography techniques are applied into the 4 images, it combines into one using these images. It offers one system to users through this procedure.

4 Conclusion

In this paper, Around-View monitoring system using Android-based image processing integrates images which are seen from top to bottom not just full-faced images after image processing with 4 wireless cameras. Unlike full-faced image, as it makes bird's-eye view image the over regular distance around vehicles can't be confirmed in real time. In order too overcome such shortcomings, it needs to seek for solution that records 4 channels on each before image processing. It is not easy to construct DVR system which obtain several wireless camera images with mobile device. Environment and equipment of this system is distinctive unlike widely spread AVM system. If this system is completed and implemented using wireless camera and portable tablet, it can overcome difficult installation and reduce the time. Further more, it is hoped that eventually this could help computer-related developers provide advanced technology information, enlarge convenience for drivers and minimize traffic accident rate.

Later I plan to develop a System designed through this study and to test it using wireless camera, Android Os-tablet in real vehicles. If the technology works, interworking and scalability with other application could be upgraded.

Acknowledgment

This work was supported by the Brain Busan 21 Project in 2014. And this work was supported by the Nurimaru R&BD Project(Busan IT Industry Promotion Agency) in 2014.

References

1. Naver Dictionary, "Around-View", May 2012, http://terms.naver.com/entry.nhn?docId=1656756&cid=2913&categoryId=2913
2. Wikipedia, "OpenCV", http://ko.wikipedia.org/wiki/OpenCV
3. Naver Dictionary, "Rtsp", http://terms.naver.com/entry.nhn?docId=300466&cid=43665&categoryId=43665
4. Wikipedia,"VLC", http://ko.wikipedia.org/wiki/VLC_%EB%AF%B8%EB%94%94%EC%96%B4_%ED%94%8C%EB%A0%88%EC%9D%B4%EC%96%B4

5. Jae-Hyeok Yeon, Hyo-Taek Im, Jae-Hong Park, Korea Institute of Maritime Information and Communication Sciences, 2010, "Design and implementation of RTSP-based streaming server, the dynamic"

Identification of Opportunities for Move Method Refactoring Using Decision Theory

Siriwan Boonkwan and Pornsiri Muenchaisri

Department of Computer Engineering, Faculty of Engineering, Chulalongkorn University, Bangkok, Thailand

Siriwan.B@Student.chula.ac.th, Pornsiri.Mu@chula.ac.th

Abstract. The early versions of object-oriented software system may function correctly, but sometimes its internal structure may not be well designed. Some classes may be god classes with several different behaviors. God classes have low cohesion in which some of their methods should be moved to another class. Finding a target method of a class to be moved to another class is a challenging problem. This paper proposes an approach to identity opportunity for move method using decision theory. The approach searches for candidate classes and chooses only one class with the highest coupling value which is used as decision criteria of Laplace method. A preliminary evaluation is performed on object-oriented software to demonstrate the effectiveness of the proposed approach. The results show that the proposed method can improve the design quality of the source code.

Keywords: Move Method Refactoring, Decision Theory, Software Metric, Cohesion, Coupling

1 Introduction

One of the main goals of software designers is to design a software system with low complexity. Designers intentionally spread the responsibility of software to several classes in order to decrease coupling and increase cohesion [5], [6]. Refactoring is a technique that changes the design structure of software without affecting the behavior of the software. Refactoring normally improves nonfunctional attributes of software such as software cohesion and coupling [1] increases efficiency, and reduces redundancy in which the software is easier to understand [2], [3]. Some researchers apply refactoring to improve software quality. Several studies have used the coupling and cohesion metrics [7], [8] to offer identification of opportunities for refactoring. Bavota [4] uses the Game theory to identify refactoring opportunities to indicate chances to extract class. Sales [9] has created recommendation algorithm for move method refactoring of software. Kimura [11] has proposed technique to search move method refactoring candidates by analyzing method traces of software systems. Napoli [12] proposes a technique to suggest refactoring that computes metrics by GPU for large software systems.

In general, a strongly coupled system shows impact on the change of class and will affect other classes [10]. A lowly cohesive system shows a class which has several responsibilities. A class with strong coupling and low cohesion needs refactoring. Moving method refactoring may be applied when a class has several behaviors or

K.J. Kim (ed.), *Information Science and Applications*,
Lecture Notes in Electrical Engineering 339, DOI 10.1007/978-3-662-46578-3_104

when there are too much dependency among classes and highly coupled [1]. Identifying moving method refactoring opportunities should decide from a weak design of source code. Finding a target method of a class to be moved to another class is a challenging problem. Decision theory is a decision making technique that will choose the best option out of several options. Decision making consists of three phases: finding possible options for making decision; finding events for measuring each option; and selecting option with Laplace criteria decision making. A Laplace criterion is a part of the decision making under uncertainty, which provides equally weight for each option.

This paper proposes an approach to identity opportunity for move method using decision theory. The approach searches for candidate classes and chooses only one target class with the highest coupling value used as Laplace criteria. The candidate class that has called the method target will be definitely regarded as one option. The Laplace decides on metrics used to measure the coupling of source code in each option. The paper can be classified as follows. Section 2 reviews related work. Section 3 proposes metric used in identifying opportunities of move method refactoring. Section 4 offers move method refactoring using decision theory. Section 5 is application of Laplace to Move Method refactoring. Finally, Section 6 is the paper's conclusion and future work.

2 Related Work

Giuseppe Pappalardo and Emiliano Tramontana [7] have proposed novel metric for identifying extract class refactoring opportunity. The metric is used to measure correlation of methods in a class based on a calling interaction between parameters and attributes. The researchers has separated method into two sets which are weakly interact method and strength interact method. After they got strength interact method, they has extracted method from this current class into the new one.

Nikolaos Tsantalis and Alexander Chatzigeorgiou [8] have proposed an identifying move method refactoring opportunities to solve Feature Envy bad smell problem. They used an algorithm to measure a distance between system entities (attribute/method) and class's extracts a list of behavior preserving refactoring by depending on set of preconditions measurement. As a result, this methodology has measured the effect of refactoring suggestion based on novel entity placement metric which is considered that an entity has organized on appropriate class.

Vitor Salesy, Ricardo Terrayz, Luis Fernando Miranday, and Marco Tulio Valente [9] have proposed the new way to use move method refactoring on the basis of dependencies set of method. The researcher has described a comparison between dependencies method similarities in the class with using algorithm to compute the similarities for creating dependency set to move.

Shuhei Kimura, Yoshiki Higo, Hiroshi Igaki and Shinji Kusumoto [11] have proposed dynamic analysis technique for checking possible refactoring. Analyzing traces during program process collected and analyzed data which came from the practical processing of source code. The researcher has separated method in interactive class and counted calling from a method to the other one. If there is an unusual call or an unusual one, system will mark on it to declare that this method needs a move method refactoring.

3 Metrics for Identification to Move Method

In this part, details regarding measurement of coupling metrics will be described. The relationships of methods and variables have accessed to a method of another class and accessed to the variables and methods within the class are measured. We deploy both existing metrics and the proposed new metrics in this study. The existing metrics: Call-based Interaction between Method (CIM) [5] and Relative Method Coupling (RMC) [12] are used to measure of coupling between classes and methods.

We present new metrics: Interaction coupling between classes (ICBC) and Interaction method coupling (IMC).

ICBC is a measurement of coupling between classes. If the value of ICBC is 0, it indicates that there is no call to the target method from another class. If the value of ICBC is greater than 0, it indicates that coupling of class C_i and class C_j have increased.

$$ICBC = \frac{Interaction(C_i, C_j) + Interaction(C_j, C_i)}{2} \quad (1)$$

Where $Interaction(C_i, C_j)$ is the number of calls from class C_i to variables and methods of class C_j. $Interaction(C_j, C_i)$ is the number of calls from class C_j to variables and methods of class C_i .

IMC is a measurement of coupling between classes and coupling between method m and other methods within class. The IMC has a value in the range of [0, 1]. If the value of IMC is 1, the metric indicates that there is no call to a target method within class. If the value of IMC is 0, the metric indicates that there is no call to a target method in another class.

$$IMC = \frac{call(C_j, m)}{call(C_i, m) + call(A_{C_i}, m) + call(C_j, m)} \quad (2)$$

Where m is method in the class C_i . $call(C_i, m)$ is the number of calls that method m calls within class C_i . $call(A_{C_i}, m)$ is the number of calls that method m uses attributes within the class C_i and $call(C_j, m)$ is the number of calls that method m is called from class C_j.

4 The Proposed Approach for Move Method Refactoring

This section presents the proposed approach in details on how to apply decision making of Move Method refactoring opportunities in order to improve quality of objects-oriented software. The proposed method can help developers to select a target method that will be moved to an appropriate class. The Decision theory is used to select a candidate class that a target method must be moved from the existing class to an appropriate target class. This paper uses Laplace criteria of Decision theory to identity opportunity for move method refactoring.

4.1 Overview of the proposed approach

Figure 1 shows an overview of the proposed approach which consists of three parts. The C# source code is input of the approach. The first part, Parser uses the Abstract Syntax Tree (AST) to represent the structure of source code (i.e., correlation between method and method calls). The second part consists of: "Finding candidate classes" and "Choosing the target class by Laplace criteria". The third part is Move method refactoring, which moves the target method to an appropriate class. The output is the improved internal structure of C# source code.

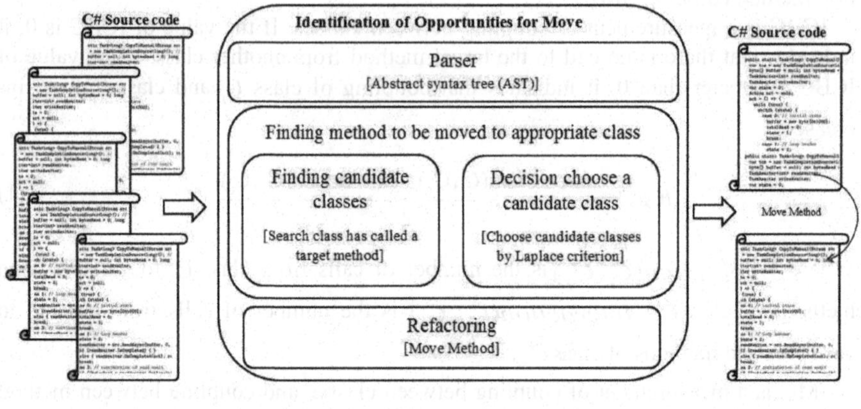

Fig. 1. Overview of the proposed approach

4.2 Finding method to be moved to appropriate class

Finding the target method which will be moved to appropriate target class is divided into two principal parts.

1. Find candidate classes which have at least one call to a target method.
2. Choose a target class from candidate classes
   (a) Measure CIM, RMC, ICBC and IMC of each candidate class.
   (b) Choose the appropriate target class from candidate classes by Laplace criterion.

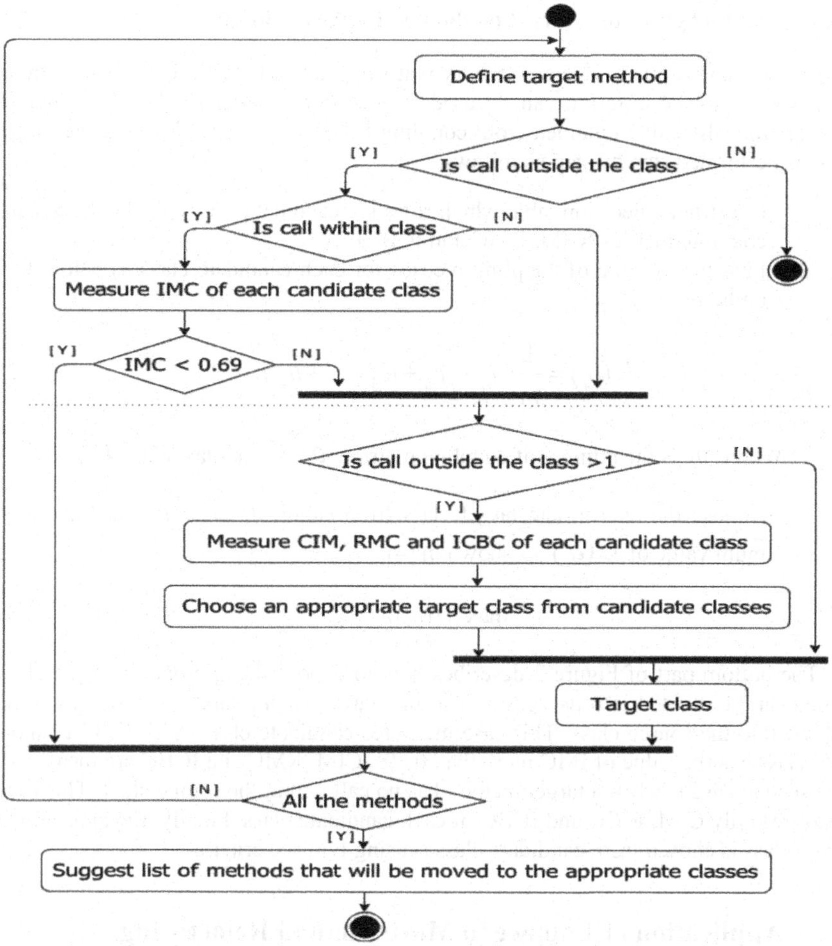

Fig. 2. Overview Activity Diagram of the Identification of Opportunities for Move Method Using Decision Theory

Finding candidate classes

Candidate classes are classes which send messages to the target method. These are three kinds of candidate classes.

- The first kind is that the target method is called from only one candidate class and there is no call to a target method within the class.
- The second kind is that the target method is called from many candidate classes and there is no call to a target method within the class.
- The third kind is that the target method is called from many candidate classes and there is some calls to a target method within the class.

The top part of Figure 2 describes how to identify opportunities for Move Method refactoring, which finds the appropriate target class for the target method. The first step is specifying the target method from the list of methods of classes. The second step is finding candidate classes that has some calls to a target method

Choose a target class for a target method by Laplace criteria

Laplace is one criteria of decision theory that can select the best option from various options. A Laplace criterion can be used for decision making in choosing suitable answer from the value obtained from coupling between classes. Choosing the target class is performed step by step as follows.

1. Construct a decision table which rows are candidate classes and columns are coupling metrics (CIM, RMC and ICBC).
2. Find the average of coupling metrics for each candidate class as below formula.

$$U(a_i) = \frac{1}{m}(u_{i1} + u_{i2} + u_{i3} + ... + u_{in}) \tag{3}$$

Where m is the number of coupling metrics. a_i is candidate class i.

3. Choose the appropriate target class from candidate classes with the maximum value of $U(a_i)$ as shown in (4).

$$\max\{U(a_i)\} \tag{4}$$

The bottom part of Figure 2 describes how to choose the appropriate target class which can be divided into two cases. The first case, when a target method has some calls within the source class. This case measures candidate class by IMC. If a candidate class has the value of IMC more than 0.69[1], CIM, RMC and ICBC are measured. The second case is when a target method has no call within the source class. This case measures only CIM, RMC and ICBC of each candidate class. Finally, the appropriate target class is chosen from candidate classes using Laplace criteria.

5 Application of Laplace to Move Method Refactoring

The proposed approach is applied to C# source code of Online Ordering system[2]. Table 1 shows target methods, source classes, candidate classes and number of calls. Table 2 shows IMC of candidate classes which call target method CusShipping. Table 3 shows CIM, RMC, ICBC, and $U(a_i)$. Class OrderDetail is chosen to be the target class since it has the highest value of $U(a_i)$ for method CusShipping of Table 3 in CC#7. Table 4 shows all target classes that the target methods are suggested to be moved to. Table 5 shows the experiment results in which LCOM[3] value decreases 22.67%, TCC[3] value increases 3.28%, CBO[3] value increases 25.00%, and RFC[3] decreases 21.44%. The experiment results indicate that the cohesion is increased and the coupling is decreased. It suggests that applying the move method refactoring can improve the quality of the software.

---

[1] 0.69 is a ratio between the number of calls that the target method calls within its class and the number of calls that the target method calls other classes.

[2] http://www.codeproject.com

[3] LCOM is abbreviation of Lack of Cohesion, TCC is Tight Class Cohesion, CBO is Coupling Between Objects, and RFC is Respond For a Class.

Table 1. The number of call method between source classes and candidate classes

CC#	Target Method	Source Class	Candidate Class	#access source	#access candidate
1	LogUser	Administrator	SearchFacade	0	2
2	LogUser	Administrator	UserManager	0	1
3	LogUser	Administrator	Log	0	3
4	ByKeyword	SearchFacade	UserManager	1	2
5	ByKeyword	SearchFacade	KeywordSet	1	3
6	CreditCardInfo	Customer	UserManager	0	1
7	CusShipping	Order	OrderDetail	1	4
8	CusShipping	Order	Genre	1	2
9	CusShipping	Order	ShippingInfo	1	1
10	DisplayProduct	Product	OrderDetail	2	3
11	DisplayProduct	Product	ShippingCart	2	4
12	GetGenreCus	Genre	ShippingCart	2	1
13	GetGenreProduct	Genre	ShippingInfo	2	2
14	GetGenreProduct	Genre	Customer	2	1

CC = Candidate class

Table 2. IMC measurement of method CusShipping

CC#	Source Class	Candidate Class	IMC
7	Order	OrderDetail	0.800
8	Order	Genre	0.710
9	Order	ShippingInfo	0.750

Table 3. Decision table for method CusShipping

CC#	Events			$U(a_i)$
	CIM	RMC	ICBC	
7	0.440	0.200	0.200	0.280
8	0.220	0.150	0.100	0.157
9	0.110	0.033	0.050	0.064

Table 4. Suggestion methods for move to appropriate target class

Target Method	Source Class	Target Class
CreditCardInfo	Customer	UserManager
LogUser	Administrator	Log
ByKeyword	SearchFacade	KeywordSet
CusShipping	Order	OrderDetail

Table 5. Coupling and Cohesion measurement of Online Ordering System

Class	Move Method Refactoring							
	Before				After			
	LCOM	TCC	CBO	RFC	LCOM	TCC	CBO	RFC
Administrator	1.00	0.33	0	3.00	0	0.33	0	2.00
SearchFacade	1.00	0.33	1.00	5.00	0	0.33	2.00	5.00
UserManager	0	0.67	2.00	7.00	2.00	0.67	2.00	6.00
Customer	0	0.67	0	3.00	0	0.33	0	2.00
Log	0	1.00	1.00	6.00	0	1.00	0	4.00
KeywordSet	0	1.00	1.00	6.00	0	1.00	0	4.00
Genre	1.00	0.33	1.00	8.00	1.00	0.33	1.00	7.00
ShippingInfo	2.00	0.67	2.00	3.00	1.00	1.00	2.00	3.00
Order	3.00	0.67	0	4.00	0	1.00	0	1.00
OrderDetail	0	0.67	2.00	11.00	0	0.33	0	8.00
Product	1.00	0.67	2.00	8.00	1.00	0.67	2.00	8.00
ShoppingCart	0	0.33	0	6.00	2.00	0.33	0	5.00
Mean	0.75	0.61	1.00	5.83	0.58	0.63	0.75	4.58
Std.dev.	1.05	0.46	0.85	2.10	0.89	0.46	0.76	1.95
Percentage change of Mean value					↓22.67	↑3. 28	↓25.00	↓21.44

6 Conclusions and Future work

The paper has proposed the new approach to find opportunity for move method refactoring using Laplace of decision theory. The approach searches for candidate classes and chooses only one class with the highest coupling value which is used as decision criteria of Laplace method. Two new metrics are presented. A preliminary evaluation is performed on object-oriented software to demonstrate the application of the proposed approach. In future work, we plan to use this theory on additional conditions, such as decision making under risk for best choices in software improvements. Furthermore, we plan to use more empirical researches with other systems and various experiment settings.

References

1. M. Fowler, K. Beck, J. Brant, W. Opdyke, and D. Roberts, Refactoring: Improving the Design of Existing Code. Addison-Wesley (1999).
2. Mika Mantyla, Jari Vanhanen and Casper Lessenius. A Taxonomy and an Initial Empirical Study of Bad Smells in Code. Proceeding of the International Conference on Software Maintenance (ICSM'03) IEEE (2003).
3. Dennis, A., B. H. Wixom and D. Tegarden. System Analysis and Design: An Object-Oriented Approach with UML. John Wiley & Sons, Inc. (2002)
4. G. Bavota, R. Oliveto, A. D. Lucia, G. Antoniol, and Y. G. Gu´eh´eneuc, "Playing with refactoring: Identifying extract class opportunities through game theory," ICSM, (2010).
5. G. Bavota, A. D. Lucia, A. Marcus, and R. Oliveto, "A two-step technique for extract class refactoring," ASE, pp. 151–154. (2010)
6. D. Poshyvanyk, A. Marcus, R. Ferenc, and T. Gyim?thy, "Using information retrieval based coupling measures for impact analysis," Empir. Softw. Eng., vol. 14, no. 1, pp. 5–32, (2008).
7. G. Pappalardo and E. Tramontana, "Suggesting Extract Class Refactoring Opportunities by Measuring Strength of Method Interactions," (2013).
8. N. Tsantalis, S. Member, and A. Chatzigeorgiou, "Identification of Move Method Refactoring Opportunities," vol. 35, no. 3, pp. 347–367, (2009).
9. V. Sales, R. Terra, L. F. Miranda, and M. T. Valente, "Recommending Move Method Refactorings using Dependency Sets," no. Section VII, pp. 232–241, (2013).
10. V. S. Bidve and A. Khare, "A Survey of coupling measurement in object," vol. 2, no. 1, pp. 43–50, (2012).
11. S. Kimura, Y. Higo, H. Igaki, ànd S. Kusumoto, "Move code refactoring with dynamic analysis," 2012 28th IEEE Int. Conf. Softw. Maint., pp. 575–578, Sep. (2012).
12. P. Joshi and R. K. Joshi, "Microscopic coupling metrics for refactoring," Conf. Softw. Maint. Reengineering, p. 8 pp.–152, (2006).

Requirements Engineering for Cloud Computing Using i*(iStar) Hierarchy Method

Sandfreni
School of Electrical Engineering and Informatics
Institut Teknologi Bandung
Bandung, Indonesia
sandfreniaz@gmail.com

Nabila Rizky Oktadini
School of Electrical Engineering and Informatics
Institut Teknologi Bandung
Bandung, Indonesia
nabilarizky@students.itb.ac.id

Kridanto Surendro
School of Electrical Engineering and Informatics
Institut Teknologi Bandung
Bandung, Indonesia
surendro@gmail.com

Abstract— **Cloud computing gets increasingly established in higher education as an option for modelling the quality of services and education system. Requirement engineering is a process to formulate user system needs, manifested by identifying the stakeholders in cloud computing at university which are used in operational activities where such operations require services that allow network access from anywhere and anytime, e.g., networks, servers, storage, applications and services. The adoption of cloud computing in a university course is highly related to the readiness of the University. Using iStar hierarchy make a great combination for improving the quality of software. Design requirement engineering framework through iStar hierarchy approach in the implementation of requirements engineering.**

Keywords—Cloud Computing, Requirements Engineering, Hierarchy iStar.

Introduction

University is an institution that has a strategic position and role in the achievement of the education objectives as a macro in which it repairs as well as produces qualified human resources. In the implementation of activities at the College, it includes educational activities, teaching, research and service to the community, a college is required to contribute much to the development of a country's public life. The activities required fulfilling an academic system that can improve the quality of education. Therefore, the presence of cloud computing technology can provide a service that is easy to use by all academics in college with cost efficient. Requirement engineering is the first step in using system academic college based cloud computing. There is a process of understanding the needs of users system in requirement engineering, which is realized by means of identifying what are involved in the system and how to communicate the needs of the College academic system. The application of requirements engineering on an information systems development process has become a necessity because this stage is an important and necessary stage in order to result a good quality information system and has corresponding expectation functions. One of the methods that still continue to be developed until now and is often used is the iStar that concern with stakeholders in the system and the relation between them. This research utilize hierarchy iStar approach to accommodate Multi-Stakeholder Distributed System (MSDS) in university environment, the iStar approach alone has limitation in defining the requirement of multiple stakeholder [1].

Literatur review

Cloud Computing

Cloud computing is a model for enabling ubiquitous, convenient, on-demand network access to a shared pool of configurable computing resources (e.g., networks, servers, storage, applications, and services) that can rapidly be provisioned and released with minimal management effort or service provider interaction [3]. Model of cloud computing consist of four deployment models, three service models and five service attributes.

K.J. Kim (ed.), *Information Science and Applications*,
Lecture Notes in Electrical Engineering 339, DOI 10.1007/978-3-662-46578-3_105

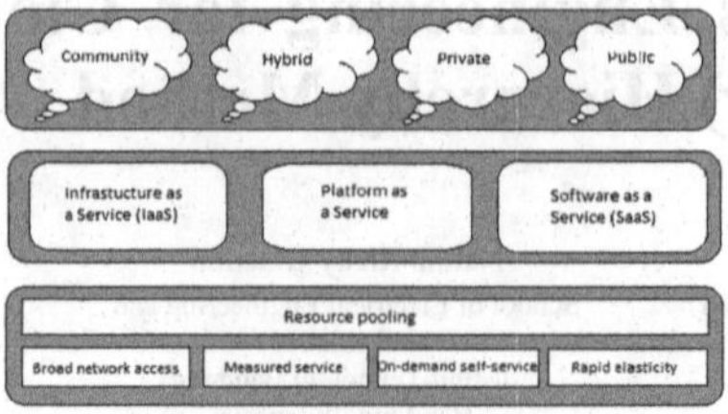

Figure 1. *Cloud Computing Architecture*

Cloud computing can be implemented by providing components such as servers, hardware, and networking required. Cloud computing users can install applications that use the infrastructure. Users also can choose how to use cloud computing services offered by vendors as needed.

Requirements Engineering

The following, broader definition is one of the most long-standing, and comes from a DoD software strategy document dated 1991:

Requirements engineering "involves all life-cycle activities devoted to identification of user requirements, analysis of the requirements to derive additional requirements, documentation of the requirements as a specification, and validation of the documented requirements against user needs, as well as processes that support these activities" (DoD 1991).

While this definition covers the identification, analysis, development and validation of requirements, it omits to mention the vital role that requirements play in accepting and verifying the solution (usually called verification rather than validation.) A more recent definition, given in the context of software engineering, suffers the same defect, but emphasizes the goal-oriented (or purpose-oriented) nature of requirements engineering, and hints at the importance of understanding and documenting the relationships between requirements and other development artefacts:

Requirements engineering is the branch of software engineering concerned with the real world goals for, functions of, and constraints on software systems. It is also concerned with the relationship of these factors to precise specifications of software behavior, and to their evolution over time and across software families (Zave 1997).

To conclude, requirements engineering is the subset of systems engineering concerned with discovering, developing, tracing, analyzing, qualifying, communicating and managing requirements that define the system at successive levels of abstraction [9].

iStar (i*)

The ***i**** framework proposes an agent-oriented approach to requirements engineering centering on the intentional characteristics of the agent. Agents attribute intentional properties (such as goals, beliefs, abilities, commitments) to each other and reason about strategic relationships. Dependencies between agents give rise to opportunities as well as vulnerabilities. Networks of dependencies are analyzed using a qualitative reasoning approach. Agents consider alternative configurations of dependencies to assess their strategic positioning in a social context [2]. i* offers a requirements modelling framework to support the "early phase" of requirements engineering. It is intended to assist in the understanding of the organizational environment (of some potential system), in the exploration of alternate system proposals and how they would fit into various work settings, and in the analysis of the impact of alternatives system arrangements on organizational participants [3].

In *i**, the central conceptual modeling construct is the actor. It is an abstraction which is used to refer to an active entity that is capable of independent action. Actors can be humans, hardware and software, or combinations thereof. Actors are taken to be inherently autonomous - their behaviors are not fully controllable, nor are they perfectly knowable [3].
Dependency represents the intension relationships between two actors. Type of dependency fall into four categories [2]:

- **A goal** is a condition or state of affairs in the world that the actor would like to achieve. It is expressed as an assertion in the representation language. How the goal is to be achieved is not specified, allowing alternatives to be considered.
- **A task** specifies a particular way of doing something. When a task is specified as a subcomponent of a (higher) task, this restricts the higher task to that particular course of action.

- **A resource** is an entity (physical or informational) that is not considered problematic by the actor. The main concern is whether it is available (and from whom, if it is an external dependency).
- **A soft goal** is a condition in the world which the actor would like to achieve, but unlike in the concept of (hard-) goal, the criteria for the condition being achieved is not sharply defined a priori, and is subject to interpretation.

The i* modeling consist of:

- Strategic Dependency (SD), represents the relationship of dependency among actors, which are each actor's strategies to identify:
  a. Their requirements
  b. The way to achieve the requirements
  c. Other actors whom they depend to achieve their requirements
- Strategic Rationale (SR), modeling events inside the actor. This model represents how the actor achieve their goal.

Hierarchy in Requirement Engineering Model

The purpose of Requirements Engineering specified from the requirements or requirements that reflect the goals of a system and also the needs of all stakeholders who have a relationship. Requirements engineering is the foundation for the success of the development of a system from start to the end and an important factor for the quality of the system and the productivity of the product of a system.

With the multiple layers of abstract system to the bottom layer of the most detailed, tracing requirements is needed to track the relationship between the requirements of each layer (Hull et al, 2011). There are several variations on iStar framework, such as GRL and Tropos, but there is an element of the definition of inheritance that is not yet complete, despite the fact that the inheritance is already included in the framework since its first definition. Several studies using inheritance, but the studies were not explicitly provided guidelines of this concept or its use (Lopez, 2008).

In the Multi-Stakeholder Distributed System (MSDS), in which each element is created, owned, and implemented by the stakeholders themselves, there are limitations of iStar when defining needs of diverse stakeholders. One of the problems that arise in the modeling of the MSDS there is a need to use inheritance in order to create a hierarchy of actors (Lopez, 2009). In an organization, the actor passed down the same organizational goals and they do not have a goal in itself, so there is no competition between actors (Yu, 1996).

By having a goal, then the actors have a responsibility to meet these goals. By knowing the responsibilities of an actor, it can be seen needs. However, the responsibilities within an organization may be delegated from the actor to other actors in the hierarchy below it (Chopra, 2011). Once the organizational structure is formed, then there are goals that can be mapped according to the hierarchy of actors who have these goals (Rahwan et al, 2006).

Research Methodology

The perspectives and experiences of the researcher influence a research design. These views and experiences are used to guide how and for whom the research problem is being investigated, what research approach and methods of gathering suitable research data will be used [10].

Author will focus on iStar hierarchy model to improve the quality of system in university according to some related researches of cloud computing and then evaluate it. Then, author will make a modeling process, the dependency relationship between actors in each business processes represents by strategic dependency modelling and also strategic rationale to determine what internal processes that occur in an actor.

Discussion

iStar method will be applied to a cloud computing university and will develop an information system of the University that has been measured using the index of satisfaction of Users of the system.

To get the needs of the information system to be developed, it will be elaborated beforehand the purpose of this research:

1. Design the stages in the process of Requirements Engineering to identify the needs of the technology, the role and the user in optimizing the adoption of cloud computing
2. Improve the capability of institutions to conduct surveillance, creating stability, predictabilities, and versatility as an educational organization.

3. Maintain and improve the quality of the University in order to make it as a world-class University.

In this study, there is a case that is not optimal in adopting cloud computing at University because it is not clearly known to every need required both from the technology side, roles and users. Requirements engineering process begin with stakeholder goals and then followed by elicitation, modeling, and analysis process. Through these processes, goals are transformed into specification and the function of the system.

In process elicitation, each business processes in university environment supported by existing system need to be re-examined and the actor involved in the system identified.

In modeling process, the dependency relationship between actors in each business processes represents by strategic dependency modelling the model exhibit the relationship between the actors both in the same and different level. In addition strategic dependency modeling we need to construct strategic rational modeling to identify internal process inside the actor. In i-star modeling represents in four categories, goal, task, resource, and soft goal.

Study Case iStar Hierarchy Modelling

We use application provisioning process in university environment as the study case. Application provisioning process involves three divisions, business owner division, IT application developer division and IT infrastructure division. Business owner division is the user that utilize application to support their business. IT application help business owner division to develop the application. And IT infrastructure provide infrastructure resources, such as power, server and network for the application. Business owner division initiate the development of application to support their business. They invite application developer to help them develop the application. Application developer define the requirement for their application, such as memory, cpu, hard disk, networking, operating system requirement. Application developer will inform infrastructure division about the requirement and ask them to provide the resources. Infrastructure division will check the availability of resources and allocate the resource if possible.

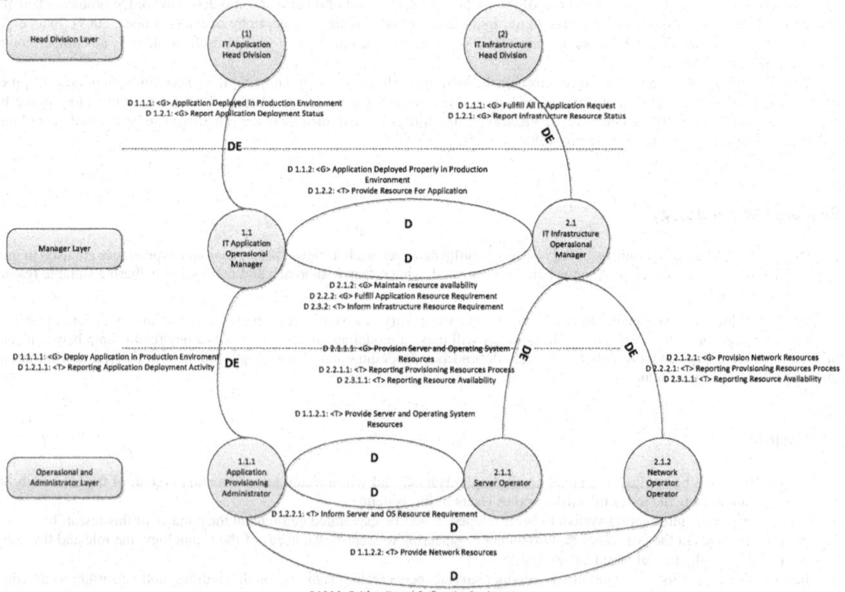

Figure 2. *Strategic Dependency "the appearance hierarchy in application provisioning process"*

In addition to the strategic dependency we also need to build strategic rationale to determine what internal processes that occur in an actor. In modeling iStar, the internal process within the actor occur in goal, task, resource, and softgoal. There is a sequential process in strategic rationale. Within use the activity principle in Goal Based workflow and implement it in the task, so in every task decomposition can be changed to explain the sequential in iStar. In this below there are strategic rationale diagram to Application Provisioning Manager, Server Operator, Network Operator.

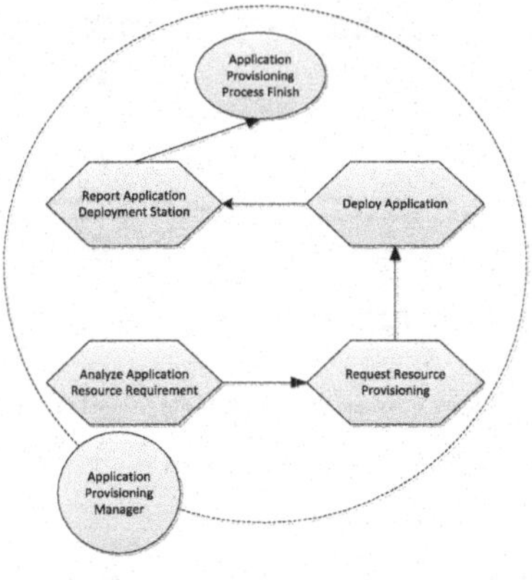

Figure 3. *Strategic Rationale Formulation "Application Provisioning Manager"*

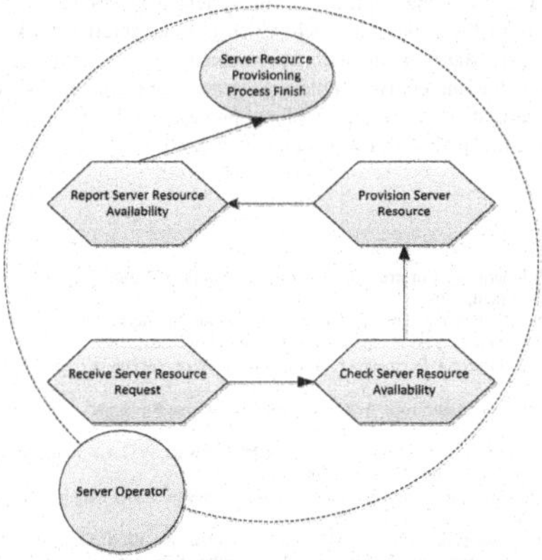

Figure 4. *Strategic Rationale Formulation "Server Operator"*

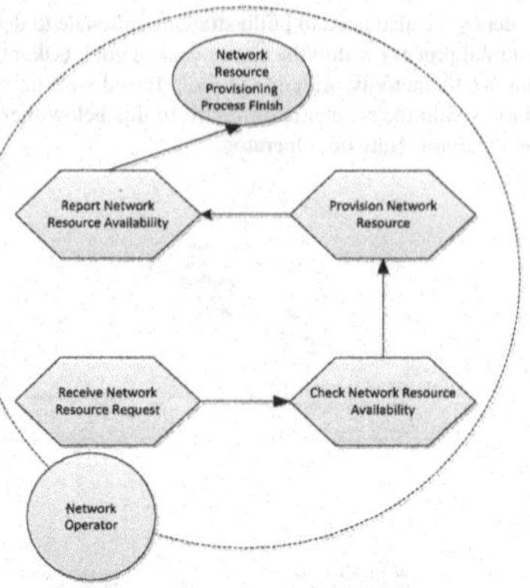

Figure 5. *Strategic Rationale Formulation "Network Operator"*

Conclusion

Requirement engineering is one of the ways to identify objectives and analyze the needs of an information system. By using the method of iStar we can figure out the purpose of the requirement engineering approach with the use of which we can know the hierarchical description of needs specific and easier to do a search so that it can understand the reasons for the use of such a system as well as the functions of the system are derived through the goal. iStar methodology approach bring social understanding into the requirement engineering process, as the implication, the system developed efficiently and meet the stakeholder needs. This research utilizes iStar hierarchy methodology approach which has characteristics like iStar methodology and also support Multi-Stakeholder Distributed System (MSDS) in university environment.

iStar hierarchy methodology help the university to build new systems or refine the existing systems with precise software requirement specification. In the end consistent system development process will help the universities to increase their capability as educational organization and improve the quality as a world-class university.

References

[1] Lopez, L, "A Complete Definition of the Inheritance Construct in i*". *Proceeding of the ER 2009 PhD Colloquium, affiliated to the 28th International Conference on Conceptual Modelling (ER-2009)*. Brazil, 2009.
[2] Yu, Eric: Social Modeling and i*. *Conceptual Modeling: Foundations and Apllications*. Springer, 2009.
[3] Yu,Eric. S: *Modelling Strategic Relationships for Process Reengineering*. PhD thesis, Canada. 1996.
[4] Anwer, S., Ikram, N., "Goal Oriented Requirement Engineering: A Critical Study of Techniques", XIII ASIA PACIFIC SOFTWARE ENGINEERING CONFERENCE (APSEC'06), 2006.
[5] Lapouchnian, A., "Goal-Oriented Requirements Engineering: An Overview of the Current Research", Department of Computer Science University of Toronto, 2005.
[6] Hui, B., Liaskos, S.m Mylopoulos, J., "Requirements Analysis for Customizable Software: A Goals-Skills-Preferences Framework". Proc. International Conference on Requirements Engineering (RE'03), Monterey, USA, September 2003
[7] M. Teruel, E. Navarro, and V. López-Jaquero, "Comparing Goal-Oriented Approaches to Model Requirements for CSCW, "Evaluation of Novel Approaches to Software Engineering, pp. 169 184, 2012.
[8]A. Van Lamsweerde, Goal-oriented requirements engineering: a guided tour, vol. 249, no.August. IEEE Comput. Soc, 2001, pp. 249 – 262.
[9] Hull Elizabeth, Jackson Ken, Jeremy Dick, "Requirements Engineering", Third edition, Springer, 2011.
[10] Janssen, M. & Cresswell, A. (2005) An Enterprise Application Integration Methodology for E-Government. Journal of Enterprise Information Management, 18(5), pp.531-547.

Power Consumption Profiling Method based on Android Application Usage

Hiroki Furusho*[1], Kenji Hisazumi[2], Takeshi Kamiyama[3], Hiroshi Inamura[3], Tsuneo Nakanishi[4], and Akira Fukuda[5]

[1]Graduate School of Information Science and Electrical Engineering (ISEE), Kyushu Univ., Fukuoka, Japan
[2]System LSI Research Center, Kyushu Univ., Fukuoka, Japan
[3]Research Laboratories, NTT DOCOMO Inc., Kanagawa, Japan
[4]Department of Electronics Engineering and Computer Science, Fukuoka Univ., Fukuoka, Japan
[5]Faculty of ISEE, Kyushu Univ., Fukuoka, Japan

Abstract. In this paper, we propose a method for collecting essential data to profile energy consumption of applications running on Android OS. Existing power-estimation methods are unable to account for all possible usage patterns, since developers can only prepare a limited number of profiling test cases. Our proposed method analyzes the power consumption using a log collected during an application use on a smart phone of a particular user. In our method, the logging code that tracks application usage data for the user is automatically embedded into the application. Our method uses this usage information to estimate power consumption, and provide developers with helpful hints for decision making and application tuning. In this paper, we analyzed the power consumption of an open-source application based on the log collected and estimated the power-saving effects of the application. The power consumption of the application was reduced by tuning the application according to the data derived from the results of the analyses; thus confirming that the method provides valid information to determine power-saving techniques.

1 Introduction

Reducing the energy consumed by Android smart phones is a crucial task for application developers. While this problem can be addressed at various levels, it is important to reduce the energy consumption of individual applications that vary significantly on their behavior. The simplest method for reducing the energy consumption of smart phone applications is to eliminate problems: excessive create instances, loop statement errors, communication process errors and bugs.

Energy-profiling methods can identify the points at which the applications consume excessive energy and to determine methods to reduce their overall consumption.

* furusho@f.ait.kyushu-u.ac.jp

K.J. Kim (ed.), *Information Science and Applications*,
Lecture Notes in Electrical Engineering 339, DOI 10.1007/978-3-662-46578-3_106

Existing methods estimate energy consumption of the entire smart phone by using data obtained from the OS (e.g. CPU time, amount of access to file systems, traffic, *etc.*). However, these methods are unable to identify specific points where excessive energy is consumed or account for all possible usage patterns and environments, developers can only prepare limited profiling test cases.

This paper proposes a method to analyze and visualize the energy consumption of applications running on Android OS depending on the actual application usage data obtained from the smart phone of a user. In addition, the profiling method analyzes the energy consumption tendencies of each usage.

The method consists of four phases: 1) Weaving logging codes into an application to automatically obtain OS resource consumption and estimate energy consumption, 2) Distributing the application to users, 3) Gathering data from the application, and 4) Visualizing the data from various viewpoints. Developers can determine methods to reduce energy consumption by analyzing the visualized usage data. We evaluate profiling results of an open-source application to which applied our profiling method is applied.

The structure of this paper is as follows. In Section **2**, we discuss related work. Section **3** provides an overview of the proposed method and the requirements for log collection technique used for profiling. Sections **4** explains parameter collection. Experimental results are reported in Section **5**. Section **6** concludes with a summary and describes future issues.

2 Related Work

Most smart phone power-analyzing employ model-based estimation. The basic form of the power consumption model can be represented by the linear equation (1) shown below:

$$P_{estimate} = \sum_{m \in M} C_m \cdot V_m \tag{1}$$

$P_{estimate}$, M, C_m and V_m represent estimated power, a set of the factors related to power consumption (CPU time, data communication, access to storage, display *etc.*), usage of a factor $m \in M$, and its coefficient, respectively. C_m is calculated by regression analysis of resource usage and measured power consumption.

Several researchers present the methods that use a power model to estimate the power consumption of the entire device, using operating time of each part of the device as parameter [1][2]. However, it is difficult to identify the contribution of an application to the total power consumption because smart phones can run several processes simultaneously.

An estimation method using values obtained from a Linux process file system [3] overcomes the issue [4] because the process file system records the device usage (hereinafter referred to as "resource usage") for each process separately. Mittal et al. proposed an energy consumption profiling method for the CPU, wireless communication (3G, Wi-Fi) and display [5]. The display energy is consumed by the application because the interface of the application itself usually occupies the display of the smart phone.

3 Approach

3.1 Overview

This section describes the profiling process of our proposed method. First, developers embed the logging code into an application. Resource logging should have minimum overhead and is automatically applied to application during development. The details will be described in Section 4. Second, the developers distribute the application. The application assembles the resource consumption from the Android system and creates log data. Finally, the log data is collected in the profiling server where the developers can view the profiling results.

Using our method, developers can obtain profiling information from unexpected test cases, and usage information from various users.

3.2 Selective log collection method

This section explains the collection method of the resource consumption of a smart phone. First, we describe the properties and overhead of our method. Second, we propose a selective log collection method.

Log collection code is embedded into the application to accumulate resource consumption data from the smart phone. Next, the resource consumption data is sent to a server. Therefore, the profiling method increases the power consumption and run time of an application. The profiling method must reduce overhead as much as possible since the logging process can affect the accuracy of the estimate and the operation of the smart phone.

The logging process consists of reading resource consumption data from the process file system in the device storage, writing log into the device storage and wirelessly communicating the data to the server. The profiling method should minimize these overheads while it collects resource consumption information.

Our profiling method collects the resource consumption from each important component of the application, such as Activity and Service classes of the Android OS. An Activity class provides the GUI for the functions of an application. The sequence of the activities in Activity class indicates a rough usage pattern.

To fulfill the above requirements, the authors embed the logging code using AspectJ [6]. AspectJ is an extended Java programming language for aspect-oriented programming [7]. Aspect-oriented programming is a cording technique to modularize the pieces of the application. Using AspectJ, developers are able to reduce the labor of the embedding source code.

4 DEFINITION OF USAGE LOG

Our profiling method estimates power consumption each component of an application. This section describes the usage log for profiling.

Activity class that provides an application interface. The Activity class executes callbacks according to the condition of the application. Resource consumption in the process file system indicates the total consumption since the time of boot. For example, when measuring the power consumed for displaying the application, we must collect log information before the display is turned on and after it is turned off. AspectJ includes an additional object named aspect, which is not a part of Java. The aspect object contains the conditions of the embedding point (pointcuts) in the source code and the embedding codes (advice). When an application is generated, Java bytecode, which generated from the advice, into the application. The proposed method uses three types of pointcuts: the calling of a method defined in source code, the calling of a method in the Android Library, and the assignment of a variable.

Service class runs longer than Activity class and performs background processing. Our profiling method collects the resource consumption of every method in the Service class.

5 EVALUATION

We evaluated our profiling method for an open source application.

5.1 Environment

Target application of our experiment was a Twitter client named Crowdroid. We embedded the logging code into Crowdroid and collected usage log of 51 hours from thirteen subjects. Subjects used the Galaxy Nexus smart phone, on the condition that it communicates only via Wi-Fi.

The usage log contains usage time, method name, CPU time and communication time. In addition, we included the tweet number in the usage log as a Crowdroid-specific parameter.

The proposed method estimates the amount of power delivered by the smart phone battery when each part is running. Assuming that the battery supplies a constant power voltage, we used the battery current consumption data as an indication of the power consumption. As related work described in [4], we created a power consumption model of the Galaxy Nexus using multiple regression analysis with its parts usage and current consumption of Galaxy Nexus as related work [4]. The Wi-Fi sending/receiving parameter was changed from throughput to communication time because both parameters are nearly equivalent. We did not verify estimation accuracy of the power consumption model, but assumed that it had an error of about 10%, determined in [4]. The proposed method can collect all the resource consumption data indicated by [4], but we did not measure the storage access because Crowdroid scarcely accesses storage. We also did not measure the wireless LAN activity because the wireless LAN is shared by many applications and processes simultaneously.

The current consumption is calculated by the power model as shown in equation (1) in Section 2. Table1 shows the parameters and coefficients. Parameters used in the evaluation are shown as follows:

Table 1. Parameters and coefficients of the current-consumption model

Parameters		Coefficients	
V_{offset}(screen *etc.*)	Display time [s]	C_{offset}	$1.648e^{-1}$
V_{CPU}	Frequency [kHz] × CPU use rate [%]	C_{CPU}	$1.910e^{-7}$
V_{WLAN}	Communication time [s]	C_{WLAN}	$2.005e^{-1}$

Display time is the difference between the times before and after displaying an application screen. The CPU use rate is the quotient of user-mode CPU time and the total CPU time of all modes in the process file system. Communication time is the run-time of the method executed when receiving or transmitting data.

5.2 Overhead of log collection process

We measured the log collection time to determine its effect on the operation of the smart phone. We added a function to record the run-time data of the log collection process in the proposed method. We embedded the altered logging code into a simple test application that displayed a preset character string on the screen. The log-collection process has three steps: reading the processes file, obtaining the run time, and writing the log file. We performed these processes 10 times and calculated the average run time.

The average runtime of reading the processes file, obtaining the run time, and writing the log file is $27.7[ms]$, less than $1[ms]$ and $5.1[ms]$, respectively. The log-collection process is performed both before and after the application interface is displayed. If the user switches screens approximately every 10 seconds, the log-collection process scarcely affects the operation because the log-collection time ratio is less than 1%.

5.3 Tuning the Activity class

We analyzed the current consumption from the logs of the all subjects. Figure 1 shows the total current consumption of the Activity class instances of the Crowdroid application.

Activity 2 consumes the most current of all the Activity instances; its main function is to display the tweets the user is following. Activity 2 is always displayed when Crowdroid starts. We found that the display of the smart phone and other Activity class account for most of the power consumption by Activity 2. However, we excluded Activity 2 from our optimization because the energy consumed by the display is proportional to the usage time, which is beyond the control of the developer. Upon investigation, we found that improving the CPU was beyond the scope of our study. Finally, we decided to focus on the communication device.

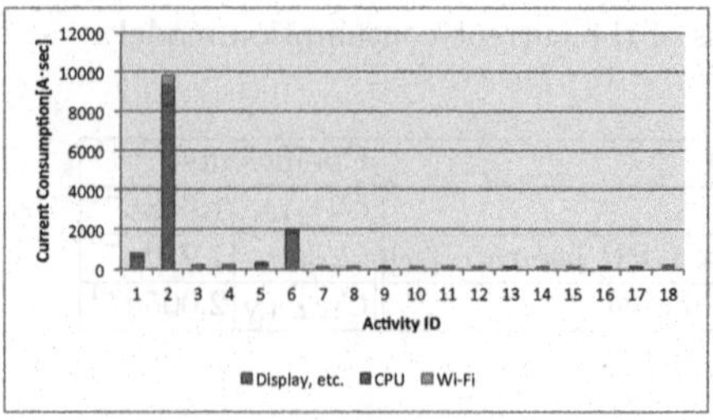

Fig. 1. Aggregate current consumption for each Activity class

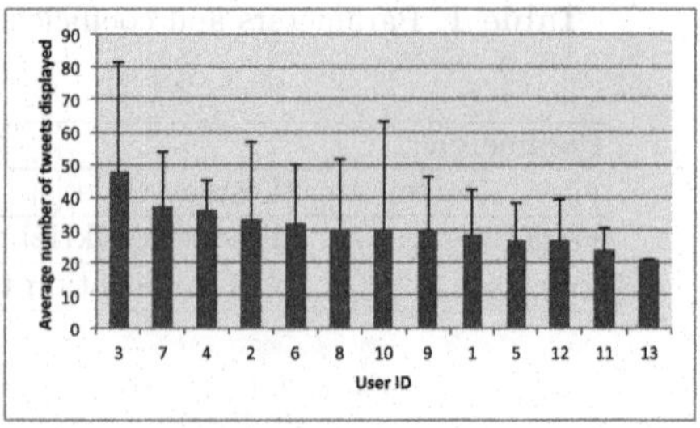

Fig. 2. Average tweet count of Activity 2

Figure 2 shows the mean and variance of the number of displayed tweets for each subject. Most subjects had received more than 21 tweets as an additional download with an unmodifiable value. If download count is reduced, energy of connect to server is also reduced. If the download count could be reduced, the energy of connecting to the server would also be reduced.

We estimated the effect of reducing the power consumption by suppressing these additional downloads. In this simulation, x is the number of tweets that can be determined from the log. Figure 3 shows the time of communication to the Twitter server, which is a modified number of downloads. The solid line represents the linear approximation curve. The time of downloading x tweets from the Twitter server is represented as $T_{opt} = 45.507x + 938.87[ms]$ using least squares method. The communication time for an optimal number of tweets can be estimated from this equation. The optimized current consumption for downloading tweets can be represented by equation (2), shown below:

$$P_{optW} = P_{WLAN}\frac{T_{opt}}{T} \tag{2}$$

P_{optW}, P_{WLAN} and T represent optimized current consumption for communication to the Twitter server, the current consumption calculated from the log, and the communication time in the log, respectively.

Figure 4 shows a comparison of the current consumption estimated from the log and simulation results. Optimizing the number of communications reduces the amount of current consumed because of communication by approximately 20%. Though the effect of the improvement is about 2% of the overall power consumption of the smartphone, this case is easy to apply because usability is scarcely affected. By visualizing the relationship of power consumption and average usage of a user in real environment, the proposed method supports the detection of power saving. We specifically mentioned this process of tuning in this section.

5.4 Tuning the Service class

We performed an experiment to analyze the Service and Activity classes. We collected a usage log over a month from two subjects. Figure 5 shows the current

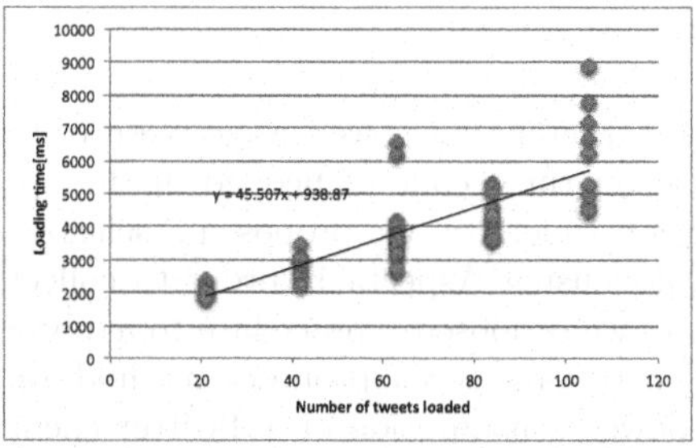

Fig. 3. Measurement of data communication time

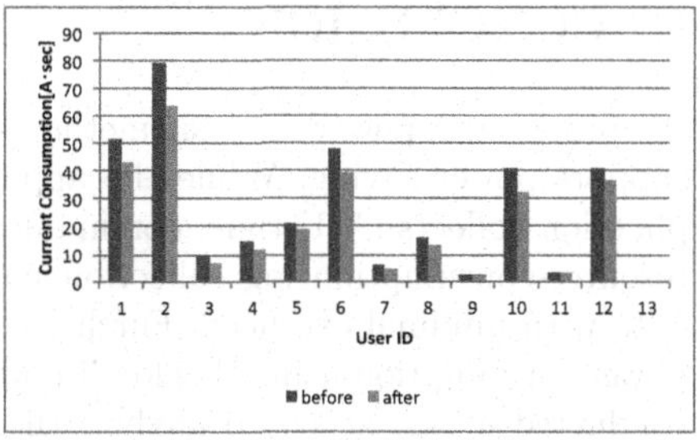

Fig. 4. Simulation of the optimized number of tweets

consumption of the top ten measurement targets. Unfortunately, the authors were removed from the study because of a flaw in the measurement data; hence, the current consumption for data communication is omitted from Figure 5.

Module B is Activity 2 in Figure 1, but module A consumes more current than module B. Module A serves as the notification function for new messages. Module

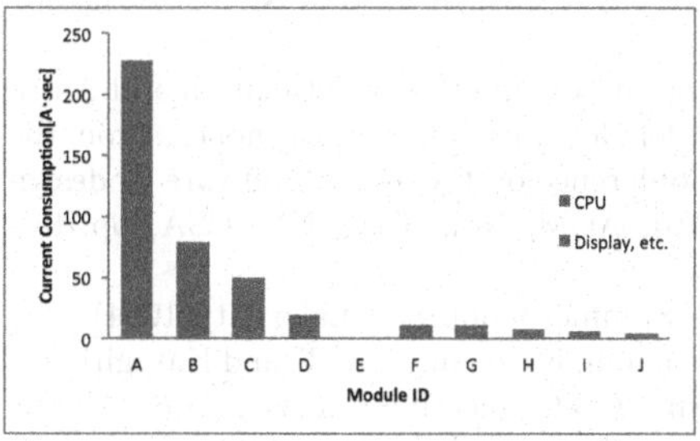

Fig. 5. Current consumption analysis of Activities and Services

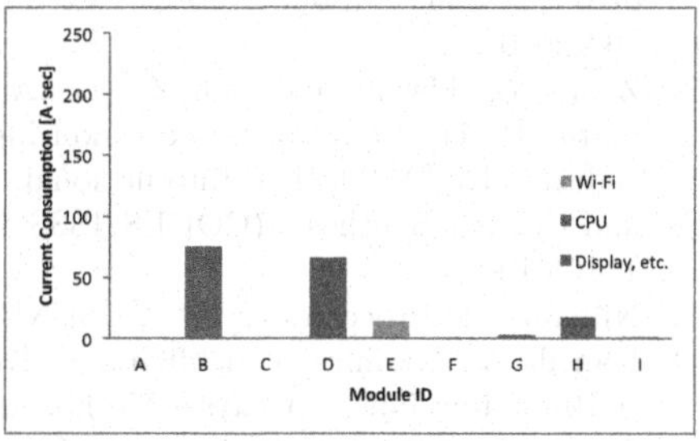

Fig. 6. Current consumption after optimization

A is executed at the time of Crowdroid startup and at regular intervals thereafter. The user can disable the notification function by changing the settings; however, in reality, Crowdroid calls module A regardless of the settings. Our subjects stated that they did not use the notification function. Thus, we changed the source code of Crowdroid to call module A according to the settings defined by the user. We collected the usage log of the altered Crowdroid application for over a month (Figure 6). Module A, C and F which runs in notification process was suppressed. The proposed method detects power consumption, which developers have missed.

6 CONCLUSION

We proposed a power analysis method based on actual power usage in order to support power saving. While the application runs, code embedded in the application collects CPU time, communication time, *etc.* We proposed a selective resource-consumption log collection method using AspectJ in order to collect data with minimal overhead. Furthermore, we proposed a procedure to analyze power consumption using this log. In order to evaluate the proposed method, we conducted a case study to analyze the power consumption of a Twitter client named Crowdroid. We found that it was possible to identify potential high power consumption areas, and devise a power-saving strategy using the analysis information. We estimated the effect of the strategy using the log. We also found that the Service class that performs background processing consumes a huge amount of power. We modified Crowdroid to suppress the activation of unnecessary modules, and verified it with simulation.

Verification of the validity of the simulation is part of our future work.

References

1. Cignetti,L.T., Komarov, K. and Ellis, C.S.:Energy estimation tools for the Palm, Proc. the 3rd ACM International Workshop on Modeling, Analysis and Simulation of Wireless and Mobile Systems (MSWIM '00), ACM, New York, NY, USA, pp.96–103(2000).
2. Zhang, L., Tiwana, B., Qian, Z., et al:Accurate Online Power Estimation and Automatic Battery Behavior Based Power Model Generation for Smartphones, Proc. the eighth IEEE/ACM/IFIP International Conference on Hardware/Software Codesign and System Synthesis (CODES/ISSS '10), ACM, New York, NY, USA, pp.105–114(2010).
3. Killian, J.T.:Processes as Files, USENIX Summer Conf. Salt Lake City(1984).
4. Kaneda, Y., Okuhira, T., Ishihara, T., Hisazumi, K., Kamiyama T. and Katagiri, M.: A Run-Time Power Analysis Method using OS-Observable Parameters for Mobile Terminals, 2010 International Conference on Embedded Systems and Intelligent Technology (ICESIT 2010), Vol.1, pp.39–44 (2010).
5. Mittal, R., Kansal, A. and Chandra, R.:Empowering Developers to Estimate App Energy Consumption, Proc. the 18th Annual International Conference on Mobile Computing and Networking (Mobicom '12), ACM, New York, NY, USA, pp.317–328(2012).
6. The Eclipse Foundation, The AspectJ Project, available from(http://www.eclipse.org/aspectj/) (accessed 2013-05-04).
7. Kiczales, G., Lamping, J., Mendhekar, A., et al:Aspect-Oriented Programming, Proc. the European Conference on Object-Oriented Programming (ECOOP), Vol. LNCS 1241, Springer Verlag (1997).

Evergreen Software Preservation: The Conceptual Framework of Anti-Ageing Model

Jamaiah H. Yahaya[1], Aziz Deraman[2], Zuriani Hayati Abdullah[3]

[1,3] Faculty of Information Science and Technology, Universiti Kebangsaan Malaysia, Bangi, Selangor, Malaysia
[2] School of Informatics & Applied Mathematics, Universiti Malaysia Terengganu, Kuala Terengganu, Terengganu, Malaysia

jhy@ukm.edu.my, a.d@umt.edu.my, zha.ukm@gmail.com

Abstract. The symptom of degradation in term of quality is observed as the indicator of ageing phenomenon in software system. In human and living creators, ageing is an inescapable manifestation for every living creature on earth. In human being, this phenomenon of delaying the ageing process is normally known as anti-ageing. We try to understand and learn the process of ageing in software through understanding the human ageing process. Unlike human ageing, software ageing can be delayed by identifying factors that influence the ageing. Ageing in software is occurring when the software is degraded in term of its quality, user's satisfaction and dynamic. Previous studies indicated that software ageing factors possibly will be classified into some categorization such as cost, technology, human, functionality and environment. Our past experiences in software quality and certification motivate us to the development of software anti-ageing model and its related areas which are the ageing factors and rejuvenation index. This paper presents the background issues in software ageing which includes software quality and certification, and focus further on the conceptual framework of software anti-ageing model and preliminary formulation of anti-ageing model.

Keywords: software anti-ageing, software ageing, software quality, evergreen software, software certification

1 Introduction

Software ageing in computer science discipline has been introduced earlier when researchers investigated the ageing in software system such as in operating system. Smooth performance degradation has been also called software ageing and is a consequence of the exhaustion of system resources, such as system memory or kernel structures, the accumulation of round off errors, database deadlocks, and the contention for a pool of limited software resources. It also refers to accumulation of errors during the software execution, which are ultimately results in crash or hanging failure [2]. Degradation of software performance is characterized by the software age. Thus, since

K.J. Kim (ed.), *Information Science and Applications*,
Lecture Notes in Electrical Engineering 339, DOI 10.1007/978-3-662-46578-3_107

then, many efforts have been devoted to characterize and mitigate the software ageing phenomenon, that is, the accumulation of errors occurring in long-running operational software. As a result, a significant body of knowledge has been established and an international community of researchers in the area of Software Aging and Rejuvenation (SAR) [1]. Former research by Parnas identified two types of software ageing in application software, which are caused by the results of the changes that have been made and the failure of the software or software to adapt to dynamic environment [3].

In current fast growing technology demand, software engineers are becoming a technology savvy in order to cope with the rapid changes in the environment. Failure to adapt with dynamic changes will results the relevance and vital of software getting lesser to its environment which is called a phenomenon of getting old and age. The characteristics of a software which initially must be built with the capability of modifiable and scalable, thus will give flexibility and enable it to stay young and relevant in their operating environment [7][8][16]. The process of delaying the ageing in software is called anti-ageing process or rejuvenation. It can be done by detecting and classifying the ageing factors that may cause the ageing and implement the reverse action to convey the anti-ageing process.

This paper starts with the research background in software ageing, software quality and certification. Later, a discussion on conceptual framework and the initial formulation of anti-ageing model will be deliberated, and concludes with a conclusion.

2 Background Works

Previous studies indicated that the relevance of the software throughout its life span depended on the quality of the software [19]. So, it is believed that software ageing is closely related to the quality and certification of the software in the specific operating environment [16]. In other words, software can be prevented from ageing and stays young with the assurance of good quality throughout its life cycle. The assurance of good quality can be achieved through certification process [19]. The following section discusses issues and state-of-the-art in software quality and certification.

2.1 Software Quality and Certification

The awareness of the software quality has been increased in most industrial sectors including software sector. Quality by definition is a subjective concept because quality is in the eyes of the beholder. Different people see quality in different views and perspectives. One way to view quality is through user's perspective which to assess product or services and relates it in customer's satisfaction level [5]. On the other hand, software quality in technical perspective can also be measured by three categories of measurements which are: internal measures, external measures and quality in use measures [6]. Internal measures is the process of evaluating on static measures of intermediate product, external measures evaluate on the behavior of the code while the quality in used evaluates the basic set of quality in used characteristics which may affect the software in certain operating environment.

For the last forty years, several software quality models have been developed and among them are the well-known such as McCall model (1977), Boehm model (1978), FURPS model (1987), ISO9126 (1991), Dromey model (1996), Systematic Quality model (2003), PQF model (2007), and SQuaRE (2011). These models demonstrate quality characteristics of a software in term of efficiency, maintainability, usability, reliability, functionality and portability [7][13]. In recent years, human aspect is a new element of software quality measurement that has been included in PQF model which are not introduced in earlier models. Measuring software product quality by reckoning the human aspect that relates to the user's perspective and expectations are recognized and become a challenge today [7][17].

The term certification in general is the process of verifying a property value associated with something, and providing a certificate to be used as proof of validity. Software certification is the extended quality process intended to ensure and guarantee the quality standard of a software based on certain quality benchmark and accordance to the country standard. Results from certification will provide a valuable recognition on the quality of the software organization which can support the buoyancy and trustworthiness of the organization.

In Korea, there is a certification program which call good software to ensure that all software products comes in Korean's industry will be tested and verified according to good software standard[20]. K-model can be applied to small and medium sized business or project for measuring the quality values of the underlying processes. Our research group has developed a certification model based on end product quality approach named as SCM-prod model and a tool, SoCfeS, support tool for certification process [7][19]. This model has been tested successfully in the several business environments in Malaysia. The results from the case studies reported a significant satisfaction of software's developer, manager, and stakeholder, who feel more confidence in using this model for assessing and certifying software product. The certification exercises can be repeated several times during the life cycle of the products and therefore continuous quality is maintained and will delay the ageing process of software. We believe the certification exercises are useful to ensure quality and to maintain sustainable quality and preserve evergreen of the software operating in the environment. Evergreen is defined in general as having an enduring and lasting freshness, success, or popularity. In this scope of research evergreen software can be defined as the enduring and everlasting freshness, success and popularity of the software in its' operating environment.

2.2 Software Ageing and Anti-Ageing Phenomenon

Formerly, software ageing is referred to a phenomenon in long-running software system that shows an increasing of failure rate in which the occurrence of a progressive degradation in software performance and may lead to undesirable hangs and crashes [4]. The accumulation of software errors and failure to perform as user intended such as hang or crush also considered as ageing process [8-9]. The characteristics were described and identified as follows: memory bloating/leaks, shared- memory-pool latching, unreleased file locks, accumulation of un-terminated threads, file-space fragmentation, data corruption/round off accrual, thread stacks bloating and overruns[2][8][10]. Most of these studies focused on the ageing factors of software systems.

Software ageing can also be understood as similar to biological system of human being [3][11]. By using two examples such as human ageing and software life cycle, applications software can be implied, and view as a category of human evolution. This analogy is appropriate because it creates certain realization about the software. First, the application software exists inside a given environment. Furthermore, much like their biological evolution where they progress and adapt to their environment and later they grow old. Finally, the life cycle is a series of stages which a living thing passes from the beginning of its life until its death. So similarly in software life cycle, we can imply that software system and applications eventually die [12] after certain time in their environment.

However, the causes of software ageing are different from the biological organism such as humans. The human gets older when the time passes by which can be measured by number of years. Contrary with the nature and human, software will not subject to weakness or physical deterioration, thus it will not get old along with time [12]. Based on our initial investigation done in Malaysia we discover that software may experience ageing as early as approximately two to four years after being used. Therefore, we cannot determine the age of the software by numbers and years. It can be claimed that it does not matter how long the application software has been used as long as the software in the good quality and dynamic with environment changes, the software will stay young and healthy [16]. Even though in some circumstances, software ageing is inevitable but by understanding the factors of software ageing, in some ways may help to prevent the occurrence of ageing earlier. Those factors will be explained in more detail the next section.

Anti-ageing is a process to delay, prevent, and retard the ageing process from occurs premature. In human biological ageing and anti-ageing addressed by Klatz [14] indicates that anti-ageing factors for human are by practicing a healthy lifestyle, such as avoid eating unhealthy food, avoid drinking alcohol, stop smoking, stay slim, regular exercises and stress reduction management. Software anti-ageing may include prevention and rejuvenation actions such as adaptive, corrective, preventive, and perfective can be the anti-ageing action for application software.

3 The Conceptual Framework

A conceptual framework refers to "a theoretical structure of assumptions, principles, and rules that holds together the ideas comprising a broad concept." The design of this conceptual framework is based on the scope of software ageing, anti-ageing and software quality for application software.

Previous study has revealed that software ageing is closely related to software product quality. The effective and practical approach for managing ageing of the software is through software quality and certification process [16]. This was discussed in section 2.1. Figure 1 shows the conceptual framework of software anti-ageing that consists of main components: software quality models, software certification models, ageing factors, and the algorithm and formulas. The underlying theories in software quality and certifications were carried out to derive the reliable and supportive ageing factors as shown in Fig.1.

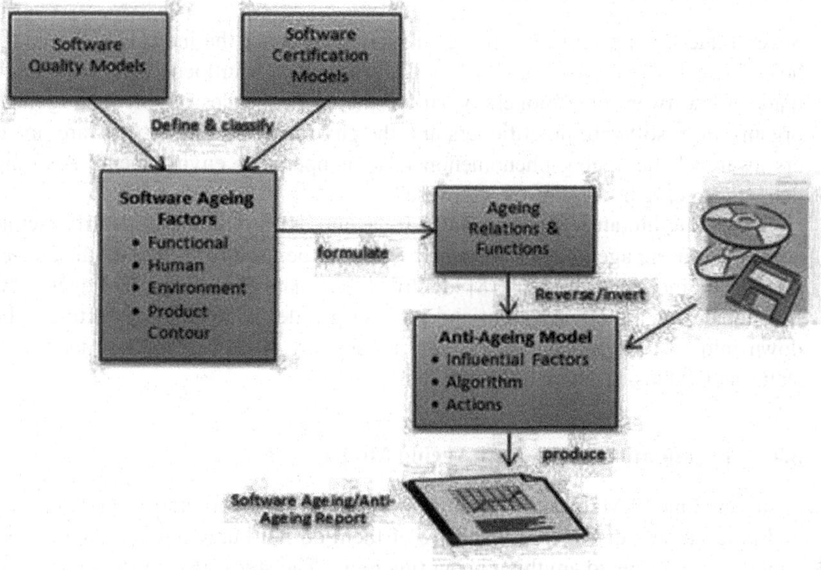

Fig. 1. The conceptual framework of software anti-ageing model

3.1 Software Quality

Software quality is closely related to the occurrence of software ageing. The more good quality of the software the less of software failure or software ageing will occur. Software quality can be measured by a number of variable which can be categorized by external variables and internal variable. The quality attributes of software product need to be known, recognized and classified to absorb and learn the underlying factors associated with ageing and anti-ageing.

3.2 Software Certification

Based on our previous works in software quality and certification indicated that certification practices in the software operating environment is essential to ensure the software stays young and healthy. Software certification models and attributes are essential to be investigated and studied to understand the contributing factors and features of software anti-ageing process and model. This research focuses on application software where application software is a program that is designed to perform specific task for end users. Application software can be used as a business tool that supports and assists the business process thus, it is crucial to study the phenomenon of software ageing and anti-ageing towards application software. It is different from previous works where their focuses are on software systems [1][[2][12][15].

3.3 Software Ageing and Factors

Several questions need to be answered such as what are the indicators of the ageing, how it affects the performances of the software, does it influence the working performance of software practitioners, what are the consequences of software ageing to an organization, software practitioners and the environment, and do software practitioners aware of the ageing phenomenon in their operating environment? Also, how to delay the ageing process of the software?

Systematic literature study on software ageing is carried out to identify factors that contribute to the ageing process which also may be used to formulate the anti-ageing factors and model. Aziz et al [18] identifies four main classification of ageing factors: functional, human, environment and product profile. Each of these factors is broken down into metrics and measures that can support in term of measuring the ageing status and values based on quantitative measures.

3.4 Formulation of the Anti-Ageing Model

A survey done in Malaysia with 50 respondents from information technology experts and practitioners discovered that 64% of them do not know anything about software ageing or never heard anything about this topic. The study also showed that only 30% of the respondents have experienced in software ageing related to the software they used, 20% of them never experienced while 50% were uncertainty. This findings reveal that the ageing and anti-ageing issues in software domain is still new but shows a consideration importance to the respondents.

In general mathematical definition, a "relation" is a relationship between sets of information. In this scope of study software ageing can be established as a set of several identified factors. Based on our recent findings 18], the ageing factors of the application software have been identified and may be formulated, as in:

$$\text{Software Ageing (SAS)} \sim f(\text{human, environment, functional, product}) \quad (1)$$

$$\text{SAS} \sim f\,(\text{h, e, f, p}) \quad (2)$$

$$y = k.x \quad (3)$$

$$\text{SAS} = k.f\,(\text{h, e, f, p}) \quad (4)$$

The above relations or equations (1) and (2) are derived from the ageing factors where software ageing score (SAS) can be derived given the values of factor human (h), environment (e), functional (f) and product (p). Each factors associated with ageing mentioned above are broken down into measurable metrics which can be assigned and transformed into numerical values. Thus, we say that given a value of all the factors, we know the computed value of SAS. In this case, SAS is the ageing score and can be mapped and transformed into an ageing status of specific software.

The ageing relation (equation 2) is converted to a function (a well-behaved relation) which means that given a value, we know exactly where to go, given an x we get only and exactly one y (see equation 3). Thus, The SAS function shown in (4) indicates that there is a constant value of k to normalize the linear equation of ageing function. The k value is unknown yet at this stage and still under investigation by our research group. There is a possibility that the k value varies for h, e, f and p. This will be an interesting investigation to come up with the possible k through an empirical study.

Therefore, to develop the anti-ageing model, the relationships between ageing factors are relevant and been explored further. As we understand anti-ageing means the reverse action or inversion of ageing. Thus, we possibly will formulate the anti-ageing model and equation as the reverse or inversion function or negation operator of ageing as shown in (5).

$$\text{Anti-Ageing-Score (AAS)} = k.f\,(h, e, f, p) \qquad (5)$$

; Where k is the constant to represent the negation operator.

The formulation discussed in this paper is a preliminary discovery and finding of this research project. Further works need to be carried out in detail to formulate and normalize the relationships and verify the relevance of the formula as well as the practicality of the measurements in real environment.

4 Conclusion

The initial works related to software ageing and anti-ageing in application software have been presented in this paper. The symptom of degradation in term of software quality is observed as the indicator of ageing factors in application software. The ageing phenomenon was observed and experienced through series of software certification practices carried out by this research group. Recently, we have identified software ageing associated factors and they may be considered as the influential factors of ageing. The classifications of factor are defined as functional, human, environment and product profile. Further study and exploration are needed to confirm the correlation between factors and formulate the anti-ageing equations and model. The anti-ageing model proposed in this paper is the preliminary work and it is valuable to prevent ageing and to preserve the software young and evergreen in the environment.

Acknowledgements

This research is funded by the Fundamental Research Grant Scheme, Malaysia Ministry of Higher Education.

References

1. Cotroneo, D., Natella, R., Pietrantuono, R. and Russo, S. Software Aging and Rejuvenation: Where We Are and Where We Are Going, 2011 IEEE Third Int. Work. Softw. Aging Rejuvenation, no. 30, pp. 1–6, Nov (2011)

2. Cassidy, K.J., Gross, K.C. and Malekpour, A. Advanced pattern recognition for detection of complex software aging phenomena in online transaction processing servers, Proc. Int. Conf. DependableSyst. Networks, pp. 478–482 (2002)
3. Parnas, D.L. Software Aging Invited. IEEE, pp. 279–287 (1994)
4. Zhao, J., Trivedi, K.S., Wang, Y. and Chen, X. Evaluation of software performance affected by aging. 2010 IEEE Second Int. Work. Softw. Aging Rejuvenation, vol. 3, pp. 1–6, Nov (2010)
5. Jin, H. and Zeng, F. Research on the definition and model of software testing quality, Proc. 2011 9th Int. Conf. Reliab. Maintainab. Saf., pp. 639–644, Jun (2011)
6. Suryn, W., Bourque, P., Abran, A. and Laporte, C. Software product quality practices - quality measurement and evaluation using TL9000 and ISO/IEC 9126. Intermag Eur. 2002 Dig. Tech. Pap.2002 IEEE Int. Magn. Conf., pp. 156–160 (2003)
7. Yahaya, J.H., Deraman, A., Baharom, F. and Hamdan, A.R. Software Certification from Process and Product Perspectives. *IJCSNS International Journal of Computer Science and Network Security*, 9(3), March (2009)
8. Sachin Garg, K. S. T., Aadvan Moorsel, Vaidyanathan, K. .A Methodology for Detection and Estimation of Software Aging. Software Reliability Engineering, Proceedings. The Ninth International Symposium (1998)
9. Grottke, M., Li, L., Vaidyanathan, K.. and Trivedi K. S. Analysis of Software Aging in a Web Server. *IEEE Trans. Reliab.* 55(3) pp. 411–420, Sept (2006)
10. Zheng, P., Xu, Q. and Qi, Y. An Advanced Methodology for Measuring and Characterizing Software Aging. 2012 IEEE 23rd Int. Symp. Softw. Reliab. Eng. Work., pp. 253–258, Nov (2012)
11. Sustainment, S. Geriatric Issues of Aging Software. *The Journal of Defense Software Engineering*. pp 4-7 Dec (2007)
12. Constantinides, C. and Arnaoudova, V. Prolonging the aging of aoftware systems. *Encyclopedia of Information Science and Technology* [Online]. Second Edition (8 Volumes) (2009)
13. ISO/IEC 25010. Systems and Software Engineering – Systems and Software Quality Requirements and Evaluation (SQuaRE) (2011)
14. Klatz, R. Definition of Anti-Aging Medicine. *Academic Journal Article*, Generations , 25(4), Winter (2002)
15. Hanmer, R. Software rejuvenation. Proceedings of the 17th Conference on Pattern Languages of Programs. ACM (2010)
16. Yahaya, J. H. & Deraman, A. Towards a Study on Software Ageing for Application Software: The Influential Factors. *IJACT: International Journal of Advancements in Computing Technology*, 4(14), pp. 51-59 (2012)
17. Yahaya, J.H., Deraman, A., Hamdan, A.R and Jusoh, Y.Y. User-Perceived Quality Factors for Certification Model of Web-Based System. International Journal of Computer, Information, Mechatronics, Systems Science and Engineering Vol:8 No:5, pp. 576-582 (2014)
18. Deraman, A., Yahaya, J.H., Zainal Abidin, Z.N. and Mohd Ali, N. Software Ageing Measurement Framework Based on GQM Structure. *Journal of Software and Systems Development*, 2014 (2014):1-12 (2014)
19. Yahaya, J. H., Deraman, A. & Hamdan, A. R. Continuosly Ensuring Quality through Software Product Certification: A Case Study. Proceedings of the International Conference on Information Society (i-Society 2010), London, UK, 28-30 June (2010)
20. Hwang, S.M. Quality Metrics for Software Process Certification based on K-model. 2010 IEEE 24th International Conference on Advanced Information Networking and Applications Workshops, pp 827-830 (2010).

A Framework Analysis for Distributed Interactive Multimedia System on Location-based Service

Shan Liu and Jianping Chai

School of Information Engineering
Communication University of China
Beijing, 100024, China

Email: liushan@cuc.edu.cn

Abstract—This paper presents a system solution to distribute and share school multimedia resources effectively. Especially, the paper presents the importance and feasibility of the location-based service in personalized, distributed and interactive multimedia system for digital campus. We focus on building a wireless roaming network platform in mobile applications, together with the characteristics of reliability, security, scalability and other design principles, such as object-oriented design patterns. Our goal is to achieve the multimedia content distribution and sharing information in the wireless roaming network by designing system function modules and terminal server application modules. The results demonstrate that the designed system can provide a channel for external communication for school management level, and a platform for internal resource sharing. The designed system can also provide a channel for independent study and information exchange platform for school staff and visitors.

Keywords—location-based service; multimedia system; content distribution; wireless roaming

1 Introduction

With the development of new media technologies and widely spread use of new mobile media, such as smart phones and tablets computers, the people's daily life have greatly changed. Recently, the concepts of digital city, digital home, digital campus, etc. have become more and more popular [1]. In addition, the information and communication technology together with the location-based services enable us to acquire the personalized information and resources conveniently [2].

Recently, the users tend to get various kinds of services in a certain range and access to a variety of resource using mobile devices. Especially for the campus, how to make the teachers and students teaching and learning effectively has become more and more important. Our work aims to set up a personalized, distributed and interactive multimedia system, which will be open to all the teachers and students on campus. Our goal is to merge the teaching and learning resources together that the teachers and students can obtain the corresponding resources according to their needs

K.J. Kim (ed.), *Information Science and Applications*,
Lecture Notes in Electrical Engineering 339, DOI 10.1007/978-3-662-46578-3_108

through the system. It can even become a learning exchange platform where the teachers and students can share information on campus.

1.1 Mobile Positioning Technology

The mobile positioning technology makes use of mobile communications network technology, by measuring and computing the wireless signal that received via mobile terminal, to acquire the location information of the mobile device or a particular person. In this way, we can provide various services related to the location information, which is also called "Location-Based Service" (LBS) [3].

Nowadays, the mobile positioning technology has been widely used in our daily life, and can be categorized into three main types:

Mobile device based positioning system: A mobile device based positioning system also can be named as forward link positioning system. As shown in Fig. 1(A), a mobile device can receive different wireless signals when the mobile device covered by different base station. All these wireless signals will carry some characteristic information such as strength of the field and propagation time, which related to the location information of the mobile device. Use these information, we can obtain the location formation between the mobile device and base stations by computing.

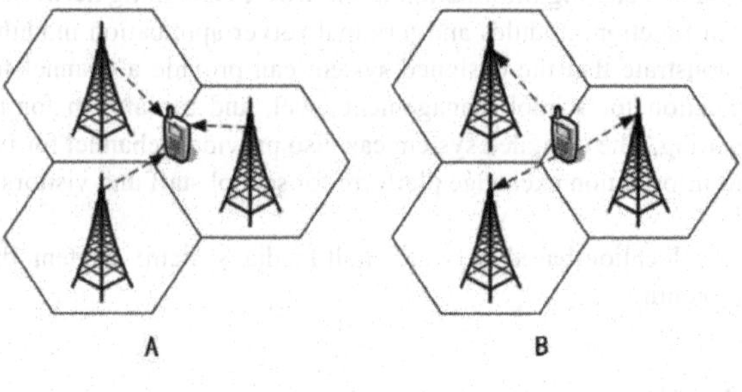

Fig. 1. Comparison between two mobile positioning systems.

Network based positioning system: A network based positioning system also can be named as reverse link positioning system. As shown in Fig. 1(B), we can see that the base stations around the mobile device will test the signals sending from the mobile device initiatively at this time. All the signals will be transmitted to the information-processing center and processed there. Then we can obtain the location formation between the mobile device and base stations by computing.

GPS device based positioning system: A GPS based positioning system integrates the GPS receiver into the mobile device, and locate the mobile device by GPS system. The integration of the GPS receiver will lead to the following problems: larger volume, more energy consumption, higher integration cost and longer time for first-time positioning.

Compared with above technologies, the mobile device based positioning system and the GPS device based positioning system will be more precise, but need more effort on reconstruction to add essential software and hardware. While the network based positioning system can be used directly to achieve the similar result without doing any reconstruction on mobile device.

1.2 Distributed Multimedia System

The multimedia mainly refers to the integration of the medium such as video, audio, image and text. In the modern society, the multimedia has been commonly used to display and share the information, exchange the ideas and feelings in our daily life.

The distribution of the multimedia content over the Internet mainly uses the content distribution network (CDN) technology, which builds the overlay network between the applications and Internet [4]. The CDN technology makes up many defects of current IP network, which provides a flexible and efficient content distribution service to multimedia applications [5]. This technology has been widely used in the applications of multimedia network and has brought convenience to the large-scale video and audio content distribution over the Internet.

2 System Architecture

2.1 System Design Principle

A distributed interactive multimedia system requires higher reliability, safety, expandability and interactivity to satisfy the property of integration, hierarchy and association.

Reliability: In order to avoid the system failures and data errors, which usually lead to system shut down or system crash, it requires the system to improve the disturbance resisting capacity and failure control ability.

Safety: To ensure the system running in the safety mode, we need to consider the security issues in the design stage. Thus we need to have an effective mechanism to authenticate the users and prevent illegal user login. In addition, we need to separate the users with different authority levels depending on their types, which is important to protect the system from unexpected damage.

Expandability: An appropriate system architecture with developing technology need to be considered in the design stage to make the system meet the subsequent expansion needs.

Interactivity: The system is designed to serve for customers, thus a convenient and reliable interaction is required to provide good service.

Maintainability: A concise and simple man-machine interface is required for the maintenance personnel to do the maintenance job.

2.2 System Architecture Overview

The system platform uses multimedia content as the sources of information, and uses the computer network as information dissemination channels, by which the sources of information are distributed to the user terminal. The information sources and system equipment, together with the management of terminal users are maintained and consolidated by the system maintenance personnel in the background management module. The user's terminal requests the multimedia content via the applications, which the server has already prepared.

The whole system is separated with three different layers: central server, subnetwork and user group. The central server provides service to terminal user and administrator, also dispatches the database and controls the network. The sub-network is a wireless local area network mainly consists of routers, which accesses the backbone network and working as a data bridge. The user group mainly consists of applications in user's terminal. The system architecture is shown in Fig. 2.

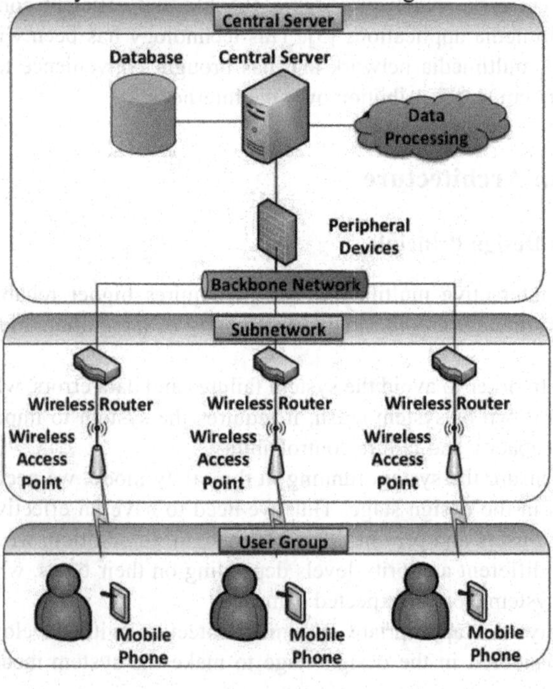

Fig. 2. System architecture.

2.3 System Architecture Design

The system is divided into the central server module and terminal application module. The central server module consists of the background management module and terminal access module, which the background management module including the function modules of the system maintenance personnel. The detail structure of the central server module and terminal application module are shown in Fig. 3 and Fig. 4, respectively.

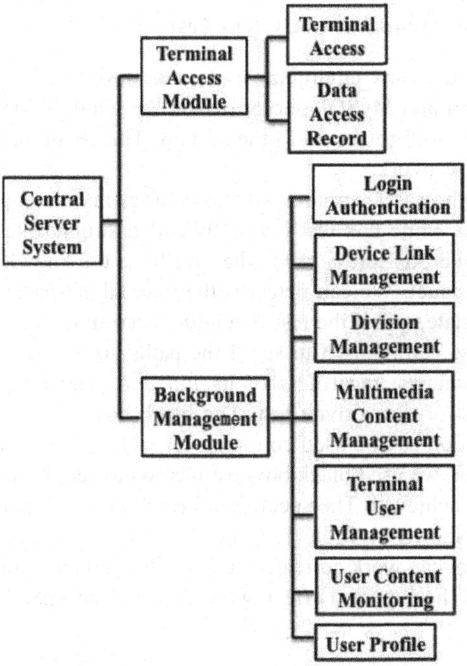

Fig. 3. Structure of central server module.

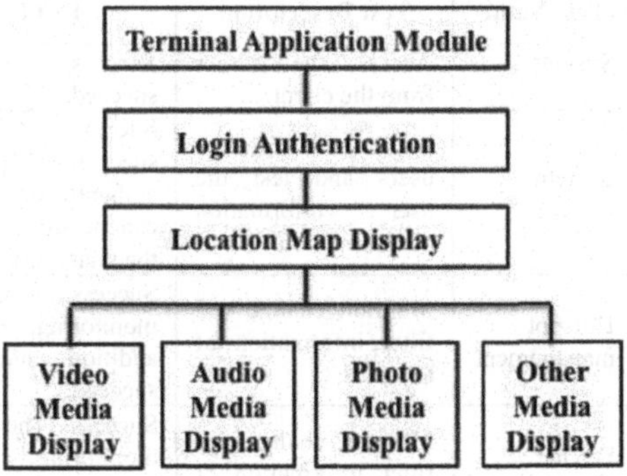

Fig. 4. Structure of terminal application module.

3 Evaluation

3.1 Background Management System Test

Description of system test environment: In our designed system, the background management system and MySQL server run on the windows desktop server, and the Chrome browser is used to access to the system. The client and server will have access to the campus network.

Results of system test: There are two software-testing methods, white box test and black box test. The white box test is usually called structure testing or logical drive test, where the white box test is used when we have the knowledge of internal work process and the product. We can detect if the internal action of the product working under the normal state, and if the test is running according to the internal structure of the program. It also can be used to see if the paths are able to work correctly in the determined requirements, regardless of its function. The black box test is usually called function test or data driven test. The black box test is usually used to detect whether each function can be used correctly, regardless of its internal operation principle. In our system, we used black box test method to see if our goal of system function design can be achieved. The specific information of test performance and results for each case are shown in Table 1. From the test we can see that the background management system can work correctly in each case and the various function modules are able to realized the expected effect, which also can accept a large number of client accesses.

Table 1. Description of test performance and results

Case No.	Test Name	Test Description	Test Result
01	System access	Access the server from the client.	Success. Access succeed.
02	System login	Login with different users and test the user information authentication.	Success. User login succeed and the authentication operation can be used to warning the unauthorized user .
03	Hotspot management	Monitor, change and filter information of hotspots.	Success. The action of monitoring, filtering, addition and deletion succeed.
04	Device management	Monitor, change and filter information of devices.	Success. The action of monitoring, filtering, addition and deletion succeed.
05	Multimedia management	Monitor, change and filter multimedia resources.	Success. The action of monitoring, filtering, addition and deletion succeed.
06	Binding of hotspot and multimedia	Bind and unbind of hotspot and multimedia.	Success. The action of binding and unbinding succeed.
07	User	Monitor, change and	Success. The action of

	management	filter information of users.	monitoring, filtering, addition and deletion succeed.
08	User profile management	Monitor and record the user's action and provide password reset..	Success. The user's action is recorded and the password can be reset.
09	Stability test	Test the system stability.	Success. The system keeps running in stable states and can respond immediately.
10	Stress test	Increase the connections of server and test the system response.	Success. The response of the server is stable when increase the connections.

3.2 Terminal Application Test

Description of terminal application test environment: The background management system run on the Tomcat server with windows operation system and the MySQL server run on the same computer. The terminal application software is installed on the Android mobile phone and access the campus local area network through a wireless router.

Results of terminal application test: The test also uses black box test method. We will test each function of the terminal applications in different cases. The results of test performance and results are listed in Table 2. Form the test result we can see that the terminal applications can provide the reliable services and can run smoothly in the mobile phone.

Table 2. Description of test performance and results

Case No.	Test Item	Test Description	Test Result
01	System Access	Login the application to access the server.	Success. Access succeed.
02	Map Resources	Map display, hotspots display and position display.	Success. The map and hotspots can be displayed and the position is correct.
03	Video Media	Request to video media display and download video resources.	Success. All the video resources display correctly and can be downloaded.
04	Audio Media	Request to audio media display and download audio resources.	Success. All the audio resources display correctly and can be downloaded.
05	Photo Media	Request to photo media display and download photo resources.	Success. All the photo resources display correctly and can be downloaded.
06	Other Media	Request to other media display and download resources.	Success. All the other resources display correctly and can be

			downloaded.
07	Access Record	Record the connections information, user login and logout information and the access information of multimedia resources.	Success. All the actions including the connections information, user's action and access information of multimedia can be recorded.

4 Conclusion

In this paper we have introduced an approach to design a location-based service in personalized, distributed and interactive multimedia system for digital campus. The study of the key technologies and the modules design provide us an overview of the system structure and the main work of this paper. We have analyzed the system characteristic and demand, proposes a feasible scheme. The study of the key technologies on the system implementation helped us to choose a suitable development technology to develop the system. The analysis of the system based on the demand of the design, system structure, system function and database structure has been pointed out. The Java Web technology and Android application development technology are used to realize each parts of the system. Finally, we have evaluated on system performance. Based on the system performance and results, we can conclude that system can provide a communication channel for the school management, a platform for sharing resources in digital campus. Future work will expand our work to larger network environment and more users.

References

1. Y. Wang, "Survey of Digital Campus Construction in China," *Modern Distance Education on Research*, vol. 4, no. 112, pp. 39-50, 2011.
2. J. Reed and T. Rappaport, "An Overview of the Challenges and Progress in Meeting the E-911 Requirement for Location Service," *IEEE Communication Magazine,* April 1998, pp. 30-37.
3. Z. Fan, P. Deng and L. Liu, *Wireless Positioning in Cellular Network.* Electronic Industry Press, Beijing, 2002.
4. J. Song and H. Tao, "The Development and Application of Content Distribution Network Technology," *Television Technology,* vol. 6, pp. 75-77, 2005.
5. H. Yin, T. Zhan and C. Lin, "Multimedia Network: From Content Distribution Network to Future Internet," *Chinese Journal of computers,* vol. 6, pp. 1120-1130, 2012.
6. X. Zhang, G. Peng and W. Wei, "Research of Marine Traffic Information Sharing Technology Based on WEB," *Institute of navigation China 2007 Annual Symposium of excellent papers,* 2007, pp. 44-48.
7. Z. Zhao and G. Li, "A New Architecture Based on B/S Structure and C/S Structure," *Computer Application,* vol. 8, 2004, pp. 35-38.

An Empirical Validation of Coupling and Cohesion Metrics as Testability Indicators

Amos Orenyi Bajeh, Shuib Basri, Low Tang Jung

Computer and Information Sciences Department.
Universiti Teknologi PETRONAS
Bandar Seri Iskandar, 31750 Tronoh, Perak, Malaysia
amos_g02040@utp.edu.my, bajehamos@unilorin.edu.ng,
Shuib_basri@petronas.com.my, lowtanjung@petronas.com.my

Abstract. The measurement of the quality attributes of software is a management approach to building quality software product. Testability is a measure of the capability of the software to be subjected to testing. It is a desirable quality attribute of software product because it ensures the reliability of software. This paper presents an empirical validation of coupling and cohesion metrics as indicators of the testability quality attributes of Object-Oriented (OO) software from High-Level Design (HLD) perspective. Open-source OO software samples are used for the empirical analysis and three test case complexity metrics are taken as the measure of the complexity of testing the software sample. The results of the empirical validation showed that the coupling metrics and some of the cohesion metrics investigated are good indicators of the testability of OO software design.

Keywords: object-oriented software, testability, metrics, coupling, cohesion

1 Introduction

Measurement is one of the fundamental aspects of any engineering discipline and software engineering is no exception [1]. Software measurement can be carried out during the different phases of development but predominantly, it is carried out on High-Level Design (HLD) and Low-Level Design (LLD). The measurement of HLDs has the benefit of equipping developers with early information indicating the need to fine-tune designs before implementation; this is more beneficial since the task of reviewing software design after implementation cost more in terms of budget and programmer's effort, and it is error prone.

Testability as defined in the ISO/IEC9126 standard [2, 3] is the capability of the software to be tested. The IEEE standard glossary [4] defined testability as "the degree to which a system or component facilitates the establishment of test criteria and the performance of tests to determine whether those criteria have been met. Testability is one of the desirable quality attributes of software products because it determines the

K.J. Kim (ed.), *Information Science and Applications*,
Lecture Notes in Electrical Engineering 339, DOI 10.1007/978-3-662-46578-3_109

complexity involved in software functional verification and validation exercise. Thus, the measurement of the testability of software designs can aid developers in making informed design decisions that will lead to developing testable software. This study presents an empirical validation of Object-Oriented (OO) HLD coupling and cohesion metrics as indicators of the testability quality attribute of OO software.

The other parts of this paper are organized as follows: Section 2 presents related works in the area of OO software testability measurement. Section 3 discusses the OO HLD metrics for coupling and cohesion, and the research method used. Section 4 presents the results and discussion of the empirical analysis. The paper is concluded in Section 5 with some discussion on future direction.

2 Testability Metrics

Studies have been conducted in the area of software testability measurement. [5, 6] showed that the testability of software designs are affected by class interactions in UML class diagram. Concurrent class interactions in class diagram can indicate part of software design that needs to be structurally modified and improved in order to reduce the effort that will be required for testing the software.

Bruntink and Deursen [7] reported a study on the use of OO metrics for predicting testability. The first nine (9) OO metrics described in Table 1 were studied. Out of the nine metrics, only the inheritance and encapsulation metrics: DIT, NOC, NOF and NOM are measurable from OO HLDs. The coupling and cohesion metrics, FOUT and LCOM, are measurable from LLDs. Thus there is a need to investigate HLD coupling and cohesion metrics as testability indicators in order to be able to estimate the testability of OO software from HLD-this is the focus of this paper.

Table 1. OO Metrics

OO Metric	Description
DIT (Depth of Inheritance Tree)	: is the distance from a class to the root class of inheritance hierarchy i.e. the number of ancestors of the class
FOUT (Fan Out)	: is the count of the number of other classes used by a class' methods.
LCOM (Lack of Cohesion Of Methods)	: is the measure of the interrelatedness of the components (attribute and methods) of a class. A normalized metric with value ranging between [0, 1]: 1 for a non-cohesive class and 0 for a perfectly cohesive class.
LOCC (Line of Code per Class)	: is the count of the total number of lines of code in the methods of a class, excluding blank lines and comments.
NOC (Number Of Children)	: is the count of the number of immediate children of a class in inheritance hierarchy
NOF (Number Of Fields)	: is the count of the number of attributes (data components) in a class.
NOM (Number of Methods) / NOO (Number Of Operations)	: is the count of the number of methods (operations or functions) in a class.
RFC (Response For Class)	: is the sum of the number of methods in a class and the number of other classes' methods invoked by the class.
WMC (Weighted Method per Class)	: is the sum of the McCabe's cyclomatic complexities of all the methods in a class.
CBO (Coupling Between Objects)	: this is the number of classes to which a given class is coupled with or connected with.

A generic testability measurement model was proposed by Khan and Mustafa [8] using regression analysis approach. The model comprises of two OO design mechanisms: encapsulation and inheritance; and coupling as the determinant factors of testability (equation 1). In order to use this model, specific metrics that measure encapsulation, inheritance and coupling attributes of OO software design have to be employed. The model implies that the higher the value of testability (*Test.*) of an OO software, the more difficult to test the software in question, and vice versa.

$$Test. = -0.08 * Encap + 1.12 * Inher + 0.97 * Coup \quad (1)$$

The model in equation 1 was empirically validated by Kout et al. [9]. The validated model, depicted in equation 2, uses NOO, DIT and CBO metrics (described in Table 1) as the specific OO metrics for encapsulation, inheritance and coupling respectively.

$$Test. = -0.08 * NOO + 1.12 * DIT + 0.97 * CBO \quad (2)$$

The coupling metric, CBO, in the model is also measurable from OO LLDs.

In order to be able to measure testability from OO HLDs, there is a need to first, establish the relationship between OO HLD metrics and testability. This implies that the HLD metrics have to be empirically validated as good testability indicators. Thereafter measurement models, such as Eq. 2, can be developed to measure the testability of OO software from the HLD perspective. This study presents an empirical validation, focusing on OO HLD coupling and cohesion metrics.

3 Methodology

The approach to the empirical validation in this study involves the use of open-source OO software products and the use of statistical analysis technique for the identification of the relationship between OO HLD coupling and cohesion metrics, and the complexity of the unit test cases of the software products.

3.1 Coupling and Cohesion metrics

The OO HLD metrics for the measurement of coupling between classes and the cohesion of classes investigated in this study are defined in this section:

The HLD coupling metrics are:

- Number of Association (NAssoc): this is the total number of entities (classes or interfaces) that a given class is in association with in a software design. This metric was initially proposed by Genero et al [10, 11] as the class-scope metric NAS. The metric is a HLD metric that measures an aspect of the CBO metric in terms of the direction of coupling; it measures the classes used as data type in the measured class.
- Import Coupling (IC): This metric is the count of the number of classes that a class used or depends upon for its operations. It is equivalent to the NDepOut metric proposed by Genero et al. [10, 11].

The HLD Cohesion metrics are:

- Cohesion Among Methods of a Class (CAMC) [12]: this metric computes the relatedness among the methods of a class by measuring the attribute-method relationship in the class based upon the parameter list of the methods of the class. The formal definition of this metric is:

  $$CAMC = \frac{a+k}{k(l+1)}$$

  where a is the summation of the 1s in a parameter occurrence matrix which is a matrix of k number of rows and l number of columns. K is the number of methods in a class and l is the number of unique data types of the parameters of all the methods in the class. The parameter-occurrence matrix is generated by placing a 1 in the cell Cij where the method i has at least one parameter of type j and 0 otherwise. Thus:

  $$a = \sum_{i=1}^{k}\sum_{j=1}^{l} C_{ij}$$

- Similarity-based Class Cohesion (SCC): this metric was proposed by Al Dallal and Briand, and it was defined to measure the other class elements relationships that are not measured by the other cohesion metrics defined above. It measures the method-method, method-attribute, attribute-attribute, and method invocation direct and transitive interactions identified using the class and communication diagrams defined by the Unified Modeling Language (UML). The metric is formally defined as follows:

  $$SCC = \begin{cases} 0 & \text{if } k = 0 \text{ and } l \geq 1 \\ 1 & \text{if } k = 1 \text{ and } l = 1 \\ MMIC & \text{if } k > 1 \text{ and } l = 1 \\ \dfrac{k(k-1)[MMAC(C) + 2MMIC(C)] + l(l-1)AAC(C) + 2lkAMC(C)}{3k(k-1) + l(l-1) + 2lk} \end{cases}$$

  where $MMIC$, $MMAC$, AAC and AMC are four cohesion interaction types which are Method-Method Invocation Cohesion, Method-Method through Attribute Cohesion, Attribute-Attribute Cohesion and Attribute-Method Cohesion respectively [15].

- Normalized Hamming Distance (NHD): this metric was proposed by Counsell et al. as a more robust alternative to CAMC. It uses the same parameter-occurrence matrix used in CAMC but computes the average parameter agreement between each pair of methods in a class. The parameter agreement between methods is the number of entries that matches in the corresponding rows in the parameter-occurrence matrix. NHD is formally defined as follows:

  $$NHD = 1 - \frac{2}{kl(k-1)} \sum_{j=1}^{l} x_j(k - x_j)$$

  where x_j is the number of 1s in the jth column of the parameter-occurrence matrix.

- Scaled Normalized Hamming Distance (SNHD): this is a scaled value of NHD. It measures the closeness of the NHD to the maximum value of NHD compared to the minimum value. A detailed discussion of this metric is also provided in [13, 14]

3.2 Test Case Metrics

In [7] the difficulty involved in developing a test case for OO classes is measured by the size of the test case. Two metrics used in the study are the number of Lines of Code (LOC) and the number of JUnit 'assert' methods invoked in the test case (NOTC). The metrics are collected from the JUnit test cases for the classes of the sample software used in the study. It is assumed that the more the LOC, the more the effort put into testing the class. Also, 'Assert' methods are provided by the JUnit testing framework for the comparison of the expected values of a test and the actual values of the class been tested. Thus, the higher the number of 'assert' methods invoked, the more complex the test case. In this study, these metrics are used as the size and complexity of the test cases, but renamed to reflect that they are test case measurement. The LOC for the test cases is designated as tLOC and the number of 'assert' method invocation is designated as tAssert. The letter 't' signifies test case.

3.3 Hypotheses Formulation

Having defined the OO HLD metrics for OO classes and the size and complexity metrics for test cases, the hypotheses formulation involves relating the HLD metrics to the test case metrics. Thus the hypotheses tested are as follows:

H1: There is a significant nonzero positive correlation between the Coupling metrics and the test case metrics
Increase in coupling makes testing more difficult. The more the number of entities (classes) a class is coupled with, the more the number of lines of code required to test the class because more test cases will be required to test the coupling relationships [16]. This implies that there should be positive correlation between the coupling metrics and the test case metrics if there is any relationship between them.

H2: There is a significant nonzero negative correlation between the cohesion metrics and the test case metrics
A cohesive class implies that the methods of the class perform related functions while a non-cohesive class comprises of methods that perform unrelated functions. Thus for a non-cohesive class, more complex tests have to be conducted in order to test the unrelated functionalities of the class [16]. This implies that as the cohesion increases the complexity and size of the test case should reduce and vice versa. Therefore a negative correlation should be observed between the cohesion metrics and the test case metrics.

3.4 The Software Sample and Data Collection

The OO open source software, Ant (version 1.9.2) and jfreeChart (version 1.0.15) are used as case study for the empirical analysis. The choice of the software is informed by the fact that they are used for empirical analysis in the area of testability metrics in the literature and their classes have test cases developed using the JUnit testing framework. Bruntink and van Deursen [7] used Apache Ant as one of the software in their empirical study while Kout et al. used both apache Ant and jfreeChart for their empirical analysis [9]. Table 2 presents the software and the test cases collected for the empirical analysis.

Table 2. Software sample description

Sample Software	LOC	Number of classes	No of classes with test cases
Apache Ant	195,986	878	207 (24%)
jfreeChart	185,074	623	320 (51%)

The metrics are automatically measured from the classes of the software samples and their test cases. Borland Together tool (v. 12.6) [17], an eclipse plugin tool, was used to measure the test case metrics, tLOC from the JUnit test cases of the sample software. tAssert and the coupling metrics (NAssoc and IC) are measured using the SDMetric tool[18], a software design metric measurement tool. The SDMetric tool was extended to measure the tAssert metric by using the in-built xml feature of the tool to define the metric. The cohesion metrics, CAMC, NHD, SNHD and SCC, are measured from the classes of the software by using the tool developed by Dallal et al. [15].

4 Results and Discussion

The descriptive statistics of the HLD metrics measured from the classes in the software samples are given in Table 3. Table 4 present the correlation analysis results.

The results in Table 4 show significant nonzero correlation between the OO HLD metrics and the test case metrics. The empirical data supports the **H1** hypothesis for the two coupling metrics, NAssoc and IC with *pval*s less than 0.05 (i.e. above 95% confidence level). Thus, coupling have impact on the testability of OO software and the number of association and dependency metrics can serve as good indicators of the testability of OO software from the HLD perspective. This relationship observed between the IC metric and test case metrics agrees with the result obtained by Bruntik et al. [7]: the FOUT metrics, by definition is an import coupling that counts the number of classes used by a given class; this metric is the same has the IC coupling except that the FOUT metric is measurable from the LLDs and thus measures more granular and detail information at the method-scope level than the IC metric which measures information at the class-scope level. All the cohesion metrics have significant correlation with the test case metrics. The empirical analysis supports the **H2** hypothesis for CAMC and SCC metrics only since they have negative correlation coefficients which are the expected outcome of the investigation. CAMC has a stronger correlation than SCC in the two software samples, implying that CAMC can be a better indicator of testability than SCC. On the other hand, the empirical analysis does not support the **H2** hypothesis for the NHD and the SNHD metrics since they have positive correlation coefficients. Thus the result of this study shows that the Number of Association and Dependency metrics and the CAMC metric for OO HLDs are good indicators of the testability of OO software products at the HLD stage of software development.

Table 3. The description statistics of the metrics

Software sample		NAssoc	IC	CAMC	NHD	SNHD	SCC
Apache Ant	**Min**	0	0	0	0	0	0
	Max	10	27	1	1	1	1
	Mean	0.75	2.08	0.37	0.68	0.40	0.09
	Median	0	0	0.31	0.73	0.33	0.03
	Std.	1.43	4.28	0.23	0.22	0.38	0.20
jfreeChart	**Min**	0	0	0.06	0	0	0
	Max	4	29	1	1	1	1
	Mean	o.24	2.22	0.3	0.72	0.49	0.06
	Median	0	0	0.28	0.74	0.56	0.02
	Std.	0.61	4.02	0.15	0.16	0.38	0.12

Table 4. The correlation analysis result (with pval in parentheses)

OO HLD Metric	Ant		jfreeChart	
	tLOC	**tAssert**	**tLOC**	**tAssert**
NAssoc	**0.382453** (1.29E-08)*	**0.372904** (3.14E-08)	**0.171408** (0.002091)	**0.154846** (0.005505)
IC	**0.36919** (4.4E-08)	**0.426759** (1.44E-10)	**0.471871** (3.79E-19)	**0.438932** (1.68E-16)
CAMC	**-0.32968** (1.23E-06)	**-0.23143** (0.000793)	**-0.25413** (4.15E-06)	**-0.21038** (0.00015)
NHD	0.319814 (2.63E-06)	0.202851 (0.003375)	0.313648 (9.8E-09)	0.308446 (1.76E-08)
SNHD	0.08028 (0.250191)	0.137831 (0.047648)	0.208887 (0.000167)	0.162346 (0.00359)
SCC	**-0.16201** (0.019692)	**-0.13989** (0.044386)	**-0.1296** (0.020393)	-0.10349 (0.064457)

5 Conclusion

The measurement of the quality attributes of software from the design phase can provide early information to software developers to aid them in taking design decisions that will yield good quality software product at implementation phase. OO HLD metrics need to be empirically validated to show that they can serve as indicators of external quality attributes such as testability. This paper presented an empirical study for the validation of OO HLD coupling and cohesion metrics as testability indicators. Two HLD coupling metrics (NAssoc and IC) and four HLD cohesion metrics (CAMC, NHD, SNHD and SCC) were considered for the empirical analysis. Data were collected from two open-source OO software products that have JUnit test cases developed for their classes. The NAssoc and IC metrics for coupling metrics prove to be suitable indicators of testability while the cohesion metric, CAMC, is shown to be a better indicator of testability than other cohesion metrics investigated.

For the future work, more open source software samples will be collected for more empirical validation of these HLD metrics in order to further verify the results of this study. Thereafter an objective OO HLD testability measurement model can be proposed using the validated OO HLD metrics. Also, the formal definition of the NHD and SNHD metrics can be studied in order to explain the positive correlation coefficient observed in this empirical analysis.

Acknowledgment

This work is carried out under the Universiti Teknologi PETRONAS (UTP) STIRF 2014 grant: 0153AA-C64.

References

1. F. T. Sheldon, K. Jerath and H. Chung, "Metrics for maintainability of class inheritance hierarchies," Journal of Software Maintenance and Evolution: Research and Practice, vol. 14, pp. 147-160, 2002.
2. ISO/IEC, "ISO/IEC 9126-1 Standard, Software Engineering, Product Quality, Part 1: Quality Model," Geneva, 2001.
3. ISO/IEC, "Software Engineering---Software Product Quality Requirements and Evaluation (SQuaRE), ISO/IEC25000," 2005.
4. IEEE Computer Society Professional Practices Committee, "Guide to the Software Engineering Body of Knowledge (SWEBOK-2004 Version)," IEEE Computer Society, Los Alamitos, CA, 2004.
5. B. Baudry, Y. Le Traon and G. Sunyé, "Testability analysis of a UML class diagram," in Software Metrics, 2002. Proceedings. Eighth IEEE Symposium on, 2002, pp. 54-63.
6. B. Baudry and Y. L. Traon, "Measuring design testability of a UML class diagram," Information and Software Technology, vol. 47, pp. 859-879, 2005.
7. M. Bruntink and A. Van Deursen, "Predicting class testability using object-oriented metrics," in Source Code Analysis and Manipulation, 2004. Fourth IEEE International Workshop on, 2004, pp. 136-145.
8. R. A. Khan and K. Mustafa, "Metric based testability model for object oriented design (MTMOOD)," ACM SIGSOFT Software Engineering Notes, vol. 34, pp. 1-6, 2009.
9. A. Kout, F. Toure and M. Badri, "An empirical analysis of a testability model for object-oriented programs," ACM SIGSOFT Software Engineering Notes, vol. 36, pp. 1-5, 2011.
10. M. Genero, M. Piattini and C. Calero, "Early measures for UML class diagrams," L'Objet, vol. 6, pp. 489-515, 2000.
11. M. Genero, "Defining and validating metrics for conceptual models," Computer Science Department, 2002.
12. J. Bansiya, L. Etzkorn, C. Davis and W. Li, "A class cohesion metric for object-oriented designs," Journal of Object-Oriented Programming, vol. 11, pp. 47-52, 1999.
13. S. Counsell, S. Swift and J. Crampton, "The interpretation and utility of three cohesion metrics for object-oriented design," ACM Transactions on Software Engineering and Methodology (TOSEM), vol. 15, pp. 123-149, 2006.
14. K. Kuljit and H. Singh, "An Investigation of Design Level Class Cohesion Metrics," The International Arab Journal of Information Technology, vol. 9, pp. 66-73, 1 January, 2012.
15. J. Al Dallal and L. C. Briand, "An object-oriented high-level design-based class cohesion metric," Information and Software Technology, vol. 52, pp. 1346-1361, 2010.
16. S. Mouchawrab, L. C. Briand and Y. Labiche, "A measurement framework for object-oriented software testability," Information and Software Technology, vol. 47, pp. 979-997, 2005.
17. Borland Together Tool, Available: http://www.borland.com/products/together/. accessed 5th January 2014.
18. SDMetric Design quality metric tool for UML models. Available: http://www.sdmetrics.com/.

Software Defect Prediction in Imbalanced Data Sets Using Unbiased Support Vector Machine

Teerawit Choeikiwong, Peerapon Vateekul

Department of Computer Engineering, Faculty of Engineering
Chulalongkorn University, Bangkok, Thailand

Teerawit.Ch@student.chula.ac.th, Peerapon.V@chula.ac.th

Abstract. In the software assurance process, it is crucial to prevent a program with defected modules to be published to users since it can save the maintenance cost and increase software quality and reliability. There were many prior attempts to automatically capture errors by employing conventional classification techniques, e.g., Decision Tree, k-NN, Naïve Bayes, etc. However, their detection performance was limited due to the imbalanced issue since the number of defected modules is very small comparing to that of non-defected modules. This paper aims to achieve high prediction rate by employing unbiased SVM called "R-SVM," our version of SVM tailored to domains with imbalanced classes. The experiment was conducted in the NASA Metric Data Program (MDP) data set. The result showed that our proposed system outperformed all of the major traditional approaches.

Keywords: software defect prediction; imbalanced issue; threshold adjustment

1 Introduction

Software defect is any flaw or imperfection in a software product or process. It is also referred to as a fault, bug, or error. It is considered as a major cause of failures to achieve the software quality, and the maintenance cost can be very high in order to amend all of the defects in the software production stage. Hence, it is very important to detect all of the bugs as early as possible before publishing the program. Software defect prediction is recognized as an important process in the field of software engineering. It is an attempt to automatically predict the possibility of having a defect in the software. Note that the detection can be in many levels, i.e., method, class, etc.

Data miners have been aware of the software defect issue and proposed several works in defect prediction applying conventional classification techniques[1, 2, 3, 4, 5, 6]. Although their experimental results were reported showing a promising detection performance, all of them were evaluated based on accuracy, which is not a proper metric in this domain. For example, assume the percentage of defected modules is only 10%, while the remaining modules (90%) are non-defected. Although the detection system incorrectly classifies all modules as non-defected ones, the accuracy is still 90%!

K.J. Kim (ed.), *Information Science and Applications*,
Lecture Notes in Electrical Engineering 339, DOI 10.1007/978-3-662-46578-3_110

For the sake of comparison, it is necessary to evaluate the system based on standard benchmarks. NASA Metrics Data Program (MDP) [7] is a publicly available and widely used data set. There are 12 projects described by many software metrics, such as McCabe's and Halstead's metrics. It is common in the software prediction benchmark, including MDP, that the number of defected modules is very small comparing to the non-defected ones. In the MDP benchmark, an average of the defected modules in all of the projects is 20%, and the PC2 project has the lowest percentage of defected modules with only 2.15%. This circumstance is referred to as "imbalanced data set," which is known to tremendously drop classification performance. There were many works [1, 2, 3, 4, 5, 6] using the MDP benchmark; however, most of them ignore the imbalanced issue, so their reported results should be impaired.

In this paper, we aim to propose a novel defect prediction system by applying a support vector machine technique called "R-SVM," our previous work [8] that is customary to induce SVM in the imbalanced training set. In R-SVM, the separation hyperplane is adjusted to reduce a bias from the majority class (non-defect). The experiment is conducted based on the MDP benchmark and, then, the result is compared to several classification techniques: Naive Bayes, Decision Tree, k-NN, SVM (Linear), and SVM (RBF) in terms of the measures including PD, PF, F1, and G-mean.

The rest of paper is organized as follows. Section 2 presents an overview of the related work. Section 3 describes the proposed method in details. The description of the data sets and tools are found in Section 4. Section 5 shows the experimental results. Finally, this paper is concluded in Section 6.

2 Related Work

2.1 Software Metrics

Measurement is known as a key element in engineering process. It can be used to control the quality and complexity of software. For building one product or system, we use many software measures to assess the quality of product and also to define the product attribute like in the MDP benchmark. These software metrics provide various benefits, and one of the major benefits is that it offers information for software defect prediction. Currently there are many metrics for defect prediction in software.

McCabe's cyclomatic complexity [9] and Halstead's theory [10] are ones of the well-known software metrics. They can be applied into any program languages and were introduced in order to describe as "code features," which are related to software quality. McCabe's metric is used to show the complexity of a program. From a program source code, it can directly measure the number of linearly independent paths. Moreover, it can be used as an ease of maintenance metric. Halstead's theory is widely used to measure complexity in a software program and the amount of difficulty involved in testing and debugging of software development. In addition, there are also other metrics, e.g., the Chidamber and Kemerer metrics [11] and Henry and Kafura's information flow complexity.

Table 1 shows the features used in the MDP benchmark, which does not only depend on McCabe's and Halstead's metrics, but also the statistics of program codes, Line Count.

Table 1. The software metrics used as features in the MDP benchmark.

McCabe (4 metrics)	$v(G)$	cyclomatic_complexity		
	$ev(G)$	essential_complexity		
	$iv(G)$	design_complexity		
	loc	loc_total(one line = one count)		
Line Count (5 metrics)		loc_blank		
		loc_code_and_comment		
		loc_comments		
		loc_executable		
		branch_count		
Halstead (12 metrics)	N_1	num_operands	V	volume: $V = N * \log_2 \mu$
	N_2	num_operators	D	difficulty: $D = 1/L$
	μ_1	num_unique_operands	E	effort: $E = V / \widehat{L}$
	μ_2	num_unique_operators	T	prog_time: $T = E/18$ seconds
	B	error_est	L	level: $L = V^*/V$ where $V^* = (2 + \mu_2^*)\log_2(2 + \mu_2^*)$
	N	length: $N = N_1 + N_2$	I	content: $I = \widehat{L} * V$ where $\widehat{L} = \frac{2}{\mu_1} * \frac{\mu_2}{N_2}$

2.2 Prior Works in Defect Prediction

In the field of software defect prediction, there were many trials in applying machine learning techniques to the MDP benchmark. Menzies et al. [1] applied Naïve Bayes to construct a classifier to predict software defects. There was an investigation on the feature selection using information gain. The proposed system obtained 71 % of the mean probability detection (PD) and 25 % of the mean false alarm rate (PF). Bo et al. [2] proposed a system called "GA-CSSVM". It built around SVM and used Genetic Algorithm (GA) to improve the cost sensitive parameter in SVM. The result showed that it achieved a promising performance in terms of AUC. Seliya et al. [3] introduced an algorithm called "RBBag". It employed the bagging concepts into two choices of classifiers: Naïve Bayes and C4.5. The result showed that RBBag outperformed the classical models without the bagging concept. Moreover, RBBag is more effective when it applied to Naïve Bayes than C4.5. Unfortunately, all of these works discard the imbalanced issue, so their reported results should be limited.

Shuo et al. [4] was aware of the imbalanced issue in the software defect prediction. There was an investigation on many strategies to tackle the imbalanced issue including resampling techniques, threshold moving, and ensemble algorithms. The result showed that AdaBoost.NC is the winner, and it also outperformed other conventional techniques: Naïve Bayes and Random Forest. Furthermore, a dynamic version of AdaBoost.NC was proposed and proved to be better than the original one.

Recently, support vector machine (SVM) have been applied in the field of software defect prediction showing a good prediction performance. Gray et al. [5] employed SVM to detect errors in the MDP data set. The analysis from the SVM results revealed that if a testing module has a large average of the decision values (SVM scores), there is high chance to found defects in it. Elish et al. [6] compared SVM to eight traditional classifiers, such as DT, NB, etc., on the MDP data set. The experiment demonstrated that SVM is the winner method. Thus, this is our motivation to apply a method built around SVM called "R-SVM" to detect the software errors.

2.3 Assessment in Defect Prediction

In the domain of binary classification problem (defect vs. non-defect), it is necessary to construct a confusion matrix, which comprises of four based quantities: True Positive (TP), False Positive (FP), True Negative (TN), and False Negative (FN) as shown in Table 2.

Table 2. A confusion matrix.

	Predicted Positive	Predicted Negative
Actual Positive	TP	FN
Actual Negative	FP	TN

These four values are used to compute Precision (Pr), Probability of Detection (PD), Probability of False Alarm (PF), True Negative Rate (TNR), F-measure [11], and G-mean [12], which is a proper metric that frequently used to tackle the class imbalance problem as shown in Table 3.

Table 3. Prediction Performance Metrics

Metrics	Definition	Formula
Precision (Pr)	a proportion of examples predicted as defective against all of the predicted defective	$\frac{TP}{TP+FP}$
Probability of Detection (PD), Recall, TPR	a proportion of examples correctly predicted as defective against all of the actually defective	$\frac{TP}{TP+FN}$
Probability of False Alarm (PF), FPR	a proportion of examples correctly predicted as non-defective against all of the actually non-defective	$\frac{FP}{TN+FP}$
True Negative Rate (TNR)	a proportion of examples correctly predicted as non-defective against all of the actually non-defective	$\frac{TN}{TN+FP}$
G-mean	the square root of the product of TPR (PD) and TNR	$\sqrt{(TPR)\cdot(TNR)}$
F-measure	a weighted harmonic mean of precision and recall	$\frac{2\times \text{Pr}\times \text{Re}}{\text{Pr}+\text{Re}}$

3 Our Proposed Method

3.1 Support Vector Machine (SVM)

Support Vector Machine (SVM) [13, 14] is one of the most famous classification techniques which was presented by Vapnik. It was shown to be more accurate than other classification techniques, especially in the domain of text categorization. It constructs a classification model by finding an optimal separating hyperplane that maximizes the margin between the two classes. The training samples that lie at the margin of the class distributions in feature space called support vectors.

The purpose of SVM is to induce a hyperplane function (Equation 1), where $\vec{w}$ is a weight vector referring to "orientation" and b is a bias.

$$h(\vec{w},b)=\vec{w}\cdot x+b \tag{1}$$

The optimization function to construct SVM hyperplane is shown in (2), where C is a penalty parameter of misclassifications.

$$\underset{w,b,\xi}{Minimize} \quad \frac{1}{2}w^T w+C\textstyle\sum_{i=1}^{n}\xi_i$$
$$subject\ to \quad y_i\left(w^T\phi(x_i)+b\right)\geq 1-\xi_i \quad ,\xi_i\geq 0 \tag{2}$$

In a non-linear separable problem, SVM handles this by using a kernel function (non-linear) to map the data into a higher space, where a linear hyperplane cannot be used to do the separation. A kernel function is shown in (3).

$$K(x_i, x_j) \equiv \phi(x_i)\phi(x_j) \quad (3)$$

Unfortunately, although SVM has shown an impressive result, it still suffers from the imbalanced issue like other traditional classification techniques.

3.2 Threshold Adjustment (R-SVM)

Although SVM has shown a good classification performance in many real-world data sets, it often gives low prediction accuracy in an imbalanced scenario. R-SVM [8] is an our earlier attempt that focuses to tackle this issue by applying the threshold adjustment strategy. To minimize a bias of the majority class, it translates a separation hyperplane in (1) without changing the orientation $\vec{w}$ by only adjusting b. After the SVM hyperplane has been induced from the set of training data mapped to SVM scores, L. The task is to find a new threshold, θ, that selected from the set of candidates thresholds, Θ, which gives the highest value of a user-defined criterion, *perf* (e.g., the F_1 metric):

$$\{\theta \in \Theta | \theta = \max(perf(L, \Theta))\} \quad (4)$$

To avoid overfitting issue, the output θ is an average of the thresholds obtained from different training subsets. Finally, the SVM function is corrected as below:

$$h^*(x_i) = h(x_i) - \theta \quad (5)$$

Fig. 1 shows how "shifting" the hyperplane's bias downward in the bottom graph corrects the way SVM labels the three positive examples misclassified by the original hyperplane in the bottom graph (note that the hyperplane's orientation is unchanged).

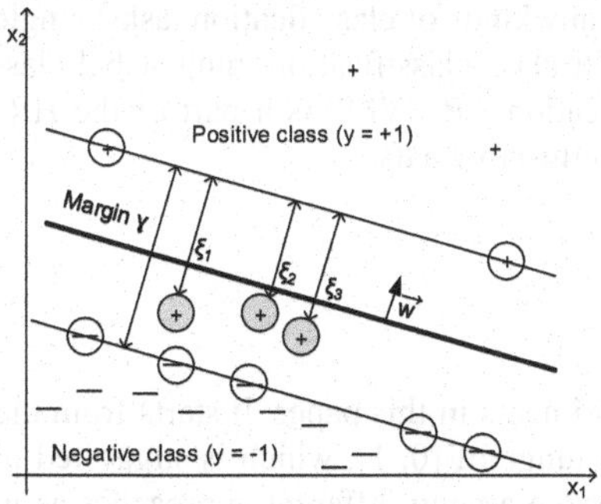

(a) SVM hyperplanes before threshold adjustment.

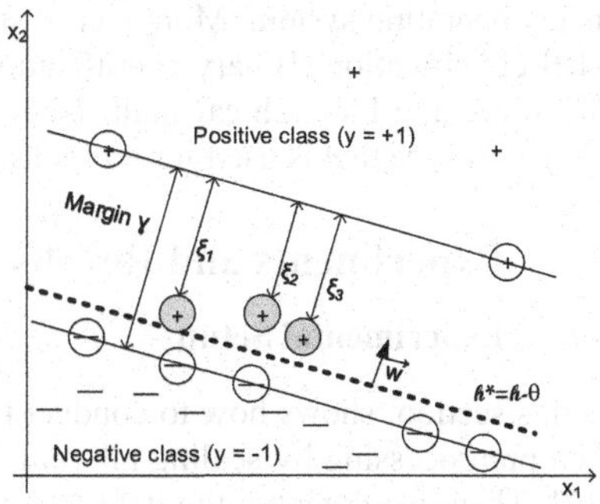

(b) SVM hyperplanes after threshold adjustment.

Fig. 1. SVM hyperplanes before (a) and after (b) threshold adjustment. The classification of three examples is corrected.

However, the enhanced hyperplane can reach an overfitting issue since the adjustment is based on just a single training data set. To increase a generalizability of the model, the resampling concept is applied to R-SVM (R stands for "resampling"). Moreover, the under-sampling concept is also employed in order to speed up the adjustment process.

4 Data Set and Tools

4.1 Data Set

The benchmark used in this experiment is the NASA Metric Data Program (MDP) [7] that comes from the NASA IV&V Facility MDP Repository, which contains a series of real software defect data from NASA spacecraft software. In this benchmark, there are 12 projects (data sets). The defect statistics is shown in Table 4. From the statistics, it has shown that the MDP data set suffers from the imbalanced issue. An average percentage of the defects is 20%, and the minimum percentage is only 2.15% in the PC2 data sets.

Table 4. Defect Statistics for Each Data Set.

Name	CM1	JM1	KC1	KC3	MC1	MC2	MW1	PC1	PC2	PC3	PC4	PC5	Avg.
#Samples	327	7782	1183	194	1988	125	253	705	745	1077	1287	1711	1448
#Attributes	38	22	22	40	39	40	38	38	37	38	38	39	36
Defect Class	42	1672	314	314	46	44	27	61	16	134	177	471	277
Non-Defect Class	285	6110	869	158	1942	81	226	644	729	943	1110	1240	1195
%Defect	12.84	21.49	26.54	66.53	2.31	35.20	10.67	8.65	2.15	12.44	13.75	27.53	20

4.2 Tools

WEKA [15] is a popular machine learning software for data mining tasks. It is a product of University of Waikato in New Zealand and was first implemented in 1997. It supports several data mining process, such as preprocessing, regression, classification, and so on. All comparison methods in this experiment are carried out in Weka.

HR-SVM [16] is our SVM software that handles the imbalanced issue by using the threshold adjustment strategy. It is built on top of LIBSVM and HEMKit, and can run in any operating system. Moreover, it supports any kinds of classification tasks: single label classification (Binary classification), multi-class classification, multi-label classification, and hierarchical multi-label classification. "R-SVM" is a part of the HR-SVM software that is used for multiclass and multi-label data.

5 Experiments and Results

5.1 Experimental Setup

In this section, shows how to conduct the experiments in this paper. It starts from the data preprocessing by scaling all values into a range of [0, 1], which is suggested in [17]. Then, we compare the detection performance among different approaches as in the following steps. Note that all experiments are based on 10 fold-cross validation and measured using PD, PF, F1, and G-mean.

- Step1: find the baseline method which is the winner of the famous traditional classifiers: Naïve Bayes (NB), Decision Tree, k-NN, and SVM.
- Step2: find the best setting for R-SVM whether or not the feature selection is necessary.
- Step3: compare R-SVM (Step2) to the baseline method (Step1) along with a significance test using unpaired t-test at a confidence level of 95%

5.2 Experimental Results

The comparison of the conventional methods. In order to get the baseline for each data set, five classifiers: Naïve Bayes (NB), Decision Tree (DT), k-NN, SVM (Linear), and SVM (RBF), were tested and compared in terms of PD, PF, F1, and G-Mean (Table 5 – 8). For each row in the tables, the boldface method is a winner on that data set. From the result, k-NN and NB showed the best performance in almost all data sets, while the standard SVM gave the worst performance since it cannot detect any errors resulting 0% detection in many data sets. For Table 7 – 8, it is interesting that F1 and G-mean unanimously showed the same winners. Since F1 and G-mean are suitable metrics for imbalanced data, we selected the winner as a baseline using F1 and G-mean as summarized in Table 9.

Table 5. Prediction performance: PD

Name	Prediction model			
	DT	*k-NN*	*NB*	*SVM*
CM1	0.190	0.143	**0.262**	0.000
JM1	0.228	**0.333**	0.189	0.090
KC1	**0.354**	0.417	0.322	0.048
KC3	0.333	0.139	**0.389**	0.028
MC1	0.065	0.304	**0.311**	0.000
MC2	0.523	**0.545**	0.386	0.364
MW1	0.148	**0.259**	0.000	0.037
PC1	0.213	**0.426**	0.350	0.000
PC2	0.000	0.000	**0.125**	0.000
PC3	0.261	0.336	**0.903**	0.000
PC4	**0.525**	0.492	0.375	0.198
PC5	0.482	**0.495**	0.206	0.161
Avg.	0.277	0.324	0.318	0.077
SD	0.172	0.166	0.219	0.112

Table 6. Prediction performance: PF

Name	Prediction model			
	DT	*k-NN*	*NB*	*SVM*
CM1	**0.091**	0.147	0.130	0.000
JM1	0.091	0.176	0.056	**0.020**
KC1	0.113	0.181	0.124	**0.028**
KC3	**0.101**	0.146	0.114	0.000
MC1	**0.005**	0.011	0.088	0.000
MC2	0.185	0.198	**0.086**	0.111
MW1	0.044	0.115	0.000	**0.009**
PC1	0.032	0.053	0.071	**0.002**
PC2	**0.007**	0.021	0.080	0.000
PC3	**0.066**	0.077	0.744	0.000
PC4	0.059	0.085	0.048	**0.004**
PC5	0.143	0.187	0.053	**0.033**
Avg.	0.078	0.116	0.133	0.017
SD	0.054	0.066	0.196	0.032

Table 7. Prediction performance: F1

Name	Prediction model			
	DT	*k-NN*	*NB*	*SVM*
CM1	0.211	0.133	**0.244**	0.000
JM1	0.292	**0.337**	0.271	0.155
KC1	0.424	**0.435**	0.386	0.289
KC3	0.375	0.156	**0.412**	0.054
MC1	0.102	**0.341**	0.122	0.000
MC2	0.561	**0.571**	0.500	0.478
MW1	0.195	**0.233**	0.000	0.067
PC1	0.271	**0.430**	0.331	0.000
PC2	0.000	0.000	**0.053**	0.000
PC3	0.303	**0.357**	0.254	0.000
PC4	**0.554**	0.486	0.447	0.324
PC5	**0.519**	0.498	0.307	0.259
Avg.	0.317	0.332	0.277	0.136
SD	0.178	0.171	0.155	0.164

Table 8. Prediction performance: G-mean

Name	Prediction model			
	DT	*k-NN*	*NB*	*SVM*
CM1	0.416	0.349	**0.477**	0.000
JM1	0.455	**0.524**	0.422	0.298
KC1	0.560	**0.585**	0.531	0.420
KC3	0.547	0.344	**0.587**	0.167
MC1	0.255	**0.549**	0.533	0.000
MC2	0.653	**0.662**	0.594	0.576
MW1	0.376	**0.479**	0.000	0.192
PC1	0.454	**0.635**	0.570	0.000
PC2	0.000	0.000	**0.339**	0.000
PC3	0.494	**0.557**	0.481	0.000
PC4	**0.703**	0.671	0.598	0.444
PC5	**0.643**	0.634	0.442	0.395
Avg.	0.416	0.499	0.465	0.208
SD	0.455	0.191	0.166	0.213

The comparison of R-SVM with and without feature selection. In this section, we intend to provide the best setting for R-SVM by testing whether or not the feature selection can improve the prediction performance. The F1-results in Table 10 illustrate that the feature selection should not be employed into the system since it tremendously decreased the performance. This should be because the number of attributes in the data sets is already small, so it cannot be further reduced.

Table 9. The winner of the baseline method for each data set in terms of F1 and Gmean.

Name	Winner	F1	G-mean
CM1	NB	0.244	0.477
JM1	k-NN	0.337	0.524
KC1	k-NN	0.435	0.585
KC3	NB	0.412	0.587
MC1	k-NN	0.341	0.549
MC2	k-NN	0.571	0.662
MW1	k-NN	0.233	0.479
PC1	k-NN	0.430	0.635
PC2	NB	0.053	0.339
PC3	k-NN	0.357	0.557
PC4	DT	0.554	0.703
PC5	DT	0.519	0.643
Avg.	-	0.374	0.562
SD	-	0.149	0.099

Table 10. A comparison of R-SVM between With and Without Feature Selection.

Name	F1 of R-SVM	
	With	*Without*
CM1	0.000	0.354**
JM1	0.008	0.411**
KC1	0.847	0.853
KC3	0.900	0.891
MC1	0.000	0.126*
MC2	0.333	0.574*
MW1	0.000	0.456**
PC1	0.000	0.392**
PC2	0.000	0.095*
PC3	0.000	0.384**
PC4	0.155	0.573**
PC5	0.845	0.846
Avg.	0.257	0.496
SD	0.379	0.264

Table 11. Comparison prediction performance measures between R-SVM and the baseline method. the boldface method is a winner on that dataset.

Name	PD		PF		F1		G-mean	
	Baseline	R-SVM	Baseline	R-SVM	Baseline	R-SVM	Baseline	R-SVM
CM1	0.262	**0.405**	**0.091****	0.130	0.244	**0.354**	0.477	**0.593****
JM1	0.333	**0.587****	**0.020****	0.301	0.337	**0.411****	0.524	**0.616****
KC1	0.354	**0.952****	**0.028****	0.777	0.435	**0.853****	**0.585****	0.461
KC3	0.389	**0.956****	**0.101****	0.722	0.412	**0.891****	**0.587**	0.508
MC1	**0.311**	0.174	**0.005****	0.030	**0.341***	0.126	**0.549****	0.409
MC2	0.545	**0.591**	**0.086****	0.346	0.571	**0.574**	**0.662**	0.657
MW1	0.259	**0.481**	**0.009****	0.075	0.233	**0.456***	0.479	**0.667****
PC1	0.426	**0.492**	**0.002****	0.096	**0.430**	0.392	0.635	**0.667**
PC2	0.125	0.125	**0.007****	0.033	0.053	**0.095**	0.339	**0.348**
PC3	**0.903****	0.463	**0.066****	0.135	0.357	**0.384**	0.557	**0.633****
PC4	0.525	**0.729****	**0.004****	0.130	0.554	**0.573**	0.703	**0.796****
PC5	0.495	**0.950****	**0.033****	0.764	0.519	**0.846**	**0.643****	0.458
Avg.	0.411	0.575	0.038	0.295	0.374	0.496	0.562	0.568
SD	0.197	0.282	0.038	0.293	0.149	0.264	0.099	0.131

* and ** represent a significant difference at a confidence level of 95% and 99%, respectively.

The comparison of R-SVM and the baseline methods. In this section, we compare R-SVM to the baseline methods, which are obtain from the first experiment as shown in Table 5-8. In Table 11 shows a comparison in terms of PD, PF, F1, and G-mean. All of the metrics give the same conclusion that R-SVM outperforms the baseline methods in almost all of the data sets. From 12 data sets, R-SVM *significantly* won 5, 3, and 5 on PD, F1, and G-mean, respectively. On average, F1-result of R-SVM outperforms that of the baselines for 32.62%, especially for the KC3 data set showing 116.26% improvement. Hence, this illustrates that it is effective to apply R-SVM as a core mechanism to early detect erroneous software modules.

6 Conclusion

Early defect detection is an important activity in the software development. Unfortunately, most of the prior works discarded the imbalanced issue, which is commonly found in the field of defect prediction, and it has known to severely affect the prediction accuracy. To tackle this issue, we proposed to employ our version of SVM called "R-SVM," which reduces a bias of the majority class by using the concept of threshold adjustment. The NASA Metric Data Program (MDP) was selected as our benchmark. It comprises of 12 projects (data sets). In the experiment, we compared R-SVM to five traditional classification techniques: Naïve Bayes, Decision Tree, k-NN, SVM (Linear), and SVM (RBF). The results showed that R-SVM overcame the imbalanced issue and significantly surpassed those classifiers.

References

[1] Menzies, T., Greenwald, J., Frank, A.: Data Mining Static Code Attributes to Learn Defect Predictors. In: IEEE Transactions on SE, vol. 33(1), pp. 2-13 (2007)

[2] Bo, S., Haifeng, L., Mengjun, L., Quan, Z., Chaojing, T.: Software Defect Prediction Using Dynamic Support Vector Machine. In: 9th International Conference on Computational Intelligence and Security (CIS), 2013, pp. 260-263. China (2013)

[3] Seliya, N., Khoshgoftaar, T.M., Van Hulse, J.: Predicting Faults in High Assurance Software. In: 2010 IEEE 12th International Symposium on High-Assurance Systems Engineering (HASE), pp. 26-34. San Jose, CA(2010)

[4] Shuo, W., Xin, Y.: Using Class Imbalance Learning for Software Defect Prediction. In: IEEE Transactions on Reliability, vol. 62(2), pp. 434-443 (2013)

[5] Gray, D., Bowes, D., Davey, N., Sun, Y., Christianson, B.: Software defect prediction using static code metrics underestimates defect-proneness. In: The 2010 International Joint Conference on Neural Networks (IJCNN), pp. 1-7. Barcelona (2010)

[6] Elish, K.O., Elish, M.O.: Predicting defect-prone software modules using support vector machines. In: Journal of System Software. vol. 81(5), pp. 649-660 (2008)

[7] NASA IV & V Facility. Metric Data Program, http://MDP.ivv.nasa.org/.

[8] Vateekul, P., Dendamrongvit, S., Kubat, M.: Improving SVM Performance in Multi-Label Domains: Threshold Adjustment. International Journal on Artificial Intelligence Tools (2013)

[9] McCabe, T.J.: A Complexity Measure. Software Engineering, In: IEEE Transactions on SE, vol. 2(4), pp. 308-320 (1976)

[10] Halstead, M.H.: Elements of Software Science. Elsevier Science Inc., (1977)

[11] Chidamber, S. R., Kemerer, C. F.: A metrics suit for object oriented design. In: IEEE Transactions on SE, vol. 20, pp. 476-493 (1994)

[12] Kubat, M., Matwin, S.: Addressing the curse of imbalanced training seta: One-sided selec-tion, pp. 179-186, 1997

[13] Han, J., Kamber, M.,: Data Mining: Concepts and Techniques, 2 ed., s.l.: Morgan Kaufmann, 2006.

[14] Cortes, C., Vapnik,V.: Support-Vector Networks. Machine Learning, pp.273-297(1995)

[15] WEKA, http://www.cs.waikato.ac.th.nz/ml/weka.

[16] Vateekul, P., Kubat, M., Sarinnapakorn, K.: Top-down optimized SVMs for hierarchical multi-label classification: A case study in gene function prediction. Intelligent Data Analysis (in press)

[17] Hsu, C.W., Chang, C.C., Lin, C.J.: A practical guide to support vector classification. Department of Computer Science and Information Engineering, National Taiwan University, (2003)

Evaluation of Separated Concerns in Web-based Delivery of User Interfaces

Tomas Cerny[1], Lubos Matl[1], Karel Cemus[1], and Michael J. Donahoo[2]

[1] Computer Science, FEE, Czech Technical University,
Charles Square 13, 12135 Prague 2, Czech Republic,
{tomas.cerny, matllubo, cemuskar}@fel.cvut.cz
[2] Computer Science, Baylor University, Waco, TX, 76798, US,
jeff_donahoo@baylor.edu

Abstract. User Interfaces (UI) play a significant role in contemporary web applications. Responsiveness and performance are influenced by the UI design, complexity of its features, the amount of transmitted information, as well as by network conditions. While traditional web delivery approaches separate out presentation of UI in the form of Cascading Style Sheets (CSS), a large number of presentation concerns are left tangled together in the structural description used for data presentations. Such tangling impedes concern reuse, which impacts the description size as well as caching options. This paper evaluates separation of UI concerns from the perspective of UI delivery. Concerns are distributed to clients through various resources/channels, which impacts the UI composition at the client-side. This decreases the volume of transmitted information and extends caching options. The efficacy is demonstrated through experiments.

Keywords: Separation of concerns, user interface, networking

1 Introduction

A User Interface (UI) defines the visual part of a computer application. Its design must not only consider usability and development efforts, but the UI must be performant and highly responsive to address growing demands in Rich Internet Applications (RIAs). The responsiveness to user requests has many influencing factors. Besides the UI design, it is influenced by page feature complexity, page size, and mostly by network conditions and HTTP delivery.

Traditional UI design approaches describe a particular UI page combining various concerns [1, 5] together. For instance, data presentation descriptions consider concerns, such as data structure, individual field presentation, user input validation, field layout, data values binding, security, etc. Although single-location, multi-concerns descriptions are easy to read to see the global picture, such an approach brings multiple disadvantages. When a single concern changes, other "tangled" concerns [1] distract designers from the modification since no explicit boundary exists in the description. The tangling further limits concern

K.J. Kim (ed.), *Information Science and Applications*,
Lecture Notes in Electrical Engineering 339, DOI 10.1007/978-3-662-46578-3_111

reuse [5]. Consider a situation when a page varies based on context. Certain variations can be solved through page conditionals, although many of the variations require designing a new, similar page. For instance a novel page description might be needed when field presentation changes with the context, the user input validation is user-sensitive, layout changes with the screen resolution, etc.

Separating UI concerns make it possible to describe each concern individually and reuse them across different data types presented in the UI. Such reuse could reduce the size of UI descriptions. Consequently, from the perspective of web delivery, providing clients UI concerns separately supports reuse and thus reduces the amount of transmitted information. Furthermore, since each concern is provided to clients separately, the client may cache a given concern for a particular time span, which is not possible in the conventional approach that delivers tangled concerns to clients.

This paper evaluates the idea of separated concern delivery in UIs for data presentations in Section 2. UI concerns are provided through multiple channels and combined at the client-side. The impact on transmission size, UI responsiveness and performance is considered and compared to traditional UI delivery in Section 3. Next, we evaluate extended UI caching options from the perspective of client-side caching and content-delivery networks (CDN). We also evaluate the impact on the server-side considering the server CPU usage. Related works are presented in Section 4. Finally, our conclusion is given in Section 5.

2 Related Work

Many approaches [1] consider Model-Driven Development (MDD) [3] with the expectation that a model captures all sorts of information at a single location. In MDD the UI is generated from the model [1]. It reduces design efforts for usual UIs, but it has limitations. When dealing with context-aware UIs, the runtime generation might be performance inefficient. MDD does not effectively address cross-cutting concerns [7] since no generalized mechanism to address multi-model integration exists [2]. Similar to MDD, Domain-specific languages (DSLs) describe UIs [1], although they fail to address cross-cutting concerns [1].

Approaches addressing cross-cutting concerns are Generative Programming (GP) [7] and Aspect-Oriented Programming (AOP) [5]. GP suggests describing concerns separately in a DSL and combining them together through a configuration to offer multiple result variants. Its operates at compile-time, which might be inefficient for context-aware UIs. The AOP operates at runtime. An example AOP adaption to UI is given in [1]. The target UI presentation is described through templates; concerns integrate through an AOP-based transformation of the audited data considering the context.

The Google Web Toolkit (GWT)[3] partially addresses UI responsiveness. It separates the UI presentation and data. It is similar to our approach, but with limited separation of concerns. It uses Java for the UI description, compiling into JS. The resulting presentation has cacheable and uncacheable JS parts and calls

[3] GWT - http://gwtproject.org, AngularJS - http://angularjs.org; November 2014

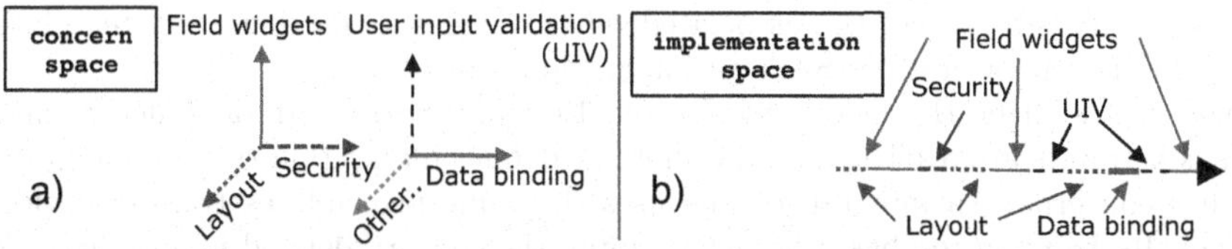

Fig. 1. Demonstration of UI cross-cutting concerns collapsing into a single dimension

Web-resource for data. The UI delivery is similar to conventional approaches. Similar to MDD, GWT faces issues with cross-cutting concerns and restated information. For instance, all the UI states are combined together at compile-time [2], which may result in a bloated UI description size.

AngularJS web framework[4], similarly to GWT, suggests data separation into another channel, although the presentation is left tangled together. In addition to GWT, it provides templating mechanism for content resolution, but it does not separate out the structure, context, or the presentation and layout templates.

3 Separation of Concerns in UI Data Delivery

Conventional programming languages, including object-oriented, provide various decomposition mechanisms. Such decomposition aims to separate concerns - information that impact the code of a particular program - into logical sections. Based on Kiczales et. al. [5], these languages lack the ability to effectively address cross-cutting concerns, which are aspects of a program that affect other concerns. One of the approaches that effectively address cross-cutting concerns is Aspect-Oriented Programming (AOP).

Recent work on separation of concerns in UI [1] considers the AOP-based UI design (AUI). The goal of such an approach is to reduce UI development and maintenance efforts. Since AOP allows describing various concerns separately, it supports concern reuse across various data type presentations. The demonstration in Fig. 1*a*) shows possible cross-cutting concerns in the UI. Each of these concerns impacts the UI data presentation and can be reasoned independently, in separate dimensions. When capturing these concerns in the implementation space, they collapse and tangle together into a single dimension [6], Fig. 1*b*).

The AUI design considers the application data model to be the main source of *join points* [8] that influence data presentation and UI concern integration [3] showing that it is possible to derive from such *join points* a variety of data presentations, and it is not specific to a particular platform. The data model is the subject of code-inspection [1, 4] for this reason. The obtained *join point* structure represents structural properties of system data, structured by fields. The *join point structure* can further be extended by runtime context (e.g., logged user, geo-location, rights, etc.).

To derive the UI for data elements, the *join point* structure is queried for its properties. A set of rules that perform the selection of integration templates uses an AOP-based mechanism consisting of *pointcuts* and *advices*. The *pointcut* specifies a query to *join points* of a particular field and context. When a *pointcut* finds a match, the associated *advice* suggests an integration template for the field presentation. Such an integration template uses the target UI language to describe the field presentation. References to the data structure and its constraints are resolved. Furthermore, it defines integration rules to integrate other concerns. An integration rule uses the same principle as transformation rules. Its *pointcut* resolves whether to apply the concern defined in an *advice*. A target layout is integrated through a layout template.

Through the separation of concerns, the AUI supports reuse. [1] shows reduction of 32% of the UI code in a large application. The UI code reductions are considerable from the development perspective, although the UI description that is delivered to clients is being generated and is equivalent to the conventional approaches. Although AUI supports concern reuse, it does not take into account the perspective of UI delivery. The separation of concerns collapses into a single, tangled description before it is delivered to the client-side, and thus the separation becomes lost.

An extension to the AUI approach that maintains the concern separation at the client-side is suggested by [2]. The perspective of UI delivery may be considered independent of AUI. The main difference for the web-delivery is that various concerns are delivered through multiple resources/channels.

The client requests an HTML page that should display particular data, e.g., multiple forms. The delivered page HTML description consists of page elements, but the forms are not described. Instead instructions on how to assemble and embed them are provided, and the required concerns are requested from other channels. The HTML description is accompanied with a JS library that contains the weaver. It takes the instructions and requests other concerns for the particular data elements to derive the aimed presentation for particular context. The design sketch for the description is provided by Fig. 2.

The particular application data definition from the data model determines the presentation structure as well as its constraints and validation rules [3]. This defines the structural concern, although the information must be provided to clients in a readable format. To avoid manual restatement of the structure information in a particular output format, the content can be derived through code-inspection [4, 3].

Besides the data structure, the application runtime context should be taken into account. For instance, guest users can see less fields that registered users, the input validation differ in given contexts; field presentation may be influenced by user location; etc. The application context may restrict existing structure or override its properties using the Annotation Driver Participant Pattern [6]. The client receives the Context-aware Structural Information (CSI) from a server in a machine readable-format (JSON, XML).

Different data and contexts give different CSIs, although for given context and data type the CSI does not change. The CSI determines the structure of the data presentation. It is structured by data fields and for each field gives detailed information of its properties, constraints, validation rules, etc. The CSI is used by the weaver to determine presentation for each particular field.

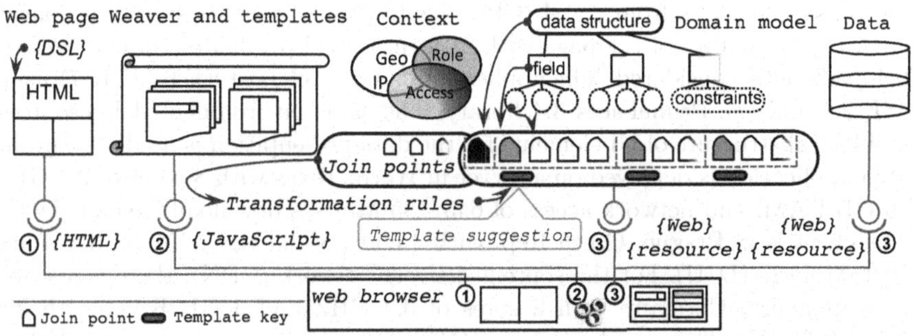

Fig. 2. Architecture of the concern-separating approach of UI data delivery. The web browser requests an HTML page (1) that is supplemented with a JavaScript library (2) responsible for the data presentation composition. The HTML page indicates data to show and the library (3) requests the CSI and data values and produces its presentation.

Similar to AUI integration templates, a set (or sets) of templates is delivered through to the clients as a JS library with templates. Template content can be resolved in the context of CSI, although the question to answer is how to select a particular template for given field. AUI [1] suggest defining small a number of generic rules that perform the selection. These rules could be part of the client-side weaver and work directly with the CSI, although it would raise its complexity at the client-side and at the same time the rule could not be based on information that was not provided from the server in the CSI. Instead the template selection executes at the server-side, and the suggested template identifier accompanies the field in CSI.

Layouts are defined through templates and delivered as a JS library. The weaver processes the CSI to determine the presentation structure and the content of field templates. Next, the weaver determines layout template and decorates the result. The data values are requested in parallel to CSI, integrated to the assembled component, and embedded at the instructed location.

The separated concerns are maintained at the client-side, and all JS sources can be cached. The weaver can cache in HTML5 LocalStorage the CSI for a particular time span and context. The weaver can further consider local context. When multiple sets of templates are provided, switching the page from read/write modes, resulting in changes to layout or presentation template set, does not require reloading other concerns than data values. Requesting the same page for a different data instance in the same context requires only loading new data values, although this is influenced by the context-awareness. The data and CSI can be even machine processed and reused by native clients, such as Android.

4 Experiments

Out experiment evaluates the Concern-Separating Approach (CSA) from the perspective of page load time, specifically caching and volume of transmitted information related to data presentations. A fragment of an existing production-level application using the conventional approach is compared with the CSA. The person profile pages from ACM-ICPC[4] contest managements system is considered. First, a subset of the page with 21 form fields is evaluated; next a 42-form field version is considered. The application is built on Java EE 6 (JDK 7) using JSF 2.1 and the PrimeFaces 3.4 library. The same environment, UI resources, and logical structure of UI elements for the data presentation is used. The evaluated application is deployed on a server in Waco, Texas with 8 cores of 2.4 GHz, 16 GB RAM, and network access of 645/185 Mbits/s download/upload (D/U). The client is in Prague, Czech Rep. with 4 cores of 2.3 GHz, 16 GB RAM and 10/6 Mbits/s (D/U). For the content-delivery network (CDN), there is a server in Nuremberg, Germany with 2 cores of 3.4 GHZ and 3 GB RAM with 200 Mbits/s (D/U). The round-trip time (RTT) between client and server is 150 ms and client and CDN is 20 ms. 50 measurements are made, and the average with the standard deviation are provided in Tables 1-3.

The first evaluation with the shorter conventional page (Table 1, Row *a*) has a main document size of 74.4 KB, which compresses to 9.2 KB through gzip. In total the page calls 10 requests (including static resources), the transmission has 218 KB, and no caching is involved. Tables 1-3 show load times for Chrome$^{37.0.2062.122}$/Firefox$^{32.0.2}$/Opera$^{24.0.1558.53}$ web browsers (including JS processing). The CSA (Table 1, Row *b*) makes 15 requests. The main document size reduces to 3.3 KB (1.3 KB compressed), although it additionally loads a JS library (3.3 KB compressed), and data (1 KB compressed) with CSI (4 KB compressed) from web-resource. The transmission has 218 KB. Notice that additional resources load in parallel. The processed size of UI is considerably smaller (Table 1, last column). Row *c* gives the percentage improvement of page load times at around 10%. The data presentation description volume reduces by 61%.

The longer, 42-field conventional page has a main document size of 98.9 KB (compressed 11.1 KB), as shown in Table 1 *d*. The CSA in Table 1 *e* requires 23 requests since it requests information from 6 data instances. The main document size is 5.3 KB (compressed 1.6 KB), the CSI 6.5 KB, and data 2.7 KB. Row *f* shows around 10% improvement in page load time. The data presentation information volume improves by 62%.

The Table 1 *a*-*f* shows improvement in the page load times and reduction in the processed UI description volume. The compressed transmission is equivalent, although both sides work with the original size. The number of requests in CSA grows, because multiple data elements are involved for CSI and value requests.

When we consider browser cache, the load times drop even further. The CSA has the advantage of concern separation, making it possible to cache the templates and CSI. The conventional approach delivers the "tangled" main document, which is 9.2/11.1 KB for the shorter/longer version (Table 2, Rows *a*,*d*). The volume of data presentation description is the same as the uncached version.

The cached CSA transmits the main document and data values. It gives 2.4/4.2 KB in 3/7 requests for the shorter/longer version (Table 2 *b*,*e*). There is a 15-20% improvement in page load time. The RTT is the dominant factor. The UI data presentation volume improves by 93% and 62-74% in the transmission.

[4] ACM-ICPC Contest management system, http://icpc.baylor.edu, December 2014

Table 1. Page load measurements

Row	Requested page	Cache	$Chrome_{\sigma}$ [ms]	$Firefox_{\sigma}$ [ms]	$Opera_{\sigma}$ [ms]	Requests	Transmission compressed	Processed UI size
a	Shorter conventional	No-cache	1637_{199}	1573_{420}	1540_{200}	10	218 KB	74.4 KB
b	Shorter CSA		1419_{114}	1417_{187}	1402_{60}	15	218 KB	29 KB
c	% change		13%	10%	9%	33%	0%	61%
d	Longer conventional	No-cache	1691_{87}	1992_{367}	1669_{195}	10	220 KB	98.9 KB
e	Longer CSA		1560_{92}	1772_{89}	1500_{90}	23	223 KB	37.1 KB
f	% change		8%	11%	10%	57%	1%	62%

Table 2. Page load measurements

Row	Requested page	Cache	$Chrome_{\sigma}$ [ms]	$Firefox_{\sigma}$ [ms]	$Opera_{\sigma}$ [ms]	Requests	Transmission compressed	Processed UI size
a	Shorter conventional	Cache	573_{21}	659_{49}	517_{12}	1	9.2 KB	74.4 KB
b	Shorter CSA		456_{29}	552_{92}	446_{28}	3	2.4 KB	4.3 KB
c	% change		20%	16%	14%	67%	74%	94%
d	Longer conventional	Cache	657_{21}	858_{105}	607_{49}	1	11.1 KB	98.9 KB
e	Longer CSA		526_{39}	593_{84}	519_{48}	7	4.2 KB	7.2 KB
f	% change		20%	31%	15%	86%	62%	93%

Table 3. Simulation download evaluation

Row	Cache	$\text{Page load}_{\sigma}$ [ms]					
		Shorter conv.	Shorter CSA	% change	Longer conv.	Longer CSA	% change
a	No-cache	1352_{66}	1182_{101}	13%	1379_{135}	$1213ms_{92}$	12%
b	Cache	312_{15}	353_{15}	12%	381_{31}	404_{63}	6%
c	CDN	631_{28}	511_{22}	19%	771_{26}	600_{17}	22%

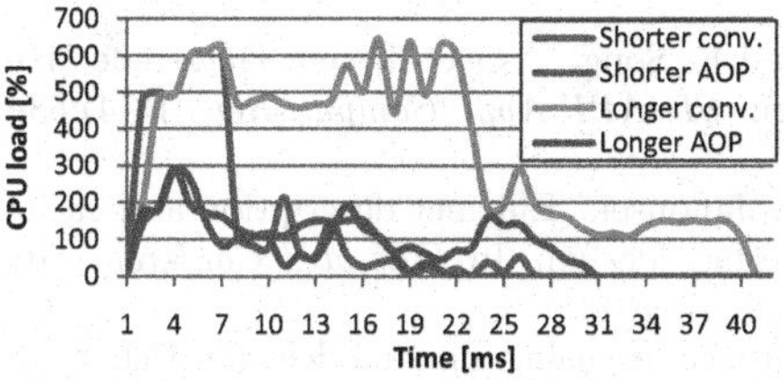

Fig. 3. Server stress-test CPU (no cache)

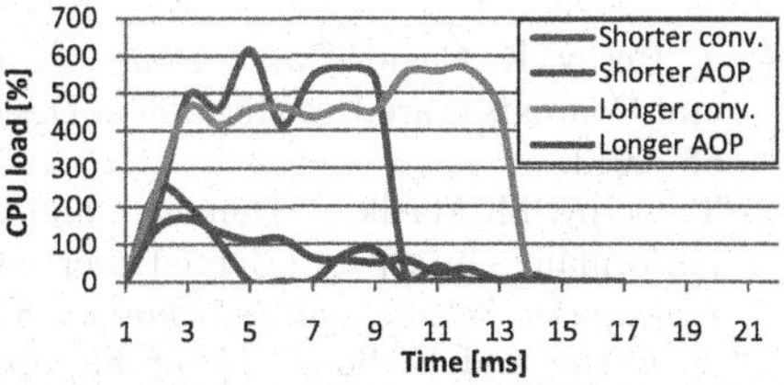

Fig. 4. Server stress-test CPU (cache)

A simulation involving browser request traces considers the network impact. This does not consider UI construction nor resource decompression. Results in Table 3 show simulation with no cache (*a*), cache (*b*) and CDN (*c*). Although *a* and *c* show improvements comparable with web browser results, *b* is worse. The explanation is the dominant RTT, which in the conventional approach occurs once but in CSA multiple times and multiple data element requests are made. Notice this experiment only involves network communication.

The server-side impact evaluation uses also the browser request traces simulation and stresses the server with 100 clients loading a particular page simultaneously. The server CPU load is measured for all four cases, not cached and cached. An Unix tool "sysstat" samples the CPU every second during the stress test. The results in Fig. 3 and 4 show that CPU load is higher and longer for the conventional approach, which match with the higher information volume transmission for the approach and delegated UI composition to clients.

5 Conclusion

This paper provides evaluation of separated concern, web-based delivery in UIs. Data presentations are provided to clients through multiple resources/channels. This approach brings clients the ability to reuse particular concerns and thus reduce the amount of delivered information. Clients gain the ability to cache given concerns for a certain amount of time and request only the concern(s) that changes. The server-side has to process less data, and its concern weaving responsibility is delegated to clients. The evaluation results confirm the expectations based on the approach.

The limitations for our approach include that it does not address page-flow, applied solely to data presentations, and thus it must build on the top of another UI framework. To improve the resource delivery, it is possible to aggregate CSI-related resources into a single request for multiple data elements and deliver it at once. Separation of platform-specific and platform independent channels is considered for future work.

6 Acknowledgments

This work was supported by the Grant Agency of the Czech Technical University in Prague, grant No. SGS14/198/OHK3/3T/13

References

1. T. Cerny, K. Cemus, M. J. Donahoo, and E. Song. Aspect-driven, Data-reflective and Context-aware User Interfaces Design. *SIGAPP Appl. Comput. Rev.*, 13(4):53–65, 2013.
2. T. Cerny, M. Macik, J. Donahoo, and J. Janousek. Efficient description and cache performance in aspect-oriented user interface design. In *Federated Conference on Coumputer Science and Information Systems*, 2014.
3. T. Cerny and E. Song. A profile approach to using uml models for rich form generation. In *Information Science and Applications (ICISA), 2010 International Conference on*, pages 1–8, 2010.
4. R. Kennard, E. Edmonds, and J. Leaney. Separation anxiety: stresses of developing a modern day separable user interface. *Proceedings of the 2nd conference on Human System Interactions*, pages 225–232, 2009.
5. G. Kiczales, J. Irwin, J. Lamping, J.-M. Loingtier, C. V. Lopes, C. Maeda, and A. Mendhekar. Aspect-oriented programming. In *IECOOP'97-Object-Oriented Programming, 11th European Conference*, volume 1241, pages 220–242, 1997.
6. R. Laddad. *AspectJ in Action: Enterprise AOP with Spring Applications.* Manning Publications Co., Greenwich, CT, USA, 2nd edition, 2009.
7. M. Schlee and J. Vanderdonckt. Generative programming of graphical user interfaces. In *Proceedings of the working conference on Advanced visual interfaces*, AVI '04, pages 403–406, New York, NY, USA, 2004. ACM.
8. M. Stoerzer and S. Hanenberg. A Classification of Pointcut Language Constructs. In *SPLAT'05: Workshop on Software-Engineering Properties of Languages and Aspect Technologies*, 2005.

Separating out Platform-independent Particles of User Interfaces

Tomas Cerny[1] and Michael J. Donahoo[2]

[1] Computer Science, FEE, Czech Technical University,
Charles Square 13, 12135 Prague 2, Czech Rep.,
tomas.cerny@fel.cvut.cz

[2] Computer Science, Baylor University, Waco, TX, 76798, US,
jeff_donahoo@baylor.edu

Abstract. User Interfaces (UIs) visualize a wide range of various underlying computer application concerns. Such orthogonal concerns present in even the simplest UIs. The expectation of support for users from various backgrounds, location, different technical skills, etc. serves to increase concern complexity. Nowadays users typically remotely access to applications from a variety of platforms including web, mobile or even standalone clients. Providing platform-specific support for multiple UIs further increases the concern complexity. Such a wide-range of concerns often results in a significant portion of the UI description being restated using platform-specific components, which brings extended development, and maintenance efforts. This paper aims to separate out the platform-independent particles of UI that could be reused across various platforms. Such separation supports reduction of information restatement, development and maintenance effort. The platform-independent particles are provided in a machine-readable format to support their reuse in platform-specific UIs.

Keywords: Separation of concerns, user interface, platform-independence

1 Introduction

Conventional UI designs describe all the displayed information and concerns [8] tangled together [9]. Usually the description can be found at a single location or logically divided to fragments. Although such monolithic design may be easier for the developer in the initial design, it does not provide enough flexibility to support UI variations for a particular context or situation. The tax for the "all at single-location" description is that each UI variation is treated as a different UI, restating information.

For example, consider a UI page that presents application data, while considering multiple concerns [1] [8]. It presents data fields in a given order, following a particular layout, where each field has a specific label and a widget. The user input validation and constraints apply in widget configuration. Furthermore, security concerns enforce access control, and the description binds to a given data

K.J. Kim (ed.), *Information Science and Applications*,
Lecture Notes in Electrical Engineering 339, DOI 10.1007/978-3-662-46578-3_112

instance. Such a description is easy to read from the global perspective, although there is no explicit separation suggesting which part of the description deals with presentation, which part is dedicated to layout, and so on.

When the system context changes, the UI must reflect the change. Consider the case when the user's access role changes; some fields should disappear and certain validation criteria may drop or emerge. This situation may lead to additional conditionals applied to the UI description, increasing its complexity. Next, the user changes the screen size (extends window or rotates the screen). The application layout should extend, although layout is usually tangled with the rest of the page elements, and this may lead to the introduction of a new page, considering and restating the same data, constraints, validation rules, conditionals, etc. Furthermore, if the user decides to relocate from computer to cellphone to finish the work, the application may be accessed through web interfaces, or in order to support native UI widget features and offline mode, the application provides a native client for given mobile platform. Such a client application must restate all page concerns, data, conditionals, etc.

A concern separating UI design approach [1, 2] does not provide the ability to find all the UI-related information within a single description or location. Instead, it is divided into independent stripes. In order to build the whole picture, all stripes must be integrated. This seems worse for a simple UI, but the big advantage is the supported reuse, support for variability [9] and consequently the possibility to distribute the stripes separately within different delivery channels to clients. Furthermore, it is possible to divide these stripes onto platform-independent and platform-specific.

This paper addresses such difficulties when supporting various clients and particular contexts. It suggests separating out the particles of the UI that are platform-independent from those that are platform-specific. The independent part is provided in a standard, machine-readable format to support its reuse across different platforms, including web, standalone and mobile platforms. Client prototypes of such an UI approach are implemented and evaluated for three different platforms.

This paper is organized as follows. Section 2 provides related work. The separation of the platform-independent particles from the UI is described in Section 3. Three platform prototypes are described in Section 4. Finally, Section 5 presents our conclusions.

2 Related Work

Existing research ranges from Model-Driven Development (MDD) [3], concerns separation through Aspect-Oriented Programming (AOP) [1], Generative-Programming (GP) [4], proprietary formats or basing on Domain-Specific Languages (DSL) [9]. Existing prototypes such as the User Interlace Protocol (UIP) [9] suggest basing on abstract UI descriptions and adding the native-client specifics to the description. Finally, a few contemporary UI development frameworks such as Google Web Toolkit (GWT) [6] or AngularJS [5] suggest separating out data values to machine-readable formats.

The idea of MDD [3] suggests that the model is the central source of information, and the rest of the system is generated from the model. This usually works for application prototypes [1], although most production systems build on code-based approaches. MDD brings multiple benefits, such as supporting platform independence and transforming information captured at the model-level to multiple locations to reduce the information re-definitions and restatements. [3] aims to derive UIs from UML models and suggests industrial standards for persistence/constraints and input validation, as well as presentation specifics at the model-level. Such an approach allows deriving rich UIs for data presentations considering certain number of variations and conditionals; however, the approach is limited by the nature of MDD. It does not address cross-cutting concerns and does not apply to code-based design.

Separation of cross-cutting concerns is the domain of AOP [8] and GP [4]. These approaches suggest describing given concerns through independent components, often involving DSL descriptions. These concerns are woven to core components to extend their behavior. AOP has well-understood concern integration mechanisms and operates at runtime. An approach considering AOP for UI concern separation is provided by [1, 2]. Compared to MDD, it operates at the code-level and at runtime considering application context. Concerns such as presentation, layout, data binding, input validation, etc. are considered. The approach simplifies the design of context-aware applications [9].

A unified description of the UI is considered by UIP [9] in a DSL format called Abstract UI (AUI). This description is platform-independent and processed by a server-side application called the UIP server. The UIP server uses a core component called the Concrete UI (CUI) generator, which takes the AUI and context including the consideration of a target platform specified by the given client (C#, iOS, Web, etc.). Based on the selected target platform, the CUI generator interprets the AUI and context, and produces the platform-specific CUI that it streams to a particular client. These proprietary clients interpret the received CUI at the platform-specific environment using native components. The goal of UIP is to address context-aware UIs. From the perspective of [3], UIP reinvents the standards and from the perspective of [1], it uses custom DSL, which demands restating information from the application data model, increasing possible errors. A novel platform requires changes to the server-side, which does not naturally scale. Next, each data element must have its custom AUI, which must correlate with the actual application data structure. To avoid the necessity of manual definition of AUI, the approach of [1] can generate the AUI through code-inspection of the application data model.

MetaWidget [7], similar to the AOP approach [1], suggests applying code-inspection to data model, deriving various types of presentation. Unfortunately, this does not address cross-cutting concerns, changing context, and requires a one-to-one mapping preventing the use of templates.

From the perspective of development frameworks and technology, HTML5 and CSS3 suggest responsive web design (RWD). RWD allows the presentation to adjust to screen size and makes the UI presentation reflect the resolution. This can be adopted, for instance, for layouts. Notice that it considers only a subset of layouts. For example, it is non-trivial to make a custom order of fields displayed at the page, and it may require absolute positioning, which become impractical from the development and maintenance perspective.

GWT [6] is a web development framework that transforms the UI description to JavaScript representations and separates application data values from the presentation to a separate stream, requested through web resources. This makes the data values separable and easy to machine process.

AngularJS [5], which is similar to GWT, separates out data values through web-resources; next it defines a templating mechanism that allows decorating data. On the other hand the mechanism does not have natural support for recursive templating.

3 Separating out Platform-Independent UI Particles

In order to reduce restatements of recurring information across platform-specific UIs that present data, it is necessary to classify various types of information. Platform-independent, model-level description of UIs is researched in [3], and it suggests that the application data model is the main driver for data presentations, although basic data structural information is not sufficient. The suggestion from [3] is to extend data descriptions with various types of profiles. For instance, the data structural information is accompanied with constraints, input validation, and field semantics for the presentation or security. [1] then shows that such extensions are practical for use in code-based applications and can be derived by code-inspection. [1, 2] suggests to treat the information accompanying the data model not only as data structure, but as AOP join points that are used to determine a particular field presentation. In addition the application runtime context is considered together with the structural information, which together produce a join point model from which can be derived the context-aware UI. The advantage of [1] is that code-inspection applied to data models derive such information, avoiding manual work. [1] uses join points in a way that applies AOP-based transformation querying field join points and determines field presentation through templates. Next, it decorates the result through a layout, determined by a template.

The above approach can be considered from the perspective of platform-independence. As suggested in [2, 3], the join point model can be treated as platform-independent. The application context extends the information received from the data model and is not specific to a particular platform. Similar to GWT or AngularJS, data values are not specific to a particular UI platform.

On the other hand, presentation and layout templates use the target UI language components and thus contain platform-specific elements. The integration of templates with the join point model and data values must be considered by a processor that has knowledge of a particular platform and is platform-specific and located at the client-side.

In order to separate these particles, the platform-independent part should be provided in a platform-independent format. Web-resources provide standard formats that are understood across platforms, and thus they naturally fit the goal. Platform-specific clients then provide presentation and layout templates and the processor.

Providing the entire join point model to clients can be impractical, mostly when considering that it can consist of internal information or information not relevant to the UI. Thus some kind of filter should exist. At the same time, it might not be known ahead of time, which join points are used to determine the presentation at the client-side. When considering that all the different clients perform the same "join point model"-to-"presentation" transformation, it essentially becomes repetition. Thus the transformation can be partially performed at the server-side, considering the unfiltered join point model.

The question is how to connect the transformation with templates that are at the client-side in a platform-specific format. It is possible to perform the "join point model"-to-"presentation template" transformation in a way that it considers an expected set of templates with a unique identifier. The result of the transformation gives each data field a template identifier and accompanies it in the provided join point model. A filtered fraction of the join point model is provided through a Web-resource, separating out information not relevant to given user access role, context, or information not relevant to UI, such us data identifier, version lock, business key, etc.

Security is considered at the server-side for both the provided join point model and values for the requested data instance and in a given context (user, location, time-frame, etc.). Thus only data that would be part of a conventional secured-view are provided to users, but in a machine-readable format.

The client-side is responsible to provide the expected set of presentation templates by supplying platform-specific components. It determines the layout and implements the processor. The processor requests the data values and join point model, giving the data structure from the server-side. Next, it populates the local presentation templates with received information. Additionally it is possible for the client-side to use multiple sets of presentation templates that are changed based on local context (readable forms, editable form, data table, report, list, etc.). It is also possible to decorate the data presentation with wizard-like components that collapse the presentation into multiple panels.

4 Platform-Specific UI Clients

Implementation of a platform-specific client requires implementing a processor that requests and interprets the join point model and data values received from the server-side. The platform-specific sets of presentation templates are defined, reflecting the expected set of template identifiers. These templates consider native components that provide expected functionality, capture constraints, input

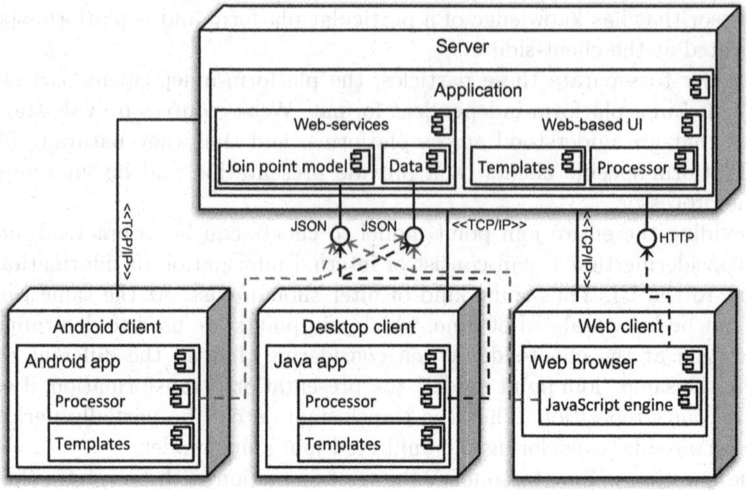

Fig. 1. Sample of deployment diagram considering three heterogeneous clients

validation, etc., given by the join point model. The advantages of platform-specific features include increasing UI usability (e.g., touch-based element selection). The appropriate presentation template is selected based on the suggested template identifier, provided by the join point model. Local context may influence the set selection. Templates for layouts are defined complying with the expected and supported screen-sizes. The processor further requests and embeds data values and has the ability to post them to the server-side.

Heterogeneous clients interpret the server-side, platform-independent information. No matter the particular processed data element, the same presentation templates apply for a particular platform. It is possible to consider generic layouts and reuse them across various data types. The size of the application data-model does not influence the size or complexity of clients and provided templates. The client application processor and templates are the same for an application providing a single data element or for an application with hundred of various data element types. This implies that changes to server-side data structures are automatically reflected by the processor at the client-side, since the provided join point model reflects it and existing templates are reused.

On the other hand if the system extends to provide a novel, unexpected constraint or data type, such a change needs to be reflected across all platform-specific clients. Even though such an extension is rare since the join point model [3] based on the industrially-standardized set of constraints and validations, it should be considered ahead of time in the design. There is no prevention mechanism that disallows clients to choose custom presentation templates and not to follow the suggestion given by the server-side.

Designing a platform-specific client becomes simplified since information is reused and restatements are reduced. For a demonstration, we implemented three clients of different platforms. The deployment diagram is shown in Fig. 1. The web-based client UI is shown in Fig. 2, an Android mobile client UI is in Fig. 3

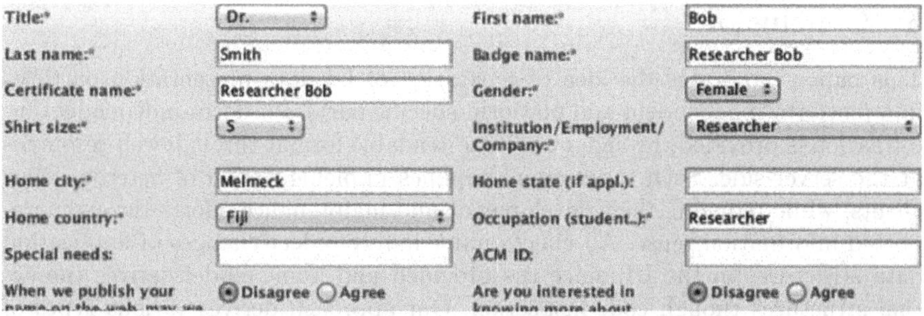

Fig. 2. Web-based UI

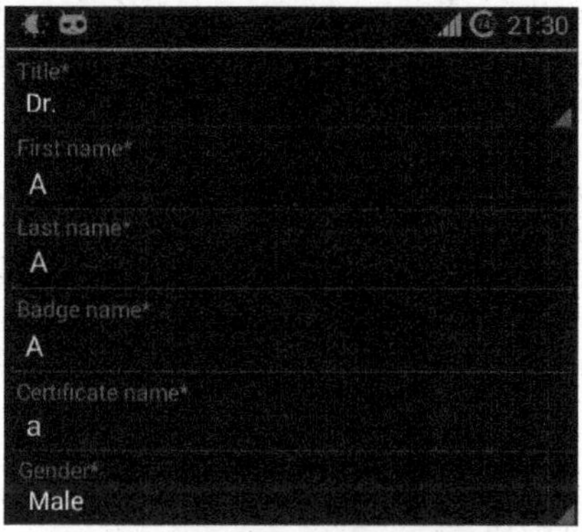

Fig. 3. Android-based UI

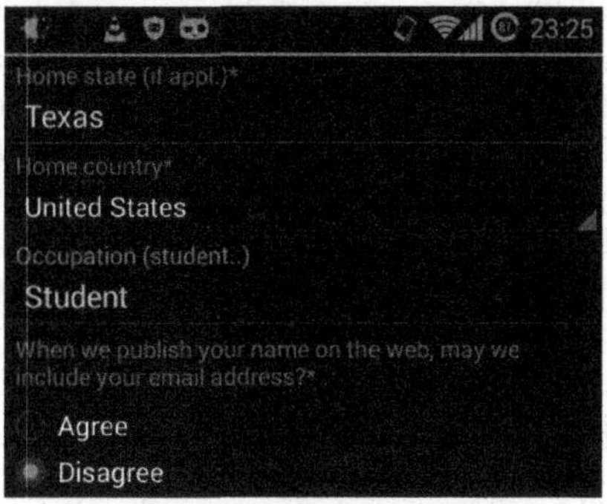

Fig. 4. Android-based UI

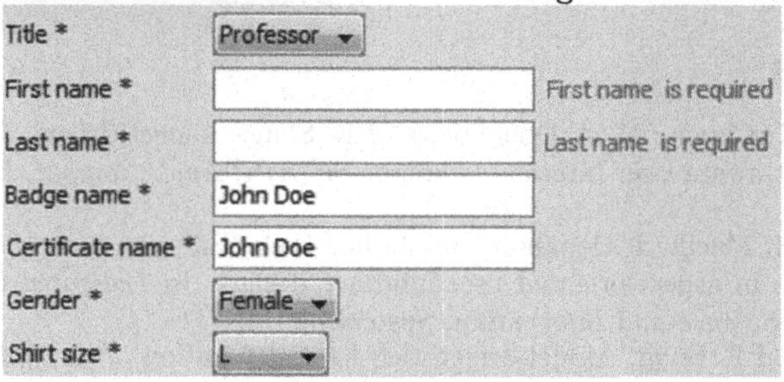

Fig. 5. Java Swing-based UI

and 4 and a standalone, Java Swing client UI is in Fig. 5. They all use the same application server, data model, domain business rules and services to provide context-aware data presentation and data manipulation. The web-based client is different from other clients as it loads the weaver and templates from the server-side in the form of a JavaScript library. Various sorts of data presentations can be derived ranging from forms, read-only forms, tables, lists, etc. Templates designed for a particular platform can be reused across different applications on the same platform. The platform-independent data presentations adjust to the server-side application and its join point model. It allows caching the join point model for a particular time span. The application page-flow is left for custom, non-automated implementation.

5 Conclusion

This paper elaborates the idea of separation of UI data presentation particles into platform-independent and platform-specific parts. Platform-independent information is provided through a machine-readable format through web-resources at the server-side. Such separation simplifies implementation of heterogeneous clients, while reducing their development and maintenance efforts through supported information reuse. All clients immediately reflect changes of application data structures in the UI, since the provided join point model derives the actual structures though code-inspection. Our approach derives various types of presentations using native components, thereby increasing usability.

The limitation of the approach is that it applies to data presentations and thus page-flow is left for custom, non-automated implementation, although the page-flow provided in a platform-independent format is left for future work. Layouts are considered as a specific particle of heterogeneous clients, although platform-independent generalization is possible but also left for future work. Future work will also consider integration with AngularJS and evaluation in a large application. The approach will be considered for Service-oriented architecture and middleware interaction.

6 Acknowledgments

This work was supported by the Grant Agency of the Czech Technical University in Prague, grant No. SGS14/198/OHK3/3T/13

References

1. T. Cerny, K. Cemus, M. J. Donahoo, and E. Song. Aspect-driven, Data-reflective and Context-aware User Interfaces Design. *SIGAPP Appl. Comput. Rev.*, 13(4):53–65, 2013.
2. T. Cerny, M. Macik, J. Donahoo, and J. Janousek. Efficient description and cache performance in aspect-oriented user interface design. In *Federated Conference on Coumputer Science and Information Systems*, 2014.
3. T. Cerny and E. Song. Model-driven rich form generation. *Information-An International Interdisciplinary Journal*, 15(7, SI):2695–2714, July 2012.
4. K. Czarnecki and U. W. Eisenecker. Components and generative programming (invited paper). *SIGSOFT Softw. Eng. Notes*, 24(6):2–19, Oct. 1999.
5. B. Green and S. Seshadri. *AngularJS*. O'Reilly Media, Inc., 1st edition, 2013.
6. R. Hanson and A. Tacy. *GWT in Action: Easy Ajax with the Google Web Toolkit*. Manning Publications Co., Greenwich, CT, USA, 2007.
7. R. Kennard, E. Edmonds, and J. Leaney. Separation anxiety: stresses of developing a modern day separable user interface. *Proceedings of the 2nd conference on Human System Interactions*, pages 225–232, 2009.
8. G. Kiczales, J. Irwin, J. Lamping, J.-M. Loingtier, C. V. Lopes, C. Maeda, and A. Mendhekar. Aspect-oriented programming. In *IECOOP'97-Object-Oriented Programming, 11th European Conference*, volume 1241, pages 220–242, 1997.
9. M. Macik, T. Cerny, and P. Slavik. Context-sensitive, cross-platform user interface generation. *Journal on Multimodal User Interfaces*, 8(2):217–229, 2014.

Implementing Personal Software Process in Undergraduate Course to Improve Model-View-Controller Software Construction

Wacharapong Nachiengmai[1] and Sakgasit Ramingwong[2]

Department of Computer Engineering Faculty of Engineering, Chiang Mai University, Chiang Mai, Thailand

[1]aj.moocm@gmail.com, [2]sakgasit@eng.cmu.ac.th

Abstract. This research attempts to implement the concepts of Personal Software Process (PSP) with Model View Controller (MVC) software construction in an undergraduate software engineering course. Exercises were redesigned to support the MVC framework based on the traditional PSP exercises. The findings proposed are a guide for instructors to improving both personal and team performance in MVC software construction.

Keywords: Personal Software Process, PSP. , Model-View-Controller, MVC.

1 Introduction

It could be said that software quality depends on the quality of its construction process [1]. Undeniably, the engineers who produce software can be another important quality factor. No matter how optimal the software process is, if the personal process is inefficient, it would be difficult to produce satisfactory software that meets customer's requirements. Thus, it is necessary to focus on the individual performance improvement in parallel with the quality control processes.

Personal Software Process (PSP^{SM}) is a practical tool that is usually used for measuring and improving individual performance[2]. There is evidence which shows that this framework is still adopted and implemented in the modern software industry[3].

MVC is a framework that aims to assist engineers to produce a new software quickly and more conveniently. It was designed to separate the software into parts. Each section can be developed independently. The engineers in MVC team can work on their own without need to wait for their colleagues' progress. As a result, the software can be produced quickly.

This study investigates an implementation of PSP in MVC software construction. The aims of this study are to create a guideline for applying the PSP's methodology to improve personal performance of software engineers or students who implement MVC framework.

K.J. Kim (ed.), *Information Science and Applications*,
Lecture Notes in Electrical Engineering 339, DOI 10.1007/978-3-662-46578-3_113

2 Related Work

2.1 Personal Software Process

PSP is a technique that aims to improve individual quality of software engineers[4]. In PSP, accumulated development data is recorded and reviewed in a systematic order. Five main KPIs, i.e. Size estimation Accuracy, Effort Estimation Accuracy, Product Quality, Early Defect Removal and Productivity are used for performance evaluation. As a result, engineers can accurately estimate their development efforts based on their own historical data. There are several studies that report the application of PSP in some university's courses that are related to programming [5-8]. The aim of these studies is to develop students to improve their competitiveness.

As aforementioned, the traditional PSP focuses on the performance process improvement for individuals. Therefore, the process of PSP was designed for the solo programming style. Yet, many researchers attempt to apply the PSP in modern software processes. One research investigates the use of PSP in university's programming course, which implements Pair Programming [9]. The result of this study shows that PSP can also help improve the personal technique efficiency just like the traditional PSP. There is some other research which applies PSP into the modern software processes, such as Scrum and other AGILE development[10],[11]. They try to add PSP process elements into typical SCRUM process framework and found that PSP provides additionally useful and effective guidance to software developers. Some researchers even implement PSP in actual industry environment [12, 13]. And some research claims that PSP helps controlling over-budget and delay problem[14].

2.2 M-V-C Software Architecture

The main purpose of the MVC that is different from another architecture is that it focuses on the separation of the coding from the user interface[15]. The benefits of this include that the developers can develop the software in a distributed fashion. Development and modification of one part usually do not impact directly on other parts. In addition, the preparation and editing in each section can be done by a specialist. Thus, the developer may not need to have expertise in all aspects to develop high-quality software.

The MVC software architecture is divided into main 3 parts: Model, View and Controller. Each section is responsible for each software development aspect. Firstly, Model is responsible for the database management of software. Secondly, View designs the user interface to meet the customers' requirement in terms of both of data inputting and rendering. Finally, Controller serves to analyze and interpret the business logic and the computational core of the software by coordination and return the processed results to the View and the Model part.

Another advantage for MVC-based software is the software is developed under the concept of the component-based software development (CBSD). Therefore, some code is developed to be reusable for a new software project [16, 17]. There is a research that tries to compare MVC software architecture with Presentation-Abstraction-Control

(PAC) with Object-Z in the formal perspective[18]. They show engineers always implement MVC in the case that the project can separate between input and output sections completely. Another study compares the implementation of MVC in ASP.NET and Java JSP projects [19]. The result of this study show that is implementing the ASP.Net framework is easier than JSP framework to achieve the MVC major concept.

3 Research Objectives and Design

3.1 Research Objectives

The main purpose of this research is to create a guideline on how PSP can be used for training and measuring individual performance of building software in the MVC framework. In the past, it is difficult to detect weaknesses of the students even after they passed the course. A system that could help the students to monitor their own process would greatly increase the efficiency of the class.

3.2 Research Design.

In this experiment, PSP is implemented in a Programming Workshop course for the Bachelor of Science, Software Engineering, Science and Technology Department, North - Chiang Mai University, Thailand. The participants include the third-year students who recently passed 4 programming related subjects, i.e., Fundamental of programming, Object-oriented Programming, Software Construction and Evolution, and Component-based Software Development. Examples of the tools that were introduced in this course include Java JSP and PHP with CodeIgniter.

1) Duration: This study involves 9 days of programming workshops. In each day, the students spent over 8 hours, separated into two phases. In the morning, a Post-Exercise Discussion is organized. The students share their experiences on how to complete the exercises. Then, before lunch, there is a lecture on software quality assurance and their next exercise. Finally, in the afternoon, the students are allowed to spend 4 hours on their assignment.

2) Role: Eighteen students participated in this experiment. They were assigned into three groups of six students based on their expertise from previous courses. Students who were responsible for Model, View and Controller roles were called M-Team, V-Team and C-Team respectively. For each exercise, random members from the three groups were selected to form a new team.

3) Tools and Computer Language: Based on their experiences, the students all build their software under the MVC pattern using CodeIgniter framework, which is the framework for building software in Web-based applications with PHP language by a MVC pattern. For PSP tool, the students implement with traditional PSP Student Workbook.

4) Exercises: There are 8 exercises that were redesigned to support the work of the MVC pattern. Some exercises were modified from the original PSP course. Table 1 displays the comparison of exercises in tradition PSP course and this experiment.

Table 1. Workshop Exercises

Exercise#	Traditional PSP.	MVC-PSP.	PSP Level.
0	Line Counter	Line Counter	-
1	Mean and S.D. calculator	Mean and S.D. calculator	0
2	Part and line counter	Part and Line counter	1.0
3	Linear regression	Listing of employee data	2.0
4	Relative size ranges	Students grade online report	2.1
5	Numerical integration 1	Mini Horoscope	2.1
6	Numerical integration 2	Mini Thai-English Dictionary	2.1
7	Significance of correlation	Mini Memo	2.1
Final Report	Final report presentation	Final report presentation	-

5) Document Redesign: Since all documents on PSP are designed for solo programming, special templates needed to be developed for students who worked on Model and View part of the development.

Table 2 introduces two new PSP documents for MVC project. These include Database Relational Specification Template (DST) for Model-Team. This template shows the detail and relationship of the entire database in the project. The User Interface Specification Template (UST) is developed for View-team in order to design the user interface.

Table 2. Document Design for MVC-PSP

Team	Documentations
Model	• Functional Specification Template : FST • Database Relational Specification Template : DST*
View	• User Interface Specification Template : UST* • Operational Specification Template : OST
Controller	• Logic Specification Template : LST • Functional Specification Template : FST

*new documents

6) Pre-Exercise Lecture: In the early morning of the class, there is a lecture on the importance of the software quality assurance and personal improvement. In order to highlight the importance of communication, 15 minutes are allocated to each team for retrospection. After that, there is another session for explaining the detail of the new exercises.

7) Post-Exercise Evaluation: Starting on day 3 of the class before the daily lecture, the lecturer spend about 30 minute for every team to discuss their exercises. The trainer also showed the past exercise's statistics and suggests how to improve the team's performance in the new exercise. Fig.1 illustrates the cycle of this MVC-PSP course.

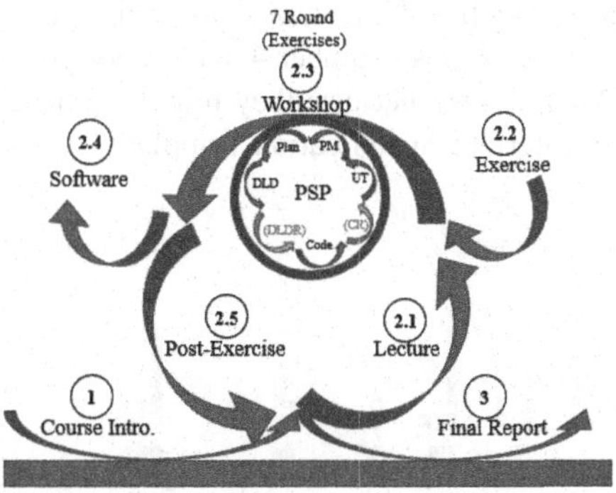

Fig. 1. MVC-PSP Course Design

4 Results

The performance analysis of this experiment is based on four indexes including Time Estimation Accuracy (percent of error), Size Estimation Accuracy (percent of error), Defect Density (defect per KLOC) and Productivity (LOC per hour). The analysis is performed on each group as well as the overall picture.

4.1 Overall

Fig. 2 illustrates the results of the 4 indexes of PSP. All of them show satisfactory positive trends. The error in both time and size estimation seem to be continuously decreasing. The Defect Density also tends to decrease. This suggests the improving quality of the software and the process. Additionally, the Productivity tends to increase during the last assignments.

4.2 Model Team

Model team's result seem to be similar to the overall's. The Time Estimation of this team became much more accurate during the last exercises.

In this team, the students produce the highest defect density especially in the second exercise because they have no experience for retrieving data from a text file to array format. Later when they can solve their problem, the defect density value just decreased continuously.

The error in time and size estimation become increasingly close to zero. This is an advantage from reusable software components. The students can reuse some code in the new exercises. This improves their productivity successively.

4.3 View Team

The View team was significantly different from another team in terms of Size Estimation. This team did not encounter problems in the first exercise since they are comparatively small. However, in later exercises, the software become larger. The View-team's members had problems with estimating the project time using PROBE C of PSP calculation (calculating based on average program size). They had 983.024 and 1185.721 percent of error in exercise number 4 and 5 respectively. In later their trend is better and straight down to zero because they found the benefit from planning and design phrase with User Interface Specification Template.

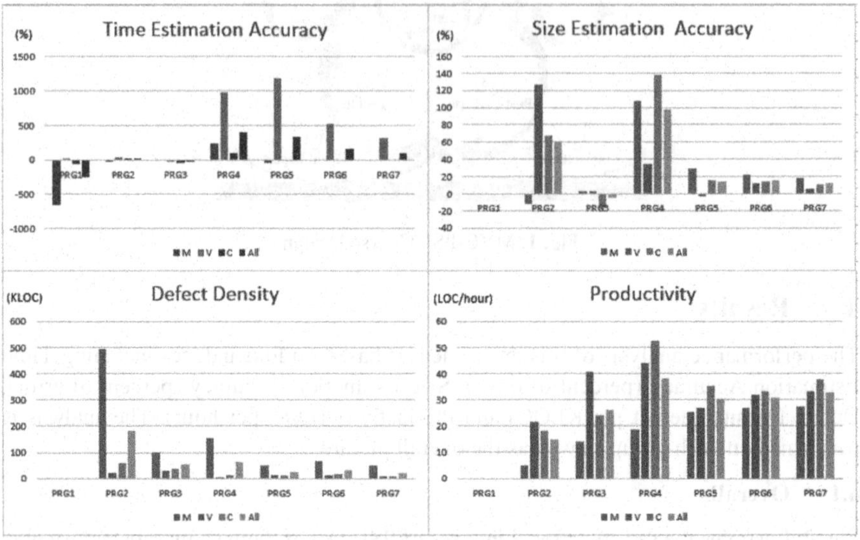

Fig. 2. Overall MVC-PSP results

The Defect Density of the View team is the lowest among the three groups. This is likely because they have a powerful tools which can convert to HTML code automatically. Most of the defects involve mistakes in naming the variables that are later used by another team. The introduction of UST greatly reduce this problem.

4.4 Controller Group

Although expected to be the most erroneous group, the estimating precision of the controller teams is satisfactorily high, despite the fact that their responsibilities on logic and calculation are the most difficult task amongst the three groups. This is arguably due to the benefits from using PROBE A, the estimation based on the linear regression of historical proxy data.

5 Conclusion

This research implements the concept of PSP and MVC in an undergraduate programming course. Parts of the Traditional PSP process are revised to support MVC Framework. The results suggest that the PSP has helped to improve students' personal performance, as seen from all of the PSP's indexes, i.e., time estimation accuracy, size estimation accuracy, defect density and productivity. However in MVC software project, there are several other factors that may affect overall performance such as the communication between team members, team selection methodology, personal workload, etc. Other advance techniques such as Team Software Process (TSP) could be implemented to further improve the performance of the MVC software team.

References

[1] F. Baharom, J. Yahaya, A. Deraman, and A. R. Hamdan, "SPQF: Software Process Quality Factor," in *Electrical Engineering and Informatics (ICEEI), 2011 International Conference on*, 2011, pp. 1-7.

[2] W. S. Humphrey, "The personal process in software engineering," in *Software Process, 1994. 'Applying the Software Process', Proceedings., Third International Conference on the*, 1994, pp. 69-77.

[3] J. Krishnamurthy and E. Nyshadham, "Quality Market: Design and Field Study of Prediction Market for Software Quality Control," in *System Sciences (HICSS), 2011 44th Hawaii International Conference on*, 2011, pp. 1-9.

[4] W. Hayes, "Using a Personal Software Process to improve performance," in *Software Metrics Symposium, 1998. Metrics 1998. Proceedings. Fifth International*, 1998, pp. 61-71.

[5] P. Abrahamsson and K. Kautz, "The personal software process: experiences from Denmark," in *Euromicro Conference, 2002. Proceedings. 28th*, 2002, pp. 367-374.

[6] J. Kamatar and W. Hayes, "An experience report on the personal software process," *Software, IEEE,* vol. 17, pp. 85-89, 2000.

[7] S. K. Lisack, "The Personal Software Process in the classroom: student reactions (an experience report)," in *Software Engineering Education & Training, 2000. Proceedings. 13th Conference on*, 2000, pp. 169-175.

[8] L. A. Williams, "Adjusting the instruction of the personal software process to improve student participation," in *Frontiers in Education Conference, 1997. 27th Annual Conference. Teaching and Learning in an Era of Change. Proceedings.*, 1997, pp. 154-156 vol.1.

[9] R. Guoping, Z. He, X. Mingjuan, and S. Dong, "Improving PSP education by pairing: An empirical study," in *Software Engineering (ICSE), 2012 34th International Conference on*, 2012, pp. 1245-1254.

[10] J. A. Stark and R. Crocker, "Trends in software process: the PSP and agile methods," *Software, IEEE,* vol. 20, pp. 89-91, 2003.

[11] R. Guoping, S. Dong, and Z. He, "SCRUM-PSP: Embracing Process Agility and Discipline," in *Software Engineering Conference (APSEC), 2010 17th Asia Pacific*, 2010, pp. 316-325.

[12] M. Morisio, "Applying the PSP in industry," *Software, IEEE,* vol. 17, pp. 90-95, 2000.

[13] K. El Emam, B. Shostak, and N. H. Madhavji, "Implementing concepts from the Personal Software Process in an industrial setting," in *Software Process, 1996. Proceedings., Fourth International Conference on the*, 1996, pp. 117-130.

[14] I. Etxaniz, "Software Project Improvement through Personal Software Process in a R&D Center," in *EUROCON, 2007. The International Conference on "Computer as a Tool"*, 2007, pp. 413-418.

[15] U. B. Information Services and Technology. (June 2004, June 24, 2014). *Model-View-Controller: A Design Pattern for Software*. Available: https://ist.berkeley.edu/as-ag/pub/pdf/mvc-seminar.pdf

[16] F. Stallinger, A. Dorling, T. Rout, B. Henderson-Sellers, and B. Lefever, "Software process improvement for component-based software engineering: an introduction to the OOSPICE project," in *Euromicro Conference, 2002. Proceedings. 28th*, 2002, pp. 318-323.

[17] Z. Xinyu, Z. Li, and S. Cheng, "The Research of the Component-Based Software Engineering," in *Information Technology: New Generations, 2009. ITNG '09. Sixth International Conference on*, 2009, pp. 1590-1591.

[18] A. Hussey and D. Carrington, "Comparing the MVC and PAC architectures: a formal perspective," *Software Engineering. IEE Proceedings- [see also Software, IEE Proceedings]*, vol. 144, pp. 224-236, 1997.

[19] F. A. Masound, D. H. Halabi, and D. H. Halabi, "ASP.NET and JSP Frameworks in Model View Controller Implementation," in *Information and Communication Technologies, 2006. ICTTA '06. 2nd*, 2006, pp. 3593-3598.

Part VIII

Personalized Care Recommendation Approach for Diabetes patients Using Ontology and SWRL

Benjamas Hempo[1,*], Ngamnij Arch-int[1], Somjit Arch-int[1], Cherdpan Pattarapongsin[2]

[1]Semantic Mining Information Integration Laboratory (SMIIL)
[1]Department of Computer Science, Faculty of Science,
[1]Khon Kaen University, Khon Kaen, Thailand
[2] Nongbuarawae Hospital, Chaiyaphum Province, Thailand, 36250

benjamas_h@kkumail.com, {ngamnij, somjit}@kku.ac.th, cherdpan_p@yahoo.com

Abstract. Although there are many health organizations that publish documents of detection and self-care recommendation for diabetes patients, this information is too general and inconsistent with each patient's conditions. This research proposed the personalized care recommendation approach for diabetes patients using ontology and SWRL. The objective is to develop the diabetes knowledge-based ontologies (DKOs) expressed in Web Ontology Language (OWL) for describing the patient profile, the general self-care practices for diabetes patients, complication, sign, symptom, and disorder of the diabetes patients. The DKOs are mapped and incorporated with rules created by using the Semantic Web Rule Language (SWRL) for inferring new knowledge or new relationships according to each patient condition. The semantic rules enable the semantic recommendation for personalized care of the diabetes patient corresponding to each patient's condition.

Keywords: Personalized care, Diabetes patient recommendation, SWRL, Self-care practices, Ontology, Diabetes knowledge-based ontologies

1 Introduction

Diabetes is the silent disaster for people in the World. The latest report from the International Diabetes Federation-IDF found that more than 371 million people around the World have diabetes and the number is still increasing steadily. 80% of these people are in underdeveloped and developing countries. In Thailand, the latest report in 2009 from the Ministry of Public Health found that more than 3.5 million people in Thailand have diabetes and approximately 8,000 people per year die from diabetes. In the next 8 years, the number of diabetes patients will increase to 4.7 million people. The diabetes patients usually risk complications of diabetes, such as, foot complications, kidney disease, high blood pressure, stroke and so on. Early detection and correct treatment of diabetes can decrease the risk of developing these complications of diabetes. Although there are many health organizations that publish the documents of

K.J. Kim (ed.), *Information Science and Applications*,
Lecture Notes in Electrical Engineering 339, DOI 10.1007/978-3-662-46578-3_114

detection and self-care practice recommendation for diabetes patients, this information is too general and inconsistent with each patient's conditions. The treatment recommendations should vary depending on the cause and severity of each patient's individual condition.

This research proposes the personalized care recommendation approach for diabetes patients using ontology and SWRL. The objective is to develop an approach to recommend the best practices suitable for each diabetes patient who has different factors or conditions. The approach focuses on food and exercise recommendation, as well as the adjustment of patient life style to prevent the risk of complications. In the case of the patient with a complication, the approach provides the nursing care guideline and discharge plan suitable for each diabetes patient to take care of themselves. This research develops the diabetes knowledge-based ontologies (DKOs) expressed in Web Ontology Language (OWL) [1] for describing the patient profile, the general self-care practices for diabetes patients, complication, sign, symptom, and disorder of the diabetes patients. These ontologies are being mapped to form the integrated diabetes ontology and incorporated with the Semantic Web Rule Language (SWRL) rules [2] for inferring the new knowledge or new relationships. These semantic rules enable the semantic recommendation for personalized care of the diabetes patient corresponding to each patient condition.

The remainder of the paper is structured as follows: Section 2 presents a literature review. Section 3 presents methodology Section 4 presents experiment and evaluation results. Section 5 presents conclusion and discussion

2 Related Works

Many research studies attempted to develop the recommendation system for diabetes patients. For example, the work of Chen R-C, Huang Y-H, Bau C-T, and Chen S-M [3] developed the diabetes medication recommendation system, based on domain ontology. The ontology knowledge described the drugs' nature attributes, type of dispensing and side effects, as well as the patients' symptoms. The system utilizes SWRL and Java Expert System Shell (JESS) to induce potential prescriptions for the patients. This system is able to analyze the symptoms of diabetes as well as to select the most appropriate drug from related drugs. The work of Al-Nazer A., Helmy T., and Al_Mulhem M. [4] developed a framework that provides food recommendations for people to reduce the risks for people who have long term diseases such as diabetes and high-blood-pressure. The semantic framework uses the personalization techniques based on integrated domain ontologies, and pre-constructed by domain experts, to recommend the relevant food that is consistent with people's needs. In the context of semantic knowledge management, many researches, such as [5], enable interoperation between different ontologies, ontology mediation is especially important to enable sharing and reuse of data.

However, most research works focus on developing the systems for health-care professionals to analyze the symptoms of diseases, to plan the treatment, and to select the most appropriate drug for each patient, leaving aside providing the appropriate self-care practice recommendations corresponding to each patient's condition. Although, the researches of [3, 4] use the ontology-based approach to represent knowledge framework, those approaches still lack the use of OWL restriction properties to implement ontology mapping.

3 Research Methodology

3.1 Diabetes Knowledge-based Ontologies Design

The Diabetes Knowledge-based Ontologies (DKOs) are local ontologies extracted from different diabetes databases consisting of the databases of diabetic patient profile, diabetes sign and symptom database, diabetes complications database and diabetes self-care practice database. Examples of DKOs are described below.

3.1.1 The DKO of diabetic patient profile

The DKO of diabetic patient profile is an ontology designed to store the diabetic patient profile consisting of classes and properties as shown in Figure 2. For instance, the *Patient* class contains patient instances as members of class and has the datatype properties as *pf:pid, pf:HN, pf:firstname, pf:lastname, pf:address, pf:gender* and *pf:diabetType.* The *pf:Patient* class has the object properties as *pf:hasCongenital_disease, pf:has_GeneralExam,* and *pf:hasLabExam.* The *pf:PhysicalExam* has the *pf:visitDate, pf:weight, pf:height, pf:SBP* (Systolic Blood Pressure), and pf:DBP (Diastolic Blood Pressure) as the datatype properties. The *pf:LabExam* has the *pf:visitDate, pf:FBS* (Fasting Blood Sugar), *pf:HbA1C* (glycated haemoglobin) as the datatype properties and has the *pf:hasUA*, and *pf:hasLipidProfile* as object properties to store the urinalysis or urine test (to measure the amount of ketone, glucose and microalbuminuria), and lipid blood test (to measure total cholesterol, triglyceride, HDL, and LDL), respectively.

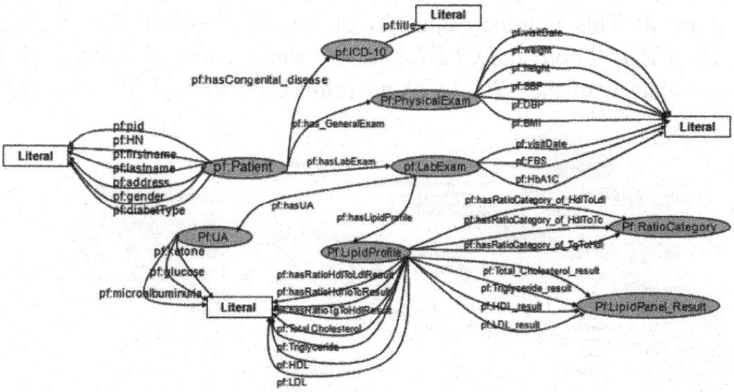

Figure 2. The *DKO of diabetic patient profile.*

3.1.2 The DKO of diabetes complication knowledge

The DKO of diabetes complication knowledge is an ontology designed to store the diabetes complication knowledge which describes sign and symptom of each diabetes complication condition as shown in Figure 3. The ontology also stores the factors obtained from lab examination result that affects the complication condition. The main classes of this ontology consist of the *CPG2:Complication*, *CPG2:SignAndSymptom*, and *CPG2:Lab_Examination_Result*. The *CPG2:Complication* and *CPG2:SignAndSymptom* can contains their own subclasses. The *CPG2:Complication* class relates with the *CPG2:SignAndSymptom* class through the *CPG2:hasSignAndSymptom* property which is an inverse property of *CPG2:signAndSymptomOf* property. The *CPG2:Lab_Examination_Result* class consists of Datatype properties indicating the factors that affect the diabetes complication, such as *CPG2:hasHDL_Level*, *CPG2:hasTotalCholesteral_Level*, *CPG2:hasLDL-Level*, *CPG2:hasTriglycerides_Level*.

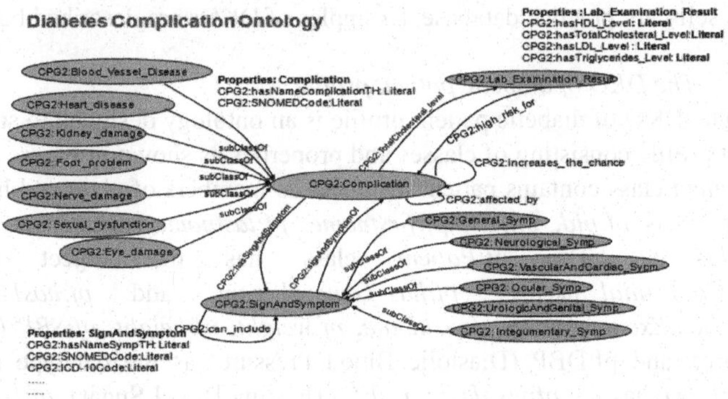

Figure 3. The DKO of the diabetes complication.

3.1.3 The DKO of self-care practice knowledge

The DKO of self-care practice knowledge is an ontology used to store the self-care practice knowledge of the diabetic patient when they have the complication as shown in Figure 4. This ontology consists of several main classes, such as, the *CPG1:Exercise*, *CPG1:Food*, and *CPG1:Lifestyle* which contains its own subclasses. These classes relate with the *CPG2:Complication* class through the Object properties.

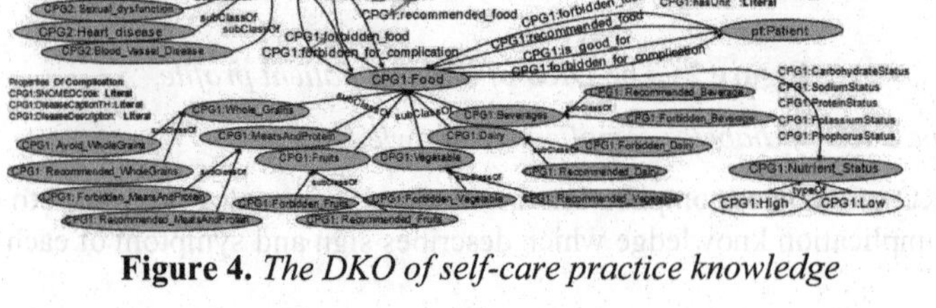

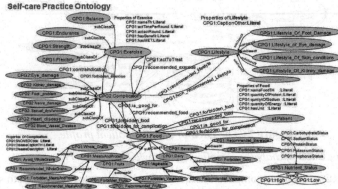

Figure 4. *The DKO of self-care practice knowledge*

For example, the *CPG2:Complication* class relates with the *CPG1:Exercise* through the *CPG1:forbidden_exercise*, and the *CPG1:recommended_exercise* properties. The DKO of self-care practice knowledge can also recommend foods and forbidden food, as well as life style, for each patient with a different complication condition. Example in Figure 5 illustrates the schema level and instance level of class relationship and instance relationship, respectively.

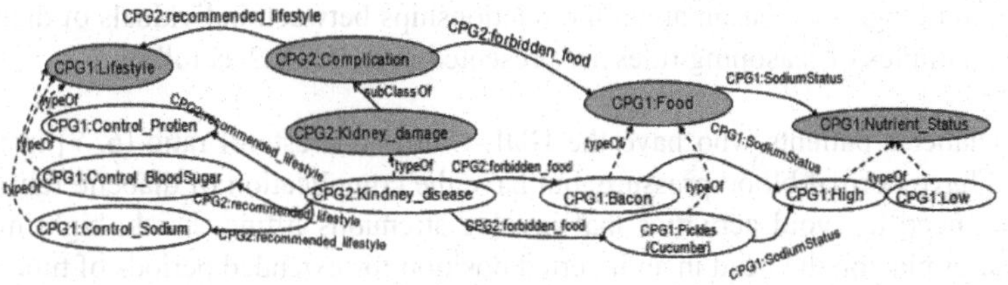

Figure 5. Example of schema level and instance level of self-care ontology.

3.3 The Design of OWL Property Characteristics

This section illustrates examples of the built-in OWL properties, such as the *owl:SymmetricProperty*, *owl:TransitiveProperty*, and *owl:inverseOf* to enhance the property characteristics and infer new instances relationship as follow.

- *owl:SymmetricProperty*

In the DKO of diabetes complication ontology, we can specify instance relationship between the *CPG2:Myocardial_Infarction* and the *CPG2:Atherosclerosis,* which are instances of the *CPG2:Heart_Disease* and the *CPG2:Blood_Vessel_Disease* classes, respectively, through the *CPG2:relatedTo* property of *CPG2:Complication* class. The relationship can be defined in a statement, (*CPG2:Myocardial_Infarction*, *CPG2:relatedTo*, *CPG2:Atherosclerosis*). If the *CPG2:relatedTo* is defined as the *owl:SymmetricProperty*, the inference engine can infer that (*CPG2:Atherosclerosis*, *CPG2:relatedTo*, *CPG2:Myocardial_Infarction*).

- *owl:TransitiveProperty*

The *CPG2:relatedTo* can also be defined as *owl:TransitiveProperty*. If we define the two statements as (*CPG2:Myocardial_Infarction*, *CPG2:relatedTo*, *CPG2:Atherosclerosis*) and (*CPG2:Atherosclerosis*, *CPG2:relatedTo*, *CPG2:HighBloodPressure)*, the inference engine can infer that (*CPG2:Myocardial_Infarction*, *CPG2:relatedTo*, *CPG2:HighBloodPressure)*.

- *owl:inverseOf*

The *CPG1:forbidden_exercise* property can be designed as an inverse properties of the *CPG1:contraindication* property. Hence, if there is a statement (*CPG2:Complication*, *CPG1:forbidded_exercise*, *CPG1:Exercise*), the inference engine can infer the following statement: (*CPG1:Exercise*, *CPG1:contraindication*, *CPG2:Complication*).

3.3 Reasoning Rule Creation

Once the integrated diabetes ontology has been generated as a result of ontology mapping process, the integrated diabetes ontology can be augmented with reasoning rules to provide an additional layer of expressivity and to support personalized semantic queries. This research used SWRL to define reasoning rules in addition to the ontology mapping process. These reasoning rules are expressed over the integrated terminologies to reason about the relationships between individuals of diabetes classes. Examples of reasoning rules are presented in RL1–RL2 as follow.

Rule1: Diabetes patients who have the HDL-C/total cholesterol ratio [6, 7] in ideal level and have normal blood pressure, but have the complication of diabetic retinopathy. They need to avoid activities that involve strenuous lifting; harsh, high-impact activities; or placing the head in an inverted position for extended periods of time.

```
pf:Patient(?pt) ∧ pf:hasLabExam(?pt, ?LabEx) ∧ pf:hasLipidProfile(?LabEx, ?LipNo) ∧
pf:hasRatioCategory_of_HdlToTc(?LipNo, pf:IDEAL) ∧ CPG2:Eye_damage(?Eyedamage)∧
CPG2:has_Complication(?pt, ?Eyedamage) ∧ pf:has_PhysicalExam(?pt, ?PhyExam) ∧
pf:has_DBP(?PhyExam, ?dbp) ∧ swrlb:lessThan(?dbp, 80) ∧
CPG1:prohibit_exercise(?Eyedamage, ?prohibit_exercise) ∧
CPG1:recommended_exercise(?Eyedamage, ?recomExercise)
→
CPG1:prohibit_exercise(?pt, ?prohibit_exercise) ∧ CPG1:recommended_exercise(?pt, ?recomExercise) ∧
CPG1:recommended_other_Exercise(?pt, "Patients with diabetes and active PDR should avoid activities that involve strenuous lifting; harsh, high-impact activities; or placing the head in an inverted position for extended periods of time")
```

Rule2: Diabetes patients who have high blood pressure more than 180/110 mm Hg and have the complication of peripheral neuropathy (a result of nerve damage). They need to limit the exercise to light intensity exercises or non-weight bearing activities which have the metabolic equivalent less than 3.0, such as swimming, yoga and stretching, because the peripheral neuropathy often causes weakness, numbness and pain, usually in patients' hands and feet.

```
pf:Patient(?pt) ∧ pf:hasLabExam(?pt, ?LabEx) ∧ pf:hasLipidProfile(?LabEx, ?LipNo) ∧
pf:hasRatioCategory_of_HdlToTc(?LipNo, pf:IDEAL) ∧ CPG2:has_Complication(?pt,
CPG2:Periheral_nerurropthy) ∧ pf:has_PhysicalExam(?pt, ?PhyExam) ∧ pf:has_SBP(?PhyExam, ?sbp) ∧
swrlb:greaterThanOrEqual(?sbp, 180) ∧ CPG1:Exercise(?LowMEtsExercise) ∧
CPG1:has_METs(?LowMEtsExercise, ?met) ∧ swrlb:lessThan(?met, 3) ∧
CPG1:Exercise(?HighMEtsExercise) ∧ CPG1:has_METs(?HighMEtsExercise, ?met2) ∧
swrlb:greaterThanOrEqual(?met2, 3) ∧ CPG1:LifestyleFoot_Problem(?LifeFoot) ∧
CPG2:Foot_Problems(?footProblem)
→
CPG1:recommended_OtherLifestyle(?pt, ?LifeFoot) ∧ pf:high_risk_of(?pt, ?footProblem) ∧
CPG1:prohibit_exercise(?pt, ?HighMEtsExercise) ∧ CPG1:recommended_exercise(?pt, ?LowMEtsExercise) ∧ CPG1:recommended_other_Exercise(?pt, "Especially Yoga lower rate of Hypertension And lower rate of High blood sugar")∧ CPG1:recommended_Other(?pt, "Periheral nerurropthy affects your feet,toes,toes,toes,arms")∧ CPG1:recommended_other_Exercise(?pt, " A moderate-intensity weight-bearing exercise program for a person with type 2 diabetes and peripheral neuropathy ")
∧ CPG1:recommended_other_Exercise(?pt, "You have a stress test before staring an exercise program")
```

4 Experiment and Evaluation

For the research experiment, the DKOs were developed by an ontology engineer using ontology editor tools such as Protégé 3.4.4 [8]. The reasoning rules were created using SWRL through the SWRLTab interface of Protégé and run under Pellet[9] and Jess reasoner . The system experimental and evaluation was designed using the 50 diabetes patients' records from the North-East hospital, Thailand. Each patient record with a different patient condition was extracted into DKO of diabetes patient profile and integrated with other DKOs to form the integrated diabetes ontology through the ontology mapping process. The integrated diabetes ontology was augmented with the reasoning rules to generate the ontology-based personalized care for diabetes patients (OPCD) through the Protégé's supporting tools. The OPCD was searched for each patient recommendation using several search criteria, such as patient id, HN, patient first name, or patient last name. The experimental outcomes return the semantic recommendation for personalized care of each diabetes patient. To evaluate the recommendation system, we have consulted the physician or nurses who take care of the diabetes patient to evaluate the correctness of system, and used the precision and recall values in order to determine the retrieval efficiency. Precision is the ratio of the number of relevant records retrieved to the total number of irrelevant and relevant records retrieved. Recall is the ratio of the number of relevant records retrieved to the total number of relevant records in the ontology. Precision and recall values are calculated from Equations (1) and (2) as follows:

$$Precision = \frac{A}{A+C} \times 100\% \qquad (1)$$

$$Recall = \frac{A}{A+B} \times 100\% \qquad (2)$$

where A is number of relevant system recommendation conformed to the physician recommendation. B is number of relevant system recommendation but not recommend to the users. C is number of irrelevant system recommendation not conformed to the physician recommendation.

Precision and Recall values can be used to calculate F-measure which is defined as a harmonic mean of precision and recall, as shown in Equation (3) below:

$$F-measure = 2 \times \left(\frac{precision \cdot recall}{precision + recall} \right) \qquad (3)$$

Evaluation of the system recommendation efficiency for 50 patient conditions shows that most system recommendation corresponded to the physician recommendation with the Precision value of 0.93 the Recall value of 0.97, and the average of overall efficiency or F-measure was 0.95. These results were in very high levels.

5. Conclusion and Discussion

This research develops an approach to recommend the best practices suitable for each diabetes patient who has different factors or conditions. The diabetes knowledge-based ontologies are developed and mapped to form the integrated diabetes ontology. The research also develops reasoning rules to enhance semantic inference of the integrated diabetes ontology to form the ontology-based personalized care for diabetes patients. The evaluation results show that the precision of system recommendation was less than 1 due to incomplete data stored in ontologies resulting in some imprecise information. However, the efficiency of system recommendation shows the F-measures are rather high. This result indicates the efficiency of the structural design and mapping of ontologies retrieval from diabetes databases. The system can respond very well to the needs of patient condition. For the next future work, we intend to implement the OPCD Web application for both patients and physician or nurses to be used for searching DKOs by themselves. We also intend to extend reasoning rules to enhace the integrated diabetes ontology to provide additional expressivity and to support more personalized semantic queries.

6 Acknowledgements

The researchers would like to thank Mrs. Petcharat Butakhiew, a nurse at the Nursing Service Division, Srinakarin Hospital, Khon Kaen Province, Thailand, who gave us valuable guidance in diabetes patients' practices. Thanks are also extended to Non-gburawae Hospital, Chaiyaphum Province, Thailand for data support.

References

1. Owl: web ontology language guide, http://www.w3.org/TR/owl-guide
2. A Semantic Web Rule Language Combining OWL and RuleML, http://www.w3.org/Submission/SWRL/
3. Rung-Ching,C., Yun-Hou, H., Cho-Tsan B., Shyi-Ming C.: A recommendation system based on domain ontology and SWRL for anti-diabetic drugs selection, vol.39, Issue 4, March 2012, pp. 3995–4006. Expert Systems with Applications
4. Ahmed, A., Tarek, H., Muhammed, A.: User's Profile Ontology-based Semantic Framework for Personalized Food and Nutrition Recommendation. ANT/SEIT 2014: 101-108.
5. Arch-int N. and Arch-int S., "Semantic Ontology Mapping for Interoperability of Learning Resource Systems using a rule-based reasoning approach", Expert Systems with Applications, Vol. 40 (18), pp. 7428–7443, 15 December 2013.
6. Mehdi, H.S., Byron J.H., Michael, S.L.: Association of Triglyceride–to–HDL Cholesterol Ratio With Heart Rate Recovery, Diabetes Care, vol. 27, no. 4, pp. 936-941 (2004)
7. The Triglyceride/HDL Cholesterol Ratio, http://www.docsopinion.com/2014/07/17/triglyceride-hdl-ratio/
8. Stanford. (1999). : The Protégé Ontology Editor and Knowledge Acquisition System , http://protege.stanford.edu/ ,
9. Evern,S., Bijan,P., Bernardo, C.G., Aditya, K., Yarden K.: Pellet: A practical OWL-DL reasoned,Web Semantics: Science, Services and Agents on the World Wide Web, vol. 5 (2), pp. 51-53(2007)

An Experimental Study for Neighbor Selection in Collaborative Filtering*

Soojung Lee

Gyeongin National University of Education
155 Sammak-ro, Anyang, Republic of Korea
sjlee@gin.ac.kr

Abstract. Similarity is a most critical index in collaborative filtering-based recommdender systems, which suggest items to their customers by consulting most similar neighbors. Current popular similarity measures may mislead the user to unwanted items in certain cases, due to their inherent properties. This study suggests a novel idea to significantly decrease such occurrences by enforcing qualifying conditions to neighbors using some simple criteria, to make consultations for their ratings. From extensive experiments, the proposed idea is found to substantially improve prediction performance of collaborative filtering based on existing similarity measures. This result is noticeable considering that such improvements are achieved by simply consulting only those neighbors satisfying the given criteria, without adopting any sophisticated similarity measure.

Keywords: Recommender system, Similarity measure, Collaborative filtering, Memory-based collaborative filtering, Content-based filtering

1 Introduction

Collaborative Filtering (CF) has been popularly used in recent commercial systems to recommend items to online customers. The items are selected based on the history of preferences of other customers. This method has been successfully utilized in various systems such as the Tapestry system, GroupLens, Video Recommender, Ringo, Amazon.com, the PHOAKS system, and the Jester system [1].

Among several types of the algorithms used for collaborative recommendation, the memory-based method refers to the history of ratings made by other users called *neighbors*. Hence, who will be selected as neighbors and how many and reliable are their ratings are most important issues regarding the quality of the system. The usage of the memory-based CF methods is to predict the rating of a yet unrated item for a user, thus recommending those items with the

* This research was supported by Basic Science Research Program through the National Research Foundation of Korea(NRF) funded by the Ministry of Education, Science and Technology(2012R1A1A3012320)

K.J. Kim (ed.), *Information Science and Applications*,
Lecture Notes in Electrical Engineering 339, DOI 10.1007/978-3-662-46578-3_115

highest predicted ratings to the user. Several methods have been developed in literature, incorporating various techniques including machine learning [3] and clustering [8] as well as utilizing domain knowledge [10].

Predicting the rating of an item for a user is usually made in two different ways, *user-based* and *item-based*. The former incorporates ratings given to the item by those users who are most similar to the user, whereas the latter refers to the ratings given by the user, of those items similar to the item. Therefore, similarity calculation is one of the most critical aspects of CF systems, which greatly influences recommendation performance.

Most common approaches to measure similarity between two users or two items are correlation-based and vector cosine-based. Examples of the former approaches are the Pearson correlation and its variants, constrained Pearson correlation and Spearman rank correlation [1]. However, these popular approaches may result in poor performance when the number of ratings given by a user is scarce, which is referred to as the *data sparsity problem*, or when the number of co-rated items are few. These problems are particularly evident for a new user or item entered to the system, known as *new user/item problem*. Detailed analysis on traditional similarity measures can be found in [2].

This paper first investigates the effect of these problems to the recommendation performance and then suggests a simple strategy to improve the performance. Through extensive experiments, it is verified that the proposed idea greatly enhances the performance of rating predictions.

2 Proposed Approach

2.1 Motivation

The motivation of our study comes from the conjecture that neighbors of low similarity with the active user or those having fewer co-rated items would contribute negatively to the recommendation performance. In order to verify this conjecture, differences between the real ratings of items given by the active user and the ratings predicted by the system for the same items are calculated. The calculation is done separately for each interval of Jaccard index [6] as well as for that of similarity. Jaccard index represents the ratio of the number of co-rated items between two users, which is calculated by $\frac{X \cap Y}{X \cup Y}$, where X and Y denote the number of ratings made by the two users, respectively. The number of co-rated items is more purely reflected on the index than on the similarity measure, which enables investigation of direct effect of the number on performance. Prediction of ratings is made based on the well known Resnick's formula [9].

Figure 1 pictures the proportion of differences between the actual and the predicted ratings for each Pearson similarity or Jaccard interval between two users. These experiments were conducted on the MovieLens dataset[1]. For instance, rating prediction made from those neighbors with less than 0.1 of similarity with the active user leads to the rating difference of zero with 28.4%, that of one with

[1] http://www.movielens.org

42.6%, and so on. Obviously, rating difference of zero indicates perfect match betwee the actual and the predicted ratings, which is to be pursued. The proportion of zero difference tends to increase with the greatest rate among all, as similarity increases. However, rating difference of one outnumbers that of zero overall, especially with the lower similarity. Hence, it seems that the effect of similarity is revealed mostly onto the rates of difference of zero and two, but not onto the other rates. This phenomenon can also be observed in the figure of Jaccard index but rather abruptly for the index>0.35. This is because the user pairs with such index are relatively rarely found in our experiment.

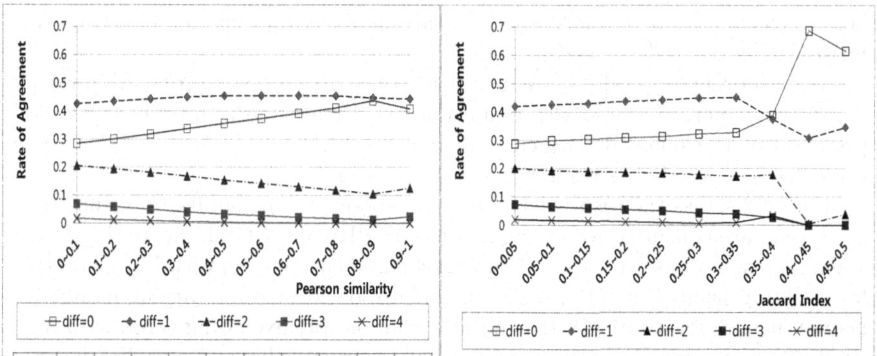

similarity	0~0.1	0.1~0.2	0.2~0.3	0.3~0.4	0.4~0.5	0.5~0.6	0.6~0.7	0.7~0.8	0.8~0.9	0.9~1
avgdiff	1.1076	1.0472	0.9874	0.9271	0.8727	0.8297	0.7833	0.7412	0.6904	0.764

Jaccard	0~ 0.05	0.05~ 0.1	0.1~ 0.15	0.15~ 0.2	0.2~ 0.25	0.25~ 0.3	0.3~ 0.35	0.35~ 0.4	0.4~ 0.45	0.45~ 0.5
avgdiff	1.116	1.074	1.052	1.021	0.999	0.96	0.95	0.941	0.318	0.423

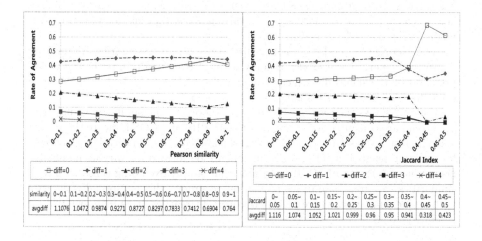

Fig. 1. Proportions of differences between the actual and the predicted ratings with respect to Pearson similarity (left) and Jaccard index intervals (right).

In the figure, the average difference, denoted by 'avgdiff', between the actual and the predicted ratings is also presented. This measure steadily decreases with the increasing similarity or Jaccard index, but slightly increases at the end. This rather strange result, especially for the similarity, is worth further investigation, because it contradicts to the common belief that neighbors with higher similarities are more reliable. Notice that the sudden drop of avgdiff in case of Jaccard index is obviously due to the abrupt increase of zero-difference rate as well as the large decrease of the other rates at the highest indices. This result along with the average difference for Jaccard index> 0.4 shown in the table allows for larger belief on Jaccard index than on similarity when the number of co-rated items exceeds some threshold.

Therefore, from all these experimental findings, it is conjectured that excluding neighbors with low or highest similarities or low Jaccard indices might enhance the mean accuracy of predicted ratings, which motivated further experiments.

2.2 Experimental Results

To implement the proposed idea, we focused on two most common similarity measures, Pearson correlation (PRS) and cosine similarity (COS), and conducted the experiments on the MovieLens dataset which has been popularly used in the related study and publicly open for research purpose. This dataset consists of 100,000 ratings of 1 to 5 scales made by 943 users on 1,682 movies. Each user has rated at least 20 movies. 80% of each user's ratings were used for training, i.e., for calculation of similarity and Jaccard index, and prediction of ratings was made on the rest 20% of testing set.

The accuracy of prediction is measured through several metrics in literature [5]. In this paper, we use MAE (Mean Absolute Error) to estimate the prediction precision of the methods, which is defined as $MAE = \sum_i |r_i - r'_i|/N$, for N predictions r'_i for the corresponding real ratings r_i.

Figure 2 shows the results when only neighbors having similarity values as in the legend are consulted for rating prediction. PRS and COS indicate the results when no restrictions are imposed on Pearson and cosine similarities, respectively, for selecting neighbors. These experiments are performed for varying number of nearest neighbors (top NN). It is notable that simply excluding those with high similarities can greatly improve prediction accuracy. This is especially the case with cosine similarity, where MAE improvement reaches at as high as 0.023 for similarity<0.98.

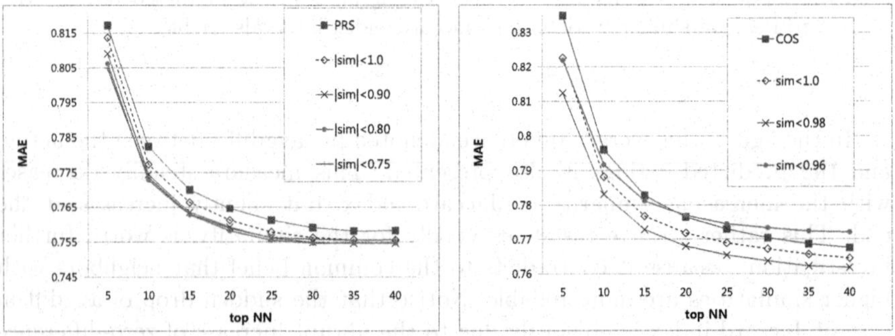

Fig. 2. MAE results with the given similarity restrictions: Pearson correlation (left) and cosine similarity (right)

The reason for such improvement is because Pearson and the cosine similarities produce results which are barely understandable, in particular cases, as discussed in [7]. For instance, when there is only one co-rated item between two users, Pearson correlation and the cosine similarity become 1, -1, or zero-divide. Furthermore, such values are also yielded when all the ratings of the common items given by each user are the same, respectively. Another problem of Pearson correlation is that it does not take the variance of ratings into account, which may cause absurd results. To illustrate this, assume that there are only two items co-rated by two users. For instance, $r_{u,i} = 5, r_{u,j} = 4, r_{v,i} = 5$, and $r_{v,j} = 1$ for two co-rated items i and j. Assuming that the rating scale allowed in the system is from 1 to 5, Pearson correlation becomes one in this case.

Considering the discussion above, it is presumed that excluding those neighbors having similarities of extreme values such as 1 and -1 for consultation of ratings can enhance the prediction accuracy. Actually, this assumption is verified in Figure 2, where the improvement is within approximately 0.003 to 0.012 in terms of MAE. However, it is notable that exclusion of neighbors having similarities lower than one further improves the accuracy in both PRS and COS; observe that consulting neighbors with absolute similarities<0.8 in case of PRS and with similarities<0.98 in case of COS seems to be the upper bound of improvement. These results suggest the adoption of more reliable scheme than similarity itself be required in collaborative filtering-based recommender systems. Contrary to our expectations, excluding the neighbors having very low similarities yields no improvement of MAE, whose detailed results are omitted here due to space constraints.

In our next experiments, restrictions based on Jaccard index are made for selecting neighbors whose results are shown in Figure 3. As previously, PRS and COS denote the results with no restriction. The trend is that consulting neighbors with higher Jaccard indices only yields better accuracy in both types of similarities. However, this is the case up to a certain point where no more improvement is achieved, as in the case of Jaccard≥0.10. This is obvious since the number of neighbors to consult gradually drops as the index increases. Therefore, limiting neighbors according to the Jaccard index is better to be personalized to each user, considering his/her number of neighbors. Note that much higher accuracy is achieved than in the case of similarity restrictions, up to 0.02 more accurate for Pearson and 0.03 for cosine similarity.

3 Conclusions

This paper demonstrates that cautious inclusion of neighbors under some criteria when consulting ratings yields much better prediction performance in collaborative filtering. The criteria is made based on two indices, similarity and Jaccard index. Besides measuring prediction quality, it is also worthwhile to evaluate the idea in terms of recommendation quality such as F1 metrics [4]. Moreover, the experiments were conducted on a dataset using a small range of integer-valued ratings, so future works on larger-scaled dataset such as Jester would be desirable.

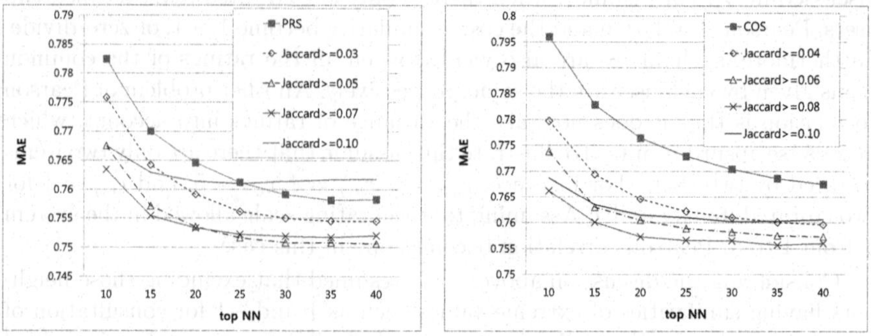

Fig. 3. MAE results with the given Jaccard restrictions: Pearson correlation (left) and cosine similarity (right)

References

1. Adomavicius, G., Tuzhilin, A.: Toward the Next Generation of Recommender Systems: A Survey of the State-of-the-art and Possible Extensions. IEEE Trans. Knowledge & Data Engineering 17(6), 734-749 (2005)
2. Ahn, H.J.: A New Similarity Measure for Collaborative Filtering to Alleviate the New User Cold-starting Problem. Information Sciences 178(1), 37-51 (2008)
3. Bobadilla, J., Ortega, F., Hernando, A, Alcala, J.: Improving Collaborative Filtering Recommender System Results and Performance Using Genetic Algorithms. Knowledge-Based Systems 24(8), 1310-1316 (2011)
4. Gao, M., Wu, Z., Jiang, F.: Userrank for Item-based Collaborative Filtering Recommendation. Information Processing Letters 111(9), 440-446 (2011)
5. Herlocker, J.L., Konstan, J.A., Terveen, L.G., Riedl, J.T.: Evaluating Collaborative Filtering Recommender Systems. ACM Trans. Information Systems 22(1), 5-53 (2004)
6. Koutrica, G., Bercovitz, B., Garcia-Molina, H.: FlexRecs: Expressing and Combining Flexible Recommendations. In: the 2009 ACM SIGMOD International Conference on Management of Data, pp. 745-758, ACM (2009)
7. Lee, S.: A Similarity Measure Using Rating Ranges for Memory-based Collaborative Filtering. Korea Association of Information Education 17(4), 375-382 (2013)
8. Nilashi, M., Jannach, D., Ibrahim, O. b., Ithnin, N.: Clustering- and Regression-based Multi-criteria Collaborative Filtering with Incremental Updates. Information Sciences 293, 235-250 (2015)
9. Resnick, P., Iacovou, N., Suchak, M., Bergstrom, P., Riedl, J.: GroupLens: An Open Architecture for Collaborative Filtering of Netnews. In: the 1994 ACM Conference on Computer Supported Cooperative Work, pp. 175-186, ACM (1994)
10. Tang, X., Deng, W., Liu, J.: A Personalized Recommendation Method Based on Comprehensive Interest. Int'l Journal of Advancements in Computing Technology 5(5), 157-164 (2013)

Transforming e-Procurement Platforms for PEPPOL and WCAG 2.0 Compliance

The anoGov-PEPPOL Project

José Martins[1], João Barroso[1,2], Ramiro Gonçalves[1,2], André Sousa[1,2], Miguel Bacelar[1], Hugo Paredes[1,2]

[1] Universidade de Trás-os-Montes e Alto Douro, Vila Real, Portugal
[2] INESC TEC (formerly INESC Porto), Porto, Portugal
jmartins@utad.pt, jbarroso@utad.pt, ramiro@utad.pt, mbacelar@utad.pt

Abstract: The increase in the complexity inherent to public administrative activities, including public procurement, has led ANO to develop a public e-procurement platform in order for Portuguese public entities to perform their contracts and acquisitions.
Given the introduction of European PEPPOL standards and the requirement for all Portuguese e-procurement platforms to be WCAG 2.0 level A compliant, ANO company established a consortium with the UTAD University in order to improve their platform and develop the required features. By using a specially designed stage-based work methodology, the R&D project was carried out with success and all initial goals were achieved.

Keywords. e-procurement, Web accessibility, anoGov, PEPPOL.

1 Introduction

The evolution of society has led to a significant increase in the complexity inherent to public administrative activities, including public procurement. In Portugal, with the creation of legal rules and regulations directed at transforming all public procurement activities to digital platforms, ANO (a Portuguese IT/IS company) created a public e-procurement platform for public entities and organizations to perform their contracts and acquisitions.

Despite ANO's initial effort to develop an efficient and quality based e-procurement platform, a new requirement appeared and ANO had to integrate this platform with Europe's PEPPOL standard while at the same time, complying with Portuguese legal requirements to be accessible according to WCAG 2.0 level A standard. This integration would allow for ANO to expand its operation to all European countries and public entities. This was the initial context for the creation of a R&D project between ANO and the University of Trás-os-Montes e Alto Douro (UTAD), named anoGov-PEPPOL,

K.J. Kim (ed.), *Information Science and Applications*,
Lecture Notes in Electrical Engineering 339, DOI 10.1007/978-3-662-46578-3_116

in order to analyze, plan, execute and test a set of new features that would completely transform anoGov platform e-procurement platform and take it to another level.

In order to achieve acceptable levels of success a new work methodology was defined and followed during the entire anoGov-PEPPOL project. This methodology was composed of five different stages that ranged from initial requirement analysis to final functional, technical and accessibility testing. By making use of the defined work methodology, both project teams were able to achieve reasonable improvements in their own abilities and knowledge. They were also able to simultaneously improve the complexity and quality of the anoGov public e-procurement platform, thus proving the success of the inherent R&D project.

The present paper is divided into five sections. We begin by presenting an introduction to the paper (section 1) and a compact e-procurement background characterization from technical, functional and accessibility perspectives. In the third section, a brief anoGov-PEPPOL project description is made in order to present the project, its goals and aims in a simple and efficient manner. The work methodology inherent to the referred R&D project is described in section 4 and we conclude by presenting some final considerations in the fifth and final section.

2 Public e-Procurement – The Portuguese Perspective

The complexity inherent to public administrative activities has raised several issues to both governments and citizens. In order to solve these problems, information technologies and systems have been used, including for conducting transactions between government and suppliers (e-procurement), thus achieving an increase in quality and cost efficiency [1].

2.1 e-Procurement Characterization

According to Corsi [2], e-procurement refers to the use of electronic mechanisms, typically Web based, to conduct transactions between entities and suppliers. In order to fully satisfy both entities and their suppliers, e-procurement platforms were forced to support all stages of the purchasing process: 1) initial requirement identification, 2) negotiation, 3) payment, and 4) contract management.

By extrapolating the e-procurement concept, one can perceive that it can be characterized according to whoever intervenes in the inherent transactions. This said, e-procurement can represent a B2B, B2C or B2G purchase or sale of supplies or services through the Internet and other online information systems [3].

When analyzing e-procurement from a public or governmental perspective, one should acknowledge that it concerns a process in which governments and public entities use Internet-related information systems to agree on the acquisition of products or services (contracting) or to purchase products or services (purchasing). From a functional point of view, e-procurement encompasses several key functionalities, such as electronic ordering and bidding, reverse auctioning and integrated automated procurement systems [1, 4].

In 2006 the Portuguese Government decided to create a set of legal regulations towards the compliance of the European eGov Action Plan, a decisive change in the way in which public entities performed their purchases occurred. Despite the initial visible changes in the Portuguese public procurement, it was only when the DL-18/2008 law was published that a legal obligation for performing all public procurement contracts through Web hosted electronic platforms came into effect [5, 6].

In order to regulate e-Procurement in Portugal, the national Government created a public entity, named CEGER that holds the powers to not only regulate, but also to apply penalties to those who fail to comply with the existing legal regulations [7].

At the present time, there are eight e-Procurement platforms in Portugal that operate concurrently in both public and private sector procurement activities, thus ensuring that all organizations with operation in the country can perform their purchases and sales in an electronic manner [8].

2.2 Portuguese e-Procurement Web Accessibility Requirements

In the opinion of W3C [9], Web accessibility refers to the creation of Web interfaces that might be perceived, understood and operated in the same manner by all people, including those with some sort of incapacity or disability.

When in 2010, the Portuguese National Institute of Statistics promoted a census study aimed on analyzing the Portuguese population, it was possible to acknowledge that almost 10% of the Portuguese population has some sort of incapacity or disability [10], thus assuming a direct relation with the European statistics [11].

The Portuguese concerns with Web accessibility started fifteen years ago and led to the publication of several laws and regulations towards the dissemination and application of the topic. Examples of these publications were the Resolution of the Ministers' Council 110/2003 - where the government presented the "National Program for the Inclusion of the Disabled Citizens in Society" [12], and the Resolution of the Ministers' Council 155/2007 – that forced the public websites to make themselves accessible to the majority of the disabled citizens [13].

Despite the relevant efforts towards the application of Web accessibility standards and the best practices in all public electronic platforms, at this time no law or regulation has been created in order to force privately-controlled Portuguese websites to become accessible to all users, including those with some sort of incapacity of disability [8, 14].

In terms of Portuguese public e-procurement, when the Portuguese Government published the 18/2008 law [6], it brought a set of mandatory requirements that e-procurement platforms, in which there was a compliance requirement with W3C's "Level A" Web Content Accessibility Guidelines (WCAG 1.0) or with a corresponding more updated regulation.

With the mandatory requirements for Web accessibility in mind, all the e-procurement platforms producers began to make an effort towards making their platforms more accessible and easy to use. However, as it can be seen by analyzing the studies that focused on the evaluation of the accessibility levels of Portuguese entities, both public and private, the levels for Web Content Accessibility Guidelines compliance fall short from what one would expect, thus allowing for the existance of poor Web accessibility levels [8, 14].

3 The anoGov-PEPPOL Project

Given the need to comply with Portuguese legal requirements for Web accessibility in e-procurement platforms, a partnership between ANO and the University of Trás-os-Montes e Alto Douro (UTAD) was created. This partnership aimed at allowing for a much needed knowledge transfer between the two entities, especially the advanced know-how concerning Web accessibility and usability that UTAD has developed over the last ten years.

3.1 Project Characterization

Over the recent years there has been a substantial technological evolution associated with how public and private sector organizations perform their operations and how they relate among themselves. [15-17]. The adoption of new Internet-based business models will certainly generate positive business revenues, thus the potential for e-business is high enough for motivating enterprises to go online [18-20].

When in 2005 the European Commission (EC) analyzed the full spectrum of electronic public procurement initiatives existent in the EU space, it identified a need for the creation of a set of standards and systems in order for all organizations from every EU member-state to be able to participate in the European e-Procurement community in an equal manner. With this in mind, in 2008 the EC created the Pan-European Public Procurement Online (PEPPOL) project that introduced mandatory standards and compliances that organizations would have to follow in order to be able to provide goods or services to any public entity in the European space [21].

Considering the above, the present paper presents a detailed description of a R&D project (anoGov-PEPPOL), performed by the University of Trás-os-Montes e Alto Douro (UTAD) , in association with ANO, which aimed to adapt ANOs electronic procurement platform (anoGov) in order for it to be compliant with the standards and technology indications that PEPPOL incorporated, thus taking a strong stand towards the development of products that are able to be traded internationally.

Adating anoGov e-procurement platform to PEPPOLs standards also transformed this software into a more innovative and competitive solution for the European market. With this in mind, ANO (who prides on being a very innovative enterprise) established a partnership with UTAD, aiming to take advantage of this University's background and current research activities on the topics of Web accessibility & usability and e-Business/e-Commerce.

The anoGov-PEPPOL R&D project was active between September 2011 to August 2014 , during which, a set of activities have been analyzed, planned and performed through a very interactive work and collaboration methodology between both institutional entities. Due to this effort it was possible to not only adapt anoGov platform to PEPPOL but also to make this a more usable and accessible internet platform.

3.2 Work Methodology

The work and collaboration methodology that was followed during this project can be characterized by a sequence of stages (composed by events, decisions or actions) that allowed for a mutual agreement on all the performed steps. During the first stage, both the ANO and UTAD teams analyze a given requirement individually and in detail in order to obtain a full understanding of its functional and technical characteristics, as well as what kind of tasks might be associated with it. After reaching conclusions on the analyzed requirement, both teams schedule a live meeting where they would share notes and opinions.

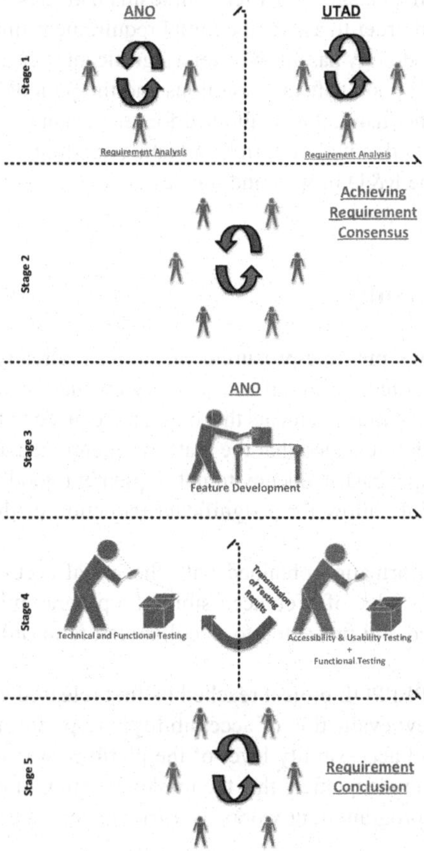

Fig. 1. -anoGov-PEPPOL project work methodology.

After the teams from UTAD and ANO performed an individual analysis to a given requirement, they meet and shared the results of their previous work and discussed the strategy through which they were going to approach the specific requirement. When the work strategy definition occurred (achieved from a consensual agreement), a task schedule was designed in order to ensure alignment between ANO's team (responsible for the development of all features) and UTAD's team that would be responsible for the functional testing, as well as for the accessibility and usability tests.

During the third stage, ANOs team underwent technical efforts towards developing the agreed feature, always incorporating in their developments the concerns and suggestions that UTAD's team was transmitting over time. When this development stage reached its end, both teams started a testing stage where all feature details were analyzed and tested from a technical, functional, accessibility and usability perspective. Every time UTAD's team was faced with some sort of error, it transmitted its feedback to ANO's team in order for them to carry out the necessary corrections or adjustments.

In the fifth and final stage, and after both teams had finished their tests, a live meeting was schedule in order to assure the initial requirement finalization.

This work methodology has allowed for a significant improvement in both the efforts need for developing good software solutions and the periods of time and costs needed to do so. On the other hand, it also allowed for an exchange of knowledge between the University and ANO that was very rich and understanding, which would not have been possible without the R&D project and the designed and implemented work methodology.

4 Project Results

Given that anoGov's public e-procurement platform already existed, when anoGov-PEPPOL project started, an initial accessibility evaluation was performed, aiming to set the basis for all the improvements that ought to be made in the context of this project. This initial evaluation revealed that the platform mentioned above was far from accessible, thus assuming a clear tendency to not implement good-practices in terms of programming that should allow for a significant reduction of the accessibility errors detected.

Based on the information obtained with the initial accessibility testing combined with UTAD's knowledge of Web accessibility, a programming manual of good-practices was developed and presented to 's collaborators in order to improve their skills and results.

When anoGov-PEPPOL project reached its first released version in a "development environment", a new evaluation of accessibility was performed and a set of improvements to the overall accessibility level of the platform was determined. This allowed for the overall acknowledgement that the implementation of best practices for Web accessibility-related programming works as a trigger for accessibility improvements of websites.

The improvement of the anoGov platform accessibility levels was a reality throughout the remaining accessibility tests performed during the duration of the project. This fact was proven when the WCAG 2.0 compliance level A was reached when the project ended and the final version of the platforms was achieved.

Besides assuming the requirement for achieving a WCAG 2.0 level A compliant public e-procurement platform, the teams associated with the project also intended to ensure that a minimum level of simplicity and ease of use was available for all platform users. In order to ensure this reality at the same time as all feature developments, several usability manual tests [22, 23] were performed, using both IT/IS experts and simple administrative personnel. The feedback obtained from these usability tests allowed for design and functional changes during the process of feature development and also allowed for significant savings in time and effort (compared to having to perform all changes after the full feature development).

5 Final Considerations

The anoGov-PEPPOL project was initially defined as a project that would allow for a customization of an existing public e-procurement platform to integrate with Europe's PEPPOL program. In addition to this initial goal, ANO, in association with UTAD University, defined extra goals by assuming that ANOs public e-procurement platform should also be WCAG 2.0 level A compliant (ensuring that it was accessible to a wide range of users) and easy and simple to use.

By assuming a stage-based work methodology the R&D project was executed without any schedule problems or deviations and with all initial goals achieved. This work methodology, created specifically for the project, proved to be the correct one for this type of development.

With the execution of this project, and in addition to the platform developments that completely integrated it with PEPPOL and made it WCAG 2.0 level A compliant, ANO also seized the opportunity to improve the technical abilities of its collaborators and their standards of good-practices, and UTAD also acknowledged a wide range of know-how on how to develop and manage public e-procurement platforms.

Acknowledgements

This paper was supported by the project anoGov PEPPOL, Nº 21.578, financed by the European Regional Development Fund, through the North Operational Programme by the Systems of Incentives for Research and Technological Development.

6 References

1. Moon, M., *E-procurement management in state governments: diffusion of e-procurement practices and its determinants.* Journal of Public Procurement, 2005. **5**(1): p. 54-72.
2. Corsi, M., *E-procurement overview*. 2006, OECD.
3. Farmer, D., et al., *Procurement, Principles and Management*. 2008: Pearson Education.
4. Ndou, V., *E-government for developing countries: opportunities and challenges.* The Electronic Journal of Information Systems in Developing Countries, 2004. **18**.

5. EU, *Digital Agenda: eGovernment Action Plan to smooth access to public services across the EU* E. Commission, Editor. 2006.
6. Sócrates, J., *Decreto-Lei n.º 18/2008*, P. Government, Editor. 2008.
7. santos, F., et al., *Despacho Nº 32639-a/2008*, P. Government, Editor. 2008.
8. Goncalves, R., et al. *Portuguese web accessibility in electronic public procurement platforms*. in *Information Systems and Technologies (CISTI), 2010 5th Iberian Conference on*. 2010.
9. W3C, *Web Content Accessibility Guidelines 2.0*, W.W.W. Consortium, Editor. 2008.
10. INE, *Census 2011*. 2011, Instituto Nacional de Estatística.
11. EU *People with disabilities*. 2014.
12. Barroso, J., *Resolução do Conselho de Ministros 110/2003*, P. Government, Editor. 2003.
13. Sócrates, J., *Portuguese Ministers Council Resolution N.155/2007*, P. Government, Editor. 2007.
14. Gonçalves, R., J. Martins, and F. Branco, *A Review on the Portuguese Enterprises Web Accessibility Levels – A Website Accessibility High Level Improvement Proposal.* Procedia Computer Science, 2014. **27**(0): p. 176-185.
15. Koussouris, S., G. Gionis, and F. Lampathaki, *Transforming traditional production system transactions to interoperable eBusiness-aware systems with the use of generic process models.* International Journal of Production Research, 2009. **48**(19): p. 5711-5727.
16. Trkman, P., *The critical success factors of business process management.* International Journal of Information Management, 2010. **30**(2): p. 125-134.
17. Gonçalves, R., et al., *Enterprise Web Accessibility Levels Amongst the Forbes 250: Where Art Thou O Virtuous Leader?* Journal of Business Ethics, 2013. **113**(2): p. 363-375.
18. Morais, E., A. Pires, and R. Gonçalves, *e-Business Maturity: Constraints associated with their evolution.* Journal of Organizational Computing and Electronic Commerce, 2011.
19. Jones, P., et al., *The proposal of a comparative framework to evaluate e-business stages of growth models.* International Journal of Information Technology and Management, 2006. **5**(4): p. 249-266.
20. Wall, B., H. Jagdev, and J. Browne, *A review of eBusiness and digital business - applications, models and trends.* Production Planning & Control, 2007. **18**(3): p. 239-260.
21. Mondorf, A., D. Schmidt, and M. Wimmer. *Ensuring Sustainable Operation in Complex Environment: the PEPPOL Project and its VCD System.* in *MCIS*. 2010. Tel Aviv, Israel.
22. Bento, M. and J. Lencastre. *Avaliação da Usabilidade do Protótipo Multimédia "Alfa E Beta"*. in *Atas do II Congresso Internacional TIC e Educação*. 2012.
23. Rubin, J. and D. Chisnell, *Handbook of usability testing: howto plan, design, and conduct effective tests*. 2008: John Wiley & Sons.

Analyzing Developer Behavior and Community Structure in Software Crowdsourcing

Hui Zhang[1], Yuchuan Wu[2] and Wenjun Wu[3]

State Key Laboratory of Software Development Environment,
Beihang University, Beijing, China
{[1]hzhang, [2]wuyuchuan, [3]wwj}@nlsde.buaa.edu.cn

Abstract. Recently software crowdsourcing has become an emerging development paradigm in software ecosystems. At present, most research efforts on software crowdsourcing focus on modelling its competitive nature from aspect of incentive mechanism and developer decision using game theory. Few work has been done on analyzing the impact of developer behavior and community structure on software crowdsourcing practices. Based on social network modeling and analysis methodology, this paper studies a popular community of software crowdsourcing named as TopCoder. We discover that the online activities of TopCoder users can be characterized as a temporal bursty pattern. Such a user behavior leads to the similarity in the participants of TopCoder contests occurred within the consecutive time intervals. Furthermore we introduce a competition social network to study the influence of the cooperation and competition between developers on their rating in the TopCoder community by analyzing competition social network. In addition to the community-wide developer network, this paper extends the modeling method to examine topologic characteristics of TopCoder projects and reveal the correlation between these structural features and the outcome quality of the projects.

KEYWORDS

software crowdsourcing, behavior, social network, centrality, competition

1 Introduction

With the increase in software system complexity, many companies open up their platforms to embrace the collaborative effort from online developers to develop software. Software crowdsourcing is a promising approach to allow organizations to outsource software development tasks to a virtual and on-demand workforce. By tapping into the collective intelligence of participants, organizations aim to acquire the high quality software with a reduced cost while stimulating the community to focus on certain skills or knowledge in various software ecosystems.

Currently, major software crowdsourcing platforms including Apple's App Store, TopCoder [1], and uTest [2], demonstrate the advantage of crowdsourcing in terms of software ecosystem expansion and product quality improvement. Apple's App Store is an online IOS application market [3], where developers can directly deliver their creative designs and products to smartphone customers. These developers are motivated to contribute innovative designs for both reputation and payment by the micro-payment mechanism of the App Store. Another software crowdsourcing example – TopCoder, creates a software contest model where programming tasks are posted as contests and the developer of the best solution wins the top prize. Following this model, TopCoder has established an online platform to support its ecosystem and gathered a virtual global workforce with more than 250,000 registered members and nearly 50,000 active participants.

Many researchers used game theory to study the impact of TopCoder's prize structure and reputation system on the TopCoder community members, and analyzed the principal factors such as project payment and requirements on the quality of the outcome in the competition. But few efforts have been made to analyze behavior of individual developer, developer competitive interactions and community structure in TopCoder programming contests. More important, it is essential to study whether there is any correlation between community structure and the quality of the product of TopCoder programming contests.

Social network analysis (SNA) has widely employed to study the evolution and structure of a wide variety of online groups and communities, including blogsphere, Twitter, online forums, social networking sites, and so forth. From a social network perspective, the TopCoder community includes two major entities, one entity is set of developers in the community, the other is the set of contests organized by TopCoder. For the relationship between these two sets of entities, there are two kinds of social network models: one is the participatory network describing the relationship between developers and contests, which can be modeled with the bipartite graph. The other is the competition network describing the relationship between developers competing with each other. This goal of this paper is to

K.J. Kim (ed.), *Information Science and Applications,*
Lecture Notes in Electrical Engineering 339, DOI 10.1007/978-3-662-46578-3_117

study the topologic properties of both networks in the TopCoder community and discover the relation between these structural properties and major indicators in crowdsourcing including community ranking and project quality.

This paper is organized as follows: Section 2 presents an overall statistical analysis of the TopCoder crowdsourcing community from the contest and developer perspective. Section 3 examines the bursty developer behavior pattern and its impact on the temporal closeness of nodes in the bipartite graph model of TopCoder contests. . Section 4 introduces a competition network model to study the influence of cooperation and competition relationship among TopCoder developers on their community ratings. Section 5 studies the topological properties of both the bipartite graph and competition network of TopCoder projects and evaluates these primary properties that can affect the quality of the solutions delivered by these projects. Related work is described in Section 6. Conclusions and future work are presented in Section 7.

2 Overall Statistical Analysis

The software development in the TopCoder community divide the project into the eight phases including conceptualization, requirement specification, UI prototype, architecture and component design, component implementation, testing and deployment. At each phase, TopCoder organizes multiple programming contests to accomplish the tasks. At present, there are totally 14751 contests and 40726 submissions on the TopCoder site. Table 1 presents the overall statistics of the TopCoder contests, including the category of the contest, the numbers of the contests, the numbers of the submissions, the reward and score of the contest. Table 2 shows the staticstics of the TopCoder developers, including their submissions and average award that every developer has won.

Table 1. Contest Analysis

	Tasks	Submissions	score	reward	duration
Architecture	807	1.86	88.72	2013.6	19.12
Assembly	2871	1.87	88.43	1948	16.97
Bug Hunt	358	3.82	/	495.7	5.48
Code	1422	3.77	78.33	1154.1	10.62
Component Des	2999	2.42	88.82	983.3	16.85
Component Dev	3117	3.74	89.28	876.6	18.45
Conceptualization	288	3.24	81.85	1657.6	20.72
Content Creation	99	3.38	77.32	1313.3	18.94
Copilot Postings	526	3	68.54	231.8	12.21

Table 2. Developer Analysis

Worker statistics	Count
Submissions(mean)	10.9
Submissions(median)	2
Wins(mean)	58.60%
Wins(median)	58.30%
Mean reward sought	446
Median reward sought	429

2.1. From the Perspective of the Contest

Figure 1 displays the number of the new generated contests every month from January 2004 to July 2014. During the ten years, the number of the new contests generated in general is on the rise. Especially from 2011 to 2013 the growth line is very steep. Then a decline occured in the number of the contests from 2013 until 2014 and went back to a trend of rising sharply. This shows that the enthusiasm of the developers to participate in the contest vary differently in months of the same year.

As Table 1 shows, there are great difference between different categories of contest. The contests of the Component Development, Component Design, Assembly type have the top three contests amount, because almost every software have these phases. And the top three kinds of contest with most awards is Architecture, Assembly and Conceptulization, maybe they are hard to implement. Bug Hunt, Code, Component Development have most submissions, indicating the most intense competition enthusiasm or a small degree of difficulty.Some contest bonuses only less than ten dollars, and some have more than $5000. Figure 2 shows the CDF function of the awards the five type contests with most submissions. Some awards such as 500, 1000, 1500, 2000, 3000 are common, the award distribution of Archivetecture and Assembly contest is higher. The awards of Component Development and Component Design Contests are under 1500 yuan. The discretion of the awards and the difficulty of the contest should have a direct relationshipThe period of different type contest is not the same, the Conceptualization contest has longest average period, and it has most awards.

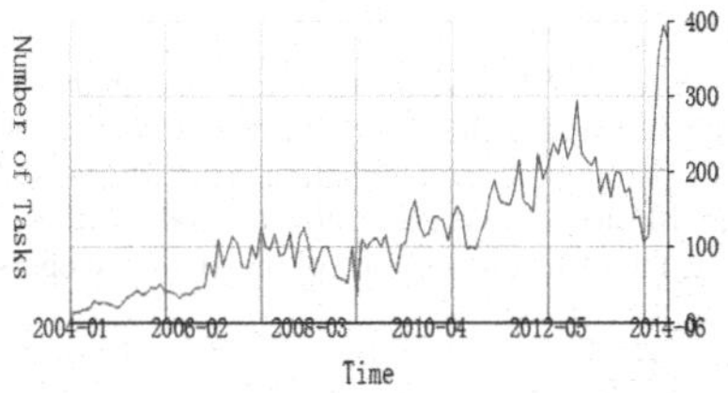

Figure 1. The Distribution of number of the new contests

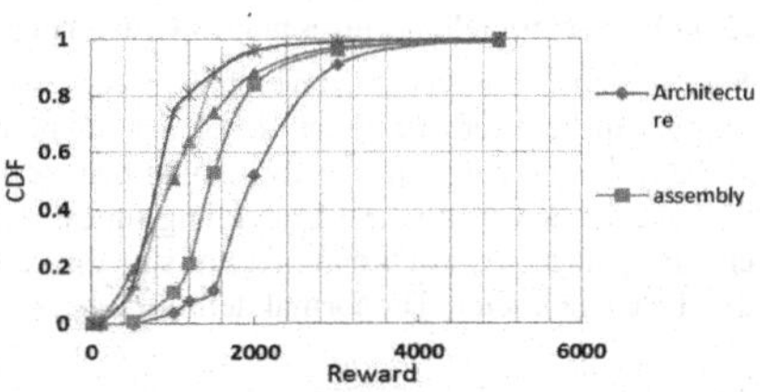

Figure 2. The CDF of Awards of five types Contest

2.2. From the Perspective of the Developer

The distribution of the submissions per developer meets the long tail distribution, each worker's average submission number is 10.9, the median is 2. Figure 3 displays the CDF distribution of the average awards of all the developers, the blue line represents developers with more than 50 submissions, and the red line represents developers with at least one submission. Visible developers with more submissions choose higher bonus than the average, such as $ 500, $ 700. In contrast, almost 50% of all the developers choose the the average award less than $100.

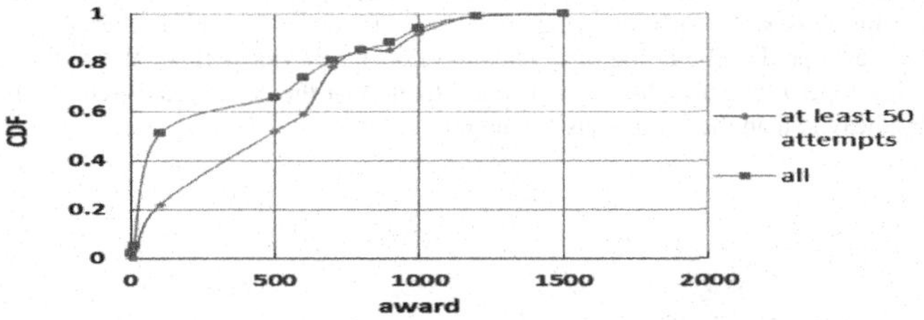

Figure. 3. The CDF of developers's Awards selection

3 TopCoder Developer Behavior

In this section, we study developer behavior pattern on their earnings, and the developer behavior pattern is related to the contest sequence, after modeling the developer behavior pattern appropriately, find each possible influencing factors, analyze its relation with rating.

3.1. Developer Behavior Pattern

When we study the online activity of TopCoder developers and after analyzing the data, we found that the online activities of TopCoder developers demonstrate a short temporal continuity of their participation in contests. Within a relative short time interval when a developer has enough free time and energy, he actively engages with contests in the TopCoder community. And for the rest of the time, he becomes idle and doesn't commit any contribution to TopCoder projects. Figure 4 presents a typical example of such an on-off activity in the TopCoder Specification Contests of the developer with id 20437508, X-axis: 274 contests sorted in the time order, Y-axis: where attend or not.

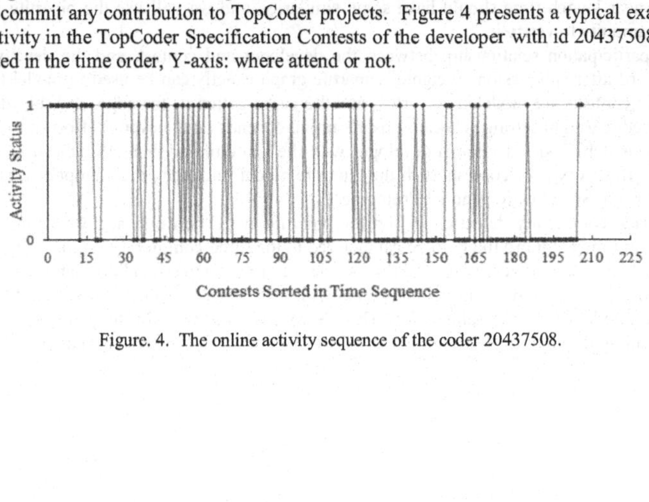

Figure. 4. The online activity sequence of the coder 20437508.

3.2. Behavior modeling and Kmeans Clustering Analysis

There are 3097 component development contests and 6685 registered developers. Firstly we study the relationship between the times developer attending the contest and their rating. As can be seen from Figure 5, there are no obvious correlation between the participation times and the rating. Then choose the 101 developers with the most attending times, range from 105 to 16, and use vector to model the user behavior pattern, regard the number of successively times as the feature, every one of the 101 developers has a behavior vector. The formal definition is as follows:

$$V = < v_1, v_2, ..., v_{17} >$$

where v_1 is the number of successively not attending more than 1000 times, v_2 is the number of successively not attending between 301 and 1000 times, v_3 is the number of successively not attending between 51 and 300 times, v_4 is the number of successively not attending between 16 and 50 times, v_5 is the number of successively not attending between 7 and 15 times, v_i is number of successively not attending $(12 - i)$ times, $i \in [6,11]$, v_i is number of successively attending $(i - 11)$ times, $i \in [12,17]$. Since the total attending times of developer has no effect on his rating, so each feature should divide by developer's total attend times.

Since developers demonstrate continuous behavior mode of participating in the contest, we further analyze the impact of this behavior pattern on developer's rating. We apply the K-Means [4] clustering to the 101 vectors, set three class, the results are as Table 3 shows, different kinds of user behavior patterns really produce different rating, and developers in one clustering class tend to have similar ratings, in different clustering class seem to have ratings with big difference. Using the clustering model of this situation, we can also predict the rating of a given developer behavior pattern. We further analyzed the correlation of developer rating and the vertex feature, found that the continuous attending feature affects developer rating heavily than the continuous not attending feature.

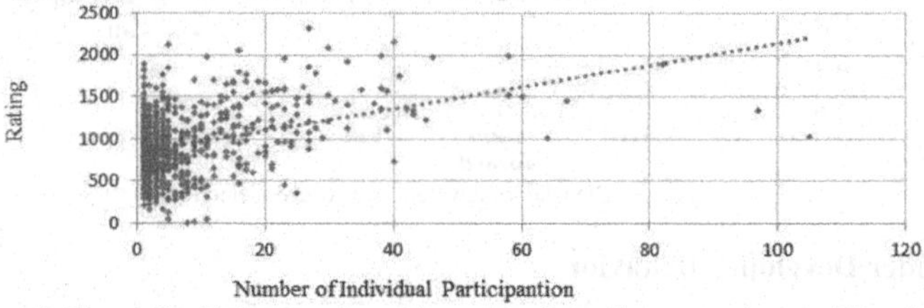

Figure. 5. The relationship between deveper's number of attending contest times and his rating.

Table 3. K-Means Clustering Result With 3 Class

Class	0	1	2
Number	54	8	39
Average rating	1161.38	1571.75	1423.23

3.3. Verification from Bipartite Graph of the TopCoder Community

In the above part we found developers prefer to consecutively participate in the contest, think about it, the contests with near launch time should have some similarity, and we will use the bipartite graph of the TopCoder community to verify it.

There are participation relationship between the developer and contest, and the developer will get points and award after submission. Weighted bipartite graph exactly can be used to model the situation, developers and contests are modeled as vertex, and the participation as the edge, and the edge weight is the point. The edge weight setting is because user's rating depends on the overall effect of the points from every attended match, TopCoder ranks developer with the Elo rating system. We construct the bipartite networks for different type of contests with the bipartite model, including the Component Development, Component Design, Specification and UI Prototype.

Then we made community detection [5] on the bipartite graph network, and we found that contests with near launch time are more likely to be categorized in the same community. In the scatter plot shown in Figure 6, the abscissa axis represents the time sequence of the contests and ordinate axis shows all the community class of the bipartite graph, and points with near abscissa values are always have the same ordinate value, marked with the same color. This shows that contests with temporal similarity tend to have more common developers, thus undoubtedly the consequence of the behavior pattern of developers.

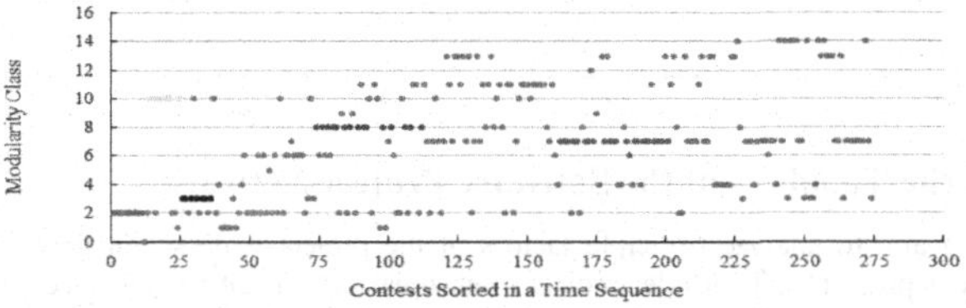

Figure. 6. Relationship between Contest community and time sequence

4 Competition Network and Community Rating

In this section, we model the competitive relationship among TopCoder developers based on social network analysis (SNA) methodology. A competition network is introduced to describe the structure properties of community-wide competition and its impact on the quality of the outcomes of crowdsourcing projects.

4.1. Competition Network Model

The competition network is a directional weighted graph G=(W,E), where W is a set of vertices that represent TopCoder developers , E is a set of directional edges representing the competition relationship between these developers. Usually, participant u regards participant v a competitor when u and v show up in the same contest. To measure the intensity of their competition, we define a weight metric as follows:

$$w_{u,v} = \frac{\sum_{c \in C} \frac{x_{u,c} x_{v,c}}{rank(c,v)}}{\sum_{c \in C} x_{u,c}}$$

C is the contest collection, u and v denotes two developers. When u delivers a submission to a contest c, x(u,c)=1, otherwise, x(u,c)=0. rank(c,v) is defined as the ranking of the software solution delivered by the participant v in the contest c. This factor can indicate the significance of the developer v's presence on the developer u in terms of competition. The higher is the rank of v, the more competitive influence presented by the developer v on developer u. These weights are not necessarily symmetric and are normalized to avoid biases because of the disparity among the number of the submissions delivered by developers.

We mainly choose five network centrality indicators, including Weighted Degree, Betweenness Centrality, Authority Hub[14], PageRank[15], Local Clustering Coefficient.

4.2. Multiple Linear Regression Analysis and Correlation Analysis

We perform a correlation and multiple linear regression analysis of the competition network for the TopCoder Component Design contest. , In the regression, the centrality feature is regarded as the independent variable, and the developer community rating and average points are regarded as the dependent variable. Table 4 shows that all correlation coefficients with the exception of the Closeness are of statistical significance (p-value at most 0.05), which suggests that there is a correlation between the independent and dependent variables. The positive correlation coefficient of the weighted in-degree indicates that the developers who have more competitors have higher rating in the community. As the weighted out-degree of a developer node represents the competition from other developers, the negative correlation coefficient of the weighted out degree apparently confirms the common sense that a relatively weak developer who have to face more competitors tend to have lower rating in the TopCoder community. The result also shows that local clustering coefficient is negative, indicating that close competition from a developer's neighbors lowers his community rating. Furthermore, as both the HITS algorithm and PageRank algorithm are very effective to identify the most important node, the positive correlation between these metrics and community rating enables us to predict developers with high rating score using the hits and the PageRank algorithm.

Table 4. The Pearson correlation of the centrality and rating.

Centrality metric	Correlation	Sig.
InDegree	.419	.000
OutDegree	-.489	.000
Closeness	-.189	.670
Betweenness	.123	.000
Authority	.466	.000

Hub	.472	.000
PageRank	.520	.000
ClusteringCoefficient	-.239	.000

5 Analysis of the Topology of the Software Project Network

In this section, we aim to analyze principle factors of the project-wide social network that affect the quality of TopCoder project. A TopCoder project often breaks down into a sequence of contests ranging from requirement analysis to implementation and testing. Therefore, interaction between contests and developers in the development process of each project can be formalized again with the bipartite graph model and competition network model.Every project consists of a sequence of TopCoder contests organized by the project's manager. On the platform of TopCoder, the outcomes of all these contests are evaluated by a review team with members selected from experienced developers in the TopCoder community. Thus the quality of the outcome from every contest can be quantized as a score ranging from zero to one hundred.

We selected 59 TopCoder projects for this study. For each project, its quality is measured by calculating the average scores of its relevant contests. Based on the same analytic approach as specified in Section 4, we run the regression analysis to find out the potential correlation between the topological characteristics and the quality of TopCoder projects.

Three global graph metrics of a project network are positively associated with the quality of the project outcome including weighted average degree, average cluster coefficient and network diameter. It is clear that the size of the competition network has positive impact on the quality of a crowdsourced project because more participants can bring in better solutions for the requirement of customers. The first half of Table 5 illustrates the statistical significance of the correlation between the topological metrics of the bipartite graph and the quality of projects. And the second half of Table 5 exhibits the statistical significance of the correlation between the topological metrics of the competition graph and the quality of projects.

The average weighted degree of the bipartite graph model is positive correlated with the project quality, illustrates that the more powerful ability the participated developers have, the higher quality of the software project, and the graph density of the bipartite graph is positively related to project quality, revealing that developers can also play in a variety of different types of the contest in favor of software quality assurance, Positive correlation between the average degree, network diameter, graph density, average cluster coefficient of the competition network and the software project quality, showing the more and stronger of the rival of the competitor, the higher the quality of the software project. The connected components amount is negative correlated with the software project quality, that the more closely linked between the competitors and the more intense of the competition, the higher quality software project to be produced.

We choose two representative software projects from that 59 TopCoder projects, the first one is "24 hour Omicron Breeding Studio" with high quality, and the second one is "Studio Replatforming" with low quality. In Figure 7 we visualize them with our defined bipartite graph, competition network model, contests and developers are separated by node size in the bipartite graph, nodes in different community have different color, nodes with higher degree are bigger in the competition network. One can see that the average degree, density, network diameter, average cluster coefficient of the first project is higher than the second one, consistent with our above conclusion.

Table 5. The Pearson correlation of the network topology and software quality

model	topology metric	Correlation	Sig.
Bipartite Network	Ave Degree	.428	.015
	Density	.462	.015
	Ave path length	.117	.041
Competition Network	Ave Degree	.541	.043
	Diameter	.236	.187
	Density	.494	.129
	Connected Component	-.312	.179
	Ave Clustering Coefficient	.560	.070

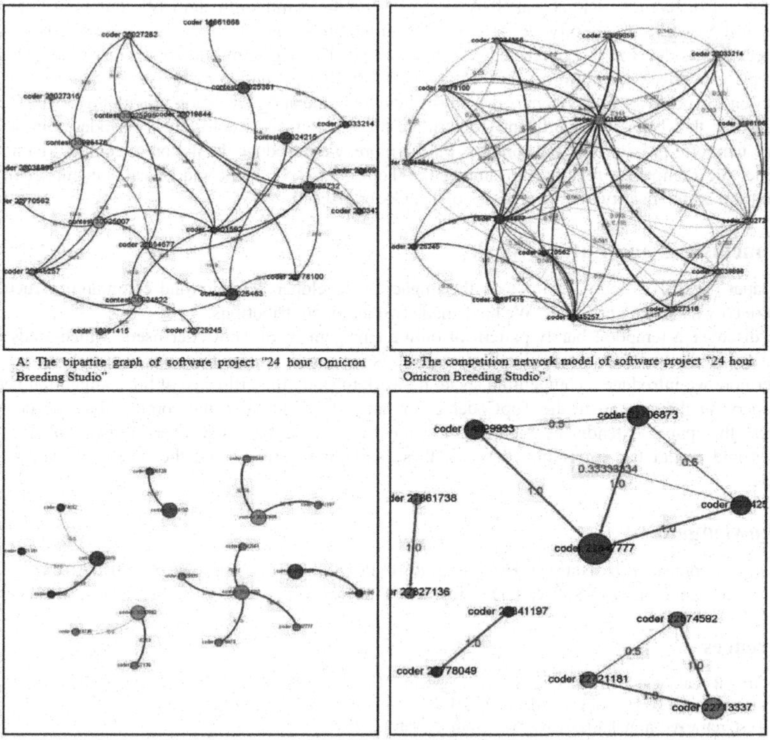

A: The bipartite graph of software project "24 hour Omicron Breeding Studio"

B: The competition network model of software project "24 hour Omicron Breeding Studio".

C: The bipartite graph model of software project "Studio Replatforming".

D: The competition network model of software project "Studio Replatforming".

Figure. 7. The bipartite graph, competition network model of two software project

6 Related Work

Many scholars have conducted in-depth research on how to take advantage of collective intelligent to develop high quality software in a cost-effective way. Archak and Sundararajan [6] use the game theory to analyze the competitiveness of the crowdsourcing software, especially in the price structure optimization. Archak [7] extends this method to the study of the influence of the reputation system of TopCoder, and analyzes the main factors in a competition, including the project payment and quality demand. Similarly, other scholars try to establish the business model [8] [9] of crowdsourcing software, and use the auction theory to construct a strategy model for the participants to easily involved in. Bacon [10] introduces a new kind of software evaluation pattern, driven by the market mechanism of providing rewards for developers, testers and bug reporters.

At present, few studies have been done about the crowdsourcing software with the social network analysis method, but researchers have published many papers on social network analysis of open source software communities. Yongqin Gao [11] tried to take the open source community as a cooperative network in his doctoral thesis, every developer and project is a node of the network. And he analyzed the experimental data they collected from SourceForge, and obtained the statistical and topological information of open source software developer cooperation network.

In [12], the author applies social network analysis to CVS data, graphing network measurements such as degree distribution, clustering coefficient in modules, weighted clustering coefficient, and connection degree of modules for various projects at different time periods in the histories of Apache, Gnome, and KDE. He concluded that both the module network and the developer network exhibit small-world behavior. Weiss [13] uses a similar approach to determine how communities add new developers (preferential attachment) and that developers migrate through communities.

The major difference between our SNA of crowdsourcing community and OSS community lies in the nature of the two kinds of software development paradigm. As an OSS community is mostly based on volunteering, its developer network tends to be more stable, long-term and collaborative ties where the ties TopCoder developer networks tend to be short-term, transit and competitive. Therefore, our study focuses on the behavior of developers driven by their free time and motivation of winning prizes. And we find out that this behavior significantly determine the topological structure of TopCoder contest graph through temporal proximity of the contests. Furthermore, we introduce the notion of competition network to model the competing relationship among the TopCoder developers and reveal the impact of such competition on the quality of the outcomes of TopCoder project.

7 Conclusions and Future Work

This paper uses SNA approach to study the impact of developer behavior and community structure on software crowdsourcing practices. We have made two major contributions:

We discover a temporal bursty pattern of online engagement of TopCoder users, which leads to the similarity in the participants of TopCoder contests occurred within the consecutive time intervals. Furthermore we introduce a competition social network to study the influence of the competition between developers on their rating in the TopCoder community. In addition to the community-wide developer network, this paper extends the modeling method to examine topologic characteristics of TopCoder projects and reveal the correlation between these structural features and the outcome quality of the projects.

Acknowledgements

This work is sponsored by National High-Tech R&D Program of China (Grant No. 2013AA01A210) and the State Key Laboratory of Software Development Environment (Funding No. SKLSDE-2013ZX-03).

References

[1] Karim R. Lakhani, David A. Garvin, Eric Logstein, "TopCoder: Developing Software through Crowdsourcing," Harvard Business School Case 610-032, 2010.
[2] uTest, retrieved from https://www.utest.com/ on Jan 12, 2013.
[3] Jan Bosch, From Software Product Lines to Software Ecosystems, SPLC '09 Proceedings of the 13th International Software Product Line Conference.
[4] Hartigan, J. A., Wong, M. A. (1979). "Algorithm AS 136: A K-Means Clustering Algorithm". Journal of the Royal Statistical Society, Series C 28 (1): 100–108. JSTOR 2346830.
[5] Blondel V D, Guillaume J L, Lambiotte R, et al. Fast unfolding of communities in large networks[J]. Journal of Statistical Mechanics: Theory and Experiment, 2008, 2008(10): P10008.
[6] Archak N, Sundararajan A. Optimal design of crowdsourcing contests[J]. ICIS 2009 Proceedings, 2009: 200.
[7] Nikolay Archak, Money, "Glory and Cheap Talk: Analyzing Strategic Behavior of Contestants in Simultaneous Crowdsourcing Contests on TopCoder.com," Proceedings of the 19th international conference on World Wide Web, 2010.
[8] Dominic DiPalantino, Milan Vojnovi´c, "Crowdsourcing and All-Pay Auctions," EC '09 Proceedings of the 10th ACM conference on Electronic commerce.
[9] John Joseph Horton, Lydia B. Chilton, "The Labor Economics of Paid Crowdsourcing," EC '10 Proceedings of the 11th ACM conference on Electronic Commerce, Pages 209-218.
[10] David F. Bacon, Yiling Chen, David Parkes, and Malvika Rao, "A Market-Based Approach to Software Evolution," 24th ACM SIGPLAN conference companion on Object oriented Programming Systems Languages and Applications, October 25–29, 2009, Orlando, Florida, USA.
[11] Y. Gao, V: Freeh and G. Madey. Analysis and modeling of the open source software community. NAACSOS, Pittsburgh, 2003.
[12] L. Lopez-Fernandez, G. Robles, and J. M. Gonzalez-Barahona. Applying social network analysis to the information in cvs repositories. In Proceedings of the First International Workshop on Mining Software Repositories (MSR 2004), Edinburgh, UK, 2004. URL citeseer.ist.psu.edu/710559.html.
[13] M. Weiss, G. Moroiu, and P. Zhao. Evolution of open source communities. Open source systems, pages 21{32, 2006.
[14] Brin, S.; Page, L. (1998). "The anatomy of a large-scale hypertextual Web search engine". Computer Networks and ISDN Systems 30: 107–117. doi:10.1016/S0169-7552(98)00110-X. ISSN 0169-7552. Edi
[15] Kleinberg, Jon (December 1999). "Hubs, Authorities, and Communities". Cornell University. Retrieved 2008-11-09

A Semantic Similarity Assessment Tool for Computer Science Subjects Using Extended Wu & Palmer's Algorithm and Ontology

Chayan Nuntawong[1], Chakkrit Snae Namahoot (✉)[1] and Michael Brückner[2]

[1]Department of Computer Science and Information Technology, Faculty of Science, Naresuan University, Phitsanulok, Thailand
chayan.nuntawong@gmail.com, chakkrits@nu.ac.th

[2]Department of Educational Technology and Communication, Faculty of Education, Naresuan University, Phitsanulok, Thailand
michaelb@nu.ac.th

Abstract. This paper presents a process model and a system for calculating the correspondence between the content of courses in Computer Science with the standard of The Thailand Qualifications Framework for Higher Education (TQF: HEd). The aim is to improve the curriculum of universities in Thailand by meeting the standards and decreasing the duration of the operation to be more convenient. We designed an ontology of courses and TQF: HEd, and then we developed the system as a web application that can map the information from two ontologies using the extended Wu & Palmer's algorithm and WordNet. At last, the system summarizes the consistency of the course descriptions with the body of knowledge of TQF: HEd. Tests with sample data show that this method indicates whether the course description is consistent with the standards or not. It also indicates the relative importance of each part of the body of knowledge in teaching of this subject as well.

Keywords: Ontology · Semantic Similarity · Qualifications Framework · Web Application

1 Background and Problems

The higher education management in Thailand has had some problems relating standards of curriculum and teaching because the curriculum and teaching design in each discipline of most universities vary. This leads to the situation that most students try to enter prestigious universities where they expect the best teaching quality. However, each university can receive only a limited number of students, so many students who not admitted must chose to study at other universities that may still have problems in accordance with the teaching standard. This is a problem of inequality in education and can affect suitability of careers after graduation as well.

K.J. Kim (ed.), *Information Science and Applications*,
Lecture Notes in Electrical Engineering 339, DOI 10.1007/978-3-662-46578-3_118

The Office of the Higher Education Commission (OHEC), Ministry of Education, has developed a Thailand Qualifications Framework for Higher Education (TQF: HEd) [1] as a framework for each university in Thailand to improve the curriculum and contents to provide teaching being consistent and more homogeneous. This is particularly useful for students, and every university in the country is bound to follow the guidelines. Nevertheless, the implementation of the curriculum of subjects in each university is still complicated, consumes a lot of time, resources and experts (manpower). Moreover, some universities that improved the curriculum have still shown differences from TQF: HEd, e.g. in terms of teaching, course syllabus and course descriptions. This might have been caused by the difficulty of management as mentioned above. And even if the courses were updated according to the TQF: HEd, we also found differences in the emphasis in the teaching of the course content that varies with university. Moreover, in 2015 Thailand is entering the ASEAN Economic Community (AEC) [2], which collaborates in education for relationship and integration and in developing the ability of students to enter the career market in ASEAN [3] by encouraging the students to transfer credits from its university to other universities in member states. In practice, there are problems with the standard curriculum and teaching that may vary other ASEAN countries as well. In the past, there were no tools or technology helping to improve the curriculum with ease and shorten the process in a more convenient way.

To deal with this situation, we have developed the concept for a solution and a system that can determine the consistency between curriculum and TQF: HEd using data from Computer Science curricula. We use the principle of ontologies and applied semantic similarity algorithm of Wu & Palmer together with WordNet to obtain consistency of curriculum and the TQF: HEd standard that can help improve the curriculum process and solve problems in operation, as discussed above.

2 Related Work

Ontologies have been used a lot in the educational field. Brinson et al. [4] have studied the expertise, certifications and education in cyber forensics, and then have designed the ontologies used in the cyber forensics courses. Jiang et al. [5] have developed a computer networks course ontology that was designed from lesson topics to the various details that students must learn in each lesson. Bagiampou and Kameas [6] have presented an ontology regarding software engineering courses. It was developed in the form of Use Case Diagrams (UCD) that used existing methods with the collaboration of specialist software engineers to develop an ontology. In addition, Yao and Zhang [7] have designed knowledge representation using an ontology design for a MIS (Management Information System) course using Protégé 2000 and OWL.

In terms of Qualifications Framework. Nuntawong and Namahoot [8] have studied the structure of TQF: HEd and designed an ontology for this framework in Computer Science. Then they proposed a Structure-based Ontology Mapping technique for comparison between Computer Science curriculum ontology and TQF: HEd ontology. It was shown that a Structure-based Ontology Mapping can exhibit similarities between two ontologies and suggest improvements for a more consistent curriculum. There were still problems in the accuracy of comparisons. Other Ontology Mapping techniques have been made available in a study by Liu et al. [9], covering Structure-based, Syntax-based or Instance-based techniques. A number of Ontology Mapping tool have also been published by Gawich et al. [10], for example Smatch, Ontobuilder, Prompt, and the VSMCOS Tool.

Regarding semantic similarity techniques, one of the most popular is the Wu & Palmer's algorithm [11]. Much research has followed this principle for ontology mapping, e.g. the study of Wongkalasin and Archint [12] that determined the similarity of two different ontologies, and combined them to create a domain ontology for an E-library search system. Wongkalasin and Archint [13] also used a semantic similarity for Ontology Mapping in the E-Learning domain using the principle of Wu & Palmer for reference a meaning. In addition, the study of Shafa'amri and Atoum [14] relating a method to increase the efficiency of Ontology Mapping for Semantic Web by separating the structure of RDF is used in the design of ontologies, and considered a similar structure. They used two methods: string matching and semantic matching with Wu & Palmer's algorithm in the part of the comparison process. Even with the development of WordNet::Similarity by Pederson et al. [15], which is a tool that helps to measure the similarity of two words or concepts using vary semantic similarity principles, see Wu & Palmer, Leacock & Chodoroc [16] and Lin [17]. That tool can be both word comparison, and sentence comparison that the system will cut each word in sentences then determine the similarity of each word. But this tool cannot determine the semantic similarity of whole meaning in the sentence.

It can be seen that the semantic similarity techniques mentioned above considered only the similarity between words. They cannot be used for checking the curriculum and Qualifications framework data that are characterized by sentences that consist of several words, which may be written in various ways but have the same meaning. Therefore, in this research we have adopted guidelines from the related work for determining the semantic similarity as required to consider the meaning of the words in a sentence together with an extended Wu & Palmer's algorithm and WordNet [18] in order to have a consistent overview of the curriculum and TQF: HEd.

3 Methodologies and System Design

3.1 Ontology Design

In this paper, we use an ontology for TQF: HEd relating Computer Science (O_{TQF}), see [8] adapted for semantic similarity of curricula. The root of this ontology has a set of classes about courses in the qualifications framework, and each course class comprises the "Knowledge Area" class defined in TQF: HEd by 14 classes, e.g. Algorithms and Complexity, Intelligent Systems and Software Engineering. Each Knowledge Area class has a related "Body of Knowledge" class.

For a curriculum ontology (O_{CC}), the system imports the curriculum data from users that consists of the course description in English (only for international use). Then the system converts the data to an ontology that can be used for comparison with O_{TQF} as well.

3.2 System Architecture

The system architecture that show in Fig. 1, as follows:

1. Import computer science course containing course description (CD) of each subject in English language and convert to ontology (O_{CC}).
2. Retrieve body of knowledge (BK) in knowledge area (KA).from O_{TQF} and course description from O_{CC} which already compared to WordNet ($CD_{WordNet}$)
3. Calculate semantic similarity values between $CD_{WordNet}$ and BK (see section 3.3)
4. Display similarity assessment results of subject.

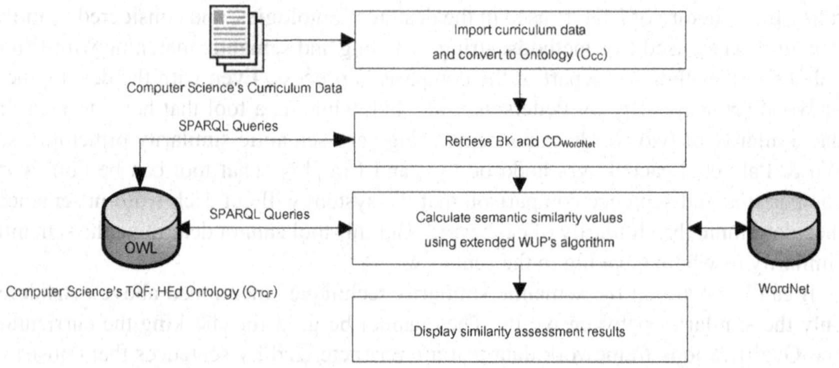

Fig. 1. The system architecture

3.3 Semantic Similarity Calculation

This section explains the calculation process of semantic similarity between $CD_{WordNet}$ and BK as follows:

1. The system creates a matrix with size m × n, where m is the number of course descriptions and n is the number of bodies of knowledge that have been normalized using WordNet. This process involves word segmentation, e.g. deleting all prefixes, prepositions and conjunctions in order to determine the similarity between each word in the matrix.
2. Calculate the similarity using Wu & Palmer's algorithm [11] as in formula 1.

$$Sim_{wup}(CD_{WordNet}, BK) = \frac{2 * Depth(LCS)}{Depth(CD_{WordNet}) + Depth(BK)} \quad (1)$$

3. Extended Wu & Palmer's algorithm for calculating average values from each matrix using formula 2.

$$Sim_{matrix} = \sum_{i=1}^{m} \sum_{j=1}^{n} \left\{ \frac{\left\{ \frac{Sim_{wup}(CD_{WordNet'}BK)_{i,j}}{n} \right\}}{m} \right\} * 100 \qquad (2)$$

4. Finally, the system selects the topic that has the maximum average value of semantic similarity from the matrix in each body of knowledge and uses it for estimate the weight values.

Table 1 shows an example of a calculation from matrix size m × n between some topic in course description of Artificial Intelligence (A.I.) subject, one of core courses from Computer Science's program, Naresuan University and three topics of Body of Knowledge in "Intelligent Systems" Knowledge Area from TQF: HEd.

The full details of course description of A.I. subject in this example is as follows: "Meaning, history and branches of artificial intelligent, intelligent agent concept, fundamental issues of artificial intelligent, development and application of artificial intelligence, machine learning, neural network, game playing, problem solving and search strategies, knowledge representation, knowledge based reasoning, expert systems and smart systems."

Table 1. An example of a calculation from matrix size m × n between topics "Problem Solving and Search Strategies" from course description of A.I. subject and the Body of Knowledge in "Intelligent Systems" from TQF: HEd

Segmentation of Course Description	Sim_{wup} of each body of knowledge					
	fundamental	issue	search	strategy	Knowledge based	reasoning
problem solving	0.54	0.63	0.82	0.72	0.73	0.88
search	0.57	0.82	1	0.55	0.66	0.82
strategy	0.43	0.7	0.55	1	0.82	0.52
Sim_{matrix}	61.50%		77.33%		73.83%	
…	…		…		…	
Sum of weight values (%)	41.67%		33.33%		25%	

For example, we want to calculate the average value of semantic similarity from the matrix in Table 1 that contains the words or phrases "problem solving", "search", "strategy" from the course description and words "search", "strategy" from the Body of Knowledge. We then calculate with Wu & Palmer's algorithm from formula 1, and get:

- Semantic similarity of "problem solving" with "search" is 0.82 and "strategy" is 0.72.
- Semantic similarity of "search" with "search" is 1 and "strategy" is 0.55.
- Semantic similarity of "strategy" with "search" is 0.55 and "strategy" is 1.

Then, we calculate the average value of this matrix by formula 2 and obtain the average value of the semantic similarity in this matrix as 77.33%.

Finally, we calculate the frequency of the most consistent part of each topic, which shows the course description in this subject have the weight values: "Fundamental Issues" 41.67%, "Basic Search Strategies" 33.33%, and "Knowledge based Reasoning" 25%.

4 Testing and Results

The semantic similarity tool was developed with JSP. Fig. 2 shows results of the semantic similarity assessment of course descriptions in A.I. subject and Intelligent Systems knowledge area. The results shows that the A.I. subject meets standards of TQF (i.e., no need to change course descriptions) and the three bodies of knowledge have approximate weight values of 42%, 33% and 25%, respectively. So, the A.I. subject at Naresuan University mostly emphasized fundamental issues. Also we have tested this subject with five other universities and found that U1, U2, U3, U4 and U5 emphasized basic search strategies (40%), knowledge based reasoning (46.15%), fundamental issues (50% and 41.67%) and basic search strategies (54.55%), respectively.

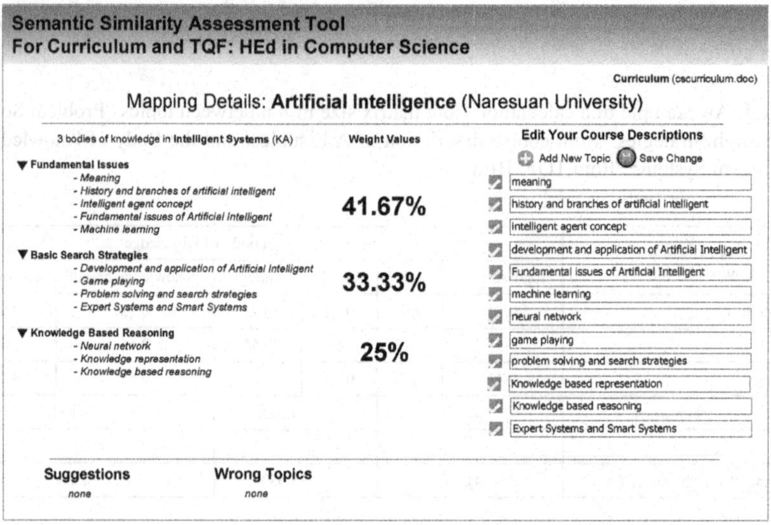

Fig. 2. Example of the system

The results indicate that the A.I. subject of these universities has the highest importance in teaching bodies of knowledge. However, we found that some weight values from some bodies of knowledge are very low. That may have several reasons, including the choice of words in the written descriptions above that using terms that are not consistent with the definitions in TQF: HEd. Thus calculation with Wu & Palmer's algorithm results in low semantic similarity values. Another reason might be that descriptions contain unusual names or words that do not appear in WordNet.

However, the results of this test show that the patterns in written descriptions of each course influence the consistency with the body of knowledge in TQF: HEd.

5 Conclusions and Further Work

In this paper, we presented the scope of the application of the principle finding semantic similarity, to be used to determine the consistency of the course by comparison with Body of Knowledge in TQF: HEd in the form of ontology. We extended Wu & Palmer's algorithm and WordNet to find the semantic similarity between two ontologies. Then, we tested with sample data of course descriptions from six universities in Thailand. The results show that the system can demonstrate the consistency of the subject with the Body of Knowledge in Knowledge Areas from TQF:HEd, which can be used in the overall assessment of the consistency of the subject, and that it can estimate the weight of importance of teaching in various fields of the Body of Knowledge. However, this test still faces the problems that the course description is not fully consistent with some elements of TQF:HEd because some words in the written description are not consistent with the definition of TQF:HEd and the usedof WordNet, which is not covered by a specific term in computer science. Also, the test only uses data from the course descriptions. If the information used in the design of course ontology has more details and is more comprehensive than this, it may determine semantic similarity of each word with greater resolution.

For the future work, we will add more features in order to process the calculation all of the courses and subjects at once. In addition, it is necessary to include more course information from other universities in Thailand and ASEAN member countries for a better comparison (nationally and internationally). Most importantly, this work should be based on a terminology of Computer Science, other than WordNet, being internationally accepted and applied.

6 References

1. Ministry of Education, http://www.mua.go.th/users/tqf-hed/news/FilesNews/FilesNews2/news2.pdf.
2. AEC Blueprint, http://www.thai-aec.com/aec-blueprint
3. Office of Higher Education Commission, Bureau of International Cooperation Strategy, http://inter.mua.go.th/main2/files/file/KM%2029_6_55.ppt
4. Brinson. A., Robinson. A. and Rogers. M.: A cyber forensics ontology: Creating a new approach to studying cyber forensics, pp. 37-43, Digital Investigation, Volume 3, Supplement (2006)
5. Jiang. L., Zhao. C. and Wei. H.: The Development of Ontology-Based Course for Computer Networks. In: Computer Science and Software Engineering 2008 International Conference, pp. 487 – 490 (2008).
6. Bagiampou. M. and Kameas. A.: A Use Case Diagrams ontology that can be used as common reference for Software Engineering education. In: Intelligent Systems (IS), 6th IEEE International Conference, pp. 35-40 (2012)

7. Yao. Z. and Zhang. Q.: Protégé-Based Ontology Knowledge Representation for MIS Courses. In: International Conference on Web Information Systems and Mining, pp. 787-791 (2009)
8. Nuntawong. C. and Namahoot. C. S.: An Analysis of Curricula in Computer Science Using Structure-based Ontology Mapping. In: The 9th National Conference on Computing and Information Technology (NCCIT 2013), Thailand. pp. 855-860 (2013)
9. Liu. X., Cao. L. and Dai. W.: Overview of Ontology Mapping and Approach. In: Broadband Network and Multimedia Technology (IC-BNMT) 2011 4th IEEE International Conference, pp. 592-595 (2011)
10. Gawich. M., Badr. A., Ismael. H. and Hegazy. A.: Alternative Approach for Ontology Matching. In: International Journal of Computer Applications (0975-8887), Vol. 49-No.18, pp. 29-37 (2012)
11. Wu. Z. and Palmer. M.: Verb semantics and Lexical Selection. In: Proceeding of 32nd annual meeting of the association for computational linguistics (ACL), Las Cruces, US, pp. 133-138 (1994)
12. Wongkalasin. S. and Archint. N.: Semantic Ontology Merging for Digital Library Domain. In: The 5th National Conference on Computing and Information Technology (NCCIT 2009), Thailand. pp. 288-293 (2009)
13. Wongkalasin. T. and Archint. N.: Ontology Mapping on E-Learning Domain with WordNet based on Semantic Similarity Measure. In: The 5th National Conference on Computing and Information Technology (NCCIT 2009). Thailand. pp. 218-224 (2009)
14. Shafa'amri, K. H. and Atoum. J. O.: A Framework for Improving the Performance of Ontology Matching Techniques in Semantic Web. In: International Journal of Advanced Computer Science and Applications (IJACSA), Vol. 3, pp. 8-14 (2012)
15. Pedersen, Patwardhan, and Michelizzi.: WordNet::Similarity - Measuring the Relatedness of Concepts. In: Proceedings of the Nineteenth National Conference on Artificial Intelligence (AAAI-04), pp. 1024-1025 (2004)
16. Leacock, C. and Chodorow, M.: Combining local context and WordNet similarity for word sense identification.In. MIT Press. pp. 265-283 (1998)
17. Lin, D.: An Information-theoretic definition of similarity. In: Proceeding of the International Conference on Machine Learning (1998)
18. Princeton University, http://wordnet.princeton.edu

Comparison of Architecture-Centric Model-Driven Web Engineering and Abstract Behavioural Specification in Constructing Software Product Line for Web Applications

Daya Adianto[1], Maya Retno Ayu Setyautami[1], and Salman El Farisi[1]

Universitas Indonesia, Formal Method in Software Engineering Lab, Faculty of Computer Science, Depok, West Java, 16424, Indonesia
fmse@cs.ui.ac.id

Abstract. We investigated how a software product line (SPL) for Web application is realized by following an established Web application development methodology called Architecture-Centric Model-Driven Web Engineering (AC-MDWE). The development process is done by using Abstract Behavioural Specification (ABS), which is an executable modelling language that provides SPL-related features. We created a case study by implementing a product line for E-commerce Web applications. The product line is realisable using ABS with modifications to the original AC-MDWE process. ABS can provide several benefits such as control during product generation, feature traceability, and preserving integrity of core assets. However, it is still not ready for creating production-level Web application and lack of readability in the generated artefacts.

Keywords: software product line, architecture-centric model-driven web engineering, abstract behavioural specification, ABS, AC-MDWE, model-driven development

1 Introduction

One of crucial parts in creating a software product line (SPL) is assembling a set of core assets which used for creating variety of products in the product line [1]. The core assets may include software artefacts such as source code, templates, and documentations. The development of products in the product line will use the core assets to derive new products.

An example of methodology that has been applied in creating an SPL is Architecture-Centric Model-Driven Web Engineering (AC-MDWE) [2]. It follows a two-track iterative processes to create a set of core assets that consists of Web application model, templates, model transformation rules, and a code generator. Once the core assets are complete, a new Web application in the same domain can be generated by creating its model using rules stated in the core assets. The model then can be transformed into deployable Web application using the code generator.

K.J. Kim (ed.), *Information Science and Applications*,
Lecture Notes in Electrical Engineering 339, DOI 10.1007/978-3-662-46578-3_119

In other hand, there exists a modelling language called Abstract Behavioral Specification (ABS) that provides object-oriented modelling notation and a set of SPL-supporting features such as feature modelling, delta modelling, product line configuration, and product selection [3].

As an example of SPL methodology, AC-MDWE has been used successfully in creating Web application products in commercial setting[1]. However, it is not yet defined on how to realize variation points in its core assets. At its current state, it is projected that the core assets may get bigger as new features are introduced into the product line. ABS, however, allows developers to control how features are included in the product line.

The paper is structured as follows. Section 2 gives overview on how SPL is realized by following AC-MDWE methodology. Section 3 gives overview about ABS. Section 4 describes a ABS-based MVC framework for Web application. Section 5 describes the experiment based on case study creating SPL for E-commerce Web application. Section 6 discusses the experiment. Section 7 concludes the paper with a summary and a brief mentions of future work.

2 Architectural Centric Model-Driven Web Engineering (AC-MDWE)

AC-MDWE is an extension to Architecture-Centric Model-Driven Software Development [4] that applied in Web application engineering. It consists of two iterative processes namely domain architecture track and application track.

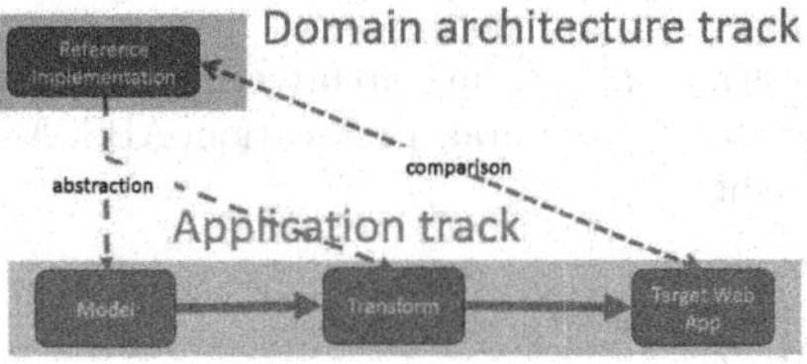

Fig. 1. AC-MDWE two track development processes

Domain architecture track produces a working Web application called reference implementation. Reference implementation consists of artefacts that are essentials in order to run the Web application on a server.

Application track divided into three subprocesses, which are (1) reference implementation modelling, (2) template and model transformation rules preparation, and (3) Web application generation via code generator.

The code generator takes the model of reference implementation, the templates, and model transformation rules as the input to generate a Web application artefacts that comparable to the reference implementation. New products

[1] Personal communication with one of AC-MDWE authors

can be generated if its model follows the pattern in reference implementation's model.

Whenever there are new features being requested, there are two possible options. First, if the new features are projected to be available in every subsequent products, the implementation is inserted into existing reference implementation. Consequently, the model of reference implementation, templates, and model transformation rules have to be updated as well.

If the new features are specific to certain products, they only have to be implemented in the products. The codes do not have to be integrated back into the reference implementation.

3 Abstract Behavioral Specification (ABS)

ABS is an executable, object-oriented modelling language produced by HATS project [5]. ABS is also supplied with a tool built on Eclipse plug-in. ABS tools[2] can automatically compile ABS models into other programming languages, such as Java and Maude. ABS combines functional, object-oriented, and concurrent programming in a single language [6]. ABS has in common with implementation-oriented languages by maintaining a Java-like syntax and a control flow close to an actual implementation [9]. The complete specification of ABS syntax and semantics can be seen in ABS Language Specification [8].

ABS is comprised of several layers of language that model a system, they are:

1. Core ABS Language
   Core ABS, simply called ABS, is the main layer of ABS that specifies core behavioural modules based on object oriented modelling. Each object of system modelled in class and each class must implement one or more interface. Core ABS syntax similar to Java syntax, excludes inheritance and overloading [9]. Moreover, code reuseability in ABS is achieved by delta modelling mechanism [10] that will be explained in the next section.
2. Micro Textual Variability Language
   Micro Textual Variability Language (μTVL) is an ABS language that defines a feature model of the system by providing boolean or integer constraints of features. It can be obtained from the feature diagram. ABS tool is completed with μTVL checker that helps the developer evaluate whether the product is valid or not.
3. Delta Modeling Language
   Delta Modeling Language (DML) is an ABS language that supports ABS to represent delta oriented programming (DOP) [10] to develop Software Product Line (SPL). In DOP, feature model, that has been described in the previous section, is related to delta modules that defines the modification of core modules. Delta in ABS modifies the Core ABS to different codes to generate various products.

[2] http://tools.hats-project.eu/

4. Product Line Configuration Language
   Product Line Configuration Language (CL) is an ABS language that manages the relationship between feature model and delta. Each features can have one or more deltas and each delta may also relate to one or more features. CL defines that relationship to describe when a feature is chosen, which delta will be applied to Core ABS.
5. Product Selection Language
   Product Selection Language (PSL) is an ABS language that defines the products that will be generated based on the features. To create a product, we define the product name and the list of features that is required. ABS tool help the verification of the selected product to guarantee that the product complies to the constraints in μTVL.

4 ABS MVC Framework

ABS MVC Framework is a web application framework based on Abstract Behavioral Specification (ABS) modelling language. This framework was built to help software developer to adopt ABS for web application engineering. Furthermore, this framework can help software developer separating the business logic, data and presentation layer using Model-View-Controller (MVC) design pattern to make the code cleaner and more reusable.

Like other web application framework, we need a web application server to make our web application accessible in the network. Unfortunately, we cannot use existing web server like Tomcat or Glassfish to run this framework (although we can convert the ABS Code into Java Code). Because of this reason, we have been built a light web application server (not production ready) to run our web application in the network so that we can access it using our web browser.

The interaction between client, server, and the ABS MVC framework can be seen in Fig. 2.

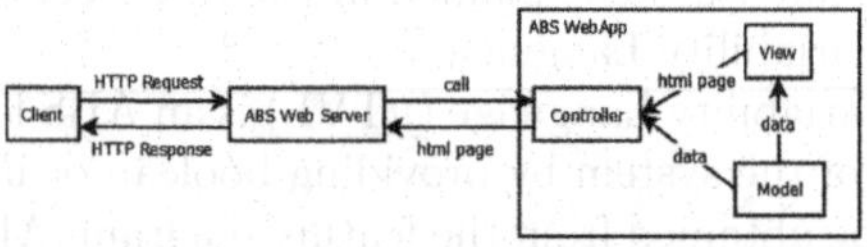

Fig. 2. ABS MVC Framework on its application server

5 Case Study and Experiment

5.1 Case Study

The products in the product line are E-commerce web applications. The feature diagram for common and varying features can be seen in Fig. 3.

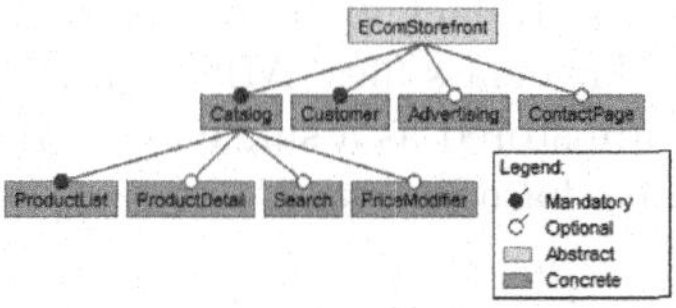

Fig. 3. Feature model of the case study

5.2 Product Line Setup

The product line is initialised as an ABS project using ABS plugin on Eclipse Luna (4.4) IDE with Modeling package[3]. The project folder is structured as follows.

The root folder contains an Ant[4] build script and 3 subfolders: (1) base, (2) productline, and (3) products. The base folder contains the base product model that follows the proposed ABS MVC framework folder convention. The productline folder contains ABS feature model, product line configuration, and product selection. The products folder is initially empty and will contain folders where each folders represent a product in the product line.

5.3 Base Product Model

We created a base product model of the Web application. The base product model consists of ABS modules and classes that represent the components in the Web application. The components of Web application are modelled as MVC components. Model and Controller components are represented as ABS classes while the View components consist of HTML templates written using Thymeleaf[5] template engine.

For the case study, we created ABS models of the Model and Controller components. The view components are represented as HTML template. We also created supporting ABS classes for containing sample data.

5.4 Initial Product Line

The product line configuration is initialised with the name of the product line and common features present in the base product. Since there are no products except the base product, the product selection is left empty at this step.

5.5 Base Product Generation

Product generation is done using ABS code generator that invoked using Ant build script. In this case study, we use ABS-to-Java code generator to transform

[3] http://www.eclipse.org/downloads/ packages/eclipse-modeling-tools/lunasr1
[4] http://ant.apache.org/
[5] http://www.thymeleaf.org/

the base product model to artefact of base product. The artefact consists of ABS models compiled to Java class files, ABS standard library, and the HTML templates. The contents are stored as a single Java Archive (JAR) file that can be deployed to custom Java-based server.

5.6 New Products Generation

We propose a convention-based approach in adding new features and product varieties into the product line. The approach is explained as follows:

1. Create a new directory in products directory named with the new product's name.
2. Create deltas that will be required to implement features present in the new product.
3. Store each deltas in its respective module's directory in the product directory.
4. If the new product requires new view template, create a new HTML template and store it in the view folder.

The steps to update the feature model, product line configuration, and product selection are as follows:

1. Update the feature model by adding new features into the feature model while also adhering to existing constraints, if any.
2. Update the product line configuration to reflect the addition of new features in the feature model.
3. Define the order of delta applications for every new features in the product line configuration.
4. Declare new product varieties with its required features in the product selection.

The generation of new product is done using the same code generator by passing additional argument containing the name of target product. The result is a JAR file containing the compiled Java codes and view templates which can be deployed to the custom Java-based server.

6 Discussion

6.1 Comparing Reference Implementation and Base Product

AC-MDWE suggested that we create a reference implementation as the basis for further products in the product line. Reference implementation contains artefacts written in commonly used technologies in Web development, e.g. Java, JSP, HTML, XML.

In contrast, ABS as a modelling language is not yet able to create a working reference implementation for Web application. Instead, we have to create a base product model that representing a Web application. The model then transformed into artefacts written in a target language. In our experiment, we use ABS-to-Java code generator to transform the base product model into Java-based artefacts.

6.2 Subsequent Products Generation

AC-MDWE does not use feature model in its product line. Features in the Web application are directly implemented in the reference implementation. In cases of adding new features, AC-MDWE suggested that the implementation of new features might included back into the reference implementation. The changes in the reference implementation may also propagate to the core assets in the product line due to the necessities of updating the model, templates, and transformation rules.

ABS can bring improvement to the process of how new features included in the product line. New features are represented in the feature model as optional features that may or may not present in any products in the product line. The implementation details of a feature is contained as one or more deltas. The product line configuration specifies the required deltas in order to implement a feature. The product selection defines the products in the product line and its features. Products generation is handled by code generator which uses the information in the feature model, deltas, product line configuration, and product selection on how to transform the base product model to a target product model and generate its artefacts.

In our experiment, we defined 2 products where the deltas and HTML templates for each products are self-contained in its respective product folder. The feature model, product line configuration, and product selection is added into existing files in productline folder. The folder containing base product model is not modified at all.

7 Conclusions and Future Work

This paper has demonstrated how a software product line for Web application is achieved by following AC-MDWE methodology using ABS. The core assets consist of base product model and initial product line configuration. New products can be added into the product line by creating a set of deltas and views in separate folder from the base product model. Feature specifications, order of delta applications, and variety of products are later added into the product line configuration and product selection.

We have to tailor AC-MDWE processes to adapt into ABS modelling process. Since ABS is a modelling language, it is currently not possible to create a reference implementation on ABS. Instead, we replace the reference implementation with base product model, which later can be transformed into artefact of Web application using ABS code generator.

While ABS can introduce benefits into AC-MDWE such as control during product generation, feature traceability and preserving integrity of core assets, we are aware that the current solution is not yet ready for production-level Web applications. The custom Java-based server is not yet able to handle sessions and serve non-ABS resources.

For future work, we would like to create an ABS-based template engine for creating HTML pages. Furthermore, we are considering to extend an existing

Java-based server stack such as Apache Tomcat by adding module for handling Web application components modelled in ABS. We would also like to see the possibilities to modify ABS language so that the feature model of new products can be separated from the original feature model.

8 Acknowledgement

This research is partially supported by research grant for project entitled "Development of Highly Adaptable and Reliable Software Development Tool to Support Requirement Changes and Variation of User Needs" led by Ade Azurat.

We are grateful to the members of Formal Method in Software Engineering lab at Faculty of Computer Science Universitas Indonesia for their feedback. The first author would also like to thank Faculty of Computer Science Universitas Indonesia for their support with the Teaching Assistant Graduate Student Scholarship.

References

1. Linda M Northrop. Sei's software product line tenets. *IEEE software*, 19(4):32–40, 2002.
2. Eban Escott, Paul Strooper, Jorn Guy Suß, and Paul King. Architecture-centric model-driven web engineering. In *Software Engineering Conference (APSEC), 2011 18th Asia Pacific*, pages 106–113. IEEE, 2011.
3. Reiner Hähnle. The abstract behavioral specification language: a tutorial introduction. In *Formal Methods for Components and Objects*, pages 1–37. Springer, 2013.
4. Markus Völter, Thomas Stahl, Jorn Bettin, Arno Haase, Simon Helsen, and Krzysztof Czarnecki. Model-driven software development: Technology. *Engineering, Management. Wiley*, 5:6, 2006.
5. HATS(Highly Adaptable and Trustworthy Software using Formal Models), `http://www.hats-project.eu`
6. Wong, Peter Y.H, et al. "The ABS tool suite: modelling, executing and analysing distributed adaptable object-oriented systems." *International Journal Software Tools Technology Transfer*. Springer: 2012.
7. Einar Broch Johnsen, Reiner Hähnle, Jan Schäfer, Rudolf Schlatte, and Martin Steffen. Abs: A core language for abstract behavioral specification. In *Formal Methods for Components and Objects*, pages 142–164. Springer, 2012.
8. HATS. The Abs Language Specification for ABS version 1.2.0, 2013. `http://tools.hats-project.eu/download/absrefmanual.pdf`
9. Johnsen, Einar B, et al.: ABS: A Core Langage for Abstract Behavioral Specification. In: Proceedings of the Formal Methods for Components and Objects. Springer (2011)
10. Schaefer, Ina, et al: Delta-oriented Programming of Software Product Lines. In: 14th Software Product Line Conference, pp.77-91. Springer (2010)

Prediction of Query Satisfaction Based on CQA-Oriented Browsing Behaviors

Junxia Guo[1], Hao Han[2], Cheng Gao[1], Takashi Nakayama[2] and Keizo Oyama[3,4]

[1]College of Information Science and Technology, Beijing University of Chemical Technology
Beijing, China
gjxia@mail.buct.edu.cn

[2]Department of Information Science, Faculty of Science, Kanagawa University
Kanagawa, Japan
han@kanagawa-u.ac.jp
nakayama@info.kanagawa-u.ac.jp

[3]National Institute of Informatics
Tokyo, Japan
oyama@nii.ac.jp

[4]The Graduate University for Advanced Studies (SOKENDAI)
Tokyo, Japan

Abstract. Browsing satisfaction with community-based websites has been studied mainly based on webpage content. In this paper, we exploratively analyze the factors affecting the browsing behaviors of client users to predict the query satisfaction level in a Community-based Question Answering (CQA) website. The experiment's results show that different categories of information are affected by different factors, and explain that considering the factors of browsing behaviors could improve the prediction accuracy of query satisfaction.

Keywords: Web Browsing Behavior · Community-based Question Answering · Query Satisfaction

1 Introduction

Different from general search engines, community-based websites could serve more effectively for subjective or specific searching cases by providing flexible inside search functionalities (site-specific search engine). However, inappropriate query results, which cannot satisfy user requests, are not rare. The reasons that these site-specific search engines have a low level of effectiveness are not always immediately evident [1].

The content quality and query relevance are important aspects of the searching mechanism of community-based websites. Here, we focus our analysis on Community-based Question Answering (CQA) data and browsing logs. CQA websites, such as

K.J. Kim (ed.), *Information Science and Applications*,
Lecture Notes in Electrical Engineering 339, DOI 10.1007/978-3-662-46578-3_120

Yahoo! Chiebukuro (Japanese Yahoo! Answers), are well structured and have a similar record format, which could be utilized for the study of webpage content quality. Furthermore, the browsing behaviors of client users could provide context-oriented information for the optimization of search functionalities in CQA websites. Client-side Web logs, called Web browsing logs, consist of a series of continuous records, such as webpage accessing and information searching.

In this paper, we present a query satisfaction modeling framework that is based on the integration of CQA data and browsing logs. The query satisfaction level is exploratively analyzed, and a novel context-oriented approach is developed to predict users' query satisfaction based on studying diverse features extracted from CQA data and browsing logs.

The rest of this paper is structured as follows. In Section 2, we introduce the data set of users' browsing behaviors. In Section 3, we present the factors used to assess the satisfaction levels of CQA-oriented queries in detail. We conduct a comprehensive experimental analysis of the satisfaction level of different categories in Yahoo! Chiebukuro, and give the primary reasons for low satisfaction level or dissatisfaction in Section 4. In Section 5, we present an overview of the related work. Finally, we conclude this paper and discuss future work in Section 6.

2 Data Set of Users' Browsing Behaviors

Net view data and Yahoo! Chiebukuro data are used as our data set. Net view data is a browsing log of webpage access, collected and provided by Nielsen for research and investigation. It contains 81,168,263 records of webpage access made by 24,498 panel users from different locations and with different ages, genders, educations, marriage status, jobs, and incomes. The personal information, such as the user name or ID parameters in URL, is deleted beforehand in order to protect the personal information of panel users. Each record contains the panel user's ID, access time, dwell time (in seconds), access URL, and referrer URL (if it has one). The referrer URL is the previous Web resource accessed by the user and the resource from which the current URL was found in the entry log. The referrer URL is used to track consecutive streams of events in the logs.

Since the browsing logs of different users are relatively independent, we divide the data according to the panel user ID, and we sorted it by access time. If these continuous records are well mapped in a kind of tree structure, each user's browsing logs are composed of multiple trees, or a forest structure. In a forest, each browsing record is a node. Then, we filter out those records related to query processes, and separate them into independent query processes/sessions according to the session characteristics [2] by matching the referrer URL and access URL.

We selected Yahoo! Chiebukuro as the experimental CQA website. After the data pre-processing, there are 120,182 query processes related to Yahoo! Chiebukuro. A simple XPath pattern based extractor is employed to extract the text content and non-text content from Yahoo! Chiebukuro webpages. Here, text content contains the main text portions of webpages, such as the text of posted questions and answers. Non-text content contains related statistical data, such as the length of text (the number of words) and the grade of the answerer.

3 Factors of Satisfaction Level

The assessment of the user's query satisfaction level could be considered a type of classification task. Thus, we use machine-learning techniques to establish the prediction framework for satisfaction level. To take advantage of these techniques, one of the most important procedures is selecting the most effective factors used in the assessment and prediction. Based on the analyses of users' browsing logs and related CQA webpages, we classify the effective factors into three categories. They are the CQA page non-text features, query correlative features, and the features of user browsing behavior.

Yahoo! Chiebukuro webpages use the same information structure, including posted questions, answers, user related information (e.g., grade), answer voting information, answer evaluation, and so on. According to our manual analysis and the existing researches about CQA data, we select 12 features to represent the basic CQA page non-text of a Chiebukuro webpage, as shown in Table 1.

We selected 4 query correlative features based on the analysis of the abovementioned separated, independent query processes. They are "value of relevance between query keywords and the text of the question and best answer," "value of relevance between query keywords and the text of question," "value of relevance between query keywords and the title of question," and "length of query keywords (number of words in query keyword excluding postpositional particles)." The value of these relevance values are calculated by TF-IDF based on the Japanese morphological analyzers, Juman[1]. A word is given a higher TF-IDF value if it appears frequently on one Yahoo! Chiebukuro webpage and rarely appears on other webpages.

Table 1. CQA page non-text features on the Chiebukuro webpage

Name	Type
Question length (number of words in the question's text)	Integer
Best answer length (number of words in the best answer's text)	Integer
Average answer length (average number of words in the answers to a question)	Float
Best answerer grade (user grade of designated best answerer)	Enum (0-7)
Average answerer grade (average user grade of answerers)	Enum (0-7)

[1] http://nlp.ist.i.kyoto-u.ac.jp/EN/index.php?JUMAN

Max graded answer (the highest graded answer)	Enum (0-7)
Has a thankful comment (comment is given to answer/answerer by asker, or not)	Yes/No
Times marked useful (number of clicks on "Like" button)	Integer
Rating (asker's rating)	Enum (0-5)
Answers number (number of answers)	Integer
Viewers number (number of webpage viewers)	Integer
Duration from submission to best answer (interval days between posted answer and best answer)	Float

In addition, similarly, 7 features of user browsing behavior are selected, as shown in Table 2. Here, the "average dwell time on webpage" is used to measure the browsing speed, which reflects whether users carefully read information on webpage or not (degree of interest). These features can reflect the users' activities from different views.

Table 2. Features of user browsing behavior

Name	Type
Average dwell time on webpage $(\frac{\text{dwell time (seconds)}}{question\ length + answer\ length})$	Float
Clicked recommended link on webpage (recommended webpage link is clicked, or not)	Yes/No
Clicked link in answer (link embedded in answer is clicked, or not)	Yes/No/NULL
Is it the end of query (query process ends or not)	Yes/No
Launch the same or similar query (another query with similar or the same query keywords is requested, or not)	Yes/No
Sum of browsing pages (sum of browsed webpage during query process)	Integer
Times user queried in search engine (number of requests sent to search engine in query process)	Integer

4 Experiment and Analysis

To predict a user's satisfaction with a query, we need training data with labeled satisfaction scores. For less subjectiveness, a multiple check mechanism was employed. Three assessors were requested to independently label 1500 records with the following questions. Fleiss' Kappa [3] was adopted to measure the agreement among these three assessors.

1. "What do you think about the information on this page? (Good, Normal, Not good, and Not sure)." Through this question, we seek to know if the answers on this webpage could help the asker on a certain level.
2. "To what level do you think the searching keyword matches the question and best answer on this page? (Well matched, Matches to a certain level, Does not match, and Not sure)." This question is used to learn whether they are unsatisfied with the webpage because of low relevancy.
3. "Has the information on this webpage become out-of-date?" This question is designed to identify the valueless information that is out-of-date at the time of the query.
4. "Do you think this webpage satisfied the user in the query process? (Satisfy, Satisfy to a certain level, Does not satisfy much, Dissatisfy, and Not sure)."

The assessors gave the results after a full consideration. After removing the records labeled "Not sure" in Question 4, there are 1,486 valid records left for further analysis and experiments. The value of Fleiss' Kappa is 0.25, which interprets as a "fair agreement" among assessors [3]. Considering the consistency of the labeled results, we convert a four-class classification problem into a two-class classification problem. The answer "satisfy to a certain level" is merged into "satisfy" and "does not satisfy much" is merged into "dissatisfy." Then, the value of Fleiss' Kappa becomes 0.477, which interprets as a "moderate agreement." This indicates that the labeled results of two-class classification reflect consistency and reliability.

Every learning algorithm tends to suit some problem types better than others. For example, Naïve Bayes classifiers (probabilistic classifiers) have strong independence assumptions between the features, and AdaBoost (Adaptive Boosting) is sensitive to noisy data and outliers. Thus, our experiments utilize Weka[2], among which, C4.5, SVM (Support Vector Machines), Naïve Bayes, Logistic, and AdaBoost are employed with features that are explained in Section 3. Here, we use different sets of features to observe the influence and effectiveness of the features of user browsing behavior. Table 3 shows the results *without* the features of user browsing behavior. Table 4 shows the results *with* the features of user browsing behavior. These results clearly explain the importance of the features of user browsing behavior, especially in the C4.5 and AdaBoost models.

[2] http://www.cs.waikato.ac.nz/ml/weka/

Table 3. Classification results without the features of user browsing behavior

Method	Precision	Recall	F-Measure	ROC Area
C4.5	0.555	0.578	0.565	0.617
SVM	0.577	0.598	0.583	0.627
Naïve Bayes	0.6	0.623	0.606	0.687
Logistic	0.556	0.555	0.555	0.635
AdaBoost	0.552	0.6	0.57	0.652

Table 4. Classification results with the features of user browsing behavior

Method	Precision	Recall	F-Measure	ROC Area
C4.5	0.776	0.835	0.803	0.858
SVM	0.729	0.745	0.727	0.764
Naïve Bayes	0.702	0.743	0.721	0.849
Logistic	0.697	0.685	0.691	0.782
AdaBoost	0.784	0.845	0.811	0.834

On a CQA website, the questions are usually classified into different categories, and there are 16 main categories on Yahoo! Chiebukuro. Each different category has different properties, which could reflect the varied information quality and bring about different satisfaction levels with the query processes. There are 246,559 Yahoo! Chiebukuro webpages that were found in the query processes. As shown in Table 5, we find that the satisfaction rates vary between different categories. The "Internet and PC" category has the highest rate and "News, Politics, and Affairs" has the lowest satisfaction rate. The satisfaction rates of other categories vary from around 30 % to 40 % because the information of different categories has different properties, user groups, and time sensitivity.

Table 5. Satisfaction rates of the 16 Yahoo! Chiebukuro categories

Category	Num.	Percent	Category	Num.	Percent
Entertainment, Hobbies	58846	35.3	Career	9343	37.5
Live Guide	27128	35.8	Country, Region, Travel	9654	35.8
Health, Fashion, Beauty	25125	38.3	Business, Economics	9397	36.7
Internet, PC	22205	40.4	Manners, Ceremonies	4450	37.1

Sports, Outdoors, Car	15731	33.3	Yahoo! JAPAN	3563	33.6
Life and Love, Relationships	20988	30.1	News, Politics, Affairs	4521	22.4
Child, Schools	14886	34.2	Computer Technology	660	36.8
Culture, Science	13892	31.6	Others	6098	35.6

Note. "Num." = the number of webpages and "Percent" = percentage of users that are satisfied.

5 Related Work

Ranking constitutes the core retrieval function of search engines and provides lists of top content items, typically in response to a query. In the context of CQA, ranking must take into account factors beyond relevance, such as category and publication time. Approaches that combine different rankings using different criteria are of particular interest. Chelaru et al. analyzed and detected the sentiment in queries for query recommendation and trend analysis (discovering unknown topics that trigger both positive and negative opinions among the users of a search engine) based on sentiment classifiers [4].

More recent query log analyses have been carried out to understand the query behavior in terms of search interface and content. In our study, we apply similar methodological techniques, such as analyzing session characteristics [2]. Kumar et al. carried out a large-scale user behavior study on search and toolbar data using logs from the Yahoo! system. They proposed a classification of browsing page views based on content and search. They found that around half of the page views belong to the content categories and a third to the communication category [5]. Cheng et al. carried out a user study to characterize the main types of page views, and they presented a method to predict the search intent of users by considering their browsing behavior. They employed a machine learning approach to rank queries using toolbar data [6]. Aikawa et al. defined a subjective/objective question classification task based on experiments on Yahoo! Chiebukuro, and they employed Naive Bayes with n-gram or maximal repeats features [7].

However, little research has been dedicated to understanding the satisfaction and difficulties that users encounter on search engines. We believe a deep analysis of context-oriented query satisfaction with search engines is required to address this.

6 Conclusion and Future Work

In this paper, we gave an exploratory prediction of the query satisfaction levels of Yahoo! Chiebukuro based on the analysis of the effectiveness of browsing behavior features, and we presented the satisfaction rates of different categories through experiments based on the data set of 120,182 query processes made by 24,498 panel users. We also explained the possible reasons for this difference and discussed the significance of features based on prediction results.

In future work, we would like to conduct further analyses of the reason(s), especially the time sensitive factors. We will develop practical applications, such as calculating the value of "shelf life" of Web information, for further time-oriented optimization in search engines.

Acknowledgements. This work was partially supported by the Fundamental Research Funds for the Central Universities grants ZY1317 (China) and funded by the National Institute of Informatics and Kanagawa University under joint research grants (Japan).

References

1. Ohshima, H., Jatowt, A., Oyama, S., Nakamura, S., Tanaka, K.: Towards Improving Web Search: A Large-Scale Exploratory Study of Selected Aspects of User Search Behavior. Proceedings of the 10th International Conference on Web Information Systems Engineering, pp. 379–386 (2009)
2. Guo, J., Gao, C., Xu, N., Lu, G., Han, H.: Analyzing Query Trails and Satisfaction based on Browsing Behaviors. Proceedings of the 10th Web Information System and Application Conference, pp. 107–112 (2013)
3. Sim, J. and Wright, C. C.: The Kappa Statistic in Reliability Studies: Use, Interpretation, and Sample Size Requirements. Physical Therapy 85(3), pp. 257–268 (2005)
4. Chelaru, S., Altingovde, I.S., Siersdorfer, S., Nejdl, W.: Analyzing, Detecting, and Exploiting Sentiment in Web Queries. ACM Transactions on the Web 8(1): 6, (2013)
5. Ravi Kumar and Andrew Tomkins. A Characterization of Online Browsing Behavior, Proceedings of the 19th International Conference on World Wide Web, pp. 561–570, 2010.
6. Zhicong Cheng, Bin Gao and Tie-Yan Liu. Actively Predicting Diverse Search Intent from User Browsing Behaviors, Proceedings of the 19th International Conference on World Wide Web, pp. 221–230, 2010.
7. Aikawa, N., Sakai, T., Yamana, H.: Community QA Question Classification: Is the Asker Looking for Subjective Answers or Not? IPSJ Transactions on Databases Vol. 4(2), pp. 1–9 (2011)

The Competition-based Information Propagation in Online Social Networks

Liyuan Sun[1], Yadong Zhou[2], Xiaohong Guan[1,2]

[1] CFINS, Tsinghua University, Beijing 100084, China
[2] MOE KLINNS Lab, Xi'an Jiaotong University, Xi'an 710049, China

Abstract. Information propagation in Online Social Networks is of great interest to many researchers. Common issues on the existing models include complete network structure requirement, topic dependent model parameters, isolated spreading assumption and so on. In this paper, we study the characteristics of information propagation for multiple topics based on the data collected from Sina Weibo (one of the most popular microblogging services in China). According to the observations, we propose a Competition-based Multi-topic Information Propagation (CMIP) model without the existence of network structure. From a perspective of topic competition, we treat the information propagation as two stages: the gain and the loss of user attention resources. Simulation results validate the model and verify that this model effectively generates the single-peak and succession phenomenon in the propagation of information and is consistent with observations.

Keywords: Online social networks (OSNs), Information propagation, Topic competition, Multiple topics, Topic dynamics

1 Introduction

Online social networks (OSNs), like Facebook, Twitter, Youtube and Sina Weibo [3], have rapidly developed into an important media platform. The propagation of information in OSNs is convenient and immediate that accordingly creates both opportunities and challenges on information security management and business development. Therefore, analyzing, modeling and predicting the propagation of information is of great interest to researchers.

Various models, such as Linear threshold (LT) model [1], independent cascades (IC) model [2] and their extended models [3, 4], are applied to characterizing the propagation of information in social networks as a propagation based on a static graph structured underlying network. However, the absence of a complete underlying network structure and the insufficiency of the required neighborhood information of an OSN give rise to the restricted applications of these models. Non-graph based approaches, such as epidemic models [5–7], focus on the temporal dynamics with no assumption about the existence of a certain graph

[3] http://weibo.com, one of the most popular microblogging services in China.

K.J. Kim (ed.), *Information Science and Applications*,
Lecture Notes in Electrical Engineering 339, DOI 10.1007/978-3-662-46578-3_121

structure, are issued that the models parameters are heavily topic dependent. Therefore, the generalization and prediction of these non-graph approaches are unreliable. Besides, the assumption that each piece of information spreads isolatedly [8–10] is also questionable. Such issues have been noticed by researchers [11, 12] and many efforts are made to improve the accuracy of the above models. Nevertheless, the propagation of multiple topics has only been studied recently [13].

In this paper, we develop a generative model based on the mechanisms of resource competition rather than topic propagation. To be more specific, we consider users and users attention as resources. Therefore, the process of topic propagation is accordingly regarded as a resource competition between different topics. From the perspective of a user, the behavior of choosing a topic can be abstracted to joining and leaving a topic, while from the perspective of a topic it means gaining or losing attention of the participating users. Therefore, we attempt to answer the following questions in this paper: how various topics evolve and compete for user attentions in an OSN, how they gain and lose attentions of participating users, and what the determining factors are in this competition.

Towards this end, the major contributions of this paper are listed in three aspects as follows. First, we propose a competition-based multi-topic model which takes the gain and the loss of user attention into account. This model is not limited by the necessary of network structure and the neighborhood information, and it considers the interactions between topics in the form of resource competitions. To the best of our knowledge, no such models have been found to study multiple topics. Second, we measure the actual Sina Weibo data and find that daily number of total active users who post messages are stable which gives evidence to our model. Third, we demonstrate that our model can produce similar fluctuations to what is observed in the actual data by means of simulations.

The rest of the paper is organized as follows. Section 2 presents the data we used in this paper and some observations from the data. Section 3 presents the model we proposed. Section 4 presents simulation experiments and the analysis of results. Finally, Section 5 presents discussions and our conclusions.

2 Data Description and Some Observations

2.1 Dataset Collection and Preprocess

Sina Weibo is one of the most popular microblog platforms in China. Similar to Twitter, Sina Weibo has a limited length of 140 characters text. It allows its users to disseminate short information at anytime and anywhere by their mobile devices. This convenience makes it become an important platform for people to acquire information, exchange opinions and connect with others. We collected a dataset from March 1 to May 31, 2013 by using Sina API. In total, it consists of 16.08 million microblogs from 68,754 different users. The data also includes the create time of microblogs.

To get the information of topics, we take advantage of topic hashtag #, which is used in pairs in a microblog to refer to certain pieces of news or events (e.g. #Ya'an earthquake# is a topic about earthquake happened in Ya'an). Thus, we identify all the phrases labeled by topic hashtags. However, some microblogs also talk about certain news or events but don't use topic hashtags in their contents. To find all related microblogs, we search the whole data using those identified phrases. If one microblog contains more than one phrase, we classify it to the first appearing phrase. Then we discard phrases which exhibit nearly uniform volume over time. To observe the complete lifetime of a phrase, we only keep phrases that first appeared after March 2. We order the phrases by their total volume, focus on 80 most frequently mentioned phrases that appear in March 2013 and generate the topic dataset used in the following studies.

2.2 Daily Number of Active Users

We first measure the number of active users who post messages and plot it in Figure 1. From March 1 to May 31, 2013, the number of active users in every day is between 30,000 to 40,000 and periodic in seven days. Although Sina Weibo is reported to have a growing number of new users every day, the daily number of active users is almost stable. Therefore, if we take users as resources, the total amount of user resources can be seen as a constant.

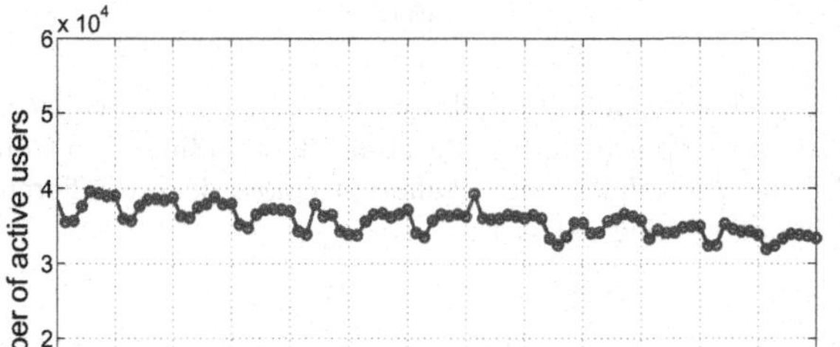

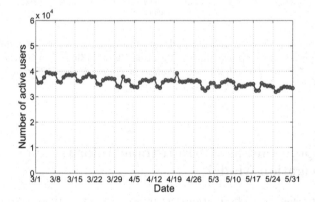

Fig. 1. Daily number of active users in Sina Weibo from March 1 to May 31, 2013. It is periodic in seven days but relatively stable, which is between 30,000 and 40,000.

2.3 Dynamics of Topic Popularity

The dynamics of topic popularity presents a single-peak and succession phenomenon. In Figure 2, we illustrate the daily number of participants for each topic, i.e. topic popularity. Because of the limited space, 30 topics are shown here. Different topics are distinguished by different colors. Generally speaking, one topic rises quickly to the peak then falls gradually. This rising and falling patterns are also observed in other researches [14]. Because of limit attention [15], users can only join finite topics in each time unit. User attentions also transfer among topics over time. A user loses his interest in an old topic when he is attracted by a new topic. Therefore, the dynamics of topic popularity is determined by the preference of user attentions.

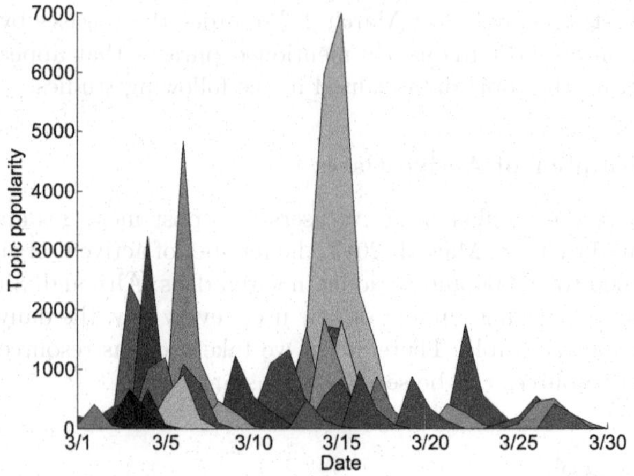

Fig. 2. Dynamics of topic popularity in days. Because of the limit space, only 30 topics are shown. These topics appear during March 2013, including "315 Consumers' Day", "Celebrity Li Yuchun's Birthday" etc.. Different colors denote different topics. They show similar rising and falling patterns.

3 Competition-based Multi-topic Information Propagation Model

According to the above observations, we present a competition-based multi-topic information propagation model (CMIP) in this section. The spreading process of topics can be described as over time topics changing, New topics continuously appear and old topics gradually disappear. A user participates in a topic at some time via posting or forwarding microblogs. After a period, this user loses his interest in the topic and then engages in other topics. It is to say, latter topic win the topic competition by drawing users attention to itself from former topics.

For the description convenience, we use urns and balls to illustrate our model. Let balls denote user attention and urns denote topics, so that users involving in different topics are imitated as balls in different urns. In our model, time runs in discrete periods $t = 1, 2, ..., T$ where T is the end of observations, and there is a collection of S users (i.e. balls), each of which joins a single topic (i.e. an urn) in one time unit. S balls distribute in K_t urns which are labeled as $1, 2, ..., K_t$, where K_t is the number of topics at time t.

In each time unit t, a user will take two stages of action: the stage of leaving (i.e. defection) and the stage of joining (i.e. recruitment), respectively corresponding to that one ball is chosen from an urn based on rule R_g and put into an urn based on rule R_p (see Figure 3). We assume that one user can only appear in one topic at a time because of limit user attention [15]. Let $U_{i,t}, i = 1, 2, ..., K_t$ denote the number of users in topic i at time t. We have $\sum_{i=1}^{K_t} U_{i,t} = S$.

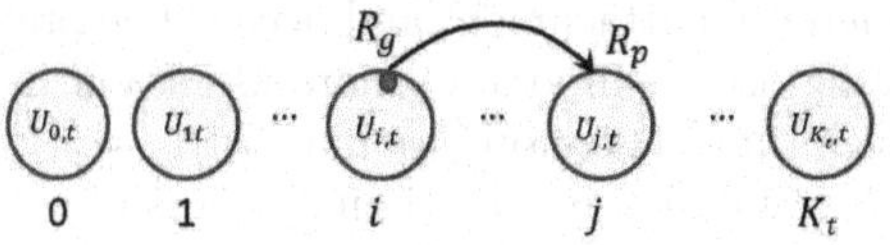

Fig. 3. Illustration of our two-stage model at time t. Each circle denotes an urn, labeled from 1 to K_t. A ball in urn i is chosen under rule R_g, and put into urn j under rule R_p.

The evolvement of CMIP model include topic arrival and popularity increase. First, we consider that topics arrive one by one, so let $\{N_t, t \geq 0\}$ denote the process of topic arrival. During time period $[0, T]$, topic i starts in time τ_i and ends in time κ_i, where $i = 1, 2, ..., N_T$. $B_i = \kappa_i - \tau_i$ denotes how long the topic lasts. Let $\{K_t, t \geq 0\}$ denote the number of topics at each time, and $K_0 = 0$. We define continuity function $w(x, y)$ as following

$$w(x,y) = \begin{cases} 1 & 0 < x < y \\ 0 & \text{otherwise} \end{cases} \tag{1}$$

Then, we have

$$K_t = \sum_{i=0}^{N_t} w(t - \tau_i, B_i) \tag{2}$$

Equation (2) with (1) means topic i is alive only if $t - \tau_i < B_i$, otherwise it disappears. Second, we consider how each topic gains and losses user resources. Let $P[R_g = i, R_p = j, t]$ denote the probability of getting a ball from topic i and putting it into topic j at time t $(i, j = 1, 2, K_t)$. The number of users $U_{i,t}$ might increase 1, decrease 1 or keep the same at each time unit, so it follows:

$$\begin{aligned} P[U_{i,t} = x] = & P[U_{i,t-1} = x - 1] \cdot P[R_g \neq i, R_p = i, t] \\ & + P[U_{i,t-1} = x + 1] \cdot P[R_g = i, R_p \neq i, t] \\ & + P[U_{i,t-1} = x] \cdot P[R_g \neq i, R_p \neq i, t] \end{aligned} \tag{3}$$

where $U_{i,t} \neq 0$. For simplicity, we assume the two stages of losing and gaining resource are independent, so that $P[R_g = i, R_p = j, t]$ in Equation (3) equals $P[R_g = i, t] \cdot P[R_p = j, t]$, where $P[R_g = i, t]$ denote the probability of getting a ball from topic i, and $P[R_p = j, t]$ denote the probability of putting a ball into topic j. The format of $P[R_g = i, t]$ and $P[R_p = j, t]$ determine the evolution of topics.

The probability of getting a ball from an urn $P[R_g = i, t]$ is proportional to the existing time of this topic, because users are easier to leave those topics that have lasted for a long time. Therefore, we have topic i loses a user with probability proportional to $f(t - \tau_i)$, i.e. $P[R_g = i, t] \propto f(t - \tau_i)$, where τ_i is the time when i is first produced, and $f(t - \tau_i) = t - \tau_i$.

The probability of putting a ball into an urn $P[R_p = j, t]$ refers to at least two factors [14]: preferential attachment and decay. The preferential attachment indicates more existing users will attract more users. On the other hand, users' interests decrease with time. Therefore, we have $P[R_p = j, t] \propto U_{j,t-1} \cdot \theta_j(t - \tau_j)$, where $\theta_j(t) = \theta_{j,0} \cdot t^{-1}$ [14], $\theta_{j,0}$ is the original fitness of topic j.

4 Experiments

In this section, simulation experiments are given to demonstrate the effectiveness of CMIP model.

4.1 Experimental Setups

Initialization. We set the total user resource S as a constant. In the simulation, the parameter S is set to be 20,000 based on previous observations on real data, which means that topics can spread to a maximum number (20,000) of users.
Topic Arrival and Departure. We assume the arrival of new topics is a Poisson Process, which is common in describing physical phenomenon. Therefore, the interval of two consecutive topics follows exponential distribution with a rate parameter λ. Besides, we assume the origin fitness of topic $\theta_{i,0}$ follows uniform distribution in simulations. Moreover, each topic lasts for fixed time steps.

4.2 Results

In this section, we validate the dynamics of temporal topic popularity generated by our model and compare it with that in the actual data. We find through simulations that our competition model produces fluctuations that are similar to what is observed in real data (Figure 2). Figure 4 shows the results of a typical simulation of the model with total resource $S = 20,000$, topic arrival rate $\lambda = 300$. We can see the temporal topic popularity of each topic first rapidly increase to the peak then decrease. In real data, trending topics also present a single-peak look. Therefore, simulation results match the real data well. However, the peak of topic in real data is sharper than that in simulations. This is due to the assumption that topic popularity only changes by one in each time unit in our model, so that even resources are continuously attracted by one topic, this topics quickest increase is linear. This can be improved by extending the assumption, e.g. allowing one user follows more than one topic at the same time. Besides, the decrease of some topics in the model is too sharp, due to the fixed existing duration of topics in simulations.

Although our model is simple and rough, it reflects basic characters of topic diffusion and provides a new view of modeling the dynamic of topics. We consider a more accurate analysis of this topic competition model as an interesting open question.

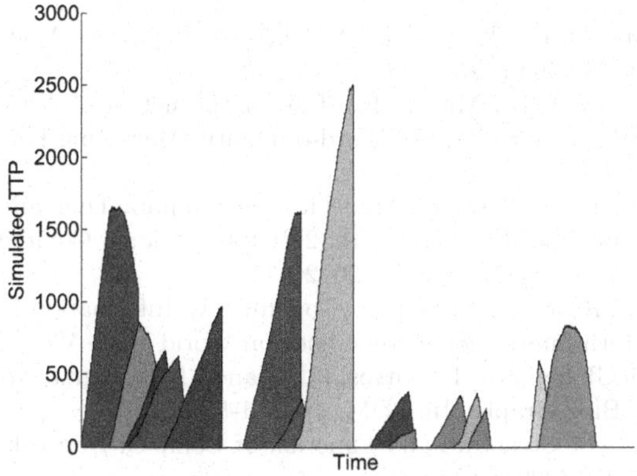

Fig. 4. Dynamics of Temporal Topic Popularity generated by our model. Similarly, different colors denote different topics and top 15 highest volume topics are shown. Although it's not accurate, our model can effectively generate the single-peak and succession phenomenon for topics.

5 Conclusions

A competition-based multi-topic information propagation model is proposed in this paper. Different from the analysis of network structure and user profile, we study the information propagation in online social networks from the perspective of topic competition and treat them as two stages: the gain and the loss of user attentions. Simulation results demonstrate that this model could effectively generate similar fluctuations of topic propagation with that observed in the actual data. In the future, we will improve our model from three aspects: (1) extending the rules of leaving and joining a topic, (2) considering different types of users since some opinion leaders have more influence than normal users, (3) considering different types of topics to build a more accurate model for predicting. At the same time, more theoretical analysis will be given.

Acknowledgments. We really appreciate the valuable discussions with Don Towsley and Zhili Zhang. We also thank Xiaoxiao Sun and Lili Liu for providing the dataset used in this paper. This work was supported in part by the National Natural Science Foundation (61221063), 863 High Tech Development Plan (2012AA011003) and 111 International Collaboration Program, of China. Yadong Zhou was supported by National Natural Science Foundation (61202392) and Specialized Research Fund for the Doctoral Program of Higher Education (20120201120023).

References

1. Granovetter, M.: Threshold Models of Collective Behavior. American Journal of Sociology. pp. 1420-1443 (1978)
2. Goldenberg, J., Libai, B., Muller, E.: Talk of the network: A Complex Systems Look at the Underlying Process of Word-of-mouth. Marketing Letters. vol. 12, pp. 211-223 (2001)
3. Guille, A., Hacid, H.: A Predictive Model for the Temporal Dynamics of Information Diffusion in Online Social Networks. In: 21st International Conference Companion on World Wide Web. pp. 1145-1152 (2012)
4. Gruhl, D., Guha, R., Liben-Nowell, D., Tomkins, A.: Information Diffusion through Blogspace. In: 13th international conference on World Wide Web. (2004)
5. Leskovec, J., McGlohon, M., Faloutsos, C., Glance, N.S., Hurst, M.: Cascading Behavior in Large Blog Graphs. In: SDM. pp. 551-556 (2007)
6. Newman, M.E.: The Structure and Function of Complex Networks. SIAM review, vol. 45, pp. 167-256 (2003)
7. Hethcote, H.W.: The Mathematics of Infectious Diseases. SIAM review, vol. 42, pp. 599-653 (2000)
8. Shen, H.W., Wang, D., Song, C., Barabsi, A.L.: Modeling and Predicting Popularity Dynamics via Reinforced Poisson Processes. arXiv preprint arXiv:1401.0778 (2014)
9. Myers, S.A., Zhu, C., Leskovec, J.: Information Diffusion and External Influence in Networks. In: 18th ACM SIGKDD, pp. 33-41 (2012)
10. Yang, J., Leskovec, J.: Modeling Information Diffusion in Implicit Networks. In: ICDM, pp. 599-608 (2010)
11. Guille, A., Hacid, H., Favre, C., Zighed, D.A.: Information Diffusion in Online Social Networks: A survey. ACM SIGMOD Record, vol. 42, pp. 17-28 (2013)
12. Myers, S.A., Leskovec, J.: Clash of the Contagions: Cooperation and Competition in Information Diffusion. In: ICDM, pp. 539-548 (2012)
13. Xu, P., Wu, Y., Wei, E., Peng, T.Q., Liu, S., et al.: Visual Analysis of Topic Competition on Social Media. IEEE Transactions on Visualization and Computer Graphics, vol. 19, pp. 2012-2021 (2013)
14. Leskovec, J., Backstrom, L., Kleinberg, J.: Meme-tracking and the Dynamics of the News Cycle. In: 15th ACM SIGKDD, pp. 497-506 (2009)
15. Weng, L., Flammini, A., Vespignani, A., Menczer, F.: Competition among Memes in a World with Limited Attention. Scientific Reports, vol. 2 (2012)

e-Learning Recommender System for Teachers using Opinion Mining

Anand Shanker Tewari[1], Anita Saroj[2], Asim Gopal Barman[3]

[1]CSE Department, NITPatna, Patna, India
e-mail:anand@nitp.ac.in
[2]CSE Department, NIT Patna, Patna, India
e-mail:anitasaroj786@gmail.com
[3]ME Department, NIT Patna, Patna, India
e-mail:asim@nitp.ac.in

Abstract: In recent few years e-learning has evolved as one of the better alternative of the classroom approach. e-learning has crossed the geographical boundaries and now it is in the reach of every learner who is using internet. But merely presence of the e-learning websites does not make sure that all the content is very effective for the learners. Generally for learning any subject the learner has to traverse many websites for various topics because no single website provide all the best content about the subject at a single place. So we have to analyze the learner's reviews about the website subject content in order to deliver all the best content at a single place. In this paper we proposed a new e-learning recommender system named as A^3. It analyzes the learner's opinions about the subject contents and recommends the teachers, who have uploaded the tutorial on to the website to change the particular portion of the subject topic which is difficult to understand by the learners using opinion mining, not the complete topic. By this system after some time all the best content about the subject will be available at the single place.

Keywords: Opinion mining. Recommender System. User reviews. Feature Extrac tion.

1 Introduction

With the fast expansion of internet and other related technologies, e-learning has also become very popular among students and other learners. The main goal of the e-learning system is to provide knowledge to the student and other learners in a more better and effective way. The mental aptitude of every human being is not same, some of them are above average, majority of them are average and some of them are below average. So it is necessary to take learner's opinion about the subject content in order to make it more lucid for majority of the people. Learners' reviews will contain the information about the subject topic relevance. In this paper our recommender system (A^3) will collect the learners 'opinions about the particular topic of a subject. With the

K.J. Kim (ed.), *Information Science and Applications*,
Lecture Notes in Electrical Engineering 339, DOI 10.1007/978-3-662-46578-3_122

help of the learners opinions using opinion mining and feature extraction A^3 will find out the particular portion of the topic which is difficult to understand by the majority of the learners inside a complete tutorial. AfterwardA^3 will recommend teachers to replace only that portion (harder) of the topic, with the new one and wait for the learner's opinion about the new change portion of the topic. After some cycles of $A^3$the e-learning website will have the best e-material of the particular topic on which it is applied. Now that topic can be easily understood by the majority of the learners. This paper is organized as follows. Section 2 discusses e-learning, Section 3 gives an introduction of the opinion mining, Section 4 gives an overview of feature extraction, section 5 gives detail working of our e-learning recommender system (A^3). Section 6 gives practical implementation and the conclusion is given in Section 7.

2 e-Learning

In the 1960s, Stanford University psychology professors Patrick Suppes and Richard c. Atkinson introduced e-learning with using computers to teach mathematics and reading to young children in elementary schools [1]. With the expansion of the IT related services e-learning has reached to almost every part of the globe. One of the greatest advantage of e-learning is that any of the learner can study from anywhere and at anytime. e-learning provides interactive, user-friendly, and prompt platform for the learner [2].e-learning uses text, pictures, animation, sound etc to teach learners[3]. e-learning is also becoming very useful for people with physical disabilities[4]. For the success of any e-learning system, it has to fulfill few conditions such as flexibility, which allows the system to change itself according to the needs of every user (learner and tutor) [5], ease of use and interactivity [6].

3 Opinion Mining

Opinion mining comes under the data mining. It analyses individual opinions such as orientations of the opinions [7, 8, 9, 10, 11]. Our recommender system, A^3 analyses the individual reviews of learner about the topics of a subject. It uses feature based opinion mining which focuses on sentence level to discover learners opinion about various topics of a subject. The main cause of using feature based opinion mining is to find out features of a topic and opinion words which express opinions and then find out the polarity of each opinion word i.e. positive or negative [12].

4 Feature Extraction

Topic features are usually nouns or noun phrases in review sentences and opinion words are adjectives. A^3 uses Stanford part-of-speech tagger, which parses each review and produces part-of-speech tag for each word i.e. noun, verb, adjective etc. Each review sentence is stored in the database along with the part-of-speech tagging information. By this way A^3 extracts the feature of the learner reviews about the topic. For example, if our tutorial is of Data Structure and topic is Link List then feature extraction process will find out features that are good or bad in this e-learning tutorial. Features in case of Link List tutorial are explanation of topic, algorithmic part, programming part etc. For example user has given the review after reading the tutorial of link list is:

Algorithm is hard to understand.
Example 1.

This sentence will pass through the Stanford part-of-speech tagger. Tagger will tag each word in the sentence, result is shown in the figure 1, when tagger is applied to example 1.

Output
run: Loading default properties from trained tagger taggers/left3words-wsj-0-18.tagger Reading POS tagger model from taggers/left3words-wsj-0-18.tagger ... done [0.6 sec]. **Input:** Algorithm is hard to understand. Output of one 1 :**Algorithm/NNP is/VBZ hard/JJ to/TO understand/VB ./.** BUILD SUCCESSFUL (total time: 1 second)

Figure 1. Output of Stanford part-of-speech tagger.

Stanford part-of-speech tagger uses the standard tags from the University of Pennsylvania (Penn) Treebank Tag-set. It is given in table 1.

Table 1. Penn. Treebank Tag-set

Tag	Description	Tag	Description
CC	coordinating, conjunction	PRP$	possessive, pronoun
CD	numeral, Cardinal	RB	adverb
DT	determiner	RBR	adverb, comparative
EX	existential there	RBS	adverb, superlative
FW	foreign word	RP	particle
IN	preposition or conjunction, subordinating	SYM	symbol
JJ	adjective or numeral, ordinal	TO	“to” as preparation or infinitive marker
JJR	adjective, Comparative	UH	interjection
JJS	adjective, superlative	VB	verb, base form
LS	list item marker	VBD	verb, past tense
MD	modal auxiliary	VBG	verb, gerund or present participle
NN	noun, common, singular or mass	VBN	verb, past participle
NNS	noun, plural, common	VBP	verb, not 3rd person singular, Present tense
NNP	proper, noun, singular	VBZ	verb, 3rd person singular,

			present singular
NNPS	Proper, noun, plural	WDT	WH-determiner
PDT	pre-determiner	WP	WH-pronoun
POS	genitive marker	WP$	Possessive, WH-pronoun
PRP	personal, pronoun	WRB	WH-adverb

It is clear from the figure 1 that in the sentence given in example 1 feature is 'Algorithm' and adjective is 'hard'.

5 Working of A^3 Recommender System

Any of the e-learning website that is using our recommender system (A^3) must have the provision to take learners reviews about the tutorial uploaded by the teacher. Generally any of the topic tutorial is divided in many of the subtopics. So it is very obvious that some of subtopics are written so well that any of the learner can understand it very easily but few of the subtopics are harder or not to understood by many of the learners. For e-learning website that is designed to be used by thousands of learners, the website will get enormous amount of reviews for a particular topic. So it becomes very tough for the subject teacher, who have uploaded the learning material on to the website to read each and every review and make changes accordingly in to the subject topic content so that it becomes more lucid for the learners. A^3 will recommend teachers those subtopics that needs improvement in tutorial, so that it becomes easy for the learners to understand. A^3 will work as follows:

1. A^3 will collect all the reviews about each and every tutorial in the COLLECTOR. It stores the reviews in the database topic wise.

2. Stored reviews are sent to the CLASSIFIER. Classifier will classify each review in to any of the three categories. i.e. positive, negative or neutral using SentiWordNet.

3. For every topic there is NEGATIVE block, it will collect all the negative classified reviews of the topic along with the date timestamp.

4. When the number of reviews of any topic will become greater than 10 then COMPARATOR block count the number of total reviews of topic (TR) and total number of negative reviews (NR) of any topic from the NEGATIVE block. Now TR is divided by the NR, if the result is in between 1.0 and 2.0. i.e. $1.0 \leq$ result ≤ 2.0, then all the negative reviews are made available to the TOPIC EXTRACTOR block. If the value of the result is greater than 2.0 (result > 2.0) than it is assume that tutorial is good and does not need any type of modification for learners.

5. FEATURE EXTRACTOR block will extract the features from the negative reviews using Stanford part-of-speech tagger, stores features in the database

and made it available to the RECOMMENDER block . These features are actually subtopics that are harder to understand by the learners.

6. Now RECOMMENDER block of A^3 will generate recommendations to the teacher who have uploaded the tutorial. Recommendation will contain the name of all the subtopics that needs improvement from the learner's point of view.

7. Now teacher will again upload only the recommended subtopics with better illustration. Again new cycle of A^3 will start. Recommendation cycle will terminate based on the result of the COMPARATOR block.

The A^3 recommendation system is represented by block diagram in the figure 2.

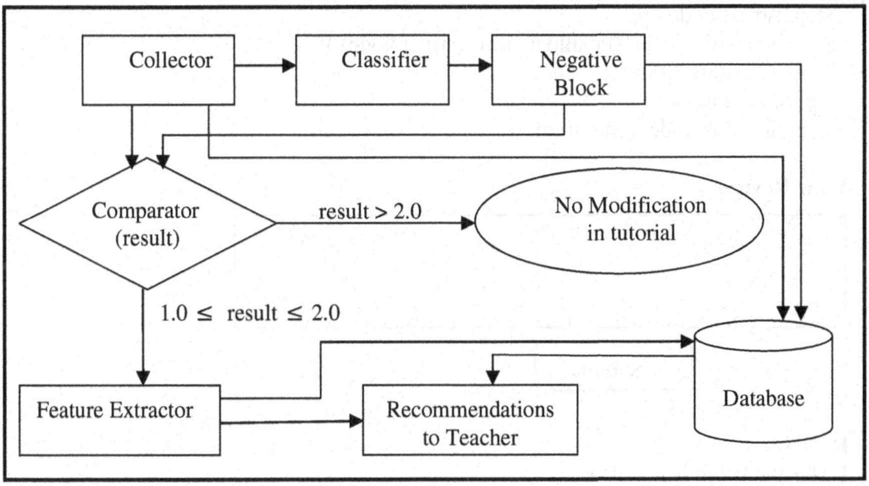

Figure 2. Block diagram of A^3 recommendation system

6 Practical Implementation

The teacher will upload the tutorial topic in the e-learning website that is using A^3 recommender system. The uploaded learning material is visible on to website web-page. Learners will write their reviews about the tutorial in the same webpage as shown in figure 3.

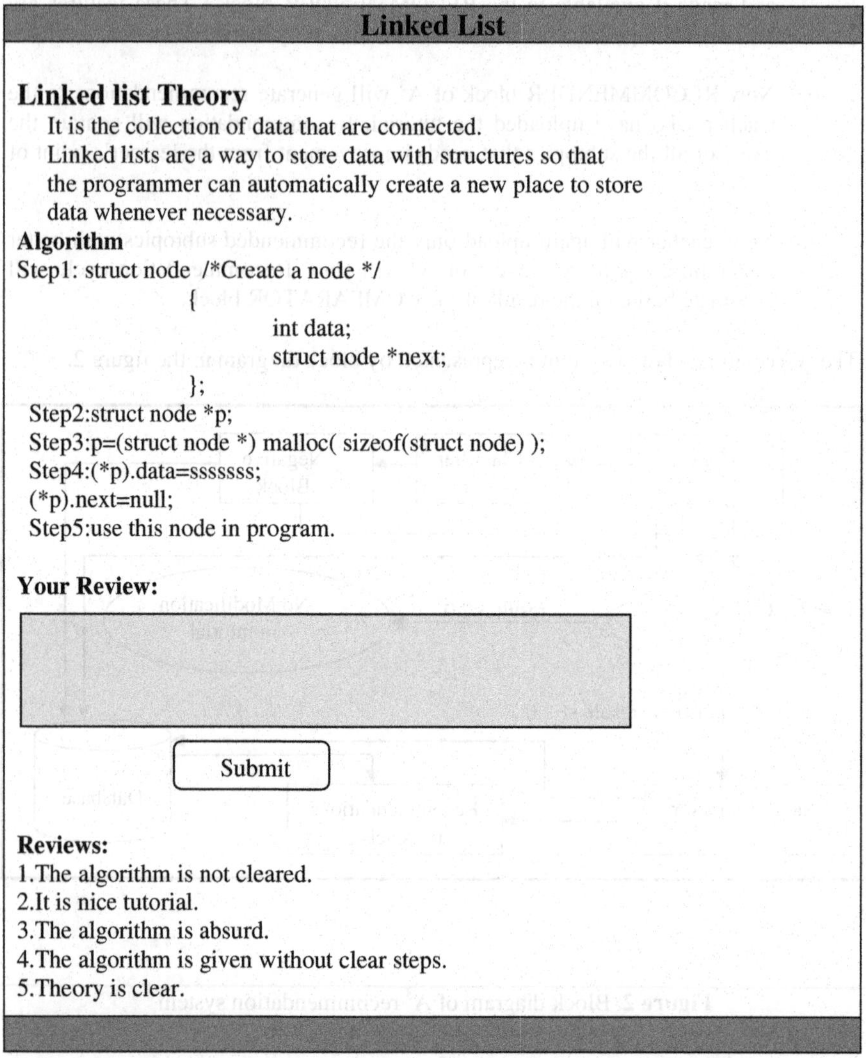
Linked List
Linked list Theory
It is the collection of data that are connected.
Linked lists are a way to store data with structures so that the programmer can automatically create a new place to store data whenever necessary.
Algorithm
Step1: struct node /*Create a node */
{
int data;
struct node *next;
};
Step2:struct node *p;
Step3:p=(struct node *) malloc(sizeof(struct node));
Step4:(*p).data=sssssss;
(*p).next=null;
Step5:use this node in program.
Your Review:
Submit
Reviews:
1.The algorithm is not cleared.
2.It is nice tutorial.
3.The algorithm is absurd.
4.The algorithm is given without clear steps.
5.Theory is clear.

Figure 3. Webpage of e-learning website.

A^3 will store information about all the teachers that are uploading learning material on to the e-learning website in *teacher information table* shown in table 2.

Table 2. Teacher Information table.

Field Name	Description
TiD	Teacher unique id
SiD	Subject Teacher Dealing with
Email	Email

Table 3. contains information about all the topics of the subject, known *as Subject Topic table.*

Table 3. Subject Topic table.

Field Name	Description
SiD	Subject unique Id
TopicID	Topic unique Id

Table 4. stores the tutorial of every topic of a subject, known as *Tutorial table.*

Table 4. Tutorial table.

Field Name	Description
TopicID	Topic unique Id
Content	Content of Topic
Date	Date of creation of content / content change

Table 5, stores all reviews of a topic and its classification done by the CLASSIFIER, known as *Review table.*

Table 5. Review table.

Field Name	Description
TopicID	Topic unique Id
ReviewId	Review unique Id
ReviewContent	Content of every review
Review date	Review Date
ReviewClass	Positive / Negative / Neutral

Feature Extractor table, shown in table 6, contains subtopics of all the negative classified reviews

Table 6. Feature Extractor table.

Field Name	Description
TopicID	Topic unique Id
SubtopicName	Name of subtopic that needs improvement

Recommender table, shown in table 7, is used to generate recommendations to the teacher.

Table 7. Recommender table

Field Name	Description
TopicID	Topic unique Id
SubtopicName	Name of subtopic that needs improvement
TiD	Teacher unique id
Date	Date of Recommendation Generation
RFlag	Extra information related to recommendation

By using table 2 to table 7, A^3 recommender system will generate recommendations for teacher about the topics uploaded by him and needs better explanation.

7 Conclusion

A^3 recommender system analyses the problem faced by the learners in understanding the topic. A^3 uses opinion mining to finds out the particular portion of the topic i.e. subtopic where the learner is facing problem. Then after it finds out concern teacher who is dealing with the topic and generate recommendations for him. Recommendation actually contains the name of the subtopics that needs better explanation for learners. Teacher will change only those subtopics for which recommendation is generated. e-learning website that is using A^3 recommender system will have best tutorial for the learners.

References

1. Naidu, S. (2005). Evaluating distance Association of Australia, 9-11 Schenk, P. Rogers, & G. A. Berg (Eds.), Naidu, S., Oliver, M. & Koronios (1999). Group, Inc.: Hershey PA.
2. B. Khan, Web-based Instruction, Englewood Cliffs, New Jersey: Education Technology Publication, 1997.
3. R. Clark & R. Mayer, E-learning and the Science of Instruction: proven guidelines for consumers and designers of multimedia learning, San Francisco: Jossey-Bass, 2003.
4. W. C. Tseng, "Analyzing web accessibility of Chinese online learning web sites, "Journal of National Taipei Teachers College Education, vol. 17(1), pp. 271-298, 2004.
5. L. C. Seng, T. T. Hok; "Humanizing E-learning" 2003 International Conference on Cyber worlds, 2003; Singapore.
6. H. Giroire, F. Le Calvez, G. Tisseau; "Benefits of knowledge-based interactive learning environments: A case in combinatorics"; Proceedings of the Sixth International Conference on Advanced Learning Technologies, 2006. pp:285-289.
7. G. Vinodhini, R.M. Chandrasekaran, Sentiment analysis and opinion mining: a survey, Int. J. Adv. Res. Comput. Sci. Software Eng. 2 (6) (2012).
8. M. Chau, J. Xu, Mining communities and their relationships in blogs: a study of online hate groups, Int. J. Hum Comput Stud. 65 (1) (2007).

9. A. Reyes, P. Rosso, D. Buscaldi, From humor recognition to irony detection: the figurative language of social media, Data Knowl. Eng. 74 (2012) 1–12.
10. F. Salvetti, S. Lewis, C. Reichenbach, Automatic opinion polarity classification of movie reviews, Colorado Res. Linguist. 17 (1) (2004).
11. T.S. Raghu, H. Chen, Cyberinfrastructure for homeland security: advancesininformation sharing, data mining, and collaboration systems, Decis. Support Syst. 43 (4) (2007).
12. Padmapani P. Tribhuvan, S.G. Bhirud, Amrapali P. Tribhuvan, "A Peer Review of Feature Based Opinion Mining and Summarization", (IJCSIT) International Journal of Computer Science and Information Technologies, Vol. 5 (1) , 2014.

Part IX
IWICT

Evaluating Heuristic Optimization, Bio-Inspired and Graph-Theoretic Algorithms for the Generation of Fault-Tolerant Graphs with Minimal Costs

Matthias Becker, Markus Krömker, and Helena Szczerbicka

Modeling and Simulation Group, University Hannover, Welfengarten 1, 30167 Hannover
xmb|hsz@sim.uni-hannover.de
http://www.sim.uni-hannover.de

Abstract. The construction of fault-tolerant graphs is a trade-off between costs and degree of fault-tolerance. Thus the construction of such graphs can be viewed as a two-criteria optimization problem. Any algorithm therefore should be able to generate a Pareto-front of graphs so that the right graph can be chosen to match the application and the user's need. In this work algorithms from three different domains for the construction of fault-tolerant graphs are evaluated. Classical graph-theoretic algorithms, optimization and bio-inspired approaches are compared regarding the quality of the generated graphs as well as concerning the runtime requirements.
As result, recommendations for the application of the right algorithm for a certain problem class can be concluded.

Keywords: fault-tolerant graphs, bio-inspired algorithm, graph optimization

1 Introduction

The construction of a graph for a given set of nodes is a task needed in many application areas, such as traffic, transportation, communication or sensor networks. The set of nodes to be connected may represent cities that have to be connected by roads or by energy or communication lines. The resulting network ideally should reveal properties such as connectivity (all nodes are part of the network), minimal costs concerning the total length of edges and in many cases also fault-tolerance, so that the network is still able to server its purpose while one or more connections are temporarily broken.

Regarding the literature about fault-tolerant networks two different fields are concerned. In graph theory, k fault-tolerance means that k arbitrary edges can be removed and the graph is still connected, which in turn implies that the node degree is $k + 1$. That definition is often too strict for practical problems, since even for the lowest level of fault-tolerance, each node is connected by at least

K.J. Kim (ed.), *Information Science and Applications*,
Lecture Notes in Electrical Engineering 339, DOI 10.1007/978-3-662-46578-3_123

two edges. In practice it is often sufficient, if a majority of nodes is connected in a fault-tolerant way, and it is eligible that outer nodes for from the center are connected by only one connection.

As consequence, most of the graph-theoretic algorithms for construction of fault-tolerant networks have limited practical use due to the anticipated definition of fault tolerance. Often it is too expensive to connect nodes in a fault-tolerant way, that lie at the edge of the covered area. Thus we use a more practical definition of fault-tolerance as given in [1], that offers a more fine grained definition of fault-tolerance, also taking the lengths of the edges into account

Nevertheless some modified versions of graph-theoretic algorithms can be used for our purpose as well as we use an agent-based approach and an optimization algorithm for the construction of fault-tolerant graphs. In a previous work [2] it has been shown, that fault-tolerant network construction can be accomplished using the (k, t)-spanners algorithm [3] by setting the desired fault-tolerance $k = 1$ and varying the stretch factor t. Although originally designed for construction of fault-tolerant networks using the graph-theoretic definition of fault-tolerance, the spanner algorithm can be used to construct a variety of networks, that reveal desired properties also in the sense of the practical definition of fault-tolerance. Further algorithms are the β-Skeleton algorithm [4,5], the k-Yao algorithm [6–8] and an agent-based slime mold simulation [9]. By variating the parameters of the two graph-theoretical algorithms other algorithms are included as special cases, e.g. setting $\beta = 1$ results in generation of Gabriel Graphs.

2 Problem Statement

All algorithms presented here have as input a set of nodes S and calculate as result a graph $G(S, E)$, where E is the set of edges. Further notation needed is given below:

- Graph G
- S: Set of nodes of G
- E: Set of edges e of G
- length(e): weight of edge e representing the distance of two nodes
- MST(G): Minimum Spanning Tree of G
- N: Number of nodes of G

The quality of the found solutions is mainly imposed by the costs and the fault-tolerance. In order to make results comparable, both measures are normalized. The costs are normalized to the costs of the minimal spanning tree, which is the graph for the given set of edges with minimal cost. The fault-tolerance is a real number between one and zero and can be informally formulated as percentage of the length of edges, that can be taken away and the graph still remains connected. Note that this definition is different to the graph-theoretic notion of fault-tolerance, where the fault-tolerance can only be expressed as an integer number. Then, fault-tolerance of one means, that one arbitrary connection can be removed (this is equivalent to our definition of fault-tolerance) and fault-tolerance of k means, that k arbitrary edges can be removed with the graph still being connected. The reason of using the length related definition of fault-tolerance is, that it is a much more coarse measure, with which practical applications can be described more detailed. There, often it is too expensive to connect an outer node in a fault-tolerant way. Using the graph-theoretic definition of fault-tolerance would result in a fault-tolerance of zero, while the length-related definition of fault-tolerance might still by high, when the not-fault-tolerant connection is only a small part of the network. In the following we present the notation needed to calculate the relevant measures, analogous to [1]:

Total length of all connections:

$$\mathrm{TL}(G) = \sum_{e \in E} \mathrm{length}(e)$$

Cost of the network relative to the minimum cost (of the MST):

$$\mathrm{Cost}(G) = \frac{\mathrm{TL}(G)}{\mathrm{TL}(MST(G))}$$

Fault Tolerance FT that takes into account the length of edges, since a long connection is more prone to fault:

$$\mathrm{FT}(G) = \frac{\sum_{e \in E | (G \setminus \{e\}) \text{ isConnected}} \mathrm{length(e)}}{\mathrm{TL}(G)}$$

In the literature, a quality measure relating fault-tolerance and the length of the graph is called the *efficiency*:

$$\mathrm{Efficiency}(G) = \frac{\mathrm{FT}(G)}{\mathrm{Cost}(G)}$$

3 Literature Review

In graph-theory spanner algorithms are used to construct a connected graph and (k, t)-spanner algorithms have been developed for construction of fault-tolerant spanners [10]. In previous work it has been shown, that fault-tolerant network construction can be accomplished using the (k, t)-spanners algorithm [2] by varying the t parameter. Although originally designed for construction of fault-tolerant networks using the graph-theoretic definition of fault-tolerance, the spanner algorithm can be used to construct a variety of networks, that reveal desired properties also in the sense of the practical definition of fault-tolerance.

Further algorithms for construction of spanners are the β-Skeleton algorithm [4, 5] and the k-Yao algorithm [6–8]. Both algorithms are not especially designed to achieve fault-tolerance, however the resulting graphs can show the desired properties if the parameters of the algorithms are tuned in that sense.

Recent approaches with bio-inspired algorithms are also concerned with network construction. Slime mold based computing raised attention in scientific renowned journals [1, 11, 12] lately. In vitro experiments using a real slime mold as well as simulations of slime based on the slime molds behavior have been done for building networks or finding a shortest path through a maze. In [1] both, a real slime mold *Physarum polycephalum* and computer simulations based on a tube model, have been used in order to construct a fault-tolerant and efficient transport network for the Tokyo rail system. The natural slime mold as well as the simulated slime mold generate networks that are similar to the existing rail system of Tokyo connecting Tokyo with its surrounding cities. While the quality of solutions generated by real slime mold show considerable variations, the networks constructed by tubular simulations show a very regular structure in their quality that is correlated with one parameter. In another approach [9] an agent based simulation of *Physarum polycephalum* turned out to better approximate the characteristics of the real slime mold's networks. These previous studies had the character of a feasibility study concerning the question, what kind of technical problems can be solved by slime molds (respectively simulations thereof) at all. Apart from feasibility it is of course the question whether those bio-inspired approaches are competitive, compared with well-known classical algorithms computing those problems. The question of computational complexity has been tackled in [13]. The results of that study suggested, that the β-skeleton representative has the fastest growing runtime complexity ($O(N^3)$ and is the first one to reach its limits, at approximately 400 nodes. The 6-Yao-algorithm shows better runtime behavior owed to the theoretical complexity ($O(N^2)$) and is feasible for problems up to a size of approx. 1500 nodes. The slime mold algorithm using the matrix method during runtime suffers from updating the edge count. While having better runtime than the β-skeleton algorithm (from approx. 300 nodes on) the runtime is always higher that that of the 6-Yao-algorithm. Only when using the Medial Axis Transform on the result of the slime mold algorithm, the slime mold algorithm shows the best runtime characteristics. Having more constant overhead, for small sized problems it is slower than the other two algorithms, however the runtime grows very slowly, so that the runtime becomes better than the runtime of the β-skeleton algorithm for approx. 300 nodes and better than the Yao graph for over 1100 nodes.

4 Goal-Oriented Graph Construction

In the previous works, the properties "fault-tolerance" and "cost" of the graphs were more or less by-products of the algorithms. Although those algorithms can be adjusted by parameter tuning in such a way, that the resulting graph indeed form a Pareto-front of fault-tolerant graphs with little costs, it seems reasonable to use an algorithm that work more goal-oriented.

Therefore we use a heuristic optimization algorithm for construction of graphs with hopefully exactly the desired properties. First experiments seemed critical concerning the runtime, since the goal function evaluation is very time intensive.

Calculating the cost of the graph and its connectedness is not crucial, however the calculation of the degree of fault-tolerance is expensive. In each optimization step, whether a new edge is added or removed, for calculation of the normalized fault-tolerance, for each it has to be checked, whether the graph is still connected if the edge is remove. However these calculations could be speed-up, so that the heuristic optimization approach turned out to be a viable solution. Starting with the MST, the greedy algorithm does a search step which can consist of adding or removing an edge. After the search step, the goal function value is calculated, and it is decided whether to do that step or not, depending on whether the goal function value increases. During our experiments it turned out, that the weighting between costs and fault-tolerance does not influence the results too much. As consequence in the following experiments we chose to weight the two goal criteria equally.

5 Experiments and Results

The algorithms are run on the same sets of random nodes, with increasing number of nodes. For up to one hundred nodes, the obtained graphs are evaluated for their degree of fault-tolerance and costs.

In Figures 5, 4, 3 and 2 the results are presented. It can be seen that the optimization algorithm produces mostly the best results. 'Best' means the produced networks have a high fault tolerance, between 90 and 100 % and at the same time the least cost, mostly around 1.5 times the length of the minimal spanning tree. All five algorithms produce good results concerning fault-tolerance, however often at much higher costs than the optimization algorithm. The slime mold algorithm also has relatively low costs however at the expense of low fault-tolerance, or even graphs that are not connected. For less nodes, slime mold and 6-Yao produce some graphs with less than 80% fault-tolerance.

Clearly the optimization algorithm has the least normalized costs around 1.5, then the slime mold algorithm follows with costs mostly above 1.5. Both Yao algorithms produce graphs with costs around or above 2, and the β = 1-skeleton producing even more costly solutions. This separation of the slime mold algorithm and the optimization algorithm with fault-tolerant and cheap graphs on the one hand and the graph-theoretic algorithms with also fault-tolerant but much more expensive graphs on the other hand is increasing with larger number of nodes to be connected.

In a second set of experiments the runtime of the algorithms has been evaluated in order to find out which algorithms are best suited for large problems. In fig. 1 the runtime results for node sets of up to 1500 nodes are given. Each point in the graph is the average of 30 runs. The variance is always low, thus not shown here. It can be seen that the hill climbing algorithm asymptotically has the worst performance, together with the β-skeleton algorithm, which has the theoretical complexity of $O(N^3)$. The β-skeleton algorithm could only be evaluated for 400 nodes in a reasonable amount of time. The slime mold algorithm also has increasing runtime so that this algorithm is not viable with large number of nodes. The best algorithms concerning the runtime are the Yao-algorithms.

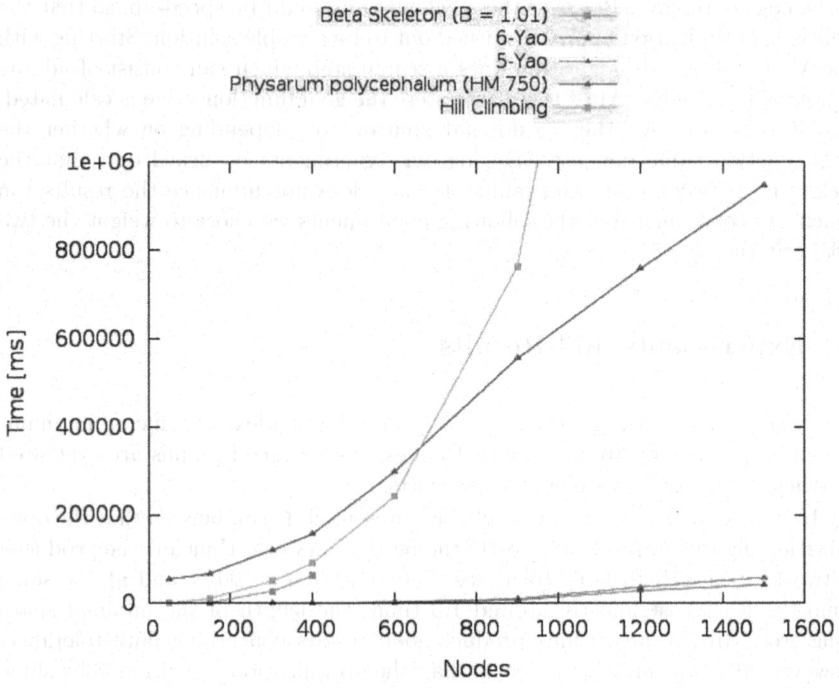

Fig. 1. Runtime

6 Conclusion

In this work the goal-oriented generation of fault-tolerant graphs with low costs by a greedy optimization algorithm has been evaluated, since other algorithms connect nodes to a graph, however have the property of fault-tolerance only as by-product. The expectation, that better graphs will be generated by application of a goal oriented algorithm, turned out to be true: the costs of graphs generated in such a way are much lower than using other algorithms, while the fault-tolerance is comparably high. However it turned out that the optimization approach is not viable for very large sets of nodes due to the rapidly increasing runtime, then, graph-theoretic approaches are much faster. We conclude that if possible, the goal-oriented optimization approach should be used whenever viable, since the properties of the graph can be determined much more detailed than using other approaches. Not only cost and fault-tolerance can be considered, but also other important measures such as the spanning ratio. Of course when this approach is not applicable due to runtime restrictions, graph-theoretic approaches are the last resort to get a result at least, even if it is not optimal. In future work one might consider to generate a suboptimal graph using a Yao algorithm and as second step apply optimizations on that graph.

References

1. A. Tero, S. Takagi, T. Saigusa, K. Ito, D. Bebber, M. Fricker, K. Yumiki, R. Kobayashi, and T. Nakagaki, "Rules for biologically inspired adaptive network design," *Science*, vol. 327, no. 5964, p. 439, 2010.
2. M. Becker, W. Sarasureeporn, and H. Szczerbicka, "Comparison of bio-inspired and graph-theoretic algorithms for design of fault-tolerant networks," in *ICAS 2012, The Eighth International Conference on Autonomic and Autonomous Systems*, 2012, pp. 1–7.
3. J. Gudmundsson, G. Narasimhan, and M. Smid, *Encyclopedia of Algorithms*. Berlin: Springer-Verlag, 2008, ch. Geometric spanners, pp. 360–364.
4. D. Kirkpatrick and J. Radke, "A framework for computational morphology," *Machine Intelligence and Pattern Recognition. Computational geometry*, p. 217248, 1985.
5. P. Bose, L. Devroye, W. Evans, and D. Kirkpatrick, "On the spanning ratio of gabriel graphs and beta-skeletons," *SIAM Journal on Discrete Mathematics*, vol. 20, no. 2, pp. 412–427, 2006.
6. X.-Y. Li, P.-J. Wan, and Y. Wang, "Power efficient and sparse spanner for wireless ad hoc networks," in *Computer Communications and Networks, 2001. Proceedings. Tenth International Conference on.* IEEE, 2001, pp. 564–567.
7. A. C. Yao, "On constructing minimum spanning trees in k-dimensional spaces and related problems," Stanford, CA, USA, Tech. Rep., 1977.
8. A. C.-C. Yao, "On constructing minimum spanning trees in k-dimensional spaces and related problems," *SIAM Journal on Computing*, vol. 11, no. 4, pp. 721–736, 1982.
9. M. Becker, "Design of fault tolerant networks with agent-based simulation of physarum polycephalum," in *Evolutionary Computation (CEC), 2011 IEEE Congress on*, june 2011, pp. 285 –291.
10. G. Narasimhan and M. Smid, *Geometric Spanner Networks.* New York, NY, USA: Cambridge University Press, 2007.
11. W. Marwan, "Amoeba-Inspired Network Design," *Science*, vol. 327, no. 5964, p. 419, 2010.
12. T. Nakagaki, H. Yamada, and A. Toth, "Intelligence: Maze-solving by an amoeboid organism," *Nature*, vol. 407, 2000.
13. M. Becker, F. Schmidt, and H. Szczerbicka, "Applicability of bio-inspired and graph-theoretic algorithms for the design of complex fault-tolerant graphs," in *Proceedings of the 2013 IEEE International Conference on Systems, Man, and Cybernetics*, ser. SMC '13. Washington, DC, USA: IEEE Computer Society, 2013, pp. 2730–2734. [Online]. Available: http://dx.doi.org/10.1109/SMC.2013.465

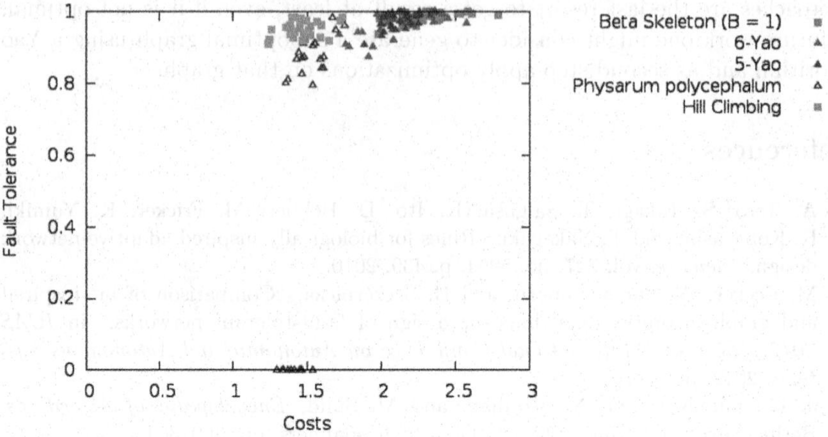

Fig. 2. 40 Nodes

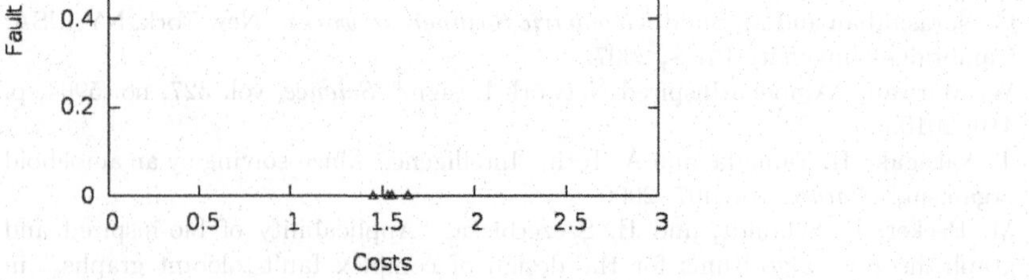

Fig. 3. 60 Nodes

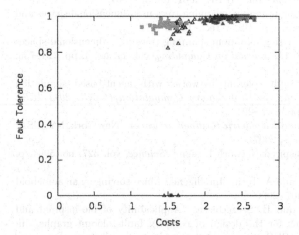

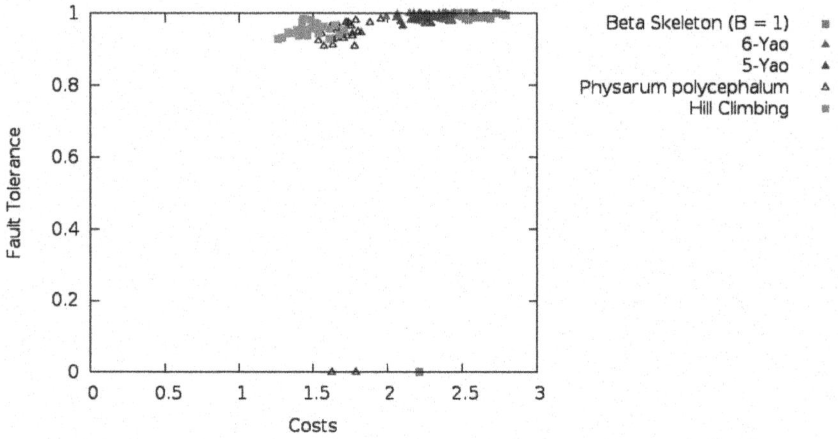

Fig. 4. 80 Nodes

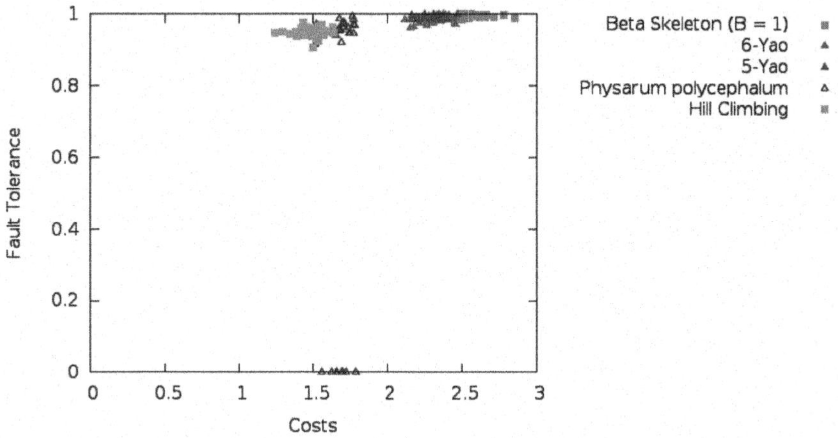

Fig. 5. 100 Nodes

Part X
ICWCIA

Real-time Night Visibility Enhancement Algorithm Using the Similarity of Inverted Night Image and Fog Image

Jae-Won Lee[1], Bae-Ho Lee[2], Yongkwan Won[2], Cheol-Hong Kim[2], Sung-Hoon Hong[2*]

[1] Department of Electronics and Computer Engineering, Chonnam National University, Korea
777kamja@naver.com

[2] School of Electronics and Computer Engineering, Chonnam National University, Korea
{bhlee, ykwon, chkim22, hsh}@jnu.ac.kr

Abstract. In this paper, we propose improved night visibility enhancement algorithm based on haze removal method. The proposed method use new haze removal method in place of the conventional method. This de-haze method is very good and faster than traditional method. It also uses a further Contrast Limit Adaptive Histogram Equalization (CLAHE) for sharpening the image. The proposed method can be applied to any application that uses a visible light camera, and it is appropriate to apply a black box, vehicle camera, and cell phone camera, since it is possible that real-time processing.

Keywords: Night Vision Enhancement, Night Visibility Enhancement, Dark Images, Haze Removal

1 Introduction

Black box, vehicle camera, surveillance camera and camera applications must be able to ensure a good quality to the user in a variety of environments. In General, however, the visible light camera does not guarantee a good image quality during the nighttime. So, many night visibility enhancement algorithms were proposed. As an exemplary method, the method of using the brightness curve, the method that emphasizes reflection component by the filter processing, and the method is such that the haze removal process in inverted video. Drago proposed a method to extend the brightness of the dark region by using the logarithmic transformation curve is similar to gamma correction [1]. This method shows image quality degradation and color noise due to excessive image enhancement. The method using brightness conversion curve based on fuzzy techniques was proposed by Cai, but this method does not shows good enhancement [2]. And method using adaptive brightness transformation curve to average brightness considering human visual characteristic was proposed by Cheng [3]. However, this method does not show too good enhancement effect. Dong found that inverted dark image is similar to haze image [4]. Based on this, they were applied to each inversed R, G, B component de-haze method based on dark channel prior [5]. This method shows good night visibility enhancement effect, but it is difficult to real-time processing because it requires a very large amount of calculation. Dong's meth-

K.J. Kim (ed.), *Information Science and Applications*,
Lecture Notes in Electrical Engineering 339, DOI 10.1007/978-3-662-46578-3_124

od improves night image stability and effectively compared to other methods. However, as mentioned above, this method very slow.
For this reason, we propose improved night visibility enhancement method to solve this problem using new fast haze removal method with local histogram equalization. The proposed method is algorithm that improves the way of Dong's. Proposed method improves the visibility of night image using proposed new haze removal formula derived from haze model. Our algorithm is possible effective real-time image enhancement because new haze removal operation is faster than He's haze removal operation. And we use local histogram equalization for sharpening of enhanced image. In this paper, we use Contrast Limit Adaptive Histogram Equalization (CLAHE) for local histogram equalization [6]. Our Proposed method is to ensure a reliable and effective performance and high speed than other methods.

2 Proposed night visibility enhancement algorithm

Proposed method's idea was derived from Dong's theory [4] that inversed luminance image is similar to haze image. Dong used He's haze removal method based on dark channel prior. The de-hazing method is effective way of removing the fog, but this method is very slow to apply real-time operation due to matting operation. In order to solve long operation time problem, Dong are using the relationships between video frames for reducing time to obtain main parameter. Through this, the overall processing time approximately four times reduced. Nevertheless, there is a difficulty in t complete real-time operation and a disadvantage that a still image can be applied. So, we propose improved night visibility enhancement method. This method is combined with new proposed haze removal for real-time operation and Contrast Limit Adaptive Histogram Equalization (CLAHE) [6] for image sharpness.

Fig. 1. Similarity of the inverted night image and haze image. (a) Example of inverted night image 1, (b) Example of inverted night image 2, (c) Example of haze image 1, (d) Example of haze image 2

Fig. 1 shows the similarity between the inversed of nighttime image and haze image. Fig. 1 (a) and (b) are examples of inverted nighttime images, and Fig. 1 (c) and (d) are examples of haze images. Night image and fog image which is inverted can be seen is very similar. It is possible to enhance an image by using such a similarity using haze removal method. Dong has been used the RGB color coordinate system in order to improve nighttime images, but we use the YCbCr color coordinate system. The proposed method improve Y signal of night image by applying night image enhancement algorithm based on new haze removal methods, and adjust the color difference signals Cb, Cr using improved rate of Y signal. Also, our method can obtain a more sharpened image using the local histogram equalization. Inverted luminance signal, as in equation (1) can be obtained by a value obtained by subtracting the input luminance signal at 255 is the maximum brightness value.

$$\overline{I_{dark}}(x) = 255 - I_{input\_y} \tag{1}$$

Inversed dark image can be expressed as equation (2) using a model of the haze image that widely used to describe the formation of a haze image [5].

$$\overline{I_{dark}}(x) = \overline{I_{org}}(x)t(x) + I_{\infty}(1 - t(x)) \tag{2}$$

where $\overline{I_{dark}}(x)$ is the x-th pixel of inversed luminance component of dark image from camera. $\overline{I_{org}}(x)$ is x-th pixel of inversed luminance component of enhanced dark image. I_{∞} is the brightest pixel value in the inversed luminance signal. t(x) is transmission rate and indicates the degree of visibility of the image. Therefore, the visibility enhanced image $\overline{I_{org}}(x)$ may be recovered by using I_{∞} and t(x) obtained from $\overline{I_{dark}}(x)$.

From equation (2), the transmission rate t(x) and $\overline{I_{org}}(x)$ the enhanced luminance component can be calculated using equation (3) and (4)

$$t(x) = \frac{\overline{I_{dark}}(x) - I_{\infty}}{\overline{I_{org}}(x) - I_{\infty}} \tag{3}$$

$$\overline{I_{org}}(x) = \frac{\overline{I_{dark}}(x) - I_{\infty}}{t(x)} + I_{\infty} \tag{4}$$

In this paper, I_{∞} set to the maximum brightness value, and transmission rate t(x) is assumed by equation (5).

$$\tilde{t}(x) = 1 - \frac{\alpha \overline{I_{org}}(x)}{I_{\infty}}, \quad 0 < \alpha < 1 \tag{5}$$

Through the above process, we can get estimated value of transmission rate and enhanced image. Finally, we can obtain quadratic equation such as equation (6) that obtainable enhanced image using the two estimated values.

$$\widetilde{I_{org}}(x) = I_{\infty} - \widetilde{\overline{I_{org}}}(x) = I_{\infty} - \frac{I_{\infty}(\alpha + 1)}{2\alpha}\left(1 - \sqrt{1 - \frac{4\alpha \overline{I_{dark}}(x)}{I_{\infty}(\alpha + 1)^2}}\right) \tag{6}$$

Parameter α is determine the degree of image enhancement. It is possible to obtain a large effect of enhancement by using α close to 1.0.

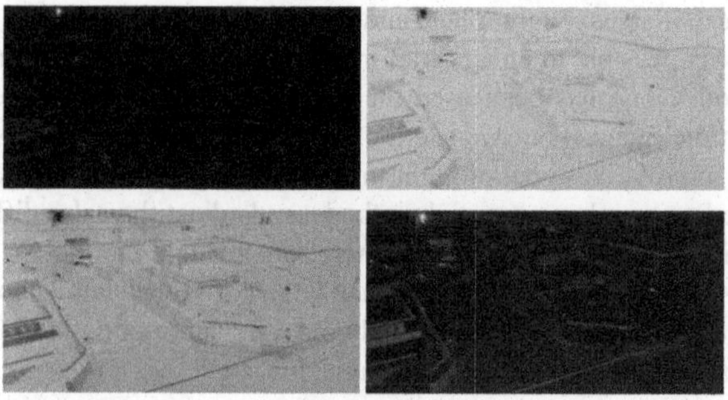

Fig. 2. The procedure of proposed night visibility enhancement method. (a) input Y image, (b) inverted Y image, (c) enhanced inverted Y image, (d) enhanced Y image.

Fig. 2 shows enhancement process of our proposed method. Fig.2 (a) is luminance image obtained through the YCbCr color conversion. The inverted grayscale image is similar to the haze image shown in Fig.2 (b) is obtained through invert operation. Fig.2 (c) shows improved image by our new haze removal method. Fig.2 (d) shows enhanced luminance image obtained by re-inverted operation.

In addition, we use CLAHE for sharpness of image. Fig 3 shows the effect of our method when used with the CLAHE. Fig 3 (b) shows result of CLAHE, the result is not good for night visibility enhancement. However, in our method, CLAHE helps to enhance the sharpness of the image.

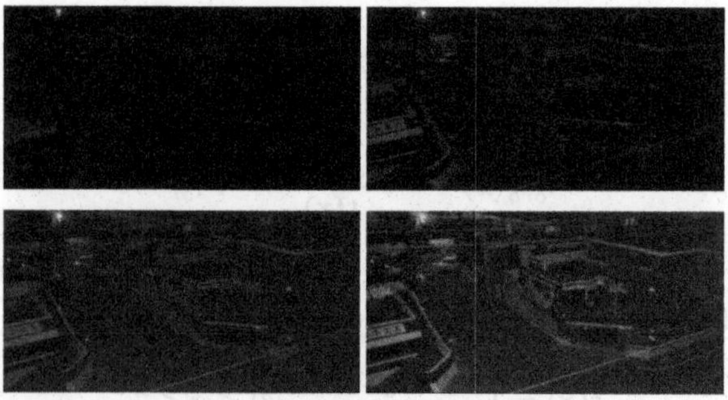

Fig. 3. Comparison result of between CLAHE using or not. (a) input Y image, (b) CLAHE-ed image, (c) Brightness enhancement Y image, (d) Brightness and sharpness enhancement Y image with CLAHE method.

As previously mentioned, our proposed night image enhancement method uses the YCbCr color coordinate system. Therefore, when a color image is restored, the color is not enough if the original color difference signals are used. Fig 4 (b) shows color image is restored using the original color difference signal. This image is color information is very low. So we obtain an increased color difference signal by using the equation (7), reconstruct color image that color information is sufficient by using obtained improved color difference signal. Improved color difference signal may be obtained through the brightness improvement ration of the luminance signal. Fig 4 (c) shows restored image color information is sufficient.

$$Cb_{out}(x) = Cb_{in}(x) \times A(x)$$
$$Cr_{out}(x) = Cr_{in}(x) \times A(x) \quad (7)$$
$$A(x) = (\ Y_{in}(x) + abs(\ Y_{in}(x) - Y_{out}(x)\ )\ )/Y_{in}(x)$$

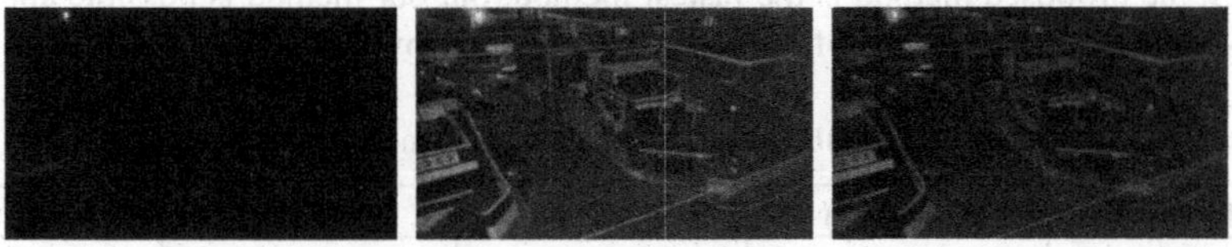

Fig. 4. The example of color component enhancement. (a) Input color image, (b) Enhanced image using original color component, (c) Enhanced image using enhanced color component.

The degree of improvement depends on the value of α. If alpha is low, there is little improvement, if alpha is high, the improvement effect is higher. Fig 5 shows the improved results according to α.

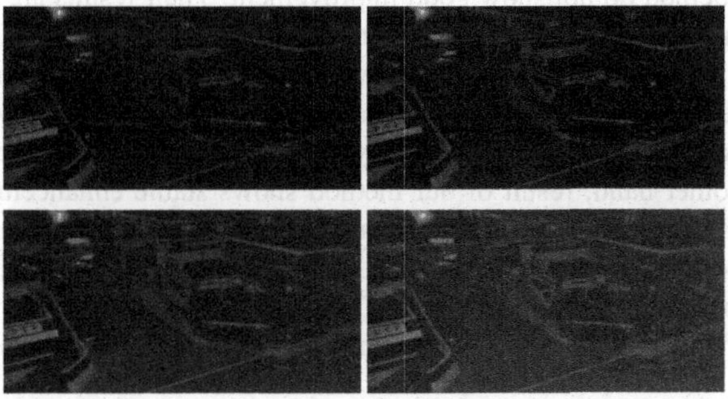

Fig. 5. The result of night visibility enhancement due to parameter α. (a) α = 0.2, (b) α = 0.4, (c) α = 0.6, (d) α = 0.8

3 Experimental Results

In order to demonstrate the performance of the proposed method, we analyze the objective and subjective results with other four methods. To quantitatively evaluate, we were compared with each execution time obtained using MATLAB code and CAF (Comprehensive optimal quality Assessment Function) used in the Cheng;s paper[3]. CAF is quantitative assessment using IE (Information Entropy), average brightness value (AG), average contrast (AC) as equation (8). CAF is a high value indicates that the more the image is much improved.

$$CAF = IE^{\alpha} \times AC^{\beta} \times NGD^{\gamma} \quad \left(\alpha = 1, \beta = \frac{1}{4}, \gamma = \frac{1}{2}\right) \tag{8}$$

$$NGD = \frac{(127.5 - dist(127.5 - AG))}{127.5}$$

In table 1, our result of execution time is not fastest method. In experimental of C code, however, our method shows 41.11 fps for image sizes 1280x960 without CLAHE method, and 26.39 fps for image sizes 1280x960 with CLAHE method. Although the proposed method is the fastest method, but our method is possible apply to real-time applications. CAF of the proposed method shows second high value.

Table 1. Quantitative measurement

Method	Time(sec)	CAF_in	CAF_out	CAF_ratio
Clahe	0.065	13.4	17.887	1.33
Cheng	0.030	13.4	16.357	1.22
Drago	0.052	13.4	26.182	1.96
dong	6.571	13.4	22.895	1.71
proposed	0.130	13.4	23.272	1.74

Fig 6, 7 and 8 shows night visibility enhancement results for night image. CLAHE and Cheng's method do not show good improvement. Their results are still dark. The result of Drago's method shows very strong enhancement result. However, color noise appears around the image due to excessive enhancement. Although fast, it is difficult to use in applications requiring stability. Dong's method shows a steady enhancement result, but it is difficult to apply to real-time application because it is very slow. On the other hand, result of our method shows stable enhancement effect and there is no color noise too. Fig. 9 show results for backlight image. The overall image is dark due to backlight. CLAHE and Cheng's method does not show sufficient brightness enhancement results. Drago's, Dong's, and our proposed methods improve visibility of image, but our method shows most reliable and clear results than others.

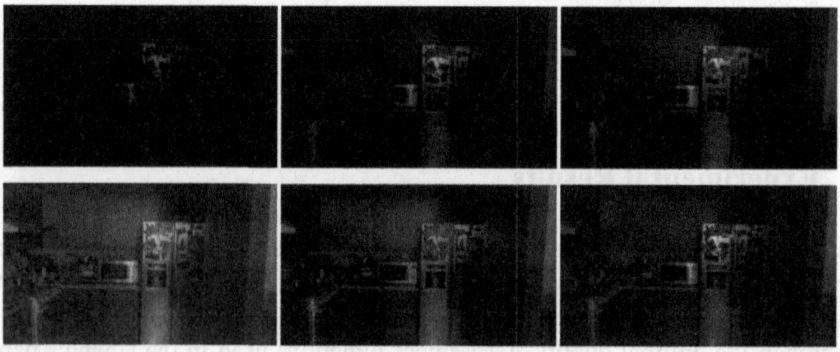

Fig. 6. The result of night visibility enhancement of various method for nighttime image 1. (a) Original image, (b) CLAHE method, (c) Cheng's method, (d) Drago's method, (e) Dong's method, (f) Proposed method

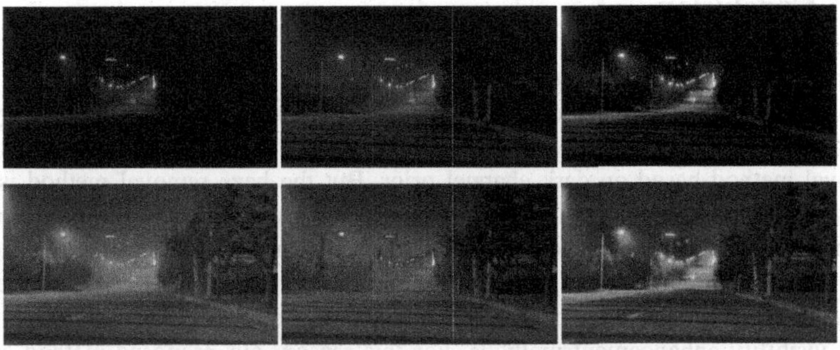

Fig. 7. The result of night visibility enhancement of various method for nighttime image 2. (a) Original image, (b) CLAHE method, (c) Cheng's method, (d) Drago's method, (e) Dong's method, (f) Proposed method

Fig. 8. The result of night visibility enhancement of various methods for nighttime image 3. (a) Original image, (b) CLAHE method, (c) Cheng's method, (d) Drago's method, (e) Dong's method, (f) Proposed method

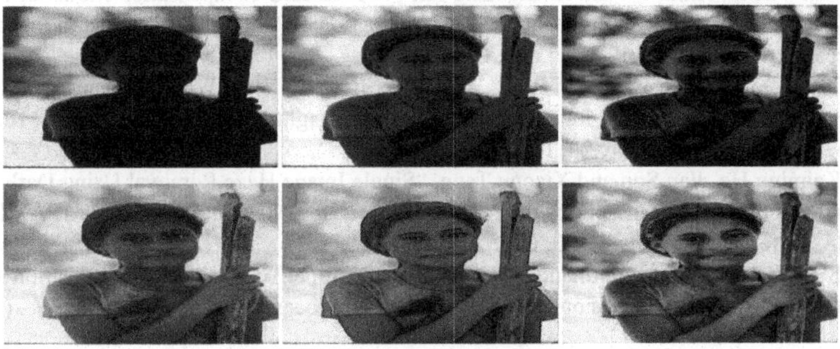

Fig. 9. The result of night visibility enhancement of various methods for back light image 1. (a) Original image, (b) CLAHE method, (c) Cheng's method, (d) Drago's method, (e) Dong's method, (f) Proposed method

4 Conclusion

We proposed improved night visibility enhancement method based on using haze removal method on inversed dark images. This method based on Dong's theory that inversed dark image is similar to haze image. In Dong's paper, he used He's haze removal method based on dark channel prior. But the haze removal method is very slow because of the complex matting operation. We propose new haze removal method for night visibility enhancement using brightness transformation model based on haze model and contrast limit adaptive histogram equalization. In experiment results, our method is very fast and shows good enhancement quality compared to traditional night visibility enhancement methods. For this reason, our proposed method can be used in light visible camera applications that require real-time processing.

Acknowledgment

This research was financially supported by the Ministry of Education, Science Technology (MEST) and National Research Foundation of Korea (NRF) through the Human Resource Training Project for Regional Innovation (No.2012H1B8A2025531).

References

1. F. Drago, K. Myszkowski, T. Annen and N. Chiba, "Adaptive Logarithmic Mapping For Displaying High Contrast Scenes", Computer Graphics Forum, Volume 22, Issue 3, PP. 419–426, September 2003
2. Limei Cai, Jiansheng Qian and Xuzhou, "Night Color Image Enhancement Using Fuzzy Set", 2nd International Congress on Image and Signal Processing 2009(CISP '09), PP. 1-4, Oct. 2009
3. Jiaji Cheng, Xiafu Lv and Zhengxiang Xie "A Predicted Compensation Model of Human Vision System for Low-light Image", 3rd international Congress on Image and Signal Processing(CISP2010), PP. 605 – 609, Oct. 2010
4. Xuan Dong, Guan Wang, Yi Pang, Weixin Li, Jiangtao Wen, Wei Meng, and Yao Lu, "A Fast Efficient Algorithm for Enhancement of Low-Lighting Video," in Proc. ICME, pp.1-6, Barcelona, Spain, July 2011.
5. Kaiming He, Jian Sun and Xiaoou Tang, "Single Image Haze Removal Using Dark Channel Prior", IEEE Transactions on Pattern Analysis and Machine Intelligence (PAMI), Vol 33, Issue 12, PP. 2341 – 2353, Dec. 2011
6. S. M. Pizer, E. P. Amburn and J. D. Austin et al, Adpative Histogram Equalization and Its Variations. Computer Vision, Graphics and Image Processing No. 39, pp. 355-368, (1987)

A Service Protection mechanism Using VPN GW Hiding Techniques

PyungKoo Park[1], HoYong Ryu[1], GyungTae Hong[2], SeongMin Yoo[2], Jaehyung Park[3], JaeCheol Ryou[2]

[1] Network SW Research section, ETRI, Korea
parkpk@etri.re.kr
[2] Dept. of Computer Engineering, Chungnam Natl. Univ., Korea
[3] School of Electronics and Computer Engineering, Chongnam Natil. Univ., Korea

Abstract. The recent supply of smartphone and the late change in the internet environment had lead users to demand safe services. Especially, the VPN technology is being researched as a key technology for providing safe services within the cloud environment or the data-center environment. However, the openness of IP is a critical threat to the VPN technology, which provides service via sharing IP address of its gateway. The exposed IP address is venerable to many kinds of attacks, and thus VPN gateway and its service are also venerable to these threats. This paper proposes a VHSP mechanism, which prevents exposure of IP address by assigning temporal IP address for the VPN gateway and its services. VHSP assign temporal IP address per-user bases. Moreover, this paper had verified performance of VHSP and original VPN in various conditions.

Keywords: VPN; IKE; Hiding address

1 Introduction

Recently, as the use of mobile devices, such as smartphone and tablet PC, become common place, user begin to demand service in cloud network to be safely delivered. As a result, VPN technology, which is for safely servicing at remote location while in cloud or mobile environment, is gaining attention [1-2]. This VPN technology is categorized into SSL VPN or IPSec VPN based on layer the technology provides security at. These two categories are methods for data communication from a user device to a VPN gateway via a secure tunnel. When doing the data communication via secure tunnel, the IP of VPN gateway must be disclosed to the user in order to configure VPN tunnel. However, this disclosed address of VPN gateway has problem of being exposed defenselessly to the outsiders. The exposed address is venerable to attack such as flooding and DDoS (Distributed Denial-of-Service), and thus cause problem in stability of the service [3-4]. Especially, when the data protected by VPN tunnel is accessing the service server by using the exposed address, it become harder to detect the attack.

This paper proposes VHSP (VPN GW hiding mechanism for service protection), in order to protect VPN gateway. VHSP prevents exposure of the address in the IPSec

K.J. Kim (ed.), *Information Science and Applications*,
Lecture Notes in Electrical Engineering 339, DOI 10.1007/978-3-662-46578-3_125

VPN environment, by assigning different temporal IP address of the VPN gateway and different temporal IP address of the service server per each user. VHSP does not require expansion on existing devices, and can provide the address protection mechanism without performance degradation by only expanding VPN gateway.

The rest of this paper is organized as follows, section 2 analysis existing VPN technology, section 3 explains IKE protocol, which is often used for configuring IPSec VPN tunnel, section 4 explains about the system plane and the control plane that are needed for proposed VHSP mechanism and IKE expansion, section 5 shows result of performance evaluation of VHSP and concludes in the section 6.

2 Model Related Works

VPN technology is a tunneling technology to encrypt and protect the data transmitted between each node. VPN technology is categorized into SSL VPN or IPSec VPN based on OSI 7 layer the tunneling technology is applied. IPSec VPN has inconveniency of making user to use separate IPSec VPN client in order to access the VPN gateway, but it can encrypt and decrypt all the application programs that uses IP. Moreover, it is advantage on performance over SSL VPN [5]. In case of the SSL VPN, Internet browser supports SSL protocol, the user does not need to install separate program to use the SSL VPN. However, SSL VPN can only be applied to applications and normally it is only applied to the web applications. Moreover, in the VPN gateway's viewpoint, SSL VPN is comparable lesser in performance, and is hard to provide robust data encryption [6-7].

A. SSL VPN

SSL VPN is a VPN servicing method that uses SSL layer located between OSI's 4th and 5th layer. Following figure shows the structure of the SSL protocol.

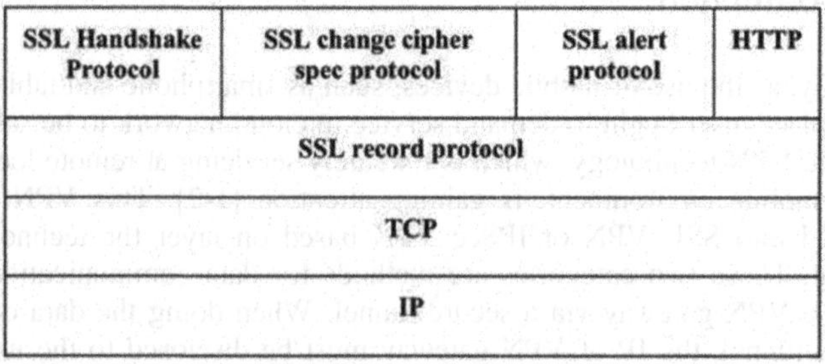

Fig. 1. SSL protocol stack

As shown in the figure 1, SSL is a protocol that operation above the transport layer. Because most of internet browser provides it, SSL VPN had big strength of being able to provide service to user without installation of client software. This feature provides benefit to user with high mobility, such as user of mobile environment. However, it is difficult to apply SSL VPN to application that does not service based on web, and SSL VPN is lesser in performance compare to the IPSec VPN [6].

B. IPSec VPN

IPSec VPN is a VPN service method that is serviced at network layer of the OSI layer structure. Due to its characterizing of being provided at network layer, the biggest characteristic of IPSec VPN is the fact that it can support all the application that uses IP. Following figure shows structure of the IPSec protocol.

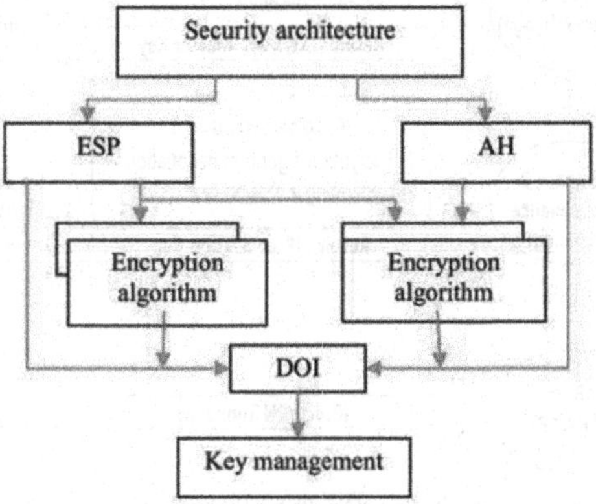

Fig. 2. IPSec protocol stack

As shown in the figure 1, IPSec protocol is composed of various modules, such as encryption algorithm, DOI(Domain of Interpretation), key management, and packet encapsulation module (ESP, AH). Through the use of IPSec VPN, user can be provided with data confidentiality with very robust data encryption. In addition, IPSec VPN provides various modes (such as main mode, aggressive mode, etc) based on service providing method, exhibits high efficiency, and supports automated key management mechanism (IKE) for the safe usage of the key.

3 Backgrounds

IPSec VPN technology requires management protocol that can pre-configure SA (Security Association) with status information between the VPN gateway and the VPN client, which utilize VPN service. Among the many Management protocol for IPSec VPN, IKE (Internet Key Exchange) is the most generally used protocol. IKE negotiates SA between two VPN nodes, and generates encryption key based of the negotiation information. The generated keys are safely shared via sharing algorithm like DH (Diffie-Hellman), and are used for protecting data. Following figure shows configuration procedure of IKE protocol.

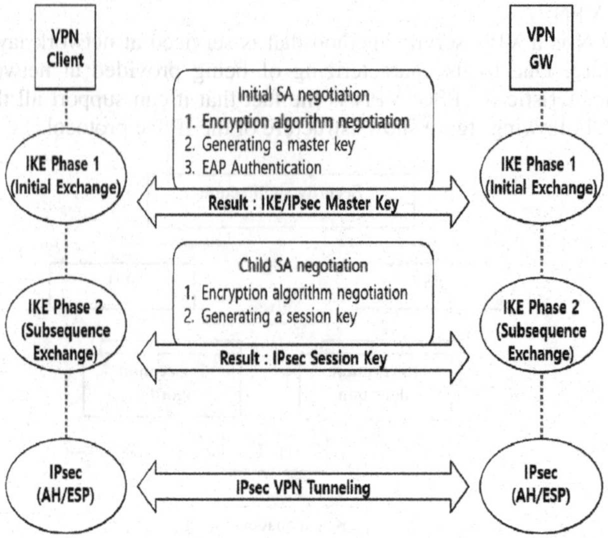

Fig. 3. IKE protocol procedures

IKE protocol is composed of phase 1, which form secured channel, and phase 2, which generate and configure SA to IPSec engine in data plane for actual IPSec communication. In phase 1, IKE_SA_INIT and IKE_AUTH are used to configure secured channel. IKE_SA_INIT process exchange DH key and trade nonce value and required parameters for negotiation and configuration of IKE SA. After the IKE_SA_INIT process is finished, messages are encrypted and protected via configured IKE SA. In IKE_AUTH process, ID and authentication information is exchanged for the mutual authentication and negotiate parameters needed for the phase 2. Phase 2 is generated and configured SA to IPSec engine in data plane. It exchanged CREATE_CHILD_SA and INFORMATIONAL_EXCHANGE information. CREATE_CHILD_SA negotiate CHILD SA in order to protect IPSec communication. INFORMATIONAL_EXCHANGE exchange information regarding errors or events that occurred during the process of the protocol. IPSec tunnel is created based on the SA information generated through phase 1 and phase 2, then IPSec VPN communication is executed [12-13].

4 VHSP : VPN GW Hiding mechanism for service Protection

In order to protect major servers such as IPSec VPN gateway and service servers from attack via exposed address, VHSP interlocks address of servers with user authentication, and changes address every time user connect. Following figure shows conceptual design of the VHSP service.

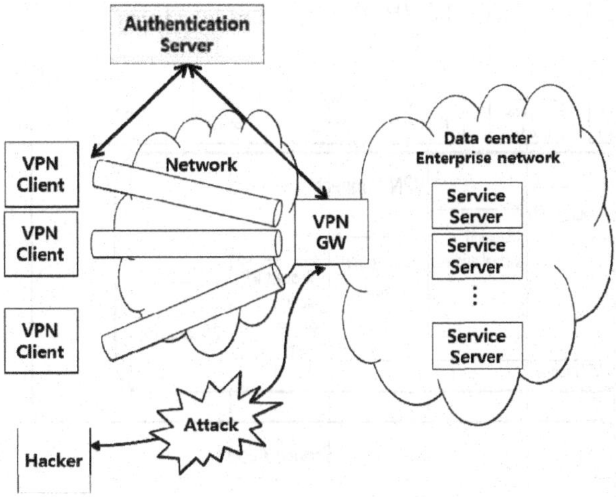

Fig. 4. VHSP conceptual service

VHSP requires expansion on IKE protocol in order to make IKE protocol work with virtual interface management technique and in order to make IKE protocol to rightfully configure IPSec tunnel for VHSP.

A. VHSP System Architecture

The VHSP system is composed of the IKE engine module, the IPSec interface module, Ingress/Egress mapping module, and the virtual interface modules, which is for dynamically configuring hiding address within the data plane. Following figure is the VHSP's system architecture.

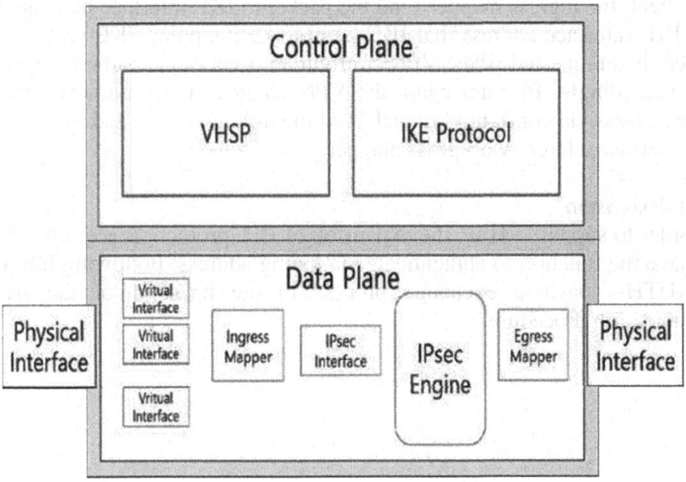

Fig. 5. VHSP system architecture

Following figure shows the VHSP system's service flow.

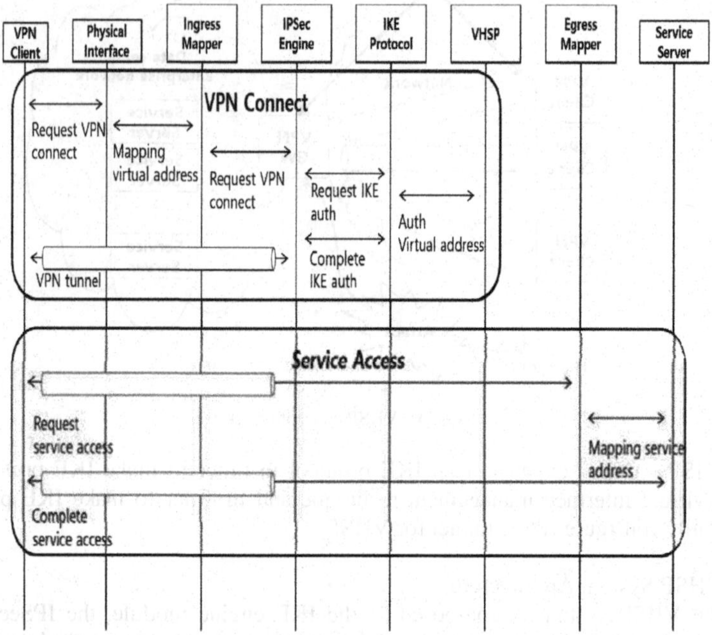

Fig. 6. VHSP system service flow

When IKE VPN connection request is sent to the hiding address that was assigned to the client, the ingress mapper send the packet to IKE interface by changing address to the IKE interface address that IKE knows. At this point, IKE protocol thinks that the client had requested IPSec VPN configuration via NAT, and configures the IPSec tunnel accordingly. In other hand, the VPN client can request service using hiding service address via configured tunnel. And the hiding service address is re-mapped to the real service address via Egress mapper.

B. IKE Extension

In order to support VHSP, the expansion of IKE protocol is required. IKE protocol must have the function to authenticate via hiding address. Following figure shows the IKE_AUTH's payload exchange process in the phase 1 of the IKE protocol communication procedure.

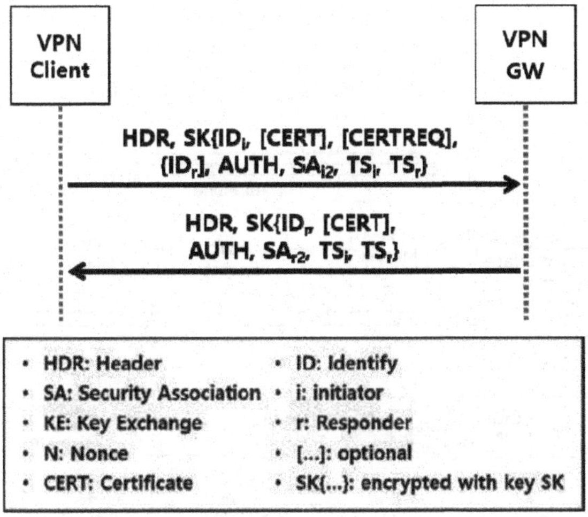

Fig. 7. IKE_AUTH message extension

IKE protocol exchange ID information and authentication information for the mutual authentication via IKE_AUTH's payload. During this process, the client and gateway IP is included in the ID information. In order to finish the mutual authentication and generate authentication information, each node's ID information must match. However, since the client uses the hiding gateway address in the VHSP system, the mutual authentication cannot successfully finished. There for VHSP IKE protocol had been modified to check the client's use of virtual Interface and to allow the mutual authentication to succeed with the right hiding address. The VHSP uses VHSP module in the control plain in order to allow IKE protocol to authenticate hiding address. While processing IKE_AUTH's payload, The IKE protocol send client information to VHSP. This procedure are not executed every for every VPN communication. It is only executed when the IPSec VPN is connected for the 1st time or when the periodic rekey process is executed. It has little effect on the performance.

C. Data forwarding mechanism

This data forwarding mechanism part describe how VHSP system receive packet and how hiding address are changed to the real address. Receiving and transmitting procedures undergoes same process for address mapping. Only difference between two procedures is the direction of packet. Following figure shows VHSP system's data forwarding mechanism.

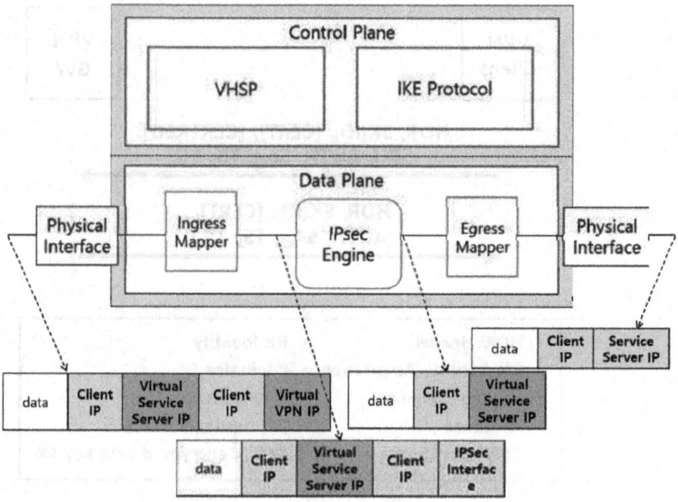

Fig. 8. VHSP data forwarding

VPN client requests the access to allowed service. At this point, the hiding VPN gateway address and the hiding service address are used. VPN gateway receives packet via real interface, and change the hiding VPN address to real VPN address through Ingress mapper. The packet with real VPN address is received by IPSec engine and undergoes encryption/decryption according to defined policy. Then, the packets real address is change to the hiding address through Egress mapper. The service servers receive packets with real address, and it can't know whether the packet transfer was via VPN or not.

5. Implementation and Evaluation

A VPN gateway, a service server, and VPN clients were prepared for the evaluation of VHSP, and the service server was stationed at the VHSP domain. The test measure and compare the "packet transmission rate" and the "transmission duration" in normal IPSec VPN environment and in VHSP's IPSec VPN Environment.

Table 1. VHSP traffic input rate

TCP traffic	UDP traffic
TCP 64byte	UDP 64byte
TCP 1,000byte	UDP 1,000byte

The service server was set to only reply with the exact same packet as the received packet. VPN client generates random 64, 1000 byte for each TCP and UDP, and send those packet to the service server. Each 100,000 date were sent and RTT (round-trip time) was compared.

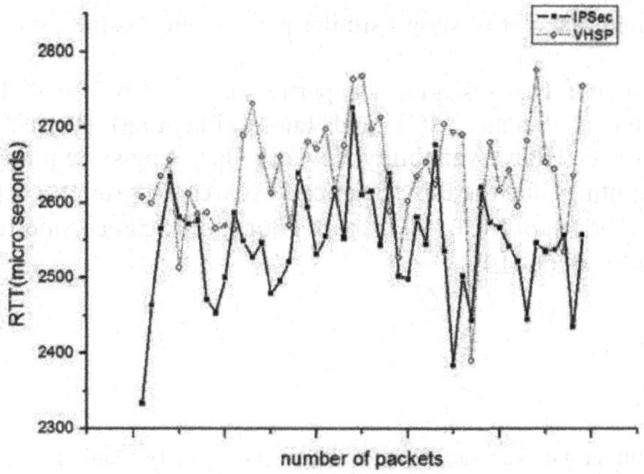

Fig. 9. TCP traffic RTT

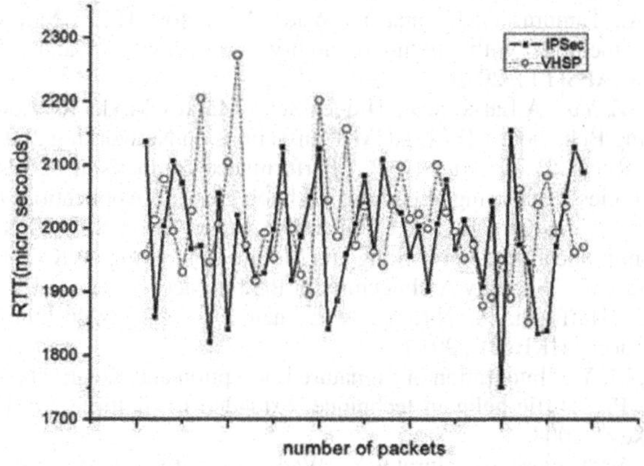

Fig. 10. UTP traffic RTT

As seen in the figure 9 and 10, VHSP exhibit only the within micro-second difference compare to existing VPN gateway. This result supports that the service via hiding address can be provided with almost no performance loss.

5 Conclusion

This paper proposed VHSP to solve the problem caused by address exposure of existing VPN. VHSP prevent address exposure by assigning temporal VPN gateway address and service server address per user. Moreover, the performance evaluation of VHSP confirmed that VHSP shows similar performance as the existing VPN gateway.

Acknowledgment : This research was partly supported by the IT R&D program of MSIP (Ministry of Science, ICT and Future Planning) [Project No. 10043380, Development of the High Availability Network Operating system for supporting Non-Stop Active Routing] and the Basic Science Research Program through NRF (National Research Foundation) of MOE (Ministry of Education, Science, and technology [Grant No. 2012R1A1A4A01004195].

References

1. Jung, Y., "Tunnel gateway satisfying mobility and security requirements of mobile and IP-based networks", Communications and Networks, 2011
2. Palomares, D., Palomares, D., Migault, D., Migault, D., Velasquez, W., Velasquez, W., et al., "High Availability for IPsec VPN Platforms: ClusterIP Evaluation", Availability, Reliability and Security (ARES), 2013
3. Dewan, P., Naidu, D. K. S., & Durham, D. M., "Denial of service attacks Using Internet Key Exchange protocol" Consumer Communications and Networking Conference, 2008
4. Kuboniwa, A., Tamura, T., Huruta, Y., Wada, Y., Satou, H., & Motono, T., "IPsec-GW redundancy method with high reliability", Information and Telecommunication Technologies (APSITT), 2010
5. Y. Xie and S. Z. Yu, "A Large-Scale Hidden Semi-Markov Model for Anomaly Detection on User Browsing Behaviors," IEEE/ACM Transactions on Networking 2009.
6. Kotuliak, I., Rybar, P., & Truchly, P., "Performance comparison of IPsec and TLS based VPN technologies", Emerging eLearning Technologies and Applications (ICETA), 2011.
7. Mao, H., Zhu, L., & Qin, H., "A Comparative Research on SSL VPN and IPSec VPN", Wireless Communications, Networking and Mobile Computing (WiCOM), 2012.
8. Kent, S., & Seo, K., "Security Architecture for the Internet protocol (RFC 4301)", 2005
9. Kaufman, C., Hoffman, P., Nir, Y., & Eronen, P., "RFC 5996-Internet Key Exchange protocol Version 2 (IKEv2)", 2010
10. Ding, Y., & Li, Y., "Integration of Signature Encryption and Key Exchange", 2008
11. Biswas, G. P., "Diffie-hellman technique: extended to multiple two-party keys and one multi-party key", 2008
12. Kuboniwa, A., Tamura, T., Huruta, Y., Wada, Y., Satou, H., & Motono, T., "IPsec-GW redundancy method with high reliability", Information and Telecommunication Technologies (APSITT), 2010
13. Jiang, Z., & Xie, Y., "Study and implement of VPN penetrating NAT based on IPSec protocol", Transportation, Mechanical, and Electrical Engineering (TMEE), 2011

Networking Service Availability Analysis of 2N Redundancy Model with Non-Stop Forwarding

Dong-Hyun Kim[1], Jae-Chan Shim[2], Ho-Yong Ryu[2], Jaehyung Park[3], Yutae Lee[4]

[1] Department of Computer Engineering, Pusan National University, Busan 609-735, Korea
[2] Network SW Platform Research Section, Electronics and Telecommunications Research Institute, Daejeon 305-700, Korea
[3] Electronics and Communication Engineering, Computer and Information Chonnam National University, Gwangju 500-757, Korea
[4] Department of Information and Communication Engineering, Dongeui University, Busan 614-714, Korea
ylee@deu.ac.kr

Abstract. This paper focuses on availability of a networking service application with 2N redundancy model and non-stop forwarding. The redundancy model is a form of resilience that ensures service availability in the event of component failure. In the model there are at most one active service unit and at most one standby service unit. The active one is providing the service while the standby is prepared to take over the active role when the active fails. Non-stop forwarding is another method to increase the availability of networking service, which can continue forwarding the service packets based on its copy of the forwarding information until the failed component is recovered. We design our analysis model using Stochastic Reward Nets, then analyze the relationship between an availability and non-stop forwarding using SPNP.

Keywords: Stochastic reward nets, SPNP, 2N redundancy, non-stop forwarding, availability, networking service applications

1 Introduction

Various complex systems are required to be dependable in use and one important aspect of a system's dependability is availability [1]. Availability is defined as the probability of the system being available for use at a given point in time. The need for high availability in networking service applications is very stringent. Achieving high availability for an application is not always an easy task, especially when this high availability is quantified by having the services available 99.999% of the times (or five 9's) [2]. For example, 99.999% availability is about five minutes and fifteen seconds of service denial over the span of one year.

The concept of redundancy models is typically used to provide high level of availability [2]. Redundancy models can increase the availability, using a duplicate device when a system failure occurs. In the Availability Management Framework

K.J. Kim (ed.), *Information Science and Applications*,
Lecture Notes in Electrical Engineering 339, DOI 10.1007/978-3-662-46578-3_126

(AMF) specification [3], several redundancy models have been defined: the No redundancy, 2N, N+M, N-Way, and N-Way Active redundancy models [3]. This paper focuses on 2N redundancy model, where there are at most one active Service Unit (SU) and at most one standby SU. The active SU is providing its service while the standby is continuously ready to take over the active role in case of a failure.

Non-Stop Forwarding (NSF) is a feature that allows networking service applications to continue forwarding the service packets even in the event of a network failure. This is done by separating the control and the data plane, having one process involved in building the routing table and another process in forwarding the packets. Thus, NSF also improves network availability, making a service application more resilient to failures. NSF is performed within a specified amount of time, which can be set by an administrator or calculated by the service application [4].

This paper considers the availability of a networking service application with 2N redundancy model and NSF. The availability analysis of any application is based on analyzing the various states that the system undergoes during its lifespan. This analysis mainly focuses on capturing the failures that cause the system to switch to a faulty state, and the repairs that shift the system back to a healthy state [2]. Since the occurrence of failures is erratic by nature, stochastic models have been used to analyze the availability. Markovian models have been extensively used for this purpose because of their expressiveness and their capability of capturing the complexity of real systems [5][6]. The Markov reward model, an extension of Markovian model, associates a reward rate with each state of the Markov chain, thus enriching the model with more capabilities of capturing the system performance and dependability. A major drawback of using Markovian models is the large number of states that need to be represented in the model [5]. As an alternative, this paper designs its analysis model using Stochastic Reward Nets (SRNs). SRNs, an extension of Stochastic Petri Nets [7], can be automatically transformed into Markov reward models that, in turn, are solved to obtain the availability using tools such as Stochastic Petri Net Package (SPNP) [2].

The objective of this paper is to construct an analysis model of 2N redundancy with NSF function using SRNs and to analyze the relationship between an availability and NSF using SPNP.

2 2N Redundancy Model

The concept of redundancy models is typically used to provide high level of availability [2]. Redundancy models can increase the availability, using duplicated devices when a system failure occurs. There are various redundancy models. AMF supports the 'No redundancy,' 2N, N+M, N-Way, and N-Way Active redundancy models [3]. These redundancy models vary depending of the number of SUs that can be active and on standby for the Service Instants (SIs), each of represents the workload assigned to a SU, and how these assignments are distributed among the SUs [8].

In this paper, we focus on 2N redundancy model. In the model, at most one SU is active and at most one SU is on standby for all SIs. The components in the active SU execute the service, while the components in the standby are prepared to take over the active role if the active fails.

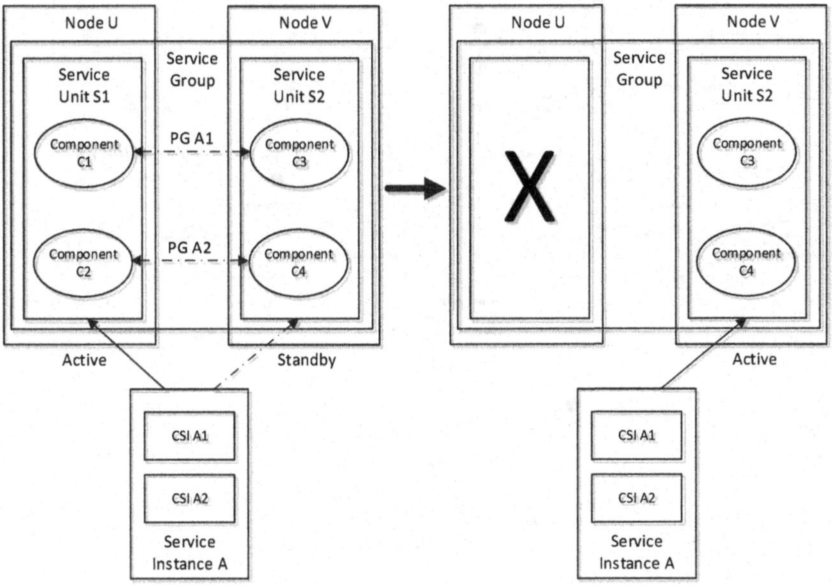

Fig. 1. Example of 2N redundancy model with two SUs on different nodes [3]

Fig. 1 shows an example of the 2N redundancy model with two SUs on different nodes, which collaborate to provide and protect an SI. After a fault that disables Node U, the SU S2 on Node V will be assigned to be active, as shown in Fig. 1. The two SUs may even reside on the same node [3].

3 SRN Model

Fig. 2 presents the SRN model for the 2N redundancy with NSF we are analyzing. Initially, the two SUs, C_1 and C_2, of the model are assigned different states, where C_1 is active and C_2 is on standby. Transition *Tiaf* fires at a rate of *λiaf*, signaling that a failure has occurred to C_i in its active state, and places a token in the *C_i_Active_faulty* state. When a SU transits from being active to being faulty, the fault is still not isolated yet, and therefore the other SU cannot transit to the active state until the faulty SU is cleaned up [2]. This constraint is captured by guards *Giaa* and *Gisa_aa*. When transition *Tiac* fires, signaling that C_i is cleaned up, the recovery can be executed. When C_i is in the standby state and the other SU is not in the active nor in the active_faulty state, transition *Tisa_aa* from standby to active is enabled. When a SU is in the standby state, a failure can still occur. This is captured by transition *Tisf*.

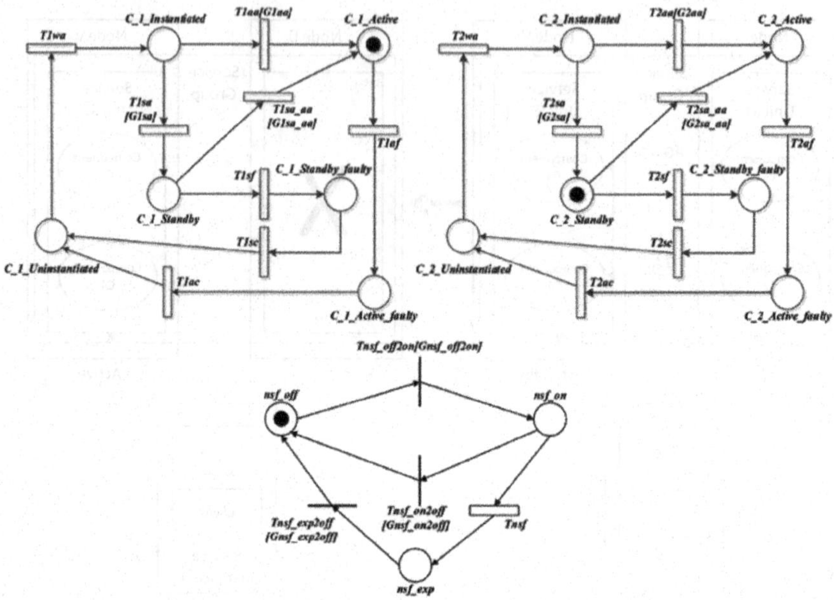

Fig. 2. SRN model of 2N redundancy with NSF

For NSF, we add three places and one token: *nsf_off*, *nsf_on*, and *nsf_exp*. The token is in the place *nsf_off* when the system operates normally, that is, one of the two SUs is in the active state. The token is in *nsf_on* when both SUs are not in the active state and the NSF period is not over. The token is in *nsf_exp* when both SUs are not in the active state but the NSF period is over.

Table 1 describes the true conditions of the guards used in this SRN model.

4 Numerical Results

Table 2 illustrates the rates we used in our analysis. This section also considers the system without NSF function, which corresponds to the case of λnsf being infinity. Table 3 depicts the calculated availability, where we assume that the values of λisa_aa, λisa, and λiaa are equal for simplicity. Note that the average switching time from standby to active is the reciprocal of λisa_aa and the average NSF period is the reciprocal of the NSF expiry rate λnsf.

Table 1. Guard conditions

Parameter	**Guard condition**
Gisa_aa	When the other SU is not in active nor active faulty state
Gisa	When the other SU is in active or active faulty state
Giaa	When the other SU is in active or active faulty state
Gnsf_off2on	When both SUs are not in active state
Gnsf_on2off	When one of the SUs is in active state
Gnsf_exp2off	When one of the SUs is in active state

Table 2. Parameters' values

Parameter	**Value (1/sec)**	**Description**
λiaf	1/43200	Active failure rate (MTTF = 12 hours)
λisf	1/360000	Standby failure rate (MTTF = 100 hours)
λiac	1/0.027	Cleanup rate for an active component
λisc	1/0.023	Cleanup rate for a standby component
λiwa	1/0.542	Instantiate rate
λisa_aa	1/0.900 ~ 1/0.400	Switching from standby to active assignment rate
λisa	1/0.900 ~ 1/0.400	Standby assignment rate
λiaa	1/0.900 ~ 1/0.400	Active assignment rate
λnsf	1 ~ infinity	NSF expiry rate

It can be seen from Table 3 that, as the average NSF period become longer, even the larger switching time from standby to active can achieve the availability of 99.999%. For example, in the case without NSF function (that is, in the case with NSF period of zero), the average switching time larger than 500 ms does not achieve the five nines. On the other hand, when the average NSF period is larger than 200 ms, the average switching time of 500 ms achieves the five nines. In case with average NSF period of one second, even the average switching time of 800 ms achieves the availability of 99.999%.

Table 3. Expected service availability

Average NSF period	Average switching time from standby to active	Availability rate
0 ms	400 ms	99.9990116 %
	500 ms	99.9987801 %
100 ms	400 ms	99.9992066 %
	500 ms	99.9989812 %
200 ms	500 ms	99.9991265 %
	600 ms	99.9989096 %
300 ms	600 ms	99.9990307 %
	700 ms	99.9988205 %
400 ms	600 ms	99.9991276 %
	700 ms	99.9989277 %
500 ms	700 ms	99.9990170 %
	800 ms	99.9988207 %
600 ms	700 ms	99.9990926 %
	800 ms	99.9989050 %
700 ms	700 ms	99.9991574 %
	800 ms	99.9989780 %
800 ms	800 ms	99.9990418 %
	900 ms	99.9988630 %
900 ms	800 ms	99.9990982 %
	900 ms	99.9989262 %
1,000 ms	800 ms	99.9990982 %
	900 ms	99.9989827 %

5 Conclusion

This paper focused on the availability of a networking service application with 2N redundancy model. To improve the availability, we have added the NSF function to the 2N redundancy model. We designed our analysis model using Stochastic Reward Nets, then analyzed the relationship between an availability and nonstop forwarding using SPNP. From numerical results, we can see that, as the average NSF period become longer, even the larger switching time from standby to active can achieve the availability of 99.999%.

Acknowledgments. This research was partially supported by the IT R&D program of MSIP(Ministry of Science, ICT and Future Planning) / IITP(Institute for Information & Communications Technology Promotion) [10043380, Development of The High Availability Network Operating System for Supporting Non-Stop Active Routing] and by Basic Science Research Program through the National Research Foundation of Korea(NRF) funded by the Ministry of Education(NRF-2013R1A1A4A01013094).

References

1. M. Neil and D. Marquez, "Availability modeling of repairable systems using Bayesian networks," Engineering Applications of Artificial Intelligence, Vol. 25, pp. 698-704, 2012.
2. A. Kanso, F. Khendek, A. Mishra, and M. Toeroe, "Integrating legacy applications for high availability: a case study", 2011 IEEE 13th International Symposium on High-Assurance Systems Engineering, pp. 83-90, 2011.
3. Service Availability Forum, Application Interface Specification. Availability Management Framework SAI-AIS-AMF-B.03.01.
4. IP Routing: OSPF Configuration Guide, CISCO IOS Release 15S, CISCO Systems, San Hose, CA, U.S.A.
5. G. Ambuj and L. Stephen, "Modeling and analysis of computer system availability," IBM Journal of Research and Development, Vol. 31, No. 6, pp. 651-664, Nov. 1987.
6. A. Wood, "Availability modeling," Circuits and Devices Magazine, IEEE, Vol. 10, No. 3, pp. 22-27, May 1994.
7. J. Muppala, G. Ciardo, and K. Trivedi, "Stochastic reward nets for reliability prediction," Communications in Reliability, maintainability and Serviceability, Vol. 1, No. 2, pp. 9-20, July 1994.
8. P. Salehi, P. Colombo, A. Hamou-Lhadj, and F. Khendek, "A model driven approach for AMF configuration generation," 6th International Workshop, SAM 2010, Oslo, Norway, LNCS 6598, pp. 124-143, 2011.

Property Analysis of Classifiers for Sleep Snoring Detection

Tan Loc Nguyen, Young Y. Lee, Su-il Choi and Yonggwan Won

School of Electronics and Computer Engineering
Chonnam National University, Gwangju, Korea

nguyentanloc032003@gmail.com, {clonelee, sichoi, ykwon}@jnu.ac.kr

Abstract. Sleep snoring has become a serious concern for long term healthcare, as well as an indicative diagnosis to other critical diseases. Recently, many researches show that the snoring sound can be analyzed in different domain by various classifiers. In this paper, a comparison study with various classifiers has been provided for analyzing the advantages and the characteristics of the classifiers for sleep snoring detection. As the results, correlational filter multilayer perceptron neural network (f-MLP) and support vector machine (SVM) classifiers achieved the better generalization performances with the classification rate over 96% for the time domain snoring data set. Besides, the filtered data as the output of filter layer in f-MLP also provided a high discriminative feature set which made most of the classifier succeeded in their works. One important observation was that the ordinary multilayer perceptron (o-MLP) could not generalize with the time domain input data.

Keywords: sleep snoring detection, classification, correlation filter, multilayer perceptron neural network, Bayesian classifier, support vector machine.

1 Introduction

Snoring is a respiratory sound which originates during sleep. It is a result of the oscillatory motion of the tissues in the throat. Tissues of the human body, in sleep, are relaxed that may cause constrictions along the upper airway, and breathing triggers mechanical oscillations of the tissues such as soft palate or tongue around the constriction [1]. It is a typical inspiratory sound, even though a small expiratory component can be heard or recorded, especially from the in obstructive sleep apnea syndrome (OSAS) patient, with different spectral features.

In the last 20 years, the snoring problem has been one of serious concerns and entered the realm of clinical medicine. It is a prevalent symptom, and about 50% of the adult population snore frequently. It has been also reported as a risk factor for the development of diseases such as ischemic brain infraction, systemic arterial hyper-arterial hypertension, coronary artery disease, and sleep disturbance [2]. In recent years, several studies have also shown the relationship between snoring and OSAS, which is usually

K.J. Kim (ed.), *Information Science and Applications*,
Lecture Notes in Electrical Engineering 339, DOI 10.1007/978-3-662-46578-3_127

associated with loud and heavy snoring sound [2]. On the other hand, investigation into the sleep sounds also provides information about breathing abnormalities, OSAS or other pathologies, such as upper airway resistance syndrome, and supports health assessment [1]. Snoring is also one of the main factors causing the sleep disruption for sleep partners or even snorers.

For last several decades, many works have been performed for snoring detection. The detector usually contains a pre-processing block and a following classifier. The pre-processing procedure is usually composed of one or more steps of filtering, segmentation, feature extraction and normalization for the sleep sound to be a suitable form for classification process [3]. Based on the discriminative characteristics of pre-processed data, the classifiers will be trained to be able to separate the snoring state from the normal respiratory ones. They used several methods for classification such as robust linear regression with principal component analysis (PCA) on the sub-band energy distributions [1], AdaBoost classifier using acoustic features from time and spectral domains [3], hidden Markov model (HMM) based method [4], fuzzy mean clustering [5], wavelet-based spectral analysis [6], genetic algorithm and support vector machine (SVM) [7], etc. Beside the time domain based classification, most of the classification method used the frequency information extracted from the acoustic signals of sleep snoring sounds.

In this research, various classifiers having different characteristics, such as the ordinary multilayer neural network (o-MLP), correlational filter multilayer neural network (f-MLP), Bayesian classifier, Support Vector Machine SVM), were used to examine the snoring data with its various feature transformations. The result indicates the strength of each classifier as well as the characteristics of snoring sound in different feature domains.

2 Literature Reviews

Nowadays, the recognition problems can be solved with many standard and popular classifiers such as Bayesian classifier, multilayer perceptron neural networks, SVM, and etc. Each of them has their own properties with different pros and corns for a certain data set.

Bayesian classifier was designed based on statistical Bayesian decision theory which quantifies the tradeoffs between various classification decisions. This classifier requires the conditional density estimation function as well as prior knowledge for input data. In general, classifying decision is made to minimize the overall risk which is computed by maximizing the weighted posterior probability according to its decision penalties [8].

Besides, multilayer perceptron neural network (MLP) has recently been rediscovered as an important alternative to various standard classification methods. MLP is actually a graph of connected units representing a mathematical model which mimics the functioning of human nervous system. Those units are called processing units or simply neurons which are interconnected by unidirectional or bidirectional arcs called weights. MLP takes input data by a set of input units and delivers its output via a set of output units. These sets are also called input and output layers, respectively. The remaining units are called hidden units which are responsible for performing nonlinear feature mapping and computing the network logic. The MLP classification performance is characterized by several factors, for example activation function, network architecture, learning algorithm, etc [9].

In addition, a modified neural network, f-MLP, has recently introduced to provide a superior performance to frequency information dominant data classification. By integrating a feature extraction module to the ordinary MLP structure, f-MLP can simultaneously train both the hidden layer of correlational filter coefficients and the weight parameters of other layers using back-propagation learning algorithm. The correlational filter layer is implemented to extract the frequency information and to create a new feature set in power spectrum domain after filtered by correlational operation with filter coefficients. The new created feature set is then fed into the next higher layer of the network. The f-MLP performance also depends on a number of factors such as network topology, activation function, etc [10].

Furthermore, SVM has also been used widely in many application domains due to its dynamic characteristic configurations and mature optimization techniques. SVM is theoretically a sound approach for controlling model complexity. It scans and picks important instances to construct the discrimination surface by maximizing the margin between categorized data instances. The discrimination surface can be normally linear or nonlinear by applying various kernel functions, implicitly mapping their inputs into high-dimensional feature spaces. SVM can also perform multiclass classifications in various ways, either by an ensemble of binary or by extending the margin concepts. The performance of SVM is highly affected by the optimization techniques including regularization conditions, kernel function options, etc [11].

3 Data set description

3.1 Raw data collection

Randomly chosen 15 participants who have the sleep snoring symptoms were asked to wear a neck band which is equipped with a small recording device with a condenser microphone during overnight sleep. The microphone was uni-directional device and faced toward the neck, which can minimize the noise disturbance from the environment. The recorded signals were digitized with the sampling frequency of 8 kHz and 16-bit resolution as mono channel raw data [10].

3.2 Pattern Data Set

Firstly, the snoring periods were determined by hearing examination throughout entire length of collected raw data. Then, 1024 samples were removed from the beginning and from the end of each snoring periods to achieve only the central clear snoring zones in order to have a gap between its neighbor non-snoring zones. The window of 1024 points was used to randomly collect non-repeated segments from those zones. This procedure was also repeated for the other non-snoring periods without necessity of achieving clear zone in the same breathing sound raw data. Afterward, 1000 samples from each class were uniformly selected from each recording raw data. When the segmentation procedure was completed for 15 participants, 1000 samples for each class were taken from the total 15 sets of segmented data. Fig. 1 shows the procedure of segmentation for each participant raw data.

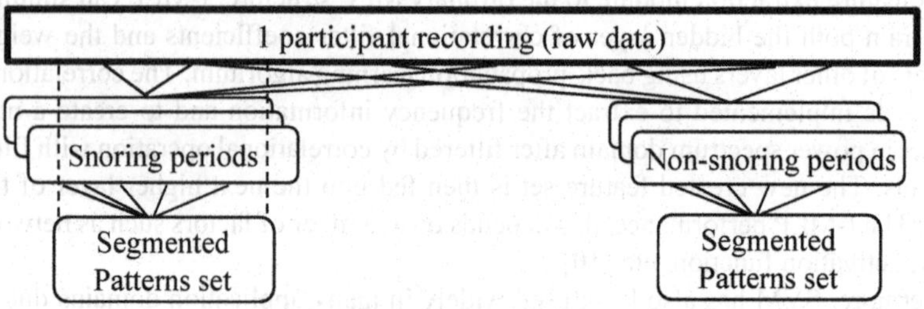

Fig. 1. Segmentation Procedure

To analyze the characteristics of the snoring data as well as the performance of classifiers, the data which input to each stages of f-MLP with some particular conditions were examined. The size of these data was as same as the time domain snoring data. The Fig. 2 shows the demonstration of where the dataset were taken.

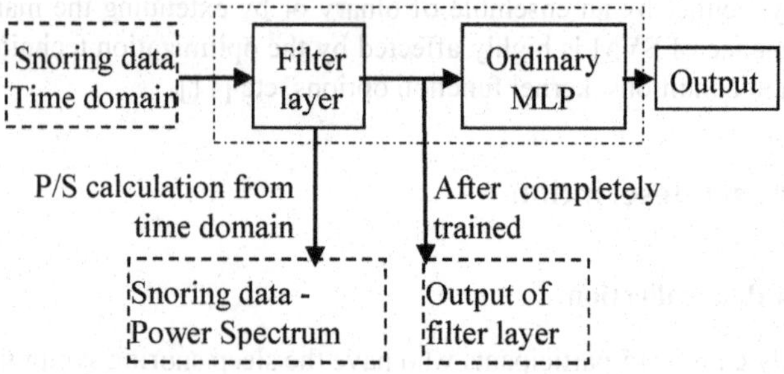

Fig. 2. Diagram of dataset generation

From the original snoring in time domain (time domain dataset), the power spectrum analysis was used to achieve the snoring data in power spectrum domain (P/S dataset). This is the double sided power spectrum transformation and the calculated dataset had the dimension as same as the time domain snoring dataset.

Furthermore, after completely trained by f-MLP, the output of filter layer (filtered data), was taken to investigate this layer's function. The output from this layer is the power spectrum calculation form filtered snoring signals by the correlational filter operation with the filter coefficients. Therefore, this dataset also has the dimensionality as same as the original time domain snoring dataset.

In each of the dataset, training all of the classifiers was done with the 70%, while the rest 30% unseen data was used for testing these classifiers.

4 Experimental results

4.1 Experimental Set-up

In the experimental studies, several classifiers, such as the f-MLP, ordinary multi-layer neural network (o-MLP), Bayesian classifier, support vector machine (SVM), were trained by the training set and tested with the testing set. Note that each classifier has their own properties with different pros and corns for a certain data set

The f-MLP consisted of 2 layers of the filter layer and the output layer, which has only a single hidden layer of the correlational filter. The filter layer had the structure of 1024 units with complex number architecture. Output layer had only 2 units which represent the 2 class problem. Other network configurations and training criterion were the same as in [10].

The o-MLP was also trained and tested with same data set for comparison purpose. However, the size of the neural network (i.e., the number of parameter such as weights and biases) and the training conditions cannot be fairly compared with those of f-MLP. Thus, nearly optimal conditions for the network size and training process were exhaustedly found, and the network of a single hidden layer with 45 hidden nodes was chosen [10].

In addition, the Bayesian classifier and Support Vector Machine (SVM) were also used for performance comparison purposes. The Bayesian classifier used the Gaussian distribution to establish the class relative frequency distribution for the prior probability, while the SVM used the 3rd order polynomial kernel function with the sequential minimal optimization for training procedure.

4.2 Classification Results

To avoid overtraining with the neural networks and to get time efficiency, cross-validation with the test data set was applied for the stopping criterion additional to the number of maximum training epochs and pre-defined minimum sum-squared-error (SSE).

Table I shows the comparison results of the averaged 3 repeated experimental studies for both f-MLP and o-MLP, Bayesian classifier and SVM. Note again that, for each trial, 1,000 training data for each category (snoring and non-snoring) were randomly retrieved from the pattern data pool of several thousands. And then 70% was used for training and 30% was used for testing.

5 Analysis

The o-MLP achieved good training (more than 96%) for most of the dataset. However, the testing accuracy for the time domain dataset was failed with only 81.3%. This indicated the failure in generalization of the ordinary MLP for the classification with the patterns which has the discrimination power by the hidden frequency information in time domain dataset.

Most of the classifiers provided a good generalization and highest testing accuracy with the filtered dataset among all given dataset. This result specifies that the correlational filter layer of f-MLP well acts as a feature extraction module by providing a better feature set which has more discriminant power than the original feature set.

Finally, based on the high ability in frequency information extraction, f-MLP has achieved the best performance in snoring detection compared to o-MLP, Bayesian classifier and SVM.

Table 1. Performance comparison table.

Signal Domain	Accuracy	o-MLP	f-MLP	Bayesian	SVM
Time Domain dataset	Training	98.6	97.7	70.8	100
	Testing	81.3	97.2	68.3	94.7
Power spectrum dataset	Training	96.2	N/A	68.3	79.8
	Testing	96.7		65.8	78.8
Filtered dataset	Training	97.4	N/A	88.3	97.7
	Testing	96.8		87.3	95.67

6 Conclusion

Various classifiers including Bayesian classifier, o-MLP, f-MLP and SVM have been trained to detect the snoring sounds in this study. The experimental results show that the f-MLP and SVM which can provide the good generalization capability achieved good performance in time domain snoring data. Only the o-MLP was successful in classify the power spectrum data which directly transformed from the time domain data. Finally, the filtered dataset achieved from the output of filter layer in f-MLP can provide more discriminative feature set for most of the classifiers to achieve good performances.

Acknowledgement. This work was supported by the National Research Foundation of Korea (NRF) grant funded by the Korea government (MSIP) (NRF-2013R1A2A2A04016782).

References

1. Cavusoglu M, Kamasak M, Erogul O, Ciloglu T, Serinagaoglu Y, Akcam T (2007) An efficient method for snore/nonsnore classification of sleep sounds. Physiological measurement 28 (8):841-853. doi:10.1088/0967-3334/28/8/007

2. Dalmasso F, Prota R (1996) Snoring: analysis, measurement, clinical implications and applications. The European respiratory journal 9 (1):146-159
3. Dafna E, Tarasiuk A, Zigel Y (2013) Automatic detection of whole night snoring events using non-contact microphone. PloS one 8 (12):e84139. doi:10.1371/journal.pone.0084139
4. Lee HK, Lee J, Kim H, Ha JY, Lee KJ (2013) Snoring detection using a piezo snoring sensor based on hidden Markov models. Physiological measurement 34 (5):N41-49. doi:10.1088/0967-3334/34/5/N41
5. Beeton RJ, Wells I, Ebden P, Whittet HB, Clarke J (2007) Snore site discrimination using statistical moments of free field snoring sounds recorded during sleep nasendoscopy. Physiological measurement 28 (10):1225-1236. doi:10.1088/0967-3334/28/10/008
6. Hossen A, Jaju D, Al-Ghunaimi B, Al-Faqeer B, Al-Yahyai T, Hassan MO, Al-Abri M (2013) Classification of sleep apnea using wavelet-based spectral analysis of heart rate variability. Technology and Health Care: official journal of the European Society for Engineering and Medicine 21 (4):291-303. doi:10.3233/THC-130724
7. Maali Y, Al-Jumali A (2012) Automated Detecting and Classifying of Sleep Apnea Syndrome Based on Genetics-SVM. International Journal of Hybrid Intelligent Systems 9 (4):2003-2210. doi:10.3233/HIS-2012-00157
8. Duda RO, Hart PE, Stork DG (2001) Pattern classification. In., 2nd edn. Wiley, New York, pp 20-63
9. Aggarwal CC (2014) Data Classification: algorithms and applications. In: Chapman & Hall/CRC data mining and knowledge discovery series. CRC Press, Taylor & Francis Group, Boca Raton, pp 205-243
10. Nguyen TL, Won Y Performance Comparison of Multilayer Neural Networks for Sleep Snoring Detection. In: Proceeding of the 4th International Conference on IT Convergence and Security, Beijing, China, 2014. pp 427-429
11. Aggarwal CC (2014) Data classification: algorithms and applications. In: Chapman & Hall/CRC data mining and knowledge discovery series. CRC Press, Taylor & Francis Group, Boca Raton, pp 187-204

An Innovative Detection Method Integrating Hybrid Sensors for Motorized Wheelchairs

Sanghyun Park[1], Hyunyoung Kim[2], Jinsul Kim[3*], Taeksoo Ji[4], Myoung Jin Lee[5]

School of Electronics and Computer Engineering, Chonnam National University, 77 Yongbong-ro, Buk-gu, Gwangju 500-757, Republic of Korea

[1]sanghyun079@gmail.com, [2]hyuny.kim1@gmail.com, [3]jsworld@jnu.ac.kr, [4]tji@jnu.ac.kr, [5]mjlee@jnu.ac.kr

Abstract. In this paper, we illustrate an innovative detection method integrating hybrid sensors for motorized wheelchairs. By recognizing its surrounding circumstance, it can prevent accidents as well as provide convenience for users. Its sensors are designed to sense obstacles so that the technology is able to control motors to avoid collision or turning over. Using the 8 infrared distance sensors detect 360° of an obstacle. In addition, the controlling of motor's speed based on the distance can avoid obstacle collision safely.

Keywords: Wheelchair Control System, Infrared Sensor, Safety Drive, Triangulation method

1 Introduction

Wheelchairs are essential for the disabled or elderly people who have walking difficulties. Comparing to the hand-operated wheelchairs, the joystick of electric wheelchairs allows its users to use less power to control and more intuitive manipulation interfaces. However, when users are not familiar with the electric wheelchair or a suddenly change in high approaches or terrain can lead the wheelchair to be overturned which is the potential cause of an additional fatal accident. Wheelchairs are being continuously developed in a wide variety of technologies [1], which are demonstrated by a stair climbing wheelchair, a self-standing up wheelchair after overturn, and a wheelchair controlled by other body parts except hands [2][3]. In this paper, our research interests are limited to a control platform which automatically control an electric wheelchair using sensors. By allowing the electric wheelchair to automatically control motors according to obstacles, accidents can be prevented as well as more usable interface can be available.

K.J. Kim (ed.), *Information Science and Applications*,
Lecture Notes in Electrical Engineering 339, DOI 10.1007/978-3-662-46578-3_128

2 Related Work

Intelligent Powered Wheelchair [4] is an autonomous navigation system which detects obstacles and is able to notify the users to avoid them (see Fig. 1). Electric wheelchair consists of laser sensors and a sound buzzer enables the electric wheelchair to detect the nearby barrier and emit a sound so that the user can move it in a different direction.

Fig. 1. Intelligent Powered Wheelchair

This wheelchair can be controlled by the voice and touch screen. The electric wheelchair is developed to guide users to avoid obstacles by detecting obstacles and making sounds. However, if the user is not able to stop the wheelchair at the moment, there is a high risk of accident. Hence, because the dedicated touch screen wired to the system, it cannot communicate with currently widely used smartphones. To overcome these limitations, this paper proposes an automatically controlled electric wheelchair based on sensors and motor controllers.

3 A Sensor-Based Smart Wheelchair Control Platform

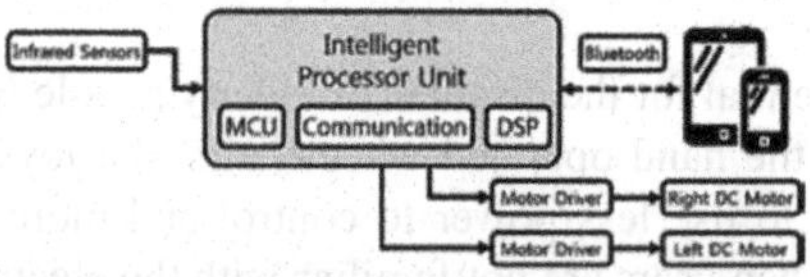

Fig. 2. Sensor-based control platform structure

Smart wheelchair control platform is infrared sensors which recognize neighboring objects including obstacles and keep the intelligent processor unit informed the distance and the location of them. Fig. 2 illustrates how the platform is designed. The processor unit communicates with infrared sensors in real time and concludes the current situation the user is facing. According to a circumstantial estimation logic, it controls to motors which motorize the wheelchair. The smartphone provides the remote monitoring and controlling interfaces by being connected to the unit via Bluetooth.

3.1 Measuring the position of obstacles using Infrared distance sensors

An infrared distance sensor is used to detect an obstacle in real time. Figure 3-(a) an infrared distance sensor, an obstacle have a maximum distance and the minimum distance range. Equation (1) is used to detect the distance of the obstacle by using 2 sensors

IR LED and Position sensitive IR Detector. "*a*" is the distance between the sensor and the obstacle, and "*c*" is the distance between the IR LED and PSD.

"*x*" is value that reflected the distance of an obstacle through the center of lens on PSD sensor. "*b*" the distance between the lens with IR LED and PSD. Using triangulation method obtains the distance of obstacle "*a*". Thus, the equation (1) can calculate the distance of the obstacle and from that we can control the motor of electric wheelchair due to distance. Figure 3-(b) is graph measuring of the obstacle based on the equation (1), "*a*" area shows when the obstacle is recognized in the maximum detection distance. If "*a*" area is recognized as an obstacle, the electric wheelchair will prepare to set up limited maximum speed of motor to avoid obstacle. If "*b*" area is recognized as an obstacle in minimum detection distance, the electric wheelchairs should stop motor to avoid collide with obstacle. As shown in the figure, at the "*b*" area the value of the IR distance sensor is constant, then the user can recognize obstacles through a Smartphone. After user clicks the button "OK" and the electric wheelchair can move over obstacle. Area "*c*" is the maximum detection distance of obstacles, now the maximum motor speed limit movement of the electric wheelchair is released.

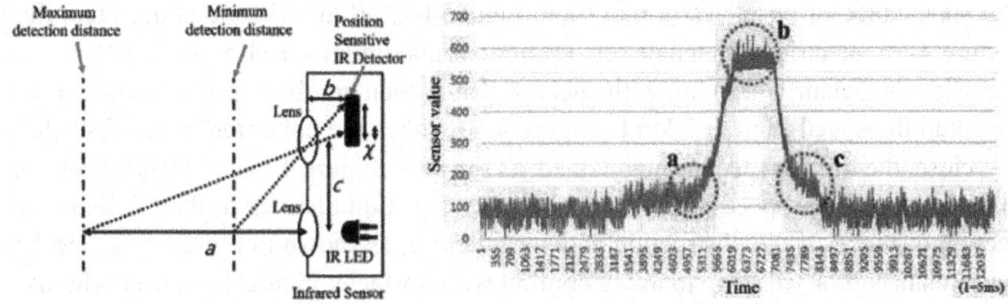

(a) (b)

Fig. 3. Using infrared distance sensors to measure distance with obstacle

$$\frac{b}{a} = \frac{x}{d}, \qquad a = \frac{bd}{x} \tag{1}$$

3.2 Motor Control According to Obstacles

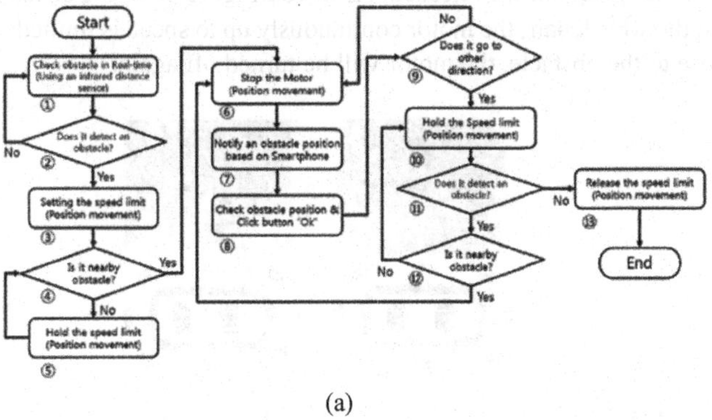

(a)

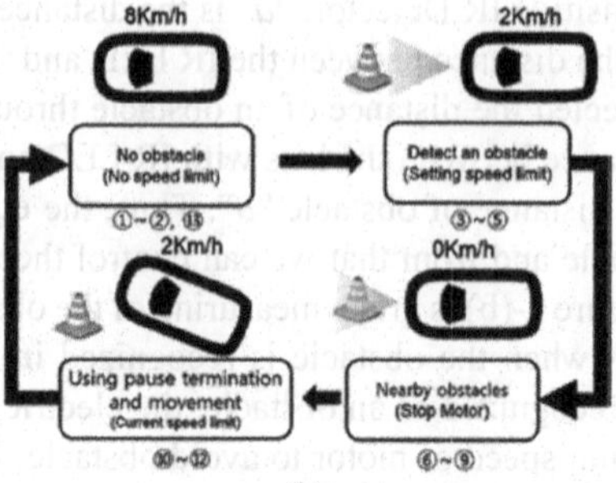

(b)

Fig. 4. Checking obstacle and controlling motor using sensors

Using eight infrared sensors, the wheelchair discovers obstacles in full 360 degrees in real time. Figure 4-(a) shows a flow chart of the motor control system according to the obstacle detection, Figure 4-(b) refers to the situation in accord with the flow chart. When the electric wheelchair moves, according to Figure 4-① using the sensor to check the appearance of obstacles in real time. Figure 4-②, if an obstacle is detected, then Figure 4-③, electric wheelchair sets up limited maximum speed. For example, when electric wheelchair is moving with speed 8 Km/h, then an obstacle is detected, it will be setup the speed limit is 2 km/h. Figure 4-④, check the condition if the obstacle is not close, the limited speed is maintained, as shown in Figure 4-⑤. Electric wheelchair will move with the speed limit from maximum detection distance to the minimum detection distance. If the obstruction is detected close, as shown in Figure 4-⑥, electric wheelchair motor will stop to avoid conflict with obstacle. Figure 4-⑦, user who uses the smart phone will receive information the location of the obstacle, then the user can easily determine the location of obstacles. In Figure 4-⑧, when user have determined the location of obstacles, they press the button "OK" to turn on the motor to continue moving. Figure 4-⑨ check the condition, if electric wheelchair moves toward the other place, then in Figure 4-⑩ the maximum speed limit will be maintenance, if wheelchair do not move to another place, the motor will be turn off again. If the obstacle is not recognized as shown in Figure 4-⑪ leave out the maximum detection distance, then in Figure 4-⑬, the speed limit is released. However, Figure 4-⑫, the obstacle is detected not close to the wheelchair, the motor continuously up to speed is limited, and if wheelchair is close to the obstacle, the motor will be turned off again.

Fig. 5. Infrared Distance Sensors are attached to electric wheelchair

Figure 5 shows the infrared distance sensors are attached to detect obstacle with full 360° degrees the real time.

4 Conclusion

In this paper, we developed an innovative detection method integrating hybrid sensors to control safely the smart wheelchair. Eight infrared distance sensors enabled the control unit to check the obstacles nearby the wheelchair in real time, and limit the speed of the motors based on the measured distance to avoid crashing into the obstacles. Using infrared distance sensors detect a distance with an obstacle in real time, according to the maximum detection distance and the minimum detection distance to control the speed of the motor. By controlling the moving speed of the motor according to the detected distance, a wheelchair can stop when it is near obstacles to prevent the users from getting hurt. Further, by limiting the maximum speed, the user is able to recognize the obstacle, so they can control the electric wheelchair to prevent the risk of collision.

In the future, a research on the system that responds quickly in a variety environments will be conducted. Lastly, a variety of research will be continued to implement systems which can be easily used by people with any kinds of disabilities.

Acknowledgment

This work (Grants No.C0218369) was supported by Business for Cooperative R&D between Industry, Academy, Research Institute funded Korea Small and Medium Business Administration in 2014. Also, it was supported by Korea Evaluation Institute of Industrial Technology, Korea, 2014.

References

1. Wallam, Fahad and Muhammad Asif.: Dynamic finger movement tracking and voice commands based smart wheelchair. International Journal of Computer and Electrical Engineering. vol. 3, no. 4, pp. 497-502. (2011)
2. Srivastava, Preeti, S., Chatterjee and Ritula Thakur.: Design and Development of Dual Control System Applied to Smart Wheelchair using Voice and Gesture Control. IASTER's-International Journal of Research in Electrical & Electronics Engineering. vol. 2, no. 2, pp. 1-9. (2014)
3. M. AL-Rousan and K. Assaleh.: A wavelet-and neural network-based voice system for a smart wheelchair control. Journal of the Franklin Institute. vol. 348, no. 1, pp. 90-100. (2011)
4. Honore, W., et al.: Human-oriented design and initial validation of an intelligent powered wheelchair. RESNA Annual Conference. (2010)
5. David Sanders., Ian Stott., Jasper Graham-Jones., Alexander Gegov and Giles Tewkesbury.: Expert system to interpret hand tremor and provide joystick position signals for powered wheelchairs with ultrasonic sensor systems. Industrial Robot: An International Journal. vol. 38, no. 6, pp. 585-598. (2011)

Part XI
IWSATA

Road Weather Information transmission method for digital multimedia broadcasting

Lee, SangWoon

Abstract Many countries gather information for the road surface condition related weather. But there are no proper way to give the gathered information to the drivers on the roads. In this paper a method to transmit the road information especially related with weather condition. The proposed method is testified via terrestrial digital multimedia broadcasting system, and can be deployed for other digital mobile broadcasting system for example DAB, HD Radio, etc.

Keywords TPEG, TTI, RWI, DMB

Lee, SangWoon
Dept of Multimedia, Namseoul University, 91 Daehak-ro Seonghwan-eup, Cheonan,, South Korea
e-mail: Quattro@nsu.ac.kr

■ Funding for this paper was provided by Namseoul University.

1. Introduction

Traffic accidents can be happened by the fog or ice on the road surface related weather. A research reported that weather plays a role in 24 percent of all crashes, and the annual economic costs of these deaths and injuries is estimated at $42 billion in America.[1] In this paper a road weather information transmission method based on TPEG is proposed and suggests ways to deliver the road weather information to the drivers and it be helpful to prevent and reduce accident which can be happened because of bad road conditions. TPEG is an international protocol to provide TTI service and a bearer and language independent protocol which use broadcasting channels like DAB, FM Radio data radio channel (DARC), digital video broadcasting (DVB), internet and others digital transmission networks. In this paper, a new application which is interoperable with TPEG protocol is proposed. In section 2, the concept of proposed method including the structure of TPEG are presented, In section 3, the implementation the proposed road weather information transmission system is described, In section 4 experiments for the verification of the proposed method are provided. Finally, a brief conclusion is given in section 5.

K.J. Kim (ed.), *Information Science and Applications*,
Lecture Notes in Electrical Engineering 339, DOI 10.1007/978-3-662-46578-3_129

Most of the existing research on road weather information has focused the survey on the effects and damage analysis by the weather on the road and road weather gathering systems. American National ITS Architecture regarding the Weather Information Processing and Distribution Market Package illustrated processes and distributes the environmental information collected from the Road Weather Data Collection market package. This market package uses the environmental data to detect environmental hazards such as icy road conditions, high winds, dense fog, etc. [2] Many transportation agencies have increasingly used advanced technologies known as road weather information systems (RWIS) in their operations. [3]

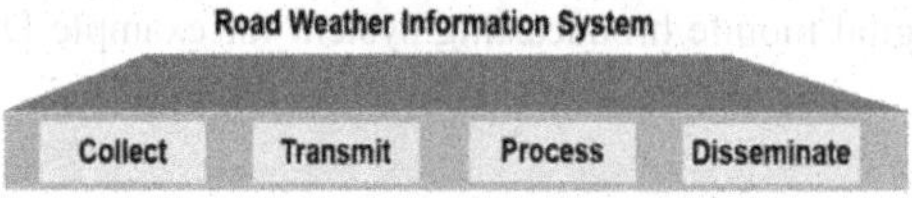

Fig. 1 Road Weather Information System (RWIS) Functions

For the road weather information service, the time needed to sending the information to the customers is very important and short time delay is desirable. And a research shows the current average data age to the service provider and customers is 55min to 24 hours. [2]

Though many administrations gather the road weather information and try to service to the public, it is needed to study more effective way of the transmission of the information. An example of road weather information service using website in Alaska area, the drivers can get the road weather information for the road to go before departure or on the way by selecting the point on the map, but it seems not easy to get the information while driving in cars. [5]

2. Proposed Method

In this study, to give the road weather information to the drivers in the car, international standard scheme named TPEG is applied. TPEG series applications are legislated as ISO technical standards. TPEG includes the common formats and frame structures of the TPEG itself besides of several application protocols, and another new application can be easily added into the existing TPEG protocols with compatibility. Therefore, all the TPEG applications have the same frame structure and synchronization scheme. From the OSI 7 layer point of view, TPEG is defined as layer 3 for synchronization, layer 4, 5 and 6 for the frame structure, layer 7 for the application itself. [6]

Any TPEG application has same frame structure and each frame includes synchronization word (Sync Word), the length of service frame (Field Length), the header CRC, the frame type indicator and the service frame. The header CRC is two bytes long and calculated on 16 bytes including the sync word, the field length, the frame type and the first 11 bytes of the service frame, which provides error detection and protection for the synchronization elements and not for the data within the service frame. [7]

Each service frame comprises service IDs, the encryption indicator and the component multiplex comprising one or more component frames. Each transport frame may be used by only one service provider and one dedicated service which supports a mixture of applications. The service component multiplex is a collection of one or more component frames, the type and order of which are freely determined by the service provider. Each component frame comprises the service component identifier, the length of the component data (Field Length), the component header CRC and the component data. The Identifier of a declarative structure is uniquely defined within each application. The same number may be used in different applications for completely different purposes. Within an application a common identifier designates a common structure. The design of an application may use declarative structures to implement placeholders or to change the composition of elements in a fixed structure.

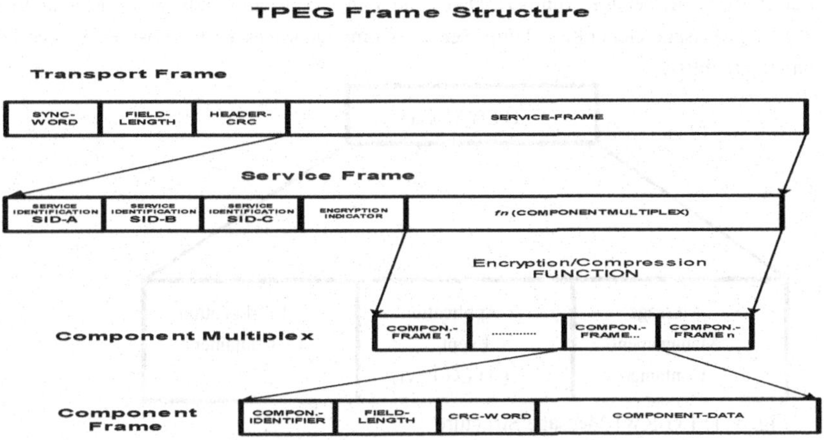

Fig. 2 TPEG Frame Structure

In any TPEG application, the management data, application information and location information components should be contained in a TPEG-Message. A message management component is necessary for the decoders to recognize when the message is received and to make decision for the validity time of the message, etc. The application information is the main data of the message as traffic or traveler information and be in an event container. One TPEG message contains one to one information for a road section or a point as a location and location container is used to give the location information. [6].

3. Implementations

3.1 TPEG Road Weather Information Application Design

The purpose of road weather information transmission is to give the road weather information to the end users via wireless multimedia network. The various road or road surface conditions of wet, iced, windy, snow and other status of road section or point can be transmitted to the driver on the road with car navigation system, smartphone or other receivers via digital broadcasting media or communication networks. The data types are numerical data, still images or videos from weather (meteorological) satellites, i.e. all kinds of multimedia data could be included. Also the road weather information transmission method is designed fit for the international TPEG standards. And it can be a new TPEG application service with the name of TPEG-RWI (Road Weather Information) according to the naming rule of TPEG application. [8]

3.1 TPEG Road Weather Information message component Design

The TPEG-Message container which delivers road weather information message includes message management container, location container by the TPEG message structure as Figure 3. For the message management container, several elements contained. MID (Message Identifier), VER(Version Number), Dates and Times, Start and Stop Times, Message Expiry Time, Message Generation Time, Service Component Reset and Cross-Reference Information are those.

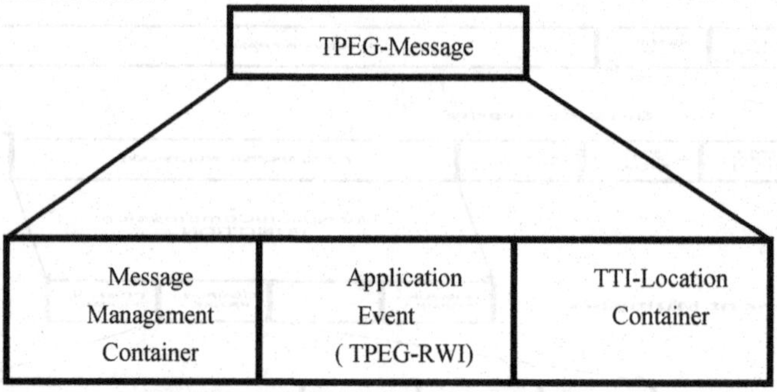

Fig. 3 TPEG-RWI Message Structure

All the message management elements and location elements are predefined in the TPEG frame structure and related specifications, the main design of TPEG-RWI is for the status container. This standard use two location reference method regard to TPEG-decoder without digital map and maintain the compatibility of existing location reference method in TPEG-RWI transmission. The status information describe follow the hierarchical structure, it guarantee the TPEG decoder compatibility by specification expansion and component addition. The status container has several classes, the highest class exist two set in the present version as define the status container, and adding or sub-class of the class of future expansion is possible only with the addition of an identifier.

RWI application comply with syntax structure and signification coding in TPEG Part2 - Syntax, Semantics and Framing Structure.[7] The service component id (scid) is allocated dynamically by the TPEG-SNI application. The CRC check of the n* road weather message is two bytes long, and is based on the ITU polynomial $x^{16} + x^{12} + x^5 + 1$. The CRC is calculated from all the bytes of the data. To proposed road weather information transmission method using TPEG was realized as commercial service system. The system configuration is as Figure4.

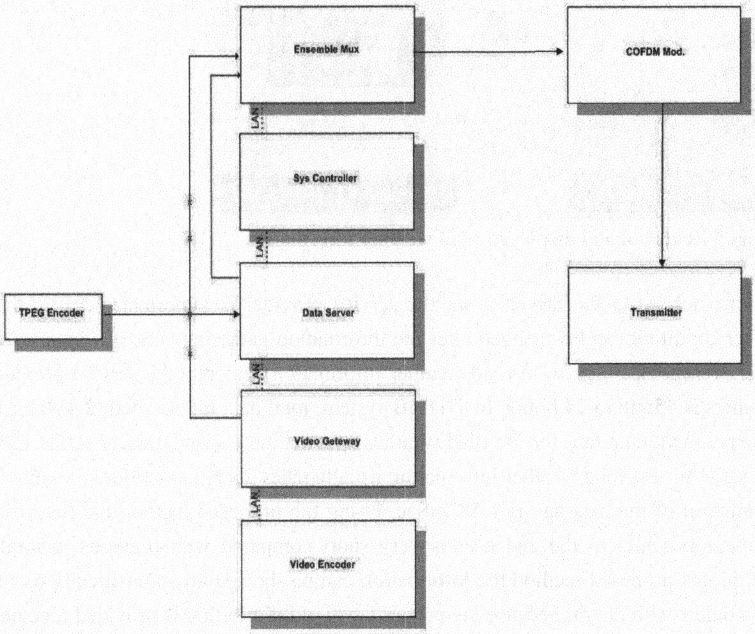

Fig. 4 Road weather information transmission system structure

4. Field Experiments and Evaluation

For the verification of the proposed road weather information transmission method, a terrestrial digital multimedia broadcasting system (T-DMB) is used as a transmission system. [8] Road weather condition information data gathered by the road weather information systems on a highway are transferred to a special road weather information server. Via the server a new TPEG road weather information encoder implemented according to the proposed method TPEG-RWI application, generates TPEG-RWI messages. These messages are inputted to the data server and Ensemble multiplexer of the T-DMB system for the COFDM modulation and transmission. The transmitted road weather information data coded as TPEG-RWI and multiplexed with other video and audio signal the information is received and decoded in a user terminal. This terminal has T-DMB receiver and TPEG decoder including TPEG-RWI application decoder also. In this experiment a car navigation type receiver is used and the results shows the proposed method works well. The transmission system structure and the receiver display shows the received road weather information are showed in Figure 5.

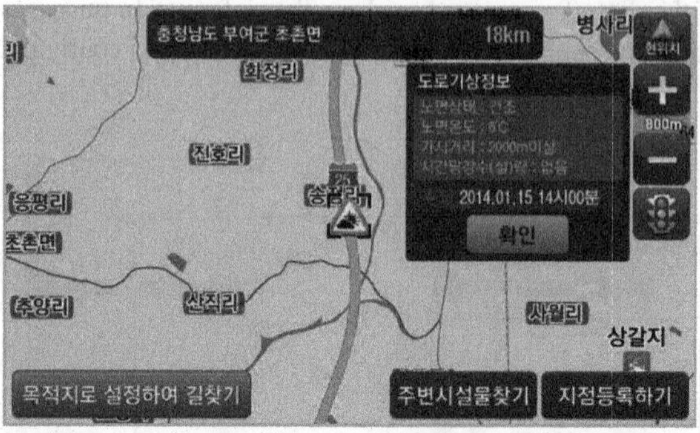

Fig. 5 Received and displayed road weather information

The time needed to the drivers from the service provider is important, because the road weather condition can be changed after the information gathering. The previous work suggested average data age from road weather information center to the service provider and customers is 55min to 24 hours. In T-DMB system, total data rate is about 2.4Mbps, and in this experiment data rate for the road weather information transmission is set as 128Kbps. A message for the road weather information usually takes 20Byte, therefore the time for the transmission of the message is 1.587mSec. Using the proposed method the time from the transmission center to the end user is very short compared with previous method. And more, in this proposed method the information can be showed up automatically on the display monitor with alarm, because the proper location information is provided together with road information. This method can be used to reduce the traffic accident which can happen from the road weather conditions.

5. Conclusion

In this paper, new method for road weather information transmission via digital broadcasting network. In this method the international traffic and traveler information standard scheme named TPEG was adopted and a new road weather information application was proposed as a name of TPEG-RWI. The various road or road surface conditions of wet, iced, windy, snow and other status of road section or point can be transmitted to the driver on the road with a transmission time of 1.587mSec for a message. And automatically displayed on the display or alarmed by sound for the drivers to know the forward road condition. According to the transmission and receiving field experiments, the proposed method works well via terrestrial digital multimedia broadcasting system. This proposed method also can be deployed to any other digital transmission media including broadcasting and communication networks. For the commercial service, enough numbers of road weather information system should be installed operated for the information gathering.

References

1. A Road Weather Research Agenda, Results of the Road Weather Policy Forum, Novem ber 8-9, 2010]
2. Clarus Concept of Operations, October 2005, Publication No. FHWA-JPO-05-072]
3. State Liability and Road Weather Information Systems (RWIS), Jaime Rall, National C onference of State of Legislatures, April, 2010]
4. Managing Traffic Signals During Storms by Ahmed Abdel-Rahim and C.Y. David Yan g, November/December 2012, Public Roads Magazine Vol. 76 • No. 3
5. http://www.dot.state.ak.us/iways/roadweather
6. Wei-feng, Dai Xi, and Zhu Tong-yu, Research of Dynamic Location Referencing Meth od Based on Intersection and Link Partition, World Academy of Science, Engineering a nd Technology Vol:2 2008-10-26
7. ISO/TS 18234-2, Traffic and Travel Information (TTI) – TTI via Transport Protocol Ex pert Group (TPEG) data-streams – Part 2: Syntax, Semantics and Framing Structure (S SF)].
8. ISO/TS 18234-1, Traffic and Travel Information (TTI) – TTI via Transport Protocol Ex pert Group (TPEG) data-streams – Part 1: Introduction, Numbering and Versions (TPE G-INV/102)],
9. Sang-Hee Lee, Kang-Hyun Jo, Design and Implementation of Interactive TPEG-UCM Application Service Using T-DMB, International Journal of Multimedia Technology, S ept. 2012, Vol. 2 Iss. 3, PP. 72-82

Color Overlapping Method of Projection System on LEDs

Yongseok Chi, Youngseop Kim, Je-Ho Park, Kyoungro Yoon, Cheong-Ghil Kim, Yong-Hwan Lee

Display and Optics R&D Lab, ACTS Co., Ltd.,
Dept. of Electrical and Electronics Engineering, Dankook University,
Dept. of Computer Science and Engineering, Dankook University,
School of Computer Science and Engineering, Konkuk University,
Dept. of Computer Science, Namseoul University,
Dept. of Smart Mobile, Far East University

Abstract. This paper proposes a primary color overlapping method for increasing the brightness of projection system using light-emitting diodes (LEDs). The proposed approach drives the pulse width modulated signals of red, green and blue LEDs. The proposed color overlapping method synthesizes secondary colors of yellow and cyan from primary colors of RGB. By the proposed method, brightness of the projected image is improved by bout 30%, compared to the conventional non-color overlapping method in a project system with LEDs.

Keywords: Color Overlapping, Secondary Color, Projection System, Light-Emitting Diode

1 Introduction

The light-emitting diode (LED) has well-known advantages for portable projection systems. It provides a solid-state light source and is compact, reliable, and does not require the use of mercury. Moreover, its lifetime is longer than that of a standard lamp light source. Therefore, the LED is very environmentally friendly. Recently, the development of high-brightness LEDs has resulted in LEDs being adopted as the lighting sources of projection systems [1].

K.J. Kim (ed.), *Information Science and Applications*,
Lecture Notes in Electrical Engineering 339, DOI 10.1007/978-3-662-46578-3_130

Technically, an LED is designed to operate with a forward driving current under both the continuous waveform driving method and the pulse width modulation (PWM) driving method [1]. Fig. 1 shows the brightness and efficiency characteristics of typical LEDs. It is noted that the brightness is not proportional to the current and the efficiency decreases as the current increases. This problem becomes serious when an LED is used for a projection system, where the LED is used in the very high current mode [2, 3].

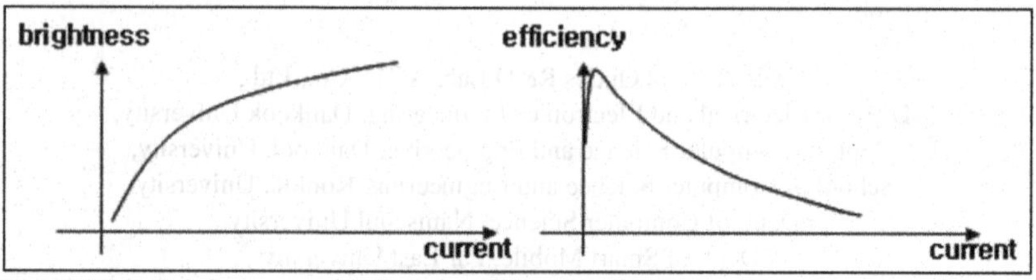

Fig. 1 LED Brightness and Efficiency Characteristics

In this paper, the design of a new LED driving method is proposed. Whereas traditional driving methods use red (R), green (G), and blue (B) colors [3-5], the proposed driving method uses R, G, B, yellow (Y), and cyan (C) colors. The driving efficiency of the LEDs is improved by increasing the duty ratio of the secondary colors (Y, C, and magenta (M)).

In Section 2, the proposed method is described. Experimental results and conclusions are given in Sections 3 and 4, respectively.

2 Proposed Method

A color overlapping method is proposed for providing more efficient and higher-brightness illumination than conventional LED projection systems in which the colors do not overlap. The driving efficiency of the LEDs is improved by increasing the duty ratio of the secondary colors (such as Y and C) obtained from color synthesis between the primary colors (R, G, and B).

The driving pulse wave of the LEDs (R, G, and B) is shown in Fig. 2. Y is produced from R and G by synthesis. C is produced from G and B by synthesis. Each frame is composed of a sequence of several colors and the color sequence consists of primary and secondary colors. For instance, when one frame (of duration 1/60 Hz = 16.67 ms) is composed of a sequence of five colors, as shown in Fig. 2, the available timeslot for each color in the sequence is 3.33 ms. The percentage of this 3.33 ms that consists of a secondary color is called the overlap duty ratio.

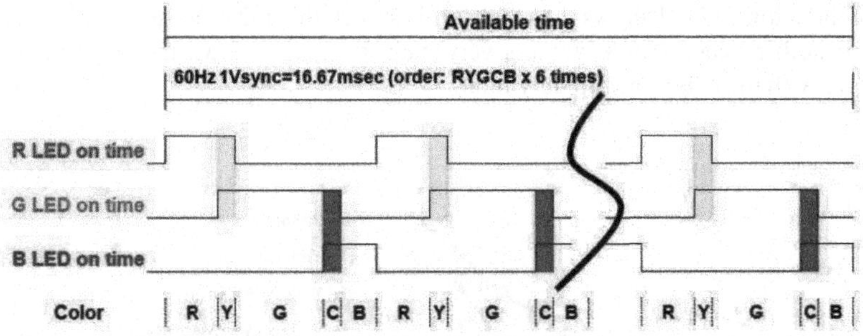

Fig. 2 Waveform of Primary Color Overlapping Scheme

Several possible examples of the proposed color overlapping are shown in Fig. 3. The brightness (in ANSI lumens) of the projector with increasing duty ratio of the generated secondary color (Y, C, and M) for the frames at a frequency of 60 Hz is measured, and the optimal combination is found, as described in Section 3. It is noted that, due to the limits of the frame frequency, the duty ratio of the primary color decreases as the duty ratio of the secondary color generated from primary color synthesis increases. This may cause the result of bad color linearity. Therefore, the secondary color ratio is a very important factor in image quality.

	Red LED on time (Duty)	Green LED on time (Duty)	Blue LED on time (Duty)	Remark
No color overlap	Red Duty	Green Duty	Blue Duty	Seconadry color Duty (Y,C,M,W) : 0~45%
Overlap Yellow	Duty = Red Duty + Yellow Duty	Duty = Green Duty + Yellow Duty	Blue Duty	
Overlap Yellow & Cyan	Duty = Red Duty + Yellow Duty	Duty = Green Duty + Yellow Duty + Cyan Duty	Duty = Blue Duty + Cyan Duty	
Overlap Yellow & Magenta	Duty = Red Duty + Yellow Duty + Magenta Duty	Duty = Green Duty + Yellow Duty	Duty = Blue Duty + Magenta Duty	
Overlap Yellow & Cyan & Magenta	Duty = Red Duty + Yellow Duty + Magenta Duty	Duty = Green Duty + Yellow Duty + Cyan Duty	Duty = Blue Duty + Cyan Duty + Magenta duty	
Overlap Cyan	Red Duty	Duty = Green Duty + Cyan Duty	Duty = Blue Duty + Cyan Duty	
Overlap White	Duty = Red Duty + White Duty	Duty = Green Duty + White Duty	Duty = Blue Duty + White Duty	

Fig. 3 Examples of Overlapping Primary Colors

3 Experimental Results

To find the optimal combination of overlapping primary colors among all the combinations, experimental results of brightness measurements (in ANSI lumens) for four cases are shown: Y; Y and C; Y and M; Y, C, and M. It is noted that our experiments for all the combinations have shown that one of these cases shows the best result of all the combinations. Fig. 4 shows the results. It is shown that the proposed color overlapping method increases the brightness very well. It is also shown that the brightness increases with the color overlap duty ratio.

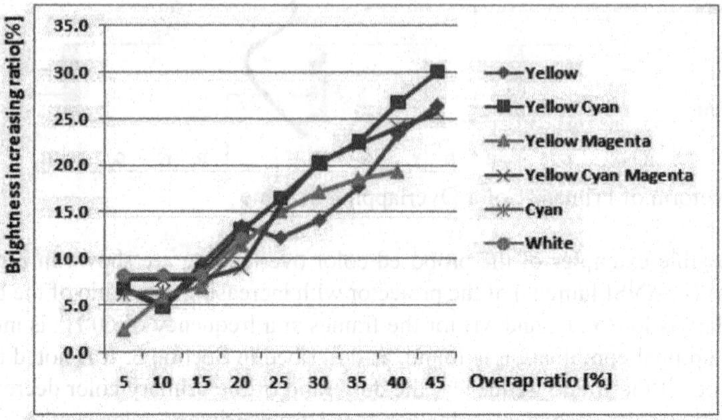

Fig. 4 Brightness Increasing vs. Color Overlapping Ratio

For the color overlap duty ratio up to 45%, the brightness increase ranges up to 30%. When the color overlap duty is fixed, the maximum increase in brightness among all possible color overlapping combinations is obtained from the YC overlap.

The total power consumption of each of the color overlapping methods is shown in Fig. 5. As can be seen, the YC overlap demands the highest current. However, compared to the increasing rate of brightness, the current increasing rate of YC overlapping is known to be limited. Hence, it is concluded that the YC overlap is the most efficient way to increase the brightness among all the color overlapping combinations

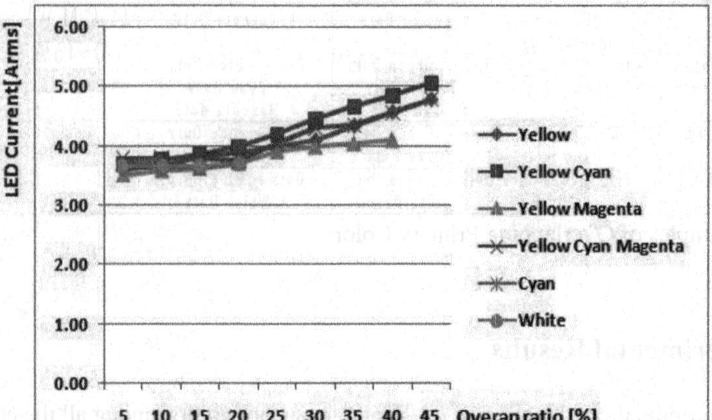

Fig. 5 Power Consumption of Overlap Types

4 Conclusion

This paper proposes a high-brightness LED projection system, which utilizes the secondary color overlapping method. The experimental results show that the secondary colors (Y, C, and M) may increase the efficiency of LEDs in a high-brightness projection system that is compact and has low power consumption, which demonstrates the merits of this projector design. The proposed method improves brightness by 30%, when the color overlap is composed of R, G, B, Y, and C sequential colors

Acknowledgements

This research was supported by the ICT R&D Program of MSIP/IITP [2014, 3D Smart Media/Augmented Reality Technology, KCJR Cooperation International Standardization (I5501-14-1007)]

References

1. T. Nonaka, M. Matsuda, T. Hase, "Additive Color Mixing Model based on Human Color Vision for Bayer-type Pixel Structures", IEEE 10th International Symposium on Consumer Electronics, 2006.
2. K. Kurahashi, "Visual Color Shifts in Spatial Array of Three Primary Colors", Journal of Institute of Television Engineers of Japan, vol.40, no.5, pp.392-397, 1986.
3. W. Kunzman, G. Pettitt, "White Enhancement for Color Sequential DLP", SID Conference Proceedings, pp.121-124, 1998.
4. L.A. Yoder, "Introduction to Digital Micromirror Device (DMD) Technology", Texas Instruments Application Report, DLPA008, 2008.
5. L.J. Hornbeck, "DLP(Digital Light Processing) fur ein Display mit Mikrospiegel-Ablenkung", Fernseh-und KinoTechnik, pp.555-564, Berlin, Germany, 1996.
6. G.S. Pettitt, A. DeLong, A. Harriman, "Colorimetric Performance Analysis for a Sequential Color DLP Projection System", SID, 1996.
7. M. Kawashima, K. Yamamoto, K. Kawashima, "Display and Projection Device for HDTV", IEEE Transactions on Consumer Electronics, vol.34, pp.100-110, 1988.

A System Design for Audio Level Measurement based on ITU-R BS.1770-3

Sang Woon Lee, Hyun Woo Lee, and Cheong Ghil Kim

Abstract Nowadays, inter-program level jumps in digital broadcasting programs have been a major source of sound nuisance. Therefore, control of loudness becomes one of the most important audio issues. In early days of digital audio, the level of a given piece of audio was determined by measuring sample-peak level. Later, the concept of perceived loudness was introduced. In this paper, we introduce a system design for audio level measurement based on ITU-R BS.1770-3 in order to measure true-peak audio level. The system board is designed with TMS320C6727 Floating Point Digital Signal Processor (DSP).

Keywords Audio Level Measurement, Perceptive Loudness, Digital Broadcasting Program, Sound Nuisance.

S.-W. Lee
Dept. of Multimedia, Namseoul University, 91 Daehak-ro, Seonghwan-eup, Cheonan, Chungnam, South Korea
e-mail: lejunee@gmail.com

H.-W. Lee
Cudo Communication Inc. , 1467-80 Seocho-dong, Seocho-gu, Seoul, South Korea
e-mail: xenodigm@cudo.co.kr

C.-G. Kim(✉)
Dept. of Computer Science, Namseoul University, 91 Daehak-ro, Seonghwan-eup, Cheonan, Chungnam, South Korea
e-mail: cgkim@nsu.ac.kr

K.J. Kim (ed.), *Information Science and Applications*,
Lecture Notes in Electrical Engineering 339, DOI 10.1007/978-3-662-46578-3_131

1. Introduction

Most everyday listening situations involve reproduced sound which is a combination of music and speech. The loudness of the sound, perceived by the listener, will not only depend on his or her volume setting, but also on the particular source and format of the audio material [1]. Especially, the loudness control in television broadcasting has posed a problem for broadcasters and audiences for many years. Program normalization practices based on measuring the level – rather than the loudness – of the program has resulted in loudness jumps occurring both between programs and between channels [2].

Since the early days of digital audio, the most common way of determining the level of a given piece of audio has been to measure sample-peak level. After then, another solution was introduced; level should be measured by how loud the listener perceives a given piece of audio - in other words, Perceived Loudness in combination with a new, improved way of measuring peaks called True-peak is the solution to the problem [3].

In this paper we introduce a system design for audio level measurement based on ITU-R BS.1770-3 [4] in order to measure true-peak audio level. The system board is designed with TMS320C6727 Floating Point DSP.

In the next section, we review the algorithm to measure audio program loudness and true-peak audio level. Section 3 describes the system design for audio level measurement based on ITU-R BS.1770-3. The last section concludes this paper with a summary and suggestion about future works.

2. Background

2.1 Algorithm

ITU-R BS.1770-3 specifies the multichannel loudness measurement modeling algorithm. The algorithm consists of following stages:

i. "K" frequency weighting;

ii. mean square calculation for each channel;

iii. channel-weighted summation (surround channels have larger weights, and the LFE channel is excluded);.

iv. gating of 400 ms blocks (overlapping by 75%), where two thresholds are used;

v. the first at –70 LKFS;

vi. the second at –10 dB relative to the level measured after application of the first threshold.

Figure 1 shows a block diagram of the various components of the algorithm. Labels are provided at different points along the signal flow path to aid in the description of the algorithm. The block diagram shows inputs for five main channels (left, centre, right, left surround and right surround); this allows monitoring of programs containing from one to five channels. For a program that has less than five channels some inputs would not be used. The low frequency effects (LFE) channel is not included in the measurement [3].

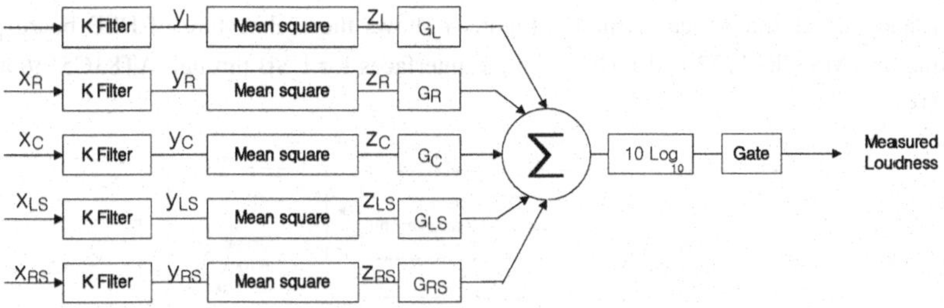

Fig. 1 Simplified block diagram of multichannel loudness algorithm

2.2 Measurement and Control Technology

In [4] we introduce measurement and control technology based on loudness. The purpose was to investigate and develop the measurement system that could replace the current ones with those for the broadcasting volume measurement and automatic control based on loudness in compliance with the international standards such as ITU-R, EBU, and so on. Figure 2 illustrates a conceptual sample workflow for the stress relaxation and evaluation of broadcasting sound and environmental noise. In a normal living condition, television and other noise sources could cause stress on audiences. The workflow begins with volume measurement of television and stress assessment of the audiences with consideration with television loudness and other noise sources and room reverberation. After that it controls the loudness of television, eventually does stress relaxation.

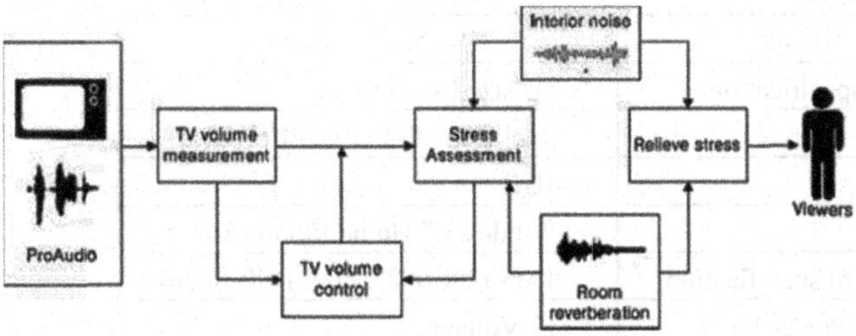

Fig. 2 Conceptual workflow for measurement

3. Implementations

This section introduces the design architecture for audio level measurement based on ITU-R BS.1770-3 using TMS320C6727 DSP. It is 32-/64-bit floating-point DSP at 300 MHz; the CPU can achieve a maximum performance of 2400 MIPS/1800 MFLOPS by executing up to eight instructions in parallel each cycle with efficient memory system. External Memory Interface

(EMIF) supports for 100-MHz SDRAM and flash memory. The main application areas are professional audio and emerging audio. The details of its specification on hardware are shown in Table 1.

Figure 3 shows the configuration of the system consisting of DSP board for loudness measurement and digital audio measurement. Figure 4 shows the architecture of DSP board [6] consisting of TMS320C6727 and EMIF with the interfaces for USB through AT83C5136 and SDRAMs.

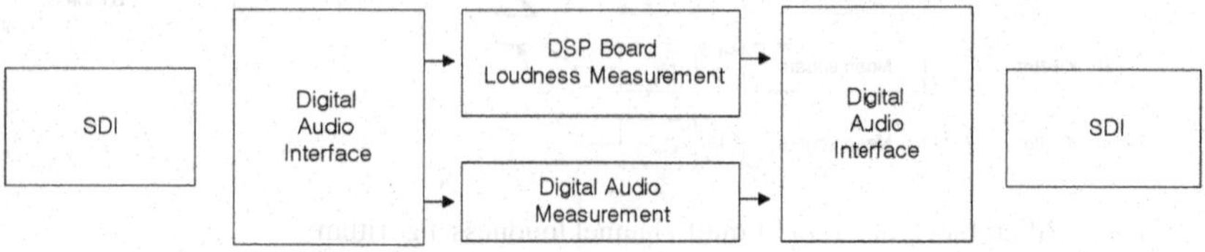

Fig. 3 Overview of system components

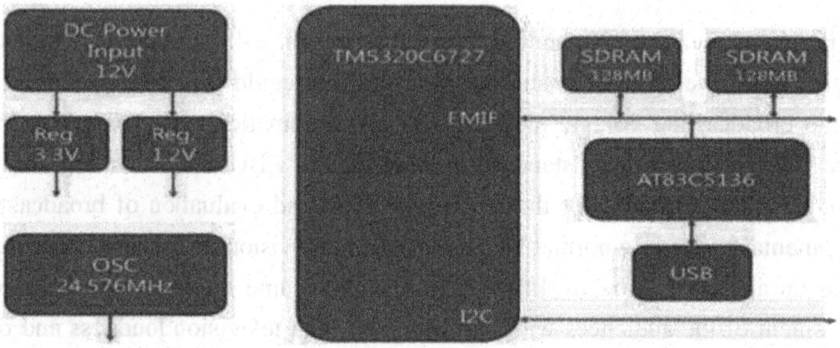

Fig. 4 Architecture of DSP board

Table 1. System parameters

Items	Specification
System specification	2U sized server type
CPU	Intel Core i5-4440 CPU(3.1GH)
Memory	4GB
OS	Windows 7 Home Edition x64
DSP board specification	TMS320C6727 기반 DSP Board
DSP DC Power Input	12 Voltage
DSP Interface	USB 2.0
DSP DC Power Input	12 Voltage
DSP Interface	USB 2.0
DSP Oscillator	24.576MHz

4. Conclusion

Currently, control of loudness becomes one of the most important audio issues. And the concept of perceived loudness was introduced rather than measuring sample-peak level. In this paper we introduce a system design for audio level measurement based on ITU-R BS.1770-3 in order to measure true-peak audio level. The system board is designed with TMS320C6727 DSP with special features for audio applications. Our future work will introduce its details of implementation and performance evaluation.

Acknowledgement

This work was supported by the ICT R&D program of MSIP/IITP. [2014-044-055-002, Loudness Based Broadcasting Loudness and Stress Assessment of Indoor Environment Noises]

References

1. Esen Skovenborg and Soren H. Nielsen, Evaluation of Different Loudness Models with Mu sic and Speech Material. Convention Paper at the 117th Convention, Audio Engineering So ciety, 2004.
2. Eel co Grimm, Esben Skovenborg, and Gerhard Spikofski, Determining an Optimal Gated Loud ness Measurement for TV Sound Normalization, in Proc. of the 128th Convention of Audio Engineering Society, May 2010.
3. tc electronic, What Is Loudness And Why Is It Important? at http://www.tcelectronic.com/l oudness/loudness-explained/
4. Recommendation ITU-R BS.1770-3, Algorithms to measure audio program loudness and tr ue-peak audio level, 2012 08.
5. Sangwoon Lee, BongjinBaek, and CheongGhil Kim, A Study on Audio Levels and Loudne ss Standard for Digital Broadcasting Program, IT Convergence and Security (ICITCS) 2014, Page(s):1 - 3, 16-18 Oct. 2014
6. Cudo Communication Inc. at http://cudo.co.kr

A Prototype Implementation of Contents Sharing System on Networked Heterogeneous Devices

Cheong Ghil Kim and Dae Seung Park

Abstract This paper introduces a prototype design and implementation of contents sharing service application on networked heterogeneous devices. This is because users need to share their media contents and play them continuously by switching from one device to another with the wide spreading mobile devices. For example, N-Screen services allow users to manage their active sessions among multiple devices having different capabilities and architectures. In this paper, the basic sharing configuration consists of three devices: PC, Smartphone, and Tablets. The application was implemented with RESTful web services and an open service platform. The prototype was successfully implemented and its performance was simulated on sample 3D video with different file sizes to minimize load latency when changing devices. The latency was decreased greatly on Smartphone from 1.6 to 0.9 seconds being compared with the one before.

Keywords N-screen, RESTful Web Services, Android, Continuous Play, Open API.

C.-G. Kim(✉)
Dept. of Computer Science, Namseoul University, 91 Daehak-ro, Seonghwan-eup, Cheonan, Chungnam, South Korea
e-mail: cgkim@nsu.ac.kr

D.-S. Park
Dept. of Computer Science, Namseoul University, 91 Daehak-ro, Seonghwan-eup, Cheonan, Chungnam, South Korea
e-mail: eosono@naver.com

K.J. Kim (ed.), *Information Science and Applications*,
Lecture Notes in Electrical Engineering 339, DOI 10.1007/978-3-662-46578-3_132

1. Introduction

Our computing environment is moving rapidly form PC toward mobile devices such as smart phones or tablet pc, and so on. The wide spreading popularity of smart devices including smart TV and IPTV has emphasized the importance of seamless display feature between various heterogeneous devices. Here, users may need to access and share the contents at anytime, anywhere and on any of their devices [1].

As for examples of contents sharing system, N-screen [2] is a technology aimed to provide seamless computing environment by supporting synchronized data or program or display; DLNA [3] is a popular standard for content sharing in home networks. Especially, N-Screen services allow users to manage their active sessions among multiple devices having different capabilities and architectures.

This paper introduces a prototype design and implementation of contents sharing service application on networked heterogeneous devices. The basic sharing configuration consists of three devices: PC, Smartphone, and Tablets. This work is an extended version of the previous work by updating DB structure and including Tablets. The application was implemented with RESTful web services and an open service platform as before. For the performance evaluation, the operation was evaluated by measuring the delay time while switching between different devices.

This paper is organized as follows. We briefly introduce the architectural overview of contents sharing system in Section 2. Then, we introduce the system implementations and the performance evaluation is made in Section 3. The conclusion is covered in Section 4.

2. Architectural Overview

The core element of N-Screen Service is a platform that mediates the use of content or services on multiple devices [1]. This service is initiated by the concept of 3-screeen in which this service makes contents available on the screen of three TV, personal computer, such as a mobile phone based on the integration of the network.

In general, two methods are known as the way of synchronizing media for multiscreen service. So the continuous playing scheme for video is possible by either sharing time or frame information through Cloud. As shown the Fig. 2, the proposed system allows the continuous playing feature between multiple devices such as PC, Smartphone, and Tablets. Here, a separate external DB server is designed to upload videos.

Fig. 3 shows the software architecture of the proposed system based on RESTful web services. REST stands for Representational State Transfer is an abstract model of the web architecture to guide the redesign and definition of the HTTP and URIs. The technologies that make up this foundation include the Hypertext Transfer Protocol (HTTP), markup languages such as HTML and XML, web-friendly formats such as JSON, and Uniform Resource Identifier (URI). REST is an architectural style for networked applications, which consists of several constraints to address separation of concerns, visibility, reliability, scalability, flexibility, performance, and so on. The destination of REST is how to determine a fine definitive Web program to drive forward. Choosing a hyperlink on the Web page, the program could make another Web page return to the user and run further [1.6]. Figure 2 shows the part of program codes which process DB on four protocols.

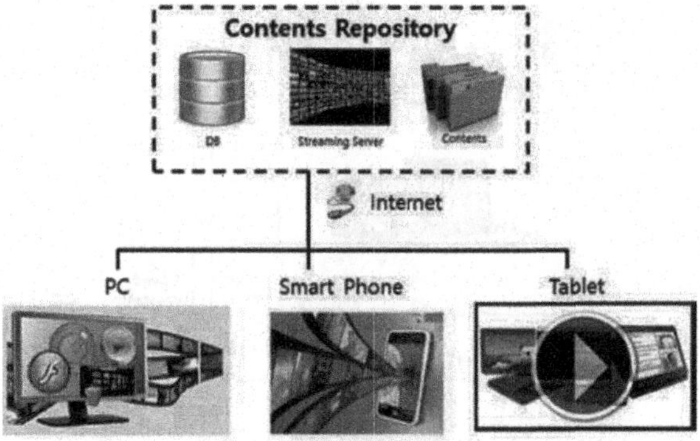

Fig. 1 Simplified block diagram of multichannel loudness algorithm

REST stands for Representational State Transfer is an abstract model of the web architecture to guide the redesign and definition of the HTTP and URIs. The technologies that make up this foundation include the Hypertext Transfer Protocol (HTTP), markup languages such as HTML and XML, web-friendly formats such as JSON, and Uniform Resource Identifier (URI). REST is an architectural style for networked applications, which consists of several constraints to address separation of concerns, visibility, reliability, scalability, flexibility, performance, and so on. The destination of REST is how to determine a fine definitive Web program to drive forward. Choosing a hyperlink on the Web page, the program could make another Web page return to the user and run further [4]. Figure 2 shows the part of program codes which process DB on four protocols.

```
if($func == "SET")
{
    $result2 = mysql_query("insert into player values('', '" . $member .
     "', '" . $name . "', '" . $src . "', '" . $timestamp . "')");
    echo "Success";
}
else if($func == "GET")
{
    $result = mysql_query("select * from player");
    while($row = mysql_fetch_array($result)){
        echo $row['serial'] . ":" . $row['member'] . ":" . $row['name'] .
         ":" . $row['src'] . ":" . $row['timestamp'] . "₩n";
    }

}
else if($func == "UPDATE")
{
    $result2 = mysql_query("update player set name='" . $name . "', src='" .
    |$src . "', timestamp='" . $timestamp . "' where serial=" . $serial);
    echo "Success";
}
else if($func == "DELETE")
{
    $result2 = mysql_query("delete from player where serial=" . $serial);
    echo "Success";
}
```

Fig. 2 Simplified block diagram of multichannel loudness algorithm

Figure 3 shows the initial DB architecture using MS-SQL 2012 [1] and Figure 4 shows the new version using MySQL with HTTP which enables a unified interfaces for all devices through Apache and PHP.

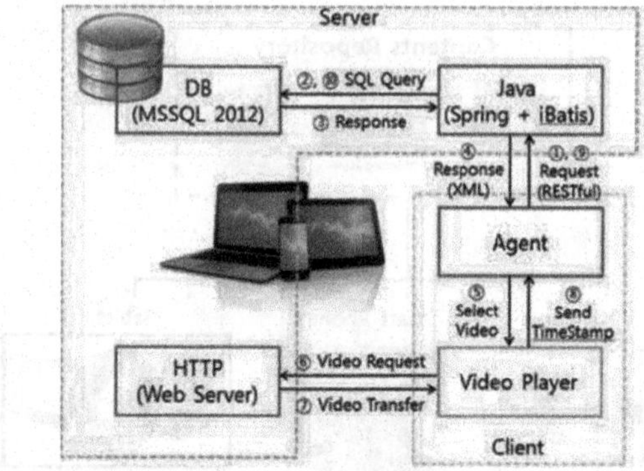

Fig. 3 MS-SQL based DB

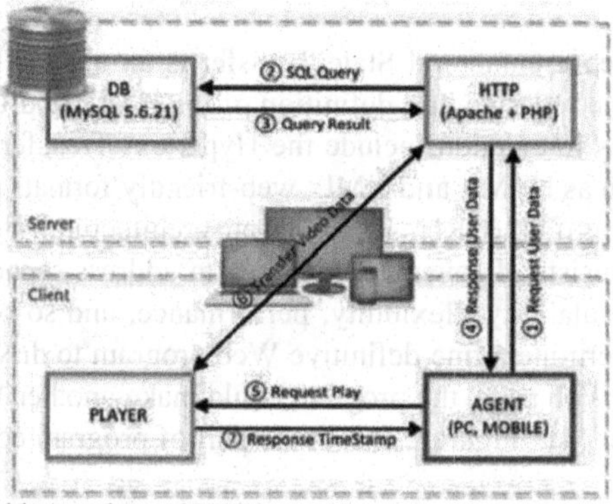

Fig. 4 MySQL DB with Apache and PHP

3. Experimental Results

For the performance evaluation, this paper measure the latency time for continuous playing between different devices consisting of PC, Smartphone, and Tablets. The technical details are shown at Table 1. These devices have better hardware performance than those used in [1] and may influence on the performance improvement greatly.

In [1], the latency time for Smartphone was around 1.6 seconds; however, in this work, it was decreased to less than one seconds. This improvement comes from the amendments on DB and the enhancements on hardware specifications as shown in Table 1. Figure 5 shows the load latency on three different file sizes. They are denoted as Small, Mid, and Big, respectively. In case of Tablets, its CPU specification is A9 Dual Core 1.0GHz which is lower than Smartphone; therefore, its latency time takes long.

Table 1. System parameters

Device	Specification
Smartphone	CPU: Qualcomm Snapdragon 600 Processor 1.7GHz Quad Core CPUs Memory: 2GB DDR2 RAM OS: Android 4.4 Kitkat
3D Tablet	CPU: MTK6577 A9 Dual Core 1.0GHz Processor GPU: 3D Power VRSGX 531 Operation System: Android 4.0 Nand Flash: 8-32G (optional) RAM: 1G LPDDR2
PC	CPU: i3-3240 (3.40GHz) RAM: 16GB DDR3 STORAGE: SAMSUNG SSD OS: Windows 8.1 x64

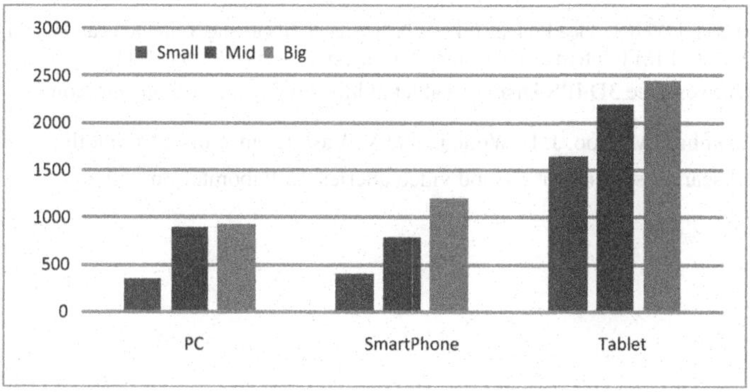

Fig. 5 Load latency on various file sizes

4. Conclusion

This work introduces contents sharing scheme in which the basic sharing configuration consists of three devices: PC, Smartphone, and Tablets. The application was implemented with RESTful web services and an open service platform. The prototype was successfully implemented and its performance was simulated on sample 3D video with different file sizes to minimize load latency when changing devices. The latency was decreased on Smartphone from 1.4 to 0.9 seconds being compared with the one before.

Acknowledgement

This work was supported by the ICT R&D program of MSIP/IITP. [2014(I5501-14-1007), 3D Smart Media/Augmented Reality Technology, Korea-China-Japan-Russia Cooperation International Standardization]

References

1. C. Suppatoomsin, A. Ampawasiri, and C. G. Kim, Implementation of a Continuous Playing Sys tem based on RESTful Web Services, Int'l Conference on International Conference on Information Science and Applications (ICISA 2014), May 2014.
2. C. Yoon, T. Urn, and H. Lee, "Classification of N-Screen Services and its Standardization", 2012 14th International Conference on Advanced Communication Technology (ICACT), p p597-602, 19-22 Feb. 2012.
3. V. D. Mai and Y. Kim, Using DLNA Cloud for Sharing Multimedia Contents beyond Hom e Networks, ICACT2014, pp. 54-57, February 16~19, 2014.
4. H. Li, RESTful Web Service Frameworks in Java, Signal Processing, Communications and Computing (ICSPCC), 2011 IEEE International Conference on, pp. 1 - 4, 14-16 Sept. 2011
5. Truly Glasses Free 3D IPS Display Tablet at http://www.masterimage3d.com/
6. N. Narasimhan, M. Doo, J. F. Wodka., and V. Vasudevan, CollecTV intelligence: A 3-scree n 'social search' system for TV and video queries, CollaborateCom, 2010

Erratum to: Finding Knee Solutions in Multi-Objective Optimization Using Extended Angle Dominance Approach

Sufian Sudeng, Naruemon Wattanapongsakorn and Sanan Srakaew

Erratum to: Chapter 79 in: K.J. Kim (ed.), *Information Science and Applications*, Lecture Notes in Electrical Engineering 339, DOI 10.1007/978-3-662-46578-3_79

Sanan Srakaew was not listed among the authors.

The online version of the original chapter can be found under
DOI 10.1007/978-3-662-46578-3_79

S. Sudeng (✉) · N. Wattanapongsakorn · S. Srakaew
Department of Computer Engineering, King Mongkut's University of Technology Thonburi, Bangkok 10140, Thailand

K.J. Kim (ed.), *Information Science and Applications*,
Lecture Notes in Electrical Engineering 339, DOI 10.1007/978-3-662-46578-3_133

Printed in the United States
By Bookmasters